MW00561559

THE LEGUMINOSAE

*Publication of this volume
has been made possible in part
by grants from the
Natural History Council
and the
College of Agricultural and Life Sciences
University of Wisconsin-Madison*

THE
LEGUMINOSAE

A Source Book
of Characteristics, Uses,
and Nodulation

O. N. Allen &
Ethel K. Allen

The University of Wisconsin Press

Published in the United States of America, Canada, and Japan by
The University of Wisconsin Press
114 North Murray Street
Madison, Wisconsin 53715, USA
ISBN 0–299–08400–0

Published in Europe and the British Commonwealth (excluding Canada) by
Scientific and Medical Division
Macmillan Publishers Ltd
London and Basingstoke, UK
ISBN 0–333–32221–5

Copyright © 1981
The Board of Regents of the University of Wisconsin System
All rights reserved

First printing 1981

Printed in the United States of America

For LC CIP information see the colophon

To our own remarkable symbiosis

A spadeful of soil is a kingdom in itself.

Contents

Illustrations

Preface

This volume incorporates data that have been accumulated during a 45-year survey of nodule incidence in the family Leguminosae. In its present form, the work has grown beyond its original conception. In order to evaluate the scope of nodulation within the Leguminosae we found it essential to learn the extent of the family, and, because there exists no contemporary taxonomic treatise of this vast family in its entirety, we have provided herein comprehensive descriptions of all the known genera.

Both authors were introduced to and schooled in the intricacies of legume-*Rhizobium* symbiosis during the years 1927–1930 under the expert tutelage of Professor I. L. Baldwin and Professor E. B. Fred, of the Agricultural Bacteriology Department, University of Wisconsin, Madison, who pioneered much of the research in this general subject in the United States. A 15-year residence in the Hawaiian Islands, 1930–1945, and subsequent visits afforded us the opportunity to examine a vast tropical leguminous flora, the symbiotic status of which was then unknown. We were especially favored by the then active tropical plant importation program of the Experiment Station of the Hawaiian Sugar Planters' Association, and had free access to their various arboreta and experimental areas in the company of expert botanists. South Pacific and Southeast Asian species, as well as little-known literature pertinent to our interest in nodulation and practical uses of tropical legumes, were readily available there.

This fortunate introduction to tropical Leguminosae in a most favorable habitat enabled us to conduct field examinations and seedling tests of over 150 different species (35 genera) of the Hawaiian flora. From our work the following generalizations were deduced (Allen and Allen 1936a,b): (1) the older textbook statement that all legumes have nodules and the ability to fix nitrogen was incorrect; (2) nodulation was least prevalent among species of Caesalpinioideae; (3) woody tropical species were as readily and effectively nodulated as tender herbaceous temperate zone legumes; (4) corky periderm and more extensive vascular systems were evident in histological nodule sections of older woody species, especially in perennial nodules; (5) strains of rhizobia from members of all three subfamilies were usually cross-infective; (6) cross-inoculation was readily attained with established members of the so-called cowpea group; (7) within the same genus, e.g., *Cassia*, the nodulability of some species was consistently negative while that of others was consistently positive.

Thus began a hobby, which later became almost an obsession, leading us to search national and foreign herbaria and even the most obscure literature for indications of and references to nodulation. Permission to dig in fields and botanical gardens during our foreign travels enabled us to confirm older reports of nodulated leguminous species, to fill the blanks in our lists, and concurrently to establish warm friendships for the exchange of ideas, as well as of nodules, cultures, and seeds.

Our enthusiasm for the survey was spurred, furthermore, by the botanical appeal of this remarkable family — its enormous size and its global distribution, its beauty and its usefulness — and by our curiosity about the extent of the symbiotic phenomenon, especially in the tropics.

The invitation to return to the Bacteriology Department at the University of Wisconsin in 1947 provided us during the next 30 years with access to superb libraries, laboratory facilities, and climatically controlled plant-growing chambers where the newly isolated exotic rhizobial strains and seeds in our sizable collection could be tested and studied.

After the reports of the preliminary survey were published (Allen and Allen 1947, 1961), it became obvious that within the family structure there were patterns to the symbiotic voids that seemed to be linked to tribal and generic phylogeny. Thanks to the fine cooperation and the generosity of rhizobiologists and botanists near and far afield, this global survey now includes nodulation data for fully 50 percent of the genera and 20 percent of the species known to exist. Regional quests have yielded much new information about the leguminous flora of Argentina, Australia, Costa Rica, Hawaii, New Zealand, the Philippines, Puerto Rico, Trinidad, Venezuela, Singapore, and especially since 1967, of South Africa and Zimbabwe. An active exchange of cultures and nodules between most of the investigators, and also personal verification in several of these areas, give credibility to these reports. These are discussed, and credit is duly accorded, under the respective generic synopses.

A global census of nodulation in the Leguminosae has been a long-felt need, not only to reveal the potential of little-known species, but to indicate gaps in our knowledge for future workers. The ultimate goal is to find or develop new or more efficient nitrogen-fixing systems. Now that both the urgency and the multidisciplinary appeal of the problem are being recognized, it seems an especially advantageous time for work in this crucial area to go forward. Just as crop breeders today use huge collections of seeds of both wild and cultivated species from which to select germ plasm for improved crops, so too must *Rhizobium* workers launch a parallel effort in quest of new symbiotic nitrogen-fixing systems within the Leguminosae. The present scope of documented nodulability, as set forth in the appendices, portends a vast potential for future study. At least half of the genera in this family have never been screened for symbiotic performance. Moreover, in the rain forests of Colombia and Guiana legumes are said to constitute more than 50 percent of the trees; they are also the main component in humid forests of Nigeria and the Congo Basin. It is doubtful that the oft-quoted global estimate of 70 million metric tons of nitrogen fixed per year has included any consideration of the fixed nitrogen pool of dominant tropical forest legumes, for virtually no quantitative data exist for these nodulated giants. The rapid landscape changes that now accompany the exploitation of tropical regions make it urgent that legume:rhizobia symbiosis in natural ecosystems be studied and evaluated posthaste before these important resources are forever destroyed.

Introduction

Nitrogen, in its elemental gaseous form, N_2, constitutes four-fifths of the world's atmosphere. This is a virtually inexhaustible supply, yet very few plants and no animals can assimilate nitrogen in its free form. Because nitrogen is the essential constituent of the proteins necessary for cell protoplasm, all organisms are dependent on having it available in a form which they can utilize. Most plants derive their nitrogen from the mineralization of soil organic matter and plant residues. To obtain high yields of crops used for human and animal food, the naturally occurring supply is augmented by application of synthetic nitrogen fertilizers. The cost of nitrogenous fertilizers, however, is already high, and they will become even more expensive because of diminishing supplies, greater demand, and increasing energy costs. Also, the use of high-analysis N-fertilizer is aggravating other already serious problems, such as nitrogen losses from the soil, nitrate movement into the subsoil, and disturbances in the water quality of our lakes and streams. In addition, the nutritional demand that will be placed on our supplies of fixed nitrogen by an estimated doubling of the global population to exceed 7 billions by the year 2000 carries grave implications. The need to conserve fixed nitrogen supplies, moreover, extends beyond the basic need for food and fodder for all living things, whether they be economically important or purely ornamental and esthetic. Modern ecology in recent years has shown this to be one of the primary focal points of vegetational adaptation. Living plants and ecosystems are organized to just that end—to obtain and preserve usable nitrogen.

Biological nitrogen fixation, particularly of the symbiotic type, plays a crucial ecological role in maintaining adequate nitrogen resources in the plant world. Quite distinctive in this respect are the numerous members of the giant family Leguminosae which can thrive without any fixed nitrogen or with a minimal supply from the soil. Here, through the agency of specific bacteria (*Rhizobium* species) which invade the root hairs and establish a mutually beneficial association inside their cortical root swellings or nodules, free air nitrogen is converted into fixed nitrogen for eventual plant protein assimilation and storage. This plant-bacteria collaboration is so intimate that one may conceive of the association as a new form of life. The complexities of nodule tissues, and the high degree of specialization for gaseous and enzyme regulations and vascular transport, indicate a long-standing co-evolution between the two partners. Thus, this select group of plants developed an efficient means for meeting its nitrogen requirements and thereby obtained an evolutionary advantage over most other living organisms. The great evolutionary and ecological success of the Leguminosae speaks for itself.

The recently intensified studies of the tropical grain legumes and their associated rhizobia stem from the pressing need of densely populated nations to maximize their productivity of high protein food crops without depleting their fixed nitrogen resources. Aspects of the extensive role leguminous plants have in these agricultural programs throughout the world are treated by Sornay (1916), Pieters (1927), the International Institute of Agriculture, Rome (1936), Whyte et al. (1953), Heath et al. (1973), Vincent et al. (1977), National Academy of Sciences (1979), and Summerfield and Bunting (1980).

THE FAMILY LEGUMINOSAE

This enormous plant family, with a worldwide distribution, has a currently estimated 16,000 to 19,000 species in about 750 genera. In economic importance it is second only to the grasses, Gramineae; in size, only to the Orchidaceae and the Compositae. Taxonomists conventionally have divided the family into three clearly distinct subfamilies, Mimosoideae, Caesalpinioideae, and Papilionoideae; division has been based mainly on floral differences (Taubert 1894), a concept adhered to in the generic synopses in the present volume. Although some recent taxonomic restructuring has accorded full family status to each of the three subdivisions, as Mimosaceae, Caesalpiniaceae, and Fabaceae in Order Leguminales (Hutchinson 1964), this is a matter of choice, for, whatever their rank, the distinctions between the three basic groups are clear and universally accepted. The current trend for the elevation of subtribes to full tribal status (Hutchinson 1964; Polhill 1976) has served to improve the demarcation of previously ill-defined or controversial genera, and many such revisions are included here.

In order to resolve these issues, an International Legume Conference was held at the Royal Botanic Gardens, Kew, England, in August 1978. Since this book was then in production, we could not benefit from or adjust to any new taxonomic decisions that were promulgated at that conference. Precirculated drafts for discussion indicated, however, that some additions as well as deletions may become necessary in the future, both in recognition of genera and in their placement into tribes; but this would involve only minor restructuring and will not affect the nodulation data.

Characteristics of the family

Trees, shrubs, woody vines, and annual or perennial herbs. Leaves usually
  alternate and compound—bipinnate, simply pinnate, or palmate, rarely
  simple. Inflorescence variously racemose, in simple racemes, panicles,
  spikes, or heads. Flower structure varies to the extent that 3 subfamilies
  are recognized; corolla typically 5-parted; stamens 3–many, mostly 10,
  free, or united by their filaments in various ways; pistil single, simple,
  free. Fruit characteristically a pod* (legume), dehiscent or indehiscent.

*The word "pod" is used throughout in a morphological sense, not always in a strictly terminological sense.

Characteristics of the subfamilies

I. MIMOSOIDEAE: Trees, shrubs, woody vines, a few perennial herbs. Leaves pinnate or bipinnate, often with glands on the rachis; leaflet pairs many, some with sleep movements, or touch-sensitive; [in certain groups, e.g., *Acacia*, characteristic broad modified petioles (phyllodes) simulate leaflets lost after the seedling stage]; stipules in some genera reduced to spines. Flowers regular (actinomorphic), crowded into globose heads, or cylindric spikes, rarely racemose, some inflorescences monoecious; calyx lobes 5, usually valvate; petals 5, equal, valvate in bud, usually united above the base; stamens mostly 10 or multiples thereof, free, joined at the base, or united, forming a tube. Fruit a pod, straight, curved, or spirally twisted, usually 2-valved, dehiscent, some breaking transversely into 1-seeded segments.

II. CAESALPINIOIDEAE: Trees, shrubs, rarely scandent, rarely herbs. Leaves pinnate or bipinnate, rarely simple; some with translucent dots. Flowers mostly irregular (zygomorphic), usually in showy racemes or panicles; calyx lobes 5, overlapping or separate; petals usually 5, sometimes rudimentary or absent, slightly unequal, the upper petal distinctive and innermost in bud; stamens free or joined, 10 or fewer, some may be staminodes; anthers dehiscing by lateral slits or terminal pores. Fruit a pod, sometimes indehiscent, some with winged sutures (samara), or fleshy and drupelike.

III. PAPILIONOIDEAE: Trees, shrubs, herbs, annual or perennial. Leaves mostly palmately 3- or more- foliolate, or odd or evenly pinnate, not bipinnate, rarely simple; some with tendrils or spines. Flowers very irregular (papilionaceous), in 1- to many-flowered, terminal or axillary inflorescences; calyx tubular, regularly 5-toothed or lobed; petals 5, unequal, overlapping; standard uppermost and outermost; the 2 wings lateral, intermediate; the 2 keel petals lowermost, inside, usually joined along the lower margin, hiding the ovary and stamens; stamens 10, rarely fewer, usually united into 2 bundles (diadelphous) of 9 + 1 or 5 + 5, or in 1 tubular bundle (monadelphous), rarely free (Sophoreae). Fruit a variously shaped pod, straight, curved, winged, or moniliform, usually 2-valved and dehiscent, some transversely jointed (loments), ripe joints separating.

Uses

Many species in Mimosoideae and Caesalpinioideae are valuable for their timber, dyes, tannins, resins, gums, insecticides, medicines, and for fibers; in tropical areas, where they abound, they are, in addition, among the world's handsomest flowering trees, vines, and shrubs. Numerous members of the Papilionoideae, moreover, are economically important, especially in temperate areas, as edible and highly nutritional crops for human and animal consumption, for forage, fodder, ground cover, green manures, and erosion control, and as major honey sources. As pioneer plants in arctic regions they form the hub of an efficient nitrogen source for the entire ecosystem (Allen et al. 1964).

Nodulation

The ecological uniqueness of Leguminosae centers about the tubercles, or nodules, of their root systems. These were depicted in the earliest printed botanical illustrations, in the famous herbals of Fuchs (1542), Bock (1556), Dalechamps (1587), and Malpighi (1679) (reproduced in Fred et al. 1932). So characteristic, so seemingly constant was their occurrence on leguminous plants that Wydler (1860) considered them of diagnostic value for taxonomic identification, and most botanists to this day are of the general opinion that legumes and nodules are always associated. As the present survey will show, however, the ability to nodulate seems to be consistently absent within certain sections of the family, so that, although the ability to nodulate is not an ubiquitous attribute of the Leguminosae, its occurrence does appear sometimes to be of taxonomic value in defining sections, genera, or tribes. Yet, even though a species is capable of participating in this remarkable symbiotic relationship, nodulation cannot occur unless the compatible rhizobia are present in the rhizosphere.

THE GENUS *Rhizobium*

The symbiotic process

The soil-improving properties of leguminous plants were recognized for centuries in many parts of the world, but it was not until the mid-19th century that chemists and physiological botanists applied scientific methods to an inquiry of the sources of nitrogen available to green plants. Experiments by Hellriegal and Wilfarth (1888) established beyond doubt that root nodules bore a causal relationship to the assimilation of free nitrogen. Beijerinck, in 1888, isolated pure-culture nodule bacteria from several leguminous species and proved, via Koch's postulates, that they were the causative microsymbionts. The agronomic significance of this monumental insight was soon realized in the production of annual food crops, such as peas and beans, and the planting of pasture legumes for the support of meat animals. Since those early days, as a natural consequence, reliable procedures have been developed commercially for producing effective cultures of root-nodule bacteria appropriate for a particular legume. These are grown on agar, in liquid media, or peat-based material, for use as inocula for coating seeds to insure that the proper types of root-nodule bacteria are present in the rhizosphere during seed germination. Recently, the International Biological Programme (IBP) and the Agency for Industrial Development (AID) have published instructions on how to produce rhizobial inocula for promising native legume crops, an important consideration in trying to improve the protein diet in underdeveloped areas (see Vincent 1970).

To make cultures more readily available to research workers and agronomists, the IBP published a *World Catalogue of Rhizobium Cultures* (Allen and Hamatová 1973), which contains the listing of about 3,000 strains, their host source, symbiotic responses, and general characteristics, as well as the institutions or reference collections where they are preserved. This catalogue is now being revised and enlarged, in the hope that it may lay the foundation for the development of a world bank of *Rhizobium* cultures.

In the free-living state, rhizobia are almost ubiquitous soil inhabitants that occur more abundantly along the rhizoplane and in the rhizosphere of legumes than of other plants. They are aerobic non-spore-forming Gram-negative short rods, readily isolated by plating out surface-sterilized crushed nodules onto suitable nutrients, such as yeast-extract mannitol agar. The recognized *Rhizobium* species have distinguishing physiological characteristics of fast or slow growth, acid or alkaline production, copious or scant gum formation, and serum zone production or absence thereof in litmus milk. The one characteristic which validates all of them as members of the genus *Rhizobium* is their ability to form root nodules on a leguminous host.*

The categorical statement that rhizobia are unable to fix N_2 except in symbiosis must now be revised. Using highly sensitive methodology, proof of positive fixation, albeit quantitatively small, by *Rhizobium* in culture media was recently established (McComb et al. 1975; Pagan et al. 1975). However, for all practical purposes the statement is still valid, for in natural ecosystems both partners are essential for N_2-fixation.

Rhizobia are designated in several ways, according to different sets of circumstances imposed when the plant partners are changed. Thus the descriptive designations are relative, and apply only to the prevailing partnership interaction. The same plant with a different rhizobial strain, or a different plant with the same strain, may evoke quite different or even opposite results.

Among the various designations commonly used to describe the rhizobia:plant association are the following:

Rhizobial attributes	*Plant responses*
noninfective; invasive	resistant; susceptible
inefficient; efficient	noneffective; effective
specific; versatile	selective; promiscuous

An effective symbiosis implies a beneficial plant response. Microbiologists have tended to regard the rhizobia as the controlling partners in the expression of the symbiotic relationship. The rhizobial strain has therefore been designated as effective or ineffective on the test host variety. The genetic characters and physiology of the host plant are also determinative in symbiosis. Infectiveness implies compatibility; according to whether the range of susceptible hosts is narrow or broad, the rhizobia are regarded, respectively, as specific or versatile, and the hosts as discriminating or promiscuous.

Cross-inoculation grouping

Early investigators regarded all rhizobia as a single species capable of nodulating all leguminous plants. As attention focused more critically on

Rhizobium symbiosis with *Parasponia* (*Trema*), elm family, Ulmaceae (Trinick and Galbraith 1976; Akkermans et al. 1978), is claimed as an exception.

Introduction

differences in the infectiveness of various strains, it became apparent that the ability of a given strain to produce nodules on certain plants and not others tended to be specific. Extensive cross-testing of many different strains on multiple hosts led to the establishment of bacteria-plant cross-inoculation groups. As defined by Fred et al. (1932), these are "groups of plants within which the root nodule organisms are mutually interchangeable."

With few exceptions, particular strains of rhizobia are able to nodulate only plants within a certain group of legumes. Species of *Rhizobium* are therefore designated by their host range (Table 1).

Table 1. Designated species of *Rhizobium*

Microsymbiont	Growth rate	Cultural reaction	Host affinities	Plant group
R. meliloti	fast	acid	*Melilotus, Medicago, Trigonella*	Alfalfa
R. trifolii	fast	acid	*Trifolium*	Clover
R. leguminosarum	fast	acid	*Pisum, Lathyrus, Vicia, Lens, Cicer*	Pea
R. phaseoli	fast	acid	*Phaseolus* (temperate zone spp.)	Bean
R. lupini	slow	alkaline	*Lupinus, Ornithopus*	Lupine
R. japonicum	slow	alkaline	*Glycine*	Soybean

The numerical size of some of the legume host groups is now much too unwieldy for comprehensive reciprocal cross-testing. The interchange within such a group may be imperfect, or gradations in compatibility may be encountered, in which case "preferred hosts" or "subgroups" may be indicated.

One cross-inoculation group, the versatile so-called cowpea miscellany, comprises semitropical and tropical vine, herb, shrub, and tree species from various genera of all three subfamilies. Isolates from each of these hosts are able to nodulate cowpea, *Vigna unguiculata* (L.) Walp. (= *V. sinensis* (L.) Endl. ex Hassk.), a notably promiscuous recipient. The reciprocal crosses are invasive, but not always effective. These broad-spectrum microsymbionts, commonly called cowpea rhizobia, have not yet been designated with a species name.

The concept of cross-inoculation groupings drifted into disorder between 1940 and 1950. It culminated with J. K. Wilson's (1944a) campaign to abandon the entire group classification, mainly on the basis of his own data purporting to prove that group specificities do not exist. Yet Wilson's own illustrations of collar-type and crotch-type "nodules" and some other root malformations (Wilson 1939a) nullify his conclusions that many intergroup crosses were attained.

The apparent dual membership of the species of *Phaseolus*, some in the bean group, others in the cowpea group, was cited as a strong argument against the cross-inoculation group concept. This anomaly has now been resolved by modern taxonomic evidence which splits the temperate *Phaseolus* species from the tropical ones, consigning the latter to *Vigna*, hence to the cowpea group. Rhizobia, here, proved to have a more discriminating taxonomic judgment than plant taxonomists, and thus we have a fine example of how "symbiotaxonomy" can provide valuable supporting evidence for phylogenetic kinships.

Recently, the validity of the cross-inoculation concept has been reinforced by highly sophisticated techniques of immunofluorescent staining, enzymology, and immunochemistry. Much of the work emphasizes the root hair surface and phytohemaglutinins or lectins as the recognition factor for rhizobial acceptance, binding, and invasion.

We had envisioned cross-inoculation groups as separate adaptative peaks rising at different times and to different intergrading levels from a common, all-purpose legume progenitor, in the manner of mountain peaks from an ocean floor (Allen and Allen 1947). Norris (1956, 1958a) formulated this idea much more precisely and logically in his theory of the development of group specificities. In his view, cowpea rhizobia, the ancestral type, symbiosed promiscuously with the many primitive woody leguminous genera that thrived on the leached acid soils of the tropics during the Upper Cretaceous and Lower Tertiary periods. Furthermore, the evolutionary advances to the herbaceous and eventually the annual plant habit, and the adjustment to calcareous soils and to the temperate habitats necessitated by vast climatic changes in relatively recent times, must have provoked adaptation in the rhizobial partner also. These evolutionary changes are now expressed in specificities or preferential selection. This may explain why the most recently evolved genus, *Trifolium*, is so highly selective that certain African species cannot symbiose with rhizobia of European *Trifolium* species and vice versa (Norris and Mannetje 1964).

THE ROOT NODULE PHENOMENON

Morphogenesis and nodule types

Nodule initiation, development, and function entail intricate biochemical and biophysical processes, many of which are still not fully understood. Nodule initiation involves (1) rhizobia-host recognition factors at loci on root hair surfaces; (2) rhizobial entry and progressive invasion via infection threads into the root cortex; (3) the evocation of meristematic loci with active cell proliferation accompanied by hypertrophy and usually polyploidy of the invaded cells; (4) release of the microorganisms from the infection threads; (5) rapid bacterial and plant cell division; and (6) differentiation of the tissue into nodule zones with the central bacteroid zone as the site of hemoglobin, essential in the fixation process, and the peripheral vascular strands as the transport system linked with the parent root. Each of the stages in nodule development has been examined in some detail in older reviews by Allen and Allen (1950, 1954, 1958) and by Nutman (1956).

In the juvenile stage all nodules are small and spherical. They are not visible as outgrowths of the root cortex until about a week after their inception. The shape of the nodule is governed by the extent and location of the meristem, which is the site of active cell division and tissue differentiation. Hemispherical peripheral meristems produce spherical nodules, such as those of *Arachis*, *Desmodium*, *Glycine*, *Lotus*, and *Vigna*. Elongated cylindrical finger-forms result from the distal growth of apical meristems. If for some reason portions of this tip grow at an unequal rate or if cleavage of the meristem occurs, bifurcated, digitate, palmate, or coralloid nodules result. Such nodules are typical of *Caragana*, *Onobrychis*, *Trifolium*, and *Vicia*. In those types of nodules, the vascular connecting tissue between the mother root cortex and the nodule outgrowth is constricted and rather narrow. Divided lateral meristems are responsible for the nodules found on *Lupinus* species. Horizontal growth in opposite directions within these nodules produces a hypertrophy which tends to surround the parent root.

Proposals have been published attempting to correlate nodule types with leguminous tribal classification (Spratt 1919; Corby 1971), but because there are so few morphological shapes and so very many Leguminosae subdivisions, this appears to be a nearly impossible task. It can be stated, however, that the plant usually determines the shape of the nodule, for example, round nodules on *Alysicarpus vaginalis* (L.) DC. and branched nodules on *Alysicarpus longifolius* W. & A., both caused by the identical rhizobial strain.

Nodules of most temperate-zone species are annual, being lysed or shed as the plant comes to fruition and the foliage drops off. However, 4–6-year-old perennial nodules are reported for *Wisteria* and *Caragana* in temperate climates (Jimbo 1927; Allen et al. 1955). In autumn their growth becomes dormant, to be renewed each year at the tips, giving such nodules a beaded appearance.

Tropical nodules tend to have a perennial existence by virtue of their often lenticelled corky periderm, annual ring growth at the vascular connection, continuously prolonged meristematic activity, and extensively developed vascular systems. *Sesbania grandiflora* Poir. is a good example (Harris et al. 1949).

The mere presence of nodules is no guarantee of benefit to the plant. Effective symbioses are indicated by large pinkish nodules attached to the upper laterals and on the tap root and a dark green foliage. Ineffective symbioses are indicated by the presence of very numerous small greenish white nodules scattered distally along secondary and tertiary roots, with nitrogen-hunger symptoms often reflected in unthrifty yellowish foliage.

The absence of nodules does not necessarily indicate inability on the part of a plant to enter into an effective symbiosis. Negative findings may reflect an unfavorable temporary situation. There are four common explanations for their absence concomitant with natural plant growth. (1) Shedding of nodules may be a consequence of drought, of flooding, or of clipping the plant's foliage. If these events have occurred at an early stage, new nodules usually will form again when plant growth is renewed. Autumnal shedding in herbaceous perennials is a seasonal, annual occurrence. (2) Unfavorable soil type and soil pH, insufficient solar radiation, and temperature extremes all adversely affect rhizobial populations and nodule formation. (3) The absence of compatible rhizobia in the rhizosphere obviously precludes infections. For example, because American soils were initially devoid of soybean rhizobia, the first soybeans introduced into the United States from Japan failed to nodulate until a soil inoculation program was adopted, using, at the outset, soil imported from Japanese soybean fields. (4) At the genetic level, a resistant or noninvasive line may result from plant breeding and selection, but this can be corrected in trials with better adapted strains.

False nodulation

It is crucial in nodulation surveys to distinguish clearly between true rhizobial nodules and other anomalous root hypertrophies that may, on casual inspection only, simulate nodules. Nematode galls, crown galls, and "false nodules" are the types of growths most commonly mistaken for true rhizobial nodules. Nematode galls caused by microscopic eelworms, *Heterodera radicicola*, are hard, knotty, irregular enlargements of the root cylinder itself. Crown galls produced by a bacterial plant pathogen, *Phytomonas tumefaciens*, are tumors near the upper crown of the root. "False nodules" or collar-type hypertrophies, devoid of bacteria but filled with starch, occur mainly at root axils in greenhouse tests with incompatible inocula (Allen and Allen 1940b). This may be due to some rhizospheral stimulation by the noninvasive rhizobia. Such collar-type and crotch-type nodules may explain some incongruous cross-inoculation data (Wilson 1939a). Pseudonodulation may be a consequence of spraying plants with growth regulatory chemicals such as halogen-substituted benzoic acid derivatives (Allen et al. 1953); such hypertrophies are blunt, deformed modified rootlets. Pocket knife bisections of any of the above anomalous structures readily show that these lack the characteristic color, texture, zonal tissue organization, and cortical outgrowth of rhizobial nodules.

Inherent inability to form nodules

The inability to nodulate is seemingly determined by basic phylogenetic factors of the host plant and is, for example, a constant characteristic of the majority (over 60 percent) of the genera in Caesalpinioideae studied. Possible factors involved here are (1) the presence of physical and morphological barriers to invasion, such as wiry, dark-colored rootlets and sparse thick-walled root hairs which are obvious, macroscopically, on most of the non-nodulating species; (2) the presence of cell constituents containing tannins, flavonoids, quinones, or other phenolic compounds which exert an antibiotic or physicochemical restriction to rhizobial invasion and growth (Allen and Allen 1976); (3) the absence of lectins, phytohemaglutinins, or essential enzymes from the root hairs, thereby accounting for the failure in recognition or binding between the rhizobia and the root surface; (4) the absence of disomatic cells in the cortical root tissue, which, according to Wipf and Cooper (1940), are essential for the initial meristematic activity which culminates in nodule formation; if nodulation failure can be linked to the absence of polyploidy, then a means for overcoming this deficiency in the plant may be sought by inducing it artificially; (5) a relationship between the chromosomal number and the presence of nodules; hypothetical phyletic lines within the Leguminosae as determined by chromosomal evidence were diagrammed by Turner and Fearing (1959). If the nodulation data of our survey are superimposed over the tribes listed for Caesalpinioideae, there is good correlation between failure to nodulate in the chromosomal lines of Amherstieae, Bauhinieae, Cassieae, and Eucaesalpinieae with a base number of $x = 7$ and ready nodulation in the lines of Amherstieae, Cassieae, Cynometreae, Sclerolobieae, and Swartzieae with a base number of $x = 8$.

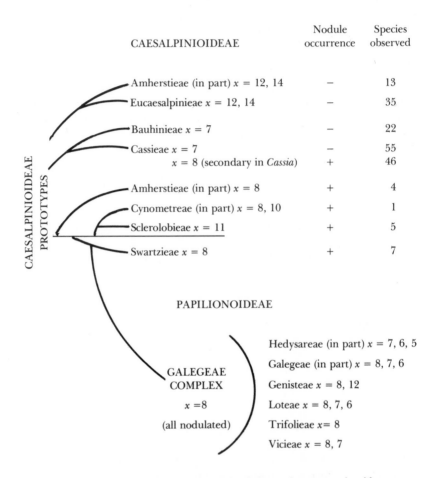

Hypothetical phyletic lines as determined from chromosomal evidence (modified from Turner and Fearing 1959), correlated with our nodulation data as reported in the present study.

Although this correlation is not absolute, it may explain why different species within the same genus, e.g., *Cassia*, have opposite nodulating capabilities. It suggests that plant breeding and grafting experiments between such species in the same genus or tribe might evoke the nodulation response, and could eventually lead to the recognition of the particular chromosome that carries the gene controlling the ability to nodulate. This becomes all the more plausible in view of the fact that botanists agree that Papilionoideae arose from some primitive 8-chromosome stock and that the incidence of nodulability is predominant within this taxon.

Ectotrophic mycorrhiza

Ectotrophic mycorrhiza, rare in Leguminosae, have been reported for some tropical African species of *Afzelia, Anthonotha, Brachystegia, Gilbertio-dendron, Julbernardia,* and *Parmacrolobium,* all in Caesalpinioideae, tribe Amherstieae, and all nonnodulated. Apparently this alternative type of symbiosis accommodates nutritional plant needs, but whether nitrogen fixation is involved has not been stated. Root hairs in the true morphological sense were not observed on such plants either in their natural habitat or in cultivation (Jenik and Mensah 1967).

Overview of the Study

This source book is designed to meet the need for a comprehensive presentation of generic diagnoses of the Leguminosae and to serve as a review of the uses and value of legumes and as a survey of nodulation scope. Our basic objectives have been (1) to present a complete alphabetical listing of all the genera, describing each taxonomically within its proper tribe and subfamily in accordance with accepted systems of classification; (2) to describe the importance, uses, and potential of prominent species against the backdrop of the genus to which they belong; and (3) to tabulate the known nodulated and nonnodulated species in each surveyed genus and to identify the geographical areas from which the reports emanated. The data are complete through 1977, and in some instances later work has been incorporated.

Most of this encyclopedic information is, understandably, not original, but, rather, has been collected from many sources, some now out of print and others difficult to locate. It is regrettable that for many genera little else besides the botanical description can be provided. It is hoped that the inclusion of those genera here will prompt workers to add such peripheral information as is available in regional reports or will tempt them into future botanical explorations.

To verify nomenclature and to eliminate synonymy and duplication, *Index Kewensis* and the *Gray Herbarium Index* (Harvard University, 1968) were consulted. We deemed it essential in the species lists to give the plant name as originally published in nodulation reports, and, consequently, have sometimes had to include corrections because of inaccurate identification, rules of priority, or *nomina conservanda* decisions. In such cases the rejected name, in parentheses, accompanies the accepted name, e.g., *Albizia lophantha* (Willd.) Benth. (= *Acacia lophantha* Willd.), *Macroptilium lathyroides* (L.) Urb. (= *Phaseolus lathyroides* L.).

The subfamily and tribal positions were structured according to Taubert's classification in Engler and Prantl (1894), which, at the start of this project, was the only comprehensive treatise available on the global distribution of the family. The enormous increase in botanical knowledge about Leguminosae during the first half of this century is shown in the updated information contained both in Hutchinson (1964) and in the revised seventh edition of Willis (1966). The numerical size of the family has grown apace, having nearly doubled since Taubert's time. Even since Hutchinson's treatment of the assemblage, the number of newly named genera and species has increased approximately 10 percent (Table 2).

Table 2. Estimated size of family Leguminosae

Subfamily	Taubert (1894)		Hutchinson (1964)		Present study	
	Genera	Species	Genera	Species	Genera	Species
Mimosoideae	31	1,000	56	2,800	66	2,900
Caesalpinioideae	90	1,000	152	2,800	177	2,800
Papilionoideae	308	8,000	482	12,000	505	14,000
Total	429	10,000	690	17,600	748	19,700

THE GENERIC SYNOPSES

The accounts of the genera are constructed according to the following format: In addition to the generic name, preceded by an asterisk if it is conserved, the heading for each entry includes the designation of the subfamily, the tribe, and the subtribe when appropriate. Directly below the generic name are given generic numbers. The first number is that assigned by Taubert (1894) in Engler and Prantl's *Die natürlichen Pflanzenfamilien*; a number enclosed in parentheses designates a position change or a new genus inserted by Harms in his supplements to Engler and Prantl (Nachträge I–IV, 1897, 1900, 1908a, 1915a). The number following the slant line is the four-digit generic number in Dalla Torre and Harms' *Genera Siphonogamarum* (1900–1907), included for the convenience of the many herbaria which still base their arrangements upon this system. If the genus is not numbered, it was more recently described.

Next follows the etymology of the generic name, obtained wherever possible from the author's original description, otherwise by translation of the Greek or Latin word elements; unfortunately, some generic etymologies are simply not available or not fathomable.

The type species or lectotype is given as a separate entry.

The actual generic diagnosis opens with the size, i.e., the number of species in the genus, the plant habit, the taxonomic description, followed by original or pertinent references (if no references are given, the source is Taubert).

The remainder of each entry treats the following: (1) the tribal position, if controversial, or if changed since Taubert; (2) endemic geographical areas and basic ecological habitats; (3) uses, economic importance, distinguishing or singular characteristics of certain species; here it should be noted that expressions of timber density are just approximations and are included for comparative purposes, since information on the prescribed moisture content was seldom available; (4) any pertinent information or discussion about the nodulation reports, such as nodule descriptions, histology, cross-inoculation; and (5) known nodulated and nonnodulated species, areas of examination and authorities for the reports (we have taken the liberty of correcting botanical species authorities where designations were obviously erroneous); these are listed in alphabetical tabular form, unless the reports are monospecific or relatively few. Some records represent only one account deemed indisputable, authored recently by highly qualified workers. Many records consist of multiple entries from various regions, and to conserve space, many multiple observations are designated "widely reported, many investigators." It is impossible to evaluate the credibility of some early unverified observations of exotic species in distant lands, but even so, such reports can provide a working base for future regional screenings.

Negative data are included, since in conservation programs — e.g., afforestation, reforestation, roadside plantings, land reclamation — the choice of a suitable nodulating rather than a nonnodulating species has economic and ecological advantages.

THE NODULATION SURVEY DATA

Following the generic synopses are three appendices in tabular form, listing the genera, alphabetically arranged, but segregated according to subfamily. These appendices enumerate the species in each genus and the number of nodulated and nonnodulated species reported. In the total counts we have omitted synonyms. Blank spaces indicate the voids in our knowledge. Each appendix ends with a computation of the total numbers of genera and species reported within the respective subfamily.

Criteria for verifying nodulation

There are certain essential requirements that should be met before reporting nodulation of a plant under study:

(1) Correct identification of the host plant. This is best accomplished with the aid of competent local botanists, both in the field and in botanical garden nurseries, by the skilled use of field guide books of the regional flora, and by authentication via reference collections in established herbaria. It is desirable to collect and prepare several duplicate voucher specimens, with roots and nodule attachments if possible, for deposit at one or more universities or national herbaria for future reference and proof of validity. Methods for collecting and preserving plant specimens are given in Willis (1957) and in Appendix V of Vincent's *IBP Handbook* (1970).

(2) Collection of nodule specimens. Collection of whole nodules should be accompanied by records of their size, shape, location on the root system, relative abundance, color, texture, plant habit and habitat, and the date of collection for future collation in the event of seasonal cycles in nodule occurrence. For ease in handling and to avoid wounding the nodule, it is desirable to include a snip of root attachment. Rapid transfer of specimens to dry vials or plastic bags and the briefest possible storage time at low temperatures are recommended.

(3) Isolation of rhizobia from the nodule. Inasmuch as a concurrent goal is to build up a rhizobial collection, the isolation of the microsymbiont, its purification, its maintenance, and demonstrated proof (via Koch's postulates) that it, and it alone, is the causative microorganism are all vital requirements. This demands impeccable bacteriological procedures.

(4) Plant tests with the isolate. This calls for a supply of host seeds and aseptic plant-growth equipment. In the event that conspecific seeds are unavailable, cross-testing with another species in the same genus or with plant

species of closely related genera in the same tribe may be indicative. Unless the host seeds are surface-sterilized, the authenticity of nodulation by the isolate under test is dubious. It must be the responsibility of each investigator to present proof positive that the reported results are obtained with *bona fide* *Rhizobium* species and that the results are reproducible.

(5) Maintenance of a reference collection of rhizobial isolates. Although not an essential requirement in determining the ability of a particular legume to be nodulated, a collection of organisms previously shown to be infective is of value for comparative purposes. Subcultures of the parent stock cultures should be tested periodically for infectivity and strain efficiency and as controls in subsequent experiments, especially those involving noninvasive mutants.

Nodulation occurrence in monotypic genera

To counter the criticism that a judgment of nodulation based on only one species in a genus does not constitute a fair sampling, it must be remarked that of 748 genera, fully 237, i.e., 31.5 percent, have only a single species. The following table shows the disposition of these monotypic genera within the subfamilies and the extent of nodulation reports concerning them.

Table 3. Nodulation data within monotypic genera

Subfamily	Number of monotypic genera	Genera reported	Nodulated	Nonnodulated
Mimosoideae	19	3	2	1
Caesalpinioideae	66	10	5	5
Papilionoideae	152	37	36	1
Total	237	50	43	7

The 50 genera reported represent 21 percent of the total monotypic genera in the family, 13 percent of the total genera reported, but only 1.6 percent of the species total herein recorded. The status of one monotypic genus, *Amherstia* subfamily Caesalpinioideae, is still in doubt owing to conflicting reports, which have been omitted here.

Incidence of nodulation: summarization

The data in the appendices are summarized in Table 4. Upon analysis, this compilation shows that 48 percent of the leguminous genera have been examined for nodulation, and of these, 86 percent were found to be nodulated. Of the genera examined in Mimosoideae and Papilionoideae, 83 percent and 95 percent, respectively, had nodulated species, whereas in Caesalpinioideae only 40 percent of the genera examined evinced an ability to nodulate.

At the species level, 15 percent of the family have been examined, of which 91 percent were recorded with nodules. Among Mimosoideae and Papilionoideae this registers 90 and 98 percent, respectively; among Caesalpinioideae, only 30 percent of the species reported were nodulated.

Table 4. Summary of the nodulation survey data in the appendices

Subfamily	Number of genera	Number of genera reported				Estimated number of species	Number of species reported			
		+	+/−	−	Total		+	+/−	−	Total
Mimosoideae	66	18	8	5	31	2,900	351		37	388
Caesalpinioideae	177	13	13	39	65	2,800	72	6[a]	180	258
Papilionoideae	505	241	14	14	269	14,000	2,416		46	2,462
Total	748	272	35	58	365	19,700	2,839	6[a]	263	3,108

[a]One species each of *Amherstia, Copaifera, Eperua, Hymenaea, Mora,* and *Saraca.*

It is significant not only that every tribe has been reported, but that evidence of symbiotic ability is claimed for at least one genus in each tribe. In the subfamily Caesalpinioideae, the lesser ability to nodulate apparently follows tribal patterns. Tribe Eucaesalpinieae seems most resistant to symbiotic associations, with only one, and that unconfirmed, positive report for the monotypic genus *Colvillea,* as against negative data for all 35 species examined in 13 other genera.

The following 10 genera had the largest number of species reported for nodulation performance: in Mimosoideae, *Acacia* (217); in Papilionoideae, *Indigofera* (194), *Crotalaria* (145), *Trifolium* (141), *Astragalus* (102), *Tephrosia* (95), *Desmodium* (76), *Vicia* (65), and *Aspalathus* (60); in Caesalpinioideae, *Cassia* (99).

Suggested reading for other aspects of symbiotic biological nitrogen fixation

In this volume we have not attempted to cover all aspects of symbiotic biological N_2 fixation, for this broad topic has been intensively studied in recent years as new instruments and new techniques in biochemistry and microbiology have been developed. We have found helpful the following excellent reviews:

Döbereiner, J., Scott, D. B., Burris, R. H., and Hollaender, A., eds. 1978. *Limitations and Potentials of Biological N_2 Fixation in the Tropics.* Basic Life Sci. 10. 398 pp. New York: Plenum Press.

Hollaender, A., ed. 1977. *Genetic Engineering for Nitrogen Fixation.* Basic Life Sci. 9. 538 pp. New York: Plenum Press.

Newton, W. E., and Nyman, C. J., eds. 1974. *Proceedings of the 1st International Symposium on Nitrogen Fixation,* 2: 313–717. Seattle: Washington State University Press.

Newton, W., Postgate, J. R., and Rodriquez-Barrueco, C. 1977. *Recent Developments in Nitrogen Fixation.* New York: Academic Press. 622 pp.

Nutman, P. S., ed. 1976. *Symbiotic Nitrogen Fixation in Plants.* Int. Biol. Progr. 7. 584 pp. Cambridge: Cambridge University Press.

Quispel, A., ed. 1974. *The Biology of Nitrogen Fixation.* Front. Biol. 33. 769 pp. New York: American Elsevier.

San Pietro, A., ed. *Photosynthesis and Nitrogen Fixation, Part B.* Methods in Enzymology 24. 526 pp. New York: Academic Press.

Acknowledgments

We are deeply sensible of the many kindnesses extended by staff members of the Steenbock Library, University of Wisconsin–Madison, not only in their perseverance in locating obscure literature, but for providing a "home away from home" office space for convenience, concentration, and safe-keeping of our records.

The enlarged scope of the nodulation survey was due in recent years to a regular exchange of unpublished observations of regional nodulation reports. In this regard, warm thanks are given to H. D. L. Corby, Salisbury, Zimbabwe; N. Grobbelaar, Pretoria, South Africa; Jorge León, Turrialba, Costa Rica; Delia de Rothschild, Buenos Aires, Argentina; B. C. Park, United States botanist in Libya, and W. R. C. Paul, Paradeniya Garden, Sri Lanka.

For identification of legumes on field excursions we are indebted to botanists John J. Koranda, Alaska; John Parham, Fiji; G. C. Lugod, Los Baños, Philippines; Nola Hannon, Sydney, and N. C. W. Beadle, Armidale, Australia; and Chia Huang, Taipei, Taiwan.

True appreciation is expressed here to the late Sir Arthur Hill, Director, Kew Gardens, England, for permission to examine root systems of cultivated exotic legumes; to Rothamsted Experimental Station, Harpenden, England, for accommodating our needs for laboratory experiments; to the staffs of Foster Gardens, Honolulu, and the several arboreta in Hawaii; to the nursery personnel at Bogor Botanic Garden, Java, who willingly helped with digging and root examinations in 1977; also to Joe C. Burton, the Nitragin Company, Milwaukee, Wisconsin, who often shared the task of rhizobial isolations, maintenance, and plant tests of exotic strains, and who generously gave us the photographs of nodulated roots for this book.

We thank Kenneth B. Raper, Bacteriology Department, University of Wisconsin–Madison, and Philip S. Nutman, Head of the Soil Microbiology Department, Rothamsted Experimental Station, Harpenden, England, for encouraging publication of the book.

For support toward funding the publication we acknowledge with gratitude the College of Agricultural and Life Sciences, and the Natural History Council, University of Wisconsin, Madison.

We are grateful to W. C. Frazier, University of Wisconsin–Madison, for useful suggestions, and especially to A. G. Norman, University of Michigan, for much needed editorial criticism of the Introduction.

We appreciate the assistance of James A. Lackey, botanist at the Smithsonian Institution, for taxonomic advice on Phaseoleae, and the devoted help of Alex Lasseigne, botanist, University of Wisconsin Herbarium, in the final decisions concerning structuring and illustrations.

Acknowledgments

To Hugh Iltis, University of Wisconsin botanist, for free access to his personal reprints, rare books, herbarium collections, and for the many stimulating discussions throughout the past 15 years, special thanks.

For his editorial assistance, especially in refining the Greek and Latin derivations, credit and thanks are due Robert J. Henry.

We are much obliged for, and gratefully acknowledge, the fine perceptiveness, the invaluable suggestions, and the preciseness of Elizabeth A. Steinberg, chief editor, the University of Wisconsin Press.

Last, but actually first, to Professor I. L. Baldwin and Professor E. B. Fred for launching us on this happy venture, true gratitude.

ILLUSTRATIONS

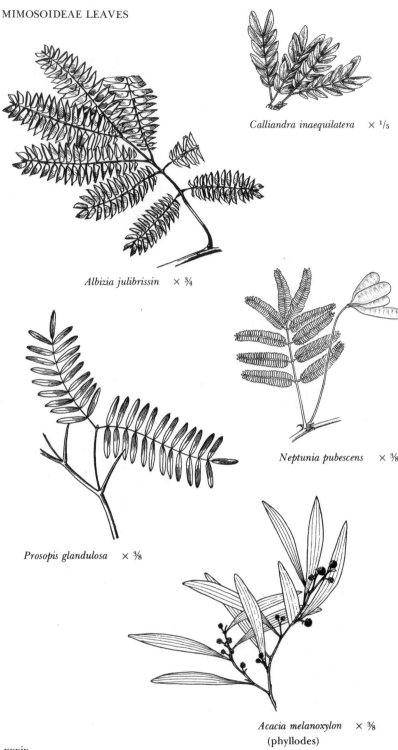

Calliandra inaequilatera × ⅕

Albizia julibrissin × ¾

Neptunia pubescens × ⅜

Prosopis glandulosa × ⅜

Acacia melanoxylon × ⅜
(phyllodes)

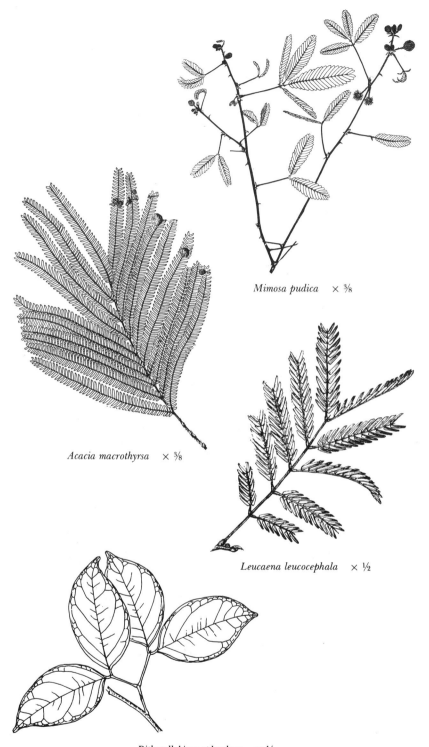

Mimosa pudica   × ⅜

Acacia macrothyrsa   × ⅜

Leucaena leucocephala   × ½

Pithecellobium splendens   × ⅕

Colospermum mopane × ⅜

Cercis canadensis × ⅜

Piliostigma thonningia × ⅜

Monopetalanthus richardsiae × ½

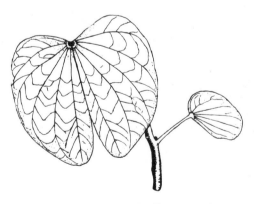

Bauhinia monandra × ½

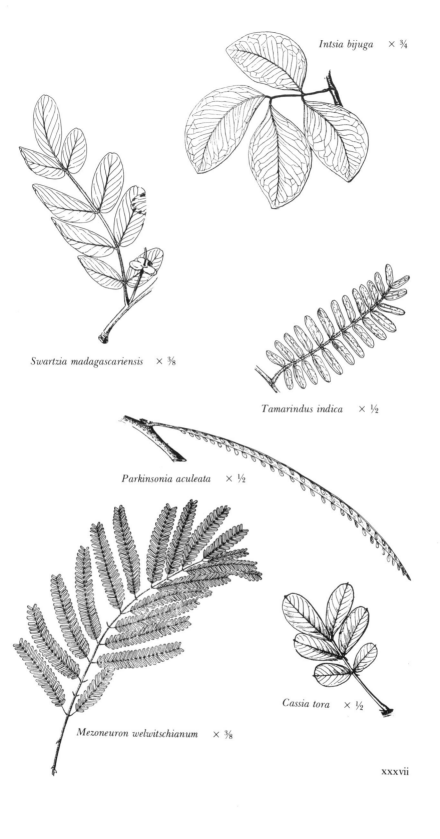

Intsia bijuga × ¾

Swartzia madagascariensis × ⅜

Tamarindus indica × ½

Parkinsonia aculeata × ½

Mezoneuron welwitschianum × ⅜

Cassia tora × ½

xxxvii

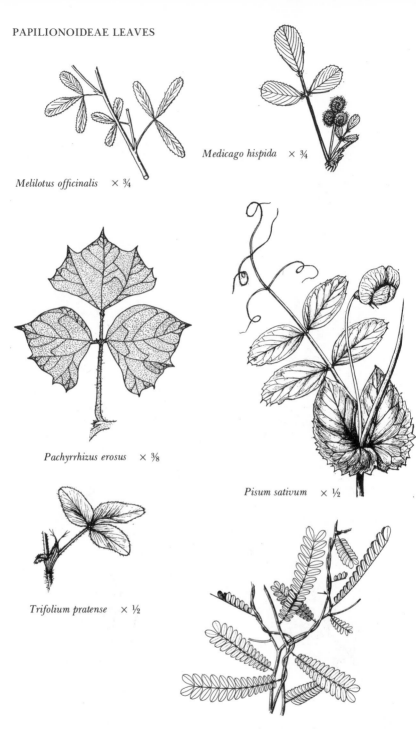

Melilotus officinalis × ¾

Medicago hispida × ¾

Pachyrrhizus erosus × ⅜

Pisum sativum × ½

Trifolium pratense × ½

Abrus precatorius × ½

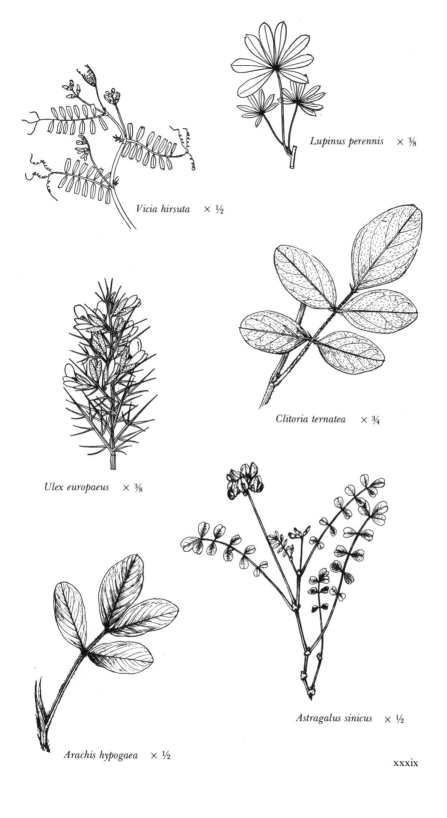

Vicia hirsuta × ½

Lupinus perennis × ⅜

Ulex europaeus × ⅜

Clitoria ternatea × ¾

Arachis hypogaea × ½

Astragalus sinicus × ½

xxxix

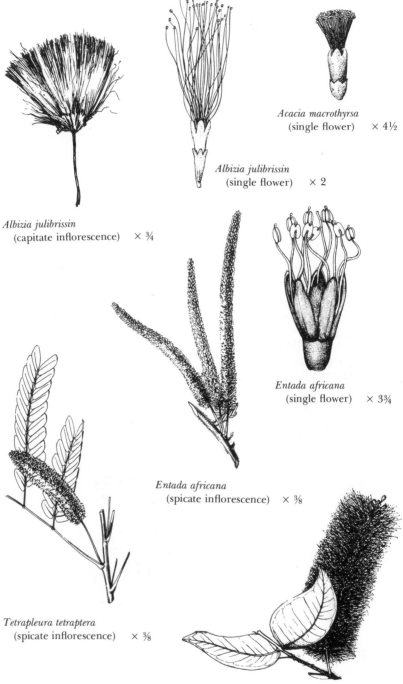

Acacia macrothyrsa
(single flower)   × 4½

Albizia julibrissin
(single flower)   × 2

Albizia julibrissin
(capitate inflorescence)   × ¾

Entada africana
(single flower)   × 3¾

Entada africana
(spicate inflorescence)   × ⅜

Tetrapleura tetraptera
(spicate inflorescence)   × ⅜

Inga laurina
(spicate inflorescence)   × ½

Leucaena leucocephala
(single flower) × 4½

Leucaena leucocephala
(globose inflorescence) × ½

Xylia ghesquierei
(globose inflorescence) × ½

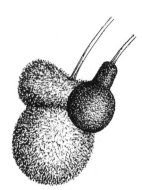

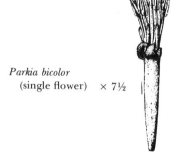

Parkia bicolor
(single flower) × 7½

Parkia bicolor
(inflorescences) × ¾

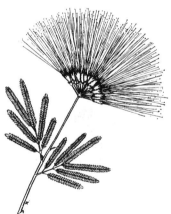

Calliandra brevicaulis
(semiglobose inflorescence) × ⅜

Albizia falcataria
(single flower) × 15

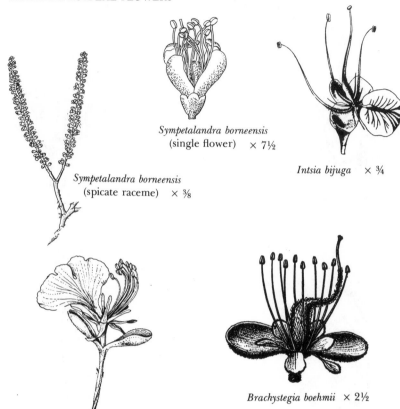

Sympetalandra borneensis
(single flower)   × 7½

Sympetalandra borneensis
(spicate raceme)   × ⅜

Intsia bijuga   × ¾

Berlinia orientalis   × ½

Brachystegia boehmii   × 2½

Bauhinia monandra   × ¾

Mezoneuron angolense   × 1½

Swartzia madagascariensis × 1½

Tamarindus indica × 1½

Cassia tora × ½

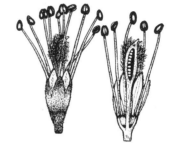

Erythrophleum guineense × 4½

Cordyla africana × ¾

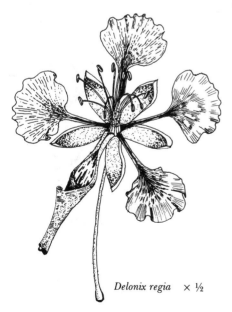

Delonix regia × ½

xliii

Trifolium pratense
(capitate inflorescence)   × ½

Clitoria ternatea   × ¾

Trifolium pratense
(single flower)   × 3

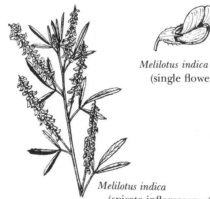

Melilotus indica
(single flower)   × 3

Melilotus indica
(spicate inflorescence)   × ½

Tephrosia noctiflora
(single flower)   × 2¼

Tephrosia noctiflora
(axillary raceme)   × ½

Mucuna macrocarpa   × ¼

xliv

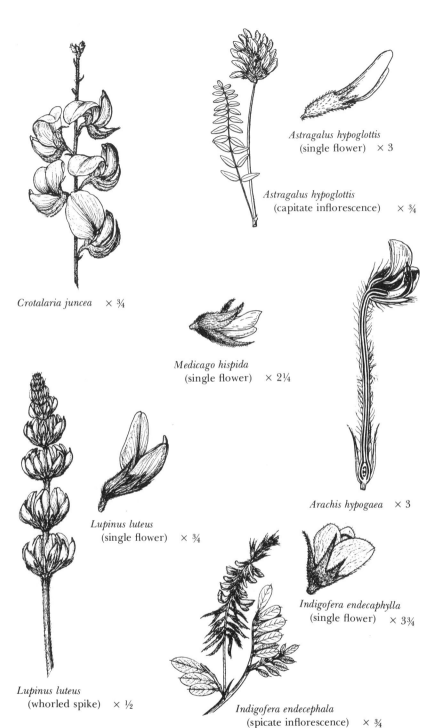

Crotalaria juncea × ¾

Astragalus hypoglottis
(single flower) × 3

Astragalus hypoglottis
(capitate inflorescence) × ¾

Medicago hispida
(single flower) × 2¼

Arachis hypogaea × 3

Lupinus luteus
(single flower) × ¾

Indigofera endecaphylla
(single flower) × 3¾

Lupinus luteus
(whorled spike) × ½

Indigofera endecephala
(spicate inflorescence) × ¾

Acacia hebeclada × ¼

Dichrostachys glomerata × ¼

Acacia nilotica × ¼

Parkia singularis × ½

Albizia julibrissin × ½

Xylia ghesquierei × ½

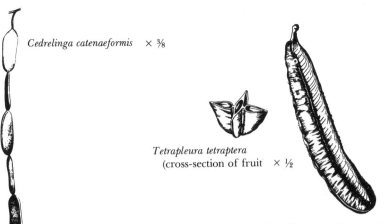

Cedrelinga catenaeformis × ³⁄₈

Tetrapleura tetraptera
(cross-section of fruit × ½

Tetrapleura tetraptera × ½

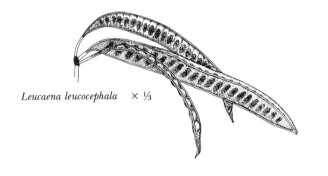

Leucaena leucocephala × ⅓

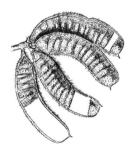

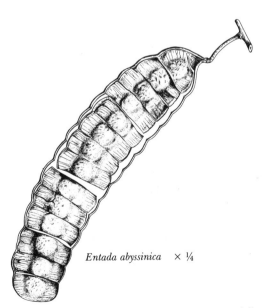

Mimosa pigra × ½

Entada abyssinica × ¼

xlvii

CAESALPINIOIDEAE PODS

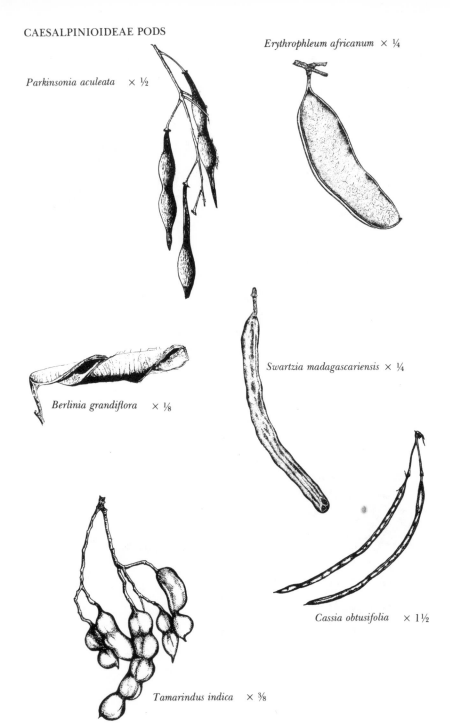

Parkinsonia aculeata × ½

Erythrophleum africanum × ¼

Berlinia grandiflora × ⅛

Swartzia madagascariensis × ¼

Cassia obtusifolia × 1½

Tamarindus indica × ⅜

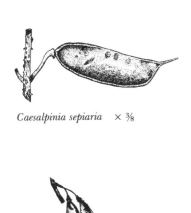

Caesalpinia sepiaria × ⅜

Afzelia quanzensis × ¼

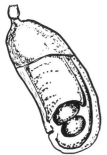

Cercis canadensis × ⅜

Sindora beccariana × ⅓

Cassia javanica × ⅓

Hymenaea courbaril × ½

Burkea africana × ⅜

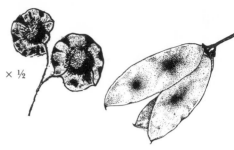

Pterocarpus vidaliana × ½

Lonchocarpus latifolius × ½

Desmodium velutinum × ¾

Lablab purpureus × 2¼

Melilotus officinalis × 2¼

Alysicarpus zeyheri × 2¼

Medicago hispida × 2¼

Pisum sativum × ¾

Ormocarpum bibracteatum × ¾

Astragalus bisulcatus × 2¼

1

Crotalaria natalita × ¾

Baptisia australis × ¾

Erythrina baumii × ¾

Medicago sativa × 2¼

Cicer arietinium × ¾

Tipuana speciosa × ½

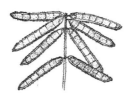

Mucuna puriens × ¾

Calopogonium mucunoides × ¾

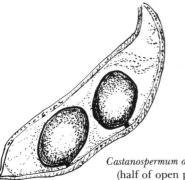

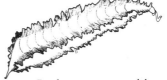

Psophocarpus tetragonolobus × ½

Castanospermum australis
(half of open pod) × ½

li

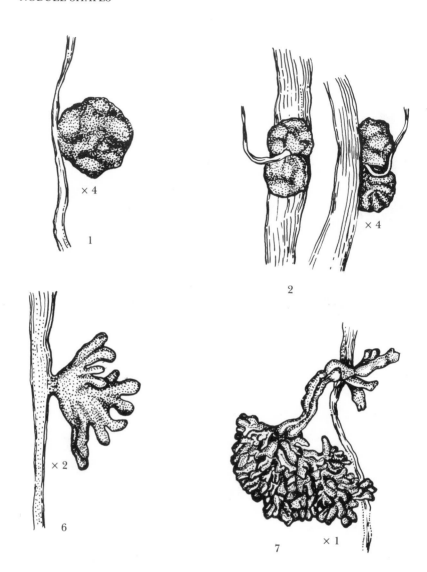

1. Globose: *Glycine, Vigna, Erythrina, Sesbania*
2. "Siamese": *Aeschynemone, Arachis*
3. Elongate: *Albizia, Tephrosia, Vicia*
4. Bifurcate: *Leucaena, Medicago*
5. Fan-shape and palmate: *Acacia, Crotalaria*
6. Coralloid: *Albizia moluccana*
7. Perennial branched clusters: *Swartzia trinitensis*
8. Perennial globose-apical: *Wisteria sinensis*, after (A) 1, (B) 2, and (C) 4 years

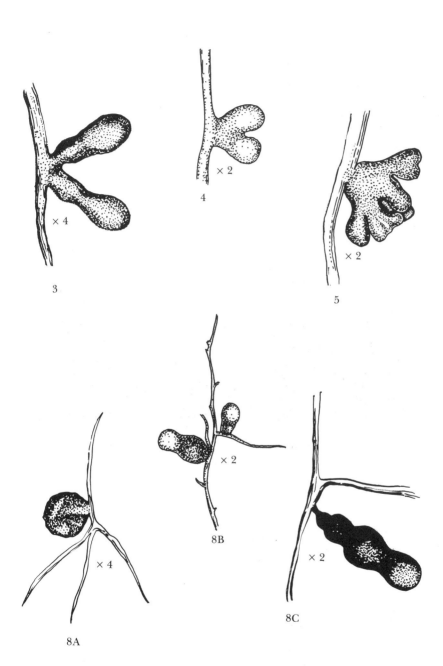

× 4

4

× 2

3

× 2

5

8A

× 4

8B

× 2

8C

× 2

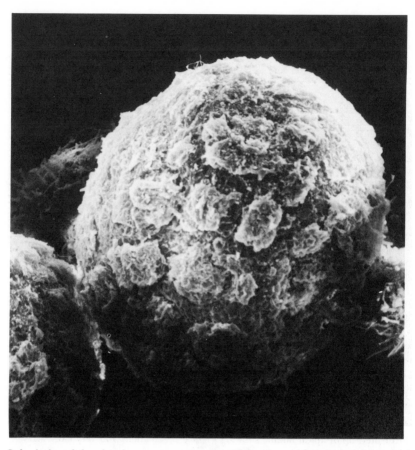

Spherical nodule, showing remnant patches of former epidermis and interstitial fissures of ruptured surface caused by internal pressures during nodule tissue proliferation. (× 63)

Nodule surfaces, showing features that enhance surface area and maximize intake of atmospheric nitrogen. These scanning electron photomicrographs were obtained from gold-palladium-coated nodule preparations viewed on a Jelco JSM–U3 scanning electron microscope at the Scanning Electron Microscope Facility of the Department of Entomology, University of Wisconsin–Madison.

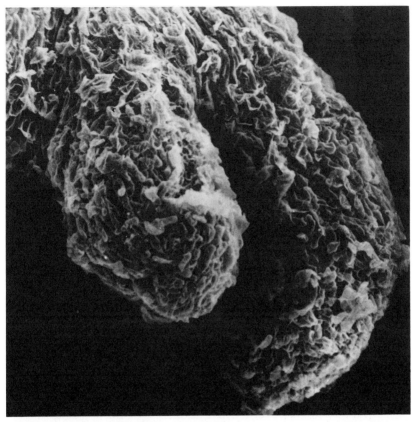

Bifurcated tip of apical nodule showing surface reticulations and very loose arrangement of outer parenchyma cells. (× 38)

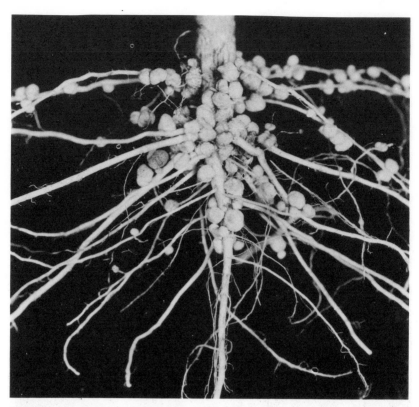

Nodulated root system: *Glycine max.*

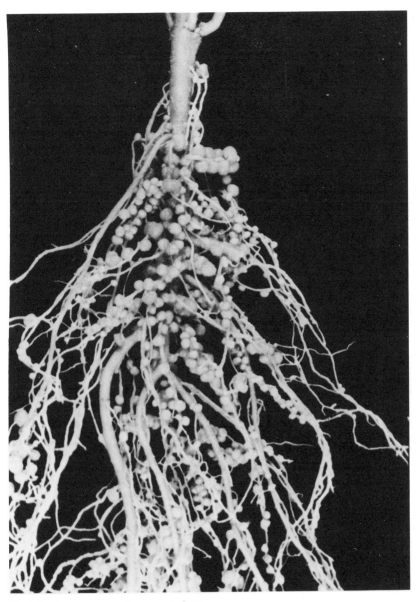

Nodulated root system: *Lotus corniculatus*.

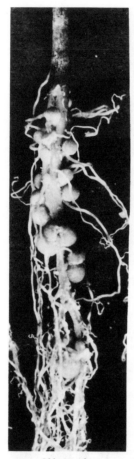

2½ months

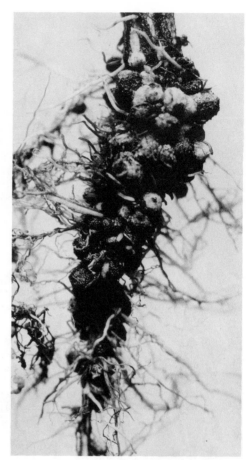

12 months

Nodulated root system: *Sesbania grandiflora.*

Nodulated root system: *Medicago* sp.

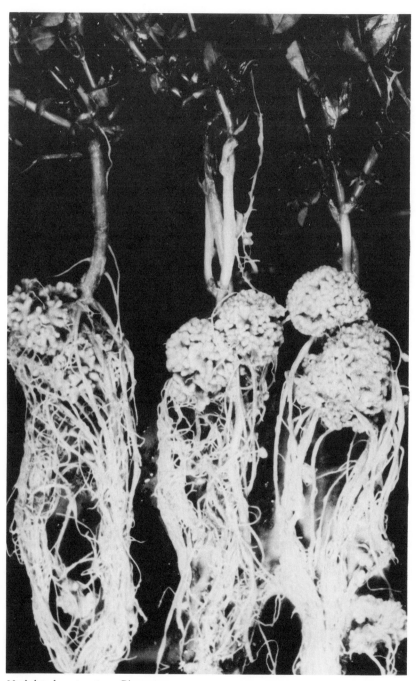

Nodulated root system: *Pisum* sp.

Nodulated root system: *Cajanus cajan*.

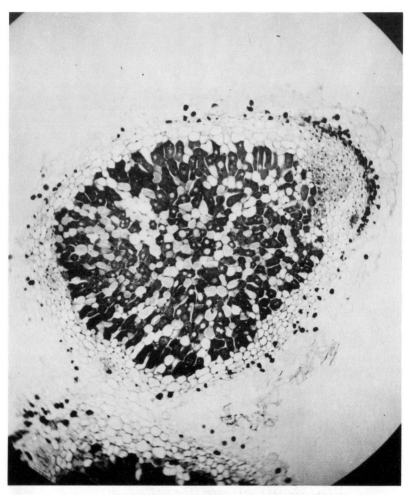

Longitudinal section of *Leucaena leucocephala* nodule: meristem apical. (× 48)

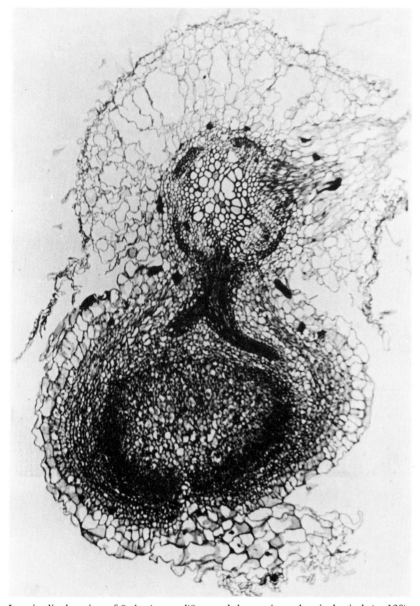

Longitudinal section of *Sesbania grandiflora* nodule: meristem hemispherical. (× 130)

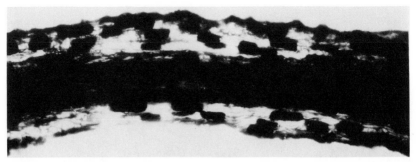

Tannin in root cortex of *Cassia tora*. (× 67)

THE LEGUMINOSAE

From Greek, *habros*, "delicate," "pretty," "soft," in reference to the beauty of the flowers and also to the softness of the leaves.

Type: *A. precatorius* L.

12–15 species. Shrubs, undershrubs, or lianas, annual or perennial, slender, dextrorsally twining, tendrils absent. Leaves paripinnate, subsessile or short-petioled; leaflets usually small, many-paired, opposite, usually with a small bristle at apex of rachis; stipels minute and slender, or reduced to inconspicuous protuberances; stipules small, linear-lanceolate, caducous. Flowers small, pink or white, subsessile or short-pedicelled, in axillary or terminal racemes or fascicled at nodes; bracteoles 2, small, caducous; calyx campanulate, subtruncate; teeth very short or obsolete, upper 2 subconnate; standard ovate, acute, short-clawed, somewhat adherent to staminal tube; wings oblong-linear, long-clawed, narrower and shorter than the broad, curved keel; stamens 9, united in a sheath split above, vexillary stamen absent; anthers uniform; ovary subsessile, few- to many-ovuled; style short, curved, glabrous; stigma terminal, capitate. Pod linear or oblong, flat, thinly subseptate, 2-valved; seeds spherical, hard, lustrous, often bright red and black. (Bentham and Hooker 1865; Amshoff 1939; Boutique 1954; Verdcourt 1970a)

The tribal positioning of this genus has been a problem for many decades. De Candolle (1825a) reviewed the possible alignment of the genus with *Vicia* on the basis of foliar characters, with *Glycine* relative to similarities in ligneous, climbing stems and pod characteristics, and with *Phaseolus* relative to mode of seed germination. In a later publication, De Candolle (1825b) placed *Abrus* in the tribe Phaseoleae. Bentham (1864) considered *Abrus* "intermediate between the tribes Vicieae, Phaseoleae, and Dalbergieae." Subsequently, Bentham and Hooker (1865) and Taubert (1894) placed the genus in the tribe Vicieae. In the ensuing years, various lines of evidence corroborated by the chromosome data of Senn (1938b) and the disk electrophoretic study of seed globulin proteins (Boulter et al. 1967) have supported the removal of *Abrus* from Vicieae to a position in, or near, the tribe Phaseoleae. In 1964, Hutchinson accommodated *Abrus* to the monogeneric tribe Abreae intermediate between Glycineae and Vicieae.

Abrus species are widespread and pantropical. They inhabit roadsides and thickets, more commonly in moist to wet than in dry areas, and occur between sea level and 300-m altitude.

The type species is commonly known as false or Indian licorice, jequirity, crab's-eye, weather vine, jumbee bean, rati, and prayer bead. Its coarse, twining stems often reach lengths of 6–8 m under favorable conditions. This objectionably aggressive growth has contributed to its disfavor as a cover crop on rubber plantations. However, its delicate foliage and attractive, small pink or white flowers borne in axillary or terminal racemes have ornamental beauty. The leaves are sensitive to light changes and droop from a horizontal to a vertical position at night and during storms. Several early publications described this species as a predictor of meteorological changes (Kew Bulletin 1890; Altamirano 1905).

3

Each pod of the type species contains 3–5 shiny scarlet seeds marked with a prominent enamellike black spot surrounding the hilum. The seeds were used by goldsmiths of East Asia in early times as standard weights, their weight being a remarkably constant 0.113 g. Jequirity seeds were used to weigh the famous Koh-i-noor diamond (Watt 1908).

The seeds are poisonous. Reputedly, half a seed chewed before swallowing is lethal to man; 60 g of seed ingested by a horse results in death within 18 hours; bovine animals, goats, and dogs are more resistant. Lethal symptoms resemble those of snake venom. Jequirity seed necklaces, brooches, and pins are commonly sold as costume or novelty jewelry. Careless handling of these items near the mouth and eyes portends potentially dangerous consequences, since release of the toxic principle is enhanced as a result of the seed coat's having been bored or pierced (Ransohoff 1955).

The occurrence of 9 distinct cell layers in the seed coat accounts for its impermeability to water and other solvents and is probably the basis for claims that seeds with an intact seed coat may be swallowed with impunity. Two pigmented materials, a scarlet one from the outermost layer of cells and a yellow one from the innermost cell layer, are released only from seeds with ruptured seed coats (Sarkar 1914). It is said that the seed may be eaten as a pulse after sufficient cooking, since the poisonous property is destroyed at 65°C. Nonetheless, seed ingestion is not recommended.

The toxic substance of the seed is abrin, a toxalbumin of 2 fractions, a globulin and an albumose, with chemical and pharmacological properties resembling those of ricin, a toxic protein of the castor bean, *Ricinus sanguineus*. Abrin, however, is a much stronger irritant to the conjunctiva; moreover, animals immunized against abrin are susceptible to ricin poisoning and vice versa (Steyn 1934). Abrin ranks among the most potent toxic principles known. A lethal dose is within the narrow range of 0.001–0.002 mg/kg of body weight (Quisumbing 1951). Lethal action is immediate when abrin is administered subcutaneously and also may result from application of an infusion of a bruised seed to the conjunctiva. Many instances of the criminal use of jequirity seeds for poisoning humans and animals in India were reported by Watt (1908).

The roots and leaves yield about 1.5 percent and 9–10 percent glycyrrhizin, $C_{42}H_{62}O_{16}$, respectively; hence the common name "false licorice."

NODULATION STUDIES

Nodules of *Abrus* species are small and spherical. Cultural and biochemical characteristics of isolates from *Abrus* nodules resemble those of the slow-growing soybean-lupine-cowpea type of rhizobia (personal observation). Results from reciprocal plant-infection tests with *A. precatorius* L. and other species have confirmed kinships within the cowpea miscellany (Allen and Allen 1936a; Wilson 1944a,b).

In general, nodulation data support botanical criteria in the removal of the genus *Abrus* from Vicieae to a position within, or near, Phaseoleae, inasmuch as (1) nodules on *Abrus* species have spherical meristems in contrast to apical meristems of nodules on plants in Vicieae; (2) cultural and biochemical characteristics of rhizobia from *Abrus* species do not conform with those of the fast-growing, acid-forming *Pisum-Lathyrus-Vicia* type rhizobia of Vicieae; (3) species of *Abrus* are pantropical, whereas those of Vicieae occur exclusively within the temperate zone; and (4) rhizobia from *Abrus* species lack ability to nodulate *Lathyrus*, *Lens*, *Pisum*, and *Vicia* species but are compatible with plant members of the so-called cowpea miscellany.

Nodulated species of *Abrus*	Area	Reported by
fruticulosus Wall. ex W. & A.	S. Africa	Grobbelaar et al. (1967)
laevigatus E. Mey.	Java	Keuchenius (1924)
precatorius L.	Widely reported	Many investigators
ssp. *africanus* Verdc.	Zimbabwe	Corby (1974)
pulchellus Wall.	Zimbabwe	Corby (1974)
schimperi Hochst. ex Bak.		
ssp. *oblongus* Verdc.	Zimbabwe	Corby (1974)

Acacia Mill.   Mimosoideae: Acacieae
11 / 3446

From Greek, *akis*, "point," "barb," in reference to the thorns; ancient Greek name for an Egyptian tree.

Type: *A. arabica* Willd.

800–900 species. Trees, medium-sized to gigantic, or shrubs, erect, often woody climbers, rarely undershrubs, deciduous, sometimes evergreen, very rarely herbs, unarmed or armed with spinescent stipules or internodal thorns. Leaves bipinnate, or in many species reduced after the seedling stage to a leaflike petiole termed a phyllodium; phyllodia of diverse shapes, often with a gland at the base of the blade; leaflets usually small, many-paired; petiole often glandular; stipules, if not spiny, small and deciduous, or absent. Flowers bisexual or polygamous, sessile or short-pedicelled, usually yellow or white, in stalked, showy, globose heads, cylindric spikes, or short racemes, the inflorescences solitary, paired, or fascicled in leaf axils or in terminal paniculate racemes; bracts often 2, scalelike; bracteoles stalked, bladelike, inconspicuous, usually subtending each flower; calyx campanulate or funnel-shaped; lobes 5, 4, rarely 3; petals as many as the divisions of the calyx, very rarely none, somewhat joined in the lower half, rarely free; stamens numerous, usually more than 20, yellow or white, exserted, free or connate only at the base; filaments filiform; anthers small, sometimes with an apical gland; ovary sessile or stalked, 2- to many-ovuled; style long, slender; stigma minute, terminal. Pod very variable, oblong or linear, straight, curved, moniliform, or variously contorted, flat, convex, or cylindric, membranous, leathery, or woody, 2-valved, dehiscent or indehiscent; seeds numerous, somewhat flattened, arranged transversely or lengthwise. (Bentham 1864; Bentham and Hooker 1865; Baker 1930; Gilbert and Boutique 1952a; Isely 1973)

Representatives of this genus are widespread; foliferous members are quite cosmopolitan in tropical and subtropical regions. Only about 7 species are native in the southern United States; phyllodineous forms are almost entirely indigenous to Australia and the Pacific islands. Habitats range from arid areas of low or seasonal rainfall to moist forests and river banks. Species are found in all soil types.

Acacia is the largest mimosoid genus and the second largest in the family, yet only about 75 species have economic value; of these, only 50 are cultivated. Many form small, spiny, slow-growing scrub that serves to prevent wind and rain erosion. *A. armata* R. Br., *A. glaucescens* Willd., *A. pycnantha* Benth., and *A. suaveolens* Willd. are favored for seashore plantings because they tolerate salt spray and control sand dunes. Other members are impressive for their enormous size; as cited by Menninger (1962), a specimen of *A. galpinii* Burtt Davy in northwestern Transvaal measured about 71 m high, was 26 m in girth at a distance of 1 m above the ground, and had a crown diameter of approximately 60 m. In the area was evidence that considerably larger trees had been destroyed by hurricanes or fires.

Some species are among the most ornamentally attractive in the leguminous family because of their fluffy balls of fragrant yellow flowers and their lacy, feathery foliage. *A. dealbata* Link, florists' mimosa, is a typical example. *A. albida* Del., *A. caffra* (Thunb.) Willd., *A. galpinii*, and *A. tortilis* (Forsk.) Hayne excel in stark, majestic beauty in their native habitats. The picturesqueness of *A. pendula* A. Cunn. in the dry areas of Queensland and New South Wales is unforgettable. *Acacia* flowers are often fragrant; among the honey- to violet-scented blossoms used in the manufacture of commercial perfumes are those of *A. cavenia* (Mol.) H. & A., and *A. farnesiana* (L.) Willd. whose flower oils provide the foundation essences of Cassie Ancienne and Cassie Romaine.

The foliage of many acacias is grazed and is economically important as cattle food; however, it is important to bear in mind that the pods and leaves of some members contain considerable amounts of cyanogenetic glucosides, which may yield free hydrocyanic acid (Rimington 1935a,b; Steyn and Rimington 1935; Webb 1948; Watt and Breyer-Brandwijk 1962) and alkaloids (Raffauf 1970; Mears and Mabry 1971) toxic to livestock.

Other members of the genus are best known for the dangerous thorns borne on trunks and branches. The tenacious thorns of *A. detinens* Burch. are in inconspicuous hooked pairs; hence the specific name meaning "to detain"; those of *A. karroo* Hayne measure 17–18 cm long. *A. tortilis*, the umbrella thorn, is armed with two types of spines, one long, straight, and white, the other small, curved, and brown. *Acacia* thickets are generally impenetrable barriers and can impale game.

Habitation of the swollen, stipular thorns of various xerophytic acacias by ant colonies is one of the best-known examples of myrmecophily. About 9 neotropical species share this obligatory relationship in combination with about 5 ant species of the genus *Pseudomyrmex*; these associations are not restricted interspecifically (Janzen 1966, 1967). In each mutualistic relationship, the enlarged stipular thorns serve as the domicile for the ants, which feed upon the enlarged foliar nectaries and modified leaflet tips, known as

Beltian bodies; in exchange, the ants, which are among the most ferocious of the ant world, protect the host acacia against phytophagous insects and other predators. Janzen (1967) postulated that should *Pseudomyrmex ferruginea*, the ant associate of bull's horn acacia, *A. cornigera* Willd. disappear abruptly from the area, only individual shoots of the acacia would survive in isolated sites and only under unusual circumstances. Obligatory myrmecophily has not been recorded amid the large assemblage of phyletically diverse Australian acacias. Absence of spinelike stipules and extrafloral nectaries has been offered as an explanation (Brown 1960).

Historians are in general agreement that the "shittah trees" of Biblical times were *Acacia* species and, accordingly, supplied the "shittim wood" used to build the Ark of the Covenant and the Altar and Table of the Tabernacle. It is most probable that *A. seyal* Del. and *A. tortilis* were the source, since these timber trees are believed to have been the principal, if not the only, sizable ones in the Arabian desert during that period and are today common in the Sinai peninsula (Moldenke and Moldenke 1952). "Seyal" in Arabic means "torrent"; thus as a specific name it alludes to the abundance of the species in ravines through which water rushed during rainy seasons. *A. arabica* Willd. may also have been an ancient source of wood.

Acacia wood is coarse grained, with a density of 640–800 kg/m$_3$, is highly durable, and responds satisfactorily to finishing treatments, but it is worked with difficulty. Wood of the small-sized, slow-growing, spiny species serves for small objects, such as smoking pipes, implement handles, and furniture. It is especially valued for fuel and charcoal, since these trees are often common in areas where the fuel supply is limited. Wood of the large trees is used for paneling, boats, and musical instruments. Hawaiian mahogany, the wood of *A. koa* Gray, a monarch of forests in the Hawaiian archipelago at one time, was prized for surfboards, canoes, and bowls. *A. glaucescens*, *A. melanoxylon* R. Br., and *A. nigrescens* Oliv. are among the major timber trees today.

Sykes et al. (1954) appraised the black wattle, *A. mearnsii* De Wild. (= *A. mollissima* Willd. = *A. decurrens* Willd.), as "the most important source of vegetable tannin material in the world." The average yield is 35–39 percent of the air-dry weight of the extracts, but the yield from the bark of good strains may be higher (Bryant 1946; Sykes et al. 1954). Within the last decade, more than 136,000 metric tons of bark and extract, equivalent to two-thirds of the world's total, were being exported yearly from South Africa (Purseglove 1968). Yields in excess of 15 percent are generally obtained from *A. albida*, *A. arabica*, *A. binervata* DC., *A. bussei* Harms, *A. catechu* Willd., *A. cyanophylla* Lindl., *A. decurrens*, *A. decurrens* var. *mollis*, *A. longifolia* Willd., *A. pycnantha*, *A. saligna* Wendl., *A. seyal*, and *A. subulata* Brenan (Greenway 1941; Bryant 1946; Sykes et al. 1954).

Acacia species are the source of gum arabic, an article of international commerce since the first century A.D. The term "arabic" had its origin with reference to the port of Aden, Arabia, the major trading outlet of early times. *A. arabica* was considered for many years the prime source of this completely water-soluble, mucilaginous gum, but more recently emphasis has shifted to about 24 other members of the genus that yield commercial grades (Mantell

1947; Howes 1949; Glicksman and Schachat 1959). Currently, *A. senegal* (L.) Willd. (= *A. verek* Guill. & Perr.), a tree 5–7 m tall, with a life-span of 25–30 years, is the principal source. This is especially true in the Sudan, where 90 percent of the marketed gum is supplied by this species and less than 10 percent comes from *A. seyal*. Today the best grade is known as Kordofan gum, so named from a province in the Sudan. El Obeid, also in the Sudan, is the most important market and auction site of gum arabic in the world. Privately and commercially owned "gum-gardens" of *A. senegal* and other species are maintained in various parts of North Africa and India.

Gum arabic normally exudes from acacia stems and branches as soft drops or tears, which later become firm and hard, or it may be artificially obtained by tapping or wounding (Mantell 1947). It is produced only by trees in an unhealthy condition; moreover, improved gum yields follow lessened vitality of the trees. It remains to be decided whether its formation is a result of a pathological condition due to an infection or the result of an unfavorable environmental condition.

The gum is marketed either as tears or as a transparent, amorphous powder varying in color from white, the highest grade, to shades of yellow to pale rose. Solubility in water, high viscosity, and reduced surface tension are among the simple and desirable properties that account for the world demand of gum arabic. Moreover, it is nontoxic, odorless, tasteless, colorless, does not crystallize, and under various conditions, it is an emulsifying agent, a protective colloid, stabilizer, adhesive or binder, sensitizer, and finishing agent. Uses of the gum are multifold in industries engaged in the production of food, pharmaceuticals, medicines, cosmetics, adhesives, paints, polishes, and inks, and in lithography, photography, textile-sizing, and other processes (Glicksman and Schachat 1959).

Nodulation Studies

Nodules on greenhouse seedlings are light colored, round, and smooth surfaced; those on mature shrubs and trees in their native habitats are multilobed, dark brown, and rugose with corky to woody surfaces (Allen and Allen 1936a). The perennial "Mimosoideae type" nodule, as described by Spratt (1919), is denoted by the presence of protective corky parenchyma, tannin inclusions, and abundant, well-developed vascular tissue. Histological studies of nodules from approximately 20 *Acacia* species showed typical morphogenesis and inner tissue maturation (Dawson 1900b; Wendel 1918; Lechtova-Trnka 1931; Rothschild 1970; personal observation).

Acacia rhizobia are monotrichously flagellated, slow growing, and produce alkalinity without serum zone formation in litmus milk. The results of Wilkins (1967) in Australia and of Habish (1970) in the Sudan support, in principle, the concept of ecological adaptation of native rhizobia strains to high temperature and arid conditions. Host-infection studies define *Acacia* species within the conglomerate cowpea miscellany (Burrill and Hansen 1917; Walker 1928; Allen and Allen 1936a, 1939; Wilson 1939a; McKnight 1949a; Lange 1961; Habish and Khairi 1968, 1970).

Nodulated species of *Acacia*	Area	Reported by
abyssinica Hochst. ex Benth.	S. Africa	N. Grobbelaar (pers. comm. 1965)
ssp. *calophylla* Brenan	Zimbabwe	Corby (1974
acinacea Lindl.	Australia	Harris (1953)
acuminata Benth.	Australia	Lange (1959, 1961)
	Ryukyus	Ogimi (1964)
adunca A. Cunn.	S. Africa	Grobbelaar & Clarke (1974)
alata R. Br.	Australia	Lange (1959)
albida Del.	Widely reported	Many investigators
anceps DC.	Australia	Harris (1953)
aneura F. Muell.	Australia	Beadle (1964)
	S. Africa	Grobbelaar et al. (1964)
arabica Willd.	Malaysia	Wright (1903)
	Hawaii, U.S.A.	Allen & Allen (1936a)
	Sudan	Habish & Khairi (1970)
arenaria Schinz	Zimbabwe	Corby (1974)
armata R. Br.	Widely reported	Many investigators
aroma Gillies ex H. & A.	Argentina	Rothschild (1970)
aspera Lindl.	S. Africa	Grobbelaar & Clarke (1974)
ataxacantha DC.	Zimbabwe	Corby (1974)
auriculaeformis A. Cunn.	Hawaii, U.S.A.	Allen & Allen (1936a)
baileyana F. Muell.	Widely reported	Many investigators
berteriana Spreng.	France	Prillieux (1879)
bidentata Benth.	Australia	Lange (1959)
biflora R. Br.	Australia	Lange (1959)
blakelyi Maid.	Australia	Lange (1959)
bonariensis Gillies ex H. & A.	Argentina	Rothschild (1970)
borleae Burtt Davy	Zimbabwe	Corby (1974)
brachybotrya Benth.	S. Africa	Grobbelaar & Clarke (1974)
brachystachya Benth.	S. Africa	Grobbelaar & Clarke (1974)
burkei Benth.	S. Africa	Grobbelaar et al. (1967)
	Zimbabwe	Corby (1974)
buxifolia A. Cunn.	Australia	Beadle (1964)
bynoeana Benth.	France	Lechtova-Trnka (1931)
	Australia	Beadle (1964)
caffra (Thunb.) Willd.	S. Africa	Grobbelaar et al. (1967)
calamifolia Sweet	Australia	Beadle (1964)
cana Maid.	Australia	Beadle (1964)
cardiophylla A. Cunn.	S. Africa	Grobbelaar & Clarke (1974)
catechu Willd.	Hawaii, U.S.A.	Allen & Allen (1936a)
cavenia (Mol.) H. & A.	Argentina	Rothschild (1970)
	S. Africa	Grobbelaar & Clarke (1974)
celastrifolia Benth.	Australia	Lange (1959)
chariessa Milne-Redhead	Zimbabwe	Corby (1974)
cognata Domin	S. Africa	Grobbelaar & Clarke (1974)

Acacia

Nodulated species of *Acacia* (cont.)	Area	Reported by
colletioides (A. Cunn.) Benth.	Australia	Lange (1959); Beadle (1964)
complanata A. Cunn. ex Benth.	Australia	Norris (1959b)
confusa Merr.	Hawaii, U.S.A.	Allen & Allen (1936a)
	Japan	Asai (1944)
	Philippines	Bañados & Fernandez (1954)
	Ryukyus	Ogimi (1964)
constricta Benth.	N.Y., U.S.A.	Wilson (1939a)
(*cornigera* Willd.)	Great Britain	Dawson (1900b)
= *spadicigera* Cham. & Schltr.		
cultriformis A. Cunn.	N.Y., U.S.A.	Wilson (1945a)
	Ryukyus	Ogimi (1964)
cunninghamii Hook.	Australia	McKnight (1949a); Norris (1959b)
cyanophylla Lindl.	Australia	Lange (1959)
	S. Africa	Roux & Warren (1963)
cyclops A. Cunn.	Australia	Lange (1959)
davyi N. E. Br.	S. Africa	Grobbelaar et al. (1967)
dealbata Link	Widely reported	Many investigators
deanei (R. T. Bak.) WCMcG.	Australia	Beadle (1964)
	S. Africa	Grobbelaar & Clarke (1974)
decora Reichb.	S. Africa	Grobbelaar & Clarke (1974)
decurrens Willd.	Malaysia	Wright (1903)
	Australia	Benjamin (1915)
	Hawaii, U.S.A.	Allen & Allen (1936a)
diptera Lindl.	Australia	Lange (1959)
discolor (Lam.) Willd.	Australia	Hannon (1949)
doratoxylon A. Cunn.	Australia	Beadle (1964)
drummondii Lindl.	Australia	Lange (1959, 1961)
ehrenbergiana Hayne	Sudan	Habish & Khairi (1970)
elata A. Cunn. ex Benth.	Australia	Hannon (1949)
eremophila Fitzg.	Australia	Lange (1959)
ericifolia Benth.	Australia	Lange (1959)
erinacea Benth.	Australia	Lange (1959)
erubescens Welw. ex Oliv.	Zimbabwe	Corby (1974)
	S. Africa	Grobbelaar & Clarke (1975)
esterhazia MacKay	U.S.D.A.	Kellerman (1910)
estrophiolata F. Muell.	Australia	Beadle (1964)
extensa Lindl.	Australia	Lange (1959)
exuvialis Verdoorn	Zimbabwe	Corby (1974)
falcata Willd.	Australia	Benjamin (1915); Norris (1956)
farnesiana (L.) Willd.	Widely reported	Many investigators
filifolia Benth.	Australia	Lange (1959)
fimbriata A. Cunn. ex G. Don	Australia	Bowen (1956)
fleckii Schinz	Zimbabwe	Corby (1974)
	S. Africa	Grobbelaar & Clarke (1974)
flexuosa H. & B.	Venezuela	Barrios & Gonzalez (1971)

10

Nodulated species of *Acacia* (cont.)	Area	Reported by
galpinii Burtt Davy	S. Africa	Grobbelaar et al. (1967)
	Zimbabwe	Corby (1974)
genistoides (F. Muell.) Benth.	Australia	Lange (1959)
georginae F. M. Bailey	S. Africa	N. Grobbelaar (pers. comm. 1963)
gerrardii Benth.	Zimbabwe	Corby (1974)
	S. Africa	Grobbelaar & Clarke (1975)
giraffae Willd.	S. Africa	Grobbelaar et al. (1964); Roux & Marais (1964)
gladiiformis A. Cunn.	Australia	Beadle (1964)
glaucescens Willd.	Hawaii, U.S.A.	Allen & Allen (1936a)
glaucoptera Benth.	Australia	Lange (1959)
goetzei Harms		
ssp. *goetzei*	Zimbabwe	Corby (1974)
ssp. *microphylla*	Zimbabwe	Corby (1974)
grandicornuta Gerstner	Zimbabwe	Corby (1974)
	S. Africa	Grobbelaar & Clarke (1975)
granitica Maid.	Australia	McKnight (1949a)
hakeoides A. Cunn.	Australia	Beadle (1964)
harpophylla F. Muell. ex Benth.	Australia	Norris (1959b)
harveyi Benth.	Australia	Lange (1959)
hastulata Sm.	Australia	Lange (1959)
hebeclada DC.	Zimbabwe	Corby (1974)
ssp. *hebeclada*	S. Africa	Grobbelaar & Clarke (1972)
hereroensis Engl.	Zimbabwe	Corby (1974)
heterophylla Willd.	Germany	Morck (1891)
	Great Britain	Dawson (1900b)
(*hispidissima* DC.)	Germany	Lachmann (1891)
=*pulchella* R. Br.		
homalophylla A. Cunn.	Australia	Beadle (1964)
horrida (L.) Willd.	Hawaii, U.S.A.	Allen & Allen (1936a)
horridula Meissn.	Australia	Lange (1959, 1961)
huegelii Benth.	Australia	Lange (1959)
intsia Willd.	W. India	V. Schinde (pers. comm. 1977)
jonesii Muell. & Maid.	S. Africa	Grobbelaar & Clarke (1974)
juniperina Willd.	Australia	Benjamin (1915); Norris (1956, 1958a)
karroo Hayne	S. Africa	Mostert (1955)
	Zimbabwe	Corby (1974)
kauaiensis Hillebr.	Hawaii, U.S.A.	Allen & Allen (1936a)
kempeana F. Muell.	S. Africa	Grobbelaar & Clarke (1974)
kirkii Oliv.	Zimbabwe	Corby (1974)
ssp. *kirkii*	Zimbabwe	W. P. L. Sandmann (pers. comm. 1970)

Nodulated species of *Acacia* (cont.)	Area	Reported by
koa Gray	Hawaii, U.S.A.	Allen & Allen (1936a)
var. *hawaiiensis* Rock	Hawaii, U.S.A.	Allen & Allen (1936a)
var. *lanaiensis* Rock	Hawaii, U.S.A.	Allen & Allen (1936a)
koaia Hillebr.	Hawaii, U.S.A.	Allen & Allen (1936a)
kraussiana Meissn. ex Benth.	S. Africa	Grobbelaar & Clarke (1975)
latifolia Benth.	Widely reported	Many investigators
leptoneura Benth.	S. Africa	Grobbelaar & Clarke (1974)
leucophloea (Roxb.) Willd.	W. India	V. Schinde (pers. comm. 1977)
ligulata A. Cunn.	Australia	Harris (1953); Beadle (1964)
linearis Sims	Australia	Benjamin (1915)
lineata A. Cunn.	France	Lechtova-Trnka (1931)
linifolia Willd.	Ill., U.S.A.	Burrill & Hansen (1917)
	Australia	Hannon (1949); Norris (1956)
longifolia Willd.	Ill., U.S.A.	Burrill & Hansen (1917)
	Ryukyus	Ogimi (1964)
	Argentina	Rothschild (1970)
var. *floribunda* F. Muell.	Ill., U.S.A.	Burrill & Hansen (1917)
	Fla., U.S.A.	Carroll (1934a)
var. *sophorae* F. Muell.	Australia	Harris (1953)
luederitzii Engl.		
var. *luederitzii*	Zimbabwe	Corby (1974)
var. *retinens*	S. Africa	Grobbelaar & Clarke (1975)
lunata Sieb.	France	Clos (1896)
	Australia	Benjamin (1915)
macrantha H. & B.	Hawaii, U.S.A.	Allen & Allen (1936a)
macrothyrsa Harms	Zimbabwe	Corby (1974)
mearnsii De Wild.	Zimbabwe	Corby (1974)
melanoxylon R. Br.	Widely reported	Many investigators
mellei Verdoorn	S. Africa	Grobbelaar & Clarke (1975)
mellifera Benth.	Sudan	Habish & Khairi (1968, 1970)
ssp. *detinens* (Burch.) Brenan	S. Africa	Grobbelaar et al. (1967)
	Zimbabwe	Corby (1974)
microbotrya Benth.	Australia	Lange (1959)
	S. Africa	N. Grobbelaar (pers. comm. 1966)
mollissima Willd.	Australia	Ewart & Thomson (1912)
	S. Africa	Byl (1914)
	Ryukyus	Ogimi (1964)
mooreana Fitzg.	Australia	Lange (1959)
myrtifolia Willd.	Australia	Harris (1953); Lange (1961)

Nodulated species of *Acacia* (cont.)	Area	Reported by
nebrownii Burtt Davy	S. Africa	Grobbelaar & Clarke (1972)
	Zimbabwe	Corby (1974)
neriifolia A. Cunn. ex Benth.	S. Africa	Grobbelaar & Clarke (1974)
nervosa DC.	Australia	Lange (1959)
nigrescens Oliv.	Zimbabwe	Corby (1974)
nigricans R. Br.	Australia	Lange (1959)
nilotica (L.) Willd. ex Del.		
ssp. *kraussiana* (Benth.) Brenan	S. Africa	Grobbelaar et al. (1967)
	Zimbabwe	Corby (1974)
obliqua A. Cunn. ex Benth.	S. Africa	N. Grobbelaar (pers. comm. 1963)
obscura DC.	Australia	Lange (1959)
orfoto Schweinf.	Sudan	Habish & Khairi (1970)
oswaldii F. Muell.	Australia	Beadle (1964)
oxycedrus Sieb.	Australia	Hannon (1949); Norris (1956)
parramattensis Tindale	Australia	R. W. McLeod (pers. comm. 1962)
pennata (L.) Willd.	W. India	V. Schinde (pers. comm. 1977)
pentadenia Lindl.	Australia	Lange (1959)
permixta Burtt Davy	Zimbabwe	Corby (1974)
	S. Africa	Grobbelaar & Clarke (1975
podalyriaefolia A. Cunn.	Hawaii, U.S.A.	Pers. observ. (1940)
	Australia	McKnight (1949a)
	Ryukyus	Ogimi (1964)
polyacantha Willd.	Sudan	Habish & Khairi (1970)
ssp. *campylacantha*		
(Hochst. ex A. Rich.) Brenan	S. Africa	Grobbelaar & Clarke (1972)
	Zimbabwe	Corby (1974)
pravissima F. Muell.	Ryukyus	Ogimi (1964)
prominens A. Cunn. ex G. Don	Ryukyus	Ogimi (1964)
pubescens (Vent.) R. Br.	Australia	Bowen (1956)
pulchella R. Br.	Australia	Lange (1959)
pumila Maid.	Australia	Benjamin (1915)
pycnantha Benth.	France	Clos (1896)
	Australia	Harris (1953)
	Ryukyus	Ogimi (1964)
reficiens Wawra		
ssp. *reficiens*	S. Africa	Grobbelaar & Clarke (1972)
rehmanniana Schinz	Zimbabwe	Corby (1974)
	S. Africa	Grobbelaar & Clarke (1975)
restiacea Benth.	Australia	Lange (1959)
rhetinodes Schlechtd.	Australia	Harris (1953)
richii Gray	Hawaii, U.S.A.	Allen & Allen (1936a)
rigens A. Cunn.	Australia	Beadle (1964)

Nodulated species of *Acacia* (cont.)	Area	Reported by
robusta Burch.	Hawaii, U.S.A.	Allen & Allen (1936a)
	S. Africa	Roux & Marais (1964)
ssp. *clavigera* (E. Mey.) Brenan	Zimbabwe	Corby (1974)
ssp. *robusta*	Zimbabwe	Corby (1974)
rostellifera Benth.	Australia	Lange (1959)
rubida A. Cunn.	Australia	Beadle (1964)
	S. Africa	N. Grobbelaar (pers. comm. 1965)
salicina Lindl.	Australia	Harris (1953); Beadle (1964)
saligna Wendl.		
(= *Mimosa saligna* (Labill.)	France	Laurent (1891)
scorpoides A. Chev.	Hawaii, U.S.A.	Allen & Allen (1936a)
semperflora [?]	Ill., U.S.A.	Burrill & Hansen (1917)
senegal (L.) Willd.	Sudan	Habish & Khairi (1968)
var. *leiorhachis* Brenan	Zimbabwe	Corby (1974)
var. *rostrata* Brenan	Zimbabwe	Corby (1974)
	S. Africa	Grobbelaar & Clarke (1975)
seyal Del.	Sudan	Habish & Khairi (1968)
sieberana DC.	Ryukyus	Ogimi (1964)
	S. Africa	Roux & Marais (1964)
	Sudan	Habish & Khairi (1970)
var. *vermoesenii* (De Wild.) Keay & Brenan	Zimbabwe	Corby (1974)
var. *woodii* (Burtt Davy) Keay & Brenan	S. Africa	Grobbelaar et al. (1967)
	Zimbabwe	Corby (1974)
(*sophorae* R. Br.) = *longifolia* Willd. var. *sophorae* F. Muell.	Australia	Harris (1953)
spadicigera Cham. & Schltr.	Hawaii, U.S.A.	Allen & Allen (1936a)
spathulata (F. Muell.) Benth.	Australia	Lange (1959)
spinescens Benth.	Australia	Harris (1953)
squamata Lindl.	Australia	Lange (1959)
stenoptera Benth.	Australia	Lange (1959)
stricta Willd.	Germany	Lachmann (1891)
strigosa Link	Australia	Lange (1959)
stuhlmanii Taub.	S. Africa	Grobbelaar & Clarke (1972)
	Zimbabwe	Corby (1974)
suaveolens Willd.	Australia	Benjamin (1915); Norris (1959b)
subcaerulea Lindl.	Australia	Lange (1959)
sulcata R. Br.	Australia	Lange (1959)
swazica Burtt Davy	S. Africa	Grobbelaar & Clarke (1975)
tamminensis Pritz.	Australia	Lange (1959)

Nodulated species of *Acacia* (cont.)	Area	Reported by
tennispina Verdoorn	S. Africa	Grobbelaar & Clarke (1975)
tetragonocarpa Meissn.	Australia	Lange (1959)
tetragonophylla F. Muell.	Australia	Lange (1959); Beadle (1964)
tortilis (Forsk.) Hayne		
ssp. *heteracantha* (Burch.) Brenan	S. Africa	Grobbelaar et al. (1967)
	Zimbabwe	Corby (1974)
ssp. *spirocarpa* (Hochst. ex A. Rich.) Brenan	Zimbabwe	Corby (1974)
triptera Benth.	Australia	Beadle (1964)
tucumanensis Griseb.	Argentina	Rothschild (1970)
uncifera Benth.	S. Africa	Grobbelaar & Clarke (1974)
urophylla Benth.	Australia	Lange (1959)
verticillata Willd.	Hawaii, U.S.A.	Allen & Allen (1936a)
	Australia	Harris (1953)
	S. Africa	Grobbelaar & Clarke (1974)
victoriae Benth.	Australia	Beadle (1964)
villosa Willd.		
var. *glabra*	Java	Keuchenius (1924)
visco Lorentz ex Griseb.	Argentina	Rothschild (1970)
visite Griseb.	S. Africa	Grobbelaar et al. (1964)
volubilis F. Muell.	Australia	Lange (1959)
welwitschii Oliv.		
ssp. *delagoensis* (Harms) Ross & Brenan	S. Africa	Grobbelaar & Clarke (1972)
	Zimbabwe	Corby (1974)
xanthophloea Benth.	S. Africa	E. R. Roux (pers. comm. 1964)
	Zimbabwe	Corby (1974)

Examinations of the following species to date have not revealed nodules:

Nonnodulated species of *Acacia*	Area	Reported by
adenocalyx Brenan & Exell	Zimbabwe	Corby (1974)
cambagei R. T. Bak.	Australia	Beadle (1964)
excelsa Benth.	Australia	Beadle (1964)
glomerosa Benth.	Venezuela	Barrios & Gonzalez (1971)
greggii Gray	Ariz., U.S.A.	Martin (1948)
loderi Maid.	Australia	Beadle (1964)
pentagona (Schum.) Hook. f.	Zimbabwe	Corby (1974)
schweinfurthii Brenan & Exell		
var. *schweinfurthii*	Zimbabwe	Corby (1974)
silvicola Gilb. & Bout.	Zaire	Bonnier (1957)
stenophylla A. Cunn.	France	Naudin (1897b)
	Australia	Beadle (1964)
suffrutescens Rose	Ariz., U.S.A.	Martin (1948)

Acrocarpus Wight ex Arn.
91 / 3548

Caesalpinioideae:
Eucaesalpinieae

From Greek, *akros*, "apex," "at the tip," and *karpos*, "fruit," in reference to the terminal position of the pods on the branches.

Type: *A. fraxinifolius* W. & A.

3 species. Trees, 20–30 m tall, erect, often buttressed, unarmed, deciduous, leafless during anthesis. Leaves large, bipinnate; leaflets ovate, acuminate; stipules small, triangular, caducous. Flowers large, scarlet, opening before the leaves, in dense, simple or branched, spikelike racemes in defoliated axils, or 2–3 at apices of branches; bracts small, caducous; calyx tubular-campanulate; lobes 5, subequal, short, lanceolate, deciduous; petals 5, narrow, imbricate in bud, subequal; stamens 5, free, exserted, alternate with the petals; filaments long; anthers oblong-linear, versatile; ovary stalked, many-ovuled; style very short, inflexed; stigma terminal, small. Pod long-stalked, flat, linear-lanceolate, compressed, narrow-winged along the ventral suture; seeds 2–many, obovate, compressed, small, flat, brown, narrowed at base. (Bentham and Hooker 1865)

Members of this genus occur throughout tropical Indo-Malaya and the Sikkim Himalayas, in rain forests.

The type species, commonly known as pink cedar or shingle tree, is often referred to as the largest tree in the rain forests of India, where it reaches 60 m or more in height, is clear of branches up to 50 m, and has a girth of about 9 m. In drier areas, it is a medium-sized tree that branches near the base. The tree can be easily grown from the tiny seeds, which average 57/g and germinate readily. The gorgeous scarlet red flowers are borne in dense, spikelike clusters that bestow a flamelike beauty on the tree before the foliage is renewed. Young leaves are bright red for a short time (Menninger 1962). *A. fraxinifolius* W. & A. has been planted as shade for tea in Java and for coffee in India, but its slow growth is a disadvantage. The hard, durable wood is esteemed in western Sumatra for construction purposes. Use of the wood by the Javanese is generally limited to tea boxes and furniture (Burkill 1966).

Adenanthera L.
22 / 3459

Mimosoideae: Adenanthereae

From Greek, *aden*, "gland," and *anthera*, "anther"; the anthers are tipped with a deciduous gland.

Type: *A. pavonina* L.

12 species. Trees, erect, tall, often to about 25 m, unarmed, deciduous. Leaves bipinnate; leaflets small, elliptic-oblong, alternate, many-paired; stipules small. Flowers yellow becoming orange, or white, hermaphroditic or polygamous, usually 5-merous, mostly fragrant, small, short-pedicelled, in slender, dense axillary, spikelike racemes, or paniculate at apices of shoots; calyx small, campanulate, short-toothed; petals valvate in bud, equal, joined

below the middle, or free; stamens 10, free, short-exserted; anthers with a deciduous, apical gland; ovary sessile, many-ovuled; style filiform; stigma small, terminal. Pod narrow, linear or falcate, compressed or turgid about the seeds, 2-valved, valves entire, somewhat convex, leathery, curved or much twisted at dehiscence; seeds vivid scarlet, or red and black, many, hard, thick, lens-shaped, sometimes separated by semicomplete partitions continuous with the endocarp, usually surrounded by a thin pulp. (Bentham and Hooker 1865)

Representatives of this genus are common in moderately dry areas of the Old World and the New World, Australia, and the South Pacific islands.

A. pavonina L. is commonly known as the bead tree throughout tropical areas because its attractive seeds are widely used for necklaces and other decorative ornaments. The narrow range in weight of the seeds is 0.26–0.32 g; in earlier times they were known as Circassian seeds of commerce, used throughout Asia in weighing diamonds, gold, and silver (Burkill 1966).

In many areas throughout the South Pacific *A. pavonina* is common along roadsides, but in others it is a cultivated ornamental. The trees reach a height of about 20 m, with slightly buttressed trunks. They have been used extensively as shade for coffee, clove, and rubber trees in Malaysia and Indonesia. All plant parts have served various uses in native medicine (Quisumbing 1951; Watt and Breyer-Brandwijk 1962; Burkill 1966). Native use of plant parts as a soap substitute is based on the presence of a considerable amount of lignoceric acid, a 24–carbon straight-chain saturated fatty acid, in the seed.

The sapwood is light brown and the heartwood reddish; hence the name "red sandalwood" in many tropical areas. The wood is strong, hard, and durable, and its use is reserved mostly for the making of cabinets and fine furniture. The red dye and gum exuded from the wood are of no commercial importance. The wood of *A. intermedia* Merr., known as tanglin or ipil-tanglin in the Philippines, resembles that of ipil, an *Intsia* species. This magnificent tree yields wood with a fine, glossy texture and a straight grain, which is resistant to the attack of termites and beetles. The heartwood of young trees is a bright yellow that becomes reddish brown in large, old trees and darkens to a rich chocolate color on exposure. Use of this high-quality wood is limited to flooring, house furnishing, wheels, and paving blocks.

NODULATION STUDIES

Nodules were not found on greenhouse plants of *A. pavonina* examined by Lechtova-Trnka (1931) at the Muséum d'Histoire Naturelle, Paris, nor by the authors (1961) on hundreds of seedlings grown in diverse soil types in Hawaii, with and without rhizobial inocula mixtures, or on tree specimens in the Philippines (personal observation 1962). Similarly, nodules were absent on *A. intermedia* examined by Bañados and Fernandez (1954) and by the authors (1962) in the Los Baños area, Laguna, Philippines, and on *A. bicolor* Moon, an indigenous tree in Singapore (Lim 1977). However, Lim and Ng (1977) reported the presence of sparse brown nodules on *A. pavonina* in Singapore.

Adenocarpus DC.   Papilionoideae: Genisteae, Spartiinae
219 / 3680

From Greek, *aden*, "gland," and *karpos*, "fruit," in reference to the sticky glandular pods.

Lectotype: *A. complicatus* (L.) Gay

15–20 species. Shrubs, unarmed; branches alternate, rigid, silky-pubescent. Leaves palmately 3-foliolate; leaflets oblong-lanceolate, acute; stipules small, caducous, or none. Flowers orange yellow, in dense, terminal racemes or clusters; bracts and bracteoles small, caducous, or foliaceous, persistent; calyx tubular, pubescent, prominently bilabiate, usually with 2 upper lobes free, lower 3 joined, dentate, glandular-punctate; standard orbicular; wings obovate-oblong; keel incurved, subequal to standard, short-beaked; stamens all joined into a tube; anthers dimorphic, alternately long and short; ovary sessile, many-ovuled; style upcurved; stigma capitate. Pod linear, compressed, viscous with glandular tubercles, 2-valved, dehiscent, many-seeded. (Bentham and Hooker 1865)

Hutchinson (1964) transposed this genus to Laburneae trib. nov., but Polhill (1976) retained it in Genisteae *sensu stricto*.

Members of this genus are native to the Canary Islands, the Mediterranean area, and the mountainous regions of Spain and tropical Africa. They thrive in light forests and scrub prairies.

The leaves and, to a lesser extent, the stems and seeds of 8 species examined yielded the following alkaloids: adenocarpine, $C_{19}H_{24}N_2O$, decorticasine, $C_7H_{12}N_2O$, and sparteine, $C_{15}H_{26}N_2$ (see Harborne et al. 1971).

NODULATION STUDIES

Information about *Adenocarpus* nodules is nil, except for Clos's statement (1896) that nodules of *A. intermedius* DC. are large, and Dawson's comment (1900b) that infection threads were not observed in histological sections of *A. decorticans* Boiss.

Nodulated species of *Adenocarpus*	Area	Reported by
complicatus (L.) Gay	S. Africa	Grobbelaar & Clarke (1974)
(= *intermedius* DC.)	France	Clos (1896)
decorticans Boiss.	Great Britain	Dawson (1900b)
foliolosus DC.	S. Africa	N. Grobbelaar (pers. comm. 1966)

Adenodolichos Harms   Papilionoideae: Phaseoleae,
(425) / 3911   Phaseolinae

From Greek, *aden*, "gland," thus a glandular or sticky plant resembling *Dolichos* species.

Lectotype: *A. rhomboideus* (O. Hoffm.) Harms

15 species. Herbs, subshrubs, or shrubs, erect or twining. Leaves alternate or opposite or subwhorled, pinnately 3-(5-) foliolate; leaflets glandular

beneath; stipels usually present; stipules ovate-acuminate, somewhat persistent. Flowers purple, blue, red, or white, pedicelled, in few-flowered, axillary racemes or panicles; rachis swollen at the base of pedicels; bracts present; bracteoles 2; calyx cupular; tube shorter than the lobes; lobes 5, upper 2 united to the middle or higher, oblong-lanceolate, subequal, lower lip 3-lobed, longer; petals clawed; standard obovate-orbicular, with 2 callosities at the base, longer than the wings and the upturned, boat-shaped, apically truncate keel; stamens 10, diadelphous; 5 anthers dorsifixed, 5 basifixed; ovary short-stalked, 2-ovuled; style bent inward at right angle, laterally compressed, villous on the inside above; stigma terminal, oblique. Pod oblique-oblong, glandular, dehiscent, 1–2-seeded; seeds subglobose. (Harms 1908a; Wilczek 1954)

Representatives of this genus are distributed throughout tropical eastern and central Africa. They flourish in cleared forests and wooded savannas up to 1,000-m altitude.

Adenodolichos species are usually among the first plants to appear after the burning of bush areas in many parts of tropical Africa. The pods of *A. punctatus* (P. Micheli) Harms are eaten as a vegetable; the roots have medicinal properties. The coarse stems of several species are used as roofing ties in hut construction.

NODULATION STUDIES

Corby (1974) observed nodules on *A. punctatus* in Zimbabwe.

Adesmia DC.   Papilionoideae: Hedysareae, Patagoniinae
(330) / 3800

From Greek, *a*-, "without," and *desme*, "bundle," in reference to the free stamens.

Type: *A. muricata* (Jacq.) DC.

230–250 species. Herbs, annual or perennial, erect or decumbent, shrublets or shrubs, 3–4 m tall, unarmed or with glandular-haired or spiny leafstalks. Leaves paripinnate or imparipinnate; leaflets numerous, entire or dentate, ovate, hairy. Flowers usually yellow, sometimes streaked with red, in axillary or terminal racemes; bracts small; bracteoles absent; calyx campanulate; lobes 5, subequal, or lowest lobe a little longer; standard orbicular, longer than the keel; wings oblique-oblong; keel obtuse, acute, or beaked; stamens 10, all filaments free, or 2 dilated at base and adnate to the claw of the standard; anthers uniform; ovary sessile, 2- or more-ovuled; style filiform; stigma small, terminal. Pod linear, flattened, usually straight along upper suture, wavy along lower suture, segmented, segments flat or convex, breaking away entirely, or only from the upper margin, often covered with bristly hairs, 2-valved or indehiscent; seeds orbicular. (De Candolle 1825c; Bentham and Hooker 1865; Burkart 1952, 1964a, 1967)

19

Burkart (1967) accepted in principle Hutchinson's positioning (1964) of this genus in Adesmieae trib. nov. but preferred to regard its position as more closely aligned to the tribe Sophoreae than to Hedysareae.

Representatives of this large genus are found almost exclusively in eastern Argentina, Bolivia, Peru, and in the dry chaparral of central Chile. Some species are drought-resistant, others thrive in areas of 500-cm annual rainfall; a few species occur up to 3,000-m altitude.

Adesmia is presumably the largest genus of Argentinian leguminous plants. All members serve as good ground cover and have potential in soil erosion control. *A. bicolor* (Poir.) DC., *A. latifolia* (Spreng.) Vog., and *A. muricata* DC. show good vitality after continuous grazing and are useful in pastures, but in general, although *Adesmia* foliage has a high protein content, it is not relished by cattle.

NODULATION STUDIES

Nodules of *Adesmia* species, examined histologically in detail by Rothschild (1967a), were spherical and occurred singly, or in pairs, in the axils of lateral rootlets. Rothschild called the paired nodules "siamese" because both shared a common main vascular connection of the root lateral.

Nodulated species of *Adesmia*	Area	Reported by
bicolor (Poir.) DC.	Argentina	Burkart (1952); Rothschild (1967a, 1970)
capitellata (Clos) Hauman	Argentina	Rothschild (1967a, 1970)
glomerula Clos var. *glomerula*	Argentina	Burkart (1964a)
latifolia (Spreng.) Vog.	Argentina	Rothschild (1967a, 1970)
muricata DC.	S. Africa	Grobbelaar et al. (1967)
punctata (Poir.) DC.	Argentina	Rothschild (1970)
retusa Griseb.	Argentina	Burkart (1964a)

Aenictophyton A. Lee   Papilionoideae: Bossiaeae

From Greek, *ainiktos*, "baffling," and *phyton*, "plant," in reference to difficulties in classification of the genus.

Type: *A. reconditum* A. Lee

1 species. Shrub, small, perennial. Leaves sometimes absent, or few, 1-foliolate, on a short petiole with 2 awl-shaped stipules; leaflets linear to narrow-elliptic. Flowers yellowish with brown markings, short-pedicelled, in terminal racemes; bracts and bracteoles present; calyx funnel-shaped, sparsely pubescent; teeth 5, subequal, upper 2 joined for greater part of their length than the lower ones; petals long-clawed; standard broad-orbicular;

wings and keel obtuse; stamens monadelphous, all joined into a sheath open adaxially; filaments alternately longer and shorter; anthers large, equal, sub-circular, alternately dorsifixed and basifixed; ovary short-stalked, glabrous, 6–8-ovuled; style short, bent; stigma terminal. Pod flattened, oblong, tapered at base and apex; seeds with small, white, hooded aril. (Lee 1973, Polhill 1976)

The plant has characters most closely aligned with the tribe Genisteae, subtribe Bossiaeinae, except for its terminal racemose inflorescence, a charac-ter of the tribe Galegeae. Both Lee (1973) and Polhill (1976) positioned this new genus in the tribe Bossiaeae.

This monotypic genus is endemic in Australia, in desert zones of the Northern Territory and Western Australia. It is found in deep, red, sandy soil.

The foliage is grazed by cattle.

Aeschynomene L.   Papilionoideae: Hedysareae,
324 / 3793   Aeschynomeninae

From Greek, *aischynomenos*, "modest," "ashamed," in reference to the sensi-tiveness of the leaves to touch and temperature; the species are sometimes called American sensitive plants.

Type: *A. aspera* L.

150–250 species. Herbs, erect or procumbent, undershrubs or shrubs, very rarely twining, or small to medium trees up to 10 m tall. Leaves paripin-nate or imparipinnate, alternate or subfascicled, short-petioled; leaflets small, 2–80, entire or serrate, sometimes alternate or subopposite on the same rachis, generally asymmetric at the base, subsessile, linear, usually minutely glandular-punctate and sensitive; rachis often ending in a short point; stipels none; stipules paired, some appendaged at the base, membranous, foliaceous, or leathery, persistent or caducous. Flowers orange to yellow, often suffused with purple or red, small, conspicuous, handsome, solitary or in short, simple or branched, usually lax, axillary racemes or panicles, rarely umbellate, sel-dom terminal; bracts small, caducous or persistent, often stipulelike; brac-teoles paired, appressed to the calyx, usually caducous; calyx short, campanu-late; lobes 5, subequal or bilabiate, upper lip entire or bilobed, lower lip entire or trilobed; petals commonly short-clawed; standard diversely shaped, usually orbicular, emarginate; wings obliquely obovate or oblong, unilaterally ap-pendaged at the base, subequal to the standard, free or slightly adherent to each other; keel petals falcate, incurved, or deeply bent, joined; stamens 10, unequal, joined two-thirds or one-half of their length into a tube which splits longitudinally along 1 or both sides at anthesis, usually forming 2 groups of 5 stamens each; anthers similar, dorsifixed or subbasifixed; ovary linear, stalked, stalk inserted in the center of the receptacle, 2- to many-ovuled; disk short, annular; style incurved, not bearded, usually glabrous and persistent; stigma small, terminal. Pod subsessile or long-stalked, linear or elliptic, nar-

21

row, compressed, rarely torulose, 1–18-jointed, deeply indented along the lower margin only, indehiscent or rarely dehiscent by the lower suture, breaking transversely into 1-seeded segments, each segment flat or slightly convex, rectangular or semiorbicular, rarely elliptic; seeds light brown to black, smooth, small, kidney-shaped. (Bentham and Hooker 1865; Amshoff 1939; Léonard 1954a, b; Rudd 1955, 1959)

Species are chiefly tropical, numerous in the Americas, less common in Africa, Asia, Australia, and Pacific areas. Approximately one-half of the species are hydrophytes, occurring in rice paddies, flooded areas, and marshes, and along lake and river margins. They often dominate as pioneer plants in tropical freshwater, swamp-forest, and waterlogged sites and sometimes form floating islands with grasses and herbs in association. Other species occur in dry waste areas, savannas, on rocky hillsides, and on sandy beaches.

Only a few species are cultivated as ornamentals. *A. elaphroxylon* (Guill. & Perr.) Taub. is planted along watersides in South America and Malaysia (Rudd 1955, 1959) for the beauty of its conspicuous orange yellow flower racemes. However, this species, called the ambatch or pith tree, is more renowned for its economic and ecological value. A profuse formation of adventitious roots usually occurs on the lower portion of the bole above the water level and also on the prostrate stems and submerged parts. In native areas, this massive root system becomes a barrier to floating vegetative flotsam and results in an accumulation of sudd which settles and fills inundated areas.

A. aspera L., in tropical Asia, and *A. elaphroxylon*, in tropical Africa, are the two most important sources of ambatch wood, a whitish, soft, spongy pith used in the manufacture of helmets, art paper products, and artificial flowers. Ambatch wood has a density of 110–190 kg/m³; thus it is one of the lightest woods known. The wood of *A. hispida* Willd., a Cuban species, is said to have a density of only 40 kg/m³, or about a third that of balsa wood, *Ochroma lagopus*, or about one twenty-fifth that of water. It serves native uses for floats, fish nets and lines, boat poles, sandals, sun and pith helmets, and rafts and canoes.

A. americana L. and *A. indica* L. merit mention as good forage and green-manuring plants. The former species has had extensive usage as soil cover on teak estates in marshy areas of Java, as green manure and for hay in tropical America and Indonesia, and recently has undergone extensive trials in the southern United States. Before it becomes woody, *A. indica* is particularly suitable for cattle browse. *A cristata* Vatke, *A. fluitans* Peter, *A. indica*, and *A. nilotica* Taub. provide important fodder on the infertile sandy soils and in the river-fringe areas of the Western Province of Zambia (Verboom 1966). The crude protein content of young shoots of *A. fluitans* ranges from 14.8 to 26.6 percent. This species is a hardy, valuable graze crop, and its shoots are readily browsed by cattle and eaten as a vegetable by the natives. Roots and shoots are produced in abundance from the nodes of its long, trailing stems.

NODULATION STUDIES

The occurrence of both stem nodules and root nodules on certain species is a phenomenon unique to this genus.

22

Presumably, Jaensch (1884) observed stem nodules on *A. elaphroxylon* but did not realize their significance. The protuberances shown in his drawing (plate V, figure 6), labeled *lysigenen gummigang*, appear similar to those depicted by Jeník and Kubíková (1969).

A comprehensive understanding of the histology and function of stem nodules on *Aeschynomene* species is lacking. Histological studies by Hagerup (1928) of the stem nodules on *A. aspera* growing along the banks of the Niger River showed bacteria-filled host cells. Since root nodules were not present, Hagerup speculated that nitrogen assimilation was not possible in underwater roots; therefore stem nodules were an alternative adaptation.

Suessenguth and Beyerle (1935) observed only stem nodules on preserved specimens of *A. paniculata* Willd. from Bahia, Brazil, and *A. indica* from Sudan, but both stem and root nodules were present on specimen material of *A. indica* from Malawi. Hundreds of uniformly hemispherical nodules, 1 mm in diameter and arranged in vertical rows, covered the entire stem area of *A. paniculata* between the water roots and the aerenchymous roots of the stem region. Young nodules were yellowish. Older ones were darker and contained a needle-thin pore. Nodules were never present on adventitious water rootlets.

Nodules borne on submerged, lower stem parts of *A. indica* were superficially similar to those borne on the roots (Arora 1954). Both types were alike in size and shape, occurred only at the juncture of emerged rootlets, were similar in internal structure, and contained hemoglobin. Rhizobial infection of each type was initiated through ruptured tissue and not through root hairs. The major difference was in origin. The root nodules were endogenous in origin, i.e., they arose within the pericycle, and thus originated like those on the peanut (Allen and Allen 1940a); the stem nodules were exogenous and arose from cortical tissue.

Root systems of *A. elaphroxylon* from various localities of Ghana lacked nodules, but the bark of stems and the root spurs near the water surface bore a profusion of regularly and closely spaced hemispherical protuberances (Jeník and Kubíková 1969). Preserved stem segments, 7–8 cm long, kindly sent to the authors by Professor J. Jeník, bore abundant nodules containing masses of pleomorphic rods unlike rhizobia in size, morphology, and staining properties. An inquiry into the identity of the causal agent, morphogenesis, and nitrogen-fixing properties of these hypertrophies is inviting.

Round and ovoid nodules, 0.5 cm in diameter, were described by Klebahn as early as 1891 on interwoven, fibrous roots of *A. elaphroxylon* from the black mud of the Upper Nile. The bacteroidal areas contained numerous starch granules amid simple rod-shaped and biscuit-shaped bacteria enclosed within a corky parenchyma. Infection threads were observed only in young nodules. Klebahn deduced that the lack of the proper organisms in the soil accounted for the absence of nodules on seedlings cultured subsequently in Cairo.

Information on the cultural and biochemical characteristics of rhizobial isolates from *Aeschynomene* species is nil, but the sparse evidence from host studies prompts the conclusion that the rhizobia are members of the slow-growing group. *A. americana* was placed in the cowpea miscellany by Carroll (1934a). A single strain from this species was effective on *Stylosanthes mucronata* and *Vigna sinensis*; another strain from *A. rhodesiaca* Harms was effective on *Phaseolus lathyroides* (Sandmann 1970a). Both strains were ineffective on *Stylosanthes guianensis*.

Stem nodules of *Aeschynomene*	Area	Reported by
aspera L.	Mali	Hagerup (1928)
elaphroxylon (Guill. & Perr.) Taub. (= *Herminiera elaphroxylon* Guill. & Perr.)	Ghana	Jeník & Kubíková (1969)
evenia W. F. Wright var. *serrulata* Rudd	Venezuela	Barrios & Gonzalez (1971)
filosa Mart.	Venezuela	Barrios & Gonzalez (1971)
indica L.	India	Arora (1954)
	Japan	Pers. observ. (1962)
paniculata Willd.	Brazil	Suessenguth & Beyerle (1935)

Root nodules of *Aeschynomene*	Area	Reported by
abyssinica (A. Rich.) Vatke	Zimbabwe	Corby (1974)
americana L.	Widely reported	Many investigators
var. *javanica*	Java	Keuchenius (1924)
aphylla Wild	Zimbabwe	Corby (1974)
aspera L.	Mali	Hagerup (1928)
	Sri Lanka	W. R. C. Paul (pers. comm. 1951)
bracteosa Welw. ex Bak.	Zimbabwe	Corby (1974)
brasiliana (Poir.) DC. var. *brasiliana*	Venezuela	Barrios & Gonzalez (1971)
elaphroxylon (Guill. & Perr.) Taub.	Upper Nile	Klebahn (1891)
	S. Africa	Grobbelaar & Clarke (1974)
evenia W. F. Wright var. *evenia*	Venezuela	Barrios & Gonzalez (1971)
var. *serrulata* Rudd	Venezuela	Barrios & Gonzalez (1971)
falcata DC.	Java	Keuchenius (1924)
	S. Africa	Grobbelaar et al. (1967)
filosa Mart.	Venezuela	Barrios & Gonzalez (1971)
fluitans Peter	Zambia	Verboom (1966)
fulgida Welw. ex Bak.	S. Africa	Grobbelaar & Clarke (1974)

Root nodules of *Aeschynomene* (cont.)	Area	Reported by
gazensis Bak. f.	Zimbabwe	Corby (1974)
grandistipulata Harms	Zimbabwe	Corby (1974)
hispida Willd.	S. America	Suessenguth & Beyerle (1935)
histrix Poir.		
var. *histrix*	Venezuela	Barrios & Gonzalez (1971)
var. *incana* (Vog.) Benth.	Venezuela	Barrios & Gonzalez (1971)
indica L.	Widely reported	Many investigators
inyangensis Wild	Zimbabwe	Corby (1974)
laxiflora Boj.	S. America	Suessenguth & Beyerle (1935)
mediocris Verdc.	Zambia	Verdcourt (1973)
megalophylla Harms	Zimbabwe	Corby (1974)
micrantha DC.	S. Africa	Grobbelaar et al. (1964)
mimosifolia Vatke	Zimbabwe	Corby (1974)
minutiflora Taub. ex Engl.	Zimbabwe	Corby (1974)
nilotica Taub.	Zimbabwe	Corby (1974)
nodulosa (Bak.) Bak. f.	S. Africa	Grobbelaar & Clarke (1972)
var. *nodulosa*	Zimbabwe	Corby (1974)
var. *glabrescens* Gillett	Zimbabwe	Corby (1974)
nyassana Taub.	S. Africa	Grobbelaar & Clarke (1972)
	Zimbabwe	Corby (1974)
paniculata Willd.	S. America	Suessenguth & Beyerle (1935)
pratensis Donn. Smith		
var. *caribaea* Rudd	Venezuela	Barrios & Gonzalez (1971)
puertoricensis Urb.	Puerto Rico	Dubey et al. (1972)
rehmannii Schinz	S. Africa	N. Grobbelaar (pers. comm. 1973)
var. *leptobotrya* (Harms) Gillett	S. Africa	Grobbelaar et al. (1967)
rhodesiaca Harms	Zimbabwe	Corby (1974)
rudis Benth.	Argentina	Rothschild (1970)
	Venezuela	Barrios & Gonzalez (1971)
schimperi Hochst. ex A. Rich.	Zimbabwe	Corby (1974)
schliebenii Harms		
var. *mossambicensis* (Bak. f.) Verdc.	Zimbabwe	Corby (1974)
sensitiva Sw.	Trinidad	DeSouza (1966)
trigonocarpa Taub.	Zimbabwe	Corby (1974)
viscidula Michx.	Ga., U.S.A.	Pers. observ. (U.W. herb. spec. 1970)

Affonsea A. St.-Hil.   Mimosoideae: Ingeae

1 / 3435

Named for Martin Affonso DeSouza, Portuguese botanist.

Type: _A. juglandifolia_ A. St.-Hil.

7–8 species. Trees or shrubs, unarmed. Leaves once pinnate; leaflets large, few pairs; petiole glandular, often winged between leaflet pairs; stipules persistent. Flowers sessile, large, silky-haired, in loose spikes or interrupted racemes; calyx tubular-campanulate; lobes 5 or more; corolla tubular; petals 5, united below the middle, valvate; stamens very numerous, equal; filaments united into a tube at the base, free and long-exserted above; anthers small, verrucose, 4-lobed; carpels usually 3, rarely 2, 5, or 6, free, villous; ovules numerous; styles separate, filiform; stigmas capitellate. Pods linear, straight. (Bentham and Hooker 1865)

The pluricarpellate condition in _Affonsea_ occurs in only 2 other closely related mimosoid genera, _Archidendron_ and _Hansemannia_. It is considered to be an ancestral relict from the more primitive Rosaceae.

Affonsea species inhabit tropical eastern Brazil.

Afgekia Craib   Papilionoideae: Galegeae, Robiniinae

Type: _A. sericea_ Craib

1 species. Shrub, scandent, unarmed. Leaves imparipinnate; leaflets 15–17, oblong, silky; stipels present; stipules conspicuous, somewhat persistent. Flowers pedicelled, disposed singly on long rachis of axillary racemes; bracts silky, conspicuous, imbricate in bud; calyx bilabiate, upper lobe shorter, lower lobe longer than the calyx tube; petals clawed; standard large, hairy, with 1 callosity at the base and 2 higher up; wings spear-shaped; keel hairy, blunt, incurved; stamens 10, vexillary stamen united only at the middle with the others; anthers uniform, longitudinally dehiscent; ovary stalked, 2-ovuled; style inflexed, not coiled, bearded on upper half; stigma small, terminal. Pod oblong, turgid, not septate, valves woody, dehiscent; seeds 2, ellipsoid, shiny, with elongated, persistent funicle and long, pale green hilum half encircling the seed. (Craib 1927)

This monotypic genus is endemic in eastern Thailand. The plants occur on savannas at 300-m altitude.

This floriferous shrub is a fine ornamental climber for tropical parks and botanical gardens.

Name indicates that the genus is indigenous only in Africa and resembles
Ormosia.

Type: *A. laxiflora* (Benth.) Harms

6 species. Shrubs or trees, ranging from untidy trees, 6–10 m tall usually
in savannas, to giant monarchs in rain forests. Leaves, imparipinnate or abruptly pinnate; leaflets 7–13, alternate or subopposite, lanceolate or ovate-
elliptic; stipels setaceous; stipules rapidly caducous. Flowers small, many,
cream or greenish white, in copious, lax, terminal panicles or racemes; calyx
campanulate; teeth 5, upper 2 subconnate, usually incurved, lower lip
3-dentate; standard suborbicular; wings obliquely obovate-oblong; keel petals
free or scarcely joined; stamens 10, free, unequal, 1–2 often abortive; anthers
versatile, elliptic; ovary subsessile or short-stalked, 2- or many-ovuled; style
slender, short, curled above; stigma incurved, lateral. Pod flat, oblong or
broadly linear, slightly woody, with a narrow wing more conspicuous on the
ventral than on the dorsal suture, sometimes slightly constricted between the
seeds; seeds 1–7, lens-shaped, flat. (Harms 1908a, 1915a)

The questionable status of this genus was reviewed recently by Knaap-
van Meeuwen (1962), who concluded that *Afrormosia* should be reduced to
Pericopsis.

Representatives of this genus occur in tropical Africa, especially in the
Ivory Coast, Ghana, Cameroon, and the Congo Basin. They grow on open
savannas and in semiwoodlands and rain forests.

Afrormosia wood has many desirable qualities, such as high strength,
durability, dimensional stability, and ease of working. The timber of *A. elata*
Harms, now preferably named *Pericopsis elata*, is the most often mentioned in
the timber trade. This rain-forest giant usually reaches a height of 50 m, with a
trunk diameter of 1.5 m. The wood is known as kokrodua and also as African
teak, although it lacks the oily finish, becomes darker rather than lighter in
color upon exposure, and is heavier than true teak, *Tectona grandis*. It serves a
variety of uses but is especially favored for shipbuilding, interior trim, furniture, decorative veneer, and flooring (Kukachka 1960, 1970; Titmuss 1965;
Rendle 1969).

NODULATION STUDIES

Nodules were reported on *Pericopsis elata* (as *A. elata*) in Yangambi, Zaire
(Bonnier 1957), and on *Pericopsis angolensis* (as *A. angolensis* (Bak.) Harms) in
Zimbabwe (Corby 1974).

Named for Adam Afzelius, 1750–1837, Swedish botanist, student of Linnaeus, and professor at Uppsala.

Type: *A. africana* Sm.

30 species. Seldom shrubs, mostly trees, unarmed, ranging from 10 to 20 m tall with short, irregular trunks to large, erect trees, 35 m or more tall with girths of 6 m or more; boles straight and cylindric with first branches often 20 m or more from the ground; limbs massive; crowns broad, dense or open. Leaves paripinnate; leaflet pairs few, leathery; petiolules short, twisted. Flowers large, decorative, sweet-scented, often reddish pink, in simple or paniculate racemes in upper leaf axils, or on branch tips; bracts ovate, concave, caducous; bracteoles large, concave, completely enveloping the young flower bud, later reflexed, deciduous; calyx tubular; sepals 4, unequal, overlapping; petal 1, fully developed, long-clawed, the others rudimentary or absent; fertile stamens 7–8, free or partly united at the base; filaments long but shorter than the petal; anthers oblong, longitudinally dehiscent; staminodes usually 2; ovary long-stalked, stalk joined to the calyx tube wall, many-ovuled; style long, exserted; stigma dark, small, subcapitate. Pods large, asymmetric, oblique-oblong, kidney-shaped or semicircular, flattened, valves 2, thick, woody, subindehiscent, transversely septate between the 4–9 or more seeds, usually filled with a thin pulp; seeds oval or oblong, shiny, black or red, partially enveloped by a fleshy, bright yellow, orange, or red aril. (Bentham and Hooker 1865; Harms 1913; Baker 1930; Léonard 1950, 1952, 1957; Phillips 1951)

All species of *Pahudia* are treated as synonyms of *Afzelia* by Léonard (1950).

Members of this genus are widely distributed in tropical Asia and Africa. They thrive in secondary and high forests, and are usually more common in semideciduous than in evergreen forests; species often occur in tidal river areas.

A. africana Sm., *A. bella* Harms, *A. bipindensis* Harms, *A. bracteata* Vog. ex Benth., *A. pachyloba* Harms, *A. palembanica* Bak., and *A. quanzensis* Welw. are valuable timber trees in various parts of tropical Africa. The specific epithet *quanzensis* is often spelled *cuanzensis*, to denote its common occurrence along the banks of the Cuanza River, Angola.

The seeds are used for necklaces, curios, and ornaments (Burkill 1966); those of *A. quanzensis* are favored by hornbills, and the fallen flowers, both fresh and dried, are relished by cattle and game (Coates Palgrave 1957). Some reports allude to the poisonous properties of the seeds of *A. africana* (Dalziel 1948). The outer seed layers of *A. palembanica* and *A. quanzensis* yield a lipoxanthin yellow dye that is used to stain mats and native cloth.

Afzelia wood is of moderately good quality and resembles mahogany, but is of no importance in the timber trade (Rendle 1969). *A. quanzensis* is commonly known as Rhodesian or pod mahogany, *A. africana* as mahogany bean. The sapwood is commonly straw-colored and easily distinguished from the reddish brown heartwood. The wood is durable, hard, resistant to decay and termite attack, and normally fine or straight grained with a uniform coarse texture, but not easy to work. Tools used on this wood need constant oiling to prevent them from being clogged by its thick gum. *Afzelia* wood serves for making dug-out canoes, furniture, paneling, general carpentry, and heavy construction. The wood of *A. quanzensis* is quite useful for purposes where small shrinkage is desired. It takes a high polish, bends easily, and is favored for musical instruments and ornamental drums. The bark of most species yields tannin.

NODULATION STUDIES

Two species, *A. africana* examined in Ghana (Jeník and Mensah 1967), and *A. quanzensis* in Zimbabwe (Corby 1974), lacked nodules. The specimens examined in Ghana were from a riverine forest and also from seeds collected in Sierra Leone. Root hairs on *A. africana* in the true morphological sense were not observed, but ectotrophic mycorrhizal hyphae were conspicuously present. J. F. Redhead was credited by Jeník (personal communication 1968) with observing ectotrophic mycorrhizae also on *A. bella* in western Nigeria.

Airyantha Brummitt   Papilionoideae: Sophoreae

Named for H. K. Airy Shaw, Kew Herbarium, in recognition of his services particularly to Malesian botany.

Type: *A. borneensis* (Oliv.) Brummitt

2 species. Woody climbers, erect shrubs or small trees, up to 8 m tall; young branches and inflorescences densely brown-hairy. Leaves 1-foliolate, petioled, pulvinate with paired stipels at the base; blade leathery, elliptic-oblanceolate, acuminate. Flowers white, yellowish, or pale blue, on short pedicels in short racemes or in sessile fascicles in leaf axils, or clustered into leafless terminal panicles; bracteoles 2, conspicuous, oblong to broad-ovate; calyx lobes 5, free or variously connate; eventually the entire calyx splits off, leaving a collar just above the base of the calyx tube; petals subequal; standard pubescent on outer surface; stamens 10, free; filaments pubescent; ovary tomentose. Pod twisted into a spiral, semicircle, or circle, constricted between the 2–7 flat seeds, valves velvety brown, dehiscent; seed color changes from purplish black to red at maturity. (Brummitt 1968a)

This genus is closely allied to *Baphia* and *Baphiastrum*.

The type species occurs in Borneo and the Philippines; *A. schweinfurthii* (Taub.) Brummitt, the other species, is indigenous to tropical western and central Africa.

Named for Il Sig. Cavalier Filippo degl' Albizzi, 18th-century Italian naturalist of Florence and member of a noble family of Tuscany; in an expedition to Turkey, he collected many seeds including those of the first described species, *A. julibrissin* Durazz. The generic name was misspelled for about 100 years. As Little (1945) explained, the original spelling with a single *z* by Durazzini in 1772 was deliberately changed in 1884 to *Albizzia* by Bentham, who preferred the latter spelling. Correct spelling with a single *z* followed Lawrence's recommendation in 1949.

Type: *A. julibrissin* Durazz.

150 species. Shrubs or trees, rarely lianas, unarmed, deciduous; some tree species reach a height of 50 m, with wide-spreading branches and flat or umbrella-shaped crowns; foliage usually feathery in appearance. Leaves bipinnate; leaflets small, in many pairs, or large and in few pairs; petioles and main rachis usually with more or less conspicuous nectaries or jugal glands; stipules setaceous, or obsolete, rarely large and membranous. Flowers yellow, white, pink, tinged with green or purple, small, regular, in globose, solitary or fascicled heads, or cylindric spikes; flowers subtended by linear-lanceolate to broad-ovate bracts; calyx campanulate or tubular, dentate or briefly lobate; petals 5, inconspicuous, united to the middle or beyond; stamens numerous, usually more than 15, elongated, white, rose, red, or purple, conspicuous, united to the middle or beyond into a tube; anthers small; ovary lanceolate or linear-oblong, several-ovuled; style threadlike, usually longer than the stamens; stigma small, terminal. Pod often papery-thin, flat, pendulous, red-brown or straw-colored, broadly linear, straight or almost so, nonseptate, 2-valved, tardily dehiscent to indehiscent; pods of some species remain on the trees for many months; seeds ovate or oblong, thick, with filiform funicle. (Bentham and Hooker 1865; Baker 1930; Standley and Steyermark 1946; Gilbert and Boutique 1952; Brenan 1959)

Albizia is a large and complicated pantropical genus; species are most numerous in the Old World tropics. They are closely allied to, and often mistaken for, *Acacia* species and vice versa. The plants are well adapted to poor soils and are found up to 1,600-m altitude. They are common in low bush, secondary forests, along sandy river beds, and in savannas. Some species are fairly salt-tolerant and occur in coastal areas. The trees thrive on lateritic alluvial soils and sandy mining areas.

Most species are rapid growing, shallow rooted, and easily uprooted by wind storms. Burkill (1966), in describing *A. falcata* (L.) Backer as "one of the quickest-growing trees of Malaysia," cited growth records of trees which reached heights of 16 m in less than 3 years, 33 m in 9 years, and 45 in 17 years. This species has been utilized in Malaysia for afforestation and is in current production on timber plantations in Hawaii.

Most of the species have value in soil improvement and as shade on tea and coffee plantations (Jackson 1910). Especially esteemed for these reasons are *A. chinensis* (Osb.) Merr., *A. falcata*, *A. moluccana* Miq., *A. odoratissima* (Willd.) Benth., and *A. stipulata* (DC.) Boiv. Fosberg (1965) regarded the 2 last-named species as synonyms of *A. falcataria* (L.) Fosberg; they are planted as ornamentals in parks, roadsides, and gardens.

Everett (1969) cited *A. julibrissin* Durazz., silk tree, one of the few warm-temperate species, as hardy in California. It was introduced into cultivation in the United States about 1745 and is now found from Virginia into Louisiana. It serves as wildlife cover, browse, and as an ornamental because of its elegant, finely divided foliage and large pink or pink and white floral pompons. The light brown, papery, flat pods persist on the tree for a long time.

Albizia wood varies greatly in quality and durability because of the rapid growth of the trees. The density varies between 544 and 800 kg/m^3 (Titmuss 1965). Some specimens are soft and lack strength. The first Javanese factory to manufacture matches used the white wood of *A. falcata* (Burkill 1966). Wood resembling walnut is obtained from *A. acle* (Blanco) Merr., *A. amara* (Roxb.) Boiv., *A. lucida* Benth., *A. mollis* Boiv., *A. odoratissima*, and *A. thompsoni* Brandis. *A. lebbeck* (L.) Benth., frequently misspelled *lebbek* (see Lawrence 1949), was exported in large shipments to Europe as East Indian walnut from the Andaman Islands at the start of the 20th century. Watt (1908) remarked that it was "taxed in Burma at a higher rate than teak-wood, a circumstance that indicates the local value put on this timber." Agricultural and industrial instruments and appliances, furniture, picture frames, and cartwrights' work are made from this wood. It is hard, heavy, and tough, has a medium to coarse texture, and is easy to work. The wood is known in the timber trade as siris or kokko; the species is known in Hawaii and the Philippines as woman's tongue because the pods remain on the tree for 3–5 months and make a clattering noise in the wind.

The sawdust of some species, namely, *A. ferruginea* (Guill. & Perr.) Benth., *A. gummifera* (J. F. Gmel.) C. A. Smith, *A. lebbeck*, and *A. zygia* (DC.) Macb., is aromatic, peppery, and irritating to the nasal passages. The roots of several species have a disagreeable odor and are not desirable in proximity to water supplies.

The saponin content in the bark of *A. saponaria* (Lour.) Bl. is sufficiently high to be used commercially as soap in Malaysia and the Philippines. The bark of this species and several others has fish-stupefying and insecticidal properties. A preparation from the bark of *A. anthelmintica* Brongn. is a tapeworm remedy in Namibia and Ethiopia. Other medicinal uses of plant parts and chemical substances of *Albizia* species are described and documented by Watt and Breyer-Brandwijk (1962) and Dalziel (1948).

Gum exudes from the trunks of *A. amara*, *A. lebbeck*, *A. odoratissima*, *A. procera* (Roxb.) Benth., and *A. stipulata*. It is not of high quality, but in Burma and Sri Lanka it is mixed with other gums as an extender (Howes 1949).

NODULATION STUDIES

Albizia species have promise as soil-improvement plants because they are abundantly nodulated. The nodules in many species are dichotomously branched, and give evidence of a perennial growth. Rhizobial isolates produce a moist, slightly opaque growth on agar slants. Colonies are raised and mucilaginous. Tyrosinase is produced from asparagine. Alkaline reactions are produced in litmus milk without the production of a serum zone (Allen and Allen 1939).

Albizia species apparently have a kinship with the cowpea miscellany as judged by plant-infection studies using *A. lophantha* (Willd.) Benth. (Maassen and Müller 1907), *A. lebbeck* (Allen and Allen 1939), and *A. distachya* (Vent.) Macb. (Lange 1961).

Nodulated species of *Albizia*	Area	Reported by
acle (Blanco) Merr.	Hawaii, U.S.A.	Allen & Allen (1936a)
adianthifolia (Schum.) W. F. Wight	S. Africa	Grobbelaar et al. (1964)
	Zimbabwe	Corby (1974)
amara (Roxb.) Boiv.		
ssp. *sericocephala*	Zimbabwe	Corby (1974)
anthelmintica Brongn.	Zimbabwe	Corby (1974)
antunesiana Harms	Zimbabwe	Corby (1974)
brevifolia Schinz	Zimbabwe	Corby (1974)
carbonaria Britt.	Costa Rica	J. León (pers. comm. 1956)
chinensis (Osb.) Merr.	Hawaii, U.S.A.	Allen & Allen (1936a)
distachya (Vent.) Macb.	Australia	Lange (1961)
ealensis De Wild.	Zaire	Bonnier (1957)
falcata (L.) Backer	N.Y., U.S.A.	Wilson (1939a)
forbesii Benth.	Zimbabwe	Corby (1974)
glaberrima (Schum. & Thonn.) Benth.		
var. *glabrescens* (Oliv.) Brenan	Zimbabwe	Corby (1974)
gummifera (J. F. Gmel.) C. A. Smith		
var. *gummifera*	Zimbabwe	Corby (1974)
harveyi Fourn.	Zimbabwe	Corby (1974)
julibrissin Durazz.	Widely reported	Many investigators
katangensis Willd.	Hawaii, U.S.A	Allen & Allen (1936a)
lebbeck (L.) Benth.	Widely reported	Many investigators
lebbekoides (DC.) Benth.	Hawaii, U.S.A.	Allen & Allen (1936a)
lophantha (Willd.) Benth.	Germany	Maassen & Müller (1907)
(= *Acacia lophantha* Willd.)		
moluccana Miq.	Widely reported	Many investigators
odoratissima (Willd.) Benth.	Malaysia	Wright (1903)
	Hawaii, U.S.A	Allen & Allen (1936a)
petersiana (Bolle) Oliv.		
ssp. *evansii* (Burtt Davy) Brenan	Zimbabwe	Corby (1974)
procera (Roxb.) Benth.	Hawaii, U.S.A.	Allen & Allen (1936a)
retusa Benth.	Hawaii, U.S.A.	Allen & Allen (1936a)
saponaria (Lour.) Bl.	India	Parker (1933)
	Hawaii, U.S.A.	Allen & Allen (1936a)
schimperana Oliv.		
var. *amaniensis* (Bak. f.) Brenan	Zimbabwe	Corby (1974)
var. *schimperana*	Zimbabwe	Corby (1974)
stipulata (DC.) Boiv.	Widely reported	Many investigators
tanganyicensis Bak. f.	Zimbabwe	Corby (1974)
versicolor Welw. ex Oliv.	Zimbabwe	Corby (1974)
zimmermannii Harms	Zimbabwe	Corby (1974)

Type: *A. insignis* (Benth.) Endl.

12 species. Trees, large, unarmed. Leaves imparipinnate or 1-foliolate above; leaflets opposite or subopposite, large, few, leathery; stipels absent; stipules small. Flowers showy, white, in simple axillary or terminal panicles; bracts small, caducous; bracteoles absent; calyx turbinate, entire in bud, splitting into 2–5 segments during anthesis; petals 5, unequal, overlapping, ventral petal exterior, broad, others oblique; stamens uniform, numerous, indefinite in number; filaments slender; anthers uniform, dorsifixed, large, linear, acuminate, longitudinally dehiscent; ovary stalked, incurved; stigma small, terminal. Pod thick, ovoid, with 1 large seed. (Bentham and Hooker 1865)

Aldina species are indigenous to Venezuela, Guiana, and northern Amazon areas.

The wood of these rather rare trees is hard, heavy, coarse, unattractive, and lacks commercial importance (Record and Hess 1943).

Alepidocalyx Piper   Papilionoideae: Phaseoleae, Phaseolinae

From Greek, *a-*, "without," *lepis*, "scale," in reference to the absence of bracteoles around the calyx.

Type: *A. parvulus* (Greene) Piper

3 species. Herbs, perennial, with small globose tubers; stems erect or twining above. Leaves pinnately 3-foliolate, pallid; leaflets ovate or lanceolate, rather thick; stipels present. Flowers violet; bracteoles none; calyx lobes not longer than the tube; petals long-clawed; standard with a transverse callus; keel curled or coiled; stamens diadelphous. Pod linear, compressed, short-beaked, several-seeded. Otherwise as in *Phaseolus*. (Piper 1926)

This genus is intermediate between *Phaseolus* and *Minkelersia* (Piper 1926).

Representatives of this genus are native to Mexico, New Mexico, and Arizona. They grow in conifer forests on mountains up to 2,500-m altitude.

Alexa Moq.   Papilionoideae: Sophoreae
140 / 3600

Type: *A. imperatricis* (Schomb.) Bak.

9 species. Trees, tall, unarmed. Leaves imparipinnate; leaflets alternate or subopposite, large, leathery; stipels absent; stipules caducous. Flowers large, white, yellow, orange, or scarlet, in lateral, pendent racemes on leafless

twigs; bracts minute; bracteoles absent; calyx large, thick, closed in bud; teeth short, sinuous; petals thick, subequal; standard ovoid, reflexed above, emarginate or bilobed at apex; wings and keel petals erect, oblong, similar, free; stamens free; anthers oblong-linear, dorsifixed, versatile; ovary long-stalked, several-ovuled; style incurved; stigma terminal, minute. Pod long, large, flattened, woody, velvety, 2-valved, nonseptate within, but spongy between the orbicular, thick seeds, spirally twisting upon ripening. (De Candolle 1825a; Bentham and Hooker 1865; Amshoff 1939)

Alexa species are limited to the wet lands of Guiana and the Brazilian Amazon Basin.

The wood of several species "appears suitable for interior construction and carpentry but with little, if any, commercial possibilities" (Record and Hess 1943).

NODULATION STUDIES

Nodules were not present on 9 seedlings of *A. imperatricis* (Schomb.) Bak. examined at 3 sites in Bartica, Guyana (Norris 1969).

Alhagi Tourn. ex Adans.   Papilionoideae: Hedysareae,
315 / 3783   Euhedysarinae

From Arabic, *al-hāj*, "camel thorn."
Type: *A. maurorum* Medik.
5 species. Shrubs, low, up to about 1 m tall, intricately branched; twigs green, rigid, thickly beset with axillary and terminal spines; spines generally hard, 12–25 mm long, sharp. Leaves obovate-oblong, small, scalelike, simple, entire; stipules small, free. Flowers small, yellow, pink, red, or lavender, solitary or few, short-pedicelled in axillary racemes; rachis rigid, apically spinescent; bracts minute; bracteoles absent; calyx campanulate; lobes 5, short, subequal; standard obovate, short-clawed; wings equal to the standard, falcate-oblong, free, at first folded, later reflexed; keel petals incurved, obtuse, subequal to the standard and wings; stamens 10, vexillary stamen free; anthers uniform; ovary subsessile, many-ovuled; disk present; style filiform, smooth, incurved; stigma terminal, small. Pod linear, cylindric, thick, small, stalked, smooth, often constricted between the seeds, septa subdouble but joints continuous, indehiscent; seeds 1–5, subreniform. (Bentham and Hooker 1865)

Members of this genus are native to Asia Minor, the North African Sahara, the Mediterranean region, central Asia, and the Himalayas. They occur in waste areas on alkaline soils and are drought-resistant.

All species are effective in lessening windblown soil erosion. *A. camelorum* Fisch., *A. graecorum* Boiss., and *A. maurorum* Medik., all commonly called camel thorn, serve as camel fodder in desert areas. During the hot summer months the stems and branches of the last-named species exude a sweet, gummy substance that hardens into reddish brown lumps; these lumps are collected by Arab nomads who consider them a delicacy. The gum also serves medicinally as a laxative and expectorant (Carmin 1950). At one time it was believed that this substance was the "manna" or "mannah" of Biblical accounts (Moldenke and Moldenke 1952), and one of the earlier proposed names for the genus was *Manna* D. Don; however, this idea no longer prevails.

Presumably, *A. camelorum* was introduced into California "in packings of shipments of date cuttings from Africa, or in impure alfalfa seed from Turkestan" (Jepson 1936). The species is now established in the Colorado desert and the San Joaquin Valley of the United States, where it is a pest because of its aggressiveness, hardy revegetation from deep rootstocks, and copious, pungent spines. Considerable money has been spent toward its eradication.

NODULATION STUDIES

Nodules were observed on roots of *A. maurorum* examined in Israel (Carmin 1950) and in Iraq (Khudairi 1957). In Iraq the plants most commonly occur in growth associations with *Prosopis stephaniana*. The nodules on both species are perennial; however, their rhizobial isolates differ from each other and are unlike strains from cultivated legumes (Khudairi 1969).

Alistilus N. E. Br.   Papilionoideae: Phaseoleae, Phaseolinae

From Latin, *ales*, "winged," and *stilus*; the style is winged.

Type: *A. bechuanicus* N. E. Br.

1 species. Herb, perennial, procumbent or prostrate, not twining. Leaves pinnately 3-foliolate, on long petioles; leaflets ovate, densely ciliate, terminal leaflet usually 3-lobed, deltoid-acute, laterals rhomboid, 2-lobed; stipels present; stipules ovate, acute, persistent. Flowers few, on erect, long peduncles in axillary racemes; bracts small, caducous; bracteoles absent; calyx hairy, campanulate, subbilabiate; teeth 5, shorter than the calyx tube, ovate, obtuse, upper lip entire, lower lip 3-lobed, lowest lobe longest, acute; standard sessile, orbicular, apex briefly bilobed with 2 inflexed auricles and 2 callosities at the base; wings oblong, pouched or eared near the base, broad-clawed; keel oblique-ovate, obtuse, incurved, subsessile; stamens 10, vexillary stamen joined to others only at the middle, kneed at the base; anthers uniform, ovoid, dorsifixed near the base; ovary linear, hairy, few-ovuled; style abruptly bent at a right angle, smooth, flattened or narrowly and membranously winged on each side, persistent; stigma terminal, truncate, with a thin ring of hairs at the base. Pod flat, linear, slightly falcate, 5-seeded. (Brown 1921; Phillips 1951)

The genus is closely related to *Dolichos* and *Lablab* (Brown 1921; Verdcourt 1970c).

This monotypic genus is endemic in Botswana and Transvaal.

NODULATION STUDIES

Nodules were observed on *A. bechuanicus* N. E. Br. in South Africa (N. Grobbelaar personal communication 1974).

Alysicarpus Desv.   Papilionoideae: Hedysareae, Desmodiinae
(340) / 3810

From Greek, *halysis*, "chain," and *karpos*, "fruit"; the pods are moniliform, with indehiscent, 1-seeded joints in chain formation.

Type: *A. bupleurifolius* (L.) DC.

30 species. Herbs, some woody, erect or prostrate, annual, biennial, or perennial, glabrous or loosely hairy, sometimes glandular-pubescent. Leaves 1-foliolate, very rarely pinnately 3-foliolate, linear-lanceolate; stipels 2; stipules dry, striate, acuminate, free or connate, opposite the leaf, long-persistent. Flowers small, on short, usually paired pedicels, in slender, terminal, or rarely axillary racemes; bracts scarious, mostly deciduous; bracteoles lacking; calyx deeply cleft, rigidly glumaceous, persistent; tube campanulate, short; lobes lanceolate, longer than the tube, stiff, dry, 2 upper lobes often connate halfway up or almost to the apex; standard obovate or orbicular, sessile or narrowed into a claw; wings oblique-oblong, adhering to the slightly incurved, obtuse keel; keel often appendaged on 1 or both sides with a membrane; stamens 10, diadelphous; anthers kidney-shaped; ovary sessile or short-stalked, few- to many-ovuled, sometimes glandular-pubescent; style filiform, incurved; stigma terminal, capitate. Pod erect, somewhat compressed but not flat, constricted between the seeds into 1-seeded segments, segments ovate, globose or truncate at each end, usually even-sided, convex or turgid, sometimes hairy, indehiscent; seeds suborbicular or globose. (Bentham and Hooker 1865; Schindler 1926; Léonard 1954a)

Members of this genus are indigenous to tropical Africa and Asia. Species were introduced into, and are now widespread throughout, Polynesia, Australia, and tropical America. They are a common weed in grassy savannas, sunny wastelands, and along roadsides.

A. glumaceus (Vahl) DC. and *A. vaginalis* (L.) DC., both commonly called Alyce clover, are cultivated as summer graze for domestic animals but preferably are used as hay after harvesting and storage. The plants do not produce large quantities of foliage because of their 1-foliolate habit. Ingestion of immature foliage is said to cause mucous diarrhea in horses. *A. vaginalis* is considered a good cover crop in Malaysia to prevent erosion on clay soils of rubber plantations (Burkill 1966). Native medicinal uses of leaf and root extracts of *A. rugosus* (Willd.) DC. are made for coughs and fever, and of *A.*

zeyheri Harv. & Sond. as a treatment for impotence (Watt and Breyer-Brandwijk 1962). The latter species is used as a snakebite remedy in some African communities (Uphof 1968).

NODULATION STUDIES

Nodules on *A. longifolius* W. & A. were tender and variously branched, similar to those on *Crotalaria* species, whereas those on *A. vaginalis* were round with rugged surfaces, as is the type common on roots of *Vigna sinensis* (Allen and Allen 1936a). Results from diverse plant-infection tests warrant the inclusion of *Alysicarpus* species in the cowpea miscellany (Allen and Allen 1936a, 1939; Wilson 1945a; Sandmann 1970a).

Nodulated species of *Alysicarpus*	Area	Reported by
belgaumensis W. F. Wight	W. India	V. Schinde (pers. comm. 1977)
bupleurifolius (L.) DC.	Sri Lanka	Wright (1903)
glumaceus (Vahl) DC.	Kenya	Bumpus (1957)
	Zimbabwe	W. P. L. Sandmann (pers. comm. 1970)
longifolius W. & A.	Hawaii, U.S.A.	Allen & Allen (1936a)
ludens Wall.	Java	Keuchenius (1924)
monilifera DC.	W. India	Bhelke (1972)
nummularifolius DC.	Java	Keuchenius (1924)
	Philippines	Bañados & Fernandez (1954)
ovalifolius (Schum.) J. Léonard	Zimbabwe	Corby (1974)
rugosus (Willd.) DC.	Hawaii, U.S.A.	Allen & Allen (1936a)
	Sri Lanka	W. R. C. Paul (pers. comm. 1951)
ssp. *perennirufus* J. Léonard	Zimbabwe	Corby (1974)
ssp. *rugosus*	Zimbabwe	Corby (1974)
tetragonolobus Edgew.	W. India	Bhelke (1972)
vaginalis (L.) DC.	Widely reported	Many investigators
zeyheri Harv. & Sond.	Zimbabwe	Sandmann (1970a)
	S. Africa	Grobbelaar & Clarke (1975)

Amblygonocarpus Harms   Mimosoideae: Adenanthereae
(21a) / 3458

From Greek, *amblys*, "blunt," *gonia*, "angle," and *karpos*, "fruit"; the pods are tetragonal.

Type: *A. andongensis* (Welw. ex Oliv.) Exell & Torre

1 species. Tree, tall, bark rough, stems glabrous. Leaves bipinnate, eglandular; pinnae pairs 2–5; leaflets 10–20 per pinna, alternate or subopposite, short-petioluled, small, obovate-elliptic, usually notched at the apex. Flowers fragrant, white to yellow, on slender pedicels in solitary or paired, axillary,

dense, spikelike racemes; bracts small, quite persistent; calyx campanulate; tube very short; lobes 5, dentate, valvate in bud; petals 5, rarely 6, oblong, pointed, valvate, free; stamens 10, rarely 12, fertile, free; anthers exserted, uniform, without an apical gland; ovary short-stalked, several- to many-ovuled; style filiform; stigma small. Pod purplish, smooth, straight, oblong, obtusely tetragonal, rhomboid in cross section, short-stalked, incompletely septate within, indehiscent; seeds 6–8, hard, brown, smooth, glossy, oval, slightly flattened but not winged. (Harms 1897; Baker 1930; Gilbert and Boutique 1952a; Brenan 1970)

In contrast to *Tetrapleura* species, in *Amblygonocarpus* the entire width of the pod valve is thickened and the anthers are eglandular, even in bud (Brenan 1970). The prominent 4 ribs on immature pods resemble those of mature *Tetrapleura* species and possibly account for the confusion in nomenclature.

This monotypic genus is endemic in tropical Africa, from Ghana to the Sudan, and in East Africa. The plants occur in deciduous and open forests, moist woodlands, grassy savannas, and in sandy soil from sea level to 1,370-m altitude.

The tree reaches heights up to 25 m and girths up to 2 m, with a clean, straight bole and a flat, open crown. The red brown heartwood and gray sapwood are very hard, have a density of 960 kg/m³, and take a fine polish (Eggeling and Dale 1951). The pulverized pods are used medicinally in some native areas as a dressing for chronic ulcers (Dalziel 1948).

NODULATION STUDIES

This species lacked nodules in Zimbabwe (Corby 1974).

Amburana Schwacke & Taub.   Papilionoideae: Sophoreae
(129a) / 3588

The Brazilian name, *amburana*, refers to the amberlike texture and hue of the wood.

Type: *A. cearensis* (Allem.) A. C. Smith

2 species. Trees, medium-large, 10–20 m tall, unarmed. Leaves im-paripinnate; leaflets 11–25, alternate, elliptic, strongly pinnately nerved. Flowers yellowish white, small, fragrant, in long, lax, axillary racemes; bracts early deciduous; calyx tubular, rimmed with 5 short teeth; petal 1, rounded-cordate, short-clawed, base cordate; stamens 10, free, subequal; anthers short, rounded, dorsifixed; ovary long-stalked, stalk mostly adnate to the calyx tube, linear, hooked, 2-ovuled; style very short, recurved; stigma small, terminal. Pod narrow-oblong, dark brown, short, dry, leathery, 2-valved, 1–2-seeded, rounded at both ends, flat, except over the rugose seeds, dehiscent at apex; seed ovoid, with a basal papery wing.

Amburana was originally positioned as 61a in Caesalpinioideae (Taubert 1894). Subsequently it was equated with *Torresea* and assigned its present position in Papilionoideae (Harms 1897). The name *Torresea* is now excluded because of its confusion with *Torresia* family Gramineae (Smith 1940; Kukachka 1961a).

Both species range from tropical and subtropical Brazil to northeastern Argentina. They are abundant in high forests and regions along rivers not subject to inundation.

A. acreana (Ducke) A. C. Smith, the larger of the 2 species, reaches a height of 30 m, with a trunk diameter of 1 m. The wood is known as ishpingo in the timber trade and is valued for durability and decorativeness in carpentry, cooperage, and furniture. The wood of both species is yellowish brown with an orange hue, and has a waxy texture and the fragrant scent and taste of coumarin or vanilla. This last property accounts for the South American vernacular names "cumarú" and "cumaré" for these species. A vanilla-scented, volatile oil extracted from the seeds of *A. acreana* is used in the same manner as that from tonka bean (*Dipteryx* sp.) to perfume soap and to flavor snuff and tobacco (Record and Hess 1943; Kukachka 1961a).

Amherstia Wall.   Caesalpinioideae: Amherstieae

65 / 3519

Named for Lady Amherst, 1762–1838, wife of Lord Amherst, artist, collector, and patron of botany, and her daughter, Lady Sara Elizabeth. Lord Amherst was at one time British Governor of Burma.

Type: *A. nobilis* Wall.

1 species. Tree, evergreen, erect, often 15 m or more tall, with a girth of 1–2 m. Leaves paripinnate; leaflets 4–7 pairs at first reddish bronze, limp, becoming large, thinly leathery, glossy dark green; stipels absent; stipules axillary, lanceolate, deciduous. Flowers vermillion blotched with yellow, odorless, 20–30, on very long red pedicals in lax, axillary and terminal, pendent racemes often 1 m long; bracts caducous; bracteoles 2, red, persistent; calyx tube elongated, narrow; sepals 4, large, petaloid, red, imbricate in bud, unequal, 2 lowest usually connate; petals 5, the upper petal innermost, broad, showy, lateral 2 lanceolate-obovate, lower 2 rudimentary; stamens 10, red, upper stamen free, short, other 9 united, 4 short, 5 long; anthers versatile, longitudinally dehiscent, ovary stalked, stalk adnate to calyx tube, 4–6-ovuled; style thin. Young pod yellow green, red-veined; mature pod smooth, brown, elongated, falcate, flat, woody, 2-valved, dilated along upper suture, containing 1–6, transverse, compressed seeds; seeds often infertile. (Bentham and Hooker 1865; Backer and Bakhuizen van den Brink 1963)

This monotypic genus is endemic in Burma and the East Indies. The plants grow in teak-forest areas. Cultivation of *Amherstia* is limited almost entirely to botanical gardens and tropical greenhouses; plants are usually propagated from cuttings. They require protection from wind, afternoon sun, and high humidity, and prefer a rich, deep, loamy, well-drained, calcareous, almost neutral soil in a humid climate below 500-m altitude.

This tree, commonly called pride of Burma and queen of the flowering trees, is considered to be "the most glorious of all the flowering trees in the entire world" (Menninger 1962).

According to Menninger (1962), this species "was first brought to flower in captivity by Louisa Lawrence, wife of William Lawrence, the great surgeon . . . in the Lawrencian hot-house soon after its discovery in Burma in 1837." Having enjoyed her triumph, Mrs. Lawrence presented the plant to Kew Gardens, where it died soon thereafter.

NODULATION STUDIES

Wright (1903) listed this species among 54 leguminous plants bearing nodules in the Royal Botanic Gardens, Peradeniya, Sri Lanka; however, plants examined in Singapore (Lim 1977) and in the Philippines (personal observation 1962) were not nodulated.

Amicia HBK   Papilionoideae: Hedysareae,
321 / 3790   Aeschynomeninae

Named for Jean Baptiste Amici, Italian scientist of the early 19th century.

Type: *A. glandulosa* HBK

8 species. Herbs, annual, biennial, and perennial, or shrubs, some gland-dotted. Leaves paripinnate, palmately or pinnately 4-foliolate; leaflets obovate or obcordate; stipels absent; stipules usually large, deciduous. Flowers yellow, rarely violet, large, in short, lax, pedicelled, axillary racemes; bracts and bracteoles broad, herbaceous; calyx deeply 5-cleft, very unequal, 2 upper lobes large, broad, laterals very small, lower oblong, shorter than upper ones; standard obovate-oblong, clawed, emarginate; wings obovate or narrow, sometimes short; keel blunt, incurved, subequal to standard; stamens monadelphous, joined into a tube split above; anthers elliptic, dorsifixed; ovary sessile, multiovuled; style filiform; stigma small, terminal. Pod lomentaceous, compressed, linear, arcuate, with 2–several, indehiscent, quadrate joints. (Bentham and Hooker 1865)

Members of this genus occur from Mexico to Argentina, in the tropical and subtropical Andes. The plants flourish in high prairies and damp alpine meadows up to 3,000-m altitude.

A. andicola (Griseb.) Harms, *A. fimbriata* Harms, and *A. medicaginea* Griseb. are browsed by cattle in Argentina.

Ammodendron Fisch. ex DC.   Papilionoideae: Sophoreae
145 / 3605

From Greek, *ammos*, "sand," and *dendron*, "tree."
  Type: *A. argenteum* (Willd.) Ktze.
  8 species. Shrubs; foliage silvery-white pubescent. Leaves paripinnate; leaflets 1–2 pairs; rachis tip spinescent; stipels lacking; stipules small. Flowers small, violet or yellow, in terminal racemes; bracts small, caducous, often absent; bracteoles absent; calyx tube short, campanulate; lobes subequal, upper 2 joined; standard ovoid, reflexed; wings oblique-oblong; keel petals incurved, free, overlapping at the back; stamens 10, free; ovary sessile; style awl-shaped, incurved; stigma small, terminal. Pod linear or curved, flattened, winged along both sutures, indehiscent, 1–2-seeded; seeds subcylindric, oblong. (De Candolle 1825a,b; Bentham and Hooker 1865)
  Representatives of this genus are native to western and central Asia, especially Turkestan and along the Caspian Sea. They grow in sandy soil on plains, hillslopes, seashores, and in deserts; some are found in quicksands.
  The absence in *Ammodendron* species of matrine alkaloids, which appear to be restricted to *Sophora* and related genera of the tribe Sophoreae, together with the presence of sphaerocarpine, one of the simple ammodendrine-hystrine alkaloids characteristic in *Genista* species, prompted Mears and Mabry (1971) to suggest the inclusion of *Ammodendron* in the tribe Genisteae.
  Nodulation studies offer promise in establishing tribal kinships.

Ammothamnus Bge.   Papilionoideae: Sophoreae
144 / 3604

From Greek, *ammos*, "sand," and *thamnos*, "shrub."
  Type: *A. lehmannii* Bge.
  4 species. Shrublets or shrubs, small, silky-pubescent. Leaves imparipinnate; leaflets small, numerous; stipels absent; stipules small, awl-shaped. Flowers white, in simple, terminal racemes; bracts setaceous; bracteoles inconspicuous; calyx tubular; lobes short, broad, subequal; standard ovoid; wings broad-oblong, subfalcate; keel petals blunt, slightly incurved, fused at the back; stamens 10, free or joined only at the base; anthers equal, versatile; ovary sessile, multiovuled; style short; stigma small, terminal. Pod linear, 2-valved, continuous within, twisted when ripe, densely tomentose; seeds ovoid. (Bentham and Hooker 1865)
  Members of this genus are scattered throughout Turkestan, Syria, and the far-eastern Soviet Union. They flourish in sandy, arid areas.
  Various quinolizidine alkaloids, namely, ammothamnine, matrine, pachycarpine, and sophocarpine, have been isolated from *A. lehmannii* Bge. (Sadykov and Lazur'evskii 1943b) and *A. songoricus* (Schrenk) Lipsky (Yakovleva et al. 1959). The roots of *A. lehmannii* yield a red dye with excellent dyeing properties for silk (Sadykov and Lazur'evskii 1943a).

From Greek, *a-*, "without," and *morphos*, "shape," i.e., deformed, in reference to the imperfectly developed flowers having only 1 petal.

Type: *A. fruticosa* L.

20–25 species. Subshrubs or shrubs, large, bushy, rarely herbs; foliage heavily scented. Leaves imparipinnate; leaflets entire or dentate, numerous, crowded, small, gland-dotted; stipels minute; stipules small, awl-shaped, caducous or absent. Flowers irregular, imperfect, small, bluish violet, whitish, or dark purple, densely clustered in long, single or branched, terminal, spikelike racemes; bracts and bracteoles narrow, awl-shaped, caducous; calyx inversely conical, usually gland-dotted; lobes 5, subequal, lowest lobe longer, persistent; petal 1; standard obcordate, erect, narrowed into a claw, incurved, enveloping the stamens and style; wings and keel petals absent; stamens 10, monadelphous; filaments elongated, long-exserted beyond the standard, united below into a tube; ovary sessile, 2-ovuled; style slender, recurved, glabrous or pubescent; stigma terminal, capitate. Pod short, oblong, often recurved and asymmetric, usually gland-dotted, indehiscent, 1–2-seeded; seeds glossy, oblong, subreniform. (Bentham and Hooker 1865; Palmer 1931)

Members of this genus are indigenous to North America, ranging from southern Canada to northern Mexico. They are well represented in Texas and California. Although plants are found along stream beds, primarily they grow in dry and sunny areas, on prairies, plains, and hillsides up to 1,600-m altitude. Species are commonly called false indigo, spicebush, or river locust.

These species are useful as windbreaks in border and buffer strips because of their dense growth, firm anchorage, shrubby habit, and immunity to insect pests. *A. canescens* (Nutt.) Pursh, *A. fruticosa* L., and *A. microphylla* Pursh provide wildlife food and are suitable for landscape planting, soil cover, and erosion control. The type species is a fine ornamental bush. *A. canescens*, leadplant, was regarded as an indicator of lead ore in southern Wisconsin and northern Illinois about 100 years ago, when lead and zinc mining there was at its peak. Claims that this species is a zinc-indicator plant were discounted by Cannon (1960a).

In the late 1930s, certain color tests considered specific for rotenone indicated the presence of that insecticide in the roots, stem bark, and seed of *A. fruticosa*, giving rise to the hope that the species might serve as a source of that insecticide during World War II; however, neither rotenone nor any rotenoid compounds proved to be present. Amorphin, $C_{33}H_{40}O_{16}$, a glycoside, and its aglycone, amorphigenin, $C_{22}H_{22}O_7$, which presumably accounted for the aforementioned false color tests for rotenone, were isolated from seed extracts by Acree et al. (1944). The resinous pustules of the seed pods yield amorpha, a toxic principle for aphids, cinch bugs, cucumber beetles, and a repellent against cattle flies (Brett 1946; Roark 1947).

NODULATION STUDIES

Rhizobia from *Amorpha* nodules are monotrichously flagellated short rods that produce moist to gummy colonies on yeast-extract mannitol and calcium glycerophosphate agars, and alkaline reactions without serum zones in litmus milk (Bushnell and Sarles 1937).

Nodules were described on *A. canescens*, *A. fruticosa*, and *A. microphylla* by Bolley as early as 1893. Limited plant-infection tests by Maassen and Müller (1907) and Burrill and Hansen (1917) prompted consideration of *A. canescens* and *A. fruticosa* as a select rhizobia-plant group, a concept that prevailed until the late 1930s. Later studies (Bushnell and Sarles 1937; Wilson 1939a, 1944b; Appleman and Sears 1943; Wilson and Chin 1947) evince a limited host-range profile of *Amorpha* species and rhizobia, yet a close affinity within the cowpea miscellany. Histological studies of *A. fragrans* Hort. nodules revealed nothing unusual (Lechtova-Trnka 1931).

Nodulated species of *Amorpha*	Area	Reported by
canescens (Nutt.) Pursh	U.S.A.	Many investigators
elata Bouché	N.Y., U.S.A.	Wilson (1939a)
fragrans Hort.	France	Lechtova-Trnka (1931)
fruticosa L.	Widely reported	Many investigators
var. *angustifolia* Pursh	Wis., U.S.A.	Bushnell & Sarles (1937)
glabra Desf.	Wis., U.S.A.	J. C. Burton (pers. comm. 1970)
herbacea Walt.	Wis., U.S.A.	Bushnell & Sarles (1937)
microphylla Pursh	U.S.A.	Many investigators

Amphicarpaea Ell.
386 / 3860

Papilionoideae: Phaseoleae, Glycininae

From Greek, *amphi-*, "both," and *karpos*, "fruit"; in reference to the two kinds of fruits formed by the two kinds of flowers. The spelling *Amphicarpaea* is conserved over the more common alternate spelling, *Amphicarpa*.

Type: *A. bracteata* (L.) Fernald

3–4 species. Herbs, annual or perennial, twining, climbing. Leaves pinnately 3-foliolate; leaflets elliptic-ovate, petioluled; stipels present; stipules persistent, membranous, lanceolate. Flowers 2 types: (1) *chasmogamous*: papilionaceous, white, blue, violet, or red, in axils of upper leaves in long-peduncled, lax, axillary racemes; bracts striate, ovate to narrow-lanceolate, conspicuous, persistent; bracteoles setaceous or obscure; calyx campanulate or cylindric; tube oblique and gibbous at the base; lobes 5, subequal, upper 2 shorter and united completely or nearly so; petals small, subequal; standard oval, erect, distinctly clawed, slightly inflexed and eared above the claw, enfolding the other petals; wings falcate-obovate, adherent to the keel petals; keel petals slightly shorter than the wings, oblong, somewhat obtuse, incurved, long-clawed, loosely joined to each other; stamens diadelphous; anthers uniform; ovary subsessile or obscurely short-stalked, many-ovuled; style filiform, inflexed, glabrous or slightly villous; stigma small, globular, pubescent; (2) *cleistogamous*: imperfect flowers borne on lower, creeping branches; calyx cuneiform; petals rudimentary, scalelike, or absent; stamens free, slightly longer than the petals; anthers 2–5, fertile; ovary with short, inflexed style. Pod of chasmogamous flowers linear or falcate, thin-walled, compressed, several- to many-seeded, 2-valved, valves coiling after dehiscence; seeds compressed, subglobose; pod of cleistogamous flowers subterranean, fleshy, ellipsoid, indehiscent, usually 1-seeded. (Bentham and Hooker 1865; Shively 1897; Hauman 1954c; Turner and Fearing 1964)

Representatives of this genus are distributed throughout temperate and tropical North America, northern and central Asia, the Himalayas, Japan, and tropical Africa. They thrive in grassy and wooded areas. Asian and American species are not readily distinguishable.

The closely related members of the genus cause confusion in identification and delineation. Turner and Fearing (1964) considered that only 3 species, each confined to a separate continent, were valid: *A. africana* (Hook. f.) Harms, in the cool, higher mountainous areas of Africa up to 2,300-m altitude; *A. edgeworthii* Benth., in sunny localities, in open bush valleys, and on hills and slopes of the Himalayas and eastern Asia; and *A. bracteata* (L.) Fernald, in cool, moist, wooded habitats of North America.

The subterranean fruits and roots of *A. bracteata*, hog peanut, are edible and nourishing, and were an important food of the American Indians, especially in the Missouri Valley (Medsger 1939). It was once cultivated in southern areas of the United States. The seeds serve as food for wild game birds and rodents. All members of the genus are important in soil improvement, as soil cover, and in erosion control.

NODULATION STUDIES

Spherical nodules on *A. comosa* Pitch. were described by Schneider (1892), and on *A. monoica* (L.) Ell. by Bolley (1893). Schneider (1892) termed the microsymbionts of *Amphicarpaea* species *Rhizobium dubium* sp. nov., thus separate and distinct from other species of rhizobia.

Lenticels were observed in the corky surface tissue which developed from a phellogen layer in the cortex of *A. comosa* nodules (Schneider 1893a). Because these lenticels were located above the vascular bundles, Schneider surmised that they served for the interchange of gasses between the nodule interior and exterior.

Because of the inability of rhizobial strains from *A. monoica* to nodulate plant species of 21 diverse genera, Burrill and Hansen (1917) considered the species highly selective. Later plant-infection studies (Conklin 1936; Bushnell and Sarles 1937; Wilson 1939a) discounted this exclusiveness by showing plant-infection kinships within the cowpea miscellany.

Nodulated species of *Amphicarpaea*	Area	Reported by
bracteaea (L.) Fernald	Widely reported	Many investigators
(= *monoica* (L.) Ell.)		
var. *comosa* (L.) Fernald	Minn., U.S.A.	Schneider (1892)
(= *comosa* Pitch.)		
var. *pitcheri* (T. & G.) Fassett	Wis., U.S.A.	Bushnell & Sarles (1937)
(= *pitcheri* T. & G.)	N.Y., U.S.A.	Wilson (1939a)
edgeworthii Benth.	Japan	Pers. observ. (U.W. herb. spec. 1971)

Amphimas Pierre ex Harms

(125a) / 3568a

Papilionoideae: Sophoreae

From Greek, *amphi*, "all around," and *himas*, "strap"; all petals have narrow, straplike lobes.

Type: *A. ferrugineus* Pierre ex Pellegr.

4 species. Trees, tall, up to about 40 m, branches stout, densely pubescent, deciduous, unarmed. Leaves imparipinnate; leaflets 7–23, alternate or subopposite, lanceolate to narrow-obovate, acuminate or mucronate, entire, with numerous lateral nerves; stipels filiform, often large, entire; stipules foliaceous, caducous. Flowers cream or yellow, many, small, hermaphroditic or sometimes unisexual, on short, jointed pedicels, in long, lax, terminal, racemose panicles; bracts and bracteoles narrow, caducous; calyx briefly campanulate, 10-ribbed; lobes 5, shallow, equal, shorter than the tube, valvate, deltoid, both surfaces hairy; petals 5–6, imbricate, linear, each deeply cleft to form 2 ribbon-shaped, fleshy lobes; stamens 10, free, subequal, slightly exserted, thick at the basal insertion on the disk; filaments filiform; anthers ovoid or rounded, dorsifixed, caducous; ovary elliptic, long-stalked, 1–2-ovuled; style oblique; stigma minute, capitate, concave. Pod long-stalked, very compressed, narrow, oblong-elliptic, papery, prominently veined, 2-valved, broadly winged along both margins, 2–1-seeded; seeds oblong-reniform, sessile, with thin spongy aril. (Harms 1908a; Baker 1930; Pellegrin 1948; Wilczek 1952)

Amphimas occupies a transitional position between the advanced Caesalpinioideae and the primitive genera in Papilionoideae (Harms 1908a, 1915a; Wilczek 1952; Hutchinson 1964). The isolation of afrormosin from the wood of *A. pterocarpoides* Harms is supportive evidence for a closer alliance with Papilionoideae, since the occurrence of flavonoids is apparently restricted to this subfamily (Harborne et al. 1971).

Amphimas species are native to tropical West Africa. They flourish in dense, humid forests or rain forests in Zaire, and in drier, high forests in Liberia.

Decoctions of a red sticky resin from the bark of *A. pterocarpoides* are used medicinally in Liberia for treating dysentery, and in the Ivory Coast as a remedy for anemia. An alkaloid in the bark serves as a fish and rat poison (Kerharo and Bouquet 1950). The wood is hard and heavy but not very durable. It is used locally for minor construction.

Amphithalea E. & Z.

184 /3644

Papilionoideae: Genisteae, Lipariinae

From Greek, *amphithales*, "blooming on both sides."

Type: *A. ericifolia* (L.) E. & Z.

15 species. Shrubs, small, erect, usually silky-haired, often heathlike. Leaves simple, ovate, obovate, lanceolate, or linear, flat, or margins recurved; stipules none. Flowers small, blue purple, rose, or white, with tip of the keel dark-colored, subsessile, usually paired in leaf axils, sometimes in terminal leafy spikes, rarely in crowded heads; bract 1; calyx tubular or campanulate,

usually hairy; lobes 5, subequal, ovate-lanceolate, acuminate, sometimes short, obtuse, upper 2 broader or united above; standard obovate or orbicular, notched or bilobed; wings oblong, long-clawed; keel shorter than the wings, straight, blunt, pouched or spurred at the base, long-clawed; stamens diadelphous, vexillary stamen free from the base, others connate above; filaments linear, glabrous; anthers alternately shorter and versatile, longer and basifixed; ovary sessile or short-stalked, 1–7-ovuled; style subcylindric, incurved, villous below; stigma terminal. Pod ovate-oblong, often acute, hairy, compressed, 2-valved, nonseptate within, 1–4-seeded. (Bentham and Hooker 1865; Phillips 1951)

Polhill (1976) assigned this genus to the tribe Liparieae in his revision of Genisteae *sensu lato.*

Members of this genus are confined to South Africa, mainly in the western and southwestern region of the Cape Province. They grow in flats, hills, and mountainous areas.

NODULATION STUDIES

In South Africa *A. intermedia* E. & Z. was nodulated (N. Grobbelaar personal communication 1978).

Anadenanthera Speg.   Mimosoideae: Eumimoseae

From Greek, *an-*, "without," *aden*, "gland," and *anthera*, "anther," referring to what is now considered a questionable characteristic of the genus, its lack of anther glands.

Lectotype: *A. peregrina* (L.) Speg.

2 species. Trees, up to 37 m tall, elegant; trunks unarmed or armed at the base with conical projections; crown umbrellalike; branches contorted; twigs more or less pubescent, often warty. Leaves large, feathery, bipinnate; pinnae pairs 7–35; leaflets in 20–80 pairs, narrow, sessile, opposite; basal petiolar gland usually present. Flowers small, sessile, white, cream, or flesh-toned, 30–50, in globose heads or in fascicles of 1–7 on slender peduncles in leaf axils; calyx campanulate, 5-toothed; petals 5, free or somewhat coherent; stamens 10, 2–3 times length of the petals, free; anthers with or without glands; ovary sessile or nearly so. Pod 10–30 cm long, straight, flattened, glabrous, oblong to linear with thick margins, constricted but continuous within, not pulpy, dehiscent along 1 suture only, valves remaining attached to other suture, leathery; seeds 8–16, flat, suborbicular-elliptic, glossy, with a rim or sharp margin. (Altschul 1964)

The 2 species of this genus were formerly in section *Niopa, Piptadenia.* Segregation was made primarily on the occurrence of distinctly globose inflorescences in contrast to spicate inflorescences of *Piptadenia* species (Altschul 1964).

Both species are native to tropical South America and the West Indies. They thrive in riverside forests, savannas, and on rocky slopes up to about 2,000-m altitude.

The timber of both species is hard, heavy, fine grained, and in general use for house and heavy construction purposes. *A. colubrina* (Vell.) Brenan yields an angico-type gum, one of the Brazilian gum arabics, used for mucilage.

The seeds of both species were the source of cohoba, or yopo, a snuff used in aboriginal witchcraft and ceremonial rites in the West Indies. The ceremonial use was described as early as 1496 by Ramon Pane, who was with Columbus on his second voyage (Stromberg 1954). The snuff, inhaled through a bifurcated tube fitting the nose, produced hallucinogenic effects. A quest for the causal agent in the seeds led to the isolation of the alkaloid bufotenine, $C_{12}H_{16}N_2O$, N, N-dimethyltryptamine, and several indole bases (Stromberg 1954; Fish et al. 1955; Pachter et al. 1959). Intoxication in human subjects using the 2 alkaloids and snuff made from *A. peregrina* (L.) Speg. seeds was not confirmed in experiments conducted by Turner and Merlis (1959).

NODULATION STUDIES

Campêlo and Döbereiner (1969) described pink coralloid nodules on secondary roots of *A. peregrina* (reported as *Piptadenia peregrina*) in Brazil. The rhizobia were slow growing, and produced small colonies and alkaline cultural reactions. This plant was not nodulated by rhizobial strains from *Clitoria racemosa*, *Dolichos lablab*, *Phaseolus vulgaris*, and *Vicia graminea*, but was nodulated by strains from *Mimosa caesalpiniaefolia*.

Anagyris L.   Papilionoideae: Podalyrieae
155 / 3615

From Greek, *an-*, "without," and *gyros*, "ring," "circle," in reference to the straight keel and pod.

Type: *A. foetida* L.

2 species. Shrubs. Leaves petioled, palmately 3-foliolate; stipules joined into 1, leaf-opposed. Flowers quite large, yellow, pedicelled, several in short, axillary or apical racemes; bracts stipulelike, or small and deciduous; bracteoles absent; calyx campanulate, persistent, subbilabiate; teeth sharply indented, subequal; standard shorter than the wings, emarginate and outcurving at the apex, sides folded, not reflexed; wings oblong, long-clawed; keel longer than the wings, straight, blunt, claws and apex free; stamens free; ovary short-stalked, multiovuled; style filiform; stigma small, terminal. Pod stalked, compressed, linear, constricted and septate between the seeds, dehiscent; seeds few, kidney-shaped. (Bentham and Hooker 1865)

Both species are native to the Mediterranean area, from Tenerife to Saudi Arabia.

The leaves are sold throughout Greece and Dalmatia as *herba Anagyris*, a purgative. The seeds have emetic properties. The heavily scented foliage is poisonous to cattle.

Four alkaloids, anagyrine, cytisine, methylcytisine, and sparteine, occur in plant parts of *A. foetida* L. (Mears and Mabry 1971). The occurrence of *Baptisia*-type lupine alkaloids in *Anagyris* and in species of 3 other Northern Hemisphere genera of this tribe, *Baptisia*, *Piptanthus*, and *Thermopsis*, suggests a closer phyletic kinship to the tribe Genisteae than to the Southern Hemisphere members of the Podalyrieae (Turner 1971).

NODULATION STUDIES

Nodules were observed on *A. foetida* in France by Naudin (1897b). Rhizobial isolates from nodulated plants in Kew Gardens, England, were culturally and physiologically akin to fast-growing, acid-producing rhizobia (personal observation 1936).

Anarthrophyllum Benth.   Papilionoideae: Genisteae,
206 / 3666   Crotalariinae

From Greek, *an-*, "without," *arthron*, "joint," and *phyllon*, "leaf"; the leaves lack petioles.

Type: *A. desideratum* (DC.) Benth.

15 species. Shrubs or subshrubs, much-branched, generally silky-haired, often forming low cushions, xerophytic. Leaves rigid, sometimes sharp-pointed; leaflets 1–5, nonpetioled, entire or spinelike, united at the base; stipules often sheathlike, persistent, small, or large and leafletlike. Flowers red or yellow, solitary or in small umbels, terminal; calyx campanulate, bilabiate, upper lip 2-parted, much smaller than the lower 3-parted lip; petals blunt; standard short, exterior pubescent; wings oblong; keel blunt, incurved, folded on each side, clawed; stamens 10, joined into a tube, split above; anthers alternately long and basifixed, short and dorsifixed; ovary hairy, oblong, subsessile, few-ovuled; style smooth, incurved; stigma terminal, capitate. Pod leathery, ovoid or oblong, compressed, nonseptate, 2-valved, spirally dehiscent, 1- to several-seeded. (Bentham and Hooker 1865; Burkart 1952; Soraru 1973)

Polhill (1976) assigned this genus to the tribe Crotalarieae in his reorganization of Genisteae *sensu lato*.

Representatives of this genus occur in Argentina and Chile, southward into Patagonia. They flourish in desolate, dry, cold areas in the Andean Mountains; the plants usually grow on sandy soils.

The spineless species are browsed by sheep and goats. The heavier bushes are used as windbreaks and for firewood (Burkart 1952).

From Portuguese, *andira*, the Brazilian name for these trees.

Type: *A. racemosa* Lam. ex J. St.-Hil.

35 species. Trees, unarmed, usually large; branchlets slightly tomentellous. Leaves imparipinnate, rarely 3-foliolate; leaflets usually opposite, 7–15, narrow-oblong, slightly acuminate, at first pubescent underneath, later glabrous, terminal leaflet usually long-stalked and obovate; stipels setaceous, awl-shaped, or none. Flowers small, pink or violet, handsome, fragrant, subsessile, usually densely clustered in large, spreading, terminal or lateral panicles; bracts and bracteoles small, caducous; calyx cupular; tube broad, blunt; teeth minute, pubescent; petals 5, all clawed; standard rounded, not appendaged; wings oblong, almost straight, free; keel petals similar to the wings, free, overlapping at the back but not joined; stamens diadelphous, anthers versatile; ovary long-stalked, 2–4-ovuled, rarely 1-ovuled; style short, incurved; stigma small, terminal. Pod drupaceous, long-stalked, pendulous, glabrous, thick, ellipsoid or obovate, green, fleshy outside, hard or woody inside, indehiscent, 1-seeded. (Bentham and Hooker 1865; Amshoff 1939)

Andira species occur primarily in tropical North and South America, and the West Indies; they are sparingly represented in tropical West Africa. Common in moist to wet coastal forests, *Andira* species are abundant along stream banks, in swampy areas, and in savannas.

Andira inermis HBK, the best-known and most widely distributed species, shares such common names as partridgewood, cabbage bark, and angelin with several other members of the genus. This attractive tree often exceeds 35 m in height; the trunk is usually flanked with buttresses up to 3 m. The ragged unsightly bark is suggestive of that on cabbage palms and has an unpleasant cabbagelike odor. It has fish-stupefying properties and was the source of "worm bark" used as a vermifuge in early European medicine.

Because of their majestic stature and fragrant, showy flowers, which serve as a source of nectar for bees, hummingbirds, and butterflies, medium-sized *Andira* species are used as garden and park ornamentals. Some specimens serve as shade plants on coffee plantations. The wood of *A. inermis* is reddish brown to black, hard, heavy, very resistant to attack by fungi and insects, and has an average density of 800 kg/m^3 (Titmuss 1965). It does not feature in world markets because it is not available in quantity. It is used for furniture, cabinets, boat and bridge construction, carts, and small turnery articles, such as billiard cues, butts, umbrella handles, and canes (Record and Hess 1943).

The bark of *A. galeothiana* Standl. and *A. vermifuga* Mart. also has native uses as a fish poison, narcotic, vermifuge, and emetic.

The bark and seeds of *A. inermis* and *A. retusa* HBK yield the alkaloids berberine, and andirine or angelin (*N*-methyltyrosine). A crystalline substance found in trunk cavities of *A. araroba* Aguiar is the source of Goa or Bahia powder, from which Chrysarobin U.S.P. is prepared (Merck 1968). This yellow brown microcrystalline fungicide is used to treat skin diseases, especially ringworm.

A. laurifolia (Sw.) Willd., a xerophytic shrub of South American savannas, attracts attention as a "hypogeous tree" for its root-branches that radiate 5 m or more about the trunk and from which foliage shoots arise.

Nodulation Studies

Nodules were reported on seedlings of *A. inermis* in Hawaii (Allen and Allen 1936a) and in Trinidad (DeSouza 1966). Rhizobial colonies are moist and mucilaginous (Allen and Allen 1939). All strains produced alkaline reactions on media containing various carbohydrates with the exception of dextrose; alkaline reaction in litmus milk was accompanied by clear, distinct serum zones within 6 weeks' incubation. The strains showed an infective-host range within the cowpea miscellany. Nonavailability of seeds prevented testing of the strains on *A. inermis* seed.

Root systems of *Andira surinamensis* (Bondt.) Splitg. examined in Venezuela lacked nodules (Barrios and Gonzalez 1971).

Androcalymma Dwyer   Caesalpinioideae: Cassieae

From Greek, *aner, andros*, "man," and *kalymma*, "cowl," alluding to the resemblance of the deflexed anther to a drawn cowl.

Type: *A. glabrifolium* Dwyer

1 species. Tree, large, up to 30 m tall, with a trunk diameter about 0.5 m. Leaves imparipinnate; leaflets ovate-elliptic, alternate, 3–5, acuminate, leathery, the apical leaflet largest. Flowers very small, perfect, short-stalked in many-flowered, terminal cymes; bracts minute, very caducous; calyx fleshy; sepals 5, free, overlapping, pubescent; petals 5, overlapping, elliptic, equal, slightly longer than sepals; stamens 4, free, equal; filaments clavate, thick, abruptly narrowed at apex; anthers basifixed, erect in bud, inflexed at anthesis, subversatile, opening by 2 apical pores; ovary sessile, pubescent; style very short; stigma small. Pod not seen. (Dwyer 1957a; Koeppen 1963)

This monotypic genus is endemic in the upper Amazon Basin, Brazil.

Androcalymma heartwood is golden or russet brown; the sapwood is somewhat lighter in color. The hard, heavy wood has large pores which give it a coarse texture. The occurrence of sclerotic parenchyma, uncommon in leguminous wood, is an unusual diagnostic characteristic that permits rapid identification (Koeppen 1963). This feature is shared with species of *Martiodendron*, also in Cassieae (Koeppen and Iltis 1962).

From Greek, *ankylos*, "curved," in reference to the bent calyx.

Type: *A. oligophyllus* (Bak.) Bak. f.

10–14 species. Shrubs, 1–2 m tall, or trees, up to 20 m tall, unarmed. Leaves imparipinnate; leaflets few, petioluled, alternate or subopposite, large, somewhat acuminate. Flowers yellow or white, on short-jointed stalks in racemes developed either among new foliage or on old wood; bracts minute; calyx conical or tubular, straight or bent in the middle, minutely toothed or subentire; petals 5, clawed, unequal, inserted in mouth of calyx tube; standard suborbicular; wings and keel petals linear-oblong, equal in length; stamens 10, monadelphous, inserted in mouth of calyx tube; filaments free, flattened; anthers oblong, subbasifixed; ovary tomentose, short-stalked, stalk adnate to side of calyx tube, multiovuled; style very short; stigma terminal. Pod yellow or orange, short-stalked, fleshy, elongated, beaked, sometimes longitudinally ridged, narrowly constricted between the 2–7, narrow-oblong, turgid seeds, indehiscent. (Taubert 1896; Harms 1897, 1915a; Baker 1929; Pellegrin 1948)

Members of this genus are scattered throughout tropical West Africa, especially Nigeria, Zaire, and Cameroon. They flourish in rain forests and wooded areas.

The bright yellow wood of *A. oligophyllus* (Bak.) Bak. f. and *A. zenkeri* Harms is used locally for construction purposes.

Antheroporum Gagnep.   Papilionoideae: Galegeae,
                                             Tephrosiinae

From "anther" and Greek, *poros*, "pore."

Type: *A. pierrei* Gagnep.

2 species. Trees. Leaves imparipinnate; leaflets 5–9; stipels and stipules none. Flowers many, rose or violet, in upper axillary or terminal, fascicled racemes; bracts small, persistent; bracteoles 2, subtending the calyx; calyx tube campanulate; lobes 4, upper lobe truncate-emarginate, others triangular, equal, shorter than the tube; petals long-clawed, subequal; standard obcordate; wings oblong, narrow, joined to the ovate, oblique keel petals; stamens monadelphous or tardily diadelphous; anthers orbicular, uniform or nearly so, longitudinally dehiscent; ovary puberulent, 2-ovuled; style short, awl-shaped, apex glabrous; stigma pointed. Pod stalked, oblong, thick, woody, large, compressed, 2-valved, dehiscent, 1-seeded; seed flat, spherical, with a subapical hilum. (Gagnepain 1915)

The anthers dehisce "by a split lengthwise," not by a terminal pore, as stated in the original description (Hutchinson 1964). Gagnepain (1915) considered this genus a member of the tribe Dalbergieae, closely akin to *Fissicalyx*, which does have anthers with terminal pores.

Both species are indigenous to Vietnam and Thailand.

The seeds contain small amounts of rotenone (Jones 1942).

From Greek, *anthos*, "flower," and *nothos*, "false"; probably in reference to the large bracteoles which seem to be a part of the perianth.

Type: *A. macrophylla* Beauv.

15 species. Trees, medium to large; branchlets rusty-haired. Leaves paripinnate; leaflets few, not glandular-punctate, often pubescent below, large, symmetric, elliptic, coriaceous, midrib not curved, apical pair largest; petiolules stout, not twisted; stipels none; stipules paired, interpetiolar, free or connate, caducous. Flowers small, white or yellow, clustered along rusty-pubescent rachis of erect, terminal or axillary lax panicles; bracts minute, caducous; bracteoles 2, large, thick, concave, opposite, valvate in bud, later spreading, persistent; calyx tube cup-shaped, shallow; sepals 4–5, free, sub-equal or 2 adaxial ones larger and joined; petals small, 2–6, usually only 1 well-developed, broadly bilobed; fertile stamens usually 3, rarely 4–5; anthers opening by slits; staminodes usually 6; ovary short-stalked; style coiled; stigma capitate. Pods variable, usually pubescent, elongated or short, thick and broad or irregular, usually diagonally ridged, usually not winged, margins thick, tardily dehiscent; seeds thick, irregularly shaped. (Léonard 1955, 1957; Voorhoeve 1965)

These species were originally included in *Macrolobium*, which is now reserved entirely for the distinctive American elements (Cowan 1953). Within recent years the excluded African species of *Macrolobium* have been accommodated to *Anthonotha*, *Gilbertiodendron*, *Isomacrolobium*, *Paramacrolobium*, and *Pellegriniodendron* (Keay et al. 1964; Voorhoeve 1965).

Representatives of this genus occur in equatorial West Africa from Sierra Leone to the Congo Basin. They flourish in riverine and rain forests and savanna woodlands.

Most of the tree species have hard, tough wood suitable for local construction purposes. The high forest tree, *A. fragrans* (Bak. f.) Exell & Hillc., reaches a height of about 30 m, with a trunk diameter of 1 m.

Peyronel and Fassi's account (1960) of ectotrophic mycorrhiza on the roots of *A. macrophylla* Beauv. is one of the rare reports of this alternate type of symbiosis in tropical leguminous plants. Apparently, it is restricted to members of the tribe Amherstieae.

Anthyllis L.   Papilionoideae: Loteae
230 / 3691

From Greek, *anthyllis*, a plant name used by Dioscorides.

Type: *A. vulneraria* L.

30–50 species. Herbs or shrubs, evergreen or deciduous, annual or perennial. Leaves imparipinnate, 5- to many-, rarely 1–3-foliolate; leaflets narrow-elliptic to obovate, 1 or both surfaces silky-haired; stipules minute and caducous, or absent. Flowers yellow, white, crimson, or purple, usually in dense, axillary heads or clusters, rarely in fascicles or singly in bract axils; bract size varies, usually small and setaceous, or absent, some palmate, larger than the calyx; calyx tubular, campanulate, often inflated after anthesis; lobes 5, subequal or upper 2 longer and partly connate; petals long-clawed, 4 lower ones often joined to the base of the staminal tube; standard ovate, abrupt or 2-eared at the base; wings ovate, obtuse; keel petals obtuse or apiculate, shorter than the wings, incurved, swollen on each side, usually adnate to the staminal tube; stamens at first all connate into a closed tube, later upper stamens partly or entirely free; anthers uniform; ovary stalked, rarely subsessile, 2- to many-ovuled; style glabrous; stigma terminal. Pod usually stalked, ovoid, short-linear, straight, falcate, or arcuate, equal in length to, and usually included within, the persistent calyx, continuous within or transversely septate, indehiscent or tardily 2-valved. (Bentham and Hooker 1865; Rehder 1940; Komarov 1971)

Members of this genus are distributed throughout Europe, North Africa and western Asia; there are about 20 species in the Mediterranean area. The plants are common in meadows and pastures.

The species are well adapted to dry soils, seed profusely, and serve as ground cover, soil-binders, and fodder plants, especially for sheep. The type species, known as sand clover or kidney vetch, has been cultivated in Europe since about 1858 as a forage, meadow, and prairie plant. In ancient medical lore this herb was called woundwort because of its use to heal wounds.

NODULATION STUDIES

Roots of *A. henoniana* Coss. ex Batt. lacked nodules in Libya (B. C. Park personal communication 1960). Prevailing arid conditions may have accounted for these negative data.

Our knowledge of rhizobia from *Anthyllis* species is limited almost entirely to strains from *A. vulneraria* L. Nodules were described on this species by Treviranus as early as 1853. Of historical interest is the fact that rhizobia of this species were included among those given the name *Rhizobium radicicola* by Hiltner and Störmer in 1903.

Anthyllis nodules are of the *Phaseolus* type. Vascular strands are prominent in the cortical surface (Štefan 1906). The rhizobia are monotrichously flagellated rods. Bacteroid forms are swollen at one end and resemble clover bacteroids.

Early rhizobial studies led to the consideration of a *Lotus-Anthyllis* plant group, restrictive *inter se* (Fred et al. 1932). Wilson's results (1939a) showed that seedlings of *A. vulneraria* were nodulated by rhizobia from diverse leguminous species. Subsequent studies (Jensen 1963, 1964; Abdel-Ghaffer and Jensen 1966; Jensen 1967; Jensen and Hansen 1968) confirmed and extended Wilson's conclusion that promiscuous symbiotic kinships prevail for these species.

Nodulated species of *Anthyllis*	Area	Reported by
cornicina L.	S. Africa	Grobbelaar & Clarke (1974)
jacquinii Kerner	Austria	Pers. observ. (U.W. herb. spec. 1971)
lotoides L.	S. Africa	Grobbelaar & Clarke (1974)
montana L.	S. Africa	Grobbelaar et al. (1967)
polyphylla Kit.	Hungary	Pers. observ. (U.W. herb. spec. 1971)
vulneraria L.	Widely reported	Many investigators

Antopetitia A. Rich.   Papilionoideae: Hedysareae, Coronillinae

Named for Dr. Anton Petit, botanical research worker in Ethiopia.

Type: *A. abyssinica* A. Rich.

1 species. Herbs, annual; stems long, slender, pubescent. Leaves imparipinnate; leaflets subsessile, 5–11, small, narrow, oblanceolate; stipels lacking; stipules 2, minute, glandlike. Flowers yellow orange, small, long-stalked, 2–8, in axillary umbels; bracts slender; bracteoles minute, filiform, caducous; calyx tubular, pubescent; lobes 5, acutely deltoid, subequal, ciliate; all petals very long-clawed, subequal in length; standard rounded; wings oblong; keel blunt, straight; stamens diadelphous; anthers uniform; ovary linear, stalk attached laterally to the calyx tube; ovules several; filiform, incurved; stigma enlarged, terminal. Pod recurved along dorsal suture, with 3–5 distinct, spherical, 1-seeded segments, each 2-valved, dehiscent, dorsal suture persistent; seeds globose. (Léonard 1954a)

This monotypic genus is endemic in tropical Africa. It occurs primarily in mountainous regions of Ethiopia, Nigeria, the Sudan, and Zimbabwe. The plants grow along paths and roadsides up to 3,400-m altitude.

NODULATION STUDIES

Corby (1971) reported branched nodules on *A. abyssinica* A. Rich. in Zimbabwe.

Aotus Sm.   Papilionoideae: Podalyrieae
174 / 3634

From Greek, *a-*, "without," and *ous, otos,* "ear," in reference to the absence of bracteoles.

Type: *A. villosa* (Andr.) Sm.

15 species. Shrubs, small, several species only 0.3 m tall; branches pubes-

cent, often wandlike. Leaves simple, narrow-linear, sometimes whorled; surface above shiny, margins recurved; stipules absent. Flowers small, yellow or mixed with purple, borne in axillary, short-pedicelled clusters of 3, or in short, terminal racemes; bracts small, soon deciduous; bracteoles absent; calyx pubescent; lobes subequal or upper 2 broader and joined in an upper lip; petals long-clawed; standard usually yellow, orbicular, longer than the oblong wings; keel incurved, often purple; stamens free; ovary sessile or stalked, silky-villous, 2-ovuled; style filiform; stigma terminal, minute. Pod small, ovate, flat or turgid, 2-valved, longer than the persistent calyx, 1–2-seeded. (Bentham 1864)

Aotus species are confined to Australia. They are well distributed in New South Wales, Queensland, Victoria, and Western Australia and are widespread in Tasmania. The plants grow in tablelands, wooded valleys, and heath and coastal areas.

All species are good ground cover.

NODULATION STUDIES

Infusions of crushed, elongated, unbranched nodules of *A. villosa* (Andr.) Sm. were noninfective on alfalfa, clover, peas, vetch, and garden beans (Ewart and Thomson 1912). Cowpea plants responded ineffectively to a rhizobial strain inoculum from *A. lanigera* A. Cunn. (McKnight 1949a). Lange (1961) obtained sporadic nodulation of cowpea, soybean, *Phaseolus lathyroides*, and imported lupine species by inoculation with strains from 3 *Aotus* species; however, nodulation of native *Lupinus* species was consistently more favorable. Significantly effective symbiosis was obtained with *Lupinus digitatus* Forsk. as the host plant (Lange and Parker 1961).

Nodulated species of *Aotus*	Area	Reported by
lanigera A. Cunn.	Queensland, Australia	McKnight (1949a)
preissii Meissn.	W. Australia	Lange (1959)
tietkinsii F. Muell.	W. Australia	Lange (1959)
villosa (Andr.) Sm.	N.S. Wales, Australia	Ewart & Thompson (1912)
	W. Australia	Benjamin (1915); Lange (1961)

Apaloxylon Drake de Castillo
(45c) / 3494b

Caesalpinioideae:
Cynometreae

From Greek, *hapalos*, "tender," and *xylon*, "wood."

Type: *A. madagascariensis* Drake del Castillo

2 species. Trees with trunks cylindric, straight, ash gray, barren of twigs except at the top. Leaves paripinnate; leaflets small, 15–18 pairs, deciduous. Flowers short-stalked, 3–4 in axillary racemes, developing after leaf-fall; bracts and bracteoles caducous; calyx concave-tubular with annular disk; se-

pals 4, ovate, white, imbricate; petals none; stamens 10, yellow, unequal; anthers very small, 2-lobed, dorsifixed, dehiscent by lateral slits; ovary short-stalked, linear, asymmetric; style elongated. Pod samaroid, asymmetric-oblong, indehiscent, 1-seeded; seed pendulous from top of pod.

This genus was positioned by Harms (1908a), who regarded it as closely allied to *Bathiaea*. Viguier (1949) described the second species, *A. tuberosum* R. Vig.

Both species are native to Madagascar. They grow on rocky ground.

Aphanocalyx Oliv.   Caesalpinioideae: Cynometreae
45 / 3493

From Greek, *a-*, "not," and *phaneros*, "visible," in reference to the obsolete calyx.

Type: *A. cynometroides* Oliv.

3 species. Trees, tall, unarmed. Leaves short-petioled; leaflets 1 pair, sessile, hemielliptic, 2–3-nerved, leathery; stipules linear, very caducous. Flowers white, congested in short axillary racemes; bracts membranous, caducous; bracteoles 2, opposite, sepaloid, enclosing the bud, persistent; calyx rudimentary or reduced to minute teeth; petal 1, posterior, obovate with cuneiform base, slightly longer than the bracteoles; the rest usually rudimentary or only 1 other petal somewhat developed; stamens 10, all fertile; filaments slender, smooth, free or slightly joined at the base; anthers small, dorsifixed, longitudinally dehiscent; ovary densely brown-pilose, short-stalked; style filiform; stigma terminal, capitate. Pod flattened, asymmetric-oblong, valves 2, woody, dehiscent; seeds flattened, disk-shaped. (Oliver 1871; Léonard 1952, 1957)

Members of this genus are indigenous to tropical West Africa; they occur sporadically in dense forests up to 1,400-m altitude.

These trees reach a height of 24–45 m, with long, clear trunks measuring 50–70 cm in diameter. The trunks and branches exude a brown gum.

*Apios Fabr.   Papilionoideae: Phaseoleae, Erythrininae
398 / 3874

From Greek, *apios*, "pear," in reference to the shape of the tubers.

Type: *A. americana* Medik.

10 species. Herbs, perennial, climbing, arising from slender stocks having tuberous enlargements. Leaves imparipinnate; leaflets 3–9, usually 5–7, ovate-lanceolate, acuminate, petioluled; stipels minute; stipules small. Flowers small, several, pedicelled, purple, brownish red, or white, in axillary, rarely terminal, peduncled racemes; rachis nodose; bracts linear to oblong, cadu-

cous; bracteoles 2, small, linear, caducous; calyx campanulate to turbinate, somewhat 2-lipped, 2 upper lobes broad, obtuse, ovate, somewhat connate, 2 lateral lobes short, triangular, acute, the lowest lobe longest, triangular-lanceolate; standard short-clawed, ovate, obovate, orbicular, or heart-shaped, reflexed, obscurely eared at the base; wings shorter than the standard, obovate or linear-oblong, short-clawed, slightly eared, adherent to the keel; keel petals longest, short-clawed, strongly incurved, becoming coiled; stamens diadelphous; ovary sessile; ovules numerous; style smooth, slender, thickened and incurved above, not barbate; stigma terminal. Pod linear, subfalcate, smooth, flat, the 2 valves spirally twisting after dehiscence; seeds several to many. (Bentham and Hooker 1865; Wilbur 1963)

Apios species are native to eastern and central North America and eastern Asia. The type species is indigenous to the United States. The plants are common in thickets, low, damp areas, and along ponds and marshes.

The tuberous rootstocks of the type species, known as groundnut, wild bean, bog or Indian potato, are sweet and edible raw, boiled, or roasted. As many as a dozen moniliform tubers may occur along the fibrous lateral roots. Early American historians proclaimed the groundnut one of the most important plants in Martha's Vineyard, where it saved the lives of many Pilgrims during their first year in New England. According to Asa Gray (*fide* Medsger 1939), had civilization started in America, *A. americana* Medik. "would have been the first edible tuber to be developed and cultivated." The tubers on *A. priceana* Robins. are solitary and very large.

NODULATION STUDIES

Nodules were recorded on *A. americana*, as *A. tuberosa* Moench, by Harrison and Barlow (1906) and Edwards and Barlow (1909). Neither report specified root-nodule location relative to tuber formation. Holm (1924) noted the absence of root hairs on secondary roots of mature plants; his drawings showed nodular swellings only on tuber rootlets.

Four rhizobial strains isolated from *A. americana* nodules were not tested on the host because of nonavailability of seed, but they produced nodules on cowpea plants (Carroll 1934a); accordingly, the species was considered a member of the cowpea miscellany. In confirmation, Bushnell and Sarles (1937) described the rhizobia as monotrichously flagellated rods with cowpea-type, slow cultural growth.

Apoplanesia C. Presl   Papilionoideae: Galegeae, Psoraliinae
241 / 3704

From Greek, *apo*, "free," and *plane*, "distinctly"; the keel petals are not united.

Type: *A. paniculata* C. Presl

1 species. Shrub, erect, or small tree, up to 10 m tall, unarmed; foliage glandular-punctate. Leaves imparipinnate; leaflets subopposite, numerous, entire, small, black-glandular-punctate; stipels and stipules none. Flowers

white, many, small, borne in axillary and terminal panicled racemes; calyx membranous; lobes obtuse, subequal, enlarged after flowering, foliaceaous, 3-ribbed, reticulate; petals subequal, short-clawed, free; standard obovate-oblong, reflexed; wings oblique-linear; keel petals free, obtuse, oblong-spathulate, undulate; stamens 10, joined at the base, sheath split above; anthers uniform; ovary sessile, 1-ovuled; style filiform, glabrous; stigma oblique, capitate. Pod semiorbicular, flat, leathery, boat-shaped, apiculate, rugose, half included in the calyx, conspicuously glandular-punctate, indehiscent, 1-seeded. (Bentham and Hooker 1865; Rydberg 1919a)

This monotypic genus is endemic in Mexico, Guatemala, and Venezuela. The plants grow in dry, lightly forested areas on hillsides or plains.

Although unimportant commercially, *Apoplanesia* wood is hard, resilient, resistant to decay, and easy to work. These qualities explain its widespread use among the Indians of Central America in making bows, and for turnery and small cabinets (Record and Hess 1943).

The dense masses of white flowers, each bearing a yellow spot at the base of the standard, add ornamental beauty to the trees. The bark yields a yellow dye.

Aprevalia Baill.   Caesalpinioideae: Eucaesalpinieae
104a / 3562

Type: *A. floribunda* Baill.

1 species. Tree; branches twisted. Leaves bipinnate; pinnae pairs 3–5; leaflets narrow-ovate, 5–7 pairs and an apical leaflet. Flowers medium-large, greenish yellow, in many-flowered terminal and lateral racemes on twigs, blooming before leaves unfold; bracteoles deciduous; calyx subcampanulate; lobes 5, thick, valvate, sometimes 2 connate; petal 1, small, clawed; stamens 10, in 2 rows, unequal; filaments slender; anthers large, dorsifixed, long-exserted; ovary sessile, tomentose; ovules many, about 30–40; style slender, inflated above; stigma slightly thickened. Pod not described. (Taubert 1894)

This monotypic genus is endemic in Madagascar.

**Apuleia* Mart.   Caesalpinioideae: Cassieae
77 / 3532

The spelling *Apuleia* Mart. was conserved over *Apuleja* Mart.

Type: *A. praecox* Mart. (= *A. leiocarpa* (Vog.) Macbr.)

2 species. Trees, tall, unarmed, deciduous. Leaves imparipinnate; leaflets 5–11, large, elliptic, alternate, leathery; stipules minute or none. Flowers small, white, fragrant, often polygamous, in axillary corymbs on leafless

branches; bracts minute; bracteoles none; calyx tube short, turbinate; sepals 3, thick, imbricate; petals 3, basally narrowed, imbricate; stamens 3, anthers erect, linear-oblong, basifixed; filaments apically narrowed; ovary short-stalked, stalk adnate to staminal tube; style thick, incurved; stigma terminal, truncate. Pod flattened, oblong, thinly leathery, upper suture narrowly winged, indehiscent; seeds 1–2, ovate, transverse, compressed. (Bentham and Hooker 1865)

Both species are native to tropical America; they are common in Venezuela, and from Brazil into Argentina. The plants usually grow on well-drained clay soil.

In the Amazon Basin, *A. praecox* Mart. is commonly a gigantic tree 50 m tall, with a trunk diameter of 1–1.5 m, and nearly half of the lower trunk clear of branches; yet in subtropical regions it seldom exceeds 25 m in height. The wood finishes smoothly, is durable, and has a density of 800–960 kg/m³. The freshly cut heartwood is golden yellow, but acquires a reddish, or coppery, hue upon exposure. Deeply colored specimens smell slightly rancid (Record and Hess 1943). The yellow color is due to the presence of apuleins, oxygenated flavones (Braz and Gottlieb 1968; Braz et al. 1968). Hardness and durability of the wood may be related to its high siliceous content (Koeppen and Iltis 1962). The wood has local use for heavy construction, flooring, and vehicle shafts but does not feature in export trade.

Apurimacia Harms   Papilionoideae: Galegeae, Robiniinae

Named for Apurímac, Peru, a province in the Andes, where the collections were made.

Type: *A. michelii* (Rusby) Harms

4 species. Shrubs and undershrubs. Leaves imparipinnate; leaflets small, 9–19 pairs; stipules linear-lanceolate, caducous. Flowers many, paired, in elongated, racemelike panicles; bracteoles small, at the base of the calyx; calyx short, cup-shaped, 2 upper lobes joined into a broad bifid lip, lateral and lowermost lobes deltoid or lanceolate; standard suborbicular or obovate, short-clawed, not eared; wings oblique-oblong, obtuse, clawed, appendaged; keel petals joined along dorsal edge, clawed, appendaged; stamens 10, vexillary stamen united with others only at the middle; ovary short-stalked, 4–6-ovuled; style incurved; stigma minute. Pod woody, compressed, oblanceolate, 2-valved, few-seeded. (Harms 1923a)

Representatives of this genus are native to the tropical and subtropical northern Andes in Peru and Bolivia.

From Greek, *a-*, "without," and *rhachis*, "spine," "midrib," in reference to the absence of erect branches.
    Type: *A. hypogaea* L.
    9–19 species. Herbs, some woody at the base, annual or perennial, erect or prostrate. Leaves paripinnate, rarely 3-foliolate, lowermost leaves sometimes reduced to stipules; leaflets entire, usually in 2 pairs, oblong-obovate, obtuse or apically short-pointed, upper surfaces smooth, lower surfaces usually covered with a light down; stipels absent; stipules somewhat elongated and linear, partly adnate to the base of the petiole, apices free. Flowers showy, yellow, yellowish white, or orange, sometimes with red striations, solitary or 3–several in short spicate racemes in axils of lower leaves, subtended by linear, striate bracts and bracteoles; calyx conspicuously bilabiate; calyx tube filiform, greatly elongated, resembling a peduncle; lobes membranous, upper 4 connate, lower lobe slender, distinct; petals inserted at apex of the calyx tube; standard suborbicular, without basal auricle, short-clawed; wings oblong, free, usually with a basal auricle; keel incurved, beaked; stamens 10, occasionally 9, monadelphous, all joined into a tube thickened at base, vexillary stamen sometimes aborted; globose anthers borne on longer filaments usually alternate with oblong anthers on shorter filaments, dorsifixed; ovary subsessile at base of the tube, 1–7-ovuled, borne on a short gynophore which elongates, grows downward, and penetrates the soil with the ovary at the apex. Pod subterranean, reticulate, thick, constricted but not septate between the seeds, indehiscent; seeds usually 1 in wild forms, 2–6 in cultivated forms, ovoid to oblong, fleshy. (Bentham and Hooker 1865; Chevalier 1933; Burkart 1939a; Hermann 1954; Purseglove 1968)
    Differentiation of species is fraught with difficulty; hence the number varies widely (Hermann 1954; Purseglove 1968). Only *A. hypogaea* L., *A. monticola* Krap. & Rigoni, and *A. pusilla* Benth. are annual; all others are perennial (Leppik 1971). Several decades ago, 12 commercially important varieties were in cultivation only in the United States (Loden and Hildebrand 1950). Recent coverages of varieties and cultivars have been given by John et al. (1954), Bunting (1955), Krapovickas and Rigoni (1957), Gibbons et al. (1972), and Maeda (1973a,b).
    Arachis species occur in tropical and subtropical regions. They are native to South America, but have been introduced into many areas.
    A. hypogaea, peanut, groundnut, goober, the only species in cultivation, is not known in the wild state. It is a warm-season plant, preferring a rainfall range from 50 to 100 cm a year, and is best suited to well-drained, friable, loam soils containing adequate amounts of phosphates, potash, and calcium; propagation is feasible from cuttings, but the plant is usually grown from seed.
    Investigators generally agree that Brazil was the original distribution center of the peanut. Leppik (1971) recently referred to evidence supporting

the distribution of about 15 species from the Amazon River through Brazil, Bolivia, Paraguay, and Uruguay into northern Argentina. Mendes (1947) credited specific origin of the peanut to the region now known as Mato Grosso, Brazil. According to Woodroof (1966), peanuts were cultivated as early as 950 B.C. by Indians in Brazil and Peru. Pods have been found in coastal Peruvian excavations dated about 800 B.C. The first record of the plant in North America was from a cave in Mexico dated about the time of Christ (Purseglove 1968). Evidence suggests that the peanut was being cultivated in the West Indies in pre-Columbian times. Dissemination of the peanut throughout the world is generally attributed to the voyages of the Spaniards and the Portuguese during the 16th and 17th centuries. The Magellan Expedition carried the distinctive Peruvian form into the Far East between 1519 and 1521 (Sornay 1916). Introduction of the peanut into the United States probably occurred via slave ships (Ferris 1922). Cultivation of the peanut in the environs of Jamestown, Virginia, was mentioned by Thomas Jefferson in 1781.

Importance of the peanut is multifold. Its fresh foliage is relished by hogs and cattle, produces high-quality hay, and has value as a green manure and for soil improvement. The flowers furnish rich nectar for bees. The plant is best known for its nutritive and easily digestible seeds, or nuts, that are eaten raw or roasted. They are a rich source of vitamins of the B complex, especially thiamin, riboflavin, and nicotinic acid, and are a cheap source of protein. One gram of peanuts contains about 5.8 food calories.

Peanuts rank second to soybeans in commercial importance as a source of high-quality vegetable oil characterized by the presence of arachidic and lignoceric acids as well as glycerides of oleic and linoleic acids. Major culinary uses are made of the oil in food recipes, salad dressings, deep-fat frying, the canning of sardines, in margarine, certain cheeses, in confectionery, and as a diluent for adding flavor to meat extracts. Industrially, the oil is an important ingredient in leather dressings, furniture polishes, paints, varnishes, lubricants, and insecticides. In its highly refined state the oil is a lubricant for fine watches. In lesser developed countries it is used as fuel and illuminant, although efficiency in these categories is poor. Medicinally, it is prescribed as a laxative and is an ingredient of liniments and ointments. The *British Pharmacopoeia* recognizes the oil as a substitute for olive oil. In the cosmetic trade, peanut oil is a constituent of soaps, shaving creams, pomades, and lotions.

Peanut cake, the residue following the expression of the oil, contains 7–9 percent nitrogen, about 1.4 percent phosphoric acid, and 1.25 percent potassium oxide. The cake has fertilizer value in rice paddies, and for vegetable crops, sugar cane, and bananas. It is also a satisfactory ingredient in mixed feed for cattle and work animals. Ardil, or Sarelon, an artificial fiber, is made from the cake protein. The husks, or shells, are used in the manufacture of furfural, xylose, cellulose, plastics, and mucilage (Merck 1968). The fuel-value of about 3 metric tons of peanut shells is equivalent to 1 metric ton of coal.

NODULATION STUDIES

Nodule formation and host response

Presumably, nodules were first noted on *A. hypogaea* by Poiteau in 1853. Bentham's depiction of nodulated roots of *A. glabrata* Benth. in Martius' *Flora Braziliensis*, published in the mid-19th century, warrants mention here, inasmuch as plant-root examination is not a conventional practice of taxonomic botanists.

Nodules appear on peanut roots about the time the third set of leaves is forming, but their beneficial effect on plant growth is not apparent until the 5th week (Allen and Allen 1949a). Entry of rhizobia solely through the ruptured sites of emerged lateral rootlets instead of root-hair invasion, and the presence of considerable reserve food material in the fleshy persistent cotyledons, account for these two conditions, respectively.

Peanut nodules are morphologically simple, spherical, 1–5 mm in diameter, with a broad basal connection indicative of continued pericambial activity. Their sites are invariably in the axils of lateral roots. Because of the tetrarch pattern of the xylem, this characteristic is particularly conspicuous on tap roots from which laterals emerge in an orderly alignment of 4 vertical rows (Allen and Allen 1940a).

Certain conclusions may be drawn from the inoculation studies to date: (1) The peanut is susceptible to nodulation by rhizobia from diverse species of tropical leguminous plants; however, the fixation of high levels of nitrogen is limited to relatively few rhizobial strains (Allen and Allen 1940a; Sarić 1964; Habish and Khairi 1968; Dadarwal et al. 1974). (2) The low level of nitrogen response, so frequently noted in peanut nodulation studies, is undoubtedly due to the lesser number of viable rhizobia per unit of nodule tissue (Staphorst and Strijdom 1972).

Nodule histology

Invasion of root tissue by rhizobia is unique. Normal lateral root hairs are rare or entirely absent (Pettit 1895; Richter *fide* Reed 1924; Allen and Allen 1940a). Tufted clusters or rosettes of thick-walled hairs, as noted by Waldron (1919) and Reed (1924), are present in root axils; however, these hairs do not participate in the infection process. Instead, rhizobia gain entry into pericyclic cells of the root stele through ruptured cortical tissue at the site of emerged rootlets. Initiation of nodule meristems occurs entirely within the endodermis, adjacent and in juxtaposition to the protoxylem strand from which the rootlet emerged. Thus, peanut nodules are endogenous and until emergence remain within the root endodermis, surrounded by a peripheral periderm layer whose origin is also pericyclic. Nodule emergence from the site of origin simulates rootlet emergence in its ability to push or digest its way through the ruptured cortex of the mother root (Allen and Allen 1940a).

In other respects, morphological development through maturation and senescence follows the normal pattern. Cytologically, the bacteroid zone displays three unusual aspects. (1) Because of the absence of infection threads, rhizobia are disseminated throughout the inner tissue into newly formed cells solely by passive transmission during mitosis. (2) Infection is not accompanied by host cell hypertrophy; the latter is attributable to the presence of diploid cells only (Kodama 1967). Tetraploidy, regularly observed by Wipf and Cooper (1938) in nodules of red clover, vetch, and pea, was confirmed by Kodama (1967) in apical type nodules of these and many other species, but not in species with spherical nodules. A positive correlation between the presence of infection threads and the stimulation of polyploid cell divisions was noted earlier (Wipf and Cooper 1940). Since *Arachis* nodules lack both infection threads and tetraploid cells and, moreover, are spherical, future studies should be aimed toward correlating these events in other species. (3) The most conspicuous inclusions in infected cells are spherical, plastidlike bodies, 3μ in diameter, which become so numerous they obscure the rod-shaped rhizobia (Allen and Allen 1940a). In fresh preparations these bodies were assessed to be nonviable and in microchemical tests to be aleurone, a proteinaceous plant product described earlier in peanut cotyledons (Godfrin 1884) and in peanut nodules (Lecomte 1894). Similar "spherules" were regarded as large involution forms of rhizobia (Lechtova-Trnka 1931). Inouye and Maeda (1952) called these bodies "bacteroid-balls" and suggested that they were aggregations of several rhizobial cells possibly united by a mucoidal excretion of the bacteria. Recently Staphorst and Strijdom (1972) identified these spheroplastlike cells as bacteroids by means of a fluorescent antibody technique. These bodies made their appearance in peanut nodules as young as 5 days, were limited to nodules only of *Arachis* species, and were present in effective and also ineffective associations.

Nodulated species of *Arachis*	Area	Reported by
(*diogoi* Hoehne) = *glabrata* Benth. var. *hagenbeckii* (Harms) Hermann	Australia	Norris (1959b)
duranensis [?]	India	Dadarwal et al. (1974)
erecta [?]	S. Africa	Staphorst and Strijdom (1972)
glabrata Benth.	Brazil	Bentham (1859–1870)
	India	Dadarwal et al. (1974)
hypogaea L.	France	Poiteau (1853)
f. *nambyguarae* (Hoehne)	Brazil	Seppilli et al. (1941)
Hermann	S. Africa	Staphorst and Strijdom (1972)
marginata Gardn.	India	Dadarwal et al. (1974)
prostrata Benth.	E. Africa	Horrell (1958)
	India	Dadarwal et al. (1974)
villosa Benth.	Argentina	Burkart (1952); Rothschild (1970)
	India	Dadarwal et al. (1974)
villosulicarpa Mart. ex Hoehne	S. Africa	Staphorst and Strijdom (1972)

From Greek, *archi-*, "first," and *dendron*, "tree," in reference to the primitive floral and fruit characters.

Type: *A. vaillantii* F. Muell.

30–33 species. Shrubs about 2 m tall to trees about 30 m tall, unarmed; branchlets gray to brown, often lenticellate and hollow relative to their myrmecophilous nature. Leaves bipinnate; pinnae 1–4 pairs; leaflets opposite, 2–7 pairs, usually large or variable in shape and size, oblique at the base; rachis frequently extending beyond the uppermost leaflet pair in the form of a thread; rachillar glands usually near base of petiole and invariably present between adjacent leaflets; stipules small, caducous. Flowers large, simple or compound, racemose to pseudoumbellate, axillary or cauliflorous; bracts small, usually caducous, occasionally glandlike; calyx large, cupular-campanulate, truncate, 4–5 lobed; petals 4–5, white to greenish to yellow orange, membranous or fleshy, united almost to the apex; stamens many, indefinite, united for most of length, conspicuously exserted; anthers small; ovaries 2–15 per flower, free, sessile, many-ovuled; styles long; stigmas inconspicuous or small, knob-shaped. Pod more or less fleshy, straight, curved or twisted, deeply lobed along 1 or both sutures, often bright-colored, not pulpy, valves thick-walled, contorted after tardily dehiscing; seeds blackish, transverse, ovoid to ellipsoid, aril lacking. (Bentham and Hooker 1865; de Wit 1952)

Mohlenbrock's comprehensive reviews (1963b, 1966) treated this genus as a section of *Pithecellobium*. The presence of multiple ovaries in these species is a primitive feature which some botanists interpret as an ancestral relict from Rosaceae.

Members of this genus are native to Papua New Guinea, the Moluccas, the Philippines, the Solomon Islands, and Queensland, Australia. They thrive in marshy, swampy areas and dense rain forests from near sea level to 1,600-m altitude.

Plant parts of *A. vaillantii* F. Muell. are generally considered poisonous. The pods and bark have a hot, nauseous, acrid taste and are more toxic than the beans and leaves; some form of a saponin-type compound is suspected as the active principle (Webb 1948).

Arcoa Urb.   Caesalpinioideae: Eucaesalpinioideae

Named for Count G. Arco, in honor of his contributions to wireless telegraphy.

Type: *A. gonavensis* Urb.

1 species. Shrub or small tree, with 2 kinds of short, lateral, spurlike branches, 1 bearing foliage, the other, flowers. Leaves bipinnate, clustered;

leaflets small, numerous, oblong. Flowers polygamous or dioecious, yellow, slightly fragrant, subsessile in clustered spikes; bracts small, triangular, obtuse or semilunar; calyx cupular, with inner disk hairy; lobes 5, slightly imbricate, unequal; petals 5–6, oblong, equal, overlapping; stamens 12, short, free; filaments free, as long as the anthers; anthers versatile, longitudinally dehiscent; ovary oblong, narrow; stigma sessile, broad. Pod small, short-stalked, subcylindric, oblong, fleshy, indehiscent, pulpy between the 1–6 obovate seeds. (Urban 1929)

The genus has botanical affinity with *Gleditsia*; the pods resemble those of *Tamarindus.*

This monotypic genus is endemic in Haiti, on the Island of Gonave.

*Argyrolobium E. & Z.   Papilionoideae: Genisteae, Spartiinae
212 / 3673

From Greek, *argyros,* "silvery," and *lobion,* "pod."

Lectotype: *A. argenteum* (Jacq.), E. & Z. (= *A. sericeum* (Spreng.) E. & Z.)

70 species. Herbs, undershrubs or small shrubs, silky-haired, some with simple branches from tuberous rootstocks. Leaves palmately 3-foliolate; leaflets obovate or linear, some large and mucronate; stipels none; stipules free, sometimes foliaceous. Flowers yellow, often tinged with red, solitary, paired, or many, in leaf-opposed or terminal short racemes, fascicles, or sub-umbels; cleistogamous and chasmogamous flowers in many species; bracts and bracteoles small; calyx tube short, campanulate, deeply bilabiate, often hairy; lobes 5, usually longer than the tube, upper 2 free or connate, lower 3 connate, 3-dentate; standard suborbicular, subsessile; wings free, obovate, short-clawed, usually with lunate folds between the upper veins; keel short, incurved, not beaked; stamens monadelphous; anthers alternately long and basifixed, short and versatile; ovary sessile or short-stalked, silky-haired, multi-ovuled; style incurved, glabrous; stigma terminal. Pod linear, flattened, sometimes subtorulose, silky-haired, eglandular, 2-valved, continuous or septate between the many seeds, dehiscent. (Bentham and Hooker 1865; Harvey 1894; Phillips 1951; Polhill 1968a).

Cleistogamy within this genus was described by Harms (1909, 1917a). Hutchinson (1964) grouped *Argyrolobium* with *Lupinus* to constitute Lupineae trib. nov.; however, this positioning was not considered entirely satisfactory by Polhill (1968a); in 1976 he included this genus in Genisteae *sensu stricto.*

Members of this genus are scattered throughout South Africa, the highlands of tropical Africa, the Mediterranean area, and India. Mostly xerophytes, they thrive in upland grasslands, deciduous woodlands, and arid regions.

The species serve primarily for ground cover, wildlife protection, and erosion control.

NODULATION STUDIES

Cultural and biochemical characterization of *Argyrolobium* rhizobia, their host affinities, and histological studies of *Argyrolobium* nodules merit study in the light of Hutchinson's alignment (1964) of the genus with *Lupinus* in Lupineae trib. nov.

Nodulated species of *Argyrolobium*	Area	Reported by
adscendens Walp.	S. Africa	Grobbelaar & Clarke (1972)
calycinum Jaub. & Spach	U.S.D.A.	L. W. Erdman (pers. comm. 1950)
candicans E. & Z.	S. Africa	Grobbelaar & Clarke (1972)
collinum E. & Z.	S. Africa	Grobbelaar & Clarke (1975)
eylesii Bak. f.	Zimbabwe	Corby (1974)
humile Phill.	S. Africa	Grobbelaar & Clarke (1975)
lanceolatum E. & Z.	S. Africa	Grobbelaar & Clarke (1972)
lancifolium Burtt Davy	S. Africa	Grobbelaar & Clarke (1975)
megarrhizum Bolus	S. Africa	Grobbelaar & Clarke (1975)
molle E. & Z.	S. Africa	Grobbelaar et al. (1964)
patens E. & Z.	S. Africa	Grobbelaar & Clarke (1972)
pauciflorum E. & Z.	S. Africa	Grobbelaar et al. (1967)
var. *semiglabrum* Harv.	S. Africa	Grobbelaar et al. (1967)
pilosum Harv.	S. Africa	Grobbelaar & Clarke (1975)
pumilum E. & Z.		
var. *verum*	S. Africa	Grobbelaar & Clarke (1972)
rupestre Walp.	S. Africa	Grobbelaar et al. (1967)
ssp. *rupestre*	Zimbabwe	Corby (1974)
sericeum E. & Z.	S. Africa	Grobbelaar et al. (1967)
speciosum E. & Z.	S. Africa	Grobbelaar et al. (1964)
stipulaceum E. & Z.	S. Africa	U. of Pretoria (1955–1956)
tomentosum (Andr.) Druce	S. Africa	Grobbelaar et al. (1967)
	Zimbabwe	Corby (1974)
transvaalense Schinz	S. Africa	Grobbelaar & Clarke (1972)
tuberosum E. & Z.	S. Africa	Grobbelaar et al. (1967)
	Zimbabwe	Corby (1974)
uniflorum Harv.	Libya	B. C. Park (pers. comm. 1960)
	S. Africa	N. Grobbelaar (pers. comm. 1963)

The Latin adjectival suffix, *aria*, denotes possession; the conspicuous aril is a distinguishing character.

Type: *A. robusta* Kurz

1 species. Tree, up to 16 m tall. Leaves imparipinnate; leaflets opposite; stipels present. Flowers white, in terminal panicles; calyx wide, upper 2 teeth somewhat larger; standard suborbicular; wings and keel petals very similar, short-clawed, free, falcate; stamens 10, free, unequal, fertile; anthers versatile; ovary shortly and stoutly stipitate, 2-ovuled; style filiform, reflexed; stigma lateral. Pod oblong, terete, fleshy to leathery, dehiscent; seeds 1–2, large, oblong, black, enclosed by a fleshy, red aril, cotyledons thick. (Kurz 1877; Taubert 1894)

This monotypic genus is endemic in Burma.

Arthrocarpum Balf. f.   Papilionoideae: Hedysareae,
                                                                    Stylosanthinae

From Greek, *arthron*, "joint," and *karpos*, "fruit."

Type: *A. gracile* Balf. f.

2 species. Shrubs, up to about 9 m tall. Leaves imparipinnate; leaflets narrow-elliptic, few pairs, under surfaces with distinct reddish purple reticulation between the lateral veins; stipels absent; stipules persistent. Flowers sessile, orange yellow, in unbranched, terminal or axillary spikes; bracteoles 4, whorled, persistent; calyx tubular, bilabiate; lobes 5, the upper lip consisting of the 2 upper and 2 lateral lobes, the lower lip consisting of the ventral lobe; standard orbicular, larger than the keel, tapering into a claw, not appendaged; wings oblique-oblong; keel petals narrow, incurved, obtuse, subequal; claws of the wings and keel petals adherent to the staminal sheath; stamens 10, united into a tube slit dorsally; ovary sessile, many-ovuled; style long, straight, curved upward near the tip. Pod compressed, silky-haired, jointed, segments several, subelliptic; endocarp spongy; indehiscent; seeds narrow obovoid. (Balfour 1882, 1888; Gillett 1966c)

This genus has closer affinities with *Chapmannia* (Gillett 1966c) than with *Ormocarpum* (Taubert 1894). Characters distinguishing *Arthrocarpum* from *Ormocarpum* are explicitly defined by Gillett (1966c).

Both species occur in Somalia and Socotra. They grow in sandy soil.

Species of *Arthrocarpum*, *Chapmannia*, and *Pachecoa* are the only members of the Leguminosae known to have a dark red network of star-shaped, tannin-filled mesophyll cells on the underside of the leaflets (Gillett 1966c). This pattern is more evident in old herbarium specimens than in fresh material.

A. gracile Balf. f. has ornamental properties; the other species, *A. somalense* Hillc. & Gillett, a low shrub common on red, sandy soil in Somalia, is grazed by animals.

From Greek, *arthron*, "joint," *kleos*, "glory," and *anthos*, "flower"; the fruit is jointed, the flowers are beautiful.

Type: *A. sanguineus* Baill.

20 species. Shrubs. Leaves pinnately 3-foliolate, leaflets variable in size and shape; stipels none; stipules 2, small. Flowers large, in axillary racemes; bracts short, usually in 2 vertical ranks; bracteoles 2, small, persistent; calyx campanulate; lobes 5, appearing 4-lobed, 2 upper lobes joined, lower 3 smaller; all petals clawed; standard ovoid, erect or reflexed, shorter than the wings and keel; wings usually falcate and adherent to the beaked keel; stamens 10, diadelphous, unequal, 5 longer than others; anthers oblong, subdorsifixed; ovary stalked, slender, many-ovuled; style filiform, incurved; stigma small, thickened. Pod flattened, elongated, with many, 1-seeded, oblong, membranous segments, long-stalked in the persistent calyx, the apical segment long-pointed; seeds subreniform. (Taubert 1894)

Arthroclianthus species are limited to New Caledonia.

All members of the genus have ornamental value.

From Greek, *arthron*, "joint," and Spanish, *saman*, "rain tree"; species resemble those of *Samanea*, except that the pod breaks transversely into 1-seeded joints.

Type: *A. pistaciaefolia* (Willd.) Britt. & Rose

10 species. Trees or shrubs, unarmed. Leaves bipinnate; pinnae pairs 2–30; leaflets small and many-paired or large and few-paired; petiole and rachis with or without glands. Flowers usually perfect, in globose heads, these axillary or clustered at ends of branches, solitary or paired, fascicled or grouped in short racemes; calyx campanulate-turbinate; lobes 5, short, valvate in bud; petals 5, joined at the base into a tube; stamens numerous; filaments long, exserted, all connate at the base; ovary sessile or stalked; ovules several to many; style filiform. Pods narrow-oblong, falcate or circinate, margins smooth, nearly straight and regularly indented or constricted between the thickened, leathery, 1-seeded, indehiscent segments that separate and break away from the margins at maturity; seeds lack an aril. (Burkart 1949b; Gilbert and Boutique 1952a)

This genus was established to define a South American species of *Samanea* having compressed, linear, septate pods that ultimately break into 1-seeded segments at maturity (Britton and Killip 1936). Over the years about 6 members of *Pithecellobium*, *Mimosa*, and *Samanea* have been transferred into *Arthrosamanea*; however, its generic status still awaits general acceptance.

In South America members of this genus occur in tropical Venezuela, Colombia, and Brazil; in Africa, from Sierra Leone to Zaire and Uganda.

They are commonly found along river banks and lagoons and on river islands.

The wood is strong and of good quality but not available in marketable quantities.

Artrolobium Desv.   Papilionoideae: Hedysareae, Coronillinae

From Greek, *arthron*, "joint," and *lobion*, "pod."

Lectotype: *A. scorpioides* (L.) Desv. (= *Ornithopus scorpioides* L.)

5 species. Herbs, annual, or dwarf shrubs. Leaves imparipinnate, 3–9-foliolate, terminal leaflet usually largest; stipules membranous, usually joined, opposite leaf axils. Flowers small, yellow, clustered in heads on slender, axillary or terminal stalks; bracts caducous; bracteoles absent; calyx campanulate; lobes 5, equal, very short or truncate; petals clawed; standard and wings obovate; keel short, sharply incurved; stamens 10, diadelphous; filaments somewhat dilated; ovary sessile, several-ovuled; style glabrous, incurved at right angle. Pod sessile, linear or circinately curved, jointed, ridged but not constricted between the segments, indehiscent.

The genus is a segregate derived, in part, from elements of *Coronilla* and *Ornithopus* (Hutchinson 1964).

Species of this genus are native to southern Europe, eastward to the Caucasus Mountains, and to North Africa. They thrive in dry, open habitats.

NODULATION STUDIES

Nodules were present on greenhouse plants of *A. vaginalis* (Lam.) Desv., labeled *Coronilla vaginalis*, in Kew Gardens, England (personal observation 1936).

Aspalathus L.   Papilionoideae: Genisteae, Crotalariinae
202 / 3662

From Greek, *aspalathos*, the name of a sweet-scented shrub; only a few species, however, have a perceptible odor.

Type: *A. chenopoda* L.

245 species. Shrublets or shrubs up to 2 m tall, decumbent or erect; mostly xerophytes with heath habits, fleshy, spiny, hairy or glabrous. Leaves commonly 3-foliolate or fascicled or tufted on a leaf tubercle, rarely solitary; leaflets sessile, entire, flat, broad or ericoid, pinoid, sometimes sharp-pointed; stipules none. Flowers yellow, rarely white, red, or purple, crowded in terminal racemes or spikes, or solitary or few-fascicled in leaf axils; bracts and bracteoles often foliaceous; calyx campanulate, hairy or glabrous; lobes 5,

subequal, upper 2 shorter and broader, lowest lobe longer than others, ovate, linear, or lanceolate, sometimes pungent, rarely pilose; standard usually orbicular and short-clawed; wings oblong or falcate, with linear claw, often with transverse ridges; keel blunt, incurved, sometimes slightly beaked, pouched, short, with a linear claw; stamens monadelphous, staminal tube split above; anthers unequal, alternately shorter and versatile, longer and basifixed; ovary usually short-stalked, glabrous or hairy, 2- to several-ovuled; style smooth, linear, incurved; stigma minutely capitate, terminal. Pod oblique-ovate or linear-lanceolate, acute or blunt, glabrous or silky, turgid or compressed, 1- to several-seeded. (Bentham and Hooker 1865; Phillips 1951; Dahlgren 1960, 1961, 1963a, 1965a)

Polhill (1976) assigned this genus to the tribe Crotalarieae in his reorganization of Genisteae *sensu lato*.

All representatives of this large genus are native to South Africa. They are abundant in the southwestern districts of the Cape Province. The plants grow on hillsides, in flats, in sandy areas, and among sclerophyll vegetation.

The dried leaves of many species are used for bush tea, a substitute for conventional tea. The plants are good soil-binders.

NODULATION STUDIES

Cultural and biochemical studies of *Aspalathus* rhizobia are lacking. If one accepts Wilson's spelling (1939b) *Aspalanthus* as an orthographic error, a rhizobial strain from *A. sarcodes* Vog. & Walp. incited nodules on *Crotalaria grantiana* but not on *C. verrucosa*.

Nodulated species of *Aspalathus*[a]

abietina Thunb.
acicularis E. Mey. ssp. *acicularis*
acuminata Lam. ssp. *acuminata*
albens L.
alopecurus Benth.
angustifolia (Lam.) Dahlgr. ssp.
    angustifolia
araneosa L.
argyrophanes Dahlgr.
aspalathoides (L.) Dahlgr.
asparagoides L. ssp. *rubro-fusca* (E. & Z.)
    Dahlgr.
astroites L.
biflora E. Mey.
bracteata Thunb.
capensis (Walp.) Dahlgr.
carnosa Berger
cephalotes Thunb. ssp. *cephalotes*;
    ssp. *violacea* Dahlgr.
cheonpoda L.
chortophila E. & Z. ssp. *chortophila*

ciliaris L.
cinerascens E. Mey.
commutata (Vog.) Dahlgr.
contaminatus Druce
cordata (L.) Dahlgr.
    (= *Borbonia cordata* L.)
crassisepala Dahlgr.
crenata (L.) Dahlgr.
divaricata Thunb. ssp. *divaricata* Dahlgr.
ericifolia L. ssp. *ericifolia*; ssp.
    minuta Dahlgr.
flexuosa Thunb.
gerrardii Bolus
hispida Thunb. ssp. *albiflora* (E. & Z.)
    Dahlgr.; ssp. *hispida*
hystrix L. f.
juniperina Thunb.
lactea Thunb. ssp. *adelphea* (E. & Z.)
    Dahlgr.
laeta Bolus
laricifolia Berger ssp. *canescens* (L.)
    Dahlgr.

Nodulated species of *Aspalathus* (cont.)

linearis (Burm. f.) Dahlgr. ssp. *linearis*
linguiloba Dahlgr.
macrantha Harv.
microphylla DC.
muraltioides E. & Z.
nigra L.
opaca E. & Z. ssp. *opaca*
retroflexa L. ssp. *retroflexa*
rubens Thunb.
salteri (L.) Bolus
sarcodes Vog. & Walp.
sericea Berger ssp. *sericea*

setacea E. & Z.
spicata Thunb.
    ssp. *neglecta* (Salt.) Dahlgr.;
    ssp. *spicata* Dahlgr.
spinescens Thunb. ssp. *lepida* (E. Mey.)
    Dahlgr.
spinosa L. ssp. *spinosa*
stenophylla E. & Z. ssp. *garciana* Dahlgr.
teres E. & Z. ssp. *teres*
triqueta Thunb.
ulicina E. & Z. ssp. *ulicina* Dahlgr.
uniflora L. ssp. *uniflora*

[a]With the exceptions of *A. contaminatus* Druce in South Africa (Strijdom personal communication 1969), and *A. sarcodes* Vog. & Walp. in New York (Wilson 1939b), all other species were reported nodulated in South Africa by N. Grobbelaar and his associates (1967, 1972, 1975).

Asphalthium Medik.   Papilionoideae: Galegeae, Psoraliinae

From Greek, *asphaltos*, "asphalt," in reference to the pungent, pitchlike odor of the plants.

Type: *A. bituminosum* (L.) Ktze. (= *Psoralea bituminosa*)

2 species. Herbs, perennial, gland-dotted. Leaves usually pinnately 3-foliolate, sometimes palmately 3-foliolate; leaflets entire or margins toothed; stipules lanceolate. Flowers blue violet, in heads or short spikes on long, slender peduncles; bracts ovate-acuminate; calyx campanulate, not gibbous; lobes 5, sharply spine-tipped, unequal, lowest lobe longest; standard narrow, arrow-shaped, with a short, broad claw, and small, basal lobes; wings shorter, semisagittate with long, lanceolate-arcuate, basal lobes; keel petals shortest, rounded, joined at the apex; stamens all joined at the base, 1 free above, others alternately long and short, united to near apex. Pod ovoid, with long, sword-shaped beak which eventually breaks away, pericarp joined to ripe seed, indehiscent. (Rydberg 1919a)

Botanists formerly regarded this genus congeneric with *Psoralea*.

Plants of this genus are found in southern Europe, North Africa, and Asia Minor. They grow on poor, sandy soil.

A. bituminosum (L.) Ktze. (= *Psoralea bituminosa*) is commonly known as asphalt clover. All plant parts were formerly used pharmaceutically as *herba trifolii bituminosi* to relieve stomach cramps and intermittent fever.

NODULATION STUDIES

Both *A. bituminosum* and *A. acaulis* (Stev.) Ktze. were reported nodulated as *Psoralea* species. Nodules were observed on the former in France by Naudin (1897d) and on the latter in France (Laurent 1891), in the United States (Wilson 1939a), and in South Africa (Grobbelaar et al. 1967). Presumably, *A. acaulis* is nodulated by heterogeneous strains of rhizobia (Wilson 1939a).

Astragalus L.   Papilionoideae: Galegeae, Astragalinae
298 / 3766

From Greek, *astragalos*, "anklebone," used as a form of dice; the rattle of the seeds in the dry pods resembles the sound of dice in a throwing cup.

Type: *A. christianus* L.

1,500–2,000 species. Herbs, undershrubs, or shrubs, mostly perennial, unarmed; shrubby species often with petioles hardened, spinescent, persistent. Leaves usually imparipinnate, rarely palmately 3-foliolate or 1-foliolate; leaflets entire; stipels none; stipules free or adnate to petiole, or leaf-opposed. Flowers small, purple, white, or pale yellow, racemose or spicate, rarely in umbels or solitary, on axillary peduncles or as a scape from the rootstock; bracts small, caducous, papery; bracteoles usually minute or absent; calyx tubular to campanulate; lobes subequal, shorter than tube; petals long-clawed; standard erect, ovate-oblong, or fiddle-shaped; wings oblong, gibbous, linear-clawed; keel shorter than wings, obtuse, clawed, sometimes with the 2 segments connate; stamens diadelphous; anthers uniform; ovary sessile or short-stalked, many-ovuled; style straight or curved, smooth; stigma small, terminal. Pod sessile or stalked, membranous to leathery, mostly turgid, usually 1-celled, often imperfectly divided into 2 cells by a membrane from the dorsal suture, 2-valved, dehiscent or indehiscent, several- to many-seeded; seeds kidney-shaped. (Bentham and Hooker 1865; Rydberg 1929; Barneby 1964; Komarov 1965)

This genus, the largest one in the Leguminosae, is currently grouped into more than 100 subdivisions. The North American species are chiefly perennial herbs or subshrubs with a thick stem base or caudex at or below the soil surface from which tufted, often stemless, herbaceous growth arises perennially. Old World species are more often shrubby. Many species in western and central Asia are spiny and develop phyllodes or modified leaves in xerophytic habitats.

Rydberg (1929) advocated the segregation of the North American astragali into 28 genera, but most botanists have favored retention as a single genus. However, the distinctiveness of genus *Oxytropis* is now accepted. Old World species of *Astragalus* and *Oxytropis* have a haploid chromosome number of 8 (Senn 1938b; Darlington and Wylie 1955). New World astragali have haploid numbers of 11 and 12 and rarely 13 and 14 (Vilkomerson 1943;

Turner 1956). Accordingly, *Oxytropis* species with a chromosome number of 8 are deemed more closely related to Old World astragali than are the two groups of Old and New World astragali to each other (Ledingham 1957, 1960). Barneby (1964) recognized 384 species and 184 varieties of *Astragalus* in North America. *Astragalus* is the largest genus of the Russian flora; Komarov (1965) recognized 889 species.

Representatives of this genus are cosmopolitan outside the tropics and Australia; southwestern Asia is the largest center of distribution. They are common in the northern half of the Northern Hemisphere, extending into the circumpolar Arctic Zone, and in mountains of South America and Asia Minor. Species occur in prairies, steppes, and semidesert areas. Most species are sun-loving and prefer a light, porous soil; many are drought-resistant.

Despite the importance suggested by the size of the genus, comparatively few species are significant agriculturally. Species acceptable in some areas are viewed unfavorably in others. Despite the attractive connotations of the name milk vetch, cattlemen are advised to be cautious in selecting astragali for forage. Species that bear large, inflated seed pods and those that emit a sickening, garlicky odor should be rejected as forage. Goodding (1939) commented favorably on *A. alpinus* L., *A. lonchocarpus* Torr., and *A. tenellus* Pursh; *A. arenarius* L. and *A. glycyphyllos* L. are recommended for forage purposes in Poland (Zimny 1964); *A. hypoglottis* L. is favored in the European Alps; *A. hamosus* L. is well accepted in the Mediterranean area; *A. chinensis* L. f., *A. cicer* L., *A. falcatus* Lam., and *A. sinicus* L. have been used with impunity as fast-growing pasture forage and intercycle green manure crops in Oriental rice culture, especially in Taiwan and Japan. *A. glycyphyllos* and *A. hypoglottis* are highly suitable species for calcareous soils. The short-lived African perennial, *A. abyssinicus* Steud. & Hochst., has shown promise in Kenya. About the turn of the century, Warren (1909) commented that *A. crassicarpus* Nutt. was being eaten so ravenously on midwestern prairies of the United States that it was being exterminated by stock. *A. arrectus* Gray, *A. cicer*, *A. dauricus* DC., *A. mortoni* Nutt., and *A. rubyi* Greene & Morris are useful in soil erosion control.

A. alopecurioides L. in France, *A. aristatus* L'Hérit. in the Alps, and *A. boeticus* L. and *A. echinus* C. A. Mey. in Asia Minor, are valued as ornamentals.

It is said that the large, proteinaceously rich, unripe seed pods of *A. caryocarpus* Ker-Gawl. and *A. mexicanus* DC. may be eaten in the same manner as green peas. According to Medsger (1939), quantities of beans from the former species, known as ground plum or buffalo pea, were presented as gifts by the American Indians to the members of the Lewis and Clark Expedition during their westward journey in 1804.

The importance of the astragali may be considered in four categories: as a source of a gum, as indicators of selenium and of uranium, and as locoweeds.

(1) As a source of gum tragacanth. Tragacanth gum has had a long and fascinating history. It is the best-known carbohydrate gum of the insoluble type. Theophrastus described it in the third century b.c. (Mantell 1947). It is one of the oldest drugs listed in *Materia Medica* and has been officially recognized in every edition of the *United States Pharmacopeia* since 1820.

In cold water the gum swells and forms a thick, transparent, soft, adhesive jelly but does not dissolve. After a swelling period of several hours, followed by agitation with additional water, it forms a strong mucilage. The insoluble portion, bassorin $(C_{11}H_{20}O_{10})_n$, a substance similar to pectin, constitutes 60–70 percent of the gum. The soluble portion, tragacanthin, consists of a ring containing 3 molecules of glucuronic acid and 1 molecule of arabinose with a side chain of 2 molecules of arabinose (Norman 1931). Tragacanth differs from soluble gum arabic of *Acacia* species in being more opaque, less brittle, less glassy, and duller in luster (Mantell 1947). Asia Minor, particularly Iran and Turkey, and India provide most of the world supply. The annual consumption in the United States between 1929 and 1940 averaged 120.8 metric tons.

This gum is obtained from grazing wounds, or by tapping or excising. It occurs as a result of an alteration, or metamorphosis, of the soft juicy cells in the pith portion primarily of older stems and branches. Plants are tapped after the second year and have a profitable gum life of about 7 years.

The gum is used in the dressing of yarn, the sizing of textiles and art paper, in color paste for wallpaper, for polishing sewing threads, in rope- and twine-finishing, and in mixtures for waterproofing and stiffening felt materials. Medicinally, it imparts firmness to pill masses and troches; it is a constituent of glycerine, shaving and dental creams, jellies, lotions, nail polishes, emollient skin preparations, depilatories, permanent-wave fixers; and it serves as a lubricant for surgical instruments and catheters.

The following species of *Astragalus* are among the major sources of gum tragacanth:

adscendens Boiss. & Haussk.	*hamosus* L.
brachycalyx Fisch.	*heratensis* Bge.
brachycentrus Fisch.	*kurdicus* Boiss.
cerasocrenus Bge.	*leioclados* Boiss.
creticus Lam.	*microcephalus* Willd.
cylleneus Boiss. & Heldr.	*myriacanthus* Boiss.
echidnaeformis Sirj.	*parnassi* Boiss.
elymaiticus Boiss. & Haussk.	*prolixus* Sieb.
eriostylys Boiss. & Haussk.	*pycnocladus* Boiss. & Haussk.
globiflorus Boiss.	*strobiliferus* Royle
gossypinus Fisch.	*stromatodes* Bge.
gummifer Labill.	*verus* Oliv.

(2) As selenium-indicator plants. The role of various astragali in the detection and mapping of seleniferous areas has been the subject of comprehensive reviews by Moxon et al. (1938), Beath et al. (1939, 1940), Trelease and Beath (1949), and Cannon (1957, 1960a,b, 1964, 1971). Selenium was discovered in 1817 by Berzelius in the refuse of a sulphuric acid factory. The general chemical behavior of this nonmetallic element is closely related to tellurium and sulphur, especially the latter. It is about as rare as gold and bromine (Trelease and Beath 1949). The world production of selenium in 1969 was about 614.5 metric tons; about 200 metric tons are used annually in manufacturing processes in the United States.

Astragalus species are the only leguminous representatives of the so-called Beath selenium indicators; moreover, they are the most important members in the group (Trelease and Beath 1949). Since 1934, when Beath et al. directed attention to the vigorous growth of certain astragali in seleniferous areas, about 30 species have been recognized as selenium indicators (Trelease and Beath 1949). Among them are the following species:

albulus Woot. & Stand.	†*osterhouti* M. E. Jones
argillosus M. E. Jones[a]	†*pattersoni* Gray
†*beathii* Porter	*pattersoni* Gray
†*bisulcatus* (Hook.) Gray	var. *praelongus* (Sheld.) M. E. Jones[e]
confertiflorus Gray[b]	†*pectinatus* Dougl.
confertiflorus Gray	*pectinatus* Dougl.
var. *flaviflorus* (Ktze.) M. E. Jones[c]	var. *platyphyllus* M. E. Jones[f]
†*crotalariae* (Benth.) Gray	†*preussii* Gray
†*diholcos* (Rydb.) Tidest.[d]	*racemosus* Pursh
†*moencoppensis* M. E. Jones	*sabulosus* M. E. Jones
†*oocalycis* M. E. Jones	†*toanus* M. E. Jones

† Used in geobotanical prospecting techniques (Cannon 1971).

[a–f] The revised names of these species according to Barneby (1964) are respectively: [a]*A. flavus* Nutt. ex T. & G. var. *argillosus* (Jones) Barneby, [b]*flavus* Nutt. ex T. & G. var. *candicans* Gray, [c]*flavus* Nutt. ex T. & G. var. *flavus* Barneby, [d]*bisiculatus* (Hook.) Gray, [e]*praelongus* Sheld., and [f]*nelsonianus* Barneby.

An explanation for selenium absorption is not well defined. Its essentiality as a microtrophic element (Trelease and Trelease 1938, 1939) and the possibility that its absorption has a genetic basis (Beath et al. 1939) have both been proposed. Other species of *Astragalus* remain free of selenium, or contain very low concentrations, or are not stimulated by it; moreover, some species are poisoned by minute concentrations. These striking physiological differences were shown in an experiment wherein *A. racemosus* Pursh grown in solution and sand culture was stimulated by selenium, as selenite, in concentrations from 0.33 to 9 ppm, whereas *A. crassicarpus* was injured severely by 0.33 ppm (Trelease and Trelease 1939).

Intensity of the pungent, garliclike odor of selenium-absorbing astragali is directly proportional to the amount of the element they contain. The accumulating power varies according to the species and the plant's phase of growth and physiological condition; in turn, the supply power of a soil depends upon the nature and concentration of the selenium compounds dissolved in the soil solution (Trelease and Beath 1949). Of 92 specimens of *A. racemosus* collected at various stages of development and on various types of soil, 44.6 percent of the plants contained 10–79 ppm selenium; 32.6 percent, 80–639 ppm; and 17.4 percent, 640–5,119 ppm. One plant yielded 15,000 ppm. Only 4.3 percent of the specimens lacked selenium (Trelease and Trelease 1939).

Selenium-accumulating astragali have accounted for enormous losses of cattle and sheep by acute and chronic poisoning. Horses are less affected because of their discriminating taste for food.

(3) As uranium-indicator plants. Deserving of special mention here is the

use of *A. pattersoni* Gray and *A. preussii* Gray in prospecting for uranium in the sedimentary carnotite deposits of the Colorado Plateau, where selenium was an accompanying element (Cannon 1957, 1960b). The latter species was indicative of radioactive ground as far east as Texas. Eighty-one percent of the ore that occurred within 10 m of the surface was located in endemic areas of these plants. Plants served to detect about 10 ore bodies that were missed in previous drilling procedures (Cannon 1964). The discovery of the ore body in Poison Canyon, near Grants, New Mexico, was attributed to the presence of *A. pattersoni* (Cannon 1964). Cannon (1971) also cited the use of *A. garbancillo* Cav. to locate uranium ore in Peru, and of *A. declinatus* Willd. as an indicator of hydrothermal copper-molybdenum deposits in the Soviet Union.

(4) As locoweeds. Various astragali are known as locoweeds, a designation not reserved solely for these plants. The word "loco" is a Spanish adjective meaning "crazy"; locoism is the malady. Fraps and Carlyle (1936) isolated a toxic base, locoine, from *A. earlei* Greene, but this compound was not chemically defined. Although not caused by selenium, locoism is often confused with selenium poisoning because the symptoms are similar and the malady occurs in the same localities (Trelease and Beath 1949).

Many locoweeds are unpalatable, but some animals, especially horses, tend to become addicts; in this respect, locoism resembles a drug habit. The effects of the toxic principle are cumulative. Symptoms differ with the animal affected, but among the most common are hallucinations, delirium, defective vision, slowness of gait, sunken eyeballs, and lusterless hair. One of the first indications in sheep is difficulty in herding.

The following astragali are known as locoweeds but not as selenium indicators: *A. bigelovii* Gray, *A. convallarius* Greene (= *campestris* Gray), *A. diphysis* Gray, *A. earlei*, *A. hornii* Gray, *A. lentiginosus* Dougl., *A. mollissimus* Torr., *A. nothoxys* Gray, *A. tetrapterus* Gray, and *A. wootoni* Sheld.

NODULATION STUDIES

Astragali rhizobia are small monotrichously flagellated coccoid rods. Detailed physiological studies (personal observation), based on rhizobial strains from 18 *Astragalus* species, delineated 2 cultural groups, a condition that occurs also among *Lotus* rhizobia. One group resembled clover and pea rhizobia in the production of abundant mucoid rapid growth on mannitol yeast-extract mineral salts agar and media containing glucose, maltose, galactose, and dextrin (Asō and Ohkawara 1928; Itano and Matsuura 1934; Bushnell and Sarles 1937; Chen and Shu 1944; Ishizawa 1953). Strains from 15 species in this category strongly reduced 0.05 percent selenite. In contrast, strains from *A. sinicus* L., *A. americanus* (Hook.) M. E. Jones, and *A. alpinus* L. produced scant cultural growth similar to slow-growing rhizobia, and reduced selenite only slightly. All strains grew poorly on media containing xylose as a carbon source, and produced alkaline reactions, generally with distinct serum zones, in litmus milk.

Young nodules on *Astragalus* species are ovoid and later become finger-shaped; those on *A. glycyphyllos* become rosette-shaped (Lechtova-Trnka 1931). Cortical cell stimulation opposite the triarch protoxylem strands follows rhizobial invasion in roots of *A. sinicus* (Inouye et al. 1953). Infection threads commonly appear bulged and contorted. Upon release and establishment in the bacteroid area, the rhizobia become pear-shaped and later spherical. Infected host cells commonly undergo a sixfold to eightfold increase in volume. The morphogenesis and development of nodules on 3 species of *Astragalus* were described in detail by Lechtova-Trnka (1931). Ishizawa and Toyoda (1955) observed bacteroidal forms in effective nodules, but only small rod-shaped rhizobia in ineffective nodules. The correlation between rhizobial morphology and effectiveness has since been confirmed for many other genera. The transformation to pleomorphic bacteroidal forms is now known to be an essential prelude for nitrogen fixation (see Quispel 1974).

Nodulated *Astragalus* species occur in geographical habitats ranging from frigid to warm, and from wet to arid. Extensive studies conducted throughout several decades at the Plant Introduction Nursery at Kirovsk on the Kola Peninsula within the Arctic Circle have emphasized the importance of leguminous plants for soil improvement and the maintenance of reindeer herds in arctic regions. Of interest here is the recording of nodules on astragali species introduced into the botanical garden at Kirovsk (Roizin 1959). Typical rhizobia were isolated from *A. frigidus* (L.) Bge. and *A. subpolaris* Boriss. & Schischk.

Eight *Astragalus* species were nodulated abundantly in various Alaskan sites ranging from Umiat and Point Barrow along the Arctic Coast to Matanuska Valley (Allen et al. 1964). The plants were growing in a 30–50-cm soil layer overlying permafrost. The nodules were bifurcated, and their red interiors evidenced leghemoglobin. Effective symbiosis was obvious in the thriftiness and luxuriant green foliage. *A. alpinus* and *A. umbellatus* Bge. were frequently found effectively and copiously nodulated scant centimeters above the perpetual ice layer in tundra sites about 5 km south of Point Barrow, at 71° 18′ north latitude, the northernmost location in the United States. Isolates from these nodules showed various degrees of effectiveness as inocula for *A. cicer*.

Absence of nodules on *A. alopecias* Pall., *A. ammodendron* Bge., *A. campyloninchus* Fisch. & Mey., *A. chirvensis* [?], *A. filicaulis* Fisch. & Mey. and *A. nothoxys* Gray, examined in Arizona by Martin (1948), and on *A. lanigerus* Desf. and *A. leucacanthus* Boiss., examined in Libya by B. C. Park (personal communication, 1960) is believed to be due to unfavorable ecological or seasonal circumstances.

Results of plant-infection tests in earlier years prompted the conclusion that symbiotic associations within this genus were exclusively *inter se* (Bushnell and Sarles 1937; Chen and Shu 1944; Ishizawa 1954). The results of tests conducted by Wilson (1939a, 1944b) and Wilson and Chin (1947) were to the contrary. Their results with *A. rubyi* do not merit consideration, since this species is now accepted as *Oxytropis riparia* Litv. (Barneby 1964). Nonetheless, host plant:rhizobial relationships of 18 astragali (personal observation) showed marked versatility. Notably, all astragali rhizobia did not nodulate all *Astragalus* species tested. Five strains nodulated 2 species of *Phaseolus*, 3 strains nodulated alfalfa and sweet clover, and 3 strains nodulated cowpea. All strains were noninfective on *Lupinus, Glycine, Trifolium, Pisum, Lathyrus,* and

Vicia species, and *Phaseolus lunatus.* In reciprocal plant-infection tests, rhizobial strains from alfalfa, red clover, and bean produced nodules on *Astragalus* species. Moreover, upon reisolation, these nodules yielded rhizobia, which again nodulated their respective original host plants.

Nodulated species of *Astragalus*	Area	Reported by
aboriginum Richards.	Alaska, U.S.A.	L. J. Klebesadel (pers. comm. 1961)
adpressus Labill.	N.Y., U.S.A.	Wilson (1939a)
agrestis Dougl. ex Hook.	N.Dak., U.S.A.	M. D. Atkins (pers. comm. 1950)
alopecuroides L.	France	Lechtova-Trnka (1931)
alpinus L.	Widely reported	Many investigators
americanus (Hook.) M. E. Jones	Alaska, U.S.A.	Allen et al. (1964)
annularis Forsk.	Libya	B. C. Park (pers. comm. 1960)
arenarius L.	Poland	Zimny (1964)
aristatus L'Hérit.	European Alps	Kharbush (1928)
armatus Willd.	U.S.D.A.	L. T. Leonard (*Rhizobium* collection 1956)
asper Jacq.	S. Africa	N. Grobbelaar (pers. comm. 1966)
atropilosulus (Hochst.) Bge. ssp. *burkeanus* (Harv.) Gillett var. *burkeanus*	Zimbabwe	Corby (1974)
austrinus Small	Ariz., U.S.A.	Martin (1948)
berytheus Boiss.	Israel	M. Alexander (pers. comm. 1963)
bisulcatus Gray	N.Y., U.S.A.	Wilson (1944b)
boeticus L.	Israel	M. Alexander (pers. comm. 1963); Hely & Ofer (1972)
	S. Africa	Grobbelaar & Clarke (1974)
brachycarpus Bieb.	N.Y., U.S.A.	Wilson (1939a)
brazoensis Buckl.	Tex., U.S.A.	Ball (1909)
callichrous Boiss.	Israel	Hely & Ofer (1972)
canadensis L.	Widely reported	Many investigators
caryocarpus Ker-Gawl.	Widely reported	Many investigators
ceramicus Sheld. var. *filifolius* (Gray) Hermann (= *Phaca longifolia* Nutt.)	Nebr., U.S.A.	Warren (1909)
chaborasicus Boiss. & Haussk.	U.S.A.	Burton et al. (1974)
chamissoni [?]	Argentina	Rothschild (1970)
chinensis L. f.	N.Y., U.S.A.	Wilson & Chin (1947)
cicer L.	Widely reported	Many investigators
coulteri Benth.	N.Y., U.S.A.	Wilson (1944b); Wilson & Chin (1947)
crassicarpus Nutt.	Widely reported	Many investigators
cruciatus Link	Libya	B. C. Park (pers. comm. 1960)
	Israel	Hely & Ofer (1964)
decumbens Gray	N.Dak., U.S.A.	Bolley (1893)

Nodulated species of *Astragalus* (cont.)	Area	Reported by
drummondii Dougl.	N.Dak., U.S.A.	M. D. Atkins (pers. comm. 1950)
echinus C. A. Mey.	N.Y., U.S.A.	Wilson (1944b); Wilson & Chin (1947)
eucosmus Robins.	Alaska, U.S.A.	Allen et al. (1964)
falcatus Lam.	Widely reported	Many investigators
flexuosus (Hook.) Dougl.	N.Dak., U.S.A.	Bolley (1893)
frigidus (L.) Bge.	Arctic U.S.S.R.	Roizin (1959)
galegiformis L.	S. Africa	N. Grobbelaar (pers. comm. 1966)
glycyphyllos L.	Widely reported	Many investigators
gracilis Nutt.	Wis., U.S.A.	Burton et al. (1974)
gummifer Labill.	U.S.D.A.	L. T. Leonard (pers. comm. 1950)
hamosus L.	France	Clos (1896); Lechtova Trnka (1931)
	Israel	Hely & Ofer (1972)
harringtonii Cov. & Standl.	Alaska, U.S.A.	Allen et al. (1964)
hornii Gray	N.Y., U.S.A.	Wilson (1939a)
hypoglottis L.	European Alps	Kharbush (1928)
kentrophyta Gray	Utah, U.S.A.	Pers. observ. (U.W. herb. spec. 1971)
kralikii Coss ex Batt.	Libya	B. C. Park (pers. comm. 1960)
lentiginosus Dougl. var. *fremontii* (Gray) S. Wats.	Calif., U.S.A.	Pers. observ. (U.W. herb. spec. 1971)
(*lepagei* Hult.) = *aboriginum* Richards.	Alaska, U.S.A.	Allen et al. (1964)
leptocarpus T. & G.	Tex., U.S.A.	Pers. observ. (U.W. herb. spec. 1971)
leucophyllus T. & G.	Wis., U.S.A.	Burton et al. (1974)
lindheimeri Engelm. ex Gray	Tex., U.S.A.	Pers. observ. (1960)
lonchocarpus Torr.	Ariz., U.S.A.	Martin (1948)
(*lotoide* Lam.) = *sinicus* L.	Japan	Hosoda (1928)
membranaceus Bge.	N.Y., U.S.A.	Wilson (1939a); Wilson & Chin (1947)
menziesii Gray	N.Y., U.S.A.	Wilson (1939a); Wilson & Chin (1947)
mexicanus DC.	Ark., U.S.A.	Pers. observ. (U.W. herb. spec. 1971)
miser Dougl.	Wis., U.S.A.	J. C. Burton (pers. comm. 1971)
missouriensis Nutt.	N.Dak., U.S.A.	Pers. observ. (U.W. herb. spec. 1971)
mollissimus Torr.	Okla. U.S.A.	L. T. Leonard (pers. comm. 1940)
mongholicus Bge.	N.Y., U.S.A.	Wilson (1939a)
monspessulanus L.	S. Africa	N. Grobbelaar (pers. comm. 1955)
mortoni Nutt.	N.Y., U.S.A.	Wilson (1944b)
neglectus (T. & G.) Sheld.	Ontario, Canada	Pers. observ. (U.W. herb. spec. 1971)

Nodulated species of *Astragalus* (cont.)   Area   Reported by

nigrescens Nutt.	Calif., U.S.A.	Pers. observ. (U.W. herb. spec. 1971)
nuttalianus DC.	N.Y., U.S.A.	Wilson & Chin (1947)
var. *trichocarpa* T. & G.	Tex., U.S.A.	Pers. observ. (U.W. herb. spec. 1971)
odoratus Lam.	S. Africa	Grobbelaar et al. (1967)
(*onobrychis* Polle) = *hypoglottis* L.	N.Y., U.S.A.	Wilson (1944b); Wilson & Chin (1947)
orbiculatus Ledeb.	U.S.A.	Burton et al. (1974)
pectinatus (Hook.) Dougl.	N.Y., U.S.A.	Wilson (1944b); Wilson & Chin (1947)
podocarpus C. A. Mey.	Italy	Cappelletti (1928)
ponticus Pall.	S. Africa	N. Grobbelaar (pers. comm. 1966)
purshii Dougl.	N.Dak., U.S.A.	M. D. Atkins (pers. comm. 1950)
reflexistipulatus Miq.	Japan	Ishizawa (1954)
robbinsii Gray	S. Africa	N. Grobbelaar (pers. comm. 1966)
(*rubyi* Green & Morris) = *Oxytropis riparia* Litv.	N.Y., U.S.A.	Wilson (1939a)
sealei Lepage	Alaska, U.S.A.	Allen et al. (1964)
semibilocularis DC.	N.Y., U.S.A.	Wilson (1944b); Wilson & Chin (1947)
sikkimensis Benth. ex Bge.	S. Africa	N. Grobbelaar (pers. comm. 1966)
siliquosus Boiss.	U.S.A.	Burton et al. (1974)
sinicus L.	Japan; China	Many investigators
stipulatus D. Don	U.S.A.	Burton et al. (1974)
striatus Reiche	N.Dak., U.S.A.	M. D. Atkins (pers. comm. 1950)
subpolaris Boriss. & Schischk.	Arctic U.S.S.R.	Roizin (1959)
(*succulentus* Richards.) = *crassicarpus* Nutt.	N.Y., U.S.A.	Wilson (1944b); Wilson & Chin (1947)
sulcatus L.	S. Africa	Grobbelaar & Clarke (1974)
triphyllus Pursh	N.Dak., U.S.A.	Pers. observ. (U.W. herb. spec. 1971)
umbellatus Bge.	Arctic U.S.S.R.	Kriss et al. (1941)
	Alaska, U.S.A.	Allen et al. (1964)
utahensis T. & G.	N.Y., U.S.A.	Wilson (1944b); Wilson & Chin (1947)
verus Georgi	N.Y., U.S.A.	Wilson (1944b); Wilson & Chin (1947)
vexilliflexus Sheld.	N.Dak., U.S.A.	M. D. Atkins (pers. comm. 1950)
wootoni Sheld.	N.Y., U.S.A.	Wilson (1944b); Wilson & Chin (1947)
yukonis M. E. Jones	Alaska, U.S.A.	Allen et al. (1964)
zingeri Kurzch.	U.S.S.R.	Pers. observ. (U.W. herb. spec. 1971)

Ateleia (Moc. & Sessé ex DC.) Dietr.   Papilionoideae:
129 / 3587   Sophoreae

From Greek, *ateles*, "incomplete"; the flowers have only 1 petal.

Type: *A. pterocarpa* (Moc. & Sessé ex DC.) Dietr.

17 species. Shrubs or small trees, unarmed; young branches often densely pubescent; lenticels conspicuous. Leaves imparipinnate, 5–28-foliolate; leaflets small, alternate or subopposite; stipels none; stipules very minute or absent. Flowers small, white or yellow, few to many, in axillary or terminal racemes 5–20 cm long; bracts minute; bracteoles absent; calyx cup-shaped, truncate or minutely 5-toothed; petal 1, long-clawed, hood-shaped, margin undulate; stamens 6–10, rarely 11, free or joined at the base; filaments shorter than the petal, alternately subequal; anthers uniform, ellipsoid, dorsifixed; ovary short-stalked, 2-ovuled; style reduced; stigma ovate, peltate. Pod samaroid, small, semiorbicular, flat, membranous, 2-valved, indehiscent, upper suture narrow-winged, usually 1-seeded; seed kidney-shaped, compressed. (Mohlenbrock 1962c; Rudd 1968a)

Ateleia is considered a primitive, papilionaceous genus having strong affinities with the Caesalpinioideae, especially to the tribe Swartzieae.

Members of this genus are widespread in tropical and subtropical Mexico, the West Indies, and Central and South America. They thrive in mountainous regions, swampy forests, or open areas.

The tree species rarely reach 10 m in height, with trunks only 20 cm in diameter. Accordingly, the hard, heavy, compact wood is available only in small sizes. The thick, yellowish sapwood is used primarily for implement handles and for other purposes requiring considerable strength. The heartwood is poorly developed (Record and Hess 1943).

Atylosia W. & A.   Papilionoideae: Phaseoleae, Cajaninae
(417) / 3895

From Greek, *a-*, "without," and *tylos*, "knob," in reference to the blunt, unbeaked, keel petals.

Lectotype: *A. candollei* W. & A.

36 species. Herbs or shrubs, trailing, twining or erect. Leaves pinnately 3-foliolate, occasionally subpalmate; leaflets entire, lower surface often with orange or yellow glandular dots; stipels absent; stipules small or absent. Flowers yellow, axillary, in sessile or stalked pairs or fascicles, or in short, lax, panicled racemes; bracts broad, membranous, early caducous, often not seen; bracteoles absent; calyx deeply bilabiate; lobes elongated, falcate, acuminate, upper 2 nearly or entirely joined, lowermost lobe longest; standard broad, recurved, with 2 basal ears, often with 2 minor callosities inside; wings obliquely obovate or oblong; keel incurved, but not beaked, blunt; stamens diadelphous; anthers uniform; ovary sessile; ovules 3–7; style incurved, upper half thicker, often hairy; stigma small, terminal. Pod sessile within the calyx, oblong or elongated, sometimes tomentose, compressed with transverse

or oblique indentations or grooves between the seeds, 2-valved, valves leathery or thin, septate; seeds ovate or orbicular, hilum oblong. (Bentham 1864; Bentham and Hooker 1865)

Representatives of this genus inhabit tropical Asia, Australia, India, and Madagascar; only 1 species occurs in tropical Africa. The plants grow along seashores, in fallow fields, open heaths, brushwood, grassy wild areas, and young forests.

A. cinerea F. Muell., *A. grandiflora* F. Muell., *A. marmorata* Benth., *A. pluriflora* F. Muell., *A. reticulata* Benth., and *A. scarabaeoides* (L.) Benth. are valuable ground cover in Queensland and Northern Territory, Australia. *A. barbata* W. & A. and *A. scarabaeoides* are useful green manures for tea estates in Java.

NODULATION STUDIES

Atylosia nodules are round, large, and occur singly or in aggregates (Lechtova-Trnka 1931; Sandmann 1970b). Host cell hypertrophy is rather meager, but infection threads are conspicuous. Following release from the threads, the rhizobia retain their short rod-shapes and tend to prevail in the center of the host cells (Lechtova-Trnka 1931).

A rhizobial strain from *A. scarabaeoides*, regarded as a member of the cowpea miscellany by Chen and Shu (1944), was not infective on *Astragalus sinicus*. Two strains in Zimbabwe produced opposite reactions on *Vigna sinensis*; the highly effective one was ineffective on *Centrosema pubescens*, the ineffective one was effective on *Phaseolus lathyroides* (Sandmann 1970a).

Nodules have been reported on *A. barbata* in France (Lechtova-Trnka 1931), on *A. lineata* W. & A. in western India (V. Schinde personal communication 1977), and on *A. scarabaeoides* in China (Chen and Shu 1944), in Sri Lanka (W. R. C. Paul personal communication 1951), and in Zimbabwe (Sandmann 1970a).

Aubrevillea Pellegr.   Mimosoideae: Piptadenieae

Named for Professor A. Aubréville, noted French forester.

Type: *A. kerstingii* (Harms) Pellegr.

2 species. Trees, deciduous, unarmed, up to 50 m tall, with trunk diameters of 1 m; buttresses narrow, flared up to 3 m from base; bole straight, cylindric; crown dense, rounded, brilliant red when the young leaves first emerge, deep green at maturity. Leaves bipinnate; pinnae 4–8 pairs; leaflets many-paired, small, sessile, narrow-oblong. Flowers bisexual, in long, spikelike panicles or lateral racemes; bracts caducous; calyx cupular; lobes 5, short, valvate before flowering; petals 5, lanceolate, united in lower part into a tube with the base of the filaments; stamens 8–10, rarely 6–7; filaments united at the base and adherent to corolla tube; anthers dorsifixed, without glands; ovary oblong, stalked, 5–7-ovuled; style short. Pod samaroid, oblong, flat,

finely reticulate, papery, indehiscent; seeds 1–3, kidney-shaped, flat. (Pellegrin 1933; Gilbert and Boutique 1952a)

The genus, native to tropical West Africa is well represented in Nigeria, Sierra Leone, and the Congo Basin. The plants grow in humid, high forests.

The wood is a pleasant gray brown color with a violet tinge and has a striped interlocking grain (Voorhoeve 1965).

Augouardia Pellegr.   Caesalpinioideae: Cynometreae

Named for Monseigneur Augouard, Archbishop of Brazzaville.

Type: *A. letestui* Pellegr.

1 species. Tree. Leaves paripinnate; leaflets 4–5 pairs, opposite or subopposite, subcoriaceous, entire, elliptic, acuminate, without translucid dots, lateral nerves anastomosed, others joined in a marginal nerve; stipules caducous or absent. Flowers white, subsessile or on short pedicels, in more than 2 rows, in dense terminal panicles; bracts caducous; bracteoles 2, imbricate, suborbicular, keeled, enclosing the flower bud; calyx shortly funnel-shaped, base thickened; sepals 4, oval, blunt, subequal, imbricate in bud; petals none; stamens 3–4, fertile, joined at the base; anthers inflexed, dorsifixed; staminodes 4–5, free, awl-shaped; ovary sessile, 2-ovuled; style elongated, recurved; stigma small, capitellate. Pod not described. (Pellegrin 1924; Léonard 1957)

This genus, established by Pellegrin (1924) and positioned in Caesalpinioideae, was segregated by Hutchinson (1964) along with others to form a new subfamily, Brachystegioideae.

This monotypic genus is endemic in equatorial West Africa.

Austrodolichos Verdc.   Papilionoideae: Phaseoleae, Phaseolinae

Name means "Australian *Dolichos*."

Type: *A. errabundus* (Scott) Verdc.

1 species. Herb, perennial, slender, semiprostrate, straggling, or twining. Leaves pinnately 3-foliolate, petioled; leaflets papery, prominently pinnately veined; stipels lanceolate, persistent; stipules oblong-lanceolate, persistent. Flowers mauve or pink, solitary, seldom 2, on slender pedicels on short, axillary peduncles; bracts small, ovate, early caducous; bracteoles 2, large, elliptic-ovate, conspicuous, silky-pubescent, foliaceous, persistent; calyx campanulate, densely pubescent; 2 upper lobes subconnate into a blunt bilobed lip, lateral lobes ovate, obtuse, lowermost lobe deltoid, subacute; petals

clawed, glabrous; standard rounded-obcordate, eared at the base; wings oblique-oblong, with a small tooth on the lower side near the claw; keel oblong, blunt, neither beaked nor incurved; stamens diadelphous; filaments unequal, vexillary filament with a small, raised wing at the base, apices slightly dilated; anthers uniform; ovary linear, about 11-ovuled; disk short-tubular, slightly lobed; style slender, curved, glabrous, not inflated; stigma small, neither capitate nor penicillate. Pod (immature) linear, velvety-haired; seeds not described. (Verdcourt 1970c)

This monotypic genus is endemic in northern Australia, in the Darwin area. The plants usually occur in semishade on gravelly sand.

Baikiaea Benth.   Caesalpinioideae: Amherstieae
56 / 3507

Named for Dr. William Balfour Baikie, West African traveler and Royal Navy Surgeon of the Niger Expeditions in 1854 and 1857.

Type: *B. insignis* Benth.

10 species. Trees, unarmed, slow-growing, medium to large, up to 30 m tall with girths 1–1.25 m; boles usually clear up to 7 m; tap root exceptionally long. Leaves imparipinnate or paripinnate; leaflets usually alternate, 1–5 pairs, petioluled, oblong-elliptic, usually large, entire, generally lacking pellucid dots, leathery, deciduous; stipules small, short, scalelike. Flowers usually very large, cream or orange, nodding, few or many, in terminal or axillary racemes or panicles; bracts and bracteoles usually small, caducous; calyx turbinate or campanulate; lobes 4, posterior lobe largest, linear or oblong, margins membranous, subvalvate; petals 5, unequal, uppermost petal inside and largest, 2 laterals smaller, 2 lower ones narrow, exceeding the calyx in length, obovate-spathulate or broad-oblanceolate, long-clawed, sometimes villous; stamens 10, vexillary stamen free, others joined at the base; filaments longer than the petals, alternate filaments stout, silky-pilose below, others thinner, glabrous; anthers linear, versatile, longitudinally dehiscent, incurved and narrowed after dehiscence; ovary flattened, long-stalked, stalk adnate to the calyx tube, densely villous, 1- to several-ovuled; disk absent; style elongated, glabrous; stigma terminal, small, globular. Pod flat, asymmetric at the base, woody, conspicuously covered with dense, brown or blackish, felty hair, 2-valved, valves spirally twisted during dehiscence; seeds few, large, elliptic or suborbicular, dark red. (Léonard 1952, 1957)

Baikiaea species are widespread in tropical Africa. They thrive in savannas, rain forests, and swamp areas.

Baikiaea flowers are among the most beautiful of the Leguminosae; those of *B. insignis* Benth., which measure about 25 cm in diameter, are said to be the largest (Menninger 1967). The blossoms generally open in the evening and fade the following afternoon. *B. plurijuga* Harms is established only on the Kalahari sands of Zimbabwe (Coates Palgrave 1957), a formation that once covered most of Africa, but now occurs only in isolated patches. Germination of the seed is not difficult; however, in Florida, Menninger (1967) was unsuccessful in culturing plantlets of *B. plurijuga* beyond 15 cm tall. The seed pods explode with a resounding noise resembling that of pistol shots.

The wood of *B. plurijuga* is marketed as Rhodesian chestnut or Rhodesian teak and is exported in considerable quantities. Perhaps the most important indigenous tree exploited commercially in Zambia and Zimbabwe, this species is highly esteemed (Rendle 1969). The seasoned wood has a density of 900 kg/m³ and is valued for its exceptional strength, durability, stability, and resistance to fungi, termites, and borers (Titmuss 1965). The wood is difficult to work and dulling of tool edges is common, but the finished product takes a high polish and its reddish brown luster is attractive. The grooved parquet floor laid in 1952 in the rebuilt Corn Exchange, London, is made of Rhodesian teak (Coates Palgrave 1957). The use of the wood for furniture is limited because of its weight.

NODULATION STUDIES

Roots of *B. plurijuga* growing in Zimbabwe lacked nodules (Corby 1974), as did those of *B. insignis* in Singapore (Lim 1977).

Bakerophyton Hutch.   Papilionoideae: Hedysareae,
                                          Aeschynomeninae

Name means "Baker's plant"; named for E. G. Baker, 1864–1949, specialist in the botany of tropical Africa.

Type: *B. lateritium* (Harms) Hutch.

3 species. Herbs, annual; stems slender, prostrate or straggling. Leaves imparipinnate; leaflets several-paired, oblong-linear, median nerve and lateral nerve spread at a right angle, upper part finely toothed; stipules ovate-lanceolate, not appendaged, serrulate, persistent. Flowers many, on very short, filiform pedicels in axillary or pseudoterminal racemes, axis of inflorescences very zigzag; bracts 3-lobed, persistent; bracteoles lanceolate, persistent; calyx bilabiate; lobes elliptic-oblong, ciliate above, upper lip 2-toothed, lower lip 3-toothed; standard oblong-rectangular; wings obovate to elliptic, free; keel petals united, bidentate at the top; stamens 10, united in a sheath, open above; anthers dimorphic, 5 larger and basifixed, alternate 5 much smaller and dorsifixed; ovary 2-ovuled; disk minute; style very short, glabrous, persistent. Pod enclosed in the bracteoles and persistent calyx, with 1–2 suborbicular, warty joints, indehiscent. (Léonard 1954a; Hutchinson 1964)

Léonard (1954a) considered *Bakerophyton* a monospecific subgenus of *Aeschynomene* consisting only of *A. lateritia* Harms. Hutchinson (1964) elevated this subgenus to generic status based on the occurrence of dimorphic anthers in contrast to uniform anthers characteristic of other *Aeschynomene* species. Restricted habitat of *A. lateritia* Harms in conjunction with this morphological feature prompted Maheshwari (1967) to transfer *A. neglecta* Hepper and *A. pulchella* Planch. to *Bakerophyton* as *B. neglectum* (Hepper) Maheshwari and *B. pulchellum* (Planch. ex Bak.) Maheshwari, respectively.

Members of this genus are native to tropical Africa. They flourish on wooded savannas and along roadsides, mostly on sandy soil.

Nodulation Studies

Small, round nodules are depicted on roots of *B. lateritium* (Harms) Hutch. (= *A. lateritia* Harms) in plate 23, of Léonard's treatment (1954a) of *Aeschynomene*.

Balisaea Taub.   Papilionoideae: Hedysareae,
(329a) / 3798   Aeschynomeninae

Named for Campos der Serra do Balisa, Brazil, the site of its discovery (Harms 1897).

Type: *B. genistoides* Taub.

1 species. Herb, perennial. Leaves imparipinnate; leaflets alternate, rarely opposite, setaceous; stipels absent; stipules small, persistent. Flowers pale orange, few, in elongated, lax, leaf-opposed racemes; bracts stipulelike; bracteoles 2, persistent; calyx subcampanulate; lobes 5, lanceolate; petals subequal; standard rounded, short-clawed; wings broad-ovate, oblique, apex blunt; keel petals subrectangular, connate from apex to middle, curved and beaked toward apex; stamens 10, united into a sheath split above; anthers alternately long and basifixed, short and dorsifixed; ovary stalked, 1–2-ovuled; style bearded along the inner side; stigma terminal. Pod linear, flat, slightly constricted between the seeds. (Harms 1897)

This monotypic genus is endemic in Balisa, Brazil. The plants grow in open fields of mountainous areas.

Baphia Lodd.   Papilionoideae: Sophoreae
152 / 3612

From Greek, *baphe*, "dye," in reference to the red dye obtained from the wood.

Type: *B. nitida* Lodd.

60–100 species. Shrubs, scrambling or climbing, or trees up to 25 m tall. Leaves 1-foliolate, large, elliptic or lanceolate; petioles with upper and lower or contiguous pulvini; stipels minute, usually absent; stipules small, poorly developed, early caducous. Flowers white, yellow, or purplish pink, solitary or few, in short, axillary or subterminal racemes, sometimes panicled or fascicled in leaf axils; bracts small; bracteoles linear-suborbicular, large and deciduous or shorter than the calyx and semipersistent; calyx ovoid to globose, apex short-toothed, split to the base at anthesis into 2 reflexed lobes, or along 1 side and spathelike; petals subsessile or short-clawed; standard broad, sometimes emarginate, often with yellow or orange blotch at the base; wings oblique-oblong; keel petals incurved, incompletely fused at the back; stamens 10, free; anthers uniform; ovary subsessile, 2–6-ovuled; style long, incurved; stigma small, terminal. Pod linear-oblong or lanceolate, plano-compressed, acuminate, leathery to woody, sometimes pubescent, continuous within or slightly filled, beaked, 2-valved, dehiscent along both sutures; seeds few to several, lenticular, flat, brown. (Bentham and Hooker 1865; Harms 1913; Pellegrin 1948; Phillips 1951; Brummitt 1968a,b)

Representatives of this genus are found in tropical West Africa, primarily in the Congo Basin; they also occur, though restricted in distribution, in Madagascar, Natal, and East Africa. They thrive in humid forests, dense evergreen and semievergreen thickets, and along water courses.

Baphia shrubs are generally used as ornamentals, hedges, and live fences because of their ease of propagation from cuttings and their profusion of flowers. *B. racemosa* Walp. is called violet tree because of its fragrance. The hard, close-grained, fine-textured wood of *B. nitida* Lodd. was the original camwood of commerce; at present, it is more commonly known as African sandalwood. Its uses are limited to small carpentry items, such as umbrella handles, walking canes, table legs, small beams, and pillars. The heartwood of all species yields a red dye. *B. nitida* replaced the Brazil woods (*Caesalpinia* species) as the first dye wood exported from West Africa, but was in turn replaced by American logwood, *Haematoxylon campechianum* L. (Dalziel 1948). The water-insoluble, alkali-soluble, turkey-red-colored dye was used by many African natives to mark faces in ceremonial dances, and as a paint for images used in religious rites; thus, the tree was considered sacred. A dye paste made with fats and oils served medicinally to treat sprains, swollen joints, and skin diseases.

NODULATION STUDIES

Nodules were observed on an unidentified *Baphia* species in Yangambi, Zaire (Bonnier 1957) and on *B. massaiensis* Taub. ssp. *obovata* (Schinz) Brummitt in South Africa (Grobbelaar and Clarke 1975), but not on the latter in Zimbabwe (Corby 1974). Trees of *B. nitida*, an introduced species in Singapore, bore moderate numbers of elongated, yellowish brown, smooth-surfaced nodules (Lim 1977).

The Latin suffix, *astrum*, denotes incomplete resemblance or inferiority to the allied genus *Baphia*.

Type: *B. brachycarpum* Harms

12 species. Scandent shrubs or trees, up to 8 m tall. Leaves 1-foliolate, large, elongated, simple; petioles thick, short, and jointed at both ends; branches, petioles, and pedicels densely rusty-pubescent; stipels small, near petiole apex; stipules lanceolate, caducous. Flowers several or many, in densely hairy, axillary and terminal racemes or racemelike panicles; bracts lanceolate or ovate-lanceolate, acute; bracteoles 2, broadly ovate-lanceolate, acuminate, usually reflexed, simulating small ears at the base of the calyx, velvety tomentose; calyx ovoid, large, oblique, densely pubescent, closed in bud but irregularly split or divided down 1 side in flower; petals glabrous, about the same length as the calyx; standard subsessile, rounded, emarginate; wings long-oval, blunt, sessile; keel petals oblique, blunt, free; stamens 10; filaments free, glabrous, longer than the anthers; anthers narrow, attached near base; ovary subsessile, densely yellow brown villous, 1–2-ovuled; style curved; stigma minute. Pod short, broad, turgid or linear-oblong, incurved, narrowed toward the base, beaked at the tip, velvety tomentose, 2-valved, constricted between the seeds, tardily dehiscent; seeds 1–2, oblong, scarlet or purplish black. (Harms 1913, 1915a; Pellegrin 1948)

Baphiastrum differs from *Baphia* in having shorter and broader pods that are 1–2-seeded, free keel petals, and long anthers.

Members of this genus occur in tropical West Africa, especially in Gabon, Nigeria, and the Congo Basin.

Baphiopsis Benth. ex Bak.   Caesalpinioideae: Swartzieae
118 / 3576

The Greek suffix, *opsis*, denotes resemblance to *Baphia*.

Type: *B. parviflora* Benth. ex Bak.

2 species. Shrubs or trees, up to 15 m tall, unarmed. Leaves 1-foliolate, entire, broad, elliptic, acuminate, petioled; stipules caducous or absent. Flowers white, small, in short, axillary umbels or racemes on older branches; bracteoles 2, opposite, striate, caducous; calyx tube very short, oblong, membranous, entirely closed at first, splitting irregularly when expanded; petals 6, oblong, subequal, imbricate, same length as the calyx; stamens numerous, 14 or more, subequal, hypogynous; filaments of varying lengths, not exserted; anthers oblong, basifixed; ovary sessile, smooth, linear, 2–6-ovuled; style filiform, apex hooked; stigma small, capitate. Pod thick, semiorbicular, upper suture curved, the other straight, apiculate, 1–2-seeded; seeds very large. (Pellegrin 1948; Gilbert and Boutique 1952b)

The differences between *Baphiopsis* and *Baphia* species so greatly outnumber the similarities that these 2 genera are positioned in different sub-families.

Both species of this genus are forest trees, native to tropical Africa, from southern Nigeria to Tanzania.

Baptisia Vent.   Papilionoideae: Podalyrieae
158 / 3618

From the Greek verb *baptizo*, "dye," alluding to the use of some of the species as a dye substitute for indigo.

Type: *B. alba* (L.) Vent.

35–50 species. Herbs, deep-rooted, perennial; stems erect, sometimes declined, stout-succulent or slender-firm, slightly or firmly ribbed, arising annually from stout, thick, woody rhizomes; foliage usually blackening upon drying. Leaves digitately 3-foliolate, occasionally bifoliolate or unifoliolate; leaflets usually sessile, sometimes short-petioled, obovate to lanceolate, apex acute or obtuse; stipules usually minute, lanceolate to ovate-cordate, cadu-cous. Flowers yellow, cream, white, or blue, in terminal or axillary racemes, rarely solitary, on slender pedicels; bracts setaceous-lanceolate to ovate-acuminate, caducous; bracteoles usually absent; calyx campanulate, per-sistent, slightly bilabiate, upper lip entire, notched, or bilobed; lower lip deeply trilobed; petals 5, equal in length, caducous; standard reniform-orbicular, not longer than the wings, sides reflexed, sometimes eared; wings oblong, straight; keel petals partly joined, straight or slightly incurved; sta-mens 10, free; filaments equal; ovary stalked, multiovuled; style incurved; stigma minute, terminal. Pod short- or long-stalked in the persistent calyx, ovoid or subcylindric, papery or leathery, inflated and beaked, nonseptate, usually many-seeded. (Bentham and Hooker 1865; Larisey 1940a)

Representatives of this genus are common in North America, especially the eastern and southeastern areas of the United States. They thrive in dry woods and plains, and are well adapted to sandy and gravelly alluvial soils.

The leaves and pod of *Baptisia* species often turn black upon maturity. Because the sap turns slate blue or black upon exposure to air, names that suggest various indigo properties have been given to some species. An in-ferior indigo dye from *B. tinctoria* (L.) R. Br. was used in earlier days in mountain communities of the southern United States to color homespun textiles.

The handsome lupinelike inflorescences have ornamental value. Dried roots of wild forms were used medicinally by the Indians as emetics and cathartics. Various species are blamed for livestock losses (Pammel 1911; Kingsbury 1964); however, toxicity generally occurs only when large quan-tities of fresh herbage are eaten. All species are good ground cover in sunny locations because of their bushy habit and extensive root systems but are not considered acceptable range indicators because they tend to decrease on an overstocked area. Their forage value is poor because of their unpalatable, bitter taste. *B. leucantha* T. & G. is regarded as poisonous to cattle. *B. bracteata* Muhl. is credited with value as an indicator of mineral lead deposits (Cannon 1960a; Nicolls et al. 1964–1965), since it thrives on old lead-diggings.

The genus *Baptisia* exemplifies the better understanding of taxonomic relationships within a plant group resulting from chemical studies. Flavonoids (Horne 1966; Harborne et al. 1971) and alkaloids (Cranmer and Mabry 1966; Mears and Mabry 1971) have played the most prominent roles. The former have served in identifying species, in designating specific groups indicative of an evolutionary sequence, for the validation of hybrids, and in analyzing population structures, or patterns, within the genus. The alkaloids have been less helpful infragenerically but have pointed to phyletic relationships at the generic level. Three major alkaloids present in *Baptisia* species are anagyrine, cytisine, and methylcytisine (Cranmer and Mabry 1966). Because tricyclic alkaloids are not only the predominant ones in *Baptisia* and *Thermopsis* species, but are also common to members of the tribe Genisteae, Cranmer and Turner (1967) were of the opinion that members of *Baptisia* have a closer phyletic relationship to Genisteae than to the Southern Hemisphere element of the Podalyrieae.

NODULATION STUDIES

The placement of *B. tinctoria* in the cowpea cross-inoculation group by Burrill and Hansen (1917) was confirmed by Conklin (1936). *B. leucantha* and *B. leucophaea* Nutt. were added to this category by Bushnell and Sarles (1937). Rhizobia from these species were monotrichously flagellated, produced slow, scant growth on various media, an alkaline reaction without serum zone formation in litmus milk, and no gelatin liquefaction. Wilson (1939a) reported nodulation of *B. australis* (L.) R. Br. by rhizobial strains from *Cassia, Crotalaria, Desmodium, Stizolobium*, and *Vigna* species of the cowpea miscellany. A rhizobial strain from *B. australis* was reciprocally infective on species of these genera; it also nodulated species in the tribes Trifolieae and Loteae, and some tribes in the subfamily Mimosoideae.

Nodulated species of *Baptisia*	Area	Reported by
alba (L.) Vent.	Wis., U.S.A.	Bushnell & Sarles (1937)
australis (L.) R. Br.	U.S.A.	Many investigators
(= *minor* Lehm.)	S. Africa	Grobbelaar et al. (1967)
bracteata Muhl.	Kans., U.S.A.	Warren (1909)
leucantha T. & G.	Wis., U.S.A.	Bushnell & Sarles (1937)
leucophaea Nutt.	Wis., U.S.A.	Bushnell & Sarles (1937)
tinctoria (L.) R. Br.	U.S.A.	Many investigators

Named for J. B. G. Barbier, contributor to *Materia Medica*, author of several treatises on pharmacology (De Candolle 1825b).

Type: *B. pinnata* (Pers.) Baill.

1 species. Shrub, erect or scandent, woody, 1–4 m tall; branches slender, covered with dense, brown hairs. Leaves imparipinnate; leaflets numerous, usually about 15, thin, entire, pubescent below; stipels long-subulate; stipules subulate-acuminate. Flowers large, scarlet, each subtended by a pair of bracteoles, grouped 2–3 along rachis in axillary and terminal racemes; bracts and bracteoles subulate-acuminate, the latter persistent; calyx tubular, large, densely pubescent; lobes 5, long, setaceous; petals long- and slender-clawed; standard oblanceolate; wings oblong-elliptic, joined to and shorter than the obtuse keel; keel subequal to the standard; stamens diadelphous; anthers uniform; ovary sessile, many-ovuled; style elongated, hairy along the upper inner surface; stigma small, terminal. Pod straight, flat, linear, 2-valved, septate between the transverse-oblong seeds. (Bentham and Hooker 1865; Rydberg 1924)

Hutchinson (1964) created a new tribe, Barbiereae, to accommodate this genus and *Genistidium*.

This monotypic genus is endemic in Mexico, the West Indies, and tropical South America. It occurs in wet forests or shady thickets.

This shrub is an attractive ornamental, but it is somewhat uncommon.

Barklya F. Muell.　　　　　Papilionoideae: Sophoreae
123 / 3581

Named for Sir Henry Barkly, 1815–1898, once Governor of Victoria, Australia, and Cape Colony, South Africa; patron of botany.

Type: *B. syringifolia* F. Muell.

1 species. Tree, small to medium, 7–20 m tall; foliage dark, evergreen. Leaves 1-foliolate, heart-shaped, long-petioled; stipules small, ovate, deciduous. Flowers small, orange yellow, on short pedicels in dense, axillary or terminal racemes up to 20 cm long; bracts minute; bracteoles none; calyx campanulate; lobes short, blunt; petals subequal, obovate, erect, free, long-clawed; standard broader than others; stamens 10, free, longer than petals; anthers arrow-shaped; ovary stalked, several-ovuled; style short; stigma terminal. Pod stalked, oblong-lanceolate, flat, valves thin and barely separating, transversely reticulate; seeds 1–2, flat. (Bentham 1864; Bentham and Hooker 1865)

This monotypic genus is endemic in tropical and subtropical eastern and northern coastal areas of Queensland. Often found in rain forests, the plants require rich, loamy soil.

This handsome, slow-growing, floriferous tree is known as Lilac Barklya in Australia and as the gold blossom tree in Florida and California, where it was introduced as an ornamental.

Batesia Spruce ex Benth. Caesalpinioideae: Sclerolobieae
109 / 3567

Named for H. W. Bates, distinguished traveler in the Amazon.

Type: *B. floribunda* Spruce ex Benth.

1 species. Tree, gigantic, unarmed. Leaves imparipinnate, large; leaflets 9–11, large, leathery; rachis ridged or slightly winged; stipules short, triangular, intrapetiolar. Flowers yellow, in large, terminal panicles; bracts and bracteoles narrow, early caducous; calyx campanulate; lobes 5, imbricate; petals 5, subequal, ovate, uppermost innermost, imbricate; stamens 10, free; filaments villous at the base; anthers uniform, longitudinally dehiscent; ovary short-stalked, stalk oblique, free at the base of the calyx, apically dilated, subjointed; ovules few; style short, thick; stigma ciliate, concave, terminal. Pod short, subfalcate, swollen, woody, with elevated margins, ridged on the inside, dehiscent along 1 suture; seeds 2–3, suborbicular, transverse, compressed, thick, subcircular, shiny, pale red. (Bentham and Hooker 1865)

This monotypic genus is endemic in northern Brazil. The plants flourish in sandy, noninundated land and riverine forests.

The durable, coarse-textured wood is used locally for furniture and small carpentry items. The heartwood has a rich chestnut brown color with a golden luster (Record and Hess 1943).

Bathiaea Drake del Castillo Caesalpinioideae:
(45b) / 3494a Cynometreae

Named for Perrier de la Bathie; he collected the specimens from Plateau d'Ankaratra, bois à Besofotra, Madagascar.

Type: *B. rubriflora* Drake del Castillo

1 species. Tree, foliage sparse. Leaves paripinnate; leaflets 8, alternate, obovate-elongated, rather large. Flowers red, short-pedicelled, 6–8, in axillary racemes; bracts and bracteoles deciduous; calyx tube slightly concave accommodating an annular disk; lobes 5, obovate, imbricate; petals 5, obovate, subequal, upper petal innermost; stamens 10, unequal, posterior stamen shorter; filaments elongated, slender; anthers ellipsoid, opening by means of a side slit; ovary stalked; ovules few, arranged in 2 rows; style filiform, elongated; stigma small, terminal. Pod samaroid, asymmetric, indehiscent, anterior suture thicker; seed with thick cotyledons, pendulous from top of fruit. (Harms 1908a)

This monotypic genus is endemic in Madagascar.

Named for Captain Baudouin, in recognition of his service to botany in the exploration of New Caledonia.

Type: *B. sollyiformis* Baill.

2–6 species. Trees, small, unarmed. Leaves entire, obovate, 1-foliolate; stipules caducous, striate. Flowers few, in slender, axillary racemes; bracts 2, very small; calyx depressed cone-shaped; lobes 5, subequal, lanceolate, smooth, imbricate; petals similar to calyx lobes but thinner; stamens 10, inferior, free, subequal, inversely pyramidal, thickened and truncate at apex; anthers basifixed, turned inwardly, arrow-shaped, 2-celled, opening by apical pores; ovary short-stalked, free, villous, few- usually 3-ovuled, subhorizontal or obliquely descending; style slender, slightly thickened at apex. Pod stalked, fleshy, often woody, drupaceous, falsely obliquely septate by the thickened endocarp between the 2–5 transverse seeds. (Baillon 1866)

Baudouinia species are native to Madagascar.

Baueropsis Hutch.   Papilionoideae: Hedysareae,
                                                        Euhedysarinae

This genus was originally described and named *Bauerella* Schindl. in honor of Ferdinand Bauer, 1760–1826, botanical illustrator of Robert Brown's collection of Australian plants deposited at the British Museum (Schindler 1926). *Bauerella* was changed to *Baueropsis* because *Bauerella* Borsi had priority for a genus in the family Rutaceae.

Type: *B. tomentosa* (Schindl.) Hutch.

1 species. Shrub or shrublet; branches wandlike. Leaves 1-foliolate; stipels none; stipules present. Flowers papilionaceous, in axillary heads, usually 3 in each bract axil; calyx tube narrow; lobes 5, unequal; standard oblong, clawed, not eared or calloused; wings spurred, incurved, narrow, obtuse, slightly adnate to the keel; keel petals long-clawed, semiobovate, obtuse, lower margins joined; stamens diadelphous, free parts filiform; anthers uniform, narrow-ovoid, dorsifixed near the base; ovary short-stalked, 1-ovuled; disk absent; style long, apically incurved and thickened; stigma small, terminal. Pod not described. (Schindler 1926; Hutchinson 1964)

This monotypic genus is endemic in Australia.

Named by Linnaeus because the bilobed leaves symbolized to him two Swiss botanists, the Bauhin brothers, John, 1541–1613, and Caspar, 1560–1624. John contributed various publications, but is especially remembered for *Historia Plantarum*, published about 40 years after his death. Caspar's two great contributions were *Prodromas* and *Pinax*. He is also remembered for his promotion of natural affinities as the basis of taxonomy, his list of botanical synonyms of the era, and his insistence that a plant should have two names. This last-named thesis paved the way for the binomial system of Linnaeus.

Type: *B. divaricata* L.

550–575 species. Trees, small, or shrubs, erect, straggling, or high climbing, with circinate tendrils; stems sometimes flattened with undivided tendrils at the base of the racemes. Leaves simple, 2- to many-nerved, entire, or 2-lobed apically, usually deeply cleft, almost 2-foliolate, some resembling the imprint of a cloven hoof, midrib usually bristly between the lobes; stipules various, often small, caducous. Flowers large, showy, fragrant, purple, pink, white, or yellow, few, often bibracteolate, in simple, terminal or axillary racemes, corymbs, or panicles; calyx short-turbinate, campanulate, or elongated, limb of the calyx before anthesis entire, or contracted apically and 5-dentate, in anthesis the limb variously cleft, spathaceous, or with 5 valvate lobes, often reflexed, occasionally glandular; petals 5, erect or spreading, subequal, broad, imbricate, usually long-clawed; stamens 10 or less, some reduced to staminodes, free or connate; filaments linear, usually flattened, lanceolate, sometimes scaly or hairy at the base, sterile filaments shorter than the fertile ones; anthers ovate, oblong, or linear, versatile, longitudinally dehiscent; ovary stalked, rarely sessile, stalk free or joined to receptacle, 2- to many-ovuled; disk present or absent; style slender, short; stigma small, sometimes peltate, oblique. Pod stalked, oblong, linear, or ovate, flat, leathery, subfleshy, membranous or woody, indehiscent or 2-valved, valves usually elastic, sometimes septate between the seeds; seeds flat, ovate or round, compressed. (Bentham and Hooker 1865; Standley and Steyermark 1946; Phillips 1951)

In his revision of Malaysian Bauhinieae, de Wit (1956) elevated several sections of *Bauhinia* to generic rank. The presence or absence of tendrils was considered a valid distinguishing basis for segregation.

Representatives of this pantropical genus are distributed throughout the Old World and the New World. They are native to India, Sri Lanka, southern China, and tropical Africa. Three species are endemic in Australia. Only about 50 species are cultivated. The plants flourish in second-growth forests, thickets, moist areas, savannas, and on river banks.

The importance of *Bauhinia* species lies in their ornamental use in parks and gardens, and along avenues and roadsides. *B. acuminata* L., *B. monandra* Kurz, *B. purpurea* L., *B. tomentosa* L., and *B. variegata* L. are especially prized for this purpose. The flowers of some species are orchidlike, thus the common name "poor man's orchid" for *B. variegata*, and pink, or park orchid tree, for *B. monandra*. The last-named species is also known as the St. Thomas tree in the Virgin and Hawaiian islands, and as butterfly Bauhinia in Puerto Rico.

B. fassoglensis Kotschy ex Schweinf. and *B. retusa* Roxb. are exemplary of the arboreal species that yield a clear, light-colored gum. The seeds of *B. esculenta* Burch. are rich in protein and in a golden yellow, transparent oil that has a pleasant odor and a bitter, nutty taste. The shell and kernel oil yield is about 50 percent of their weight. Cosmetic properties of the oil are similar to those of almond or apricot oil. The seeds were used at one time in Europe as a culinary substitute for the almond. The boiled pods of *B. petersiana* Bolle are eaten in some African areas. The seeds of this species have been used as a substitute for coffee; for this reason it was known to early settlers of Zambia as Zambesi coffee (Coates Palgrave 1957). The leaves of *B. malabarica* Roxb. are used for flavoring fish and meat in the Philippines. Plant decoctions of *Bauhinia* species have been used as medicines to treat abdominal and liver troubles, fevers, common colds, coughs, and abscesses. They have also served as vermifuges. The properties are astringent.

The bark of various species has been used for cordage, roofing small huts, and making native cloth. By weight it yields about 18–20 percent tannin. The heartwood of *B. tomentosa* is almost black; the sapwood is pale pink. The wood is fine grained, easily worked, hard, heavy, strong, and finishes with a medium to high luster. The wood of *B. variegata* is often termed mountain ebony. In Asia, the name "chrysanthemum wood" is applied to *B. championii* Benth. because the cross section is so figured. It is a favorite wood in the Orient for making chests; however, commercial use of *Bauhinia* wood is negligible because of the small supply.

Nodulation Studies

The occurrence of black, or dark-colored, wiry roots typical of nonnodulated caesalpinoid species characterizes *Bauhinia* species. The reports of nodulation on senescent roots of field plants of *B. microstachya* (Raddi) Macb. (Rothschild 1970) and of their occurrence on only 1 plant among the many unnamed *Bauhinia* species examined (Lechtova-Trnka 1931) are viewed with skepticism.

Nonnodulated species of *Bauhinia*	Area	Reported by
acuminata L.	Philippines	Bañados & Fernandez (1954)
	Hawaii, U.S.A.	Pers. observ. (1962)
benthamiana Taub.	Venezuela	Barrios & Gonzalez (1971)
bidentata Jack	Singapore	Lim (1977)
binata Blanco	Hawaii, U.S.A.	Allen & Allen (1936b)
candicans Benth.	Argentina	Rothschild (1970)
carronni F. Muell.	N. S. Wales, Australia	Beadle (1964)
corymbosa Roxb.	Hawaii, U.S.A.	Pers. observ. (1939)
cumingiana (Benth.) Vill.	Philippines	Bañados & Fernandez (1954); Pers. observ. (1962)
diphylla Buch.-Ham.	W. India	V. Schinde (pers. comm. 1977)

Nonnodulated species of *Bauhinia* (cont.)	Area	Reported by
excisa (Griseb.) Hemsl.	Trinidad	DeSouza (1966)
galpinii N. E. Br.	India	Parker (1933)
	Hawaii, U.S.A.	Pers. observ. (1962)
	Zimbabwe	Corby (1974)
kirkia Oliv.	S. Africa	Grobbelaar & Clarke (1975)
kochiana Korth.	Singapore	Lim (1977)
kuntheana Vog.	Guyana	Norris (1969)
macrantha Oliv.	Zimbabwe	Corby (1974)
malabarica Roxb.	Philippines	Bañados & Fernandez (1954)
megalandra Griseb.	Trinidad	DeSouza (1966)
monandra Kurz	Hawaii, U.S.A.	Allen & Allen (1936b)
	Philippines	Bañados & Fernandez (1954); Pers. observ. (1962)
pauletia Pers.	Costa Rica	J. León (pers. comm. 1956)
	Trinidad	DeSouza (1966)
petersiana Bolle	Zimbabwe	Corby (1974)
purpurea L.	France	Naudin (1897b)
	Java	L. G. M. Baas-Becking (pers. comm. 1941)
	Philippines	Bañados & Fernandez (1954)
racemosa Lam.	W. India	V. Schinde (pers. comm. 1977)
reticulata DC.	Singapore	Lim (1977)
tomentosa L.	Hawaii, U.S.A.	Pers. observ. (1939)
	Java	L. G. M. Baas-Becking (pers. comm. 1941)
	Philippines	Bañados & Fernandez (1954)
ungulata L.	Hawaii, U.S.A.	Pers. observ. (1962)
variegata L.	Hawaii, U.S.A.	Allen & Allen (1936b)
	S. Africa	Grobbelaar et al. (1964)

Baukea Vatke   Papilionoideae: Phaseoleae, Glycininae
392 / 3867

Named for H. Bauke, 1852–1880, plant collector.
   Type: *B. maxima* (Boj.) Baill. (= *B. insignis* Vatke)
   1 species. Shrub, arborescent or climbing. Leaves pinnately 3-foliolate; leaflets entire; stipels present; stipules sessile, narrower above, awned. Flowers handsome, large, yellow, in axillary racemes; bracts and bracteoles cadu-

cous; calyx elongated, campanulate; 2 upper lobes connate into 1 2-toothed lobe, ciliate, lower lobes longer, pointed; standard broadly obovate-oblong, reflexed, auricles inflexed at the base; wings oblong, adherent to and shorter than the oblong, blunt keel; stamens diadelphous; anthers basifixed, uniform; ovary stalked, villous, 2–3-ovuled; style awl-shaped; stigma globose, terminal, glabrous. Pod 2–3-seeded, short-stalked, 2-valved.

This monotypic genus is endemic in northwestern Madagascar.

Behaimia Griseb.   Papilionoideae: Dalbergieae,
360 / 3831   Lonchocarpinae

Type: *B. cubensis* Griseb.

1 species. Tall shrub or tree, unarmed. Leaves imparipinnate; leaflets in few pairs, narrow, leathery; stipels absent. Flowers small, in simple or branched, terminal or axillary racemes; bracts small; bracteoles 2, small; calyx acute at the base, bilabiate; teeth 5, short, 2 upper ones joined to the middle; petals long-clawed; standard orbicular, reflexed above the claw, not eared; wings erect, spathulate-oblong; keel petals free, obtuse, oblique-semicordate, slender-clawed; stamens 10, diadelphous, 9 connate or 7 connate and other 2 more or less free; anthers versatile, oblong-elliptic; ovary sessile, tomentose, few- to 6-ovuled; style awl-shaped; stigma minute, capitate. Pod sessile, elliptic-oblong, flat, very thin to almost membranous, transversely reticulate, not winged, indehiscent, both ends acute, usually 1-seeded; seed flat, kidney-shaped. (Bentham and Hooker 1865; León and Alain 1951)

This monotypic genus is endemic in Cuba.

The hard, tough, heavy wood has a fine texture and a low luster, but it is difficult to work. Its uses have been limited to railway ties and general carpentry (Record and Hess 1943).

Belairia A. Rich.   Papilionoideae: Hedysareae,
(318a) / 3787   Aeschynomeninae

From Greek, *belos*, "dart," "stinger," and the verb *haireo*, "grasp," in reference to the sharp spines.

Type: *B. spinosa* A. Rich.

5 species. Shrubs or small trees, armed with awl-shaped, spiny stipules. Leaves paripinnate; leaflets in few pairs, or 3-foliolate, rigid, small. Flowers yellow, solitary or subfascicled, on long, slender pedicels, at old nodes or clustered on short twigs; bracts caducous; bracteoles membranous, small, ovate, persistent; calyx obliquely turbinate or campanulate; lobes 4–5, very short; petals longer than the calyx, erect, acute; standard trapezoid, the 4

anterior petals subequal, narrower and somewhat longer, free, linear-lanceolate; stamens usually free, exserted; anthers uniform, ovate; ovary linear, stalked, smooth, 2–3-ovuled; style filiform, smooth, incurved; stigma minute. Pod stalked, oval or elongated, plano-compressed, with conspicuous venation, indehiscent; seeds ovoid to kidney-shaped, flattened, smooth. (Bentham and Hooker 1865; Harms 1908a; León and Alain 1951)

This genus was positioned as number 130 in the tribe Sophoreae by Taubert (1894). Urban (1900) recognized the close similarity of *Belairia* with *Pictetia* tribe Hedysareae. In subsequent revisions and corrections of the Engler and Prantl system, Harms (1908a) assigned the generic number 318a to follow *Pictetia*.

Members of this genus are indigenous to Cuba. They are common on rocky sites and poor savanna soils; the root systems in basaltic soils are very short and thick.

The trunk bark breaks away in large red flakes. The wood is hard and of good quality, but because it is available only in small sizes, it is used primarily for posts and small turnery articles (Record and Hess 1943).

Bembicidium Rydb. Papilionoideae: Galegeae, Robiniinae

From Greek, *bembix*, "a top," and *idion*, a diminutive suffix, in reference to the small, turbinate flower buds.

Type: *B. cubense* Rydb.

1 species. Shrub, low, unarmed; branches rusty-haired. Leaves paripinnate; leaflets entire, leathery, wrinkled, veinless above, midrib prominent beneath; petiole and rachis broadly winged, wings discontinuous; rachis extended beyond the uppermost leaflets; stipels obsolete; stipules lanceolate, persistent. Flowers purplish, solitary, axillary; calyx turbinate, as broad as long, 2 lips broad, subequal, acute; petals subequal; standard obovate, emarginate, short-clawed; wings and keel petals equal in length and shape, blades oblique-lanceolate, rounded at apex, eared at base, claws short, straight; blades of keel petals united at middle only; ovary short-stalked, linear, many-ovuled; style glabrous, bent inward at the base, not hooked at the apex; stigma minute, terminal. Pod not described. (Britton 1920; León and Alain 1951)

This monotypic genus is endemic in Cuba. The plants grow in mountain woods.

Type: *B. benoistii* Maire

1 species. Herb, prostrate, glaucous, annual; stems numerous. Leaves sessile or subsessile, imparipinnate, 6–9 foliolate; leaflets 3–4 pairs, entire, lowest 2 stipuliform. Flowers 2–4, umbellate, in axils of small, 5-foliolate leaves; calyx lobes 5, equal, long-ciliate; petals glabrous, exserted; standard pale purple; wings white, eared at the base; keel longer than other petals, dark purple at apex, much-curved above the claw, with an oblique, straight, obtuse beak; stamens diadelphous; alternate filaments thickened below the anthers; ovary glabrous, abruptly narrowed into an oblique style; stigma truncate. Pod long-exserted, smooth and cylindric, becoming more or less flattened dorsiventrally, swollen at intervals but not jointed, septate by thinly membranous partitions, each segment 1-seeded, indehiscent; seeds in 2 rows, much-compressed, black-speckled. (Maire 1924)

This monotypic genus is endemic in Morocco.

Bergeronia M. Micheli
362 / 3833

Type: *B. sericea* M. Micheli

1 species. Tree, up to 15 m tall, trunk fluted, about 0.3 m in diameter. Leaves imparipinnate, 9–13 foliolate; leaflets, elliptic-oval or lanceolate, entire, reticulately nerved, pubescent above and especially below; stipels none; stipules short, deciduous. Flowers rose lilac, solitary or paired on short pedicels in simple axillary racemes; bracts small, caducous; bracteoles 2, setaceous, subtending the calyx, caducous; calyx tubular-campanulate, silky-pubescent; lobes 5, short, acute, lowest lobe somewhat longer than the upper ones; petals clawed, silky-pubescent ouside; standard broad-orbicular, erect; wings, free, oblong; keel petals straight, blunt, shorter than the wings, connate at the back; stamens 10, diadelphous; anthers elliptic-oblong, glabrous, versatile, uniform; ovary sessile, linear, pubescent, 1- to several-ovuled; style slightly incurved, glabrous; stigma globose, terminal, small. Pod linear, curved, plano-compressed, densely pubescent, filled between the oblong-reniform seeds, breaking into indehiscents segments when ripe. (Burkart 1952)

This monotypic genus is endemic in Paraguay and northern Argentina. It grows along river banks and on forested islands at the confluence of the Paraguay and Paraná rivers.

This tree is noteworthy for its beautiful flowers and the shade it provides. The wood, which requires care in seasoning, has poor durability and a low luster; the reddish yellow heartwood emits a disagreeable odor (Record and Hess 1943).

Named for Andreas Berlin, 18th-century Swedish botanist and student of Linnaeus.

Type: *B. acuminata* Soland. ex Hook. f.

15 species. Trees, rarely shrubs, unarmed, evergreen, usually 25–40 m tall, with trunk diameters 1 m or more and crowns dense, obconical; bases of trunks usually ridged, seldom buttressed. Leaves paripinnate, 2–6 paired; leaflets large, leathery; stipules various, large, foliaceous, subpersistent or small to inconspicuous, caducous. Flowers white, handsome, fragrant, large, borne on stout stalks in stout terminal racemes or panicles, often corymbose; bracts small, leathery, ovate, caducous; bracteoles 2, large, obovate or spathulate, enclosing each flower bud, spreading or deciduous at anthesis; calyx tube short or elongated; lobes 5, papery, subequal, free, linear, imbricate; petals 5, upper petal large, suborbicular, much longer, crinkled, long-clawed, hooded and folded in bud, lower petals smaller, shorter, linear, narrower; stamens 10, rarely 5, usually all fertile, diadelphous, exserted; filaments long, slender; anthers oblong, uniform, longitudinally dehiscent; ovary short-stalked, stalk adnate to the calyx tube, 4–8-ovuled; style filiform; stigma terminal, capitate. Pod large, flat, leathery to woody, upper sutures often dilated and ridged, usually at a right angle to the stout stalk; pods usually bursting violently and curling spirally on dehiscence, releasing round, flat seeds. (Bentham and Hooker 1865; Hauman 1952)

Members of this genus are widespread in tropical Africa and are especially common throughout Liberia and Nigeria. They thrive in high forests or low bush areas on well-drained soil. Some species favor swamp and riverine sites.

B. confusa Hoyle and *B. grandiflora* (Vahl) Hutch. & Dalz. are the principal sources of *Berlinia* wood, which is sometimes sold under the name "rose zebrano" (Rendle 1969). Differentiation between the wood of *Berlinia* species is fraught with difficulty (Voorhoeve 1965; Kukachka 1970). In general, the wood is similar to white oak in durability, density, and usefulness. In recent years it has attracted attention as a decorative hardwood and has been gaining popularity for fine interior joinery (Rendle 1969).

NODULATION STUDIES

Root systems of *B. grandiflora* examined in Yangambi, Zaire, by Bonnier (1957) lacked nodules.

From Latin, *bis*, "twice," and *serrulatus*, "serrate," in reference to the serrations of the crest on each side of the flattened pod, resembling a 2-edged saw.

Type: *B. pelecinus* L.

1 species. Herb, annual, pubescent, spreading, weakly erect, with slender branches. Leaves imparipinnate; leaflets numerous; stipels absent; stipules small, free, papery. Flowers blue or yellowish, with blue tips, 2–11, small, on slender peduncles, in short, axillary racemes; calyx campanulate; teeth 5, subequal; petals short-clawed; standard ovoid, erect; wings falcate, oblong, free; keel petals short-clawed, subequal to the wings in length, obtuse; stamens 10, diadelphous, only 5 fertile; anthers uniform; ovary sessile, many-ovuled; style short, thickened, incurved; stigma terminal, capitate. Pod linear, dorsiventrally compressed, 2–4 cm long, sutures forming a central line between the sinuate-dentate crests or outer margins of the 2 flattened, indehiscent valves; pod divided inside by a narrow longitudinal septum connecting the sutures, each half containing about 8–10, small, kidney-shaped seeds. (Bentham and Hooker 1865; Gillett 1964)

This monotypic genus is endemic in the Canary Islands and the Mediterranean areas of Europe and North Africa.

NODULATION STUDIES

Nodules were observed on the roots of *B. pelecinus* L. in the Jardin de Botanique de Toulouse, France (Clos 1896), in Spain (Buendia Lazaro 1966), and in South Africa (Grobbelaar et al. 1967).

Bolusanthus Harms   Papilionoideae: Sophoreae
(147a) / 3607a

Named for Dr. Harry Bolus, 1834–1911, Cape of South Africa botanist and founder of the Bolus Herbarium, Cape Town.

Type: *B. speciosus* (Bolus) Harms

1 species. Tree, slender, 10–13 m tall; branches subglabrous. Leaves imparipinnate, on long petioles; leaflets 7–15, opposite or alternate, asymmetric at the base, lanceolate or oblong-lanceolate, acuminate, silky-haired when young. Flowers blue violet, several to many, in pendent, lax, terminal racemes; bracts and bracteoles small, awl-shaped, deciduous; calyx campanulate-cupular, silky; lobes 5, ovate-lanceolate, subequal, upper 2 broad-ovate and connate higher up, other 3 acute, lanceolate; petals clawed; standard suborbicular, very broad, rounded, emarginate; wings oblong, obtuse; keel petals appendaged on 1 side, similar, oblong, as long as wings, free or nearly so; stamens 10, free, subequal, glabrous; anthers elliptic, dorsifixed; ovary short-stalked, 4–5-ovuled, silky-haired; style smooth, glabrous, curved, narrowed toward the base; stigma small, hairy, capitate. Pod short-stalked, oblong, linear, flat, reticulate, not winged, narrowed toward apex, tardily dehiscent; seeds up to 5. (Harms 1908a)

Harms (1908a) considered the genus closely allied to *Calpurnia* and assigned it number 147a in conformance with the system devised by Taubert (1894).

This monotypic genus is endemic in subtropical South Africa, northeastern Transvaal, and Zimbabwe. It is easily grown from seed, and usually occurs in open areas on granitic, basaltic, or limestone soils. The plants are slow growing and intolerant of heavy soils. They are considered drought-resistant in some areas and withstand light frosts.

Because the floral habit and hue markedly resemble that of vine wisterias, this attractive ornamental is called tree wisteria. It is one of the most handsome specimens of the wild flowering flora of subtropical South Africa (Palmer and Pitman 1961).

Bolusanthus wood is white, hard, durable, and considered valuable, but it is not available in large quantities. The heartwood is resistant to white ants and borers. Its use is reserved for furniture, veneers, and small carpentry items. Natives in various areas use decoctions of plant parts for medicinal purposes.

NODULATION STUDIES

B. speciosus (Bolus) Harms was nodulated in experimental studies by rhizobial strains from *Albizia julibrissin*, *Baptisia australis*, *Desmodium canadense*, and *Stizolobium deeringianum* (Wilson 1939a), and in natural habitats of South Africa (Grobbelaar et al. 1967) and Zimbabwe (Corby 1974).

Bolusia Benth.   Papilionoideae: Galegeae, Tephrosiinae
267 / 3732

Named for Dr. Harry Bolus, 1834–1911, Cape of South Africa botanist and founder of the Bolus Herbarium, Cape Town.

Type: *B. capensis* Benth.

4 species. Herbs, perennial, or undershrubs, many-stemmed, silky-pubescent, often with a long tap root. Leaves simple or palmately 3-foliolate; leaflets linear, elliptic, or obovate; stipules ovate or lanceolate, acuminate, sometimes foliaceous and coarsely dentate. Flowers 1–few, greenish yellow to pink, on leaf-opposed peduncles; bracts and bracteoles present; calyx short-campanulate, deeply 5-lobed; lobes subequal, ovate-lanceolate, 2 upper lobes broader, more or less connate; standard very broad, emarginate, sometimes hooded and distinctly keeled; wings falcate-obovate, free, clawed; keel long, linear, spirally contorted, spurred at the base; stamens diadelphous, spirally coiled inside the keel; anthers alternately elongated-linear and basifixed with filaments short and oblong-elliptic and dorsifixed with filaments long; ovary sessile, many-ovuled; style glabrous; stigma terminal, capitate. Pod oblong, turgid, many-seeded; seeds horseshoe-shaped. (Baker 1929; Phillips 1951)

Polhill (1976) reclassified this genus in the tribe Crotalarieae, with affinities to Genisteae *sensu lato*.

An endemic African genus, *Bolusia* consists of 2 species in dry parts of the northern Cape Province and 2 in Namibia.

NODULATION STUDIES

Nodules have been observed on *B. rhodesiana* Corb. in South Africa (N. Grobbelaar personal communication 1963) and in Zimbabwe (Corby 1974).

Borbonia L.   Papilionoideae: Genisteae, Crotalariinae
193 / 3653

Named for Gaston de Bourbon, Duke of Orleans and son of Henry IV of France; he was a patron of botany.

Type: *B. cordata* L.

12–13 species. Shrubs or shrublets, rarely decumbent, glabrous, occasionally villous. Leaves simple, entire or dentate, with many, rigid, longitudinal veins, usually sharp-pointed at the apex, and heart-shaped at the base; stipules lacking. Flowers yellow, often becoming reddish with age, solitary or grouped in short racemes or subumbels, axillary or terminal; bracts and bracteoles often setaceous; calyx campanulate or tubular-campanulate; lobes 5, subequal, ovate or lanceolate, acute or pungent; petals usually hairy; standard subsessile, suborbicular, villous outside, clawed; wings oblique or obovate, short-clawed or with a long, linear claw; keel incurved, obtuse, saccate or gibbous, often with a fold on each side; stamens unequal, all connate into a sheath split above, staminal tube adnate to claw of the wings and keel; anthers alternately short and versatile, elongated and basifixed; ovary sessile, few- to many-ovuled, usually glabrous; style curved, awl-shaped; stigma small, terminal. Pod linear or lanceolate, compressed, longer than the calyx, oblique-acute, 2-valved, valves leathery, continuous inside. (Bentham and Hooker 1865; Phillips 1951)

The incorporation of all *Borbonia* species into *Aspalathus* has been proposed by Dahlgren (1963c, 1968).

Members of this genus are limited to South Africa, mostly in the southwestern area of the Cape Province. They usually occur in flats and on dry hillslopes.

Decoctions of plant parts have served as remedies for respiratory catarrh, bronchitis, asthma, and as a diuretic and astringent.

NODULATION STUDIES

Grobbelaar and Clarke (1972) observed nodules on *B. lanceolata* L. in South Africa.

Named for Bossieu de la Martinière, French botanist on the expedition of La Pérouse, a French navigator; all members of the expedition perished in a shipwreck near the Island of Vanikoro, in the Solomon Island group, 1788.

Type: *B. heterophylla* Vent.

40 species. Shrubs, 2–3 m tall, rarely shrublets; branches cylindric, flattened, or 2-winged. Leaves alternate or opposite, simple, entire or dentate, glabrous, often glaucous, or reduced to scales; stipules minute, brown, lanceolate, or absent. Flowers yellow, red, or streaked, solitary or in small clusters, axillary; bracts 2, 3, or more, imbricate, dry, rigid, at base of pedicel, outermost bract minute, persistent, innermost bract longer, deciduous; bracteoles similar; calyx lobes 5, 2 upper ones short, broad, separate or somewhat connate into an upper lip, lower 3 equal, very small; petals clawed; standard orbicular, about 2 times longer than calyx, reflexed; wings oblong; keel petals obtuse, broader than wings; stamens united into a sheath split above; anthers uniform, dorsifixed, equal, oblong; ovary several-ovuled, often ciliate; style awl-shaped, incurved; stigma terminal, small. Pod sessile or short-stalked, flat, linear-oblong, not winged, continuous or septate inside, valves flat, thin, edges thickened, dehiscent, opening along both sutures. (Bentham and Hooker 1865; Polhill 1976)

The genus originally encompassed 3 sections differentiated by leaf arrangement, calyx, and pod characters. The 6 species with opposite leaves, upper calyx lobes obtuse, and long-stalked pods in section *Oppositifoliaea* were segregated by Hutchinson (1964) to constitute *Scottia* (q.v.), a genus not recognized by Polhill (1976).

Representatives of this genus are distributed throughout Australia, especially in New South Wales, Victoria, Western Australia, and Tasmania. They thrive in gullies, in rocky, dry areas and sandstone soils, and often in rather arid zones.

The ornamental beauty of various members of this genus is greatly enhanced by the occurrence of brilliantly colored papilionaceous flowers on leafless branches. The pendent coral red flowers of *B. walkeri* F. Muell. account for its common name "coral pea."

Ingestion of the foliage is considered deleterious to livestock and causes a malady known as peg leg, which results in a high percentage of deaths (Webb 1948). After recovery, animals walk stiff-legged with joints swollen to an abnormal size.

Nodulation Studies

Creamy brown, ovoid nodules measuring 5–12 mm long by 2–4 mm wide were observed on *B. cinerea* R. Br. by Ewart and Thomson (1912). Cultural characteristics and plant-infection tests of 7 species prompted Lange (1959, 1961) to regard the rhizobia as members of the slow-growing lupine and soybean type and the plant species as members of the cowpea miscellany.

Nodulated species of *Bossiaea*	Area	Reported by
†*aquifolium* Benth.	W. Australia	Lange (1959)
cinerea R. Br.	Victoria, Australia	Ewart & Thomson (1912)
†*dentata* Benth.	W. Australia	Lange (1959)
eriocarpa (R. Br.) Benth.	W. Australia	Lange (1959)
heterophylla Vent.	Queensland, Australia	G. D. Bowen (pers. comm. 1957)
laidlawiana Tov. & Morris	W. Australia	Lange (1959)
linophylla R. Br.	W. Australia	Lange (1959)
ornata (Lindl.) Benth.	W. Australia	Lange (1959)
preissii Meissn.	W. Australia	Lange (1959)
prostrata R. Br.	Australia	Harris (1959)
pulchella Meissn.	W. Australia	Lange (1959)
webbii F. Muell.	W. Australia	Lange (1959)

†Segregated as *Scottia* spp. (Hutchinson 1964); Polhill (1976) does not recognize this change.

Bowdichia HBK   Papilionoideae: Sophoreae
133 / 3591

Named for Thomas Edward Bowdich, 1790–1824; he traveled extensively in Africa as a plant collector.

Type: *B. virgilioides* HBK

2 species. Trees, medium to large, unarmed. Leaves imparipinnate; leaflets numerous, 5–21, coriaceous; stipels none. Flowers attractive, lilac blue or white, densely clustered in lax, terminal panicles; bracts and bracteoles small; calyx teeth valvate; petals often with curled margins; standard broad-orbicular; wings obovate or broad-oblong, longer than the standard; keel petals oblong, free, smaller than the wings; stamens 8–10, free, slightly unequal; anthers versatile; ovary stipitate, many-ovuled; style filiform, apex inflexed; stigma terminal, capitate. Pod oblong-linear, 6–8 cm long, flattened, membranous, narrowly winged along the upper suture, wine red or orange, indehiscent; seeds small, black, hard, oblong, compressed, transverse. (Bentham and Hooker 1865)

Both species are native to tropical South America, specifically northern Brazil, Guiana, and Venezuela.

The sizes of these trees vary within a wide range. In dry wooded areas on plains and slopes of the coastal range in central and northern Brazil, *B. virgilioides* HBK is a small tree about 5 m tall, but specimens found in forests along the Rio Negro and the lower Amazon are commonly 50 m in height, with trunk diameters in excess of 1 m.

Bowdichia wood is hard, tough, strong, without odor or taste, and has a density of 993 kg/m³. The heartwood is a dull chocolate to reddish brown, and sharply demarcated from the whitish sapwood. It is difficult to work and highly resistant to decay. Despite its very coarse texture and irregular and interwoven grain, the wood can be finished smoothly. It is used for heavy construction in Brazil but is not considered suitable for furniture (Record and Hess 1943). The corky bark is of no commercial value. Pounded stems of *B. virgilioides* apparently are used by natives of Guyana as fish poison (Menninger 1962).

NODULATION STUDIES

Nodules were lacking on the whitish roots of *B. virgilioides* growing on the savannas of Venezuela (Barrios and Gonzalez 1971).

Bowringia Champ. ex Benth.   Papilionoideae: Sophoreae
150 / 3610

Named for Dr. Bowring and his son, J. O. Bowring.

Type: *B. callicarpa* Champ. ex Benth.

2 species. Shrubs or small lianas, slender-stemmed, glabrous. Leaves large, 1-foliolate, rounded at the base, long-acuminate, prominently nerved on both surfaces; stipels absent; stipules very small. Flowers white, few, in short, axillary racemes or panicles; bracts very small, caducous; bracteoles minute, somewhat persistent; calyx cup-shaped to campanulate, membranous, truncate; lobes 5, short, subequal; petals short-clawed; standard orbicular, apically notched, reflexed; wings falcate-oblong; keel petals similar to the wings, subequal or as long as the wings, slightly connate dorsally; stamens 10, free or slightly joined at the base; anthers oblong, very small; ovary stalked, many-ovuled; style awl-shaped; stigma small, terminal. Pod stalked, thick, ovoid or subglobose, membranous to thinly leathery, 2-valved, dehiscent; seeds 1–2, brown or vivid scarlet, glossy, oblong or globose, strophiole large, cup-shaped. (Harms 1913; Pellegrin 1948; Toussaint 1953; Hutchinson and Dalziel 1958)

The type species is native to Hong Kong, southern China, Indochina, and Borneo. The native tropical African species, *B. mildbraedii* Harms, occurs in Gabon, Cameroon, and Congo.

From Greek, *brachys*, "short," and *semeia*, "standard."

Type: *B. latifolium* R. Br.

16 species. Shrubs, or undershrubs, low, usually less than 1 m tall; stems hoary, leafless, except for small scales, branching dichotomously, ends spinescent, or with leaves alternate or opposite. Flowers red, sometimes yellowish green or nearly black, terminal or axillary, solitary or several together, or crowded on short, basal scapes; bracteoles usually none; calyx lobes 5, subequal, linear-lanceolate, 2 upper lobes often connate high up; standard shorter and narrower than the wings, often quite small; wings narrow, oblong; keel usually longer and broader than the wings, incurved, and turned uppermost on the recurved pedicels; keel petals connate dorsally; stamens 10, free; ovary sessile or stipitate, villous, several-ovuled, often surrounded by a cup-shaped, inner disk; style filiform; stigma small, terminal. Pod ovoid to elongated, turgid, valves leathery, often hairy. (Bentham and Hooker 1865)

The genus is limited to Australia. Its members are well distributed in Western Australia and Northern Territory, in mountain, river, and coastal areas.

These species are attractive. The type species, commonly called Swan River pea, has bright red flowers; the leaves are dark green with silvery undersides.

NODULATION STUDIES

Nodules were first ascribed to a member of this genus on plants grown in the greenhouse of the Muséum d'Histoire Naturelle, Paris (Lechtova-Trnka 1931). Histological sections of nodules of *B. lanceolatum* Meissn. displayed typical bacteroid tissues and hypertrophy of invaded cells.

Lange (1959) observed nodules on field-excavated root systems of *B. lanceolatum* and 4 other species, *B. aphyllum* Hook., *B. latifolium* R. Br., *B. praemorsum* Meissn., and *B. sericeum* (Sm.) Domin in Western Australia. The erroneous inclusion of *B. aphyllum* in the *Bossiaea* listing was corrected by Lange by hand in the reprint sent to the authors. Isolates from nodules of 3 species showed characteristics of the slow-growing cowpea, lupine, and soybean type of rhizobia. Affinity with members of the cowpea miscellany was shown in plant-infection tests (Lange 1961).

From Greek, *brachys*, "short," and *stege*, "shelter," in reference to the bracteoles sheathing the flower buds.

Type: *B. spiciformis* Benth.

70 species. Trees, medium to large, rarely shrubs, unarmed, mostly deciduous; bark thick, fibrous. Leaves paripinnate, usually red, bronze, or pink,
pilose and limp when newly emerged; leaflets diversely shaped, 2–70 pairs,
sessile, often pellucid-punctate, papery or leathery; stipules display great variation in size, shape, position, the presence of a basal auricle, and duration.
Flowers small, greenish yellow, in short, dense, simple or panicled, axillary or
terminal racemes; bracts concave, caducous; bracteoles 2, opposite, valvate,
sepaloid, often pubescent, completely enveloping the flower in bud, later
reflexed, caducous; calyx very short, narrow, lined with a lobed, glandular
disk bearing 2–10 floral segments called tepals, i.e., sepals and petals much
reduced and indistinguishable; stamens usually 10, sometimes 9–12, or rarely
up to 18, all fertile, joined briefly at the base, exserted; anthers dorsifixed,
elliptic-subglobular, longitudinally dehiscent; ovary stipitate, oblong, tapering
at both ends, usually densely villous; ovules 4–10; style sigmoid or bent at a
right angle to the ovary; stigma small, terminal or lateral, turbinate-capitate.
Pod flat, oblong, boat-shaped, obliquely beaked, woody, upper suture winged
on each side, elastically dehiscent with both valves rolling up and twisting;
seeds 1–8, discoid, flat, glossy. (Bentham and Hooker 1865; Harms 1913;
Baker 1930; Pellegrin 1948; Hoyle 1952; Léonard 1957)

Although differentiation between *Brachystegia* species is at best fraught
with difficulty, some members are more difficult to distinguish than others.
Striking morphological differences occur in the growth of the same species in
different soils, altitudes, and rainfall areas (Topham 1930). White (1962)
referred to this genus as the most difficult taxonomically throughout Africa
because of the complicated patterns of variations in definitive characters.
Many of the species are remarkably tropophytic (Hoyle 1952). Nonetheless,
Hutchinson (1964) designated this genus as the type of the new subfamily
Brachystegioideae in family Caesalpiniaceae, order Leguminales.

Members of this genus are widely distributed throughout tropical Africa.
Species flourish in rain forests, woodland savannas, gravelly escarpments,
near streams and lagoons, and in intermediate and fringing forests. They
grow mostly on well-drained soils, and tend to be dominant in woody, vegetated areas, especially in Malawi, where they are ecologically important in
water and soil conservation (Topham 1930).

B. spiciformis Benth. is "perhaps the best known of Central African trees"
(Coates Palgrave 1957). The bright-colored new spring leaves of this handsome tree are often mistaken for flowers. The seeds are ejected 10 m or more
upon dehiscence of the elastic pods.

B. leonensis Hutch. & Burtt Davy is said to be the largest tree species in

Liberia (Voorhoeve 1965), where it reaches a height of 45 m, with a trunk diameter up to 1.8 m. Old specimens may have thick, heavy buttresses 1.8 m up from the base and straight, cylindrical boles, clear of branches up to 30 m. Dalziel (1948) described a felled tree measuring slightly more than 58 m in height, with a girth of about 7 m above buttresses 5.3 m from the ground.

Brachystegia wood lacks value as general purpose timber (Coates Palgrave 1957; Rendle 1969; Kukachka 1970). The slash of most species is hard, dense, fibrous, and bright red in color when freshly cut but becomes darker on exposure. It exudes a small amount of clear red sap. The sapwood is thick, usually about 12 cm, white, and distinct from the unequally colored heartwood. The wood is brown, coarse, medium hard, with a density of about 750 kg/m³, rough, blunts cutting edges quickly, is difficult to work, fairly gummy, does not splinter, is resistant to impregnation, and is rather variable. The wood of *B. boehmii* Taub. is susceptible to borer, termite, and fungal attack; that of *B. spiciformis* is slow to season, tends to twist after treatment, warps easily, splits readily, but is used to make excellent smelter poles as fuel for refining copper in Zambia, where it is known as miombo (White 1962). The name "okwen" is given to the wood of *B. nigerica* Hoyle & A. P. D. Jones and *B. kennedyi* Hoyle (Rendle 1969), used mainly for firewood, charcoal manufacture, as poles for the construction of huts and, to a limited extent, as a decorative veneer and for flooring.

The species do not furnish a good paper pulp. The fibrous bark is commonly beaten into bark cloth used for native clothing and for making rope, baskets, and sacks. *Brachystegia* gum is dark and of poor quality and low solubility (Howes 1949). Yields of tannin from the bark range from 3 to 13 percent. Decoctions of the bark and roots serve various native medicinal purposes. The seeds of *B. appendiculata* Benth. are edible.

NODULATION STUDIES

Corby (1974) did not find nodules on 7 species examined in Zimbabwe: *B. allenii* Burtt Davy & Hutch., *B. boehmii*, *B. glaucescens* Burtt Davy & Hutch., *B. manga* De Wild., *B. microphylla* Harms, *B. spiciformis*, and *B. utilis* Burtt Davy & Hutch.

The inference of nodule occurrence on *B. boehmii* and *B. spiciformis* by Sandman (1970b) and the inclusion of these species in the "cowpea-group (including strain-specific legumes)" was disclaimed by Corby (personal communication 1972). Also, *B. laurentii* (De Wild.) Louis ex Hoyle lacked nodules in Yangambi, Zaire (Bonnier 1957). This species symbioses with ectotrophic mycorrhiza, a phenomenon considered rare in the tropics and thus far noted only among members of a few genera in the Caesalpinioideae (Fassi and Fontana 1962a).

Named for Dimitrie Brandza, 1846–1895, Romanian botanist, author of research on the therapeutic properties of the Gentianaceae.

Type: *B. filicifolia* Baill.

1 species. Tree, unarmed, up to 10 m tall. Leaves bipinnate; pinnae 6–9
pairs; leaflets small, asymmetric at the base, eared; petiole jointed at the base
bearing a cup-shaped gland above the joint; stipules minute. Flower buds
subglobose; flowers small, racemose-cymose, crowded at the tips of twigs on
1-year-old wood; calyx obconical, with an inner, glandular, scalloped disk;
lobes 4, imbricate, upper lobe broader; petals 5, rarely 4 or 6, subequal,
longer than calyx, long-clawed, imbricate; stamens 10; filaments free, incurved in bud, later long-exserted; anthers dorsifixed, ellipsoid, with a glandular connective; ovary stalked; ovules 8–12 in 2 rows; style slender, incurved
at the apex; stigma apical. Pod stalked, very large, oblong or obovate, flat,
each suture somewhat sinuate and thickened, often empty at the base; seeds
few, on slender placenta, separated by incompletely formed endocarp septa.
(Baillon 1869)

This monotypic genus is endemic in Madagascar and the Seychelles.

It is a handsome tree with hard red wood.

Brenierea Humbert Caesalpinioideae

Named for M. J. Brenière of the Agricultural Service of Madagascar; he
collected the material and made possible the complete study of the only
species.

Type: *B. insignis* Humbert

1 species. Tree, small, 6–8 m tall, xerophytic; branchlets flattened into
cladophylls, closely dotted with glandlike scales. Leaves short-petioled,
2-foliolate, borne singly on lateral nodes of the cladodes; leaflets sessile, small,
obovate, entire, caducous; stipules abortive. Flowers subsessile, minute,
5-merous, sometimes 4–6, slightly zygomorphic, in short, axillary, spiciform
clusters above leaf scars; each with a minute bract and 2 bracteoles; calyx
gamosepalous, ovoid, scaly, 5-dentate above; petals lacking; stamens 5, opposite calyx lobes, alternating with 5 narrow, linear staminodes; anthers versatile, dorsifixed, longitudinally dehiscent; ovary subsessile, 2-ovuled, compressed laterally; stigma capitate, sessile. Pod flat, short-stalked, elliptic-
orbicular, 2-seeded, nonseptate, dehiscent (?); seeds suborbicular-triangular,
flat, cotyledons thick. (Humbert 1959)

Despite the occurrence of distinctive characters, Humbert (1959) re-
frained from positioning this genus in a tribe of Caesalpinioideae, for it
seemed unrelated to any other heretofore described.

This monotypic genus is endemic in Madagascar. The plants grow on
rocky, calcareous or siliceous soil.

Named for Adolphe T. Brongniart, 1801–1876, French botanist.

Lectotype: *B. podalyroides* HBK

60 species. Shrubs or small trees, unarmed, usually silky-villous, rarely glabrous. Leaves imparipinnate; leaflets entire, numerous, rounded, petioluled; stipels minute, awl-shaped or setaceous when present; stipules herbaceous, persistent or deciduous, sometimes large. Flowers large, brownish yellow, violet, or flesh-colored, pedicelled, solitary or in 2–7-flowered umbels in leaf axils, rarely in terminal racemes; individual flowers closely subtended by foliaceous, persistent or deciduous, paired bracteoles resembling stipules or bracteoles sometimes reduced to hair tufts; bracts ovate-lanceolate or setaceous; calyx subbilabiate; lobes 5, subequal, upper 2 united about two-thirds their lengths, lower 3 free to near the base or united higher up; petals subequal, clawed; standard broad, orbicular or broad-obovate, short-clawed; wings oblique-lanceolate, free, somewhat falcate, claw short, fleshy, eared at the base; keel petals oblique-obovate or broad-lunate, incurved, blunt, tapering to the base, blades united above; stamens diadelphous; anthers uniform, or alternate ones shorter; ovary sessile or stalked, enclosed at the base by a crenate ringed callus or cup; style filiform, incurved; stigma small. Pod short-stalked, flat, silky-haired, elongated, with an inner filling of cellulose, 2-valved, valves leathery, with a narrow wing margin along the upper suture, several-seeded. (Bentham and Hooker 1865; Standley 1920–1926; Rydberg 1923)

Representatives of this genus are scattered throughout Mexico, Central America, and the Andes. They prefer dry areas. *B. minutifolia* S. Wats. occurs in southwestern Texas at altitudes of about 1,000 m on dry, alkaline soils and limestone gravel.

Brownea Jacq.   Caesalpinioideae: Amherstieae
(70) / 3524

Named for Dr. Patrick Browne, native of County Mayo, Ireland. Browne's medical studies led to a consuming interest in botany that resulted in his becoming a pioneer naturalist in Jamaica, 1746–1755. Following publication of his *Natural History of Jamaica* in 1756, he became Curator of the Oxford Botanical Garden. Browne classified about 1,200 plants of Jamaica.

Type: *B. coccinea* Jacq.

30 species. Shrubs, mostly trees, unarmed, evergreen, small to medium, 10–15 m tall; a few stout-branched species up to 20 m tall. Leaves paripinnate, new leaves lax and often bright pink or purplish, splashed with white, becoming stiff and green with maturity; leaflets 4–18 pairs, leathery, often large; stipules foliaceous, caducous. Flowers tightly packed in globular heads, 10–22 cm in diameter, rose, red, seldom white, showy, the outermost ring of flowers opening earliest; with successive openings the flower resembles a huge rosette, borne on terminal branches, sometimes in lower trunk areas; bracts

usually large, numerous, colored, caducous; bracteoles 2, colored, connate into a floral sheath longer than the calyx, cleft on 1 side; calyx tubular-campanulate; lobes 4, petaloid, subequal, imbricate; petals 5, ovate to oblong, clawed, imbricate, uppermost petal inside, broader; stamens 10–11, free or slightly connate in the lower half; anthers uniform, oblong, longitudinally dehiscent; ovary pubescent, stalk short and adnate to the calyx tube, many-ovuled; style filiform; stigma dilated, capitate. Pod oblong, compressed, straight or falcate, woody or leathery, 2-valved, upper suture often thickened, elastically dehiscent; seeds few, transversely arranged, ovate, compressed. (Pittier 1916a)

Members of this neotropical genus are native to northern South America and the West Indies.

B. ariza Benth. (= *B. princeps* Linden), *B. coccinea* Jacq., and *B. latifolia* Jacq. (= *B. speciosa* Reichb.) are handsome ornamentals (Menninger 1962). *B. grandiceps* Jacq., Rose of Venezuela, is often termed the queen of the genus. It is cultivated in botanical gardens around the world and considered by some people the most beautiful flowering tree of America. The flower heads of *B. capitella* Jacq. are equally attractive; the stamens normally protrude 7 cm or more beyond the petals.

Brownea wood is neither of good quality nor available in quantity. Trunks of some species are often hollow and inhabited by ants. The sapwood is brownish gray, odorless, tasteless, hard and heavy, and has a density of about 800 kg/m^3. (Record and Hess 1943). The heartwood is often not normally developed, dark brown to black, and durable but not attractive. The wood finishes smoothly but splinters easily.

NODULATION STUDIES

Nodules were lacking on botanical garden specimens of *B. coccinea* in Hawaii (personal observation), and field specimens of *B. grandiceps* in the Philippines (Bañados and Fernandez 1954; personal observation 1962; J. Davide personal communication 1963), and on *B. latifolia* in Trinidad (De-Souza 1966). In Singapore, Lim (1977) observed smooth globose nodules on *B. ariza*, but none on *B. coccinea*, *B. capitella*, and *B. crawfordii* W. Wats., all introduced species.

Browneopsis Huber   Caesalpinioideae: Amherstieae
(70a) / 3524a

The Greek suffix, *opsis*, denotes resemblance to the genus *Brownea*.

Type: *B. ucayalina* Huber

3 species. Trees, small or large, up to 30 m tall, unarmed. Leaves paripinnate; leaflets 3–4 pairs, opposite or subopposite, oblong, short-petioluled; common petiole jointed at the base. Flowers subsessile, in densely crowded heads covered with bracts, on trunks and branches; bracts increasing

in size from below upwards, lowermost bract very short, intermediate bract larger and broadly rounded, all minutely hairy; bracteoles absent; calyx tube shallow, disklike, fleshy; lobes 4, petaloid, free or more or less connate; petals 3–4, rudimentary, linear or elongated-obovate; stamens 12–15; filaments united below into a tube split on the upper side; anthers oblong, dorsifixed; ovary multiovuled, stalked, stalk adnate to the calyx tube, linear; style elongated; stigma capitate. Pod long-stalked, linear, slightly curved, short-beaked, dehiscent. (Harms 1908a; Britton and Rose 1930)

Members of this genus differ from *Brownea* species in having more stamens and rudimentary petals, and in the absence of a sheath formed by connate bracteoles (Pittier 1916a).

Browneopsis species are indigenous to Central America and the Amazon region of Brazil.

Brya P. Br.   Papilionoideae: Hedysareae,
319 / 3788   Aeschynomeninae

Type: *B. ebenus* DC.

8 species. Shrubs, erect, prickly, or small trees, deciduous, seldom over 8 m tall, with trunks about 20 cm in diameter. Leaves imparipinnate; leaflets 1–many, but usually 3-foliolate, clustered with scarcely a common petiole; stipels absent; stipules spinescent and persistent, or inconspicuous and caducous. Flowers yellow or orange, small, in short, few-flowered, axillary or terminal cymes or fascicles; bracts and bracteoles small, persistent; calyx lobes subequal or lower lobe longer; standard suborbicular, clawed; wings falcate-oblong; keel incurved, blunt; stamens all connate into a sheath split above; anthers uniform; ovary sessile, or short-stalked, 2- or few-ovuled; style filiform, incurved; stigma terminal, very small. Pod sessile or short-stalked, with 1–2 broad, flat, indehiscent joints, upper joint usually sterile, lower suture curved; seeds kidney-shaped. (Bentham and Hooker 1865; León and Alain 1951)

Brya species are native to the West Indies. They are common in thickets and pastures, on limestone soil, rocky plains, hills, and in arid areas. *B. chrysogonii* León & Alain, *B. buxifolia* Urb., *B. ebenus* DC., and *B. subinermis* León & Alain, are well distributed in Cuba.

B. ebenus, a hardwood of considerable value, is known as Jamaican, or West Indian ebony, cocuswood, or grandillo in the timber trade. In local areas it is called torchwood because inflammable resins in the wood ensure a durable flame for native torches (Garratt 1922). The rich brown-to-black, durable heartwood finishes smoothly with high luster. It is popular for cutlery handles, musical instruments, inlays, and jewelry boxes (Record and Hess 1943; Titmuss 1965).

NODULATION STUDIES

B. ebenus was reported nodulated in Malaysia (Wright 1903).

Bryaspis Duvign. Papilionoideae: Hedysareae,
Aeschynomeninae

From the Greek verb _bryo_, "be full of," and _aspis_, "shield"; numerous flower bracts shield the flowers.

Type: _B. lupulina_ (Planch. ex Benth.) Duvign.

1 species. Herb, woody, erect or straggling; stems smooth, straw-colored. Leaves imparipinnate; leaflets 2–3 pairs, thin, glabrous, asymmetric, petiolule of terminal leaflet longer than those of laterals; stipels deciduous; stipules broad-ovate, conspicuously veined, not adnate to the petiole, persistent. Flowers small, yellow, crowded in flattened conelike racemes resembling those of hops, _Humulus lupulus_; bracts very large, membranous, persistent, at the base of each flower, overlapping each other along the entire length of the inflorescence, concealing the flowers and later the pods; calyx funnel-shaped, bilabiate, upper lip 2-toothed, lower one 3-toothed; standard suborbicular, base narrowed into a long claw; wings oblong, not appendaged; keel petals straight, partly joined along the rear marginal fringe; stamens all connate, sheath split above or later in 2 bundles of 5; anthers uniform; ovary short-stalked, 2-ovuled; disk inconspicuous; style incurved; stigma small, terminal. Pod very small, hard, with 2 rounded reticulate joints. (Duvigneaud 1954a)

This monotypic genus is endemic in tropical West Africa and the Sudan. The plants grow in moist places.

Buchenroedera E. & Z. Papilionoideae: Genisteae,
203 / 3663 Crotalariinae

Named for W. L. V. Buchenroeder, South African botanist and friend of Ecklon and Zeyher.

Lectotype: _B. alpina_ E. & Z.

23 species. Shrubs, small, much-branched, or stems simple, from an underground rootstock, densely silky-pubescent. Leaves palmately 3-foliolate, petioled; leaflets linear-lanceolate, often hooked at the apex; stipules leaflike. Flowers rose, white, yellow, or purplish, crowded in terminal spikes or heads, rarely solitary. Bracts foliaceous; bracteoles absent; calyx campanulate or urceolate; lobes very short, subequal, ovate and acuminate; petals smooth or tomentose; standard suborbicular or ovate, long-clawed, claw broad, linear; wings oblong, often transversely ridged, with long, linear claw; keel plano-convex, obtuse, shorter than wings and standard, with a long, linear claw, sometimes pouchlike; stamens all connate into a sheath split above; anthers unequal, alternately short and versatile, long and basifixed; ovary sessile, several- to many-ovuled, hairy; style incurved; stigma terminal. Pod oblique-acute, ovoid, turgid, scarcely longer than the calyx, 1- to few-seeded. (Bentham and Hooker 1865; Phillips 1951)

The genus is closely akin to _Aspalathus_ (Dahlgren 1963b). Polhill (1976) assigned this genus to the tribe Crotalarieae, segregated from Genisteae _sensu stricto_.

Members of this genus are scattered throughout South Africa, from Natal to the Uitenhage district.

114

NODULATION STUDIES

Two species are reported nodulated in South Africa: *B. lotonoides* Scott-Elliot (Grobbelaar and Clarke 1975) and *B. tenuifolia* E. & Z. var. *pulchella* (E. Mey.) Harv. (Grobbelaar and Clarke 1972).

Burkea Benth.   Caesalpinioideae: Dimorphandreae
34 / 3474

Named for Joseph Burke, British botanist; he collected plants in Africa with Charles Zeyher.
   Type: *B. africana* Hook.
   1 species. Tall shrub with thick branchlets or tree, unarmed, deciduous, usually 10–12 m tall, occasionally up to 20 m or more, and a girth of 2 m, with thick branches, a straight bole of 6–7 m or more, and a flattened crown. Leaves abruptly bipinnate; pinnae few-paired, opposite; leaflets elliptic or elliptic-oblong, petioluled or subsessile, small or medium in size, alternate; petioles often rusty-tomentose; stipules filiform, early caducous. Flowers small, creamy white, fragrant, in elongated, simple or panicled spikes; bracts small; calyx tube short, campanulate; lobes 5, rounded, oblong, equal; petals 5, subequal, obovate or elliptic, obtuse, imbricate, more than 2 times longer than the calyx; stamens 10, subequal, shorter than the petals; filaments linear, uniform, oblong-oblanceolate, longitudinally dehiscent, connective pointed, bearing an inflexed, sessile apical gland; ovary sessile or shortly stipitate, 1–2-ovuled, densely villous; style short, thick; stigma truncate or concave, terminal. Pod oblong-lanceolate, flat, reticulate, compressed, brittle, sub-coriaceous, indehiscent (?), persistent on trees after new leaves appear, 1-seeded; seeds compressed, suborbicular. (Bentham and Hooker 1865; Wilczek 1952)
   This monotypic genus, dominant and codominant in Zambia, is present throughout Africa as far north as Ethiopia and west to Nigeria. The plants commonly inhabit dry sandy soil in savannas and woodlands up to 1,500-m altitude.
   The bark and fruit yield tannin and have fish-stupefying properties. The stems yield a yellowish to brownish red, translucent, soluble gum used to treat dysentery (Howes 1949). Decoctions of the bark and leaves are used medicinally as mouthwashes against scurvy and in treating trachoma (Coates Palgrave 1957).
   B. africana Hook., known commonly as the wild seringa or Rhodesian ash, is a graceful ornamental admired for its shaggy white appearance when in flower and for its red, bronze-colored autumn foliage (Palmer and Pitman 1961). It is a good garden tree. The reddish brown wood is of good quality with a handsome grain and takes a good polish, but is difficult to tool and does not take nails well. It has a density of 707 kg/m^3. It is used for implement

handles, durable posts, buildings, uprights, bridge construction, heavy-duty and parquet flooring, wagon naves, furniture, fuel, and charcoal.

NODULATION STUDIES

Specimens of this species examined in Zimbabwe lacked nodules (Corby 1974).

Burkillia Ridley   Papilionoideae: Galegeae, Tephrosiinae

Named for I. H. Burkill, 1870–1965, Director of the Gardens, Straits Settlements, Malaysia, and author of *A Dictionary of the Products of the Malay Peninsula.*

Type: *B. alba* Ridley

1 species. Shrub. Leaves simple, petioled, large, lanceolate or elliptic. Flowers numerous, small, in axillary racemes; calyx campanulate; lobes 5, unequal, 2 upper lobes very short, lower 3 longer, bristly; standard broad, round, clawed; keel obcuneate, widely bilobed; stamens 10, the upper 5 shorter than the lower 5; anthers uniform; ovary cylindric, 2-ovuled; style long, filiform; stigma capitate. Pod elongated, compressed, narrowed at the base, wider and beaked at the apex, 2-seeded. (Ridley 1925)

This monotypic genus is endemic in Malaysia.

Burtonia R. Br.   Papilionoideae: Podalyrieae
169 / 3629

Named for D. Burton, collector for the Royal Botanic Gardens, Kew, England.

Type: *B. scabra* R. Br.

12 species. Shrubs, rarely undershrubs, erect, branches smooth or pubescent, some rigid, heathlike. Leaves simple, or palmately or pinnately compound; terminal leaflet sessile; stipules minute or none. Flowers yellow, orange red, or purple blue, solitary in the upper leaf axils or in terminal racemes; bracts usually small; bracteoles small, usually near the middle of the pedicel; calyx tube very short; lobes 5, longer than the tube, valvate, upper 2 broader than the lower 3; petals short-clawed; standard orbicular or kidney-shaped, longer than other petals; wings oblong or falcate, smaller than the broad, blunt keel; stamens free; ovary sessile or short-stalked; ovules 2, small,

on long, thick, folded funicles; style incurved, broad at base. Pod small, ovoid or subglobular, turgid; seeds 1–2, small. (Bentham 1864)

The genus is closely related to *Gompholobium*.

Representatives of this genus are limited to Australia, primarily in the northern and western areas.

NODULATION STUDIES

B. conferta DC. is nodulated in the South-West Province of Western Australia (Lange 1959).

***Bussea* Harms**   Caesalpinioideae: Eucaesalpinieae
(102a) / 3559a

Named for Dr. W. C. O. Busse, 1865–1933; he collected specimens of the type species in East Africa in 1901.

Type: *B. massaiensis* (Taub.) Harms

4–6 species. Shrubs and trees, medium to large, up to 35 m tall, with trunk diameters of 75 cm, unarmed. Leaves bipinnate; leaflets opposite or alternate, long-pointed, asymmetric, glabrous; stipules filiform, caducous. Flowers usually yellow, bright, fragrant, on short pedicels in axillary or terminal paniculate racemes; bracts caducous; calyx oblique, large, broad, short, obliquely cupular; lobes 5, triangular-ovate, imbricate in bud, subequal, subacute, margins hyaline; petals 5, broad-clawed, spathulate, crinkly, unequal, 1 smaller and narrower than others, margins crenate, lower dorsal area villous; stamens 10, uniform; filaments densely villous; anthers oblong, basifixed, longitudinally dehiscent; ovary short-stalked, 2–3-ovuled; style sparsely pilose; stigma peltate-capitellate. Pod compressed, thick, woody, not winged, lance-shaped, narrowing to the base, 2-valved, valves splitting from the top and recurved after opening, shortly apiculate on side near apex; seeds 2–3, yellowish brown near apex of the pod, flat. (Harms 1908a; Wilczek 1952)

Members of this genus occur in Gabon, Ghana, Liberia, Sierra Leone, and Zaire. They thrive in lowland forests.

B. occidentalis Hutch., common in the northern areas of Liberia, is perhaps the most prominent member of the genus. The ornamental value of this species lies in its long flowering period and abundance of fragrant, bright yellow flowers covering the evergreen crowns of the trees in forest areas.

Bussea wood is hard, heavy, tough, strong, and difficult to work (Voorhoeve 1965). It is suitable only for rough unfinished work. Its availability is too limited for export.

The seeds of *B. occidentalis* are roasted and eaten by natives of tropical West Africa (Dalziel 1948).

Named for John Stuart, Earl of Bute, 18th-century patron of botany; he was the Confidential Advisor to the mother of King George III. Stuart maintained a botanical garden on the grounds of Kew House. Later the king combined Kew House and his Richmond Lodge to establish the Kew Gardens of today.

Type: *B. monosperma* (Lam.) Taub.

3 species. Shrubs, climbing, or erect trees, medium-sized, 7–15 m tall, slow-growing, deciduous, tomentose; trunks usually crooked; branches irregular, rough; bark gray. Leaves pinnately 3-foliolate, soft, silky-haired, becoming leathery; stipels present; stipules small, caducous. Flowers silky-pubescent, showy, yellow, orange, or red, in dense, pendent, fascicled or paniculate racemes; bracts small, caducous, bracteoles at the base of the calyx, narrow, caducous; calyx large, campanulate; lobes short, deltoid, upper 2 connate into a broad, entire or emarginate lip; petals subequal in length; standard ovate, acute, recurved; wings falcate, adherent to the keel petals; keel petals much-incurved, obtuse or acute, semilunar; stamens diadelphous; anthers dorsifixed, uniform; ovary sessile or short-stalked, 2-ovuled; style elongated, beardless, incurved; stigma terminal, small or truncate. Pod oblong or broad-linear, about 10 cm long, densely hairy when young, yellow brown at maturity, achenelike, lower part flat, empty, thin, winglike, indehiscent, upper part thick, 2-valved, dehiscent, 1-seeded; seed plano-compressed, obovate. (Bentham and Hooker 1865; Prain 1908)

Butea species are indigenous to tropical India, Sri Lanka, Burma, and Malaysia; they are cultivated in Florida. The plants thrive on black loam in India and on the reclaimed saline soils of the Punjab. They are common in teak forests and on plains and hills up to 1,000-m altitude.

Most of the year the type species is an unsightly, twisted, and distorted tree, but its common name, "flame of the forest," becomes meaningful when its stiff, bright orange and vermilion flowers encompass its crown (Cowen 1965). Because the keel of the flowers has a pronounced beak, the tree is also called the parrot tree. The 3 leaflets are considered emblematic of the Hindu Trinity; hence the tree is esteemed throughout India. The scentless flowers contain much nectar; they are pollinated by birds.

Uses of plant parts of *B. monosperma* (Lam.) Taub. are varied (Watt 1908; Burkill 1966; Everett 1969). It is one of the most important hosts for lac insects. The bark is a source of Pala fiber, a cordage. In addition to their use as ornaments, the flowers yield a brilliant, but not fast, yellow dye that is used decoratively in holy Hindu festivals. The leaves serve as fodder for buffaloes and elephants. The flowers and leaves possess astringent, diuretic, and aphrodisiac properties. The seeds, ground into a paste with honey, are used medicinally as a purgative and vermifuge; mixed with lemon juice they are used as a rubefacient. The seeds contain about 18 percent oil, called moodoga oil, that is effective against hookworms; the seeds are also a source of small amounts of a resin, large amounts of albuminoids, and lipolytic and proteolytic enzymes. The gum exuded from the stems, Butea or Bengal kino, is astringent. The fibrous roots are used in making rope sandals. The dirty white, soft wood, which is durable under water, is used for well curbs and fuel, and to make water dippers, sacred utensils, and charcoal. The type species has considerable value in the recovery of saline soils.

Nodulation Studies

No nodules were present on *B. monosperma* examined in western India (V. Schinde personal communication 1977).

Cadia Forsk.
122 / 3580

Papilionoideae: Sophoreae

Type: *C. purpurea* (Piccioli) Ait.

5 species. Shrubs, erect, small, or trees with slender branches; ornamental. Leaves imparipinnate; leaflets small, numerous or large and few; stipels absent; stipules small. Flowers white, rose, or purple, large, handsome, solitary in leaf axils or in few-flowered racemes; bracts small; bracteoles absent; calyx broad-campanulate; lobes 5, broad, subequal; petals uniform, free, obovate or suborbicular, short-clawed, uppermost petal outermost; stamens 10, free, subequal, shorter than the petals; anthers uniform, linear, versatile; ovary subsessile or short-stalked, many-ovuled; style incurved, awl-shaped; stigma small, terminal. Pod linear, stipitate from the persistent calyx, acuminate, compressed, leathery, continuous within, 2-valved; seeds orange red, numerous, compressed, ovate-orbicular. (Bentham and Hooker 1865; Baker 1930)

Cadia species are native to tropical East Africa, Madagascar, and Arabia.

Caesalpinia L.
102 / 3559

Caesalpinioideae: Eucaesalpinieae

Named for Andreas Caesalpini, of Arezzo, 1519–1603, Italian botanist, Professor of Medicine in Pisa, and Chief Physician to Pope Clement VIII. Caesalpini is especially remembered for *De Plantis Libri*, published in 1583, one, if not the first, of the theoretical books on botany of the Renaissance. His taxonomic system was based solely on reproductive structures.

Type: *C. brasiliensis* L.

200 species. Shrubs, tall climbers, small and tall trees, mostly armed with spines or curved, hooked, sharp thorns, rarely unarmed. Leaves bipinnate, lacy, attractive; leaflets few to many, opposite, rarely alternate, small or large, herbaceous or leathery, usually without translucid glandular dots; stipels rarely present; stipules large, conspicuously leafy or minute. Flowers yellow, red, or variegated, showy, handsome, medium to large, in lax, multiflowered, paniculate, axillary or terminal racemes; bracts small or large, caducous; bracteoles none; calyx short-campanulate; lobes 5, short but longer than the calyx tube, imbricate, lowest lobe outermost in bud, concave or boat-shaped, often

larger; petals 5, orbicular-oblong, subequal or uppermost petal smaller and usually long-clawed; stamens 10, declinate, free; filaments linear, red or yellow, alternately long and short, villous or glandular at the base; anthers uniform, dorsifixed, longitudinally dehiscent; ovary sessile or short-stalked, glabrous, pubescent or glandular, free from calyx; ovules 2–many in 2 rows; style filiform; stigma terminal, generally concave, ciliate. Pod very variable, often prickly, pubescent or glandular, ovate, lanceolate, or falcate, or flat and smooth, straight, linear or inflated, beaked, sutures thickened, often filled between the seeds, 2-valved or indehiscent; seeds 1–8, transverse, ovate, orbicular, or globose. (Bentham and Hooker 1865; Baker 1930; Amshoff 1939; Wilczek 1952; Brenan 1967)

The genus is one of ancient origin. Fossil plants closely resembling *Caesalpinia* species were ascribed to the Tertiary period. This genus, called *Poinciana* in some localities in keeping with the Linnean synonym (1753), is not to be confused with the Linnean synonym *Poinciana* (1760) for *Delonix*.

Members of this large genus are distributed throughout the tropics and subtropics, primarily in America and Asia. They are now widely dispersed and are common on dry, sandy uplands, rocky areas, and seashores.

Caesalpinia species are garden ornamentals and hedge plants. The beauty of *C. pulcherrima* (L.) Sw., whose showy red flowers are borne in long spikes, is reflected in its common names: pride of Barbados, paradise flower, Spanish carnation, peacock's crest, flower tree, and others. This semi-drought-resistant species flowers in favorable habitats when only 8 months old. The yellow-flowered variety, *aurea*, is equally beautiful. *C. peltophoroides* Benth., false Brazilwood, is an avenue, park, and garden ornamental of unsurpassed beauty and form. This medium-sized, much-branched, rapid-growing unarmed tree thrives under varied soil and climatic conditions.

Most species are rich sources of tannin used for tanning leather and in the preparation of inks and dyes. The pods of *C. brevifolia* Baill. yield up to 60 percent tannin. *C. coriaria* (Jacq.) Willd., a shrub native to semiarid regions in Mexico and South America, is now widely cultivated in India. The S-shaped pods, the divi-divi of commerce, yield about 50 percent tannin, mostly as ellagitannin. The pods of *C. brevifolia* and *C. tinctoria* HBK are the algarobilla of commerce, a rich source of gallic acid used medicinally as a urinary astringent and in the manufacture of gallic esters, inks, dyes, and pyrogallol. *C. paraguariensis* (Parodi) Burk., *C. sappan* L., *C. sepiaria* Roxb., and *C. spinosa* (Mol.) Ktze. are also commercial sources of tannin. *C. sepiaria* is known as Mysore thorn in India and various tropical areas as "wait-a-bit" bush because of its sharp, recurved spines. It forms impenetrable thickets throughout Malaysia, the Philippines, and Japan.

Plant-part decoctions are used in folk medicine to treat intermittent fever, and as an abortifacient, emmenagogue, and general tonic. Bonducin, $C_{20}H_{28}O_8$, an amorphous, white, bitter glycoside, is obtained in generous quantities from the seed cotyledons of *C. bonduc* Roxb., *C. bonducella* Flem., and *C. crista* L. (Heckel and Schlagdenhauffen 1886; Tummin Katti 1930; Tummin Katti and Puntambekar 1930). Bonducin is sometimes referred to as "poor man's quinine" because it is used as a substitute for quinine in the treatment of intermittent fever. The seeds of *C. bonducella*, called fever, nikker, or nikar nuts, are gray, round, smooth, and very hard. Buoyancy of the seed accounts, in part, for this species' being widely dispersed pantropically by ocean currents. The seeds are used as talismans, for leis and beads, and by children as marbles; they yield oils used in cosmetics and in medicinal preparations (Godbole et al. 1929; Ghatak 1934). *C. praecox* R. & P., a shrubby tree of Argentina and Chile, exudes a golden yellow gum brea that contains about 80 percent arabin, is completely soluble in water, and is an acceptable substitute for gum arabic (Howes 1949).

C. echinata Lam. is the national tree of Brazil. The name "Brazil" had its origin in the Portuguese words "brésil" or "brazil," which mean bright red, resembling glowing coals, and were used to describe the color of a *Caesalpinia* wood abundant in this area. The name "brésil" was first applied to *C. sappan*, an evergreen shrub of India and Malaysia highly prized by Portuguese traders as a commercial red dye source. Following the discovery in 1500 of what is known today as Brazil, the Portuguese found that a dominant tree in this region, *C. echinata*, was the best commercial source of this prized dyewood. The trees commonly reached heights of 30–35 m, with trunk diameters of 1 m. The wood became a monopoly of the Portuguese crown from 1623 until the middle of the 19th century. During this period private exploitation of this species in Brazil was prohibited.

The red dye of *C. echinata*, brazilin, was used extensively for about 200 years in the textile and lacquer industries, and in the early 1900s it was acclaimed as a nuclear stain of biological materials and as an indicator in acid-base titrations. It resembles hematoxylin in requiring ripening, or oxidation, whereupon its color becomes an intense red; mordanting is necessary to achieve permanency. The red dye becomes purplish with alkalies and yellowish with acids (Uphof 1968). Commercial dye wood trade prevailed until about the middle of the 19th century, but with the synthesis of coal-tar dyes, the market collapsed. The isolation and structure of brazilin, $C_{16}H_{14}O_5$, were reported by Perkin et al. (1926–1928) and by Pfeiffer et al. (1930). The ± form of brazilin was synthesized in 1963 by Dann and Hofmann.

The wood of *C. crista* is known as Fernambuco, or Pernambuco, and that of *C. braziliensis*, as Brazilwood. These woods are generally uniform, very hard, compact, and strong, and have a rich, vinous, red color when seasoned (Record and Hess 1943). The wood of *C. sappan* is as hard as ebony; however, it is available neither in quantity nor in acceptable sizes for commercial purposes. Its use is limited to small turnery items and inlays; the wood is especially prized for violin bows.

NODULATION STUDIES

Presumably, factors other than the probable absence of compatible rhizobia in the rhizosphere account for the lack of nodules on *Caesalpinia* species. The black-pigmented, wiry, fibrous roots of *Caesalpinia* species bear a marked resemblance to those of nonnodulated *Gleditsia* and *Cassia* species (q.v.).

Nonnodulated species of *Caesalpinia*	Area	Reported by
brevifolia Baill.	France	Naudin (1894)
coriaria (Jacq.) Willd.	Philippines	Bañados & Fernandez (1954)
	Venezuela	Barrios & Gonzalez (1971)
crista L.	Hawaii, U.S.A.	Allen & Allen (1936b)
decapetala (Roth) Alst.	Zimbabwe	Corby (1974)
	S. Africa	Grobbelaar & Clarke (1974)
exostemma Moc. & Sessé	U.S.A.	Leonard (1925)
japonica S. & Z.	Japan	Asai (1944)
mexicana Gray	Java	Pers. observ. (1977)
paraguariensis (Parodi) Burk.	Argentina	Rothschild (1970)
(= *melanocarpa* Griseb.)	U.S.A.	Leonard (1925)
pulcherrima (L.) Sw.	U.S.A.	Leonard (1925)
	Hawaii, U.S.A.	Allen & Allen (1936b)
	Java	L. G. M. Baas-Becking (pers. comm. 1941)
	Philippines	Bañados & Fernandez (1954)
var. *aurea*	Java	L. G. M. Baas-Becking (pers. comm. 1941)
sappan L.	Hawaii, U.S.A.	Allen & Allen (1936b)
sepiaria Roxb.	Hawaii, U.S.A.	Allen & Allen (1936b)
spicata Dalz.	W. India	V. Schinde (pers. comm. 1977)
spinosa (Mol.) Ktze.	S. Africa	Grobbelaar & Clarke (1974)

Cajalbania Urb.   Papilionoideae: Galegeae, Robiniinae

Name refers to Province Pinar del Río near Cajálbana, Cuba, where the first specimens were collected.

Type: *C. immarginata* (Ch. Wright) Urb.

1 species. Shrub. Leaves imparipinnate; leaflets opposite, sometimes sub-alternate, numerous, 9–17; rachis not winged; stipels minute; stipules long-acuminate, basally lanceolate, connate below, intrapetiolar. Flowers bright purple, axillary or 1- to few-clustered from nodes of old wood, pedicels jointed below the calyx; calyx bell- to cup-shaped, base oblique, margin sub-bilabiate; lobes short, posterior lobe much smaller; petals very unequal; standard short-ovate, neither notched nor eared; wings oblong-lanceolate, longer than standard but shorter than the keel, free from the keel and short-clawed; keel petals short-clawed, very long, oblong, acute, united above but

free at apex; stamens diadelphous, 1 free, others unequal, anterior ones longer; anthers ovate, equal; ovary linear, stalked, smooth, multiovuled; style straight; stigma terminal, minutely flattened. Pod not described. (Urban 1928)

According to León and Alain (1951), this species is identical with *Suavallella immarginata* (Ch. Wright) Rydb.

This monotypic genus is endemic in Cuba. It occurs in pine forests.

Cajanus DC.   Papilionoideae: Phaseoleae, Cajaninae
414 / 3892

From the Malayan *kachang*, "bean," "pea."

Type: *C. cajan* (L.) Millsp. (= *C. indicus* Spreng.)

2 species. Shrublets, short-lived, perennial, erect, much-branched; young stems angled, whitish gray-tomentose. Leaves pinnately 3-foliolate, spirally arranged, with long, awl-shaped, subpersistent stipules; leaflets lanceolate, entire, acute at both ends, both surfaces pubescent, underside usually dotted with minute, yellow, resinous glands. Flowers large, yellow and purple, in pairs or scattered along the rachis of erect, axillary, peduncled racemes; bracts caducous; bracteoles none; calyx campanulate; lobes 5, acute, 2 upper lobes united into a 2-toothed lip, lower lip much smaller, 3-dentate; standard large, orbicular, reflexed, with 2 inflexed auricles at the base; wings oblique-obovate, equal in length to the incurved, obtuse keel; stamens diadelphous; anthers uniform; ovary subsessile, 4–7-ovuled; style thickened above middle, dilated below the oblique, terminal stigma. Pod linear, broad, hairy, oblique-acute, beaked, compressed, 2-valved, diagonally depressed between the seeds with transverse lines, but scarcely septate within; seeds usually brown but variable in size, shape, and color, usually round or oval, with a small, lateral, oblong, white hilum. (Bentham and Hooker 1865; Amshoff 1939)

The genus is represented in the tropics and subtropics of both hemispheres. Both species are widely cultivated in pure and mixed stands.

Two varieties of *C. cajan* (L.) Millsp. are recognized: the yellow-flowered, early-maturing variety, *flavus* DC., and the perennial, late-maturing variety, *bicolor* DC., whose flower standards are red or purple or streaked with these colors. The former variety was known as early as 1687; the latter was recognized in 1800 (de Sornay 1916). These varieties cross readily, and over 100 cultivars are recognized in India (Purseglove 1968). *C. kerstingii* Harms is a lesser-known wild species in West Africa.

The origin of *C. cajan*, known throughout the world as pigeon pea, Angola pea, Congo pea, dhal, no-eye pea, red gram, and by many other vernacular names, is still uncertain. Purseglove (1968) construes this plant as a native of Africa, where it was cultivated before 2000 B.C.; from Africa, it spread along trade routes into other areas. Today it ranks as the second most important pulse crop in India.

The pigeon pea is remarkable for its adaptability to diverse climates and soil types. It is moderately drought-resistant, yet its growth is enhanced by 60 cm of rainfall per annum or more; however, it does not do well in waterlogged soils, nor is it suitable to the humid wet tropics. It may be propagated easily from seed, but it does not transplant well. The root system is deeply penetrating. Calcium is a major element requirement. It is a prolific seeder. Krauss (1927) reported the production of 6,460 seeds weighing 1,150 g from 1,430 pods borne the first year on a single plant of a Hawaiian hybrid. Campbell and Gooding (1962) cite Bowles' postulation of a potential yield of about 897 kg of green pods per hectare from a Grenada long-podded selection; however, the normal yield is much less.

The usefulness of this short-lived plant is exceptional. The young green pods are eaten as a vegetable. The ripe seeds, a good source of vitamin B, are eaten cooked and are canned commercially. In India, the edible seeds in the form of a split pulse are known as dal or dhal. Field plants provide excellent fodder and browse, good quality hay, and silage. Pigeon pea is planted as a green manure and cover crop, for live fences and windbreaks, as temporary shade for young cacao and tea, and as a prop-plant for vanilla. The leaves are hosts for silkworm culture and lac insects. The deep root systems and the reseeding habit of these plants help prevent soil erosion. Natives in some areas use the stems as fiber for thatched roofs, for baskets, and as a source of charcoal. Leaf and stem decoctions have been used medicinally in India to treat sore throat, and as a diuretic, antidysenteric, and laxative.

NODULATION STUDIES

Diverse reports confirm the occurrence of nodules on the pigeon pea grown in widely different geographical areas. The nodules tend to be large, cylindrical to fork-lobed, and copious on the central taproot system. Infected cells in the bacteroid area of effective nodules are radially elongated and constitute from 60 to 70 percent of the total number of cells. Uninfected cells are considerably smaller and contain numerous starch grains. Leghemoglobin is abundantly present. The rhizobia are markedly elongated; bacteroidal forms are not produced. Longevity of the nodule is enhanced by the presence of a layer of thick-walled, pitted sclerid cells in the cortex (personal observation; Arora 1956c).

Nodules are produced on this species by a wide spectrum of rhizobial strains from species of other genera, and vice versa (Allen and Allen 1939; Wilson 1939a). Ineffective responses are uncommon.

From the Greek prefix, *kalli*, "beautiful," and *aner, andros*, "man," in reference to the prominent, attractive stamens.

Type: *C. inermis* (L.) Druce

150 species. Shrubs, low, straggling, much-branched, rarely perennial herbs or small trees, unarmed. Leaves bipinnate; leaflets small, numerous and in many pairs, or larger and in few pairs or 1, the single pair of pinnae often 3-foliolate, usually larger toward the apex of the pinna; stipules lanceolate, nervose, membranous, leaflike or hardened, sometimes bristly, persistent, often crowded at the base of the petiole. Flowers showy, red or white, polygamous, 5–6-parted, peduncled, in simple, axillary heads or in terminal, fascicled heads often 8 cm in diameter resembling pompons or powder puffs; calyx campanulate, 5-dentate or lobate, rarely deeply cleft; sepals valvate; petals 5, small, united halfway; stamens monadelphous, numerous, 10–100, long-exserted, bright red, yellowish green, or white, borne on the edge of the disk lining the calyx; anthers minute, often glandular-hairy; ovary sessile or subsessile, many-ovuled; style long, slender, often extending beyond the anthers; stigma small, terminal. Pod flat, linear, narrowed at the base, nonseptate within, margins thick, riblike, 2-valved, valves elastically dehiscent from the apex and sharply recurved after dehiscence; seeds 8–fewer, obovate or orbicular, compressed. (Bentham and Hooker 1865; Standley and Steyermark 1946)

Members of this large genus are distributed throughout tropical and subtropical America. Although it is considered native to the New World, some species are present in India, Madagascar, and West Africa. *Calliandra* species are well represented in the southwestern United States, Mexico, and the warm regions of South America. There are about 10 species in Argentina. *C. chilensis* Benth. is a xerophytic shrub endemic in Chile. *Calliandra* species occur in sandy washes and canyons in Colorado and Texas and on rocky igneous or limestone soil. In Mexico, various species inhabit dry forested mountain slopes up to 1,500-m altitude; others thrive in moist thickets.

All species are attractive; many are cultivated as ornamentals. Collectively, the species are called powder puff trees, or shrubs. *C. grandiflora* Benth., *C. inaequilatera* Rusby, and *C. tweedii* Benth. are especially noteworthy; their large flowers measure up to 9 cm in diameter. The first-named species is known as cabellos de ángel, or angel's hair, in Latin America, in reference to the long, silky stamens. Flowers of *C. haematoma* Benth., lehua haole of Hawaii, are used in making attractive garland leis.

Aside from their ornamental value, all species have soil-binding properties, and some serve as forage. *C. eriophylla* Benth., a low, densely branched, spineless perennial shrub, sometimes called false mesquite, is valued in the western United States as browse for livestock and deer; the seeds are relished by wild fowl.

In Mexico, astringent, viscid plant extracts from the roots of *C. grandiflora* have been used as remedies for eye diseases, diarrhea, and indigestion. Formerly, the bark of *C. houstonii* (L'Hérit.) Benth. was marketed in Europe as pambotana bark, as a substitute for quinine, and as an antiperiodic. The roots of *C. anomala* (Kunth) Macb. are used in Mexico and Central America to retard the fermentation of tepache, an alcoholic beverage made from pulque.

Calliandra wood is hard, heavy, strong, medium textured, and easy to work but not highly lustrous. Minor uses are made of it for small carpentry, implement handles and frames, and for fuel. In Mexico, the tannin extracted from the wood of *C. anomala* is used for tanning hides.

NODULATION STUDIES

Both nodulated and nonnodulated *Calliandra* species are documented. Information substantiating lack of nodules on the nonnodulated species is sparse, but is noted below as a matter of record. In-depth host-specificity studies with diverse rhizobial strains are merited.

Nodulated species of *Calliandra*	Area	Reported by
affinis Pitt.	Venezuela	Barrios & Gonzalez (1971)
foliolosa Benth.	Argentina	Rothschild (1970)
grandiflora Benth.	Hawaii, U.S.A.	Pers. observ. (1977)
guildingii Benth.	Trinidad	DeSouza (1966)
haematacephala Hassk.	Java	Pers. observ. (1977)
haematoma Benth.	Hawaii, U.S.A.	Pers. observ. (1940)
inaequilatera Rusby	Singapore	Lim (1977)
selloi (Spreng.) Macb.	Argentina	Rothschild (1970)
surinamensis Benth.	Java	Pers. observ. (1977)
tweedii Benth.	Argentina	Rothschild (1970)

Nonnodulated species of *Calliandra*	Area	Reported by
eriophylla Benth.	Ariz., U.S.A.	Martin (1948)
humilis Benth.	Ariz., U.S.A.	Martin (1948)
parvifolia (H. & A.) Speg.	Argentina	Rothschild (1970)

Calophaca Fisch.   Papilionoideae: Galegeae, Astragalinae
294 / 3762

From Greek, *kalos*, "beautiful," and *phakos*, "legume."

Type: *C. wolgarica* Fisch.

5 species. Shrubs or shrublets. Leaves imparipinnate; leaflets 3–13 pairs, entire, ovate-elliptic, leathery; stipels none; stipules large, membranous to leathery. Flowers few, large, mostly yellow, capitate and crowded or scattered on long-peduncled, axillary racemes; bracts and bracteoles usually caducous; calyx tubular; lobes 5, imbricate, lanceolate, acute, subequal, or upper 2 connate high up; petals clawed; standard ovate or suborbicular, sides folded backwards; wings obovate-oblong, subfalcate, free; keel shorter, incurved; stamens diadelphous; anthers round, uniform; ovary sessile, many-ovuled;

style glabrous, filiform, bent; stigma small, terminal. Pod oblong-cylindric, hairy, stipitate-glandular when young, nonseptate, 2-valved, the valves curling at maturity; seeds few, dark brown, kidney-shaped. (Bentham and Hooker 1865; Komarov 1971)

Members of this genus occur from the southern Soviet Union to the mountains of central and northern Asia. They are usually found in dry places.

These large-flowered shrubs have ornamental value. In some areas the seeds serve as feed and the bark as fiber and coarse cordage.

Calopogonium Desv. Papilionoideae: Phaseoleae, Galactiinae
403 / 3879

From Greek, *kalos*, "beautiful," and *pogon*, "beard"; the stems and pods of the type species have long, stiff, spreading, brownish hairs.

Type: *C. mucunoides* Desv.

12 species. Herbs, tall, climbing, perennial, sometimes fruticose vines. Leaves pinnately 3-foliolate, large; leaflets sometimes lobed; stipels present. Flowers small or medium, blue or violet, subsessile, fascicled on nodes along rachis of axillary racemes; bracts linear, hairy, caducous; bracteoles small, caducous; calyx campanulate; lobes 5, 2 upper lobes distinct or semiconnate into a 2-toothed lobe, lower 3 lanceolate; standard obovate, with greenish yellow blotch and 2 inflexed basal auricles; wings narrow, adherent to the keel; keel petals shorter than wings, blunt, oblong; stamens diadelphous; anthers dorsifixed, uniform; ovary sessile, hairy, several- to many-ovuled; style filiform, somewhat incurved, not barbate; stigma capitate, terminal. Pod linear, flat, 2-valved, elastically dehiscent, septate between the 5–10, globose, orbicular, compressed seeds. (Desvaux 1826; Bentham and Hooker 1865; Amshoff 1939)

Members of this genus are scattered throughout tropical America, the West Indies, and from Mexico to Argentina. They thrive in moist thickets, bushy, rocky areas, and light forests.

Most of these species are effective in erosion control. *C. mucunoides* Desv. is widely used as a green manure on plantations in Java and the Philippines. This vigorous creeping perennial roots at the nodes and forms dense mats, 50–60 cm thick, endures high rainfall, and thrives on a wide range of well-drained soils. It has promise in grass mixtures on newly cleared areas because of its tolerance for sun. Growth on coconut plantations has the disadvantage of being so dense that locating fallen coconuts is difficult; also, the matty growth provides a lush harbor for snakes (Joachim 1929). *C. orthocarpum* Urb. is more drought-resistant than *C. mucunoides* and is preferable in many areas because of its aggressiveness, profuse seeding ability, and acceptable palatability as hay. *C. caeruleum* (Benth.) Hemsl. is equally aggressive but not palatable.

NODULATION STUDIES

Nodulation has been reported for *C. mucunoides* by many workers. Nodules have also been observed on *C. caeruleum* in Taiwan and the Philippines by the authors (unpublished data 1962), in Trinidad by DeSouza (1966), in Papua New Guinea by M. J. Trinick (personal communication 1968), and in Singapore by Lim (1977), and on *C. orthocarpum* in Puerto Rico by Dubey et al. (1972).

Rhizobia from *C. mucunoides* are pleomorphic rods, 1.2μ by 0.7μ with rounded ends. Growth on solid media is scant. Colonies are circular, raised, smooth, moist, and transparent to translucent. Acid formation on bromthymol-blue glucose agar is slight. Glucose is utilized more readily than mannitol (Aquino and Madamba 1939).

Plant-host affinities of rhizobia from *Calopogonium* species are akin to members of the cowpea miscellany (Aquino and Madamba 1939; Ishizawa 1954; Galli 1958; personal observation 1963).

Calpocalyx Harms
(19a) / 3455

Mimosoideae: Adenanthereae

From Greek, *kalpis*, "urn," in reference to the urnlike calyx.

Type: *C. dinklagei* (Taub.) Harms

6–11 species. Woody climbers, or evergreen trees. Leaves bipinnate; pinnae 1 pair; leaflets several pairs, 7–10 cm long; conspicuous, flat glands on petiole and between leaflets of some species. Flowers orange brown to reddish yellow, slightly fragrant, small, sessile, in erect, peduncled spikes, axillary or congested at apices of branches; flowers cauliflorous on some species; young flowers catkinlike; calyx campanulate-cylindric, shortly 5-dentate, valvate; petals 5, adnate at base; stamens 10; filaments free, inserted at the base of the petals; anthers with a caducous, apical gland; ovary subsessile, many-ovuled; style slender; stigma acute, terminal. Pod short-stalked, falcate, thick, woody, often 20–28 cm long by 5–9 cm wide near top, slightly winged on the ventral edge, sides smooth or striate; on some species the pods open forcibly, scattering the seeds; seeds 1–3, ellipsoid or irregularly shaped, up to 4 cm long by 2.5 cm wide by 1 cm thick, slightly compressed. (Harms 1897; Baker 1930; Pellegrin 1948)

Members of this genus inhabit tropical West Africa. *Calpocalyx* species are usually understory trees that are seldom dominant, but somewhat common in evergreen forests near water.

The large members of this genus reach heights of 30 m and more and have boles up to 1 m in diameter that are often curved and forked at lower levels but commonly straight, cylindrical, and free of branches 20 m or more above the ground. The trunks are often buttressed with thin, sharp butt flares or root spurs and with spreading surface roots. The crowns are usually small.

The wood of *C. brevibracteatus* Harms, one of the better-known species, is reddish brown, variegated, hard, heavy, and attractive when finished, but it is very difficult to work. Its uses are limited to house posts and implement handles. The softer, lighter-weight wood of *C. aubrevillei* Pellegr. warps easily and is subject to fungus and insect damage. The evaporated leachate of the burnt ashes is a native source of salt in Liberia, where it occurs in rich stands (Voorhoeve 1965). Minor medicinal uses are made of plant-part decoctions.

Calpurnia E. Mey.   Papilionoideae: Sophoreae
147 / 3607

Named for Calpurnius, an imitator of Virgil; the plants are closely allied to and resemble those in the genus *Virgilia*.

Lectotype: *C. intrusa* (R. Br.) E. Mey.

6 species. Erect shrubs or small trees up to 6 m tall; branches angular, tawny, glabrous or pubescent. Leaves imparipinnate; leaflets numerous, 3–many pairs, opposite or alternate, blunt or notched at apex, usually thinly appressed-pubescent on both surfaces; stipels absent; stipules small or inconspicuous. Flowers yellow, few to many, pendent, in axillary and terminal racemes; bracts small; bracteoles absent; calyx campanulate; lobes 5, short, broad, upper 2 smaller and partly connate; standard suborbicular, bilobed, erect or slightly recurved, long-clawed, deeply channeled; wings falcate-oblong, obtuse, long-clawed; keel petals incurved, obtuse, long-clawed, joined along the back; stamens 10, free or connate at base; anthers small, dorsifixed, uniform; ovary stalked, 6- or more-ovuled; hairy or glabrous; style arcuate, short, awl-shaped; stigma small, terminal, capitate. Pod stalked, flat, broad-linear, membranous, compressed transversely between 2–6 seeds, narrowly winged along the upper suture, indehiscent; seeds transverse, compressed, oval-oblong. (Bentham and Hooker 1865; Phillips 1917, 1951)

Representatives of this genus are native to tropical Africa, South Africa, and India. They grow in mountain forests.

When in bloom, *C. aurea* Benth. and *C. floribunda* Harv. closely resemble *Laburnum anagyroides*, European laburnum; these species are cultivated as ornamentals and as shade plants on coffee plantations. *C. floribunda*, *C. silvatica* E. Mey., *C. subdecandra* (L'Hérit.) Schweik., and *C. villosa* Harv. are among the most attractive members of the genus; *C. subdecandra* grows up to an altitude of 2,100 m. The bitter leaf infusions of *C. intrusa* (R. Br.) E. Mey. are applied to sores on cattle to drive out maggots (Watt and Breyer-Brandwijk 1962). The aerial portion of this species yields oroboidine, $C_{20}H_{27}N_3O_3$ (Goosen 1963; Gerrans and Harley-Mason 1964). The role of this alkaloid has not been defined.

NODULATION STUDIES

C. aurea (N. Grobbelaar personal communication 1968), *C. intrusa*, and *C. villosa* are nodulated in South Africa (Grobbelaar and Clarke 1972).

From "calyx" and Greek, *tome*, "a part left after cutting"; the upper part of the calyx drops when the flower opens. The genus is alternately spelled *Calicotome*.

Type: *C. villosa* (Poir.) Link

5 species. Shrubs, low, erect, deciduous or evergreen; branches spinescent. Leaves palmately 3-foliolate, petioled; leaflets small, sessile; stipules absent or inconspicuous. Flowers yellow, axillary, solitary or in short, umbellate fascicles amid leaf clusters or ebracteate racemes; bracteole broad, 3-partite, at apex of pedicels; calyx turbinate, membranous, truncate; teeth 5, short, apical portion breaking away as the flower opens, leaving a cuplike remnant; standard ovate, erect, longer than the obovate-oblong wings and the blunt, incurved keel; stamens monadelphous; anthers alternately short and versatile, long and basifixed; ovary sessile, many-ovuled; style incurved, glabrous; stigma capitate, oblique. Pod narrowly linear-oblong, compressed, thickened or 2-winged on the upper suture, continuous inside, dehiscent, several-seeded. (Rehder 1940; Rothmaler 1949)

Polhill (1976) included this genus in Genisteae *sensu stricto*.

Species of this genus are common in the Mediterranean area, particularly in North Africa. They flourish in dry, rocky, evergreen scrub areas. *C. spinosa* Link has become naturalized in the environs of Palmerston North, New Zealand.

The species lack outstanding importance. However, *C. villosa* (Poir.) Link, a dominant shrub in devastated areas of Israel, "plays an important role in the successive stages leading to arboreal climax communities" (Zohary 1968).

NODULATION STUDIES

Nodules have been reported on *C. intermedia* Boiss. in Libya (B. C. Park personal communication 1960), on *C. spinosa* in South Africa (Grobbelaar and Clarke 1974), and on *C. villosa* in Israel (M. Alexander personal communication 1963).

*Camoensia Welw. ex Benth.   Papilionoideae: Sophoreae
131 / 3589

Named for Luis de Camoëns, 1524–1580, celebrated Portuguese epic poet and soldier, author of *The Lusiades*; he sailed with Vasco da Gama on his voyages of discovery.

Type: *C. maxima* Welw. ex Benth.

2 species. Shrubs, climbing or robust woody forest vines reaching 20 m or more tall. Leaves palmately 3-foliolate; leaflets elliptic-acuminate, large, leathery, petioluled; stipels awl-shaped. Flowers large (those of *C. maxima* Welw. ex Benth. very large), fragrant, 6–10, in erect, stoutly peduncled, axillary racemes; bracts and bracteoles short, caducous; calyx turbinate-campanulate; lobes 5, thick, imbricate in bud, reflexed at anthesis, posterior 2

connate; petals 5, long-clawed, free, white with yellow crinkled margins; standard large, broad-orbicular, with yellow blotch in the center; other 4 small, oval to wedge-shaped; stamens 10, monadelphous, free above, united below; anthers uniform, linear, dorsifixed; ovary long-stalked, 2–3-ovuled, sometimes pubescent; style long, thin, involute in bud; stigma terminal, small, capitate. Pod broad, linear, compressed, thick, leathery, 2-valved; seeds 3–4, obovoid, transverse, compressed. (Bentham and Hooker 1865)

Both species are native to tropical West Africa, but have been introduced into many tropical areas as ornamentals.

The exceedingly fragrant, tubular, individual flowers of *C. maxima* are often said to be the most handsome of the subfamily. Dehiscing pods expel the seeds with considerable force.

NODULATION STUDIES

Nodules were absent on an unidentified *Camoensia* species introduced into the greenhouses at Villa Thuret, France (Naudin 1897a).

Campsiandra Benth.   Caesalpinioideae: Sclerolobieae
114 / 3572

From Greek, *kampe*, "bending," and *aner, andros*, "man," in reference to the long, wavy stamens.

Type: *C. comosa* Benth.

3 species. Trees, medium to large, unarmed. Leaves imparipinnate; leaflets alternate, several, large, oblong, leathery; stipules small, narrow, caducous. Flowers yellow or rose red, showy, borne in short, terminal, panicled racemes; bracts small, deciduous; bracteoles absent; calyx campanulate; lobes 5, short, imbricate; petals 5, oblong, subequal, imbricate; stamens 12–20, free; filaments colored red, pink, or white, long-exserted; anthers ovate, versatile, opening by longitudinal splits; ovary short-stalked, linear, many-ovuled; style filiform; stigma small, terminal. Pod large, flat, oblong, straight or falcate, leathery, 2-valved; seeds flat, large, with brittle coat. (Bentham and Hooker 1865; Amshoff 1939)

Members of this tropical American genus are mostly confined to the Amazon Basin, along streams and lakes.

The type species is native to Guyana and is synonymous with *C. surinamensis* Kleinh. of Surinam; the other 2 species, *C. angustifolia* Spruce and *C. laurifolia* Benth., are common in regions of Brazil along the Amazon. The last-named species reaches a height of 30 m, with a straight cylindrical trunk 60 cm in diameter and free of branches 12–15 m above the ground. The wood has a density of 900–1,100 kg/m^3; it is highly durable and finishes attractively, but the supply is very limited. Its use is reserved for heavy construction purposes (Record and Hess 1943). The seeds are used in some areas as a source of starch. In Brazil the roots and leaves are used medicinally to counteract fevers (Uphof 1968).

Camptosema H. & A.
408 / 3884

Papilionoideae: Phaseoleae, Diocleinae

From the Greek verb *kampto*, "bend," and *semeia*, "standard," in reference to the strongly reflexed standard.

Type: *C. rubicundum* H. & A.

12–15 species. Shrubs, or shrublets, climbing or suberect, perennial. Leaves pinnately 3-foliate, rarely 1-, 5-, or 7-foliolate; leaflets elliptic-oblong, subcoriaceous, smooth; stipels present; stipules caducous. Flowers brilliant red, handsome, pedicelled, in axillary, fasciculate racemes; bracts deciduous; bracteoles small, often deciduous; calyx tubular; lobes 4, upper 2 joined into 1, laterals shorter, lowermost longest; standard ovate-oblong, appendaged at base with inflexed auricles; wings oblong, free or slightly adherent to the oblong keel; stamens 10, pseudomonadelphous, vexillary stamen free at base, connate with others in the middle; anthers elliptic, uniform; ovary stalked, with basal, tubular disk, many-ovuled; style awl-shaped, glabrous, incurved; stigma small, terminal. Pod stalked, linear, leathery, 2-valved, falsely partitioned between the small, compressed, elliptic seeds, tardily dehiscent. (Bentham and Hooker 1865; Burkart 1970)

Representatives of this genus are widespread in South America, primarily in Brazil, but also throughout central northern Argentina and the lowlands of the Uruguay River. They thrive in small wooded areas.

Several species serve as ornamental vines in parks and gardens. *C. rubicundum* H. & A. is a highly prized introduction in Italy and California.

Campylotropis Bge.

Papilionoideae: Hedysareae, Desmodiinae

From Greek, *kampylos*, "bent," and *tropis*, "keel."

Type: *C. chinensis* Bge.

65 species. Shrubs or shrublets; stems usually pubescent. Leaves 3-foliolate, rarely 1-foliolate; leaflets elliptic to ovate; stipels usually absent; stipules 2, persistent. Flowers mostly rosy or purple, few, in axillary and terminal racemes; bracts small, 1-flowered; bracteoles 2, early caducous; calyx campanulate; lobes 5, upper 2 partly joined; standard rounded or ovate, acuminate; wings subequal, subfalcate, blunt, somewhat joined to the keel by a large basal auricle; keel inflexed at a right angle, apex pointed; stamens diadelphous, filaments slender; ovary short-stalked, 1-ovuled; style filiform, base pilose; stigma small, terminal. Pod small, flattened, lenticular, with pointed apex, indehiscent, 1-seeded. (Schindler 1924; Ricker 1946)

The genus was formerly a section in *Lespedeza*.

Members of this genus are indigenous to eastern Asia and Taiwan, especially in the provinces of Yünnan and Szechwan in China. Species occur in bushlands and along roadsides.

Generic name is the Latinized version of *kanavala*, the vernacular Malabar Indian name for a plant species, presumably *Canavalia maritima* (Aubl.) Thou. (Sauer 1964).

Type: *C. ensiformis* (L.) DC.

50 species. Herbs, twining, or prostrate, slender, annual or stout, tall, woody, perennial climbers. Leaves pinnately 3-foliolate; leaflets variously shaped, mostly ovate and acuminate, often large, entire, papery or leathery, undersurfaces usually hairy; stipels generally minute, caducous; stipules small, wartlike or inconspicuous, caducous. Flowers generally purple or violet, varying from reddish to bluish and from dark to pale, sometimes white, with yellow markings near the petal bases, often resupinate, 2–6, pedicelled at swollen nodes on short or long peduncles in axillary thyrses; bracts small; bracteoles 2, caducous, subtending each flower; calyx campanulate, tubular at the base, bilabiate, upper lip large, truncate or 2-lobed, lower lip smaller, minutely 3-toothed or entire; standard large, broad-obovate, reflexed, calloused along the midvein above the claw; wings free, narrow, subtwisted or falcate, eared above the claw; keel wider than the wings, incurved, free at the base, bluntly beaked, sometimes gibbous, eared above the claw; stamens monadelphous or vexillar stamen free at the base and connate with the others in the middle; anthers uniform; ovary sessile, usually hairy, many-ovuled; style incurved or folded with the keel, beardless; stigma small, capitate, terminal. Pod broad-linear, flattened, sometimes swollen, compressed between the seeds, with a prominent ridge or rib along the upper suture, endocarp papery, white, separating, 2-valved, valves leathery, often thinly filled between the seeds, dehiscent, sometimes explosively, or indehiscent; seeds 4–15, large, ovate or orbicular, oblong to elliptic, strongly to moderately compressed, usually various shades of reddish brown, mottled, hilum linear. (De Candolle 1825b; Bentham and Hooker 1865; Piper and Dunn 1922; Piper 1925; Sauer 1964)

Sauer (1964) expressed the opinion that this genus probably diverged from other Phaseoleae during the Cretaceous period.

Members of this genus are widely distributed throughout the tropics and subtropics of the Old World and the New World and the islands of the West Indies and South Pacific. They inhabit sandy and rocky coastal seashores, woodlands, and thickets.

Two species extensively cultivated are *C. gladiata* (Jacq.) DC., sword bean, in the Far East, especially India, and *C. ensiformis* (L.) DC., jack bean, in the southwestern United States. They are grown as green manure and as cover and rotation crops favored for their rapid and heavy growth, semi-drought-resistance, deep-rooted habit, and shade tolerance. For jack bean Krauss (1911) reported yields of 34–43 metric tons of green foliage per hectare in Hawaii. The best seed-crop yield was 1,420 kg/ha. Krauss cited the composition of the leaves, in percent, as follows: water, 76.81; ash, 2.7; cellulose, 6.36; fat, 0.48; nonnitrogenous matter, 8.44; and nitrogenous matter, 5.21. Nutritionally the foliage appears acceptable, but cattle find it coarse and unpalatable. Various opinions prevail regarding digestibility of the foliage and the presence of poisonous properties in the mature pods and seeds. The seeds of this species have a record of toxicity in Zimbabwe, but the toxic principle is not known. Orrù and Fratoni (1945) eliminated hydrocyanic acid, a cyanogenetic glucoside, and urease, as possible causes. Young seeds and pods of the jack bean are cooked and eaten as vegetables, but once they have matured and become hard, they are not delectable.

All species are excellent soil-binders and soil-improvement plants. *C. maritima* is a common pantropic strand plant. Dissemination of the seed of this species along the beaches of many South Pacific Islands is attributed to ocean currents. Throughout Polynesia the orchid-colored maunaloa flowers of this and other *Canavalia* species are strung into ornamental garland leis.

The enzyme, urease, was first crystallized by Sumner (1926a,b) from jack bean meal. This enzyme hydrolyzes urea to ammonium carbonate and is used as a standard laboratory reagent for determining the urea content of urine and blood. It occurs rather generally throughout the plant kingdom and is present in some members of each subfamily; however, *C. ensiformis* and *C. obtusifolia* (Lam.) DC. have the largest known urease content of any higher plant (Damodaran and Sivaramakrishnam 1937; Granick 1937). Recent reviews by Varner (1960), Reithel (1971), and Bailey and Boulter (1971) discuss the properties, chemistry, and function of urease.

The occurrence of canavanine, $C_5H_{12}N_4O_3$, an amino acid, in seeds of *C. ensiformis* is of related interest (Kitagawa and Tomiyama 1929), yet not unique, except for its occurrence only in papilionaceous species; it is lacking in Mimosoideae and Caesalpinioideae surveyed to date (Turner and Harborne 1967; Bell 1971). Canavanine and urease are present concurrently in 15 genera, a biochemical coincidence that merits further inquiry. The role and significance of this amino acid in chemotaxonomy are reviewed by Bailey and Boulter (1971), Bell (1958, 1971), and Birdsong et al. (1960).

Concanavalin A, recognized about the turn of the century in seed extracts of jack bean meal, was isolated in crystalline form and described by Sumner and Howell (1936). The clumping of various animal erythrocytes in high dilution was described. The characterization and reactions of this phytohemagglutinin attracted considerable attention in the years that followed. In keeping with evidence that concanavalin A reacts with certain polysaccharides to form precipitates, much as a specific antibody reacts with an antigen (So and Goldstein 1967), Daniel and Wisnieski (1970) showed the usefulness of the reaction in identifying and characterizing polysaccharide constituents of numerous mycobacterial culture filtrates. Preliminary studies by Burger and Noonan (1970) showed a preferential agglutination of tumor cells by concanavalin A. By covering sites on transformed cells with monovalent concanavalin A, the growth pattern of transformed fibroblasts was restored to that of normal cells. Reversibility of reaction was possible. This technique holds promise for future research with tumor tissue.

NODULATION STUDIES

Young spherical nodules on *Canavalia* species become cylindrical and often apically bilobed at maturity. In a detailed study of nodule morphogenesis and development on *C. gladiata*, Narayana and Gothwal (1964) observed effective nodules only on secondary and tertiary rootlets. The first nodules were evident on the 6th day; the maximum number was obtained within 30 days. Infection threads in the root hairs were branched. Cells in the apical meristem were progressively invaded by extensions and further branching of the threads. Only about 50 percent of the cells in the bacteroid area were infected, and only invaded cells displayed hypertrophy. Ten to 12 vascular bundles observed in cross sections of the nodule cortex attested to an ample vascular supply. Seasonal degeneration ultimately emptied the inner bacteroid area, whereupon only the hollow outer cortex remained. Ineffective white nodules and false, or atypical, collar-type nodules at the base of lateral roots were also observed. The latter, described earlier by Narayana (1963a), were composed of hypertrophied parenchyma and lacked any evidence of infection.

Reciprocal host-infection tests signify the inclusion of *Canavalia* species within the cowpea miscellany (Richmond 1926a; Carroll 1934a; Allen and Allen 1936a, 1939; Wilson 1939a; Burton 1952). The need for careful rhizobial strain selection to obtain a highly effective level of nitrogen fixation is evident from the prevalence of mediocre and ineffective responses.

Nodulated species of *Canavalia*	Area	Reported by
campylocarpa Piper	Hawaii, U.S.A.	Allen & Allen (1936a)
cathartica Thou.	Hawaii, U.S.A.	Pers. observ. (1940)
ensiformis (L.) DC.	Widely reported	Many investigators
gladiata (Jacq.) DC.	Widely reported	Many investigators
var. *gladiata*	S. Africa	Grobbelaar & Clarke (1974)
lineata (Thunb.) DC.	Japan	Asai (1944)
maritima (Aubl.) Thou.	Hawaii, U.S.A.	Pers. observ. (1940)
	Australia	Norris (1959b)
virosa (Roxb.) W. & A.	Zimbabwe	Corby (1974)
(= *ferruginea* Piper)	S. Africa	Grobbelaar et al. (1964)
	S. Africa	Grobbelaar & Clarke (1974)

Caragana Lam.   Papilionoideae: Galegeae, Astragalinae
293 / 3761

Generic name is the Latinized version of *karaghan*, the Mongolian name for these plants.

Lectotype: *C. arborescens* Lam.

80 species. Shrubs or small trees, up to 6 m tall, hardy, deciduous. Leaves paripinnate, often fascicled; leaflets small, 2–18 pairs, entire; rachis usually persistent, becoming lignified and spine-tipped; stipels absent; stipules awl-shaped, membranous, later hardened and spinescent. Flowers yellow, rarely white or pink, peduncled, solitary or in few-flowered fascicles at older nodes or in axils at the base of young shoots; bracts and bracteoles often awl-shaped;

calyx tubular or campanulate, swollen at the back; lobes subequal, upper 2 usually smaller; standard ovate or suborbicular, upright, sides recurved, narrowed into a claw; wings oblong, obliquely truncate, long-clawed; keel straight, blunt; stamens diadelphous; anthers uniform; ovary subsessile, several-ovuled; style straight or slightly curved; stigma terminal, small. Pod linear, cylindric or inflated, tip usually curved and pointed, nonseptate, 2-valved, twisting during dehiscence; seeds ellipsoid or subglobose. (Bentham and Hooker 1865)

Members of this genus are native to northeastern Europe, the Soviet Union, central Asia, the Himalayas, and China; species have been introduced into many regions in northern latitudes. They grow on stony slopes and in gullies and thickets.

The vast native range of the Siberian pea tree, *C. arborescens* Lam., which encompasses about 620,000 square miles (over 160 million hectares), extends across Siberia from about 77° to 120° east latitude and from 48° to 60° north longitude (Moore 1965). Its growth vigor in the United States diminishes south of Nebraska. Because of its remarkable cold- and drought-resistant properties, it is cultivated extensively for home shelter-belt purposes and as a windbreak hedge plant on the plains and prairies of Canada and the northern United States (Cram 1952). The seeds are food for wildlife, and in some areas the young pods serve as a vegetable. The leaves yield an azure dye, and the bark provides a fiber for ropes. It is especially valued as a soil-improvement plant and as supplementary food for the maintenance of reindeer herds within the Arctic Circle on the Kola Peninsula (Roizin 1959).

NODULATION STUDIES

Rhizobia were isolated by Beijerinck from *C. arborea*, = *C. arborescens*, as early as 1888. Maassen and Müller's conclusion (1907) that rhizobia from *C. frutescens* Medik. nodulated only their homologous host went unchallenged until Wilson (1939a) reported nodulating promiscuity by strains from this species with a vast range of plants in diverse genera and tribes. Considerable cultural and physiological variation among 14 strains from *C. arborescens* was noted by Gregory and Allen (1953); however, in general, all belonged to the slow-growing cowpea-soybean-lupine type of rhizobia. Host-infective patterns were quite uniform, although various examples of nonreciprocal cross-inoculation were observed. Of special interest was the finding that caragana rhizobia reisolated from nodules formed by them on *Trifolium pratense* retained the ability to form nodules on *Caragana arborescens*.

Caragana nodules are among the few perennial types that have been studied histologically. Nodule morphogenesis is similar to that of nodules of herbaceous species (Lechtova-Trnka 1931; Allen and Allen 1950, 1954); however, as maturation progresses, the basal areas develop a woody texture and a dark brownish black surface. During seasonal cessation of active nodule function, the vascular bundles *per se* terminate and are transformed into rootlets by continued procambial differentiation after their outward extension from the nodules (Allen et al. 1955). This type of rootlet development differs from that of *Sesbania* nodules wherein rootlet meristems are initiated within the endodermis of vascular bundles and grow outwardly at right angles to the bundle axis (Harris et al. 1949).

Throughout the life of a *Caragana* nodule the volume of tissue functionally active in nitrogen fixation remains more or less constant; accordingly, as the nodule becomes larger, the ratio of this volume to total nodule mass becomes smaller. On 1-month-old nodules the ratio of functional bacteroidal tissue to total nodule mass is about 1:1; in 2-month-old nodules, the ratio is about 1:2; in 6-month-old nodules the functional bacteroidal tissue constitutes only about 20 percent of the nodule mass. Of particular interest is the coexistence of juvenile and senescent tissues in close proximity for long periods. Growth equilibrium, development, and function of the nodule do not appear unbalanced during its existence.

Nodulated species of *Caragana*	Area	Reported by
arborescens Lam.	Widely reported	Many investigators
aurantiaca Koehne	Arctic U.S.S.R.	Roizin (1959)
(*chamlagu* Lam.)	Japan	Asai (1944); Ishizawa
= *C. sinica* (Buc'hoz) Rehder		(1954)
frutescens Medik.	Widely reported	Many investigators
var. *pendula* Dipp.	Arctic U.S.S.R.	Roizin (1959)
frutex (L.) K. Koch	Ill., U.S.A.	Appleman & Sears
		(1943)
maximowicziana Kom.	Arctic U.S.S.R.	Roizin (1959)
microphylla Lam.	Arctic U.S.S.R.	Roizin (1959)
pekinensis Kom.	U.S.A.	Wilson (1939a)
sophoraelia Tausch	Arctic U.S.S.R.	Roizin (1959)
tragacanthoides Poir.	France	Lechtova-Trnka
		(1931)
	Arctic U.S.S.R.	Roizin (1959)
turkestanica Kom.	Arctic U.S.S.R.	Roizin (1959)

Carmichaelia R. Br.   Papilionoideae: Galegeae, Robiniinae
282 / 3748

Named for Captain Dugald Carmichael, Scottish army officer and botanist of the 18th century; he collected plants in New Zealand.

Type: *C. australis* R. Br.

41 species. Mostly shrubs, a few small trees, often leafless during flowering; xerophytes with rushlike or flattened stems, with minute scales at the nodes. Leaves, when present, imparipinnate with 3–many, obcordate, small, scalelike leaflets; stipels absent; stipules small, membranous. Flowers rose, white, or lilac-streaked, short-stalked in short, solitary or clustered racemes in nodes of branches; bracts small, membranous; bracteoles small, inserted on the pedicel or adnate to the calyx; calyx tubular or campanulate; teeth subequal, imbricate in bud; standard orbicular, narrowed into a short claw; wings oblong, free, often shorter than the standard; keel incurved, blunt; stamens

10, 1 free, often shorter than the other 9 united into a sheath; anthers uniform; ovary short-stalked, many-ovuled; style incurved, glabrous; stigma small, terminal. Pod flat, ovate or elliptic-oblong, acuminate, sutures thick, valves separating from the margin during dehiscence; seeds few, flat. (Allan 1961)

This genus is native to New Zealand and Lord Howe Island. Its members thrive in alluvial shady areas, stream sides, grasslands and forest margins, coastal areas, and limestone and rocky places.

Nodulation Studies

Milovidov's (1928a) descriptions of nodules on *C. australis* R. Br. and of the isolates which he considered the causative organism have historical significance only. His designation of the isolates as *Bacterium radicicola* forma *carmichaeliana* lacked a sound basis. Thin and inconspicuous infection threads in nodules of *C. australis* were observed by Dawson (1900b); however, Lechtova-Trnka (1931) depicted rather bizarre strands. Instead of rhizobial alignment in single file, the rods were transported 5 to 10 abreast in wide multibranched and multibulged, finger-shaped strands that protruded in all directions. The rhizobia were released into the invaded cells in the form of microcolonies or mucoid aggregates by means of a conidial-like pinching off of the thread tips, or by constrictions within the bulged strands. These observations merit confirmation.

Kew Gardens, England, specimens of *C. subulata* Kirk were well nodulated (personal observation 1936). Nodules on *C. australis* were reported by Dawson (1900b), Milovidov (1928a), and Grobbelaar and Clarke (1974) in England, France, and South Africa, respectively. Also, all of the following *Carmichaelia* species in New Zealand are nodulative (R. M. Greenwood personal communication 1975): *C. aligera* Simpson, *C. angustata* Kirk, *C. arborea* Druce, *C. exsul* F. Muell., *C. flagelliformis* Col. ex Hook. f., *C. grandiflora* Hook. f., *C. kirkii* Hook. f., *C. monroi* Hook. f., *C. odorata* Col. ex Hook. f., *C. orbiculata* Col. ex Hook. f., *C. ovata* Simpson, *C. petriei* Kirk, *C. robusta* Kirk, *C. violacea* Kirk, *C. virgata* Kirk, and *C. williamsii* Kirk.

Carrissoa Bak. f.   Papilionoideae: Phaseoleae, Cajaninae

Named for Luiz Wittnich Carrisso, 1886–1937, Portuguese botanist with the mission to Angola and founder of *Conspectus Florae Angolensis*.

Type: *C. angolensis* Bak. f.

1 species. Shrublet; rootstock thick, elongated, woody; stems several, 15–20 cm long. Leaves simple, entire, linear-lanceolate, pinnately nerved,

glandular underneath, at first silvery pubescent; stipels none; stipules linear-lanceolate, reddish. Flowers short-stalked, solitary, axillary; calyx tube linear, short, bilabiate; upper 2 sepals connate, shorter; petals short-clawed, sub-equal; standard ovate, emarginate, yellow with red streaks; wings asymmetric, elliptic or oblong; keel rounded dorsally; stamens diadelphous; ovary densely pubescent; style pubescent; stigma terminal. Pod flattened, elliptic-oblong, dehiscent, usually 2-seeded. (Baker 1932)

The genus is allied to *Eriosema* and *Rhynchosia*.

This monotypic genus is endemic in Angola.

Cascaronia Griseb.   Papilionoideae: Galegeae, Astragalinae
302 / 3770

From Spanish, *cáscara*, "bark," in reference to the gray, cracked bark of the trunk.

Type: *C. astragalina* Griseb.

1 species. Tree, 5–25 m tall, with a trunk diameter up to 1 m; bark thick, gray, deeply grooved, corky, and resinous. Leaves imparipinnate; leaflets alternate and also opposite, 5–13 pairs, elliptic and lanceolate-oblong, sticky, glandular on under surface; stipels none; stipules caducous. Flowers yellow, about 1 cm long, borne in axillary, peduncled racemes; bracts minute, cadu-cous; bracteoles none; calyx turbinate-campanulate, very glandular, weakly bilabiate; lobes 5, upper 2 longer, semiunited, lower 3 free, triangular; corolla membranous; standard obovate; wing and keel petals free, obliquely clawed; stamens diadelphous, unequal; anthers convergent, uniform; ovary stipitate, oblong, glandular, 2–3-ovuled, narrowed into a slender, slightly incurved style; stigma terminal. Pod flat, oblong-elliptic, narrowly winged on dorsal side, reticulate and gland dotted, glabrous, shortly stipitate, indehiscent, 1–2-seeded; seeds flat. (Burkart 1943; Hutchinson 1964)

The preceding description accommodates Hutchinson's observation (1964) of diadelphous stamens in contrast to monadelphous stamens (Burkart 1943). Burkart (1943) deemed this genus more appropriately positioned in the tribe Dalbergieae because of fruit characters.

This monotypic genus is endemic in Argentina. It grows near rivers and in cleared forests.

The bark exudes a red gum of no commercial importance. The red, hard, heavy wood has a fine texture and is of fair quality, but the supply is limited. It is used primarily for fuel and in the turnery of casual objects. Plant parts, especially the pods and flowers, have the citrus fragrance of rue, *Ruta graveolens*.

Throughout ancient times the word "cassia" denoted any second-class *Cinnamomum* bark marketed as a substitute for, or as an adulterant of, Chinese cinnamon, *Cinnamomum cassia*, whose bark and leaves were articles of commerce because of their aromatic and purgative properties. Linnaeus used the name in the generic sense for similar reasons.

Lectotype: *C. fistula* L.

Over 600 species. Trees, shrubs, and herbs. Leaves paripinnate, rarely reduced to phyllodes; leaflets usually broad, glabrous or hairy, 1–many pairs; petioles often glanduliferous, some sensitive to touch; stipules present. Flowers actinomorphic, yellow, white, pink, or orange, showy, attractive, often large, solitary in leaf axils or few to many, in axillary and terminal racemes, simple or panicled, erect or pendent; bracts and bracteoles various; calyx tube short or absent; lobes 5, subequal, ovate or lanceolate, sometimes acuminate, overlapping; petals 5, spreading, overlapping, sometimes long-clawed, subequal or lowermost petal larger, uppermost petal innermost; stamens 10, perfect and subequal, or very unequal with upper 3 imperfect, rudimentary or absent, or only 5; filaments short or long, linear or rarely dilated; anthers uniform or lower ones larger, dorsifixed or basifixed, opening by a basal or terminal pore or a longitudinal slit; ovary sessile or stalked, few- to many-ovuled; style long or short; stigma terminal, small, truncate. Pod linear, flat or cylindric, rarely winged or quadrangular, membranous, woody or leathery, nonseptate or with transverse corky compartments or pulpy septations between the seeds, elastically dehiscent, the valves twisting, or indehiscent; seeds few to many, ovate, compressed. (Bentham 1871; Steyaert 1952; Irwin 1964; Irwin and Barneby 1976)

The genus *Cassia* was originally defined by Linnaeus in the first edition of his *Species Plantarum*. In 1871 Bentham appraised the genus as "an excellent instance of a large, widely distributed, much varied, but well-defined group." Over the years certain sections have been given generic consideration, notably the division of the American members into 28 separate genera (Britton and Rose 1930). Most botanists, however, still favor the retention of Bentham's infrageneric organization of *Cassia* into 3 subgenera, *Fistula*, *Senna*, and *Lasiorhegma*; within this framework, *Cassia* is the fourth largest genus in family Leguminosae, the largest in subfamily Caesalpinioideae, and among the 25 largest genera of dicotyledonous plants.

Representatives of this large genus are pantropical; only a few species grow in temperate regions. They occur in moist or dry thickets, thinly forested hillsides, waste or cultivated areas, and along roadsides.

Cassia species have varied economic uses, but their importance is disproportionate to the enormous size of the genus. Species in subgenus *Fistula* undoubtedly have the greatest popularity because of their floriferous, ornamental beauty. Well known in parks, gardens, and along avenues in all warm countries are the shower trees, so called because their fallen petals carpet the ground beneath them. Their beauty is enhanced by their long flowering periods, handsome form, and spreading, open crowns. The following species are among the most popular:

C. fistula L.	Golden shower, Indian laburnum
C. grandis L. f.	Pink or coral shower
C. javanica L.	Rainbow shower
C. leiandra Benth.	Bronze shower
C. nodosa Buch.-Ham.	Pink and white shower
C. siamea Lam.	Kassod tree, Siamese cassia

Cassia wood is of reasonably good quality but lacks commercial timber value. Its local uses are mostly for posts, beams, fuel, and small carpentry items. The bark of *C. auriculata* L., tanners' cassia, yields about 20 percent tannin.

In general, *Cassia* species have not proved highly satisfactory as browse plants, green manures, and soil cover because of their slow growth, low yield, susceptibility to fungal and insect attacks, slow decomposition, and toxic and unpalatable properties. Primary use has been for shade on tea and coffee plantations.

Natives of various tropical areas use decoctions of plant parts to treat ringworm, eye diseases, rheumatism, and a wide variety of other ailments (Watt and Breyer-Brandwijk 1962). The following appraisal of *C. fistula* by Sir Hans Sloane, an early 18th-century medical botanist in Jamaica, is of historical interest (Pertchick and Pertchick 1951): "The Pods or Canes are us'd to purge the Belly of Choler, and the Blood of vicious Humours, being pulped, and to cool the Kidneys, and generally thought proper in Diseases of the Breast." However, none of these medicinal uses rivals that of senna leaves for cathartic purposes. *C. acutifolia* Del., *C. angustifolia* Vahl, and *C. obovata* Collad. are commercial sources of Alexandrian, Arabian, and Italian senna, respectively. The laxative property is said to be due to the presence of emodin, $C_{15}H_{10}O_5$, and related glucosides.

NODULATION STUDIES

Discontinuity of nodulating ability within *Cassia* becomes obvious at the subgeneric level when nodulation data are superimposed upon Bentham's infrageneric organization of the genus. Paucity of symbiobility within subgenera *Fistula* and *Senna* is in marked contrast to nodule prevalence in subgenus *Lasiorhegma*. Although conflicting reports of nodulation for a few species in *Fistula* and *Senna* present an anomaly, the negative evidence outweighs the positive. Indeed, it is quite likely that *C. wrightii* Gray, section *Chamaecrista*, will eventually be found nodulated under growth conditions more favorable than those in arid Arizona (Martin 1948). This assumption is prompted by the report of nodules on *C. leptadenia* Greenm. in Zimbabwe (Corby 1974), the only other species in section *Chamaecrista* heretofore reported nonnodulated (Martin 1948).

Information pertinent to nodule morphology, histology, and rhizobial characterization relative to the 4 nodulated species in *Fistula* and *Senna* (see table below) is lacking. Nodules on species in section *Chamaecrista* have apical meristems that become dichotomously branched with age (personal observation). Rhizobia from these nodules are slow-growing strains of the cowpea-soybean-lupine type. Host-infection studies of the members of the *Chamaecrista* complex studied to date, i.e., *C. chamaecrista* L. (Burrill and Hansen 1917; Wilson 1939a) as *C. fasciculata* Michx. (Bushnell and Sarles 1937), *C. mimosoides* L. (Allen and Allen 1936a), and *C. nictitans* L. (Conklin 1936), and 1 species in subgenus *Senna*, *C. pilifera* Vog. (Allen and Allen 1936a), have verified these plants and their symbionts as members of the cowpea miscellany.

Why certain *Cassia* species are nodulated and others are not has long been an enigma. Significant physicochemical factors operative in *Fistula* and *Senna* militate against nodule formation. Root hairs are uncommon; when present, they are sparse and thick walled. Heterogeneous rhizobia have the ability to enter root hairs of *C. grandis* L. f. and *C. tora* L., 2 nonnodulating species, but they become immobilized in the basal area of the hairs (personal observation). Neither infection threads nor rhizobial entry into cortical cells has been observed. Plant parts yield antibacterial compounds (Fowler and Srinivasian 1921; Itano and Matsuura 1936; Wasicky 1942; George et al. 1947; Robbins et al. 1947; Anchel 1949; Hauptmann and Nazario 1950). Roots of nonnodulating species are commonly dark brown to black and wiry. Cortical root tissue is extremely narrow and rootlet production is meager. Anthraquinones and their derivatives are rare in Leguminosae, but they occur abundantly in *Cassia* species (Harborne et al. 1971; Tiwari and Yadava 1971). Simple phenolic compounds, tannins, quinones, and derivatives are easily demonstrated microchemically in the overlapping cortical root cells of *C. fistula*, *C. grandis*, and *C. tora* (see p. lxiv). The assumption that these cell layers present a physicochemical barrier stems from their role in thwarting nematode gall formation (personal observation), a unique function that accounts for the success of *C. tora* in many areas as a nematode-trap crop.

Favorable returns from *C. occidentalis* and *C. tora* as green manures (Leonard and Reed 1930) gave rebirth to the idea that bacterial symbiosis may occur within root tissue without forming nodules (Fehér and Bokor 1926; Friesner 1926). This assumption became untenable after all attempts to demonstrate microorganisms within plant tissues of *C. tora* were negative (Allen and Allen 1933). The following statement was issued by the International Institute of Agriculture (1936): "As a rule, all leguminous plants of *Cassia* genus, without nodules, are harmful to the main crops with which they are associated. Their use either as a cover crop or as a shade is not recommended."

In recognizing *Fistula*, *Senna*, and *Lasiorhegma* as subgenera, Bentham (1871) deduced that these groups "were differentiated at a time when the configuration of land and water was very different from what it is now, and had been well fixed before the areas they occupied had become broken up by the intervention of apparently insurmountable obstacles. . . ." His remark that some of the subordinate groups "are exclusively limited to one hemisphere, showing the possibility of their formation since the interposition of impassable barriers," is especially apropos of those members of section *Chamaecrista* that are almost exclusively indigenous to the Western Hemisphere.

Paucity of nodulation within *Fistula* and *Senna* in contrast to its prevalence in the 3 sections of *Lasiorhegma* is construed here as evidence that nodule-forming ability within *Cassia* is a relatively recent physiological adaptation. Moreover, the question arises whether requisite genetic ingredients for symbiosis ever existed within the primitive components of *Fistula* and *Senna*.

Chamaecrista has had a particularly varied taxonomic history relative to its status as a section or as a generic segregate (Senn 1938a,b; Irwin and Turner 1960; Irwin and Barneby 1976). The data presented in this synopsis suggest, in principle, the concept of symbiotaxonomy (Norris 1965), a term analogous to chemotaxonomy in connoting phylogenetic relationships of leguminous species relative to rhizobial affinities. The consistent symbiobility within *Chamaecrista* merits consideration as evidence for the elevation of section *Chamaecrista* to subgeneric or generic status (Corby 1974; Allen and Allen 1976).

Cassia: Subgenera *Fistula* and *Senna*

Nonnodulated species	Area	Reported by
abbreviata Oliv.		
ssp. *abbreviata*	Zimbabwe	Corby (1974)
ssp. *beareana* (Holmes) Brenan	Zimbabwe	Corby (1974)
aculeata Pohl.	Venezuela	Barrios & Gonzalez (1971)
acutifolia Del.	France	Lechtova-Trnka (1931)
alata L.	Hawaii, U.S.A.	Allen & Allen (1936b)
	Philippines	Bañados & Fernandez (1954)
	Trinidad	DeSouza (1966)
antillana (Britt. & Rose) Alain	Hawaii, U.S.A.	Pers. observ. (1962)
artemisoides Gaud.	U.S.D.A.	Leonard (1925)
	Australia	Beadle (1964)
bacillaris L. f.	Philippines	Bañados & Fernandez (1954)
bauhinioides Gray	Ariz., U.S.A.	Martin (1948)
bicapsularis L.	U.S.D.A.	Leonard (1925)
	Venezuela	Barrios & Gonzalez (1971)
biflora L.	Trinidad	DeSouza (1966)
circinnata Benth.	Australia	N. C. W. Beadle (pers. comm. 1963)
(*coluteoides* Collad.)	Zimbabwe	Corby (1974)
= *pendula* HBK ex Willd. var. *pendula*		
corymbosa Lam.	U.S.D.A.	Leonard (1925)
	S. Africa	Grobbelaar et al. (1964)

143

Nonnodulated species in *Fistula*
and *Senna* (cont.)

	Area	Reported by
covesii Gray	Ariz., U.S.A.	Martin (1948)
desolata F. Muell.	Australia	Beadle (1964)
emarginata L.	U.S.D.A.	Leonard (1925)
eremophila A. Cunn.	Australia	Beadle (1964)
fistula L.	Hawaii, U.S.A.	Allen & Allen (1936b)
	Philippines	Bañados & Fernandez (1954)
	Brazil	Campêlo & Döbereiner (1969)
floribunda Cav.	Zimbabwe	Corby (1974)
	S. Africa	Grobbelaar & Clarke (1974)
fruticosa Mill.	Trinidad	DeSouza (1966)
glauca L. f.	Hawaii, U.S.A.	Allen & Allen (1936b)
italica (Mill.) Lam.	Zimbabwe	Corby (1974)
ssp. *arachoides* (Burch.) Brenan	S. Africa	Grobbelaar et al. (1967)
javanica L.	Java	Muller & Frémont (1935)
	Hawaii, U.S.A.	Pers. observ. (1962)
	S. Africa	Grobbelaar et al. (1964)
	Brazil	Campêlo & Döbereiner (1969)
(*laevigata* Willd.) = *floribunda* Cav.	Germany	Morck (1891)
	Java	Steinmann (1930); Muller & Frémont (1935)
	Hawaii, U.S.A.	Allen & Allen (1936b)
latifolia G. F. W. Mey.	Guyana	Norris (1969)
leptocarpa Benth.	Ariz., U.S.A.	Martin (1948)
marilandica L.	U.S.D.A.	Leonard (1925)
marksiana Domin	Hawaii, U.S.A.	Pers. observ. (1962)
medsgeri Shafer	Ill., U.S.A.	Burrill & Hansen (1917)
	U.S.D.A.	Leonard (1925)
moschata HBK	Hawaii, U.S.A.	Allen & Allen (1936b)
	Philippines	Bañados & Fernandez (1954)
	Venezuela	Barrios & Gonzalez (1971)
multijuga Rich.	Java	Steinmann (1930); Muller & Frémont (1935)
	Trinidad	DeSouza (1966)
nodosa Buch.-Ham.	Hawaii, U.S.A.	Allen & Allen (1936b)
obovata Collad.	S. Africa	Mostert (1955)
obtusifolia L.	Widely reported	Many investigators
oracle [?]	N.Y., U.S.A.	Wilson (1939a)
petersiana Bolle	Zimbabwe	Corby (1974)
phyllodinea R. Br.	Australia	Beadle (1964)
pleurocarpa F. Muell.	Australia	Beadle (1964)

Nonnodulated species in *Fistula* and *Senna* (cont.)	Area	Reported by
quadrifoliolata Pitt.	Venezuela	Barrios & Gonzalez (1971)
saeri Britt. & Rose	Venezuela	Barrios & Gonzalez (1971)
sericea Sw.	Venezuela	Barrios & Gonzalez (1971)
siamea Lam.	Java	Muller & Frémont (1935)
	Hawaii, U.S.A.	Allen & Allen (1936b)
	Philippines	Bañados & Fernandez (1954)
	S. Africa	Grobbelaar et al. (1964)
singueana Del.	Zimbabwe	Corby (1974
sophera L.		
var. *schinifolia* Benth.	S. Africa	Grobbelaar & Clarke (1974)
sturtii R. Br.	Australia	Beadle (1964)
timorensis DC.	Java	Steinmann (1930)
tomentosa L. f.	U.S.D.A.	Leonard (1925)
	Java	Steinmann (1930)
tora L.	Widely reported	Many investigators
wislizeni Gray	Ariz., U.S.A.	Martin (1948)

Nodulated species	Area	Reported by
leiandra Benth.		
var. *guianensis* Sandw.	Guyana	Norris (1969)
pilifera Vog.	Hawaii, U.S.A.	Allen & Allen (1936a)
sophera L.	S. Africa	Grobbelaar et al. (1967)
speciosa Schrad.	Brazil	E. S. Lopes (pers. comm. 1966)

Conflicting reports	Nodulated	Nonnodulated
didymobotrya Fresen.	Sri Lanka (W. R. C. Paul pers. comm. 1951)	Java (Steinmann 1930; Muller & Frémont 1935) Zimbabwe (Corby 1974)
grandis L. f.	Sri Lanka (Wright 1903)	Hawaii, U.S.A. (Allen & Allen 1936b)
hirsuta L.	Sri Lanka (Wright 1903)	Sri Lanka (Bamber & Holmes 1911; Holland 1924)
	Java (Keuchenius 1924)	Java (Steinmann 1930; Muller & Frémont 1935) Philippines (Bañados & Fernandez 1954) Trinidad (DeSouza 1966)

Conflicting reports in *Fistula*
and *Senna* (cont.)

	Nodulated	Nonnodulated
occidentalis L.	Java (Keuchenius 1924)	U.S.D.A. (Leonard 1925)
	Philippines (Bañados & Fernandez 1954)	Java (Steinmann 1930; Muller & Frémont 1935)
		U.S.S.R. (Lopatina 1931)
		Japan (Itano & Matsuura 1936)
		Hawaii, U.S.A. (Allen & Allen 1936b)
		Argentina (Rothschild 1970)
		Venezuela (Barrios & Gonzales 1971)
		Zimbabwe (Corby 1974)

Cassia: Subgenus *Lasiorhegma*
Section *Absus*

Nodulated species	Area	Reported by
absus L.	Widely reported	Many investigators
hispidula Vahl	Trinidad	DeSouza (1966)

Section *Apoucouita*

Nodulated species	Area	Reported by
pteridophylla Sandw.	Guyana	Norris (1969)

Section *Chamaecrista*

Nodulated species	Area	Reported by
bauhiniaefolia Kunth	Venezuela	Barrios & Gonzalez (1971)
biensis (Stey.) Mend. & Torre	S. Africa	Grobbelaar & Clarke (1975)
	Zimbabwe	Corby (1974)
calycioides DC.	Venezuela	Barrios & Gonzalez (1971)
capensis Thunb.		
var. *keiensis* Stey.	S. Africa	Grobbelaar & Clarke (1972)
chamaecrista L.	U.S.A.	Burrill & Hansen (1917); Leonard (1925); Wilson (1939a)
	Puerto Rico	H. D. Dubey (pers. comm. 1970)
chamaecristoides Collad.	Mexico	Pers. observ. (1970)

Nodulated species in *Chamaecrista*

(cont.)	Area	Reported by
comosa Vog.	S. Africa	Grobbelaar et al. (1967)
cultrifolia HBK	Venezuela	Barrios & Gonzalez (1971)
diphylla L.	Puerto Rico	Kinman (1916)
falcinella Oliv.		
var. *parviflora* Stey.	Zimbabwe	Corby (1974)
(*fasciculata* Michx.)	Widely reported	Many investigators
= *chamaecrista* L.		
fenarolii Mend. & Torre	Zimbabwe	Corby (1974)
flexuosa L.	Venezuela	Barrios & Gonzalez (1971)
glandulosa L.	Mexico	Pers. observ. (1970)
gracilior (Ghesq.) Stey.	Zimbabwe	Corby (1974)
gracilis Kunth	Venezuela	Barrios & Gonzalez (1971)
hochstetteri Ghesq.	Zimbabwe	Corby (1974)
katangensis (Ghesq.) Stey.	Zaire	Steyaert (1952)
kirkii Oliv.	S. Africa	Grobbelaar & Clarke (1975)
var. *kirkii*	Zimbabwe	Corby (1974)
leptadenia Greenm.	Zimbabwe	Corby (1974)
leschenaultiana DC.	Java	Keuchenius (1924); Steinmann (1930); Muller & Frémont (1935)
	Sri Lanka	Paul (1949)
	Taiwan	Pers. observ. (1962)
lucesiae Pitt.	Venezuela	Barrios & Gonzalez (1971)
mimosoides L.	Widely reported	Many investigators
nictitans L.	U.S.A.	Burrill & Hansen (1917); Shunk (1921); Leonard (1923); Conklin (1936)
parva Stey.	Zimbabwe	Corby (1974)
patellaria DC.	Java	Keuchenius (1924); Steinmann (1930); Muller & Frémont (1935)
	Trinidad	DeSouza (1966)
	Argentina	Rothschild (1970)
polytricha Brenan		
var. *polytricha*	Zimbabwe	Corby (1974)
pumila Lam.	Java	Van Helten (1915); Muller & Frémont (1935)
	W. India	Y. S. Kulkarni (pers. comm. 1972)

Nodulated species in *Chamaecrista* (cont.)

	Area	Reported by
quarrei (Ghesq.) Stey.	Zimbabwe	Corby (1974)
rotundifolia Pers.	S. Africa	U. of Pretoria (1955–56); N. Grobbelaar (pers. comm. 1963)
	Australia	Norris (1959b)
	Zimbabwe	Corby (1974)
serpens L.	Venezuela	Barrios & Gonzalez (1971)
simpsoni Pollard	U.S.D.A.	Leonard (1925)
swartzii Wickstr.	Puerto Rico	Dubey et al. (1972)
tagera L.	Venezuela	Barrios & Gonzalez (1971)
wittei Ghesq.	Zimbabwe	Corby (1974)
zambesica Oliv.	Zimbabwe	Corby (1974)

Nonnodulated species	Area	Reported by
wrightii Gray	Ariz., U.S.A.	Martin (1948)

Cassia in perspective

Subgenera	Species	Habitat	Habit
Fistula	15	Tropics	Trees
Senna	270	Tropics and subtropics	Trees, shrubs, few herbs
Lasiorhegma			
Section *Absus*	100	Tropical N. and S. America	Trees, shrubs
Section *Apoucouita*	6	Tropical S. America	Trees
Section *Chamaecrista*	270	Tropics and temperate N. and S. America	Shrubs, herbs

Nodulation profile of *Cassia*

Subgenera	Nodulated	Non-nodulated	+/− reports	Total
Fistula & *Senna*	4	50	4	58
Lasiorhegma				
Section *Absus*	2			2
Section *Apoucouita*	1			1
Section *Chamaecrista*	37	1		38
Total	44	51	4	99

From Latin, *castanea*, "chestnut," and *spermum*, "seed," in reference to the marked resemblance of the large, shiny, brown, round seeds to those of *Castanea* species, family Fagaceae.

Type: *C. australe* A. Cunn.

2 species. Trees, tall, glabrous, evergreen; foliage glossy, dense, dark green. Leaves large, imparipinnate; leaflets large, 9–15, short-petioluled, elliptic, tapering, leathery; stipels absent. Flowers large, orange to reddish yellow, in short, loose racemes in the axils of old branches; bracts minute; bracteoles none; calyx thick, large, colored; teeth broad, very short; standard obovate-orbicular, narrowed into a claw, recurved; wings and keel petals shorter than the standard, free, subsimilar, erect, oblong; stamens 10, free; anthers linear, versatile; ovary long-stalked, many-ovuled; style incurved; stigma terminal, blunt. Pod elongated, subfalcate, turgid, leathery to woody, 2-valved, valves hard, thick, spongy inside between the 2–5, large, globose, chestnut brown seeds. (Bentham and Hooker 1865)

The status of the other species, *C. brevivexillum* Domin (= *C. australe* A. Cunn. var. *brevivexillum* F. M. Bailey) is subject to question.

Both species are native to the tablelands of northeastern Australia, Queensland, and New South Wales. The plants were introduced into Sri Lanka about 1874, and are now somewhat common in India and the East Indies.

C. australe, commonly known as the Moreton Bay or Australian chestnut, or blackbean, is a handsome evergreen tree that often reaches a height of 40–45 m, with a trunk diameter of 1 m. It provides excellent shade and is cultivated as an ornamental in Java and India.

The edibility of the roasted seed of *C. australe*, often equated with that of the European chestnut, has been overestimated. Some writers rate its edibility about equal to that of acorns, or as acceptable only under dire circumstances of need and hunger. The unpleasant purgative effects and indigestibility of the fresh seeds have been attributed to their high saponin content of about 7 percent. Purportedly, ingestion of the seed has been fatal to horses, cattle, and pigs. At one time Australian cattle-owners advocated the eradication of the trees from range areas. The astringency of fresh seeds is reduced or removed by soaking and roasting, although even after such treatment ill effects are known to occur. Toxic substances have not been defined (Webb 1948; Menninger 1962; Watt and Breyer-Brandwijk 1962).

Blackbean timber is resistant to wood-rotting fungi and to termites. The sapwood is subject to beetle attack. The sapwood and heartwood are sharply distinct. This resembles walnut, is well known on the market, and is in general demand, but the supply is limited. It has a density of about 800 kg/m³, is beautifully grained, almost black, strong, durable, and takes a good polish. The sawdust produces an irritation of the nasal mucosa. The wood is used in inlays, high-class furniture, cabinets, panels, umbrella handles, beamed ceilings, plywood, and carved objects, such as jewel and glove boxes (Titmuss 1965; Everett 1969).

NODULATION STUDIES

Greenhouse specimens grown in potted soil throughout a 3-year span and examined periodically lacked nodules (personal observation). However, this finding should be regarded as inconclusive, inasmuch as the seeds were not treated with heterogeneous rhizobial strains, the substratum was not native rhizospheric soil and the large amounts of reserve nutrients in the persistent seed cotyledons may have deterred a symbiotic association.

Catenaria Benth.   Papilionoideae: Hedysareae,
(337)   Desmodiinae

From Latin, *catena*, "chain," in reference to the jointed pod.

Type: *C. caudata* (Thunb.) Schindl.

1 species. Small shrub. Leaves pinnately 3-foliolate; leaflets lanceolate, terminal leaflet larger; stipels and stipules awl-shaped, spinescent. Flowers yellow or greenish white, in terminal and axillary racemes; bracts awl-shaped, persistent, often striate; calyx turbinate-campanulate; lobes 4, upper lobe bidentate; standard elliptic-oblong, erect, obtuse or subcordate at the base, short-clawed; wings spurred, crinkled; keel petals abruptly narrowed into a claw; stamens diadelphous, vexillary stamen free except at the base; ovary sessile, several-ovuled; disk thick, tubular; style glabrous; stigma terminal, capitate. Pod sessile, flat, linear-oblong, about 5-jointed, segments equal, oblong, 1-seeded, covered with brown hooked hairs; seeds flat. (Schindler 1924)

This genus was formerly a section in *Desmodium* (Schindler 1924).

This monotypic genus is endemic in semi-tropical India, China, Japan, and Taiwan. It occurs in waste places and thickets at low altitudes.

NODULATION STUDIES

C. caudata (Thunb.) Schindl., reported as *Desmodium caudatum* DC., was observed nodulated in Japan (Asai 1944). It is grown as a fodder plant in Taiwan.

From Greek, *kathormion*, "necklace"; the moniliform pod resembles a string of beads.

Type: *C. moniliforme* (DC.) Merr.

15 species. Trees, mostly tall, up to about 25 m, or shrubs, unarmed, young plants sometimes spinescent. Leaves bipinnate; pinnae 2–8 pairs, opposite; leaflets several to many pairs, oblong-lanceolate; rachis sometimes glandular between the upper pinnae pairs. Flowers white, in rounded heads or pseudoumbels, peduncled, 1–3, in leaf axils; calyx tubular-funnel-shaped; sepals 4–5, united, short-toothed; petals 4–5, united below; stamens numerous, 10–20 or more; filaments united below into a slender tube, long-exserted; ovary slightly stalked, many-ovuled; style slender. Pod compressed, linear-oblong, falcate or spirally curved, margins lobed, moniliform, eventually breaking at the joints into leathery or hard, 1-seeded, rounded segments, indehiscent; seeds compressed. (Hutchinson and Dalziel 1958; Mohlenbrock 1963a; Burkart 1964b; Brenan and Brummitt 1965; Brenan 1970; Brummitt 1970)

Cathormion was formerly in the *Pithecellobium* complex (Kostermans 1954). Brenan (1970) considers the distinction between *Cathormion* and *Samanea* very tenuous.

Members of this genus are found in tropical South America, West and East Africa, Malaysia, and northern Australia. They grow in freshwater swamp forests and along river banks.

The pale, yellow brown streaked wood has the properties of *Pithecellobium* wood and is used for planks and veneer.

NODULATION STUDIES

Nodules occur on *C. leptophyllum* (Harms) Keay in Zimbabwe (H. D. L. Corby personal communication 1977).

Caulocarpus Bak. f.   Papilionoideae: Galegeae, Tephrosiinae

From Greek, *kaulos*, "stem," and *karpos*, "fruit"; the stems bear fruit repeatedly.

Type: *C. gossweileri* Bak. f.

1 species. Shrublet, to about 1.6 m tall, rootstock woody; stems numerous, angular, grayish-pubescent. Leaves sessile, palmately 3–5-foliolate; leaflets cuneate-oblanceolate, apex blunt; stipels none. Flowers in short, terminal racemes; calyx tube short; lobes subequal, triangular; standard spreading, exterior densely hairy, longer than the keel; wings oblong; stamens diadelphous; anthers uniform; ovary stalked, linear, many-ovuled; style awl-shaped, somewhat flat; stigma small, terminal. Pod stalked, oblong, compressed, not membranous, about 7-seeded. (Baker 1926)

The genus is closely allied to *Lamprolobium* and *Tephrosia* (Baker 1926).

This monotypic genus is endemic in Angola.

Name refers to cedar, *Cedrela*, because of similarities in the bark and wood.

Type: *C. catenaeformis* Ducke

1 species. Tree, large, handsome, unarmed. Leaves bipinnate; pinnae pairs 2; leaflet pairs 2–4, large, ovate-elliptic, thick-veined. Flowers small, greenish white, sessile, in few-flowered heads, in large, terminal and axillary inflorescences; calyx short; lobes 5, minute, triangular; petals 5, equal, joined at the base; stamens many, all joined below into a tube, free above, long-exserted. Pod pendulous, stalked, elongated, flat, twisted and constricted between the 1-seeded joints, segments about 6, oblong, submembranous, thickened but not hard around the seeds, sutures prominent, indehiscent, joints break apart at maturity; seeds large, flat, suborbicular, centrally located in each joint. (Ducke 1930)

This monotypic genus is reported only from the humid primeval forests of the Amazon region along the Trombetas River near Óbidos and the Tapajóz and Tocantins rivers in Brazil.

This tree is one of the most colossal of the Amazon region, where it reaches a height of 50 m, with a trunk diameter of 1.85 m at 1.5 m from the ground. The coarse-textured, firm, tough wood has limited uses for general construction and inexpensive furniture (Record and Hess 1943). It is commonly called cedrorana, or false cedar.

Cenostigma Tul.　　　　　　　　　　Caesalpinioideae: Sclerolobieae
105 / 3563

From Greek, *kenos*, "empty," "exposed," and "stigma."

Type: *C. macrophyllum* Tul.

6 species. Trees, clothed with feltlike, stellate hairs. Leaves paripinnate or imparipinnate; leaflets leathery; stipels absent; stipules small. Flowers large, yellow, in lax, terminal or branched racemes; bracts small, deciduous; calyx tube short; lobes 5, imbricate, lowermost boat-shaped, larger; petals 5, subequal, obovate, imbricate; stamens 10, free, bent downward; filaments woolly at the base; anthers uniform, longitudinally dehiscent; ovary subsessile, linear-oblong, tomentose, free, few-ovuled; style filiform, club-shaped above; stigma small, terminal. Pod oblanceolate, flat, leathery to woody, conspicuously veined, 2-valved, thinly filled between the transverse, orbicular, compressed seeds. (Tulasne 1843; Bentham and Hooker 1865)

Members of this genus are indigenous to Mato Grosso, Brazil, and Paraguay.

Centrolobium Mart. ex Benth.
356 / 3827

Papilionoideae: Dalbergieae,
Pterocarpinae

From Greek, _kentron_, "spur," and _lobion_, "pod," in reference to the stout lateral spine on the prickly pod.

Type: _C. robustum_ (Vell.) Mart. ex Benth.

5 species. Trees, unarmed, deciduous, up to about 30 m tall, with trunk diameters often 1 m, buttressed; bark grayish. Leaves large, imparipinnate, 7–21-foliolate; leaflets opposite or subopposite, entire, oblong-ovate, apex acuminate or blunt, pubescent or glabrous, conspicuous red glands on the undersides, membranous to subcoriaceous; stipels none; stipules large, caducous, acute, deltoid to orbicular. Flowers yellowish or purplish, medium-sized, in large, densely flowered, terminal panicles; bracts ovate or narrow; bracteoles narrow, caducous; calyx subturbinate-campanulate, pubescent, broad, basally blunt; lobes 4, subequal, uppermost lobe broader, emarginate, obtuse or bidentate, basal lobes acute; standard broad-ovate or orbicular; wings oblong or oblique-obovate; keel petals joined at the back, subsimilar to the wings; stamens 10, monadelphous, all connate into a sheath split above; anthers versatile; ovary sessile or short-stalked, 2–3-ovuled; style filiform, usually persistent as a stout spine, incurved; stigma small, terminal. Pod subsessile, large, samaroid, base thick, prickly with many weak spines, a stout lateral stylar spine subtending the falcate-oblong wing at the top, indehiscent; seeds 1–3, oblong-subreniform, septa transverse. (Bentham and Hooker 1865; Rudd 1954; Dwyer 1965)

Representatives of this genus are native to eastern Venezuela, Panama, and northeastern Brazil.

Centrolobium trees are called porcupine trees in some areas because of their prickly samaroid pods, which resemble a chestnut bur with the wing attachment of a gigantic maple seed. The spines often measure 2.5 cm in length; the stout spur at the base may reach 7.5–10 cm in length. Several species are popular avenue trees in cities of Brazil because of their rapid growth, attractive flowers, and oval-shaped crowns.

The wood, known as canarywood, porcupine wood, or zebrawood in the timber trade, is yellow to orange red with purple streaks that usually change to reddish brown upon exposure. It takes a satiny finish and is highly esteemed for furniture, cabinetwork, flooring, and boatbuilding. The wood is not commonly exported because the supply is only about equal to the local demand (Record and Hess 1943; Rudd 1954).

*Centrosema Benth.
(384) / 3858

Papilionoideae: Phaseoleae, Phaseolinae

From Greek, _kentron_, "spur," and _semeia_, "standard"; the outer surface of the standard petal has a short, basal spur.

Type: _C. brasilianum_ (L.) Benth.

50 species. Shrubs or herbs, prostrate or climbing. Leaves petioled, pinnately 3-foliolate, sometimes 1–5–7-foliolate or palmately 3–5-foliolate; leaflets entire, ovate to elliptic; stipels filiform or linear-lanceolate; stipules

153

persistent, striate. Flowers showy, resupinate, white, violet, rose, or bluish, 1–several, on single or paired peduncles in leaf axils; lower bracts paired, stipulelike, upper bracts united, rigid, striate; bracteoles larger, striate, pairs united, appressed to the calyx; calyx campanulate, short, bilabiate; lobes 5, subequal, upper lip bidentate, emarginate, lower lip tridentate, center lobe longest; petals clawed; standard broad, flat, short-spurred, pouched dorsally above the short, incurved, folded claw; wings oblong, oblique, or falcate, spurred at the base, slightly longer than the broad, obtuse, incurved keel petals; stamens 10, vexillary stamen free or slightly coherent with others above the base; anthers uniform; ovary subsessile, multiovuled; disk present; style incurved, dilated near the apex, barbate around the terminal stigma. Pod subsessile, linear, compressed, subseptate, both sutures thickened, with a longitudinal rib near each margin of the 2 valves or narrowly winged along the lower suture, sometimes pointed, dehiscent; seeds transversely oblong or globular, thick or flat, aril short. (Bentham and Hooker 1865; Wilczek 1954)

Species of this genus are indigenous to tropical and subtropical America. They have been widely introduced elsewhere and are easily naturalized in sandy or well-drained loam soil.

Centrosema species are generally useful as soil-binders, green manures, and forage, but they do not withstand trampling. In Java several species are cultivated as ground cover under teak and as hedge plants. In Malaysia and the Philippines, *C. pubescens* Benth. is a desirable intercycle crop on rubber, coconut, and palm oil plantations because of its low, twining habit, dense green cover, shade tolerance, and drought resistance (Beeley 1938; Whyte et al. 1953; Watson 1957). In recent decades this species has become increasingly important as a legume component on well-drained, coastal and subcoastal soils in tropical and subtropical pastures in Australia (Walsh 1958; Bowen 1959a). Other beneficial properties of this species are its leafy foliage, lack of woody growth up to 18 months, prolific seeding ability, and persistence as a solo crop, or in mixtures.

NODULATION STUDIES

Bowen (1959a) attributed the following desirable aspects of nitrogen fixation to *C. pubescens*: (1) continued nodulation throughout the entire growth span and continued trace infections even during dormant and winter phases of plant growth; (2) continuous increase in nodule weight until the cessation of plant growth; and (3) highly efficient translocation of fixed nitrogen from nodules to upper plant parts, as evinced by minute residual amounts of nitrogen in underground parts. Later, Bowen and Kennedy (1961) noted a heritable variation in nodulation responses among stable seed lines. Such a diversity would condition greatly the agronomic and nodulating value of a species.

Wilson's data (1939a) prompt the conclusion that *C. pubescens* and *C. virginianum* (L.) Benth. are nodulated promiscuously by rhizobia from diverse genera. However, Ishizawa (1954), Galli (1958), and Bowen (1959b) showed that cross-inoculation kinships were confined exclusively to the cowpea miscellany.

Nodulated species of *Centrosema*	Area	Reported by
angustifolium (HBK) Benth.	Venezuela	Barrios & Gonzalez (1971)
brasilianum (L.) Benth.	Venezuela	Barrios & Gonzalez (1971)
pascuorum Mart.	Venezuela	Barrios & Gonzalez (1971)
plumieri (Turp.) Benth.	Sri Lanka	W. R. C. Paul (pers. comm. 1951)
	Philippines	Bañados & Fernandez (1954)
	Australia	Bowen (1959b)
	Trinidad	DeSouza (1966)
pubescens Benth.	Widely reported	Many investigators
venosum Mart.	Venezuela	Barrios & Gonzalez (1971)
virginianum (L.) Benth.	Java	Keuchenius (1924)
	U.S.A.	Wilson (1939a)
	Australia	Bowen (1959b)
var. *latifolium* (L.) Benth.	Puerto Rico	Dubey et al. (1972)

Ceratonia L.

74 / 3529

Caesalpinioideae: Cassieae

From Greek, *keras*, "horn," in reference to the long, slender, curved pod.

Type: *C. siliqua* L.

1 species. Tree, evergreen, small to medium, up to 15–20 m tall, with a trunk diameter of 30 cm or more; bark gray, thin, smooth but flaky on old trees; crown broad, round, dense. Leaves paripinnate; leaflets 2–5 pairs, elliptic, at first red brown and hairy becoming leathery, dark green, glossy; stipules minute, caducous. Flowers polygamous or dioecious, small, apetalous, usually 30–50, in lateral or axillary short racemes on older branches; perennial renewal of the racemes from endogenous buds at the same location produces warty mounds on the bark; bracts and bracteoles small, scalelike, caducous; calyx turbinate; sepals 5, short, imbricate, deciduous; *male flowers*: reddish to yellow; calyx disk with 5 fertile stamens; filaments slender; anthers ovate, versatile; odor disagreeable; *female flowers*: greenish; ovary short-stalked on center of the disk, curved; ovules numerous; style short; stigma enlarged, peltate; *bisexual flowers*: stamens and pistil on the same disk, disk probably nectariferous. Pod pendent, brownish violet, thick, leathery, linear-oblong, curved, ranging from 8 to 25 cm long and from 2 to 3 cm broad, thickened along both sutures, indehiscent, with an agreeably flavored, mucilaginous, sweet pulp between the 5–15 obovoid-lenticular seeds. (Bentham and Hooker 1865; Coit 1951)

This monotypic genus is endemic in the eastern Mediterranean region, extending into Asia Minor and Arabia; it is common in Israel.

The carob tree or St.-John's-bread does occur in the wild state, but it is also extensively cultivated for fodder. Best growth occurs in deep, heavy, well-drained loams, a warm but not hot climate, and at low elevations near the sea. Its drought resistance is explained, in part, by its snakelike, deeply penetrating root system.

According to De Candolle (1908), the carob "grew wild in the Levant, probably on the southern coast of Anatolia and in Syria, perhaps also in Cyrenaica. Its cultivation began within historic time. The Greeks diffused it in Greece and Italy, but it was afterwards more highly esteemed by the Arabs who propagated it as far as Moroco and Spain." In 1854 about 8,000 plants, grown from seed in Spain, were imported into the United States and distributed throughout the southern states.

The carob tree has been the subject of various historical reviews (Mantell 1947; Coit 1951; Moldenke and Moldenke 1952; Wheeler 1955). The Egyptians, at the time of the Pharaohs, fed their cattle on it and made wine from the pods. The tree is not mentioned in the Old Testament, but in his history of plants, written in the fourth century B.C., Theophrastus commented on its cultivation at Rhodes. Early Greek writers mentioned it as a source of food. In the New Testament the carob tree is referred to by its Greek name. Dioscorides, in the first century A.D., praised the carob in his *Materia Medica* as a laxative and diuretic. The finding of carob pods in the storehouses of Pompeii is evidence that they were known to the Romans as early as 79 A.D. It is generally accepted by philologists that the "husks" of Jesus' parable of the Prodigal Son in the King James Bible, Luke 15:16, were carob pods (Moldenke and Moldenke 1952). Some commentators are of the opinion that the "locusts" eaten by John the Baptist, Matthew 3:4, were carob pods and not insects. An error is believed to have occurred whereby a transcriber substituted the Hebrew *g* for the *r* in the word "cherev"; this changed the word in translation to "locust" from "carob" (Moldenke and Moldenke 1952). Accordingly, the tree was known as "Johannis brodbaum," or St.-John's-bread, in the Middle Ages.

It is said that the seeds were the original carat weights used by jewelers and apothecaries and that the word "carat" is derived from "girit," the Arabic name for the tree. Prior to 1913 the carat was equivalent to 205.3 mgm, but since then it has been standardized to 200 mgm, or 3,086 grains.

Pods of the carob tree were the only available food for the cavalry horses in Wellington's Peninsular Campaign during the Napoleonic Wars, and also for Allenby's campaign in Palestine in World War I (Howes 1949). Coit (1951) states that the rural people of southern Greece subsisted largely on carob pods during World War II, after the country had been stripped of livestock and most other food. *Ceratonia* is planted extensively in Mediterranean foothills as a forest-forage tree and for fodder.

Carob pods appear in the open markets of southern Europe as a cheap confection. In some parts of Austria and Germany, the roasted pods are blended with chicory as a substitute for coffee. In Sicily the pods are an ingredient in the commercial manufacture of alcohol. Stripped of seeds, the pods are made into a nutritious feed cake for cattle. Profarin, or trefarin, a yellow high-protein baking flour, is prepared from the seeds (Watt and

Breyer-Brandwijk 1962). A highly nutritious, easily digested, beige-colored carob powder processed from seeded pods is an acceptable substitute for chocolate in baking and candymaking (Logan 1960).

Carob-seed gum is a manogalactan consisting of galactan, mannan, pentosans, proteins, nitrogen, cellular tissue, and mineral matter (Knight and Dowsett 1936). It has a molecular weight of 310,000 (Rol 1959), and a viscosity equal to that of tragacanth; its emulsifying properties are more efficient than those of tragacanth but inferior to those of *Acacia* gum. In 1939, the United States imported more than 1,800 metric tons of carob-seed gum from Great Britain, France, and the Mediterranean area (Howes 1949); in 1955, the total importations exceeded 5,880 metric tons (Rol 1959).

The gum has current and potential use in foods, cosmetics, pharmaceuticals, photographic film emulsions, water emulsion paints, lithograph and writing inks, metal and shoe polishes, and adhesives. It is also used for flavoring tobacco before curing, sizing textiles, leather-finishing and tanning, and as a rubber latex creaming and thickening agent, ceramic binder and plasticizer, and match-head and striking binder (Coit 1951).

Other uses of the carob tree are less significant. Despite its attractiveness and its evergreen shade, it is unsuitable for avenue purposes because the roots bulge sidewalks and pavements, and the pod litter is objectionable. The wood is moderately heavy, tough, and reddish brown when properly aged; its texture resembles that of cherry, but it has no commercial importance. As firewood, it should be used only in stoves because of its tendency to spark.

NODULATION STUDIES

Numerous examinations of seedlings and older plants have shown this species to be nonnodulated in France (Naudin 1897b), in the United States (Leonard 1925; Coit 1951; Wheeler 1955), and in South Africa (Grobbelaar et al. 1964).

Cercidium Tul.   Caesalpinioideae: Eucaesalpinieae
97 / 3554

From Greek, *kerkidion*, "weaver's shuttle," in reference to the fancied resemblance of the pods to a weaver's shuttle.
    Type: *C. praecox* (R. & P.) Harms (= *C. spinosum* Tul.)
    8–10 species. Shrubs, tall, or small trees, up to 8 m tall; branches stout, usually tortuous, often grooved, armed at the nodes with short, slender, straight spines; bark bright green. Leaves bipinnate, small, early deciduous, petioled; pinnae 1–2 pairs; leaflets small, oblong or obovate, 2–4 pairs; petiolar gland absent; rachis cylindric; stipels none; stipules inconspicuous or none. Flowers yellow, showy, in short, lax, few-flowered, axillary racemes or corymbs, pedicels slender; bracts small, membranous; bracteoles minute or absent; calyx short-campanulate; lobes 5, subequal, acute, linear, valvate in bud, reflexed at maturity; petals 5, oblong or orbicular, imbricate, long-

clawed; upper petal broader and longer clawed than others, slightly auriculate and claw conspicuously glandular at the base, often red-spotted and inside others; stamens 10, free, exserted, equal, inserted with petals on margin of disk; filaments slender, pilose below; anthers ovoid, uniform, versatile; ovary short-stalked, smooth, many-ovuled; style slender, infolded in bud; stigma minute, terminal. Pod short-stalked, linear-oblong, compressed or somewhat turgid, slightly constricted between the seeds, woody, membranous or leathery, 2-valved, valves venulose, tardily dehiscent, 4–5-seeded; seeds ovoid, compressed, longitudinal. (Bentham and Hooker 1865; Johnston 1924)

This genus was formerly congeneric with *Parkinsonia* (Johnston 1924).

Members of this genus are widespread in tropical and subtropical America, ranging from the southwestern United States to Chile. They are well adapted to arid and semiarid areas and to a wide range of soils. The plants are found in gullies and up to 1,400-m altitude.

Cercidium species are all known as paloverde, or green tree, in the southwestern United States and Mexico. They have little value but are ornamentally attractive in full flower. The bright green bark renders the plant conspicuous during the leafless period. Their use as forage is limited mostly to periods of drought, when other forage food is scarce. The plants provide good cover for wildlife. The flowers are a source of wild honey. The pods are moderately nutritious and fairly palatable but are eaten sparingly by livestock. The Indians used the ground seeds of *C. torreyanum* (S. Wats.) Sarg. as meal for cakes and in the preparation of a beverage.

The wood of *C. floridum* Benth. is soft, close grained, and has a density of 500 kg/m^3. The largest and most widely distributed species, *C. praecox* (R. & P.) Harms, has hard, moderately heavy but brittle wood of no commercial value. It finishes smoothly, is easy to work, odorless, and tasteless, but perishable. The wood burns quickly and gives off an unpleasant odor. The greenish semitransparent gum which exudes periodically from the stem of this species has local use in soapmaking when dissolved in alkali (Record and Hess 1943).

Nodulation Studies

Nodules were absent on *C. torreyanum* in its native environment in Arizona (Martin 1948). Also, greenhouse specimens of *C. floridum* examined in Wisconsin were devoid of nodules (personal observation). The roots were black and wiry, a condition that is encountered frequently among other nonnodulating caesalpiniaceous species.

Cercis L.   Caesalpinioideae: Bauhinieae
71 / 3526

From Greek, *kerkis*, the ancient name for the Judas tree, *C. siliquastrum* L.

Type: *C. siliquastrum* L.

5–7 species. Shrubs or trees, up to 10 m tall, unarmed, deciduous. Leaves on long petioles, deciduous, simple, entire, broad, heart-shaped to kidney-shaped, palminerved; stipules ovate, small, scalelike or membranous, caducous. Flowers rose lilac, in sessile or short-stalked fascicles or clustered racemes on old wood prior to leaves unfolding; bracts small, scalelike, sometimes imbricate at base of racemes; bracteoles minute or absent; calyx turbinate-campanulate, oblique, purplish; lobes 5, short, broadly triangular; petals 5, small, subequal, clawed, oblong-ovate, vaguely papilionaceous, upper 3 smaller than lower 2, upper petal smallest and enclosed in bud by the wing petals and encircled by the keel petals; stamens 10, free, declinate; filaments short, persistent; anthers fertile, short, uniform, versatile, longitudinally dehiscent; ovary short-stalked, free, many-ovuled; style filiform; stigma noblike. Pod slightly stalked, oblong to broad-linear, flat, acute at both ends, 2-valved, narrowly winged along upper suture, tardily dehiscent; seeds usually many; obovate, compressed, transverse, reddish brown. (Bentham and Hooker 1865; Rehder 1940; Wilbur 1963)

Species of this genus are represented in the north temperate regions of Asia, Japan, the United States, southern Europe, and northern Mexico. All species grow well in rich, moist loam or sandy soils, on limestone hills, along streams, or as forest undergrowth.

The lineage of this genus extends back to the Tertiary period (Axelrod 1958). Specimens closely resembling *Cercis* species have been identified among the rare fossil records of leguminous species. Bible legends and traditions maintain that *C. siliquastrum*, the European Judas tree or redbud, was the tree on which Judas Iscariot hanged himself; however, no mention of a tree is made in Matthew 27:5.

Cercis is one of the few woody leguminous genera occurring in the North Temperate Zone. Its members are highly ornamental in spring when, before the new leaves appear, the branches and upper parts of the trunk display profuse rose-lilac blossoms. This habit of cauliflory is rare among temperate zone trees. The Asian species, *C. racemosa* Oliv. & Hook., is perhaps the most showy for its pendulous racemes of pink blossoms. *C. canadensis* L., the American Judas tree or American redbud, is the official tree of the state of Oklahoma. The flowers of *C. chinensis* Bge., Chinese redbud, the largest member of the genus, are a fair source of honey and are relished in salads in some areas for their sweet acid flavor; the bark has certain antiseptic properties.

Cercis wood is of good quality, has a density of about 750 kg/m^3, and a coarse texture. It is easy to work and finishes smoothly. The wood does have limited use in turnery and cabinetmaking, but it is neither in great demand nor available in quantity (Record and Hess 1943).

Nodulation Studies

Reports are in agreement that nodules are lacking on *C. siliquastrum* (Lachmann 1858; Morck 1891; Clos 1893; Wilson 1939c; Grobbelaar et al. 1964) and on *C. canadensis* (Buckhout 1889; Harrison and Barlow 1906; Leonard 1925; Erdman and Walker 1927; Wilson 1939c; personal observation). The listing of a rhizobial culture by Buchanan (1909a) as "*Cercis* bacteria" is inexplicable and is in all likelihood erroneous. Asai's report (1944) of nodules on *C. chinensis* in Japan merits confirmation; it is likely that he misinterpreted coralloid mycorrhizal growths for leguminous nodules.

Chadsia Boj.
263 / 3728

Papilionoideae: Galegeae, Tephrosiinae

Named for Captain Chads of the Royal Navy; he invited Bojer to accompany him on an expedition to Madagascar and the Comoro Islands in 1835.

Type: *C. flammea* Boj.

18 species. Shrubs, erect or climbing, or lianas. Leaves imparipinnate; leaflets usually more than 3, with numerous parallel veins; stipels absent; stipules awl-shaped. Flowers large, scarlet red or orange yellow, borne singly, fascicled, or in racemes at leafless nodes on old wood, along stems or branches; bracts minute; bracteoles absent; calyx broad, oblique, somewhat dorsally inflated; lobes 5, unequal, linear, upper 2 joined, lowermost lobe longest; standard lanceolate with long-pointed apex; wing petals shorter than the standard, the apical margins joined and acuminate, adherent above the base to the keel petals; keel petals falcate, beaked, longer than the other petals; stamens 10, monadelphous, vexillary stamen free only at the base, then joined with others above into a tube; anthers uniform; ovary subsessile, linear, many-ovuled; style filiform, smooth, short, with a small, terminal stigma. Pod elongated, flattened, beaked by a persistent style, 2-valved, dehiscent; seeds oblong. (Bojer 1843; Bentham and Hooker 1865)

Members of this genus are native to Madagascar.

Chaetocalyx DC.
317 / 3785

Papilionoideae: Hedysareae, Aeschynomeninae

From Greek, *chaite*, "long flowing hair"; the calyx is beset with glandular setae.

Type: *C. scandens* (L.) Urb.

11 species. Vines, herbaceous, slender, twining; stems herbaceous or slightly woody, slender, subterete, striate. Leaves imparipinnate, 5–17-foliolate; leaflets oblong, elliptic, subacute or emarginate, few, entire; stipels absent; stipules paired, deltoid to lanceolate, entire to setose-ciliate or laciniate. Flowers yellow, medium-sized, few, in axillary or terminal racemes, panicles, or fascicles, seldom solitary; bracts and bracteoles integrate, the former slightly smaller; calyx campanulate; tube gibbous or symmetric, glabrous or pubescent, usually beset with glandular setae; lobes 5, subequal; petals yellow, sometimes with red or violet; standard outwardly pubescent or glabrous, spathulate, obovate or suborbicular, clawed, slightly longer than the oblong wings and keel petals; stamens 10, subequal, about as long as the keel; filaments glabrous or slightly pubescent, free from apex to about midlength, then connate to form a sheath usually split along the vexillary side and sometimes also the carinal side; anthers dorsified, ellipsoid; ovary sessile or short-stalked, 6–16-ovuled, glabrous to densely pubescent; style filiform, glabrous; stigma small, terminal, capitate. Pod lomentaceous, elongated, laterally compressed, reticulate-striate or subterete and longitudinally striate, 6–16-segmented, inconspicuously constricted between the 1-seeded joints, surface smooth or with glandular setae; seeds sublustrous, smooth, kidney-shaped,

reddish brown. (De Candolle 1825a,b; Bentham and Hooker 1865; Amshoff 1939; Rudd 1958)

This genus is represented in tropical southern Mexico, and from Central America to northern Argentina. Its members occur in moderately moist localities up to about 2,000-m altitude, along stream banks and roadsides, or in thickets and forests.

Nodulation Studies

Native plants of *C. longiflora* Gray in Brazil and *C. latissiliqua* (Poir.) Benth. ex Hemsl. in Costa Rica lacked nodules. *Chaetocalyx* was introduced into Australia as a promising pasture crop, but all attempts to induce nodulation by inoculation with 84 rhizobial strains from 30 different legume genera, imported native soils, and top grafting experiments proved unsuccessful (Diatloff and Diatloff 1977). The yellow pigmentation of the plant roots was a factor in the nonnodulating character and also in the effectiveness of the plants as a trap crop in nematode-infested soils.

Chapmannia T. & G.   Papilionoideae: Hedysareae,
331 / 3801   Stylosanthinae

Named for Dr. Alvan Wentworth Chapman, 1809–1899, American botanist; he contributed much to the knowledge of the botany of Florida.

Type: *C. floridana* T. & G.

1 species. Herb, erect, viscid. Leaves imparipinnate; leaflets entire, few; stipels absent; stipules awl-shaped. Flowers small, yellow, in short, long-peduncled, terminal racemes; bracts small, 2-stipulate; calyx membranous, wide, cylindric, narrowed at the base, shortly stipelike; 4 upper lobes connate, lowermost lobe free, small; standard suborbicular; wings oblique-obovate; keel incurved, blunt, subequal to the standard; stamens monadelphous; anthers oblong, alternately basifixed and dorsifixed; ovary sessile, many-ovuled; style elongated, filiform; stigma minute, terminal. Pod linear, cylindric, upper suture almost straight, lower suture sinuate; segments ovoid, 1-seeded, surface with soft glandular bristles or spines, striate lengthwise; seeds subreniform. (Bentham and Hooker 1865; Gillett 1966c)

This monotypic genus is confined to Florida.

Old leaves show a beautiful dark red reticulation on the under surfaces due to tannin in large, star-shaped cells of the mesophyll. Closely related species in *Arthrocarpum* and *Pachecoa* also have this unique reticulation (Gillett 1966c).

Chesneya Lindl.   Papilionoideae: Galegeae, Astragalinae

Named for F. R. Chesney, British soldier and explorer, leader of the 1835 expedition to the Tigris and Euphrates.

Lectotype: *C. rytidosperma* Jaub. & Spach

18 species. Herbs, usually perennial, woody only at the base; suberect or decumbent, acaulescent or short-stemmed, silky-pubescent, eglandular. Leaves imparipinnate; leaflets entire, 1–7 pairs; stipels none; stipules folia-ceous. Flowers large, red, yellow, or violet, 1–3, on elongated, axillary pedun-cles; calyx tubular, gibbous at the base, subbilabiate, upper lip 2-toothed, lower lip longer, tridentate; standard hairy, ovate, apically notched, slightly longer than the obtuse keel and wings; stamens diadelphous; anthers uni-form; ovary sessile, not glandular; style curved, pubescent below; stigma capitate, papillary. Pod sessile, oblong-linear, eglandular, pubescent, com-pressed, nonseptate, valves twisted at maturity; seeds kidney-shaped, flat-tened, wrinkled or pitted. (Komarov 1971)

This genus, formerly a section of *Calophaca*, is now considered generically distinctive on the basis of herbaceous habit and eglandular ovaries.

Species of this genus are widespread, ranging from Asia Minor to central Asia, and from the Soviet Union to Mongolia and eastern India. They occur on stony and gravelly mountain slopes.

Some species have ornamental value.

Chidlowia Hoyle   Caesalpinioideae: Amherstieae

Named for Chidlow Vigne, silviculturist, Gold Coast Forest Service; he suggested that his specimens represented a new genus.

Type: *C. sanguinea* Hoyle

1 species. Tree, glabrous, up to 25 m tall, unarmed. Leaves paripinnate; leaflets 4–6 pairs, opposite or subopposite, not glandular-punctate, lowest ovate, middle pairs oblong-elliptic, uppermost pair obovate-elliptic, under surfaces dull, leathery; stipules small, caducous. Flowers wine red, in elon-gated, narrow, pendent, spiciform panicles, 1–2 together on old wood, some-times terminal on young branches, pedicels very short; bracts and bracteoles minute, caducous; calyx campanulate; teeth 5, very short, equal, rounded, open in bud; disk fleshy, adnate to the calyx tube, bearing petals and stamens; petals 5, sessile, subequal, strongly imbricate, much longer than the calyx, united at the base; stamens 10, equal, free; filaments glabrous, filiform, very long; anthers uniform, longitudinally dehiscent; ovary stalked, many-ovuled; style filiform; stigma small, terminal. Pod oblong-linear, acute at both ends, smooth, large, up to 60 cm long by 6 cm broad, valves leathery-woody, elastic-ally dehiscent along both sutures; seeds, as many as 15, flat, suborbicular. (Hoyle 1932; Pellegrin 1948; Léonard 1957)

This monotypic genus, endemic in tropical West Africa, is especially widespread in Liberia, Sierra Leone, the Gold Coast, and the Ivory Coast. It is often dominant in humid forests.

The dull, gray brown, moderately heavy, hard wood lacks commercial importance. Chidlowine, an alkaloid with hypotensive and antianemic properties extracted from the leaves, was patented in France (*Chemical Abstracts 57,* #15256-d 1962). The relation of this alkaloid to triacanthine found in the leaves of *Gleditsia triacanthos* is not clear (Raffauf 1970).

Chloryllis E. Mey.   Papilionoideae: Phaseoleae, Phaseolinae
(423) / 3912

From Greek, *chloros*, "pale green," in reference to the green flowers.

Type: *C. pratensis* E. Mey.

1 species. Herb similar in habit to *Dolichos* species except for the following characters: flowers green; upper calyx lip round; keel short, boat-shaped; style flattened at base, upper half slender, cylindric, hairy; and pod oblong, margins warty.

This genus has had a varied taxonomic background. Taubert (1894) designated *Chloryllis* a section of *Dolichos*; Harms (1908a) segregated it as a monotypic genus; recently, Verdcourt (1970c) construed it as a subgenus of *Dolichos*.

This monotypic genus is endemic in South Africa.

Chordospartium Cheesem.   Papilionoideae: Galegeae,
(282c)   Robiniinae

From Greek, *chorde*, "string," and *spartos*, the name of a broom plant, presumably in reference to the slender branches.

Type: *C. stevensonii* Cheesem.

1 species. Tree, small, up to about 8 m tall; branches drooping, rushlike; branchlets slender, grooved, leafless, with small scales at the nodes. Flowers lavender, about 20, in short, dense, solitary or fascicled racemes, peduncles woolly-hairy; bracts and bracteoles small; calyx cup-shaped, pilose; teeth 5, minute, subequal or upper tooth smaller; standard orbicular, reflexed, short-clawed, longer than the wings; wings falcate, free; keel incurved, blunt, equal to the standard; stamens diadelphous; anthers uniform; ovary sessile, silky-pubescent, few-ovuled; style slender, incurved at apex, bearded along the inner margin; stigma minute, terminal. Pod short, turgid, rhomboid-ovoid, indehiscent, mostly 1-seeded. (Harms 1915a; Allan 1961)

This monotypic genus is endemic in the South Island, New Zealand. It is usually found along tributaries of the Wairau, Awatere, and Clarence rivers.

NODULATION STUDIES

Nodules were observed on this species in New Zealand (R. M. Greenwood personal communication 1975).

From Greek, *chora*, "place," and *zema*, "drink." The name alludes to its discovery in Australia in 1792 by the French botanist Labilladière; after frustrating experiences with salt springs, he found this plant growing over a freshwater spring.

Type: *C. ilicifolium* Labill.

16 species. Shrubs or undershrubs, erect, up to about 1 m tall, or prostrate. Leaves usually alternate, simple, entire or with pungent-pointed lobes; stipules small and bristly, or absent. Flowers orange or red, borne in loose, terminal or rarely axillary racemes; bracteoles 2, small, deciduous; calyx lobes subequal, imbricate, upper 2 usually broader and united above; petals clawed; standard orbicular or kidney-shaped, emarginate, longer than the oblong wings; keel shorter than the wings, straight and obtuse or with an erect point, rarely incurved, often purplish red; stamens free; ovary sessile or stalked, hairy, many-ovuled; style short, incurved; stigma terminal, often oblique. Pod ovoid, turgid or compressed, nonseptate within, containing many, shiny seeds. (Bentham 1864; Bentham and Hooker 1865)

Members of this genus are widely distributed in Western Australia and Queensland, but are not common in New South Wales. They inhabit river areas, gravelly plains, and coastal areas.

Several species are called flame pea. They are good shrubbery border plants, long blooming, showy, and hardy. The type species, *C. ilicifolium* Labill., has thorn-pointed leaves and the heath habit of English holly (*Ilex aquifolium*).

Nodulation Studies

Seedlings of *C. ilicifolium* were nodulated by rhizobial strains from 30 of the 32 diverse leguminous species tested by Wilson (1939a). Therefore, he categorized the genus as symbiotically promiscuous. *C. parviflorum* Benth. was assigned to the cowpea miscellany by Bowen (1956), as were *C. aciculare* (DC.) C. A. Gardn., *C. cytisoides* Turcz., *C. dicksonii* R. Grah., *C. ericifolium* Meissn., *C. ilicifolium*, and *C. reticulatum* Meissn. by Lange (1961) on the basis of slow-growing cultural characteristics of the rhizobia and restricted, not versatile, cross-symbiobility.

Christia Moench Papilionoideae: Hedysareae, Desmodiinae
(342) / 3812

Lectotype: *C. vespertilionis* (L.) Bakh. f.

12 species. Herbs, prostrate, or shrubs, small, erect. Leaves pinnate, 1–3-foliolate; leaflets often broader than long; stipels present; stipules free, awl-shaped or striate. Flowers white or purplish, small, paired along rachis of simple or branched terminal racemes; bracts early caducous; bracteoles ab-

sent; calyx campanulate, enlarged after anthesis, longer than the pod; lobes broad, acute, subequal; standard obovate or obcordate, narrowed into a claw, basal auricles and callosities lacking; wings oblique-oblong, shorter than and adherent to the blunt, incurved keel; stamens 10, diadelphous; anthers uniform; ovary with 2–8 ovules; style awl-shaped, inflexed and glabrous above; stigma terminal, capitate. Pod subsessile or stalked, articulate, deeply constricted between ovate joints, joints folded back upon each other within the persistent calyx, indehiscent; seeds orbicular or subglobose. (Bentham and Hooker 1865; Meeuwen et al. 1961)

Christia species are scattered throughout Indochina, Malaysia, and northern Australia. They occur on dry, grassy areas, in sandy soils, and along roadsides.

The type species is cultivated in gardens of Malaysia for its ornamental, odd-shaped leaves. *C. reniformis* DC. has shown promise as a cover crop.

NODULATION STUDIES

Nodules were reported on *C. vespertilionis* (L.) Bakh. f. as *Lourea vespertilionis* (L.) DC., its synonym, growing in greenhouse tests in the United States (Wilson 1939a), in experimental field plots in the Philippines (Bañados and Fernandez 1954), and on the shrubs introduced into Singapore (Lim 1977). In Wilson's tests, 10 rhizobial strains from members of the cowpea miscellany were cross-infective on this species.

Chrysoscias E. Mey.   Papilionoideae: Phaseoleae, Cajaninae

From Greek, *chrysos*, "gold," and *skia*, "shadow," in reference to the flower color.

Lectotype: *C. erecta* (Thunb.) C. A. Smith

6 species. Shrublets, virgate, semierect, scandent, woody, silky-pubescent, gland-dotted. Leaves pinnately 3-foliolate; leaflets rigid, narrow-oblong, margins sometimes revolute, lower surfaces gland-dotted; stipels absent; stipules ovate, subpersistent. Flowers golden yellow, few, in axillary, peduncled umbels or solitary; bracts small, caducous; bracteoles none; calyx tube very short, deeply cleft; lobes 5, subequal, foliaceous, lanceolate, as long as or longer than the tube, upper 2 usually more or less connate; standard large, clawed; wings heart-shaped at the base; keel with scalelike appendages; stamens 10, diadelphous; anthers uniform; ovary 2-ovuled. Pod turgid, pilose, dehiscent; seeds kidney-shaped, black. (Harvey 1861–1862; Smith 1932)

Members of this genus were formerly a section of *Rhynchosia*.

Species of this genus are indigenous to South Africa. They thrive in the mountains of the Transvaal and southwestern Cape Province.

The Latin name for the chick pea.

Type: *C. arietinum* L.

14 species. Herbs, annual or perennial, erect, 0.5–1 m tall, often with glandular hairs, sometimes almost leafless. Leaves usually imparipinnate, rarely paripinnate or trifoliolate; leaflets small, 3–8 pairs, edges dentate or deeply incised; petiole ending in a terminal tendril or spine or odd leaflet; stipels absent; stipules foliaceous, oblique, sometimes dentate. Flowers small, white, violet, or mixed, solitary or few, peduncled, in axillary racemes; bracts small; bracteoles absent; calyx oblique, tube swollen at the base, bilabiate; lobes 5, subequal or upper 2 shorter; standard obovate, narrowed into a broad claw; wings free, oblique-obovate; keel broad, obtuse or subacute, incurved, dilated; stamens diadelphous; filaments dilated above; anthers uniform; ovary sessile, 2- or more-ovuled; style filiform, incurved, glabrous above; stigma terminal. Pod sessile, ovate or oblong, turgid, 2-valved, dehiscent; seeds 1–4, irregularly obovoid, cotyledons thick, hilum small. (Bentham and Hooker 1865; Tutin et al. 1968)

Cicer species are native to the Mediterranean region and western Asia. They require clay-loam soil.

The type species, commonly called chick pea, garbanzo bean, or gram, was well known to the ancient Egyptians, Hebrews, and Greeks. Today it is still one of the most important pulse crops in many areas, particularly in India, where over 8 million hectares annually are devoted to its cultivation. The average yield is about 675 kg/ha (Purseglove 1968). It is also cultivated on a large scale in the Middle East, Spain, Argentina, and Taiwan, where it is valued as a soil-renovating crop. Chick pea is sparingly cultivated in the United States.

The root system is extensive. A light-textured soil and a cool climate are required for high yields. Heavy rains, sandy or wet soils, and a hot humid climate are not tolerated. This drought-resistant crop may be grown in pure stands or in admixtures with cereals (Howard et al. 1915; Shaw and Khan 1931).

Chick peas are eaten fresh, dried, boiled, and roasted. The cooked seeds are used as ingredients of soups and salads, or ground into flour. Roasted roots have served as a coffee substitute. The cooked green pods and tender shoots are consumed as a vegetable; the herbage contains appreciable quantities of oxalic acid and is not good forage. *Cicer* hay has been reported toxic to horses (Pammel 1911).

NODULATION STUDIES

The presence of nodules on *C. arietinum* L. was reported by Wydler as early as 1860, and confirmed by Naudin (1897b) and numerous later workers in Europe, India, and the United States. Host-specificity studies by Simon (1914) erroneously placed this species in the pea cross-inoculation group. Rasumowskaja (1934) and Raju (1936) concluded that the plants and the rhizobia of chick pea were highly selective *inter se* and therefore indicative of a separate plant group. Plant tests by Sanlier-Lamark and Cabezas de Herrera (1956) were generally confirmatory except for some unilateral compatibility between alfalfa plants and cicer rhizobia.

In the Soviet Union, nodulation on *C. chorassanicum* Popov was reported by Popov (1929) and Rasumowskaja (1934).

Culturally, isolates from chick pea nodules produce acid reactions in media containing sugars, an alkaline reaction with succinic acid, and excellent growth on substrates containing succinic and malic acids, erythritol, and dulcitol. Colonial growth is moderately fast, moist, and raised (Raju 1938b).

The yield of field-grown chick peas in Israel was nearly doubled by inoculations of the seeds with effective homologous rhizobial strains (Okon et al. 1972).

The detailed histological study of chick pea nodules by Arora (1956a) traced their development from root-hair invasion through maturity and senescence. A unique feature was the passage of the infection thread through differentiated, elongated cells in the cortex. These cells could be recognized after nodule maturity. Ineffective and false nodules devoid of bacteria were also described.

Cladrastis Raf.   Papilionoideae: Sophoreae

146 / 3606

From Greek, *klados*, "branch," and *thraustos*, "brittle."

Type: *C. lutea* (Michx.) K. Koch (= *C. tinctoria* Raf.)

5 species. Trees, deciduous, medium to large, occasionally 20 m tall, with trunks up to 1 m in diameter; branches zigzag. Leaves imparipinnate, petioled; winter buds 4 together and all enclosed within the inflated, hollow base of the petioles, terminal bud lacking; leaflets alternate, 7–15, entire, membranous, petioluled; stipels present or absent; stipules absent. Flowers showy, fragrant, white, yellow, pink, in large, terminal, long, often drooping, paniculate racemes; bracts caducous; bracteoles absent; calyx cylindric-campanulate, enlarged on the upper side, oblique-obconic at the base, puberulent; lobes 5, imbricate in bud, broad, short, triangular, upper 2 nearly completely united; petals clawed; standard orbicular, emarginate or entire, recurved; wings oblique-oblong, 2-auricled; keel petals free, dorsally overlapping, slightly incurved; stamens 10, free; filaments distinct, slightly incurved above, glabrous; anthers versatile, quadrate-elliptic; ovary linear, short-stalked, few-, usually 3–6-ovuled; style awl-shaped. Pod linear, lanceolate, plano-compressed, short-stalked, walls thin and membranous, winged or not, tipped with remnants of the persistent style, tardily dehiscent, 4–6-seeded; seeds short-oblong, compressed, brown. (Takeda 1913; Rehder 1940; Ohwi 1965)

This Asian-American genus is now considered distinct from *Maackia*, a closely related but exclusively Asian genus. Early botanists regarded the genera synonymously. Segregation was effected on the basis of *Cladrastis* species bearing alternate leaflets and buds covered, or hidden, by the inflated base of the leaf petioles, in contrast to *Maackia* species having opposite leaflets and exposed, scaled buds (Takeda 1913; Ohwi 1965).

Four species are found in Southeast Asia, 1 in North America. Only the type species is indigenous to the south-central United States, particularly North Carolina, Kentucky, and Tennessee. The plants grow on limestone ridges and cliffs, and along river banks.

C. lutea (Michx.) K. Koch, commonly called American yellowwood, reaches a height of 20 m, with a trunk diameter up to 1 m. It is ornamentally attractive, with fragrant, creamy white, *Wisteria*-like flowers and bright yellow autumn foliage. The wood is yellow but turns brown upon exposure; it is very hard, heavy, strong, and close grained. The wood is not considered especially attractive, although it finishes smoothly. It is used mainly for fuel, implement handles, and gunstocks (Record and Hess 1943). The heartwood and roots yield a clear yellow dye.

C. platycarpa (Maxim.) Makino and *C. sinensis* Hemsl. are known as Japanese and Chinese yellowwood, respectively. These trees often reach heights of 25 m and more. The white or pinkish flowers of the latter species are borne in upright branched panicles. A yellow spot adorns the bottom of each standard; the pods are winged (Everett 1969). *C. shikokiana* (Makino) Makino (= *Sophora shikokiana*) is a minor ornamental in Japan.

NODULATION STUDIES

Evidence to date points to the conclusion that *Cladrastis* species lack nodules. Naudin (1897a) commented about the absence of nodules on introduced greenhouse specimens of this genus in France, but he did not designate the species. Lopatina (1931) failed to find nodules in the Soviet Union on *C. amurensis* (Rupr. & Maxim.) Benth., a species now assigned to *Maackia*. Convincing reports negate the occurrence of nodules on *C. lutea* in the United States (Wilson 1939c) and on *C. platycarpa* and *C. shikokiana* in Japan (Uemura personal communication 1971). Wilson's conclusion was arrived at from experimental tests involving rhizobial isolates from species of 32 leguminous genera. Because *C. lutea* is a self-pollinating species, Wilson incorporated this information along with other data purporting to show a correlation between nonnodulation and self-pollination.

A more extensive inquiry into the nodulation status of *Cladrastis* species is warranted. It is one of the few genera of the subfamily Papilionoideae for which only negative observations have been reported.

Clathrotropis Harms   Papilionoideae: Sophoreae
(136a) / 3596

From Greek, *klathron*, "lattice," and *tropis*, "keel," in reference to the opening between the 2 incompletely joined keel petals.

Type: *C. nitida* (Benth.) Harms

4 species. Trees, small to medium, unarmed. Leaves very large, im-

paripinnate; leaflets 5–7; stipules small. Flowers white, often with a purple blotch, fragrant, borne in large, spreading, terminal panicles; calyx unequally 5-toothed; petals subequal in length, short-clawed; standard round, not appendaged; keel petals partly joined dorsally; stamens free, unequal; ovary sessile, few-ovuled; style filiform; stigma small, terminal. Pod compressed, oblong, large, often about 17 cm long, beaked at the apex, sutures thickened, valves woody or leathery, elastically dehiscent, usually 1-seeded; seeds compressed, irregularly shaped. (Harms 1908a; Ducke 1933; Amshoff 1939)

The members of this genus constituted a section of *Diplotropis* until they were segregated and elevated to generic rank (Harms 1908a).

Species of this genus occur in the central and northern Amazon region of northern Brazil, Guyana, and Trinidad. They flourish in wet tropical forests.

The timber of *C. brachypetala* (Tul.) Kleinh., known as blackheart, *C. grandiflora* (Tul.) Harms, *C. macrocarpa* Ducke, and *C. nitida* (Benth.) Harms, known as acapú, has been examined for commercial possibilities but does not show great promise. In general, the wood is hard, dark brown to black, coarse textured, and fairly straight grained, but varies in durability and has a low luster. It is not considered attractive. Its uses are limited to heavy construction purposes.

NODULATION STUDIES

Nodules of *C. brachypetala* sent to the authors' laboratory from Trinidad by DeSouza (1963, 1966) were large, oval to elongated, with pink interiors. Rhizobial isolates were typical of the cowpea type, as judged from cultural and physiological reactions. The nodules on *C. macrocarpa* in Bartica, Guyana, were numerous, brown, and 6–10 mm in length (Norris 1969).

Cleobulia Mart. ex Benth.   Papilionoideae: Phaseoleae,
411 / 3888   Diocleinae

Type: *C. multiflora* Mart. ex Benth.

3 species. Shrubs or climbing shrublets, tall. Leaves pinnately 3-foliolate; stipels present; stipules absent. Flowers small, purple, densely clustered along the rachis of elongated, axillary racemes; bracts and bracteoles small, caducous; calyx lobes 5, upper 2 connate into 1 entire lobe, lateral lobes smaller; standard suborbicular, with basal, inflexed auricles; wings small, free, not longer than the calyx; keel blunt, incurved; vexillary stamen free at base, connate at the middle with the others; anthers uniform; ovary subsessile, many-ovuled; style smooth, apically dilated-truncate; stigma attached to the dorsal surface. Pod broad-linear, flattened, 2-valved, filled between the seeds. (Bentham and Hooker 1865; Maxwell 1977)

Members of this genus are native to Brazil.

*Clianthus Soland. ex Lindl.　　　　　Papilionoideae: Galegeae,
(285) / 3753　　　　　　　　　　　　　　　　　　Coluteinae

From Greek, *kleos*, "glory," and *anthos*, "flower," in reference to the exceptional beauty of the inflorescence.

Type: *C. puniceus* (G. Don) Soland. ex Lindl.

8 species. Herbs, semiviny, woody, or undershrubs, smooth or densely
hairy. Leaves imparipinnate; leaflets numerous, entire, ovate; stipels absent;
stipules broad, awl-shaped, herbaceous. Flowers large, red or white, pendent,
in short, peduncled, axillary and terminal, umbellike racemes; bracts long-
acuminate; bracteoles subpersistent; calyx campanulate, usually hairy; lobes
5, subequal or upper 2 basally broader, connate above; petals short-clawed,
subequal, glabrous; standard acuminate, reflexed, base with dark blotch, swollen; wings obtuse, falcate-lanceolate, shorter than the standard; keel incurved, beaked, equal to or longer than the standard; stamens diadelphous;
anthers uniform, dorsifixed; ovary stalked, glabrous, many-ovuled; style awl-
shaped, incurved, inner margin bearded above; stigma minute, terminal. Pod
linear turgid, oblong with pointed tip, 2-valved; seeds small, numerous,
kidney-shaped. (Bentham and Hooker 1865; Black 1948; Allan 1961)

Representatives of this genus are indigenous to Australia and New Zealand. They grow on rocky ridges, sand hills, and in scrub areas.

All species have ornamental properties and are widely cultivated in gardens. The type species, indigenous to New Zealand, is known as parrot's-bill
or parrot's-beak because of the beak-shaped keel. The large red or white
flowers of *C. speciosus* (G. Don) Aschers. & Graebn., Sturt pea, the indigenous
Australian species, are spectacular because of the shiny, purplish black blotch
at the base of the standard.

Nodulation Studies

Nodules on *C. formosus* (G. Don) Ford & Vick. were described as flat or
cylindrical, 6–7 mm long, and sometimes notched at the apex (Beadle 1964).
Host-infection studies of 2 species by Wilson (1939a, 1944b) and Wilson and
Chin (1947) showed promiscuity within the cowpea miscellany.

Nodulated species of *Clianthus*	Area	Reported by
breviolata [?]	N.Y., U.S.A.	Wilson (1944b)
formosus (G. Don) Ford & Vick.	N.S. Wales, Australia	Beadle (1964)
puniceus (G. Don) Soland. ex Lindl.	Germany	Lachmann (1891)
	France	Clos (1896)
speciosus (G. Don) Aschers. & Graebn.	S. Australia	Harris (1953)
(= *dampieri* A. Cunn.)	N.Y., U.S.A.	Wilson (1939a)

Climacorachis Hemsl. & Rose　　　　Papilionoideae: Hedysareae,
(324a) / 3794a　　　　　　　　　　　　　　Aeschynomeninae

From Greek, *klimax*, "ladder," and *rhachis*, "spine," "midrib"; the axis of the
inflorescence is zigzag.

Type: *C. mexicana* Hemsl. & Rose

2 species. Shrubs, branches slender, low, wiry, somewhat hairy. Leaves pinnate; leaflets small, 9–20 pairs; stipules peltate, striate, caducous. Flowers 2–12, yellow, streaked with purple, bibracteoleate, in short, zigzag, pilose racemes; calyx bilabiate, the upper lobe broad, rounded; standard orbicular; keel blunt, shorter than the wings; stamens in 2 bundles of 5 each; ovary stalked, sometimes pubescent. Pod oblong, obtuse, 2–4-seeded, not jointed or constricted between the seeds. (Rose 1903; Harms 1908a)

This genus is closely related to *Aeschynomene*, but the pods are very different.

Both species occur in western Mexico, up to altitudes of 1,500 m.

Clitoria L.   Papilionoideae: Phaseoleae, Phaseolinae
383 / 3857

From Greek, *kleitoris*, "clitoris"; the small keel suggests the mammalian clitoris.

Type: *C. ternatea* L.

70 species. Herbs, rarely erect, suffrutescent, perennial, sometimes large, erect or scandent shrubs with long, sinistrorsally twining branches, rarely trees. Leaves imparipinnate, 3- or more-foliolate, rarely unifoliolate; leaflets 1–4 pairs, large, ovate or oblong-lanceolate, petioluled; stipels present; stipules striate, persistent. Flowers large, resupinate, white, purplish blue, or red, axillary, solitary or paired or in short, few-flowered racemes; bracts small, ovate-lanceolate, persistent, paired, stipulelike; bracteoles 2, usually large, striate, appressed to the base of the calyx, persistent; calyx tubular, glabrous inside, bilabiate; 2 upper lobes subconnate, lower lip tridentate with the middle lobe longest; standard large, much longer than the other petals, erect, emarginate, rim wavy, not eared, narrow at the base; wings falcate-oblong, adherent at the middle to the keel, long- and slender-clawed; keel shorter than the wings, incurved, acute, with a slender claw; stamens diadelphous, vexillary stamen free or slightly joined below with the others; anthers uniform, alternately dorsifixed and basifixed; ovary short-stalked, linear, many-ovuled; disk papery, lobed; style compressed, elongated, incurved, bearded inside lengthwise above. Pod linear, stalked, compressed or convex, somewhat thickened on the upper or both sutures, beaked with the persistent style, 2-valved, valves twisting on dehiscence, with or without membranous septa between the 2–many, globose or compressed seeds. (Bentham and Hooker 1865; Baker 1929; Amshoff 1939; Wilczek 1954; Hutchinson and Dalziel 1958)

The genus is represented throughout warm and temperate areas of both the Old World and the New World. One species is endemic in Australia, 4 species in Argentina. They flourish along roadsides, in grassy savannas, in dry or moist thickets, and in pine forests. The plants grow up to 2,000-m altitude in Mexico.

Several species, especially *C. mariana* L., *C. racemosa* G. Don, a fast-growing, blue-flowered tree, *C. rubiginosa* Juss. ex Pers., a fragrant, mauve-flowered vine, and the type species, are cultivated along fences and trellises as ornamentals. *C. laurifolia* Poir. (= *cajanifolia* Benth.) is popular as a contour hedge on terraces of tea plantations to prevent soil erosion and as a cover crop on teak estates. The foliage is unpalatable for stock, but pigs eat the fleshy roots (Whyte et al. 1953). *C. rubiginosa* is regarded highly as a forage vine in the Philippines. The royal blue flowers of *C. ternatea* L. yield a dye similar to litmus as an indicator of acidity and alkalinity; it is used to dye mats in Malaysia. This species is a good soil-binder because of its rhizomatous roots and vining stems. It was once cultivated in Sri Lanka but not regarded highly as a green manure and forage crop. The foliage and pods are grazed by livestock. Natives in some areas consume the green pods as a vegetable.

The seeds contain a fixed oil, starch, tannic acid, and a bitter acid resin, and have been used in folk medicine as an emetic, diuretic, and cathartic. Root and leaf decoctions have been used medicinally to reduce swellings, as poultices, and as purgatives (Quisumbing 1951).

C. brachycalyx Benth., a deciduous tree of about 12–15-m height in Guyana, *C. fendleri* Rusby, a fairly large tree in northern Colombia, and *C. hoffmanseggii* Benth., a medium-sized tree in the Amazon region, have similar wood. The sapwood is yellow and not easily differentiated from the heartwood. The wood is moderately hard, heavy, coarse textured, straight grained, and splits readily. Because it is stringy and lacks durability and luster, it is not marketed and has no commercial possibilities (Record and Hess 1943).

Nodulation Studies

C. ternatea, Kordofan or butterfly pea, is perhaps the species most studied by rhizobiologists. Joshi's early work (1920) showed that rhizobia from this species did not produce nodules on *Cicer*, *Pisum*, and *Phaseolus* species. Carroll's placement (1934a) of this species in the cowpea miscellany was confirmed by Allen and Allen (1936a). In a plant-infection study, 29 of the 54 strains of rhizobia from 28 leguminous plants of Hawaii produced ineffective responses on plantlets of *C. ternatea*, and 21 strains incited effective reactions. Only strains from *Albizia lebbek* and *Cytisus scoparius* failed to form nodules (Allen and Allen 1939). Wilson's results (1939a) were confirmatory in principle.

Nodulated species of *Clitoria*	Area	Reported by
australis Benth.	Australia	Bowen (1956)
biflora Dalz.	W. India	Bhelke (1972)
guianensis (Aubl.) Benth.	Venezuela	Barrios & Gonzalez (1971)
laurifolia Poir.	Puerto Rico	Dubey et al. (1972)
(= *cajanifolia* Benth.)	Sri Lanka	Holland (1924)
mariana L.	N.C., U.S.A.	Shunk (1921)
racemosa G. Don	Philippines	Pers. observ. (1962)
	Brazil	Campêlo & Döbereiner (1969)
rubiginosa Juss. ex Pers.	Philippines	Pers. observ. (1962)
ternatea L.	Widely reported	Many investigators

The Greek suffix, *opsis*, denotes resemblance to *Clitoria*.

Type: *C. mollis* Wilczek

1 species. Shrublet, erect; bark fibrous. Leaves pinnately 3-foliolate; leaflets ovate, subcoriaceous, petioluled; petiole short; rachis much longer than petiole; stipels persistent; stipules not extended below the insertion, persistent. Flowers large, on short pedicels, in dense, many-flowered, axillary racemes shorter than the leaves; bracts and bracteoles small, persistent; calyx fleshy, exterior tomentose, bilabiate; tube exceeds the lobes; lobes 5, imbricate, subequal, 2 upper lobes connate into a bidentate lip, lower lip tridentate, middle lobe slightly longer; all petals long-clawed; standard broad, slightly longer than the other petals, tapered at the base; wings oblanceolate, without callosities, narrowed at the base, not adnate to the keel; keel petals short, erect, oblanceolate, beaked; stamens 10, diadelphous; anthers apiculate at the summit, 5 dorsifixed on short filaments alternate with 5 basifixed on long filaments; ovary linear, long-stalked, 2- to many-ovuled; disk lobed; style long, erect or slightly curved, glabrous above, pubescent toward the base; stigma terminal, puberulent-glandular. Pod stalked, narrowed at both ends, with persistent style. (Wilczek 1954)

This monotypic genus is endemic in Africa, ranging from Zaire to Sudan. The plants thrive in humid savannas.

Cochlianthus Benth.   Papilionoideae: Phaseoleae, Erythrininae

399 / 3875

From Greek, *kochlion*, "coiled," and *anthos*, "flower," in reference to the coiled keel.

Type: *C. gracilis* Benth.

2 species. Herbs, slender, climbing, subglabrous, leaves and flowers becoming blackish when dry. Leaves pinnately 3-foliolate, petioled, membranous, stipellate. Flowers medium-large, pendent, in axillary, slender racemes; rachis nodose; bracts and bracteoles minute, caducous; calyx campanulate, often clothed with appressed, silky hairs; lobes short, 2 upper ones connate into 1, lateral 2 smaller, lowest lobe longest, lanceolate; petals subequal in length; standard broad-ovate, with 2 inflexed ears; wings oblong, slightly longer than the standard; keel narrow, linear, shorter than the wings, curved, coiled, snaillike; stamens diadelphous; anthers uniform; ovary very short-stalked, multiovuled; style filiform, peltate-dilated; stigma large, terminal. Pod linear, incurved, compressed, inconspicuously septate, 2-valved; seeds squarish. (Bentham and Hooker 1865)

Both species occur from India to southwestern China.

From Greek, *kodarion*, "little fleece," in reference to bracts being caducous
while flowers are still in bud.

Type: *C. gyrans* (L. f.) Hassk.

2 species. Shrublets or erect shrubs, small. Leaves 3–1-foliolate; terminal
leaflet large, oval-oblong, laterals very small or almost completely reduced;
upper leaf surfaces, tops of stems, rachis of inflorescences, and pods hairy;
stipels and stipules present. Flowers few, overlapped with large, caducous
bracts while in bud, in terminal or axillary racemes or panicles; bracteoles
none; calyx very small, campanulate, subbilabiate, minutely 5-toothed;
standard ovoid, not eared; wings short-clawed; keel petals obtuse, long-
clawed, lower margin sharply incurved; stamens diadelphous, vexillary stamen connate only at the base; filaments alternately long and short; anthers
ovoid, dorsifixed; ovary linear; disk absent; style short, inflexed at right angle,
glabrous, dilated above; stigma terminal, slightly oblique-capitate. Pod subsessile, linear, compressed, inconspicuously jointed, joints subquadrate, dehiscent
along entire length of 1 suture, not breaking up into joints; seeds kidney-
shaped, attached aril bipartite. (Schindler 1924)

Both species were originally included within the genus *Desmodium*. Segregation was based on distinctive dehiscence of the pods (Schindler 1924).

Both species occur in Indo-Malaya and the Philippines. They flourish in
grassy fields, brushwood, and teak forests in Java.

The type species is commonly known as the telegraph plant because the 2
lateral leaflets gyrate by short, jerky motions resembling semaphores. Its
flowers are at first lilac but soon turn orange. The attractive pink lilac flowers
of *C. gyroides* (Roxb.) Hassk. later become blue violet.

These species are useful on rubber and tea plantations as cover crops and
green manure and as temporary shade on young coffee plantations; they are
palatable to stock.

Nodulation Studies

Plants of *C. gyrans* (L. f.) Hassk. (as *Desmodium gyrans*) lacked nodules
when grown in the University Botanical Gardens, Cambridge, England, but
those grown later in native soil imported from Calcutta, India, were nodulated (Dawson 1900b). Nodule sections and stained smears of pure cultures
showed exceptionally large rhizobia containing deeply stained bodies.
Nodules on *C. gyrans* (as *D. gyrans*) have also been reported in Europe by
Štefan (1906) and in Java by Keuchenius (1924).

Nodules on *C. gyroides* (as *Desmodium gyroides*) have been observed in Java
(Keuchenius 1924), in Sri Lanka (W. R. C. Paul personal communication
1949), in Japan (Ishizawa 1953), and in South Africa (B. W. Strijdom personal
communication 1970). Rhizobia from *C. gyroides* were slow-growing forms
with plant-infection affinities akin to members of the cowpea miscellany
(Ishizawa 1953, 1954).

Coelidium Vog. ex Walp.   Papilionoideae: Genisteae,
186 / 3646   Lipariinae

From Greek, *koilos*, "hollow," in reference to the frequently concave leaves.
    Type: *C. ciliare* Vog. ex Walp.
    15 species. Shrubs, small, much-branched, silky-haired, ericoid; branches
sometimes terminally spinescent. Leaves sessile, simple, entire, linear or oblong, usually hairy, occasionally scattered, twisted or wrinkled, margins rolled
inward; stipules none. Flowers small, purple, rose, or yellow, subsessile, clustered into terminal leafy heads or 2-flowered on long, axillary peduncles;
bracts absent; bracteoles 1, rarely 2; calyx tubular-campanulate, hairy; lobes
5, deltoid, subequal, upper 2 partly joined; standard elliptic or suborbicular,
reflexed, claw broad, short; wings oblong, long-clawed, sometimes eared at
the base; keel oblong, straight, blunt, pouched or spurred on each side; stamens monadelphous, all connate into a short staminal tube sometimes adnate
to the lower part of the calyx tube; filaments slender, free above; anthers
alternately shorter and versatile, longer and basifixed; ovary sessile or short-
stalked, villous, 1-ovuled; style incurved; stigma terminal. Pod lanceolate or
ovate, usually beaked, 2-valved, villous, 1-seeded. (Bentham and Hooker
1865; Phillips 1951)
    Polhill (1976) assigned this genus to the tribe Liparieae in his revision of
Genisteae *sensu lato*.
    Representatives of this genus are limited to South Africa, mostly in the
southwestern districts of the Cape Province. They inhabit rocky, hilly, scrub
localities.

Collaea DC.   Papilionoideae: Phaseoleae, Diocleinae

Named for Luigi Aloys Colla, 1766–1848, botanist and member of the Ac-
cademia di Scienze dei Turin, Italy.
    Type: *C. speciosa* DC.
    3 species. Shrubs or shrublets, erect, rarely twining; rhizome woody.
Leaves usually 3-foliolate, villous; stipules small, oval, soon deciduous. Flowers medium to large, purple, in axillary racemes or subumbel; calyx campanulate; lobes 4, upper lip entire, broader, lower lobe longest; standard ovate or
rounded, narrowed with margins inflexed or eared at the base; wings obovate
or oblong; keel similar but subequal, straight; stamens diadelphous, 1 united
only in the lower half with the others; anthers equal, short, dorsifixed; disk
forming a short sheath; ovary subsessile, often hairy; ovules numerous; style
linear, incurved; stigma small, truncate. Pod sessile, linear, compressed,
leathery, filled between oblong seeds, 2-valved, dehiscent. (De Candolle
1825a,b,c; Burkart 1971; Lackey 1977a)
    This genus was formerly a section in *Galactia* (Bentham and Hooker
1865). Lackey (1977a) included it in the subtribe Diocleinae.
    Collaea species are native to tropical and subtropical America.
    Species of this genus merit cultivation for their ornamental beauty.

Named for the Cologan family, Port Orotavo, Tenerife; they extended great hospitality to scientific visitors.

Type: *C. procumbens* Kunth

15 species. Herbs, climbers, sometimes trailing, perennial, unarmed; rootstocks hard, sometimes tuberous. Leaves pinnately 3-foliolate, occasionally 1- or 5-foliolate; stipels present; stipules small, striate. Flowers violet, red, or purple, attractive, axillary, subsessile, solitary or fascicled, or in umbels or short racemes; bracts and bracteoles persistent, lanceolate and striate, linear, or setaceous; calyx tubular; lobes 4, uppermost lobe very broad, lower lobe longer; standard obovate, narrowed into a claw, erect, sides reflexed; wings oblique-oblong, slightly adherent to the keel; keel petals incurved, obtuse, shorter than the wings; stamens diadelphous; anthers uniform or alternate ones smaller; ovary stalked, few- to many-ovuled; style incurved, awl-shaped; stigma terminal, capitate. Pod linear, septate between the compressed, orbicular or subquadrate seeds. (Rose 1903; Fearing 1959; Turner and Fearing 1964)

Cologania is most closely related to *Amphicarpaea*, but is readily distinguished by its slender, tubular calyx.

Members of this genus range from tropical America into Argentina, with Mexico the center of proliferation; few species occur in Central and South America. In the United States their presence is restricted to the southwestern border region, particularly the pine forests of Arizona. Species grow up to about 2,400-m altitude, on rocky hillsides, in moist thickets, along roadsides, and in open dry areas.

The ornamental beauty of *Cologania* species has not been fully appreciated; specimens are only occasionally found in botanical gardens. *C. capitata* Rose, *C. grandiflora* Rose, and *C. hirta* (Mart. & Gal.) Rose are among the most attractive (Rose 1903).

NODULATION STUDIES

Nodules were present on University of Wisconsin herbarium specimens of *C. angustifolia* Kunth from Zacatecas, Mexico, and *C. broussonetti* (Balb.) DC. from Durango and Michoacán, Mexico (personal observation 1970).

Colophospermum Kirk ex J. Léonard   Caesalpinioideae: Cynometreae

From Greek, *kolophon*, "summit," "top," and *sperma*, "seed," in reference to germination of the seeds above ground while still in the pod.

Type: *C. mopane* Kirk ex J. Léonard (= *Copaifera mopane* Kirk. ex Benth.)

1 species. Shrub or tree, up to about 20 m tall, unarmed; trunk gray brown; bark deeply fissured. Leaves 2-foliolate; leaflets sessile, asymmetric, pellucid-punctate, leathery, with a minute, sessile, foliar appendage between their inner margins; the leaflets fold together and hang downward during mid-

day heat; petiole short, stout; stipules caducous. Flowers small, in 2 vertical rows, white or pale green, in slender, pendent racemes or panicles; bracts minute, caducous, rarely large and persistent; bracteoles none; calyx tube short or absent; sepals 4, alternate 2 outermost, covering the inner 2 in bud; petals absent; stamens 20–25, free, equal in length, exserted; filaments filiform, glabrous; anthers uniform, oblong, dorsifixed, longitudinally dehiscent; ovary sessile or stalked, stalk hairy, oblique, sometimes glandular, compressed, 1-ovuled; style cylindric, often attached laterally on the ovary. Pod stalked, oblique-elliptic or falcate-ovate, pendent, compressed or turgid, thinly leathery, sparsely glandular, indehiscent, 1-seeded; seeds yellowish, kidney-shaped, resin-dotted, sticky, wrinkled, cotyledons corrugated. (Leónard 1949, 1957)

This monotypic genus is endemic in tropical Africa; it is especially common in the interior Transvaal, Zimbabwe, Malawi, and Mozambique. The trees flourish on sandy soil, low-lying alluvium flood plains, and in grasslands. They are tolerant of low rainfall and poorly drained, neutral or slightly acid soils.

Mopane woodlands, where these mopane, butterfly, or turpentine trees are dominant, occur in many low-lying parts of Africa. These are distinctive, savanna-type vegetation belts. Mopane scrub provides browse and fodder to cattle and wild game, especially elephants. The nutritious leaves contain about 12 percent crude protein and have a turpentine odor; however, the milk, butter, and meat of the animals that consume the foliage are not tainted even when large amounts are ingested (Palmer and Pitman 1961).

These trees supply Rhodesian ironwood in the timber trade. The sapwood is yellowish and the heartwood is reddish brown to almost black. The hard, heavy, durable, and termite-resistant wood is used in heavy construction, for fence posts, mine props, bridge pilings, and small handicraft articles. Mopane areas are often struck by lightening during thunderstorms and are considered a dangerous refuge because the wood burns when green.

Mopane trees are subject to defoliation by caterpillars of moths that have a wing span in the adult stage of about 12 cm. These hairless, black, spotted mopane worms are roasted or sun-dried by the natives, who relish them for their nutty flavor. The trees are also host to a small, irritating but nonstinging mopane bee, a *Melipona* species, that is attracted to the moisture of the victim's eyes and nose. Attacks are most vicious during midday, and repellents afford little relief.

Natives in various areas use wood extracts to treat inflamed eyes and root extracts as a tapeworm remedy. The bark is used for fiber, is a source of copal, and yields 5–8 percent tannin that is used for staining leather. The roots contain a resin. The seeds and pods yield a balsam of no commercial value (Coates Palgrave 1957; Palmer and Pitman 1961).

Nodulation Studies

Specimens of *C. mopane* Kirk ex J. Léonard examined in South Africa by Grobbelaar and Clarke (1972) and in Zimbabwe by Corby (1974) lacked nodules.

From *koloutea*, an ancient Greek name of a leguminous plant.

Type: *C. arborescens* L.

27 species. Shrubs, much-branched, unarmed, or rarely spiny, deciduous. Leaves imparipinnate, rarely 3-foliolate; leaflets 4–6 pairs, small, oval, entire; stipels none; stipules small. Flowers yellow, orange, brownish red, or almost black, in few-flowered, long-stalked, axillary racemes; bracts and bracteoles very small or absent; calyx campanulate, slightly bilabiate; lobes 5, subequal or upper 2 shorter; standard suborbicular with 2 swellings above a short claw; wings falcate-oblong, short-clawed; keel broad, much incurved, blunt, claws of petals long and connate; stamens diadelphous; anthers uniform; ovary stalked, many-ovuled; style incurved, longitudinally bearded on inner side near apex, apex inflexed or involute; stigma large, thick and prominent below apex, inserted obliquely on inner edge of style and surrounded by hairs. Pod stalked, very inflated, walls papery, indehiscent or opening at the beak; pods burst when squeezed; seeds kidney-shaped. (Bentham and Hooker 1865; Baker 1929; Rehder 1940; Tutin et al. 1968; Komarov 1971)

Colutea species are distributed throughout central and southern Europe, principally around the Mediterranean, and from northeastern Africa into Asia Minor and western Asia. They thrive in sunny localities on poor dry soil.

These ornamentals have been introduced into northern Europe, Argentina, and the United States for their attractive flowers and decorative pods, which often turn red when mature. The genus is commonly called bladder senna because the leaves of some species, notably *C. arborescens* L., have cathartic properties similar to senna (*Cassia* spp.). Seed germination is slow and irregular, but these shrubs grow rapidly. The rarer species are often grafted on *C. arborescens* or propagated from cuttings for horticultural stock.

C. istria Mill. has value in desert agriculture in the Negev area not only as a pasture plant relished by sheep and goats, but also as a component of bush dams to diminish erosion by filtering soil and plant debris from flood waters (Koller and Negbi 1955). Other drought-resistant species serve well for shelter belts and gully windbreaks. *C. arborescens* is one of the few shrubs in the Himalayan valleys of northern India yielding good wood for fuel, implements, and small furniture.

Nodulation Studies

Accounts of nodulation have been reported for *C. arborescens* by many workers, for *C. cilicica* Boiss. & Bal. in New York by Wilson (1939a), and for *C. persica* Boiss. in France by Lechtova-Trnka (1931). Seedlings of *C. arborescens* and *C. cilicica* were versatile in symbiotic performance with members of the cowpea miscellany and also with *Trifolium* and *Vicia* species (Wilson 1939a).

Broad, prominent, multibulged infection threads were observed in nodule sections of *C. arborescens* and *C. persica* (Lechtova-Trnka 1931). Presumably, the globular masses of rhizobia within the host cells of the bacteroid area were enclosed within the membranes now recognized to be the site of nitrogen fixation. Information concerning cultural and physiological characteristics of *Colutea* rhizobia is sparse.

Named for Sir Charles Colville, distinguished Scottish officer under Wellington in the Napoleonic Wars and mid-19th-century Governor of Mauritius. The trees in Mauritius were raised from seeds brought from Madagascar in 1824 by the botanist Wenzel Bojer (Menninger 1962).

Type: *C. racemosa* Boj.

1 species. Tree, unarmed, deciduous, up to 15 m tall; trunk thick, straight, bark gray, horizontally ridged, with loose thin scales which peel; underbark reddish; branches thick, spreading, studded with corky nobs. Leaves bipinnate, fernlike, long; pinnae opposite, 15–25 pairs; leaflets small, numerous, in spiral rows, opposite, sometimes alternate, 18–30 pairs; stipules minute, caducous. Flowers burnt orange to scarlet, showy, densely crowded in pendent, long, cylindric or cone-shaped racemes clustered 10–12 at the tips of branches; each flower usually about 2 cm long; bracts membranous, colored, soon caducous; bracteoles none; calyx tube short, 1 side swollen; sepals 5, rarely 4, thick, induplicate-valvate; petals 5, imbricate, uppermost petal inside, very broad, lateral ones smaller, obovate, lower ones outside, narrow; stamens 10, free, deflexed; filaments thick, hairy at base; anthers large, uniform; ovary subsessile, free, many-ovuled; style filiform; stigma small, terminal. Pod straight, elongated, turgid, 2-valved; seeds transverse, oblong, small. (Bentham and Hooker 1865)

This monotypic genus is endemic in Madagascar. The type species, commonly known as Colville's glory, has been introduced as a handsome ornamental for parks and gardens in India, the West Indies, Hawaii, and Florida. It is best suited to moist or low, dry areas.

NODULATION STUDIES

This species was reported nodulated in the Royal Botanic Gardens, Peradeniya, Sri Lanka (Wright 1903); however, in view of its taxonomic position in the tribe Eucaesalpinioideae and the scant information conveyed, Wright's report merits confirmation.

Condylostylis Piper   Papilionoideae: Phaseoleae, Phaseolinae

From Greek, *kondylos,* "knuckle," in reference to the shape of the style.

Type: *C. venusta* Piper

2 species. Herbs, twining. Leaves pinnately 3-foliolate; each leaflet 3-nerved from the base; stipels, stipules, and bracteoles striate-nerved. Flowers purple or white, in few-flowered, axillary racemes; rachis with pedicellar glands; calyx campanulate; teeth broad, obtuse, short; standard orbicular, thick, eared at the base; wings long-clawed, oblong, constricted below the middle; keel petals long-clawed, lower portion broad, constricted and twisted above the middle, beak bottle-shaped; stamens diadelphous, the free stamen thickened at the base and bent at a right angle; anthers oblong; style with a

globose enlargement slightly above the middle, constricted and smooth near the apex and tipped with a spathulate appendage, bearded about the stigma; stigma roundish, lateral. Pod linear, short-beaked, thin, slightly compressed, dehiscent, 6–8-seeded; seeds cylindric, hilum more than half as long as the seed. (Piper 1926)

This genus is closely related to, but readily distinguished from, *Phaseolus* and *Vigna* by the form of the style and the peculiar keel.

Both species are native to Central and South America and are especially common in Costa Rica, Belize, and Colombia at altitudes up to 1,600 m.

Conzattia Rose   Caesalpinioideae: Eucaesalpinieae
(97a)

Named for Cassiano Conzatti, 1863–1951, Director of the Normal School, Oaxaca, Mexico, plant collector, and author of *Los Generos Vegetalis Mexicano*.

Type: *C. arborea* Rose

3 species. Tall shrubs or medium-sized trees, unarmed, up to 15 m tall, with a trunk diameter up to 0.6 m, and with a spreading crown. Leaves large, bipinnate; pinnae 10–15 pairs; leaflets oblong, acute, about 20 pairs; stipules minute. Flowers yellow, small, many, in long, slender racemes clustered near the ends of branches; calyx tube campanulate, very short, shorter than the lobes; lobes valvate, later reflexed, subequal; petals 5, equal, separate; stamens 10, erect; filaments free, glabrous above, hairy at the base; ovary white-woolly. Pod flat, broad-linear, acute at both ends, narrowly winged on each suture, dehiscent, few-seeded. (Rose 1909; Harms 1915a; Britton and Rose 1930)

The genus is closely akin to *Cercidium*.

Members of this genus are indigenous to Mexico, in the states of Oaxaca, Puebla, and Michoacán.

The wood of *C. sericea* Standl. is lightweight, firm, and finishes smoothly, but it is susceptible to decay and to insects. The species lacks commercial importance (Record and Hess 1943).

*Copaifera L.   Caesalpinioideae: Cynometreae
(42) / 3490

From *copai*, the name used by the Tupi Indian tribes of the Amazon for the resin of the copaiba tree, and the Latin verb *fero*, "bear," "produce."

Type: *C. officinalis* (Jacq.) L.

25 species. Trees, shrubs or undershrubs. Leaves paripinnate or imparipinnate; leaflets opposite or alternate, 1–many pairs, rarely unifoliate, emarginate or acuminate at the apex, usually gland-dotted; stipels absent; stipules small, deciduous. Flowers small, in 2 vertical rows, sessile or on short pedicels in terminal or subterminal, racemose or paniculate inflorescences; bracts small, oval, oblique; bracteoles 2, caducous, usually absent; calyx tube short; lobes 4, unequal, 1 larger than the other 3, imbricate or subvalvate, on short, disk-bearing papery receptacle; petals none; stamens 8–10, free, usually alternately long and short; filaments glabrous; anthers dorsifixed, longitudinally dehiscent; ovary sessile or short stalked, 2–7-ovuled; style filiform; stigma terminal, truncate. Pod somewhat stalked, usually leathery, suborbicular or elliptic, compressed or dilated, 2-valved, dehiscent or subdehiscent; seeds 1–2, with or without a short aril. (Amshoff 1939; Pellegrin 1948; Léonard 1949, 1952, 1957)

Copaifera is primarily a tropical South American genus; however, 5 species do occur in Africa. The plants thrive in humid forests and moist districts, especially along the Amazon and Congo rivers.

Copaifera trees are better known as producers of copal or copaiba balsam than as sources of timber. The resins, obtained either directly from living trees or in a semifossilized state from the ground where the trees are growing, or have grown, are characterized by their hardness and relatively high melting points. Commercial varieties of copaiba balsam range from the yellowish brown, thick, malodorous product of *C. reticulata* Ducke to the fluid, transparent, and less strongly scented product of *C. multijuga* Hayne (Everett 1969). Most copal from Brazil is obtained from *C. guyanensis* Desf., *C. langsdorfii* Desf., *C. martii* Hayne, *C. multijuga*, *C. officinalis* (Jacq.) L., and *C. reticulata*. Almost all of the copal supply from Pará in the Amazon area is obtained from *C. guyanensis* and *C. reticulata*. *C. demeusei* Harms is the principal copal source in Zaire. These copals are used mainly in varnishes, lacquers, paints, and to a limited extent in medicines (Howes 1949).

C. ehie A. Chev., *C. mildbraedii* Harms, and *C. salikounda* Heckel are tall, majestic forest trees that often reach a height of 40 m or more, with straight, cylindrical boles about 1 m in diameter above the buttressed base and free of branches for the first 25 m. The density of copaiba wood varies from 700 to 900 kg/m^3. The wood is hard, tough, usually straight and fine grained, highly durable, and may or may not be odorous. It is used for heavy construction, bridges, shipbuilding, and furniture (Record and Hess 1943).

NODULATION STUDIES

Information regarding nodulation of *C. officinalis*, the only member of the genus examined, is inconsistent. Wright (1903) reported nodules present on specimens growing in Sri Lanka; however, specimens examined in Trinidad (DeSouza 1966) and in Venezuela (Barrios and Gonzalez 1971) lacked nodules. The roots of plantlets are dark brown to black, quite typical of those found on nonnodulating members of the Caesalpinioideae.

Corallospartium Armstr.   Papilionoideae: Galegeae, Robiniinae
(282a) / 3749

From Greek, *korallion*, "coral," and *spartos*, the name of a broom plant, in reference to the stiff branching habit of this shrub.

Type: *C. crassicaule* (Hook. f.) Armstr.

1 species. Shrub, rigid, up to 2 m tall; branches thick, erect, yellowish, deeply grooved, nearly leafless at maturity. Leaves small, rare, linear- or ovate-oblong, fugacious. Flowers creamy, pedicels pilose, densely fascicled; calyx woolly, campanulate-turbinate, bluntly 5-toothed, teeth subequal; standard large, broad, reflexed, short-clawed; wings falcate, oblong, obtuse, eared near the base, shorter than keel petals; keel incurved, subequal to standard, oblong, obtuse; stamens diadelphous; ovary few-ovuled, villous; style silky-haired at apex and base. Pod small, deltoid, winged dorsally, villous, short-beaked, thinly and reticulately 2-valved, 1-, or rarely 2-seeded. (Harms 1900; Allan 1961)

Hutchinson (1964) included this genus in Carmichaelieae trib. nov.

This monotypic genus is endemic in South Island, New Zealand. It occurs on mountain and subalpine grassland, scrub, and rocky areas.

NODULATION STUDIES

Nodules were observed on this species in New Zealand (R. M. Greenwood personal communication 1975).

Cordeauxia Hemsl.   Caesalpinioideae: Eucaesalpinioideae
(102b)

Named for Captain H. E. S. Cordeaux; he supplied Kew Gardens with the plant specimens.

Type: *C. edulis* Hemsl.

1 species. Shrub, low to 3 m tall, deep-rooted, evergreen; lower branches dense, straight, broomlike, hard. Leaves paripinnate; leaflets usually oval-oblong, 4-paired, leathery, undersides profusely dotted with reddish, scalelike glands; stipules none. Flowers few, yellow, in corymbs at apex of branches; calyx short; lobes 5, blunt, valvate, glandular; petals 5, subequal, clawed, spoon-shaped; stamens 10, free; basal half of filaments hairy; anthers versatile; ovary short-stalked, 2-ovuled, densely glandular; stigma obtuse. Pod leathery, compressed-ovoid, curved, apex beaked, 2-valved, dehiscent, usually 1-seeded; seeds ovoid, endosperm lacking, cotyledons thick. (Harms 1915a)

In Hemsley's opinion (1907) the genus seemed related to *Schotia* in the tribe Amherstieae, but Harms (1915a) reassessed the generic characteristics and assigned it to position 102b near *Caesalpinia*.

This monotypic genus is endemic in Somalia, Malawi, and Ethiopia. It grows on sandy soils of pH 7.8–8.4 up to 800-m altitude, in desert areas with mean annual rainfall of 10–20 cm and a mean annual temperature of 30°C.

The seeds, called yeheb or yebb nuts, are regarded as a possible commercial source of oil, starch, and sugar (Marassi 1939). They lack alkaloids and glucosides, are low in protein, calcium, and iron, but contain 36–41 percent carbohydrates and 10–12 percent fat. Their consistency resembles that of cashews. The steamed or stewed nuts are relished by the native peoples. The leaves are infused for tea (Greenway 1947). The leaves yield a magenta stain, cordeauxione, used to dye cloth. This pigment is the only naphthoquinone found in the Leguminosae (Harborne et al. 1971). When the leaves are touched, they stain the hands red. Somalis use about 200 g of pulverized leaves to dye 10 m of 90 cm-wide calico. When the leaves are grazed by goats, the dye is deposited in their bones, staining them bright orange.

The bush can survive extreme drought (2 years without rainfall, as during 1974–1975), but it is in danger of extinction within a few years if no steps are taken to protect it from overgrazing and to set aside some of the crop for seeding purposes.

NODULATION STUDIES

According to Dr. Marco Paolo Nuti of Pisa University (personal communication 1978) the roots penetrate the ground to 3 m. They show small secondary rhizomes near the surface and root nodules on the younger roots. Rhizobia are not recorded to have been isolated.

Cordyla Lour.   Caesalpinioideae: Swartzieae
115 / 3573

From Greek, *kordyle*, "club," in reference to the long-stalked ovary that projects beyond the stamens.

Type: *C. africana* Lour.

5 species. Trees, usually tall, unarmed. Leaves deciduous, imparipinnate; leaflets alternate or subopposite, petioluled, glabrous or slightly pubescent, usually 9–10 pairs, ovate-oblong, marked with many translucid dots or lines; stipules caducous, linear-lanceolate. Flowers in short, dense, softly woolly-tomentose, axillary racemes or subfascicled at nodes on older twigs before the leaves reappear; bracts and bracteoles minute, linear, caducous; calyx turbinate, entire in bud, splitting into 3–5 lobes as the flower expands, often somewhat leathery; petals none; stamens many, often 100–300, more or less in 4 ranks, free or slightly connate at the base forming a ring in the throat of the calyx tube, the stamens of the 2 exterior rows fertile and with long filaments, anthers dorsifixed and longitudinally dehiscent, the stamens of the 2 interior rows staminoidal, sterile, anthers absent, much shorter and often united *inter se* in groups of 2–4; ovary long-stalked, many-ovuled, in 2 rows; style short, conical; stigma small, terminal. Pod large, drupaceous, ellipsoid, oblong, or cylindric, stalked, apiculate, pericarp leathery, brittle and rigid when dry, pulpy inside; seeds usually 4, sometimes 6, subelliptic. (Baker 1930; Milne-Redhead 1937; Capuron 1968b)

Members of this genus are widespread in tropical West Africa, Sudan, Zimbabwe, Natal, and Mozambique. They occur in hot rift valleys, fertile alluvial soil, and on rocky hillsides.

The yellow fruit of *C. africana* Lour. has soft, edible pulp, rich in ascorbic acid (Carr 1957). It is similar in size to a mango or a lemon, and is locally called bush mango. Elephants and other wild animals relish the fruit of *C. richardii* Planch. ex Milne-Redhead, but its unpleasant odor renders it less popular for human consumption.

The wood of these tree species is yellowish, durable, and suitable for general carpentry, joinery, and plywood. Trunks of large, old, hollow trees are prized for making drums and dug-out canoes.

NODULATION STUDIES

In Zimbabwe, *C. africana*, the only species of the genus examined to date, lacked nodules (Corby 1974).

Corethrodendron Fisch. & Basiner	Papilionoideae:
314 / 3782	Hedysareae, Euhedysarinae

From Greek, *korethron*, "broom," and *dendron*, "tree."
Type: *C. scoparium* Fisch. & Basiner
1 species. Shrub, gray-pubescent. Lower leaves imparipinnate; leaflets entire; upper petioles leafless, spinescent; stipels absent; stipules 2, united, leaf-opposed, caducous. Flowers purple (?), in axillary, long-stalked racemes; bracts small, caducous; bracteoles minute; calyx lobes subequal, upper 2 convergent; standard broad-obovate, short-clawed; wings short; keel blunt, incurved, subequal to standard; stamens diadelphous; anthers uniform; ovary stalked, multiovuled; style filiform, incurved; stigma small, terminal. Pod linear, subcylindric, jointed, segments ovate, indehiscent after breaking away; seeds kidney-shaped. (Bentham and Hooker 1865)
This monotypic genus is endemic in central Asia.

Cornicina Boiss.	Papilionoideae: Loteae

From Latin, *corniculus*, "little horn," in reference to the shape of the pod.
Type: *C. hamosa* (Desf.) Boiss.
1 species. Herb, annual, stems ascending to erect, hairy. Leaves sessile, imparipinnate; leaflets narrow, elliptic, up to 5 pairs, basal pair in lieu of stipules. Flowers yellow, 8–10, in axillary, solitary, peduncled heads; calyx long-tubular, curved, persistent, not inflated; lobes 5, unequal, awl-shaped; standard truncate, notched at the base, abruptly narrowed into a claw; sta-

mens diadelphous; filaments free above and widened; style glabrous. Pod reflexed and up-curved, with a long beak exserted from the calyx, not winged, pericarp hard, transversely septate, several-seeded, indehiscent. (Hutchinson 1964)

The genus is closely related to and often included in *Anthyllis*.

This monotypic genus is endemic in the Iberian Peninsula and adjacent northwestern Africa.

Coronilla L.   Papilionoideae: Hedysareae, Coronillinae
306 / 3774

From Latin diminutive of *corona*, meaning "little crown," in reference to the flower heads.

Type: *C. varia* L.

55 species. Herbs, annual or perennial, or low shrubs; stems ascending, usually glabrous, sometimes silky-pubescent. Leaves imparipinnate, rarely 1–3-foliolate; leaflets small, 3–12 pairs, bluish green, entire, oblong; stipules large and leaflike or small, free or connate. Flowers umbellate, yellow, rose, rarely purple, white, or streaked, pedicelled, on long axillary peduncles; bracts small; bracteoles absent; calyx short-campanulate, more or less 2-lipped; teeth 5, short, subequal, ovate-oblong, 2 upper teeth somewhat connate; petals 5, long-clawed; standard suborbicular; wings obliquely obovate-oblong; keel incurved, beaked; stamens 10, diadelphous; alternate or all filaments dilated above; anthers uniform; ovary sessile, many-ovuled; style incurved, awl-shaped, glabrous; stigma small, capitate. Pod lomentaceous, straight or curved, cylindric, 4-angled or ridged longitudinally, transversely jointed but not constricted between the 1-seeded, subcylindric, indehiscent segments; seeds transversely oblong. (Bentham and Hooker 1865; Gams 1924)

Members of this genus are well represented in the Mediterranean region and also in northern and central Europe, northern Asia, and Africa. They inhabit roadsides, waste areas, drylands, grasslands, and scrub on calcareous soils.

C. varia L., crownvetch, is important as a weed suppressor and as a cover plant on highway embankments and strip-mined areas, and in erosion control (Ruffner and Hall 1963). It is a hardy, long-lived, deep-rooted, semireclining herbaceous perennial that is spread by seed or by rhizomes. Nonpalatability of the forage and inability of the plants to reestablish themselves after being cut or plowed are the major shortcomings in the use of crownvetch. However, in the Soviet Union this species is cultivated for fodder.

C. emerus L., *C. scorpioides* (L.) K. Koch and *C. varia* have ornamental value and are often found on banks and hillsides in gardens and parks. The forage value of *C. emerus* and *C. scorpioides* is diminished by the presence of coronillin, $(C_7H_{12}O_5)_x$, a toxic glucoside, in the leaves and flowers. The alkaloid cytisine, $C_{11}H_{14}N_2O$, is found in the seeds of *C. varia* (White 1951). Native medicinal uses of plant parts are for diuretic purposes and as cardiac tonics.

Nodulation Studies

Crownvetch rhizobia are monotrichously flagellated short rods. They are slow growing, and produce small, compact colonies on solid media and an alkaline reaction in litmus milk without serum zone formation.

Early investigators (Maassen and Müller 1907) reported that crownvetch rhizobia symbiosed only with the homologous host species. This opinion prevailed for about 3 decades (Fred et al. 1932). Wilson's results (1939a) with *C. emerus, C. glauca* L., and *C. varia* as macrosymbionts lacked confirmation *in toto*, but nonetheless indicated a restricted susceptibility range to rhizobia. These results were confirmed in principle by Schreven (1972). Crownvetch nodulation was obtained only with rhizobia from crownvetch and sainfoin, *Onobrychis vicifolia.* The presence of crownvetch rhizobia in 32 of 57 soil samples collected in 7 provinces of the Netherlands is noteworthy.

Young globular nodules of *Coronilla* species evolve into clustered, bifurcated shapes. The nodule meristems are localized apically. Rhizobial invasion and dissemination in 3 species of *Coronilla* were described by Lechtova-Trnka (1931). Rhizobia within the infection threads were rod-shaped, but upon release into the host cytoplasm of *C. glauca* and *C. scorpioides* they assumed swollen club and pear shapes. Enlarged spherical bacteroids were dominant in nodules of *C. iberica* Bieb. Each spherule contained 4–8 minute unstained areas that were interpreted to be vacuoles. Electron microscopic studies would probably determine whether these structures are groupings of rhizobia within a membrane (Goodchild and Bergersen 1966) or a distinctive type of bacteroid *per se.*

Nodulated species of *Coronilla*	Area	Reported by
cappadocica Willd.	S. Africa	N. Grobbelaar (pers. comm. 1965)
coronata L.	U.S.S.R.	Lopatina (1931)
cretica L.	Widely reported	Many investigators
emerus L.	N.Y., U.S.A.	Wilson (1939a)
	S. Africa	Grobbelaar & Clarke (1974)
glauca L.	Widely reported	Many investigators
iberica Bieb.	France	Lechtova-Trnka (1931)
montana Scop.	S. Africa	N. Grobbelaar (pers. comm. 1965)
pentaphylla Desf.	France	Naudin (1897b)
repanda Boiss.	Israel	M. Alexander (pers. comm. 1963)
scorpioides (L.) K. Koch	Widely reported	Many investigators
(*vaginalis* Lam.)	Kew Gardens, England	Pers. observ. (1936)
= *Artrolobium vaginalis* (Lam.) Desv.		
valentina L.	Libya	B. C. Park (pers. comm. 1960)
	S. Africa	Grobbelaar & Clarke (1974)
varia L.	Widely reported	Many investigators

Coroya Pierre   Papilionoideae: Dalbergieae, Pterocarpinae
(350a) / 3821a

Type: *C. dialoides* Pierre

1 species. Tree, small; young twigs pubescent, later smooth. Leaves imparipinnate; leaflets alternate, 3–6, ovate to elliptic, terminal leaflet largest; stipels absent; stipules caducous. Flowers very small, densely clustered in pubescent, terminal, cymose panicles; bracts and bracteoles lanceolate, deciduous; calyx lobes 5, pubescent, front lobe larger; standard obcordate, larger than the other petals; keel petals entirely free, elliptic, blunt; stamens diadelphous; anthers ellipsoid, dorsifixed; ovary stalked, hairy, 1-ovuled; style very short. Pod not described. (Harms 1908a)

Plant characters resemble those of *Dalbergia* and *Pterocarpus*.

This monotypic genus is endemic in Vietnam.

Corynella DC.   Papilionoideae: Galegeae, Robiniinae
(275) / 3740

From Greek *koryne*, with Latin diminutive suffix; name means "little club." The name was recorded *Corynitis* Spreng. (=*Corynella* DC.) by Taubert (1894); priority of *Corynella* was later recognized by Harms (1897).

Lectotype: *C. polyantha* (Sw.) DC.

6 species. Low shrubs. Leaves paripinnate; leaflets leathery; stipels minute; stipules awl-shaped, firm, erect, sometimes spinescent. Flowers purplish, fascicled on short, axillary branches before leafing, or in clusters on older nodes; bracts small; bracteoles none; calyx campanulate, subbilabiate; lobes 5, very short or long and awl-shaped, upper 2 connate above; petals short-clawed; standard suborbicular, reflexed; wings free, oblique-oblong, conspicuously eared, claws curved; keel petals slightly incurved, obtuse, longer than the wings and standard; stamens 10, subequal, diadelphous; filaments subequal; anthers uniform; ovary stalked, many-ovuled; style glabrous, thickened above, incurved, strongly hooked at the apex; stigma subterminal. Pod linear, lanceolate, slightly stalked, plano-compressed, many-seeded. (De Candolle 1825c; Bentham and Hooker 1865; Rydberg 1924)

Members of this genus are native to the West Indies.

Coursetia DC.   Papilionoideae: Galegeae, Robiniinae
278 / 3744

Named for George Dumont de Courset, author of *Botaniste Cultivateur*, Paris, 1802; his botanical garden displayed plants of central and southern Europe.

Type: *C. tomentosa* (Cav.) DC.

25 species. Shrubs or small trees, up to about 6 m tall, unarmed. Leaves paripinnate or imparipinnate; leaflets numerous, entire, thin, oval or lanceo-

late, petioluled; stipels rudimentary or absent; stipules awl-shaped, persistent, sometimes spiny. Flowers small, white tinged with rose or purple, some with yellow centers, in axillary or terminal racemes, tomentose or glandular; bracts small, caducous; bracteoles none; calyx campanulate or turbinate, about as broad as high; lobes 5, long, subequal, upper 2 joined above midway; petals equally long; standard broad-orbicular, short-clawed, edges reflexed, sometimes with 2 callosities; wings oblique-oblanceolate, with a basal ear, short-clawed; keel obtuse to acuminate, short-beaked, with a small basal ear, longer clawed than the wings; stamens diadelphous, vexillary stamen free or connate with others in the middle; anthers uniform; ovary sessile, several-ovuled; style stiff, inflexed at base, hairy above the middle; stigma capitate. Pod sessile or short-stalked, slender, linear, flattened, some covered with rust-colored hairs, 2-valved, margins usually constricted around the seeds and somewhat torulose but continuous within, lower portion sometimes narrow and empty, dehiscent; seeds subcircular, many, black. (De Candolle 1825c; Bentham and Hooker 1865; Rydberg 1924)

Representatives of this genus are indigenous to the southwestern United States, Mexico, and from Central America to Brazil and Peru. They thrive in canyons, on dry rocky slopes, and in thickets and woodlands up to 1,000-m altitude.

The plants of this genus lack commercial possibilities. The foliage is reasonably good browse.

C. glandulosa Gray and *C. microphylla* Gray are hosts for lac insects of the genus *Tachardia*. The transparent yellow or orange gum produced on the branches by the insect infestation is sold in pharmacies in Mexico under the name "goma Sonora." This gum is dissolved in water for cough syrups; it was once thought to be a remedy for tuberculosis. It is also an ingredient of a high-grade varnish. *C. madrensis* M. Micheli, *C. mollis* Robins. & Greenm., and *C. polyphylla* Brandegee are shrubs frequently found in Mexico.

The tallest tree species, *C. arborea* Griseb., is widely distributed in the West Indies, Panama, and Venzuela. Its wood is of good quality, hard, and heavy, and has a density of about 960 kg/m^3. It is straight grained, fine textured, and lustrous, but it is not available in quantity or in large pieces. The heartwood is not differentiated from the sapwood. The bright yellow color of the wood renders it attractive for small turnery objects (Record and Hess 1943).

NODULATION STUDIES

Round, medium-sized, brown nodules were observed by DeSouza (1963, 1966) on mature trees of *C. arborea* in Trinidad.

From Latin, *cracca*, the name of a vetch. *Cracca* Benth. is conserved over *Benthamantha* and not to be confused with the homonyms *Cracca* L. or *Cracca* (Riv.) Medik., synonyms for *Tephrosia* and *Vicia*, respectively.

Lectotype: *C. glandulifera* Benth.

10 species. Perennial herbs or low, slender shrubs, unarmed. Leaves imparipinnate; leaflets many, small, oval; stipels minute; stipules filiform, free, setaceous. Flowers small, yellow or white, in short, axillary racemes; bracts awl-shaped or setaceous, caducous; bracteoles absent; calyx campanulate; lobes 5, subequal, subulate-acuminate; upper 2 united to the middle; petals subequal; standard orbicular or kidney-shaped, clawed, sides reflexed; wings oblong-obovate, clawed, free, reflexed; keel petals obovate, beaked, inflexed, united to apex; stamens diadelphous; ovary sessile, many-ovuled; style inflexed, upper part bearded inside; stigma capitate. Pod linear, flat, transversely impressed between the seeds, septate within, 2-valved; seeds numerous, tetragonal. (Bentham and Hooker 1865)

Cracca species are native to tropical and warm temperate America. They flourish in thin forests, moist or dry thickets, and open areas up to 2,000-m altitude.

Cracca species lack special interest (Standley and Styermark 1946). Most of them are considered weeds, but *C. glabrescens* Benth. is occasionally used as a forage plant (Burkart 1952). *C. edwardsii* Gray is heavily grazed in Arizona.

Nodulation Studies

Round to elongated nodules were observed on *C. caribaea* (Jacq.) Benth. in Puerto Rico (Dubey et al. 1972). A rhizobial strain from *C. cathartica* Rydb. was listed in the *Rhizobium* culture collection of the United States Department of Agriculture in the 1930s (L. T. Leonard personal communication 1935). Nodules were lacking on *C. edwardsii* (as *Benthamantha edwardsii* Rose) examined by Martin (1948) in Arizona.

Craibia Harms & Dunn   Papilionoideae: Galegeae,
(257b)   Tephrosiinae

Named for William G. Craib, Assistant for India, of the Kew Gardens staff, and later Professor of Botany at Aberdeen, Scotland.

Lectotype: *C. brevicaudata* (Vatke) Dunn

10 species. Trees, or shrubs, usually with golden brown or black woolly hairs, especially on leaf rachis, calyx, and ovaries; youngest branches sparsely

pilose. Leaves imparipinnate, rarely simple; buds conspicuous, often with broad scales; leaflets alternate, few; stipels usually absent. Flowers mostly white or pink, showy and fragrant, in terminal or axillary racemes, or paniculate; pedicels short, usually about 5 mm; bracts and bracteoles caducous; calyx usually cupular; lobes broad, shorter than the tube, upper pair joined; standard round, short-clawed; wings and keel petals oblong; stamens diadelphous; anthers similar; ovary sessile or short-stalked, 2–6-ovuled, lower part usually empty; disk absent; style inflexed; stigma small, capitate. Pod short-stalked, oblong-obovate, flat, beaked, tapering at the base, valves leathery, twisting, dehiscent; seeds 1–3, asymmetrically oval, black or dark brown, hilum short, surrounded by a short, white, cupular strophiole. (Dunn 1911a; Gillett 1960a)

This genus, positioned by Harms (1915a), is closely allied to *Millettia*.

Members of this genus are indigenous to equatorial Africa. They grow in rain forests, grasslands, rocky gullies, and hillsides up to 2,000-m altitude.

These slow-growing trees reach 15 m in height; they have exceptional beauty when in full bloom. Their strongly scented, creamy white flowers are conspicuous against the light green foliage. The wood of *C. grandiflora* (M. Micheli) Harms, which is almost white, is used for furniture but not on a commercial scale. The sap of this species is considered to have aphrodisiac properties. The fruit of *C. affinis* De Wild. is edible.

NODULATION STUDIES

Nodules on *C. brevicaudata* (Vatke) Dunn ssp. *baptisarum* were observed in Zimbabwe by Corby (1974).

Cranocarpus Benth.   Papilionoideae: Hedysareae,
345 / 3815   Desmodiinae

From Greek, *kranion*, "skull," and *karpos*, "fruit," in reference to the helmet-shaped pod.

Type: *C. martii* Benth.

2 species. Shrubs. Leaves 1-foliolate, large; 2 small lateral leaflets sometimes present; stipels 2, awl-shaped; stipules free, setaceous-acuminate. Flowers yellow or white, in axillary racemes; pedicels solitary, bibracteolate; calyx oblique; teeth subequal in length, upper 2 broader; standard ovate, reflexed, long-clawed; wings slanted, oblong, free; keel hood-shaped, strongly inflexed; stamens monadelphous, connate into a sheath open above; ovary short-stalked, 1–2-ovuled; style slender, incurved; stigma terminal, small. Pod stalked, compressed, helmet-shaped, glandular-haired, upper suture indented in the middle, lower suture arched, indehiscent; seeds subreniform. (Bentham and Hooker 1865)

Schindler (1924) preferred the positioning of this genus in Hedysareae, subtribe Aeschynomeninae, on the basis of pod structure and monadelphous stamens. Hutchinson (1964) placed *Cranocarpus* in Lespedezeae trib. nov.

Both species occur in Brazil.

From Greek, *kraspedon*, "edge," "border," and *lobion*, "pod"; the pod has a narrow ventral margin.

Type: *C. schochii* Harms

1 species. Shrub, scandent. Leaves 3-foliolate. Flowers violet, small, short-pedicelled in rounded heads or long spikes; calyx cupular, pubescent; teeth subequal, lower lanceolate-ovate, lateral deltoid, acute, uppermost broad, emarginate; standard ovate or obovate-suborbicular, slightly eared at the base, short-clawed; wings oblique, narrow-oblong, obtuse, base shortly dentate; keel suberect, apex downcurved, base denticulate; stamens 10, diadelphous; ovary short-stalked, 5–8-ovuled, linear, pubescent; style smooth, curved; stigma minutely capitulate. Pod sessile, flattened, linear-lanceolate, apiculate, papery, pubescent, with a narrow ventral margin; seeds 3–5. (Harms 1921a)

This monotypic genus is endemic in western China. The plants occur up to 2,000-m altitude.

Cratylia Mart. ex Benth.   Papilionoideae: Phaseoleae,
409 / 3885   Diocleinae

Lectotype: *C. hypargyrea* Mart. ex Benth.

5–8 species. Shrubs, tall, climbing. Leaves large, pinnately 3-foliolate; leaflets acuminate; stipels present; stipules small. Flowers large, white to rose violet, loosely spaced in long, axillary, erect racemes on nodose rachis; bracts caducous; bracteoles small, caducous; calyx campanulate, often silky-pilose outside; lobes 5, upper 2 wide, obtuse, joined into 1, entire or emarginate, lower 3 triangular, free; petals clawed; standard large, orbicular, reflexed, without ears or basal callosities; wings ovate, free, reflexed; keel petals about the same length as the wings, oblong, broad, obtuse, short, inflexed; stamens 10, fertile, vexillary stamen dilated at the base and joined at the middle with the other 9; anthers oval, uniform; ovary stalked, linear, many-ovuled, with a tubular, basal disk; style incurved, glabrous or with dorsal hairs; stigma terminal, capitate, glabrous. Pod oblong-linear, compressed, flattened, 2-valved; seeds oblong, compressed, transversely positioned, and thinly separated within the pod. (Bentham and Hooker 1865, Burkart 1943)

Members of this genus are native to South America, from northern Argentina to northern Brazil.

These high-climbing shrubs are of no particular use, but *C. floribunda* Benth. merits consideration as a cultivated ornamental.

NODULATION STUDIES

Nodules were observed on *C. floribunda* in Brazil by Galli (1958). Rhizobia from this species lacked ability to form nodules on species of 17 genera, including members of the cowpea miscellany. The rhizobia were slow growing, produced scanty growth on solid media, an alkaline reaction in bromcresol purple milk, and acid reactions on mannose and rhamnose media (Galli 1959).

From Greek, *krotalon*, or Graeco-Latin, *crotalum*, "rattle," in reference to the castanet sound of the loose seeds when the dry pods are shaken.

Type: *C. laburnifolia* L.

550 species. Herbs, annual or perennial, or shrubs, mostly erect, rarely spinescent. Leaves unifoliate or palmately 3-foliolate, rarely 5- or 7-foliolate, sessile or on long petioles; stipels absent; stipules usually present, free from the petiole, sometimes conspicuous, decurrent along the stem. Flowers usually yellow, occasionally streaked with red or purple, often showy, in terminal or leaf-opposed racemes, rarely solitary, often opening in the afternoon; bracts small, rarely foliaceous; bracteoles small, rarely none; calyx often obliquely campanulate; lobes 5, subequal, usually free, rarely with the upper 2 forming a lip and the lower 3 more or less connate; standard large, orbicular, rarely ovate, longer than the wings and keel, often callous above the short claw; wings obovate or oblong; keel scythe-shaped, incurved, beaked; stamens 10, monadelphous, all joined below into a sheath split above; anthers unequal, small and versatile alternate with long and basifixed; ovary sessile or short-stalked, 2- to many-ovuled; style strongly incurved, with linear pubescence above. Pod subcylindric, globose or oblong, turgid or inflated, continuous within, 2-valved, dehiscent; seeds small, several to many, rarely only 1, loose and usually rattling in the dry, leathery pod at maturity. (Bentham and Hooker 1865; Amshoff 1939; Senn 1939; Standley and Steyermark 1946; Phillips 1951; Munk 1962; Backer and Bakhuizen van den Brink 1963)

According to Polhill (1968b, 1976) the lectotype is *C. lotifolia* L. Both he and Hutchinson (1964) assigned the genus to the tribe Crotalarieae, segregated from Genisteae *sensu lato*.

Representatives of this large genus are well distributed in tropic and warm areas of both hemispheres. There are about 400 species in Africa, 35 in Indo-Malaya, and 15 in Australia. They thrive on dry, sandy soils, stony hills, grassy banks, and sandy coastal areas. A few species occur in moist meadows and valleys.

C. alata Buch.-Ham., *C. anagyroides* HBK, *C. ferruginea* Grah., *C. goreensis* Guill. & Perr., *C. intermedia* Kotschy, *C. juncea* L., *C. laburnifolia* L., *C. sericea* Retz. (= *C. spectabilis* Roth), and *C. usaramoensis* Bak. f. have widespread use as green manures and fodder crops. Tall shrub species have value as nurse crops for coffee and cacao and for terracing tea. *C. sericea* is a favorite in tung and citrus groves to provide shade, control weeds, and as a turn-under crop for soil enrichment. *C. juncea*, sunn hemp, is perhaps the fastest-growing of all *Crotalaria* species and, accordingly, has been found satisfactory as a green-manuring crop in rotation with potatoes, tobacco, and wheat, and especially for soil-improvement purposes in sugar cane and pineapple culture. The foliage is optimum for turn-under in 8–10 weeks. The highest yields of maize and rice in Java field trials followed the turn-under of *C. anagyroides* and *C. juncea* (Goor 1954).

The stems and bark of *C. burhia* Buch.-Ham., *C. cannabina* Schweinf., *C. intermedia*, *C. juncea*, *C. retusa* L., and *C. saltiana* Andr. are sources of a strong fiber. The bast fiber of *C. juncea*, sunn hemp, is rated second to that of jute, *Corchorus capsularis*, in tensile strength and durability; however, its strength is not equal to that of *Cannabis sativa*. In India, sunn hemp fiber is used for cot strings, rope, sacking, floor matting, fish nets, marine cordage, and twine. The fiber has good pulping characteristics (Nelson et al. 1961). Because of its high cellulose and low ash content, the bast fiber is also used in the manufacture of cigarette paper and high-quality tissue paper (White and Haun 1965). During World War II it was in great demand for making camouflage nets.

Maladies of cattle, sheep, and horses resulting from ingestion of *Crotalaria* foliage are generally termed some form of acute or chronic crotalism. Presumably, this term was first applied to a disease of horses that was characterized by excessive elongation of the hooves and sensitivity to pressure and warmth and was accentuated by pain and locomotor impediment (Stalker 1884). Crotalism was known in early days throughout the central plains of the United States as the Missouri river bottom disease; it was so common that farmers did not know from week to week whether a team would be available for farm work (Pammel 1911). *C. sagittalis* L. was proved to be the cause. Numerous other *Crotalaria* species have been held responsible for the poisoning of livestock, poultry, and swine (Theiler 1918; Steyn 1934; Webb 1948; Watt and Breyer-Brandwijk 1962; Verdcourt and Trump 1969).

Monocrotaline, $C_{16}H_{23}NO_6$, was isolated from seeds, leaves, and stems of *C. spectabilis* by Neal et al. (1935). This initial isolation of a specific alkaloid from a *Crotalaria* species followed in the wake of one acute and three chronic cases of poisoning in cattle (Becker et al. 1935). In the ensuing years more than 40 pyrrolizidine-type alkaloids, characteristically causing severe liver damage, were isolated from diverse *Crotalaria* species (see Harborne et al. 1971). While all of these alkaloids are hepatotoxic, it appears that monocrotaline, retrorsine, $C_{18}H_{25}NO_6$, and retronecine, $C_8H_{13}NO_2$, are the most potent. Cox et al. (1958) provided unequivocal proof of the stability and toxicity of monocrotaline by isolating it from cirrhotic livers of horses that had died from eating *C. spectabilis*.

Various *Crotalaria* species tolerate heavy metals. *C. cobalticola* Duvign. & Plancke flourishes on rubbish and excavation sites of old copper mines in the Katanga Province, Zaire. Because these soils contain about 2,000 ppm cobalt, Duvigneaud (1959) termed this species a cobaltophyte. Plant tissue showed 500 ppm cobalt on a dry weight basis, an equivalent of 2 percent cobalt on a dry ash basis. Inasmuch as excessive cobalt interferes with iron metabolism essential for chlorophyll formation, chlorosis has been used to advantage in cobalt prospecting. Relatedly, *C. florida* Welw. ex Bak. var. *congolensis* (Bak. f.) Wilczek was termed a manganophile because of its growth on manganese-rich soil. *C. cornettii* Taub. & Dewevre, *C. dilolensis* Bak. f., *C. françoisiana*

Duvign. & Timp., and *C. peschiana* Duvign. & Timp. are cuprophytes. These species have practical value in copper prospecting (Duvigneaud and Timperman 1959).

NODULATION STUDIES

Crotalaria rhizobia are monotrichously flagellated, slow growing, and produce an alkaline reaction without serum zone formation in litmus milk. Nodules develop apically and become cylindrical with branched pink tips. In old specimens the tips develop multilobes. Bacteroidal areas abound in leghemoglobin; ineffective nodules are uncommon. Host cell infection in the nodule approaches 100 percent (personal observation 1938; Rothschild 1963). Apparently spread of infection occurs by means of host cell mitotic divisions (Arora 1956b).

Early infection tests indicated that *C. juncea* was host-specific, whereupon it was designated as group XIV of the 18 distinctive rhizobia:plant groups then recognized (Walker 1928); however, this conclusion no longer prevails. Consistent reciprocal crossings between cowpea and 22 *Crotalaria* species were reported by Carroll (1934a); furthermore, precipitin tests showed close kinships of seed proteins of 15 *Crotalaria* species with cowpea but not with pea and clover (Carroll 1934b). Subsequent results confirm the inclusion of *Crotalaria* species in cowpea miscellany (Allen and Allen 1936a, 1939; Conklin 1936; Burton 1952). Considerable promiscuity in symbiotic affinity was shown by Wilson (1939a, 1944b, 1945a).

Nodulated species of *Crotalaria*	Area	Reported by
abbreviata Bak. f.	Zimbabwe	Corby (1974)
acicularis Buch.-Ham. ex Benth.	Java	Keuchenius (1924)
agatiflora Schweinf.	S. Africa	Grobbelaar et al. (1967)
alata Buch.-Ham.	Widely reported	Many investigators
alexandri Bak. f.	Zimbabwe	Corby (1974)
allenii Verdoorn	S. Africa	Grobbelaar & Clarke (1975)
anagyroides HBK	Widely reported	Many investigators
anisophylla Welw. ex Bak.	Zimbabwe	Corby (1974)
anthyllopsis Welw. ex Bak.	Zimbabwe	Corby (1974)
argyrea Welw.	S. Africa	N. Grobbelaar (pers. comm. 1974)
assamica Benth.	Hawaii, U.S.A.	Allen & Allen (1936a)
astragalina Hochst.	Hawaii, U.S.A.	Allen & Allen (1936a)
australis Bak. f.	S. Africa	Grobbelaar & Clarke (1974)
axillaris Dryand.	Fla., U.S.A.	Carroll (1934a)
barkae Schweinf.	Zimbabwe	Corby (1974)
barnabassii Dinter ex Bak. f.	Zimbabwe	Corby (1974)
bequaertii Bak. f.	Zimbabwe	Corby (1974)
brachycarpa Burtt Davy	S. Africa	Grobbelaar & Clarke (1975)

Nodulated species of *Crotalaria* (cont.)	Area	Reported by
bracteata Roxb.	Philippines	Bañados & Fernandez (1954)
(*brownei* Bert. ex DC.)		
=*saltiana* Andr.	Sri Lanka	Paul (1949)
	Trinidad	DeSouza (1963)
burhia Buch.-Ham.	Pakistan	A. Mahmood (pers. comm. 1977)
burkeana Benth.	S. Africa	Grobbelaar & Clarke (1975)
capensis Jacq.	S. Africa	Grobbelaar & Clarke (1972)
caudata Welw. ex Bak.	Zimbabwe	Corby (1974)
cephalotes Steud. ex A. Rich.	S. Africa	Grobbelaar et al. (1967)
chirindae Bak. f.	Zimbabwe	Corby (1974)
cleomifolia Welw. ex Bak.	Zimbabwe	Corby (1974)
collina Polh.	Zimbabwe	H. D. L. Corby (pers. comm. 1973)
comosa Bak.	S. Africa	Grobbelaar & Clarke (1974)
cunninghamii R. Br.	Hawaii, U.S.A.	Allen & Allen (1936a)
	Australia	Beadle (1964)
cylindrostachys Welw. ex Bak.	Zimbabwe	Corby (1974)
damarensis Engl.	S. Africa	Grobbelaar et al. (1967)
	Zimbabwe	Corby (1974)
deserticola Taub. ex Bak. f.	Zimbabwe	Corby (1974)
dissitiflora Benth.	Australia	Beadle (1964)
distans Benth.	S. Africa	Grobbelaar et al. (1967)
	Zimbabwe	Corby (1974)
doidgeae Verdoorn	S. Africa	Grobbelaar & Clarke (1975)
dura Wood & Evans	S. Africa	Grobbelaar & Clarke (1975)
erecta Pilg.	S. Africa	U. of Pretoria (1954)
eremicola Bak. f.	S. Africa	Grobbelaar & Clarke (1975)
eriocarpa Benth.	Mexico	Pers. observ. (U.W. herb. spec. 1971)
falcata Vahl ex DC.	Widely reported	Many investigators
filicaulis Welw. ex Bak.	Zimbabwe	Corby (1974)
filipes Benth.	W. India	Bhelke (1972)
flavicarinata Bak. f.	Zimbabwe	Corby (1974)
fulva Roxb.	Hawaii, U.S.A.	Allen & Allen (1936a)
gazensis Bak. f.		
ssp. *gazensis*	Zimbabwe	Corby (1974)
glauca Willd.	Zimbabwe	Corby (1974)
globifera E. Mey.	S. Africa	Grobbelaar & Clarke (1972)
goreensis Guill. & Perr.	Widely reported	Many investigators
grantiana Harv.	Widely reported	Many investigators
heidmannii Schinz	Zimbabwe	Corby (1974)
hildebrantii Vatke	Widely reported	Many investigators

Nodulated species of *Crotalaria* (cont.)	Area	Reported by
hispida Schinz	S. Africa	Grobbelaar & Clarke (1975)
hyssopifolia Klotz.	Zimbabwe	Corby (1974)
incana L.	Widely reported	Many investigators
ssp. *purpurascens* (Lam.) Milne-Redhead	Zimbabwe	Corby (1974)
insignis Polh.	Zimbabwe	H. D. L. Corby (pers. comm. 1973)
intermedia Kotschy	Widely reported	Many investigators
juncea L.	Widely reported	Many investigators
kapirensis De Wild.	Zimbabwe	Corby (1974)
kipandensis Bak. f.	Zimbabwe	Corby (1974)
laburnifolia L.	Widely reported	Many investigators
ssp. *australis* (Bak. f.) Polh.	Zimbabwe	Corby (1974)
	S. Africa	Grobbelaar & Clarke (1975)
ssp. *laburnifolia*	Zimbabwe	Corby (1974)
lachnocarpoides Engl.	Zimbabwe	Corby (1974)
lachnophora Hochst. ex A. Rich.	Zimbabwe	Corby (1974)
	S. Africa	Grobbelaar & Clarke (1974)
lanceolata E. Mey.	Widely reported	Many investigators
ssp. *lanceolata*	Zimbabwe	Corby (1974)
ssp. *prognatha* Polh.	Zimbabwe	Corby (1974)
linifolia L. f.	Australia	Bowen (1956)
longirostrata H. & A.	Hawaii, U.S.A.	Allen & Allen (1936a)
longithyrsa Bak. f.	Zaire	Bonnier (1957)
lotoides Benth.	S. Africa	Grobbelaar et al. (1964, 1967)
lupulina HBK	Ariz., U.S.A.	Martin (1948)
macaulayae Bak. f.	S. Africa	Grobbelaar et al. (1967)
macrocarpa E. Mey.	S. Africa	Grobbelaar et al. (1967)
macrotropis Bak. f.	Zimbabwe	Corby (1974)
maxillaris Klotz.	Widely reported	Many investigators
maypurensis HBK	Venezuela	Barrios & Gonzalez (1971)
medicaginea Lam.	Pakistan	A. Mahmood (pers. comm. 1977)
meyerana Steyd.	S. Africa	N. Grobbelaar (pers. comm. 1974)
microcarpa Hochst. ex Benth.	Zimbabwe	Corby (1974)
monteiroi Taub. ex Bak. f.		
var. *galpinii*	Zimbabwe	Corby (1974)
(*mucronata* Desv.)	Philippines	Bañados & Fernandez (1954)
= *pallida* Ait.		
(*musijusii* [?])	Sri Lanka	Holland (1924)
= *usaramoensis* Bak. f.	Java	Keuchenius (1924)
(*mundyi* Bak.)	Widely reported	Many investigators
= *macrotropis* Bak. f.		
mysorensis Roth	Widely reported	Many investigators
nana Burm. f.	W. India	Bhelke (1972)

Nodulated species of *Crotalaria* (cont.)	Area	Reported by
natalitia Meissn.	Hawaii, U.S.A.	Allen & Allen (1936a)
var. *natalita*	Zimbabwe	Corby (1974)
var. *rutshuruensis* De Wild.	Zimbabwe	Corby (1974)
obscura DC.	S. Africa	Grobbelaar et al. (1967)
ochroleuca G. Don	Widely reported	Many investigators
oocarpa Bak.	Fla., U.S.A.	Carroll (1943a)
orixensis Willd.	W. India	V. Schinde (pers. comm. 1977)
orthoclada Welw.	S. Africa	Grobbelaar & Clarke (1974)
pallida Ait.		
var. *pallida*	Zimbabwe	Corby (1974)
pallidicaulis Harms	Zimbabwe	Corby (1974)
paulina Schrank	S. Africa	U. of Pretoria (1954)
phylicoides Wild	Zimbabwe	Corby (1974)
pilosa Mill.	Trinidad	DeSouza (1966)
	Argentina	Rothschild (1970)
pisicarpa Welw.	Zimbabwe	Corby (1974)
	S. Africa	Grobbelaar & Clarke (1975)
platysepala Harv.	Zimbabwe	Corby (1974)
podocarpa DC.	Zimbabwe	Corby (1974)
polysperma Kotschy	Widely reported	Many investigators
pseudo-eriosema Vatke	Java	Keuchenius (1924)
quinquefolia L.	Philippines	Bañados & Fernandez (1954); Pers. observ. (1962)
recta Steud. ex A. Rich.	S. Africa	U. of Pretoria (1955–1956)
	Zimbabwe	Corby (1974)
reptans Taub.	Zimbabwe	Corby (1974)
retusa L.	Widely reported	Many investigators
rhodesiae Bak. f.	S. Africa	N. Grobbelaar (pers. comm. 1963)
	Zimbabwe	Corby (1974)
rogersii Bak. f.	Zimbabwe	Corby (1974)
rotundifolia (Walt.) Poir.	Fla., U.S.A.	Carroll (1934a)
sagittalis L.	Widely reported	Many investigators
saltiana Andr.	N.Y., U.S.A.	Wilson (1944b)
schinzii Bak. f.	Zimbabwe	Corby (1974)
	S. Africa	Grobbelaar & Clarke (1975)
schlechteri Bak. f.	Zimbabwe	Corby (1974)
semperflorens Vent.	Malaysia	Wright (1903)
senegalensis (Pers.) Bacle ex DC.	Zimbabwe	Corby (1974)
sericea Retz.	Widely reported	Many investigators
(= *spectabilis* Roth)		
sessiliflora L.	Japan	Asai (1944)
shamvaënsis Verdoorn	Zimbabwe	Corby (1974)
shirensis (Bak. f.) Milne-Redhead	Zimbabwe	Corby (1974)

Nodulated species of *Crotalaria* (cont.)	Area	Reported by
silvestris [?]	N.Y., U.S.A.	Wilson (1944b)
sparsifolia Bak.	Zimbabwe	Corby (1974)
spartea Bak.	Zimbabwe	Corby (1974)
	S. Africa	Grobbelaar & Clarke (1975)
sphaerocarpa Perr. ex DC.	Hawaii, U.S.A.	Allen & Allen (1936a)
	S. Africa	Grobbelaar et al. (1967)
spinosa Hochst. ex Benth.	Zimbabwe	Corby (1974)
steudneri Schweinf.	Zimbabwe	Corby (1974)
stipularia Desv.	Argentina	Rothschild (1970)
	Venezuela	Barrios & Gonzalez (1971)
stolzii (Bak. f.) Milne- Redhead ex Polh.		
var. *stolzii*	Zimbabwe	Corby (1974)
striata DC.	Widely reported	Many investigators
stricta Schrank	S. Africa	Grobbelaar et al. (1967)
subcapitata De Wild.		
var. *subcapitata*	Zimbabwe	Corby (1974)
tetragona Roxb.	Hawaii, U.S.A.	Allen & Allen (1936a)
usaramoensis Bak. f.	Widely reported	Many investigators
valetonii Backer	Widely reported	Many investigators
valida Bak.	Zimbabwe	Corby (1974)
vallicola Bak. f.	Fla., U.S.A.	Carroll (1934a)
variegata Welw. ex Bak.	Zimbabwe	Corby (1974)
vasculosa Wall. ex Benth.	Zimbabwe	Corby (1974)
velutina Benth.	Venezuela	Barrios & Gonzalez (1971)
verrucosa L.	Widely reported	Many investigators
vestita Bak.	W. India	V. Schinde (pers. comm. 1977)
virgulata Klotz.	Zimbabwe	Corby (1974)
virgultalis Burch.	S. Africa	Grobbelaar & Clarke (1975)

Cruddasia Prain   Papilionoideae: Phaseoleae, Diocleinae
(412a) / 3890

Type: *C. insignis* Prain

1 species. Climber; stems slender, tawny-pubescent. Leaves pinnately 5-foliolate; leaflets ovate-lanceolate, apices acute, bases cuneate, margins entire, glabrous above, appressed-puberulent underneath; stipels filiform; stipules rigid, spinulose-setaceous, caducous. Flowers white or purplish, fascicled in long, axillary racemes; rachis nodose; bracts and bracteoles minute, caducous; calyx campanulate, appressed-pubescent; lobes 5, upper 2 connate into a short-bidentate lip, lower 3 broadly triangular, acute, with the laterals shorter than the central lobe; standard suborbicular, apex notched, exterior

silky; wings oblong-ovate, slightly adherent to the base of the keel; keel not beaked, equal to the wings; stamens 10, monadelphous; anthers uniform; ovary sessile, densely silky-haired, many-ovuled; style inflexed, smooth, except for a tuft of hairs around the terminal stigma. Pod about 8 cm long by 8 mm wide, flat leathery, slightly appressed-puberulent, faintly depressed but thickly filled inside between the seeds, 2-valved; seeds 10–12, suborbicular, compressed, pale greenish, smooth. (Harms 1900; King et al. 1901; Lackey 1977a)

According to King et al. (1901), *C. insignis* Prain "might with almost equal propriety be referred to the subtribe Galactieae, especially to the section *Colloea* of the genus *Galactia* itself, where also the vexillary stamen is connate with the remaining nine, or to the subtribe Diocleae, especially to the genus *Pueraria* of which it exhibits most of the characters though it differs very markedly in having 5-foliolate leaves." Harms (1900) positioned the genus in the latter subtribe; more recently Hutchinson (1964) designated the former. Lackey (1977a) has consigned it to the subtribe Ophrestiinae, all of whose members gave negative tests for canavanine.

This monotypic genus is endemic in Burma and Thailand. The plants grow at elevations up to 1,600 m.

*Crudia Schreb.   Caesalpinioideae: Amherstieae
(46) / 3495

Named for Dr. J. W. Crudy, 1753–1810 (?), collector of plants in the Bahamas.

Type: *C. spicata* (Aubl.) Willd.

50–55 species. Trees, small to large, unarmed. Leaves imparipinnate; leaflets numerous, alternate, large, oblong-elliptic, glossy above, usually leathery, dull or brownish-villous underneath, not gland-dotted, densely veined, on twisted petiolules; stipules small, narrow and caducous or large, foliaceous, and persistent, more or less joined at the base. Flowers small, in 2 rows, on long, hairy pedicels, pendent in loosely flowered, terminal or axillary racemes on young branches; bracts and bracteoles small, caducous; calyx cup-shaped, short, with disk; lobes 4, rarely 5, ovate, membranous, imbricate in bud, reflexed at anthesis; petals none; stamens 10, rarely less, free, glabrous, exserted; filaments slender; anthers ovate-oblong, versatile, longitudinally dehiscent; ovary sessile or short-stalked, stalk joined to the calyx tube, hairy, few- to several-ovuled; style filiform; stigma small, terminal. Pod flat, oblong, oblique-orbicular, rigid, leathery, sometimes rusty-haired, 2-valved, margins often thickened, dehiscent, 1–2-seeded; seeds large, compressed, orbicular-reniform. (Bentham and Hooker 1865; Léonard 1952)

Members of this pantropical genus occur mostly in tropical America. They usually thrive in moist regions, along streams, in freshwater swamps, or in evergreen rain forests.

The smaller tree species, typified by *C. harmsiana* De Wild., *C. klainei* Pierre ex De Wild., *C. laurentii* De Wild., and *C. michelsonii* J. Léonard, range in height from 2 to 20 m. *C. obliqua* Griseb. in Trinidad and *C. senegalensis* Planch. and *C. gabonensis* Pierre ex Harms in equatorial Africa are lofty trees up to 40 m tall, with trunks 1 m in diameter.

Crudia wood has not been examined extensively. Record and Hess (1943) inferred that none of the timber of the tropical American species had commercial potential. Wood of *C. obliqua* has a density of 850–960 kg/m³, a medium texture, and low luster, but is plain and unattractive. The wood of the Philippine species, *C. blancoi* Rolfe, is hard but not durable. Voórhoeve (1965) commented relative to the wood of *C. gabonensis*, "Of three logs brought to a sawmill at Bomi Hills (Nigeria), two were discarded after the first ruined the saw." Apparently the wood is little used except as bumpers on ore trucks.

NODULATION STUDIES

Nodules were absent on 10 seedlings of *C. parivoa* DC. examined in the rain forest at Belém, Pará, Brazil, by Norris (1969). The brown root systems were typical of other caesalpinaceous species that lack nodules.

Cryptosepalum Benth.   Caesalpinioideae: Amherstieae
54 / 3505

From the Greek verb *krypto*, "hide"; thus, "hidden sepals," in reference to the minute calyx being hidden by a pair of bracteoles.

Type: *C. tetraphyllum* (Hook. f.) Benth.

12–15 species. Low undershrubs, shrubs or small trees, much-branched; a few species up to about 30 m tall, with a trunk diameter of about 80 cm. Leaves paripinnate; leaflets in numerous pairs, seldom 1–2 pairs, sessile or subsessile, not glandular-punctate; stipules usually linear, caducous. Flowers small, whitish rose, in 2 rows, on slender, glabrous or pubescent pedicels, in many-flowered axillary or terminal racemes, solitary or few-fascicled; bud scales overlapping at the base of young racemes, early caducous; bracts minute; bracteoles 2, opposite, large, membranous, petaloid, concave, densely puberulent along margins and on inner sides, persistent, enclosing the young flower; calyx cupular, short, very reduced; teeth 4, glabrous, minute, rudimentary, scalelike or absent; petal 1, small, orbicular, sessile; stamens usually 3, occasionally 4–5, arising with the petal and ovary stalk from a basal disk; filaments filiform, equal; anthers oblong, elliptic, versatile, longitudinally dehiscent; staminodes 0–3, short; ovary short-stalked, stalk adnate to the side of the receptacle, 1–5-ovuled; style filiform, curved; stigma truncate, terminal. Pod small, flat, woody, ovate or quadrate, apiculate, sutures thick, elastically dehiscent, 1- to few-seeded. (Bentham and Hooker 1865; Pellegrin 1948; Léonard 1952, 1957)

Because of the difficulties in species differentiation, the genus is considered a "taxonomic nightmare." As a solution, Duvigneaud and Brenan (1966) designated two groups: (1) shrubs and small trees exemplified by *C. exfoliatum* De Wild. with aerial perennial stems and clustered axillary inflorescences; and (2) shrubs, typified by *C. maraviense* Oliv. with stems that die back or are burned off, inflorescences in a single terminal raceme on each branch, and thick rootstocks. Members of the latter group are common on fireswept savannas; their scalelike buds give rise to new growth the next season after burning.

Representatives of this genus are native to tropical Africa. They are widespread in high rainfall areas, lowland forests, and savannas.

The poor-quality wood of *C. tetraphyllum* (Hook. f.) Benth. is utilized in Nigeria along with that of *Brachystegia*, *Didelotia*, and *Tetraberlinia* species under the misleading name of African pine (Voorhoeve 1965). Bast fibers from *C. pseudotaxus* Bak. f. are used in the manufacture of cords and sacks in regions of the Congo Basin. The flowers are a good source of honey.

NODULATION STUDIES

C. maraviense examined by Corby (1974) in Zimbabwe lacked nodules.

Cullen Medik.   Papilionoideae: Galegeae, Psoraliinae

Type: *C. corylifolia* (L.) Medik.

2 species. Herbs, annual or perennial, base sometimes shrubby. Leaves pinnately 1–3-foliolate, gland-dotted; leaflets coarsely dentate. Flowers in axillary, peduncled spikes; calyx campanulate, not gibbous, glandular-punctate; lobes 5, lowest lobe longest, upper 2 somewhat connate; standard obovate, with a short, straight claw; wings straight, as long as the standard but longer than the keel petals, oblong, with prominent basal lobe, claws free; keel petals shorter than the wings, broad, obliquely lunate, united at the rounded apex, each petal slightly united with the adjacent wing at the base, claws free; stamens diadelphous, vexillary stamen free except at the base, others united; ovary 1-ovuled; style curved above; stigma capitate. Pod glandular-warty, beak short, erect, pericarp thin, adherent to the obliquely kidney-shaped seed. (Rydberg 1919a)

Both species are found from Arabia to India and Burma.

The aromatic bitter seeds of *C. corylifolia* (L.) Medik. are administered in India and Indochina as a laxative and as a tonic for stomach ache. This plant shows promise as a green manure in Sri Lanka.

NODULATION STUDIES

Both species are nodulated: *C. americanum* (L.) Rydb. (= *Psoralea americana* L.) in South Africa (Grobbelaar et al. 1967) and *C. corylifolia* in Sri Lanka (W. R. C. Paul personal communication 1951) and in western India (Y. S. Kulkarni personal communication 1972).

From Latin, *cupula*, "little cup," and Greek, *anthos*, "flower"; the large, round bracteoles are united, forming a cup below the calyx.

Type: *C. bracteolosus* (F. Muell.) Hutch.

1 species. Shrublet. Leaves simple, linear or lanceolate, 1-nerved, finely reticulate below, margins recurved; stipules free, linear-subulate. Flowers red, axillary, 1–2, on slender pedicels; bracts none; bracteoles 2, orbicular, shortly united forming a cup below the calyx; calyx broadly campanulate, silky-haired outside; lobes 5, equal; standard short-clawed, ovate, reflexed, slightly exceeding the calyx; wings shorter than the keel; keel petals about twice as long as the calyx, falcate; stamens 10, free; anthers ovoid, basifixed; ovary silky-haired, 6-ovuled; style smooth. Pod not described. (Hutchinson 1964)

This monotypic genus is endemic in Western Australia.

Cyamopsis DC.   Papilionoideae: Galegeae, Indigoferinae
238 / 3700

From Greek, *kyamos*, "bean," and *opsis*, "resemblance."

Type: *C. tetragonoloba* (L.) Taub. (= *C. psoraloides* DC.)

3–4 species. Herbs, annual, with thin, silky, appressed, medifixed, T-shaped, gray or white hairs on erect stems, up to about 3 m tall. Leaves imparipinnate, 3–7-foliolate; leaflets linear or obovate, margins entire or coarsely serrated; stipels none; stipules small, setaceous. Flowers few, small, purple or lilac, in axillary, lax racemes, solitary in the axils of bracts; bracts caducous; bracteoles none; calyx broad, oblique-campanulate, sometimes pubescent; lobes 5, narrow, 3 lower lobes longer, lowest lobe longest; petals very similar; standard obovate-elliptic, sessile; wings oblong or obovate-oblong, sessile, free; keel erect, obtuse, short, incurved, sides not appendaged; stamens monadelphous; anthers uniform, connectives apiculate; ovary sessile, hairy, few- to many-ovuled; style incurved at apex; stigma capitate. Pod linear, pubescent, 4-angled, acuminate, with membranous partitions between the seeds, 2-valved; seeds many, compressed, quadrate. (Bentham and Hooker 1865; Phillips 1951)

Members of this genus are native to the tropics of Africa and Asia. They are usually found on light sandy soil.

C. tetragonoloba (L.) Taub., Calcutta lucerne, cluster or guar bean, is probably indigenous to India; it is not known to occur in the wild state. *C. dentata* (N. E. Br.) Torre, *C. senegalensis* Guill. & Perr. (= *C. stenophylla* (Bonnet) A. Chev.), and *C. serrata* Schinz occur in the semiarid zones of Africa (Hymowitz 1972).

Guar was at one time referred to as "an old crop with a new future" (Esser 1947). Because of its bushy habit, drought resistance, and tolerance of alkaline soils, it was cultivated for decades in India and Southeast Asia for cattle fodder and soil erosion control, and as a green manure, soil-improvement crop and vegetable. Some varieties reach heights of 3–4 m and are grown as shade for ginger in tropical areas. Guar was introduced into the southwestern United States about the turn of the century as a summer forage and soil-improvement crop. Inasmuch as neither its performance as a forage crop nor its value as a cash crop was outstanding, it was more or less forgotten until World War II, when the paper, textile, and food industries needed a galactomannan gum as a substitute for imported carob bean gum (Hymowitz 1972).

Guar gum is obtained from the water-soluble fraction of flour made from grinding seed endosperms; in general, this fraction consists of about 63 percent mannose and 35 percent galactose. The mucilaginous principle is soluble in both cold and hot water, insoluble in oils, ketones, and esters, stable over a wide pH range, tasteless, odorless, nontoxic, stable to heat, and has 5–8 times the thickening power of starch. There are commercial outlets for guar gum in the food industry as a protective colloid and thickener, in the pharmaceutical industry as a stabilizer and binding agent, and in the mining industry as a flocculating agent; it is also used in the paper, cosmetics, and explosives industries (Schlakman and Bartilucci 1957; Goldstein and Alter 1959; Smith and Montgomery 1959). Consumption of guar gum in the United States exceeds 6,800 metric tons annually.

NODULATION STUDIES

C. tetragonoloba has been known for many years as a nodulated plant in diverse geographical areas. Recently nodules were observed on *C. dentata* in Zimbabwe (Corby 1974) and on *C. serrata* in South Africa (N. Grobbelaar personal communication 1974). It is generally accepted that guar is included within the cowpea miscellany of rhizobia-host-plant relationships (Richmond 1926a; Raju 1936; Wilson 1939a; Erdman 1948a; McLeod 1960). A histological study of guar nodules revealed nothing unusual (Narayana 1963b).

Cyathostegia (Benth.) Schery   Caesalpinioideae: Swartzieae

From Greek, *kyathos*, "cup," and *stege*, "shelter"; the calyx is cup-shaped.

Type: *C. matthewsii* (Benth.) Schery

2 species. Shrubs or small trees. Leaves imparipinnate, 3–13-foliolate; leaflets alternate, ovate or olbong-lanceolate, usually silky-haired; stipels none; stipules linear, lanceolate, caducous. Flowers small, 10–30, in axillary or terminal racemes; bracts and bracteoles small, caducous; calyx cup-shaped, subtruncate; teeth 5, small, deltoid, valvate in bud; petal 1, white, clawed, oblong-elliptic, sometimes sparsely pubescent; stamens 20–30; filaments united at the base, free above, variable in length, shorter than the petal; ovary short-stalked, pubescent, 1–2-ovuled; style long; stigma terminal, subcapitate. Pod not winged, flat, semiorbicular, 2-valved, beaked with the persistent style, dehiscent; seeds usually 2, reddish brown. (Rudd 1968a)

The controversial positioning of this genus in Caesalpinioideae and Papilionoideae has been reviewed by Rudd (1968a).

Both species are native to Peru and southern Ecuador. The plants grow in xerophytic deciduous woods in inter-Andean valleys at 500–2,300-m elevations.

Cyclocarpa Afzel. ex Urb. Papilionoideae: Hedysareae,
325 / 3794 Aeschynomeninae

From Greek, *kyklos*, "circle," and *karpos*, "fruit," in reference to the shape of the pod.

Type: *C. stellaris* Afzel. ex Urb.

1 species. Herb, erect or prostrate, annual, unarmed, glabrous, usually less than 15 cm tall; stems naked, slender. Leaves short-petioled, paripinnate; leaflets small, 2–4 pairs, sensitive; stipels none; stipules medifixed, ovate-lanceolate, thin. Flowers small, 1–4, yellow, in axillary, umbellike racemes; bracts acuminate, persistent; bracteoles small, membranous; calyx bilabiate, upper lip entire or 2-parted, lower lip 3-parted; petals subequal in length; standard round, reflexed, emarginate, short-clawed; wings oblong, unequally-sided at base, finely denticulate in upper part; keel oblique-obovate, short-clawed; stamens diadelphous, in 2 bundles of 5, or connate into a slit tube; anthers equal; ovary subsessile, linear, flat, ridges with minute warts, 10–14-ovuled; style arcuate; stigma terminal, small. Pod linear, flattened, coiled into 1–1.5 spirals, with 8–10 segments, subdeltoid, sutures persistent after joints are shed, dehiscent. (Léonard 1954a; Steenis 1960a)

This monotypic genus is endemic in Malaysia, tropical Africa, and northern Queensland. It occurs on savannas and cleared land, and often in cultivation with *Hyparrhenia* grass in the lower Congo Basin.

NODULATION STUDIES

This species was nodulated in Zimbabwe (Corby 1974).

Cyclolobium Benth. Papilionoideae: Dalbergieae,
351 / 3822 Pterocarpinae

From Greek, *kyklos*, "circle," and *lobion*, "pod."

Type: *C. brasiliense* Benth.

6 species. Shrubs or small to medium trees. Leaves 1-foliolate; stipels 2; stipules small. Flowers purple, medium-sized, in solitary or fascicled, axillary or lateral racemes; bracts small, narrow; bracteoles minute, inconspicuous, caducous; calyx campanulate; teeth subequal or the upper 2 united higher up; standard suborbicular, flat, spreading; wings oblong; keel petals oblong, short, obtuse, slightly joined at the back; stamens diadelphous; anthers versatile; ovary stalked, many-ovuled; style filiform, incurved; stigma terminal, truncate. Pod stalked, round or sickle-shaped, membranous, upper suture narrow-winged, slightly thickened over the 2–3 oblong, transverse seeds, indehiscent. (Bentham and Hooker 1865)

Representatives of this genus are widespread from southeastern Brazil northward through the coastal forests, in the lower Amazon region, and Guiana.

C. vecchii Samp. reaches a height of about 20 m, with a trunk diameter up to 50 cm. Members of the genus have been little explored for commercial purposes. The wood has a low luster, a mild licorice odor, a straight grain, and a fine texture (Record and Hess 1943). It is hard, heavy, compact, not difficult to work, and finishes smoothly, but it is not available in sufficient quantities to justify marketing.

Cyclopia Vent.   Papilionoideae: Podalyrieae
160 / 3620

From Greek, *kyklos*, "circle," and *pous*, "foot," in reference to the circular depression at the base of the calyx.

Type: *C. genistoides* Vent.

15–20 species. Shrubs, erect, young parts often silky-haired. Leaves palmately 3-foliolate, rarely 1-foliolate, sessile; leaflets narrow, linear to ovate, sessile or short-stalked, glabrous or pubescent, margins often revolute; stipules minute or absent. Flowers large, bright yellow, stalked or sessile, solitary in leaf axils; bracteoles 2, rarely 3, at the base of the pedicels; calyx tube shallow, truncate and intruse at the base; lobes subequal, rounded or pointed, slightly longer than the tube; standard suborbicular, folded at the base, short-clawed; wings oblong, falcate, transversely folded; keel incurved, beak blunt with a small triangular pocket; stamens 10, free or slightly united at the base; filaments dilated; ovary sessile or short-stalked, many-ovuled; style filiform, arcuate; stigma minute, terminal. Pod compressed, oblong, continuous inside, valves leathery, black, several-seeded. (Bentham and Hooker 1865; Kies 1951; Phillips 1951)

Cyclopia species are confined to South Africa, mainly in the southwestern districts of Cape Province. They thrive on rocky hills and open places.

The leaves and flowers of several *Cyclopia* species are used as bush tea, a palatable substitute for *Thea*. Common names for species used in this manner are bostee, boertea, or honey tea (Kies 1951); the leaves of some species are marketed commercially.

Nodulated species of *Cyclopia*	Area	Reported by
falcata (Harv.) Kies	S. Africa	Grobbelaar & Clarke (1975)
genistoides (L.) R. Br. var. *genistoides*	S. Africa	Grobbelaar & Clarke (1972)
maculata (Andr.) Kies	S. Africa	Grobbelaar & Clarke (1975)
montana Hofm. & Phill. var. *glabra*	S. Africa	Grobbelaar & Clarke (1972)

From Greek, *kylix*, "cup," in reference to the shape of the disk at the base of the ovary to which the anthers are attached.

Type: *C. gabunensis* Harms

2 species. Trees, very large. Leaves bipinnate; pinnae opposite, 1–few pairs; leaflets alternate or subopposite, 1–2 pairs, ovate-elliptic, long acuminate. Flowers yellowish green, on apically jointed pedicels in elongated, spikelike, paniculate racemes; bracts minute, caducous; calyx shortly campanulate; teeth 5, minute; petals 5, equal, oblong-lanceolate, acute, free, valvate; stamens 10, free; filaments affixed to a conspicuous, cup-shaped basal disk; anthers tipped by a small, deciduous gland; ovary oblique-oblong, pilose, stalked, many-ovuled; style short, slender; stigma minute, cuplike. Pod very long, usually about 0.75 m, narrow, straight or curved, covered with a rusty mat or scales, margins thick; seeds many, long, thin, longitudinally positioned in the pod, winged all around, wing unequally heart-shaped at the base, with a long, filiform funicle at 1 end. (Harms 1897, 1908a; Baker 1930; Pellegrin 1948; Hutchinson and Dalziel 1958)

Both species are indigenous to equatorial Africa. They inhabit tropical rain forests.

C. gabunensis Harms is widespread in the rain forests of West Africa, where it generally reaches a height of 25 m, with a diameter of 1 m. An exceptional lofty specimen about 65 m tall, with a trunk girth about 12 m and bearing pods up to 1 m long and 4.3–4.75 cm broad, was recorded by Hutchinson and Dalziel (1958). The wood, known as okan in the timber trade and as denya locally, is used primarily for heavy construction. The supply is ample for local needs but insufficient for regular export. It is coarse grained, very hard, heavy, resistant to fungi, marine borers and termites, and is rated with teak and other woods of the highest durability class (Rendle 1969). The heartwood darkens from greenish brown to a rich reddish brown, but it is difficult to polish. The other species, *C. paucijuga* (Harms) Verdc., occurs in Kenya (Milne-Redhead 1951).

Cylindrokelupha Kosterm.

Mimosoideae: Ingeae

From Greek, *kylindros*, "cylinder," and *kelyphos*, "pod."

Type: *C. bubalina* (Jack) Kosterm.

According to Kostermans (1954), *Pithecellobium* should be divided into about 10 generic segregates. The 7 species of *Cylindrokelupha* are judged distinctive on the basis of trees being spineless, pod cylindric, dehiscent along both sutures, valves not twisted after dehiscence, and seeds cylindric, truncate. (Mohlenbrock 1963a)

Species of this genus are found in Malaysia and Indonesia.

From Greek, *kylistos*, "large," in reference to the calyx, which becomes much enlarged after flowering.

Type: *C. villosa* Ait.

7 species. Herbs and shrubs, climbing, branches slender, downy, some woody. Leaves pinnately 3-foliolate; leaflets resinous-punctate on lower surface; stipels absent. Flowers yellow to orange in axillary racemes; bracts membranous, caducous; bracteoles absent; calyx campanulate; lobes bladdery, blunt, later enlarged and scarious, upper 2 connate, lateral 2 short, lowest lobe largest, concave; standard suborbicular, basal ears inflexed; wings narrow; keel incurved, blunt; stamens diadelphous; anthers uniform; ovary subsessile; style threadlike; stigma terminal. Pod small, oblique-oblong, enclosed in the enlarged calyx, 2-valved, 1-seeded. (Bentham and Hooker 1865; Kurz 1877; Ali 1967)

Inasmuch as the type species, *C. villosa* Ait., was subsequently recognized to be a synonym of a species of *Rhynchosia*, it and 6 other species previously described under *Cylista* Ait. were transferred to *Rhynchosia*. Ali (1967) proposed that the name *Cylista* W. & A. be conserved for the remaining 6 species, typified by *C. scariosa* Roxb.; these differ from *Rhynchosia* species primarily in their characteristic membranous calyx. This proposal was rejected. Ali (1968) then proposed a new generic name, *Paracalyx*, as an alternate to accommodate the remaining ex-*Cylista* species. This awaits approval.

The proposed segregated species are native to tropical India, Somalia, and Socotra.

From Greek, *kymbe*, "boat," and *semeia*, "standard"; the standard and pods are boat-shaped.

Type: *C. roseum* Benth.

1 species. Vine, up to 10 m., semiwoody; stems twining, hoary-strigose. Leaves pinnately 3-foliolate; leaflets entire, broad-lanceolate to oblong, appressed-pubescent below; stipels small; stipules triangular. Flowers rose red or purplish, large, showy, clustered 2–3, fascicled-racemose along the upper third of erect, slender, axillary peduncles; rachis nodose; bracts ovate; bracteoles triangular; calyx campanulate; upper 2 lobes connate into 1 bidentate lobe, laterals acute, lower lobe lanceolate; standard oblong-ovate, membranous, with 2 minute, reflexed ears at the base; wings oblanceolate, free, eared; keel petals oblong, somewhat shorter than the wings, incurved, fused at the apex, subacuminate; stamens diadelphous, vexillary stamen free, others fused at the base, free above; anthers uniform; ovary subsessile, gray-tomentose, often 6-ovuled; style glabrous, long, down-curved, persistent; stigma truncate. Pod oblong-falcate, compressed, apiculate, thinly pithy be-

tween the seeds, canescent becoming ferruginous, finally glabrescent, dehiscent; seeds usually 4, thick, hard, dark brown, lustrous, transversely oblong, hilum wide and encircling nearly one-half of the seed. (Bentham 1840; Bentham and Hooker 1865; Maxwell 1970)

This genus is closely related to *Dioclea* and *Camptosema*.

This monotypic genus is endemic in southern Central America and Northern South America. It flourishes in moist lowlands and along river banks.

Cymbosepalum Bak.   Caesalpinioideae: Cynometreae
(35b) / 3477

From Greek, *kymbe* "boat"; 1 sepal is boat-shaped prior to full floral expansion.

Type: *C. baroni* Bak.

1 species. Tree, small, unarmed. Leaves paripinnate; leaflets 3–4 pairs, sessile, obovate, leathery, articulate at the base. Flowers profuse, in upright racemes, grouped 3–4 in leaf axils; bracts small, deltoid, acuminate, caducous; calyx tube very short; sepals 5, unequal, lowermost one largest, at first boat-shaped, later spreading, straight; petals 5, equal, oblanceolate, blunt; stamens 10, same length as petals; filaments threadlike, free; anthers oblong, versatile; ovary short-stalked, linear, 2–3-ovuled; style filiform, curved, thickened apically; stigma terminal. Pod not described. (Harms 1897)

This monotypic genus is endemic in the northern areas of Madagascar.

Cynometra L.   Caesalpinioideae: Cynometreae
35 / 3475

From Greek, *kyon*, *kynos*, "dog," and *metron*, "measure"; the flaccid pink or white young leaves resemble the tongue of a panting dog.

Type: *C. cauliflora* L.

60–70 species. Shrubs or trees, erect, unarmed, small to medium, occasionally large, up to about 50 m tall. Leaves opposite, rarely alternate, paripinnate, not pellucid-punctate; young leaves lax, pink, white, or red, later becoming deep green or brownish; leaflets pairs 1–few, entire or emarginate, glabrous, leathery, asymmetric, short-petioled, oblique, often with a basal petiolar gland, upper pair longest; stipules linear, rarely foliaceous, subpersistent. Flowers small, in more than 2 rows, in short, dense axillary or terminal racemes or panicles, or sometimes fascicled on trunks and old branches; bracts often dry, scalelike; bracteoles present at the base or middle of the pedicel; calyx tube short, turbinate or campanulate; sepals 4–5, imbricate in bud, later reflexed; petals 5, free, subequal, narrow-lanceolate, upper petal innermost; stamens 8–13, usually 10; filaments all free, long, exserted, rarely joined at the base; anthers often deeply bilobed at the base, longitudinally dehiscent; ovary usually stalked, free, few-ovuled; disk absent; style filiform; stigma terminal, truncate or capitate. Pod often obliquely attached to the thickened stalk, either flattened, oblong, woody, smooth, 2-valved, elastically dehiscent, or ovoid, turgid, wrinkled or warty, indehsicent; seeds mostly 1–2, thick, compressed. (Bentham and Hooker 1865; Amshoff 1939; Pellegrin 1948; Léonard 1951, 1952; Knaap-van Meeuwen 1970)

Members of this genus are pantropical and are distributed throughout Indo-Malaya, Africa, the West Indies, and Central and South America. Many species thrive along river banks and in mangrove swamps and rain forests, up to about 700-m altitude.

The lower branches and trunk of *C. cauliflora* L., a shrubby tree native to India and Malaysia, bear a profusion of fleshy, wrinkled, 1-seeded, semicircular pods that are similar in flavor to unripe apples. These green yellow, succulent, semiacid fruits are marketed as nam-nam and are suitable for pickling or stewing.

A number of African species, notably *C. alexandri* C. H. Wright, *C. ananta* Hutch. & Dalz., and *C. leonensis* Hutch. & Dalz., are evergreen forest trees reaching heights up to 50 m. They are flanked at the base with prominent thin buttresses, which extend some distance from the base along the ground as spreading surface roots (Voorhoeve 1965).

The wood of *C. alexandri* is said to be termite-resistant, and suitable for bridges and flooring (Eggeling and Dale 1951). *C. bauhiniaefolia* Benth., widespread in Guiana, Brazil, and Argentina, has pink sapwood and reddish brown heartwood. It is not considered good timber but is useful for small articles. The wood of *C. retusa* Britt. & Rose, a common understory tree in lowland forests of Guatemala, provides good fuel and charcoal (Record and Hess 1943). *C. ramiflora* L. is not only a useful timber tree in the Philippines for posts, house supports, and fuel, but also has horticultural value.

Nodulation Studies

Three species, *C. hankei* Harms, examined in Yangambi, Zaire, by Bonnier (1957), *C. ramiflora*, in the Philippines (personal observation 1962), and *C. cauliflora*, in Java (personal observation 1977), lacked nodules.

Cytisopsis Jaub. & Spach   Papilionoideae: Loteae
234 / 3695

The Greek suffix, *opsis*, denotes resemblance to *Cytisus*.

Type: *C. dorycniifolia* Jaub. & Spach

1 species. Shrublet, spreading, rootstock woody, thick. Leaves sessile, palmately 5–7-foliolate; leaflets linear, entire, with silky, white pubescence;

stipules absent. Flowers quite large, yellow, 1–3, at tips of short twigs; bracteoles 2; calyx long, tubular, base dilated, bilabiate above, upper lip bidentate, lower lip tridentate, narrow, and shorter; petals long-clawed, lowermost petal adnate to staminal tube; standard ovate; wings and keel blunt, slightly incurved; stamens diadelphous; filaments apically thickened; anthers uniform; ovary sessile, multiovuled; style slightly incurved, thickened above; stigma truncate. Pod linear, straight, subcylindric, silky-haired, exserted from the calyx, thinly septate within, 2-valved, dehiscent; seeds many, subglobose. (Bentham and Hooker 1865)

This monotypic genus is endemic in Asia Minor and Israel. It occurs in pine forests and on coastal hills.

Cytisus L.   Papilionoideae: Genisteae, Cytisinae
221 / 3682

From Greek, *kytisos*, Latin, *cytisus*, the name of a cloverlike fodder plant found in the Greek island Kíthira.

Type: *C. sessilifolius* L.

50 species. Shrubs, deciduous or evergreen, rarely small trees, unarmed; branches usually ribbed, stiff, green. Leaves palmately 3-foliolate, rarely 1-foliolate, sometimes minute, nearly wanting or reduced to decurrent phyllodes, crowded on older branches; stipules minute and setaceous, or none. Flowers showy, bright yellow, white, rarely purple, solitary or in axillary or leafy terminal racemes or fascicles; bracts and bracteoles small, caducous, sometimes leaflike and subpersistent; calyx campanulate, bilabiate; lobes 5, short, 2 upper lobes connate, sometimes free, lower lip 3-dentate; petals free; standard suborbicular or ovate, greatly exceeding the wings and keel, without auricles; wings oblong-obovate or falcate; keel straight or incurved, obtuse or briefly acuminate, claws free, not beaked; stamens monadelphous; anthers alternately long and basifixed, short and versatile; ovary sessile, 2-, usually many-ovuled; style incurved, glabrous above; stigma terminal, capitate or obliquely decurrent. Pod linear-oblong, compressed, continuous inside, rarely subseptate, 2-valved, dehiscent. (Bentham and Hooker 1865)

This genus is closely related to *Genista*. Polhill (1976) included it in Genisteae *sensu stricto*.

Cytisus species are widely distributed, both wild and cultivated, throughout Europe, the Mediterranean area, North Africa, and western Asia. The plants are commonly found along gravelly waysides and in sandy waste areas. Best growth is achieved in full sunlight on poor soils with good drainage.

Cytisus species are ornamentally attractive. *C. proliferus* L. f., tree lucerne, and *C. scoparius* Link, Scotch broom, are cultivated in many areas as hedges and live fences.

Few *Cytisus* species have browse value because of toxic alkaloids present in the foliage and seeds of numerous species. Anagyrine, cytisine, lupanine, *N*-methylcytisine, and sparteine of the quinolizidine series are most commonly found throughout the genus (Leonard 1953; Willaman and Schubert 1961; Mears and Mabry 1971). Loss of livestock rarely occurs, but fatalities have been recorded in Europe. Poisonings usually occur when hungry animals overindulge. Symptoms of poisoning vary from slavering, vomiting, and staggering, to motor nerve paralysis, convulsions, and death by asphyxiation. *C. proliferus* var. *palmensis* Christ (= *C. palmensis* (Christ) Hutch.) and *C. maderensis* Masf. (= *C. stenopetalus* Christ) have been mentioned as important fodder plants in the Canary Islands; and, similarly, *C. scoparius* and *C. monspessulanus* L. are eaten by sheep and cattle in New Zealand (White 1943; Whyte et al. 1953; Watt and Breyer-Brandwijk 1962). In view of the erratic occurrence and distribution of alkaloids in plant parts, and the taxonomic confusion with *Sarothamnus* and *Genista* species, caution in the use of *Cytisus* species as forage is advised. According to Fadiman (1965), smoking cigarettes made from dried flowers of *C. canariensis* Steud. (= *Genista canariensis*) induces a pleasant minor psychedelic state without undesirable residual effects.

NODULATION STUDIES

Nodules were reported on *C. nigricans* L. and *C. purpureus* Scop. as early as 1860 by Wydler, who considered nodule formation a key character in the classification of leguminous plants. Nodules were observed on *C. albus* Link by Lachmann in 1891. In the ensuing years, nodulation has been reported for one-half of the species in the genus, yet cultural and biochemical characterizations of *Cytisus* rhizobia are lacking.

Maassen and Müller (1907) considered *C. scoparius* a separate rhizobia: plant group, but later workers (Erdman and Walker 1927; Pieters 1927; Conklin 1936) assigned this species and *C. hirsutus* L. to the cowpea miscellany. This conclusion was confirmed in detailed experimentation by Allen and Allen (1939) and Wilson (1939a, 1945a). The early study by Pereira Forjaz (1929) on the presence of molybdenum, nickel, and cobalt in nodules of *C. proliferus* var. *palmensis* but not in parent roots was an experimental prelude to a better understanding of biocatalytic reactions involved in symbiotic nitrogen fixation.

Nodules on *Cytisus* species and certain other woody annuals were grouped as the "*Genista* type, subgroup 1" on the basis of external and internal structure (Spratt 1919). Nodules of this group have distinct zones of bacteroid tissue separated by sterile tissue owing to (1) a rarity of infection threads, (2) disruption of the initial, broad, hemispherical meristematic activity, and (3) localization of subsequent activity to several small peripheral areas. Accordingly, nodule surfaces become irregular and covered with a thick protective periderm. Lechtova-Trnka (1931) interpreted differences in nodules from 6 *Cytisus* species. Nodules on *C. sessilifolius* L. and *C. virescens* Wohlf. were spherical to ovoid in contrast to cylindrically elongated nodules on *C. capitatus* Jacq., *C. purgans* (L.) Boiss., and *C. scoparius*. Typical intracellular infection threads were observed in *C. scoparius* nodules; however, in those of *C. capitatus* the rhizobia were disseminated to newly formed cells only by mitotic host cell divisions. Peculiar broad, mucous, intercellular filaments that presumably meandered between adjacent host cells and often formed bulbous cul-de-sacs were depicted in nodules of *C. capitatus* and *C. purgans*. These filaments often surrounded islands of noninvaded, nonhypertrophied host cells. The authors have observed a similar anomalous condition in roots invaded and deformed by a noncompatible rhizobial strain, i.e., alfalfa roots invaded by clover rhizobia.

Nodulated species of *Cytisus*	Area	Reported by
albus Link	Germany	Lachmann (1891)
†*austriacus* L.	S. Africa	Grobbelaar & Clarke (1974)
†*borysthenicus* Grun.	U.S.S.R.	Gordienko (1960)
capitatus Jacq.	France	Lechtova-Trnka (1931)
decumbens Spach	Wis., U.S.A.	J. C. Burton (pers. comm. 1966)
elongatus (L.) Scop.	S. Africa	N. Grobbelaar (pers. comm. 1965)
fragrans Lam.	N.Y., U.S.A.	Wilson (1939a)
(= *Genista supranubia* Spach)		
hillebrandtii Briq.	Kew Gardens, England	Pers. observ. (1936)
†*hirsutus* L.	N.Y., U.S.A.	Conklin (1936)
(*laburnum* L.)	Widely reported	Many investigators
= *Laburnum anagyroides* Medik.		
nigricans L.	Europe	Many investigators
proliferus L. f.	Australia; Europe	Many investigators
var. *palmensis* Christ	France	Pereira Forjaz (1929)
purgans (L.) Boiss.	France	Lechtova-Trnka (1931)
†*purpureus* Scop.	France	Lechtova-Trnka (1931)
ramosissimus Ten.	France	Prillieux (1879)
†*ruthenicus* Fisch. ex Wolosz.	S. Africa	Grobbelaar & Clarke (1974)
sagittalis (L.) K. Koch	Germany	Lachmann (1891)
scoparius (L.) Link	Widely reported	Many investigators
var. *andreanus* Hort.	France	Lechtova-Trnka (1931)
sessilifolius L.	France	Lechtova-Trnka (1931)
†*supinus* L.	S. Africa	Grobbelaar & Clarke (1974)
(= *Genista supina* (L.) Scheele)	N.Y., U.S.A.	Wilson (1945a)
virescens Wohlf.	France	Lechtova-Trnka (1931)

†Regarded as *Chamaecytisus* species by some botanists.

Named for Hugo Gustaf Adolf Dahlstedt, 1856–1935, collector of Scandinavian plants.

Type: *D. pinnata* (Benth.) Malme

2 species. Shrubs, erect. Leaves imparipinnate; leaflets 5–7, elliptic; stipels absent. Flowers large, red, pendent in terminal panicles; bracts and bracteoles minute; calyx tubular; teeth 4, very short; petals clawed, subequal; standard oblong, straight, not eared; wing petals adnate to the keel; keel petals joined only at the apex; stamens 10, all joined to midway, free above; anthers dorsifixed; ovary stalked, hairy, multiovuled; style bent, smooth; stigma blunt. Pod large, oblong with thin brittle pericarp, indehiscent; seeds few, thick, kidney-shaped. (Malme 1905; Harms 1908a)

Both species are native to subtropical Brazil, in the environs of São Paulo and Rio de Janeiro.

D. pinnata (Benth.) Malme has insecticidal and fish-poisoning properties (Burkart 1952).

Dalbergia L. f.
350 / 3821

Papilionoideae: Dalbergieae, Pterocarpinae

Named by Linnaeus' son in honor of Nils and Carl Gustav Dalberg; they were associated with the elder Linnaeus during the last half of the 18th century. Nils was a student of Linnaeus and later a naturalist. Carl Gustav was a plant collector in Surinam.

Type: *D. lanceolaria* L. f.

100–300 species. Lianas, high-climbing, shrubs, much-branched, some climbing by means of stem tendrils, or trees, often up to 30 m tall, sometimes villous, spiny. Leaves imparipinnate, rarely 1-foliolate; leaflets alternate, small to large, oblong, broad-elliptic or obovate, few to numerous; stipels absent; stipules generally small, caducous, sometimes large, persistent. Flowers numerous, small, purple or white to yellowish, in terminal and axillary racemes, cymes, or panicles, each flower subtended by a pair of bracteoles, often deciduous; bracts small, broad, subpersistent; calyx campanulate; lobes 5, unequal, upper 2 generally larger and more obtuse than others and partly joined, lowest lobe usually longest; petals clawed; standard rounded, emarginate; wings oblong; keel petals oblong, obtuse, shorter, often eared, united at the top; stamens monadelphous with all connate into a sheath split on 1 side, or diadelphous with the vexillary stamen free, or the sheath split on both sides, making 2 bundles; anthers small, basifixed, 2-lobed, usually apically dehiscent; ovary stalked, 1- to few-ovuled, sometimes hairy; style short; incurved; stigma small, terminal. Pod flattened and samaroid in 1 group, thickened and drupaceous in another, oblong or linear, stalked, not thickened or winged at the sutures but usually harder and reticulate about the seed, indehiscent, 1–4-seeded; seeds plano-compressed, subreniform. (Bentham and Hooker 1865; Baker 1929; Amshoff 1939; Standley and Steyermark 1946; Phillips 1951; Cronquist 1954b)

Members of this large genus are distributed over tropical and subtropical areas of the Old World and the New World. They are well represented in eastern and central Africa and Central and South America. There are about 12 species in Malaysia and 12 in Zimbabwe, Transvaal, and Mozambique. *Dalbergia* species are often prominent as secondary growth, in low coastal woodlands, light forests, savannas, and in thickets along open swampy ground. The pods of some species float in water; thus the plants are easily spread along stream banks. The plants are well adapted to sandy soils and limestone escarpments; they occur up to 1,700-m altitude.

Most members of this genus are lianas; *D. junghuhnii* Benth. and *D. rostrata* R. Grah. are examples of these woody climbers. *D. arbutifolia* Bak., *D. ferruginea* Roxb., *D. lactea* Vatke, and *D. martinii* F. White exemplify the multibranched, woody, climbing shrubs.

The genus is best known for those trees that yield valuable timber. In India the wood of *D. sissoo* Roxb., a much-cultivated tree, is perhaps surpassed in importance only by teak (*Tectona*). *Dalbergia* wood is commonly called rosewood in the Americas and India because of its fragrance; Barrett (1956) defines the dalbergias as the true rosewoods. In Africa and China the wood is commonly known as blackwood because of the dark-veined heartwood. The following are among the most prominent and valuable species (Titmuss 1965; Rendle 1969; Kukachka 1970):

D. baroni Bak.	Madagascar rosewood
D. cearensis Ducke	Brazilian kingwood, violeta
D. cochinchinensis Pierre	Siam rosewood
D. cubilquitensis (Donn. Smith) Pitt.	Guatemala, or Honduras, rosewood
D. greveana Baill.	Madagascar rosewood
D. latifolia Roxb.	East Indian rosewood, Bombay blackwood, black rosewood, rosetta rosewood
D. melanoxylon Guill. & Perr.	Senegal ebony, African blackwood, Bombay rosewood, Javanese palisander
D. nigra Allem.	Bahia, Brazilian, or Rio rosewood
D. retusa Hemsl.	Nicaragua rosewood, cocobolo
D. sissoo Roxb.	Indian teakwood, Indian sissoo
D. spruceana Benth.	Amazon rosewood, Jacarandá do Pará
D. stevensonii Standl.	Honduras rosewood

Dalbergia wood was once a trade article in the Mediterranean area (Burkill 1966). The sapwood is generally white or yellowish and sharply distinguished from the chocolate brown to purple black heartwood. The grain is usually straight and very fine, the luster varies from medium to high, and the odor is roselike. The density varies from 750 to 1000 kg/m^3. The wood is durable, not difficult to work, and finishes most attractively. The wood of some species has a rather high oil content. The fine dust resulting from the working of *D. retusa* Hemsl. often causes an allergic response in susceptible individuals (Kukachka 1970).

Dalbergia wood is generally reserved for high-quality Oriental cabinets and chests, interior trim, ornamental turnery articles, and musical instruments, such as xylophone, marimba, and piano cases. It is preferred over ebony for woodwind instruments (Rendel 1969). It is used also to a lesser extent to make canes and handles for cutlery and brushes, and for inlays, steering wheels, jewelry boxes, and carved figures. The heavily scented heartwood of the vine *D. junghuhnii* is incorporated in the manufacture of incense and joss sticks used in Hindu and Chinese temples.

Various compounds of the 4-phenylcoumarin group have been isolated from the heartwood of *Dalbergia* species. These substances may account for the durable properties of *Dalbergia* wood (Krishnaswamy and Seshadri 1962). The occurrence of flavonoid compounds in the Leguminosae, particularly in *Dalbergia* and *Machaerium* species, was reviewed by Ollis (1968).

D. sissoo, known as tali in India, is a common shade and fodder source along sand and gravel river banks in the Himalayas up to 1,500-m altitude.

NODULATION STUDIES

K. S. N. Murthy (personal communication 1955) stated that nodules of *D. melanoxylon* Guill. & Perr. examined in India bore nodule rootlets. Nodules of *D. monetaria* L. f. in Puerto Rico (Dubey et al. 1972) were hard, woody, rough surfaced, and reddish brown at first, turning black with age. In Singapore, *D. oliveri* Gamble ex Prain, an introduced tree species, lacked nodules (Lim 1977).

Nodulated species of *Dalbergia*	Area	Reported by
arbutifolia Bak.	Zimbabwe	Corby (1974)
armata E. Mey.	Sri Lanka	Wright (1903)
	S. Africa	Grobbelaar et al. (1967)
boehmii Taub.	Zimbabwe	Corby (1974)
ferruginea Roxb.	Philippines	Bañados & Fernandez (1954)
fischeri Taub.	Zimbabwe	Corby (1974)
lactea Vatke	Zimbabwe	Corby (1974)
latifolia Roxb.	W. India	V. Schinde (pers. comm. 1977)
martinii F. White	Zimbabwe	Corby (1974)
melanoxylon Guill. & Perr.	India	K. S. N. Murthy (pers. comm. 1955)
	Zimbabwe	Corby (1974)
monetaria L. f.	Puerto Rico	Dubey et al. (1972)
nitidula Welw. ex Bak.	Zimbabwe	Corby (1974)
obovata E. Mey.	S. Africa	Grobbelaar & Clarke (1975)
paniculata Roxb.	W. India	V. Schinde (pers. comm. 1977)
sissoo Roxb.	Zimbabwe	H. D. L. Corby (pers. comm. 1972)
	W. India	V. Schinde (pers. comm. 1977)
sympathetica Nimmo	W. India	Bhelke (1972)

Dalbergiella Bak. f.   Papilionoideae: Dalbergieae,
                                            Lonchocarpinae

The Latin diminutive suffix denotes small *Dalbergia*.

Type: *D. welwitschii* Bak. f.

3 species. Shrubs, scandent or small trees, deciduous. Leaves imparipinnate; leaflets small, numerous, 6–13 pairs, oblong-elliptic, asymmetric at the base, opposite or nearly so, slightly pubescent underneath. Flowers white, pink, yellow, some streaked with red or violet, in many-flowered, paniculate racemes; calyx hairy, campanulate, deeply bilabiate, upper lip longer, obtuse, emarginate, lower lip deeply 3-toothed; standard orbicular, slightly emarginate, ridged on back, claw long, slender, channeled; wings oblong, long-clawed; keel petals shorter than the wings, united at apex; stamens diadelphous; anthers dorsifixed, with a blackish connective at the apex; ovary linear, sessile, 5–8-ovuled. Pod oblong, lanceolate, thin, flat, reticulate, membranous, slightly pubescent, 1-seeded. (Pellegrin 1948)

Representatives of this genus are native to tropical Africa, from Gabon to Zimbabwe. They occur in hot, dry woodlands, riverine forest, and on poor stony soil.

The profusely flowering tree *D. nyasae* Bak. f. is valued as an ornamental for parks and gardens; the wood is suitable for small carpentry items.

NODULATION STUDIES

Corby (1974) reported nodules present on *D. nyasae* in Zimbabwe.

*Dalea L.   Papilionoideae: Galegeae, Psoraliinae
246 / 3709

Named for Dr. Samuel Dale, 1659–1739, eminent English botanist and author of articles on *materia medica*.

Type: *D. alopecuroides* Willd.

150 species. Shrubs, herbs, rarely trees, annual or perennial, plant parts usually gland-dotted. Leaves imparipinnate; leaflets small, entire, rarely 3-foliolate, sometimes stipellate; stipules small, awl-shaped often glandlike. Flowers small, purplish, dark blue, white, rarely yellow, in showy, terminal or leaf-opposed spikes; bracts usually membranous, broad, concave above, glandular-dentate; bracteoles absent; calyx campanulate; lobes 5, subequal, lowest lobe sometimes longer than the others; corolla imperfectly papilionaceous; standard heart-shaped, clawed, inserted in the bottom of the calyx, sometimes appendaged with 2 inflexed ears; wings and keel straight, often longer than the standard; keel petals usually united along their lower margins; stamens 10, rarely 9, monadelphous, all connate into a sheath split above and forming a cup at the base; anthers uniform, often glandular-tipped; ovary sessile or shortly stipitate, 2-, rarely 3-ovuled; style awl-shaped; stigma small, terminal. Pod small, ovoid, membranous, sometimes conspicuously ribbed, mostly indehiscent, 1-seeded, enclosed in a persistent calyx; seeds kidney-shaped, cotyledons broad and flat.

Dalea is generally considered a genus of the Western Hemisphere, common in the dry, warm regions of the southwestern United States and extending throughout Mexico into Central America and the Andean area of western

South America. About 65 species are found in the southwestern United States. They are common on sand dunes, arroyo banks, dry and desert soils, and in alluvial bottom lands. Some species occur up to 2,000-m altitude.

All species are well adapted to poor soils. The much-branched bushes have value on sand dunes as wind-erosion-control plants. *D. alopecuroides* Willd. is used to some extent as a green manure in the midwestern United States. Many of the wild species contribute to the forage value of stock ranges; none are reported poisonous, but livestock do not find many of the species palatable, possibly because of the presence of resins and essential oils in the glandular foliage. *D. formosa* Torr. and *D. greggii* Gray, however, are relished as browse by mule deer and the Sonora white-tailed deer in Arizona. The twigs of some species are woven into baskets by some Indian tribes in the Southwest. Most of the perennial species are highly ornamental and worthy of cultivation.

NODULATION STUDIES

Nodules within this genus were first reported on *D. alopecuroides*, Wood's clover (Schneider 1892). Schneider's deduction that this species possessed rhizobia unlike those common to other plant groups was confirmed in principle 2 decades later by Whiting et al. (1926). Cultural and biochemical reactions affirmed the rhizobia as being more similar to strains of *Rhizobium trifolii* than to *R. meliloti*; however, neither alfalfa nor clover plants were receptive to *Dalea* rhizobial strains and the converse was also true.

Later, Sears and Clark (1930) showed that, although rhizobial strains from *Dalea* were infective but not effective on garden and navy bean plants, no reciprocal cross was achieved.

Six species listed in Martin's survey (1948) lacked nodules: *D. albiflora* Gray, *D. amoena* S. Wats., *D. batocaulis* [?], *D. formosa*, *D. mollis* Benth., and *D. wrightii* Gray. Presumably, arid conditions accounted for lack of nodules on these species at the time of their examination in Arizona.

Nodulated species of *Dalea*	Area	Reported by
†*alopecuroides* Willd.	Widely reported	Many investigators
†*aurea* Nutt.	N.Y., U.S.A.	Wilson (1939a); Wilson & Chin (1947)
frutescens Gray	Ariz., U.S.A.	Martin (1948)
lagopus (Cav.) Willd.	Ariz., U.S.A.	Martin (1948)
lumholtzii Robins. & Fernald	Ariz., U.S.A.	Martin (1948)
(*ordiae* Gray) = *Thornbera ordiae* (Gray) Rydb.	Ariz., U.S.A.	Martin (1948)
parryi T. & G.	Ariz., U.S.A.	Martin (1948)
pogonathera Gray	Ariz., U.S.A.	Martin (1948)
†(*scoparia* Gray) = *Psorothamnus scoparius* (Gray) Rydb.	N.Y., U.S.A.	Wilson (1939b)
(*spinosa* Gray) = *Psorodendron spinosum* (Gray) Rydb.	Calif., U.S.A.	Pers. observ. (U.W. herb. spec. 1971)
virgata Lag.	Ariz., U.S.A.	Martin (1948)

†Also reported as *Parosela*, the rejected generic name.

Named for James Andrew Ramsey, 1812–1860, Marquis of Dalhousie, British Governor General of India, 1848–1856.

Type: *D. bracteata* R. Grah.

3 species. Shrubs, subscandent or trailing, 2–3 m tall; branches slender, often gray-downy. Leaves large, 1-foliolate; stipels absent; stipules ovate-lanceolate. Flowers white, in copious, terminal panicles and axillary sub-corymbose racemes; bracts opposite, ovate or rounded, stipuliform, persistent; bracteoles similar, enclosing the flower bud; calyx campanulate, short-toothed, deltoid; standard orbicular, subsessile, emarginate; wings oblique-oblong, free; keel petals short-clawed, erect, oblique, broader than wings, slightly joined at back; stamens 10, free; anthers linear-oblong, short; ovary subsessile or short-stalked, few-ovuled; style slightly incurved; stigma small, terminal. Pod angled, oblique-oblong, linear, compressed, rigidly leathery, pointed at each end, continuous within, 2-valved, dehiscent; seeds 1–3, lens-shaped, flat, black, shiny. (Bentham and Hooker 1865; Baker 1929; Pellegrin 1948; Toussaint 1953)

Members of this genus occur in tropical West Africa and the eastern Himalayas.

NODULATION STUDIES

Nodules were lacking on *D. africana* S. Moore examined in Yangambi, Zaire (Bonnier 1957).

Daniellia Benn. Caesalpinioideae: Amherstieae
60 / 3512

Named for Dr. W. F. Daniell, 1818–1865, botanist and collector of plants in Senegal and Sierra Leone, 1841–1853.

Type: *D. thurifera* Benn.

12 species. Trees, unarmed, large, aromatic. Leaves paripinnate; leaflets petioluled, pinnately nerved, oblique at the base; stipules unequal, oval and small, or large, lanceolate and foliaceous, very caducous, encompassing the terminal bud. Flowers variously colored, in lax, short-panicled, axillary or terminal racemes; bracts and bracteoles oval-oblong, early deciduous; calyx tube narrow, with disk; lobes 4, overlapping, lanceolate, free, unequal, caducous; petals 5, imbricate, sessile, 2 generally large, 1 small and 2 very small, rarely 1 large and 4 very small or rudimentary; stamens exserted, 10, joined at the base or free, hairy, rarely glabrous; anthers linear, dorsifixed, longitudinally dehiscent; ovary long-stiped, stipe partly adnate to calyx tube, many-ovuled; style exserted; stigma terminal, capitate. Pod stalked, compressed, subfalcate, subcoriaceous, 2-valved, usually 1-seeded; seed flat, pendulous from the apex near top of 1 of the valves, lower part of pod empty; the valve with the attached seed breaks away and is dispersed; the other valve drops away later.

Cyanothyrsus, designated 60a by Harms (1897), was recognized later (Harms 1915a) to be congeneric with *Daniella*.

Members of this genus are indigenous to tropical West Africa; they occur in forests and savannas.

In general, members of this genus are huge trees reaching heights in excess of 45 m, with boles clean and cylindrical to 30 m above the ground and with trunk diameters of 1–1.5 m at waist level. The trunk base is usually flanked by large root swellings but is not buttressed.

All species yield a brown, fragrant balsam or oleoresin; *D. oliveri* Hutch. & Dalz., *D. similis* Craib ex Holl., and *D. thurifera* Benn. are the most copious producers. The fresh product is obtained by cutting out square pieces of the bark, but much of it is procured in the semifossilized state (Howes 1949). The resin is used as an adulterant of copaiba balsam, as a torch resin, or dammar, to fumigate closed areas, and as a frankincense; it is not considered of much commercial value.

Daniellia timber is marketed under various names, but the product is often termed ogea, with reference to *D. ogea* (Harms) Rolfe and closely related species. The wood is moderately coarse textured and lightweight, with an air-dry density of about 480 kg/m^3; the cut surface is apt to be woolly but can be finished satisfactorily. It is a reasonably strong timber but is not durable. The heartwood is likely to be gummy. Limited amounts are used as a decorative veneer and for plywood, but generally the wood is used for crates and for construction purposes not requiring durability and high-quality appearance (Titmuss 1965; Voorhoeve 1965; Rendle 1969).

Dansera Steenis   Caesalpinioideae: Cassieae

Named for Benedictus Hubertus Danser, 1891–1943, Dutch botanist; he collected extensively in Java.

Type: *D. procera* Steenis

1 species. Tree. Leaves simple, ovate or lanceolate; petiole short, jointed at base and apex; stipules narrow, caducous. Flowers white, small, in axillary cymes, pedicels and buds densely hairy; bracts and bracteoles small, caducous; calyx lobes 3, hairy on both sides, free, imbricate, outer lobe hooded; petals none; stamens usually 6, 2 with longer filaments; anthers basifixed, with apical pore; ovary oblong, tomentose, 1-ovuled; style smooth, apex recurved. Pod smooth, ovoid or flattened, mesocarp pulpy, exocarp hard, sutures grooved, indehiscent, 1-seeded. (Steenis 1948)

This genus was reduced to a subgenus of *Dialium* by Steyaert (1953) because of obvious similarities in floral and fruit characters; however, Hutchinson (1964) considered it distinctive.

This monotypic genus is endemic in Sumatra.

219

Named for the Reverend Hugh Davies, 1739–1821, Welsh botanist.

Lectotype: *D. acicularis* Sm.

60 species. Shrubs, or shrublets, glabrous or hairy; stems decumbent or ascending, rigid. Leaves flat or cylindric, simple, entire, stiff, horizontal or vertical, sometimes sharp-pointed, reduced to short prickles, or wanting; stipules minute or absent. Flowers attractive, often small, few, yellow to orange or red, in short, lax, axillary or lateral racemes, umbels, or short clusters, rarely solitary or terminal; bracts small, dry, scalelike, at base of pedicels; bracteoles absent; calyx 5-toothed; teeth short, equal, or upper 2 united into a truncate lip; petals slender-clawed; standard usually yellow with dark center, orbicular or kidney-shaped, often shorter than the wings and keel; keel often crimson, curved almost into a right angle, glabrous, beaked, triangular; stamens free; 5 outer filaments often flattened; ovary short-stalked, glabrous, 2-ovuled; style blunt; stigma small, terminal. Pod glabrous, more or less compressed, 8–12 mm long, triangular, acute, upper suture nearly straight, lower suture much-curved, 1–2-seeded.

Representatives of this genus are native to Australia. They are well distributed in New South Wales, Queensland, Victoria, and Tasmania, but occur chiefly in Western Australia. They are common in mountain ranges and along slopes, on tablelands, and in coastal and river areas.

Some species, when not in flower, may be mistaken for phyllodineous *Acacia* species.

Nodulation Studies

Rhizobia from *Daviesia* are slow-growing cowpea-type strains (Norris 1958a; Lange 1961). Host affinities are akin to members of the cowpea miscellany. A non-calcium-demanding strain produced effective nodules on *Pultenaea villosa* (Norris 1959b). Strains of *Rhizobium lupini* lacked ability to nodulate *D. divaricata* Benth., *D. euphorbioides* Benth., *D. horrida* Meissn., and *D. quadrilatera* Benth. (Lange 1962).

Nodulated species of *Daviesia*	Area	Reported by
aphylla (F. Muell.) Benth.	W. Australia	Lange (1959)
brevifolia Lindl.	W. Australia	Lange (1959)
cordata Sm.	W. Australia	Lange (1959)
corymbosa Sm.	S. Australia	Harris (1953)
flexuosa Benth.	W. Australia	Lange (1959)
genistifolia A. Cunn.	N.S. Wales, Australia	Benjamin (1915)
hakeoides Meissn.	W. Australia	Lange (1959)
incrassata Sm.	W. Australia	Lange (1959)
pectinata Lindl.	W. Australia	Lange (1959)
reversifolia F. Muell.	W. Australia	Lange (1959)
teretifolia R. Br.	W. Australia	Lange (1959)
ulicifolia And.	Queensland,	Bowen (1956)
(= *ulicina* Sm.)	S. Australia	Harris (1953)
umbellulata Sm.	Queensland, Australia	Bowen (1956); Norris (1958a)

Named for De Corse, plant collector in Madagascar.

Type: *D. schlechteri* (Harms) Verdc.

4 species. Herbs, perennial, creeping or climbing; rootstocks woody. Leaves pinnately 3-foliolate; leaflets thin, lanceolate; stipels lanceolate; stipules ovate-elliptic, persistent. Flowers mauve, blue, or purple, on short pedicels in axillary fascicles of 4–20, or in clustered racemes; bracts and bracteoles ovate, subpersistent; calyx campanulate; lobes 5, deltoid, upper 2 rounded, connate, lower 3 acute; petals clawed; standard broad, reflexed, with 2 basal appendages; wings oblong; keel boat-shaped, beaked apex tightly coiled; stamens diadelphous; anthers subuniform, 5 basifixed, 5 dorsifixed, pollen grains triangular; ovary narrow, many-ovuled; style thin, flexuous at base with apex thickened and strongly curved; stigma surrounded by dichotomously branched hairs. Pod narrow, linear, apex sickle-shaped, nonseptate, dehiscent; seeds ellipsoid. (Verdcourt 1970c)

The genus is closely allied to *Physostigma* and *Dolichos*.

Members of this genus are common in tropical eastern and central Africa, Transvaal, Zimbabwe, Zambia, and Madagascar. They occur on loamy and sandy soils in savanna woodlands, scrub forests, and open grasslands, climbing over grasses and small shrubs.

The beans of *D. schlechteri* (Harms) Verdc. are eaten by natives in Tanzania.

NODULATION STUDIES

Corby (1974) reported nodules present on *D. schlechteri* in Zimbabwe.

Delaportea Thorel ex Gagnep.   Mimosoideae: Adenanthereae
(22a)

Type: *D. armata* Thorel ex Gagnep.

1 species. Tree, tall; branches spiny. Leaves abruptly bipinnate; pinnae pairs 3; leaflets 5–8 pairs, sessile, large, oblong, midrib eccentric; stipules thorny. Flowers peduncled, in small, globose heads in panicles, peduncles with whorls of bracts midway; bracteoles minute; calyx tube campanulate; sepals 5, short, blunt, hairy; petals 5, alike, same length as the sepals; stamens 15–20, free; anthers tipped with a stalked gland; ovary sessile, smooth; style short; stigma minute. Pod narrow, spirally twisted or arched, woody, nonseptate, many-seeded. (Gagnepain 1911; Harms 1915a)

This monotypic genus is endemic in Laos.

(= *Poinciana* L., 1760, not 1753)

From Greek, *delos,* "evident," "clear," and *onyx,* "claw," in reference to the conspicuously long-clawed petals.

Type: *D. regia* (Boj. ex Hook.) Raf. (= *Poinciana regia* Boj.)

3 species. Trees, unarmed. Leaves bipinnate, 30–50 cm long; pinnae 10–25 pairs; leaflets small, numerous, 20–40 pairs, deciduous; stipels none; stipules inconspicuous. Flowers zygomorphic, showy, orange or red, in terminal or axillary, corymbose racemes; bracts small; bracteoles none; calyx tube disklike or shortly turbinate; lobes 5, subequal, valvate in bud; petals 5, broad, edges ruffled, long-clawed, 4 subequal, uppermost petal usually variegated; stamens 10, perfect, free; filaments long, incurved toward apex, hairy at the base; anthers longitudinally dehiscent; ovary sessile or short-stalked, many-ovuled; style short, filiform; stigma truncate, ciliate. Pod large, up to 60 cm long, pendent, linear, compressed, hard, woody, 2-valved, with broad, almost solid partitions between many, transverse, oblong seeds.

Members of this genus were originally native to Madagascar. They are now widely distributed in the tropics and subtropics, especially in Africa, the Caribbean area, and Florida.

D. elata (L.) Gamble and *D. baccal* Bak. f. are the other 2 species.

The prominence of this genus rests almost entirely on the type species, which is known commonly as the flamboyant tree of Madagascar, the peacock flower of India, the royal poinciana of the Hawaiian archipelago, and the flame tree in many other areas. This fast-growing majestic tree reaches a height of 10–18 m, prefers well-drained soil, and grows well near the sea, but is shallow rooted and does not tolerate shade. The slightly buttressed trunk supports a wide-spreading crown with a diameter 1.5–2 times the height of the tree. Because of the umbrella-shaped crown and the shallow root system, the trees should not be planted near houses.

Few leguminous trees surpass the royal poinciana in beauty and form. The first blossoms unfold before the leaves reappear. Soon the crown becomes a vivid splash of orange, red, and crimson. The faintly scented blossoms measure 7–12 cm in diameter, in large clusters. The trees remain in flower for 3–4 months, with fernlike foliage providing additional ornamental beauty. During the early growth years the tree bark is smooth, but it becomes furrowed and gnarled with age.

The tree has little value aside from its ornamental use along avenues and in parks and gardens. The huge black woody pods can serve as a minor source of fuel. In the South Pacific islands, the seeds are strung into necklaces or leis and are used for decoration on women's purses and belts. The wood is soft, coarse grained, and not durable.

Nodulation Studies

Twenty-three plants, ranging in age from 1 to 12 months and grown in native soils of Hawaii, and numerous plantlets grown in sand culture from seed inoculated with diverse rhizobia, all lacked nodules (Allen and Allen 1936b). This finding has been confirmed in principle by Bañados and Fernandez (1954), the authors (personal observation 1962) in the Philippines,

Ogimi, in the Ryukyus (1964), and Grobbelaar et al., in South Africa (1964). In western India *D. elata* was devoid of nodules (V. Schinde personal communication 1977). In Singapore very sparse white nodules were observed on roots of *D. regia* (Boj. ex Hook.) Raf. seedlings, but never on mature plants (Lim and Ng 1977).

Dendrolobium Benth.   Papilionoideae: Hedysareae, Desmodiinae

From Greek, *dendron*, "tree," and *lobion*, "pod," in reference to the thick joints of the pod.

Type: *D. umbellatum* (L.) Benth.

12–17 species. Small trees or shrubs, unarmed; young shoots usually silky-haired. Leaves pinnately 3-foliolate; stipels present; stipules striate, caducous. Flowers usually white, in dense, axillary umbels on short peduncles; bracts narrow, stipulelike; calyx narrow, campanulate, with 2 bracteoles at the base; lobes 4, upper lobe entire or 2-dentate; standard broad, narrow at the base; wings oblong, slightly joined to the straight, blunt keel; stamens monadelphous, united at the base for at least half their length; ovary sessile, few-ovuled; style glabrous, bent at a right angle but not thickened above; stigma capitate. Pod falcate, with several squarish, thick, indehiscent joints.

Dendrolobium species originally constituted a section in *Desmodium*. Schindler (1924) presented evidence of sufficient distinctiveness to warrant the segregation of about 13 species as a separate genus.

Members of this genus are native to tropical Indo-Malaya and Northern Territory, Australia. They occur along calciferous beaches and on sandy soil, in humid, grassy localities, teak forests, and rice fields.

Denisophytum R. Vig.   Caesalpinioideae

Apparently the name means "Denis' plant"; Viguier (1949) did not explain this.

Type: *D. madagascariense* R. Vig.

1 species. Tree, small, 3–4 m tall; young branches very tomentose, eglandular, deciduous. Leaves paripinnate, 6–10-foliolate; leaflets elliptic, margins parallel, eared on 1 side at the base; stipels absent; stipules small, tomentose, deciduous. Flowers yellow, pedicelled, in small, lax, multiflowered racemes; bracts tomentose, stipulelike; calyx tube very short; lobes 5, imbricate, more or less united, upper lobes apically rounded, lower ones narrow, thickened; petals 5, subobovate, free, clawed, shorter but broader than the sepals; stamens 10, all fertile, unequal; filaments villous and thick at the base; anthers

dorsifixed, opening by longitudinal slits; ovary on short stalk free from the calyx, few-ovuled; style short, cylindric. Pod short-stalked, beaked, subvesicular, glabrous, dehiscent; seeds few, small, flattened, lens-shaped. (Viguier 1949)

This monotypic genus is endemic in northwestern Madagascar. The plants grow in dry woods on gneiss.

Derris Lour.   Papilionoideae: Dalbergieae,
366 / 3838   Lonchocarpinae

From Greek, *derris*, "leather covering," in reference to the tough seed pods.
Type: *D. trifoliata* Lour.

70–80 species. Vines, woody, perennial, or shrubs, usually twining, sometimes creeping, rarely trees. Leaves imparipinnate, stipulate, without tendrils or stipels; leaflets 3–many. Flowers small, profuse, showy, violet, pink, or white, usually fascicled, in axillary or terminal racemes or panicles; bracts small, caducous; calyx subtended by 2 basal, often caducous bracteoles; tube truncate or cupular; teeth indistinct or very short, lowermost tooth often longest; standard obovate or orbicular, sometimes with inflexed basal ears; wings oblique-oblong, slightly adherent to the incurved keel above the claw; stamens monadelphous, all connate into a tube, or vexillary stamen free at the base; anthers versatile; ovary sessile or short-stalked, hairy, ovules 2–many; style slender, incurved, glabrous above; stigma small, terminal. Pod brown, leathery, orbicular or elongated, flat, rigid, narrowly winged along 1 or both sutures, indehiscent; seeds 1–6, flat, kidney-shaped or orbicular.

Representatives of this genus are native to tropical East Asia and the East Indies. They are widely distributed in the Old World tropics. The majority of the species are wild; commercially important species are cultivated.

All species tolerate a wide range of soil conditions, from fertile loams to barren areas. Well-drained soils, a well-distributed rainfall, and a tropical climate are preferable. Wild species occur as woody climbers in tidal areas, mangrove swamps, lowland forests, secondary jungles or brushwood, and along muddy banks of rivers.

The importance of *Derris* species has its origin in the primitive practice throughout Southeast Asia, South America, the Philippines, East Africa, and the South Pacific Islands of throwing soaked or crushed plant parts into streams for the purpose of stupefying fish. The fish may be eaten with impunity. *Derris* species were among the earliest plants used for this purpose.

Concurrent with the use of *Derris* species for fish-stupefication was the incorporation of macerated plant parts and decoctions in arrow poisons and in shampoos to eliminate lice. Oxley's report (1848) of the effectiveness of *Derris* root emulsions in destroying insects on young nutmeg trees in Singapore was perhaps the first record of the agricultural use of *Derris* species as an insecticide. Emulsions were used later by Chinese gardeners to control insect pests on vegetables. As early as 1877, *D. elliptica* (Roxb.) Benth. was being cultivated in Singapore for this purpose. The first exports of plant roots of this species from the Far East to the Western world began in the early part of the 20th century. The first patent relating to the use of *Derris* as an insecticide was granted in 1911 by the United Kingdom.

Chemical studies aimed at isolating the insecticidal principle began in the late 1800s. In 1902 Nagai isolated from *D. chinensis* Benth. colorless crystals melting at 163°C, which he called rotenone. This substance was identical with the colorless, crystalline compound isolated in 1892 and in 1895 by Geoffroy from *Robinia nicou*, now correctly named *Lonchocarpus nicou*. Geoffroy called his substance nicouline. Nagai's contribution was accorded little significance amid the isolation of diverse rotenoid substances during this early period. In 1933 LaForge et al. reported the structure of rotenone, $C_{23}H_{22}O_6$.

Rotenone production from members of the genus *Derris* has centered largely on *D. elliptica* (Georgi and Teik 1932; Higbee 1948), but insecticidal properties of a lesser degree (Holman 1940) are shown in plant parts of the following species: *D. amazonica* Killip, *D. chinensis*, *D. heptaphylla* Merr., *D. malaccensis* Prain, *D. philippinensis* Merr., *D. polyantha* Perk., *D. robusta* Benth., *D. thyrsiflora* Benth., and *D. trifoliata* Lour. (= *D. uliginosa* Benth.).

Species of *Lonchocarpus*, *Milletia*, *Mundulea*, and *Ormocarpum* also yield rotenone, but only *Lonchocarpus* species excel *Derris* species in commercial production.

Rotenone first makes its appearance in the secondary cortical cells opposite the protoxylem in *D. elliptica* roots between the 5th and 6th weeks of plant growth (Worsley and Nutman 1937). Thereafter it spreads in scattered groups throughout the xylem parenchyma and cortex. The rotenone-containing cells are morphologically unspecialized. An increase in number of these cells occurs as the roots advance in age.

Maximum yield of rotenone is obtained between the 18th and 25th months. Air-dry root yields at the latter time usually vary from about 880 kg/ha when *D. elliptica* is used as an intercycle crop to about 1,100 kg/ha as a solo crop (Georgi and Curtler 1929). Harvesting is not difficult, since about 95 percent of the root system lies within the top 45 cm of the soil surface. Studies in Puerto Rico showed that 81 percent of *Derris* roots were located in the top 44 cm of soil (Moore 1940). The rotenone yield from wild plants cultivated under the most favorable conditions is about 5–6 percent, whereas yields up to 13 percent are obtained from selected cultivars. Tropical Asia, the East Indies, and the Philippines are the principal sources of export. The United States imported about 2,950 metric tons of derris roots in 1940 (Purseglove 1968). The comprehensive coverages of this subject by Roark (1932) and Holman (1940) are of interest.

Derris roots yield three other insecticidal compounds less effective than rotenone (Davidson 1930), namely, deguelin, an isomer of rotenone, and the isomers tephrosin and toxicarol (Clark 1930a). Toxicity of *Derris* species to man and test animals is very low (Jones et al. 1968). Malaccol, $C_{20}H_{16}O_7$, an optically active phenol, was isolated from *D. malaccensis* in 1940 by Harper.

Other uses of *Derris* species are of comparatively minor importance. *D. el-*

225

liptica, *D. malaccensis*, *D. microphylla* (Miq.) Val., and *D. robusta* have served as green manures, shade trees, and double-cropping on rubber, cacao, coffee, kapok, and tea plantations. Natives of eastern Malaysia consume the raw and cooked leaves of *D. heptaphylla* and also use them as a culinary flavoring. The flexible stems of *D. malaccensis* and *D. trifoliata* make good cordage. The tree species, *D. polyphylla* Koord. & Val. and *D. robusta*, yield wood useful for small buildings and turnery objects. Plant-part preparations of *D. trifoliata* are used medicinally as stimulants, antispasmodics, and counterirritants. The bark of this species contains about 9.3 percent tannin.

NODULATION STUDIES

Root systems of uncultivated field specimens of *D. microphylla* in Hawaii bore abundant apical-type nodules. Rhizobial colonies on yeast-extract, asparagus, and mannitol mineral salt agars were white, translucent to opaque, and discernible within 1 week. Moderate alkalinity without serum zone formation was produced in litmus milk within 3 weeks (Allen and Allen 1936a). Two strains were effective on diverse plants of the cowpea miscellany. Ineffective reactions were obtained on *Clitoria ternatea*, *Lespedeza stipulacea*, and *Phaseolus lunatus* (Allen and Allen 1939). A strain from *D. elliptica* tested by Chen and Shu (1944) failed to nodulate *Astragalus sinicus*; these authors also regarded *D. elliptica* as allied with the cowpea group.

Lim (1977) reported the absence of nodules on *D. dalbergoides* Bak., an introduced shrub in Singapore.

Agricultural reports on the commercial growth of *Derris* species lack mention of nodule occurrence and effects of rhizobial inoculants on plant yields. Since plantings are made from rooted cuttings, immersion of the roots into a rhizobial slurry at the time of transplantation from the propagating frames to the field nurseries would be a feasible means of assuring nodulation, if seed treatment had been unsatisfactory.

Nodulated species of *Derris*	Area	Reported by
elliptica (Roxb.) Benth.	Widely reported	Many investigators
indica (Lam.) Benth.	W. India	V. Schinde (pers. comm. 1977)
microphylla (Miq.) Val.	Hawaii, U.S.A.	Allen & Allen (1936a)
philippinensis Merr.	Philippines	Bañados & Fernandez (1954)
scandens (Roxb.) Benth.	W. India	V. Schinde (pers. comm. 1977)

From Greek, *desme*, "bundle," and *anthos*, "flower," in reference to the dense flower heads.

Type: *D. virgatus* (L.) Willd.

40 species. Herbs, perennial, or shrubs, unarmed, some arborescent. Leaves bipinnate; leaflets many, small, narrow, with a sessile gland between the lowest pair; stipules bristly, persistent. Flowers white or creamy yellow, small, in axillary heads on long, upright peduncles, all perfect, or the lowest ones sterile, apetalous, with short staminodes; calyx campanulate, 5-toothed; petals 5, free, or somewhat joined; stamens 10 or 5, free, exserted, about twice as long as the petals; anthers apically eglandular; ovary subsessile, many-ovuled; style thick; stigma terminal. Pod brown, linear or curved, membranous to leathery, continuous or subseptate between lengthwise-placed, ovate, flattened seeds.

Members of this genus are native to warm regions of North and South America, especially Texas, Florida, and from the West Indies to northern Argentina. They occur along roadsides, ditches, grassy savannas, and at low elevations. *Desmanthus* species are well adapted to silty clay soils.

All species are weedy shrubs and prolific seeders; the scanty foliage is good cattle fodder. *D. virgatus* (L.) Willd. reaches a height of about 3 m and has possibilities as a hedge plant. It is cultivated for cattle graze in the West Indies, Hawaii, and Mauritius.

NODULATION STUDIES

Cultural and biochemical characteristics of *Desmanthus* rhizobia are akin to those of the slow-growing members of Group II rhizobia. Alkaline reactions with serum zone formation are produced in litmus milk. In a broad spectrum of reciprocal host-infection tests using *D. virgatus*, effective responses were obtained in combinations with *Prosopis juliflora* and *Leucaena glauca*, ineffective or no nodulation in tests with *Mimosa pudica*, and lack of nodulation in combinations with *Glycine max* and *Lupinus polyphyllus*. Wilson's data (1939a) showed considerable versatility in the nodulating performance of *D. illinoensis* (Michx.) MacM. and *D. leptalobus* T. & G.

Nodulated species of *Desmanthus*	Area	Reported by
brachylobus Benth. (= *Darlingtonia glandulosa* Benth.)	France	Clos (1896)
depressus H. & B. ex Willd.	U.S.A.	J. C. Burton (pers. comm. 1966)
illinoensis (Michx.) MacM.	U.S.A.	Many investigators
leptalobus T. & G.	U.S.A.	Wilson (1939a)
virgatus (L.) Willd.	Hawaii, U.S.A.	Pers. observ. (1938)
	S. Africa	Grobbelaar et al. (1967)
	W. India	Y. S. Kulkarni (pers. comm. 1972)

From Greek, *desmos*, "bond," "chain," and *hode*, "like," in reference to the resemblance of the jointed pod to links on a chain.

Lectotype: *D. scorpiurus* (Sw.) Desv.

350–450 species. Herbs, annual or perennial, or shrubs, erect or sub-scandent, foliage and pods profusely hooked-hairy. Leaves pinnately 3-, sometimes 5-foliolate, rarely 1-foliolate; leaflets elliptic, lanceolate, ovate, or obovate; stipels usually large; stipules often striate, dry, free or united, leaf-opposed. Flowers small, mostly purple, some pink, blue, or white, in axillary or terminal racemes or panicles; bracts striate, awl-shaped and persistent or membranous and caducous; bracteoles conspicuous and persistent, minute or absent; calyx tube short, turbinate or campanulate, bilabiate; lobes 5, upper 2 subconnate into a bidentate lip, lower lip tridentate, middle lobe much longer; standard oblong to orbicular, narrow-clawed; wings oblong, adherent to the middle of the keel, short-clawed; keel straight or incurved, long-clawed, partly connate, obtuse, rarely beaked; stamens 10, monadelphous in the group with Asian affinities, diadelphous in the American group; anthers uniform; ovary sessile or stalked, 2- to many-ovuled; style inflexed, incurved, beardless; stigma terminal. Pod lomentaceous, usually stalked, compressed, membranous or leathery, exserted from the calyx, 1 or both sutures indented between the seeds, separating into indehiscent or tardily dehiscent, 1–many, 1-seeded segments at maturity; seeds compressed, orbicular-reniform. (Bentham and Hooker 1865, in part; Schindler 1917, 1926; Standley and Steyermark 1946; Schubert 1954; Isely 1955)

Species of this large genus are widespread in the warm parts of the New World and the Old World. They are not native to New Zealand, continental Europe, and the western United States. There are both wild and cultivated species. The plants grow in dry, grassy, sandy waste areas, and in thickets providing semishade, up to 2,000-m altitude.

Two phylogenetic groups are recognized (Isely 1955). Tropical America is the center of distribution of most of the American species; the others have Asian affinities.

Members of this genus are important pioneer plants in providing ground cover, wildlife protection, and for erosion control of denuded, burned-over areas. However, desmodia usually disappear as other vegetative cover develops and shading becomes pronounced. *D. gangeticum* (L.) DC. and *D. heterocarpon* (L.) DC. are noteworthy for their heavy matty growth and weed-control value in open sandy areas. Although many species of this large genus occur intermixed in the forage and browse of native grasslands, species with forage value are few and their commercial culture is limited. The amount of foliage produced by different species varies considerably; as an example, *D. heterophyllum* (Willd.) DC. provides palatable fodder but has a low yield. Most desmodia form woody crowns and have ligneous roots that are difficult to remove. However, the most objectionable feature is the effective habit of seed dispersal by means of the hooked hairs on the pod segments. The tenacious adherence of the segments to clothing or animals has earned many species such common names as tick clover, tick trefoil, and beggarweed.

A number of *Desmodium* species have shown promise as suitable substitutes for alfalfa and clover in suptropical areas. *D. barbatum* (L.) Benth. and *D. discolor* Vog. are good wild forage on acid, calcium-deficient soils in Brazil and northern Argentina. Kaimi clover, or creeping beggarweed, *D. canum* (J. F. Gmel.) Schinz & Thell., is favored in Hawaii and Florida for its tolerance to acid soil and a wet, warm climate. The economic potentials of various desmodia under semitropical conditions in Hawaii were studied by Hoska and Ripperton (1944) and Younge et al. (1964). *D. intortum* Urb. produced over 926 kg/ha of dry matter and 141 kg/ha of protein per annum. The plant is regarded highly as fodder and graze for beef cattle. *D. latifolium* DC. is a good graze for horses in Nigeria; but Webb's mention (1948) of string halt and Chillagoe disease of horses in Australia, caused by *D. brachypodum* Gray and *D. umbellatum* (L.) DC. respectively, merits consideration before assuming that all *Desmodium* species are nontoxic.

Plant parts of various species have served in folk medicine as febrifuges, remedies for dysentery and liver diseases, and have been used in poultices and other decoctions to treat acne, ulcers, catarrh, abscesses, and eye diseases (Uphof 1968). During Cuba's wars of independence, root extracts of *D. canum* were used in hospitals to heal wounds (León and Alain 1951). The roots of *D. gangeticum* yield a lactone, $C_{16}H_{30}O_2$ (Avasthi and Tewari 1955a,b) and 7 alkaloids (Ghosal and Banerjee 1969). Various tryptamine bases and related compounds were isolated from plant parts of *D. puchellum* (L.) Benth. by Ghosal and Mukherjee (1965, 1966). Of these, hordenine, $C_{10}H_{15}NO$, has been used in experimental medicine as a sympathomimetic, and bufotenine, $C_{12}H_{16}N_2O$, as a hallucinogen.

The stems of some species are woven into baskets and prawn traps. A dye of no commercial value, but one used by natives in some areas, is obtained from desmodia flowers. The leaves of *D. oldhami* Oliv. are used in Japan for tea. The wood of *D. umbellatum* serves as fuel in Taiwan.

NODULATION STUDIES

The spherical nodules on *Desmodium* species resemble those on *Vigna* species. Infection threads were not observed as a mode of rhizobial dissemination in nodules of *D. dillenii* Darl. (Lechtova-Trnka 1931). The rod-filled bacteroid cells of these nodules, greatly enlarged and vacuolated, were interspersed between small starch-packed, noninvaded cells. Polyploid cells were lacking in the bacteroidal area of nodules of *D. fallax* Schindl. (Kodama 1967). This observation contradicts Wipf's (1939) proposal of the essentiality or correlation between ploidy and infection threads.

Results of reciprocal host-infection studies are convincing that *Desmodium* species are members of the cowpea miscellany (Burrill and Hansen 1917; Carroll 1934a; Allen and Allen 1939; Bhide 1956; Bowen 1956; Galli 1958). Symbiotic compatibility is apparently lacking with soybean and lupines. Antisera produced by cowpea and lespedeza rhizobia agglutinated desmodia rhizobia at titers of 1:2560; no agglutination was obtained in the presence of bean, clover, and soybean antisera (Carroll 1934b).

Nodulated species of *Desmodium*	Area	Reported by
adscendens (Sw.) DC.	Widely reported	Many investigators
var. *robustum* Schub.	Zimbabwe	Corby (1974)
affine Schlechtd.	Argentina	Rothschild (1970)
	Venezuela	Barrios & Gonzalez (1971)
asperum Desv.	Venezuela	Barrios & Gonzalez (1971)
(*auriculatum* DC.)	Java	Keuchenius (1924)
= *Pteroloma auriculatum* (DC.) Schindl.		
axillare (Sw.) DC.		
var. *acutifolium* (Ktze.) Urb.	Trinidad	DeSouza (1966)
var. *genuinum* Urb.	Trinidad	DeSouza (1966)
barbatum (L.) Benth.	Hawaii, U.S.A.	Allen & Allen (1936a, 1939)
var. *dimorphum* (Welw. ex Bak.) Schub.	Zimbabwe	Corby (1974)
batocaulon Gray	Ariz., U.S.A.	Martin (1948)
bracteosum (Michx.) DC.		
var. *longifolium*	Widely reported	Many investigators
cajanifolium DC.	Hawaii, U.S.A.	Allen & Allen (1936a)
canadense (L.) DC.	Widely reported	Many investigators
canescens (L.) DC.	Widely reported	Many investigators
canum (J. F. Gmel.) Schinz & Thell.	Widely reported	Many investigators
(= *supinum* (Sw.) DC.)		
(= *frutescens* (Jacq.) Schindl.)		
capitatum DC.	Java	Keuchenius (1924)
	Philippines	Bañados & Fernandez (1954)
(*caudatum* (Thunb.) DC.)	Japan	Asai (1944)
= *Catenaria caudata* (Thunb.) Schindl.		
†*ciliare* (Willd.) DC.	N.C., U.S.A.	Shunk (1921)
cinereum DC.	Zimbabwe	H. D. L. Corby (pers. comm. 1968)
cuspidatum (Willd.) Loud.	U.S.A.	Many investigators
(= *grandiflorum* (Walt.) DC.)		
diffusum (Roxb.) DC.	India	Bhide (1956)
dillenii Darl.	Widely reported	Many investigators
discolor Vog.	Widely reported	Many investigators
distortum Macb.	Sri Lanka	W. R. C. Paul (pers. comm. 1951)
dregeanum Benth.	S. Africa	Grobbelaar et al. (1967)
elegans Benth.	Java	Keuchenius (1924)
fallax Schindl.	Japan	Kodama (1967)
gangeticum (L.) DC.	Widely reported	Many investigators
glutinosum (Willd.) Wood	Widely reported	Many investigators
(= *acuminatum* (Michx.) DC.)		
(*gyrans* DC.)	Widely reported	Many investigators
= *Codariocalyx gyrans* (L. f.) Hassk.		
(*gyroides* DC.)	Widely reported	Many investigators
= *Codariocalyx gyroides* (Roxb.) Hassk.		

Nodulated species of *Desmodium* (cont.)	Area	Reported by
heterocarpon (L.) DC.	Java	Keuchenius (1924)
(= *polycarpon* (Poir.) DC.)	N.Y., U.S.A.	Wilson (1939a)
heterophyllum (Willd.) DC.	Widely reported	Many investigators
hirtum Guill. & Perr.	S. Africa	Grobbelaar et al. (1967)
illinoense Gray	Widely reported	Many investigators
intortum Urb.	Trinidad	D. I. A. DeSouza (pers. comm. 1963)
jucundum Thwait.	Malaysia	Wright (1903)
†*laevigatum* (Nutt.) DC.	U.S.A.	Many investigators
latifolium DC.	Java	Keuchenius (1924)
laxiflorum DC.	Philippines	Bañados & Fernandez (1954)
leiocarpum G. Don	Zimbabwe	Sandman (1970a)
limense Hook.	S. Africa	B. W. Strijdom (pers. comm. 1967)
var. *sandwicense* Schinde	Sri Lanka	W. R. C. Paul (pers. comm. 1951)
marilandicum (L.) DC.	U.S.A.	Many investigators
microphyllum DC.	Fiji	Pers. observ. (1962)
molle (Vahl) DC.	N.Y., U.S.A.	J. K. Wilson (pers. comm. 1940)
molliculum (HBK) DC.	Venezuela	Barrios & Gonzalez (1971)
muelleri Benth.	Australia	Bowen (1956)
natalitium Sond.	S. Africa	Grobbelaar & Clarke (1975)
nemorosum F. Muell. ex Benth.	Australia	Bowen (1956)
nicaraguensis Oerst. ex Benth.	West Indies	Pers. observ. (1958)
nudiflorum (L.) DC.	U.S.A.	Many investigators
oldhami Oliv.	Japan	Pers. observ. (U.W. herb. spec. 1971)
ospriostreblum Chiov.	Zimbabwe	Corby (1974)
ovalifolium Wall. ex Merr.	Widely reported	Many investigators
pabulare (Hoehne) Malme	Hawaii, U.S.A.	Allen & Allen (1936a)
pachyrhizum (Sw.) Vog.	Venezuela	Barrios & Gonzalez (1971)
†*paniculatum* (L.) DC.	U.S.A.	Many investigators
pauciflorum (Nutt.) DC.	Ind., U.S.A.	Pers. observ. (U.W. herb. spec. 1971)
procumbens (Mill.) A. S. Hitchc.	Philippines	Bañados & Fernandez (1954)
(*pulchellum* (L.) Benth.)	Java	Keuchenius (1924)
= *Phyllodium pulchellum* (L.) Desv.	Philippines	Bañados & Fernandez (1954)
purpureum H. & A.	Widely reported	Many investigators
racemosum DC.	Japan	Asai (1944)
repandum (Vahl) DC.	Zimbabwe	Corby (1974)
rhytidophyllum F. Muell.	Australia	Bowen (1956)
salicifolium (Poir.) DC.		
var. *salicifolium*	Zimbabwe	Corby (1974)
scorpiurus (Sw.) Desv.	Philippines	Bañados & Fernandez (1954)
sessilifolium T. & G.	Australia	Bowen (1959b)

231

Nodulated species of *Desmodium* (cont.)	Area	Reported by
setigerum (E. Mey.) Benth. ex Harv.	Zimbabwe	Corby (1974)
sintenisii Urb.	Puerto Rico	Dubey et al. (1972)
tanganyikense Bak.	Zimbabwe	Corby (1974)
tiliaefolium G. Don	India	Aggarawal (1934)
tortuosum (Sw.) DC.	Widely reported	Many investigators
trichocaulon DC.	Australia	G. D. Bowen (pers. comm. 1957)
triflorum (L.) DC.	Widely reported	Many investigators
(*triquetrum* (L.) DC.)	Java	Keuchenius (1924)
= *Pteroloma triquetrum* (DC.) Benth.		
umbellatum (L.) DC.	Java	Pers. observ. (1977)
uncinatum (Jacq.) DC.	Widely reported	Many investigators
(= *intortum* Fawc. & Rend.)		
(= *sandwicense* E. Mey.)		
varians (Labill.) G. Don	Australia	Bowen (1956)
velutinum (Willd.) DC.	Zimbabwe	Corby (1974)
virgatum Zoll.	Philippines	Bañados & Fernandez (1954)
†*viridiflorum* (L.) DC.	N.C., U.S.A.	Shunk (1921)

†*Meibomia* synonyms were listed by Shunk (1921).

Detarium Juss.   Caesalpinioideae: Cynometreae
43 / 3491

From *detar*, the Wolof name for the tree in Senegal.

Type: *D. senegalense* J. F. Gmel.

3–4 species. Trees, unarmed. Leaves pinnate; leaflets few or many pairs, usually alternate, elliptic, leathery, translucent gland-dotted; stipules inconspicuous. Flowers creamy white, small, crowded in narrow, axillary panicles; bracts and bracteoles small, early caducous; calyx with a small, disklike receptacle; calyx tube miniscule; sepals 4, minutely imbricate; petals none; stamens 10, free, alternately long and short; anthers versatile, oval; ovary sessile, free, 2–6-ovuled; style filiform; stigma small. Pod globular or disk-shaped, drupaceous, sessile, indehiscent, with fibrous pulp and 1 stony, orbicular, thick, flattened seed. (Léonard 1952)

Members of this genus occur in tropical Africa. They grow in forests and dry or moist savannas.

In forests these trees tend to be large, wide crowned, and buttressed, with trunk diameters of 1–1.5 m, whereas in open woodland areas they are commonly medium sized. All forms yield a viscous, yellowish brown oleoresin, which is collected, dried, and sold under various names in local markets. This fragrant copal is used as a pomade, a fumigant, a resin soap, a skin-disease medicine, a chewing gum purgative, and as an adhesive to mend pottery.

Detarium wood is pleasantly scented, of moderate weight, dark, and fine grained, but the timber is soft and not of commercial value. Its burns slowly and is favored as a fuel because of the agreeable odor. The bark is said to be poisonous. The fruit of *D. senegalense* J. F. Gmel. has an edible, green, sweet mesocarp. The dried brown fruits are sold in native markets. Local uses of various species are described in detail by Dalziel (1948).

Dewevrea M. Micheli   Papilionoideae: Galegeae,
(261a) / 3726   Tephrosiinae

Named for Alfred Dewèvre, 1866–1897; his plant collection from equatorial Africa was conserved in the Jardin Botanique de Bruxelles.

Type: *D. bilabiata* M. Micheli

2 species. Shrubs, climbing or small trees; branches smooth or rusty-haired. Leaves imparipinnate; leaflets 2–3 pairs, ovate, leathery; stipels none; stipules elliptic to obovate, acuminate, caducous. Flowers yellow, in large, axillary and terminal panicles; bracts and bracteoles minute, very caducous; calyx tube short, broad, bilabiate, lips entire, subequal; standard broad-ovate, reflexed when fully expanded, short-clawed with 2 callosities; wings and keel petals broad, free; keel petals somewhat imbricate, obtuse; stamens 10, diadelphous; anthers glabrous, versatile; ovary subsessile, 3–4-ovuled, surrounded at the base by a 10-lobed, raised disk; style filiform, short or long; stigma capitate, very small. Pod narrowed toward base and apex with 1–4 large seeds. (Harms 1900; Hauman 1954a)

Both species are common in equatorial West Africa. They thrive in humid forest shade along water courses and marshes.

Young leaves are cooked as a vegetable; plant parts of both species have fish-poisoning properties.

NODULATION STUDIES

Nodules were observed on *D. bilabiata* M. Micheli in Yangambi, Zaire (Bonnier 1957).

Dewindtia De Wild.   Caesalpinioideae: Amherstieae
(54a) / 3505a

Type: *D. katangensis* De Wild.

1 species. Subshrub with woody rhizome; branches erect, straight, multifascicled, renewed seasonally, glabrous or pubescent, often scaly at the base. Leaves paripinnate; newly emerged leaves reddish pink; leaflets oblong-lanceolate, 2–9 pairs with a small flat gland at the base of each petiolule; rachis channeled, slightly winged; stipules linear, very caducous. Flowers many, small, white or pink, in solitary terminal racemes up to 27 cm long, bloom-

ing before the leaves are renewed; bracteoles broad, elliptic, enclosing the flower bud; calyx tube very short, cup-shaped; sepals and petals greatly reduced to 2 awl-shaped and 2 wartlike rudimentary structures; stamens usually 6, all fertile, uniform; filaments filiform; anthers elliptic, versatile, longitudinally dehiscent; ovary subsessile, 2-ovuled; style filiform, curved at the apex. Pod not described. (Harms 1908a; Baker 1930; Léonard 1952)

The reduction in calyx and corolla is also characteristic of *Didelotia*, *Brachystegia*, and *Cryptosepalum*, all closely akin to *Dewindtia* (Harms 1908a). Léonard (1952) merged this genus with *Cryptosepalum* and renamed the species *Cryptosepalum katagense* (De Wild.) J. Léonard.

This monotypic genus is endemic in the high mountains of Katanga Province, Zaire. The plants grow in forest clearings.

Dialium L.   Caesalpinioideae: Cassieae
75 / 3530

Origin of the generic name has various interpretations. Voorhoeve (1965) refers to *Dialium* as the "Latin form of the Greek plant name 'dialion' used by Linnaeus indiscriminately for this genus." Keay et al. (1964) stated that *Dialium* is derived from the Greek word for "destroyed," in reference to the fact that the petals soon fall off. The term is also said to refer to the insecticidal properties of the crushed wood, especially that of *D. engleranum* Henriques. Gunawardena (1968) attributes the name to Dialis, the priest of Jupiter, signifying high rank.

Type: *D. indum* L.

70 species. Trees, unarmed, medium to large, often buttressed or with thick root swellings or thin butt flares; bark usually exuding red brown resin; rarely shrubs. Leaves imparipinnate, often rusty-pubescent when young; leaflets 3–21, usually in pairs, entire, elliptic, acuminate, profusely veined; stipules small, linear, caducous. Flowers yellow, white, or greenish, small, in dense axillary and terminal panicles; bracts small, cup-shaped, often silky-haired, caducous; calyx tube very short and saucer-shaped, or absent; sepals 5–7, thick, petaloid, imbricate, often pubescent, subequal, spreading or recurved at anthesis, soon deciduous; petals 1, 2, or none, rarely 5, not longer than the sepals, oblong-spathulate, entire, very caducous; stamens usually 2, sometimes 5, rarely 10, inserted at the edge of the hairy disk; filaments short, thick, straight or bent; anthers basifixed, bilobed, dehiscent by lateral slits; ovary obliquely short-stalked, hairy, 2–3-ovuled; style straight, smooth; stigma terminal, small. Pod drupaceous, small, ovoid, subglobose, or disk-shaped, indehiscent, often velvety brown or black; seeds 1–2, in pulpy endocarp. (Steyaert 1952)

Chakravarty (1969) considers the genus "a relict of an ancient group of apocarpous Leguminosae . . . or of a parallel stock," as judged by the reduction of petals and stamens and the presence of more than 1 ovary in some species.

Members of the genus are indigenous to tropical Africa, Madagascar, and Asia. There are about 10 species in the Western Hemisphere from southern Mexico to northern Brazil. *Dialium* species occur in high rainfall or inundated areas, moist, dense, semideciduous forests, and along river and lagoon banks. Some species prefer a well-drained soil, savannas, and hill slopes.

The fruits of several species have edible pulp suggestive of small plums, or tamarinds. They are sold in Malaysian and Sri Lanka bazaars primarily for use in chutneys.

The wood is hard, heavy, strong, and durable. The highly siliceous nature of some species inhibits teredine borer attack (Koeppen and Iltis 1962). The timber of young trees consists primarily of sapwood. The heartwood of mature trees may be dark mahogany brown or beautifully ripple-marked. In general, the hardness and the scant supply of the lumber limit its uses to boat and house construction, ship masts, rudders, and utility poles. The exploitation of *Dialium* wood in Malaysia around 1900 resulted in severe logging restrictions (Burkill 1966). The wood of *D. guianense* (Aubl.) Sandw., the only species occurring in North America, has a density of 900 kg/m^3; the logs do not float.

Decoctions of the bark and leaves of several West African species are employed in folk medicine for combating fevers, as a gargle for toothache and mouth sores, and in the treatment of infant diarrhea. The bark is a substitute for betel nut in Indochina (Dalziel 1948; Burkill 1966).

NODULATION STUDIES

Nodules were observed by Bonnier (1957) on *D. zenkeri* Harms but not on *D. pachyphyllum* Harms, in a forest in Yangambi, Zaire. Nodule morphology was not described. *D. engleranum* lacked nodules in Zimbabwe (Corby 1974).

***Dicerma* DC.**   Papilionoideae: Hedysareae, Desmodiinae

From Greek, *di-*, "two," and *kerma*, "coin," in reference to the 2 flat, round joints of the pod.

Type: *D. biarticulatum* (L.) DC.

3 species. Herbs, prostrate, woody at the base, or rigid shrublets; young shoots sometimes softly pubescent. Leaves pinnately 3-foliolate; leaflets oblong-lanceolate, small; stipules scarious, united, 3-forked. Flowers yellow or red, small, 2–3 in each narrow bract along the rachis of long narrow terminal racemes; calyx bibracteolate, bilabiate; lobes 5, upper 2 united almost to the apex, lower 3 acute; calyx disk tubular, thin around the ovary base; standard oblong-obovate, erect; wings oblong; keel straight, blunt; vexillary stamen joined with others at the base, free above; ovary sessile, 2-ovuled; style glabrous; stigma capitate. Pod straight, exserted, deeply indented between the 2 flat, round joints. (Schindler 1924)

Dicerma is closely related to *Desmodium* wherein it was formerly included

235

as a section. Schindler (1924) considered *D. biarticulatum* (L.) DC., *D. hispidum* Schindl., and *D. novoguineense* Schindl. sufficiently distinctive to merit their segregation into a separate genus, *Dicerma*.

Representatives of this genus are native to Burma, Papua New Guinea, and subtropical Australia.

NODULATION STUDIES

D. biarticulatum was reported nodulated in Queensland, Australia (Bowen 1956).

Dichilus DC.   Papilionoideae: Genisteae, Crotalariinae
204 / 3664

From Greek, *di-*, "two," and *cheilos*, "lip," in reference to the bilabiate calyx.

Type: *D. lebeckioides* DC.

5 species. Undershrubs, slender, erect. Leaves long-petioled, palmately 3-foliolate; leaflets linear, oblanceolate, or obovate, sometimes mucronate, pilose; stipules none or inconspicuous. Flowers yellow, nodding, solitary or 1-paired, at apices of lateral branches; bracteoles minute; calyx campanulate, bilabiate, glabrous, upper lip 2-lobed, lower lip 3-lobed; standard suborbicular or ovate, with a channeled, short linear claw; wings eared with a broad linear claw; keel obtuse, longer than the standard and wings; stamens monadelphous, all united into a sheath split above; anthers alternately shorter, dorsifixed, and longer, basifixed; ovary subsessile, many-ovuled, ciliate on the sutures; style incurved; stigma terminal. Pod linear-cylindric, sutures ciliate, compressed, thinly septate between the seeds, 2-valved.

Polhill (1976) assigned this genus to the tribe Crotalarieae in his revision of Genisteae *sensu lato*.

Members of this genus are indigenous to southern Africa. They occur in central regions of Cape Province, Natal, Transvaal, Lesotho, and Zimbabwe.

NODULATION STUDIES

Nodules were observed on *D. lebeckoides* DC. in South Africa (Grobbelaar et al. 1967) and in Zimbabwe (Corby 1974), and on *D. pilosus* Conrath ex Schinz and *D. strictus* E. Mey. in South Africa (Grobbelaar and Clarke 1972, 1975).

From Greek, *di-*, "two," *chroma*, "color," and *stachys*, "spike"; the flowers at the top and base of the spike are differently colored.

Type: *D. cinerea* (L.) W. & A.

20 species. Shrubs, or small, shaggy, and yet oddly attractive trees, usually 5–8 m tall, seldom over 12 m; trunk diameter usually about 22 cm; branches tend to intertwine, giving a matted appearance to the umbrella crown; branchlets armed with sharp, woody spines often ending in a hard, straight, sharp point. Leaves bipinnate; leaflets usually small, in numerous pairs, on some species leaflets close when a leaf is picked; stipules small. Flowers 5-parted, handsome, decorative, sessile in spikes, axillary, cylindric, solitary or paired, arranged in 2 sections: *lower* flowers neuter, white, rose, purplish, or pinky mauve; staminodes 10, long, filiform; ovary rudimentary, small; *upper* flowers bisexual, yellow or yellow-tipped; calyx campanulate, longer than the 5 lobes; lobes short-dentate, valvate; petals valvate, coherent below the middle; stamens 10, free, short-exserted; anthers crowned with a stalked gland; ovary subsessile, oblong, villous, several- to many-ovuled; style longer than the stamens; stigma terminal, truncate. Pods clustered, linear, compressed, much-twisted, leathery, indehiscent or irregularly dehiscent along sutures, nonseptate within; pods remain on trees for several months; seeds about 4, obovate, compressed, smooth.

The genus is closely allied to *Neptunia*.

Species of this genus are native to the tropics of the Old World. They are widespread in tropical Africa, Asia, and Madagascar. There is 1 endemic species in Australia. *Dichrostachys* is commonly found in the deciduous woodlands of Africa, wooded grasslands, the bush, savannas, and fringe forests. Optimum growth occurs on moist sites; the plants are usually stunted in dry, stony areas.

The 2 best-known and most prominent species, *D. cinerea* (L.) W. & A. and *D. glomerata* (Forsk.) Chiov. (= *D. nutans* Benth.), are soil-improvement plants. Crops planted on cleared areas do exceedingly well. All species are good cover plants on very dry soils. Under conditions of optimum growth *Dichrostachys* species often form thickets, an objectionable feature noticed following their introduction into the West Indies. The pods of most species are eaten by game and stock, despite the spiny branches. Seeds of some species are not digested; thus ingestion is a means of dispersal (Coates Palgrave 1957).

The sapwood is yellow to brown and beautifully marked with dark streaks. The heartwood is dark, brown, hard, tough, durable, and resistant to termite and borer damage, but the stems and tree trunks are small and usually crooked. Therefore, the usefulness of the wood is limited to small turnery articles, walking sticks, implement handles, and bows. The stems, roots, and wood of most species furnish considerable heat as fuel and charcoal (Palmer and Pitman 1961). The bark of most species yields a strong fiber suitable for cordage. Bark, leaf, and root decoctions are used in native medicines to heal skin infections, abscesses, and wounds, as remedies for catarrh, toothache, and snakebite, and for antidysenteric, diuretic, and vermifuge purposes. A low-quality gum is produced by *D. glomerata*. The pods of *D. major* Sim yield about 9 percent tannin.

Nodulation Studies

Wilson (1939a) reported nodules formed on *D. nutans* by rhizobia from *Albizia julibrissin*, *Desmodium canadense*, *Stizolobium deeringianum*, and *Vigna sinensis*. Typical nodule morphogenesis and development on *D. spicata* Domin were described by Gibson (1958).

Nodulated species of *Dichrostachys*	Area	Reported by
cinerea (L.) W. & A.		
ssp. *africana* Brenan & Brummitt		
var. *africana*	Zimbabwe	Corby (1974)
	S. Africa	Grobbelaar & Clarke (1975)
var. *lugardiae* (N. E. Br.)	Zimbabwe	Corby (1974)
Brenan & Brummitt	S. Africa	Grobbelaar & Clarke (1975)
var. *plurijuga* Brenan & Brummit	Zimbabwe	Corby (1974)
ssp. *argillicola* Brenan & Brummitt		
var. *hirtipes* Brenan & Brummitt	Zimbabwe	Corby (1974)
ssp. *cinerea*	S. Africa	Grobbelaar et al. (1967)
ssp. *nyassana* (Taub.) Brenan	S. Africa	Grobbelaar & Clarke (1975)
ssp. *platycarpa* (Welw. ex Bull) Brenan & Brummitt		
var. *platycarpa*	S. Africa	Grobbelaar & Clarke (1974)
glomerata (Forsk.) Chiov.	U.S.A.	Wilson (1939a)
(= *D. nutans* Benth.)		
spicata Domin	Australia	Gibson (1958)

Dicorynia Benth.   Caesalpinioideae: Cassieae
83 / 3539

From Greek, *di*-, "two," and *koryne*, "club," in reference to the 2 club-shaped stamens.

Type: *D. paraensis* Benth.

2 species. Trees, medium to very large, up to 50 m tall, with trunk diameters up to 2 m, unarmed; buttresses, if present, low; bark thick, reddish brown; branches, buds, and leaves with light gray or brown pubescence. Leaves imparipinnate; leaflets large, alternate, 5–13, uppermost largest; stipules early caducous. Flowers white or rosy white, small, numerous, in large, terminal thyrse; bracts and bracteoles ovate, early caducous; calyx tube

very short; sepals 5, free, unequal, concave, imbricate; petals 3, oblique-orbicular, edges crisped, short claw darker colored; stamens 2, free, unequal, club-shaped; filaments whitish, glabrous; anthers yellow, crustaceous, dehiscent by a V-shaped apical slit; ovary sessile, free from calyx tube, with brown pubescence, 3–6-ovuled; style filiform, inflexed; stigma small, terminal. Pod oblong-ovate, leathery, compressed, upper suture narrow-winged, 2–4-seeded, tardily dehiscent along upper suture; seeds orbicular to kidney-shaped, transverse, compressed.

The type species and 6 varieties constitute the Amazonian group; *D. guianensis* Amsh., the second member of the genus, has Surinam and French Guiana as its center of distribution (Koeppen 1967). The major distinctions between the 2 species are in floral structure and in range of distribution.

D. paraensis Benth. occurs primarily in the Amazonas Territory of Brazil in inundated forest areas, on lower terrain along the Rio Negro. *D. guianensis* inhabits well-drained sandy or sand-clay soils not subject to flooding.

The wood of *D. paraensis* is uncommon on the timber market and little used locally; however, that of *D. guianensis*, known as angelique, is highly regarded for its excellent quality (Kukachka 1964; Koeppen 1967). Three forms are recognized according to the color of the heartwood: angelique gris has russet-colored heartwood that becomes dull brown with a purplish cast upon drying; angelique rouge has reddish-cast heartwood that frequently shows purplish banding; and the heartwood of angelique blanc is grayish white. These forms do not differ in mechanical properties; all are superior to teak and white oak. *Dicorynia* is one of the few leguminous woods that accumulate silica. The individual silica particles measure about 2.5μ in diameter, occur in the vertical parenchyma and marginal cells of the wood rays, and constitute 67–87 percent of the ash from unextracted wood of the 3 forms. The wood is highly regarded for its resistance to marine borers, white rot fungi, and mechanical abrasion (Kukachka 1964). It is used for vats, for heavy construction, and for boat frames, piers, and similar marine uses.

Dicraeopetalum Harms   Papilionoideae: Sophoreae
(122b) / 3580b

From Greek, *dikraios*, "2-forked," and *petalon*, "petal," in reference to the bilobed shape of the petals.

Type: *D. stipulare* Harms

1 species. Tree. Leaves imparipinnate; branchlets densely covered with circular scars; leaflets 4–5 pairs, small, oblong or obovate; stipules lanceolate-subulate, joined at the base, persistent. Flowers many, in long racemes at tips of short branches; bracts lanceolate, silky-pubescent; calyx cupular, thickened at the base; teeth 5, lanceolate, upper 2 somewhat connate and subequal to the calyx tube; petals 5, inserted in the lower part of the calyx tube, imbricate, subequal, short-clawed, obovate-oblong, briefly bilobed apically or emarginate; stamens 10, inserted with petals; filaments free, exserted, glabrous;

anthers small, broad-ovate, subbasifixed; ovary short-stalked, linear, densely villous, 1- rarely 2-ovuled; style short, thick, smooth; stigma small, capitate. Pod not described. (Harms 1908a)

This monotypic genus is endemic in Somalia.

Dicymbe Spruce ex Benth. & Hook. f.   Caesalpinioideae:
107 / 3565   Sclerolobieae

From Greek, *di-*, "two," and *kymbe*, "shell," "boat," in reference to the 2 concave, shell-like bracteoles enclosing the flower in bud.

Type: *D. corymbosa* Spruce ex Benth.

7–10 species. Trees, medium to large, unarmed. Leaves paripinnate; leaflets in few pairs, leathery, long, pointed. Flowers white, large, in racemes or corymbose panicles; bracts thick, shell-like, early caducous; each flower subtended by 2, thick, valvate, leathery, persistent bracteoles which envelop the flower in bud; calyx turbinate, thick; lobes 4, oval-oblong, imbricate, often bilobed at apex; petals 5, ovate, subequal, broad, white, often with reddish claw or base, overlapping; stamens 10, free; filaments inflexed, hairy at base; anthers uniform, linear; ovary short-stalked, stalk free of calyx tube, many-ovuled; style long, involute in bud; stigma peltate. Pod compressed, transversely nerved, middle nerve absent, 2-valved, explosively dehiscent.

Members of this genus occur in South America, especially in the Amazon region of Brazil, Venezuela, and Guyana.

The trees are rather rare; thus the timber is uncommon and little used. Record and Hess (1943) described the one available specimen of *D. corymbosa* Spruce ex Benth. as moderately hard, heavy, tough, and strong, and having a rather fine texture and straight grain, but a low rating for durability. The fact that it is splintery or fibrous, as well as being difficult to work, detracts from its usefulness.

NODULATION STUDIES

Nodules were observed on 2 species, *D. altsoni* Sandw. and *D. corymbosa* in the Amatuk region of Guyana (Whitton 1962).

Didelotia Baill.   Caesalpinioideae: Amherstieae
53 / 3503

Named for Admiral (Baron) Didelot.

Type: *D. africana* Baill.

8 species. Shrubs, mostly trees, large, up to 50 m tall, with trunk diameters of 1 m or more, straight boles up to 25 m before branching, and open, deltoid crowns. Leaves paripinnate, short-petioled; leaflets 1–7 pairs, sessile or subsessile, oblique, asymmetric at the base, apex emarginate or acuminate;

stipules persistent or caducous. Flowers small, inconspicuous, in few-flowered, axillary or terminal racemose panicles; bracts minute; bracteoles paired, rounded, valvate, completely enclosing the flower bud; calyx rudimentary; sepals 4–5, minute; petals 4–5, or none, minute; stamens 4–5, free, subequal; filaments threadlike, exserted, alternating with 4, or fewer, rudimentary staminodia; anthers oblong, longitudinally dehiscent; ovary sub-sessile, elliptic; style filiform; stigma terminal. Pod yellow brown, smooth, up to 11 cm long by 3–4 cm wide, flat, oblong-lanceolate, rounded at the base, apiculate at apex, 2-valved, dehiscent; seeds 1–3, disk-shaped or elliptic, brown, smooth. (Oldeman 1964; Voorhoeve 1965)

Members of this genus are common in West Africa; there are 4 species in Liberia (Voorhoeve 1965). They occur in evergreen forests, river border areas, and on rocky slopes.

No commercial use is made of the medium-hard, heavy, attractive, brown wood because it is not readily available. The dark, sticky gum of certain species is of no special importance.

Didymopelta Regel & Schmalh.   Papilionoideae: Galegeae,
297 / 3765   Astragalinae

From Greek, *didymos*, "twin," and *pelte*, "shield"; the pods are partitioned into 2 shield-shaped compartments.

Type: *D. turkestanica* Regel & Schmalh.

1 species. Herb, annual, very small, pubescent with simple, appressed, rough hairs. Leaves imparipinnate; stipules free, awl-shaped. Flowers few, in racemes; bracts awl-shaped; calyx campanulate-tubular; teeth awl-shaped, upper 2 smaller than the lower 3; petals clawed; standard erect, bilobed at the apex; wings oblong; keel straight, blunt, longer than the wings; stamens 10, diadelphous; ovary stalked, 4-ovuled; style filiform; stigma terminal, capitate. Pod lens-shaped, on a slender stalk, divided inside longitudinally into 2 shield- or boat-shaped valves by the intrusion of the seed-bearing dorsal suture; each compartment 2-seeded.

The genus is closely related to *Astragalus*.

This monotypic genus is endemic in Turkestan.

Dillwynia Sm.   Papilionoideae: Podalyrieae
180 / 3640

Named for Lewis Weston Dillwyn, 1778–1855, English botanist and naturalist.

Type: *D. ericifolia* Sm.

15 species. Shrubs, slender, heathlike, erect. Leaves simple, narrow-linear or cylindric, with upper surface grooved; stipules absent. Flowers yel-

low or yellow and red, several, in short, axillary or terminal racemes, or clustered on short peduncles, rarely solitary; bracts small, very caducous; bracteoles 2, minute, on short pedicels slightly below the calyx; calyx lobes short, about as long as the tube, 2 upper lobes united into a lip; petals clawed; standard broader than long, kidney-shaped; wings narrow, crimson; keel acuminate or blunt, crimson; stamens free, inserted on rim of calyx disk; ovary short-stalked, 2-ovuled; style short, thick, hooked near apex, erect; stigma truncate, thick. Pod ovoid or oblong, turgid, subsessile, 2-valved; seeds 1–2, kidney-shaped. (Bentham 1864; Bentham and Hooker 1865)

Representatives of this genus are native to Australia. They are well-distributed in Victoria, Queensland, and Tasmania; there are 11 species in New South Wales. Species occur in coastal and scrub areas, tablelands, and on dry, stony hills and poor soils.

D. brunioides Meissn., *D. ericifolia* Sm., *D. phylicoides* A. Cunn., and *D. sericea* A. Cunn., all with masses of yellow and red flowers, and *D. juniperina* Sieb., with yellow and brown flowers, are showy spring ornamentals.

NODULATION STUDIES

Cultural characteristics and unilateral host-infection tests using a single rhizobial strain from *D. uncinata* (Turcz.) C. A. Gardn. conformed with reactions of slow-growing cowpea-type rhizobia (Lange 1961).

Nodulated species of *Dillwynia*	Area	Reported by
brunioides Meissn.	N.S. Wales, Australia	Hannon (1949)
cinerascens R. Br.	W. Australia	Lange (1959)
divaricata (Turcz.) Benth.	W. Australia	Lange (1959)
ericifolia Sm.	N.S. Wales, Australia	Hannon (1949)
floribunda Sm.	N.S. Wales, Australia	Hannon (1949)
	W. Australia	Harris (1953)
hispida Lindl.	W. Australia	Harris (1953)
juniperina Sieb.	N.S. Wales, Australia	R. W. McLeod (pers. comm. 1962)
peduncularis Benth.	Queensland, Australia	Bowen (1956)
uncinata (Turcz.) C. A. Gardn.	W. Australia	Lange (1959)

Dimorphandra Schott   Caesalpinioideae: Dimorphandreae
33 / 3473

From Greek, *di*-, "two," *morphe*, "form," and *aner*, *andros*, "man," in reference to the 2 kinds of stamens.

Type: *D. exaltata* Schott

25 species. Trees, unarmed, small to very large, some with immense

girth. Leaves bipinnate; pinnae pairs 1–many; leaflet pairs 2–many, usually hairy; stipules small, caducous. Flowers small, white, yellow, or red, in dense racemes or spikes; bracts small, early deciduous; bracteoles none; calyx campanulate; lobes 5, short, often pubescent; petals spathulate, free; stamens free, 5 fertile ones opposite the petals alternating with 5 staminodes; staminodes usually free, thickened at apex; ovary sessile or short-stalked; style short; stigma terminal. Pod elongated, flat, linear or falcate, woody or fleshy, 2-valved or indehiscent, containing small, cylindric or oval, hard seeds. (Ducke 1925a, 1935a)

Species of *Mora* are often confused with *Dimorphandra*; leaves of the former are pinnate, those of the latter bipinnate.

Members of this genus are common in tropical Brazil, Guiana, and Amazonian Venezuela. They occur in upland forests on moist soil or on dry savannas of white sand.

Very little is known about the woods of *Dimorphandra* species or their usefulness (Record and Hess 1943).

NODULATION STUDIES

D. davisii Sprague & Sandw., a dominant forest tree of the Amatuk region of Guyana, was reported by Whitton (1962) to be abundantly nodulated.

Dinizia Ducke   Mimosoideae: Adenanthereae

Named for José P. Diniz, sponsor of the botanical exploration of the Trombetas River, Brazil.

Type: *D. excelsa* Ducke

1 species. Tree, large, unarmed; young branches rusty-pubescent. Leaves bipinnate; pinnae 7–11 pairs, alternate; leaflets 8–10 pairs; petiolar glands small, indistinct. Flowers small, green, polygamous, borne in solitary or paired, large, terminal racemes; calyx teeth 5, short, valvate; petals 5, free; stamens 10, free, long-exserted, twisted; anthers not glandular; ovary short-stalked; style long-exserted. Pod large, 25–35 cm long by about 5 cm wide, thin, reddish brown, leathery, wrinkled and veined longitudinally, indehiscent; seeds transverse, compressed, shiny, black. (Ducke 1922)

Burkart (1943) interpreted *Dinizia* to be closely related to, but distinct from, *Mimozyganthus*; however, Ducke (1922) positioned the genus near *Stryphnodendron*.

This monotypic genus is endemic in Brazil. The plants grow in primeval forests bordering rivers.

These colossal trees, 60–65 m in height, with trunk diameters of about 2 m, are among the largest found in the lower Amazon region. The hard, heavy wood is decay-resistant, but is sparingly used because it is difficult to work (Ducke 1922; Record and Hess 1943).

Named for Diocles, considered by the ancient Greeks second only to Hippocrates in his knowledge of plants.

Lectotype: *D. sericea* HBK

30–50 species. Vines, woody or shrubby, tall, climbing. Leaves pinnately 3-foliolate; stipels setaceous; stipules sometimes present. Flowers large, many, violet, blue, or white, in clusters on sessile or stalked nodes on axillary, elongated racemes; bracts narrow; bracteoles orbicular, membranous; calyx campanulate, silky-haired inside; lobes 4 or 5, upper 2 united, lowest lobe longest; standard ovate, reflexed, with 2 inflexed, basal auricles; wings free, oblong; keel blunt, straight or incurved, short; stamens 10, pseudomonadelphous, vexillary stamen free only at the base; anthers all fertile or alternate ones small, abortive; ovary subsessile; ovules 2–many; style incurved, usually not bearded, dilated above; stigma terminal, truncate. Pod linear-oblong, compressed or turgid, leathery, 2-valved, both sutures narrow-winged or upper one thickened, dehiscent or indehiscent; seeds usually globose with linear hilum.

Species of this genus are pantropical in both hemispheres, principally in the Americas. They are abundant in Brazil and Peru, and there are 3 species in Argentina. *Dioclea* species are commonly found along the coast in moist parts of the tropics but are also widespread in moist inland thickets, rocky open areas, and light, wet forests.

The species are mostly wild forms and are of little economic importance. *D. lasiocarpa* Mart. ex Benth. and *D. violacea* Mart. ex Benth. are often cultivated in botanical gardens for their attractive, fragrant flowers. Flowers of the latter species, known in the Hawaiian Islands as maunaloa, or sea bean, are strung into garlands for leis. *D. reflexa* Hook. f., the marbles vine, is highly regarded in some parts of Africa. The spherical seeds are used in games; root decoctions serve to alleviate coronary pain; and seed and root extracts are said to have insecticidal properties. Freise (1936) reported the presence of physostigmine, $C_{15}H_{21}N_3O_2$, a parasympathomimetic alkaloid, in seeds of *D. bicolor* Benth., *D. lasiocarpa*, *D. macrocarpa* Huber, *D. reflexa*, and *D. violacea*. This alkaloid is most commonly extracted from the Calabar bean, *Physostigma venenosum*.

NODULATION STUDIES

D. guianensis Benth. in Trinidad (DeSouza 1966) and *D. violacea* in Hawaii (personal observation 1947) and in Trinidad (DeSouza 1966) were reported nodulated.

Diphyllarium Gagnep.

Papilionoideae: Phaseoleae,
Glycininae

From Greek, *di-*, "two," *phyllon*, "leaf," and the Latin suffix -*aris*, "provided with," in reference to the 2 leafy flower bracteoles.

Type: *D. mekongense* Gagnep.

1 species. Shrublet, climbing, twining. Leaves pinnately 3-foliolate,

eglandular; leaflets large, ovate; stipels and stipules persistent. Flowers rose-colored, clustered 2–3 at the nodes, at first hidden by 2 striated bracteoles, in paniculate, axillary and terminal racemes, nodes not swollen; bracts stipulelike, persistent, much smaller than the 2 bracteoles; calyx tube short, tubular-campanulate, 2 lateral teeth slightly shorter than tube, upper and lower teeth longer than tube; petals equal, short-clawed; standard orbicular. not appendaged; wings and keel similar, suberect; stamens diadelphous; anthers uniform; ovary sessile, many-ovuled; style nearly straight, subglabrous, awl-shaped; stigma acute, terminal. Pod sessile, flat, linear, entire, 2-valved; seeds orbicular-reniform, arranged lengthwise. (Gagnepain 1915)

This monotypic genus is endemic in Vietnam and Laos.

Diphysa Jacq.
276 / 3741

Papilionoideae: Galegeae,
Robiniinae

From Greek, *di-*, "two," and *physa*, "bladder"; the inflated exocarp forms 2 bladders on each side of the pod.

Type: *D. carthagenensis* Jacq.

18 species. Shrubs, or small, deciduous trees, rarely more than 12 m tall; some spinescent; branchlets often pubescent. Leaves imparipinnate; leaflets 5–25, entire, eglandular, alternate; stipels none; stipules small. Flowers yellow, usually large, showy, in short, axillary, lax racemes or fascicles, pedicels solitary; calyx campanulate, subtended by 2 caducous bracteoles, turbinate at base; lobes 5, unequal, upper 2 broad, rounded at apex, lower 3 narrow, lowest lobe longest; petals unequal; standard orbicular, reflexed, clawed, with 2 callosities above the claw; wings oblong, incurved; keel petals lunate, sometimes beaked; stamens 10, diadelphous; anthers uniform; ovary stalked, many-ovuled; style incurved, glabrous; stigma small. Pod stalked, flattened, oblong, exocarp veined, paper-thin, inflated, forming a bladder on each side of the pod, indehiscent, somewhat interrupted inside between the seeds; seeds transversely oblong, flattened.

Hutchinson (1964) designated this genus the sole member of his new tribe Diphyseae.

Members of this genus are common in Mexico and from Central America into Venezuela. They occur on rocky hillsides, in thin forests and dry or moist thickets; they are present mostly in the uplands.

The wood is similar to that of *Robinia pseudoacacia* and suitable for the same purposes but of little importance in the timber trade. The trees are often poorly formed and less than 12 m in height. *D. carthagenensis* Jacq., *D. floribunda* Peyr., and *D. robinioides* Benth. are the best-known species. The 2 last-named species, with their masses of brilliant yellow flowers, have ornamental value and are called tree *Crotalaria* in the native dialect. The wood of *D. racemosa* Rose has a disagreeable odor. The heartwood of some species yields a yellow dye. Some of the species form dense thickets and are propagated for live fences (Record and Hess 1943).

From Greek, *diploos*, "double," and *tropis*, "keel"; the wing and keel petals are similar in size and shape.

Type: *D. martiusii* Benth.

14 species. Trees, medium to large, unarmed. Leaves imparipinnate; leaflets 5–15, usually large, ovate, leathery; stipels none; stipules small. Flowers pink, violet, or white, in upper axillary racemes or terminal panicles; bracts and bracteoles small; calyx coriaceous, oblique, incurved; teeth 5, the 2 uppermost joined higher up; standard oblong or rounded, eared above the claw; wings and keel petals free, long-clawed, similar in size and shape; stamens 10, free; filaments alternately long and short; anthers ovate or oblong; ovary subsessile; style slender, incurved; stigma oblique, small. Pod leathery or woody, thick or membranous, oblong, 2-valved, indehiscent, 1- to few-seeded.

Representatives of this genus are native to tropical South America, Colombia, Venezuela, Guiana, and northern Brazil. They occur in upland forests, except *D. martiusii* Benth., which inhabits marshes and inundated regions.

The large tree species provide excellent hard, durable timber for heavy construction, furniture-making, and shipbuilding (Record and Hess 1943). In Guiana, Indians use an infusion of the bark of *D. brachypetalum* Tul. to destroy vermin (Uphof 1968).

NODULATION STUDIES

Three seedling specimens of *D. purpurea* (Rich.) Amsh. examined at 2 sites in Bartica, Guyana, by Norris (1969) lacked nodules.

Dipogon Liebm.   Papilionoideae: Phaseoleae, Phaseolinae

From Greek, *di-*, "two," and *pogon*, "beard," in reference to the arrangement of hairs on the style.

Type: *D. lignosus* (L.) Verdc. (= *Dolichos lignosus*)

1 species. Vine, scrambling, climbing, perennial, short-lived; stems woody, many; branches very long, slender, somewhat downy. Leaves alternate; leaflets rhomboid, elongated, acute; stipules entire, somewhat triangular. Flowers red purple, drooping somewhat, 3–6, in erect, solitary, axillary clusters; bracts acute, lanceolate, hairy; calyx campanulate, ciliate, bilabiate, upper lip bidentate, lower lip tridentate; standard suborbicular, reflexed, with 2 appendages; wings clawed, falcate, obtuse; keel petals partly joined, clawed, falcate, blunt; stamens 10, diadelphous; filaments slender, glabrous; anthers globose, dorsifixed, longitudinally dehiscent; ovary glabrous, linear, 4–8-ovuled; style strongly incurved at base and apex in same direction, middle part curved oppositely, apical half barbate beneath in 2 close, longitudinal lines, thicker near basal bend, narrowed toward apex; stigma terminal but not

fringed. Pod brownish, stalked, compressed, glabrous, short truncate at apex, 4–5-seeded. (Liebmann 1854)

The taxonomic status of this species has invited various opinions; however, its present identity appears justified (Freeman 1918; Verdcourt 1968b, 1970c).

This monotypic genus is widespread in South Africa and is extensively cultivated in South America and Australia.

This climbing herb is admired in many areas as an ornamental and also valued as a green manure and cover crop. The edible pods resemble snap beans.

NODULATION STUDIES

D. lignosus (L.) Verdc. was reported nodulated as *Dolichos lignosus* and shown to have versatile kinships in host-infection tests conducted by Wilson (1939a). Nodules were also observed on this species in South Africa (Grobbelaar and Clarke 1975).

Dipteryx Schreb.   Papilionoideae: Dalbergieae,
(371) / 3845   Geoffraeinae

From Greek, *di-*, "two," and *pteryx*, "wing," in reference to the 2 large, winglike lobes of the calyx.

Type: *D. odorata* (Aubl.) Willd.

10 species. Trees, unarmed, small to very large; bark and sapwood of some species coumarin-scented. Leaves opposite or alternate, paripinnate; leaflets opposite or alternate, 3–14, often gland-dotted; rachis slightly winged or flattened, with apex projected beyond the leaflets; stipels none. Flowers violet or rose, showy, in terminal panicles; bracts and bracteoles small or later absent; calyx tube short, campanulate, pinkish, coriaceous, pubescent, glandular, short-toothed, or with 2 upper lobes very large, winglike or petaloid, membranous, lower 3 joined into a small entire or 3-toothed lip; petals clawed, very unequal; standard obovate, emarginate; wings falcate or oblique-oblong, free, often emarginate; keel petals similar to wings, smaller, free or joined dorsally; stamens 10, all joined into a sheath split above; anthers ovate, versatile, 5 alternate ones shorter or abortive; ovary linear, glabrous, 1-ovuled, stalked; style straight or incurved, slender; stigma small, terminal, capitate. Pod drupaceous, elliptic, somewhat compressed, endocarp woody, with or without an oily, aromatic pericarp, indehiscent; seed 1, flat, oblong, strongly coumarin-scented or odorless. (Bentham and Hooker 1865; Ducke 1940)

Ducke's position (1934c, 1940) that the conservation of *Dipteryx* over *Coumarouna* was improper appears well founded.

Members of this genus are native to northern Brazil, Central America, Guiana, and Venezuela. They flourish along rivers and in woodlands.

247

The genus embodies two natural groups of plants that are sometimes casually referred to by the vernacular names "cumarú" and "false cumarú," respectively. One group consists of 4 species, *D. odorata* (Aubl.) Willd., *D. punctata* (Blake) Amsh., *D. rosea* (Spreng. ex Benth.) Taub., and *D. trifoliolata* Ducke, which are characterized by a leathery calyx and fragrant, coumarin-yielding seeds; these are the source of the tonka beans of commerce. The other 6 species have petaloid calyx wings and no parts of the plant are coumarin-scented (Ducke 1940). The lack of coumarin is compensated for, in part, by their ornamental beauty.

Of the aforementioned species, *D. odorata* is the major source of tonka beans and hence of coumarin, $C_9H_6O_2$, whose fragrance resembles vanilla. Venezuela is the principal export center (Pound 1938). Two commercial cultivars, the Angostura-Venezuelan type, and the Brazilian or Pará type, are recognized (Purseglove 1968). Although the demand for tonka beans is erratic and coumarin has been replaced to a certain extent by synthetic coumarin, a commercial market still exists for its use in flavoring snuff, cigarettes, cigars, cocoa, and confectionery, and as an ingredient of perfumes, liqueurs, sachet powders, and cosmetics. The beans also yield a high percentage of solid fat known as tonka butter, which is used in flavoring food. A copallike resin is obtained from leaves, stems, and fruits, and a kinolike gum is obtained from excised bark. *D. trifoliata*, the largest of the true tonka-bean trees, bears pods with a sweet, edible pericarp.

The members of this genus range from majestic trees 45 m or more in height, sustained by smooth trunks arising from huge buttressed bases, to others of only medium size. The hard, durable, medium- to fine-textured wood is difficult to work. Its limited supply is reserved for veneers and turnery objects. Inaccessibility of the timber sites and the value placed upon the fruit have hindered exploitation (Record and Hess 1943).

NODULATION STUDIES

Seedlings of *D. odorata* examined in Bartica, Guyana, by Norris (1969) and in the Bogor Botanic Garden, Java (personal observation 1977), lacked nodules.

Diptychandra Tul.   Caesalpinioideae: Sclerolobieae
112 / 3570

From Greek, *di-*, "two," *ptyche*, "fold," and *aner, andros*, "man"; the stamen filaments are twice-folded in bud.

Lectotype: *D. epunctata* Tul.

3 species. Shrubs or small trees, unarmed. Leaves paripinnate; leaflets membranous, minutely pellucid-punctate; stipels absent; stipules inconspicuous or absent. Flowers small, yellow, in loose, axillary and terminal racemes; bracts and bracteoles small, caducous or absent; calyx tube short; lobes 5, imbricate, subequal; petals 5, small, ovate, imbricate, subequal; stamens 10,

free, inserted with the petals; filaments haired at base, erect, twice-folded in bud; anthers ovate, versatile, elliptic, 2-lobed; ovary stalked, free, few-ovuled; style filiform, inflexed; stigma terminal, small or truncate, glabrous. Pod short-stalked, short or elongated, compressed, coriaceously 2-valved, 1–3-seeded; seeds ovoid or round, transverse, compressed, winged on 1 side or all around. (Tulasne 1843)

Members of this genus occur in Brazil, Paraguay, and Bolivia.

Discolobium Benth.

Papilionoideae: Hedysareae, Aeschynomeninae

329 / 3799

From Greek, *diskos*, "quoit," "disk," and *lobion*, "pod," in reference to the shape of the pod.

Type: *D. pulchellum* Benth.

8 species. Herbs, perennial, or shrubs, glabrous or sticky-pubescent. Leaves imparipinnate; leaflets many, rarely 1–3; stipels none. Flowers yellow, striated, on slender pedicels, in long-stalked, axillary, lax racemes; bracts and bracteoles small, persistent; calyx campanulate, bilabiate, upper lip blunt, emarginate or bidentate, lower lip 3-parted with lanceolate lobes; standard suborbicular, short-clawed; wings obovate, subequal to standard; keel blunt, shorter than the wings; stamens 10, connate into a sheath split on both sides, vexillary stamen and keel stamen almost free from the base; anthers uniform, elliptic, dorsifixed; ovary short-stalked, 3-ovuled, pubescent; style incurved, smooth; stigma apical, oblique. Pod short, indehiscent, divided into 3 horizontal disks, of which the middle is the largest, kidney-shaped, membranous, and reticular, containing 1 long kidney-shaped seed; the 2 outside disks i.e., the upper and lower ones, smaller and sterile.

Representatives of this genus occur in subtropical Brazil, Paraguay, and Argentina. They are common in wet localities, especially along rivers (Burkart 1943).

Distemonanthus Benth.

Caesalpinioideae: Cassieae

78 / 3533

From Greek, *dis*, "double," *stemon*, "stamen," and *anthos*, "flower," in reference to the flower with 2 conspicuous, fertile stamens.

Type: *D. benthamianus* Baill.

1 species. Tree, up to 35 m tall, with trunk about 1 m in diameter, with low, spreading buttresses, unarmed, deciduous. Leaves imparipinnate; new leaves bronze-colored; leaflets 9–11 pairs, alternate, ovate to lanceolate, acute, basally rounded, lower surface pubescent. Flowers in lax, axillary, paniculate cymes, appearing before the leaves; bracts small, narrow, caducous; brac-

teoles absent; calyx turbinate; sepals 5–4, sessile, reddish brown, imbricate in bud, upper sepal lanceolate, recurved when flowering; petals 3, white, sessile, about 1.2 cm long, the upper petal broader than the 2 laterals; fertile stamens 2, inserted at each side of the upper petal; anthers purple, linear, erect, opening with 2 pores at the apex; staminodes 3, opposite the petals; ovary short-stalked, 4–5-ovuled, densely brown-hairy; style thick, long, straight; stigma oblique, terminal. Pod thin, flat, membranous with transverse venation, indehiscent, 2–3-seeded; seeds flat, elliptic, brown with yellow margin. (Voorhoeve 1965)

This monotypic genus is endemic in tropical West Africa, from Sierra Leone to Gabon. The trees grow in rain forests.

The wood is known in West Africa as ayan or movingui, a valuable export commodity. Although this species does not yield a true satinwood, it was once marketed as Nigerian or yellow satinwood. The yellowish, hard, heavy heartwood has a density of about 720 kg/m³ and many properties similar to white oak. Ripple marks and ribbon stripes in the interlocked grain are displayed attractively in quartersawed wood used for cabinetwork and furniture. The wood accumulates silica, which in turn dulls cutting tools, but its presence may account for the resistance of the wood to fungi and borers, and, still further, for the value of the timber in underwater and underground construction (Titmuss 1965; Kukachka 1970).

Dolichopsis Hassl.   Papilionoideae: Phaseoleae, Phaseolinae
(424a)

The Greek suffix, *opsis*, denotes resemblance to *Dolichos*.

Type: *D. paraguariensis* (Benth.) Hassl.

1 species. Herb, climbing or prostrate. Leaves pinnately 3-foliolate; leaflets lanceolate; stipels present; stipules lanceolate, grooved. Flowers violet, solitary or paired at nodes in long peduncled racemes; bracts and bracteoles grooved, lanceolate or ovoid, caducous; calyx campanulate; lobes 5, upper 2 rounded, nearly completely connate, lower 3 triangular-toothed, connate to one-third length; standard orbicular, with 2 inflexed, basal auricles; wings falcate-obovate with short spur; keel sharply beaked, incurved; stamens 10, diadelphous, vexillary stamen free with filament thickened at base, others connate in 2 rings, innermost ones smaller; anthers elliptic-oblong, subbasifixed; ovary subsessile, 8–10-ovuled, hairy; style dilated in basal half, compressed and jointed, filiform above, bearded lengthwise; stigma lateral, cupshaped, barbed with short hairs. Pod compressed, oblong-curved, beaked, elastically dehiscent, subseptate between the 7–10, oblong, black seeds attached in the middle, hilum elongated.

Harms (1915a) positioned this genus adjacent to *Lablab*, a *Dolichos* segregate.

This monotypic genus is endemic in Argentina and Paraguay. The plants are found in moist and humid lowlands.

NODULATION STUDIES

Rothschild (1970) observed nodules on *D. paraguariensis* (Benth.) Hassl. in Argentina, where it has potential as a forage plant.

***Dolichos* L.**   Papilionoideae: Phaseoleae, Phaseolinae
(422) / 3910

From Greek, *dolichos*, "long," in reference to the long, slender stems; also the Greek name for a plant with edible pods.

Type: *D. vulgaris* L.

60 species. Herbs or subshrubs, annual and perennial, scandent, twining, trailing or erect; some branching from an underground woody rootstock, some with long, woody taproots; stems glabrous, rarely pilose. Leaves pinnately, rarely palmately, 3-foliolate; leaflets ovate, elliptic, usually broadest in the basal half, sometimes acuminate, sometimes 3-lobed; stipels and stipules present. Flowers purple, pink, blue, or creamy, solitary or fascicled in leaf axils, or in axillary or terminal racemes; bracts and bracteoles small, striate, caducous; calyx campanulate, bilabiate; upper lip entire or 2-toothed, lower lip 3-lobed; petals clawed; standard orbicular or obovate, reflexed, usually with 2 inflexed ears at the base; wings falcate or oblong, sometimes pouched or eared, longer than and adhering to the keel; keel petals bent, seldom beaked or incurved, not spirally twisted; stamens diadelphous; vexillary filament free, sometimes toothed at the base; anthers uniform; ovary sessile or subsessile, linear-oblong, 4- to many-ovuled, base encircled by a cupular disk; style somewhat thicker at the base, often bearded above; stigma terminal, capitate, usually ringed with hairs. Pod linear-oblong, flattened, pubescent or glabrous, 2-valved, sutures often thick; seeds compressed, hilum usually central.

Verdcourt (1968a) proposed *D. trilobus* L. as the lectotype. The various proposals to retypify *Dolichos* L. have been reviewed by Westphal (1975). The original position number 428 (Taubert 1894) was changed to 422 by Harms (1908a) because of the reorganization within the subtribe necessitated by various revisions and inclusions of new genera. As Verdcourt (1970c) stated, "The genus as at present constituted is far from homogeneous and its constituent parts are separated from each other by more characters than many other groups long retained as genera in other parts of the Phaseoleae." In keeping with this line of reasoning, fragmentation of the *Dolichos* complex has resulted in the definition of various new genera, along with the reduction of estimated species from approximately 150 to about 60.

Representatives of this genus are pantropical, mainly in southern Asia and Africa; no species is native to North America. They are wild and cultivated. The plants flourish in mountain meadows, forest fringes, or as twiners in tall grasses; many species are shade-tolerant.

Dolichos species are excellent ground-cover, green-manuring, forage, and soil-improvement plants. Various species are planted as nurse crops and for shade on tea, coffee, cacao, and coconut plantations and clove tree estates. The woody rootstocks of *D. buchananii* Harms, *D. kilimandscharicus* Harms ex Taub., *D. oliveri* Schweinf., and *D. trinervatus* Bak. have native uses in East Africa as soap substitutes because of their high saponin properties. The 2 last-named species also have fish-poisoning properties (Verdcourt and Trump 1969).

Nodulation Studies

Young *Dolichos* nodules are spherical but become irregular in shape with age. The rhizobia commonly produce alkaline reactions in carbohydrate media and in litmus milk. Host-specificity reactions denote kinships within the cowpea miscellany.

Nodules produced on *D. biflorus* L. by a rhizobial strain recommended to improve the growth of *Lablab purpureus* (= *D. lablab* L.) contained a black pigment; attempts to separate the black pigment from leghemoglobin were not successful (Cloonan 1963). Claims have not been borne out that the rhizobial strains responsible for black-pigmented nodules have promise as tracer, or marker, strains in studying the fate of inocula applied under field conditions.

Nodulated species of *Dolichos*	Area	Reported by
(*africanus* Brenan ex Wilczek) †= *Macrotyloma africanum* (Brenan ex Wilczek) Verdc.	Brazil	Norris et al. (1970)
angustifolius E. & Z.	S. Africa	Grobbelaar & Clarke (1975)
axillaris L.	Australia	Cloonan (1963)
baumannii Harms	Australia	UDALS culture list (1962)
biflorus L.	Widely reported	Many investigators
debilis Hochst.	Australia	UDALS culture list (1962)
falciformis E. Mey.	S. Africa	Grobbelaar & Clarke (1975)
hastaeformis E. Mey.	S. Africa	Grobbelaar & Clarke (1975)
hosei Craib	Sri Lanka	W. R. C. Paul (pers. comm. 1951)
(= *Vigna hosei* (Craib) Backer)	Trinidad	DeSouza (1963)
kilimandscharicus Harms ex Taub.	Zimbabwe	Corby (1974)
(*lablab* L.) †= *Lablab purpureus* (L.) Sweet	Widely reported	Many investigators
(*lignosus* L.) †= *Dipogon lignosus* (L.) Verdc.	N.Y., U.S.A.	Wilson (1939a)
linearis E. Mey.	S. Africa	N. Grobbelaar (pers. comm. 1974)
(*melanophthalmus* DC.) †= *Vigna unguiculata* (L.) Walp.	Europe	Laurent (1891)
oliveri Schweinf.	Zimbabwe	Corby (1974)

Nodulated species of *Dolichos* (cont.)	Area	Reported by
sericeus E. Mey.		
ssp. *formosus* (A. Rich.) Verdc.	S. Africa	B. W. Strijdom (pers. comm. 1967)
ssp. *sericeus*	Zimbabwe	Corby (1974)
(*sesquipedalis* L.)	France	Clos (1896)
†= *Vigna unguiculata* (L.) Walp. ssp. *sesquipedalis* (L.) Verdc.		
taubertii Bak. f.	S. Africa	N. Grobbelaar (pers. comm. 1971)
trilobus L.	Widely reported	Many investigators
(= *falcatus* Klein ex Willd.)		
ssp. *transvaalensis* Verdc.	S. Africa	Grobbelaar & Clarke (1975)
ssp. *trilobus*	Zimbabwe	Corby (1974)
trinervatus Bak.	Zimbabwe	Corby (1974)

†Revised generic name.

Dorycnium Vill.   Papilionoideae: Loteae
236 / 3697

From Greek, *doryknion*, a plant name used by Dioscorides.

Type: *D. suffruticosum* Vill.

12 species. Perennial herbs, or shrublets, smooth or hairy. Leaves sessile or subsessile, 5–4-foliolate; leaflets entire, 1–2 lowest ones stipulelike; stipules minute or absent. Flowers pink or white, with a dark red or black keel, in axillary, stalked umbels or heads, or in small terminal clusters; bracts below the heads sometimes 3–1-foliolate; bracteoles none; calyx subcampanulate; lobes 5, subequal or the lower ones longer; petals free from the staminal tube; standard ovate-oblong, broad-clawed; wings obovate-oblong; keel smaller, incurved, rather blunt, both sides weakly swollen; stamens diadelphous; all or only alternate filaments broad at the tip; anthers uniform; ovary sessile, 2- to many-ovuled; style incurved, glabrous; stigma terminal. Pod oblong or linear, cylindric or turgid, 2-valved, dehiscent, with or without septa between the 1–many, globose or flattened seeds.

Members of this genus are common in the Mediterranean region, from the Canary Islands to the Balkans. They are usually found on light, sandy soil, generally of low fertility.

D. hirsutum (L.) Ser., commonly called hairy Canary clover, is regarded in Buenos Aires as a fine ornamental introduction for gardens. It has shown promise in California (Crampton 1946) and Australia (Brockwell and Neal-Smith 1966) as a forage crop and a deterrent to soil erosion.

NODULATION STUDIES

Histological aspects of the spherical nodules of *D. hirsutum* were described by Lechtova-Trnka (1931). The rhizobia retained rod shapes within the bacteroid tissue.

Data concerning rhizobial characterization and host-infection range are sparse. The ability of *D. rectum* (L.) Ser. rhizobia to nodulate *Lotus* species and *Anthyllis vulneraria* was shown by Erdman and Means (1950). The strains symbiosed effectively with *L. corniculatus* but not with *L. uliginosus*. These host-plant kinships were confirmed by Brockwell and Neal-Smith (1966) with strains of rhizobia from various *Lotus* species, *Anthyllis vulneraria*, an unidentified *Hosackia* species, and *Ononis repens* as inocula for *D. hirsutum*.

Nodulated species of *Dorycnium*	Area	Reported by
herbaceum Vill.	Widely reported	Many investigators
hirsutum (L.) Ser.	France	Lechtova-Trnka (1931)
	Kew Gardens, England	Pers. observ. (1936)
	Australia	Brockwell & Neal-Smith (1966)
pentaphyllum Scop.	S. Africa	Grobbelaar & Clarke (1974)
rectum (L.) Ser.	Italy	Pichi (1888)
	Kew Gardens, England	Pers. observ. (1936)
suffruticosum Vill.	S. Africa	Grobbelaar & Clarke (1974)

Dorycnopsis Boiss.   Papilionoideae: Loteae

The Greek suffix, *opsis*, denotes resemblance to *Dorycnium*.

Type: *D. gerardii* (L.) Boiss.

1 species. Herb, perennial, up to 60 cm tall, woody at the base; stems ascending, numerous, slender, whiplike. Leaves imparipinnate; leaflets 5–11, alternate or opposite, oblanceolate, narrow-elliptic, entire; rachis not winged; stipels absent; stipules small, awl-shaped, deciduous. Flowers rose red, numerous, small, on long peduncles, not subtended by a leafy bract; bracts minute; calyx campanulate, scarcely inflated; teeth 5, equal; standard longer than the wings and keel, blunt, abruptly narrowed into a claw; stamens diadelphous; free parts of filaments not dilated; anthers uniform; style glabrous. Pod small, straight, ovate, usually 1-seeded, enclosed by the calyx, indehiscent. (Hutchinson 1964)

This genus was formerly a section of *Anthyllis*.

This monotypic genus is endemic in southern Europe and islands of the western Mediterranean.

Drepanocarpus G. F. W. Mey.　　　　Papilionoideae: Dalbergieae,
353 / 3824　　　　　　　　　　　　　　Pterocarpinae

From Greek, *drepanon*, "sickle," and *karpos*, "fruit," in reference to the shape of the pod.

Type: *D. lunatus* (L. f.) G. F. W. Mey.

15 species. Woody vines, scandent shrubs, or trees. Leaves imparipinnate; leaflets short-stalked, oblong or pointed, mostly alternate; stipels none; stipules often spiny, recurved. Flowers medium-sized, purple or white, abundant in short, axillary or terminal racemes or panicles; bracts small, caducous; bracteoles persistent; calyx campanulate; teeth 5, deltoid, minute; standard broad-ovate; wings oblong; keel petals joined at the back, incurved; stamens in bundles of 5 and 5, or 9 and 1; anthers versatile; ovary short-stalked, 1–2-ovuled; style filiform, incurved; stigma small, terminal. Pod thick, flat, curved into a circle or broadly sickle-shaped, leathery, indehiscent; seed 1, flat, kidney-shaped. (Bentham and Hooker 1865; Pellegrin 1948; Hauman 1954b)

Representatives of this genus are common to tropical America and the tropical coast of western Africa. They occur on moist ground, inundated areas along rivers, salt marshes, mangrove lagoons, on abandoned plantations, roadsides, and forest zones.

The economic importance of these species has not been explored. The wood of *D. paludicola* Standl., in Brazil, appeared to lack luster and to be without commercial possibilities (Record and Hess 1943). The bark of several species contains a red sap which stains the wood during cutting.

NODULATION STUDIES

Nodules from *D. lunatus* (L. f.) G. F. W. Mey. sent to the authors from Trinidad by DeSouza (1966) were club-shaped, 1–1.5 cm long, with the narrow elongated basal zone covered with a dark, lenticular, barklike epidermis; the bulged apex was cream-colored and smooth. Freehand sections confirmed that the red-pigmented, active nitrogen-fixing zone occurred only in the tip or youngest region of the nodules. General aspects of the nodules suggested perennial growth.

Droogmansia De Wild.　　　　　　　Papilionoideae: Hedysareae,
(337a) / 3807a　　　　　　　　　　　　Desmodiinae

Named for H. Droogmans, Financial Secretary General of the Department of Finances, Independent State of the Congo.

Lectotype: *D. pteropus* (Bak.) De Wild.

30 species. Shrubs, small. Leaves 1-foliolate; petiole broadly winged, both margins deeply indented just below the base of the attached leaflet; stipels 2; stipules striate, ciliate. Flowers white to purple, small, usually appearing before the leaves, on long rachis in dense or lax terminal or axillary racemes; bracts and bracteoles present; calyx bilabiate, upper lip 2-parted, lower lip 3-dentate, middle tooth of lower lip longest; standard suborbicular,

clawed; wings smaller and shorter than the keel petals; stamens diadelphous, 1 free below, then connate one-third to one-half length, again free at apex, 9 joined, often in 2 bundles; alternate filaments widely expanded; anthers uniform; ovary stalked, many-ovuled, stipe surrounded by a short disk; stigma terminal. Pod small, flat, usually long-stalked, densely hairy, 2–7-jointed, each joint 1-seeded. (Chevalier and Sillans 1952; Schubert 1954)

Harms (1908a) positioned *Droogmansia* adjacent to *Desmodium*. Schindler (1924) acknowledged this kinship but considered *Pteroloma* the closest kin. Hutchinson (1964) positioned *Droogmansia* in Desmodieae trib nov.

Droogmansia species occur in tropical Africa. They grow on rocky slopes, plateaus, burned-over brush, savannas, and forests.

NODULATION STUDIES

Nodules were observed on *D. pteropus* (Bak.) De Wild. var. *pteropus* in Zimbabwe by Corby (1974).

Dumasia DC.   Papilionoideae: Phaseoleae, Glycininae
387 / 3861

Named for M. Dumas, editor of *Annales des Sciences Naturelles*, renowned animal physiologist, and friend of A. P. De Candolle.

Type: *D. villosa* DC.

10 species. Herbs, twining; stems slender; base slightly woody. Leaves pinnately 3-foliolate; petiole longer than the ovate leaflets; stipels present; stipules narrow, setaceous or striate. Flowers yellow or violet, blackening with age, solitary or paired on long rachis of pendent, axillary racemes; bracts small, narrow, persistent; bracteoles minute, awl-shaped or lanceolate, usually caducous; calyx cylindric or tubular-campanulate, obliquely truncate, gibbous dorsally at the base; teeth or lobes inconspicuous or obsolete; petals all long-clawed; standard oval, erect, with 2 basal auricles; wings oblong, falcate-obovate, adnate to the keel; keel petals blunt, slightly incurved and longer than the wings; stamens 10, diadelphous; anthers uniform, dorsifixed; ovary substalked, oblong, many-ovuled; style dilated above the middle, inflexed above, awl-shaped; stigma terminal, glabrous below. Pod subsessile, narrow, small, compressed, oblong, constricted between the seeds, nonseptate within, 2-valved; seeds subglobose, black or blue. (De Candolle 1825c; Bentham and Hooker 1865; Phillips 1951)

Members of this genus are native to India, Malaysia, East Africa, Transvaal, Natal, and China. They occur up to 2,400-m altitude, in jungles, along forest borders, and in grassy wilds.

NODULATION STUDIES

Nodules have been observed on *D. truncata* S. & Z. in Japan (Asai 1944), and on *D. villosa* DC. in South Africa (Grobbelaar & Clarke 1975) and in Zimbabwe (Corby 1974).

Named for Professor George Dunbar, 1784–1851, of Edinburgh.

Lectotype: *D. heynei* W. & A.

15 species. Herbs, slightly woody, climbing, often tomentose. Leaves pinnately 3-foliolate; leaflets entire, undersurface with numerous, orange red, resinous dots. Flowers in leaf axils or in pairs along rachis in axillary racemes, rarely solitary; bracts broad, deciduous; bracteoles absent; calyx teeth pointed, upper 2 usually united, entire or 2-dentate; standard orbicular, erect or spreading, with inflexed, basal auricles, and 2 callosities inside; wings obliquely obovate or oblong; keel blunt, shorter, incurved; stamens diadelphous; anthers uniform; ovary sessile or short-stalked, many-ovuled; style inflexed. Pod linear, straight or falcate, beaked, compressed, 2-valved, valves thin, subseptate between the 4–11 seeds, funicle of seed expanded into a thick membrane.

Members of this genus are indigenous to southern India, Japan, and Australia.

D. heynei W. & A. occurs in the Himalayas up to 1,000-m altitude. It has shown promise as a green manure on rubber estates in Sri Lanka and Malaysia (Burkill 1966). *D. rotundifolia* Merr. is common in thickets in Taiwan.

NODULATION STUDIES

D. villosa Mak. was observed nodulated in Japan (Asai 1944), and *D. heynei* in Sri Lanka (W. R. C. Paul personal communication 1951).

Duparquetia Baill. Caesalpinioideae: Cassieae
85 / 3541

Named for the Reverend Père Victor Aubert Duparquet, 1830–1888; he collected plants in French Equatorial Africa, 1863–1887.

Type: *D. orchidacea* Baill. (= *Oligostemon pictus* Benth.)

1 species. Shrub, or small tree, up to 6 m tall; young parts rusty-pubescent. Leaves imparipinnate; leaflets 3–9 pairs, large, ovate-elliptic, acuminate, undersides pubescent, terminal leaflet larger; stipules caducous. Flowers showy pink or white, in erect, many-flowered, terminal panicles or racemes; bracts and bracteoles very small, scalelike; calyx small; sepals 4, distinct, large, outer ones anterior, coriaceous, inner sepal 2-parted; petals 5, rose-colored, 3 upper posterior ones lanceolate, unequal, red-veined, often glandular-dentate, the 2 lower petals rudimentary, linear, smaller; stamens 4 or 5, free; filaments short, flattened; anthers deeply 2-lobed, basifixed, large, dehiscing by a short, apical slit; ovary oblong, short-stalked, longitudinally 4-ridged or 4-winged, about 5-ovuled; style thick; stigma red, small, terminal. Pod elongated, longitudinally 4-winged, acuminate, glabrous, mesocarp and endocarp spongy, dehiscent, 2-seeded; seeds oblong, black. (Steyaert 1952)

This monotypic genus, endemic in tropical West Africa, occurs in savannas.

The twigs and branches serve in the making of rope.

Named for Père A. Duss, French botanist; he collected plants in the Lesser Antilles.

Type: *D. martinicensis* Krug & Urb.

10 species. Trees, tall, up to 50 m, unarmed, deciduous; trunk may measure 1 m in diameter, base buttressed; bark smooth, gray. Leaves imparipinnate; leaflets 5–25, large, alternate to subopposite; undersurface pubescent, on long rachis; stipels absent; stipules minute, caducous. Flowers pink to purple, some with greenish markings or variegated yellow and brown, in long, simple or panicled, axillary and terminal racemes; bracts and bracteoles caducous; calyx tube short, campanulate, slightly oblique, bilabiate, upper lip bidentate, lower lip tridentate; lobes 5, imbricate, subequal; standard orbicular, outer surface often pubescent; wings and keel petals straight, dorsally pubescent; stamens 10, all fertile, subequal in length, 1 subfree, 9 connate at the base; anthers versatile, ovate, small, dorsifixed; ovary short-stalked, about 4-ovuled, pubescent; style pubescent except near the incurved apex; stigma small, terminal. Pod short-stalked, leathery, ovoid to elliptic, turgid, compressed, 2-valved, margins of valves usually incurved, dehiscent; seeds 1–3, thick, cylindric with 1 blunt end, the other acute, cotyledons thick, aril red. (Rudd 1963)

Cashalia and *Vexillifera* are regarded as synonyms of *Dussia* (Rudd 1963).

Representatives of this genus are indigenous to tropical American rain forests of the Caribbean Islands, southern Mexico, Central America, Peru, and Amazonas, Brazil.

The timber of these trees is not in great demand. It is used for local construction in El Salvador, where it is known as cashal (Record and Hess 1943). *D. discolor* (Benth.) Amsh. is the principal species in the Amazon region.

The heartwood in most species is indistinguishable from the yellow sapwood; the bark, leaves, and pods usually contain a blood red sap. The sap from *D. macrophyllata* (Donn. Smith) Harms has native use as a purgative.

Dysolobium Prain   Papilionoideae: Phaseoleae, Phaseolinae
(434) / 3904

From Greek, *dysodes*, "ill-smelling," and *lobion*, "pod"; presumably the pods have a disagreeable odor.

Type: *D. dolichoides* (Roxb.) Prain

4 species. Herbs, twining, usually woody; stems brown-pubescent. Leaves 3-foliolate; leaflets ovate-rhomboid or oblong-lanceolate; stipels and stipules present. Flowers many, greenish white, violet-blotched, paired on nodose rachis of axillary racemes; bracts caducous; bracteoles inconspicuous; calyx campanulate; 2 upper lobes connate into a bidentate lip, lowermost lobe longest, lanceolate but shorter than the calyx tube; standard with 2 inflexed, basal auricles; keel much longer than the wings, beaked, sometimes distinctly curved and laterally deflexed; stamens diadelphous; anthers uniform; ovary

sessile, densely hairy, many-ovuled; style bearded below the oblique stigma. Pod cylindric, thick, hard, woody, elongated, oblong, densely hairy, endocarp papery, with double septa between the seeds; seeds sparingly or densely velvety-haired. (King et al. 1901; Harms 1908a)

These species were formerly in section *Dysolobium* of *Phaseolus*.

Representatives of this genus are native to eastern India, Malaysia, and Java. They occur in brushwood areas and along watersides.

Ebenus L.   Papilionoideae: Hedysareae, Euhedysarinae
313 / 3781

From Greek, *ebenos*, "ebony," "the black heartwood of ebony"; the meaning for this genus is unclear.

Type: *E. cretica* L.

20 species. Shrublets with a woody base, armed, small, seldom above 50 cm tall, or unarmed, perennial herbs, silky or shaggy-haired. Leaves imparipinnate or subdigitately 3-foliolate; leaflets entire; petioles spinose; stipels absent; stipules leaf-opposed, membranous, connate, divided at the apex. Flowers pink, red, or purple, on axillary, long-stalked, crowded spikes or heads; bracts broadly lanceolate or linear, sometimes colored, sometimes whorled, forming an involucre subtending the flower head; bracteoles inconspicuous; calyx tubular-campanulate, villous; lobes 5, subequal, awl-shaped, plumose, often as long as the petals; standard obovate or obcordate, slightly clawed; wings very short; keel obtuse, obliquely trimmed at apex, equal in length to standard; stamens usually diadelphous, 1 free only below the middle, then connate with the other 9 into a sheath; anthers uniform; ovary sessile, very obtuse and 1-ovuled, or oblong and 3–2-ovuled; style filiform, inflexed; stigma small, terminal. Pod flattened, obovate or oblong, membranous, included in the calyx tube, indehiscent, 1–2-seeded. (Bentham and Hooker 1865)

Representatives of this genus are native to the Mediterranean area, Iran, and India. They grow on cliffs and rocks and up to 2,700-m altitude in the Punjab-Himalayas.

Echinosophora Nakai   Papilionoideae: Sophoreae

From Greek, *echinos*, "hedgehog," and Arabic, *sufayra*, a tree of the genus *Sophora*; name signifies a prickly or thorny *Sophora* species and refers to the hard, persistent, spinescent stipules.

Type: *E. koreensis* (Nakai) Nakai

1 species. Shrub, low, with long, creeping rhizomes that branch out in the ground; shoots growing up in dense masses. Leaves imparipinnate; stipules

rigid, persistent, spinescent. Flowers yellow, in terminal racemes on new shoots; calyx oblique, pouched at the base; lobes 5; standard straight, longer than other petals, emarginate, long-clawed; wings narrowed at the base, shorter than the standard but longer than the keel petals; keel petals free; stamens 10, free, 5 shorter than others; anthers rounded, versatile; ovary stalked; style elongated; stigma pointed. Pod moniliform, 4-winged; seeds globose; shiny. (Nakai 1923)

This monotypic genus is endemic in Korea.

Echinospartum (Spach) Rothm.   Papilionoideae: Genisteae, Spartiinae

From Greek, *echinos*, "hedgehog," and *spartos*, the name of a broom plant; name signifies a thorny, spiny, *Spartium*-like species.

Type: *E. horridum* (Vahl) Rothm.

4 species. Shrubs, small, up to about 2 m; branches many, opposite, sometimes terminating in a small spine. Leaves 3-foliolate, usually opposite, caducous, sessile or short-petioled; leaflets linear; stipules persistent, adnate to leaflets. Flowers yellow, in headlike, terminal spikes or clusters, crowded in bract-axils; calyx campanulate, inflated, bilabiate, upper lip short, deeply bilobed, lower lip longer, curved, with 3 prominent lobes; all lobes equal to or longer than calyx tube; petals somewhat exserted, glabrous or pilose; stamens monadelphous; anthers alternately basifixed and dorsifixed. Pod oblong-ovoid, dehiscent, 1–3-seeded, villous, slightly exceeding the calyx. (Rothmaler 1941)

The genus was formerly a section of *Genista* (Rothmaler 1941).

Members of this genus occur in southwestern Europe and the Iberian Peninsula. They prefer calcareous and siliceous soils. The plants are usually found on exposed mountainous rocks or slopes.

Edbakeria R. Vig.   Papilionoideae: Genisteae, Crotalariinae

Named for Edmund G. Baker, 1864–1949, British botanist.

Type: *E. madagascariensis* R. Vig.

1 species. Shrublet. Leaves palmately 3-foliolate, densely villous; stipules foliaceous. Flowers yellow, few, in axillary or terminal culsters; calyx campanulate; lobes 5, triangular-lanceolate, longer than the calyx tube; standard suborbicular, shorter than the wings; keel straight; stamens monadelphous, sheath split above; anthers subglobose, uniform; ovary villous; style straight, glabrous, slender, shorter than the ovary; stigma capitate. Pod turgid, 1-seeded, tomentose. (Viguier 1949)

Polhill (1974) revised and broadened the concept of *Pearsonia* to include *Edbakeria*.

This monotypic genus is endemic in Madagascar. It occurs on sand dunes and dry seashores.

Eleiotis DC. Papilionoideae: Hedysareae, Desmodiinae
344 / 3814

The common name for this plant in India.

Type: *E. monophylla* (Burm. f.) DC. (= *E. sororia* DC.)

1 species. Herb, annual, with slender, fragile, trailing, densely tufted, glabrous stems. Leaves membranous, subglabrous, consisting usually of 1 rounded, emarginate, short-petioled leaflet, sometimes with a pair of minute, lateral, kidney-shaped or orbicular leaflets; stipels 2; stipules short, striate. Flowers very small, in copious, lax, terminal and axillary racemes, pedicels often paired, spreading, downy; bracts minute, ovate, striate, deciduous; calyx tube very short, membranous, truncate; teeth 5, very short, subequal, setaceous; petals short-clawed; standard suborbicular, emarginate, narrowed into a claw; wings oblong; keel short, obtuse, adherent to the wings; stamens diadelphous, vexillary stamen at length free, other 9 connate; anthers uniform; ovary subsessile, 1-ovuled; style short, straight, inflexed and thickened above; stigma capitate, terminal. Pod flattened, semioval, membranous, reticulate, boat-shaped with the upper suture straight, slightly dilated and 2-nerved, lower suture curved, thin, indehiscent, with 1 transversely oblong, subreniform seed. (De Candolle 1825b; Bentham and Hooker 1865)

This monotypic genus is endemic in eastern India and Sri Lanka. The plants grow on plains.

Elephantorrhiza Benth. Mimosoideae: Piptadenieae
27 / 3467

From Greek, *elephas*, "elephant," and *rhiza*, "root," in reference to the large rootstock.

Type: *E. elephantina* (Burch.) Skeels (= *E. burchellii* Benth.)

8–10 species. Shrubs, undershrubs or small trees, up to 7 m tall, unarmed, deciduous; rhizomes thickened, woody. Leaves bipinnate; leaflets small, many pairs, up to 50, midrib usually oblique; petioles without glands. Flowers short-pedicelled, creamy white, yellow, to brownish, numerous, small, uniform, bisexual or polygamous, crowded in cylindric, spikelike, axillary racemes, or on short, leafless stalks; calyx campanulate; sepals 5, short; petals 5, equal, longer than the calyx, valvate, free above but connate below the middle; stamens 10, fertile, short-exserted, not joined to the petals; anthers oblong, topped with a deciduous gland; ovary sessile, many-ovuled; style filiform, as long as stamens; stigma small, terminal. Pod large, woody, linear, straight or irregular but not twisted, thick, leathery, endocarp entire, separating from the exocarp, valves separate at dehiscence from persistent margins but not splitting into segments; seeds many, round, flat, transverse, enclosed in a thin pulp. (Bentham 1842; Brenan 1959)

Members of this genus are indigenous to tropical Africa, south of the equator, mostly in Angola, Namibia, Natal, Transvaal, and Zimbabwe. They occur in rocky areas, grasslands, deciduous woodlands, and river banks.

Foliage of the type species is grazed by wild and domestic animals; elands relish the pods. Steyn (1929) cautioned that the pods contain a principle toxic to sheep and rabbits. The roots of *E. elephantina* (Burch.) Skeels have a high tannin content and thus are useful for tanning leather, as an astringent, and as a khaki dye.

NODULATION STUDIES

The nodules are oblong or branched. Cultural and physiological reactions of the rhizobial isolates have not been reported. Host-specificity reactions, although sparse, indicate a kinship with the cowpea miscellany. One rhizobial strain from *E. suffruticosa* Schinz was highly effective on *Clitoria ternatea* (Sandmann 1970a). It is interpreted from Sandmann's data that this strain, #1073, was not infective on *Lotononis angolensis* or on *Stylosanthes guianensis*, but nodulated 5 of 9 plants of *Stylosanthes mucronata*, and all plants of *Vigna sinensis* tested.

Nodulated species of *Elephantorrhiza*	Area	Reported by
burkei Benth.	Zimbabwe	Corby (1974)
elephantina (Burch.) Skeels	S. Africa	Grobbelaar et al. (1964)
	Zimbabwe	Corby (1974)
goetzei (Harms) Harms		
ssp. *goetzei*	Zimbabwe	Corby (1974)
suffruticosa Schinz	Zimbabwe	Corby (1974)

Eligmocarpus R. Cap. Caesalpiniodeae: Cynometreae

From Greek, *heligmos*, "winding," "twisting," and *karpos*, "fruit"; in reference to the folded zigzag character of the ovary and pod.

Type: *E. cynometroides* R. Cap.

1 species. Tree, up to 15 m tall. Leaves imparipinnate; leaflets cuneiform, emarginate, 4–8 pairs; stipules obovate, small, deciduous. Flowers zygomor-

phic, in axillary, peduncled, cymose racemes; bracts and bracteoles caducous; calyx short; sepals 5, free, subequal; petals 5, unequal, posterior petal largest, imbricate in bud; stamens 10, all fertile, the 5 posterior stamens with stout filaments joined almost to the apex, bearing basifixed barbed anthers, the 5 anterior stamens smaller, glabrous, free; each anther with 2 longitudinal compartments with 4 loculi at the apex, dehiscence by means of apical pores; ovary short-stalked, surface zigzag with 2–3 folds; stigma short, blunt. Pod subquadrangular or suborbicular, upper suture channeled, lower suture crossed by 2–3 vertical grooves, indehiscent, endocarp very cartilaginous. (Capuron 1968a)

This monotypic genus is endemic in southeastern Madagascar at 150-m altitude.

Elizabetha Schomb. ex Benth. Caesalpinioideae: Amherstieae
69 / 3523

Named for Queen Elizabeth of Romania.

Type: *E. princeps* Schomb. ex Benth.

10 species. Trees, small to medium, unarmed. Leaves paripinnate; leaflets few to many pairs, small, leathery; stipules large, intrapetiolar, membranous, caducous. Flowers white, pink, or red, small, handsome, in globose-spicate racemes at the apex of shoots; bracts broad, leathery, imbricate or caducous; bracteoles leathery, colored, joined at the base, enclosing the calyx; calyx tube long; sepals 4, subequal, imbricate, petaloid; petals 5, oblong-ovate, subsessile, imbricate, subequal, uppermost petal innermost and broadest; stamens 9, free or shortly connate at the base, the 3 longer ones fertile, perfect, with blunt anthers longitudinally dehiscent, the 6 smaller ones with empty anthers or anthers lacking; filaments of the fertile anthers exserted; ovary stalked, stalk adnate to the calyx tube, multiovuled; style filiform; stigma capitate, terminal. Pod elongated, sickle-shaped, flat, leathery to woody, upper suture broader and thickened, 2-valved; seeds transverse, ovate, flat. (Bentham and Hooker 1865; Ducke 1934a, 1940c; Amshoff 1939)

Members of this genus occur in tropical South America, particularly Amazonas, Brazil, Guiana, Venezuela, and Colombia. Species are present in upland forests, along river banks, and on hillslopes.

The flowers are similar to, but smaller than, those of *Brownea*, a closely related genus. Most of the species are pollinated by hummingbirds, which are attracted to the sweet secretion in the flower buds. The wood of *E. durissima* Ducke is dense, hard, and heavy, but it is not exploited commercially (Ducke 1934a; Record and Hess 1943).

Named for Dr. Emin-Pascha; he sponsored Dr. Stuhlmann's collections of plants from the Kilamanjaro high mountain area (Taubert 1891).

Type: *E. eminens* Taub.

5 species. Herbs, robust or undershrubs. Leaves pinnately 3-foliolate, deciduous; leaflets entire, margins undulate; stipels and stipules small, striate, persistent. Flowers often large, pedicelled, 1–3 at old leaf nodes, usually blooming before leaf emergence, or in erect, axillary or terminal racemes, often paniculate; bracts stipulelike, tipped with a clavate gland; bracteoles 2, subtending the calyx; calyx tube subcampanulate; lobes 5, long, awl-shaped or linear-lanceolate, all or some with 1 or more apical glands, 2 upper lobes sometimes connate for one-third their length, 3 lower ones subequal; standard obovate, short-clawed, basal auricles inflexed; wings narrow-oblong, auricles 1–2, upper margin inflexed, lower twice folded towards the apex, clawed; keel petals falcate-oblong, equal to the wings, long-clawed; stamens 10, exserted, diadelphous, vexillary stamen free, persistent after flowering; anthers subglobose, dorsifixed; ovary villous, short-stalked, surrounded at the base by cupular disk, 2-ovuled; style long, upper half filiform, glabrous, rectangularly reflexed, lower half hairy; stigma glabrous, terminal, minute. Pod hairy, subrectangular, apex pointed, compressed, sutures slightly thickened, nonseptate, surrounded by the persistent calyx and disk, 1–2-seeded; seeds orbicular, compressed. (Taubert 1891, 1894; Hauman 1954c)

The genus was doubtfully positioned in Glycininae by Taubert (1891, 1894); after having been temporarily repositioned in the subtribe Cajaninae (Hauman 1954c; Hutchinson 1964), it was reestablished in this subtribe (Lackey 1977a).

Members of this genus are scattered from East Africa to Zaire. They flourish in high mountainous regions, wooded savannas, and marshlands.

The roots of *E. harmsiana* De Wild. and *E. polyadenia* Hauman are the ingredients of a native beverage in Zaire.

NODULATION STUDIES

Corby (1974) observed nodules on *E. antennulifera* (Bak.) Taub. in Zimbabwe.

Endertia Steenis & de Wit   Caesalpinioideae: Amherstieae

Named for Frederik Hendrik Endert, forester; he collected extensively in Indonesia, 1917–1935, and contributed greatly to our knowledge of Malaysian trees.

Type: *E. spectabilis* Steenis & de Wit

1 species. Tree, medium-sized, bole not buttressed, copiously branched; foliage dense, dark bright green; unarmed. Leaves abruptly pinnate; leaflets 3 pairs, upper pair usually largest, elliptic, acuminate; stipules small, axillary, early caducous. Flowers small, solitary, in terminal or axillary panicles, on old

wood, along branches, often on knotty excrescences; bracts and bracteoles persistent; calyx campanulate; lobes 4, imbricate in bud; petals 2–1, only lateral petals developing, white to yellow when young, orange red when old, rounded, short-clawed; stamens 2, perfect; filaments connate at the base into a ring around the base of the ovary; anthers broad-elliptic, dorsifixed, longitudinally dehiscent; staminodes absent; ovary 3–5-ovuled, oblong, hairy to woolly; style stout, recurved, glabrous; stigma small, terminal, capitellate. Pod stalked, oblong, tardily dehiscent, the valves coiling into a closed spiral ending in a short, straight beak; seeds flat, discoid. (De Wit 1947b)

This monotypic genus is endemic in Borneo. The plants grow at 1,000-m altitude. They are cultivated in the Bogor Botanic Garden, Java.

Endomallus Gagnep.   Papilionoideae: Phaseoleae, Phaseolinae

From Greek, *endo-*, "within," and *mallos*, "wool"; the ovary is densely pubescent.

Lectotype: *E. pellitus* Gagnep.

2 species. Shrubs, climbing; branches woody, hairy. Leaves pinnately 3-foliolate; stipels awl-shaped; stipules ovate-acuminate, persistent. Flowers showy, in loose, terminal, racemes or panicles; bracts and bracteoles minute or absent; calyx campanulate, hairy; lobes 5, subequal, small, upper 2 connate to the middle; petals subequal; standard with 2 scales above the base; wings shorter, eared; keel curved, obtriangular, obliquely short-beaked, not spirally twisted; stamens 10, diadelphous; anthers equal; ovary sessile, densely silky-pubescent, multiovuled; style filiform, curved, hardened above; stigma globose, terminal, capitate. (Gagnepain 1915)

Both species occur throughout Indochina.

Englerodendron Harms   Caesalpinioideae: Amherstieae
(62b)

Name means "Engler's tree," in honor of Heinrich Gustaf Adolph Engler, 1844–1930, German botanist.

Type: *E. usambarense* Harms

1 species. Tree. Leaves paripinnate; leaflets petioluled, lanceolate, in 2–4 pairs, symmetric at the base, without pellucid punctations. Flowers small to medium, clustered in velvety-haired panicles; bracteoles 2, persistent, enclosing buds and flowers; calyx tube short, cupular; lobes 6–7, rarely 4, subequal, imbricate in bud; petals 6, oblanceolate or spathulate, about twice as long as the sepals, subequal, clawed; stamens 12–13, of these 6–8 with elongated filaments and fertile, well-developed, dorsifixed anthers, alternate with 5–6 stamens with shorter filaments bearing minute, infertile anthers; ovary

short-stalked, densely villous, 5–8-ovuled; style villous at the base only; stigma terminal, minute, capitellate. Pod very short-stalked, obovate, flat, blunt, rounded at the base, sharp-pointed at the apex, woody, 2-valved; valves velvety-haired, dehiscent; seeds 2. (Léonard 1957)

The genus is closely allied to *Berlinia* (Harms 1915a), but it differs in the number of lobes, petals, and stamens.

This monotypic genus is endemic in tropical East Africa. The trees grow in rain forests up to 1,000-m altitude.

Entada Adans.   Mimosoideae: Piptadenieae
(28) / 3468

Name presumably comes from Portuguese, *d'entada*, "toothed," in reference to the prickles on the stems and leaves of some species.

Type: *E. monostachya* DC.

30–40 species. Woody lianas, often tall climbers, shrubs or trees, armed with prickles or unarmed. Leaves bipinnate; pinnae pairs several, the uppermost pair sometimes modified into a tendril; leaflets large and few or small and numerous; stipules small, setaceous. Flowers small, sessile, creamy yellow, green, red, or purple, bisexual or polygamous, in slender, axillary spiked racemes or terminal panicles, calyx campanulate; teeth 5, equal, narrow, short; petals 5, longer than the calyx, free or slightly joined, valvate; stamens 10, free or joined at the base, short-exserted; filaments linear or filiform; anthers with a deciduous, apical gland; ovary subsessile, many-ovuled; style filiform; stigma truncate, concave. Pod subsessile or long-stalked, straight or curved, often very large, flat, thin, woody, sutures thickened, persistent, continuous, valves transversely jointed, breaking away from the sutures into 1-seeded, indehiscent segments, endocarp persistent and separating from the exocarp; seeds flat, large, round. (Bentham and Hooker 1865; Baker 1930; Phillips 1951; Gilbert and Boutique 1952a; Brenan 1959, 1970)

Members of the genus are native to the Old World tropics. There are about 18 species in central and East Africa and Madagascar, 10 species in Asia, and 4 in America (Brenan 1970). They occur in lowland rain and streamside forests, coastal thickets, riverine areas, savannas, and sandy soil.

Frequent mention is made of the enormous size of the fruits of these species. Menninger (1967) credits *Entada* vines with having the largest leguminous pods; those of *E. phaseoloides* (L.) Merr. (= *E. scandens* (L.) Benth.), the St. Thomas tree, measure over a meter long by 12.5 cm in width. Quisumbing (1951) described the angled, twisted, and slightly curved stems of this woody climber as being as thick as a man's arm and the pendent pods as measuring 30–100 cm long by 7–10 cm wide. The pods of *E. abyssinica* Steud. ex A. Rich., a deciduous, unarmed, medium-sized tree, and *E. sudanica* Schweinf., a low branching, savanna tree, range from 15 to 37.5 cm in length by 5 cm wide (Eggeling and Dale 1951).

At maturity the valves of *Entada* pods split transversely into 1-seeded segments. The exocarp of the pod peels away, and the endocarp remains around the seed as an envelope. The seeds are mahogany brown and smooth, elliptically flat, and vary in number from 10 to 15. The large round seeds of *E. phaseoloides* are known as ai lavo by the Fijians; this is now their term for money, a word unknown to them when exchange was entirely by barter. Because each segment of the broken pod of *E. polystachya* (L.) DC. is the size of a calling card, this climber is called the calling card vine.

Root and bark decoctions of *Entada* species serve various medicinal purposes in their native areas (Watt and Breyer-Brandwijk 1962). *E. africana* Guill. & Perr. yields a small amount of rotenone and an inferior gum consisting of about 10 percent gum tragacanth and 90 percent gum arabic (Howes 1949). The seeds of *E. phaseoloides* are a poor substitute for coffee. Plant parts of many species have a high saponin content (Dutta 1954) and serve as native shampoo and laundry soap substitutes. The large hollowed-out seeds of some species are used as boxes for jewelry, snuff, and matches, and as ornamental objects. The large pods of some of the aforementioned species were used at one time as clubs by the police in the West Indies.

The sapwood of the tree species is white, the heartwood red. The wood is used for small carpentry.

NODULATION STUDIES

As early as 1903, Wright described nodules on *E. scandens* as "flat and grooved" with the "appearance of miniature finger-like projections from a common centre, the latter being the point of attachment of the nodule to the root."

Nodulated species of *Entada*	Area	Reported by
abyssinica Steud. ex A. Rich.	Zimbabwe	Corby (1974)
chrysostachys (Benth.) Drake	Zimbabwe	Corby (1974)
nana Harms		
ssp. *nana*	Zimbabwe	Corby (1974)
phaseoloides (L.) Merr.	Sri Lanka	Wright (1903)
(= *scandens* (L.) Benth.)	Philippines	Bañados & Fernandez (1954)
polystachya (L.) DC.	Trinidad	DeSouza (1966)

Enterolobium Mart.   Mimosoideae: Ingeae
6 / 3440

From Greek, *enteron*, "intestine," and *lobion*, "pod," in reference to the shape and appearance of the semicircular pods.

Type: *E. timboüva* Mart. (= *E. contortisiliquum* (Vell.) Morong)

8–10 species. Trees, unarmed, rapid-growing; trunks usually stout, long-boled, often buttressed, reaching lofty heights; crown wide-canopied, up to 30 m in diameter. Leaves fernlike, bipinnate; pinnae 5–15 pairs; leaflets 20–30 pairs; stipules small, inconspicuous. Flowers small, about 1 cm long, yellow-white to greenish, in long-stalked, globose, fluffy heads, peduncles solitary or subfascicled, axillary or upper ones forming a short raceme; calyx funnel-shaped, short-toothed; petals connate to the middle, valvate; stamens numerous, exserted, connate into a tube at the base; anthers small; ovary sessile, many-ovuled; style filiform; stigma small, terminal. Pod broad, kidney-shaped, curved into a circle about 7.5–18 cm in diameter, thick, compressed, hard, leathery, glossy, dark brown, mesocarp spongy becoming hardened and septate between the seeds, indehiscent; seeds transverse, compressed, numerous, 10–20, brown, about 15 mm long, with a yellow ring on each side. (Bentham and Hooker 1865)

Members of this genus are native to tropical America and the West Indies; some species have been introduced into the South Pacific Islands. They grow in lowlands.

E. cyclocarpum Griseb., commonly known as elephant's ear in many areas, is one of the best-known species. The word "elephant" refers to the gigantic size of the tree with its huge gray buttressed trunk and broad, ear-shaped pods. In Costa Rica the common name is "guanacaste," a word of Indian origin meaning "ear-tree." The Province of Cordillera del Guanacaste is so named because the tree is a dominant feature of this region.

In some areas of South America this fast-growing species reaches a height of 40 m or more, with a trunk diameter of 2 m (Barrett 1956). The wide-spreading crown provides excellent shade. This tree is an attraction in gardens and parks, but it is undesirable near houses or sidewalks because its shallow-surfaced, radiating, lateral roots often cause heaving and tilting of sidewalks and foundations. The pod litter underneath the tree may become objectionable.

E. cyclocarpum, *E. schomburgkii* Benth., and *E. timboüva* Mart. yield timber for cabinetwork, furniture, paneling, canoes, water troughs, and construction (Record and Hess 1943; Standley and Steyermark 1946). The wood ranges from soft, or spongy, to medium hard, with a density from 320 to 640 kg/m^3. The dull white sapwood merges into the reddish-tinged, walnut brown heartwood. The grain is straight, and the texture medium to coarse. The wood is easy to work, finishes smoothly, takes a high polish, and is unscented, but workmen are often allergic to the wood dust. The timber ranges in character from white pine to black walnut; that of *E. timboüva* is used as a substitute for cedar. The wood is not affected by dampness or attacked by termites, but the soft varieties lack durability and are not considered high grade.

Elephant's ear pods are a source of tannin and saponins; in some areas, the pods and bark are pulverized and used as a soap substitute. Commenting on the edibility of elephant's ear foliage and pods, Standley and Steyermark (1946) write, "Wherever there are cattle pastures, the trees are sure to be found, and the cattle and other stock seem to show a preference for the shade of the tree, probably on account of the fallen foliage and pods, both of which they eat greedily." However, Deutsch et al. (1965) attributed the death of four bovines to hepatogenous photosensitivity resulting from the experimental feeding of pods of *E. gummiferum* (Mart.) Macb. Medicinal uses of plant parts are minor. The jellylike gum, goma de caro, from the trunks of elephant's ear trees, is used in Venezuela as a folk remedy for chest ailments; in Brazil, an anthelmintic is extracted from the roots and seed coats of *E. timboüva*.

NODULATION STUDIES

Nodules have been recorded on *E. timboüva* in Czechoslovakia (Štefan 1906), in France (Lechtova-Trnka 1931), and in Argentina (Rothschild 1970), and on *E. cyclocarpum* in Hawaii (Allen and Allen 1936a, 1939). Nodules on the taproots and upper laterals commonly measure 15 mm in diameter. Rhizobial isolates are culturally and physiologically typical of the slow-growing type. Symbiosis in various degrees of effectiveness was shown with 20 leguminous species of the cowpea miscellany (Allen and Allen 1936a, 1939).

Eperua Aubl.   Caesalpinioideae: Amherstieae
61 / 3515

From Galibi, *eperu*, the Caribbean Indian name for the fruit in Guiana.

Type: *E. falcata* Aubl.

14 species. Trees, slender, unarmed. Leaves mostly paripinnate, rarely imparipinnate; leaflet pairs few, leathery; stipules intrapetiolar, small, leafy, deciduous. Flowers red, purple, or white, large, in short terminal panicled racemes or peduncles sometimes long, pendulous; bracts and bracteoles caducous; calyx turbinate-campanulate; sepals 4, thick, ovate or oblong, concave, imbricate, leathery; petal 1, sessile, showy, large, broad, with margin fringed or crinkled and 4 minute petalodia; stamens 10, free or slightly joined at the base; filaments usually long, sometimes alternate ones shorter and minus anthers; fertile anthers ovate to oblong, longitudinally dehiscent; ovary short-stalked, many-ovuled, stipe adnate to the side of the calyx tube; style elongated, curved in bud; stigma small, terminal. Pod large, elongated or falcate, plano-compressed, leathery to woody, 2-valved, explosively dehiscent; seeds few, oblong, ovate, flat, without aril. (Bentham and Hooker 1865; Amshoff 1939; Cowan 1975)

Species of this genus are native to tropical South America, notably Amazonas, Brazil, Guiana, and Venezuela. They occur in moist or swampy upland forests, along shores of forest rivers, on white sand, and lateritic "ironstone" soils. The common name for various species and the timber derived therefrom is wallaba. They are the dominant trees of the so-called wallaba forests.

E. purpurea Benth. is considered to be one of the most beautiful flowering trees in the world. The contrasting brilliance of the purple flower clusters crowning the bright green foliage of the forest trees is so intense that the

Indians say it hurts the eyes to look at them (Ducke 1940b). *E. leucantha* Benth. is a white-flowered ornamental. *E. oleifera* Ducke is a source of commercial resin for varnish (Record and Hess 1943).

Soft wallaba and Ituri wallaba are commercial timber names for *E. falcata* Aubl. and *E. jenmani* Oliv., respectively; the woods of these 2 species are indistinguishable commercially (Record and Hess 1943). The wood is durable, decidedly hard and heavy, and straight grained with a medium to coarse texture but has limited use because of its viscous exudations. Aitken (1930) considered the pulping properties satisfactory, but Record and Hess (1943) commented that the presence of a considerable quantity of resins in the wood of both species rendered them unsatisfactory for pulp production by the sulphite process. *E. falcata* grows in pure stands on white quartz sand otherwise quite useless for agriculture. It reaches a height of 25 m in 30 years. The red flowers hang in dense clusters and are pollinated by bats. Wallaba wood is used for telephone poles, fence posts, house frames, fuel, and charcoal. Because of their resinous content, untreated poles last 20–30 years (Record and Hess 1943).

Nodulation Studies

E. bijuga Mart., examined at Manaus, Amazonas, Brazil, and *E. falcata* and *E. grandiflora* (Aubl.) Benth., in Bartica, Guyana, lacked nodules (Norris 1969). The roots of each species were dark brown and covered with stubby outgrowths suggestive of mycorrhizal rootlets. Good nodulation of *E. falcata* was reported by Whitton (1962) on plants growing on white sand and lateritic "ironstone hills" of the Amatuk region of Guyana, but not on acidic volcanic soil types. In Whitton's opinion the poor soil, good illumination, and optimum aeration of the white sand were factors favoring nodulation. A further check on nodulation within this genus is warranted to clarify the conflicting reports, especially since the species are economically important in forestry.

Eremosparton Fisch. & Mey.　　　　　Papilionoideae: Galegeae,
287 / 3755　　　　　　　　　　　　　　　Coluteinae

From Greek, *eremia*, "desert," and *spartos*, the name of a broom plant.

Type: *E. aphyllum* Fisch. & Mey.

5 species. Low shrubs; stems erect, long, green, rushlike. Leaves reduced to scales at the nodes. Flowers small, violet, solitary or in lax racemes on elongated peduncles in the axils of small, leaflike scales; bracts small; bracteoles minute toward the apices of pedicels; calyx campanulate; teeth subequal, short, deltoid; standard broad, orbicular, emarginate, nude inside, short-clawed, slightly reflexed; wings falcate-oblong; keel blunt, incurved; stamens diadelphous; anthers uniform; ovary sessile, many-ovuled; style incurved, upper part bearded longitudinally on the back; stigma terminal. Pod short, thin, membranous, sickle-shaped, 2-valved, plano-compressed at first, later turgid; seeds 1–2, kidney-shaped. (Bentham and Hooker 1865)

Members of the genus are indigenous to Turkestan and southwestern Asia. They inhabit sandy deserts and the dry steppes along the Caspian Sea.

E. aphyllum Fisch. & Mey. and *E. flaccidum* Litw. yield spherophysine (Merlis 1952 Ryabinin and Il'ina 1954), a ganglion-blocking alkaloid having the same general formula, $C_{10}H_{22}N_4$, as smirnovine and smirnovinine (Matyukhina and Ryabinin 1964). Its occurrence also in *Astragalus, Galega, Smirnowia*, and *Sphaerophysa* has chemotaxonomic significance (Mears and Mabry 1971).

Erichsenia Hemsl.   Papilionoideae: Podalyrieae

(172a) / 3633a

Named for Frederik Ole Erichsen; he assisted G. H. Thiselton-Dyer in forming the collection of Western Australian plants.

Type: *E. uncinata* Hemsl.

1 species. Shrublet. Leaves 1-foliolate, sessile, revolute, rigid, tapering into a recurved hook at the apex; stipules bractlike at the base. Flowers medium, yellow with purple striations, solitary or in short racemes in leaf axils; calyx subbilabiate; lobes hairy, subequal, ovate-rounded, upper lip innermost, imbricate in bud, middle lobe of lower lip outermost; petals clawed; standard kidney-shaped; wings oblong; keel petals joined; stamens 10, free, alternately long and short; ovary sessile, 2-ovuled. Pod not seen. (Harms 1908a)

This genus was positioned between *Viminaria* and *Daviesia*.

This monotypic genus is endemic in Western Australia. The plants commonly occur along railways.

NODULATION STUDIES

Seedlings of this species treated with 2 strains of *Rhizobium lupinus* did not form nodules (Lange 1962).

Erinacea Adans.   Papilionoideae: Genisteae, Spartiinae

216 / 3677

From Latin, *erinaceus*, "hedgehog," in reference to the spiny nature of the plants.

Type: *E. anthyllis* Link (= *E. pungens* Boiss.)

1 species. Shrub dense, compact, about 30 cm tall; branches opposite or alternate, rigid, ribbed, spiny, pubescent, often leafless. Leaves uncommon, 1- or palmately 3-foliolate, narrow-oblanceolate, short-petioled, near ends of branches, silky-haired; stipules present. Flowers 1–3, pale blue violet, in axillary or subterminal clusters; bracts and bracteoles small, leaflike; calyx campanulate, membranous, inflated after blooming, bilabiate; teeth 5, short, sub-

equal, upper lip broader, bidentate, lower lip tridentate; petals small, narrow, long-clawed, claws adnate to the staminal tube; standard ovate, weakly 2-eared; wings narrow, upper margin wrinkled; keel incurved, blunt; stamens monadelphous; anthers alternately short and versatile, long and basifixed; ovary sessile, many-ovuled; style slender, incurved; stigma globose, terminal. Pod projected beyond the calyx, narrow-oblong, glandular-hairy, 2-valved, dehiscent, usually 4–6-seeded. (Bentham and Hooker 1865; Rehder 1940; Rothmaler 1949)

Polhill (1976) included this genus in Genisteae *sensu stricto*.

This monotypic genus is endemic in Mediterranean areas of southwestern Europe and North Africa. The shrubs grow on stony mountain slopes and in calcareous soil.

Eriosema DC. ex Desv.   Papilionoideae: Phaseoleae, Cajaninae
420 / 3898

From Greek, *erion*, "wool," and *semeia*, "standard"; the underside of the leaflets and the standard are usually pubescent.

Lectotype: *E. sessiliflorum* (Poir.) Desv.

140 species. Herbs or shrublets, perennial, erect or prostrate, rarely climbing. Leaves short-petioled, pinnately 3-foliolate, rarely 1- or 5-foliolate; leaflets elliptic or elliptic-lanceolate, usually hairy, resin-dotted, and prominently veined on the underside; stipels usually absent; stipules small, inconspicuous or large lanceolate, often striate, free or connate into 1 and leaf-opposed. Flowers usually yellow, reflexed, solitary, paired or fascicled along the rachis of axillary racemes; bracts usually lanceolate, caducous; calyx campanulate, oblique at the base, hairy; lobes 5, ovate or lanceolate, often long-acuminate, longer or shorter than the tube, all distinct or upper 2 shortly joined, sometimes lowest lobe longest; standard obovate or oblong, with a deeply channeled claw, usually villous, with 2 inflexed, basal ears, often glandular; wings narrow, oblong, usually shorter than keel petals, partly adnate, eared and with a linear claw; keel blunt, incurved, with a linear claw, often gibbous and glandular; stamens diadelphous; anthers uniform; ovary sessile, villous, 2-ovuled, surrounded by a small, cup-shaped disk; style slender, slightly thicker above; stigma small, capitate, terminal. Pod straight, compressed, oblique-oblong or rhomboid, usually villous, nonseptate within, 2-valved, dehiscent; seeds 2, rarely 1, obliquely transverse, flat. (Bentham and Hooker 1865; Amshoff 1939; Standley and Steyermark 1946; Phillips 1951; Hauman 1954b; Grear 1970)

The method of attachment of the funiculus to the hilum is the major difference between the South African species of *Eriosema* and *Rhynchosia* (Phillips 1951).

Species of this large genus occur in tropical Africa, America, Asia, and Australia. Thirty-eight species are distributed in South America, primarily in Argentina, Brazil, and Paraguay (Grear 1970). They are present in grassy savannas, open forests, hilly and swamp areas, and occasionally at sea level.

Several species have tuberous rootstocks that are consumed by natives in the Congo Basin. The yellow seeds of *E. psoraleoides* (Lam.) G. Don are cooked and eaten by South African natives. Decoctions of plant parts of several other species serve a variety of local medicinal purposes (Dalziel 1948). In some areas pounded leaves and seeds are used to stupefy fish.

Nodulated species of *Eriosema*	Area	Reported by
affine De Wild.	Zimbabwe	Corby (1974)
buchananii Bak. f.	Zimbabwe	Corby (1974)
burkei Benth.	Zimbabwe	Corby (1974)
chrysadenium Taub.	Zimbabwe	Corby (1974)
cordatum E. Mey.	Zimbabwe	Corby (1974)
var. *cordatum*	S. Africa	Grobbelaar & Clarke (1975)
var. *gueinzii* Harv.	S. Africa	Grobbelaar et al. (1967)
crinitum (HBK) Benth.	Venezuela	Barrios & Gonzalez (1971)
ellipticum Welw. ex Bak.	Zimbabwe	Corby (1974)
engleranum Harms	Wis., U.S.A.	Pers. observ. (1958)
	Zimbabwe	Corby (1974)
kraussianum Meissn.	S. Africa	Grobbelaar & Clarke (1972)
lebrunii Staner & Decr.	Zimbabwe	Corby (1974)
macrostipulum Bak. f.	Zimbabwe	Corby (1974)
montanum Bak. f.	Zimbabwe	Corby (1974)
nutans Schinz	Zimbabwe	Corby (1974)
parviflorum E. Mey.	S. Africa	Grobbelaar et al. (1967)
pauciflorum Klotz.	Zimbabwe	Corby (1974)
polystachyum (A. Rich.) Bak.	S. Africa	Grobbelaar et al. (1964)
psoraleoides (Lam.) G. Don	S. Africa	Grobbelaar et al. (1967)
	Zimbabwe	Corby (1974)
pumilum Verdc.		
var. *pumilum*	Zimbabwe	Corby (1974)
ramosum Bak. f.	Zimbabwe	Corby (1974)
rhodesicum R. E. Fr.	Zimbabwe	Corby (1974)
rhynchosioides Bak.	Zimbabwe	Corby (1974)
rufum (HBK) G. Don	Venezuela	Barrios & Gonzalez (1971)
salignum E. Mey.	S. Africa	Grobbelaar & Clarke (1972)
shirense Bak. f.	Zimbabwe	Corby (1974)
simplicifolium (HBK) G. Don	Venezuela	Barrios & Gonzalez (1971)
squarrosum Walp.	S. Africa	Grobbelaar et al. (1964)
villosum (Meissn.) C. A. Smith ex Burtt Davy	S. Africa	Grobbelaar & Clarke (1972)

From Greek, *erythros*, "red," in reference to the color of the flowers.

Type: *E. corallodendron* L.

108 species. Shrubs or trees, small to about 20 m tall; trunk, young branches, petioles, and petiolules often armed with blunt, conical thorns or recurved prickles; rarely herbs. Leaves pinnately 3-foliolate, often clustered at the ends of branches; leaflets broad-ovate, elliptic, often deltoid or rhomboid, entire, lateral leaflets often asymmetric, terminal leaflet largest, symmetric; stipels fleshy, glandlike, turning black upon drying, usually 1 at base of lateral leaflets, paired stipels at base of terminal leaflet; stipules small, ovate or linear, caducous or persistent. Flowers appearing before or with the first leaves, very showy, mostly red, some salmon, pink, orange, or yellow, solitary, paired or fascicled in erect, terminal racemes leafy at the base or in axillary racemes; bracts obovate to lanceolate, erect, densely pubescent, caducous; bracteoles linear, slender, small, caducous, or absent; calyx tubular-campanulate, bilabiate or spathelike with a slit down the base on the lower side, obliquely truncate; lobes 5, sometimes obsolete or whiplike; petals free or dorsally connate, membranous; standard large or long and narrow, erect or spreading, much larger and longer than the wings and keel, subsessile or long-clawed, not appendaged or eared; wings ovate or oblong, asymmetric, short-clawed, incurved, sometimes very small or absent; keel petals free or dorsally connate, longer or shorter than the wings but much smaller than the standard, sometimes eared; stamens monadelphous or subdiadelphous, vexillary stamen connate with others below or entirely free, others connate to the middle; filaments alternately short and long, glabrous; anthers dorsifixed, uniform; ovary stalked, linear, many-ovuled, usually tomentose; style incurved, apex awl-shaped, not hairy; stigma small, terminal. Pod stalked, linear, flattened or cylindric, usually moniliform, i.e., deeply constricted between the seeds, forming 2–14 segments, leathery or woody, narrowed at the base and apex, 2-valved, opening along both sutures or along the upper suture only; seeds ovoid, shiny, red, carmine, or brown, some with a black spot. (Bentham and Hooker 1865; Amshoff 1939; Standley and Steyermark 1946; Majot-Rochez and Duvigneaud 1954; Krukoff 1970; Krukoff and Barneby 1974)

This large genus is pantropical. There are about 70 American species (Krukoff 1939a,b, 1970), 32 African, and 18 Asian; 3 species are indigenous to Australia and Argentina.

The wide range of *Erythrina* habitats includes light forests, brush areas, river banks, swamp sites, and coastal regions. Some species are well adapted to dry, sandy, rocky regions. One species, *E. montana* Rose & Standl., occurs up to 2,900-m altitude.

An appreciation of the brilliant red flowers is reflected in such colorful names as tiger's claw or Indian coral for *E. indica* Lam., and cock's comb, or cock's spur coral, for *E. crista-galli* L.; *E. abyssinica* Lam., *E. caffra* Thunb., and *E. lysistemon* Hutch. are known in South Africa as Kafferboom, Kaffir boom, or lucky bean tree. The shiny red seeds of *Erythrina* species are popular tourist items as necklaces, rosaries, and good-luck charms.

E. glauca Willd., *E. indica*, *E. lithosperma* Miq., and *E. poeppigiana* (Walp.) Cook are invaluable as shade on coffee and cacao plantations. Their vernacular names, "madre arbol" and "madre de cacao," are indicative of their role as nurse trees. Most species are readily propagated from cuttings; hence they are popular as live fences and support plants. *E. indica* and *E. stricta* Roxb. serve the latter purpose for pepper plants in India; *E. corallodendron* L. is favored as a support for vanilla vines in Puerto Rico. The leaf-fall has value as green manure.

Erythrina wood is grayish white, spongy, lightweight, and strong but not durable. It is used for sieve frames, surfboards, dugout canoes, outrigger canoe floats, boxes, and small art carvings. *E. monosperma* Gaud. is a favorite canoe wood of the Polynesians. The dry wood of this species and the bark of *E. suberosa* Roxb. are used for the manufacture of composition cork.

Natives in various areas of the world have always made considerable use of *Erythrina* seeds, stems, and bark to stupefy fish, and as a narcotic, purgative, diuretic, and soporific. Crushed seeds have been useful as a rat poison. All *Erythrina* species examined to date have yielded alkaloids having a curarelike poisoning action. Among the first signs of curare poisonings are drooping of the eyelids, drowsiness, and loss of speech, followed by muscle paralysis of the neck, extremities, and diaphragm. Death usually results from respiratory failure (Craig 1955). The term "curare" was used early in the study of South American arrow poisons with reference, in general, to substances, such as those from *Strychnos* species, that cause a paralyzing action at the myoneural junction in skeletal muscle (McIntyre 1947). However, despite the curarelike action of aqueous extracts and concentrates of *Erythrina* plant parts, evidence is lacking that these were ever used as native arrow poisons.

The alkaloid, hypaphroine, $C_{14}H_{18}N_2O_2$, the betaine of tryptophane, a convulsive poison, was isolated near the turn of the century from *E. subumbrans* (Hassk.) Merr., then known as *E. hypaphorus* Boerl. (*fide* Folkers and Unna 1938; Merck 1968). Amorphous alkaloids were detected in other species, but it was not until Folkers and Major (1937) isolated erythroidine from *E. americana* Mill. and showed that this new alkaloid exhibited a curarelike action that chemical and pharmacological interest focused on *Erythrina* species.

Few genera have been examined so exhaustively for alkaloids as has *Erythrina*. After showing that 51 of 105 *Erythrina* species yielded alkaloids producing curarelike paralysis in frogs, Folkers and Unna (1939) concluded that alkaloid occurrence was characteristic of the genus. Alkaloids from the following 8 species were credited with the highest paralyzing potencies (Folkers and Unna 1938, 1939): *E. buchii* Urb., *E. coralloides* DC., *E. crista-galli, E. eggersii* Kruk. & Mold., *E. macrophylla* DC., *E. mexicana* Kruk., *E. occidentalis* Standl., and *E. suberosa*. Between 1937 and 1951, 20 comprehensive papers describing the occurrence, identification, and taxonomic host relationships of *Erythrina* alkaloids were published by Folkers and colleagues. Various aspects of this topic have been subjects of extensive reviews (Folkers and Unna 1938, 1939; McIntyre 1947; Marion 1952; Craig 1955; Boekelheide 1960).

NODULATION STUDIES

Nodules on *Erythrina* species tend to be large, spherical, and clustered on the central taproot system.

In 1936 the authors reported the nodulation of *E. indica* by rhizobia isolated from 13 other *Erythrina* species. Reciprocal testing of all species was not possible because of the lack of seed. In turn, positive reciprocal nodulation was obtained using rhizobia from *E. indica* and *Vigna sinensis*. In susequent tests (Allen and Allen 1939), 2 rhizobial strains nodulated effectively diverse tropical leguminous species of the cowpea miscellany; only *Canavalia ensiformis, Samanea saman*, and *Stylosanthes guianensis* were nodulated ineffectively, and *Phaseolus lunatus* not at all. In reciprocal tests *E. indica* was nodulated effectively by 51 of 53 strains from 27 species; only 2 strains from *Andira inermis* were noninfective. In the same year, Wilson (1939a) reported nodulation of *E. crista-galli* and *E. rubrinervia* HBK by slow-growing cowpea type rhizobia. Erdman (1948a) found that strains from *E. indica* were more effective on guar, *Cyamopsis tetragonoloba*, than were rhizobia from species of 7 other genera.

Nodulated species of *Erythrina*	Area	Reported by
abyssinica Lam.	India	Parker (1933)
	Hawaii, U.S.A.	Allen & Allen (1936a)
	Zimbabwe	Corby (1974)
arborea Small	Hawaii, U.S.A.	Allen & Allen (1936a)
berteroana Urb.	Hawaii, U.S.A.	Allen & Allen (1936a)
caffra Thunb.	India	Parker (1933)
	Hawaii, U.S.A.	Pers. observ. (1938)
corallodendron L.	Hawaii, U.S.A.	Allen & Allen (1936a)
crista-galli L.	Hawaii, U.S.A.	Allen & Allen (1936a)
fusca Lour.	India	Parker (1933)
	Hawaii, U.S.A.	Allen & Allen (1936a)
(= *ovalifolia* Roxb.)	Sri Lanka	Wright (1903)
glauca Willd.	West Indies	Pers. observ. (1958)
herbacea L.	Fla., U.S.A.	Carroll (1934a)
horrida Moc. & Sessé	Hawaii, U.S.A.	Allen & Allen (1936a)
humeana Spreng.	S. Africa	Grobbelaar et al. (1967)
indica Lam.	Sri Lanka	Wright (1903)
	Hawaii, U.S.A.	Allen & Allen (1936a)
insignis Tod.	Hawaii, U.S.A.	Allen & Allen (1936a)
latissima E. Mey.	S. Africa	Grobbelaar et al. (1967)
	Zimbabwe	Corby (1974)
livingstoniana Bak.	Zimbabwe	Corby (1974)
lysistemon Hutch.	Zimbabwe	Corby (1974)
monosperma Gaud.	Hawaii, U.S.A.	Allen & Allen (1936a)
pallida Britt. & Rose	Trinidad	DeSouza (1966)
poeppigiana (Walp.) Cook	Hawaii, U.S.A.	Allen & Allen (1936a)
(= *micropteryx* Poepp.)	Trinidad	DeSouza (1966)
reticulata C. Presl	Hawaii, U.S.A.	Pers. observ. (1940)
ribrinervia HBK	Hawaii, U.S.A.	Allen & Allen (1936a)
subumbrans (Hassk.) Merr.	Hawaii, U.S.A.	Allen & Allen (1936a)
(= *lithosperma* Miq.)	Sri Lanka	Wright (1905); Holland (1924)
umbrosa HBK	Sri Lanka	Wright (1903)
	Hawaii, U.S.A.	Pers. observ. (1940)
variegata L.	Hawaii, U.S.A.	Pers. observ. (1940)
var. *orientalis* (L.) Merr.	Philippines	Bañados & Fernandez (1954)
velutina Willd.	Hawaii, U.S.A.	Allen & Allen (1936a)
zeyheri Harv.	S. Africa	Grobbelaar et al. (1964)

From Greek, *erythros*, "red," and *phloios*, "bark," in reference to the red dye obtained from the bark.

Type: *E. suaveolens* (Guill. & Perr.) Brenan (= *E. guineense* G. Don)

15–17 species. Trees, medium to about 20 m tall, unarmed; buttresses, if present, short and blunt. Leaves bipinnate; pinnae opposite or subopposite, few, 2–4 pairs; leaflets alternate, petioluled, asymmetric, usually moderately large, leathery, acuminate; stipules very small, slender, early caducous. Flowers small, regular, on short pedicels, crowded in terminal, spikelike, paniculate racemes; bracts small or absent; bracteoles none; calyx campanulate; tube short; lobes 5, ovate or lanceolate, subequal, short, imbricate; petals 5, creamy yellow or greenish, narrow, spoon-shaped, free, equal, short but longer than the calyx, pilose outside, slightly overlapping; stamens 10, prominent, longer than the petals, free, equal or alternately long and short; filaments linear, glabrous, sometimes villous; anthers uniform, introrse, elliptic, dorsifixed, longitudinally dehiscent; ovary short-stalked, stalk free, usually 5–8-ovuled, sometimes villous; style very short; stigma obtuse, slightly distinct, terminal. Pod stalked, stalk slightly lateral, oblong, thick, ends rounded, leathery or woody, 2-valved, valves thick leathery, smooth or faintly ridged transversely, often opening along only 1 edge, pulpy between the seeds; seeds transverse, orbicular or quadrangular, compressed, usually about 6. (Bentham and Hooker 1865; Wilczek 1952)

The species display mimosoid and caesalpinoid characteristics. Bentham (1864) regarded *Erythrophleum* as a mimosoid genus. Brenan (1960) proposed conservation of *Erythrophleum* over *Erythrophloeum* and the type species as *E. suaveolens* (Guill. & Perr.) Brenan comb. nov.

Species of this genus are paleotropical. Representatives occur in Madagascar, Seychelles, tropical Africa, and eastern Asia; 1 species is endemic in Australia. They are present in moist evergreen, semideciduous, high, and secondary forests.

E. lasianthum Corb., in Mozambique, and *E. densiflorus* Merr., in the Philippines, typify the small to medium ornamental members of this genus in contrast to *E. africanum* (Welw. ex Benth.) Harms, *E. ivorense* A. Chev., and *E. suaveloens*, which are among the largest trees of African forests. The last-named species is known in many parts of Africa as sasswood, ordeal tree, or red-water tree.

The genus is one of the few caesalpinoids reported to contain alkaloids. Chemical interest in this genus originated from native uses of sasswood bark in ordeal trials, as a fish, game, and arrow poison, and as a rat eradicant. Aqueous infusions are red, owing to the high solubility of a dye substance. Bark decoctions have functioned in native medicine as a headache remedy, narcotic, febrifuge, purgative, and diaphoretic.

Five species yield alkaloids with digitalislike properties similar to those of bufotalins or bufotoxins (McCawley 1955), but *E. couminga* Baill. and *E. guineense* G. Don have been the principal sources. Few of these cardioactive compounds have been synthesized. Erythrophleine, $C_{24}H_{39}NO_5$, the most potent of the alkaloids, was originally isolated from *E. guineense* by Gallois and Hardy (1876), and later from *E. lasianthum* Corb. by Kamerman (1926). Cassaine was also ioslated from *E. fordii* Oliv. (Paris 1948). The structure of ivorine, $C_{28}H_{43}NO_5$, an alkaloid isolated earlier from *E. ivorense*, was reported in 1965 by Ottinger et al. The leaves of *E. suaveolens* yield taliflavonoloside, $C_{22}H_{22}O_{12}$, a rhamnoflavonoloside (Dussy and Sannié 1947). An amber-colored gum exudate obtained from the bark resembles gum arabic (Howes 1949).

The dark yellow sapwood is relatively narrow; the freshly cut heartwood is white, but upon exposure becomes dull reddish brown with dark venations. The wood, marketed as missanda, tali, or sasswood, has a density of 900 kg/m³; it is difficult to work, but finishes well, takes a fine polish, and is resistant to termites and marine borers (Rendle 1969). The wood is used primarily for house, bridge, and boat construction, heavy-duty flooring, wheel hubs, and carriages. It burns slowly and makes a good charcoal.

NODULATION STUDIES

E. africanum and *E. suaveolens* were reported nodulated in Zimbabwe (Corby 1974). Bonnier (1957) listed the latter species by its synonym *E. guineense* as nodulated in Yangambi, Zaire.

Etaballia Benth.   Papilionoideae: Dalbergieae, Anomalae
/ 3850

From *etabally*, the vernacular name in Guyana.

Type: *E. dubia* (HBK) Rudd (= *E. guianensis* Benth.)

1 species. Tree, up to about 30 m tall, unarmed. Leaves 1-foliolate, large, glabrous, ovate-oblong, acute-acuminate, leathery. Flowers yellow orange, small, in short, showy, densely clustered, axillary racemes; inflorescences at first conelike, later catkinlike; calyx tubular, pubescent; lobes 5, deltoid; petals 5, rarely 6, subequal, narrow, linear; stamens 10, rarely 11; filaments united in the lower half, free above and alternately long and short; anthers uniform, minute, dorsifixed, laterally dehiscent; ovary pubescent, 1–3-ovuled; style eccentric, very short, glabrous; stigma truncate. Pod laterally compressed, hard, pubescent, surface wrinkled, indehiscent, mostly 1-seeded. (Kuhlmann 1949; Rudd 1970c)

The correct placement of *Etaballia* is uncertain because of numerous anomalous characters (Rudd 1970c). Kuhlmann (1949) regarded it a transition type intermediate between Mimosoideae and Papilionoideae.

This monotypic genus is endemic in Guiana and Venezuela. The trees grow along the banks of the northern tributaries of the Amazon in Brazil.

The timber is attractive, very hard, and heavy. The sapwood is white and very broad, the heartwood, reddish brown streaked with black. The sap is red. The wood takes a good polish and is suitable for furniture, but the supply is scant (Record and Hess 1943).

From Greek, *eu*, "well," "nice," *cheilos*, "edge," "rim," and *opsis*, "appearance";
the leaf margins are attractively rolled downward.

Type: *E. linearis* (Benth.) F. Muell.

1 species. Undershrub. Leaves linear, leathery, scattered, with revolute
edges. Flowers red, solitary or paired, axillary; bracts very small, stipulelike;
bracteoles small; upper calyx lip large, bilobed, deeply cleft, cuneate-
orbicular, lower lip minute, trilobed, lobes equal, half-lanceolate; standard
orbicular, reflexed; wings short-clawed, somewhat longer than the dorsally
joined keel petals; stamens 10, free; filaments dilated below; anthers alternately long, basifixed, and short, dorsifixed; ovary short-stalked, 2-ovuled;
style setaceous, glabrous; stigma terminal, minute. Pod compressed, obliquely
orbicular-ovate, 2-seeded.

This monotypic species is endemic in Western Australia. It occurs in
sandy and stony areas.

Nodulation Studies

A rhizobial strain isolated from this species showed selective nodulating
ability for certain members of the cowpea-lupine-soybean miscellany (Lange
1959, 1961).

From Greek, *euchloros*, "verdant"; relevance of the name is not clear.

Type: *E. serpens* E. & Z.

1 species. Herb, perennial, small, prostrate, hairy; underground stems
give rise to leafy, rusty-haired branches at intervals; rootstocks woody. Leaves
sessile, simple, entire, lanceolate, silky-haired; stipels and stipules absent.
Flowers purplish, small, in dense, subcapitate, terminal racemes; calyx villous,
campanulate; lobes 5, deeply cleft, lanceolate, slightly shorter than the calyx
tube, hairy, 2 upper lobes broader than laterals and slightly joined higher up,
lowest lobe narrow; petals adnate to the calyx tube and with a linear claw;
standard obovate-spathulate, emarginate; wings oblique-obovate, blunt,
longer than the keel petals; keel plano-convex, incurved, obtuse; stamens
monadelphous, all joined into a tube split above, staminal tube adnate to the
calyx tube; filaments linear; anthers unequal, alternately short and versatile,
long and basifixed; ovary sessile, about 6-ovuled; style smooth, incurved;
stigma terminal, suboblique, minutely capitate. Pod turgid, ovoid, hairy,
2-valved, few-seeded. (Bentham and Hooker 1865; Phillips 1951; Dahlgren
1963b)

This monotypic genus is endemic in South Africa, specifically the Cape
districts and Namaqualand; it is sparingly found in sandy plains and river flats.

From Greek, *euchrestos*, "useful," "serviceable."
  Type: *E. horsfieldii* (Lesch.) Benn.
  4–5 species. Shrubs or shrublets, up to about 1.5 m tall, unarmed, rootstocks tuberous. Leaves imparipinnate, spirally arranged, long-petioled; leaflets 3–7, oblong, entire; stipels absent; stipules small, erect. Flowers white, borne singly along rachis of axillary, leaf-opposed or terminal racemes; bracts small, narrow, caducous; bracteoles minute; calyx obliquely and deeply campanulate, short-lobed, base saccate, dorsally gibbous; lobes 5, deltoid, short, 2 upper lobes connate into a truncate, emarginate lip; standard narrow, emarginate, ovate, narrowed into a long claw, not eared at the base; wings oblong, narrow, subequal to the short keel; keel petals subsimilar to the wings, obtuse, joined at the apex; stamens diadelphous; anthers small, versatile; ovary glabrous, long-stalked, 1–2-ovuled; style filiform, incurved, glabrous; stigma small, terminal. Pod stalked, oblong, thick, turgid, rather fleshy, drupelike, smooth, leathery, brittle when dry, indehiscent; seed 1, pendulous. (Bentham and Hooker 1865; Backer and Bakhuizen van den Brink 1963; Ohashi and Sohma 1970)
  Representatives of this genus are native to Japan, Taiwan, Malaysia, and the Himalayas. They flourish in cool mountainous areas and forests.
  Early records cite the seeds of *E. horsfieldii* (Lesch.) Benn. as bitter and poisonous; nonetheless, they have native uses in treating diseases of the chest, as a snakebite antidote, and as an aphrodisiac (Quisumbing 1951; Burkill 1966). The seeds are a source of the alkaloid cytisine, $C_{11}H_{14}N_2O$ (Henry 1939; Mears and Mabry 1971).
  The occurrence of cytisine in *Euchresta*, in conjunction with other evidence, prompted Turner (1971) to question its tribal position. This genus, according to Turner, stands apart from the other generic members of the subtribe "in being mostly a temperate group of the northern hemisphere." In essence, he asserted that *"Euchresta* is probably a primitive genus whose phyletic roots go back to those ancestral taxa which were intermediate to the tribes Sophoreae, Dalbergieae and Podalyrieae as these are expressed today." Turner's reasoning in the chemotaxonomic sense merits consideration. Ohashi (1973) arrived at a similar conclusion from purely taxonomic aspects; he segregated *Euchresta* from the tribe Dalbergieae into a new tribe, Euchresteae. Conceivably, rhizobial studies on *Euchresta* species could provide additional information on the kinships of the genus.

Eurypetalum Harms   Caesalpinioideae: Cynometreae
(40c)

From Greek, *eurys*, "wide," and *petalon*, "petal"; one of the petals is broader than long; the others are small and scalelike.
  Type: *E. tessmanii* Harms
  3–4 species. Trees, large, 30 m or more tall, with girths of 2 m or more,

slightly buttressed. Leaves paripinnate; leaflets large, lanceolate or oblong, 1–2 pairs. Flowers paniculate or in solitary or paired axillary racemes; bracts and bracteoles very small; calyx short, cupular; lobes 4, subequal, imbricate in bud, obtuse, glabrous or subglabrous, 1 ovate, larger than the others; petals 5, 1 broader than long, sessile, with margin wavy or toothed, other 4 scalelike, rudimentary, alternate with the calyx lobes; stamens 10; filaments awl-shaped, shortly connate and hairy at the base; anthers versatile, equal; ovary short-stalked, small, densely villous, 1–2-ovuled; style short, incurved, smooth; stigma small, capitellate. Pod stalked, flat, obliquely oblong-lanceolate, apiculate, 2-valved. (Harms 1915a)

Species of this genus are common in tropical Africa, Cameroon, and Gabon. They are primeval forest trees.

The hard, red Andzilim wood obtained from *E. batesii* Bak. f. is valued for cabinetwork.

Eutaxia R. Br.   Papilionoideae: Podalyrieae
178 / 3638

From Greek, *eutaxia*, "good order," in reference to the orderly arrangement of the leaves.

Type: *E. myrtifolia* (Sm.) R. Br. ex Ait.

8 species. Shrubs, often glabrous, low, sometimes procumbent or heath-like; branches small, sometimes spiny. Leaves small, simple, linear or lanceolate-ovate, entire, concave or with edges inrolled, in opposite pairs alternating at right angles with the pair above and below; stipules minute or none. Flowers yellow to orange, red brown, or purple, solitary or 2–4 in axils or terminal clusters; bracts small, caducous; bracteoles short, small, usually well removed from the calyx; calyx lobes 5, upper 2 somewhat united into a broad lip; petals long-clawed; standard orbicular, entire, broader and longer than the other petals; wings oblong, longer than the keel petals; keel straight, blunt; stamens 10, free; ovary narrowed to the base or stalked, pubescent, rarely glabrous, 1–2-ovuled; style filiform, incurved or apically hooked; stigma small, terminal. Pod ovoid or oblong, flattened or turgid, 2-valved; seeds 1–2, kidney-shaped. (Bentham 1864; Bentham and Hooker 1865)

Eutaxia species are limited mostly to Western Australia. They are present in rocky and sandy areas and on hills.

NODULATION STUDIES

In plant-infection tests conducted by Lange (1961), rhizobial strains from *E. densiflora* Turcz., *E. epacridioides* Meissn., and *E. virgata* Benth. produced nodules on *Phaseolus lathyroides*, *P. vulgaris*, and *Vigna sinensis*; nodulation of *Glycine hispida* and 5 *Lupinus* species was variable, and nil on peas, clover, and alfalfa. These results, in conjunction with cultural studies, prompted Lange to assign kinships of the *Eutaxia* strains with slow-growing, cowpea, lupine, and soybean rhizobia.

Nodulated species of *Eutaxia*	Area	Reported by
densifolia Turcz.	W. Australia	Lange (1959, 1961)
epacridioides Meissn.	W. Australia	Lange (1959, 1961)
microphylla (R. Br.) J. M. Black	N.S. Wales, Australia	Harris (1953); Beadle (1964)
obovata (Labill.) C. A. Gardn.	W. Australia	Lange (1959)
virgata Benth.	W. Australia	C. A. Gardner (pers. comm. 1947); Lange (1959, 1961)

Eversmannia Bge.   Papilionoideae: Hedysareae,
309 / 3777   Euhedysarinae

Named for Eduard Friedrich Eversmann, 1794–1860; he collected plants in Siberia.

Type: *E. hedysaroides* Bg.

2 species. Shrublets, branches spreading, stiff, gray-pubescent, with paired or solitary axillary spines. Leaves imparipinnate; leaflets small, ovate, rigid, 3–7 pairs; stipels absent; stipules membranous, connate. Flowers purple or pink in axillary, peduncled racemes; calyx tubular-campanulate, more deeply cleft above; teeth 5, upper 2 shorter than the lower 3; standard obovate, very short-clawed; wings very short, pointed; keel incurved, nearly as long as the standard; vexillary stamen entirely free or joined with the others at the middle; ovary subsessile, multiovuled; style slender, inflexed; stigma small, terminal. Pod linear, compressed, flexuous, lomentaceous, indehiscent, not constricted between the segments which tardily break away from the persistent dorsal suture; seeds round to kidney-shaped. (Bentham and Hooker 1865)

Both species are distributed throughout the southeastern Soviet Union, from the Volga Steppes to Iran and Central Asia.

Exostyles Schott ex Spreng.   Caesalpinioideae: Swartzieae
119 / 3577

From Greek, *exo*, "on the outside," in reference to the protruding style.

Type: *E. venusta* Schott ex Spreng.

2 species. Trees, small, unarmed. Leaves imparipinnate; leaflets alternate, thin, leathery, with notched or toothed margins; stipels awl-shaped;

stipules small, bristly, awl-shaped, caducous. Flowers rose or purplish red, in short, lax, axillary racemes; bracts and bracteoles small, stiff, lanceolate, sometimes acuminate, subpersistent; calyx elongated-turbinate, acuminate and entire in bud, later 3–4-cleft, the lobes doubled back during anthesis; petals 5, subequal, erect, imbricate, the uppermost petal innermost, the 4 others often convolute; stamens 10, equal; filaments short; anthers uniform, linear, basifixed, longitudinally dehiscent by means of slits; ovary stalked from the base of the deep receptacle, multiovuled; style thick, elongated, exserted in bud before the flowers open; stigma small, terminal. Pod flat, ovoid, leathery, 2-valved, sutures thick; seeds 1–3, ovate, flattened. (Bentham and Hooker 1865; Cowan 1968)

Both species are indigenous to the Province of Rio de Janeiro, Brazil.

Eysenhardtia Kunth Papilionoideae: Galegeae, Psoraliinae
245 / 3708

Named for Karl Wilhelm Eysenhardt, 1784–1825, Professor of Botany, University of Königsberg, Germany.

Type: *E. amorphoides* Kunth

15 species. Shrubs or small trees, about 8 m tall, unarmed, with trunks 15–20 cm in diameter, glandular-punctate, aromatic. Leaves pinnate; leaflets numerous, 10–20 pairs, small, gland-dotted; stipels minute; stipules small, awl-shaped. Flowers small, creamy white, fragrant, loosely clustered in terminal, spikelike racemes; bracts and bracteoles narrow, caducous; calyx lobes unequal, anterior lobe the longest, tube cleft more deeply between the posterior lobes; tube and style gland-dotted; petals subequal, oblanceolate or obovate, clawed, erect, all free; standard obovate, concave, broader than the oblong, obtuse, subsimilar wings and keel petals; stamens diadelphous; filaments united to midway; anthers uniform; ovary subsessile, 2–3-, rarely 4-ovuled; style thick, hooked at apex, more or less pubescent, with a gland sometimes near the apex, or eglandular; stigma large, introrsely oblique, capitate. Pod small, oblong or linear-falcate, beaked, flat, indehiscent(?); seed solitary, rarely 2, oblong, reniform. (Bentham and Hooker 1865; Rydberg 1919a)

Members of this genus are common from Guatemala to the southwestern United States, particularly Arizona, New Mexico, and Texas. They inhabit canyons, dry thickets, and dry, rocky, hilly areas, up to 2,000-m altitude.

The genus is represented in fossil records of the Madro-Tertiary Geoflora (Axelrod 1958).

The species are commonly called kidney wood, rock brush, or false mesquite. *E. platycarpa* Pennell & Safford, *E. reticulata* Pennell, *E. spinosa* Engelm., and *E. subcoriacea* Pennell are encountered frequently on dry slopes in Mexico; *E. adenostylis* Baill. is the common species in Guatemala, *E. peninsularis* Brand, in Baja California, Mexico, and *E. texana* Scheele, in Texas. The reddish brown, streaked, waxlike heartwood of the type species is of rather good quality, but it has not featured commercially because it is scarce and available only in small pieces. The wood has a rather fine texture and a moderately straight grain. It is hard and heavy, has a density of 900 kg/m³, takes a fine polish, finishes smoothly, and is resistant to decay. An outstanding feature is the beautiful, yellow to orange to peacock blue opalescence which the wood imparts in slightly alkaline liquids, especially when viewed against a black background. During the Spanish conquest of Mexico the wood was exported for medicinal uses to Europe as kidney wood, *lignum nephriticum*, a name also applied to *Pterocarpus indicus*. Miraculous cures for kidney and bladder ailments were reputedly imparted by liquids served in polished cups of this wood, or by liquids containing the wood chips.

In Mexico the wood is made into watering troughs for fowl. The foliage is relished by cattle and horses, despite its objectionable odor. The plant withstands severe abuse and sprouts freely from mutilated trunks. It has merit as forage and for erosion control. The flowers of all species are a good honey source. The wood of some species yields a yellow brown dye used to stain native cloth, but it has no commercial value.

NODULATION STUDIES

The type species, *E. amorphoides* Kunth, reported as *E. orthocarpa* S. Wats., lacked nodules in field and in greenhouse experiments in Arizona with commercial inoculants and rhizosphere soils (Martin 1948). *E. texana* is nodulated in Texas (J. C. Burton personal communication 1971).

Factorovskya Eig   Papilionoideae: Trifolieae

Named for Eliezar Factorovsky, 1897–1926, collaborator in botany and friend of A. Eig in Palestine.

Type: *F. aschersoniana* (Urb.) Eig

1 species. Herb, prostrate, annual. Leaves 3-foliolate, obovate, dentate. Flowers yellow, usually remaining closed, solitary on axillary peduncles; calyx lobes 5, triangular-lanceolate; standard suborbicular; keel blunt; stamens diadelphous; filaments not dilated at apex; ovary 2-ovuled, on short gynophore which, after flowering, forms an acute angle with the peduncle and elongates pushing the ovary 1–10 cm underground. Pod indehiscent, deeply 1–2-partitioned, each cell with 1 lenticular seed, upper portion often abortive. (Eig 1927)

This monotypic genus is endemic in western Israel and the eastern Mediterranean area. The plants grow in fallow fields.

Named for Caspar Fagel, Grand Pensionary of Holland, patron of botany.

Type: *F. bituminosa* (L.) DC.

1 species. Herb, robust, twining, soft-haired, sticky, heavy-scented, sub-woody at the base. Leaves pinnately 3-foliolate; leaflets ovate; stipels none; stipules ovate, striate. Flowers large, yellow, numerous, in long-stalked, axillary racemes; bracts ovate, caducous; bracteoles none; calyx campanulate, hairy, hairs bulbous-based; lobes 5, acuminate, longer than tube, upper 2 slightly connate; standard obovate, reflexed, with 2 basal ears and broad claw; wings shorter than the purple-tipped, incurved, obtuse keel; stamens diadelphous; anthers uniform; ovary sessile, several-ovuled, villous; style inflexed at middle, pubescent; stigma terminal, capitate. Pod oblong, hairy, swollen, 2-valved, continuous inside, subdepressed between the seeds; seeds 4–6, globose, black, with a short hilum. (Bentham and Hooker 1865; Phillips 1951)

Fagelia is regarded as a later homonym of *Bolusafra* by J. Lackey (personal communication 1978).

This monotypic genus is endemic in South Africa; it is confined to the southwestern districts of the Cape Province. The plants grow on hills and mountain slopes.

NODULATION STUDIES

F. bituminosa (L.) DC. is nodulated in its native habitats of South Africa (Grobbelaar & Clarke 1975).

Named for Alejandro Rodrigues Ferreira, 1756–1815, plant collector in Brazil.

Type: *F. spectabilis* Allem.

2 species. Trees, large, up to about 40 m tall, unarmed, with well-formed boles and brown black, ridged bark. Leaves imparipinnate; leaflets small, 3–7 pairs; stipels absent. Flowers small, many, yellow or white, in slender, terminal, racemose panicles; bracts and bracteoles small, caducous; calyx membranous, reddish, truncate; teeth inconspicuous; standard broad, suborbicular, reflexed; wings and keel petals free, subsimilar, narrow, oblong; stamens short, free; anthers uniform, ovate; ovary short-stalked, 1-ovuled; style very short, incurved; stigma small, terminal. Pod short-stalked, samaroid, 1-seeded at the base with a unilateral, membranous, transversely veined, apical wing resembling that of maple (*Acer* spp.), indehiscent; seeds oblong, subreniform, compressed. (Bentham and Hooker 1865; Burkart 1952)

Both species are distributed throughout southeastern Brazil, Bolivia, Paraguay, and northern Argentina. They inhabit forest and jungle areas.

The hard, heavy wood finishes smoothly and is attractive but does not feature in the timber trade. Its local use is reserved for durable construction purposes. The bark is used for tanning.

Fiebrigiella Harms
(324a)

Papilionoideae: Hedysareae,
Aeschynomeninae

Named for K. Fiebrig; his collection of specimens from southern Bolivia
formed the basis of Harms's description.

Type: _F. gracilis_ Harms

1 species. Herb, perennial, tender; stems hairy, thin, limp or prostrate.
Leaves imparipinnate, petioled; leaflets large, oblong-lanceolate or obovate-
oblong, membranous, 2–3 pairs; stipels none; stipules quite large, lanceolate,
acute, free. Flowers yellow or orange, 10–12, in lax, few or many-flowered,
long-peduncled, axillary racemes; bracts and bracteoles with fringed borders;
calyx cup-shaped, hirsute, lobed to the middle or more; lobes hairy, 2 upper
lobes connate and bidentate, lowest lobe longest; standard orbicular; wings
and shorter keel petals long-clawed; stamens diadelphous, in 2 bundles of 5;
anthers uniform, dorsifixed; ovary short-stalked, subsessile, smooth or with
basal or lateral hairs, 3–4-ovuled; style long, smooth; stigma minute, capitate,
smooth. Pod lomentaceous, short-stalked, linear-oblong, flat, straight or
slightly curved, apex briefly apiculate, with 3–5 1-seeded, oblong-rectangular
or semilunar, reticulate-nerved, dorsally acute, indehiscent segments; seeds
oblong, ovoid, laterally compressed. (Harms 1908a, 1915a; Burkart and Vilchez 1971)

The genus is closely akin to _Chaetocalyx_ and _Aeschynomene_.

This monotypic genus is endemic in Bolivia and Peru. It is present in
thickets on mountain slopes up to 3,000-m altitude.

Fillaeopsis Harms
(25c) /3465

Mimosoideae: Piptadenieae

The Greek suffix, _opsis_, denotes resemblance to _Fillaea_, a synonym for _Eryth-
rophleum_ (Harms 1908a).

Type: _F. discophora_ Harms

1 species. Tree, glabrous, medium-sized, up to 40 m tall, strongly
branched, malodorous, unarmed. Leaves petioled, bipinnate; pinnae 1–2
pairs, opposite; leaflets short-petioluled, large, alternate or opposite, few,
rounded at the base, acuminate. Flowers small, sessile, in terminal elongated,
paniculate spikes; bracts minute, scalelike; calyx saucer-shaped, very small;
lobes 5, short, ovoid, glabrous, obtuse, valvate in bud; petals 5, free, ovate,
valvate; stamens 10, arising from a thick cupular disk; filaments free, inserted
at the base of the disk; anthers with an early-deciduous, apical gland; ovary
short-stalked, oblong, glabrous, many-ovuled; style filiform, elongated, glabrous; stigma cupular. Pod sessile, flat, oblong-elliptic, very large, up to 40 cm
long by 15 cm broad, both ends blunt, densely reticulate, margins somewhat
thickened, subcoriaceous, tardily dehiscent; seeds large, nearly surrounded
by a membranous wing, flat, elliptic, transverse, brown, about 10, attached at
the middle by a long, slender funicle. (Harms 1908a; Baker 1930; Pellegrin
1948; Gilbert and Boutique 1952a)

This monotypic genus is endemic in tropical West Africa.

In local areas, the dense, red brown wood is called ironwood because its hardness and weight resemble teak. Its usefulness is limited to house construction, furniture, flooring, and paving blocks.

Fissicalyx Benth.　　　　　Papilionoideae: Dalbergieae, Geoffraeinae
373 /3845

From Latin, *fissus*, "deeply cleft"; the calyx is spathelike, cut to midway or more on 1 side only.

Type: *F. fendleri* Benth.

1 species. Tree, small, 8–15 m tall, deciduous; bole usually straight, covered with light brown, flaky bark. Leaves imparipinnate; leaflets subopposite; stipels absent; stipules deciduous. Flowers orange, in dense, terminal panicles; bracts very small; bracteoles small, persistent; calyx turbinate, rim entire or shortly bidentate, splitting spathaceously at anthesis; petals inserted with stamens at top of calyx tube; standard ovate; wings oblique-oblong; keel petals free, similar to and shorter than the wings; stamens monadelphous, joined into a sheath split above; anthers versatile, opening by 2 apical pores; ovary short-stalked, 2-ovuled; style slender; stigma minute, terminal. Pod flat, ovate-elliptic with broad, lateral wings, sutures inconspicuous, indehiscent, 1-seeded. (Bentham and Hooker 1865)

This monotypic genus is endemic in Venezuela and Panama.

The hard wood lacks economic importance because of its small size and the inaccessibility of the trees.

Flemingia Roxb. ex Ait.　　　　　Papilionoideae: Phaseoleae,
(421) / 3899　　　　　　　　　　　　　　　　　Cajaninae

Named for Dr. John Fleming, Scottish naturalist and Physician-General of the East India Company's Medical Establishment in Bengal; he published a catalogue of Indian medicinal plants. Nomenclatural problems concerning this genus have prevailed over the years. Usage of *Flemingia* in preference to *Moghania* by Verdcourt (1971) and Drummond (1972) has followed Rudd's proposal (1970b) to conserve the former.

Lectotype: *F. strobilifera* (L.) Ait.

40 species. Shrubs or herbs, perennial, erect, often tomentose, rarely twining. Leaves 1–3-foliolate, rarely 4–5-foliolate; leaflets ovate to lanceolate, undersurfaces red or yellow gland-dotted; stipels absent; stipules striate, caducous. Flowers small, pink, red, or purple, often mixed with green or

yellow, in simple or panicled racemes, dense spikes or globose heads; bracts narrow, deciduous, or large, concave, persistent, enclosing the flowers; bracteoles none; calyx short, campanulate; lobes 5, linear, acute, the lowest lobe longer; standard short-clawed, elliptic or orbicular, usually with 2 inflexed basal lobes; wings narrow-oblong, often adherent to the broad, incurved, blunt keel; stamens diadelphous; filament of vexillary stamen often broad; anthers uniform; ovary subsessile, 2-ovuled, often hairy; style filiform, incurved and thickened above the middle; stigma small, terminal. Pod short, oblong, turgid, often enclosed within the bracts, some with glandular hairs, 2-valved, nonseptate, 2-seeded. (Bentham and Hooker 1865; Hooker 1879; Amshoff 1939; Li 1944; Hauman 1954c)

Representatives of this genus are indigenous to tropical Asia, Africa, and northern Australia; species have been introduced into the West Indies and tropical South America. The plants are present in waste places, abandoned pastures, and along roadsides and streams.

The underground tubers of *F. vestita* Benth. and *F. rhodocarpa* Bak. contain an abundance of starch and are edible. Leaves of *F. strobilifera* (L.) Ait. are administered to children in Java as a mild vermifuge; in Trinidad this species is sometimes cultivated as a cover crop on coconut plantations. *F. lineata* Roxb. has had use in Malaysia as a green manure. The glandular hairs from the dried pods of *F. congesta* Roxb. and *F. grahamiana* W. & A. yield a purple or orange brown powder that is sold under the names "waras" or "warus" in Arabian markets. This crude powder imparts a brilliant orange color to silks, but is unsatisfactory for staining wool and useless on cotton (Burkill 1966); the purified crystalline product is flemingin, $C_{12}H_{12}O_3$ (Mayer 1943).

NODULATION STUDIES

Dawson (1900b) commented on the unusually large size of the rhizobia and the absence of infection threads in nodule sections of greenhouse plants of *F. semialata* Roxb.

Nodulated species of *Flemingia*	Area	Reported by
congesta Roxb.	Wis., U.S.A.	J. C. Burton (pers. comm. 1962)
	Papua New Guinea	Allen & Hamatová (1973)
grahamiana W. & A.	Zimbabwe	Corby (1974)
	S. Africa	Grobbelaar & Clarke (1975)
lineata Roxb.	Java	Keuchenius (1924)
semialata Roxb.	Great Britain	Dawson (1900b)
strobilifera (L.) Ait.	Hawaii, U.S.A.	Pers. observ. (1935)
	Sri Lanka	W. R. C. Paul (pers. comm. 1951)
	Philippines	Bañados & Fernandez (1954)
	Trinidad	DeSouza (1966)
	S. Africa	Grobbelaar & Clarke (1974)

Fordia Hemsl.   Papilionoideae: Galegeae, Tephrosiinae
259 / 3723

Named for Charles Ford, 1844–1927, plant collector in China, 1871–1902.
    Type: *F. cauliflora* Hemsl.
    12 species. Shrubs, woody or small trees, erect. Leaves imparipinnate; leaf buds covered with many, awl-shaped scales; leaflets many, entire, large, long, acutely pointed; stipels filiform, persistent; stipules awl-shaped, persistent. Flowers rose, purple, or white, many, on short pedicels, in large racemes usually arising directly from the stems and branches on old wood; calyx truncate, entire or obsoletely 5-lobed; lobes blunt, usually short; petals narrow-clawed; standard large, rounded; wings oblong, nearly straight; keel blunt, incurved; stamens diadelphous; filaments not dilated; anthers equal; ovary sessile, 2-ovuled; style incurved, glabrous; stigma small, terminal. Pod sessile or short-stalked, flattened, club-shaped, acutely tipped, leathery, 2-valved; seeds 2–3, dark brown, discoid, in apical region of the pod. (Dunn 1911b; Ridley 1925)
    Species of this genus are common in Malaysia, southern China, Thailand, and the Philippines.

Gagnebina Neck.   Mimosoideae: Adenanthereae
20 / 3456

Named for Abraham Gagnebin, 1707–1800, Swiss botanist.
    Type: *G. tamariscina* DC.
    2 species. Trees, unarmed. Leaves bipinnate; pinnae pairs many; leaflets small, linear, many-paired; petiolar gland large; jugal glands small; stipules small, setaceous. Flowers yellow, small, sessile, in slender, cylindric spikes, fascicled in upper leaf axils or at ends of twigs; calyx campanulate; lobes 5, short, equal; petals 5, free, oblong or linear, equal, valvate; stamens 10, free, slightly exserted; anthers gland-tipped; ovary stalked, many-ovuled; style long, filiform; stigma terminal, blunt. Pod oblong-linear, villous, becoming glabrous when mature, compressed, thick, membranously winged along both sutures, septate between the seeds, septa continuous with the endocarp, indehiscent; seeds many, compressed, transverse. (De Candolle 1825b; Bentham and Hooker 1865)
    Members of this genus are native to Mauritius and Madagascar.

Galactia P. Br.   Papilionoideae: Phaseoleae, Diocleinae
406 / 3882

From Greek, *gala*, "milk," in reference to the milky latex in the stems of some species, a rare occurrence in the Leguminosae.
    Type: *G. pendula* Pers.
    50 species. Herbs, prostrate, twining or erect shrubs, sometimes hairy. Leaves pinnately 3-foliolate, rarely 1–5–7-foliolate; stipels present; stipules small, lanceolate, often caducous. Flowers mostly small, red violet or white,

paired or fascicled, sometimes solitary, on firm, slender peduncles, in lax-flowered racemes; rachis nodose; bracts small, setaceous; bracteoles very small; calyx campanulate, often hairy; lobes lanceolate-linear, acuminate, longer than the calyx tube, upper 2 fused to form 1, lateral lobes often smaller, lowermost lobe often longest; petals subequal; standard round or ovate, short-clawed, margins scarcely inflexed at the base, short-clawed; wings narrow, obovate or oblong, eared, short-clawed, adherent to the keel; keel oblique-oblong, subequal or longer than the wings, short-clawed, not beaked; stamens diadelphous, vexillary stamen free or sometimes connate with others at the middle; anthers uniform; ovary subsessile, several- to many-ovuled, often villous; style linear, filiform, small, not barbate; stigma small, terminal. Pod linear, straight or incurved, flat, rarely convex on both sides, 2-valved, thinly septate inside between the seeds; seeds not appendaged. (Bentham and Hooker 1865; Amshoff 1939; Standley and Steyermark 1946; Phillips 1951; Burkart 1971)

Representatives of this genus are common worldwide in warm regions, but are especially prevalent in the Western Hemisphere. About 45 species occur in tropical America, 21 species in Cuba, 12 species in Argentina. Only 2 species, *G. muelleri* Benth. and *G. tenuiflora* (Willd.) W. & A., occur in Australia. *Galactia* species are present in brushwood areas, light forests, forest margins, and thickets; they prefer sandy soil. *G. spiciformis* T. & G., in many areas a beach vine, is not harmed by ocean spray.

The matlike growth of most species suggests usefulness in soil-erosion control, but none of the plants has been cultivated for any purpose and they apparently have no forage value. It is said that some species are toxic at maturity (Hoehne 1939). *Galactia* flowers are ornamentally attractive.

NODULATION STUDIES

Rhizobia from nodules of *Galactia* species have been sparsely studied. Cultural and biochemical characteristics are lacking. Rhizobial strains from *G. volubilis* (L.) Britt. produced nodules on seedlings of *Amorpha fruticosa*, *Crotalaria sagittalis*, *Wisteria frutescens* (Wilson 1944b) and *Astragalus nuttalianus* (Wilson and Chin 1947). Nine of the 12 strains of soybean rhizobia tested by Wilson (1945a) nodulated *G. volubilis* seedlings. Plants of *G. wrightii* Gray were devoid of nodules in their native Arizona field habitats and also in greenhouse tests when treated with various commercial inoculants (Martin 1948).

Nodulated species of *Galactia*	Area	Reported by
dubia DC.	Puerto Rico	Dubey et al. (1972)
jussiaeana Kunth	Venezuela	Barrios & Gonzalez (1971)
regularis (L.) BSP	N.C., U.S.A.	Pers. observ. (1948)
striata (Jacq.) Urb.	Venezuela	Barrios & Gonzalez (1971)
tenuiflora (Willd.) W. & A.	Java	Keuchenius (1924)
	Zimbabwe	Corby (1974)
volubilis (L.) Britt.	N.Y., U.S.A.	Wilson (1944b)

From Greek, *gala*, "milk"; it was once thought that feeding these plants to animals would increase lactation.

Type: *G. officinalis* L.

6–8 species. Herbs, perennial, erect, robust, glabrous. Leaves imparipinnate, very large at the base; leaflets entire, venose, 6–12-paired. Flowers blue, violet, or white, in axillary or terminal racemes; bracts narrow; bracteoles none; calyx tubular-campanulate; lobes 5, subequal; standard obovate-oblong, narrowed below into a short claw; wings oblong, slightly joined to a blunt, incurved keel; stamens 10, monadelphous; ovary sessile, many-ovuled; anthers uniform or alternately smaller, elliptic, dorsifixed; style tapering, incurved, glabrous; stigma small, terminal. Pod linear-cylindric, torulose, semierect, tipped by the remnant style, 2-valved, valves slender, obliquely striate, continuous within, sometimes constricted between the seeds, dehiscent; seeds 1–several, transversely oblong. (Bentham and Hooker 1865)

Members of this genus are present from the warm temperate regions of Mediterranean Europe to Iran. Three species are indigenous to the mountain regions of tropical East Africa.

The common name of the type species, "goat's rue," indicates its toxicity for goats. The foliage has a bitter taste, and causes vomiting and even death under some conditions. However, because of its adaptation to acid soil, it is cultivated to some extent in southern Europe for soil improvement and as a beeplant.

Pharmacologically, plant part decoctions of *G. officinalis* L. show diaphoretic, anthelmintic, antispasmodic, and galactagogue properties. The seeds and foliage contain galegine, $C_{16}H_{13}N_2O$, and various guanidine derivatives (see Harborne et al. 1971). Among these guanidine derivatives are spherophysine and smirnovine (Reuter 1965) of the general formula, $C_{10}H_{22}N_4$, which are important medicinally as ganglion-blocking agents. According to Mears and Mabry (1971), the limited occurrence of these smirnovine-type alkaloids in *Smirnowia*, *Sphaerophysa*, *Eremosparton*, and *Astragalus* (q.v.) of the tribe Galegeae has chemotaxonomic significance. This concept agrees in principle with Gillett's earlier deduction (1963a) that *Galega* has a closer kinship with *Astragalus* than with *Tephrosia* and is misplaced in the subtribe Tephrosiinae.

Nodulation Studies

Nodules on *G. officinalis* were described by Štefan (1906) as coralloid with abundant starch deposits. Spratt (1919) considered them the Viceae type, elongated with apical meristems.

Galega species provide high yields of dry matter and fixed nitrogen when they are properly inoculated. Effective symbiosis of *G. officinalis* with homologous strains and with rhizobia from *Vicia faba* and *V. sativa* were described by Hauke-Pacewiczowa (1952). Symbiosis was not obtained with strains from peas or alfalfa. Wilson's data (1939a) indicated that *G. officinalis* and *G. officinalis* var. *hartlandii* were highly promiscuous in their ability to form nodules with rhizobial isolates from many species of unrelated genera.

Nodulated species of *Galega*	Area	Reported by
albaflora Tourn.	N.Dak., U.S.A.	Bolley (1893)
bicolor Boiss.	Kola Peninsula, U.S.S.R.	Roizin (1959)
officinalis L.	Widely reported	Many investigators
var. *hartlandii*	N.Y., U.S.A.	Wilson (1939a)
orientalis Lam.	Kew Gardens, England	Pers. observ. (1936)
patula Stev.	Kew Gardens, England	Pers. observ. (1936)
var. *startlandii*	Kola Peninsula, U.S.S.R.	Roizin (1959)

Gamwellia Bak. f. — Papilionoideae: Loteae

Named for A. H. Gamwell, from whose herbarium specimens the generic determination was made.

Type: *G. flava* Bak. f.

1 species. Herb, bushy perennial, hairy throughout, woody at the base, stems spreading up to 35 cm. Leaves imparipinnate, 3-foliolate; leaflets sessile, linear-lanceolate, acute, subtended in the leaf axil by 1 pair of linear-lanceolate, leafletlike stipules. Flowers yellow, few, in short, terminal or leaf-opposed racemes; bracts and bracteoles linear; calyx campanulate, deeply cleft; lobes 5, linear-lanceolate, acute, subequal, longer than the calyx tube, 2 upper lobes slightly connate; petals longer than the calyx; standard obovate, short-clawed; wings oblong-oblanceolate, slender-clawed; keel petals smaller, oblong, long-clawed; stamens diadelphous; filaments unequal, outermost 2 shortest, next 2 longest, not apically dilated; anthers equal, dorsifixed; ovary hairy, many-ovuled; style filiform, slightly curved; stigma terminal. Pod linear-oblong, flattened, hairy, beaked, nonseptate, dehiscent, several- to many-seeded. (Baker 1935b)

The stipules may be interpreted as a second pair of leaflets (Hutchinson 1964). This genus was originally conceived of as closely akin to *Lotus*. Polhill's merger (1974) of several genera, including *Gamwellia*, into *Pearsonia* suggests a closer kinship with *Lotononis* in the tribe Crotalarieae (Polhill 1976).

This monotypic genus is endemic in Zambia. It is present in sandy and rocky areas in the open bush up to about 1,900-m altitude.

From Greek, *gaster*, "belly," and *lobion*, "pod," in reference to the ovoid pod.

Type: *G. bilobum* R. Br. ex Ait.

44 species. Shrubs, often decumbent, some erect, ornamentally attractive; young branches silky pubescent. Leaves opposite, or 3–4 whorled, simple, on short petioles, flat, or margins revolute, usually rigid; stipules spiny. Flowers yellow or mixed with purplish red, on short peduncles, in axillary heads or in elongated, lax, axillary or terminal racemes; bracts and bracteoles absent or caducous; calyx 5-lobed, 2 upper lobes broader and connate higher up; petals clawed; standard usually yellow, orbicular or kidney-shaped, emarginate, longer than the lower petals; wings oblong; keel blunt, often red, broader and usually shorter than the wings; stamens free; ovary stalked, rarely sessile, often villous, 2-ovuled; style incurved, slender; stigma small, terminal. Pod ovoid, turgid, valves leathery, continuous inside, 1–3-seeded. (Bentham 1864; Bentham and Hooker 1865)

Gastrolobium is indigenous to Western Australia. Members of this genus are present on rocky summits, in range areas, and are associated with water courses in sandy areas.

Gastrolobium species have accounted for large economic losses of livestock in Queensland and Western Australia; 16 of 30 species examined by Bennetts (1935) were toxic to sheep. Monofluoroacetic acid has been identified as the toxic principle in 3 species (McEwan 1964; Alpin 1967, 1971). Poisoning results from the ingestion of air-dried material as well as the green foliage. *G. grandiflorum* F. Muell., an attractive shrub of 1–2-m height with large, reddish-brown to yellow flowers, has been the primary subject of study. An antidote for this stable toxic compound is not known.

NODULATION STUDIES

Field-excavated roots of *G. brownii* Meissn., *G. crassifolium* Benth., *G. obovatum* Benth., *G. pulchellum* Turcz., *G. spinosum* Benth., *G. trilobum* Benth., *G. velutinum* Lindl., and *G. villosum* Benth. were observed nodulated in Western Australia (Lange 1959). Rhizobia from 4 species were culturally similar to slow-growing strains of the lupine-soybean type; nodules were produced by these strains on cowpea and lupine species in infection tests (Lange 1961).

From Greek, *geisson*, "hem," "border," and *aspis*, "shield"; large bracts border and shield the flowers.

Type: *G. cristata* W. & A.

3 species. Herbs, annual. Leaves paripinnate; leaflets few-paired, obovate; stipels none; stipules membranous. Flowers small, yellow or purple, in long pendent, axillary racemes, often hidden or enclosed by large, overlapping, asymmetric, membranous, hairy bracts that later also cover the pods;

bracteoles none; calyx deeply bilabiate, upper lip entire, lower lip 3-dentate; standard broad, short-clawed; wings oblong; keel petals incurved, joined near the apex; stamens all joined into a sheath split above and later also below; anthers uniform; ovary short-stalked, 2-ovuled; style incurved. Pod with the upper suture nearly straight, lower suture sinuate; septate within, joints 1–2, flat, densely reticulate, indehiscent.

This genus is closely related to the African genus *Humularia* (Duvigneaud 1954a).

Representatives of this genus are distributed throughout tropical India, Africa, and Malaysia. They occur in sandy countryside areas and along muddy edges of rice fields.

Nodulation Studies

Nodules were observed in western India on *G. cristata* W. & A. (Bhelke 1972) and on *G. tenella* Benth. (V. Schinde personal communication 1977).

Genista L.   Papilionoideae: Genisteae, Spartiinae
214 / 3675

From Latin, *genista*, the name of a broom plant. According to Smith (1963), the Plantagenet kings and queens of England adopted their name from the Latin, *planta genista*.

Type: *G. tinctoria* L.

80 species. Shrubs, rarely shrublets or small trees, deciduous or nearly evergreen, unarmed or spiny. Leaves small, lanceolate, simple, rarely palmately 3-foliolate, or absent; stipules small or absent. Flowers yellow, rarely white, fascicled or subsolitary, in terminal clusters or heads; bracts and bracteoles usually small, caducous, otherwise foliaceous and somewhat persistent; calyx bilabiate; lobes short, upper 2 free or slightly connate, lower 3 connate into a 3-lobed lip; standard ovate; wings oblong; keel petals oblong, slightly incurved, swollen at the sides; claws of wings and keel petals usually adnate to the staminal tube; stamens 10, monadelphous; anthers alternately short and versatile, long and basifixed; ovary sessile, 2- to many-ovuled; style apex incurved; stigma capitate, terminal. Pod linear-oblong or ovate, 2-valved, convex or turgid, nonseptate, dehiscent or indehiscent, 1- to many-seeded. (Bentham and Hooker 1865)

Species of *Genista* and *Cytisus* are distinguished with difficulty.

Members of this genus are native to temperate regions of Europe, northern Africa, the Mediterranean area, and western Asia. They occur on heaths, scrub and dry woodlands, and in sandy, desert, and littoral habitats.

The drought resistance of these species suggests their use in preventing sand-dune erosion. *G. aetnensis* DC., *G. rhodorrhizoides* Webb & Berth., and *G. sphaerocarpa* Lam. are cultivated as ornamentals. The fine green textile dye, Kendal green, is a mixture of the yellow dye from flowers of *G. tinctoria* L. and the blue dye from dried flowers of woad, *Isatis tinctoria*.

Genista species lack grazing and forage value because of the general occurrence of quinolizidine alkaloids in plant parts (Leonard 1953; Willaman
and Schubert 1961; Mears and Mabry 1971). Anagyrine, cytisine, lupanine,
N-methylcytisine, retamine, and sparteine have been found in more than 15
of the species examined (see Harborne et al. 1971).

Nodulation Studies

Genista rhizobia have received little attention; however, it appears conclusive that they closely resemble strains of *Rhizobium lupini* (Sarić 1959). Limited
plant studies suggest inclusion of *G. tinctoria* in the cowpea miscellany (Burrill
and Hansen 1917; Conklin 1936; Wilson 1939a). Oval-shaped bacteria were
described in nodules of *G. germanica* L. by Frank as early as 1879. Normal
morphogenesis and development of nodules from 3 *Genista* species were observed by Lechtova-Trnka (1931).

Nodulated species of *Genista*	Area	Reported by
anglica L.	Germany	Morck (1891)
(*florida* Asso)	France	Lechtova-Trnka (1931)
? = *cinerea* DC.		
germanica L.	Germany	Frank (1879)
hispanica L.	France	Lechtova-Trnka (1931)
monosperma Lam.	S. Africa	Grobbelaar & Clarke (1974)
pilosa L.	France	Lechtova-Trnka (1931)
praecox [?]	N.Y., U.S.A.	Conklin (1936)
radiata (L.) Scop.	S. Africa	Grobbelaar & Clarke (1974)
sagittalis L.	Yugoslavia	Sarić (1959)
sibirica Reichb.	France	Lechtova-Trnka (1931)
spachiana Webb	France	Laurent (1891)
(*supina* (L.) Scheele)	N.Y., U.S.A.	Wilson (1945a)
= *Cytisus supinus* L.		
(*supranubia* Spach)	N.Y., U.S.A.	Wilson (1939a)
= *Cytisus fragrans* Lam.		
tinctoria L.	Widely reported	Many investigators
(= *ovata* Waldst. & Kit.)	Denmark	Jensen (1964)
(= *virgata* Willd.)	Sweden	Eriksson (1873)
	France	Clos (1896)

Genistidium I. M. Johnston   Papilionoideae: Galegeae,
Tephrosiinae

The Greek *idion* is a diminutive suffix; the general habit of this small plant
resembles that of members of the Genisteae.

Type: *G. dumosum* I. M. Johnston

1 species. Shrub, unarmed, erect, small, much-branched, strigose. Leaves
3-foliolate; leaflets entire, elongated, imperceptibly veined; stipels none;

stipules awl-shaped, minute. Flowers yellow or reddish, solitary, rarely paired, in the axils of upper often 1-foliolate leaves; calyx campanulate; lobes longer than calyx tube, 2 upper lobes joined higher up; petals long-clawed; standard suborbicular; wing and keel petals eared, obtusely lunate, somewhat convergent; stamens diadelphous; anthers uniform; ovary subsessile, 4–6-ovuled; style inflexed, subcylindric, awl-shaped, completely bearded above middle; stigma minute, capitate, terminal. Pod linear, flattened, 2-valved, valves leathery, dry; seeds orbicular, compressed. (Johnston 1941)

The genus is closely related to *Tephrosia* and *Peteria*.

Plants of this monotypic genus occur at the summit of high cliffs of coarse volcanic tuff, which dominates San Antonio de los Alamos, State of Coahuila, Mexico, and in rocky limestone soil on Reed Plateau, Terlingua, Texas.

Geoffroea Jacq. Papilionoideae: Dalbergieae, Geoffraeinae
370 / 3842

Named for Claudio José Geoffroy, Paris physician, author of *Materia Medica* (*fide* Burkart 1949a).

Type: *G. spinosa* Jacq.

6 species. Trees or shrubs, armed and unarmed, eglandular. Leaves large, imparipinnate; leaflets 5–15, short-petioluled, alternate or subopposite on the same plant, oblong, obtuse or emarginate; stipels small, linear or awl-shaped, caducous; stipules awl-shaped, small, caducous. Flowers mostly yellow, small, with a fetid odor, borne in simple, axillary racemes or subfascicled near the tips of branchlets; bracts and bracteoles small, early caducous; calyx turbinate-campanulate; lobes 5, subequal, upper 2 united; standard suborbicular, eared at the base, claw abrupt, short; wings and keel petals nearly alike, erect, obtuse, free, conspicuously clawed, eared near the base, keel petals imbricate and joined at the back; stamens 10, usually diadelphous, sometimes all united into a split sheath; anthers uniform; ovary sessile or short-stalked, oblong, pubescent, 2–5-ovuled; disk absent; style large, cylindric, glabrous, slightly incurved; stigma small, capitate, terminal. Pod drupaceous, ovoid, endocarp woody, indehiscent, 1-seeded. (Bentham and Hooker 1865; Burkart 1949a)

Geoffroeeae trib. nov. was proposed by Hutchinson (1964), with *Geoffroea* as the type genus.

Species of this genus occur in temperate and tropical South America, the West Indies, and Panama.

Bark decoctions have vermifuge and purgative properties but are poisonous in heavy doses; the leaves are anthelmintic; roasted seeds of *G. superba* H. & B. are edible. *Geoffroea* wood is hard, heavy, and strong, with a rather fine texture and irregular grain; it is favored locally for ornamental boxes.

Name means "Gilbert's tree"; named for Professor G. Gilbert, Belgian botanist.

Type: *G. demonstrans* (Baill.) J. Léonard

26 species. Trees, generally very tall. Leaves paripinnate, rarely imparipinnate; leaflets usually in several pairs, opposite, rarely alternate, margin dotted with 1 or more glands; stipels absent; stipules small to large, sometimes foliaceous, caducous or persistent. Flowers red, purple, cream, or pink, in more than 2 rows, in panicles or sometimes in racemes, axillary or terminal, or solitary or fascicled on old wood; bracts generally small; bracteoles 2, velvety, opposite, conspicuous, concave, valvate, sepaloid, enclosing the bud, sometimes very large, persistent, often apically glandular; sepals 5, lanceolate, imbricate, 4 subequal, median sepal larger and 2-toothed; petals 5, middle petal largest, fanlike, fully developed, 2-lobed, clawed, other 4 small or scalelike, equal; stamens 9, 3 large, external, exserted, others lesser developed, reduced; anthers dorsifixed, longitudinally dehiscent; ovary short-stalked, unilaterally adnate to the receptacle, few- to many-ovuled; style filiform; stigma terminal. Pod up to 30 cm long by 10 cm wide, flattened, generally oblong, woody, with 1–3 prominent, longitudinal nerves; seeds several, large, flat, shiny, brown. (Léonard 1952)

This genus was created to accommodate African species formerly in *Macrolobium*, which now consists exclusively of species in tropical America.

Members of this genus occur throughout tropical West Africa. They are present in moist, swampy, riverside or rain forests.

The wood has no commercial export value. It is hard, heavy, reddish brown, and coarse grained. The supply from large trees is used locally for construction; that from the more available medium-sized trees is used for cabinetmaking and general carpentry.

NODULATION STUDIES

In 1957 Bonnier reported the lack of nodules on *G. dewevrei* (De Wild.) J. Léonard in the dense virgin forests of the Yangambi region, Zaire, where it was the dominant species; however, mycorrhiza were abundantly present on the roots. An apparent symbiosis of this species with 5 types of ectotrophic mycorrhiza was described by Peyronel and Fassi (1957). Later, upon observing nodules on this species growing in cleared areas and along roadsides where the soil had been disturbed and aeration was provided, Bonnier (1958) opined that "rhizobial symbiosis is an adaptation by the Leguminosae to a shortage of nitrogen. . . . There may be, therefore, a tendency to form mycorrhiza but no such tendency to nodulate, except where soils have been disturbed and presumably more leaching and loss of nitrogen has gone on." Several other nonnodulating genera in Caesalpinioideae have since been shown to harbor ectotrophic mycorrhiza, namely, *Afzelia*, *Anthonotha*, *Brachystegia*, *Julbernardia*, and *Paramacrolobium* (Peyronel and Fassi 1960).

Name means "Gillet's tree"; named for Abel Gillet, 1857–1927, French botanist; he collected specimens of these trees in Zaire.

Type: *G. klainei* (Pierre) Verm.

5–7 species. Trees, evergreen, medium to large, up to 45 m tall, with trunk diameters in excess of 80 cm. Leaves imparipinnate or paripinnate; leaflets numerous, 4–20 pairs, mostly alternate, entire, asymmetric, translucent-punctate; stipules linear, very caducous. Flowers white, small, many, in 2 vertical rows along rachides of axillary or terminal panicles; bracts small, very caducous; bracteoles 2, subopposite at base of pedicel, caducous; calyx cuplike, small; sepals 4, glandular, reflexed, 1 much larger than the others, subvalvate in bud, free later; petals 5, free, equal, silky-pubescent inside; stamens 10, free, alternately long and short; anthers dorsifixed, dehiscent along longitudinal slits; ovary short-stalked, glabrous, free, 2–5-ovuled; style filiform; stigma terminal. Pod stalked, flattened, oblique, apiculate, dehiscent, the 2 valves often prominently crackled or glandular-punctate, sometimes smooth; seeds flattened. (Léonard 1957)

Members of this genus were formerly included in *Cynometra*.

Species of this genus are indigenous to tropical West Africa. They are present in forests.

Gleditsia L.   Caesalpinioideae: Eucaesalpinieae
88 / 3544

Named for Professor J. D. Gleditsch, German botanist, 1714–1786. The valid Latinized name, *Gleditsia*, of Linnaeus (1753) has priority over *Gleditschia* of Scopoli (1777).

Type: *G. triacanthos* L.

12 species. Trees, 30 m or more tall, with a trunk diameter up to 1 m; bark deeply furrowed; trunk and branches usually armed with simple or branched, vicious, hard spines often 7.5–10 cm long, frequently in clusters; crown flat-topped, spreading; roots thick, fibrous. Leaves small to medium, long-petioled, simply paripinnate and bipinnate, both forms often on the same tree, deciduous; leaflets thin, small or medium in size, margins unevenly crenate; stipels absent; stipules small, inconspicuous. Flowers small, unisexual and bisexual, greenish white, on short pedicels, in axillary or lateral, simple or fascicled racemes; bracts small, scalelike, caducous; bracteoles absent; calyx turbinate-campanulate; lobes 3–5, narrow, subequal, open in bud or subimbricate; petals 3–5, sessile, subequal, imbricate; stamens 6–10, free, exserted, straight, inserted with the petals on the margin of the disk lining the calyx; anthers uniform, longitudinally dehiscent, much smaller and abortive in the female flower; ovary in male flowers minute, rudimentary or absent, in bisexual flowers subsessile, free; ovules 2–many; style short, stigma terminal, somewhat dilated, often oblique. Pod ovate-elongated, plano-compressed, broad, twisted, often pulpy inside, ovoid, indehiscent or dehiscent; seeds

many, brown, transverse, ovoid to suborbicular, flattened, attached by a slender funicle. (Bentham and Hooker 1865)

Representatives of this genus are well distributed in temperate and subtropic North and South America, Asia, and Africa. They show a preference for damp, sandy, and low alluvial soils.

The type species, commonly called honey locust, *G. texana* Sarg., the locust of southern bottomlands, and *G. aquatica* Marsh., water locust, are prominent in North America. *G. texana* is regarded by some botanists as a hybrid of the honey and water locust. *G. amorphoides* Taub., native to Argentina and Brazil, is appropriately called by the name "espina de corona Cristi," since the trunk bears pronged thorns up to 40 cm in length. The genus is represented in the southern Soviet Union by *G. caspica* Desf., in the Orient by *G. japonica* Miq., Japanese honey locust, and *G. sinensis* L., Chinese honey locust, and in Africa and Celebes by *G. africana* Welw. ex Benth. and *G. celebica* Koord., respectively.

Fast-growing honey locust has importance in soil-erosion control but only limited use as an ornamental shade tree because of its pod litter and dangerous thorns. The thornless variety of *G. triacanthos* L., *inermis*, has promise for residential planting.

The wood is strong, hard, heavy, and tough. It serves local uses for fence posts and minor heavy construction but has no commercial trade value. Pods of various species yield saponins and are used in primitive areas for laundry, for cleaning hides, and as an emetic and expectorant.

Chemical inquiries into the composition of plant parts of *G. triacanthos* have been extensive (Watt and Breyer-Brandwijk 1962). The pods, which may reach 45 cm in length, contain 12–14 percent sugar and 12–14 percent protein; they are relished by livestock. Orchards are planted in some areas of South Africa as fodder for cattle. The pods and seeds lack alkaloids, but the leaves contain about 0.5 percent of the alkaloid triacanthine, $C_8H_{10}N_4$ (Belikov et al. 1954; Khaidarov 1963), an alkaloid with hypotensive and antianemic properties similar to chidlowine present in the leaves of *Chidlowia sanguinea*. Raffauf (1970) attributes identical molecular formulae, $C_{10}H_{13}N_5$, to these 2 alkaloids. Tyramine and *N*-methyl-β-phenylethylamine are present in the air-dried leaves of the honey locust (Camp and Norvell 1966).

NODULATION STUDIES

Preclusion of nodules on honey locust was attributed by Nobbe et al. (1891) and McDougall (1921) to the inability of rhizobia to penetrate the thick, stiff root-hair walls. In 1926 the occurrence of cylindrical, hair-covered root swellings on 1–3-year-old plants was discovered independently by Fehér and Bokor in Germany and Friesner in the United States. Cultural characteristics, agglutination reactions, and plant-infection tests using isolates from these swellings were considered similar to those of rhizobia. Friesner's inference (1926) that a symbiosis existed despite the lack of nodule formation was not verified by Leonard (1939).

Interest in the symbiotic status of *Gleditsia* species was revived recently by Rothschild's account (1967b) of hair-covered, cylindrical swellings on roots of

G. amorphoides. Experimental infections of root hairs and invasion of the root cortex were obtained with mixtures of lupine- and cowpea-type rhizobia. However, Rothschild's drawings evinced intercellular cortical invasion and distribution of the bacteria markedly unlike the intracellular location of rhizobia in *bona fide* nodules.

Invasion of honey locust root hairs by rhizobia from *Erythrina*, *Glycine*, *Leucaena*, *Lupinus*, *Pisum*, and *Vigna* species was observed in the authors' laboratory; however, the infection threads did not penetrate beyond the epidermal wall. Disk assay tests using aqueous root extracts showed the presence of antirhizobial substances that were suggestive of flavones in other tests. Nodulation was not inhibited or diminished on diverse species of other leguminous genera grown in association with honey locust. It was concluded, therefore, that the nodule-inhibitory principle was root-contained, or nonfunctional if exuded into the rhizosphere (personal observation).

Nonnodulated species of *Gleditsia*	Area	Reported by
amorphoides Taub.	Argentina	Rothschild (1967b, 1970)
caspica Desf.	U.S.S.R.	Lopatina (1931)
japonica Miq.	Japan	Asai (1944)
sinensis L.	U.S.S.R.	Lopatina (1931)
	S. Africa	Grobbelaar et al. (1964)
triacanthos L.	Germany	Nobbe et al. (1891)
	Hungary	Fehér & Bokor (1926)
	Ind., U.S.A.	Friesner (1926)
	U.S.D.A.	Leonard (1939)
	Wis., U.S.A.	Pers. observ. (1950)
	S. Africa	Grobbelaar et al. (1964)
	Argentina	Rothschild (1967b)

Gliricidia HBK   Papilionoideae: Galegeae, Robiniinae
269 / 3734

From Latin, *glis*, "dormouse," and the verb *caedo*, "kill"; the powdered bark and seeds have been used as a rodent poison in the tropics.

Type: *G. sepium* (Jacq.) Steud. (= *G. maculata* HBK)

6–9 species. Shrubs or small, short-boled trees, usually less than 8 m tall, unarmed. Leaves imparipinnate; leaflets entire, 7–21, leathery, oblong, reticulate, pinnately veined, deciduous; stipels none; stipules small. Flowers large, showy, usually pink rose, often appearing before the leaves, in axillary racemes or solitary from old wood below the leaf axils; bracts small, caducous; bracteoles none; calyx briefly campanulate, as broad as long, truncate; lobes 5, short and broad or obsolete, upper 2 subconnate; standard large, orbicular, reflexed, sometimes with 2 callosities at the base, short-clawed; wings falcate-oblong, erect, free, longer than the keel; keel petals curved above, obtuse, united at the apex; stamens diadelphous; anthers uniform; ovary stalked, many-ovuled, 7–12; style glabrous, inflexed almost at a right angle, blunt; stigma small, capitate, terminal, papillose. Pod short-stalked, plano-compressed, broad-linear, 2-valved, valves leathery, nonseptate within, tardily dehiscent, twisting spirally; seeds flat, suborbicular. (Bentham and Hooker 1865; Amshoff 1939; Standley and Steyermark 1946)

Members of this genus are native to tropical America. They are common on hillsides and thickets, and are often present in mountain areas up to 2,000-m altitude. There are both wild and cultivated species.

These small, spreading, fast-growing trees are widely used as coffee and cacao shade trees and to support vanilla vines. The vernacular name of the type species in Latin America is madre de cacao, suggestive of its value as a nurse crop and shade tree for commercial cacao plantations. It is not uncommon for the branches of this small ornamental to break under the weight of the abundant floral sprays. The tree is equally attractive, despite its contorted trunk, when draped with its many slender, flat pods. *G. guatemalensis* M. Micheli and *G. meistophylla* (Donn. Smith) Pitt. are attractive shrubs in Guatemala. *G. ehrenbergii* (Schlechtd.) Rydb. and *G. lambii* Fernald are common in Mexico.

Gliricidia species are easily propagated from seeds or cuttings and are widely used as hedges, windbreaks, and live fences. Tilth and fertility of the soil underneath these trees are greatly improved from the flower and leaf-fall. According to Holland (1924), the average weight of green loppings per tree per year from the type species on tea plantations in Sri Lanka was 64 kilos.

Fallen and dried leaves emit an odor of new-mown hay because of the occurrence of coumarin compounds (Griffiths 1962). The flowers are edible and contain about 3 percent nitrogen, but the seeds and bark are poisonous and have been used to kill rats and other rodents in tropical America; however, Neal (1948) stated that Hawaii-grown plants were not poisonous. Leaves of *G. sepium* (Jacq.) Steud. are reputedly toxic to dogs and horses, but cows and goats are said to ingest them with impunity (Blohm 1962).

The light olive brown wood is hard and heavy with a coarse texture and an irregular grain. It is not easy to work but finishes smoothly and is well known for its durability. Its local use is for fence posts, agricultural implements, and small carpentry items, but it has no place in commercial channels (Record and Hess 1943).

NODULATION STUDIES

Nodulation of the type species has been observed in Sri Lanka (Holland 1924), in Hawaii (personal observation 1938), in the Philippines (Bañados and Fernandez 1954), in Costa Rica (J. León personal communication 1956), and in Trinidad (DeSouza 1966).

Glottidium Desv.   Papilionoideae: Galegeae, Robiniinae
(281)

From Greek, _glottidion_, "little tongue."

Type: _G. vesicarium_ (Jacq.) Harper

5 species. Herbs, annual, widely branched, stems 1–4 m tall, rigid, somewhat woody at the base. Leaves paripinnate; leaflets subalternate, many, narrow, oblong; stipels absent; stipules deciduous. Flowers yellow, pedicelled, each flower subtended by a caducous bract and a pair of caducous bractlets, few in slender-stalked racemes; calyx conspicuously obliquely campanulate; lobes 5, small, broadly triangular, acute, subequal; standard with a broad, emarginate, kidney-shaped blade, short-clawed, reflexed; wings short-clawed, eared at base; keel petals long-clawed, subequal to wings and standard in length, with a basal auricle; stamens diadelphous; style glabrous. Pod long-stalked, flat, oblong, short-beaked, thin-margined, tardily dehiscent; the valves separating into 2 layers, the inner layer thin, sacklike, containing 2 kidney-shaped seeds with a long, deep-set hilum, the outer layer firmer, somewhat inflated but not bladderlike. (Phillips and Hutchinson 1921; Rydberg 1924; Gillett 1963b)

Taubert (1894) regarded _Glottidium_ as a synonym of _Sesbania_. Definitive characters separating _Glottidium_ from _Sesbania_ were outlined by Phillips and Hutchinson (1921); Hutchinson (1964) maintained the distinctiveness of _Glottidium_ as a monospecific genus.

Plants of this genus thrive on depleted soils and in damp river bottoms of the coastal plain from southeastern North Carolina to Florida and west into Texas.

The species have erosion-control value. The pods contain a toxic saponin that causes intense inflammation of the gastrointestinal tract. Cases of poisoning in chickens, sheep, and cattle have been reported (West and Emmel 1950).

NODULATION STUDIES

The type species, often called bladder pod, bears copious nodules in Texas (personal observation 1938).

*Glycine Willd.   Papilionoideae: Phaseoleae, Glycininae
390 / 3864

From Greek, _glykys_, "sweet," in reference to the sweet smell or taste of the roots and leaves of some of the species.

Type: _G. clandestina_ Wendl.

10 species. Herbs, perennial, twining, except for _G. max_ (L.) Merr., an erect annual. Leaves pinnately 3-foliolate, rarely 5-foliolate, often hirsute, stipellate; leaflets entire; stipules small, free from petiole. Flowers white, blue, purplish, small, in axillary or terminal, short racemes, solitary, or in clusters along the rachis; bracteoles 2, small, subtending the calyx; calyx subcampanulate, pubescent; lobes 5, narrow, acute, upper pair slightly connate, lower 3 free; standard orbicular, base weakly eared, spreading; wings small, narrow, loosely adhering to the short, blunt, clawed keel; stamens monadelphous, or vexillary stamen later partly to completely free; anthers perfect; ovary sub-

sessile, 2–6-ovuled; style slender, beardless, incurved; stigma globose, terminal. Pod narrow, linear or oblong, subcylindric, cellulose-septate between the seeds, 2-valved, spirally dehiscent; seeds ovoid or oblong, hilum lateral, and caruncle papery thin. (Hermann 1962)

The taxonomic status of the genus *Glycine* has been a complex issue for many years. At one time 323 species, subspecies, and varieties were listed in *Index Kewensis*. Only 10 species were recognized by Hermann (1962). Further revision by Verdcourt (1966) included a proposal to adopt *G. clandestina* Wendl. as the type species to replace *G. javanica* L., a misidentified *Pueraria* species.

Members of this genus are native to lower eastern Asia and Africa, but none is native to the New World. There are both wild and cultivated species. They are well distributed in warm temperate areas and their cultivation on a wide range of soil types extends from the tropics to about 52° north latitude.

The economic importance of *G. max* dwarfs that of all other members of the genus. Reference to this cultigen in the literature is generally in such complimentary terms as the "little honorable plant," the "miracle, or wonder, bean," "China's most valuable gift to the Western world," the "orchid of agriculture," the "cow of China," and so on. *G. max* may have been derived from *G. ussuriensis* Regel & Maack, a wild species occurring throughout eastern Asia (Morse 1947). This view has not been entirely accepted (Hymowitz 1970).

Evidence points to the emergence of the soybean as a domesticate during the Chou Dynasty in China about the 11th century B.C. (Hymowitz 1970). As the dynasty expanded, the soybean was carried into southern China, Korea, Japan, and Southeast Asia. Northern China became the primary gene distribution center; Manchuria, the secondary gene center. Soybeans were made known to Europeans by the German botanist Engelbert Kaempfer in the late 17th century, but little interest was aroused (Morse 1947). Throughout the 18th century the importance of the soybean in Europe did not exceed that of a botanical curiosity. According to Morse (1947), the first mention of soybeans in American literature was by Mease in 1804: "The soybean bears the climate of Pennsylvania very well. The bean therefore ought to be cultivated." Systematic seed introductions were sponsored by the United States Department of Agriculture about 1898. The importance of the soybean in the production of oil, oil products, and as a source of flour and food products was realized in the United States in the early years of World War I. Soybean processing became an industry in the United States in 1920; oil and defatted meal were the two major products (Wolf and Cowan 1971). Between 1920 and 1970 soybean production increased 378-fold.

Domestication and dissemination of the soybean have been subjects of various reviews (Morse 1947; Ochse et al. 1961; Hymowitz 1970). Technical aspects of soybean culture were discussed in the comprehensive volume edited in 1963 by Norman. This is one of the few select crops to which a monthly journal, *The Soybean Digest*, is devoted. Various aspects of soybean utilization were expertly detailed by Schollenberger and Goss (1945). Uses of this so-called wonder crop appear endless (Markley 1950–1951; Wolf and Cowan 1971).

The rich chemical composition of the bean, the adaptability of the plant, and the ease with which it can be grown and harvested have contributed greatly to its use and demand. In 1963 soybean production in the United States was about 76 percent of the estimated world production; mainland China's production approximated only 17 percent (Pellett 1969). Again in 1970, the United States yield was 75 percent of the world total (Wolf and Cowan 1971). Daly and Egment (1966) predicted the use within the United States of 1.3 billion bushels of soybeans in 1980 as compared with 0.7 million bushels in 1963.

In the United States, where soybean is still a newcomer as a source of protein, a great potential exists for its use as food. Less than 1 percent of the crop was used for this purpose in 1971, whereas 12–13 percent of the dietary protein of the Japanese was derived from soybean products (Wolf and Cowan 1971).

NODULATION STUDIES

Soybean rhizobia are short rods with monotrichous subpolar flagella. Bacteroid forms are long, slender rods rarely branched or swollen. Cultural growth is slow and scant. Colonies are white to opalescent, with little gum formation. Pentose sugars provide better growth than hexoses. An alkaline reaction is produced in carbohydrate media and in litmus milk.

The causal organism of soybean nodules was first isolated and described by Kirchner (1895). These organisms were known as *Bacterium* (*Rhizobacterium*) *japonicum* Kirchner until 1926, when *Rhizobium japonicum* (Kirchner) Buchanan became the accepted name (Buchanan 1926). Until Leonard's report (1923), the opinion prevailed that soybean rhizobia were markedly host-specific. As evidence accumulated that cross-infectiveness between soybean rhizobia and cowpea plants, and vice versa, was the rule and not the exception, *R. japonicum* was proposed to include strains of both groups (Walker and Brown 1935). Two years later lupine rhizobia were suggested as physiologically adapted forms of the same species (Reid and Baldwin 1937). Computer studies of rhizobial strains by Graham (1964) prompted his segregation of all slow-growing rhizobia at the generic level, recognition of the nomen *Phytomyxa*, a generic name proposed by Schroeter (1886), and still further, *Phytomyxa japonicum* as the type species. This segregation was confirmed in principle by DNA base composition studies (DeLey and Rassel 1965). Less valid taxonomic changes were suggested by Moffett and Colwell (1968). Investigators have been reluctant to accept these drastic changes because so few strains were studied, host specificities of strains were not considered a coded character, nodule-forming abilities of the strains were not manifested, and taxonomic protocol was not respected. Consideration of the slow-growing, soybean-cowpea-lupine-type rhizobia as the symbiotic "ancestral" type (Norris 1956, 1958a, 1965) was challenged by Parker (1968).

Despite convincing evidence of the reciprocal infectiveness of soybean, cowpea, and lupine rhizobia on each other's host plants, interchangeability is not perfect and is often only unilateral. Accordingly, efforts to grow soybeans in various areas where cowpea- or lupine-type rhizobia prevail have been unsuccessful because of poor nodulation. Not until soil containing proper rhizobia was imported from Japan was soybean culture in the United States possible. Poor nodulation was imputed to be a major factor in failures to establish accessions of *G. wightii* (R. Grah. ex W. & A.) Verdc. (= *G. javanica*) in many areas of Australia (Nicholas 1971; Nicholas and Haydock 1971). In Malawi and Indonesia, where soybean is not endemic, 3 years were required to establish soybean rhizobia in sufficient numbers to provide good crop returns. Ever-increasing economic potential of *G. max* for industrial purposes and of *G. wightii* as a pasture crop emphasizes the need for artificial inoculation of these 2 species. The latter, a vinelike perennial, is palatable to stock, seeds profusely, and may be grown as a solo crop or in mixed crop seedings. Its common name, Rhodesian kudzu vine, denotes its acceptance in tropical and subtropical agriculture.

Nodulated species of *Glycine*	Area	Reported by
clandestina Wendl.	Australia	Benjamin (1915); Bowen (1956)
(*javanica* L.) = *Pueraria montana* (Lour.) Merr.	Widely reported	Many investigators
max (L.) Merr. (= *hispida* Maxim.)	Widely reported	Many investigators
(*sericea* (F. Muell.) Benth.) = *canescens* Hermann	Australia	Beadle (1964)
tabacina (Labill.) Benth.	Australia	Bowen (1956)
(= *koidzumii* Ohwi)	Taiwan	Pers. observ. (1962)
tomentella Hayata (= *tomentosa* Benth.)	Australia	Bowen (1956); Beadle (1964)
†*wightii* (R. Grah. ex W. & A.) Verdc. (= *javanica* auct. mult.)	Zimbabwe	Corby (1974)

† = *Neonotonia wightii* (Arn.) Lackey comb. nov. (Lackey 1977b).

***Glycyrrhiza* L.**   Papilionoideae: Galegeae, Astragalinae
301 / 3769

From Greek, *glykys*, "sweet," and *rhiza*, "root."
    Type: *G. glabra* L.
    30 species. Herbs or undershrubs, perennial with long, stout, sweet roots; stems erect, resinous-dotted. Leaves imparipinnate; leaflets paired, rarely 3, numerous, lanceolate, scaly or gland-dotted, especially underneath; stipels none; stipules minute, narrow, membranous, caducous. Flowers white, yellowish or blue, on short pedicels in peduncled, axillary spikes; bracts narrow, membranous, caducous; bracteoles absent; calyx weakly bilabiate; lobes 5, subequal or upper 2 shorter or joined at the base; petals narrow, scarcely coherent; standard narrow, oblong, erect, equal or exceeding the wings; wings oblique-oblong, acute; keel longer than the wings, blunt, sometimes acute; stamens diadelphous or monadelphous, vexillary stamen free or joined

with others into a tube split above; anthers with loculi confluent at the apex, sub-2-valved, alternate anthers smaller, with large, unequal valves more deeply open; ovary sessile, 2- or more-ovuled; style short, rigid, glabrous, curved at apex; stigma terminal. Pod oblong, ovate or short-linear, straight or arcuate, turgid or flattened, glandular or beset with curved prickles, rarely smooth, 2-valved, indehiscent or tardily dehiscent, continuous within, few-seeded; seeds 1–8, kidney-shaped or globose. (Bentham and Hooker 1865)

Representatives of this genus occur on all continents in temperate and warm areas. The majority of the species occur in the Mediterranean region of eastern Europe and in western Asia.

The roots of the major species have a characteristic sweet licorice flavor due to the triterpenoid saponin, glycyrrhizic or glycyrrhizinic acid, $C_{42}H_{62}O_{16}$. Upon acid hydrolysis this compound, which is about 50 times as sweet as sucrose, yields 2 molecules of glucuronic acid and 1 molecule of its sapogenin, glycyrrhetinic or glycyrrhetic acid, $C_{30}H_{46}O_4$, which lacks this property (Mitchell 1956).

Uses of licorice root and its extracts as mild laxatives, expectorants, and demulcents, in confectionery, in flavoring tobacco, in the manufacture of pills and troches, and for similar purposes are long established. Since glycyrrhetinic acid shares with other glycosides of sapoinins the property of foaming when shaken with water, licorice root extracts have been widely used in commercial fire extinguisher foams. In recent years, glycyrrhetinic acid has been used experimentally in the treatment of dermatoses, peptic ulcers, and Addison's disease because of its pharmaceutical properties similar to those of deoxycorticosterone (Houseman 1944; Molhuysen et al. 1950; Mitchell 1956).

G. glabra L. var. *typica* Regel & Herder and var. *glandulifera* (Waldst. & Kit.) Regel & Herder are the major sources of commercial licorice in the Western world. The former yields 6–8 percent glycyrrhizic acid, the latter, 10–14 percent (Houseman 1944). *G. malensis* Maxim. and *G. ralensis* Fisch. are the principal sources in the Orient. *G. lepidota* (Nutt.) Pursh, American licorice, and *G. echinata* L., a swamp plant in Israel and the Soviet Union, lack the sweet principle. The American species is a good soil-binding plant, spreads rapidly by rootstocks, and is shade-tolerant, but has low palatability as a forage.

Nodulation Studies

The nodules on the Australian species, *G. anthocarpa* (Lindl.) J. M. Black, growing on clay soil in arid areas subject to flooding were described as cylindrical to semicorralloid and up to 7 mm long (Beadle 1964).

Rhizobia isolated from *G. lepidota* were monotrichous short rods which elicited a very alkaline reaction and serum zone in litmus milk, meager dry, opaque growth on calcium glycero-phosphate, and scant, watery, transparent growth on yeast-extract mannitol mineral salts agars (Bushnell and Sarles 1937). These strains did not nodulate alfalfa, clover, pea, vetch, garden bean, soybean, lupine, and cowpea, or selected *Amorpha*, *Dalea*, *Robinia*, and *Strophostyles* species. Reciprocal trials were also negative. Accordingly, Bushnell and Sarles listed *Glycyrrhiza* as "unassigned to any cross-inoculation group."

Wilson (1939a), on the other hand, reported nodulation of *G. lepidota* by strains from species of diverse genera. In later tests, 1 strain from *G. lepidota* incited nodules on *Amorpha, Crotalaria,* and *Wisteria* species (Wilson 1944b) and on *Astragalus nuttalianus* (Wilson and Chin 1947). Mutual interchangeability between rhizobia and species of *Glycyrrhiza* and *Astragalus* is credible, inasmuch as these genera are closely allied phylogenetically.

Nodulated species of *Glycyrrhiza*	Area	Reported by
anthocarpa (Lindl.) J. M. Black	Australia	Beadle (1964)
echinata L.	S. Africa	Grobbelaar & Clarke (1975)
glabra L.	Germany	Morck (1891)
lepidota (Nutt.) Pursh	N.Dak., U.S.A.	Bolley (1893)
	Wis., U.S.A.	Bushnell & Sarles (1937)
	N.Y., U.S.A.	Wilson (1939a)

Goldmania Rose ex M. Micheli (25a) /3463a

Mimosoideae: Piptadenieae

Named for E. A. Goldman; he made a survey of mammals and collected plants in Mexico and Central America about 1900.

Type: *G. foetida* (Jacq.) Standl. (= *G. platycarpa* Rose ex M. Micheli)

2 species. Shrubs or small trees, unarmed; branchlets sometimes spinescent at the tips. Leaves bipinnate; pinnae 1–3 pairs; leaflets 1-, rarely 2-paired; petioles eglandular; rachis glandular between pinnae. Flowers small, in axillary spikes; calyx small, short; lobes 5, valvate; corolla greenish white, sympetalous, forming a long tube, puberulent to pubescent; stamens 10, free, equal; anthers topped with a spherical gland; ovary stalked, downy or pubescent; disk very small; style slender. Pod somewhat falcate, broad, flat, tardily dehiscent along 1 suture, not constricted between seeds or scarcely so, valves thick, leathery; seeds few, suborbicular or irregular, grayish white. (Britton and Rose 1928; Brenan 1955; Burkart 1969)

This genus of 2 subxerophilic species is geographically discontinuous. *G. foetida* (Jacq.) Standl., so named because the yellow green flowers are foul smelling, occurs in Mexico. The other species, *G. paraguensis* (Benth.) Brenan, is indigenous to Paraguay.

Gompholobium Sm. 168 /3628

Papilionoideae: Podalyrieae

From Greek, *gomphos,* "peg," and *lobion,* "pod," in reference to the appearance of the pods.

Type: *G. grandiflorum* Sm.

24 species. Shrubs, rarely undershrubs, low, diffuse, 0.3–3 m tall, gla-

brous, pubescent or with spreading hairs. Leaves simple or palmately or pinnately compound; leaflets 3–many, narrow, some with revolute margins, terminal leaflet sessile; stipules small, awl-shaped or lanceolate, or absent. Flowers yellow or red, often with green markings, solitary or few together in short, terminal racemes, rarely in upper axils; bracts and bracteoles small or absent; calyx deeply cleft; tube very short; lobes 5, lanceolate, longer than the tube, valvate, upper 2 falcate or slightly joined but not connate; petals very short-clawed, subsessile; standard orbicular, longer than the other petals; wings oblong, falcate; keel blunt, broader than the wings; stamens free; ovary short-stalked or subsessile, 4- to many-ovuled; style incurved, filiform, sometimes thicker from middle upwards. Pod ovoid or globular, often oblique and inflated, stalked; seeds 4–12, small. (Bentham 1864; Bentham and Hooker 1865)

Species of this genus are indigenous to Australia and are well distributed in Queensland, New South Wales, Victoria, and Western Australia. They are present on shores bordering bays and inlets, in hilly areas, and in ranges.

All species serve as good ground cover. *G. uncinatum* A. Cunn. ex Benth. forms a dense mat 1 m in diameter and about 0.3 m thick. The conspicuous scarlet to brick red flowers have a clear yellow patch at the base of the standard. The foliage is considered poisonous to sheep in New South Wales. Pammel (1911) mentioned it as having inebriant properties. *G. pinnatum* Sm. and *G. virgatum* Sieb. are also reputedly poisonous.

NODULATION STUDIES

Cultural characteristics of rhizobial strains from 5 of the 6 nodulated species reported by Lange (1959) conformed with those of slow-growing cowpea-type rhizobia (Lange 1961). The strains showed infective variability with *Glycine* and *Lupinus* species and *Vigna sinensis*, and were unable to nodulate alfalfa, clover, peas, vetch, and lens. All the strains nodulated *Lupinus digitatus*; however, the relationship was nonreciprocal (Lange 1962).

Nodulated species of *Gompholobium*	Area	Reported by
amplexicaule Meissn.	W. Australia	Lange (1959)
knightianum Lindl.	W. Australia	Lange (1959)
marginatum R. Br.	W. Australia	Lange (1959)
minus Sm.	S. Australia	Harris (1953)
pinnatum Sm.	Queensland, Australia	Bowen (1956)
polymorphum R. Br.	W. Australia	Lange (1959)
tomentosum Labill.	W. Australia	Lange (1959)
venustum R. Br.	W. Australia	Lange (1959)
virgatum Sieb. ex DC.	Queensland, Australia	McKnight (1949b); Bowen (1956)

From Greek, *gonia*, "angle," and *rhachis*, "spine," "midrib"; the rachis of the inflorescence is zigzag.

Type: *G. marginata* Taub.

1 species. Shrub, small, much-branched. Leaves paripinnate; leaflets obliquely ovate or oblong, asymmetric at the base, in 2 pairs, thin, leathery, reticulate; stipules early caducous. Flowers white, zygomorphic, medium-sized, borne on a zigzag rachis in few-flowered, terminal spikes; bracts suborbicular, cuplike, eventually caducous; bracteoles 2, alternating, orbicular-ovate, concave, subtending each flower; calyx subcylindric; lobes 4, blunt, imbricate, lowermost lobe outermost; petals 5, short-clawed, imbricate, obovate, uppermost petal innermost; stamens 10, 5 slightly longer; filaments free; anthers dorsifixed, uniform; ovary stalked, stalk adnate to calyx tube, many-ovuled; style filiform, shorter than ovary; stigma small, terminal. Pod awaits description.

This monotypic genus is endemic in southeastern Brazil.

The shrub is not economically important, but variety *elata* Kuhlmann, which reaches a height of about 20 m in rain forests, has heavy, durable wood that is highly esteemed for construction and railroad crossties (Record and Hess 1943).

Gonocytisus Spach   Papilionoideae: Genisteae

From Greek, *gonos*, "seed," "genitals"; reproductive organs resemble those of *Cytisus*.

Type: *G. angulatus* (L.) Spach

3 species. Virgate shrubs, unarmed; branches ribbed or angular-winged. Leaves 3-foliolate, sessile or short-petioled; stipules minute. Flowers yellow, in long terminal or axillary racemes; upper calyx tooth more deeply cleft than lateral teeth; stamens monadelphous; anthers ciliate at both ends. Pods sessile, flattened, oblong, pointed, dehiscent, 1–2-seeded. (Polhill 1976)

Polhill (1976) placed this genus in Genisteae *sensu stricto*. Members of this genus are native to countries bordering the eastern end of the Mediterranean.

Goodia Salisb.   Papilionoideae: Genisteae, Bossiaeinae
192 / 3652

Named for Peter Good, botanical explorer and collector with Robert Brown on H.M.S. *Investigator*.

Type: *G. lotifolia* Salisb.

2 species. Shrubs, 1–3 m tall; branches pubescent. Leaves 3-foliolate;

leaflets entire, ovate; stipels none; stipules, membranous, early caducous. Flowers numerous, yellow mixed with red or purple, borne in terminal or leaf-opposed racemes; bracts and bracteoles membranous, caducous; calyx more or less pubescent; lobes 5, 2 upper lobes joined into a 2-toothed lip, 3 lower ones narrow, equal; petals clawed; standard orbicular, yellow with purple mixture; wings yellow or red, falcate-oblong, shorter than the standard; keel yellow or red, broad, incurved, blunt; stamens united into a sheath split above; anthers versatile, alternately smaller; disk annular between stamens and ovary; ovary stalked, 2–4-ovuled; style blunt, incurved; stigma small. Pod long-stalked, oblong-falcate, plano-compressed, 2-valved, valves thin, continuous inside; seeds 1–4. (Bentham 1864; Bentham and Hooker 1865)

This genus was assigned to the tribe Bossiaeae by Polhill (1976) in his revision of Genisteae *sensu lato*.

Goodia species are limited to the temperate regions of Australia. They are common in scrub areas.

G. lotifolia Salisb. is a tall, much-branched, hardy, long-blooming, broom-like shrub cultivated in many beach gardens. It is reputedly poisonous to stock because of an unstable cyanhydrin principle in the leaves (Webb 1948).

Nodulation Studies

G. pubescens Sims, the second species, regarded by some botanists as a variety of *G. lotifolia*, was observed nodulated in Kew Gardens, England, by the authors in 1936. Wilson (1944b) obtained nodulation of seedlings of *Amorpha fruiticosa*, *Crotalaria sagitallis*, and *Wisteria frutescens*, grown in sterilized sandy loam in cotton-plugged containers, by using a rhizobial strain designated as *G. medicaginea* F. Muell., a synonym of *G. lotifolia*. Presumably, this was the same strain later reported to symbiose with *Astragalus nuttalianus* (Wilson & Chin 1947).

Gossweilerodendron Harms   Caesalpinioideae: Cynometreae

Named for John Gossweiler, an authority on the flora of Angola.

Type: *G. balsamiferum* (Verm.) Harms

1 species. Tree, tall; branches glabrous. Leaves imparipinnate; leaflets mostly alternate, rarely opposite, entire, oblong-obovate, gland-dotted; petiolule usually twisted. Flowers white, very small, disposed in 2 rows, on short pedicels in long spiciform, axillary or terminal racemes, short-paniculate; bracts and bracteoles minute, caducous; calyx tube cup-shaped, very short, shallow, broad; sepals 4, rarely 5, oval or suborbicular, free, imbricate, gland-dotted; petals none; stamens 10, rarely 8–9, exserted; filaments free, slender, densely hairy at the base; anthers small, oval-subquadrate, dorsifixed, longitudinally dehiscent; ovary stalked, stalk densely hairy and in-

serted at the base of the receptacle, oblique-ovoid, 1-ovuled at the apex; disk absent; style short, acute, slender, often S-shaped, hairy below. Pod samaroid, stoutly stalked, lanceolate-oblong, obliquely apiculate, glabrous, ventral side margined, the larger basal portion forming a membranous, reticulate, pellucid-dotted or striate wing with 1 seed at the thickened, pendulous apex, indehiscent. (Baker 1930; Pellegrin 1948; Léonard 1952, 1957)

This monotypic genus is present in tropical West Africa, south to the western Congo Basin. It is common in rain-forest areas of southern Nigeria and Angola.

This species, commonly known by its Nigerian name "agba," is one of the largest trees of the tropical rain forests of Africa. It may reach a height of 65–70 m, with a diameter of 2 m (Kukachka 1961b). Dalziel (1948) referred to a specimen at Sapoba, Nigeria, that measured 7 m in mid-girth and was not forked below 35 m. Excellent knot-free logs are obtained from straight, cylindrical trees that are devoid of buttresses and lower branches.

Agba wood is marketed as Nigerian cedar or pink mahogany from Nigeria and the Congo (Titmuss 1965). It is one of the lighter-weight tropical African timbers and has a density of about 512 kg/m^3 at a moisture content of 12 percent. The heartwood is light brown, the thick sapwood is paler in color. The wood is available in large sizes, seasons rapidly, is worked easily by hand and machine tools, finishes readily, and has a texture finer than mahogany (Kukachka 1961b). A copal obtained from the wood is used for illumination (Dalziel 1948). This copal imparts a slight but unobjectionable odor to the wood and may account for its resistance to fungi and subterranean termites. The wood is used for fine veneers, decoration, molding, flooring, furniture, and general construction.

NODULATION STUDIES

Nodules were recorded on this species in Yangambi, Zaire, by Bonnier (1957).

Gourliea Gillies ex Hook.   Papilionoideae: Sophoreae
143 / 3603

Named for Robert Gourlie, Scottish botanist in Chile.

Type: *G. decorticans* Gillies ex Hook. (= *spinosa* (Mol.) Skeels; *G. chilensis* Clos)

1 species. Shrub or small tree, usually less than 8 m tall; thorns in leaf axils and at tips of branches; bark on upper parts exfoliating. Leaves small, imparipinnate, usually in pairs or clusters on wood nodes; leaflets subopposite or alternate, 2–10 pairs, small, oval; stipels none. Flowers small, golden yellow, in short, clustered, axillary racemes at old leaf nodes; bracts minute; bracteoles none; calyx lobes 5, upper 2 truncate, subconnate, lower 3 smaller, narrow; petals long-clawed; standard broad, rounded, spreading; wings

oblique-obovate, margins undulate; keel petals shorter than the wings, blunt, incurved, joined at the back; stamens free or joined briefly at the base; anthers small; ovary sessile, multiovuled; style awl-shaped, incurved; stigma small, terminal. Pod ovoid or globose, drupaceous, resembling a plum, exocarp fleshy, endocarp woody, indehiscent; seeds 1–2, kidney-shaped, thick. (Bentham and Hooker 1865)

Burkart (1952) suggested the positioning of this genus in proximity to *Geoffraea*, tribe Dalbergieae.

This monotypic genus is endemic in temperate South America, Chile, Bolivia, and Argentina. The plants grow in dry open areas, often forming dense thickets.

In addition to being an attractive ornamental, *G. decorticans* Gillies ex Hook. yields fruits, known as chañar or chanal, which are a staple food in some areas, consumed by livestock, and used as an ingredient of a fermented beverage. The principal use of the nondurable, low-luster, irregularly grained wood is for implement handles and fuel.

NODULATION STUDIES

Nodules were present on this species growing in Kew Gardens, England (personal observation 1936).

Griffonia Baill.   Caesalpinioideae: Bauhinieae
72 / 3527

Type: *G. simplicifolia* Baill.

3–4 species. Shrubs, tall, climbing. Leaves short-stalked, 1-foliolate, leathery; stipules minute. Flowers nodding, showy, red to dark purple, in simple or panicled, terminal racemes; bracts and bracteoles small, calyx tubular, dilated or campanulate above; lobes 5, short, broad, subimbricate, persistent; petals 5, oblanceolate or oval, subequal, erect, imbricate; stamens 10, free, alternately long and short; filaments usually pubescent below; anthers uniform, longitudinally dehiscent; ovary oblong, flattened, long-stalked, stalk adnate to calyx tube, 1–3-ovuled; style short, persistent; stigma small, terminal. Pod long-stalked, swollen, 2-valved, leathery or thinly woody, 1- to few-seeded, apicula lateral. (Wilczek 1952)

Griffonia is preferred to the synonym, *Bandeiraea*; both were published in October, 1865.

Members of this genus are well distributed in tropical Africa. They are common in galleried forests of humid coastal areas.

The stems are used as cordage in hut construction and basketry. Various plant parts serve in native medicines as purgatives and for delousing purposes.

Gueldenstaedtia Fisch.   Papilionoideae: Galegeae, Astragalinae
295 / 3763

Named for Anton Johann Gueldenstaedt, 1745–1781, Russian botanist and noted traveler.

Type: *G. asiatica* Fisch.

10 species. Herbs, perennial, spineless, decumbent or nearly stemless; rootstock thick, often tuberous, with rosettes of leaves and flower stalks at the surface. Leaves imparipinnate or 1-foliolate; leaflets entire, oblong, often silky-haired; stipels none; stipules sometimes foliaceous, free or adnate to petiole. Flowers violet or yellow, solitary or umbellate on long, axillary peduncles; calyx campanulate; teeth 5, distinct, upper 2 broader than lower 3; standard round or obovate, erect; wings obovate-oblong, free; keel very short, straight, blunt; stamens diadelphous; anthers uniform; ovary sessile; style short, rolled inward at apex; stigma broad, ovary many-ovuled. Pod linear or ovoid, cylindric or inflated, twice the calyx length, nonseptate, 2-valved; seeds 6–8, kidney-shaped. (Bentham and Hooker 1865; Hooker 1879)

Representatives of this genus occur in eastern and central Asia and the Himalayas. They flourish on sandy soils, dry stony slopes, and fallow steppes; some species are present in alpine zones up to 3,000-m altitude.

Guibourtia Benn. emend. J. Léonard   Caesalpinioideae:
                                                Cynometreae

Named for Nicolas Jean Baptiste Gaston Guibourt, 1790–1867, Professor of Pharmacology, Paris; he published extensively on exotic woods and natural dyes.

Type: *G. copallifera* Benn. (= *Copaifera guibourtiana* Benth.)

15–18 species. Trees or shrubs. Leaves 2-foliolate; leaflets entire, often gland-dotted, rarely with a terminal leaflet; stipels none; stipules minute. Flowers small, in more than 2 rows, pedicels short or absent, in paniculate, terminal cone-shaped inflorescences; bracts small, caducous or subpersistent; bracteoles 2, small, persistent, cup-shaped, at the base of each flower; sepals 4, unequal, imbricate; petals none; stamens 8–12, usually 10, free, alternately long and short; filaments glabrous; anthers ovoid, dorsifixed, longitudinally dehiscent; ovary stalked, with stipe at the base of the calyx or sessile, 1–4-ovuled, usually 2; disk fleshy; style filiform; stigma terminal, capitellate. Pod stalked, round or elliptic, compressed or turgid, membranous or leathery, unilaterally winged, sometimes indehiscent; seeds 1, rarely 2, with or without a fleshy aril. (Léonard 1949, 1952, 1957)

Most of the *Guibourtia* species were segregated from *Copaifera*.

At least 4 species occur in tropical America; there are 11 or more in tropical Africa from Mozambique to Angola. They are common in moist forests, dense bush areas, savannas in low, hot, dry basins of the Transvaal, and along sandy stream banks.

Most of these species are tall, graceful trees. In the moist semideciduous forests from the Ivory Coast to Gabon, *G. ehie* (A. Chev.) J. Léonard and *G. pellegriniana* J. Léonard commonly reach heights of 40–50 m, with trunk girths of 3–5 m. The boles are straight, cylindrical, and are often free of branches up to 25 m (Voorhoeve 1965).

The timber is of good quality. The white-veined, reddish brown, fine-textured, heavy wood of *G. coleosperma* (Benth.) J. Léonard, a slightly buttressed tree of 25 m or more in height in Zambia, is commonly known as Rhodesian mahogany. It is used for paneling, cabinetmaking, and parquet flooring, but is generally considered too heavy for furniture. Table tops are made from the large burrs of old trees. The wood of *G. demeusei* (Harms) J. Léonard and *G. arnoldiana* (De Wild. & Th. Dur.) J. Léonard is in demand for furniture. The fine-grained, durable hardwood of *G. conjugata* (Bolle) J. Léonard and *G. sousae* J. Léonard, trees 15–20 m tall, is of high quality and resistant to termites.

The seeds contain an oil that is used in food and also as a stain. Boiled seeds, especially the red arils, are used by Rhodesian natives to make a nourishing beverage.

Several species yield copal. The copal from *G. demeusei* is marketed as Congo or Cameroon copal. Inhambane copal (Howes 1949), a product quite different from East African copal in appearance and properties and of closer resemblance to Accra copal, is obtained from *G. conjugata*. The resin from the type species is marketed as Sierra Leone copal and valued for varnishes.

NODULATION STUDIES

Nodules were not observed on *G. coleosperma* and *G. conjugata* examined in Zimbabwe (Corby 1974).

Gymnocladus Lam. Caesalpinioideae: Eucaesalpinieae
89 / 3545

From Greek, *gymnos*, "naked," and *klados*, "branch," in reference to the branches being bare of leaves most of the year. The leaves develop in the late spring and are shed in the early autumn.

Type: *G. dioica* (L.) K. Koch (= *G. canadensis* Lam.)

3 species. Trees, up to about 40 m tall, unarmed; branchlets stout; bark rough, deeply fissured. Leaves unequally bipinnate, caducous; pinnae alternate, many-foliolate; leaflets 7–15 per pinna, alternate, thin, entire, ovate, petioluled; stipules foliaceous, early deciduous. Flowers dioecious or polygamous; pistillate flowers borne in elongated, terminal racemes, anthers small and sterile; staminate flowers occur in short, terminal corymbs, ovaries

rudimentary or absent; regular flowers greenish white to purplish, symmetric, on long slender pedicels, in terminal or axillary panicles, pedicels bibracteolate near the middle; calyx tubular, elongated, lined with a thin, glandular disk; lobes 5, lanceolate, acute, subequal, erect; petals 4–5, oblong, rounded or acute, pubescent, as long as the calyx lobes or longer and twice as broad, spreading or reflexed, subequal, imbricate; stamens 10, distinct, short ter than the petals, and inserted with the petals on the margin of the disk; filaments slender, pilose, those opposite the calyx lobes longer; anthers oblong, versatile, longitudinally dehiscent; ovary sessile or short-stalked, acute, many-ovuled; style short, erect, thick. Pod sessile, broad-oblong, subfalcate, turgid or slightly compressed, woody, 2-valved, indehiscent or opening tardily; seeds 6–9, ovoid, thick, embedded in a dark, mucilaginous pulp. (Bentham and Hooker 1865; Sargent 1949)

The genus shows a disjunct distribution; 1 species is endemic in North America, 2 in Asia.

The North American species, *G. dioica* (L.) K. Koch, commonly called Kentucky coffee tree, American coffee tree, and Kentucky mahogany, is sometimes cultivated in parks and gardens of the eastern United States and Europe as an ornamental. The rate of seed germination is slow because of the hard seed coat, but propagation is easily obtained from cuttings.

The roasted seeds of *G. dioica* were used as a coffee substitute before and during the Revolutionary War and during the early settlement of Kentucky. Members of Long's expedition to the Rocky Mountains in 1820 recorded the palatability of a coffee brewed from the seeds (Medsger 1939).

The leaves and pulp of the pod of the Kentucky coffee tree contain the alkaloid cytisine, $C_{11}H_{14}N_2O$ (Pammel 1911). Kingsbury (1964) reported the loss of sheep, cattle, and horses after ingestion of the foliage and the fatal poisoning of a woman who ate the pulp of the pods.

The cherry red heartwood of the American species is sharply differentiated from the greenish white sapwood, but it is difficult to season. It is hard, heavy, has a density of about 700 kg/m^3, a coarse texture, a straight grain, and a medium luster. It is moderately easy to work, takes a high polish, and is durable. The wood is not commercially important, but it has been used for posts, railroad crossties, bridge construction, and sparingly for interior trimming (Record and Hess 1943).

The uses of the wood of the Chinese species, *G. chinensis* Baill., are similar to those of the American species. The saponins of the pods and bark are used in native preparations for washing fine fabrics. The ground seeds are mixed with various ingredients in the making of a perfumed soap (Uphof 1968).

The other Asian species, *G. burmanicus* C. E. Parkinson, is native to Burma and Assam, India.

NODULATION STUDIES

Nodules were lacking on plants of *G. dioica* examined in Germany by Morck (1891), in the Soviet Union by Lopatina (1931), and in the United States by Leonard (1925), Bushnell and Sarles (1937), and the authors (personal observation).

From Greek, *haima*, "blood," and *xylon*, "wood," in reference to the red dye obtained from the heartwood.

Type: *H. campechianum* L.

3 species. Trees, small, glabrous, or shrubs, much-branched, erect, spiny. Leaves paripinnate, lowest pair bipinnate; leaflets obovate, in few pairs; stipules thorny, or small and deciduous. Flowers small or medium, fragrant, yellow, in short, lax, axillary racemes; bracts minute; bracteoles absent; calyx tube short, shallowly campanulate, deeply cleft; lobes 5, slightly unequal, broadly imbricate in bud, or 4 lobes equal, with the other longer than the tube; petals 5, oblong, subequal, spreading, uppermost petal innermost; stamens 10, free; filaments almost straight, hairy at the base; anthers uniform, longitudinally dehiscent; ovary short-stalked, free, 2–3-ovuled; style slender; stigma small, terminal. Pod lanceolate, flat, membranous, persistent, dehiscent by a longtitudinal slit along the middle of the 2 valves but not at the sutures; seeds transversely oblong, hilum central. (Bentham and Hooker 1865; Amshoff 1939; Standley and Steyermark 1946; Backer and Bakhuizen van den Brink 1963)

Members of this genus are native to Central and South America and the West Indies.

The 2 major species, *H. campechianum* L. and *H. brasiletto* Karst. (= *H. boreale* S. Wats.), are very similar. The former, known as logwood, blackwood, bloodwood, or Campeche wood, seldom exceeds 8 m in height, has a gnarled or fluted trunk, and generally occurs in moist, swampy areas. The latter, known as Brazil wood, Nicaragua wood, or brazilette, is a spiny shrub common on rocky slopes and calcareous soils. As stated by Standley and Steyermark (1946), "If the two species were found growing together, they could scarcely be regarded as distinct, since they are at best none too easily separable, except by the wood, which seems to have distinct properties in the two species." The third species, *H. africanum* Stephens, is a little known, small, shrubby form found in the dry, rocky areas of Great Namaqualand, Namibia (Stephens 1913).

Logwood stain was used by the Indians of Central and South America to color cotton and wool cloth many years before Mexico was discovered by the Spaniards (Record and Hess 1943; Standley and Steyermark 1946). This stain, obtained from the heartwood, is the source of hematoxylin, a dye as valuable to the cytologist and histologist as methylene blue has been to the bacteriologist. *H. brasiletto* shares with several *Caesalpinia* species the distinction of being the source of the red dye, brazilein. These two coloring principles are of value in differentiating the woods.

The hematoxylin stain imparted to white paper, following the application of a few drops of concentrated ammonia to the cut surface of logwood, is bluish to black, whereas that of brazilette is at first red and later reddish violet (Record and Hess 1943).

Importation of logwood to England for extraction of the stain began about the middle of the 16th century. The environs of Campeche, Mexico, to which the specific epithet of the type species refers, and Belize were the early centers of export. The first attempts by the English dyers to use logwood extract were so unsatisfactory that importation was prohibited by an Act of Parliament in 1581 that remained in effect until 1662. Nonetheless, clandestine traffic persisted; in fact, the heartwood was in such demand that British privateers preyed upon Spanish cargoes for subsequent sale to English dyers. The potential of the crystalline derivative of logwood extract, hematoxylin, $C_{16}H_{14}O_6 \cdot 3H_2O$, was not fully realized until it became known that it requires ripening or oxidation to hematein, $C_{16}H_{12}O_6$, and that metallic salts of iron, aluminum, copper, chromium, or molybdenum are required as mordants.

Bohmer's alum hematoxylin (1865) was the first satisfactory stain using hematoxylin as a dye (Conn 1953). Heidenhain's iron-alum hematoxylin stain, introduced in 1892, was immediately acclaimed by cytologists and histologists. Today more than three dozen formulae of hematoxylin stains are widely accepted. The unexcelled merits of hematoxylin as a biological stain lie in its permanence and subtle polychrome properties in the staining of cell nuclear and chromatin structures, connective tissue, lipoids, glycogen, and bacteria in tissue. It is used extensively in staining fabrics and in inks; it also serves satisfactorily as a delicate reagent for iron or copper.

Brazilin, the colorless extract from the heartwood of *H. braziletto*, is a lower homolog of hematoxylin containing one less hydroxyl group. Upon oxidation it becomes the red dye known as brazilein, $C_{16}H_{12}O_5$; thus it differs from hematein by having one less molecule of oxygen. The uses of brazilein in cytology and histology have been similar to those of hematoxylin (Conn 1948, 1953). The complex chemistry of brazilin and hematoxylin was reviewed by Robinson (1958, 1962) and Dann and Hofmann (1963).

Antimicrobial properties of brazilette sawdust extracts were shown by Sánchez-Marroquín et al. (1958) to be effective against *Brucella*, *Shigella*, and *Staphylococcus* species.

Haematoxylon timber sizes are usually irregular and small. The wood has a density of about 960–1040 kg/m³. Inasmuch as the logs are heavier than water, they are commonly cut into 1-m lengths and transported downstream on palm barges. Only the heartwood has commercial value for its hematoxylin content. The color of freshly cut heartwood is a bright red, but it darkens upon exposure. It is mildly fragrant; that of the type species has the odor of violets. The grain is interlocked, the texture is coarse, and the wood is rather difficult to work by hand, but it can be finished to a smooth surface and takes a high polish. The thin, yellowish white sapwood is usually chipped away for fuel and small carpentry items.

NODULATION STUDIES

Roots of all plants of *H. campechianum* examined in botanical gardens in France (Lechtova-Trnka 1931), in the Philippines (Bañados and Fernandez 1954; personal observation 1962), in Trinidad (DeSouza 1966), in Hawaii (personal observation), in Singapore (Lim 1977), and in western India (V. Schinde, personal communication 1977) lacked nodules.

From Greek, *hals*, "salt," and *dendron*, "tree."

Type: *H. halodendron* (Pall.) Voss (= *H. argenteum* Fisch. ex DC.)

1 species. Shrub, up to 2 m tall, spreading, bluish-gray, armed, spines up to 6 cm long. Leaves paripinnate; leaflets 1–5 pairs on rachis terminating in a spine; stipules awl-shaped, spinose, with a broad base. Flowers purple, sometimes almost white, large, 1–3, at apex of axillary peduncles or fascicled at older nodes; bracteoles minute; calyx pouched at the back; lobes 5, short, upper 2 adherent, persistent; standard suborbicular, sides folded; wings falcate-oblong, free; keel petals incurved, obtuse; stamens diadelphous; anthers uniform; ovary stalked; style inflexed, not barbate; stigma small, terminal. Pod stalked, obovoid, inflated, beaked, yellowish, leathery, indented along the seed-bearing suture, tardily dehiscent; seeds few, smooth, subreniform, shining. (De Candolle 1825b; Bentham and Hooker 1865)

This monotypic genus is native to the salt steppes of central Asia.

In addition to ornamental attractiveness, *H. halodendron* (Pall.) Voss, Siberian sandthorn, has value in erosion control.

NODULATION STUDIES

Moshkov (1939) observed nodulation of this species in the northern Soviet Union. Presumably unaware of this, Federov (1940) concluded that the species is normally nonnodulative, but progeny of a graft of Siberian sandthorn onto *Caragana arborescens* could be induced to nodulate.

Hallia Thunb.   Papilionoideae: Hedysareae, Desmodiinae
348 / 3819

Named for Birger Martin Hall, a favorite pupil of Linnaeus.

Lectotype: *H. alata* Thunb.

9 species. Herbs or undershrubs, trailing or upright, with flattened or angular stems. Leaves simple, entire, tapering or heart-shaped at the base; stipels none; stipules 2, striate, coherent to and sometimes longer than the petiole. Flowers purplish blue, rarely white, solitary on axillary stalks jointed above the middle; bracts small, 3-lobed; bracteoles none; calyx campanulate, narrowed at the base, sometimes villous; lobes 5, subequal, long, pointed; petals 5, short-clawed; standard round or obovate; wings oblique-oblong, free or partly joined to the keel; keel petals incurved at apex, obtuse, subequal or shorter than the wings; stamens usually united into a sheath, vexillary one rarely free; anthers uniform; ovary subsessile, 1-ovuled; style inflexed and awl-shaped near the apex; stigma terminal, capitate. Pod compressed, ovoid, thinly membranous, reticulate, enclosed by the persistent calyx, 1-seeded. (Bentham and Hooker 1865, Harvey 1894)

Hutchinson (1964) positioned this genus in Psoralieae trib. nov.

Members of this genus occur in Cape Province, South Africa. They are common on mountain slopes and flats, in moist places, and among grasses.

Hammatolobium Fenzl
304 / 3772

Papilionoideae: Hedysareae,
Coronillinae

Original spelling was *Hamatolobium*, from Latin, *hamatus*, "hooked," and Greek, *lobion*, "pod."

Type: *H. lotoides* Fenzl

3 species. Herbs, perennial; rootstocks woody, stems much-branched, procumbent, silky-villous. Leaves imparipinnate, 2–3-, usually 5-foliolate, sometimes palmate, 2 lower leaflets stipulelike; stipules small, free, linear. Flowers yellow or purple, solitary or paired in axillary heads, subtended by a simple or 3-foliolate bract; calyx tubular-campanulate; lobes 5, subequal, upper 2 slightly connate at the base; petals all long-clawed; standard suborbicular; wings obliquely obovate-oblong, longer than the acute, incurved keel; stamens diadelphous; filaments dilated upwards; anthers uniform; ovary sessile, many-ovuled; style glabrous, inflexed, beaked, broadened at the apex; stigma terminal. Pod linear, lomentaceous, beaked, segments compressed or convex, finely veined; seeds arcuate. (Bentham and Hooker 1865; Tutin et al. 1968)

Species are distributed throughout southeastern Europe, Syria and North Africa.

Hansemannia Schum.
4 / 3438

Mimosoideae: Ingeae

Named for A. von Hansemann, 1826–1903, Director of the German New Guinea Company and promoter of the botanical exploration of Kaiser-Wilhelmsland, now Papua New Guinea.

Type: *H. glabra* Schum.

4 species. Trees or shrubs. Leaves bipinnate; pinnae 1 pair; rachis glandular. Flowers small, on slender-peduncled, loose racemes; calyx campanulate, shortly 4–5-toothed; petals 4–5, small, equal, acuminate, joined almost halfway upward from the base; stamens numerous, filamentous, exserted beyond the petals; ovary compound, 4–6-carpellate, each many-ovuled, each with a separate style. Pods several, exserted from the persistent calyx, thick, leathery, with margins straight or undulating, not notched or lobed, 4–6-seeded, indehiscent.

Hansemannia is closely related to *Affonsea* and *Archidendron*. All have pluricarpels, i.e., compound pistils with several styles, a characteristic of the more primitive Rosaceae (Burkart 1943).

Representatives of this genus are indigenous to Papua New Guinea.

Named by Schindler in honor of his friend Hansli Hegner, wife of Cnefelii.
Type: *H. adhaerens* (Poir.) Schindl.
1 species. Shrublet, climbing. Leaves 1-foliolate, petioled; stipels 2. Flowers small, fascicled in the axils of setaceous bracts, in pendulous, terminal racemes; calyx campanulate, subtended by 2 bracteoles; lobes 4, short, triangular, obtuse, emarginate, posterior lobe broad; standard suborbicular, clawed; wings spurred, oblong, obtuse, straight; keel obtuse, straight; stamens diadelphous; vexillary filament free, slender, others connate to two-thirds length; anthers ovoid, dorsifixed; ovary short-stalked, surrounded by a tubular disk, many-ovuled; style twisted; stigma short, lateral. Pod stalked, linear, compressed, turgid, constricted on the lower suture more than on the upper, joints unequal-sided, 4–5 times longer than broad, indehiscent; seeds linear-oblong, subterete, without aril. (Schindler 1924)

This monotypic genus is endemic in the Moluccas, eastern Malaysia, tropical Australia, and southeastern Borneo.

Haplormosia Harms   Papilionoideae: Sophoreae

Name combines Greek, *haploos*, "simple," and *Ormosia*, a generic name; the genus differs from *Ormosia* primarily in having simple leaves.
Type: *H. monophylla* Harms
2 species. Trees, briefly deciduous, small to medium, 15–20 m tall, with trunk diameters up to 80 cm. Leaves small, 1-foliolate, oblong or lanceolate, leathery, glossy; new leaves bright red, subtended by stipels. Flowers in few- to several-flowered racemes, pedicelled; calyx cupular, purplish red; lobes 5, persistent, overlapping in bud, 2 upper lobes broad, joined beyond the middle; petals 5; standard semiorbicular; wings oblique, asymmetric; keel petals lanceolate, margins joined near the apex; stamens 10, free; filaments glabrous; ovary stalked, surrounded by a disk, 2–4-ovuled; style slender, straight; stigma minute, terminal. Pod flat, obovate or oval, thinly lignified with a narrow wing along the margins, short-stalked, valves somewhat woody, dehiscent, 1-seeded; seed thick, oblong or narrow-elliptic, red. (Harms 1917b; Baker 1929; Knaap-van Meeuwen 1962)

Species of this genus are common in tropical West Africa. They are present in evergreen coastal forests bordering freshwater swamps, lagoons, and rivers.

Liberian black gum, the wood of *H. monophylla* Harms, is favored for furniture because of its figured grain, ability to take a fine polish, and resemblance to walnut, but the supply is limited. Its hardness, weight, and resistance to decay and termites contribute to its demand for construction purposes (Voorhoeve 1965).

Named for Countess Franziska von Hardenberg; her brother, Baron Karl von Hügel, 1794–1870, collected plants in Western Australia in 1833.
    Lectotype: *H. monophylla* (Vent.) Benth.
    2 species. Herbs, prostrate, trailing, glabrous, or undershrubs up to 2 m tall; roots long, carrotlike. Leaves 1-, 3-, or 5-foliolate; leaflets entire, lanceolate; stipels present; stipules striate, small. Flowers many, small, pale violet or pink with a greenish yellow spot on the standard, in pairs or 3–4 along the rachis of axillary racemes or in terminal panicles; bracts small, caducous; bracteoles none; calyx teeth shorter than the tube, upper 2 united; standard broad-orbicular, emarginate, not eared; wings obovate-falcate, adherent to the incurved, shorter keel; stamens diadelphous; anthers kidney-shaped; ovary sessile; ovules several; style short, thick, incurved, beardless; stigma terminal, capitate. Pod linear-oblong, compressed or turgid, 2-valved, hard, with or without a pithy pulp between the ovoid or oblong seeds. (Bentham 1864; Bentham and Hooker 1865)
    This genus was formerly considered congeneric with *Kennedia*. Distinctiveness was based on flower size and color, keel length, and the occurrence of the short, thick style.
    Members of this genus are native to Queensland and South Australia, and Papua New Guinea. They are present in dry sclerophyll at 650–1,200-m altitude.
    The long, carrotlike roots of *H. monophylla* (Vent.) Benth., native lilac, were used by bushmen as a substitute for sarsaparilla, *Smilax* sp. The plant foliage is reputed to cause colic in horses (Webb 1948).
    H. comptoniana (Andr.) Benth., a popular blue-flowered climber, is a garden ornamental in Western Australia.

NODULATION STUDIES

H. monophylla was reported nodulated under two synonyms, *Kennedya monophylla* (Wilson 1939a) and *H. violacea* (Schneev.) Stearn (Harris 1953; Bowen 1956). Seedlings were nodulated by rhizobia from diverse species in the cowpea miscellany and also from alfalfa, clover, vetch, and beans (Wilson 1939a). Rhizobia from *H. comptoniana* produced nodules on cowpea, soybean, and several lupine species, but lacked ability to nodulate alfalfa, clover, and peas (Lange 1961).

Hardwickia Roxb.   Caesalpinioideae: Cynometreae
40 / 3486

Named for Thomas Hardwicke, 1757–1835, British botanist.
    Type: *H. binata* Roxb.
    2 species. Trees, unarmed, briefly deciduous. Leaves paripinnate, 2- to few-foliolate; leaflets sessile, ovate, asymmetric, entire; stipules scalelike. Flowers small, yellow green, many, in slender, lax, panicled axillary or terminal racemes; bracts minute, scalelike; bracteoles absent; calyx very shallow; sepals

5, broad-ovate, subpetaloid, glandular, imbricate; petals and disk absent; stamens 10, infolded in bud, alternately long and short, all perfect; anthers dorsifixed, warty; ovary sessile, glandular, 1–2-ovuled; style filiform; stigma large, peltate. Pod flat, samaroid, oblong-lanceolate, 2-valved, 1-seeded and dehiscent at the apex, thin but not dehiscent at the base; seed pendulous, subreniform, compressed. (Hooker 1879; Knaap-van Meeuwen 1970)

Members of this genus occur in tropical western India. They are common in savanna forests and in areas of scant to moderate rainfall up to about 1,200-m altitude.

H. binata Roxb., the Anjan tree of India, reaches a height of about 35 m. Its dark red, heavy, durable wood has a density of 1400 kg/m³ and is in demand for heavy construction, bridge posts, looms, farm implements, and house-building. The bark yields a fiber used in the manufacture of paper, rope, cordage, and sails. An oleoresin obtained from the trunk serves as a wood preservative. Misra et al. (1964) isolated 4 diterpinoids from the oleoresin of *H. pinnata* Roxb.: hardwickic acid, kolavenol, kolavic acid, and kolavensis acid. Livestock are said to consume the leaves as fodder (Uphof 1968).

NODULATION STUDIES

Wright (1903) reported *H. pinnata* nodulated in Sri Lanka.

Harpalyce Moc. & Sessé ex DC.   Papilionoideae: Galegeae,
248 / 3711   Brongniartiinae

Named for Harpalyce; in Greek mythology, the exceptionally beautiful daughter of Clymenus of Argos.

Type: *H. formosa* DC.

25 species. Shrubs or small trees, erect, unarmed; branches often her-baceous, tomentose. Leaves imparipinnate; leaflets entire, rounded, petioluled, with yellow orange glandular dots or scales or lower surface; stipels absent; stipules minute. Flowers scarlet or purple, large, showy, in axillary racemes or terminal panicles; bracts and bracteoles linear, caducous; calyx tube very short, bilabiate, lips united to apex, entire, linear, upper lip bidentate, lower tridentate; lobes rounded, elongated; standard large, rounded, sessile or shortclawed; wings oblanceolate-falcate, short-clawed, with a basal ear; keel petals free, long, linear, incurved, joined above, tips blunt; stamens monadelphous; filaments joined into a sheath split above; anthers alternately long and short; ovary sessile, several- to many-ovuled; style curved, glabrous; stigma terminal. Pod sessile, oblong or linear, narrow or broad, 2-valved, septate within, or the compressed, oblong-ovate seeds are separated by spongy tissue. (De Candolle 1825b; Bentham and Hooker 1865; Rose 1903; Standley 1922; Rydberg 1923)

Representatives of this genus are well distributed in southern Mexico, Cuba, and Brazil. They inhabit dry, bush areas and hot, rocky hillsides.

The wood is available only in small sizes and so is unimportant commercially. The heartwood of *H. cubensis* Griseb. is hard, durable, attractive with yellowish brown and pinkish streaks, and takes a high polish; it is used to some extent for small turnery articles.

Haydonia Wilczek   Papilionoideae: Phaseoleae, Phaseolinae

Type: *H. monophylla* (Taub.) Wilczek

2 species. Herbs, perennial, erect or prostrate; stems more or less winged or angled, smooth; rootstocks swollen, woody. Leaves pinnately 3-foliolate, petioled, or simple, subsessile; leaflets papery; stipels present or absent; stipules persistent. Flowers few, blue, yellow, or greenish, on elongated, ribbed or winged, axillary peduncles; bracts more or less persistent; bracteoles 2, subtending the calyx; calyx campanulate, bilabiate; lobes 5, subequal, imbricate, upper 2 partly connate, lower 3 free, middle lobe narrow, longer; standard orbicular, base eared, 2 callosities above the claw; wings oblanceolate, spurred, longer than the keel and somewhat adherent to it; keel beaked, incurved; stamens diadelphous, 5 with a gland at the base of each anther alternate with 5 lacking glands; ovary linear, many-ovuled, sessile; disk cylindric, somewhat lobed; style curved, bearded on the inner surface just below the stigma. Pod cylindric, linear, straight, subcoriaceous, 2 valves convex, septate between the seeds, dehiscent; seeds quadrangular or subglobose, nonwinged. (Wilczek 1954)

This genus is closely related to, and should perhaps be merged with, *Vigna*.

Members of this genus occur in tropical Africa. They are common among grasses and on fallow land.

NODULATION STUDIES

In Zimbabwe *H. monophylla* (Taub.) Wilczek was reported nodulated as *Vigna monophylla* (Corby 1974).

Hebestigma Urb.   Papilionoideae: Galegeae, Robiniinae
(269a) / 3735

The prefix *hebe*, "downy," comes from the Greek *hebe*, "youth," "pubes"; the stigma is covered with soft hairs.

Type: *H. cubense* Urb.

1 species. Tree, small to medium, up to about 14 m tall, unarmed, deciduous. Leaves imparipinnate, subopposite or alternate, large; leaflets variable in outline, entire, alternate; stipels and stipules absent. Flowers pink or purplish, in racemes, blossoming before leaves are renewed on young

323

branches; bracts small; bracteoles none; calyx short, obliquely turbinate; lobes 5, short, broad; standard orbicular, reflexed, not eared, without callosities, short-clawed; wings oblong, free, near straight, with prominent basal ear; keel petals obovate, blunt, united near the apex, inner margin almost straight, with rounded basal auricle; stamens diadelphous, vexillary stamen joined with others at the base; anthers uniform; ovary stalked; ovules 5–9; style rectangular, inflexed, glabrous, awl-shaped; stigma terminal, villous; Pod sessile or subsessile, linear, margins flat, about 10–18 cm long by 2–3 cm wide, not winged, 2-valved, exocarp leathery, endocarp woody; seeds usually 2, ovate, compressed, black, separated by woody corky septa. (Urban 1900; Harms 1908a; Rydberg 1924)

This monotypic species is endemic in the West Indies and Cuba. It is common in mountain forests and brush areas.

The hard, heavy, durable wood lacks commercial importance but serves in rural areas for posts and props (Record and Hess 1943).

Hedysarum L.   Papilionoideae: Hedysareae, Euhedysarinae
310 / 3778

From Greek, *hedys*, "sweet," and *saron*, "broom."

Type: *H. coronarium* L.

70–100 species. Herbs, deep-rooted, annual or perennial, rarely shrubby; stems erect, leafy. Leaves imparipinnate; leaflets entire, numerous, 7–20 pairs, some finely pellucid-dotted; stipels absent; stipules free or connate, scarious. Flowers large, showy, pink, red, violet, white, rarely yellow, in axillary erect racemes; bracts scarious or setaceous; bracteoles setaceous; calyx campanulate; lobes 5, subequal, awl-shaped; standard ovate or obcordate, reflexed, narrowed at the base; wings oblong, shorter than the standard; keel petals obliquely truncate, rarely subarcuate, longer than the wings and standard; stamens diadelphous, anthers uniform; ovary subsessile, 4- to many-ovuled; style filiform, inflexed above the stamens; stigma small, terminal. Pod lomentaceous, plano-compressed, deeply indented along both margins and constricted between each seed, segments up to 8, oval, orbicular, or quadrate, readily separable but indehiscent; seeds usually flattened, kidney-shaped. (Bentham and Hooker 1865; Rollins 1940)

Species of this large genus are well distributed in temperate Europe, the Mediterranean region of North Africa, Asia Minor, Siberia, and in North America from Arizona into Canada and the Arctic regions. About 10 species occur in the western prairie areas of the United States.

They are well adapted to calcareous soils, and rocky and gravelly banks. Certain species are present up to 2,300-m altitude.

Apparently the foliage of all species is palatable and relished by livestock, but the plants are not plentiful enough to be important range and forage crops. *H. splendens* Fisch. ex DC. was introduced into Colombia for that purpose. *H. boreale* Nutt., one of the most common North American species, extends from Arizona to northern Canada.

The mature roots of *H. alpinum* L. are often 0.5 m long, branched, and thick as carrots. In autumn, from central Alaska to the Arctic Ocean, these roots are collected by Eskimos and Indians, who eat them raw, roasted, or boiled; however, many are left in the ground during winter for an early spring vegetable. The raw roots are also relished by brown bears, thus the common name, "bear root." The roots of *H. mackenzii* Richards., a closely related species in the same habitat, are considered poisonous. Sir John Richardson, the arctic explorer, reported that all of his men who mistook this for the edible *H. alpinum* became ill (Heller 1962). The roots of *H. gangeticum* L. have cathartic properties.

The type species is commonly called sulla clover, French honeysuckle, Spanish sainfoin, or Spanish esparcet. This fragrant, red-flowered perennial is a successful crop for soil improvement, hay, and honey in Italy and the Balearic Islands, but it has not proved popular in the United States.

NODULATION STUDIES

H. pabulare A. Nels., in Arizona, was reported devoid of nodules in field and greenhouse tests with rhizosphere soil (Martin 1948). A reexamination is warranted, inasmuch as the species has promise as a forage plant.

The common occurrence of nodules on *Hedysarum* species in arctic ecosytems is of special interest. The absence of *Hedysarum*, *Oxytropis*, and *Astragalus* species from the Greenland flora was attributed by Porsild (1930) to the possible absence of the proper rhizobia, not to the unfavorable climatic conditions. Porsild's comment that these genera are represented on Baffin Island, where the climate is even more severe than on Greenland, implied that these species are nodulated there; however, a direct statement to this effect was not made.

Nodulated species of *Hedysarum*	Area	Reported by
alpinum L.	Arctic U.S.S.R.	Roizin (1959)
ssp. *americanum* (Michx.) Fedtsch.	Alaska, U.S.A.	Allen et al. (1964)
boreale Nutt.	N.Dak., U.S.A.	M. D. Atkins (pers. comm. 1950)
capitatum Burm. f.	France	Clos (1896)
caucasicum Bieb.	Arctic U.S.S.R.	Roizin (1959)
coronarium L.	Widely reported	Many investigators
hedysaroides Schinz & Thell.	S. Africa	N. Grobbelaar (pers. comm. 1965)
iomiticum Fedtsch.	Arctic U.S.S.R.	Roizin (1959)
mackenzii Richards.	Alaska, U.S.A.	Allen et al. (1964)
mongolicum Turcz.	Arctic U.S.S.R.	Roizin (1959)
obscurum L.	Arctic U.S.S.R.	Roizin (1959)
var. *japonicum*	Arctic U.S.S.R.	Roizin (1959)
spinosissimum L.	Libya	B. C. Park (pers. comm. 1960)
	Israel	Hely & Ofer (1972)

Named by Schindler in honor of his friend Hansli Hegner, wife of Cnefelii.

Type: *H. obcordata* (Miq.) Schindl. (= *Uraria obcordata*; *Desmodium obcordatum*)

1 species. Undershrub, erect or straggling. Leaves 3-foliolate; the terminal leaflet wide and butterfly-shaped, the laterals smaller and much narrower; stipels and stipules present. Flowers violet, in branched racemes, 5–20 cm long, or paniculate, fascicled in axils of narrow, persistent bracts; bracteoles absent; calyx campanulate; lobes 5, narrow, acute; petals short-clawed; standard small, obovate, without ears; wings eared, obtuse; keel not eared; stamens monadelphous, vexillary stamen mostly free above; ovary short-stalked, linear, 3–5-ovuled; disk lacking; style acutely inflexed at right angle; stigma terminal, small. Pod stalked, flattened, 1–5-jointed, joints membranous, dilated, indehiscent; seeds compressed, without aril. (Schindler 1924)

This monotypic genus is endemic in Burma and Sumatra. It occurs in dry, grassy localities.

Helminthocarpon A. Rich.   Papilionoideae: Loteae
231 / 3692

The name *Vermifrux* Gillett (q.v.) has been proposed in preference to *Helminthocarpon* or *Helminthocarpus*; reasons for this proposal were clearly defined by Gillett (1966b).

Herpyza Ch. Wright   Papilionoideae: Phaseoleae, Glycininae
(391a)

From the Greek verb *herpyzo*, "creep," "crawl."

Type: *H. grandiflora* (Griseb.) Ch. Wright

1 species. Creeping herb; stems hairy, spreading, rooting at the nodes. Leaves 3-foliolate, long-petioled; leaflets obovate, entire, terminal leaflet larger, eglandular; stipels linear; stipules free, ovate, striate. Flowers blue, large, 1–3, in long, peduncled, axillary racemes; bracts awl-shaped; bracteoles persistent, subtending the calyx; calyx tubular-campanulate; lobes 5, subequal, ciliate, subulate-lanceolate; standard obovate-oblong; short-clawed, eared at the upper half of the claw; wings oblong, as long as standard; keel petals short, clawed, connate only in the middle; stamens 10, diadelphous, 1 free, 9 joined midway; anthers fertile, rounded; ovary sessile, pubescent, 2–4-ovuled; style smooth above, slender, erect; stigma terminal. Pod sessile, oblong, linear-ovate, flat, beaked, septate between the 2–3 subquadrate seeds, or ovate and 1-seeded. (Harms 1915a; León and Alain 1951)

The genus seems related to, but is distinctive from, *Teramnus* (Harms 1915a). J. A. Lackey (1977b) believes the relationship is with Diocleinae.

This monotypic genus is endemic in Cuba and Puerto Rico. It occurs in savannas and pine forests.

From Greek, *hesperos*, "western"; presumably this laburnumlike plant occurs only in the westernmost region of North Africa.

Type: *H. platycarpum* (Maire) Maire (= *Laburnum platycarpum*)

1 species. Tree, small, glabrous, unarmed; all shoots of 1 kind, all long. Leaves sessile, 3-foliolate; leaflets entire, lanceolate; stipules none. Flowers yellow, in erect, terminal racemes; bracts and bracteoles minute, early caducous; calyx campanulate, bilabiate, caducous; upper lip shortly 2-toothed, lower lip 3-toothed; standard reflexed; wings and keel petals free; standard, wings, and keel petals equally blunt; stamens 10, monadelphous; anthers alternately long and basifixed, short and dorsifixed; ovary linear, sessile, apex incurved; style minute, smooth; stigma minute. Pod leathery, tardily dehiscent, few-seeded. (Maire 1949)

Polhill (1976) included this genus in Genisteae *sensu stricto*.

This monotypic genus occurs in the Anti-Atlas Mountains, Morocco.

Hesperothamnus Brandegee

From Greek, *hesperos*, "western," and *thamnos*, "shrub," in reference to its habitat in western North America.

Type: *H. littoralis* Brandegee

6 species. Shrubs or small trees. Leaves imparipinnate, 5-foliolate; leaflets oblong-lanceolate, at first silky-pubescent; stipels linear; stipules linear-lanceolate, fugacious. Flowers large, purple, fascicled in dense or interrupted terminal racemes; bracts and bracteoles linear; calyx tube cupular; lobes 4, unequal in length, upper lip entire or inconspicuously 2-toothed, lanceolate-acuminate or ovate, 3 lower teeth lanceolate or ovate; petals subequal, clawed; standard suborbicular-obovate; wings oblong, falcate, free; keel petals incurved, obtuse, equal; stamens 10, 1 free only at the base then united with the other 9; anthers uniform, small; ovary pubescent, sessile; ovules 5; style smooth, curved; stigma small, terminal. Pod oblong, flattened, broad-linear, leathery, 2-valved, dehiscent; seeds flat, orbicular, with a broad funicle. (Brandegee 1919)

J. A. Lackey (personal communication 1978) does not include this in the tribe Phaseoleae.

Members of this genus are native to Baja California, Mexico, and the southwestern United States.

Heterostemon Desf.   Caesalpinioideae: Amherstieae

68 / 3522

From Greek, *heteros*, "different," and *stemon*, "stamen"; both fertile and sterile stamens are present.

Type: *H. mimosoides* Desf.

7–8 species. Trees or shrubs, small, unarmed. Leaves either paripinnate with leaflet pairs large and few, small and many, or imparipinnate with terminal leaflet very large; stipules leaflike. Flowers large, showy, orchidlike, resembling *Cattleya* species, bluish to purple, in short, terminal racemes, or on leafless nodes; bracts small; bracteoles connate, persistent; calyx tube long; sepals 4, subequal, petaloid, imbricate, sometimes rose red; petals 5, upper 3 ovoid, subequal, imbricate, lower 2 vestigial; stamens 9, connate into a sheath split above, lowermost 3 long, perfect, with oblong anthers, other 6 short, unequal, with anthers absent or sterile; ovary stalk adnate to calyx tube; ovules numerous; style slender; stigma terminal, capitate. Pod stipitate, large, linear or sickle-shaped, compressed, leathery, 2-valved; seeds several, transverse, flat, ovate or orbicular. (Bentham and Hooker 1865; Amshoff 1939)

Representatives of this genus are common in tropical South America, mainly in swampy forests of the Brazilian Amazon and Venezuela.

Ducke (1925b) described *H. mimosoides* Desf. as probably the most beautiful of all American leguminous plants because of its attractive foliage and abundant, large, beautiful flowers. The wood is easily worked and takes a smooth finish, but lacks commercial possibilities. All species merit consideration for cultivation as ornamentals.

Heylandia DC.   Papilionoideae: Genisteae, Crotalariinae
208 / 3668

Named for M. Heyland, an artist employed by De Candolle.

Type: *H. latebrosa* (L.) DC.

1 species. Herb, prostrate, much-branched, silky. Leaves ovate, subsessile, simple, entire; stipules absent. Flowers yellow, subsessile, small, solitary, axillary; calyx turbinate; lobes 5, lanceolate, upper 2 partly joined, lower 3 free; petals much-exserted, subequal; standard suborbicular, large, with 2 scales above the short claw; wings smaller than standard, obovate-oblong; keel petals narrow, incurved, joined dorsally, short-beaked; stamens united into a tube split above; anthers alternately short and dorsifixed, long and basifixed; ovary sessile, 2-ovuled; style elongated, filiform, abruptly incurved at the base, bearded above lengthwise; stigma terminal. Pod oblong, compressed, silky-haired, 2-valved, nonseptate within, 1–2-seeded. (De Candolle 1825b; Bentham and Hooker 1865; Hooker 1879)

The genus is closely related to *Crotalaria*.

This monotypic genus is endemic in India and Sri Lanka, mainly on tropical plains.

Nodulation Studies

Nodules were reported on this species in experimental studies in the United States (Wilson and Choudhri 1946) and on field plants in western India (Y. S. Kulkarni personal communication 1972).

From Greek, *hippos*, "horse," and *krepis*, "shoe," in reference the horseshoe-shaped pod segments.

Type: *H. unisiliquosa* L.

10–12 species. Herbs or shrubs, slender, annuals and perennials. Leaves imparipinnate; leaflets numerous, 3–10 pairs, entire; stipels none; stipules small, broad and membranous or lanceolate, sometimes absent. Flowers yellow, usually 2–10, rarely solitary, in axillary, nodding, peduncled umbels; bracts small; bracteoles none; calyx tubular-campanulate; lobes 5, subequal, 2 upper teeth somewhat joined; petals long-clawed; standard suborbicular; wings, falcate-obovate or oblong; keel petals acute, incurved; stamens diadelphous; alternate filaments broadened at apex; anthers uniform; ovary sessile, many-ovuled; style inflexed, apex awl-shaped; stigma small, terminal. Pod flat, lomentaceous, rarely subcylindric, compressed, upper margin deeply indented at each seed, segments lunate to horseshoe-shaped; seeds arched, hilum ventral. (Bentham and Hooker 1865)

This genus is closely allied to *Ornithopus* and *Coronilla*.

Representatives of this genus occur in the Mediterranean area and Canary Islands. They are commonly found in dry, sunny, calcareous, rocky regions.

Nodulated species of *Hippocrepis*	Area	Reported by
comosa L.	Kew Gardens, England	Pers. observ. (1936)
	S. Africa	Grobbelaar & Clarke (1974)
cornigera Boiss.	Libya	B. C. Park (pers. comm. 1960)
multisiliquosa L.	Great Britain	Dawson (1900a)
	Libya	B. C. Park (pers. comm. 1960)
	Israel	M. Alexander (pers. comm. 1963)
unisiliquosa L.	France	Clos (1896)
	Israel	Hely & Ofer (1972)

Hoffmanseggia Cav.   Caesalpinioideae: Eucaesalpinieae
100 / 3557

Named for J. Centurius von Hoffmannsegg, 1766–1849, German botanist. The spelling of the generic name with a single *n* has been a longtime controversy. The circumstances pertinent to this orthographic error were recently reviewed by Brummitt and Ross (1974) in their proposal to correct and conserve the name with a double *n*.

Type: *H. falcaria* Cav.

45 species. Perennial herbs or low shrubs, usually unarmed but some-

times with short spines, often very glandular; stems arise from tuberous roots or a woody base. Leaves bipinnate; pinnae 3–7; leaflets small, usually gland-dotted; stipels small when present; stipules small, sometimes spinescent. Flowers yellow or red, in lax, terminal or leaf-opposed racemes; bracts caducous; bracteoles none; calyx short-campanulate, often with black glands; lobes 5, short, subequal, oblong or lanceolate; petals 5, overlapping, oblong or obovate, claws sometimes glandular, the uppermost petal slightly larger; stamens 10, free, shorter than the petals, declinate; filaments linear, often glandular at the base and villous; anthers uniform; ovary subsessile, many-ovuled; style incurved; stigma terminal. Pod subsessile, flat, linear to oblong, straight or falcate, 2-valved, usually glandular or setaceous, elastically dehiscent; seeds transverse or oblique-ovate. (Bentham and Hooker 1865)

Members of this genus are present from southwestern North America to eastern Argentina; species also occur in South Africa. They are drought-resistant, and are most common in dry, sandy clay and rocky, limestone areas, up to 1,500-m altitude.

Aside from serving as a soil-binder, the species are of little economic importance. Some often form large patches in fields and pastures and become troublesome weeds difficult to eradicate. *H. densiflora* Benth., a low herbaceous perennial with yellow to red flowers in spikes, is commonly called Indian rushpea or hog potato; it is widespread from Kansas through Texas into Mexico and California. The tuberous enlargements of the roots are palatable to hogs and livestock. American Indians of the southwestern United States relished the roasted roots. *H. drepanocarpa* Gray is somewhat common on the dry hills of Texas, Colorado, and Arizona. The seeds serve as quail food. *H. patagonica* Speg. and *H. trifoliata* Cav. are reportedly common in the Andes foothills of southern Argentina. The root bark of *H. melanosticta* Gray has astringent properties. The stems yield 25–30 percent tannin and a red dye.

Most of the species with *Larrea* synonyms were derived from *Larrea* Ort. and are not to be confused with *Larrea* Cav. (Zygophyllacea), a conserved genus, famous as the creosote bush of the Mojave Desert and salt deserts of Arizona.

NODULATION STUDIES

Four species are reportedly nonnodulated: *H. densiflora* and *H. microphylla* Torr. in Arizona (Martin 1948), *H. falcaria* Cav. in Argentina (Gonzalez et al. 1969), and *H. burchellii* (DC.) Benth. ex Oliv. in Zimbabwe (Corby 1974).

An American Indian name for the type species, used as a source of fiber for thread.

Type: *H. macrostachya* (DC.) Rydb. (= *Psoralea macrostachys*)

11 species. Herbs, perennial, with rootstocks; stems erect, rarely prostrate or shrubby below. Leaves pinnately 3-foliolate; leaflets short-petioluled, entire, gland-dotted; stipules lanceolate or awl-shaped. Flowers purple or yellowish, in lax, long-peduncled, axillary spikes or racemes; calyx campanulate, glandular-punctate; lobes 5, lowest one longest; standard broad-obovate, basal lobes distinct, with a short, curved claw; wings as long as the standard, with a small, basal lobe, claw varying from equal to half as long as the blade; keel petals shorter than the wings, coherent at the tip, clawed, dark-purple tipped; stamens 10, vexillary stamen united with the other 9 only above the middle or only at the base; anthers alternately long and short; ovary 1-ovuled; style abruptly incurved above; stigma globose, bearded. Pod small, oblique-ovate to oval, pubescent, beak short and slender, indehiscent, included in the calyx, pericarp thin, free from the seed; seed brown or black, oval-reniform, shiny. (Rydberg 1919a)

With few exceptions, the species were segregated from *Psoralea*.

Members of this genus are native to California, Bolivia, and Peru.

NODULATION STUDIES

H. orbicularis (Lindl.) Rydb. (= *Psoralea orbicularis*) is nodulated (personal observation, University of Wisconsin herbarium specimen from California 1972).

Holocalyx M. Micheli   Caesalpinioideae: Swartzieae
121 / 3579

From Greek, *holos*, "entire," in reference to the nontoothed calyx tube.

Type: *M. balansae* M. Micheli

2 species. Trees, unarmed, medium to large, with dense foliage and fluted or irregular trunks. Leaves paripinnate; leaflets opposite or alternate, numerous, in many pairs, oblong, dentate, subcoriaceous, venation subparallel; stipules minute. Flowers greenish, small, in short axillary racemes; bracts and bracteoles small, persistent; calyx tube conical, truncate usually pubescent, entire or minutely toothed; petals 5, small, free, linear-oblong, caducous; stamens 12–10, free, glabrous; anthers small, dorsifixed; ovary stalked, pubescent; style erect, very short; stigma minute, terminal. Pod ovoid, fleshy, turgid, indehiscent, containing 1–3 large, ovoid seeds. (Burkart 1943)

Members of this genus occur in Brazil, Paraguay, and northeastern Argentina. They inhabit dry forest areas.

The wood of *H. balansae* M. Micheli, the better-known member of the genus, is strong, hard, resilient, durable, and heavy. Because of these qualities and its markings, medium luster, and ability to take a smooth finish, this wood is especially valued for decorative turnery items, furniture, and implement handles. A limited supply precludes exportation (Record and Hess 1943).

NODULATION STUDIES

Rothschild (1970) did not find nodules on seedlings and mature plants of *H. balansae* in native Argentinian habitats or on seedlings grown experimentally in soil composited from rhizospheres of nodulated plants of *Acacia*, *Erythrina*, and *Vigna* species.

Holtzea Schindl.   Papilionoideae: Hedysareae, Desmodiinae

Named for Nicholas Holtze, 1868–1913; he collected extensively in northern Australia, 1891–1913.

Type: *H. umbellata* Schindl.

1 species. Shrub; young branches somewhat grooved, sparsely hairy. Leaves petioled, 1–3-foliolate; leaflets ovate to oblong, undersurfaces silky-haired, prominently veined; stipels setaceous; stipules brown, triangular-falcate, persistent. Flowers white, 5–10, in short-stalked, axillary umbels; bracts small, caducous; bracteoles linear, persistent; calyx campanulate, silky, short, 4-parted; upper lobe broader; standard broad, short-clawed, not eared; wings short-clawed, falcate-oblong, obtuse, spurred; keel as long as the wings, long-clawed, obovate, apex apiculate; stamens diadelphous; anthers ovoid, uniform, dorsifixed; ovary short-stalked, 1–2-ovuled; disk absent; style incurved, not thickened; stigma small, terminal. Pod segments 1–2, thick, oblong, separating into 1-seeded, indehiscent (?) segments. (Schindler 1926)

The genus has close affinities with *Desmodium*.

This monotypic genus is endemic in northern Australia.

Hosackia Dougl. ex Benth.   Papilionoideae: Loteae
235 / 3696

Named for David Hosack, 1769–1835, Professor of Botany and *Materia Medica*, Columbia College, New York City. In 1801 Dr. Hosack purchased the land that became the Elgin Botanical Garden.

Type: *H. bicolor* Dougl. ex Benth.

50 species. Herbs, perennial or annual, seldom subshrubs. Leaves pin-

nately 2- to many-foliolate, seldom 3-foliolate; stipules membranous, foliaceous or minute and glandlike. Flowers yellow or red, in peduncled or sessile axillary umbels, rarely solitary, subtended by 1- or more-foliolate bracts; bracteoles often absent; calyx teeth subequal, often shorter than the tube; standard ovate or suborbicular, with an obtuse base and slender claw; wings obovate or oblong, reflexed; keel incurved, obtuse, short, subequal to the wings; stamens diadelphous; all or alternate filaments slightly dilated above; anthers uniform; ovary sessile, many-ovuled; style incurved, glabrous; stigma terminal. Pod linear, cylindric or plano-compressed, straight or arcuate, 2-valved, septate within, several-seeded. (Bentham and Hooker 1865)

Hosackia is closely allied to *Lotus*; many species have *Lotus* synonyms. European botanists usually consider *Hosackia* the American counterpart of *Lotus*, a strictly Old World genus.

Representatives of this genus are common in western North America from Mexico to British Columbia. They are found on light, dry, gravelly soils, some in arid to transition zones, others in humid transition zones on bluffs near the Pacific Ocean.

H. subpinnata T. & G., introduced into Chile, has forage potential (Burkart 1952). All species provide good ground cover; a few are used as border plants, but they are not prized as garden ornamentals.

NODULATION STUDIES

All nodules observed by the authors on the tap and primary lateral roots of young herbarium specimens were small, round to oval, and about 2 mm in diameter.

No cultural or cross-inoculation data have been reported for *Hosackia* rhizobia. Since *Hosackia* and *Lotus* species are closely akin botanically, it is reasonable to expect that their symbiotic spectra are also similar.

Histological and cytological studies by Peirce (1902) primarily concerned nodules of *Medicago denticulata*; however, parallel studies of nodules of other species included *H. subpinnata*. Since Peirce demonstrated that descriptive features of bur clover nodules were applicable to other nodules, one is permitted to conclude that morphogenesis and development of *Hosackia* nodules are in no way unusual. At an early date Peirce (1902) commented on the rosy pink color of nodular tips and on the size of rhizobia-infected host cells being double that of noninvaded host cells.

Nodulated species of *Hosackia*	Area	Reported by
americana (Nutt.) Piper	Idaho, U.S.A.	Pers. observ. (U.W. herb. spec. 1970)
(= *purshiana* Benth.)	N.Dak., U.S.A.	Bolley (1893)
glabratus Hell.	Calif., U.S.A.	Pers. observ. (U.W. herb. spec. 1970)
parviflora Benth.	Wash., U.S.A.	Pers. observ. (U.W. herb. spec. 1970)
pilosa Nutt.	Calif., U.S.A.	Pers. observ. (U.W. herb. spec. 1970)
subpinnata T. & G.	Calif., U.S.A.	Peirce (1902)

Named for A. P. Hove, Polish botanist and collector of plants for the Royal Botanic Gardens, Kew, England, during the mid-19th century.

Type: *H. linearis* (Sm.) R. Br.

10–12 species. Shrubs small, 2–3 m tall, erect, evergreen, branches sometimes tomentose, rarely thorny. Leaves simple, entire, some prickly-toothed or sharp-pointed, glabrous above, tomentose below; stipules small and setaceous, or absent. Flowers blue or violet, in axillary fascicles, rarely solitary or in short racemes; bracts and bracteoles small; 2 upper calyx lobes united into a broad, truncate or notched upper lip, lower 3 smaller, narrow, lanceolate; petals clawed; standard orbicular, emarginate; wings short, oblique-obovate, inner margin eared at the base; keel petals slightly coherent, much shorter than the standard, slightly incurved, blunt; stamens monadelphous, united into a sheath split above and sometimes also below; anthers alternately long and basifixed, short and versatile; ovary sessile or stalked, usually 2-ovuled; style thick, incurved; stigma terminal. Pod sessile or stalked, oblique-ovoid, swollen, smooth or tomentose, nonseptate, 2-valved, leathery, 2-seeded, dehiscent; seeds kidney-shaped. (Bentham 1864; Bentham and Hooker 1865)

This genus was assigned to the tribe Bossiaeae by Polhill (1976) in his reorganization of Genisteae *sensu lato*. *H. chorizemifolia* (Sweet) DC. was removed from this genus to form the monotypic genus *Plagiolobium* (q.v.).

Hovea is an Australian genus. Species are common in New South Wales, South Australia, Victoria, Queensland, and Tasmania. They flourish in dry forests, sheltered and grassy valleys, and dry rocky places. The plants prefer well-drained soils. Some species occur up to 1,500-m altitude.

Hovea species are favored horticulturally for the intensity of the color and the profusion of flowers. The grazed foliage of several species is toxic to livestock. Yields of 1.1–1.4 percent and 0.5 percent of sparteine, based on dry weight of the leaves, were obtained from *H. longifolia* R. Br. and *H. acutifolia* A. Cunn., respectively (Morrison and Neill 1949).

NODULATION STUDIES

Lange (1961) concluded from cultural and host-infection tests that isolates from nodules of *H. elliptica* (Sm.) DC., *H. trisperma* Benth., *H. chorizemifolia*, and *H. pungens* Benth. were typical of slow-growing cowpea, lupine, and soybean rhizobia. Strains from *H. chorizemifolia* and *H. pungens* nodulated *Lupinus digitatus* (Lange 1961; Lange and Parker 1961). In a later study (Lange 1962), strains from *L. digitatus* and *L. angustifolius* failed to nodulate *H. elliptica* and *H. pungens*.

Nodulated species of *Hovea*	Area	Reported by
acutifolia A. Cunn.	Queensland, Australia	Bowen (1956)
(*chorizemifolia* (Sweet) DC.) = *Plagiolobium chorizemifolium* Sweet	W. Australia	Lange (1959)
elliptica (Sm.) DC.	W. Australia	Lange (1959)
pungens Benth.	W. Australia	Lange (1959)
trisperma Benth.	W. Australia	C. A. Gardner (pers. comm. 1947); Lange (1959)

Named for Friedrich Alexander von Humboldt, 1769–1859, explorer and scientist.

Type: *H. laurifolia* Vahl

5–8 species. Shrubs or trees, erect, small, unarmed. Leaves paripinnate, short-petioled; leaflets sessile, 4–5 pairs, ovate-oblong, acuminate, leathery, with crackled venation, lowest pair usually smallest; rachis sometimes winged; stipules peltate, kidney-shaped, or semisagittate, foliaceous, persistent. Flowers orange or white fused with red, in dense axillary racemes, subsessile at the apex of shoots or at old nodes on old wood; bracts ovate or oblong; bracteoles large, persistent, colored, enclosing flower bud, but spreading at anthesis; calyx turbinate or narrow; lobes 4, oblong, colored, obtuse, subequal, posterior lobe sometimes broadest; petals either 5, sessile or thin-clawed, oblong-spathulate, subequal, or 3, uppermost broader than the laterals, lower 2 small, deformed or absent; stamens 5, fertile, alternating with 5, minute, dentate staminodes; anthers oblong, versatile, longitudinally dehiscent; ovary stalked, stipe adnate to calyx tube, linear, few-ovuled; style very long, thin; stigma terminal, club-shaped. Pod oblong, straight or curved, compressed, rigid, leathery, 2-valved; seeds 1–4, ovate, transverse, without aril. (Bentham and Hooker 1865; Hooker 1879; Backer and Bakhuizen van den Brink 1963)

Members of the genus are distributed throughout Sri Lanka, India, and Java.

Several species are highly ornamental. The flower-bearing twigs of *H. laurifolia* Vahl commonly have hollow, obconical internodes inhabited by ants; the nonflowering twigs are not myrmecophilous.

Humularia Duvign.   Papilionoideae: Hedysareae,
                                      Aeschynomeninae

Named for the resemblance of the inflorescence to that of the hop plant, *Humulus lupulus*.

Lectotype: *H. drepanocephala* (Bak.) Duvign.

40 species. Undershrubs, erect or spreading, rarely shrubs, usually with glandular-based hairs. Leaves paripinnate; leaflets rounded, generally 2–12, paired, opposite or subalternate, asymmetric at the base, principal nerve submedian to marginal; rachis terminating in a tapered point; stipels absent; stipules round, heart-shaped or unequally eared at the base, with dimensions often equal to that of the leaflets, leathery or submembranous, persistent or caducous. Flowers small to medium, yellow, orange, or reddish, often red-streaked, in axillary or terminal, usually dense, scorpioid or conelike racemes, rarely lax and zigzag; bracts 2-lobed, large, persistent, distichous, densely imbricate, more or less hiding the flowers and pods, membranous, and transparent becoming thickened and colored yellow, orange, or red; bracteoles membranous, persistent; calyx short, campanulate, bilabiate, membranous, upper lip 2-toothed, lower lip 3-toothed, generally ciliate at the top; standard clawed, fiddle-shaped; wings spathulate, eared and joined at the base; keel petals obovate, markedly incurved, partly united along the back margins;

stamens 10, united in 2 groups of 5; ovary stalked, 2-ovuled; disk short; style smooth, sometimes persistent. Pod small, compressed, often biconvex at maturity, usually deeply lobed into 2, small, semiorbicular, indehiscent segments; seeds reniform-suborbicular, strongly biconvex, brown, smooth, glossy. (Duvigneaud 1954a,b)

Members of this genus, which is especially well represented in Zaire, are common throughout equatorial Africa. Species are present in underbrush, light forests, and on sandy or lateritic soils of savannas and steppes.

The dried roots of *H. apiculata* (De Wild.) Duvign. are smoked as a kind of tobacco to promote energy and virility.

Hybosema Harms   Papilionoideae: Galegeae, Robiniinae

From Greek, *hybos*, "hump-backed," and *semeia*, "standard," in reference to the thick protuberance at the base of the standard.

Type: *H. ehrenbergii* (Schlechtd.) Harms

1 species. Shrub, unarmed, young branches pubescent. Leaves imparipinnate; leaflets 7–11-paired, small, oblong or elliptic; stipules long, pointed, linear-lanceolate, deciduous. Flowers several, in lax, short, axillary racemes on shoots, pedicels single or paired, jointed below the calyx; calyx long-cupular or tubular-campanulate, abruptly narrowed below, briefly bilabiate, upper lip broad, 2-toothed, lower lip shortly 3-toothed; corolla long-exserted, glabrous; standard obovate, apex notched, short-clawed, thickly callous at the base; wings clawed, narrow, obtuse, lanceolate, appendaged at the base; keel petals narrow, long-clawed, blunt, partly joined dorsally; stamens diadelphous; anthers small; ovary long-stalked, linear-lanceolate, glabrous, 6–8-ovuled; style incurved but not coiled at the tip, glabrous; stigma minutely capitate. Pod linear, lanceolate, broader at the apex, flat, nonseptate, 2-valved, valves woody, dehiscent, 3–5-seeded. (Harms 1923b)

The genus is closely allied to *Robinia* and *Gliricidia*.

This monotypic genus is endemic in Mexico and Central America.

Hylodendron Taub.   Caesalpinioideae: Amherstieae
(46a) / 3496

From Greek, *hyle*, "forest," "woodland," and *dendron*, "tree."

Type: *H. gabunense* Taub.

1 species. Tree, slender, up to about 20 m tall, sharply buttressed, with woody spines on the trunk and branches. Leaves imparipinnate; leaflets 9–15,

large, alternate, elongate-elliptic, entire, acuminate, asymmetric at the base, translucent-punctate on the margins only, with prominent lateral veins; stipules prominent, large, leathery, linear, striate, sheathing the apical bud, later caducous leaving an annual scar. Flowers small, greenish white, usually in racemose panicles or in short, axillary or terminal racemes; axial bracts ovate-reniform, somewhat twisted, leathery to scalelike, striate, imbricate in bud, tardily caducous; bracteoles 2, linear-filiform, subtending the calyx; calyx tube short; lobes 4, concave, oblong-lanceolate, subvalvate, free, unequal; petals none; stamens 8–10, exserted, alternately long and short; anthers dorsifixed, longitudinally dehiscent; ovary sessile, narrow, usually 10–12-ovuled; disk absent; style linear-subulate; stigma terminal, capitate. Pod samaroid, pendulous, more or less stalked, flat, thin, oblong-linear, both ends rounded, leathery or membranous, slightly curved, prominently veined transversely, proximal area flat, winglike; seeds 1–2, rarely 3–4, in the distal end or apex, indehiscent. (Léonard 1952, 1957)

This monotypic genus, endemic in tropical West Africa, is common in Nigeria and from Gabon to Zaire. It occurs in heavily forested regions.

The hard wood has local uses for fuel and carpentry.

Hymenaea L.   Caesalpinioideae: Amherstieae
49 / 3499

Name alludes to Hymen, the god of marriage; each leaf has a single pair of leaflets.

Type: *H. courbaril* L.

25–30 species. Trees, large, unarmed, resiniferous. Leaves 2-foliolate, short-petioled; leaflets sessile, asymmetric, leathery, pellucid-punctate; stipules broad-linear, very caducous. Flowers large or medium, mostly white, in short, terminal, densely corymbose panicles; bracts and bracteoles concave, very caducous; calyx tube thick-walled, campanulate; lobes 4, very imbricate in bud, leathery; petals 5, sessile, oblong-obovate, subequal, uppermost petal inside and slightly larger than the others, beset with glands; stamens 10, free, all perfect, glabrous; anthers oblong, uniform, longitudinally dehiscent; ovary on a short stalk often adnate to the calyx tube, few-ovuled; style filiform; stigma terminal, small. Pod drupelike, oblique-obovoid or oblong, short or rather long, thick or subterete, woody, indehiscent; seeds few, usually imbedded in a thick, mealy pulp, shape variable. (Bentham and Hooker 1865; Amshoff 1939)

Members of this genus are well distributed in tropical Central and South America, the West Indies, and Mexico. The majority of the species are native to the Amazon Basin. They are common in dry forests, on hillsides, along roadsides, and in coastal limestone areas.

In the West Indies the type species is called courbaril, courbaril plum, West Indian locust, and stinking toe. This handsome tree reaches a height of

40 m, with a trunk diameter up to 1.5 m (Kryn 1956). The bole is usually straight, enlarged at the base, or slightly buttressed. The crown is lofty and spreading. The dark red seeds are imbedded in a thick, sweet, yellow, mealy, edible, but unpleasantly odorous pulp.

The bark of old trees is thick and can be removed in long sheets. Amazon natives prize it for making canoes strong enough to carry 25–30 men. The stripped bark is sewn together at the ends, and the seams waterproofed with a gum; a few wooden crosspieces are then fitted in to hold the shape (Little and Wadsworth 1964). The bark is also a source of tannin. Bark decoctions are used medicinally to treat dysentery.

The wood resembles mahogany and at one time was commercially marketed, but the supply is limited. It is known in the trade as algarrobo or guapinol. Formerly it was one of the best woods available from the Antilles. The sapwood is white to gray; the heartwood is reddish brown, usually silvery-streaked, without odor or taste, heavy, tough, coarse textured, and has a density of about 750–1050 kg/m^3. It is difficult to work but finishes smoothly and is durable. It is typically used in ship construction, furnituremaking, cabinetwork, bars, gear cogs, and wheel fellies (Titmus 1965).

The tree is a source of a valuable, pale yellow to reddish gum, known in the trade as South American or Brazil copal, or anime gum. Large quantities in hardened lumps are obtained from the base of living trees. Barrelsful of so-called fossil gum are often found at the base of decayed trees. This copal-like gum is valuable in the preparation of special grades of varnishes, shellacs, and polishes, for medicinal purposes, in crockery cements, and to a minor extent for religious incense. Gum from the bark, known as soft anime, is less valuable and obtainable in smaller quantities. Other species of *Hymenaea* yielding resin in lesser amounts are *H. confertiflora* Mart., *H. intermedia* Ducke, *H. parviflora* Huber, *H. stigonocarpa* Mart., *H. stilbocarpa* Hayne, and *H. velutina* Ducke.

Hymenaea trees are not desirable for homesites because of their root spread, gum production, and malodorous fruits.

Nodulation Studies

Isolates from nodules on a large tree of *H. courbaril* L. in Hawaii gave cultural and physiological reactions typical of cowpea-type rhizobia (Allen and Allen 1939). Nodule-forming ability of the isolates was not ascertained on the host species because seeds were not available; but 2 strains produced nodules on 16 of 20 plant species of the cowpea miscellany. Maximum effectiveness was obtained with both strains in symbiosis with *Acacia koa* and with 1 strain on *Dolichos lablab*.

Large, rough-surfaced nodules observed on 10 of 15 plants of this species in the Philippines (Bañados and Fernandez 1954) confirmed rhizobial symbiobility within this genus. However, in Trinidad (DeSouza 1966) and in Venezuela (Barrios & Gonzalez 1971) nodulated specimens were not found.

From Greek, *hymen*, "membrane," and *karpos*, "fruit," in reference to the membranous fruits.

Type: *H. circinnata* (L.) Savi

1 species. Herb, annual, prostrate; stems pubescent. Leaves imparipinnate; leaflets 2–3 pairs, entire, terminal leaflet the largest; stipules minute, membranous, those of the lower leaves adnate to the petiole, those of the upper leaves obsolete. Flowers small, yellow, 2–4, in heads on long, axillary peduncles; lower bracts foliaceous, others small, setaceous; calyx campanulate; lobes 5, deep, subequal; petals free from the staminal tube; standard suborbicular, short-clawed; wings obovate; keel sharply inflexed, beaked; stamens diadelphous; filaments apically dilated; anthers uniform; ovary sessile, 2-ovuled; style inflexed; stigma terminal. Pod small, flat, coiled into a broad ring, outer margin membranous-winged, often sharp-toothed, indehiscent; seeds kidney-shaped. (Bentham and Hooker 1865)

This monotypic genus is endemic in the Mediterranean region and in western Asia.

This herb was reported nodulated in France (Clos 1896; Naudin 1897c), in Israel (M. Alexander personal communication 1963), and in South Africa (Grobbelaar and Clarke 1974); nodulation was also observed on a University of Wisconsin herbarium specimen from Dalmatia (personal observation 1970).

***Hymenolobium** Benth.   Papilionoideae: Dalbergieae,
359 / 3830   Lonchocarpinae

From Greek, *hymen*, "membrane," and *lobion*, "pod."

Type: *N. nitidum* Benth.

8–12 species. Trees, deciduous, unarmed, medium to very large, some up to 50 m tall, with trunks 3 m in diameter. Leaves imparipinnate, large; leaflets 5–49, small to large, oblong, leathery; stipels minute, usually absent; stipules linear or lanceolate, narrow, caducous. Flowers pale rose, red, or violet, fragrant, in lax, terminal panicles, appearing after leaves fall; bracts and bracteoles small, caducous; calyx campanulate, apex truncate, margin obscurely sinuate-dentate; petals clawed, subequal; standard often emarginate, orbicular, not appendaged, reflexed; wings free, oblique-oblong, larger than the free, falcate, keel petals; stamens 10, all united into a sheath split above; anthers versatile; ovary short-stalked, linear, few-ovuled; style slender, incurved; stigma small, terminal. Pod large, oblong or oblong-linear, flat, membranous, prominently veined, 2 of the veins subparallel near the margin towards the base, indehiscent, 1–2-seeded; pods of some species are highly colored and persist until the foliage is renewed; seed, usually 1, transversely oblong, plano-compressed. (Bentham and Hooker 1865)

Members of this genus are common in tropical South America, especially in upland Amazon rain forests of eastern Brazil, Guiana, and Venezuela.

All species yield a hard, strong, heavy wood that is useful for industrial construction. The wood of *H. excelsum* Ducke is known as angelim in the timber trade of Pará, Brazil (Record and Hess 1943).

NODULATION STUDIES

An unnamed species of *Hymenolobium* examined by Norris (1969) in Bartica, Guyana, was nodulated.

Hymenostegia (Benth.) Harms   Caesalpinioideae: Cynometreae
(35a) / 3476

From Greek, *hymen*, "membrane," and *stege*, "shelter," in reference to the membranous, petaloid bracteoles enclosing the flower buds.

Type: *H. floribunda* (Benth.) Harms

20 species. Shrubs or trees up to 30 m tall, with trunks 70–100 cm in diameter. Leaves paripinnate; leaflets sessile, asymmetric, base broad, tip pointed, midrib incurved, lower pair smaller; rachis winged or grooved between the 2–12 pairs; stipules variable, caducous, small, sometimes large and persistent. Flowers white, red, pink, or yellow, in solitary or fascicled, axillary or terminal racemes; bracts variable, often small, narrow; bracteoles 2, large, well-developed, ovate to linear-filiform, opposite, imbricate in bud, membranous, petaloid, usually persistent; calyx deeply cupular or turbinate, often gibbous at the base; sepals 4–5, overlapping, sometimes petaloid; petals 5, entire, 3 large, equal, 2 very small, rarely 1 or none; stamens 10, rarely 16–20, free; filaments filiform; anthers dorsifixed; ovary stalked, stalk adnate to the side wall of the calyx tube, 2–3-ovuled, sometimes 5–6-ovuled; disk inconspicuous or absent; style filiform; stigma terminal. Pod stipitate, broad, flat, leathery, 2-valved, curling upward during dehiscence, few-seeded. (Harms 1897; Pellegrin 1948; Léonard 1951)

Representatives of this genus occur in equatorial West Africa. They inhabit riverside forest areas.

The wood is durable, fine grained, and takes a high polish, but the quantity is very limited. Several members of the genus have ornamental value as avenue and garden trees.

Hypocalyptus Thunb.   Papilionoideae: Genisteae, Cytisinae
222 / 3683

From Greek, *hypo*, "under," and the verb *kalypto*, "hide," in reference to the leaflets, which fold downward along the midrib.

Type: *H. sophoroides* (Berger) Baill.

3 species. Shrub, 2–3 m tall, glabrous. Leaves palmately 3-foliolate; leaflets obcordate or obovate, folded back on the midrib; stipules free, small.

Flowers purple, borne in terminal racemes, often panicled; bracts and bracteoles bristly; calyx campanulate, broad, brown, membranous, intrusively
ringed at the base; lobes shorter than the tube, subequal, acute; standard
suborbicular or broad-obovate, reflexed, the short, oblong claw callous inside;
wings eared at the base, oblong-spathulate with an oblong claw bent at right
angle; keel long-clawed, incurved, shorter than the standard, beaked; stamens monadelphous; anthers alternately short and versatile, long and basifixed; ovary subsessile, few- to several-ovuled; style incurved, glabrous;
stigma terminal, minutely capitate. Pod linear, oblong, flat, 2-valved, upper
suture thickened, nonseptate, few- to several-seeded. (Bentham and Hooker
1865; Phillips 1951; Polhill 1976)

Polhill (1976) assigned this genus to tribe Liparieae, which he segregated
from Genisteae *sensu lato.*

Species of this genus are native to South Africa. They are present in
gorges and ravines, especially in the Cape area.

NODULATION STUDIES

Grobbelaar and Clarke (1972) observed nodules on *H. sophoroides* (Berger) Baill. in South Africa.

Indigofera L.   Papilionoideae: Galegeae, Indigoferinae
239 / 3702

From Latin, *indicus*, "indigo," and the verb *fero*, "bear," "produce," in reference to some species of the genus being the source of indigo dye.

Type: *I. tinctoria* L.

800 species. Herbs, annual or perennial, shrubs or small trees, unarmed,
usually clothed with appressed, silky, medifixed, branching hairs. Leaves imparipinnate or pinnately or palmately 3-foliolate, rarely simple; leaflets
ericoid, long-linear, entire, sessile or short-petioled; stipels sometimes
present; stipules small, awl-shaped, bristly, slightly adnate to the petiole.
Flowers small, red to purplish, sometimes greenish white with red streaks, in
lax or dense long-peduncled racemes or spikes, seldom sessile, solitary or
paired in leaf axils; bracts small, caducous; bracteoles absent; calyx small,
campanulate, sometimes almost truncate, hairy or glabrous; lobes 5, narrow,
acute, subequal or the lowest lobe longest; standard broad to ovoid, sessile or
short-clawed, rounded or obtuse at the apex, sometimes pubescent on the
back, often long-persistent; wings oblique-obovate, oblanceolate, oblong, or
linear, slightly adherent to the keel; keel petals oblique-lanceolate, tapering
into a long, broad claw, rounded, obtuse at the apex, rarely beaked, with a
short spur or pouch on each side; stamens diadelphous, vexillary stamen free,
others united more than halfway into a staminal sheath; anthers uniform,
connective apiculate, glandular; ovary subsessile, 1- to many-ovuled, usually
strigose; style incurved often penicillate; stigma capitate. Pod linear or oblong, straight, arcuate, rarely circinate, cylindric, rarely globular, 4–3-sided or
plano-compressed, transversely septate between the seeds; seeds globose or

truncate, compressed or quadrate. (Bentham and Hooker 1865; Amshoff 1939; Phillips 1951; Cronquist 1954a)

Species of this large genus are widespread in tropical and subtropical areas. There are about 10 species in the United States and Canada, 18 in Mexico, and 76 in tropical West Africa. *Indigofera* is one of the largest and most widespread genera in South Africa. Members of this genus are best adapted to semidry areas. Wild species occur along roadsides, in grassy fields, brushwood areas, and in open and waste places.

Six species have been prominent as the source of indigo, the dark blue dye of commerce that was once known as the "king of the dyestuffs":

I. arrecta Hochst. ex A. Rich.	Tropical Africa
I. articulata Gouan	Tropical Africa
(= *I. argentea* L.)	
I. longeracemosa Boiv.	Madagascar
I. suffruticosa Mill.	Tropical America
(= *I. anil* L.)	
I. sumatrana Gaertn.	Malaysia
I. tinctoria L.	Tropical Africa

Botanists currently accept *I. sumatrana* Gaertn. as a synonym of *I. tinctoria* L. (Backer and Bakhuizen van den Brink 1963). However, in the early days of indigofera culture, the Dutch introduced *I. sumatrana* from Malaysia into India as a species with a higher yield of indigo than *I. tinctoria*, which was taken from the Old World into the West Indies (Ghosh 1944). Prain and Baker (1902) referred to *I. sumatrana* as a species most probably developed under cultivation from the wild form of *I. tinctoria*. *I. arrecta* Hochst. ex A. Rich. and *I. longeracemosa* Boiv. were probably the richest sources of the dye.

Indigo occurs in the leaves as indican, indoxyl-β-D-glucoside, $C_{14}H_{17}NO_6$. Upon hydrolysis, this colorless glucoside yields glucose and indoxyl. Oxidation of the latter results in the sedimentation of insoluble blue flakes, or a blue lump, consisting of 60–80 percent indigo, $C_{16}H_{10}N_2O_2$, amid quantities of indigo red, indigo brown, indigo green, and other substances as impurities. Subsequent purification is a simple process of successive treatments with dilute acetic acid, caustic soda, and alcohol (Ghosh 1944). The bulk of the glucoside occurs in the leaves and young stems. Four-day-old air-dried leaves of *I. tinctoria* yield about 3 percent indican. Evidence is lacking that it serves any function in plant metabolism. The yield per unit of leaf weight of some species is much greater than that of others. Many species of *Indigofera* yield none.

Few natural by-products have played as prominent a role in history and in international trade as has indigo. The earliest use of indigo is unknown, but evidence is convincing that it was used about 2300 B.C. by the Egyptians as a dye for mummy cloth. Alexandria was the import center when the Roman Empire was at its zenith. In the 16th century A.D. indigo was more valued in Indian and Portuguese trade than were cloves. According to Watt (1908), the desire to secure a more certain supply of the dyestuff led to the formation in 1631 of the Dutch East India Company, and shortly afterwards to the overthrow of Portuguese supremacy in the East. In that year, after having replaced *I. tinctoria* with *I. sumatrana* as the indigo-producing species in India and Malaysia, the Dutch sponsored five vessels carrying a cargo of about 285 tons of indigo, considered equivalent in value to about 5 tons of gold (Burkill 1966).

Indigo was introduced into England and northern Europe about the middle of the 17th century. Prior to that time, the only blue dyestuff used by the dyers was obtained from the woad plant, *Isatis tinctoria*, of the family Cruciferae. The importation of indigo met strong resistance, and the woad growers and merchants faced bankruptcy. The woad industry, therefore, set about strengthening its own position and at the same time stopping the growth of the upstart indigo industry. First, it sponsored scientific experiments aimed at improving the culture of the woad plant. Second, edicts (the industry possessed great political clout) were issued, making both the importation and the use of indigo a criminal offense punishable by death. Finally, the woad industry publicized that indigo was poisonous. By 1737 the penalties for using indigo were removed, resistance to its use was lessened, and competition in trade became spirited. Accordingly, cultivation of *Indigofera* species developed in many areas other than India. Notably, the American colonists assumed leadership, and in particular, the West Indies surpassed India as the indigo center of the world.

In the late 1700s intense cultivation of *Indigofera* was abandoned in the West Indies because of the high duties imposed, and agricultural interest shifted to the culture of coffee and sugar. This paved the way for India to reestablish itself as the world's source of indigo. As a result of the activities of the English East India Company, Bengal indigo became the most sought-after dye in the world. This success, however, was relatively short lived.

From about the turn of the 19th century onward, monopoly by the English East India Company was weakened by migration of planters and cultivators. Poor labor relationships developed and misunderstandings consequently arose among all factions. In mid-century, cultivators in Bengal claimed they were subjected by European indigo planters "to a system of inhuman oppression which only finds a parallel in the annals of negro-slavery in America" (Chaklader 1905; *see* Ghosh 1944). In the wake of these many problems the Bengali drama *Nilderpan* appeared in 1860. It had "the same effect in helping the cause of the abolition of indigo-slavery in Bengal as *Uncle Tom's Cabin* in abolishing negro-slavery in America" (Ghosh 1944).

In 1856 William Henry Perkin, 1838–1907, was granted a patent for his epoch-making discovery of the dye, mauve, by oxidizing aniline, $C_6H_5NH_2$. Ironically, the age-old plant-indigo industry was toppled by a mere teenage lad; moreover, the base compound, aniline, had been prepared earlier from indigo. Perkin's announcement signaled the end of the plant-indigo industry. Indigo was synthesized in 1878 and 1879 by Baeyer. The first commercial process appeared in 1890 (Heumann 1890a,b). Synthetic indigo was first marketed on a large scale in 1898. During the 2 previous years, India exported 169,523 cwt of indigo. In the 1914–1915 period the amount was about one-tenth as much. In the late 1930s the export averaged only about 158 cwt a year (Ghosh 1944).

Other uses of *Indigofera* species are varied. *I. arrecta, I. hirsuta* L., *I. spicata*

Forsk., *I. suffruticosa* Mill., and *I. tinctoria* are suitable as soil cover and green manures. *I. hirsuta* has served as annual fodder in Florida and Brazil. *I. pilosa* Poir. and *I. subulata* Vahl ex Poir. are perennial fodders in Florida and Puerto Rico. *I. arenaria* A. Rich., *I. oblongifolia* Del., and *I. pauciflora* Del. are grazed by camels in the Sudan; *I. ovina* Harv. is relished by cattle, sheep, and goats in the Cape area of South Africa.

I. decora Lindl. is a showy pink rock garden ornamental in the cool high country of Sri Lanka. *I. miniata* Ort., a prostrate vine with beautiful salmon-colored flowers, has shown potential for soil erosion control in sandy areas. *I. tesymanni* Miq. is used as a windbreak and as shade on tea, coffee, and cacao plantations in Sri Lanka. This reddish-pink-flowered, rapidly growing tree reaches a height of about 7 m in 1 year.

The diverse medicinal uses of plant-part decoctions, as anthelminthics, febrifuges, and treatments for skin diseases and heart trouble are reviewed by Watt and Breyer-Brandwijk (1962) and Burkill (1966).

The roots of *I. simplicifolia* Lam. are used as arrow poison by the Fulani in Africa (Uphof 1968). *I. thibaudiana* DC. and *I. lespedezioides* HBK have insecticidal properties; solutions of macerated plant parts are applied as an external lavage to deter mites, lice, and fleas on domestic animals in some areas of Central and South America. In northern Brazil crushed roots and leaves of *I. lespedezioides* and *I. suffruticosa* are used to stupefy fish (Blohm 1962). *I. australis* Willd., *I. linifolia* (L. f.) Retz., and *I. boviperda* Morris. are regarded as poisonous to stock in Australia (Webb 1948).

NODULATION STUDIES

Nodulation was first reported within this genus on *I. caroliniana* Walt. (Pichi 1888). Only 7 species were reported nodulated before 1932. Of the 191 species now listed as nodulated, information on approximately 140 was obtained within the last decade.

Three species are reported lacking nodules: *I. alternans* DC. in South Africa (Mostert 1955) and *I. reticulata* Koehne and *I. sphaerocarpa* Gray in Arizona (Martin 1948). These negative reports most likely reflect unfavorable seasonal conditions at the time of examination.

The root system of *Indigofera* species consists of a long taproot with comparatively fine laterals. Nodulation occurs largely on the roots within the upper 10 cm soil layer.

Joshi (1920) provided the first description of rhizobia from *Indigofera* nodules. Detailed studies have affirmed cultural and biochemical kinships of *Indigofera* rhizobia with the slow-growing rhizobia and symbiobility of *Indigofera* species within the cowpea miscellany (Allen and Allen 1936a, 1939, 1940a; Raju 1936, Wilson 1939a; Burton 1952; Ishizawa 1954; Galli 1958).

Knowledge of the inner structure of young oval and mature elongated nodules is best gleaned from the histological studies of *I. artropurpurea* Hornem. and *I. macrostachya* Willd. (Lechtova-Trnka 1931). Cortical host cell invasion was initiated by very narrow infection threads in which the small rods, 1.5μ long, were aligned in single file. Upon release into the plant tissue, the rods elongated to 3.5μ and became slightly bulged at one end. Distribution within the cell assumed a one-directional orderly arrangement. In the bacteroid area of *I. artropurpurea* many small vacuoles were interspersed among the bacteria. In *I. macrostachys* usually only one large, irregular but confluent vacuole was seen. Marked hypertrophy of the host cells accompanied the release and multiplication of the microsymbiont. Adjacent interspersed noninvaded cortical cells contained abundant starch granules.

Aging and bacterial degeneration were denoted by loss of staining power in the older, basal zone of the nodules.

Kodama's observations (1967) substantiated hypertrophy of cells in the bacteroidal tissue of *I. pseudotinctoria* Matsum. Infected cells were tetraploid, i.e., the chromosome number was twice the 2n count, 16, for normal root tips.

Nodulated species of *Indigofera*	Area	Reported by
acutisepala Conrath	S. Africa	Grobbelaar & Clarke (1975)
adenocarpa E. Mey.	S. Africa	Grobbelaar & Clarke (1975)
adenoides Bak. f.	S. Africa	Grobbelaar et al. (1967)
alternans DC.	S. Africa	Grobbelaar & Clarke (1972)
anabaptista Steud.	W. India	Y. S. Kulkarni (pers. comm. 1978)
angustifolia L.	S. Africa	Grobbelaar & Clarke (1972)
annua Milne-Redhead	Zimbabwe	Corby (1974)
antunesia Harms	Zimbabwe	Corby (1974)
arenophila Schinz	Zimbabwe	Corby (1974)
argyroides E. Mey.	S. Africa	Grobbelaar & Clarke (1975)
arrecta Hochst. ex A. Rich.	Widely reported	Many investigators
artropurpurea Hornem.	France	Lechtova-Trnka (1931)
asperifolia Bong.	Venezuela	Barrios & Gonzalez (1971)
astragalina DC.	S. Africa	Grobbelaar et al. (1964)
	Zimbabwe	Corby (1974)
atriceps Hook. f. ssp. *alboglandulosa* (Engl.) Gillett	Zimbabwe	Corby (1974)
auricoma E. Mey.	S. Africa	Grobbelaar & Clarke (1974)
australis Willd.	Queensland, Australia	Bowen (1956); Beadle (1964)
bainesii Bak.	S. Africa	Grobbelaar & Clarke (1972)
	Zimbabwe	Corby (1974)
bifrons E. Mey.	S. Africa	Grobbelaar & Clarke (1972)
brachynema Gillett	Zimbabwe	Corby (1974)
brachystachya E. Mey.	S. Africa	Grobbelaar & Clarke (1972)
burkeana Benth.	S. Africa	Grobbelaar & Clarke (1972)

345

Nodulated species of *Indigofera* (cont.)	Area	Reported by
cardiophylla Harv.	S. Africa	Grobbelaar & Clarke (1972)
caroliniana Walt.	U.S.A.	Wilson (1945a)
cecili N. E. Br.	Zimbabwe	Corby (1974)
charlierana Schinz	Zimbabwe	Corby (1974)
circinella Bak. f.	Zimbabwe	Corby (1974)
circinnata Benth. ex Harv.	S. Africa	Grobbelaar et al. (1967)
	Zimbabwe	Corby (1974)
colutea (Burm. f.) Merr.		
var. *colutea*	Zimbabwe	Corby (1974)
comosa N. E. Br.	S. Africa	Grobbelaar & Clarke (1972)
cordifolia Heyne ex Roth	W. India	Y. S. Kulkarni (pers. comm. 1972)
coriacea Ait.		
var. *cana* Harv.	S. Africa	Grobbelaar & Clarke (1972)
var. *hirta*	S. Africa	N. Grobbelaar (pers. comm. 1973)
costata Guill. & Perr.		
ssp. *gonioides* (Hochst. ex Bak.) Gillett	S. Africa	Grobbelaar & Clarke (1975)
ssp. *macra* (E. Mey.) Gillett	Zimbabwe	Corby (1974)
cryptantha Benth. ex Harv.	S. Africa	Grobbelaar et al. (1967)
var. *cryptantha*	Zimbabwe	Corby (1974)
var. *occidentalis* Bak. f.	S. Africa	Grobbelaar & Clarke (1972)
cuneata Bak. ex Oliv.	S. Africa	Grobbelaar et al. (1964, 1967)
cuneifolia E. & Z.	S. Africa	Grobbelaar & Clarke (1975)
cylindrica DC.	S. Africa	Grobbelaar & Clarke (1975)
cytisoides Thunb.	S. Africa	Grobbelaar & Clarke (1972)
daleoides Benth. ex Harv.	S. Africa	Grobbelaar & Clarke (1972)
	Zimbabwe	Corby (1974)
decora Lindl.	Japan	Ishizawa (1953)
delagoaensis Bak. f. ex Gillett	S. Africa	Grobbelaar & Clarke (1972)
	Zimbabwe	Corby (1974)
demissa Taub.	S. Africa	Grobbelaar et al. (1964)
	Zimbabwe	Corby (1974)
dendroides Jacq.	Zimbabwe	Corby (1974)
denudata Thunb.	S. Africa	Grobbelaar et al. (1967)
digitata Thunb.	S. Africa	Grobbelaar & Clarke (1972)
dimidiata Vog.	S. Africa	Grobbelaar & Clarke (1972)
	Zimbabwe	Corby (1974)
discolor E. Mey.	S. Africa	Grobbelaar & Clarke (1972)

Nodulated species of *Indigofera* (cont.)	Area	Reported by
dissimilis N. E. Br.	S. Africa	Grobbelaar & Clarke (1972)
dosua Buch.-Ham.	U.S.A.	Wilson (1945a)
dyeri Britten var. *major*	Zimbabwe	Corby (1974)
echinata Willd.	S. Africa	Grobbelaar & Clarke (1974)
egens N. E. Br.	S. Africa	Grobbelaar & Clarke (1972)
emarginella Steud. ex A. Rich.	Zimbabwe	Corby (1974)
endecaphylla Jacq.	Widely reported	Many investigators
enneaphylla L.	Sri Lanka	W. R. C. Paul (pers. comm. 1951)
	Australia	Beadle (1964)
eriocarpa E. Mey.	S. Africa	Grobbelaar & Clarke (1975)
erythrogramma Welw. ex Bak.	Zimbabwe	Corby (1974)
eylesiana Gillett	Zimbabwe	Corby (1974)
fanshawei Gillett	Zimbabwe	Corby (1974)
filicaulis E. & Z.	S. Africa	Grobbelaar & Clarke (1972)
filiformis Thunb.	S. Africa	Grobbelaar & Clarke (1972)
filipes Benth.	S. Africa	Grobbelaar & Clarke (1972)
	Zimbabwe	Corby (1974)
flavicans Bak.	S. Africa	Grobbelaar & Clarke (1972)
	Zimbabwe	Corby (1974)
foliosa E. Mey.	S. Africa	Grobbelaar & Clarke (1975)
frutescens L. f.	S. Africa	Grobbelaar et al. (1964)
fulvopilosa Brenan	Zimbabwe	Corby (1974)
gairdnerae Hutch. ex Bak. f.	Zimbabwe	Corby (1974)
garckeana Vatke	S. Africa	Grobbelaar & Clarke (1972)
	Zimbabwe	Corby (1974)
gerardiana R. Grah.	India	Aggarawal (1934)
gerrardiana Harv.	S. Africa	Grobbelaar & Clarke (1972)
glandulosa Willd.	W. India	Y. S. Kulkarni (pers. comm. 1972)
glaucescens E. & Z.	S. Africa	Grobbelaar & Clarke (1975)
glomerata E. Mey.	S. Africa	Grobbelaar & Clarke (1972)
griseoides Harms	Zimbabwe	Corby (1974)
hedyantha E. & Z.	S. Africa	Grobbelaar & Clarke (1964)
	Zimbabwe	Corby (1974)
heterophylla Thunb.	S. Africa	Grobbelaar et al. (1967)
heterotricha DC.	S. Africa	Grobbelaar et al. (1967)
	Zimbabwe	Corby (1974)

Nodulated species of *Indigofera* (cont.)	Area	Reported by
hewittii Bak. f.	Zimbabwe	Corby (1974)
hilaris E. & Z.	S. Africa	Grobbelaar & Clarke (1972)
	Zimbabwe	Corby (1974)
hirsuta L.	Widely reported	Many investigators
hispida E. & Z.	S. Africa	Grobbelaar & Clarke (1975)
holubii N. E. Br.	S. Africa	Grobbelaar & Clarke (1972)
	Zimbabwe	Corby (1974)
humifusa E. & Z.	S. Africa	Grobbelaar & Clarke (1972)
incarnata Nakai	Japan	Asai (1944)
ingrata N. E. Br.	S. Africa	Grobbelaar & Clarke (1975)
inhambanensis Klotz.	Zimbabwe	Corby (1974)
kirilowii Maxim.	U.S.A.	Wilson (1945a)
langebergensis (L.) Bolus	U.S.A.	Wilson (1939a)
laxeracemosa Bak. f.	S. Africa	Grobbelaar & Clarke (1972)
linifolia (L. f.) Retz.	Australia	Beadle (1964)
	W. India	V. Schinde (pers. comm. 1977)
livingstoniana Gillett	Zimbabwe	Corby (1974)
lobata Gillett	Zimbabwe	Corby (1974)
longebarbata Engl.	S. Africa	Grobbelaar & Clarke (1975)
longepedicellata Gillett	Zimbabwe	Corby (1974)
longipes N. E. Br.	S. Africa	Grobbelaar & Clarke (1972)
lupatana Bak. f.	Zimbabwe	Corby (1974)
lyallii Bak.		
ssp. *lyallii*	Zimbabwe	Corby (1974)
macra E. Mey.	S. Africa	Grobbelaar & Clarke (1972)
macrostachya Willd.	France	Lechtova-Trnka (1931)
malacostachys Benth.	S. Africa	Grobbelaar & Clarke (1972)
maritima Bak.	S. Africa	N. Grobbelaar (pers. comm. 1974)
melanadenia Benth.	S. Africa	Grobbelaar & Clarke (1972)
	Zimbabwe	Corby (1974)
microcarpa Desv.	Venezuela	Barrios & Gonzalez (1971)
mimosoides Bak.		
var. *mimosoides*	Zimbabwe	Corby (1974)
var. *rhodesica* Bak. f.	Zimbabwe	Corby (1974)
mischocarpa Schlechtd.	S. Africa	Grobbelaar & Clarke (1975)
mollicoma N. E. Br.	S. Africa	Grobbelaar & Clarke (1975)

Nodulated species of *Indigofera* (cont.)	Area	Reported by
monantha Bak. f.	Zimbabwe	Corby (1974)
mucronata Spreng.	Brazil	Galli (1958)
neglecta N. E. Br.	S. Africa	Grobbelaar & Clarke (1975)
nummulariifolia (L.) Liv. ex Alst.	S. Africa	Grobbelaar et al. (1967)
	Zimbabwe	Corby (1974)
ormocarpoides Bak.	Zimbabwe	Corby (1974)
ovata Thunb.	S. Africa	Grobbelaar & Clarke (1972)
oxalidea Welw. ex Bak.	Zimbabwe	Corby (1974)
	S. Africa	Grobbelaar & Clarke (1975)
oxytropis Benth. ex Harv.	S. Africa	Grobbelaar & Clarke (1972)
oxytropoides Schltr.	S. Africa	Grobbelaar & Clarke (1972)
paniculata Vahl ex Pers. ssp. *gazensis* (Bak. f.) Gillett	Zimbabwe	Corby (1974)
parviflora Heyne ex W. & A. var. *parviflora*	S. Africa	Grobbelaar & Clarke (1972)
	Zimbabwe	Corby (1974)
pascuorum Benth.	Venezuela	Barrios & Gonzalez (1971)
paucifolia Del.	Malaysia	Anon. (1903)
pauxilla N. E. Br.	S. Africa	Grobbelaar & Clarke (1975)
podophylla Benth. ex Harv.	S. Africa	Grobbelaar & Clarke (1972)
poliotes E. & Z.	S. Africa	Grobbelaar & Clarke (1972)
pongolana N. E. Br.	S. Africa	Grobbelaar & Clarke (1975)
porrecta E. & Z.	S. Africa	N. Grobbelaar (pers. comm. 1966)
praticola Bak. f.	Zimbabwe	Corby (1974)
pretoriana Harms	S. Africa	Grobbelaar & Clarke (1972)
procumbens L.	S. Africa	Grobbelaar & Clarke (1972)
prostata Willd.	W. India	V. Schinde (pers. comm. 1977)
pseudoindigofera (Merx.) Gillett	Zimbabwe	Corby (1974)
pseudotinctoria Matsum.	Japan	Asai (1944); Kodama (1967)
psoraleoides L.	S. Africa	Grobbelaar & Clarke (1972)
purpurea Page ex Steud.	Philippines	Pers. observ. (1962)
rautanenii Bak. f.	S. Africa	N. Grobbelaar (pers. comm. 1974)
reducta N. E. Br.	S. Africa	Grobbelaar & Clarke (1972)
rehmannii Bak. f.	S. Africa	Grobbelaar & Clarke (1972)

Nodulated species of *Indigofera* (cont.)	Area	Reported by
retroflexa Baill.	Australia	McLeod (1962)
rhynchocarpa Welw. ex Bak.		
var. *rhynchocarpa*	Zimbabwe	Corby (1974)
rhytidocarpa Benth. ex Harv.	S. Africa	Grobbelaar et al. (1967)
ssp. *rhytidocarpa*	Zimbabwe	Corby (1974)
richardsiae Gillett	Zimbabwe	Corby (1974)
ripae N.E. Br.	S. Africa	Grobbelaar & Clarke (1975)
rostrata Bolus	S. Africa	Grobbelaar et al. (1967)
sanguinea N. E. Br.	S. Africa	Grobbelaar et al. (1967)
schimperi Jaub. & Spach	S. Africa	Grobbelaar et al. (1967)
	Zimbabwe	Corby (1974)
schinzii N. E. Br.	S. Africa	Grobbelaar & Clarke (1975)
sessilifolia DC.	S. Africa	Grobbelaar & Clarke (1975)
setiflora Bak.	S. Africa	Grobbelaar et al. (1967)
	Zimbabwe	Corby (1974)
sordida Benth.	S. Africa	Grobbelaar & Clarke (1975)
spicata Forsk.	Widely reported	Many investigators
spinescens E. Mey.	S. Africa	Grobbelaar & Clarke (1975)
stenophylla E. & Z.	S. Africa	Grobbelaar et al. (1967)
stipularis Link	W. India	V. Schinde (pers. comm. 1977)
stricta L. f.	S. Africa	Grobbelaar & Clarke (1975)
strobilifera Hochst.		
ssp. *strobilifera*	Zimbabwe	Corby (1974)
subcorymbosa Bak.	Zimbabwe	Corby (1974)
subulata Vahl ex Poir.	Brazil	Galli (1958)
	S. Africa	Grobbelaar & Clarke (1972)
subulifera Welw. ex Bak.		
var. *subulifera*	Zimbabwe	Corby (1974)
suffruticosa Mill.	Widely reported	Many investigators
sulcata DC.	S. Africa	Grobbelaar & Clarke (1972)
swaziensis Bolus	S. Africa	Grobbelaar & Clarke (1972)
var. *perplexa* (N. E. Br.) Gillett	Zimbabwe	Corby (1974)
tenuis Milne-Redhead	S. Africa	Grobbelaar et al. (1967)
ssp. *major* Gillett	Zimbabwe	Corby (1974)
tenuissima E. Mey.	S. Africa	Grobbelaar & Clarke (1972)
(*tettensis* Klotz.) = *schimperi* Jaub. & Spach	Australia	McLeod (1962)
tetragonoloba E. Mey.	S. Africa	N. Grobbelaar (pers. comm. 1974)

Nodulated species of *Indigofera* (cont.)	Area	Reported by
teysmanni Miq.	Sri Lanka	W. R. C. Paul (pers. comm 1951)
	W. India	Y. S. Kulkarni (pers. comm. 1972)
tinctoria L.		
(= *sumatrana* Gaertn.)	U.S.A.	Wilson (1939a)
	India	Raju (1936a)
var. *arcuata* Gillett	Zimbabwe	Corby (1974)
tomentosa E. & Z.	S. Africa	Grobbelaar & Clarke (1972)
torulosa E. Mey.	S. Africa	Grobbelaar & Clarke (1975)
tristis E. Mey.	S. Africa	Grobbelaar & Clarke (1972)
tristoides N. E. Br.	S. Africa	Grobbelaar et al. (1967)
trita L. f.	S. Africa	Grobbelaar et al. (1967)
var. *scabra* (Roth) Ali	Zimbabwe	Corby (1974)
var. *subulata* (Vahl ex Poir.) Ali	Zimbabwe	Corby (1974)
velutina E. Mey.	S. Africa	Grobbelaar & Clarke (1975)
vicioides Jaub. & Spach		
var. *rogersii* (R. E. Fries) Gillett	Zimbabwe	Corby (1974)
var. *vicioides*	S. Africa	Grobbelaar & Clarke (1975)
viscidissima Bak.		
ssp. *viscidissima*	Zimbabwe	Corby (1974)
viscosa Lam.	N. S. Wales, Australia	Beadle (1964)
welwitschii Bak.		
var. *remotiflora* (Taub. ex Bak. f.) Cronquist	Zimbabwe	Corby (1974)
wildiana Gillett	Zimbabwe	Corby (1974)
williamsonii (Harv.) N. E. Br.	Zimbabwe	Corby (1974)
woodii Bolus	S. Africa	Grobbelaar & Clarke (1972)
zeyheri Spreng.	S. Africa	Grobbelaar & Clarke (1972)

Indopiptadenia Brenan   Mimosoideae: Piptadenieae

Indo refers to habitat; the name indicates close similarities between the genus and *Piptadenia*, an exclusively tropical American genus.

Type: *I. oudhensis* (Brandis) Brenan

1 species. Trees, medium-sized, unarmed. Leaves bipinnate; rachis glan-

dular between the opposite pinnae pairs; leaflets 1-paired, large, subsessile, oblique-elliptic, pinnately veined; petiole eglandular. Flowers greenish yellow, dense, mostly in solitary, axillary spikes; calyx glabrous outside; lobes free; petals entirely free; stamens short; ovary smooth, short-stalked; disk inconspicuous. Pod flat, linear, curved, long-stalked, dehiscent along both sutures, valves leathery, continuous within; seeds brown, 15–20, flat, elongated, with a broad, membranous wing at each end, hilum central. (Brenan 1955)

This monotypic genus is endemic in India. It occurs in forests at the base of hills.

This species, restricted to India, differs from other *Piptadenia* species in having "elongated seeds, the eglandular petioles, the glandular rhachis, the unijugate leaves, and the corolla-lobes being free to the base" (Brenan 1955).

Inga Scop.   Mimosoideae: Ingeae
2 / 3436

The Brazilian vernacular name.

Type: *I. vera* (L.) Willd.

150–300 species. Shrubs or trees, unarmed, glabrous or pubescent. Leaves paripinnate; leaflets large, few pairs, asymmetric, larger toward the apex; rachis often winged, with large gland usually present between each pair of leaflets; stipules small and caducous or large and persistent. Flowers mostly white, or yellow, rather large, often tomentose, sessile or pedicelled in globose umbels or arranged spirally in elongated spikes, axillary on old branches, terminal on new growth; calyx tubular or campanulate; lobes 5, shallowly lobate; petals 5, joined halfway or higher, valvate; stamens numerous, joined at base or higher to form a tube, long-exserted; anthers small; ovary sessile, few- to many-ovuled; style awl-shaped; stigma terminal, small. Pod linear, flat and thin, or thick and fleshy, 4-sided or subterete, woody to leathery, tardily, if at all, dehiscent; pod usually opened by force of the germinating seeds rather than by dehiscence, releasing the naked embryo with a well-developed hypocotyl, defined as "viviparous germination" by León (1966); seeds covered with a pulpy, white aril, rarely nude. (Bentham and Hooker 1865; Pittier 1916b, 1929; Standley and Steyermark 1946)

Species of this large genus are common in tropical and subtropical America from Mexico to Brazil. They thrive in rain forests, along streams, on the lowlands of Panama, and on the wet highlands of Guatemala.

These fast-growing species range in habit from shrubs 3–4 m tall to medium-sized evergreen trees 25 m in height, with trunk diameters of 0.5 m. The tallest species, *I. altissima* Ducke, reaches 40 m in height.

Inga species have been used in association with coffee and cacao culture in America since pre-Columbian times. Their desirability is enhanced by the following characteristics: (1) their rapid growth for quick shade; (2) their ability to withstand drastic pruning to maintain appropriate size and amounts of shade; (3) their usefulness in maintaining soil fertility; and (4) their effectiveness in preventing erosion. Often several different species are used on the same plantation. Among the favored species are *I. edulis* Mart, *I. laurina* (Sw.) Willd., *I. oerstediana* Benth., *I. pittieri* M. Micheli, *I. speciosissima* Pitt., and *I. spuria* H. & B. at low altitudes, and *I. leptoloba* Schlechtd. and *I. punctata* Willd. at high altitudes.

The cuttings from *Inga* trees, which need constant pruning, furnish a fuel supply. *Inga* wood is attractive, but its coarse texture and susceptibility to attack by dry-wood termites and to decay in the soil limit its use for construction.

I. edulis, *I. feuillei* DC., and *I. macuna* Walp. & Duchas are prominent among the species grown as fruit trees for the edible sweet raw pulp. In parts of Panama they are called ice cream beans. The pods and large seeds of some other species are cooked as vegetables.

NODULATION STUDIES

The lack of nodule occurrence on *I. densiflora* Benth., *I. sapindoides* Willd., *I. spuria*, and *I. tonduzii* J. D. Smith in Costa Rica (J. León personal communication 1956) is believed by the authors to be a circumstantial coincidence unrelated to any sectional division within the genus.

Young nodules are usually round, white, and smooth surfaced; older forms are dark brown, rough surfaced, and more or less corky or woody in texture (Allen and Allen 1936a). Two strains of rhizobia from *I. laurina* were effective on diverse species of the cosmopolitan cowpea miscellany (Allen and Allen 1939). Culturally and biochemically the strains were typical of slow-growing rhizobial forms.

Nodulated species of *Inga*	Area	Reported by
edulis Mart.	Hawaii, U.S.A.	Allen & Allen (1936a)
	Costa Rica	J. León (pers. comm. 1956)
ferruginea Gray	Germany	Brunchorst (1885)
heterophylla Willd.	Trinidad	DeSouza (1966)
ingoides Willd.	Trinidad	DeSouza (1966)
laterifolia Miq.	Guyana	Norris (1969)
laurina (Sw.) Willd.	Puerto Rico	Cook & Collins (1903)
	Hawaii, U.S.A.	Allen & Allen (1936a)
marginata Willd.	Costa Rica	J. León (pers. comm. 1956)
oerstediana Benth.	Costa Rica	J. León (pers. comm. 1956)
pezizifera Benth.	Guyana	Norris (1969)
setifera DC.	Trinidad	DeSouza (1966)
thibaudiana DC.	Trinidad	DeSouza (1966)
	Guyana	Norris (1969)
venosa Griseb. ex Benth.	Trinidad	DeSouza (1966)
vera Willd.	Puerto Rico	Cook & Collins (1903)

From Greek, *is, inos,* "sinew," "fiber," and *karpos,* "fruit," in reference to the lined appearance of the pod.

Type: *I. edulis* Forst.

2 species. Trees, medium to 25 m tall, handsome; trunks usually deeply furrowed, commonly buttressed; crown dense. Leaves simple, entire, large, oblong-lanceolate short-petioled, pinnately nerved, leathery; stipules very small, soon caducous. Flowers white, cream, or yellow, fragrant, in axillary, simple or branched spikes resembling catkins when young; bracts small, connate with rachis, somewhat pouched; bracteoles small; calyx tubular-campanulate, bilabiate, membranous, irregularly 2–5-toothed; petals 4–6, usually 5, subequal, imbricate in bud, linear-lanceolate, upper part crinkled; stamens twice the number of petals, alternately long and short, the longer ones briefly joined to the petals; anthers small, uniform; ovary subsessile or short-stalked, 1-, seldom 2-ovuled; style very short; stigma oblique. Pod short-stalked, oblique-obovate, flattened, 2-valved, subdrupaceous, leathery, indehiscent, 1-seeded. (Backer and Bakhuisen van den Brink 1963)

Members of this genus occur in tropical America, Malaysia, and Oceania, especially in Fiji, Java, and Tahiti. They are usually found on rich valley soils and along stream borders, and are somewhat swamp-tolerant and salt-tolerant.

The type species is called Polynesian or Tahitian chestnut for the chestnutlike flavor of its edible seeds. The tree is ubiquitous throughout the South Pacific Islands. When breadfruit is scarce, the boiled or roasted seeds are a suitable substitute. Natives also permit the seeds to undergo a partial fermentation in underground pits, in the manner accorded breadfruit, for preservation as a reserve food supply. The leaves are eaten by cattle. The astringent bark decoctions are used in native medicine for intestinal disturbances. The wood is of good quality but not plentiful; it is used for construction of native furniture.

NODULATION STUDIES

Nodules were absent on *I. edulis* Forst. examined in the Philippines by J. G. Davide (personal communication 1963).

Type: *I. bijuga* (Colebr.) Ktze.

7–9 species. Trees, tall, deciduous, unarmed. Leaves paripinnate; leaflets in few pairs, entire, ovate to oblong, leathery; stipules connate at the base. Flowers rather large, white, pink, or red, pedicelled, each in the axil of a bract, in terminal racemes or corymbs; bracts and bracteoles caducous; calyx tube elongated; lobes 4, subequal, imbricate in bud; petals, 1 perfect, in front of posterior lobe, clawed, orbicular or kidney-shaped, others rudimentary, minute, or absent; fertile stamens 3, free, declinate, hairy at the base; filaments long; anthers dorsifixed; staminodes 1–7, small, on short filaments; ovary often imperfect, stalked, stalk adherent to the calyx tube, several- to many-ovuled; style elongated; stigma truncate, subcapitate. Pod oblong-linear, flat, thin, leathery to woody, transversely nerved, 2-valved, tardily dehiscent, septate between the 1–8 transverse seeds; seeds flat, round, without aril. (Backer and Bakhuizen van den Brink 1963)

Representatives of this tropical genus are common in Indo-Malaya, Polynesia, the Philippines, Madagascar, and Taiwan. They occur along sea-coasts, sandy beaches, mangrove swamp fringe areas, and mouths of tidal rivers periodically inundated; some species are present in hillside and inland forests up to 400-m altitude.

I. acuminata Merr., *I. bakeri* Prain, *I. bijuga* (Colebr.) Ktze., *I. palembanica* Miq., *I. plurijuga* Harms, and *I. retusa* Ktze. are major timber trees. *I. bijuga*, Moluccan ironwood or ipil, is one of the most sought-after woods in the Pacific and considered the "chief" of the Philippine timbers. The sapwood of this huge tree is 4–8 cm thick and creamy white; the yellow, freshly cut heartwood becomes reddish brown on exposure. Seasoned wood is chocolate brown, hard, usually straight grained, fine textured, and not difficult to work. Application of a watertight seal is necessary after seasoning to prevent exuda-tion of an oily stain. Properly finished wood takes a high polish and is most desirable for high-quality furniture, flooring, paving blocks, and house con-struction (Kukachka 1970). It is not durable for saltwater pilings. Samoans favored the wood for kava bowls.

Various indelible brown, yellow, and khaki shades of dye for staining native mats and textiles are obtained from the sap and bark of several dif-ferent *Intsia* Species.

Nodulation Studies

Nodules were not observed on *I. acuminata*, *I. bakeri*, and *I. bijuga* in the Philippines (J. G. Davide personal communication 1963). Nodules were also lacking on numerous seedlings of the last-named species in Hawaii (personal observation 1940) and on nursery plants in the Bogor Botanic Garden, Java (personal observation 1977). Norris (1959b) isolated a typical cowpea-type strain from nodules on *I. bijuga* growing in Tully, Queensland, Australia.

Isoberlinia Craib & Stapf   Caesalpinioideae: Amherstieae
(62a)

The Greek prefix, *iso*, denotes similarity to *Berlinia*.
   Lectotype: *I. dalzielii* Craib & Stapf
   7 species. Trees, medium-sized, up to 20 m tall, with trunks up to 40 cm in diameter, deciduous; crowns flat, spreading; buttresses, if present, short and rounded. Leaves paripinnate; leaflets opposite or subopposite, 3–4 pairs, medium to large, petioluled, rigid, papery or leathery, not pellucid-punctate, asymmetrically ovate; rachis not winged; stipules caducous. Flowers small to medium, usually white, sessile, in terminal or subterminal, panicled racemes, rarely in groups; bracts small, caducous; bracteoles 2, opposite, valvate, obovate or rounded, usually thick, large, enclosing the flower bud; calyx shortly tubular; sepals usually 5, rarely 6, subequal, free, linear; petals 5, subequal or the posterior median petal slightly broader than the others, clawed, sessile or subsessile; stamens usually 10, rarely 13, all fertile, free, more or less exserted; anthers oblong; ovary stalked, stalk adnate to the side of the receptacle, 4–6-ovuled; style elongated, filiform; stigma terminal. Pod woody, compressed, large, 2-valved, each valve channeled but not winged, elastically dehiscent; pods of some species dehiscing explosively; seeds 4–6, large, round, flat. (Hauman 1952; Léonard 1957)
   Members of this genus are native to tropical Africa. Abundant but confined to savanna forests, they are often found on stony ground.
   The pinkish, open-grained, nondurable wood is used primarily for small carpentry and fuel. A reddish, resinous exudate from the bark of several species lacks any known use (Dalziel 1948).

Isodesmia Gardn.   Papilionoideae: Hedysareae,
322 / 3791   Aeschynomeninae

From Greek, *isos*, "equal," and *desme*, "bundle"; the filaments are usually joined into 2 bundles of 5 each.
   Type: *I. tomentosa* Gardn.
   2 species. Shrublets, twining or climbing. Leaves imparipinnate; leaflets numerous, 7–17-foliolate; stipels none; stipules persistent, not at the base of the petiole. Flowers few, rather large, yellow, in axillary racemes; bracts stipulelike; bracteoles persistent; calyx campanulate; lobes subequal or the upper 2 partly subconnate; standard orbicular, emarginate; wings oblong, blunt; keel shorter than the wings, blunt; stamens connate into a sheath split above the keel or on both sides; anthers uniform; ovary sessile, many-ovuled; style filiform; stigma small, terminal. Pod sessile, linear, straight, planocompressed, joints quadrate, 6–8, leathery, longitudinally venose. (Desvaux 1826; Bentham and Hooker 1865)
   The status of this genus is questionable in the light of Rudd's deduction (1958) that both members of this genus, *I. tomentosa* Gardn. and *I. blanchetiana* Benth., should be transferred to *Chaetocalyx*.
   Both species have a distribution range from Brazil to Central America.

Isomacrolobium Aubrév. & Pellegr. Caesalpinioideae:
Amherstieae

The Greek prefix, *iso*, denotes similarity to *Macrolobium*; all species have *Macrolobium* synonyms.

Type: *I. leptorrachis* (Harms) Aubrév. & Pellegr.

10 species. Trees or shrubs, distinctive from *Macrolobium* by the following characteristics: petals 4–6, subequal; stamens 9, 3 large, fertile, exserted, others small, sterile; pod gibbous, curved, fissured or wrinkled, margins not distinctly ribbed. (Aubréville and Pellegrin 1958)

Species of this genus were originally included in *Macrolobium*, which is now reserved for the distinctive American elements (Cowan 1953). Some of the excluded African species were transferred to *Anthonotha*, which in turn was partly fragmented by Aubréville and Pellegrin (1958) to establish *Triplisomeris* and *Isomacrolobium*; these new taxa are regarded as synonyms of *Macrolobium* by Hutchinson (1964).

Representatives of this genus are native in equatorial Africa. They grow in forest habitats.

Isotropis Benth. Papilionoideae: Podalyrieae
167 / 3627

From Greek, *isos*, "equal," and *tropis*, "keel"; the keel is as long as the wings.

Type: *I. striata* Benth.

10 species. Herbs or undershrubs, small, usually less than 2 m tall, slender, or bushy; stems diffuse or ascending, often rusty-pubescent. Leaves sparse, simple or 1-foliolate, narrow, linear, some with a recurved, pungent point; stipules linear-falcate or minute. Flowers orange yellow and purple or solid purple, solitary or few, on long pedicels, axillary or in loose terminal racemes; calyx tube short; lobes longer than the tube, imbricate, 2 upper lobes united almost to the top; petals clawed; standard purple-streaked, orbicular, longer than the wings; wings obovate, slightly falcate; keel incurved, as long as wings; stamens free; ovary sessile, multiovuled; style incurved, filiform; stigma minute, terminal. Pod oblong to lanceolate, acute, often tomentose, more or less turgid, slightly beaked; seeds several, red to reddish brown, reticulate. (Bentham 1864; Bentham and Hooker 1865)

This genus is closely related to *Oxylobium* and *Chorisema*.

Members of this genus are indigenous to New South Wales, Queensland, Northern Territory, and South and Western Australia. Species occur on tablelands, slopes, gullies, and stream and creek banks.

The foliage of some species is toxic to livestock (Webb 1948).

Nodulation Studies

Lange (1959) observed nodules on field-excavated root systems of *I. cuneifolia* (Sm.) Domin and *I. juncea* Turcz. in Western Australia. Cultural and host-infection tests identified an isolate from *I. cuneifolia* as a member of the slow-growing cowpea-lupine-soybean type (Lange 1961; Lange and Parker 1961); however, in reciprocal plant tests, strains of rhizobia from *Lupinus digitatus* and *L. angustifolius* did not nodulate *I. cuneifolia* (Lange 1962).

Named for George Jackson, 1790–1811, English botanist.

Type: *J. spinosa* R. Br.

40 species. Shrubs or undershrubs; branches rigid, angular or winged; branchlets phyllodelike, resembling flat leaves, or spinescent, sometimes pubescent. Leaves, if present, few, 1-foliolate, or reduced to minute scales at the nodes. Flowers cream to yellow, or reddish purple and yellow, in terminal or lateral spikes, or scattered on the branchlets; bracts small, scalelike; bracteoles small, caducous or persistent; calyx deeply cleft; tube short; lobes long, valvate, 2 upper ones broader, seldom connate; petals shorter than the calyx, claws short; standard round or kidney-shaped, usually emarginate; wings oblong; keel straight, broader than the oblong wings; stamens free; ovary sessile or stalked, mostly 2-ovuled; style awl-shaped, incurved; stigma minute, terminal. Pod sessile or stalked, ovate or oblong, turgid or flat, scarcely longer than the calyx, sometimes silky-haired, 1- to few-seeded. (Bentham 1864; Bentham and Hooker 1865)

The genus is closely allied to *Gompholobium* and *Burtonia*.

Members of this genus are indigenous to Australia. About 30 species occur in southwestern Australia, New South Wales, Queensland, and Victoria. They are common in coastal highlands, slopes, and plains.

All species have ornamental value. *J. scoparia* R. Br., the golden rain tree, is especially prized as an ornamental for areas having light soils.

NODULATION STUDIES

Nodules on *J. furcellata* (Bonpl.) DC. and *J. spinosa* (Labill.) R. Br. are large and elliptical (C. A. Gardner personal communication 1947). Inclusion of *J. scoparia* in the cowpea grouping was proposed by McKnight (1949a); however, a strain tested by Norris (1959b) did not nodulate *Phaseolus lathyroides*, a plant usually receptive to rhizobia of the cowpea miscellany. *Jacksonia* species listed in Lange's initial report (1959) later yielded isolates with slow-growing cultural characteristics. In plant responses they displayed a natural grouping with cowpea, soybean, and lupine. None of the strains formed nodules on *Pisum*, *Medicago*, *Trifolium*, *Vicia*, *Lens*, or *Ornithopus* species (Lange 1961).

Nodulated species of *Jacksonia*	Area	Reported by
floribunda Endl.	W. Australia	Lange (1959)
furcellata (Bonpl.) DC.	W. Australia	C. A. Gardner (pers. comm. 1947); Lange (1959)
hakeoides Meissn.	W. Australia	Lange (1959)
horrida DC.	W. Australia	Lange (1959)
scoparia R. Br.	Queensland, Australia	McKnight (1949a); Norris (1959b)
spinosa (Labill.) R. Br.	W. Australia	C. A. Gardner (pers. comm. 1947); Lange (1959)
sternbergiana Hueg.	W. Australia	Lange (1959)
umbellata Turcz.	W. Australia	Lange (1959)

Named for Dr. Jacques Huber, 1867–1914, Brazilian botanist.

Type: *J. quinquangulata* Ducke

2 species. Trees, small to medium, up to about 16 m tall, unarmed; stems and branches 5-sided. Leaves large, bipinnate; pinnae pairs many; leaflets very numerous, small, paired; stipules large, foliaceous, pinnate, persistent on the flowering branchlets. Flowers yellow or purple, in short, terminal subcorymbose racemes at the tips of slender, nearly leafless branches; bracts long, bristly; bracteoles absent; calyx cupular or campanulate, obsoletely ribbed; lobes 5, imbricate, ovate; petals 5, subequal, imbricate, not clawed, erect, ovate; stamens 10, equal, elongated; filaments connate only in basal one-third; anthers dorsifixed; ovary sessile, many-ovuled; style filiform, spirally coiled in bud; stigma terminal, obliquely capitate. Pod woody, linear, flat, 2-valved, ribbed along both margins, elastically dehiscent from the apex to the base, septate between the 4–8, flat, oblong seeds. (Ducke 1922)

Members of this genus occur in tropical Brazil, mostly in the northern and eastern Amazon region. They are found in primeval forests on sandy soil.

The durable wood lacks commercial demand because of the small size and scarcity of the trees (Record and Hess 1943).

Jansonia Kipp.   Papilionoideae: Podalyrieae
162 / 3622

Named for Joseph Janson, F.L.S., patron of botany.

Type: *J. formosa* Kipp.

1 species. Shrub, seldom above 1 m tall, young branches silky-haired. Leaves simple, ovate-lanceolate, opposite, short-petioled, leathery, finely veined; stipules lanceolate-subulate, caducous. Flowers 4, together in sessile, terminal, recurved or nodding heads, enclosed in bud within 2 rows of 4 bracts; bracteoles absent; calyx very oblique, split above, villous outside; lobes 5, upper 2 minute, lower 3 elongated; petals glabrous, all clawed; standard minute, heart-shaped, lanceolate, reflexed; wings oblong; keel petals longer than the wings, dorsally joined; stamens free; ovary sessile, villous, multi-ovuled; style filiform; stigma small, terminal. Pod not described. (Bentham 1864; Bentham and Hooker 1865)

This monotypic genus is endemic in southwestern Australia.

NODULATION STUDIES

This is the only genus of the more than 20 in the tribe Podalyrieae for which no record of nodulability was found.

Named for Jules Bernard, former Governor of Gabon.

Type: *J. hochreutineri* Pellegr.

9–11 species. Trees, unarmed, up to 45 m tall, with trunks 1.5 m in diameter, and with low, sharp buttresses. Leaves paripinnate; leaflets small to large, few or many pairs, elliptic, tapered at each end, asymmetric at the base, lowest pair smallest, glabrous, usually densely gland-dotted; rachis not winged; stipules caducous. Flowers small, many, short-stalked, conspicuous, in lax, axillary or terminal panicles; bracts short, caducous; bracteoles 2, suborbicular, slightly keeled, valvate, persistent, enclosing each flower in bud; calyx tube flat or slightly concave; sepals 5, equal, large, free, oval, ciliate, slightly imbricate in bud; corolla reduced but present; petals 5, posterior petal well-developed, the other 4 very small; stamens 9–10; all filaments but 1 long-exserted, inflexed in bud, joined at the base, the stamen opposite the posterior petal free; anthers versatile, elliptic, longitudinally dehiscent; ovary sessile, villous, 4-ovuled; style filiform; stigma capitate. Pod short-stalked, rather large, oblong-linear, compressed, woody, margins thickened, 2-valved, valves inrolled after dehiscence; seeds 1–5, flattened, oval. (Pellegrin 1948; Hauman 1952; Léonard 1957)

Representatives of this genus are common in tropical West Africa. They are often dominant or codominant in regions with medium rainfall and along river banks subject to periodic inundation.

The cross-grained, hard wood of these species, known as Congo zebra wood, is not exploited commercially, but it is of good quality and is used for house construction and mine supports. Several species, exemplified by *J. globiflora* (Benth.) Troup., are appropriate for parks and avenues because of their fragrant flowers and graceful shapes. The bark yields a gum resin of little value. Among some native tribes, bark decoctions are used in ordeal trials and for minor medicinal purposes.

NODULATION STUDIES

Nodules were not found on *J. globiflora* in Zimbabwe by Corby (1974); however, symbiosis of *J. seretii* (De Wild.) Troup. with ectotrophic mycorrhiza in Zaire was described in detail by Fassi and Fontana (1961, 1962b). Ten distinct forms of mycorrhizae were found on rootlets in surface soil. This alternate type of symbiosis is extremely rare among leguminous species.

Kalappia Kosterm.   Caesalpinioideae: Cassieae

Kalappi is the name for *K. celebica* Kosterm. in the Malili region of Celebes.

Type: *K. celebica* Kosterm.

1 species. Tree, tall, unarmed. Leaves opposite, imparipinnate, eglandu-

lar, mostly 5-foliolate; leaflets alternate, entire, lanceolate or elliptic, acuminate, chartaceous, the terminal leaflet larger than the laterals; stipules inconspicuous. Flowers small, orange yellow, several on stout densely silky, axillary or subapical short, branched peduncles; bracts small, ovate, caducous; calyx obconical, inconspicuous; sepals 5, subequal, concave-elliptic, densely silky, imbricate; petals 5, clawed, erect, spreading, the 2 outermost laterals larger; stamens 10, arranged in 2 groups, alternately opposite the petals and sepals; 2 fertile stamens with long, stout, glabrous filaments and large basifixed horizontal apiculate anthers dehiscent by means of 2 apical pores; 2 pairs of lateral stamens slightly shorter, remainder reduced to staminodes, shorter and more slender; ovary sessile, compressed, densely silky, 3–5-ovuled; style incurved; stigma very small, terminal. Pod reddish brown, elongated, flat, smooth, narrowly winged along the curved, ventral suture, dehiscent on the straight dorsal suture, valves thin, smooth inside; seeds 1–3, disklike. (Kostermans 1952)

This monotypic genus is endemic in the vicinity of Malili, Celebes.

The tree is a forest giant yielding valuable timber that was formerly exported in quantity. The wood is highly esteemed for cabinetwork and furniture because of its beautiful grain pattern.

Kaoue Pellegr.   Caesalpinioideae: Eucaesalpinieae

From *kaoué* or *kahu*, the vernacular name in Ivory Coast.

Type: *K. stapfiana* (A. Chev.) Pellegr. (= *Oxystigma stapfiana*)

Kaoue was originally described as a monotypic genus (Pellegrin 1933). It is now generally regarded as congeneric with *Stachyothyrsus*, q.v. (Voorhoeve 1965).

*Kennedia Vent.   Papilionoideae: Phaseoleae, Glycininae
393 / 3868

Named for Lewis Kennedy, 1775–1818, nurseryman at Hammersmith, near London. The original spelling *Kennedia* is conserved over the more common subsequent spelling, *Kennedya* (Rickett and Stafleu 1959–1961; Burbridge 1963; *International Code of Botanic Nomenclature* 1966).

Type: *K. rubicunda* (Schneev.) Vent.

15 species. Herbs, perennial, prostrate or climbing, some woody, often

hairy. Leaves pinnately 3-foliolate, seldom 1- or 5-foliolate; leaflets entire; stipels small; stipules broad, striate or connate, persistent. Flowers red or violet to blackish, on axillary peduncles, in racemes, umbels, pairs, or solitary; bracts stipulelike, persistent or small, caducous; bracteoles none; calyx lobes subequal to the calyx tube, the uppermost 2 united into an entire or emarginate lip; standard obovate or orbicular, short-clawed with 2 inflexed, minute ears; wings oblique-oblong, joined to the incurved, obtuse or acute keel petals; stamens diadelphous; anthers uniform; ovary subsessile or short-stalked, many-ovuled; style filiform, inflexed upwards, rarely with an apical tooth; stigma terminal. Pod linear, compressed, terete or turgid, 2-valved, with pithy partitions between the seeds, rarely empty; seeds ovoid or oblong, laterally attached. (Bentham 1864; Bentham and Hooker 1865)

Members of this genus are native to Australia, mostly in western areas; a few species occur in New South Wales, Queensland, South Australia, and Tasmania. Species are most commonly found in coastal limestone areas, scrub-covered plains, and on podzolic soils of low fertility.

In reviewing the development of the Australian flora, Crocker and Wood (1947) surmised that in response to warm conditions the Australian continent was invaded by this element during the Miocene period. Similarly, Silsbury and Brittan (1955) commented that "the genus *Kennedya*, in view of its close relationship to the Glycines, and the fact that that Phaseoleae are not richly developed in Australia, is undoubtedly of Indo-Melanesian origin. . . . Whilst *Kennedya* species cannot be considered as typical of the Australian Element, the fact remains that they are endemic to this continent and show a considerable degree of adaptation to existing climatic and edaphic conditions."

All species have agronomic potential, especially in low rainfall, wheat-belt areas in Australia. Several forms serve as window-box ornamentals.

NODULATION STUDIES

The 11 species reported by Silsbury and Brittan (1955) were described as "herbaceous or woody perennials with well-developed root nodules" in Western Australia. In burned-over areas in South Australia (Strong 1938) and in arid regions of New South Wales, Australia (N. C. W. Beadle personal communication 1963), *K. prostrata* R. Br. was well nodulated. Two rhizobial strains of this species were maintained in the Type Collection of the Commonwealth Scientific and Industrial Research Organization, Division of Soils, Adelaide (Harris 1953).

Histological sections of nodules of *K. prostrata* showed the presence of typical bacteria and zone formations readily distinguishable from other non-rhizobial nodular structures also present on the root systems. Cultural characteristics of isolates from each species, as well as cross-inoculation tests, were considered by Lange (1961) to conform with those of the slow-growing members of the cowpea, soybean, and lupine groups. In subsequent tests by Lange (1962), however, the attempted reciprocal cross testing of rhizobia from *Lupinus digitatus* and *L. angustifolius* failed to nodulate any seedlings of 8 *Kennedia* species.

Nodulated species of *Kennedia*	Area	Reported by
beckxiana F. Muell.	W. Australia	Silsbury & Brittan (1955)
carinata (Benth.) Domin	W. Australia	Silsbury & Brittan (1955); Lange (1959)
coccinea Vent.	W. Australia	Silsbury & Brittan (1955); Lange (1959)
eximia Lindl.	W. Australia	Silsbury & Brittan (1955); Lange (1959, 1961)
glabrata (Benth.) Lindl.	W. Australia	Silsbury & Brittan (1955)
macrophylla (Meissn.) Benth.	W. Australia	Silsbury & Brittan (1955)
microphylla Meissn.	W. Australia	Silsbury & Brittan (1955)
(*monophylla* Vent.) = *Hardenbergia monophylla* Benth.	N.Y., U.S.A.	Wilson (1939a)
nigricans Lindl.	W. Australia	Silsbury & Brittan (1955); Lange (1959)
prorepens F. Muell.	W. Australia	Silsbury & Brittan (1955)
prostrata R. Br.	S. Australia	Strong (1938); Harris (1953)
	W. Australia	Silsbury & Brittan (1955)
	N.S. Wales, Australia	N. C. W. Beadle (pers. comm. 1963)
stirlingii Lindl.	W. Australia	Silsbury & Brittan (1955); Lange (1959)

Kerstania Rech. f.   Papilionoideae: Loteae

Named for Gerhard Kerstan, plant collector with the German Hindu Kush Expedition, 1935.

Type: *K. nuristanica* Rech. f.

1 species. Herb, perennial; rhizome slender. Leaves imparipinnate; leaflets elliptic, entire, multipaired; stipules thin, minute, membranous. Flowers 1–3, subumbellate, borne on short, axillary peduncles; calyx tube short, campanulate, margin minutely 5-toothed, truncate; petals free, smooth, curved, long-clawed; standard suborbicular; stamens 10, 9 connate into a tube, vexillary stamen more free, all free and broader above; anthers minute,

equal. Pod long-stalked, oblong-linear, flattened, nonseptate but sub-constricted between the 3–5 seeds. (Rechinger 1957)

The genus resembles *Hosackia*.

This monotypic genus is endemic in Nuristan, Afghanistan. Specimens are present in light oak forests at 1,400-m altitude.

Kerstingiella Harms   Papilionoideae: Phaseoleae, Phaseolinae
(422a)

Named for Dr. Kersting; he discovered the plant specimens and sent them to Harms from Togo.

Type: *K. geocarpa* Harms

1 species. Herb, annual, branches hairy, prostrate, rooting at the nodes. Leaves long-petioled, pinnately 3-foliolate; leaflets elliptic or obovate; stipels linear; stipules striate. Flowers cleistogamous, ivory or yellow, small, short-stalked or sessile, usually in pairs on short, axillary peduncles; calyx cupular, hairy, deeply cleft into linear-lanceolate lobes, the upper 2 almost completely joined; standard often pale violet, short-clawed, obovate; wings narrow, lanceolate; keel petals joined in the middle, somewhat broader and longer than the wings; stamens diadelphous; anthers oval; ovary usually 2-ovuled, short-stalked, surrounded at the base by a short, thick, sheathlike disk; style smooth, slightly curved; stigma terminal, ciliate. After fertilization, ovary stalk elongates geotropically into a carpopodium; the immature fruit is thus buried beneath the soil surface. Pod geocarpic, usually 2-jointed, thin-walled, wrinkled, indehiscent; seeds ovoid, white, black, red, or mottled. (Harms 1908b, 1915a; Stapf 1912)

A recent proposal that this genus and *Macrotyloma* are congeneric and, therefore, should be merged under the latter name (Maréchal and Baudet 1977) prompted Lackey (1978) to urge the conservation of the older legitimate epithet, *Kerstingiella*, for the combination. If Lackey's proposal is accepted, the genus will include 24 species distributed throughout the tropics of the Old World.

The genus position 422a was assigned by Harms (1915a). Species of *Arachis* and *Voandzeia* are also geocarpic, but they are not cleistogamous.

This monotypic genus is restricted to tropical West Africa, from Senegal to Nigeria.

Plants are cultivated, but the seed yield is low. Seeds are marketed for a nutritious food made by pounding the kernels and mixing with sheabutter. The flavor is comparable to that of *Phaseolus* beans.

NODULATION STUDIES

The sole reference to nodulation is contained in Harms' original description of the genus (Harms 1908b). He wrote, "In den Boden entsendet der Stengel an den Blattanzätzen zahlreiche feine oft mit Wurzelknöllchen besetzte Wurzeln."

Name means "King's tree"; named for Sir George King, 1840–1909.

Type: *K. pinnatum* (Roxb. ex DC.) Harms

4 species. Trees, large, handsome, evergreen. Leaves imparipinnate; leaflets 3–7, alternate, mostly asymmetric, lanceolate, acuminate, leathery, pellucid-punctate, glabrous; petiolules swollen, with many annular grooves; stipules very small, caducous. Flowers minute, on short pedicels, in long, axillary, lax-flowered, panicled racemes, hairy or glabrous; bracts minute, scalelike, persistent; bracteoles 1–2, minute, inserted on the receptacle or at its base; calyx tube very short; sepals 5, ovate, convex, gland-dotted, ciliate; petals none; stamens 10, infolded in bud, exserted at anthesis, hairy at base; anthers introrse, not cleft or apiculate, dorsifixed, longitudinally dehiscent; ovary densely hairy, 1-ovuled; disk distinct in young flower, connate with ovary stalk, later obsolete; style glabrous or hairy; stigma small, capitate. Pod variable, smooth, globular or ellipsoid, bulging over the seed area, otherwise flattened, sometimes apiculate, woody or leathery, 1-seeded. (Harms 1897; Knaap-van Meeuwen 1970)

Representatives of this genus are distributed throughout tropical India, Papua New Guinea, the Philippines, the Moluccas, the Solomon Islands, and Fiji. Common in evergreen forests at low and medium altitudes, they are often found on dry soil on volcanic hills.

The Philippine species, *K. alternifolium* Merr., reaches a height of 30–35 m, with a straight, cylindrical trunk about 1 m in diameter. The gray outer bark is shed in large, scroll-shaped patches revealing a red inner bark. The moderately hard wood is used for posts, beams, flooring, interior finish, and furniture. The dark green sap, which thickens to a gummy consistency, was an ingredient of an incense used by the early Spaniards in the Philippines (Brown 1921). A copaiba-type balsam resembling that from *Copaifera* species is obtained from the sapwood of *K. pinnatum* (Roxb. ex DC.) Harms in India (Harms 1915a).

Koompassia Maing.   Caesalpinioideae: Cassieae

76 / 3531

From *kumpas*, the vernacular name for these trees in Singapore.

Type: *K. malaccensis* Maing.

2 species. Trees, unarmed, deciduous, often gigantic, buttressed. Leaves imparipinnate; leaflets oblong-elliptic, alternate, 5–17, somewhat leathery; stipules small, free, early caducous. Flowers very small, many, in axillary or terminal panicles of small cymes; bracteoles small; calyx tube minuscule; sepals 5, small, equal, thinly pubescent, acuminate, imbricate; petals 5, subequal, slightly larger than the sepals; stamens 5, alternating with the petals; anthers large, basifixed, opening by means of 2 apical and 2 basal pores connected by a thin, nearly dehiscent rim; ovary sessile, 1-ovuled; style short, awl-shaped; stigma small, terminal. Pod oblong, compressed, twisted 180° at the base, encircled by a broad papery wing, 2-valved, indehiscent; seed 1, flat. (De Wit 1947a)

Members of this genus are indigenous to Borneo, the Malaysian Peninsula, and Papua New Guinea. They are usually found in river valleys, lowland forests, and on ridge slopes.

The type species and *K. excelsa* Taub. reach very great sizes, with widespreading crowns and large buttressed trunks (Corner 1952; Burkill 1966). The hard, heavy, coarse-textured wood, known as kempas in the timber trade, is used primarily for heavy construction in local areas (Titmuss 1965; Kukachka 1970).

NODULATION STUDIES

Nursery stock in Bogor Botanic Garden, Java, lacked nodules (personal observation 1977).

Kostyczewa Korsh.   Papilionoideae: Galegeae, Astragalinae
(294a) / 3762a

Named for P. A. Kostyczew, Russian plant collector.

Type: *K. ternata* Korsh.

2 species. Herbs, perennial, shrubby, short-stemmed, with woody rhizome, silky-pubescent throughout. Leaves in whorls of 3, long-petioled; leaflets broad-ovate, entire, terminal leaflet longer; stipels none; stipules triangular-lanceolate, joined at the base to the petiole. Flowers solitary on axillary peduncles shorter than the leaves; calyx tubular; teeth 5, acute, short, subequal; petals scarcely longer than the calyx; standard small, obovate-lanceolate, claw long, straight; wing and keel petals similar, very long-clawed; stamens diadelphous; anthers uniform; ovary with many ovules aligned in 2 rows; style awl-shaped, hairy; stigma small. Pod stalked, linear-lanceolate, inflated or turgid, thinly septate between the 8–10 seeds, valves 2, spirally twisted at dehiscence; seeds chestnut-brown, kidney-shaped, smooth. (Harms 1908a)

Representatives of this genus are native to Turkestan and the central Soviet Union. They thrive on gravel and clay soils on mountain slopes at altitudes of 1,000–1,800 m.

Kotschya Endl.   Papilionoideae: Hedysareae,
(327)   Aeschynomeninae

Named for Theodor Kotschy, 1813–1866, Austrian botanist; he collected plants in the Sudan.

Type: *K. africana* Endl.

30 species. Herbs, robust, erect, perennial, or small trees with pubescent or glabrous stems. Leaves pinnate; leaflets 4 or more, asymmetric at the base, with prominent veins branching marginally from the base, the major nerve submarginal; rachis terminated in a short point; stipels absent; stipules not spurred, sometimes persistent. Flowers blue, yellow, or orange, reflexed, in racemes, distichous, scorpioid, strobiliform; bracts hairy, entire, smaller than the flowers, scarious, striate, brownish, subtending each flower; bracteoles inserted below the calyx, free or joined, persistent, scarious, striate; calyx bristly, veined, membranous, shorter than the corolla, bilabiate, upper lip bilobed, lower lip trilobed; petals clawed; standard orbicular-obovate, usually emarginate; wings oblong-linear or elliptic, sometimes unilaterally eared, pouched, or spurred; keel petals obovate or elliptic, slightly incurved, sometimes unilaterally eared, short, blunt, united at the back, free at the apex; stamens 10, united into a tube split above and below; anthers uniform; ovary linear, stalked, stalk inserted in the center of the receptacle, 2–9-ovuled; disk very short, inconspicuous; style inflexed; stigma small, terminal. Pod stalked, flattened, enclosed by the persistent calyx; segments 1–9, eventually separating, indehiscent; seeds smooth, kidney-shaped. (Bentham and Hooker 1865; Dewit and Duvigneaud 1954a,b)

Kotschya was formerly a subgenus of *Smithia* and so most *Kotschya* species bear *Smithia* synonyms. The absence of spurred stipules on *Kotschya* species is one of the distinctive characters separating the 2 genera (Dewit and Duvigneaud 1954a,b).

This genus is well represented in tropical Africa. About 11 species occur in southern Zaire; 10 species are found in Madagascar. They are common in cleared forests, fallow land, swamps, savanna pastures, steppes, and high plateaus; they are frequently present in the environs of the headwaters of rivers, at altitudes up to 2,000 m.

The flowers of *K. speciosa* (Hutch.) Hepper and *K. ochreata* (Taub.) Dewit & Duvign. are borne in dense little spikes resembling caterpillars. The barrel-shaped, conelike inflorescence of *K. strobilantha* (Welw. ex Bak.) Dewit & Duvign. resembles hop flowers. A tea brewed from leaves of *K. strigosa* (Benth.) Dewit & Duvign. is used in Zaire as a cure for headache and stomach pains. *K. recurvifolia* (Taub.) F. White is a tree characteristic of the high mountain slopes of Kilamanjaro. The wood is useful for minor construction and for making charcoal.

NODULATION STUDIES

The following species are nodulated in Zimbabwe (Corby 1974): *K. capitulifera* (Welw. ex Bak.) Dewit & Duvign. var. *capitulifera*, *K. scaberrima* (Taub.) Wild, *K. speciosa*, *K. strigosa* var. *strigosa*, *K. strobilantha* var. *strobilantha*, and *K. thymodora* (Bak. f.) Wild ssp. *thymodora*.

Kummerowia Schindl.   Papilionoideae: Hedysareae,
                                                            Desmodiinae

Named for Professor Kummerov of Poznan, Poland.

Type: *K. striata* (Thunb.) Schindl.

2 species. Herbs, annual, creeping, much-branched, low, up to about 25 cm tall; stems and branches usually hairy. Leaves palmately 3-foliolate; leaflets obovate-oblong, subentire; stipels none; stipules 2, large, ciliate, soft-membranous, yellow turning brown. Flowers of 2 kinds, chasmogamic (petaliferous) and cleistogamic (apetalous), both borne on the same plant, small, pedicelled, axillary, with 1 bracteolelike bract subtending the calyx; bracteoles persistent, ovate; calyx campanulate; lobes 5, pinnately nerved, ovate, subacute, lower half connate; petals pink or dark violet; standard deltoid to suborbicular; wings oblong to lanceolate; keel blunt, longer than the standard and wings; stamens diadelphous; ovary 1-ovuled. Pod sessile, 2–3-times longer than the calyx, compressed, orbicular or oblong, apex rounded or acute, indehiscent, 1-seeded. (Schindler 1912; Ohwi 1965; Komarov 1972)

This genus of questionable status consists of 2 species more commonly known as *Lespedeza striata*, Kobe lespedeza, and *L. stipulacea*, Korean lespedeza or Korean clover; these species formerly constituted the subgenus *Microlespedeza*. Their segregation was based on annual habit and differences in branching and in floral characters (Schindler 1912). Agronomists in the Western Hemisphere conventionally consider these Asian importations as *Lespedeza* species; Asian botanists generally accept them as members of *Kummerowia*.

The agronomic importance of these species as forage, hay, protein supplement, and green-manuring crops has been adequately described by Pieters (1934), Henson (1957), Henson et al. (1962), Offutt and Baldridge (1973), and others. Each species adapts well to a wide range of soil types and fertility levels but produces its best yield on well-drained, fertile soil with a pH range between 5.8 and 6.2. They are easily established in lone stands or in grass mixtures, are reasonably tolerant of drought, pests, and low soil-fertility levels, and are useful for pasture purposes, rotation patterns, and in soil-erosion control measures.

NODULATION STUDIES

Nodulation of these species as members of the genus *Lespedeza* has been known for decades. Investigators of rhizobia in the Orient have similarly reported them nodulated as *Kummerowia* species (Kubo 1939; Asai 1944).

Kunstleria Prain   Papilionoideae: Dalbergieae,
(363a) / 3835   Lonchocarpinae

Named for H. H. Kunstler; he collected these plants in Malaysia.

Type: *K. curtisii* Prain

9 species. Climbers, woody, often more than 30 m long; branches slen-

der. Leaves pinnately 1–7-foliolate, usually 3- foliolate; leaflets ovoid-lanceolate, short-petioluled, firmly chartaceous, pale green, both surfaces glabrous; stipels none; stipules small, deciduous. Flowers usually dark purple, many, small, solitary, in copious, terminal and axillary thyrsoid panicles; rachis and branches sometimes rusty-pubescent; calyx campanulate, densely pubescent; teeth lanceolate, upper 2 connate into a broad, deltoid, bidentate lip, lower 3 triangular; standard ovate-oblong, entire; keel boat-shaped; keel petals slightly coherent; stamens diadelphous, vexillary stamen free from the others but adnate to the claw of the standard; filaments alternately long and short; anthers versatile, uniform; ovary sessile; ovules few, usually 2; style slightly incurved, stigma terminal, capitate. Pod flattened, thin, elongated, membranous or leathery, sometimes densely brown-pubescent and reticulate, indehiscent, not winged; seeds few, oblong, centrally located in the pod. (Harms 1900; King et al. 1901)

Kunstleria was established to accommodate 5 Malaysian species having the habit of *Spatholobus*, but with leaflets exstipellate, flowers borne on solitary, not fascicled, pedicels, pods indehiscent, and seeds situated centrally, not terminally (King et al. 1901).

Members of this genus are indigenous to Indonesia, Malaysia, and the Philippines. They inhabit dense jungles and lowland woods.

Labichea Gaud. ex DC.   Caesalpinioideae: Cassieae
79 / 3524

Named for M. Labiche, French naval officer; he accompanied Freycinet on his world voyage and died in the passage to the Moluccas.

Type: *L. cassioides* Gaud.

8 species. Shrubs or subshrubs, bushy, glabrous to pubescent, unarmed. Leaves imparipinnate or palmately 3–5-foliolate, rarely 1-foliolate; leaflets small, shiny, rigid, usually sharp-pointed; stipules small, caducous. Flowers yellow, few, in loose axillary racemes; bracts small, caducous; bracteoles none; calyx tube very short; sepals 5 or 4, overlapping, subequal; petals 5 or 4, small, equal, free, spreading; stamens 2; filaments very short; anthers basifixed, oblong-linear, either both alike with terminal pore, or 1 beaked and longer, with pollen only at the base of the tube; ovary sessile or short-stalked, free, 2–3-ovuled; style tapered, short; stigma small, terminal. Pod short-stalked, flattened, oblong or lanceolate, often villous, 2-valved; seeds transverse, few, usually with small, globular, fleshy aril. (Bentham 1864; Bentham and Hooker 1865; Gardner 1941)

Members of this genus are indigenous to Western Australia, Northern Territory, and Queensland. They inhabit rocky sandstone ravines and dry exposed areas.

From Arabic, *lablab*, the popular name for this plant in Egypt.

Type: *L. purpureus* (L.) Sweet (= *Dolichos lablab*, *D. purpureus*, *L. niger* Medik., *L. vulgaris* Savi)

1 species. Herb, perennial, often grown as an annual, erect or climbing, up to about 6 m tall, sometimes bushy, much-branched; stems purplish. Leaves pinnately 3-foliolate, on slender petioles with a swollen, basal pulvinus; leaflets ovate, subglabrous or softly hairy, entire, oblique, on petiolules with thickened pulvini; stipels small, triangular to lanceolate; stipules reflexed, persistent. Flowers purple or white, in erect, long-stalked, stiff, axillary racemes longer than the leaves; rachis swollen at base of pedicel; bracts early caducous; bracteoles short; calyx elongated, campanulate, bilabiate; lobes 5, upper 2 connate, entire or emarginate, lower 3 small; standard orbicular, reflexed, eared at the base, with a deep-channeled claw bordered by 2 parallel callosities; wings oblique-obovate, longer than the keel; keel beaked, incurved sharply at a right angle in the middle; stamens diadelphous; anthers uniform, basifixed; ovary sessile, several-ovuled, encircled by a lobed disk; style laterally compressed, incurved at a right angle, not winged along the margins, softly barbed along the inner margin; stigma terminal, glabrous. Pod variable in size and color, rounded, compressed, obliquely oblong-falcate, beaked, upper suture often knobby, tardily dehiscent; seeds 2–6, compressed, black, brown, or buff, with a long, prominent, raised hilum, partly enveloped by white, arillar tissue. (De Candolle 1825a,b; Wilczek 1954)

This species was segregated from the *Dolichos* complex (Harms 1908a; Verdcourt 1970c).

This monotypic genus is pantropical, and occurs from sea level to 2,000-m altitude in areas of moderate rainfall.

L. purpureus (L.) Sweet, bonavist, Egyptian or hyacinth bean, presumably had its origin in Asia and from there found its way into Africa. Since antiquity it has been cultivated extensively in India, southern China, and Japan, as well as in the Sudan and Egypt. It is a dryland, hardy, drought-resistant plant that does well on poor soil. These characteristics account for its importance in soil improvement, as pasture graze and fodder, as ground cover in coffee plantations and orchards, as a companion crop with corn, wheat, sorghum, and sugar cane, and as a green-manuring crop. The young pods and the boiled or parched beans are high in protein, the amino acid content is well balanced, and the lysine content is comparable to that of poultry eggs. In some regions the cooked beans are incorporated into feed for hogs, cattle, and goats; in Egypt the bean meal is a substitute for broad bean meal in cakes. The crop is grown primarily for hay in the southern United States, Cuba, and Puerto Rico. Introduction of this plant and its widespread cultivation in Brazil has improved the agricultural economy (Schaaffhausen 1963).

NODULATION STUDIES

Prior to 1971 the hyacinth bean was designated *Dolichos lablab* in all nodulation studies. Grobbelaar and Clarke (1974) properly used the name *L. purpureus* in reporting its nodulation in South Africa. In 1974 Corby reported nodules on subspecies *uncinatus* Verdc. Host-infection studies have positioned this species within the cowpea miscellany (Richmond 1926a; Conklin 1936; Raju 1936; Allen and Allen 1939). *L. purpureus*, reported as *Dolichos lablab* (Cloonan 1963), is one of diverse species on which nodules with predominantly purplish to black interiors have been described (Allen and Allen 1936a; Sandmann 1970c; Dobereiner 1971). Cloonan conjectured that the pigment was a melanin formed by phenolase in the nodules; attempts to separate it from leghemoglobin were unsuccessful. This unusual pigmentation appears to be specific in the association of certain rhizobial strains with particular hosts; however, an explanation for its occurrences and significance, if any, remains unknown.

Laburnum Fabr.   Papilionoideae: Genisteae, Spartiinae
217 / 3678

The Latin name for this plant.

Type: *L. anagyroides* Medik. (= *L. vulgare* J. Presl)

4 species. Shrubs or trees, up to 7 m tall, erect, deciduous, unarmed. Leaves palmately 3-foliolate, petioled, pubescent; leaflets elliptic, subsessile; stipels none; stipules minute or none. Flowers golden yellow, many, on short branchlets, in pendulous, leafless, axillary or terminal racemes; bracts and bracteoles small, early caducous; calyx tube very short, campanulate, obscurely bilabiate, lips obtuse, short, upper lip bidentate, longer, lower lip tridentate, usually ciliate; all petals free; standard orbicular or broad-ovate, upcurved, without basal ears or tubercles; wings shorter than standard, obovate; keel very short, incurved, glabrous, not beaked; stamens 10, monadelphous; anthers alternately long and subbasifixed, short and dorsifixed; ovary stalked, many-ovuled; style glabrous; stigma terminal, small. Pod long-stalked, linear, narrow, flattened, sutures thickened or slightly winged, slightly constricted between the seeds, continuous within, 2-valved, tardily dehiscent; seeds several, kidney-shaped. (Bentham and Hooker 1865; Backer and Bakhuizen van den Brink 1963)

Polhill (1976) included this genus in the Genisteae *sensu stricto*.

Members of this genus occur in western Europe, North Africa, and western Asia. They serve as cultivated ornamentals in gardens and as street and park plantings in many parts of Europe.

Laburnum wood is straight grained, moderately durable, and one of the hardest and heaviest of the European timbers. The texture is variable but usually fine. The sapwood is yellowish brown, the heartwood, darkish brown; both have distinctive bandings. The wood is of little, if any, commercial importance but occasionally appears on the market in small supplies. Its use is limited to cabinetmaking, inlays, and turnery, especially for small objects (Titmuss 1965; Rendle 1969).

All *Laburnum* species are toxic; *L. anagyroides* Medik. has long been regarded as one of the most poisonous. The quinolizidine alkaloid, cytisine, $C_{11}H_{14}N_2O$, occurs in all plant parts, especially the seeds (Klein and Farkass

1930; White 1943). This powerful poison causes nausea, convulsions, and death by asphyxiation.

NODULATION STUDIES

Nodules were reported on the type species by Simon as early as 1914, and have since been observed on the 3 other species: *L. alpinum* (Mill.) J. Presl, *L. maritimus* Bigel., and *L. pratense* L. Limited studies define *Laburnum* rhizobia as slow growing and the infection range of the host species as similar to others within the cowpea miscellany (Wilson 1939a).

Lamprolobium Benth.   Papilionoideae: Galegeae,
249 / 3712   Brongniartiinae

From Greek, *lampros*, "bright," "shining," and *lobion*, "pod"; the pod is glabrous.

Type: *L. fruticosum* Benth.

2 species. Shrubs, erect, up to about 2 m tall. Leaves imparipinnate, 3–5–7-foliolate, rarely solitary, silky-pubescent underneath; leaflets in few pairs, or only 1 leaflet; stipels none; stipules small. Flowers small, yellow, short-stalked, 1–3, axillary or terminal; bracts and bracteoles minute, caducous; calyx deep cleft; 2 upper lobes joined nearly to the top; petals almost equal to calyx; standard orbicular, short-clawed; wings oblique-oblong, free; keel blunt, incurved; stamens all joined into a sheath open above; anthers uniform, dorsifixed; ovary short-stalked, several-ovuled; style filiform, incurved; stigma terminal. Pod smooth, stalked, linear-oblong, very flat, 2-valved, septate; valves leathery; seeds oblong, transverse. (Bentham 1864; Bentham and Hooker 1865)

Polhill (1976) assigned this genus to the tribe Bossiaeae in his reorganization of Genisteae *sensu lato*.

Distribution of this genus is limited to Northern Territory and Queensland, Australia.

Lathriogyne E. & Z.   Papilionoideae: Genisteae, Lipariinae
185 / 3645

From Greek, *lathrios*, "covert," and *gyne*, "woman," in reference to the hidden ovary.

Type: *L. parvifolia* E. & Z.

1 species. Shrub, heathlike, small, silky-haired. Leaves simple, flat, entire. Flowers yellow, 2–4, in terminal heads, hidden and crowded amid the leaves; calyx lobes densely villous, subequal, longer than the tube, upper 2 connate above; petals shorter than the calyx; standard ovate; wings oblong, falcate;

keel narrowed into a slender tip or point, swollen on each side; stamens diadelphous; anthers alternately small and versatile, long and basifixed; ovary sessile, 1-ovuled; style incurved; stigma terminal. Pod densely villous, 2-valved, 1-seeded. (Bentham and Hooker 1865)

This monotypic genus is endemic in Cape Province, South Africa.

Lathyrus L.   Papilionoideae: Vicieae
380 / 3854

From Greek, *lathyros*, the name of a pea or pulse.

Type: *L. sylvestris* L.

130 species. Herbs, annual, many perennial, often climbing by means of tendrils; stems winged or angled. Leaves paripinnate; leaflets entire, few, often 2 pairs, rarely numerous pairs; petiole often dilated or phyllodelike, terminating in a branched tendril, bristle, or occasionally a leaflet; stipels absent; stipules commonly foliaceous, semisagittate, rarely entire at the base. Flowers purple, red, pink, white, or yellow, solitary or racemose on axillary peduncles, racemes often 1-sided; bracts minute, very caducous; bracteoles absent; calyx campanulate to turbinate; tube usually oblique, often inflated at the back; lobes subequal or 3 lower ones longer, upper 2 linear, lanceolate, deltoid to ovate; standard broad-obovate or obcordate-orbicular, blade distinct and reflexed, emarginate, short-clawed; wings narrow to broadly falcate-obovate or oblong, free or slightly adherent to the middle of the keel; keel petals shorter than the wings, incurved, clawed, obtuse; stamens 10, vexillary stamen free or partially connate with others, staminal tube truncate at the apex; filaments filiform or dilated above; anthers uniform; ovary subsessile or stalked, usually multiovuled, rarely few-ovuled; style inflexed, flattened dorsally, pubescent along the upper side or glabrous; stigma terminal. Pod compressed, oblong, continuous within, 2-valved, dehiscent; seeds 2–numerous, globose, angular, seldom flattened. (Bentham and Hooker 1865; Boutique 1954)

This genus is closely related to, and often compared with, *Vicia. Lathyrus* species differ from *Vicia* species in having leaflets usually fewer, larger, thicker, and more prominently veined, stems winged, style nongrooved, pubescent, and wings almost or completely free from the keel.

The genus is widespread in the temperate regions of both hemispheres. Its members grow wild and cultivated. They are well adapted to rather dry areas, yet tolerate waterlogging, grow well on poor land, and are resistant to cool weather. They occur in meadows, along seashores, lake and stream banks, roadsides, and in thickets, fields, and waste areas.

L. cicera L., *L. hirsutus* L., *L. ochrus* (L.) DC., *L. sativus* L., and *L. sylvestris* L. are prominent among the species used for soil cover, green manure, erosion control, and rehabilitation of cutover or burned-over land. In India and Java, *L. sativus* is considered one of the most economical pulses for fodder and green manure in rice fields during the cool winter period.

The flowers of *Lathyrus* species are papilionaceously handsome. *L. odoratus* L. is the well-known sweet pea of gardens and florists. *L. latifolius* L., everlasting pea, *L. maritimus* (L.) Bigel., beach pea, *L. splendens* Kell., royal perennial pea, and *L. tingitanus* L., Tangier pea, are valued ornamentals.

The slender herbaceous members of the genus are palatable to livestock and have been used successfully in various parts of the world. In general, however, *Lathyrus* species are viewed suspiciously because of a type of poisoning called lathyrism. *L. ochroleucus* Hook. is reportedly browsed in northern areas by moose, hares, and caribou. *L. ornatus* Nutt., *L. palustris* L., and *L. polymorphus* Nutt. have value as additive nutrient in hay and forage mixtures. *L. sativus* is the cheapest pulse in India, where the annual seed production is about 0.45 million metric tons from about 1.7 million hectares (Purseglove 1968). The starchy root tubers of *L. montanus* (L.) Bernh. (= *L. macrorrhiza* Wimm.) have a sweet astringent taste resembling chestnuts and mild antidiarrheal properties. In some European areas the tubers are boiled and eaten as a potato substitute, or crushed and infused with yeast to make an alcoholic drink.

Lathyrism is an intoxication resulting primarily from overindulgence in eating the seed, but the disease has been linked to the consumption of other plant parts as well. Moderate amounts of seed, consumed in well-balanced diets, are innocuous. The malady has played a significant role in the history of man since the time of Hippocrates (Stockman 1932). Outbreaks have usually occurred in periods of famine. Young adult males are more often afflicted with the disease than females, children, and elderly people. In some instances 6–7 percent of an area population has shown advanced symptoms. An epidemic of spastic paralysis occurred in India as recently as 1944–1945 (Watt and Breyer-Brandwijk 1962). Farm animals vary greatly in their susceptibility to the disease. Horses seem highly susceptible (Clough 1925). Acute fatal poisonings of pigs have occurred after they were turned into lathyrus fields; also, domestic birds have succumbed after brief feeding periods (Stockman 1929).

L. sativus, grass or Indian pea, or chickling vetch, *L. cicera*, flat-pod pea, and *L. clymenum* L., Spanish vetch, have long been considered sources of lathyrism, but ample evidence also documents the following species as toxic when consumed in excess: *L. hirsutus*, *L. latifolius*, *L. odoratus*, *L. pusillus* Ell., *L. sphaericus* Retz., *L. splendens*, *L. strictus* Grover, *L. sylvestris*, and *L. tingitanus* (Kingsbury 1964). Toxic properties become evident about the time the seeds are formed.

Symptoms in man usually appear as a paralysis of the muscles below the knees, pains in the back followed by weakness and stiffness of the legs, and progressive locomotive incoordination. Symptoms in animals resemble those of man, evidenced mainly by weakness and paralysis in the legs and loins.

Explanations for the cause of lathyrism have been extremely speculative and varied. The presence of hydrocyanic acid, a cyanogenetic glucoside, an acid salt of phytic acid, a volatile alkaloid, amino acid deficiencies, manganese poisoning, selenium interference with methionine metabolism, and other factors have all been proposed (Long 1917; Steyn 1934; Watt and Breyer-Brandwijk 1962; Kingsbury 1964).

Symptoms of lathyrism in experimental animals fall into two categories. Hyperexcitability and convulsions usually followed by death characterize the excessive ingestion of the highly toxic seed of *L. latifolius*, *L. splendens*, and *L. sylvestris*. Skeletal lesions are lacking. L-α, γ-diaminobutyric acid monochloride has been isolated and identified as the neuroactive factor from *L. latifolius* and *L. sylvestris* var. *wagneri* (Ressler et al. 1961). Concentrations in seeds of the former species ranged from 0.51 to 0.67 percent; seeds of the latter yielded 1.4 percent. Lameness, paralysis, and skeletal deformity characterize excessive consumption of the less toxic seeds of *L. hirsutus*, *L. odoratus*, *L. pusillus*, *L. strictus*, and *L. tingitanus*. A crystalline substance isolated from *L. pusillus* (Dupuy and Lee 1954) and *L. odoratus* (Schilling and Strong 1955) produced skeletal changes and muscular paralysis in the rat identical with symptoms induced by excessive ingestion of seed. This substance was shown to be β-(*N*-γ-L-glutamyl)-aminopropionitrile, $C_8H_{13}N_3O_3$ (Schilling and Strong 1955).

NODULATION STUDIES

As early as 1879 Frank described the bacterial forms in nodules of *L. pratensis* L. and *L. tuberosus* L. (= *Orobus tuberosus*) as slender, usually branched, long rods. Tschirch's description (1887) of bacteroids in these species was confirmed the following year by Beijerinck (1888). Several years later the bacteroid shape was one of the criteria for separating the rhizobia into 2 species. *Lathyrus* rhizobia were included with those from species of 9 genera to which the name *Rhizobium radicicola* was first applied (Hiltner and Störmer 1903).

The microsymbiont from *Lathyrus*, *Pisum*, and *Vicia* species was named *Rhizobium leguminosarum* Frank in the first edition of Bergey's *Manual of Determinative Bacteriology* in 1923; in the second edition, 1925, it was designated the type species.

Young nodules on *Lathyrus* species are commonly elongated because of a well-defined apical meristem; old nodules show branched tips. Numerous well-defined infection threads are usually conspicuous in the apical areas. Wipf's account (1939) of polyploidy in host cells of *Lathyrus* nodules following release of rhizobia from infection threads and their subsequent multiplication was confirmed by Kodama (1967). Usually more than 80 percent of the mature host cells in *Lathyrus* nodules are filled with bacteria. Peripheral layers of starch cells are often discernible. Juncture of the nodule with the root is delicate because secondary thickening is scant.

Susceptibility of *Lathyrus* species in cross-infection studies is a subject of broader scope than can be properly dealt with here. It suffices to say that of the numerous species known to be nodulated, 3 species constituted the nucleus of the so-called pea cross-inoculation group as early as 1907 (Maassen and Müller 1907). Comparatively few other species have been studied in the ensuing years, but the data of Carroll (1934a), Conklin (1936), Bushnell and Sarles (1937), and Wilson (1939a) confirmed mutual kinships between *Lathyrus*, *Pisum*, *Vicia*, and *Lens* species and their rhizobia. Reciprocal relationships seem to be restricted within the tribe Vicieae.

Only 2 species, *L. decaphyllus* Pursh and *L. graminifolius* (S. Wats.) T. G. White, both native to the southwestern United States, lacked nodules under field conditions in Arizona (Martin 1948). The root systems of these species merit reexamination; both have spreading, horizontal root systems, are drought-resistant, and have excellent potential in erosion control along highways and gullies.

Nodulated species of *Lathyrus*	Area	Reported by
alatus (Maxim.) Kom.	U.S.S.R.	Pers. observ. (U.W. herb. spec. 1970)
alpinus[?]	Italian Alps	Malan (1935)
angulatus L.	S. Africa	N. Grobbelaar (pers. comm. 1963)
annuus L.	S. Africa	Grobbelaar & Clarke (1974)
aphaca L.	Widely reported	Many investigators
articulatus L.	Arctic U.S.S.R.	Roizin (1959)
blepharicarpus Boiss.	Israel	Hely & Ofer (1972)
cicera L.	S. Africa	N. Grobbelaar (pers. comm. 1963)
clymenum L.	Germany	Salfeld (1896)
	S. Africa	Grobbelaar et al. (1967)
crassipes H. & A.	Argentina	Burkart (1952)
davidii Hance	Japan	Asai (1944)
eucosmus Butt. & St. John	Kans., U.S.A.	M. D. Atkins (pers. comm. 1950)
hierosolymitanus Boiss.	Israel	M. Alexander (pers. comm. 1963); Hely & Ofer (1972)
hirsutus L.	Widely reported	Many investigators
japonicus Willd.		
var. *glaber* (Ser.) Fernald	Wis., U.S.A.	Bushnell & Sarles (1937)
var. *pellitus* Fernald	Wis., U.S.A.	Conklin (1936); Bushnell & Sarles (1937)
laevigatus Gren.	Poland	Zimny (1964)
latifolius L.	Widely reported	Many investigators
†*luteus* (L.) Peterm.	Italian Alps	Kharbush (1928)
magellanicus Lam.	Australia	P. H. Graham (Thesis 1963)
maritimus (L.) Bigel.	Widely reported	Many investigators
marmoratus Boiss. & Bl.	Israel	M. Alexander (pers. comm. 1963); Hely & Ofer (1972)
†*montanus* (L.) Bernh.	Italian Alps	Malan (1935, 1938)
var. *tenuifolius*	Italian Alps	Malan (1938)
†*niger* Bernh.	S. Africa	Grobbelaar et al. (1967)
nissolia L.	Netherlands	Beijerinck (1888)
	Great Britain	Pers. observ. (U.W. herb. spec. 1970)
ochroleucus Hook.	U.S.A.	Many investigators

Nodulated species of *Lathyrus* (cont.)	Area	Reported by
ochrus (L.) DC.	Netherlands	Beijerinck (1888)
	Israel	M. Alexander (pers. comm. 1963)
	S. Africa	Grobbelaar & Clarke (1974)
odoratus L.	Widely reported	Many investigators
ornatus Nutt.	N. Mex., U.S.A.	Pers. observ. (1937)
palustris L.	Alaska, U.S.A.	L. J. Klebesadel (pers. comm. 1961)
var. *linearifolius* Ser.	Wis., U.S.A.	Bushnell & Sarles (1937)
var. *myrtifolius* (Muhl.) Gray	Wis., U.S.A.	Bushnell & Sarles (1937)
pisiformis L.	U.S.S.R.	Pärsim (1966)
pratensis L.	Widely reported	Many investigators
pseudocicera Pamp.	Israel	Hely & Ofer (1972)
pusillus Ell.	Ark., U.S.A.	Pers. observ. (U.W. herb. spec. 1970)
sativus L.	Widely reported	Many investigators
sphaericus Retz.	Italy	Pichi (1888)
	Sweden	Pers. observ. (U.W. herb. spec. 1970)
sylvestris L.	Widely reported	Many investigators
tingitanus L.	Widely reported	Many investigators
tuberosus L.	Widely reported	Many investigators
venosus Muhl.	Widely reported	Many investigators
†*vernus* (L.) Bernh.	Widely reported	Many investigators
wagneri[?]	France	Naudin (1897c)

†Also reported as *Orobus* synonyms.

Latrobea Meissn.
179 / 3639

Papilionoideae: Podalyrieae

Named for Latrobe, Governor of the ancient port colony of Macedonia.

Lectotype: *L. brunonis* Meissn.

6 species. Shrubs, heathlike; branches wandlike, glabrous, often pubescent when young. Leaves alternate or scattered, simple, linear, some concave, channeled on upper surface; stipules none. Flowers yellow or purplish, terminal or subaxillary on short branchlets, solitary in heads or umbels; bracts and bracteoles none, or sometimes small and at a distance from the calyx; calyx tube often ribbed; lobes 5, subequal; petals short-clawed; standard round to ovoid, obtuse or acuminate, longer than lower petals; wings narrow; keel straight or slightly incurved, as long as or longer than the wings; stamens free; ovary sessile or stalked, 2-ovuled; style slender or thickened at the base; stigma small, terminal. Pod flat, ovate or lanceolate, 2-valved, with 1–2 kidney-shaped, strophiolate seeds. (Bentham 1864; Bentham and Hooker 1865)

Members of this genus are indigenous to Western Australia.

NODULATION STUDIES

One rhizobial strain from a field-nodulated plant of *L. hirtella* (Turcz.) Benth. in Western Australia (Lange 1959) showed cultural characteristics and host-specificities akin to those of the slow-growing members of the cowpea, lupine, and soybean type (Lange 1961; Lange and Parker 1961).

Lebeckia Thunb.   Papilionoideae: Genisteae, Crotalariinae
200 / 3660

Named for H. J. Lebeck, collector in Indo-Malaya, 1795–1800.
    Lectotype: *L. sepiaria* (L.) Thunb.
    46 species. Woody shrubs or undershrubs, sometimes subherbaceous; stems straight, wandlike, unarmed or much-branched, glabrous, spiny, or silky-pubescent. Leaves simple or palmately 3-foliolate; leaflets linear-filiform, especially when only 1 leaflet develops, rarely elliptic or obovate; petiole sometimes jointed near the middle, rarely sessile; stipules small, usually none. Flowers small or large, yellow to orange red, few or many, in lax, terminal, often unilateral, racemes; bracts and bracteoles small, inconspicuous; calyx obliquely campanulate, glabrous or tomentose; lobes equal or subequal, shorter than the calyx tube, usually dentoid; petals glabrous, rarely hairy; standard ovate, reflexed, with a bent, linear, broad-channeled claw; wings oblong or obovate, eared, with a straight, linear claw; keel blunt, rarely acute or beaked; often gibbous, with a straight, linear claw, longer than the wings and standard; stamens monadelphous, joined into a tube split to the base; anthers alternately short and dorsifixed, longer and basifixed; ovary sessile or stalked, usually linear, several- to many-ovuled, glabrous; style incurved, glabrous; stigma terminal, capitulate. Pod linear-oblong, rarely lanceolate, flat or terete, seldom turgid or long-stalked, sometimes membranous, continuous or thinly septate, usually with a persistent style at the tip, 2-valved, several- to many-seeded. (Bentham and Hooker 1865; Phillips 1951)
    Hutchinson (1964) included this genus in Lotononideae trib. nov.; Polhill (1976), however, assigned it to the tribe Crotalarieae in his revision of Genisteae *sensu lato*.
    Members of this genus are distributed throughout Madagascar, Namibia, and South Africa, in Natal, Namaqualand, and Cape Province. The plants grow on gravelly flats and hillsides and in desert habitats.
    Several species, known as wild broom in South Africa, are used as ornamentals.

NODULATION STUDIES

Limited studies using 1 rhizobial strain from *L. simsiana* E. & Z. indicated host-infection affinities within the cowpea miscellany (Wilson 1944b; Wilson and Chin 1947). The ability of this strain to produce nodules on red clover merits confirmation (Wilson 1946).

Nodulated species of *Lebeckia*	Area	Reported by
cytisoides Thunb.	S. Africa	Grobbelaar & Clarke (1972)
linearifolia E. Mey.	S. Africa	N. Grobbelaar (pers. comm. 1974)
plukenetiana E. Mey.	S. Africa	Grobbelaar & Clarke (1972)
sericea Thunb.	S. Africa	Grobbelaar & Clarke (1972)
simsiana E. & Z.	S. Africa	Grobbelaar & Clarke (1972)
spinescens Harv.	S. Africa	N. Grobbelaar (pers. comm. 1974)
wrightii Bolus	S. Africa	Grobbelaar & Clarke (1972)

Lebruniodendron J. Léonard — Caesalpinioideae: Cynometreae

Named for P. L. Le Brun, botanist; his collection of specimens from the Congo Basin contributed toward the definition of the genus.

Type: *L. leptanthum* (Harms) J. Léonard

1 species. Tree, 20–30 m tall, with brownish scales generally persistent at the base of young branches and inflorescences. Leaves paripinnate; leaflets opposite or subopposite, 2-paired, entire, pinnately nerved, not pellucid-punctate; petiolules twisted; stipels minute, caducous; stipules joined, very caducous, beaked, intrapetiolar. Flowers small, disposed in 2 rows, in axillary or terminal elongated racemes; bracts small, scarious, caducous; bracteoles subopposite, membranous, nonpetaloid, small, caducous; calyx tube cup-shaped, very short, slender; sepals 4, free, distinctly imbricate, reflexed, unequal; petals 5, free, equal, lanceolate; stamens 10, free; anthers dorsifixed, longitudinally dehiscent; ovary flat, glabrous, long-stalked, stalk laterally adnate partway to the wall of the receptacle, 2-ovuled; disk none. Pod asymmetric, flat, apiculate, long-stalked, transversely nerved when young, dehiscent(?). (Léonard 1951, 1952, 1957)

This monotypic genus is endemic in equatorial Africa. The plants grow in mixed forests.

Lecointea Ducke — Caesalpinioideae: Swartzieae

Named for Paul LeCointe, upon whose botanical collection the genus was determined.

Type: *L. amazonica* Ducke

3 species. Trees, medium to large, unarmed; trunks usually deeply scal-

379

loped or fluted. Leaves 1-foliolate, coarsely serrate, large, leathery; stipules caducous. Flowers small, white, in solitary or clustered, often 1-sided racemes, reflexed; bracts persistent; bracteoles small, subpersistent; calyx tube long, campanulate to turbinate, obscurely 5-lobed; petals 5, imbricate, erect, clawed, 4 somewhat unequal, outermost twice as wide as others, all deciduous; stamens 10 or 9, unequal, free; anthers narrow-oblong, basifixed, shorter than the filaments, longitudinally dehiscent by slits; ovary sessile to stalked, 4–6-ovuled; style elongated, exserted in bud; stigma small, oblique, terminal. Pod thick, compressed, leathery, indehiscent, pericarp fleshy, mesocarp thin; seeds 1–2, thick, compressed, rounded or kidney-shaped. (Ducke 1922)

Members of this genus are common in tropical Peru, Brazil, and the lower Amazon region. They are usually found on sandy soils, periodically inundated.

The reddish brown wood of *L. amazonica* Ducke is attractive, has a fine texture, and finishes smoothly with a high polish, but lacks commercial possibilities because of the limited supply. Processing of the trunks is impeded by their fluted condition. The wood is heavier than water and resistant to decay (Record and Hess 1943). Game animals relish the pods.

Lemuropisum H. Perrier   Caesalpinioideae: Eucaesalpinieae

Name refers to a nocturnal, monkeylike mammal, *Lemur cattas*, which relishes the pealike seeds.

Type: *L. edulis* H. Perrier

1 species. Shrub, much-branched. Leaves paripinnate; leaflets few; stipules small, persistent. Flowers subsymmetric, 5–20, white, in terminal racemes; bracts thick, spinelike, caducous; calyx cup-shaped; sepals 5, subequal, valvate; petals 5, subequal, imbricate; stamens 10; filaments hairy near the base; ovary sessile; ovules numerous; style elongated, slender; stigma apical, covered with short papillae. Pod elongated, subcylindric, submembranous, depressed between the seeds, 2-valved, valves twisted upon dehiscence; seeds white, longitudinal, thick, kidney-shaped, hilum small. (Perrier de la Bathie 1938)

The genus is related to *Poinciana* L. and *Colvillea*.

This monotypic genus is endemic in southwestern Madagascar. It occurs in very dry localities on calcareous soil.

The seeds, which resemble hazelnuts, are suitable for human consumption and are especially relished by lemurs that abound in the area. This shrub, with particular reference to its seeds, is recommended for introduction into other areas with similar climatic conditions for the purpose of alleviating periodic suffering from famine (Perrier de la Bathie 1938).

Named for Peter Joseph Lenné, 1789–1866, General Director of the Gardens of the German Crown Prince.

Type: *L. robinioides* Klotz.

6 species. Shrubs or small trees, rarely 10 m tall, unarmed. Leaves imparipinnate; leaflets numerous, small, elliptic, membranous; stipels present; stipules awl-shaped or setaceous, caducous. Flowers small, nodding, purplish or greenish yellow, in short, slender, many-flowered axillary racemes, or in fascicles at old foliar nodes; bracts awl-shaped or setaceous; calyx campanulate; teeth 5, short, upper 2 joined; petals subequal; standard rounded, reflexed, nude inside, short-clawed; wings oblique-oblong, free, eared at the base, short-clawed; keel petals slightly lunate, blunt, united above, with a rounded, basal ear; stamens 10, vexillary stamen free at the base, bent, connate with others into a closed tube; anthers uniform; ovary short-stalked, many-ovuled; style long, spirally coiled at the apex, inner side pubescent; stigma terminal. Pod short-stalked, linear, flat, rough-surfaced, nonseptate within, 2-valved, valves thin, hard, elastically dehiscent; seeds 3–many, dark brown, lens-shaped. (Bentham and Hooker 1865; Rydberg 1924)

Representatives of this genus occur in southern Mexico, Guatemala, Panama, and Belize. They are present along sandy stream beds and in moist forests up to about 350-m altitude.

The hard, medium-textured wood lacks commercial value (Record and Hess 1943).

*Lens Mill.   Papilionoideae: Vicieae
379 / 3853

The Latin name for this pulse; it refers to the double-convex shape of the seeds.

Type: *L. culinaris* Medik.

5 species. Herbs, annual, low, erect or twining. Leaves pinnate; leaflets elliptic or obovate, entire, 2–8 pairs, usually with the terminal leaflet modified into a tendril or bristle; stipels none; stipules semisagittate. Flowers small, white, pale blue, or lilac, few, 1–3, on axillary peduncles, racemose or solitary; bracts and bracteoles absent or rudimentary; calyx campanulate; lobes 5, subequal, large, elongated, usually about twice the length of the tube; standard broad-obovate or suborbicular, emarginate, narrowed into a short, broad claw; wings oblique-obovate, adherent in the middle to the keel; keel petals shorter than the wings, somewhat acute or subrostrate; stamens diadelphous, the sheath with an oblique mouth; anthers uniform, elliptic; ovary subsessile, 2-ovuled; style incurved, slightly flattened dorsally above, pubescent longitudinally on the inner side; stigma apical. Pod compressed, 2-valved, continuous within, 1–2-seeded; seeds compressed, orbicular. (Bentham and Hooker 1865)

The type species is generally listed as *L. esculenta* Moench, the name given to it in 1794 by Moench; however, Lawrence (1949) cited evidence that the name *L. culinaris* Medik. given to it in 1787 by Medicus is legitimate and has priority. This judgment is accepted in *Flora Europaea* (Tutin et al. 1968).

Members of this genus are indigenous to the eastern Mediterranean area and western Asia. They grow well on impoverished soil and on a wide range of loamy soils but are not tolerant of a humid, tropical climate. Species are both wild and cultivated.

In early history *L. culinaris* was known to the Hebrews as *lentiles*, to the Greeks as *phakos*, and simply as *lens* to the Romans. It was one of the first plants to have been brought under cultivation. The role of this pulse in agriculture has been documented by De Candolle (1908), Szafer (1966), Harris (1967), Davis and Plitmann (1970), and many others. Lentils are mentioned in various books of the Old Testament: 2 Samuel 17:27–29; 2 Samuel 23:11; and Ezekiel 4:9. The best-known passage is the story of Esau's sale of his birthright to Jacob for bread and a pottage of lentils, Genesis 25:29–34.

L. culinaris is the only cultivated member of the genus. Zohary (1972) cited the lentil as "one of the primary domesticants that founded the neolithic agricultural revolution in the Near East arc." Four wild species are recognized: *L. ervoides* (Brign.) Grande, *L. montbretii* (Fisch. & May) Davis & Plitm., *L. nigricans* (Bieb.) Godr., and *L. orientalis* (Boiss.) Popov. *L. orientalis* manifests the closest morphological similarity to *L. culinaris*.

Lentils are eaten as a cooked vegetable, as parched seeds, in soups, and as ground flour made into cakes and bread. World acreage is estimated at about 1.4 million hectares (Purseglove 1968). Asia is the world's lentil production center (Youngman 1968). Washington and Idaho, in the United Sates, are the largest lentil-producing areas in the Western Hemisphere. Two cultivars of the type species are recognized: ssp. *macrospermae* (Baumg.) Barul., characterized by large pods, seeds, and flowers, the latter white, rarely blue; and ssp. *microspermae* Barul., with small pods and seeds and small flowers, mostly pink or blue (Purseglove 1968).

NODULATION STUDIES

As early as 1907, Maassen and Müller observed that bacterial isolates from nodules of *L. esculenta* not only shared physiological similarities with those from *Pisum*, *Vicia*, and *Lathyrus* species, but that all of these plants were nodulated by each other's microsymbiont. Burrill and Hansen (1917) confirmed these cross-inoculation affinities among the same genera. Gams (1924) depicted nodules on the roots of *L. culinaris* ssp. *nigricans* (Bieb.) Thell., a form now recognized as *L. nigricans* (Tutin et al., 1968).

Leonardoxa Aubrév.   Caesalpinioideae: Amherstieae

Named for J. Léonard, Belgian botanist, an authority on tropical flora of the Congo Basin.

Type: *L. africana* (Baill.) Aubrév.

3 species. Trees or shrubs. Leaves pinnate; leaflet pairs 2–4. Flowers red

or violet, in axillary short racemes; calyx tubular; lobes 4, broad, imbricate; petals 5, subequal; stamens 10; filaments all joined at the base; ovary stipitate, connate along 1 margin with the calyx tube. Pod oblong-elliptic, flattened, veined, surface smooth; seeds without aril. (Aubréville 1968)

The species were formerly classified erroneously with *Schotia*.

The distribution range of the species is from Guinea to Zaire.

The plants are considered ornamental for their colorful flowers.

Leptoderris Dunn
(366a)

Papilionoideae: Dalbergieae,
Lonchocarpinae

From Greek, *leptos*, "slender," "thin," in reference to the narrow petals and the similarity of the pod to that of *Derris* species.

Type: *L. trifoliolata* Hepper

20 species. Shrubs, usually climbing, or woody vines. Leaves imparipinnate; leaflets elliptic, rounded at the base, subacute at the apex; young leaflets densely tomentose on both sides becoming glabrous above and leathery. Flowers small, narrow, white, yellow, pink, or purple, fascicled on short pedicels, in dense panicles, rarely in racemes; bracts small, caducous; bracteoles small, ovate or awl-shaped; calyx narrow-campanulate, densely silky-haired; lobes 4, short, uppermost lobe bidentate; petals glabrous; standard erect, oblong to oblong-obovate; wings oblique-oblong, upper half partly joined with the slightly incurved keel; stamens 10, monadelphous or subdiadelphous; anthers versatile; ovary sessile, hairy, 1–3-ovuled; style incurved; stigma small, terminal. Pod oblong, rarely suborbicular, flat, membranous, slightly winged on upper suture, indehiscent, 1–2-seeded. (Dunn 1910; Harms 1915a; Baker 1929; Hauman 1954b)

Members of this genus are distributed throughout tropical West Africa, and from Zaire to Zanzibar. They grow in flooded lowlands, light forests, along river banks, and in high rainfall areas.

Leptodesmia Benth.
343 / 3813

Papilionoideae: Hedysareae,
Desmodiinae

From Greek, *leptos*, "slender," "thin," and *desme*, "bundle," in reference to the delicate stems which grow in tufts.

Type: *L. congesta* Benth. ex Bak.

5 species. Shrublets or perennial herbs, diffuse; stems slender, caespitose, 30–60 cm long. Leaves pinnately 1–3-foliolate; leaflets small; stipels present; stipules free. Flowers small, in terminal, short, compact, subcapitate or spikelike racemes; bracts broad, imbricate in bud, caducous; calyx deeply cleft; tube very short; lobes 5, narrow, setaceous, subequal; petals clawed; standard suborbicular, broad-clawed; wings oblique-oblong, free; keel nar-

rower, blunt; stamens 10, diadelphous; anthers uniform; ovary sessile, 1-ovuled; style filiform, smooth; stigma capitate, terminal. Pod small, ovoid, membranous, flattened, enclosed by the calyx, 2-valved, opening along the ventral suture, 1-seeded. (Bentham and Hooker 1865; Hooker 1879)

Hutchinson (1964) positioned this genus in Lespedezeae trib. nov.

Representatives of this genus occur in Madagascar and southern India.

Lespedeza Michx. Papilionoideae: Hedysareae, Desmodiinae
349 / 3820

Named for Vincente Manuel de Céspedes, Spanish Governor of the Florida Colony, 1784–1790, patron of Michaux. Michaux visited the Colony from February to May, 1788. The current spelling of the generic name presumably resulted from illegibility or a printer's error (Ricker 1934).

Type: *L. virginica* (L.) Britt.

140 species. Herbs or shrubs, erect or procumbent, perennial. Leaves pinnately 3-foliolate; leaflets entire, equilateral, petioluled, ovate-elongated, pinnately veined; stipels none; stipules linear or ovate-lanceolate, awl-shaped, setaceous, persistent. Flowers of 2 kinds: chasmogamous (petaliferous) and cleistogamous (apetalous), both often occurring on the same plants, the latter more abundant in some species than in others; chasmogamous flowers showy, papilionaceous, violet, purple, roseate, yellow, or whitish, rarely solitary, usually in few- to many-flowered, loosely arranged to compact, sessile or long-peduncled, axillary fascicles or racemes, sometimes in terminal panicles; bracts linear, small, subtending each pedicel; bracteoles 2, inconspicuous, subtending the calyx; calyx campanulate-cylindric; lobes 5, short, linear-subulate, subequal, persistent; standard oblong-obovate, clawed; wings oblong, slightly curved, clawed, free or slightly adherent to the keel; keel petals obovate, incurved; stamens diadelphous; anthers uniform; ovary sessile or stalked, 1-ovuled; style filiform, incurved; stigma small, terminal. Pod oval or orbicular, sessile or short-stalked, compressed, reticulate, tipped with the remnant style, indehiscent; seed 1, flat, rounded. (Bentham and Hooker 1865, in part; Isely 1955; Wilbur 1963)

About 125 species are indigenous to eastern Asia and 15 to the southeastern United States. In 1912 Schindler segregated the only 2 annual species, *L. striata* (Thunb.) H. & A., Kobe lespedeza, and *L. stipulacea* Maxim., Korean lespedeza or clover, to establish the genus *Kummerowia*. Differences in branching, floral characters, and growth habits were the basis for this decision. Agronomists in the Western Hemisphere usually consider these 2 Asian importations as *Lespedeza* species; Asian botanists generally consider them members of *Kummerowia*.

Lespedeza species are held in high esteem as forage, hay, protein-supplement and green-manuring crops, and in preventing soil erosion (Pieters 1934; Henson 1957; Henson et al. 1962; Offutt and Baldridge 1973). Kobe and Korean lespedezas and *L. cuneata* G. Don, sericea lespedeza, a perennial, are the 3 most important members of the genus in the United States. *L. hedysaroides* (Pall.) Kitagawa, *L. juncea* (L. f.) Pers., and *L. latissima* Nakai are other perennial species having forage value. Although generally inferior to Kobe and Korean lespedezas in adaptability, ease of culture, range of use, and palatability, these 3 species are more tolerant of drought, low fertility, and sandy soils (Henson et al. 1962; Offutt and Baldridge 1973). *L. capitata* Michx., *L. hirta* (L.) Hornem., *L. procumbens* Michx., *L. stuevei* Nutt., *L. violacea* (L.) Pers., and *L. virginica* (L.) Britt. are among those species especially valued for drought-resistance. *L. bicolor* Turcz., *L. cyrtobotrya* Miq., *L. formosa* (Vog.) Koehne, *L. japonica* L. H. Bailey, and *L. thunbergii* (DC.) Nakai have considerable potential in soil conservation, as wildlife food and shelter, and as ornamental shrubs (Henson et al. 1962).

Tryptophane-derived alkaloids having uterus-contracting or hallucinogenic properties have been isolated in Japanese laboratories from *L. bicolor* var. *japonica* (Goto et al. 1958; Morimoto and Oshio 1965; Morimoto and Matsumoto 1966).

NODULATION STUDIES

Inclusion of lespedezas in the cowpea miscellany dates back to the reports of Burrill and Hansen (1917), Tracy and Coe (1918), and Whiting and Hansen (1920), and was supported, in principle, by the findings of Wilson (1939a). Certain practical conclusions can be drawn from Erdman's study (1950) of 14 strains from 7 perennial species and 12 strains from the 2 annual species: (1) lespedeza rhizobia show marked host specificity relative to effectiveness; (2) effective plant responses of perennial species are most likely to result with rhizobial inocula from perennial species; similarly, isolates from the 2 annual species should be used homologously; and (3) highly effective inoculants should be selected strain mixtures to ensure best results.

Nodulated species of *Lespedeza*	Area	Reported by
angustifolius (Pursh) Ell.	Mass., U.S.A.	Pers. observ. (U.W. herb. spec. 1971)
bicolor Turcz.	Hawaii, U.S.A.	Allen & Allen (1936a)
	Japan	Asai (1944)
var. *japonica*	Japan	Pers. observ. (1962)
var. *rosa*	U.S.A.	Conklin (1936)
capitata Michx.	Wis., U.S.A.	Bushnell & Sarles (1937)
	N.J., Mass., U.S.A.	Pers. observ. (U.W. herb. spec. 1971)
cuneata G. Don	Japan	Asai (1944)
	Taiwan	Pers. observ. (1962)
cyrtobotrya Miq.	Japan	Asai (1944)
	S.C., U.S.A.	Erdman (1950)
daurica Schindl.		
var. *schimadae*	Kans., U.S.A.	M. D. Atkins (pers. comm. 1956)
formosa (Vog.) Koehne	N.Y., U.S.A.	Wilson (1939a)
frutescens (L.) Britt.	Mass., U.S.A.	Conklin (1936)
fruticosa (L.) Britt.	N.Y., U.S.A.	Wilson (1939a)

Nodulated species of *Lespedeza* (cont.)	Area	Reported by
hedysaroides (Pall.) Kitagawa	Kans., U.S.A.	M. D. Atkins (pers. comm. 1956)
hirta (L.) Hornem.	Widely reported	Many investigators
intermedia (S. Wats.) Britt.	N.Y., U.S.A.	Wilson (1944b, 1945a)
	Mass., U.S.A.	Pers. observ. (U.W. herb. spec. 1971)
juncea (L. f.) Pers.	Ohio, U.S.A.	L. W. Erdman (pers. comm. 1950)
latissima Nakai	Ohio, U.S.A.	L. W. Erdman (pers. comm. 1950)
leptostachya Engelm.	Wis., U.S.A.	Burton et al. (1974)
pilosa S. & Z.	Japan	Asai (1944)
procumbens Michx.	Ark., Ky., U.S.A.	Pers. observ. (U.W. herb. spec. 1970)
repens (L.) Bart.	Ala., U.S.A.	Pers. observ. (U.W. herb. spec. 1970)
	Wis., U.S.A.	Burton et al. (1974)
sericea (Thunb.) Benth.	Fla., U.S.A.	Carroll (1934a)
	Hawaii, U.S.A.	Allen & Allen (1936a)
sieboldi Miq.	France	Lechtova-Trnka (1931)
†*stipulacea* Maxim.	Widely reported	Many investigators
†*striata* (Thunb.) H. & A.	Widely reported	Many investigators
stuevei Nutt.	Wis., U.S.A.	Burton et al. (1974)
thunbergii (DC.) Nakai	Penn., U.S.A.	Schramm (1966)
villosa Pers.	Wis., U.S.A.	Burton et al. (1974)
(= *L. tomentosa* Sieb.)		
violacea (L.) Pers.	Widely reported	Many investigators
virgata DC.	Wis., U.S.A.	Burton et al. (1974)
virginica (L.) Britt.	Widely reported	Many investigators

†Also accepted as *Kummerowia* species.

Lessertia DC.
288 / 3756

Papilionoideae: Galegeae, Coluteinae

Named for Baron Jules Paul Benjamin De Lessert, 1773–1847, patron of botany; he maintained a botanical museum in Paris.

Lectotype: *L. perennans* DC.

60 species. Undershrubs or weak prostrate herbs, usually woolly-gray-pubescent, rarely glabrous. Leaves imparipinnate; leaflets small, entire, linear or elliptic, glabrous above; stipels none; stipules minute. Flowers reddish or dull purple, rarely white, in pedunceled axillary racemes, rarely solitary; bracts small; bracteoles minute or none; calyx campanulate, glabrous or hairy; lobes 5, ovate, subequal, shorter than the tube; all petals short-clawed; standard rounded, notched, erect or reflexed; wings oblong, usually eared, clawed, similar to keel petals; keel blunt, clawed, straight or incurved, often shorter than the standard; stamens diadelphous; anthers uniform; ovary subsessile or stalked, 2- to many-ovuled, sometimes pubescent; style incurved, tapering,

bearded in front below the terminal stigma. Pod ovoid or elongated, thinly membranous, flat or inflated, sub-2-valved, splitting tardily only at the apex; seeds many, kidney-shaped. (Baker 1929; Phillips 1951)

Representatives of this genus occur in tropical and southern Africa. Common on hillsides and sandy flats, they are often found near the seaside.

Several species are suspected of being poisonous to stock (Phillips 1926; Watt and Breyer-Brandwijk 1962).

Nodulated species of *Lessertia*	Area	Reported by
affinis Burtt Davy	S. Africa	Grobbelaar & Clarke (1975)
annularis Burch.	S. Africa	N. Grobbelaar (pers. comm. 1974)
benguellensis Bak.	S. Africa	N. Grobbelaar (pers. comm. 1974)
brachystachya DC.	S. Africa	Grobbelaar et al. (1967)
depressa Harv.	S. Africa	Grobbelaar & Clarke (1972)
emarginata Schinz	Zimbabwe	Corby (1974)
falciformis DC.	Zimbabwe	Corby (1974)
harveyana Bolus	S. Africa	Grobbelaar & Clarke (1972)
herbacea (L.) Druce	S. Africa	Grobbelaar & Clarke (1972)
incana Schinz	S. Africa	Grobbelaar & Clarke (1972)
pauciflora Harv.	S. Africa	Mostert (1955)
perennans DC.	Zimbabwe	Corby (1974)
var. *polystachya* (Harv.) L. Bolus	S. Africa	Grobbelaar & Clarke (1972)
rigida E. Mey.	S. Africa	Grobbelaar & Clarke (1972)
stenoloba E. Mey.	S. Africa	Grobbelaar & Clarke (1975)
stricta Bolus	S. Africa	Grobbelaar & Clarke (1975)
thodei Bolus	S. Africa	Grobbelaar & Clarke (1972)

Leucaena Benth.

12 / 3447

Mimosoideae: Eumimoseae

From Greek, *leukos*, "white," in reference to the color of the flowers.

Lectotype: *L. leucocephala* (Lam.) de Wit (= *glauca* (L.) Benth.)

50 species. Shrubs, small, erect, or trees, medium to large, unarmed. Leaves abruptly bipinnate; pinnae pairs numerous; leaflet pairs many and small, or few and larger; petiole usually with a conspicuous gland below the lowermost pinnae; stipules minute and caducous, or nearly spinescent and persistent. Flowers minute, sessile, usually bisexual, yellowish white or greenish, in many-flowered heads borne on long, axillary peduncles or terminally racemose-fasciculate; bracts 2 below the flower heads or near apex of

peduncle; calyx tubular-campanulate, small, short-dentate; lobes 5, minute, equal, triangular; petals 5, narrow, acute or rounded at apex, narrowed at base, minutely pubescent outside, free or with coherent margins; stamens 10, free, exserted; filaments filiform; anthers ovate, oblong, versatile, eglandular, often pilose; ovary stalked, many-ovuled; style long filiform; stigma small, terminal, slightly dilated. Pod short-stalked, thin, broad-linear, compressed between the transverse seeds, not septate, many-seeded; valves 2, membranous or thinly leathery, thickened at the margins, dehiscent; seeds brownish or black, ovate, compressed. (Bentham 1842; Bentham and Hooker 1865; Sargent 1949)

According to Gillis and Stearn (1974), the correct lectotype is *L. latisiliqua* (L.) Gillis, with *L. leucocephala* (Lam.) de Wit and *L. glauca* (L.) Benth. as synonyms. These names have been used interchangeably in this book.

Members of this genus are widespread in tropical and subtropical North and South America, Africa, and Polynesia. About 13 species are endemic in Mexico. Species are both wild and cultivated. They are well adjusted to a variety of soils, waste lowlands, wet and dry areas; the plants are often well established on steep rocky slopes, in rocky limestone soils, and in volcanic-developed regions.

Leucaena species other than the lectotype have attracted little attention. The wood of *L. brachycarpa* Urb., *L. guatemalensis* Britt. & Rose, *L. salvadorensis* Standl., and *L. shannoni* Donn. Smith is hard, heavy, and attractive, but it is not available in commercial quantities. The wood is used locally for fuel, charcoal, and durable heavy construction purposes. Both *L. esculenta* (Moc. & Sessé) Benth. and *L. pulverulenta* (Schlechtd.) Benth., trees of northern Mexico and southwestern Texas, reach heights of 15–20 m, with trunk diameters of 50 cm or more. The latter species is commonly called the lead tree because of its hard, dark brown wood.

L. leucocephala (= *L. glauca*) (de Wit 1961; Wilbur 1965), native to Mexico, but now pantropic, is known as koa haole throughout Polynesia and as ipilipil in the Philippines. This remarkable, deep-rooted, arborescent shrub, or small tree, was disseminated from Mexico into the Philippines and Indonesia during the days of Spanish occupation. It propagates readily from cuttings, stumps, or root collars, establishes itself readily, and coppices freely. Maximum elevation for the best cultivation is about 250 m above sea level. In conjunction with its prolific seeding habit (about 264,000 seeds per kg), the seed germination rate is very high, and, accordingly, the plants often form an almost impenetrable thicket. Because of the deeply penetrating taproot system and the right-angled radiation of the lateral roots, the plants are provided with secure anchorage and obtain nutrients from the lower soil strata (Dijkman 1950). Rapidly growing taproots of a 1-year-old plant often reach depths of more than 2 m; a taproot of a 5-year-old plant reached a depth of 5 m.

Koa haole is valuable for soil erosion control, soil and water conservation, reforestation, and land reclamation. It is a good cover, or contour hedge, on terrains subject to windblown erosion. Because of its deep anchorage and its ability to shade out undergrowth annuals, koa haole has been used successfully on coffee, tea, quinine, oil palm, coconut, rubber, and cacao plantations.

The extensive root system and leaf fall of this species contribute to its soil-improvement properties. The total nitrogen content of young leaves may reach 4 percent and ranges from 1.5 to 2.5 percent in mature leaves. Interplanting of this species in land reclamation and reforestation areas in Indonesia resulted in a 100 percent increase of teak. Bimonthly pruning and topping of 1,000 mature trees provided 36,000 kg/ha of wet leaves and twigs per annum, calculated· to have a fertilizer equivalent of 1,000 kg of nitrogen as ammonium sulfate and 100 kg of phosphoric acid as double superphosphate (*fide* Dijkman 1950). Yields are similarly impressive in Mauritius (Anslow 1957) and in Hawaii (Takahashi and Ripperton 1949), where koa haole is grown with various grasses and managed as a grass-legume pasture (Lyman et al. 1967), and is well regarded for high protein yield, vitamin A content, and palatability to livestock.

Numerous reports have favored the use of koa haole as a source of crude protein feed for beef and dairy cattle (Henke 1945, 1958; Kinch and Ripperton 1962). In Hawaii its production as green foliage and dehydrated feed was cheaper than that of alfalfa; also, its feed quality was superior. The hydrated product was also found to be superior to alfalfa in poultry-feeding experiments, as evinced by a higer carotene, vitamin A content, egg production, hatchability and a lower chick mortality (Palafox and Quisenberry 1948).

Distrust of koa haole as an all-purpose feedstuff had its basis in the use of plant parts as a fish poison and in the loss of hair and other symptoms of ill health after ingestion by nonruminants. Beef and dairy cattle show no ill effects, but depilatory effects are particularly prominent among mules, donkeys, swine, rabbits, and horses. Suspicion that the plants extracted selenium from the soil prevailed for many years because of the symptomatic similarities.

The investigation into the cause of these health abnormalities, particularly alopecia in nonruminants, gained impetus with the isolation of leucaenol by Mascré (1937). This optically inactive crystalline substance has the same empirical formula as the optically active compound, mimosine, isolated from *Mimosa pudica* (Renz 1936).

The agronomic potential of *L. glauca* as forage in the Australian tropics was explored comprehensively by Hutton and Gray (1959). Impressive dry matter and protein yields of this species in the Brisbane area were reported by Hutton and Bonner (1960). Mimosine concentrations ranging from 2 to 5 percent were observed in a world collection of 72 *Leucaena* lines analyzed by Brewbaker and Hylin (1965). Cross-breeding of low mimosine lines was recommended. Mimosine was absent in species of 10 other mimosoid genera.

Seeds of *L. leucocephala* yield 8.8 percent oil consisting of palmitic,·stearic, behenic, lignoceric, oleic, and linoleic acid (Farooq and Siddiqui 1954).

NODULATION STUDIES

Three species are nodulated: *L. leucocephala* (= *L. glauca* (L.) Benth.) in diverse geographical habitats, *L. glabrata* Rose in South Africa (N. Grobbelaar personal communication 1973), and *L. insularum* (Guill.) Dänik. var. *guamensis* in Guam (V. Cushing personal communication 1966). Nodules were recorded on the type species by Wright, in Sri Lanka, as early as 1903.

Well-nodulated plants of koa haole growing in barren, volcanic soil and also in thickly vegetated areas were commonly observed by the authors in Hawaii. Invariably the nodules were large, multilobed, and located on the upper, central taproot systems. Effective symbiosis was always evident. Histological sections of nodules depicted a highly functional type wherein about 40–60 percent of the bacteroid area was packed with rhizobia. Stone cells were usually abundant in the nodule cortical layer; accordingly, the nodules tended to be firm and persistent; however, perennial nodules were not observed.

Rhizobial isolates from nodules of *L. leucocephala* studied by the authors were more similar culturally and physiologically to the slow-growing types of cowpea, soybean, and lupine species than to fast-growing types. Observations by Galli (1959) were in general agreement. Strains resembling fast-growing rhizobia in colony characteristics and carbohydrate utilization were described by Trinick (1965); 1 of 3 strains studied in detail produced a brown to black pigment on yeast-extract mannitol agar; 2 of these strains produced nodules on some, but not all, of the alfalfa seedlings grown in bottles of nutrient agar. All of these crosses were ineffective. Reisolated strains maintained their cultural and antigenic identities and their ability to nodulate *Leucaena* after passage through alfalfa plants.

Wilson (1939a) recorded the nodulation of *L. glauca* by rhizobia from species of 9 genera. In tests conducted in Hawaii (personal observation), *L. glauca* was nodulated by strains from *Desmanthus virgatus*, *Mimosa pudica*, *Prosopis juliflora*, and 2 *Sesbania* species, but not by cowpea, soybean, and lupine strains. All plant responses were effective. In reciprocal tests, all *Leucaena* rhizobia nodulated cowpea, *D. virgatus*, and *P. juliflora* effectively, but not all strains nodulated *M. pudica* and *Sesbania grandiflora*; these latter responses were ineffective. Rhizobia from 17 species of 12 genera in the cowpea miscellany tested at the University of Pretoria (1954) failed to nodulate *L. leucocephala*. Reciprocal tests with a strain from this *Leucaena* species failed to nodulate any of these 17 plant species. Two strains of *L. leucocephala* rhizobia used by Galli (1958) were only host-specific in tests of 20 species. Two general conclusions have been drawn from the broad spectrum tests conducted by Trinick (1968): (1) *L. leucocephala* is nodulated only by fast-growing, culturally similar rhizobia from tropical species, not by slow-growing rhizobial strains; and (2) fast-growing rhizobial strains of *L. leucocephala* rhizobia nodulate, often effectively, tropical leguminous species whose rhizobia are of the slow-growing type. *L. leucocephala* failed to nodulate in New South Wales field trials, where 18 of 24 other species of tropical legumes were nodulated effectively by indigenous rhizobia in the soil (McLeod 1960).

Leucomphalos Benth.   Papilionoideae: Sophoreae
151 / 3611

From Greek, *leukos*, "white," and *omphalos*, "navel," "funicle," in reference to the conspicuous white aril of the seed.

Type: *L. capparideus* Benth.

1 species. Shrub or small tree up to 5 m tall; branches smooth, slender, cylindric. Leaves simple, ovate-elliptic, large, papery, both sides reticulate; stipels none; stipules inconspicuous. Flowers white, short-pedicelled, in axillary panicles near tips of branches; bracts and bracteoles small; calyx ovoid, short-toothed, membranous, splitting along 1 side as the flower expands; petals subequal, subsessile; standard broad-obovate; wings linear-oblong; keel petals not joined, broader than the wings; stamens free; anthers linear, longer than the filaments; ovary long-stalked, few-ovuled; style awl-shaped, incurved; stigma small, terminal. Pod long-stalked, falcate-ovate, turgid, beaked, leathery, 2-valved, dehiscent; seeds 1–2, deep red, with a thick, spongy, white aril. (Bentham and Hooker 1865; Pellegrin 1948)

This monotypic genus is endemic in southern Nigeria, Gabon, and Cameroon. It occurs in wooded areas.

Leucostegane Prain   Caesalpinioideae: Amherstieae
(47a) / 3497a

From Greek, *leukos*, "white," and *stege*, "shelter," in reference to the white calyx lobes enclosing the much shorter petals.

Type: *L. latistipulata* Prain

2 species. Trees, small; branches slender, pubescent. Leaves paripinnate; leaflets large, lanceolate-acuminate, 6 pairs; stipels small; stipules large, foliaceous, each with a strong midrib, intrapetiolar, connate at the base. Flowers clustered on rugose, woody nodes in few-flowered fascicles or racemes on old branches; bracts and bracteoles caducous; calyx funnel-shaped, white; tube fleshy, cylindric; lobes 4, narrow, obtuse, oblong, imbricate in bud, later reflexed; petals 2, small, much shorter than the calyx lobes, inserted in the calyx throat between the upper and lateral sepals; stamens only 2 perfect, opposite the lateral sepals, with large ovoid anthers, longitudinally dehiscent; staminodes 2, short, erect, sterile, between and above the fertile stamens; ovary stalked, oblong, stipe adnate to the calyx tube, exserted, oblong, pubescent along the margins; style filiform, coiled in bud; stigma oblique. Pod oblong, compressed, apiculate. (King et al. 1901; Harms 1908a)

The genus is closely related to *Saraca* and *Lysidice*.

Members of this genus are native to the Malay Peninsula. They are common in forested hill areas.

Leycephyllum Piper   Papilionoideae: Phaseoleae, Galactiinae

From Greek, *leukophyllos*, "white-leaved."

Type: *L. micranthum* Piper

1 species. Climbing shrub. Leaves trifoliolate; leaflets large, entire; stipules striate. Flowers small, yellowish, numerous, in dense racemes from

the axils of the upper leaves; calyx campanulate, upper lip short, bidentate, lower lip 3-toothed, the median tooth as long as the calyx tube, lateral ones short; standard obovate, stipitate, the upper margin incurved or hooded, the base not callose or eared, basal margins thickened; wings oblong, stipitate, the ear somewhat hooklike; keel oblong-obovate, stipitate, not eared; vexillary stamen free, its filament enlarged at the base, the other stamens united below, free above the middle; anthers oval; style curved, glabrous; stigma terminal, very oblique, minute; ovary pubescent. Pod not described. (Piper 1924c)

The genus was previously thought to be related to *Calopogonium* (Piper 1924c) but has recently been transferred to *Rhynchosia*, subtribe Cajaninae (Grear 1978).

This monotypic genus is endemic in Costa Rica.

Librevillea Hoyle   Caesalpinioideae: Amherstieae

Named for Libreville, Gabon, where the plant was first collected.

Type: *L. klainei* (Pierre ex Harms) Hoyle

1 species. Tree, about 30 m tall, trunk buttressed. Leaves imparipinnate; leaflets alternate, rarely opposite, asymmetrically elliptic, acuminate, not pellucid-punctate, pinnately nerved, finely veined; rachis grooved; stipules intrapetiolar, connate, subtruncate, persistent. Flowers small, on slender pedicels in terminal and axillary panicles; bracts minute; bracteoles sepaloid, valvate, sheathing the flower; persistent; calyx cup-shaped, with a disk; sepals 2, small, petaloid; petals absent; stamens usually 10, all fertile and joined briefly at the base; anthers subglobose, dorsifixed; ovary short-stalked, stalk free, 2–3-ovuled; style filiform; stigma very small. Pod short-stalked, oblong, flattened, hard, woody, sutures somewhat thicker, not winged, 2-valved, terminating in a short, sharp point, dehiscent; seeds 1–2, flat, oblong-elliptic. (Léonard 1957)

This monotypic genus, endemic in equatorial Africa, is especially common in Gabon and Angola.

Liparia L.   Papilionoideae: Genisteae, Lipariinae
182 / 3642

From Greek, *liparos*, "brilliant," in reference to the color of the flowers.

Type: *L. sphaerica* L.

4 species. Shrubs or shrublets. Leaves simple, rigid, leathery, sharp-pointed, palmately nerved, sessile; stipels and stipules absent. Flowers usually yellow to orange, densely clustered in nodding heads becoming black when dry; bracts large, conspicuous, imbricate, forming a conspicuous rosette; calyx short-campanulate, intruse at the base, sometimes densely pilose; lobes 5, shorter than the tube, 4 upper lobes small, acute, joined higher up until

midway, lowest lobe large, blunt, petaloid; standard elliptic-oblong, rounded or notched at the apex; wings oblong; keel petals small, narrow, acute; stamens unequal, diadelphous; filaments linear, glabrous; anthers linear or ovate, sometimes subsagittate, attached near the base; ovary sessile or short-stalked, densely villous, few-ovuled; style filiform; stigma small, terminal. Pod ovate or oblong, compressed; valves 2, leathery, nonseptate within. (Bentham and Hooker 1865; Phillips 1951; Polhill 1976)

Polhill (1976) assigned this genus to the tribe Liparieae in his reorganization of Genisteae *sensu lato* and designated *L. splendens* (Burm. f.) Bos & de Wit the type species.

Members of this genus are limited to South Africa. They are found primarily in the southwestern districts of the Cape Province. The plants grow on hillsides and mountain slopes.

Listia E. Mey. 　　　　　　Papilionoideae: Genisteae, Crotalariinae
198 / 3658

Named for Friederich Ludwig List, of Tilsit, specialist in the botany of Lithuania in the early half of the 19th century.

Type: *L. heterophylla* E. Mey.

1 species. Perennial herb with deep taproot and prostrate habit. Leaves palmately 3-foliolate, petioled; leaflets variable in size; stipules unequal. Flowers yellow in terminal racemes; bracts small; bracteoles none; calyx campanulate, glabrous; lobes shorter than the tube, 4 upper lobes fused in pairs, lowest lobe free and narrower than the others; standard ovate, broad-clawed; wings oblong, curved; keel petals pouched, incurved, longer than the standard, narrow-clawed; stamens monadelphous, connate into a sheath split to the base; anthers unequal, alternately short and versatile; longer and basifixed; ovary sessile, many-ovuled; style incurved, glabrous; stigma terminal, oblique. Pod linear, compressed, twisted, and folded repeatedly from side to side within the persistent dried keel; seeds many. (Bentham and Hooker 1865; Phillips 1951)

This species is very often confused with *Lotononis bainesii*. Polhill (1976), who regards the genus as untenable, has renamed the plant *Lotononis listii* nom. nov. (= *Listia heterophylla* E. Mey.). *Listia* is included here as a separate entry because the name appears in nodulation literature. However, high interspecificity of red strains of rhizobia from *Lotononis bainesii* and *Listia heterophylla* reinforces Polhill's position.

This monotypic genus is endemic in Namibia, South Africa, and Zimbabwe.

NODULATION STUDIES

Nodules were observed on *L. heterophylla* in South Africa (Mostert 1955), in Australia (Norris 1959c), and in Zimbabwe (Corby 1974). This species is known as yellow or hop clover in some localities.

In 1959 Norris reported that rhizobia isolated from this species produced red-pigmented colonies comparable to those of Brisbane strain CB376 from *Lotononis bainesii* (Norris 1959c). In plant tests conducted in Zimbabwe, *Listia heterophylla* was nodulated by strain CB376 and a red-pigmented strain isolated locally from *Lotononis angolensis*.

Loddigesia Sims   Papilionoideae: Genisteae, Cytisinae
223 / 3684

Named for Conrad Loddiges, 1738–1826, English botanist.

Type: *L. oxalidifolia* Sims

1 species. Shrub, small, glabrous. Leaves palmately 3-foliolate; leaflets rounded-obovate; stipules free. Flowers pale lavender or purple and white, in terminal racemes; calyx broad, intrusive at the base; teeth short, subequal; standard suborbicular, broad, callous inside on the short claw; wings incurved obtuse; keel petals incurved, obtuse, longer than the standard; stamens monadelphous; anthers alternately versatile and longer, subbasifixed and shorter; ovary substalked, many-ovuled; style incurved, glabrous; stigma terminal. Pod linear, ovate-lanceolate, compressed, sutures slightly thick, 2-valved, continuous inside, few-seeded. (Bentham and Hooker 1865)

The genus closely resembles *Hypocalyptus*, except that the standard of *Loddigesia* flowers is shorter than the wings and keel.

This monotypic genus is endemic in Cape Province, South Africa.

NODULATION STUDIES

Nodules were observed on *L. oxalidifolia* Sims in its native habitat (Grobbelaar and Clarke 1975).

Loesenera Harms   Caesalpinioideae: Amherstieae
(62a) / 3514

Named for Ludwig Eduard Theodor Loesener, 1865–1941, German botanist.

Type: *L. kalantha* Harms

4 species. Shrubs or small trees, up to 20 m tall. Leaves paripinnate; new leaves drooping, pink to red; leaflets 1–5 pairs, leathery, oblique-oblong, acuminate, midrib incurved, upper pairs large, stalks short, twisted. Flowers pink or white, in elongated, tomentose, terminal racemes or racemose panicles; bracts soon caducous; bracteoles 2, broad-ovate, not valvate but enfolding the flower bud, subpersistent; calyx funnel-shaped, slightly glandular at the base; lobes 4, imbricate, tomentose outside; petals 5, rarely 4, 3 large, subsimilar, obovate, short-clawed, the others minute, narrow, acute; stamens 10; filaments free, filiform, often hairy at the base; ovary 2–4-ovuled, oblong, velvety-haired, stalk adnate to the side of the calyx tube; stigma minute,

capitellate. Pod flattened, hard, woody, broad-oblong, base truncate, apex rounded or pointed, rusty-tomentose, pericarp dividing into 2 layers, the outermost smooth, thick, the innermost rust-colored; seeds 2–4, flat, round. (Harms 1897; Pellegrin 1948; Léonard 1951)

Representatives of this genus are distributed throughout tropical West Africa, Nigeria, Gabon, and Ivory Coast. They flourish in moist or swampy valleys, and along river banks.

Leaf and bark decoctions are said to have native uses for treating leprosy and syphilis (Dalziel 1948).

Lonchocarpus HBK
363 / 3834

Papilionoideae: Dalbergieae, Lonchocarpinae

From Greek, *lonche*, "spear," and *karpos*, "fruit," in reference to the lance-shaped pod.

Lectotype: *L. punctatus* HBK

175 species. Trees, small to medium, woody, scrambling shrubs, or lianas, unarmed; tree trunks up to 90 cm in diameter, often slightly buttressed; stems and leaf rachis often pubescent. Leaves imparipinnate; leaflets elliptic, 1–15 pairs. Flowers numerous, white, purple, or violet, in simple or branched, axillary racemes, or in terminal panicles, pedicels fascicled or paired; bracts small, caducous; bracteoles 2, caducous or persistent; calyx campanulate-truncate; lobes short or obsolete, 2 uppermost ones longer and connate; standard orbicular-obovate, often silky-pubescent, with 2 basal ears, often clawed; wings oblique-oblong, clawed, eared, slightly adherent to the keel above the claw; keel obtuse, curved or straight, eared, clawed; stamens monadelphous, or vexillary stamen free only at the base; anthers versatile; ovary sessile or stalked, silky-pubescent, 2- to many-ovuled; style incurved, filiform; stigma small, terminal. Pod oblong, linear-lanceolate, flat, thin, membranous or leathery, some minutely hairy, often prominently nerved on the margins, the upper suture sometimes laterally dilated, indehiscent; seeds 1–4, rarely numerous, flat, kidney-shaped or round. (Bentham and Hooker 1865; Pittier 1917; Amshoff 1939; Hauman 1954b)

Species of this large genus are well distributed in tropical America, the West Indies, tropical Africa, Madagascar and Australia, but are uncommon in Asia. Many species are indigenous to the Amazon Basin. They occur in forests, mangrove swamps, and coastal districts, and on river banks, open hillsides, and dry plains. Their presence is usually restricted to belts below 1,000-m altitude. Best growth is obtained on well-drained soil in a semitropical climate receiving a well-distributed rainfall.

The species are known by a wide variety of common names: lancepod, bitter wood, turtle bone, cabbage bark, waterwood, tree lilac, and others. *L. griffonianus* Dunn, *L. latifolius* (Willd.) HBK, *L. leucanthus* Burk., and *L. violaceus* (Jacq.) HBK are among those prized as ornamentals because of their medium size, dense foliage, evergreen bushcrowns, profusion of fragrant flowers, and graceful growth. *L. capassa* Rolfe, one of the "rain trees" of the northern Transvaal and Zimbabwe, is host for a species of the froghopper insect, *Ptyelus grossus*, order Hemiptera (Coates Palgrave 1957; Palmer and Pitman 1961). These insects obtain nourishment by piercing the wood, sucking up the sap, and ejecting almost pure water. Thomas Baines referred to this "rain tree" in his *Goldfields Diaries*, as "perhaps the same mentioned by Livingstone as harbouring an insect which distils water. The fluid appears perfectly pure and some of the trees have quite visible pools, reflecting the sky, beneath them" (*fide* Coates Palgrave 1957). This species has been the subject of various native superstitions. Cattle and game browse the leaves.

The leaves and roots of *L. cyanescens* Benth., a straggly shrub whose leaves turn blue black when dried, yields a blue indigolike dye used to color native textiles in Nigeria.

Lonchocarpus wood has no commercial possibilities. The reddish brown heartwood is sharply demarcated from the thick yellow sapwood. The wood is hard to heavy and has a density of 700–945 kg/m³. It is very coarse and not easy to work, but is quite durable (Record and Hess 1943). Its use is reserved primarily for fence posts, fuel, and charcoal.

The insecticidal and piscicidal properties of *Lonchocarpus* species have been of great interest from early times. Roark's comprehensive monographs (1936, 1938) are extremely revealing and have not been excelled in historical depth. As cited by Roark, de Rochefort as early as 1665 wrote of a certain wood, presumably a *Lonchocarpus* species, in the Antilles, that was cut into pieces, or beaten, and thrown into the water to stupefy fish. References to the use of *Lonchocarpus* plant parts for this purpose were frequent in the writings of South American explorers and historians during the 18th and 19th centuries. During this period toxic species of *Lonchocarpus* acquired the name "cubé," which was adopted as the trade name for the dried roots of commerce. They are also known under the following provincial names: "timbo" in Brazil, "haiari" in Guyana, "nekoe" in Surinam, and "barbasco" in Spanish-speaking areas (Holman 1940). All these names denote fish-stupefying plants and are not necessarily distinctive for *Lonchocarpus* species.

In 1892 and 1895 Geoffroy reported the isolation of nicouline, a colorless crystalline compound with a melting point of 162°C, from *Robinia nicou*, now correctly named *L. nicou* (Aubl.) DC. Nicouline is now known to be identical with rotenone, $C_{23}H_{22}O_6$.

The roots of 2 clones of *L. utilis* A. C. Smith grown at Mayaguez, Puerto Rico, yielded 14.02 and 16.5 percent rotenone (Moore 1938). The principal sources of the commercial supply of rotenone are from roots of *L. nicou*, *L. urucu* Killip & A. C. Smith, and *L. utilis* (Higbee 1948). Hermann (1947a) regarded *L. nicou* as a variety of *L. utilis*. The yield from leaves is commercially unimportant. Rotenone in the root is present mainly, if not entirely, in the root latex or sap. Roots containing abundant thick white latex are high in rotenone content. According to Holman (1940), "this fact has been successfully used in the field to distinguish roots high in rotenone from those with low rotenone content when the species to which the root belonged was unknown and chemical tests were impracticable." Between 1940 and 1945 yields were so large and profits so satisfactory that *Lonchocarpus* roots became the most important export of the Amazon region of Peru (C. M. Wilson 1945).

The total tonnage produced exceeded that of rubber. In 1946 the United States imported about 5.5 million kg of *Lonchocarpus* roots, chiefly from the Amazon region of Peru and Brazil and the Orinoco Valley of Venezuela (Higbee 1948).

Nodulation Studies

Wright reported nodules on *L. latifolius* as early as 1903. Rhizobia isolated from nodules located on the central taproot systems of *L. latifolius* and 3 other *Lonchocarpus* species were effective on *L. sericeus* (Poir.) HBK, the test species for this genus, and also nodulated cowpea, *Vigna sinensis* (Allen and Allen 1936a). Cultural, physiological, and biochemical characteristics of the strains were typical of the slow-growing strains of so-called cowpea rhizobia (Allen and Allen 1939). Two strains from *L. domingensis* (Pers.) DC. were generally effective on diverse species of the cowpea miscellany. Both strains were ineffective on *Acacia koa*, *Samanea saman*, and *Stylosanthes guianensis*, and failed to nodulate *Phaseolus lunatus* and *Stizolobium utile*. Strains of rhizobia from cowpea, soybean, *Mucuna deeringiana* (= *Stizolobium deeringianum*), and *Albizia julibrissin* nodulated plantlets of *L. discolor* Huber (Wilson 1939a). Accordingly, host-infection studies to date merit inclusion of *Lonchocarpus* species within the cowpea miscellany.

Presumably, the absence of nodules on *L. leucanthus* (Rothschild 1970) was the result of an unfavorable environment, in the light of a report of positive nodulation occurring on *L. nitidus* (Vog.) Benth. in another locality in Argentina.

Nodulated species of *Lonchocarpus*	Area	Reported by
bussei Harms	Zimbabwe	Corby (1974)
capassa Rolfe	Zimbabwe	Corby (1974)
discolor Huber	U.S.A.	Wilson (1939a)
domingensis (Pers.) DC.	Hawaii, U.S.A.	Allen & Allen (1936a)
eriocalyx Harms		
ssp. *wankieensis* Mend. & Sousa	Zimbabwe	Corby (1974)
ernestii Harms	Venezuela	Barrios & Gonzalez (1971)
latifolius (Willd.) HBK	Malaysia	Wright (1903)
	Hawaii, U.S.A.	Allen & Allen (1936a)
	Trinidad	DeSouza (1966)
nelsii (Schinz) Schinz ex Heering & Grimme	Zimbabwe	Corby (1974)
nitidus (Vog.) Benth.	Argentina	Rothschild (1970)
sericeus (Poir.) HBK	Hawaii, U.S.A.	Allen & Allen (1936a)
violaceus (Jacq.) HBK	Hawaii, U.S.A.	Allen & Allen (1936a)

Lophocarpinia Burk. Caesalpinioideae: Eucaesalpinieae

From Greek, *lophos*, "crest," and *karpos*, "fruit," in reference to the crested, winged pod; the ending *inia* signifies a close systematic relationship with *Caesalpinia*.

Type: *L. aculeatifolia* Burk.

1 species. Shrub, woody, up to 1 m tall; branches spreading, stiff, with apical spine. Leaves paripinnate; leaflets minute, 2–3 pairs, sessile, obovate; stipels very small, awl-shaped; stipules ovate, tips bristly. Flowers small, yellow, few, in short corymbs at the base of short, arrested branches; bracts minute, bracteoles absent; calyx fleshy, widely campanulate at the base; lobes 5, imbricate, lowermost lobe large, concave; petals 5, free, imbricate, uppermost petal innermost; stamens 10, free, declinate; filaments pilose; anthers dorsifixed; ovary linear, short-stalked, several-ovuled; stigma small, terminal. Pod long-stalked, exserted from the persistent calyx, incurved, 1–5-jointed, joints elongated, 4-winged by the extension of the margins, wings crested; seeds elongated, kidney-shaped. (Burkart 1957)

This monotypic genus is endemic in Argentina and Paraguay. The plants occur in xerophytic habitats.

Lotononis E. & Z.   Papilionoideae: Genisteae, Crotalariinae
197 / 3657

Name is a combination of the generic names *Lotus* and *Ononis*.

Type: *L. prostrata* (L.) Benth.

110–120 species. Shrubs, rarely shrublets or herbs, plants of various habit, woody or herbaceous, sometimes prostrate creepers, some forming a cushion, hairy or glabrous, often with a deep taproot, unarmed. Leaves usually petioled, palmately 3-foliolate, rarely 4–5- or 1-foliolate; leaflets linear or obovate-linear, glabrous or hairy; stipules minute or foliaceous, solitary or paired. Flowers small, yellow, solitary, umbellate or racemose, leaf-opposed, axillary or terminal, often on long peduncles, sometimes massed into a head; bracts and bracteoles present; calyx campanulate or narrow-turbinate, often hairy and membranous and slightly inflated; lobes 5, ovate or acuminate, 4 upper lobes joined into 2 pairs, lowest lobe longer, narrower, free, rarely subequal; standard round to oblong, claw sometimes apically bicallous; wings oblique-oblong, linear-clawed, eared at the base; keel petals fused along the back, often gibbous; stamens monadelphous, unequal, staminal tube cleft to the base, rarely joined above; anthers alternately short and dorsifixed, long and basifixed; ovary sessile, rarely stalked, linear, few- to many-ovuled; style curved, linear; stigma minutely capitate. Pod oblong to linear, compressed or swollen, smooth or hairy, 2-valved, continuous inside; seeds numerous, attached by a long funicle. (Bentham and Hooker 1865; Dümmer 1913; Phillips 1951)

Hutchinson (1964) placed this genus in Lotononideae trib. nov. in his revision of Genisteae *sensu lato*.

Members of this large genus are found primarily in South Africa. Species also occur in tropical Africa, Mediterranean Europe, southwestern Asia, the deserts of the lower Jordan Valley and Palestine, and in dry areas of Spain and Greece.

Lotononis species have value as soil-improvement and erosion-control plants. *L. bainesii* Bak., introduced into Queensland, Australia, in 1952 as a potential pasture legume, has since spread to various other Pacific areas, particularly Taiwan.

L. carnosa Benth., *L. involucrata* Benth., and *L. laxa* E. & Z. contain hydrocyanic acid in concentrations toxic to sheep and stock (Steyn 1929; Watt and Breyer-Brandwijk 1962). Decoctions of plant parts of *L. calycina* Benth., *L. ornata* Dümm., *L. rehmannii* Dümm., and *L. versicolor* Benth. are used in South African folk medicine as bronchitis remedies.

NODULATION STUDIES

Information on *Lotononis* rhizobia stems from Norris' isolation (1958b) of the well-known Brisbane strain CB376 from desiccated nodules of *L. bainesii* that were sent to him by J. P. Botha, Union Department of Agriculture, South Africa. With the exception that red-pigmented growth is produced on solid media, this strain was found typical of slow-growing rhizobia in every respect. In the ensuing years this stable pigmented strain has become known in the literature as the "red *Rhizobium*."

Norris (1958b) described the nodulation habit of *L. bainesii* as "similar to that of *Lupinus* spp. Nodules begin as a small spherical body, but by division of the meristem they grow around and enclasp the root until finally they may surround it like a bead on a string, and may be removed only by breaking them. The upper part of the strong tap-root is usually enclosed in such a mass of nodule tissue, only a few small nodules appearing on lateral roots."

The host range of *Lotononis* species for rhizobia and, correspondingly, the infective range of *Lotononis* rhizobia are highly specialized. In a test with 31 species of 21 genera, strain CB376 produced large, effective nodules only on its homologous host. Tiny ineffective nodules were produced in abundance on *Aeschynomene indica* and sparsely on *Crotalaria striata* and *C. lanceolata*, but *C. incana* and *C. intermedia* were not nodulated.

Sandmann's results (1970a) from a complicated series of plant-infection tests using both pigmented and nonpigmented *Lotononis* isolates emphasized further specificities. Only red-pigmented strains from *L. bainesii* and *L. angolensis* Welw. ex Bak. were effective on *L. bainesii*; 1 strain from each species also benefited *Listia heterophylla*. None of the 18 white-pigmented isolates from *L. angolensis*, *L. bainesii*, and *L. stipulosa* (Bak. f.) Schreib. nodulated *L. bainesii*; these isolates, plus 10 others from *Crotalaria* species, were ineffective on *L. angolensis*. The nodulation of *Listia* supports Polhill's view (1976) that the characters claimed for the segregation of *Listia* from *Lotononis* are untenable. He proposed *Lotononis listii* Polh. nom. nov. (= *Listia heterophylla* E. Mey.).

Nodulated species of *Lotononis*	Area	Reported by
angolensis Welw. ex Bak.	S. Africa	Grobbelaar et al. (1967)
azurea Benth.	S. Africa	Grobbelaar & Clarke (1975)
bainesii Bak.	S. Africa	U. of Pretoria (1955–1956)
calycina Benth.	S. Africa	Grobbelaar et al. (1964)
var. *hirsutissima* Dümm.	S. Africa	Grobbelaar & Clarke (1975)

Nodulated species of *Lotononis* (cont.)	Area	Reported by
corymbosa Benth.	S. Africa	N. Grobbelaar (pers. comm. 1974)
crumaniana Burch. ex Benth.	S. Africa	N. Grobbelaar (pers. comm. 1974)
cytisoides Benth.	S. Africa	Grobbelaar & Clarke (1975)
depressa E. & Z.	S. Africa	Grobbelaar & Clarke (1975)
divaricata Benth.	S. Africa	Grobbelaar & Clarke (1975)
eriantha Benth.	S. Africa	Grobbelaar & Clarke (1972)
	Zimbabwe	Corby (1974)
florifera Dümm.	S. Africa	Grobbelaar & Clarke (1972)
hirsuta Schinz	S. Africa	Grobbelaar & Clarke (1972)
humifusa Burch. ex Benth.	S. Africa	N. Grobbelaar (pers. comm. 1963)
involucrata Benth.	S. Africa	Grobbelaar & Clarke (1972)
laxa E. & Z.	S. Africa	Mostert (1955)
	Zimbabwe	Corby (1974)
var. *multiflora* Dümm.	S. Africa	Grobbelaar & Clarke (1972)
leobordea Benth.	S. Africa	Mostert (1955)
macrosepala Conrath	S. Africa	Grobbelaar & Clarke (1975)
magnistipulata Dümm.	S. Africa	N. Grobbelaar (pers. comm. 1974)
mucronata Conrath	S. Africa	Grobbelaar & Clarke (1975)
orthorrhiza Conrath	S. Africa	Grobbelaar & Clarke (1975)
peduncularis Benth.	S. Africa	Grobbelaar & Clarke (1972)
platycarpa (Viv.) Pichi-Serm.	S. Africa	Grobbelaar & Clarke (1975)
prostrata Benth.	S. Africa	Grobbelaar & Clarke (1972)
sericoflora Dümm.	S. Africa	Grobbelaar & Clarke (1975)
serpentinicola Wild	Zimbabwe	Corby (1974)
steingroeveriana Dümm.	S. Africa	N. Grobbelaar (pers. comm. 1974)
stipulosa (Bak. f.) Schreib.	Zimbabwe	Corby (1974)
	S. Africa	Grobbelaar & Clarke (1975)
tenella E. & Z.	S. Africa	Grobbelaar & Clarke (1975)
trichopoda Benth.	S. Africa	N. Grobbelaar (pers. comm. 1963)

From Greek, *lotos*, a name given to several fodder plants.

Type: *L. corniculatus* L.

100–120 species. Herbs or undershrublets, often woody at the base, erect or prostrate, annual or perennial. Leaves 4–5-foliolate, rarely palmate; leaflets entire, 3 crowded at the apex of the petiole, 1–2 near the stem closely resembling stipules; stipules minutely tuberculate or absent. Flowers yellow, often tinged or streaked with purple, red, or rose, umbellate on axillary peduncles, rarely solitary; bracts trifoliaceous; bracteoles usually lacking; calyx cylindric to campanulate, subbilabiate; lobes 5, subequal or lower lobes longer; all petals free from staminal tube but sometimes adnate to the base of the calyx tube; standard obovate, suborbicular or ovate-acuminate, clawed or sessile; wings obovate, adherent to the keel, sometimes eared, clawed and gibbous; keel incurved, petals usually fused along both margins, pouched on each side, beaked; stamens diadelphous; alternate filaments dilated at apex; anthers uniform; ovary sessile, many-ovuled; style inflexed; stigma terminal or lateral, sometimes capitate. Pod linear to oblong, straight or curved, cylindric, turgid, rarely plano-compressed, 2-valved, many-seeded, usually septate between the seeds; seeds subglobose or lens-shaped. (Bentham and Hooker 1865; Gams 1924; Isely 1951; Komarov 1971)

Lotus species are native to both the Old World and the New World. This large genus has 2 principal geographic centers of speciation: (1) the Mediterranean environs with a southward spread around the Sahara Desert and westward throughout the temperate areas to Asia; and (2) western North America (Meusel and Jåger 1962). Species of agronomic importance seemingly originated in Europe and adjacent areas of Africa and Asia. The genus is best represented in the north temperate regions of the Old World.

The trefoils, or deervetch, show a great diversity in form. The species are adapted to habitats ranging from sea level to over 3,000-m altitude, and from wet soils to xerophytic desert conditions. *L. humistratus* Greene, *L. rigidus* (Benth.) Greene, *L. salsuginosus* Greene, and *L. tomentellus* Greene are representative of those found on the sandy flats and the dry, rocky slopes and mesas of the western United States. *L. creticus* L., *L. cytisoides* L., and *L. halophilus* Boiss. & Sprun. typify those that grow exceptionally well in saline, sandy, seashore habitats. *L. palustris* Willd. and *L. tenuis* Waldst. & Kit. ex Willd. are common to ditches, river banks, and swampy ground. *L. alamosanus* (Rose) Gentry and *L. wrightii* (Gray) Greene are desirable for soil erosion control. *L. glinoides* Del. and *L. lanuginosus* Vent. frequent the hot desert areas of Palestine.

L. corniculatus L., birdsfoot trefoil, and *L. pedunculatus* Cav., big trefoil, are the major agronomic species. The following comment pertains to the nomenclatural dichotomy that prevails relative to the latter species and *L. uliginosus* Schk.: "The confusion cannot easily be resolved, because it is permissible to take either a broad or a narrow view of species limits in this group. In the broad view, *L. pedunculatus* Cav. *sensu lato* is the correct name for the entire *pedunculatus-uliginosus* group, including the cultivated forms. In the narrow view, *L. pedunculatus* Cav. *sensu stricto* can be used only for the species described by Cavanilles from the mountains of Spain (present distribution outside Spain and Portugal unknown), in which case *L. uliginosus* Schkuhr is the correct name for the widespread form(s). Where no qualification is given, it is not clear in what sense the name is being used" (Forde 1974). Other botanists (Chrtkova-Zertova 1966; Lainz 1969) consider the two entities separate and distinct species. This nomenclatural and taxonomic problem requires further study.

Both birdsfoot trefoil and big trefoil are valuable browse, pasture, and cover plants, but their agronomic properties are quite different. Birdsfoot trefoil is winter-hardy, drought-resistant, long lived, deep rooted, and a prolific seeder. Its foliage is leafy, fine stemmed, palatable, and nutritious, and thus the plants are especially useful for hay, silage, and pasture forage. In contrast, big trefoil is limited to regions with mild winter temperatures, spreads rhizomatously and stoloniferously, is comparatively shallow rooted, and is best adapted to acidic, poorly drained soils (Hughes 1962; Seaney 1973). In recent years big trefoil has become an important perennial addition to the grassland economy of New Zealand as a pioneer plant on peat and pumice, in hill country and areas not suitable to white clover (Barclay 1957).

NODULATION STUDIES

Hughes (1962) credited Culpeter (1650) [sic; for Culpeper? original not seen] with making the following observation regarding birdsfoot trefoil: "The roots do carry small white knots or kernels amongst the strings . . ." If one accepts these white knots as nodules and the strings as secondary or adventitious roots, this early observation of nodules is of historical interest.

Lotus nodules are generally round, 4–7 mm in diameter, and firm, and occur in abundance on the primary and secondary roots when the plants are properly inoculated. Efficient *Lotus* rhizobia are not ordinarily common in soils where these species have not grown previously. Wilson and Westgate (1942) reported a nitrogen content of 8.32–9.20 percent in effective nodules.

Early studies led to the concept that the host-infective range of certain rhizobial strains was limited only to *Lotus* species and *Anthyllis vulneraria*. With few exceptions (Erdman and Means 1950, Lynch and Sears 1950), rhizobial host-specificity studies have centered on *L. corniculatus* and *L. uliginosus* because of their forage and pasture potential. From these studies it was concluded that effective strains on the former species were generally ineffective on the latter, and vice versa. Effectively nodulated plants of the latter species reportedly fix considerably more nitrogen than the former (Erdman and Means 1950; Gavigan and Curran 1962). Gershon (1961) postulated the involvement of a number of genes in the effective reaction.

Data discrediting the distinctiveness of *Lupinus* species and *Ornithopus sativus* as one cross-infective group and *Lotus* species and *Anthyllis vulneraria* as another were shown by Jensen (1963, 1964). Normal slow-growing strains of *Rhizobium lupini* studied by him nodulated birdsfoot trefoil and *Anthyllis vulneraria* ineffectively, or semieffectively. Fast-growing strains of birdsfoot trefoil rhizobia and *A. vulneraria* nodulated *Ornithopus sativus* ineffectively. These findings were confirmed and extended in other examples of reciprocal and nonreciprocal infectiveness and effectiveness (Gregory and Allen 1953; Brockwell and Neal-Smith 1966; Jensen 1967; Jensen and Hansen 1968).

The microsymbionts of *Lotus* and *Anthyllis* species are now regarded as strains of *Rhizobium lupini*. Because their infective spectrum is rather heterogeneous, Jensen (1967) concluded we are "faced with a cross-inoculation group that begins to vie in promiscuity with the almost 'classical' cowpea (*Vigna sinensis*) group."

Nodulated species of *Lotus*	Area	Reported by
alamosanus (Rose) Gentry	Ariz., U.S.A.	Martin (1948)
americanus (Nutt.) Bisch.	Kans., U.S.A.	M. D. Atkins (pers. comm. 1950)
	U.S.D.A.	Erdman & Means (1950)
angustissimus L.	U.S.D.A.	Erdman & Means (1950)
	Australia	Brockwell & Neal-Smith (1966)
arabicus L.	Zimbabwe	Corby (1974)
argenteus Webb & Berth.	Libya	B. C. Park (pers. comm. 1960)
australis Andr.	Australia	Harris (1953); Norris (1959b)
(*caucasicus* Kupr.) = *corniculatus* L.	S. Africa	Grobbelaar & Clarke (1974)
	U.S.A.	Burton et al. (1974)
coccineus Schlechtd.	Australia	Beadle (1964)
conimbricensis Brot.	France	Clos (1896)
corniculatus L.	Widely reported	Many investigators
var. *ciliatus*	Ariz., U.S.A.	Martin (1948)
var. *japonicus* Regel	Japan	Ishizawa (1953)
crassifolius Pers.	N.Y., U.S.A.	Wilson (1944b)
creticus L.	Libya	B. C. Park (pers. comm. 1960)
	Israel	M. Alexander (pers. comm. 1963)
cruentus Court	Australia	Brockwell & Neal-Smith (1966)
cystisoides L.	Libya	B. C. Park (pers. comm. 1960)
discolor E. Mey.	S. Africa	Grobbelaar & Clarke (1975)
ssp. *mollis* Gillett	Zimbabwe	Corby (1974)
edulis L.	Libya	B. C. Park (pers. comm. 1960)
	Australia	Brockwell & Neal-Smith (1966)
formosissmus Greene	Calif., U.S.A.	Pers. observ. (U.W. herb. spec. 1970)
frondosus Freyn	U.S.A.	Burton et al. (1974)
grandiflorus Greene	Ariz., U.S.A.	Martin (1948)

Nodulated species of *Lotus* (cont.)	Area	Reported by
greenei (Woot. & Standl.) Ottley	Ariz., U.S.A.	Martin (1948)
halophilus Boiss. & Sprun.	Libya	B. C. Park (pers. comm. 1960)
	Israel	Hely & Ofer (1972)
hispidus Desf.	Scandinavia	Eriksson (1874)
	Australia	Brockwell et al. (1966)
	Calif., U.S.A.	Pers. observ. (U.W. herb spec. 1971)
	S. Africa	Grobbelaar & Clarke (1975)
humistratus Greene	Ariz., U.S.A.	Martin (1948)
junceus (Benth.) Greene	Calif., U.S.A.	Pers. observ. (U.W. herb. spec. 1971)
lamprocarpus Boiss.	U.S.D.A.	Erdman & Means (1950)
major Scop.	W. Australia	Cass Smith (1941)
	S. Africa	N. Grobbelaar (pers. comm. 1963)
maroccanus Ball	Australia	F. W. Hely (pers. comm. 1959); J. Brockwell (pers. comm. 1965)
micranthus Benth.	Calif., U.S.A.	Jepson (1925); Pers. observ. (U.W. herb. spec. 1971)
namulensis Brand	Zimbabwe	Corby (1974)
nevadensis (S. Wats.) Greene	Calif., U.S.A.	Pers. observ. (U.W. herb. spec. 1971)
nudiflorus (Nutt.) Greene	Calif., U.S.A.	Pers. observ. (U.W. herb. spec. 1971)
oblongifolius (Benth.) Greene	Calif., U.S.A.	Pers. observ. (U.W. herb. spec. 1971)
ornithopodioides L.	France	Naudin (1897c)
	U.S.D.A.	Erdman & Means (1950)
parviflorus Desf.	S. Africa	Grobbelaar & Clarke (1974)
pedunculatus Cav.	Egypt	Moustafa (1963)
peregrinus L.	Israel	M. Alexander (pers. comm. 1963); Hely & Ofer (1972)
pinnatus Hook.	Idaho, U.S.A.	Pers. observ. (U.W. herb. spec. 1971)
purpureus Webb	U.S.A.	Burton et al. (1974)
purshianus (Benth.) Clem. & Clem.	Calif., U.S.A.	Pers. observ. (U.W. herb. spec. 1971)
pusillus Medik.	Australia	G. P. M. Wilson (pers. comm. 1964)
requieni Mauri	U.S.A.	Burton et al. (1974)
rubellus (Nutt.) Greene	Calif., U.S.A.	Pers. observ. (U.W. herb. spec. 1971)
scoparius (Nutt.) Ottley	Calif., U.S.A.	Vlamis et al. (1964)

Nodulated species of *Lotus* (cont.)	Area	Reported by
strictus Fisch. & Mey.	U.S.A.	Burton et al. (1974)
strigosus (Nutt.) Greene	Calif., U.S.A.	Pers. observ. (U.W. herb. spec. 1971)
suaveolens Pers.	U.S.D.A.	Erdman & Means (1950)
subbiflorus Lag. ssp. *castellans* (B. & R.) Ball	S. Africa	Grobbelaar & Clarke (1974)
subpinnatus Lag.	France	Clos (1896)
	Calif., U.S.A.	Pers. observ. (U.W. herb. spec. 1971)
tennuifolius [?]	Scandinavia	Eriksson (1874)
tenuis Waldst. & Kit. ex Willd.	U.S.A.	Burton et al. (1974)
tetragonolobus L.	S. Africa	Grobbelaar et al. (1967)
	U.S.A.	Burton et al. (1974)
torreyi (Gray) Greene	Calif., U.S.A.	Pers. observ. (U.W. herb. spec. 1971)
uliginosus Schk.	Widely reported	Many investigators
weilleri Maire	U.S.A.	Burton et al. (1974)
wildii Gillett	Zimbabwe	Corby (1974)
wrangellianus F. Muell.	Calif., U.S.A.	Pers. observ. (U.W. herb. spec. 1971)

Luetzelburgia Harms

Papilionoideae: Sophoreae

Named for Dr. Ph. von Luetzelburg; he collected specimens in Brazil.

Type: *L. pterocarpoides* Harms

5 species. Trees, medium to large, or shrubs. Leaves imparipinnate; leaflets few, large, alternate or subopposite, smooth, leathery; stipels and stipules absent. Flowers small, on short pedicels, red or violet, in panicled racemes; bracts ovate, caducous, slightly acuminate; bracteoles paired, subtending the calyx; calyx obliquely funnel-shaped, villous; lobes 5, upper 2 somewhat joined; petals 5, oblanceolate, subequal, oblong, long-clawed, margins wrinkled or pleated; standard silky outside; stamens 10, free, or 9 with vexillary stamen aborted; filaments glabrous, alternately longer and shorter; anthers small, dorsifixed; ovary stalked, 1-ovuled; style smooth. Pod samaroid, hard, obovate-elliptic, with a narrow lateral wing and topped with a large, unilateral, oblong, spreading wing, 1-seeded. (Harms 1922)

Floral characteristics resemble those of *Myroxylon*; the winged pods are similar to those of *Tipuana*.

Members of this genus occur in Brazil, south of the Amazon Basin. They grow in upland forests.

The heartwood of *L. trialata* Ducke is yellowish brown, prominently striped, coarse, very hard, and durable (Record and Hess 1943).

From the generic name *Lupinus* and Greek, *phyllon*, "leaf," in reference to the *Lupinus*-like, palmate leaves.

Type: *L. lupinifolium* (DC.) Gillett (= *Tephrosia lupinifolia*)

1 species. Herb, slender, perennial, rootstock woody with radiating shoots giving rise to roots at the nodes. Leaves palmately 3–7-, mostly 5-foliolate, long-petioled, resembling those of *Lupinus* species; leaflets narrow-oblanceolate, apiculate, sensitive, folding during the heat of the day; stipels 2, lateral to leaflets, filiform, awl-shaped, persistent, recurved; stipules subulate-lanceolate. Normal flowers pink mauve in terminal racemes; flowers in lower lateral racemes sometimes cleistogamous apetalous; bracts small, linear; calyx campanulate, equally 5-lobed; lobes triangular, acute; standard broad-obovate, short-clawed; wings clawed, oblong, eared; keel oblique-elliptic; stamens monadelphous; free part of filaments filiform; anthers equal, ellipsoid, basifixed; ovary subsessile, several-ovuled; style bent at a right angle, slightly hairy towards the base. Normal aerial pods broad-linear, about 5-seeded, short-beaked, with spongy septa between the seeds; seeds sub-reniform, faintly reticulate, with hilum near the middle; cleistogamous flowers with pods 1–2-seeded, probably remaining buried in the soil. (Original not seen; see Hutchinson 1964)

This monotypic genus is endemic in tropical southern Africa. The plants grow in sandy places.

NODULATION STUDIES

Corby (1974) reported this species nodulated in Zimbabwe.

Lupinus L.   Papilionoideae: Genisteae, Spartiinae
211 / 3672

From Latin, *lupus*, "wolf," in reference to the belief that the plants take over the land and destroy, or exhaust, the fertility of the soil.

Type: *L. albus* L.

150 species. Herbs, annual or perennial, rarely shrubs. Leaves generally palmately 5–15-foliolate, very rarely 3- or 1-foliolate; leaflets entire, linear or oblanceolate; stems and petioles often hairy; stipules adnate to the base of the petiole. Flowers blue, violet, white, or variegated, rarely yellow, large, showy, scattered or whorled in erect, terminal racemes or spikes; bracts often caducous; bracteoles persistent, usually adnate to the base of the calyx; calyx deeply cleft; lobes 5, upper 2 connate into a bidentate lip, lower 3 into an entire or 3-toothed lip; standard erect, orbicular or broad-ovate with reflexed margins; wings falcate-oblong or obovate, connate at apex; keel incurved, beaked, enclosed within the wings; stamens 10, monadelphous; anthers alternately short and versatile, long and basifixed; ovary sessile, 2- to many-ovuled; style incurved, glabrous; stigma terminal, often bearded. Pod oblong, more or less compressed, usually silky-haired, septate between the seeds, valves thick and leathery, dehiscent; cotyledons thick and fleshy. (Bentham and Hooker 1865)

Polhill (1976) included this genus in Genisteae *sensu stricto*.

Species of this large genus are distributed throughout North and South America, southern Europe, Mediterranean regions, and North Africa. There are no representatives indigenous to Australia. Species are both wild and cultivated. They are commonly found on dry sites, stony slopes, and calcareous and sandy-loam soils. Most species prefer well-drained soils in warm climates.

Lupines were a subject of much discussion in the early literature of the Egyptians, Greeks, and Romans. A flour for bread was made from the seeds by the very poor people; the seeds also served for medicinal uses, and the plants were used widely for green-manuring and land reclamation (Daubeny 1857; Harrison 1913). Lupine meal was still a product of human consumption in the late 19th century throughout areas of Spain, Corsica, and northern Italy (Cornevin 1893). The meal and flour were prepared from cooked, dried, ground seeds from which the bitter principle had been rinsed. Use of lupines as a cover crop and soil-improvement plant reached a peak in the 17th and 18th centuries. Although their use is now rather limited, several decades ago the white, *L. albus* L., the yellow, *L. luteus* L., and the blue lupine, *L. angustifolius* L., were recommended as winter cover crops for soils of low fertility in the southern United States (McKee and Ritchey 1947).

All lupines are attractive ornamentals. *L. subcarnosus* Hook. (= *L. texensis* Hook.), Texas bluebonnet, is the state flower of Texas.

Major losses of domestic animals have resulted from the ingestion of lupine foliage under range conditions. The first reports of toxicity in the United States were to sheep in western areas (Chestnut 1899; Wilcox 1899; Chestnut and Wilcox 1901). The following general conclusions were gleaned from the voluminous literature on lupinosis:

(1) Not all *Lupinus* species are poisonous, but, in general, use of these plants for forage and fodder merits caution because toxicity is variable; moreover, taxonomic differentiation between harmless and poisonous species and varieties is not easy. The following are considered the principal toxic species: *L. alpestris* A. Nels., *L. angustifolius*, *L. argenteus* Pursh, *L. caudatus* Kell., *L. greenei* A. Nels., *L. laxiflorus* Dougl., *L. leucophyllus* Dougl., *L. leucopsis* Agardh., *L. luteus*, *L. onustus* S. Wats., and *L. sericeus* Pursh.

(2) Toxicity of a species varies seasonally and geographically. Plants in the preflowering stage are less likely to be toxic than thereafter, but much depends on the species concerned. All aerial parts of poisonous species are toxic to some extent; most poisonings have resulted from ingestion of pods and seeds. Drying does not render toxic foliage harmless. Podded hay is to be avoided as the sole ration for hungry animals.

(3) Animals vary greatly in susceptibility to lupinosis. The greatest losses have occurred among sheep; horses are less likely to be affected. Lupinosis is acute or chronic, depending on the amount of toxic principle ingested. Fatal cases generally occur when large quantities are consumed over a brief period; ingestion of the same amount over a long period may be innocuous. The poisonous properties are not cumulative. Excretion of toxic substances by the kidneys is rapid.

(4) Different symptoms in a given type of livestock often result from ingestion of different lupine species (Long 1917; Couch 1926; Steyn 1934; Kingsbury 1964).

(5) The active toxic principle of *Lupinus* species may be one or several of 25–30 quinolizidine alkaloids known to occur in about 75 members of the genus (Leonard 1953; Willaman and Schubert 1961; Mears and Mabry 1971). These alkaloids are not limited to the Leguminosae, but they are most commonly found in members of the Papilionoideae. Several of these so-called lupine alkaloids occur in a single species. Sparteine, $C_{15}H_{26}N_2$, lupanine, $C_{15}H_{24}N_2O$, lupinine, $C_{10}H_{19}NO$, spathulatine, $C_{33}H_{64}N_4O_5$, and hydroxylupanine, $C_{15}H_{24}N_2O_2$, are the most common. Of these *d*-lupanine and sparteine are apparently the most toxic (Couch 1926).

Nodulation Studies

Early workers described lupine rhizobia as a homogeneous group of polarly flagellated, slow-growing rhizobia with a host range limited to *Lupinus* species and *Ornithopus sativus*. The designation *Rhizobium lupini* was applied in 1931. Reid and Baldwin (1937) proposed consideration of lupine rhizobia as physiological adaptations of *R. japonicum*. Graham (1964) favored grouping of all slow-growing rhizobia with subpolar flagella into a single genetic species without regard for host-plant relationships and, accordingly, suggested the designation *Phytomyxa japonica*. DeLey and Rassel (1965) reasoned similarly but preferred the retention of *R. japonicum*.

Reports of fast-growing isolates from nodules produced by slow-growing rhizobia are recurrent (Burrill and Hansen 1917; Fred and Davenport 1918; Löhnis and Hansen 1921; Schönberg 1929; Leifson and Erdman 1958; Abdel-Ghaffar and Jensen 1966). Repeatedly, critical examination has shown two types of colonies, one fast growing, the other slow growing, and without exception, only the slow-growing type has had nodule-inciting ability (Eckhardt et al. 1931; Lange and Parker 1960). The isolation of only fast-growing, mucilaginous organisms from *L. densiflorus* Benth. prompted Abdel-Ghaffar and Jensen (1966) to propose provisional recognition of such isolates as *R. lupini* var. *densiflori*, and their eventual inclusion within the broad group of *R. leguminosarum sensu* Graham (1964) and DeLey and Rassel (1965). Any acknowledgment of fast-growing lupine rhizobia invites skepticism. Recent results of Jansen Van Rensburg and Strijdom (1971, 1972) are particularly relevant to the culture-purity pitfall suffered by investigators. Fast-growing contaminant cells from soybean nodules were mixed in various ratios with strains of *bona fide* fast-growing and slow-growing rhizobia and used as inocula for surface-sterilized seeds cultured in Leonard assemblies under plant-growing room conditions. The contaminant cells were retrieved with higher frequency from nodules produced by slow-growing than by fast-growing types. Various properties of the many contaminants misconstrued as rhizobia *per se* resembled the non-nodule-forming isolate from a *Cajanus cajan* nodule designated *Bacillus concomitans* by Palacios and Bari (1936b).

Lupine rhizobia are currently thought to encompass a wider host range than was formerly believed true (Lange and Parker 1960, 1961; Jensen 1963, 1967; Abdel-Ghaffar and Jensen 1966; Jensen and Hansen 1968). Jensen's conclusion (1967) that the host range of lupine strains may "vie in promiscuity with the almost 'classical' cowpea . . . group," agrees, in principle, with Norris' concept (1959c, 1965) of slow-growing rhizobia as ubiquitous and unspecialized in origin.

Young lupine nodules are ovoid to round, firmly attached to the root by a broad base of secondary tissue, and normally abound in leghemoglobin; ineffective nodules are uncommon. Rhizobia enter root hairs in a typical manner and penetrate the root cortex within typical infection threads, but infection strands are not evident in nodule tissue. Host cell divisions account for further dissemination of rhizobia (Milovidov 1926, 1928b; Schaede 1932). In actively dividing meristematic host cells, rhizobia collect at the poles during mitosis and are subsequently transferred to daughter cells. Absence of intracellular threads was recently confirmed by Jordan and Grinyer (1965) in ultrathin sections of young nodules of *L. luteus*. Old nodules commonly encircle the root in a gall-like manner. This massive crowding and fusion of adjacent nodules was attributed to divided lateral meristems and their horizontal growth in opposite directions (Allen and Allen 1954). Understandably, early observers considered nodule formation on lupines a diseased root condition.

Nodulated species of *Lupinus*	Area	Reported by
albicaulis Dougl.	Wis., U.S.A.	J. C. Burton (pers. comm. 1971)
albifrons Benth.	Calif., U.S.A.	Vlamis et al. (1964)
albus L.	Widely reported	Many investigators
angustifolius L.	Widely reported	Many investigators
arboreus Sims	Hawaii, U.S.A.	Pers. observ. (1939)
arcticus S. Wats.	Alaska, U.S.A.	Allen et al. (1964)
benthamii Hell.	Calif., U.S.A.	Pers. observ. (U.W. herb. spec. 1971)
bicolor Lindl.	Calif., U.S.A.	Pers. observ. (U.W. herb. spec. 1971)
ssp. *microphyllus* (S. Wats.) Dunn	Calif., U.S.A.	Pers. observ. (U.W. herb. spec. 1971)
ssp. *pipersmithii* (Hell.) Dunn	Calif., U.S.A.	Pers. observ. (U.W. herb. spec. 1971)
ssp. *tridentatus* (Eastw.) Dunn	Calif., U.S.A.	Pers. observ. (U.W. herb. spec. 1971)
ssp. *umbellatus* (Greene) Dunn	Calif., U.S.A.	Pers. observ. (U.W. herb. spec. 1971)
caespitosus Nutt.	Colo., U.S.A.	Pers. observ. (U.W. herb. spec. 1971)
confertus Kellogg	Calif., U.S.A.	Pers. observ. (U.W. herb. spec. 1971)
cruickshanksii Hook.	France	Laurent (1891)
densiflorus Benth.	Denmark	Jensen (1964)
diffusus Nutt.	Fla., U.S.A.	Carroll (1934a)
digitatus Forsk.	Australia	Riceman & Powrie (1952); Lange (1958)
elegans HBK (= *californicus* Hort.)	Germany	Lachman (1891); Stapp (1923)
flavoculatus Hell.	Calif., U.S.A.	Pers. observ. (U.W. herb. spec. 1971)
grayi S. Wats.	U.S.A.	Munns (1922)
hartwegii Lindl.	Widely reported	Many investigators
harvardi S. Wats.	Tex., U.S.A.	Pers. observ. (U.W. herb. spec. 1971)
heptaphyllus (Vell.) Hassl.	Argentina	Rothschild (1970)

Nodulated species of *Lupinus* (cont.)	Area	Reported by
hirsutus L.	Widely reported	Many investigators
kingii S. Wats.	Ariz., U.S.A.	Pers. observ. (U.W. herb. spec. 1971)
lepidus Dougl.	Wash., U.S.A.	Pers. observ. (U.W. herb. spec. 1971)
lobbii Gray	Calif., U.S.A.	Pers. observ. (U.W. herb. spec. 1971)
luteus L.	Widely reported	Many investigators
mexicana Cerv.	Mexico	Pers. observ. (U.W. herb. spec. 1971)
micranthus Dougl.	Calif., U.S.A.	Pers. observ. (U.W. herb. spec. 1971)
var. *bicolor* S. Wats.	Calif., U.S.A.	Peirce (1902); Pers. observ. (U.W. herb. spec. 1971)
microcarpus Sims	Calif., U.S.A.	J. C. Burton (pers. comm. 1972)
multiflorus Desr.	Argentina	Rothschild (1970)
mutabilis Sweet	Widely reported	Many investigators
nanus Dougl.	Wis., U.S.A.	Eckhardt et al. (1931)
nootkatensis Donn	Alaska, U.S.A.	Allen et al. (1964)
palaestinus Boiss.	Israel	M. Alexander (pers. comm. 1963)
perennis L.	Widely reported	Many investigators
var. *occidentalis* S. Wats.	Wis., U.S.A.	Bushnell & Sarles (1937)
pilosus L.	Australia	Adams & Riches (1930)
polyphyllus Lindl.	Widely reported	Many investigators
pubescens Benth.	Zimbabwe	Corby (1974)
pusillus Pursh		
var. *intermontanus* (Hell.) C. P. Smith	Utah, U.S.A.	Pers. observ. (U.W. herb. spec. 1971)
rivularis Dougl.		
var. *latifolius*	Calif., U.S.A.	Pers. observ. (U.W. herb. spec. 1971)
sabulosus Hell.	Calif., U.S.A.	Pers. observ. (U.W. herb. spec. 1971)
sellulus Kellogg	Calif., U.S.A.	Pers. observ. (U.W. herb. spec. 1971)
sparsiflorus Benth.	Ariz., U.S.A.	Martin (1948)
spectabilis Hoover	Calif., U.S.A.	Pers. observ. (U.W. herb. spec. 1971)
subcarnosus Hook. (= *texensis* Hook.)	Widely reported	Many investigators
succulentus Dougl.	Widely reported	Many investigators
termis Forsk.	Sudan	Habish & Khairi (1968)
vallicola Hell.	Oreg., U.S.A.	Pers. observ. (U.W. herb. spec. 1971)
ssp. *apricus* (Greene) Dunn	Calif., U.S.A.	Pers. observ. (U.W. herb. spec. 1971)
varius L.	Australia	Cass Smith (1941); Norris (1959b)
villosus Willd.	Fla., U.S.A.	Carroll (1934a)

Luzonia Elmer   Papilionoideae: Phaseoleae, Diocleinae
(413a)

Named for the largest of the Philippine Islands, on which the plants were first discovered.

Type: *L. purpurea* Elmer

1 species. Shrub, scandent, widely spreading, up to about 10 m tall; stems of yellow, thick, flexible, porous, water-filled wood. Leaves 3-foliolate, ascending, submembranous, upper surface lucid, dark green, dull green underneath; lateral leaflets asymmetric, petioluled, base rounded, apex acuminate, terminal leaflet usually larger, broad-obovate. Flowers purple, sessile, in suberect, slender, spicate inflorescences; bracts pubescent; calyx deep purple outside, inside yellowish, pubescent, bilabiate; lobes subequal, lower lip slightly broader, briefly tridentate; standard fiddle-shaped, about 3 cm long, with a yellow streak from the base of the short claw to the middle, apex rounded, 2-lobed; wings sharply bent below the middle, lower part curved, slender-clawed, upper half oblong, deflexed, apex obtuse; keel petals nearly straight, basal one-third narrow-clawed; stamens 10, monadelphous, curved, pinkish enclosed by the keel petals; filaments unequal in length; anthers erect, about one-half of them fertile and crowded in with the sterile ones; ovary sessile, pubescent; style curved, lower portion pubescent; stigma terminal, with an obscurely fringed ring. Pod and seeds not described. (Elmer 1907; Merrill 1910; Harms 1915a)

The nearly equal 2-lipped calyx, woody nature of the stems, and the slender, spicate inflorescences were deemed sufficient to differentiate this genus from *Canavalia*, with which it is closely akin (Elmer 1907).

This monotypic genus is endemic in Luzon, the Philippines. The plants grow in wooded ravines.

Lyauteya Maire   Papilionoideae: Loteae

Named for General Louis Hubert Gonsalve Lyautey, 1854–1934, French military commander in Morocco; his peaceful colonial administration ensured the safety of botanical explorers.

Type: *L. ahmedi* (Batt. & Pit.) Maire

1 species. Shrub, intricately branched. Leaves sessile, on a prominent pulvinus, usually 3-, rarely 4-, 2-, or 1-foliolate; leaflets subequal, entire; stipels and stipules absent. Flowers yellow orange, fused with purple, paired, rarely 3 or 1, peduncled; bracteoles minute; calyx not bilabiate; lobes 5, subequal, shorter than the tube; petals clawed, free; standard recurved; keel petals acute, upcurved; stamens diadelphous; filaments dilated below the anthers; anthers equal, subbasifixed; ovary sessile; style curved, filiform; stigma capitate. Pod linear, longer than the persistent calyx, subcompressed; seeds without strophiole. (Maire 1919)

Maire (1919) concluded from histological evidence that *Lyauteya* occupies a position intermediate between Genisteae and Loteae with affinities closest to *Anthyllis*. Hutchinson (1964) positioned the genus in the tribe Loteae.

This monotypic genus is endemic in North Africa. It occurs in mountainous regions between elevations of 1,150 and 1,650 m, in fissures of calcareous rocks along river beds.

Lygos Adans.

SEE *Retama* Raf.

Lysidice Hance   Caesalpinioideae: Amherstieae
67 / 3521

From Greek, *lysis*, "loosening," "setting free," and *dis*, "double," in reference to the 2 valves of the pod curling backwards upon opening.
    Type: *L. rhodostegia* Hance
    1 species. Small shrub or tree, up to about 8 m tall, evergreen. Leaves paripinnate; leaflets 4–6 pairs, smooth, shiny, leathery, with narrow, red margins, the terminal pair largest; stipules small, intrapetiolar. Flowers purplish red, attractive, fragrant, in axillary and terminal panicles on new growth; bracts large, persistent, bright pink, at base of peduncles; bracteoles 2, subtending the calyx; calyx funnel-shaped, fleshy, red; lobes 4, elongate-elliptic, obtuse, upper lobe broadest, imbricate in bud, reflexed at anthesis; petals 3, deep violet, subequal, obovate, long-clawed, inrolled in bud, lower 2 very small, rudimentary; stamens 6, white, exserted, lateral 2 fertile and with long filaments, posterior 2 with sterile anthers, anterior 2 unequal, bent, staminoidal; ovary with stalk adnate to the tube, 9–10-ovuled; style filiform, circinate in bud; stigma small, terminal. Pod large, flat, obovate, acuminate, leathery to woody, 2-valved, dehiscent; seeds flat, transversely oblong, separated by spongy septae. (Hance 1867; Rock 1920)
    This monotypic genus is endemic in southern China. It thrives at moist, low elevations. Specimens have been introduced into Cuba, Florida, and other subtropical areas.
    This small, rapid-growing, hardy ornamental is highly prized for its spectacular flowers that are borne in long-branched sprays (Menninger 1962).

NODULATION STUDIES

    Plantlets of *L. rhodostegia* Hance were nodulated in Hawaii (personal observation 1938).

From Greek, *lysis*, "loosening," "setting free," and *loma*, "border," "margin," in reference to the breaking away of the pod valves from the persistent sutures.

Type: *L. schiedeana* Benth.

30–50 species. Shrubs, or small trees, occasionally large, spreading, with short trunks and slender branches, unarmed. Leaves bipinnate; leaflets small, pairs numerous, or larger and few-paired; petiolar gland usually conspicuous; stipules foliaceous. Flowers small, white or greenish, in globose heads or cylindric spikes, solitary or fascicled in long-peduncled, axillary racemes; calyx campanulate; lobes valvate; teeth short; petals equal; stamens numerous, mostly 12–30, long-exserted, basally connate into a short tube but free from corolla; anthers minute; ovary sessile, or short-stalked, many-ovuled; style awl-shaped; stigma small, terminal. Pod linear, black or brown, generally straight, flattened, apex rounded, base narrow, submembranous, valves separating at maturity from the persistent, undivided margins, continuous within, containing brown, flat, small, ovate, compressed, transversely placed seeds. (Bentham and Hooker 1865; Britton and Rose 1928)

Lysiloma is considered a Mexican genus with extensions into Central America and the West Indies. A few species are native to the southwestern United States. Species occur in valleys and lakeside coves and on thinly forested, dry, rocky hillsides up to 1,000 m.

Various members have ornamental shade value, especially the rather small trees with broad crowns. *L. bahamensis* Benth. and *L. latisiliqua* (L.) Benth. (= *L. sabicu* Benth.), large, spreading trees, 15–20 m tall, with trunks 1 m or more in diameter, are popular avenue trees in Miami and other cities in Florida.

Lysiloma wood, with special reference to that of *L. latisiliqua*, is known in the timber trade as sabicú. It is hard, resistant to fungal and insect attacks, seasons slowly but well, and has a density of about 800 kg/m^3. The lustrous brown heartwood is sharply demarcated from the thin, white sapwood. At an earlier time sabicú was one of Cuba's most valuable exports to England and, to a lesser extent, to the United States; it was used for the making of fine furniture, cabinets, bobbins, and shuttles (Record and Hess 1943). The wood of *L. auritum* (Schlechtd.) Benth. and *L. divaricata* (Jacq.) Macb. has served as a commercial source of railway ties from Salvador. Market trade of *Lysiloma* timber is currently negligible because of the limited supply (Titmuss 1965). Bark extractions of various species with a high tannin content are used in Mexico for tanning purposes.

Nodulation Studies

Rhizobia isolated from *L. thornberi* Britt. & Rose incited nodules on *Amorpha fruticosa*, and *Crotalaria sagittalis* (Wilson 1944b), and on *Astragalus nuttalianus* (Wilson and Chin 1947).

Named for Richard Maack, 1825–1886, Russian naturalist.

Type: *M. amurensis* Rupr. & Maxim.

10–12 species. Trees, small or large. Leaf buds axillary, scaled, exposed; leaves imparipinnate; leaflets subcoriaceous; stipels none. Flowers white, in dense, erect, terminal racemes; bracts and bracteoles caducous; calyx inflated; lobes 4, short, broad, upper lobe larger; standard orbicular-obovate, reflexed, notched at the apex, thickened at the base; wings oblong, spear-shaped; keel petals incurved, semijoined dorsally; stamens 10, shortly joined only at the base; anthers versatile, quadrate-elliptic; ovary subsessile, few-ovuled, densely hairy; style awl-shaped, slightly incurved; stigma minute, terminal. Pod subsessile, narrow, membranous, flattened, winged along the ventral suture, scarcely dehiscent; seeds 1–5, oblong, flat. (Takeda 1913)

Early botanists considered *Maackia* a synonym of *Cladrastis*. Distinctiveness of *Maackia* was based on opposite leaflets and leaf buds of the next year being exposed and scaled, in contrast to *Cladrastis* species having alternate leaflets and leaf buds covered by the base of hollow petioles (Takeda 1913; Ohwi 1965).

Members of this genus are native to eastern Asia, namely Japan, Korea, Manchuria, and eastern Siberia.

Maackia species are ornamental but are valued primarily for their heavy, close-grained, dark wood, which is used for chests, furniture, heavy construction, paneling, railroad ties, and agricultural implements (Everett 1969). *M. amurensis* Rupr. & Maxim. var. *buergeri* C. K. Schneid., *M. chinensis* Takeda, *M. floribunda* Takeda, and *M. tashiroi* (Yatabe) Makino are propagated in Japan. The yellow dye obtained from the bark is of no commercial value. Nikonov (1959) showed that of 6 alkaloids present in plant parts of *M. amurensis*, cytisine and *l*-lupanine were the most important. The alkaloid content of 2.67 percent in the seeds was higher than that of the roots, bark, or leaves.

NODULATION STUDIES

Professor S. Uemura, Laboratory of Forest Soil Microbiology, Government Forest Experiment Station, Meguro, Tokyo, Japan, certified in 1970 (personal communication) the presence of nodules on *M. amurensis* var. *buergeri* in Japan. Until then, only negative records constituted nodulation information on members of this genus. Nodules were absent on the type species in the Soviet Union, reported as *Cladrastis amurensis* by Lopatina (1931). Wilson (1945b) reported the absence of nodules on plantlets of *M. chinensis* "grown from seed in flasks on sterilized soil and exposed separately to 106 diverse isolates of the root nodule bacteria."

From Greek, *machaira*, "dagger," in reference to the shape of the fruit, or to the prickly stipules.

Type: *M. ferrugineum* Pers.

150 species. Trees, erect, small to medium, scandent shrubs, or lianas. Leaves imparipinnate; leaflets alternate, few, 3–11, and large, or small and numerous, up to 70; stipels lacking; stipules often hard and spiny. Flowers blue, violet, pink, or white, small, on short pedicels in axillary or terminal racemes or panicles; bracts small, caducous; bracteoles persistent, surrounding the calyx; calyx campanulate-truncate, obtuse at base; lobes 5, unequal, short; standard ovate or orbicular, exterior silky; wings oblong to falcate; keel petals incurved, joined at the back; stamens 10, all joined into a sheath usually split on 1 or both sides, or vexillary stamen free; anthers versatile, oval, dorsifixed; ovary stalked with disk at the base, ovules 1–2; style thin, incurved; stigma small, terminal. Pod stalked, samaroid, thickened at the base around the single seed, apex narrowed into an oblong wing, network prominent, indehiscent; seeds compressed, ovate or kidney-shaped. (Bentham and Hooker 1865; Amshoff 1939)

Species of this large genus are distributed throughout tropical South America and the West Indies. They are especially abundant in Brazil. Many species favor moist habitats.

The majority of *Macherium* species are lianas, or shrubs, whose lateral shoots are sensitive to contact for climbing. All are florally attractive. The tree forms are suitable for shade and provide acceptable but variable timber (Record and Hess 1943). The wood of *M. schomburgkii* Benth. is known in Guyana as itaka, or tiger wood, and that of *M. robinifolium* (DC.) Vog., in Trinidad, as salt fish wood. Various native uses are made of the bark and its gum resin exudations. The samaroid fruit is eaten in some areas as a diuretic.

NODULATION STUDIES

Nodules were observed on *M. firmum* Benth. in Germany (Brunchorst 1885), on *M. robinifolium* in Trinidad (DeSouza 1963, 1966), and on *M. aculeata* Raddi in Venezuela (Barrios and Gonzalez 1971).

__Macroberlinia__ (Harms) Hauman

Caesalpinioideae:
Amherstieae

Name means "large *Berlinia*."

Type: *M. bracteosa* (Benth.) Hauman

1 species. Tree, up to 25 m tall, with a trunk 1 m in diameter; crown large, rounded; branches thick, angular. Leaves imparipinnate, 4–5 pairs; leaflets large, generally opposite, somewhat asymmetric, not pellucid-punctate; stipules large, silvery, caducous. Flowers large, white, solitary, sessile, in simple, stout, terminal racemes; bracts very large, enveloping the flower and bracteoles before blossoming; bracteoles 2, free, opposite, valvate,

leaflike, persistent, resembling the bracts, forming an involucre around each flower bud; calyx tubular; lobes 5, free, linear; petals 5, equal, clawed, longer than the calyx lobes, the middle petal spreading fanlike, the lateral ones oblong; stamens 10, 1 free, 9 joined at the base, all fertile, exserted; ovary stalked, lower part of stalk adnate to side of receptacle, 4-ovuled; style exserted; stigma very small. Pod large, oblong-rectangular, woody at maturity, upper suture marked with a double groove; seeds suborbicular. (Hauman 1952)

This monotypic genus is endemic in tropical West Africa. It usually grows as border along forest marshes and rivers.

The brown wood has the durability of oak and takes a fine polish. The squamous, grayish brown bark exudes a white resin from wounds.

Macrolobium Schreb. Caesalpinioideae: Amherstieae
(63) / 3517

From Greek, *makros*, "large," and *lobion*, "pod."

Type: *M. bifolium* (Aubl.) Pers.

50 species. Shrubs, small, some scandent, or large trees, up to about 30 m tall, unarmed. Leaves paripinnate or pseudoimparipinnate, in few pairs, rarely unifoliate; leaflets in 1 or many pairs, usually sessile and asymmetric, elliptic or ovoid, emarginate or abruptly pointed, sometimes punctate underneath; stipules foliaceous or minute, caducous. Flowers yellow, red, or white, small, in simple or compound, solitary or fascicled, often paired, axillary or terminal racemes; bracts minute, caducous; bracteoles 2, large, sheathlike, enclosing the bud, or free, sometimes red; calyx tube short-turbinate; lobes 4–5, narrow, subequal in length, imbricate, posterior lobe usually broader; petals 5–4, usually unequal, posterior petal largest, round, claw often eared, other petals vestigial, smaller, scalelike, or absent; fertile stamens generally 3, long, free, exserted, others abortive; anthers oblong, dorsifixed, longitudinally dehiscent; ovary subsessile or stalk adnate to the calyx tube, few- to many-ovuled; style filiform; stigma small, terminal. Pod smooth, oblong, ovoid or falcate, compressed, leathery to woody, 2-valved, with longitudinal ribs or transversely rugose, upper margin usually thick; seeds compressed, rounded or quadrate. (Bentham and Hooker 1865; Baker 1930; Amshoff 1939; Cowan 1953; Léonard 1957)

Cowan (1953) considered the American species separate and distinct from the African elements, which are closely allied with the *Berlinia* species in Africa but remote from other Amherstieae. In more recent years Keay et al. (1964) and Voorhoeve (1965) transferred the African species of *Macrolobium* to *Anthonotha*, *Gilbertiodendron*, *Pellegrinioidendron*, and *Paramacrolobium*.

Representatives of this genus are widespread in tropical America, especially from Panama to southern Brazil. Tree species are common among lowland riverine and savanna vegetations.

Some species are decorative and merit ornamental consideration. The timber of the large trees is suitable for cabinetwork and furniture but it is not available in quantity.

NODULATION STUDIES

Dark brown root systems of seedlings of an unidentified species examined at 5 sites in Manaus, Amazonas, Brazil, lacked nodules (Norris 1969).

Macropsychanthus Harms   Papilionoideae: Phaseoleae,
(409a) / 3887   Diocleinae

From Greek, *makros*, "large," *psyche*, "butterfly," and *anthos*, "flower," in reference to the large papilionaceous flowers.
    Type: *M. lauterbachii* Harms
    3 species. Shrubs or vines, tall, climbing, woody. Leaves petioled, 3-foliolate; leaflets broad, rounded, entire; stipels and stipules caducous. Flowers large, showy, pink, blue, or pale purple, several-fascicled, on stout nodes in upper half of terminal racemose panicles; calyx broad, campanulate-cylindric, externally puberulent, villous inside; lobes 5–4; petals broad- or long-clawed, subequal; standard obovate, deeply emarginate; wings narrow, falcate, apex rounded, with a rounded ear on 1 side at base of the blade; keel petals falcate-oblong, joined above; stamens 10, vexillary stamen free at the base but united with the others at the middle; anthers oblong-ellipsoid, dorsifixed; ovary stalked, linear, hairy; stigma minute, truncate. Pod linear, flattened, at first turgid and rusty-pubescent; glabrous and almost woody at maturity, apex acuminate; seeds large, thick, orbicular, with linear hilum. (Harms 1908a; Merrill 1910)
    Members of this genus occur in the Philippines, Papua New Guinea, and the Truk Islands. They grow along streams in moist forests.

Macroptilium Urb.   Papilionoideae: Phaseoleae, Phaseolinae

From Greek, *makros*, "large," and *ptilon*, "wing"; the wing petals are much larger than the standard.
    Lectotype: *M. lathyroides* (L.) Urb.
    15 species. Herbs, erect or climbing. Leaves pinnately 3-, very rarely 1-foliolate; leaflets suborbicular to elliptic, sometimes linear-oblong; stipels awl-shaped; stipules lanceolate, acuminate, not appendaged at the base. Flowers purplish crimson, deep pink, or whitish, in distant pairs on long-

417

stalked racemes; bracteoles short, awl-shaped, caducous; calyx narrow-campanulate or subtubular; lobes 5, free or upper 2 connate, acute, posterior lobe sometimes much reduced; standard ovate or orbicular, reflexed above, with 2 small, reflexed ears at the base; wings broad, obovate or suborbicular, erect, more deeply colored and larger than the standard, with 2 ears below the base of the limb, long-clawed, claws adnate to the staminal tube; keel petals broad-linear, boat-shaped, twisted spirally into a hood, apex obtuse, connate; stamens diadelphous, vexillary one sometimes dilated above; anthers uniform; ovary subsessile, few- to many-ovuled; style thickened apically, bearded inside below the apex; stigma lateral, capitate. Pod small, reflexed, subcylindric or compressed, straight or curved; valves 2, convex; seeds few to many. (Urban 1928; León and Alain 1951)

The species were formerly grouped as section *Macroptilium*, in *Phaseolus* (Piper 1926).

Macroptilium species occur in tropical and subtropical America and the West Indies. They are commonly considered weeds in waste places and along roadsides.

Nodulation Studies

M. atropurpureum Urb. and *M. lathyroides* (L.) Urb., formerly *Phaseolus atropurpureus* and *P. lathyroides*, respectively, have been observed nodulated in various areas where they have shown promise as tropical pasture plants. Each symbiosed promiscuously with a broad spectrum of rhizobia from plants in the conglomerate cowpea miscellany (Allen and Allen 1939; Norris 1967). *M. lathyroides*, phasey bean, proved so responsive as a "guinea pig" host in testing for presumptive rhizobia from legumes other than those in the tribes Vicieae and Trifolieae that Norris (1964) preferred it over cowpea, *Vigna unguiculata*. Moreover, its growth habit and development are suitable to test-tube growth methods, whereas cowpeas require bottle-jar assemblies or similar techniques.

Macrosamanea Britt. & Rose   Mimosoideae: Ingeae

Name means "large *Samanea*."

This genus of 8 species segregated from *Albizia* was the result of the revision of tropical American mimosoids by Britton and Killip (1936). More conservative botanists do not consider it a valid genus because of the tenuous basis for its segregation (Mohlenbrock 1963a; Hutchinson 1964).

All of the species are endemic in Colombia; most of them have *Pithecellobium* or *Inga* synonyms. Hutchinson (1964) regards the genus as congeneric with *Pithecellobium*.

From Greek, *makros*, "large," *tylos*, "knob," and *loma*, "border," "margin," in reference to the knobby sutures on some of the pods.

Lectotype: *M. uniflorum* (Lam.) Verdc. (= *Dolichos uniflorus*)

25 species. Herbs, annual or perennial, some with woody rootstocks; stems pubescent, scandent, prostrate, or erect. Leaves pinnately 3-foliolate, rarely 1-foliolate, undersurfaces more hairy than upper ones; stipels filiform, lanceolate, or obsolete; stipules deltoid. Flowers cream, yellow, green, rarely red, few, on short pedicels, axillary and fascicled, or pseudoracemose at ends of branches; calyx campanulate; lobes 4–5, deltoid, upper 2 in a bidentate lip; standard round to elliptic, with 2 parallel linear ears; wings narrow; keel not twisted; stamens diadelphous; anthers uniform; ovary linear-oblong, 3–13-ovuled; style filiform, smooth or slightly pubescent; stigma terminal, hairy-rimmed. Pod straight or curved, narrow or oblong, compressed, nonseptate, dehiscent; seeds compressed. (Verdcourt 1970b,c)

The genus was segregated from *Dolichos* by Verdcourt (1970b,c), and recently Maréchal and Baudet (1977) proposed the merger of *Kerstingiella* into it. Lackey (1978) has urged that, if the merger is accepted, *Kerstingiella* be conserved as the genus name.

Members of this genus are indigenous to Africa and Asia. Their habitat varies widely; some species are adapted to desert and dune areas, others to grasslands and areas bordering mangrove swamps.

M. uniflorum (Lam.) Verdc. is sometimes called the poor man's pulse crop in India. The seeds are eaten after being dried, fried, or ground into a meal.

NODULATION STUDIES

Norris et al. (1970) observed effective nodulation of *Teramnus uncinatus* in Brazil by a strain of *M. africanum* (Brenan ex Wilczek) Verdc. (reported as *Dolichos africanus*) obtained from Zimbabwe. Presumably, this is the only report of host-infection studies using rhizobia from *Macrotyloma* species; a kinship with the cowpea miscellany is implicit.

Corby (1974) observed nodules on the following species in Zimbabwe: *M. africanum*; *M. axillare* (E. Mey.) Verdc. var. *axillare*, var. *glabrum* (E. Mey.) Verdc., var. *macranthum* (Brenan) Verdc.; *M. daltonii* (Webb) Verdc.; *M. densiflorum* (Welw. ex Bak.) Verdc.; *M. oliganthum* (Brenan) Verdc.; *M. rupestre* (Welw. ex Bak.) Verdc.; *M. stipulosum* (Welw. ex Bak.) Verdc.; and *M. uniflorum* var. *stenocarpum* (Brenan) Verdc.

M. maranguense (Taub.) Verdc. in South Africa produces nodules (Grobbelaar and Clarke 1975).

Type: *M. grandiflora* (Gray) Scheff.

15 species. Trees, up to 20 m tall, with short trunks and rounded crowns. Leaves paripinnate, young leaves enclosed in prominent buds before unfolding; bud scales imbricate, cartilaginous, coarsely serrate; leaflets sessile, 2–15 pairs, mostly glabrous, elliptic, acuminate, asymmetric, midrib eccentric, flaccid, drooping, white or pink at first, later dark green, leathery; stipules long and narrow, present in buds, very early caducous. Flowers white or pink, in terminal and axillary, dense, globose racemes on stout rachis; each flower in the axil of a subpersistent, scalelike bract; bracteoles 2, inner bracteole small, outer bracteole broad-obovate, half encompassing the pedicel, caducous; calyx campanulate or tubular, deeply 4-cleft, rarely 5-cleft; lobes subequal, colored, imbricate in bud, reflexed at anthesis; petals 5, sessile, narrow-linear, subequal, free; stamens numerous, 15–80; filaments joined only at the base, long-exserted; anthers dorsifixed, longitudinally introrsely dehiscent; ovary sessile or short-stalked, 1–2-ovuled, glabrous; disk absent; style elongated, often S-shaped; stigma terminal, capitate. Pod stalked above scar left by the circumscissile calyx, flat to globular, thick, ridged, lower suture much-curved, upper suture almost straight, indehiscent, usually 1–2-seeded. (Backer and Bakhuizen van den Brink 1963; Knaap-van Meeuwen 1970)

Maniltoa was formerly a section in *Cynometra* (Harms 1897).

Representatives of this genus are native to Fiji, Papua New Guinea, and Indonesia. They occur in forests and along streams.

Many of the tree species are esteemed ornamentals in parks and gardens of the tropics because the young leaves are beautifully colored and flowering starts at a very young age. The 4 species in Fiji are useful timber trees.

NODULATION STUDIES

Introduced trees of *M. grandiflora* (Gray) Scheff. in Singapore bore elongated, whitish brown, moderately abundant nodules on red brown roots (Lim 1977). *M. scheffera* Schum. in the nursery of the Bogor Botanic Garden, Java, was devoid of nodules (personal observation 1977).

Margaritolobium Harms   Papilionoideae: Galegeae,
                                                               Robiniinae

Named for Margarita Island, Venezuela, where the specimens were collected.

Type: *M. luteum* (I. M. Johnston) Harms

1 species. Shrub. Leaves imparipinnate, 5-foliolate. Flowers small, many, yellow, on paired pedicels in racemes, blooming before leaves appear; calyx broadly cup-shaped, subtruncate, pubescent; upper teeth inconspicuous, lateral teeth broad, blunt, lower tooth broad; standard subtirangular-ovate, broad-clawed, with a dark blotch near claw; wings oblong, with a slender claw, acutely short-eared, apex blunt; keel petals acuminate, long-clawed, pubescent outside; stamens diadelphous; ovary lanceolate, short-stalked, 3–4-

ovuled; ovary and stipe silky-pubescent; style glabrous; stigma minute. Pod flat, thin, dehiscent (?). (Harms 1923b)

This monotypic genus is endemic in Margarita Island, Venezuela. It grows on dry hills.

Marina Liebm.   Papilionoideae: Galegeae, Psoraliinae
242 / 3705

From Latin, *marinus*, "marine," "of the sea."

Type: *M. gracilis* Liebm.

1 species. Herb, annual, slender, diffuse, glandular-punctate. Leaves imparipinnate; leaflets numerous, entire, very small; stipels minute; stipules broad, scarious, dentate. Flowers violet, small, few, on slender peduncles in axillary or leaf-opposed racemes; calyx lobes subequal; teeth ciliate; standard obovate-round, long-clawed; wings falcate-obovate; keel hood-shaped, shorter than the standard and wings; stamens 10, all connate into a sheath split above; anthers uniform; ovary sessile, 1-ovuled; style filiform. Pod enclosed by the calyx, membranous, indehiscent; seed subreniform. (Bentham and Hooker 1865)

This monotypic genus is endemic in Mexico.

Marmaroxylon Killip ex Record   Mimosoideae: Ingeae

From Greek, *marmaros*, "marble," and *xylon*, "wood," in reference to its characteristic appearance.

Type: *M. racemosum* (Ducke) Killip ex Record

1 species. Tree, large, pubescent, unarmed; bole often 20–25 m to the first limb, with a diameter of 45–60 cm above the buttressed base; bark rough. Leaves bipinnate; pinnae pairs many; leaflets small, membranous, numerous, asymmetric; rachis glandular between pinnae; stipules caducous. Flowers salmon pink, small, sessile, in pubescent racemes on old wood; calyx cupular, minutely toothed; lobes very short; stamens numerous; filaments connate into a long-exserted tube, white at base, yellow above. Pod linear-oblong, about 7.5 cm long, often semicircular, densely pubescent, nonseptate, but indented transversely between the seed, 2-valved, becoming twisted after dehiscence; seeds numerous. (Record 1940)

This monotypic genus, endemic in the Amazon region and Guiana, is quite rare. It usually occurs in dense forests.

The brownish yellow wood is attractively marked with irregular purplish brown streaks and patches due to a deeply colored resin. It was known in the Brazilian timber trade as angelim rajado. The durable wood is heavier than water and difficult to work, but is considered suitable for wheelcraft work, cabinetmaking, and marquetry (Record and Hess 1943).

Name means "Martius' tree," in honor of Carl Friedrich Philipp Martius, 1794–1868, renowned German botanist and editor of *Flora Brasiliensis*.

Type: *M. excelsum* (Benth.) Gleason

4 species. Trees, unarmed, medium to large; trunks buttressed or basally swollen; young branches and petioles with golden brown pubescence. Leaves imparipinnate; leaflets 5–13, alternate, ovate or oblong, apices acuminate, bases obtuse to cordate; stipules early caducous. Flowers showy, deep yellow, borne in a large terminal thyrse; bracts and bracteoles very caducous; calyx tube short; lobes 5, lanceolate, widths unequal, lowest lobe narrowest, all imbricate in bud, free and reflexed at anthesis, both surfaces pubescent; petals 5, obovate, clawed, uppermost petal widest and innermost; stamens free, 4–5, rarely 7, usually only 4 fertile; filaments short, thick; anthers unequal, elongated, narrow-sagittate, acuminate, dehiscing by 2 terminal pores; staminodes 0–5, without anthers; ovary sessile, free, 1-ovuled; style awl-shaped; stigma small, terminal. Pod samaroid, elliptic or oblong, flattened laterally, winged along both sutures, large, thinly leathery, sparsely pubescent at maturity, indehiscent, 1-seeded; seed in center of pod, flat, kidney-shaped. (Gleason 1935; Koeppen and Iltis 1962)

The absence of silica in the woods of *Martiodendron*, *Cassia*, and *Androcalymma* readily differentiates them from the highly siliceous woods of *Apuleia*, *Dialium*, and *Dicorynia*, also in the tribe Cassieae. Sclerotic parenchyma cells in the wood of *Martiodendron*, a rare feature among the Leguminosae, also occur in *Androcalymma*, but not in *Cassia* species (Koeppen and Iltis 1962).

Representatives of this genus are indigenous to tropical South America, Brazil, and Guiana. They are common in tropical rain forests and along rivers, and usually grow on well-drained clay soil.

The orange to red brown, durable wood of *M. excelsum* (Benth.) Gleason, a heavily buttressed tree of 40–45 m height, is very hard, tough, and strong, but is considered a refractory timber of little promise, except for local construction (Record and Hess 1943).

Mastersia Benth.   Papilionoideae: Phaseoleae, Galactiinae

407 / 3883

Named for Maxwell T. Masters, 1833–1907, taxonomic botanist of England.

Type: *M. assamica* Bak.

2 species. Shrubs, twining or creeping. Leaves arranged spirally, pinnately 3-foliolate; leaflets entire; stipels present; stipules early caducous. Flowers purple, fascicled, in axillary and terminal racemes; rachis joints nodose; bracts ovate, paired, caducous; bracteoles 2, basal, suborbicular, subpersistent; calyx turbinate; lobes lanceolate, longer than the tube, upper 2 joined into a broad, entire lip, lowest lobe longest; standard oval-round, very short-clawed, with 2 basal tubercles, ears lacking; wings oblique, equal to or slightly shorter than the broad, obtuse, slightly incurved keel; stamens

diadelphous; anthers uniform, linear, dorsifixed; ovary sessile, many-ovuled; style short, filiform, incurved, not bearded; stigma terminal, capitate. Pod linear-oblong, very flat, with narrow wing along upper suture, breaking into irregular segments, indehiscent; seeds many, oblong, transverse. (Bentham and Hooker 1865; Backer and Bakhuizen van den Brink 1963)

Members of this genus occur from the Himalayas to Celebes.

NODULATION STUDIES

In Java, Keuchenius (1924) observed nodulation on *M. bakeri* (Koord.) Bak., which is cultivated there as a cover crop and green manure.

Mecopus Benn.　　　　　　　Papilionoideae: Hedysareae, Desmodiinae
336 / 3806

Type: *M. nidulans* Benn.

1 species. Herb, slender, annual, or shrub, small short-lived, erect, ascending; branches long, whiplike. Leaves 1-foliolate, broadly rounded; petioles filiform, jointed, often persistent after leaflets falls; stipels 2, minute; stipules erect, linear-lanceolate. Flowers white, very small, in short-peduncled, many-flowered, terminal racemes which appear woolly-headed because of the densely-haired, fascicled pedicels that curve apically and invert the flowers; bracts awl-shaped, hooked at apex, persistent; bracteoles absent; calyx campanulate; teeth short, lanceolate, upper 2 joined; standard obovate with base narrow, ears lacking; wings falcate, adherent to the incurved, obtuse keel; stamens diadelphous; anthers uniform; ovary stalked, 2-ovuled; style curved; stigma small, capitate, terminal. Pod small, long-stalked, long-exserted, bent downward from the inverted calyx, concealed by the bracts and pedicels, jointed, joints 1–2, reticulate, compressed, indehiscent. (Bentham and Hooker 1865; Hooker 1879; Backer and Bakhuizen van den Brink 1963)

This monotypic genus, endemic in Southeast Asia, Java, and India, is common in teak forests.

Medicago L.　　　　　　　　　Papilionoideae: Trifolieae
227 / 3688

From Latin, *medica*, "Median," in reference to the introduction of alfalfa, medic or medick, into Europe from Media, an ancient country in southwestern Asia corresponding to northwestern Iran.

Type: *M. sativa* L.

50–100 species. Herbs, annual or perennial, rarely shrubs. Leaves pinnately 3-foliolate; leaflets obovate, margins short-toothed; stipules adnate to the petiole, sometimes dentate. Flowers small, yellow or violet, rarely variegated, in axillary, spikelike or capitate racemes or smaller clusters of 1–5; bracts small, or absent; bracteoles absent; calyx campanulate; tube short; lobes 5, subequal, sometimes acuminate and longer than the tube; petals free from the staminal column; standard sessile, obovate or oblong, straight or curved outwards, narrowed at the base; wings oblong, eared, clawed, longer than the obtuse keel; stamens diadelphous; anthers uniform; ovary sessile or short-stalked, many-ovuled; style awl-shaped, glabrous; stigma subcapitate, oblique. Pod longer than the calyx, spirally coiled or sickle-shaped, often covered with spines, hooked or somewhat curved, usually indehiscent, 1- to several-seeded. (Bentham and Hooker 1865)

Members of this genus, native to Eurasia and Africa, especially in the Mediterranean area, are now widespread in temperate areas.

Species prefer deep, well-drained, aerated soils of light texture and moderately high fertility, with a pH range from about 6.5 to 7.5. They are well adapted to a wide range of climatic conditions and are drought-resistant. Species thrive in open habitats, cultivated fields, and grasslands. They have deeply penetrating root systems.

Alfalfa is said to be the only forage crop which was cultivated before recorded history (Bolton 1962). It was introduced into Greece as early as 490 B.C. (Hanson et al. 1962). Iran was most likely the center or origin (Hendry 1923; Bolton et al. 1972). The word "alfalfa" is Arabicized Persian and traceable to the Iranian word "aspoasti," "horse fodder." The term "lucerne," a common name in many parts of the world, had its origin from the Lake Lucerne district, Switzerland, where the cultivation of this plant in very early times led to popularity throughout Europe.

No other forage crop has been the subject of so much research and so many scholarly reviews (Hendry 1923; Griffiths 1949; Bolton 1962; Hanson et al. 1962; Hanson 1972). These studies contain a wealth of information concerning the botany, geographical movement, cultivation, and utilization of this important forage crop.

The agricultural value of alfalfa has been accurately defined by Keuren and Marten (1972): "Alfalfa is a superior pasture legume for many classes of livestock because of its high yield, forage quality, and wide climatic and soil adaptation. It is a dependable and economical protein source for the grazing animal because it is independent of soil nitrogen. The protein is of good quality, especially important to nonruminants on pasture, such as swine, poultry, and horses. Alfalfa is an excellent source of calcium, magnesium, phosphorous, and vitamins A and D. Intake of alfalfa is usually greater than that of grasses of equal digestibility.

"Alfalfa provides greater flexibility for the livestock producer than many other pasture legumes. Because of upright growth habit and rapid recovery, it lends itself to harvest as hay, silage, or soilage, as well as pasture. This flexibility is a desirable characteristic in livestock areas where harvested feed is utilized for portions of the year. Its drought resistance also provides a more uniform seasonal growth pattern than that of most legumes and grasses."

NODULATION STUDIES

The term *Rhizobium meliloti* was applied in 1926 by Dangeard to those organisms capable of forming nodules on alfalfa and sweet clover. Growth of these rhizobia is fast but not as rapid as that of strains of *R. leguminosarum, R.*

phaseoli, and *R. trifolii*; the colony growth is buttery but not viscous, and an acid reaction in litmus milk is accompanied by serum zone formation. Young cells are peritrichously flagellated. Bacteroid shapes from nodules are variously branched and often club-shaped.

Medicago was one of several leguminous genera recognized about the turn of the century to be highly selective of rhizobial strains. The alfalfa cross-inoculation group, consisting of species only of *Medicago*, *Melilotus*, and *Trigonella*, came into existence within the early years of the 20th century, and has been unchallenged as a plant:rhizobia group in the intervening years. Strain variation and host specificity have been subjects of innumerable research papers. Studies by Burton and Wilson (1939), Burton and Erdman (1940), Vincent (1941), Jensen (1942), Purchase et al. (1951), Hely and Brockwell (1964), Brockwell and Hely (1966), Burton (1972), and Hely and Ofer (1972) provide valuable insights into these topics.

Nodulated species of *Medicago*	Area	Reported by
aculeata Willd.		
var. *aculeata* Willd.	Australia	Brockwell & Hely (1966)
arabica (L.) Huds. (= *maculata* Sibth.)	Widely reported	Many investigators
arborea L.	France	Naudin (1897c); Lechtova-Trnka (1931)
	Australia	Harris (1953)
aschersoniana Urb.	S. Africa	Grobbelaar & Clarke (1974)
blancheana Boiss.	Australia	Harris (1953)
carstiensis Wulf.	Europe	Gams (1923)
ciliaris (L.) All.	France	Pichi (1888)
coerulea Less.	Australia	Jensen (1942)
disciformis DC.	France	Clos (1896)
falcata L.	Widely reported	Many investigators
gaetula [?]	Australia	Vincent (1941); Jensen (1942)
(*hispida* Gaertn.) = *polymorpha* L.	Widely reported	Many investigators
intertexta (L.) Mill.	Germany	Wendel (1918)
	Australia	Harris (1953)
var. *ciliaris* (L.) Heyne	S. Africa	Grobbelaar & Clarke (1974)
laciniata (L.) Mill.	Sweden	Eriksson (1873)
	Libya	B. C. Park (pers. comm. 1960)
littoralis Rhode	Israel	M. Alexander (pers. comm. 1963); Hely & Ofer (1971)
var. *littoralis* Rhode	Australia	Brockwell & Hely (1966)
	Israel	Hely & Ofer (1972)

Nodulated species of *Medicago* (cont.)	Area	Reported by
lupulina L.	Widely reported	Many investigators
marina L.	Israel	M. Alexander (pers. comm. 1963)
minima (L.) Bartal.	Widely reported	Many investigators
var. *minima* (L.) Bartal.	Australia	Brockwell & Hely (1966)
murex Willd.	Australia	Vincent (1941); Jensen (1942)
orbicularis (L.) Bartal.	Widely reported	Many investigators
platycarpa Trautv.	Wis., U.S.A.	Wipf (1939)
polymorpha L.	Widely reported	Many investigators
(= *denticulata* Willd.)		
(= *hispida* Gaertn.)		
(= *lappacea* Desr.)		
(= *terebellum* Willd.)		
var. *brevispina*	Australia	Brockwell & Hely (1966)
	Calif., U.S.A.	Pers. observ. (U.W. herb. spec. 1971)
var. *polymorpha* L.	Australia	Brockwell & Hely (1966)
	Israel	Hely & Ofer (1972)
var. *vulgaris* (Benth.) Shin.	U.S.D.A.	L. T. Leonard (pers. comm. 1937)
	Australia	Brockwell & Hely (1966)
praecox DC.	Australia	Brockwell & Hely (1966)
rigidula (L.) All.	Ala., U.S.A.	Duggar (1934)
(= *gerardii* Waldst. & Kit.)	France	Clos (1896)
var. *agrestis* Burnat	Australia	Brockwell & Hely (1966)
rugosa Desr.	Australia	Brockwell & Hely (1966)
(= *elegans* Jacq.)	France	Clos (1896)
ruthenica Trautv.	Wis., U.S.A.	Wipf (1939)
sativa L.	Widely reported	Many investigators
scutellata (L.) Mill.	France	Laurent (1891)
	Australia	Jensen (1942)
soleirolii Duby	Australia	Harris (1953)
tornata Mill.	Libya	B. C. Park (pers. comm. 1960)
truncatula Gaertn.	Widely reported	Many investigators
(= *tribuloides* Desr.)		
var. *longiaculeata* Urb.	Israel	Hely & Ofer (1972)
var. *tricycla* (Negrè) Heyn	Israel	Hely & Ofer (1972)
turbinata (L.) All.	S. Africa	Grobbelaar et al. (1967)
(= *tuberculata* Willd.)	U.S.D.A.	Erdman (1948b)

From Greek, *melas*, "black," and *aden*, "gland," in reference to the dense black glands on plant parts.

Type: *M. badocana* (Blanco) Turcz.

3 species. Shrubs or subshrubs, erect, stout, up to 1 m tall, softly tomentose, glandular, scented. Leaves 1-foliolate, ovate-lanceolate, margins wavy or dentate, both surfaces softly villous and dotted with dense black glands; petiole jointed above; stipels 2, minute; stipules awl-shaped, persistent. Flowers small, in dense heads or short spikes, axillary, on very short peduncles or sessile; bracts ovate, black-gland-dotted; bracteoles absent; calyx black-gland-dotted, villous, bilabiate, deeply cleft, upper lip 4-lobed, lower lip entire, usually much longer; petals equal to or scarcely longer than the lower lip; standard narrow-obovate, not clawed; keel petals short, blunt; stamens diadelphous; anthers uniform, subbasifixed; ovary elliptic, 1-ovuled; style slender, glabrous; stigma minute, terminal. Pod small, reticulate, glandular, hispid, enclosed by the persistent calyx, indehiscent, 1-seeded.

Some authors consider *Meladenia* and *Psoralea* congeneric (Taubert 1894; Burbidge 1963).

Representatives of this genus occur in northern Australia, Papua New Guinea, and the Philippines.

Melanoxylon Schott　　　　Caesalpinioideae: Sclerolobieae
108 / 3566

From Greek, *melas*, "black," and *xylon*, "wood"; the color of the heartwood is almost black.

Type: *M. brauna* Schott

2 species. Trees, 30–35 m tall; trunks straight, smooth, with diameters usually exceeding 1 m, unbuttressed. Leaves imparipinnate; leaflet pairs numerous, large, leathery, undersurfaces rusty-tomentose. Flowers large, orange yellow, borne in showy terminal panicles; bracts minute, caducous, bracteoles none; calyx obliquely campanulate; lobes 5, thick, imbricate, outermost lobe smaller; petals 5, orbicular, equal, rounded, erect, imbricate; stamens 10, free; filaments hairy at the base; ovary sessile, free, multiovuled; style short, thick, incurved; stigma truncate-concave, fringed. Pod long-stalked, oblong-falcate, flattened, leathery to woody, pulpy between the seeds, 2-valved; seeds transverse, oblong, compressed, with an angular apical wing. (Bentham and Hooker 1865)

Members of this genus are common in tropical South America, especially in coastal rain forests from Bahia to São Paulo, Brazil.

The timber of *M. brauna* Schott is valued for its durability; it is used primarily in heavy construction and in making fine furniture. The hard, heavy wood is dark brown to blackish and rather difficult to work, but it finishes smoothly and attractively. The bark is a source for tannin and a brown blackish dye substance. Locally, the tree is known as baraúna or braúna (Record and Hess 1943).

From Greek, *meli*, "honey," and *lotos*, a name given to several fodder plants, thus honey lotus.

Lectotype: *M. officinalis* Willd.

20 species. Herbs, annual or biennial; foliage often with strong coumarin odor when dried. Leaves pinnately 3-foliolate; leaflets usually linear to elliptic-oblong, short-petioluled, serrate, veins usually ending in teeth; stipels absent; stipules lanceolate to awl-shaped, adnate to the petiole. Flowers small, many, yellow, white, or white tipped with blue, in slender, erect, axillary, spikelike racemes on short, reflexed pedicels; bracts minute, persistent; bracteoles absent; calyx short-campanulate; lobes 5, subequal, awl-shaped to lanceolate, acute to acuminate, shorter than the tube; petals free from the staminal tube, caducous; standard obovate-oblong, narrow at the base, subsessile, without basal ears; wings oblong, eared at the base, adherent to the keel; keel blunt, clawed, obtuse; stamens diadelphous, vexillary stamen free above, joined only at the middle with the other 9; filaments not dilated; anthers uniform; ovary sessile or minutely short-stalked, few-ovuled; style filiform, beardless, straight or incurved above; stigma small, terminal. Pod straight, beaked, globose or obovoid, some markedly reticulate, cross-ribbed or wrinkled, indehiscent or tardily so; seeds 1–few. (Bentham and Hooker 1865)

Species of *Melilotus*, a genus closely related to *Medicago* and *Trigonella* fall into two relatively distinct groups, one consisting of about 11 annual species common to Mediterranean Europe, Asia Minor, and bordering areas of Africa, and the other consisting of the biennial species of Europe and adjacent Asia (Isely 1954). They prefer temperate to warm climates.

The species are moderately winter-hardy, drought-resistant, and have a marked preference for alkaline soils with little to no tolerance for acid areas. They are valued as pasture forage, green manure, for soil-improvement, in soil erosion control, and as ground cover of depleted land. The roots penetrate the soil deeply and upon decomposition stimulate dynamic microbial activity that results in enhanced soil particle aggregation; thus, soil tilth is improved.

Because of the color of their flowers, *M. alba* Desr. and *M. officinalis* Willd. are commonly known as white sweet clover and yellow sweet clover, respectively. The flowers of both these species are sources of commercial honey. *M. altissima* Thuill. is used as a flavoring herb for a green cheese made in Switzerland.

Special care is required in the curing of sweet clover hay and in the preparation and feeding of sweet clover silage. Ingestion by cattle of moderately large amounts of improperly cured sweet clover hay, or silage, is responsible for a hemorrhagic malady known as sweet clover poisoning, or the sweet clover disease. This malady first made its appearance in Canada (Schofield 1924), and was soon reported in various areas of the north-central United States.

Pathology of the disease was described in detail by Roderick (1929). Initial clinical symptoms appear in 3–8 weeks and are evidenced by a delay in coagulation time of the blood due to the destruction of prothrombin (Steyn 1934; Kingsbury 1964). As feeding of the improperly cured hay continues, the clotting time becomes more prolonged. Swellings over the body, stiffness, lameness, hematuria, bloody stools, and hemorrhages are more advanced symptoms. Death may result from progressive weakness and excessive hemorrhaging. The disease is restricted almost exclusively to cattle. It is reversible in early stages by blood transfusions and the removal of spoiled hay from the diet.

In 1941 Campbell and Link isolated the hemorrhagic agent, $C_{19}H_{12}O_6$, from spoiled sweet clover hay that had been produced experimentally and from hay that had killed cattle in agricultural practice. The reduction in prothrombin level induced by 1.5 mg of the crystalline agent was equivalent to about 50 gm of the spoiled hay. Proof was presented the same year by Stahmann et al. (1941), through degradation reactions and by synthesis, that the hemorrhagic agent was dicoumarin, or dicoumarol, i.e., 3,3'-methylenebis (4-hydroxycoumarin). Chemical and physical properties of the naturally occurring agent in spoiled hay and of the synthetic compound were shown to be identical. Formation of dicoumarol has its origin in coumarin, $C_9H_6O_2$, the lactone of O-hydroxycinnamic acid. This volatile, vanilla-scented compound is present to some extent in all *Melilotus* species.

Research on dicoumarol and sweet clover sickness might well have been a closed chapter at that stage, had it not been for the acumen of Link and colleagues (Link 1945). The synthesis and characterization of many coumarins continued throughout the decade. In 1948 coumarin #42 was selected and proposed by Link for rodent control under the auspices of the Wisconsin Alumni Research Foundation. This compound became Warfarin, the world-famous rodenticide and later a foremost anticoagulant in the medical field. Link (1959) coined the word "Warfarin" by combining the first letters of the Wisconsin Alumni Research Foundation with "arin," in reference to coumarin.

In addition to the extra care required in the curing of sweet clover hay and silage to prevent the formation of dicoumarol, genetic studies aimed toward the development of low coumarin strains of sweet clover have offered promise (Brink and Roberts 1937). Smith (1948) was successful in transferring a low-coumarin gene from *M. dentata* Pers. to *M. alba*. In later studies the same gene was transferred from *M. alba* to *M. officinalis* (Webster 1955). Other studies are in progress.

NODULATION STUDIES

Melilotus nodules have apical meristems, thus oval, finger-shaped, and corolloid-shaped nodules are produced. Rhizobial forms are X-, Y-, and club-shaped.

Melilotus rhizobia are peritrichously flagellated and fast growing; they produce acid reactions on carbohydrate media and an acid reaction and a distinct serum zone in litmus milk. Other cultural and biochemical reactions are also akin to those of rhizobia from *Medicago* and *Trigonella* species. These characteristics in keeping with host-infection affinities confirm the rhizobia as strains of *Rhizobium meliloti*.

Nodulated species of *Melilotus*[a]	Area	Reported by
alba Desr.	Widely reported	Many investigators
altissima Thuill.	France	Dangeard (1926)
	Wis., U.S.A.	Wipf (1939)
	S. Africa	Grobbelaar et al. (1967)
caerulea Desr.	Germany	Lachmann (1891);
		Morck (1891)
caspia Grun.	Wis., U.S.A.	Wipf (1939)
dentata Pers.	N.Y., U.S.A.	Wilson et al. (1937)
	Wis., U.S.A.	Wipf (1939)
(*gracilis* DC.)	Wis., U.S.A.	Wipf (1939)
= *neopolitana* Ten.		
indica (L.) All.	Widely reported	Many investigators
(= *parviflora* Desf.)		
italica (L.) Lam.	Italy	Pichi (1888)
macrorrhiza Pers.	France	Vuillemin (1888)
messanensis (L.) All.	S. Africa	Grobbelaar & Clarke
		(1974)
neapolitana Ten.	S. Africa	Grobbelaar et al. (1967)
officinalis (L.) Desr.	Widely reported	Many investigators
(= *arvensis* Wall.)		
segetalis (Brot.) Ser.	Wis., U.S.A.	Wipf (1939)
speciosa Th. Dur.	Wis., U.S.A.	Wipf (1939)
suaveolens Ledeb.	Wis., U.S.A.	Wilson et al. (1937)
		Wipf (1939)
sulcata Desf.	Wis., U.S.A.	Wipf (1939)
wolgica Poir.	Wis., U.S.A.	Wipf (1939)
(= *ruthenica* Ser.)	S. Africa	N. Grobbelaar (pers.
		comm. 1966)

[a]Numerous inconsistencies appear in the literature regarding the gender of the generic name, and, in consequence, that of the specific epithets. The authors here have followed the precedent set by many botanists in considering *Melilotus* a feminine noun (Voss 1957; Wilbur 1963; Tutin et al. 1968).

Melliniella Harms   Papilionoideae: Hedysareae, Desmodiinae

Named for Lieutenant Mellin; he collected the specimens from the Togo savanna.

Type: *M. micrantha* Harms

1 species. Herb, prostrate; stems glabrous. Leaves simple, small, entire, elliptic, obtuse or rounded at the apex, subcordate at the base, glabrous above, appressed-puberulent below; stipels none; stipules dry, membranous, brownish, striate, lanceolate, acuminate. Flowers very small, violet, often paired in axils of bristly bracts or in short, dense, terminal glomerules; calyx tube very short, about 5 mm; lobes 5, sharply pointed, subequal, densely ciliate, upper ones almost free or connate at the base; petals long-clawed; standard broad-spathulate, slightly longer than the others; wings oblique-oblong, obtuse, longer than the keel petals; keel petals subacute or apically

obtuse, joined at the back; stamens 10, monadelphous; anthers uniform; ovary narrow, linear, subsessile, 5–8-ovuled; stigma small, capitate. Pod subsessile, exserted, oblong, compressed, about 1 cm long, sparsely pubescent, apiculate, slightly depressed between the seeds but not articulate, dehiscent along the ventral suture, not septate between the 5–8, small, brown, shiny, suborbicular, compressed seeds. (Harms 1914)

This genus is closely akin to *Alysicarpus*; hence the position number 280 originally assigned this genus (Harms 1915a) presumably should have been 340a. Hutchinson (1964) positioned *Melliniella* in Lotononideae trib. nov., more closely akin to Genisteae than to Hedysareae.

This monotypic genus is endemic in tropical West Africa, specifically Niger, Senegal, and Togo.

Melolobium E. & Z.   Papilionoideae: Genisteae, Crotalariinae
205 / 3665

From Greek, *melon*, "apple," "melon," and *lobion*, "pod," in reference to the melon-shaped segments of the pod.

Lectotype: *M. cernuum* (L.) E. & Z.

30 species. Shrubs or perennial herbs, often hairy, glandular, viscid, or spiny. Leaves petioled, palmately 3-foliolate; leaflets oblong or linear; stipules leafy. Flowers small, yellow, in spicate racemes; each flower subtended by leafy bracts; bracteoles often foliaceous; calyx campanulate or tubular, short, glandular or hairy, bilabiate, upper lip bipartite, larger and longer, lower lip tripartite; petals and sepals subequal in length; standard oblong to suborbicular, claw linear, channeled; wings oblong, with a linear claw; keel blunt, shorter than the wings and standard, sometimes gibbous; stamens monadelphous, connate into a sheath split above; anthers unequal, alternately short and versatile, long and basifixed; ovary sessile or short-stalked, 1- to few-ovuled, sometimes glandular or hairy; style incurved; stigma terminal. Pod linear or oblong, flattened, often constricted at intervals, usually glandular or hairy, 2-valved. (Bentham and Hooker 1865; Phillips 1951; Polhill 1976)

Polhill (1976) included this genus in the tribe Crotalarieae, segregated from Genisteae *sensu lato*.

Species of this genus are native to South Africa.

Melolobium species are pasture plants and sometimes referred to as clovers; however, this designation is improper because of their spinescence. In some parts of Africa the thorns of *M. candicans* E. & Z. are used around the bottoms of grain baskets to prevent rodent depredations (Watts and Breyer-Brandwijk 1962).

NODULATION STUDIES

The following species have been observed nodulated in South Africa by Grobbelaar and Clarke (1972, 1975): *M. adenodes* E. & Z., *M. alpinum* E. & Z., *M. candicans*, *M. microphyllum* E. & Z., and *M. obcordatum* Harv.

Named for Mendoravy, plant collector in Madagascar; his specimens provided the basis for the generic description.

Type: *M. dumaziana* R. Cap.

1 species. Tree, 15 m tall, trunk lenticellate. Leaves simple, entire, glabrous; stipules bristlelike, caducous. Flowers few, in axillary, cymose or subracemose inflorescences; pedicels appressed-pilose; bract and bracteoles inconspicuous, triangular; calyx short, sepals 5–6, free, ovate-triangular, persistent; petals 5–6, subequal, imbricate, base cuneate, apex acute; stamens 11–12, equal, free; filaments short; anthers basifixed, oblong, with apical pores transversely dehiscent; ovary sessile, compressed, transversely reticulate, papery, 2-valved, dehiscent; seeds 1–2, suborbicular. (Capuron 1968)

Despite the similarities of the anthers with those of *Cassia* and *Storckiella*, Capuron (1968) was reluctant to include this genus in the tribe Cassieae.

This monotypic genus is endemic in Madagascar. It grows in mountain forests.

Mezoneurum Desf.   Caesalpinioideae: Eucaesalpinieae
103 / 3560

From Greek, *mesos*, "middle," and *neuron*, "nerve"; the upper suture of the pod is not marginal but bordered by a broad, lengthwise wing; hence the suture gives the appearance of a central nerve.

Type: *M. glabrum* Desf.

35 species. Shrubs, tall, climbing, lianas, or rarely trees; branchlets and petioles usually armed with recurved prickles. Leaves bipinnate; pinnae and leaflet pairs opposite, numerous; leaflets small and numerous, or large and shiny; stipules small, caducous. Flowers variegated red and yellow or purplish pink, on short, thick pedicels, in terminal or axillary racemes or paniculate at apices of branches; bracts narrow, often fugacious; bracteoles none; calyx cup-shaped, oblique; tube usually short; sepals 5, orbicular, imbricate, lowest sepal outermost, largest, and concave; petals 5, orbicular, imbricate, subequal, or uppermost inner petal slightly dissimilar; stamens 10, declinate, free; filaments glabrous or hairy at the base; anthers uniform, basifixed, longitudinally dehiscent; ovary sessile or short-stalked, 1- to many-ovuled; style filiform, awl-shaped, usually clavate at apex; stigma oblique, small and terminal, or concave and ciliate. Pod oblong or lanceolate, compressed, thin-walled, rarely leathery, longitudinally and broadly winged along the thickened upper suture, indehiscent or tardily 2-valved; seeds 1–few, transverse, flat, orbicular or kidney-shaped. (Bentham and Hooker 1865; Wilczek 1952)

Representatives of this genus are common in tropical Asia, Africa, Madagascar, Australia, and the South Pacific Islands. They thrive along river banks and forest borders, in thickets and open country.

These species are not commercially important; however, in their native habitats they are used in a variety of ways. Filipinos eat the young leaves of *M. glabrum* Desf. in salads; leaf decoctions are used for asthma relief. Malaysians use leaf decoctions of *M. sumatrana* (Roxb.) W. & A. to combat diarrhea and as a vermifuge (Burkill 1966). *M. scortechinii* F. Muell., an Australian tree, is a source of barrister gum, an exudate similar to tragacanth gum. The wood of *M. kauaiense* (Mann) Hillebr., the indigenous Hawaiian tree species, was favored by the early Hawaiians as wood for spears, tapa implements, and house frames.

NODULATION STUDIES

Plantlets of *M. cucullatum* (Roxb.) W. & A. examined in the Philippines by Bañados and Fernandez (1954) and of *M. kauaiense* in Hawaii (Allen and Allen 1936b) lacked nodules.

Michelsonia Hauman   Caesalpinioideae: Amherstieae

Named for M. A. Michelson, Belgian plant explorer; he collected the specimens.

Type: *M. microphylla* (Troup.) Hauman

2 species. Trees, medium-sized. Leaves paripinnate; leaflets 7–15 pairs, subsessile, unilaterally eared. Flowers small to medium, short-stalked, solitary and sessile between 2 large, ovate, valvate bracteoles, in terminal or subterminal panicles; sepals 5, 3 free, 2 united above, ciliate; petals 5, subequal or the lateral petals smaller than the median petal; stamens 10, 1 free, 9 united at the base; ovary hairy, few-ovuled; stigma capitate. Pod oblong, apiculate, flat, with a nerve along the middle of each valve. (Hauman 1952)

Both species are native to equatorial Africa, specifically Zaire, Cameroon, and Gabon. They are dominant in shady forests.

The wood is exploited extensively for turnery objects because it is easy to work.

Microberlinia A. Chev.   Caesalpinioideae: Amherstieae

Name means "small *Berlinia*"; this genus resembles *Berlinia*, except for its smaller leaf size.

Type: *M. brazzavillensis* A. Chev.

2 species. Trees, unarmed, large, 40–45 m tall, usually with huge trunks and extensive buttresses; branchlets pubescent. Leaves paripinnate; leaflets small, in many pairs, asymmetric at the base, apex notched; rachis narrow-winged; stipules large, foliaceous, oblong-linear, free, intrapetiolar, caducous. Flowers on very short petioles, crowded in terminal panicles; bracts very large, leathery, enclosing the inflorescence, caducous; bracteoles hairy, obo-

vate, concave, leathery, enclosing the flower; sepals 4, posterior sepal a fusion of 2 and larger than other 3, imbricate, oblong-lanceolate; petals 5, usually white, subequal, oblong-spathulate, 1 larger and clawed; stamens 10, 1 free, other 9 briefly connate at the base; anthers uniform; ovary stalked, stalk joined laterally to the calyx tube, 3–6-ovuled; style slender, apically truncate. Pod oblong, woody, compressed, 2-valved, upper suture thickened and marked by 2 parallel ridges; seeds 3–4, ovate or orbicular, compressed. (Chevalier 1946; Normand 1947; Pellegrin 1948)

Both species are indigenous to equatorial West Africa.

"Zebrano," "zingana," and "African zebrawood" are trade names for the decorative yellow- and brown-striped timber of *Microberlinia* species; it is used almost exclusively for veneers and ornamental turnery (Rendle 1969).

Microcharis Benth.　　　　Papilionoideae: Galegeae, Robiniinae
280 / 3746

From Greek, *mikros*, "small," and *charis*, "grace," in reference to the delicate habit of the plant.

Type: *M. tenella* Benth.

5 species. Herbs, annuals; stems spreading, slender, usually bristly-haired. Leaves simple, subsessile, elliptic or linear-oblanceolate; stipules herbaceous, awl-shaped, sometimes persistent. Flowers very small, red, in axillary, few-flowered, lax racemes; bracts narrow; bracteoles none; calyx tube minute, campanulate, usually beset with gray hairs; lobes 5, subequal, lanceolate, about as long as the tube; standard orbicular, sides reflexed; wings oblique-obovate, longer and scarcely adherent to the blunt keel; stamens 10, diadelphous; anthers uniform; ovary sessile, linear, multiovuled; style short, smooth, tapering above; stigma capitate. Pod sessile, linear, flat, membranous, 2-valved, thinly septate between the transverse, oblong or quadrate seeds. (Bentham and Hooker 1865; Baker 1929; Phillips 1951)

Microcharis is closely akin to *Indigofera* and is sometimes regarded as a synonym (Hutchinson 1964).

Representatives of this genus are restricted to tropical Africa, specifically Angola, Guinea, Tanzania, and South Africa. They are usually present along river banks.

Milbraediodendron Harms　　　　Caesalpinioideae: Swartzieae
(115a)

Named for Gottfried Wilhelm Johannes Milbraed, German botanist.

Type: *M. excelsum* Harms

1 species. Trees, large, tall, unarmed, deciduous; buttresses small, rounded; bark gray brown, with thin, rectangular scales; crown large. Leaves

long, paripinnate; leaflets 24–32, with scattered glandular dots, alternate or subopposite, at first hairy, later smooth. Flowers few, produced in lower leaf axils, in short, panicled racemes, before leaf emergence; bracts lanceolate; bracteoles 2, awl-shaped, somewhat caducous; calyx with appressed, yellow pubescence, globose in bud, later splitting into 2–3 broad, deltoid, subequal, reflexed lobes; petals absent; stamens numerous, usually about 16, exserted, shortly connate at the base, attached to the rim of the disk; filaments long and thickened at the base; anthers small, dorsifixed; ovary stalked, oblong, thick, glabrous, 7–8-ovuled; disk thick, fleshy; style awl-shaped; stigma small. Pod drupaceous, globular to ovoid, large, outside leathery, inside filled with a fleshy, yellow brown, gelatinous pulp, indehiscent; seeds 4–5, large, transverse. (Harms 1915a; Gilbert and Boutique 1952b)

The genus is closely related to *Cordyla*.

This monotypic genus is endemic in tropical Sudan, Uganda, and Zaire.

The ripe, yellow pods of *M. excelsum* Harms are about the size of tennis balls. The yellow brown pulp of the pods, which have a fresh pea-pod flavor, is relished by elephants. These trees reach heights of about 60 m. The thick, straight trunks yield a handsome, durable timber, which has a strong broad-bean odor when freshly cut (Eggeling and Dale 1951).

Millettia W. & A.   Papilionoideae: Galegeae, Tephrosiinae
257 / 3720

Named for J. A. Millett, 18th-century French botanist; he worked in Canton, China, where the first collection of *Millettia* species was made (Dunn 1912).

Type: *M. rubiginosa* W. & A.

180 species. Trees, sometimes tall, shrubs or woody lianas, unarmed. Leaves imparipinnate; leaflets few to several pairs, oblong-lanceolate, opposite or alternate, reticulate, pinnately nerved, usually evergreen, leathery; stipels usually present; stipules small, caducous. Flowers purple, blue, rose, or white, showy, usually in lateral or terminal panicles, rarely in simple racemes; bracts and bracteoles usually caducous before flowering; calyx widely campanulate, truncate or 5-toothed; teeth generally short or nearly obsolete, upper 2 often partly or entirely joined; petals long-clawed; standard large, sometimes suborbicular with an oblong claw, spreading or reflexed, silky-haired or smooth, rarely eared or callous; wings oblong-falcate, free from the keel, often with a broad-linear claw and eared; keel petals incurved, obtuse, clawed, nonbeaked; stamens usually monadelphous, occasionally diadelphous; filaments filiform; anthers ovate, uniform; ovary usually sessile, linear, sometimes hairy, base surrounded by an annular disk or short sheath, 3- to many-ovuled; style filiform, terete, inflexed, glabrous; stigma small, terminal. Pod linear, oblong or lanceolate, flat or turgid, rigidly leathery or woody, 2-valved, tardily dehiscent, glabrous, sometimes velvety; seeds 1–few, orbicular or kidney-shaped. (Bentham and Hooker 1865; Kurz 1877; Dunn 1912; Hutchinson and Dalziel 1958; Gillett 1960a)

Millettia is closely akin to *Lonchocarpus*. Dunn's revision and review (1912) of the genus is comprehensive and instructive. Members of this large genus are common in the tropics and subtropics of the Old World. They are widely distributed throughout Burma, China, India, Japan, Malaysia, equatorial Africa, and Madagascar. A few species have been introduced into Florida and California. The plants usually occur in grasslands, along river banks, and in secondary, fringe, coastal, and humid forests.

The flowers of all species are beautifully tinted and are borne in drooping panicles resembling those of their closest relative, *Wisteria*. Among the showiest are *M. brandisiana* Ktze., *M. grandis* Skeels, *M. laurentii* De Wild., *M. ovalifolia* Kurz, *M. stuhlmannii* Taub., and *M. thonningii* Bak., which are planted as ornamentals along avenues, in parks, and in gardens in Asia and Africa (Menninger 1962; Palmer and Pitman 1961). *M. gracillima* Hemsl. in the Solomon Islands climbs on other trees, cutting into their bark until its stem is almost buried in theirs.

The wood of *M. laurentii*, known as wenge, a small tree of the Congo Basin, enjoys moderate popularity on the European continent for cabinetmaking, paneling, parquet flooring, and interior fittings. The timber is hard, dense, strong, elastic, highly resistant to fungal and insect attack, takes a good polish, and has a density of 800 kg/m³ (Rendle 1969). The wood is useful for rough construction, wagon wheel spokes, implement handles, and fuel.

The presence of poisonous principles in the plant tissue has been the primary reason for interest in this genus. These toxic substances have served native uses as arrow poisons, fish poisons, insecticides, vermifuges, and febrifuges. Among those used as sources of fish poisons are *M. atropurpurea* Benth. in Malaysia, *M. barteri* (Benth.) Dunn in tropical Africa, *M. auriculata* Bak. in India, and *M. taiwaniana* Hayata in Taiwan. The leaves of *M. sericea* Benth. are used in Malaysia to poultice sore eyes (Burkill 1966).

Rotenone, $C_{23}H_{22}O_6$, and similar isoflavones have been isolated from *M. ferruginea* Hochst. (Clark 1943), *M. pachycarpa* Benth. (Ghose and Krishna 1937), and *M. taiwaniana* (Kariyone et al. 1923); however, *Millettia* species are not profitable commercial sources of insecticides. Saponins are present in the leaves and pods of most species. Unnamed alkaloids have been isolated from *M. australis* Benth. (= *M. maideniana* F. M. Bailey) and *M. megasperma* Benth. (Willaman and Schubert 1961).

Nodulated species of *Millettia*	Area	Reported by
dubia De Wild.	Yangambi, Zaire	Bonnier (1957)
duschesnei De Wild.	Yangambi, Zaire	Bonnier (1957)
harmsiana De Wild.	Yangambi, Zaire	Bonnier (1957)
japonica Gray	Japan	Asai (1944)
stuhlmannii Taub.	Zimbabwe	Corby (1974)
usaramensis Taub.		
ssp. *australis* Gillett	Zimbabwe	Corby (1974)

From Greek, *mimos*, "mimic," "mime," in reference to the sensitivity of the leaves of many species; the leaves collapse temporarily or show "sleep movements" when touched.

Type: *M. pudica* L.

600 species. Herbs or shrubs, rarely trees, sometimes scrambling or climbing, armed with recurved spines or prickles. Leaves bipinnate, often sensitive, rarely not developed or reduced to phyllodes; leaflets small, few to many pairs; petioles usually eglandular. Flowers sessile, small, white, pink, or lilac, 4–5-parted, rarely 3–6-parted, perfect or polygamous, in globose heads or cylindric spikes, peduncles axillary, solitary or fascicled, upper ones sometimes racemose; calyx minute, tubular-campanulate; tube often longer than the calyx; lobes 4, sometimes 3, 5, or 6, narrow, usually setaceous; petals valvate, somewhat connate or free; stamens as many as or twice the number of lobes, free, long-exserted; anthers small, eglandular; ovary sessile, rarely stalked, 2- to many-ovuled, sometimes villous; style filiform, oblique; stigma small, terminal. Pod oblong-linear, plano-compressed, occasionally thickened, membranous or leathery, sometimes bristly or prickly, 2-valved, entire or segmented transversely into 1-seeded joints which separate at maturity from the persistent sutures, continuous or subseptate within; seeds flat, orbicular. (Bentham and Hooker 1865; Brenan 1959)

Species of this large genus inhabit tropical and subtropical regions, mostly in South America. The range extends from the southern United States to Argentina, with a few species endemic in Asia and Africa, and none indigenous to Australia. About 20 shrubby or arborescent species occur in North America. They are prolific in Mexico, Brazil, and Paraguay but scant or absent in the Andes. The plants generally thrive on limestone, igneous, and dry, gravelly soils. Some species form dense thickets in moist or dry open forests, or along river banks, up to 1,800-m altitude.

Many *Mimosa* species are graceful and deserve cultivation as ornamentals. The species with sensitive leaves are curiosities in many botanical gardens, parks, and greenhouses. The "mimosa" of florists is a misnomer; it is actually *Acacia decurrens* var. *dealbata*.

In general, herbaceous *Mimosa* species are weeds, and some are regarded as pests on agricultural land; but all are good soil-binders and provide wildlife cover. Forage usage is sparse, mainly because of their prickly habit and tendency to form objectionable thickets. *M. paupera* Benth. and *M. strigillosa* T. & G., 2 creeping, spineless species, have forage potential. *M. somnians* H. & B. and *M. uliginosa* Chod. & Hassl. are good bush browse despite their spines. *M. invisa* Mart. was used successfully as a green manure and cover plant for many years in Indonesia because its vigorous growth endures for 18–24 months and it is abundantly self-seeding. The development of a spineless variety improved its popularity.

Native medicinal use of plant part decoctions is limited to minor ailments. The leaves of *M. pigra* L., a marsh shrub, are infused as a bitter tonic for treatment of diarrhea in the West Indies. The roots of *M. pudica* L. yield about 19 percent tannin. The seeds have emetic properties.

The wood of small tree species is not particularly attractive and is used mostly for fuel. *M. bracaatinga* Hoehne, a slender tree in Brazil and Guatemala, is used as shade on coffee plantations, and may prove valuable for papermaking and rapid reforestation (Standley and Steyermark 1946).

Macerated roots of *M. hostilis* Benth. yield about 0.57 percent *N,N*-dimethyltryptamine, $C_{12}H_{16}N_2$, and an alkaloid with hallucinogenic properties (Pachter et al. 1959). The Pancaru Indians of Pernambuco, Brazil, made from this species a beverage which they used in mystico-religious ceremonies. Mimosine, $C_8H_{10}N_2O_4$, a hydroxamino acid of aromatic nature, was isolated from *M. pudica* by Renz in 1936. For further details concerning mimosine, refer to *Leucaena*.

NODULATION STUDIES

Three members of the genus have been reported lacking nodules, *M. flagellaris* Benth. in Argentina (Rothschild 1970), *M. lemmoni* Gray in Arizona (Martin 1948), and *M. myriadena* Benth. Guayana (Norris 1969). To what extent these negative data are related to the seasonal period of examination, peculiarities of the area, and paucity of plants observed is unknown.

Nodules on *M. pudica* develop slowly and require 3–4 weeks to reach diameters of 3–5 mm (personal observation). Nodules on *M. dysocarpa* Benth. tend to be large and lobed (Lechtova-Trnka 1931). Histological sections of nodules of this species showed that newly formed cells behind the apical meristem were interlaced with thin infection threads, which often branched 2–3 times within the individual host cells and bulged noticeably at the approach to host nuclei. The bacteroid zone consisted of hypertrophied, elongated host cells that had a vaporous appearance due to the presence of many small vacuoles in the cytoplasm and the large central vacuole in each invaded cell. Noninfected smaller interstitial cells contained small starch granules. The rhizobia retained their short rod shapes and showed little evidence of disintegration even in the basal zone. Tannin cells were scattered throughout the nodule cortex.

Descriptions of *Mimosa* rhizobia differ considerably. Leonard (1934) initially described rhizobia for *M. pudica* as organisms most readily isolated on slightly acid or neutral beef agar, tolerating reactions between pH 5.5 and 10.0, growing poorly on media commonly used to culture rhizobia, producing a trimethylamine odor on peptone media, and having a restricted host-infection range. The authors (personal observation 1939) obtained best growth of rhizobia from this species on mannitol-asparagus-potato-extract agar containing 0.3 percent $CaCO_3$. Ten days were required for the appearance of thin, dry, opaque colonies discernible by the unaided eye. An alkaline reaction was produced in a wide spectrum of carbohydrate media; litmus milk became alkaline, with no serum zone produced; tyrosinase reactions were negative. Isolates from an undesignated *Mimosa* species studied by Jensen (1967) were culturally typical of slow-growing *Rhizobium lupini* strains. On the contrary, isolates from *M. invisa* and *M. pudica* were fast-growing types that resembled the cultural morphology of *Leucaena* isolates (Trinick 1968). Similarly, isolates from *M. caesalpiniaefolia* Benth. produced rapid growth and alkaline reactions (Campêlo and Döbereiner 1969).

It is generally agreed that the host-infection range of *Mimosa* plants and their rhizobia is narrow. Although kinships within the cowpea miscellany have been shown (Carroll 1934a; Wilson 1939a; Trinick 1968), effective symbiosis seems most favored between species of other mimosoid genera such as *Prosopis*, *Leucaena*, *Desmanthus*, and *Acacia* (Ishizawa 1954; Trinick 1968; Campêlo and Campêlo 1970; personal observation).

Nodulated species of *Mimosa*	Area	Reported by
acanthocarpa Poir.	Germany	Morck (1891)
bimucronata (DC.) Ktze.	Brazil	Campêlo & Campêlo (1970)
biuncifera Benth.	S. Africa	Grobbelaar & Clarke (1974)
caesalpiniaefolia Benth.	Brazil	Campêlo & Döbereiner (1969)
camporum Benth.	Venezuela	Barrios & Gonzalez (1971)
casta L.	Trinidad	DeSouza (1966)
debilis H. & B.	Venezuela	Barrios & Gonzalez (1971)
dormiens H. & B.	Venezuela	Barrios & Gonzalez (1971)
dysocarpa Benth.	France	Lechtova-Trnka (1931)
floribunda Willd.	Venezuela	Barrios & Gonzalez (1971)
hamata Willd.	W. India	V. Schinde (pers. comm. 1977)
invisa Mart.	Widely reported	Many investigators
orthocarpa Spruce	Venezuela	Barrios & Gonzalez (1971)
pigra L.	Trinidad	DeSouza (1966)
	Zimbabwe	Corby (1974)
polycarpa Kunth	S. Africa	Grobbelaar et al. (1967)
	Argentina	Rothschild (1970)
polydactyla H. & B.	S. Africa	Grobbelaar & Clarke (1974)
pudica L.	Widely reported	Many investigators
somnians H. & B.	Venezuela	Barrios & Gonzalez (1971)
strigillosa T. & G.	Fla., U.S.A.	Carroll (1934a)
tenuiflora Benth.	Venezuela	Barrios & Gonzalez (1971)
tomentosa H. & B.	Venezuela	Barrios & Gonzalez (1971)
uliginosa Chod. & Hassl.	Argentina	Rothschild (1970)

From Greek, *mimos*, "mimic," "mime," to denote positioning in the Mimosoideae, *zygon*, "yoke," and *anthos*, "flower"; the flower heads are yoked in pairs or fours between spiny stipules.

Type: *M. carinatus* (Griseb.) Burk. (= *Mimosa carinata*)

1 species. Shrub or small tree, 3–5 m tall; trunk, if present, to about 15 cm in diameter, xerophytic. Leaves bipinnate; pinnae pairs few; leaflets numerous, minute, oblong, opposite or alternate; rachis glandular between the pinnae; stipules spiny. Flowers minute, bisexual, in sessile, axillary, globose heads, 1–4, between a pair of spiny stipules; sepals 5, free, oblong, fleshy, imbricate in bud, incurved; petals 5, oblong-lanceolate, free, valvate, slightly imbricate in the middle; stamens free, 5 long, 5 short; anthers eglandular, ellipsoid, dorsifixed; ovary short-stalked, 2-ovuled; stigma peltate, large, convex. Pod elliptic, somewhat samaroid, compressed, exocarp thinly leathery, fibrous, indehiscent, 1–2-seeded. (Burkart 1939b)

This genus is currently positioned in the monogeneric tribe Mimozygantheae, in preference to the tribe Eumimoseae. This new tribe is one of the few additions to subfamily Mimosoideae since Bentham's original conspectus set forth in 1865 in his *Genera Plantarum*.

This monotypic genus is endemic in Argentina. The plants grow in semiarid eastern areas of the northern and central territories.

Minkelersia Mart. & Gal.   Papilionoideae: Phaseoleae,
(436) / 3902   Phaseolinae

Type: *M. galactoides* Mart. & Gal.

4 species. Herbs, small, slender, creeping or climbing; rootstock tuberous. Leaves pinnately 3-foliolate; stipels present; stipules membranous-foliaceous. Flowers 1-several, on axillary, jointed peduncles in racemes; bracts 2–3, stipulelike, persistent; bracteoles minute or absent; calyx tube short; lobes 5, oblong, subequal, large, much longer than the calyx tube; standard obovate-oblong, erect, reflexed, folded, not eared at the base; wings shorter than the standard, oblong, slightly joined with the linear, spirally inrolled, sharp-pointed keel; stamens diadelphous; ovary sessile, multiovuled; style long, thicker and twisted within the keel, bearded lengthwise inside above; stigma large, oblique or introrsely lateral. Pod flat, compressed, linear-elongated, 2-valved; seeds rounded. (Rose 1897; Harms 1908a; Piper 1926)

The original generic number 424 (Taubert 1894) was changed to 436 by Harms (1908a) because of the reorganization within the subtribe necessitated by revisions and inclusions of new genera. J. Lackey (personal communication 1978) does not recognize this genus as distinctive from *Phaseolus*.

Members of this genus are indigenous to Mexico in the states of Oaxaca and Chiapas. They occur at altitudes between 1,300 and 2,900 m.

Named for C. K. F. Brisseau-Mirbel, Director of the Jardin du Roi, Paris.

Type: *M. rubiifolia* (Andr.) G. Don (= *M. reticulata* Sm.)

20–25 species. Shrubs or undershrubs, prostrate or erect, up to 2 m tall; stems mostly pubescent. Leaves opposite, sometimes whorled or alternate, simple, entire or with sharp-pointed lobes; stipules small, setaceous or none. Flowers yellow, orange, purplish red, or blue, solitary or clustered in axillary or terminal racemes; bracts and bracteoles small or none; calyx lobes subequal, imbricate, upper 2 often broader and more connate higher up; petals clawed; standard rounded or kidney-shaped; wings oblong, narrower and longer than the keel; stamens free, ovary subsessile, 2- to many-ovuled; style incurved; stigma terminal, capitate. Pod ovoid-oblong, inflated, divided longitudinally into 2 cells by a false septum projecting from the lower suture; seeds shiny, black. (Bentham 1864; Bentham and Hooker 1865)

Members of this genus are native to Australia. They are common in Queensland, New South Wales, and Victoria. Species thrive in coastal areas, sandstone tablelands, and highlands.

NODULATION STUDIES

Cultural and unilateral host-infection tests using rhizobial strains from 3 species (Lange 1961) confirmed McKnight's (1949a) original inclusion of *Mirbelia* species in the cowpea miscellany.

Nodulated species of *Mirbelia*	Area	Reported by
dilatata R. Br.	W. Australia	Lange (1959)
floribunda Benth.	W. Australia	Lange (1959)
ovata Meissn.	W. Australia	Lange (1959)
ramulosa (Benth.) C. A. Gardn.	W. Australia	Lange (1959)
rubiifolia (Andr.) G. Don (= *reticulata* Sm.)	Queensland, Australia	McKnight (1949a); Bowen (1956); Norris (1958a)
spinosa Benth.	W. Australia	Lange (1959)

Named for Johann Jakob Moldenhower, 1766–1827, Professor of Botany, Kiel.

Type: *M. floribunda* Schrad.

5 species. Trees, unarmed, deciduous, 15–25 m tall. Leaves bipinnate, paripinnate and imparipinnate; leaflets leathery, undersides rust-colored; stipules small, narrow, caducous. Flowers yellow, subpanicled in large, terminal racemes; bracts small, caducous, narrow; bracteoles none; calyx tube very short; lobes usually 5, seldom 4, subequal, valvate in bud, but later separating; petals 4–5, subequal, long-clawed, spreading, oblong-orbicular, margins fringed or crinkled, imbricate; stamens free; filaments smooth, 9 short,

441

straight, with anthers large, fertile, basifixed, lower most filament much longer, incurved, with anther small, sterile; ovary sessile, free, many-ovuled; style slender, long, thickened and club-shaped near the apex; stigma trim, truncate, ciliate. Pod oblong, flat, leathery, 2-valved; seeds ovoid, transversely arranged. (Bentham and Hooker 1865)

Representatives of this genus are present in tropical Brazil and Venezuela.

These plants are valued as park and garden ornamentals because of their rapid growth and magnificent flowering habit, which resembles *Delonix* species.

Monarthrocarpus Merr.	Papilionoideae: Hedysareae,
(337b)	Desmodiinae

From Greek, *monos*, "single," *arthron*, "joint," and *karpos*, "fruit," in reference to the nonarticulated pods.

Type: *M. securiformis* (Benth.) Merr.

2 species. Undershrubs, suberect, up to about 60 cm tall. Leaves pinnately 3-foliolate, sometimes simple; leaflets lanceolate or ovate-lanceolate, large, entire, reticulate below; stipels awl-shaped; stipules ovate-lanceolate, longitudinally nerved. Flowers small, in long, slender, terminal, fascicled racemes or subpanicles; bracts subulate-lanceolate, caducous; bracteoles absent; calyx tube short, 2 upper lobes connate above, lower 3 subcaudate-acuminate; standard orbicular-obovate, base narrowed, short-clawed; wings oblong; keel petals joined; stamens 10, diadelphous; anthers small, equal; ovary stalked, 1-ovuled. Pod stalked, flattened, narrow, oblique or sickle-shaped, tip acute, markedly reticulate, sprinkled with short, hooked hairs, nonarticulate, indehiscent; seed solitary, brown, glabrous, narrow-oblong. (Merrill 1910; Harms 1915a)

The genus closely resembles *Desmodium*, except for the 1-ovuled ovary, 1-seeded, 1-jointed pods, and the narrow-oblong seeds.

These species occur in the Philippines and the Moluccas, in forested areas.

Monopetalanthus Harms	Caesalpinioideae: Cynometreae
(45a) / 3494	

From Greek, *monos*, "single," *petalon*, "petal," and *anthos*, "flower"; the flower has only 1 petal.

Type: *M. pteridophyllus* Harms

8–10 species. Trees, medium, up to 30–40 m tall. Leaves paripinnate, fernlike, short-petioled; leaflets 1–many pairs, oblong, closely spaced, sessile, emarginate, asymmetric, subrectangular at the base, curving upward, not pellucid-punctate; rachis usually not grooved; stipels absent; stipules large, heart-shaped at the base, membranous, very caducous or persistent. Flowers conelike in bud, in more than 2 rows in axillary or terminal, panicled racemes shorter than the leaves; bracts oval, caducous; bracteoles 2, rust-velvet, opposite, sepaloid, concave, valvate, enveloping the flower, persistent; calyx tube very short or obsolete; sepals 0–5, when present, small, oblique-ovate, 2 united almost to the middle, the other 3 very similar, minutely toothed, fringed with rusty hairs; petal 1, usually white, short-clawed, emarginate or rounded at the apex, spathulate, much longer than the calyx, membranous; 1–4 rudimentary petals sometimes present; stamens 8–10, all fertile; filaments all briefly joined at the base, or 1 free and 9 joined, all slender; anthers dorsifixed, elliptic, longitudinally dehiscent; ovary obovate-oblong, short-stalked, 2–3-ovuled, densely hairy; style filiform, lower part hairy; stigma minute, capitellate. Pod small, compressed, broad-oblong, obtuse or mucronate, woody, stalked, 2-valved, dehiscent; seeds few. (Harms 1897; Baker 1930; Pellegrin 1948; Léonard 1952, 1957)

Members of this genus are indigenous to tropical West Africa. They are gregarious in rain forests along the Gulf of Guinea.

These trees tend to have straight boles, medium to high buttresses and semiopen crowns. The wood is hard, tough, and usually pinkish brown in color. Natives use trunks of large specimens for dug-out canoes.

NODULATION STUDIES

The authors are not aware that members of this genus have been examined for nodules; however, ectotrophic mycorrhizae, which are rare among Leguminosae, do occur on the root systems (Fassi and Fontana 1962a).

Monoplegma Piper   Papilionoideae: Phaseoleae, Phaseolinae

From Greek, *monos*, "single," and *plegma*, "braid," "wreath," presumably in reference to the ridge on each valve.

Type: *M. trinervium* (Donn. Smith) Piper

1 species. Vine, tall, with woody stems. Leaves palmately trifoliolate; leaflets entire, 3-nerved from the base, the 2 lateral nerves nearly as large as the midrib. Flowers in racemes, each pedicel with prominent glands at the base; calyx campanulate, 2-lipped, upper lip broad, emarginate, as long as the tube, lower lip with 3 broad, ovate lobes nearly as long as the upper lip, median lobe smallest; petals clawed; standard orbicular, emarginate, 2-eared at base, a thick narrow gland near the middle of the petal; wings spathulate, blunt and hooded at apex; keel bent, blunt at apex, as long as but broader than the wings; stamens diadelphous; anthers small; style hairy on inner side; stigma lanceolate, terminal. Pod large, woody, a small longitudinal ridge on each valve very near the ventral suture; inner layer of the pod not separating at maturity; seeds 1–2, globose, the narrow linear hilum covered with spongy

tissue and extending for three-fifths of the circumference. (Piper 1920, 1926; Rudd 1967)

The genus bears a resemblance to *Canavalia*, but the floral characters indicate a closer relationship to *Dolichos* (Piper 1920). Rudd (1967) regarded *Monoplegma* as a synonym of *Oxyrhynchus*.

This monotypic genus is endemic in Mexico, Guatemala, and Costa Rica. It is found in thickets.

Monopteryx Spruce ex Benth.   Papilionoideae: Sophoreae
154 / 3614

From Greek, *monos*, "single," and *pteryx*, "wing," in reference to the 1-winged pod.

Lectotype: *M. angustifolia* Spruce ex Benth.

3 species. Trees, unarmed, buttressed, gigantic, often up to 30–35 m tall, with trunk diameters exceeding 1 m. Leaves imparipinnate; leaflets leathery, large, few pairs, eglandular; stipels none. Flowers small, rose-colored, in terminal, panicled racemes; bracts and bracteoles small; calyx very short, bilabiate, upper lip large, enveloping the petals, lower lip short, weakly 3-toothed; petals sessile; standard ovate-suborbicular, reflexed, shorter than the free, oblong wings and the dorsally-joined keel petals; stamens 10, subequal, free; anthers linear-oblong; ovary stalked, 1-ovuled; style short, curved; stigma inflexed. Pod large, flat, winged below the seed, 2-valved, elastically dehiscent. (Bentham and Hooker 1865)

Monopteryx species are native to the north-central region of the Brazilian Amazon and Venezuela. They occur along rivers but not in inundated areas or marshlands.

The hard, coarse-textured wood has potential for furniture and heavy construction, but a marketable supply is inaccessible. The seeds contain large amounts of a light-colored, bitter oil, but are edible when roasted (Record and Hess 1943).

Monoschisma Brenan   Mimosoideae: Adenanthereae

From Greek, *monos*, "single," and *schisma*, "division"; the pod splits along 1 suture.

Type: *M. leptostachyum* (Benth.) Brenan

2 species. Trees, unarmed. Leaves opposite, bipinnate; pinnae usually opposite; leaflets 1–few pairs; petiole eglandular; rachis glandular between pinnae. Flowers in spikes; calyx glabrous; lobes 5, entirely free; petals smooth, joined at the base, free above; ovary glabrous to slightly pubescent, short-

stalked; disk inconspicuous. Pod straight or somewhat curved, valves flattened, torulose, leathery, deeply constricted between the seeds, dehiscent along 1 suture, with valves twisting; seeds brown, elliptic, flat, narrow-winged, funicle obliquely attached at the middle. (Brenan 1955)

Monoschisma is akin to *Anadenanthera*, from which it differs in having spicate, not capitate, inflorescences, winged seeds, eglandular petioles, a glandular rachis, and corolla lobes free to the base (Brenan 1955).

Both species are widespread throughout tropical South America.

Mora Schomb. ex Benth.   Caesalpinioideae: Dimorphandreae

From Tupi, *moira*, "tree," the vernacular South American Indian name for the trees and the timber.

Type: *M. excelsa* Benth.

10 species. Trees, unarmed, buttressed, gigantic, up to about 40 m tall, free of branches throughout the lower half. Leaves paripinnate; leaflets oblong-lanceolate, 3–4 pairs, glabrous; stipules small, caducous. Flowers small, sessile, white, in dense spikes in terminal panicles; bracts small, scalelike, caducous; bracteoles absent; calyx urnlike or campanulate, smooth; lobes 5, short-toothed, imbricate; petals 5, equal, oblong, imbricate; stamens 10, exserted, 5 opposite the petals with fertile, bilocular, dorsifixed, white-hairy anthers, longitudinally dehiscent; alternate anthers clavate, smooth; ovary free, short-stalked, several-ovuled; style compressed; stigma terminal. Pod oblong, woody, dehiscent; seeds 1–6, kidney-shaped, very large, soft. (Bentham 1839; Ducke 1925a; Amshoff 1939)

Species of this genus were formerly included in *Dimorphandra* as a section. *Mora* and *Dimorphandra* are frequently confused; both are commonly termed mora in forestry literature. *Dimorphandra* species are readily distinguishable by their bipinnate leaves and small, hard seeds.

Members of this genus are common in the West Indies, and tropical Central and South America. They are abundant and gregarious, forming pure stands in the so-called mora forests of Trinidad, and the Orinoco and Amazon deltas. Species are tolerant of shade and swampy soil.

M. excelsa Benth. and *M. gonggrijpii* (Kleinh.) Sandw., known as mora and morabukea respectively, are important timber trees in Guiana. The wood is heavier and stronger than white oak and has a density of 1025 kg/m^3, but it is not resistant to marine borers. Its use is reserved locally for heavy construction (Kukachka 1970). The heartwood of both these species contains an unusually high content of saponins (Farmer and Campbell 1950).

Seeds of *M. megistosperma* (Pitt.) Britt. & Rose (= *Dimorphandra oleifera*) are kidney- or mango-shaped, and measure 18 cm long by 12 cm wide. Each seed weighs approximately 110–115 gm. They are considered to be the largest dicotyledonous seeds known. Natives in many areas make use of them as a source of a dark red dye.

NODULATION STUDIES

In 1903 Wright reported nodulation of *Dimorphandra mora*, the synonym for *M. excelsa*, in Sri Lanka. In 1934 Davis and Richards also stated that nodules were abundant on root systems of *M. excelsa* at Morabella Creek, Guyana. Similarly, DeSouza (1966) confirmed their presence in Trinidad; but fresh rootlet material sent to the authors by DeSouza contained only abundant, minute, round, reddish brown, evenly spaced hypertrophies resembling nodular protuberances on *Podocarpus* species. Attempts to isolate rhizobia from these hypertrophies were unsuccessful (personal observation). Somewhat earlier, Johnston (1949) had concluded that a vesicular-arbuscular association existed within the roots of this species. This conclusion was sustained by Norris (1969), who judged that the stubby outgrowths on the purple brown roots of both *M. excelsa* and *M. gonggrijpii* in Guyana were mycorrhizal.

Morolobium Kosterm.   Mimosoideae: Ingeae

Type: *M. monopterum* Kosterm.

One of approximately 10 genera segregated from *Pithecellobium* by Kostermans (1954). Distinctiveness of the only species of this genus, *M. monopterum* Kosterm., was based on rachilla bearing 1 leaflet and seeds not being arillate.

This monotypic genus is endemic in the Moluccas.

Mucuna Adans.   Papilionoideae: Phaseoleae, Erythrininae
401 / 3877

From *mucunán*, the name given these plants by the Tupis, South American Indians of the Amazon valleys of Brazil.

Type: *M. urens* (L.) DC.

150 species. Herbs, vines, rarely shrubs, climbing, woody, tall, rarely short and erect. Leaves pinnate, 3-foliolate, usually large; leaflets asymmetric, often oblique-ovate, usually hairy, apiculate; stipels usually present; stipules small, usually caducous. Flowers large, showy, purple, orange scarlet, red, or yellowish, fascicled-racemose on axillary peduncles or subcymose at the apex of the peduncle; bracts and bracteoles small, caducous; calyx often hairy, bilabiate; lobes 5, ovate, upper lip bilobed, lower lip trilobed, longer; standard sessile, ovate-elliptic, folded inward, eared at the base, ears often upturned,

subequal to the wings; wings oblong-ovate, incurved, sessile, eared, often adherent to the keel; keel sessile, linear-oblong, incurved or beaked, often longer than and joined to the wings; stamens 10, diadelphous; anthers unequal, alternately longer and basifixed, shorter and dorsifixed, bearded; ovary sessile, few-ovuled, villous, surrounded by a poorly developed, usually basal, cupular to lobular disk; style linear, filiform; stigma small, capitate, terminal. Pod thick, ovate to oblong, often covered with brown, stinging hairs, 2-valved, septate within; seeds 1–10, large, round, discoid, with hilum more or less one-half the circumference or smaller, ovoid, with a short, raised hilum. (Bentham and Hooker 1865; Amshoff 1939; Hauman 1954c)

Species of this large genus are well distributed throughout tropical regions of the Old World and the New World. They are common to sandy beaches, coves, bushwood areas, forest borders, thickets, and moist areas.

Several of the viny species, especially *M. aterrima* (Piper & Tracy) Merr., Mauritius bean, *M. deeringiana* (Bort) Small, Florida velvet bean, and *M. utilis* Wall. ex Wight, Bengal bean, are rated highly as soil renovators, cover crops, green manures, and forages. The cooked beans are suitable as feed for cattle and swine. In early American literature these species were designated as members of *Stizolobium*, a generic name now rejected. *M. deeringiana* was widely promoted in Florida as an annual summer forage plant for interplanting with corn. *M. gigantea* (Willd.) DC. is commonly found throughout Polynesia along rocky sea coasts and in coves and valleys, where its tangled growth over bushes forms impenetrable thickets. *M. bennetti* F. Muell. and *M. novoguineensis* Scheff. are noteworthy as ornamentals for their spectacular inflorescences.

The barbed hairs, or trichomes, on the pods of several *Mucuna* species cause an intense stinging irritation and itching. Those on *M. pruriens* (L.) DC. have earned the species the common name "cow itch" or "cowhage," a corruption of "kiwach," a Hindi word meaning "bad rubbing." Ingestion by cattle has resulted in hemorrhage, emaciation, and death. According to Shelley and Arthur (1955b), the specimen of *M. pruriens* collected in 1688 by Dr. Hans Sloane is still on display in the British Museum, London, as an example of "itch powder."

The pruritis caused by contact with the pod hairs was at first thought to be a mechanical effect of the piercing spicules. After it was demonstrated that the itching properties were destroyed by boiling cowhage hairs, the responsible subtance was viewed as a member of the histamine-liberator group similar to that in bee and snake venom. This led to an extensive chemical inquiry. In the ensuing years various alkaloidal compounds were isolated from plant parts: mucunine and mucunadine (Mehta and Majumdar 1944; Santra and Majumdar 1953); prurienene and prurieninine and (Majumdar and Zalani 1953); mucuadine, mucuadinine, mucuadininine, prurienidine, in conjunction with the confirmation of prurieninine and nicotine (Majumdar and Paul 1954); 5-hydroxytryptamine (Bowden et al. 1954); and 5 new alkaloids in addition to certain previous ones (Rakhit and Majumdar 1956). Shelley and

Arthur (1955a,b) isolated, purified, and identified mucunain, a new plant proteolytic enzyme, as the active pruritogenic principle of *M. pruriens.*

Natelson's comment (1969) that *M. deeringiana* (as *Stizolobium*) is a rich source of L-dopa is of interest, in view of the promising results with this amino acid in the treatment of Parkinson's disease.

NODULATION STUDIES

Nodules were first recorded on *M. pruriens*, growing in Java, by Wright in 1903. Only *M. deeringiana* had merited consideration in plant-infection tests as late as 1932.

Information pertaining to rhizobial-host-plant relationships of *Mucuna* species is sparse, although it appears from the dominance of effective responses on *Mucuna* (as *Stizolobium*) that these species are in the cowpea miscellany (Burrill and Hansen 1917; Carroll 1934a). Also Ishizawa (1954) reported the successful nodulation of soybean, cowpea, and *Phaseolus angularis* by 3 strains of rhizobia from *M. capitata* W. & A. Rhizobia from nodules of *M. gigantea* are slow growing; colonies on yeast-extract mannitol mineral salts agar and asparagus-extract mannitol agar were not visible until about the 14th day and were not large enough to be isolated in less than 3 weeks. Reciprocal cross-inoculation tests using rhizobia and seeds of this species and of *Vigna sinensis* were positive (Allen and Allen 1936a).

Nodulated species of *Mucuna*	Area	Reported by
†*aterrima* (Piper & Tracy) Merr.	Sri Lanka	W. R. C. Paul (pers. comm. 1951)
bennetti F. Muell.	Trinidad	DeSouza (1963)
capitata W. & A.	Japan	Ishizawa (1954)
	Philippines	Pers. observ. (1962)
coriacea Bak.	S. Africa	Grobbelaar et al. (1967)
	Zimbabwe	W. P. L. Sandmann (pers. comm. 1970)
ssp. *irritans* (Burtt Davy) Verdc.	Zimbabwe	Corby (1974)
†*deeringiana* (Bort) Small	U.S.A.	Tracy & Coe (1918)
diabolica Backer ex Heyne	Java	Keuchenius (1924)
gigantea (Willd.) DC.	Hawaii, U.S.A.	Allen & Allen (1936a)
novoguineensis Scheff.	Trinidad	DeSouza (1963)
poggei Taub.		
var. *pesa* (De Wild.) Verdc.	Zimbabwe	Corby (1974)
pruriens (L.) DC.	Sri Lanka	Wright (1903)
	Java	Keuchenius (1924)
var. *pruriens*	Zimbabwe	Corby (1974)
var. *utilis* (Wall. ex Wight) Backer	Zimbabwe	H. D. L. Corby (pers. comm. 1972)
(= *utilis* Wall. ex Wight)	Ill., U.S.A.	Burrill & Hansen (1917)
rostata Benth.	Trinidad	DeSouza (1966)
sloanei Fawc. & Rendl.	Trinidad	DeSouza (1966)
urens (L.) DC.	Guyana	Norris (1969)
†*utilis* Wall. ex Wight	Hawaii, U.S.A.	Allen & Allen (1939)

†Reported as *Stizolobium* species.

*Muellera L. f. Papilionoideae: Dalbergieae, Lonchocarpinae
(365) / 3837

Named for Ferdinand Jacob Heinrich Mueller, 1825–1896, noted Australian botanist.

Type: *M. frutescens* (Aubl.) Standl. (= *M. moniliformis* L. f.)

2 species. Trees or shrubs, unarmed. Leaves imparipinnate; leaflets few, 5–9, entire, lanceolate; stipels none; stipules minute. Flowers rose or violet, paired on short peduncles along an axillary or lateral rachis; bracts and bracteoles minute; calyx campanulate to cup-shaped, puberulent, shallow; teeth 5, very short, broad or absent; petals clawed, smooth; standard broad-ovoid, without ears or callosities; wings falcate-oblong, eared at the base, joined to the incurved, blunt keel; keel petals oblong, shorter than the wings; stamens 10, monadelphous; anthers versatile; ovary subsessile, several-ovuled; glandular disk absent; style slender, incurved; stigma small, terminal. Pod thick, fleshy or corky, leathery, subcylindric, sometimes globose or constricted between the seeds, indehiscent; seeds usually numerous, large, flat, ovoid. (Standley and Steyermark 1946; Dwyer 1965)

Both species are native to tropical Mexico, and Central and South America. They inhabit moist thickets, forests, and stream banks at low elevations.

Muelleranthus Hutch. Papilionoideae: Genisteae, Crotalariinae

Named for Ferdinand Jacob Heinrich Mueller, 1825–1896, noted Australian botanist.

Type: *M. trifoliolatus* (F. Muell.) Hutch.

3 species. Herbs, prostrate, pubescent; stems slender, long, wiry, radiating from a central root, often overall hairy. Leaves palmately 3-foliolate, on long, slender petioles; leaflets broad-obovate, subsessile, small, entire; stipules free, ovate, large, persistent. Flowers 1–2, sometimes 3, brown or purplish yellow, on short pedicels on long slender leaf-opposed peduncles; bract linear, at base of pedicel; bracteoles 1, shortly below calyx, persistent; calyx obscurely bilabiate; lobes unequal, upper 2 short, lowermost lobe longest; standard obovate-orbicular, clawed, often with a dark-colored eyespot; wings usually obovate, dark-colored; keel petals yellow, subcircular or blunt at the apex; stamens 10; filaments united into a sheath split along the upper side; anthers uniform; ovary 6–8-, sometimes 10-ovuled, short-stalked; style awl-shaped, glabrous; stigma terminal. Pod small, short-stalked, linear-oblong, mottled faintly, transversely reticulate, containing globose, brownish, nonarillate seeds. (Hutchinson 1964; Lee 1973)

Polhill (1976) assigned this genus to the tribe Bossiaeae in his revision of Genisteae *sensu lato*.

Members of this genus are indigenous to northwestern New South Wales and central Australia.

From Latin, *mundulus*, diminutive of *mundus*, "trim," "neat," "elegant," in reference to the clean style.

Lectotype: *M. sericea* (Willd.)

15–20 species. Shrubs or small trees, silky-pubescent; bark corky. Leaves imparipinnate; leaflets entire, lanceolate or elliptic, glabrous above, pubescent underneath; stipels none; stipules small. Flowers rose or violet, clustered in axillary or terminal, leaf-opposed racemes; bracts small; bracteoles none; calyx campanulate, sub-2-lipped; lobes ovate, shorter than the tube, upper 2 subconnate; standard large, elliptic-obovate, spreading, silky-pubescent, with a transverse callus at the base of the short, curved claw; wings falcate-oblong, eared, clawed, slightly adherent to the blunt, incurved keel; stamens monadelphous, vexillary stamen free only at the base, bent, then joined with others into a closed tube; alternate filaments dilated; anthers uniform; ovary subsessile, hairy, many-ovuled; style subcylindric, smooth, incurved, apex inflated; stigma small, capitate. Pod linear, flattened, sutures thickened, tardily dehiscent; seeds 6–8, kidney-shaped. (Bentham and Hooker 1865; Hutchinson and Dalziel 1958)

This genus is closely related to *Tephrosia* and *Millettia*.

Representatives of this genus are widespread in the Old World tropics, East Africa, the East Indies, India, and Sri Lanka; the majority of species occur in Madagascar.

Species occur mostly in the wild state; there is, however, limited cultivation of select cultivars in East Africa as a source of commercial insecticide preparations. They are commonly called corkbush or silver bush in South Africa.

The seeds and bark of *Mundulea* species, especially *M. pauciflora* Bak., *M. sericea* (Willd.) A. Chev., *M. striata* Dubard and Dip., and *M. telefairii* Bak., have long records as fish poisons. In 1898 Greshoff isolated from bark extracts of the latter a white crystalline powder, which he assumed was derrid, the name then applied to what is now known as rotenone. The powder showed impressive toxicity values; goldfish weighing 50 gm were stupefied by 1 ppm of the powder in 10 minutes, by 1 in 5 million parts in 40 minutes, and 1 in 10 million parts in 50 minutes (*fide* Worsley 1936).

The piscicidal action of *Mundulea* seeds and bark is more rapid than that of *Lonchocarpus*, *Derris*, and *Tephrosia* species. Moreover, the danger to humans from eating fish killed by plant parts of some *Mundulea* species is considerable (Greenway 1936). In parts of Africa, strips of bark of *M. sericea* are tied around the legs of cattle at watering time in the rivers to protect them from reptiles. Allegedly, the toxic principle has sufficient potency to repel and even kill small crocodiles.

Worsley (1937a) isolated from *M. suberosa* DC. unidentified glucosides, alkaloids, and several insecticidal substances of which only rotenone was appreciably toxic to insects. Earlier a highly toxic glucoside, $C_{33}H_{30}O_{10}$, was attributed to *Mundulea* roots and bark (Pammel 1911). In 1959 Burrows et al. isolated mundulone, $C_{26}H_{26}O_6$, an isoflavone of novel complexity which they considered identical with *Mundulea* substance A isolated earlier by Narayana and Rangaswami (1955) and given the probable formula $C_{25}H_{24}O_6$. Three crystalline substances were described by Dutta (1955), of which the principal component was later termed munetone, $C_{21}H_{18}O_4$ (Dutta 1959). Other chemical substances are enumerated by Watt and Breyer-Brandwijk (1962).

Toxicity of *Mundulea* bark is variable, but it exceeds that of seeds and leaves. One bark specimen from an East African district was as toxic to aphids as a sample of *Derris* root containing 5.4 percent rotenone; a sample from another area was less than half as toxic (Worsley 1936). Debarked stems show low toxicity. Smooth bark also has higher potency than corky bark (Worsley 1937b). The conclusion that bark from trees growing in calciferous soils in dry areas is more toxic than bark from trees growing in wet acid soils seems warranted.

The wood of *M. sericea* lacks commercial importance. Its attractive yellow color, fine grain, and durable qualities resemble satinwood, *Chloroxylon*. Usage is confined to small turnery articles because of the scarcity of large pieces and the objectionable odor of the wood.

NODULATION STUDIES

Nodules on *M. sericea* in Hawaii (personal observation) were abundant on central taproot systems, finger-shaped to coralloid, and bore a dark-surfaced, rough periderm layer. Nodule interiors evidenced a rich leghemoglobin content. Old nodules were suggestive of perennial types. W. P. L. Sandmann (personal communication 1970) confirmed nodulation of this species in Zimbabwe.

Parker (1933) applied crushed nodules of *Erythrina caffra* to dying, non-nodulated, potted plants, designated as *M. suberosa*. Nodulation and vigorous plant growth were apparent within 2 months. Thus one may assume that *Mundulea* species are allied to the cowpea miscellany, since *Erythrina* species are recognized members.

Myrocarpus Allem.   Papilionoideae: Sophoreae
125 / 3583

From Greek, *myron*, "sweet oil," "perfume," and *karpos*, "fruit."
   Type: *M. fastigiatus* Allem.
   4 species. Trees, medium to large, unarmed; wood hard; bark rough, deeply fissured, resinous. Leaves imparipinnate; leaflets, 5–9, ovate, leathery, pellucid-punctate; stipels absent; stipules small. Flowers regular, fragrant, white or yellowish green, small, borne in small-clustered, axillary and terminal, spicate, catkinlike racemes; bracts small, scalelike; bracteoles absent; calyx short-campanulate; teeth 5, short, subequal, upper ones sometimes connate; petals 5, free, clawed, subequal, linear-lanceolate; stamens 10, free, glabrous,

exserted, very conspicuous; anthers small, dorsifixed, uniform; ovary stalked, linear, multiovuled; style short, inflexed; stigma small, terminal. Pod short-stalked, elongated, compressed, samaroid, winged along each suture, pericarp above the seeds turgid and conspicuously filled with resinous cells; seeds 1–2, oblong, hilum near apex. (Bentham and Hooker 1865; Burkart 1943)

Members of this genus are common to the tropical and semitropical eastern portions of Brazil, Paraguay, and Argentina.

These trees of 40–45-m heights, with trunk diameters of 1–1.25 m, have ornamental and shade value. The dark red brown wood is esteemed locally for durable, heavy construction purposes, but it is not available in quantity for export. The beautifully veined wood of *M. frondosus* Allem. is especially esteemed for exquisite furniture and small, decorative turnery articles; also, because of its spicy fragrance, it is favored for lining wardrobe cabinets. The balsam obtained from the bark is used as a medicine, a perfume, and an incense in churches.

Nodulation Studies

M. frondosus, the only member of the genus examined to date, is nodulated in Argentina (Rothschild 1970).

Myrospermum Jacq.   Papilionoideae: Sophoreae
127 / 3585

From Greek, *myron*, "sweet oil," "perfume," and *sperma*, "seed."

Type: *M. frutescens* Jacq.

1 species. Shrubs or small trees, unarmed. Leaves imparipinnate, deciduous; leaflets numerous, with scattered pellucid glandular dots and lines; stipels absent. Flowers large, showy, white, tinged with pink, in axillary racemes; calyx turbinate, incurved, apex membranous; lobes 5, short, broad; standard obovate; the 4 lower petals subsimilar, free, falcate-lanceolate, acute; stamens 10, free, persistent; filaments thin, long; anthers minute; ovary stalked; ovules 2–more; style awl-shaped; stigma small, terminal. Pod samaroid, stalked, compressed, apex hardened, long-narrowed to the base, 2-winged, wings long, broad, membranous, upper wing broader than the lower, indehiscent; seed oblong, compressed, located at the apex. (Bentham and Hooker 1865)

This monotypic genus is endemic in tropical central and northern South America and the West Indies. It thrives in dry, rocky, hillside areas.

The brown, greenish- to purplish-tinged heartwood is extremely hard and generally heavier than water. The wood is durable but difficult to crosscut and, accordingly, is used sparingly (Record and Hess 1943).

From Greek, *myron*, "sweet oil," "perfume," and *xylon*, "wood."

Type: *M. peruiferum* L. f.

6 species. Trees, unarmed, evergreen, 20–30 m tall, with trunks up to 75 cm in diameter; crowns handsome. Leaves imparipinnate; leaflets few, 3–11, alternate, oval-lanceolate, entire, glabrous, petioluled, usually large, with scattered, translucent, glandular oil dots or lines; stipels absent. Flowers whitish, medium-sized, in simple, axillary racemes or fascicled-paniculate at the base of branches; bracts and bracteoles small, caducous; calyx shallowly campanulate, subincurved; lobes 5, irregular; standard clawed, broad-orbicular; wings and keel petals subequal, free, narrow, spathulate; stamens 10, equal, free or shortly united at the base, falling with the petals; anthers long, large, uniform, acuminate, longer than the filaments; ovary long-stalked, 1–2-ovuled at apex; style short, filiform, incurved; stigma terminal. Pod samaroid, long-stalked, flat, indehiscent, both sutures winged, lower wing narrower than upper wing, long-tapered at the base, 1-seeded at the tip; seeds subreniform. (Harms 1908c; Amshoff 1939)

Representatives of this genus are indigenous to tropical South America, southern Mexico, and Central America, chiefly on the Pacific side; they are not common in the Amazon Basin.

M. pereirae (Royle) Klotz. and *M. balsamum* (L.) Harms (= *M. toluiferum* HBK), 2 closely related species, are commercial sources of Peru and tolu balsams, respectively. Peru balsam derived its name from Callao, Peru, the port of export, although the center of commercial production lay in northwestern El Salvador. This product was introduced into Europe by the Spaniards in the early 16th century. Tolu balsam entered commercial channels from Colombia and Venezuela (C. M. Wilson 1945; Howes 1949). *Myroxylon* trees are not a profitable source of balsam until about the 15th year; thereafter they are productive for 15–25 years. Methods of obtaining the balsam are drastic and often result in seriously injuring or killing the tree, inasmuch as the balsam is not normally exuded. Twenty-year-old trees yield about 3 kg of balsam annually. At one time El Salvador exported about 52,164 kg, mostly to the United States (Howes 1949).

Peru balsam contains 53–66 percent colorless, aromatic, oily liquid, and about 28 percent dark resin (Howes 1949). Its fragrance is attributed to vanillin, coumarin, cinnamic, and benzoic acids. Its uses are mostly as a fixative in perfumery, a flavor ingredient in cough mixtures and expectorants, in ointments, salves, and proprietary preparations, as a hair set and thickening agent, and as incense in churches.

Myroxylon wood varies from yellow orange to rose, is hard, with a density of 900–1090 kg/m³, has a spicy, fragrant scent, and does not respond well to staining. It is not a commercially marketed timber. In local areas it is used for interior trim, furniture, and cabinetwork.

NODULATION STUDIES

M. balsamum was nodulated in Hawaii (personal observation 1962).

Named for Bohumil Němec, Professor of Botany, Karlova University, Prague.

Lectotype: *N. atropurpurea* (Turcz.) Domin

12 species. Shrubs, small, rigid; young branches angular and softly pubescent, not spinous. Leaves simple, ovate or oblong, opposite or in whorls of 3–4, thick, leathery; stipules bristly. Flowers conspicuous, yellow, purple, or variegated, in short, dense, axillary or terminal fascicles or corymbs, rarely racemose; calyx villous; lobes 5, upper 2 joined higher up and shorter than the lower 3; petals clawed; standard broad, rounded; wings oblong, keel subequal to the wings, incurved; stamens free; ovary stalked, densely hairy, 4- or rarely 6-ovuled; style slender, dilated at the base; stigma small, terminal. Pod small, turgid, nonseptate; seeds strophiolate. (Domin 1923)

All members of this genus formerly constituted series *Gastrolobioideae* of *Oxylobium*.

Members of this genus are native to Western Australia.

NODULATION STUDIES

Three species, *N. atropurpurea* (Turcz.) Domin, *N. capitata* (Benth.) Domin, and *N. reticulata* (Meissn.) Domin, were reported nodulated as *Oxylobium* synonyms (Lange 1959).

Neochevalierodendron J. Léonard Caesalpinioideae: Cynometreae

Name means "Chevalier's new tree," in honor of Professor Auguste Chevalier, 1873–1956; he first collected the plant parts in Gabon in 1912.

Type: *N. stephanii* (A. Chev.) J. Léonard

1 species. Tree, 8–15 m tall, with a trunk 30–60 cm in diameter, grayish bark and dense foliage. Leaves paripinnate; leaflets 4, rarely 2, entire, cuneiform at the base, acuminate at the apex, uppermost pair opposite, lower 2 alternate, emarginate, without translucid dots; petioles twisted, stout; rachis channeled. Flowers disposed in more than 2 rows, quite large, on flattened pedicels, in fascicled, short, axillary or terminal racemes; bracts caducous; bracteoles ovate, petaloid, enveloping the bud, but not valvate, subpersistent to caducous; calyx funnel-shaped; sepals 4, free, imbricate, subequal; petals 5, 2 minute, 3 larger, cordate-orbicular, short-clawed, margins undulate, densely hairy inside; stamens 10; filaments connate at the base; anthers all fertile, dorsifixed, longitudinally dehiscent; ovary stalked, flattened, pubescent, stipe adnate to the side of the calyx tube, 2–4-ovuled; disk inconspicuous. Pod very large, long-stalked, 2-valved, narrowed at both ends, slightly acuminate at the apex, subglabrous, transversely nerved, valves leathery, suture slightly winged, dehiscent. (Léonard 1951, 1957)

This monotypic genus is endemic in Gabon.

Structure of the wood was described in detail by Sillans and Normand (1953).

Named for Colonel Sir Henry Collett, 1836–1902; he collected plants in Burma and Java.

Type: *N. wallichii* (Kurz) Schindl.

1 species. Herb, slender, creeping, rooting from all nodes. Leaves pinnately 3-foliolate; stipels awl-shaped, small; stipules striate, persistent, rigid. Flowers very small, solitary or 2–4 fascicled on long, slender pedicels; bracts long-stalked, embracing the calyx of each flower; bracteoles 2; calyx tubular, membranous; lobes 5, subequal, short, rounded, upper 2 partly or entirely connate, lower 3 semiorbicular; standard blue violet with a white base, obovate, glabrous, with 2 very small, inflexed, basal ears; wings free, oblong, spurred; keel blunt, straight, shorter than the oblong wings; stamens 10, diadelphous; anthers small, uniform; ovary short-stalked, 1-ovuled; style long, incurved, glabrous; stigma small, terminal. Pod sessile, small, turgid, 1-seeded, indehiscent, developing subterraneously by means of the recurved geotropic gynophore. (Collett and Hemsley 1890; Steenis 1960a; Backer and Bakhuizen van der Brink 1963)

Hutchinson (1964) positioned this genus in Lespedezeae trib. nov.

This monotypic genus is endemic in Burma and Java, in teak forests.

The Greek prefix *neo*, "new," indicates similarity with *Cracca* Benth.

Type: *N. heterantha* (Griseb.) Speg.

1 species. Herb, annual, villous, stemless, taproot thick to fleshy. Leaves whorled, imparipinnate, long-petioled; leaflets 1, 3, 5, or 7, obovate, terminal leaflet emarginate, wider and larger than the lateral ones, pubescent overall or only on the under surfaces; stipels absent; stipules linear-subulate, adnate to the base of the petiole. Flowers 1–few, on axillary peduncles bearing white pubescence and stipitate glands, 2 types: cleistogamous (apetalous) flowers with stamens 5 or 2, enclosed in a pubescent calyx, on very short peduncles; chasmogamous flowers large, violet, on peduncles not exactly axillary; calyx tube short, cupular; lobes lanceolate-subulate, upper 2 connate above; corolla not enclosed in the calyx; standard reniform-orbicular, short-clawed; wings obliquely clawed, eared on 1 side; keel petals joined near apex; stamens 10, diadelphous; ovary linear, shortstalked, many-ovuled, glandular-punctate; style persistent, inflexed and curved, glabrous below, pilose above on adaxial side; stigma capitellate. Cleistogamous pod white-pubescent, very small, ovoid to torulose, 1–3-seeded; chasmogamous pod linear, erect, compressed, torulose, covered with fine white pubescence and stalked glands similar to those on calyx and peduncles, transversely septate, several-seeded. (Burkart 1943)

This monotypic genus is endemic in Argentina and the Bolivian Andes. It occurs in very arid regions.

The fleshy, sweet roots are edible. The foliage is relished by goats (Burkart 1943).

Neodielsia Harms   Papilionoideae: Galegeae, Astragalinae
(299a) / 3766a

Named for Friederich Ludwig Emil Diels, 1874–1945, German botanist.

Type: *N. polyantha* Harms

1 species. Herbs. Leaves imparipinnate; leaflets 3–5, oblong; petioles short; stipules lanceolate, membranous. Flowers many, in elongated, slender, axillary racemes crowded at the tips of branchlets and stems; calyx tubular-cylindric, obliquely truncate; teeth 5, minute, acute; petals subequal, exceeding the calyx; standard obovate-oblong, clawed, slightly emarginate; wings and keel petals long-clawed, oblong, the latter incurved, rear margins joined; stamens 10, vexillary stamen united with others only in the middle; ovary lanceolate, long-stalked, surrounded by a short, basal disk, sometimes constricted between the 2 ovules; stigma minute, scarcely distinct. Pod long-stalked, flattened, oblong or narrow-elliptic, 1–2-seeded. (Harms 1905a, 1908a)

Harms (1905a, 1908a) expressed uncertainty about the systematic positioning of *Neodielsia*, but regarded it as closely akin to *Astragalus*.

This monotypic genus is endemic in central China.

Neodunnia R. Vig.   Papilionoideae: Galegeae, Tephrosiinae

Named for S. T. Dunn, author of a revision of the genus *Millettia* published in 1912.

Type: *N. atrocyanea* R. Vig.

4 species. Shrubs or small trees, 4–10 m tall; young branches usually pubescent. Leaves imparipinnate; leaflets 11–21, deciduous; stipules elongated, soon caducous. Flowers violet or reddish, few, opposite, in short, axillary racemes; bracteoles small, caducous; calyx campanulate or urn-shaped, pubescent; sepals long-connate, acutely 5-toothed above; petal lengths subequal; standard suborbicular, outside sometimes villous; wings and keel petals narrow, apices rounded; stamens diadelphous; ovary elongated, villous. Pod long, linear, compressed, dehiscent; seeds lens-shaped, compressed. (Viguier 1950)

The genus is closely allied to *Millettia*.

Neodunnia species are indigenous to Madagascar. They grow in dry, sandy forests and on seaside dunes.

Neoharmsia R. Vig.   Papilionoideae: Sophoreae

Named for Dr. H. Harms, 1870–1942, renowned German authority on Leguminosae.

Type: *N. madagascariensis* R. Vig.

1 species. Tree or small shrub, 3–4 m tall. Leaves imparipinnate, 9-foliolate; leaflets ovate-elliptic, large, base subcordate, apex obtuse or acuminate, deciduous. Flowers purplish red, small, on short, densely villous pedicels in terminal racemes, appearing before the leaves; bracteoles short, triangular-acute; calyx irregular, tomentose, oblique at the apex; teeth short, unequal, valvate; petals unequal, clawed, longer than the calyx; standard obovate, narrowed at the base; wings and keel very short, with rounded apex, free; stamens free; anthers dorsifixed; ovary elongated, sessile, glabrous, multiovuled; style filiform, straight; stigma minute, terminal. Pod short-stalked, straight or slightly curved, elongated, flat, reticulate, dehiscent; seeds 5–8, flattened. (Viguier 1952)

This monotypic genus is endemic in Madagascar. The plants grow on sand dunes near the sea.

Neorautanenia Schinz
(390a) / 3865

Papilionoideae: Phaseoleae, Glycininae

Type: *N. amboensis* Schinz (= *N. brachypus* (Harms) C. A. Smith; *N. edulis* C. A. Smith)

3–8 species. Herbs, erect, prostrate or climbing; stems and foliage densely velvety-tomentose or smooth; rootstocks sometimes turniplike, tubers up to 60 cm in diameter. Leaves pinnately 3-foliolate, long-petioled; leaflets elliptic, lanceolate-ovate, sometimes 3-lobed, cuneate at the base, acute at apex, often pubescent underneath; stipels present; stipules 2, lanceolate, persistent. Flowers blue, mauve, or white, in leafless, few- to many-flowered, axillary racemes; rachis swollen at the nodes; bracts caducous; bracteoles none; calyx campanulate, subbilabiate; lobes ovate-lanceolate, longer than calyx tube, upper lip bidentate with lobes joined higher up, lower lip tridentate to the middle, median lobe very long; standard suborbicular, reflexed, eared at the base, with a broad, linear, often channeled claw; wings oblong, long-clawed, spurred at the base, slightly adnate to the keel; keel blunt, incurved, long-clawed, sometimes eared; stamens 10, diadelphous or monadelphous, vexillary stamen free or partly joined; anthers equal, dorsifixed; ovary sessile, few- to several-ovuled, usually 3–8, hairy; disk present; style linear, thicker towards the base; stigma capitate, terminal, glabrous. Pod oblong-linear, sometimes densely tomentose, more or less leathery, septate within, dehiscent; seeds 3–8, dark brown, subglobose, hilum white. (Harms 1908a; Phillips 1951; Wilczek 1954; Verdcourt 1970a)

Harms (1908a) positioned this genus between *Glycine* and *Teramnus*, but Verdcourt (1970a) judged its kinship to be near *Dolichos* in Phaseolineae. Verdcourt recognized only 3 species: *N. amboensis* Schinz, *N. ficifolius* (Benth.) C. A. Smith, and *N. mitis* (A. Rich.) Verdc.

Neorautanenia is an African genus. Species occur in tropical and southern Africa. They flourish in bush and grasslands, forest clearings, and on termite mounds.

These plants are distinctive for their deeply lobed leaflets and tuberous roots. Verdcourt and Trump (1969) described *N. mitis* as "a shrubby herb with erect climbing or scrambling stems 3–8 ft. long (records of up to 36 ft. are mentioned in the literature) arising from huge conical root stock often well over a foot wide." Watt and Breyer-Brandwijk (1962) credited Dalziel with seeing an inedible 9–14 kg tuber of this species.

N. amboensis, *N. ficifolius*, and *N. mitis* are well known as fish poison plants and for the insecticidal properties of their tuberous roots (Tattersfield and Gimingham 1932; Holman 1940; Duuren and Groenewoud 1950a,b). An analysis of dry roots of *N. mitis* in 1950, cited by Verdcourt and Trump (1969), showed the presence of about 13 percent saponins and a substance indicative of rotenone. Various substances were isolated from *N. amboensis* (= *N. edulis*), of which edulin and neorautone were structurally related to an isoflavone, pterocarpin, $C_{17}H_{14}O_5$, and a rotenoid, pachyrrhizin, $C_{19}H_{12}O_6$ (Duuren 1961). Pachyrrhizin and several new isoflavones were extracted from *N. mitis* roots by Crombie and Whiting (1963). These authors alluded to a probable relationship between flavonoid chemistry and plant phylogeny, since *N. mitis* is the first plant known to yield an isoflavanone and its corresponding rotenoid. Verdcourt and Trump (1969) mentioned the potential of *N. mitis* roots in killing bilharzia-carrying freshwater snails.

NODULATION STUDIES

Nodules were reported present on roots of *N. brachypus* (Harms) C. A. Smith and *N. mitis* in Zimbabwe by Corby (1974) and on *N. amboensis* and *N. ficifolius* in South Africa by Grobbelaar and Clarke (1975).

Nepa Webb   Papilionoideae: Genisteae, Cytisinae

From Latin, *nepa*, "scorpion," in reference to the thorny branches.

Type: *N. boivinii* Webb

1 species. Shrub, low, diffuse; branches spiny, angular; branchlets very short, subopposite. Leaves reduced to short, triangular, scalelike, subspinescent, subopposite phyllodes. Flowers solitary, in phyllode axils, near ends of the thorny branches; calyx bilabiate, upper lip longer, curved, bidentate, lower lip tridentate; petals twice as long as calyx; standard reflexed, densely silky-villous outside; keel villous outside; anthers alternately basifixed and dorsifixed. Pod ovoid-rhomboid, short, slightly longer than calyx, 1–2-seeded, dehiscent. (Rothmaler 1941)

This species, formerly a section of *Ulex*, was judged distinctive for its dainty, zigzag branching habit, its soft and weakly armed twigs, open 2-lipped calyx, and densely silky standard and keel.

This monotypic genus is endemic in the Iberian Peninsula. It is common in sandy and gravelly seashore areas.

Nephrodesmus Schindl. Papilionoideae: Hedysareae, Desmodiinae

From Greek, *nephros*, "kidney," and *desmos*, "bond," "fetter," in reference to the constriction between the kidney-shaped joints of the pod.

Type: *N. sericeus* (Hochr.) Schindl.

7 species. Trees or shrubs. Leaves pinnately 3-foliolate; stipels and stipules present. Flowers few, in lax, panicled racemes; bracts minute; bracteoles 2, persistent, subtending the calyx; calyx campanulate; lobes 4; petals subequal but longer than the calyx, clawed; standard orbicular; wings and keel long, blunt, joined; keel petals abruptly clawed; stamens diadelphous; filaments alternately longer and shorter; anthers short, oval; ovary short-stalked, stalk inserted within a basal, tubular disk, several-ovuled; style incurved, filiform; stigma pointed, terminal. Pod flat, linear, jointed, segments 1-seeded, indehiscent. (Schindler 1917)

This genus is akin to *Desmodium*, *Hanslia*, and *Arthroclianthus* (Schindler 1924).

Nephrodesmus species are limited in distribution to New Caledonia.

Nephromeria (Benth.) Schindl. Papilionoideae: Hedysareae, Desmodiinae

From Greek, *nephros*, "kidney," and *meris*, "part," "portion"; the joints of the pod are subreniform.

Type: *N. barclayi* (Benth.) Schindl.

10 species. Shrubs or subshrubs, scandent or creeping, or herbs rooting at nodes. Leaves pinnately 3-foliolate; stipels and stipules present. Flowers paired, in axillary or terminal, panicled racemes; bracts broad, caducous; bracteoles absent; calyx broad, campanulate; lobes 4, posterior lobe 2-parted or entire; petals short-clawed, longer than the calyx; standard ovate; wings straight, eared, broad; keel curved sharply upward along lower margin; keel petals united at base; stamens monadelphous, vexillary stamen free above; filaments filiform, alternately long and short; anthers ovoid; ovary short-stalked, 2–3-ovuled; disk absent; style bent inward at right angle; stigma terminal, capitate. Pod stalked, compressed, joints 3–several, subreniform-orbicular, membranous, lower suture deeply constricted, only the upper suture dehiscent, segments 1–seeded. (Schindler 1924)

Elevation of this former section of *Desmodium* to generic rank was based on the distinctive dehiscence of the pod segments (Schindler 1924).

Members of this genus are native to tropical America. They are abundant in the region at the mouth of the Amazon and in tropical forests in Ecuador and the Antilles.

NODULATION STUDIES

N. axillaris (Sw.) Schindl. var. *acutifoliola* (Urb.) Schindl., reported as *Desmodium axillare* var. *acutifolium*, was observed nodulated in Trinidad (DeSouza 1966).

Named for Neptune, mythological god of the sea, in reference to the aquatic habit of some species.

Type: *N. oleracea* Lour.

10–15 species. Sprawling undershrubs or perennial herbs, some aquatic, with ascending or floating stems. Leaves bipinnate; leaflets numerous, small, often sensitive to touch and light changes; petiolar gland sometimes present; stipules membranaceous, obliquely heart-shaped. Flowers usually yellow or white, 5-parted, upper flowers perfect, lower often staminate or neuter, in globose or ellipsoid heads on long, axillary, solitary, erect peduncles; calyx campanulate, shallowly dentate-lobate; petals connate to middle or free, valvate; stamens 10, rarely 5; filaments not connate into a tube; anthers with or without an apical gland; ovary stalked, multiovuled; style filiform; stigma terminal, small, concave. Pod oblique-oblong, compressed, membranous to leathery, dorsal suture straight, ventral suture curved, subseptate within, 2-valved, dehiscent; seeds 4–20, transverse, ovate, compressed. (Bentham and Hooker 1865; Gilbert and Boutique 1952a)

Representatives of this genus are indigenous to tropical and subtropical regions of the Old World and the New World. They flourish in moist and swampy areas, on grassy plains, and on sand or salt flats. Free-floating species are common in the tranquil waters of Madagascar, Asia, and America.

The slender-stemmed terrestrial species tend to be excellent soil-binders. *N. lutea* (Leavenw.) Benth., common in the southern United States from Texas to Florida, forms dense mats on sites that are partially submerged during wet seasons. The stems of the aquatic species are thick, inflated, and spongy. *N. oleracea* Lour., perhaps the best-known member of this group, is commonly cultivated as an exotic in temperate conservatories. The leaves of this species are very sensitive. The sprouts and leaves are consumed in Vietnam as a pot herb.

Nodulation Studies

Nodules produced on adventitious water roots of *N. oleracea* were described as large as hazelnuts by Schaede (1940); however, those borne on earth roots were smaller, and fewer in number. Rhizobial infection was solely through epidermal ruptures, inasmuch as root hairs are lacking. The *Bläschenbacteroiden*, or spherical bacteroids, resembled very closely those described in peanut nodules (Allen and Allen 1940a). Structural development of the nodules on this species conformed with apical types of other leguminous species (Schaede 1940, 1962).

A single rhizobial strain from *N. plena* Lindl. nodulated *Amorpha fruticosa*, *Crotalaria sagittalis*, and *Astragalus nuttalianus* but not *Wisteria frutescens* (Wilson 1944b; Wilson and Chin 1947). Reciprocal tests were not conducted. Strains from *Robinia*, *Eysenhardtia*, and *Astragalus* species symbiosed effectively with *N. lutea* (J. C. Burton personal communication 1972).

Beadle (1964) credited significant nitrogen fixation to nodulated *N.*

gracilis Benth. and *N. monosperma* F. Muell. growing in overflow country, woodlands, treeless areas, and along water courses in eastern Australia.

Nodulated species of *Neptunia*	Area	Reported by
gracilis Benth.	Australia	Bowen (1956); Beadle (1964)
lutea (Leavenw.) Benth.	Tex., U.S.A.	Burton et al. (1974)
monosperma F. Muell.	Australia	Beadle (1964)
oleracea Lour.	Germany	Morck (1891); Pers. observ. (1936); Schaede (1940)
	Zimbabwe	Corby (1974)
(= *plena* Lindl.)	U.S.A.	Wilson (1944b, 1947)
(= *prostrata* (Lam.) Baill.)	Venezuela	Barrios & Gonzalez (1971)

Nesphostylis Verdc.   Papilionoideae: Phaseoleae, Phaseolinae

Name is an anagram of *Sphenostylis*, from which the genus was segregated.

Type: *N. holosericea* (Bak.) Verdc. (= *Sphenostylis holosericea*)

1 species. Herb, perennial, twining, up to 3 m tall, with tuberous rootstock. Leaves pinnately 3-foliolate, petioled; stipels present; stipules persistent. Flowers large, highly scented, borne singly on short stalks in leaf axils, apex of peduncle glandular; bracteoles 2, large, suborbicular, persistent; calyx campanulate; 2 upper lobes in a slightly bidentate lip, 3 lower lobes ovate-triangular; standard suborbicular, eared at the base, apex emarginate, glabrous, blue to mauve; wings deeper colored than standard, broad, obovate, clawed, base briefly spurred, lower part slightly grooved; keel white, obovate, clawed, apex slightly incurved, rounded; stamens diadelphous; vexillary filament spurred at base, all dilated at apex; anthers alternately subdorsifixed and basifixed; ovary linear, subsessile, many-ovuled; disk briefly sheathed, lobate; style cartilaginous, base slightly thickened, apex obtriangular, flattened, dilated; stigma terminal. Pod rusty-haired, linear-cylindric, 10–15 cm long, apex tapered, with calyx persistent at the base, dehiscent; seeds oblong-ovoid, dark brown. (Verdcourt 1970a)

This monotypic genus is endemic in tropical Africa.

This species, usually occurring as a climber in tall-grass areas, is worthy of cultivation for its beauty and fragrance.

NODULATION STUDIES

Closely allied species in *Sphenostylis* are nodulated; however, this segregated species has not been examined.

Named for the famous English scientist, Sir Isaac Newton, 1643–1727.

Type: *N. duparquetiana* (Baill.) Keay

15 species. Trees, unarmed, up to 40 m tall, with a trunk diameter of 90 cm; bark smooth; buttresses, if present, usually narrow, steep. Leaves bipinnate; pinnae opposite, 1–20 or more pairs; rachis often with a gland between each pair of opposite pinnae; leaflets sessile obliquely elliptic or oblong to linear, leathery, few–70 pairs. Flowers white or yellowish, small, sessile, numerous, in dense, slender, terminal spikes or spicate panicles; bracts very small, calyx campanulate, often pubescent; sepals joined, obscurely 5-toothed, pubescent; petals 5, equal, valvate; stamens 10; filaments thickened at the base, alternately long and short; anthers with or without a small, stalked, spherical, apical, caducous gland; ovary stalked, many-ovuled, densely hairy; style short, curved; stigma cupular. Pod leathery, flattened, linear or curved, valves papery, dehiscent along 1 margin; seeds flat, narrow-oblong, surrounded by a narrow, membranous wing, attached at the end by a long slender funicle. (Gilbert and Boutique 1952a; Brenan 1955, 1959)

The genus is closely allied to *Piptadeniastrum*, but it is distinguished by its glandular leaves, pubescent ovary, and the funicle attachment at the end instead of near the middle of the winged seed (Brenan 1959).

Representatives of this genus occur mostly in tropical West Africa, but a few species are present in tropical South America. They are common in lowland and upland rain forests, usually along stream borders and in riverine swamp forests.

Limited local use is made of *Newtonia* wood for cabinets, furniture, and canoes. It is more durable in water than in soil. The slash of *N. aubrevillei* (Pellegr.) Keay has the odor of old cheese, is fibrous, and exudes a sticky, honey-colored, clear gum. The stripped bark is chewed by natives as an aphrodisiac (Voorhoeve 1965). *Newtonia* wood is not exported.

NODULATION STUDIES

Plants of *N. buchananii* (Bak.) Gilb. & Bout. and *N. hildebrandtii* (Vatke) Torre var. *pubescens* Brenan examined in Zimbabwe lacked nodules (Corby 1974).

Named for William Nissole, 1647–1735, French botanist.

Type: *N. fruticosa* Jacq.

10–12 species. Vines, climbing, slender, perennial, more or less woody, rarely prostrate herbs; stems numerous, herbaceous becoming woody, smooth or bristly-pubescent; rootstocks woody. Leaves imparipinnate, 5-, sometimes 7-foliolate; leaflets elliptic to orbicular, obovate to obcordate, entire, surfaces micropuncticulate, glabrous to densely pubescent; stipels absent; stipules paired, joined at the base, deltoid to lanceolate, entire or glandular-denticulate. Flowers small, usually yellowish, some white, pink, or purple, in axillary fascicles, less commonly paniculate or racemose; bracts awl-shaped, intergrading with the stipules; bracteoles, if present, persistent and below pedicel joint at calyx base; calyx campanulate, toothed or truncate, articulate at the base; lobes 5, subequal, glabrous or pubescent; standard ovate-orbicular, short-clawed, reflexed, longer than the wings and keel, pubescent outside; wings falcate-oblong, clawed, glabrous or pubescent; stamens 10, subequal, as long as the keel, vexillary stamen free at the base, connate at middle with others into a sheath splitting along the vexillary side as the pod develops; ovary subsessile, 2–4-ovuled, less commonly 1-ovuled; style glabrous; stigma capitate. Pod samaroid, linear or spathulate, articulate, 2–5-jointed, segments short, broad, terminal segment sterile, flat, winglike, indehiscent; seeds kidney-shaped, laterally compressed, sublustrous, reddish brown. (Bentham and Hooker 1865; Rose 1899; Standley 1920–1926; Rudd 1956)

This genus is closely related to *Chaetocalyx*.

Members of this genus are distributed throughout tropical and warm temperate western America, ranging from southern Arizona and Texas southward to Argentina and Paraguay; all species occur in Mexico. They thrive in moist open locations along streams, roadsides, forest edges, on cliffs, and in canyons.

Economic uses of *Nissolia* species have received little attention. Plant parts of *N. fruticosa* Jacq. have fish-stupefying properties; they are said to be used as an antidote for snakebite in El Salvador.

Nogra Merr.   Papilionoideae: Phaseoleae, Galactiinae

Name is an anagram of the generic name *Grona* Benth., not *Grona* Lour.

Type: *N. grahami* (Wall.) Merr.

3 species. Herbs, prostrate or climbing. Leaves 1-foliolate; stipels present; stipules caducous. Flowers sometimes fascicled on the nodose rachis of axillary or terminal racemes; bracts narrow, caducous; bracteoles small, somewhat persistent; calyx lobes 5, longer than the tube, upper 2 partly joined; standard obovate or suborbicular with basal lobes minute, inflexed; wings falcate, slightly joined to the narrow, incurved, blunt-beaked keel; stamens 10, diadelphous; anthers uniform; ovary subsessile, many-ovuled; style smooth, slender, not barbate; stigma capitate. Pod slender, linear, 2-valved, filled between the seeds; seeds round, hilum small, strophiole thick. (Merrill 1935; Hutchinson 1964)

Nogra species are widespread in tropical Asia, from India to Thailand.

Notodon Urb. Papilionoideae: Galegeae, Robiniinae
(277a) / 3743

From Greek, *noton*, "back," and *odous, odontos*, "tooth," in reference to the petiole base being pouched and extended into a spur.

Type: *N. gracilis* Urb.

4 species. Shrubs or small trees. Leaves paripinnate; leaflets entire, leathery, few pairs, veinless above, deciduous; petiole base pouched and expanded into a straight or recurved spur or prickle; rachis somewhat winged, with the wing discontinuous at the sites of leaf insertion, terminating in a fine, free tip; stipels minute; stipules lanceolate, apex subulate-pointed, caducous. Flowers bluish or rose purple, in small fascicles in axils of old leaves; calyx short, campanulate, membranous; lobes 5, very short, upper 2 close together, triangular, obtuse, other 3 rather remote; standard suborbicular, with a short claw but without a callus; wings oblong, straight, short-clawed; keel petals longer than the standard and wings, united along the upper half of the lower incurved margin, other margin straight; stamens 10, diadelphous, subequal; anthers uniform; ovary short-stalked, linear, several-ovuled; style awl-shaped, arched, smooth; stigma minute, terminal, indistinct on inner side of style apex. Pod small, pointed at both ends, linear-oblanceolate. (Urban 1899; Harms 1908a; Britton 1920; Rydberg 1924; León and Alain 1951)

Representatives of this genus are indigenous in western Cuba. They flourish on rocky savanna soil.

The rich, olive brown streaked heartwood of *N. gracilis* Urb., given the name "granadillo de Cuba," is hard, heavy, easy to work, and suitable for small turnery articles (Record and Hess 1943).

Notospartium Hook f. Papilionoideae: Galegeae, Robiniinae
283 / 3751

From Greek, *notos*, "south," and *spartos*, the name of a broom plant; the species are endemic in the South Island, New Zealand.

Type: *N. carmichaeliae* Hook. f.

3 species. Shrubs or small trees; branches slender, whiplike, pendent, grooved, nearly leafless. Leaves reduced to minute scales at the nodes. Flowers pink- or purple-flushed and -veined, on slender racemes at the nodes; calyx campanulate; teeth 5, short, triangular, subequal; standard short-clawed, obcordate, reflexed; wings free, oblong-falcate, shorter than the blunt, incurved keel; stamens diadelphous; anthers nearly uniform; ovary subsessile, multiovuled; style incurved, hooked at the apex, inner margin barbate; stigma terminal. Pod short-stalked, linear, compressed or torulose, beaked, septate inside, indehiscent; seeds 10–15, brown or reddish yellow, mottled with black. (Bentham and Hooker 1865; Allan 1961)

Hutchinson (1964) positioned this genus in Carmichaelieae trib. nov.

This genus is endemic in South Island, New Zealand. The species inhabit montane river valleys and river terraces in drainage areas.

NODULATION STUDIES

R. M. Greenwood (personal communication 1975) observed nodules on *N. glabrescens* Petrie in New Zealand.

Oddoniodendron De Wild.　　　　Caesalpinioideae: Amherstieae

Name means "Oddon's tree"; Oddon collected the holotype of this tree in the lower Congo Basin.

Type: *O. micranthum* (Harms) Bak. f. (= *O. gilletii* De Wild.)

1 species. Tree, unarmed. Leaves imparipinnate; leaflets 2–5, alternate, slightly asymmetric at the base, not pellucid-punctate; petiolules grooved; stipules joined, bidentate, persistent, enclosing the leaf bud. Flowers in 2 or more rows, in panicles or sometimes racemes, axillary or terminal, solitary or many-fascicled; bracts caducous, small; bracteoles 2, valvate, opposite, persistent, sepaloid, enclosing the flower bud; calyx tube short; sepals 4, imbricate, entire, subequal, pilose inside; petals 5, equal, entire, imbricate before anthesis; stamens usually 10, sometimes 8, free, all fertile, exserted, alternate ones longer, positioned between the petals, shorter ones opposite petals; anthers dorsifixed, longitudinally dehiscent; ovary flattened, 2–5-ovuled, hairy, on a stalk inserted to the wall of the receptable near the base; style filiform, subglabrous. Pod flat, oblong, woody, the 2 valves rolling up during dehiscence, sutures not winged; seeds compressed, coats easily detached from cotyledons when dry, but becoming mucilaginous in water. (Léonard 1952, 1957)

This monotypic genus is endemic in tropical West Africa.

The hard, durable wood is used for local construction.

Oleiocarpon Dwyer　　　　Papilionoideae: Dalbergieae,
　　　　　　　　　　　　　　　　　Geoffraeinae

From Latin, *oleum*, "oil," and Greek, *karpos*, "fruit."

Type: *O. panamense* (Pitt.) Dwyer

1 species. Tree, tall. Leaves imparipinnate; leaflets large, oblong, 12–14 pairs, opposite or subopposite, with the tip of the rachis extended beyond the uppermost pair; stipels absent; stipules minute, caducous. Flowers pink to purple, quite persistent in dense, terminal panicles up to 40 cm long; bracts small, caducous; bracteoles 2; calyx campanulate, bilabiate; lobes 5, upper 2 very conspicuous, oblong, glandular-punctate, lower 3 minute; standard rounded, emarginate; wings and keel petals oblong, obtuse, asymmetrically

bilobed; stamens 10, monadelphous; filaments subequal; ovary stalked, glabrous; style slightly thick; stigma truncate, capitate. Pod stalked, elliptic, drupaceous, oily, at first with a gray green pubescence, hairs early caducous; seed 1, flat, light brown, almond-shaped. (Dwyer 1965)

Explanations for the segregation of this species from *Coumarouna* and the kinships of this genus with *Dipteryx* are reviewed and well documented by Dwyer (1965).

This monotypic genus is endemic in Panama, Colombia, and Costa Rica. It occurs in forested areas.

The beautiful flowers attract leaf-cutting ants. The oily pods are used to provide torch light (Dwyer 1965).

Olneya Gray   Papilionoideae: Galegeae, Robiniinae
274 / 3739

Named for Stephen T. Olney, 1812–1878, Rhode Island botanist.

Type: *O. tesota* Gray

1 species. Shrub or small tree, evergreen, 8–10 m tall, trunk gnarled, diameter often exceeding 45 cm; branches armed with stipular spines, small twigs unarmed; bark thin, scaly; foliage somewhat heavy. Leaves paripinnate or imparipinnate; leaflets in pairs of 5–8, small, entire, with a grayish to bluish green pubescence; stipels absent. Flowers pink, white to purplish white, in short, axillary racemes; bracts caducous; calyx lobes subequal, upper 2 connate above; standard broad-orbicular, clawed, emarginate, reflexed, with 2 broad, inflexed, basal ears and 2 callosities inside; wings oblique-oblong, free; keel broad, incurved, obtuse; stamens 10, diadelphous; anthers uniform; ovary sessile, several-ovuled; style inflexed, barbate above the middle; stigma thick, capitate. Pod convex over seeds, moniliform, glandular-hairy, 2-valved, valves thick, leathery, continuous within; seeds 1–2, broad-ellipsoid. (Bentham and Hooker 1865)

This monotypic genus is endemic in southern Arizona, California, and northwestern Mexico; in general, it is confined to the southwestern deserts of the United States.

This species, better known as desert or Sonora ironwood, palo fierro, or tesota, is a handsome flowering tree worthy of consideration as an ornamental, although it is not currently cultivated. Jepson (1936) described it as "one of the largest and finest trees of the southwestern United States." Its foliage is relished by animals and also serves as escape and refuge cover for wildlife. Mules, donkeys, and horses prefer the foliage to grass. The fleshy young pods and flowers are edible. Early settlers in the western United States used the seeds to make a beverage similar to hot chocolate (Cook 1919).

The sapwood is yellowish white; the heartwood is a rich brown and somewhat variegated. The wood is as hard as ebony, heavier than water, and durable, but tends to be brittle. Because it is scarce and available only in small pieces, the wood serves little purpose other than for fuel and small objects (Record and Hess 1943).

NODULATION STUDIES

Nodules were absent on field specimens and seedlings grown in rhizosphere soil in the greenhouse (Martin 1948). Paucity or absence of lateral roots, the deeply penetrating nature of the taproot, and the confinement of this species to desert areas are undoubtedly factors that contributed to the lack of nodules in the Arizona tests. Recently Corby (personal communication 1977) obtained good nodulation in Zimbabwe on plants grown from imported seeds.

***Onobrychis* Mill.**   Papilionoideae: Hedysareae, Euhedysarinae
312 / 3780

From Greek, *onobrychis*, a name used by Dioscorides for a leguminous plant.
   Type: *O. viciifolia* Scop. (= *O. sativa* Lam.)
   100 or more species. Herbs, mostly perennial, some are spinescent shrubs. Leaves imparipinnate, without tendrils; leaflets entire; petioles sometimes persisting to form spines; stipels absent; stipules free or connate, scarious. Flowers white, pink to purple, rarely yellow, in axillary peduncled racemes or spikes; bracts scarious; bracteoles small, subtending the calyx; calyx campanulate; lobes 5, subequal, awl-shaped; standard orbicular or obcordate, scarcely clawed and narrowed toward the base; wings short; keel truncate at the apex, obtuse, equal to or longer than the standard; stamens diadelphous, vexillary stamen free at the base, later connate with others into a closed tube; anthers uniform; ovary sessile, 1–3-ovuled; style filiform, inflexed; stigma small, terminal. Pod sessile, compressed, semiorbicular, not jointed, margins usually toothed, sides strongly veined, prickly or pitted, indehiscent; seeds 1–3, broad-reniform or transversely oblong. (Bentham and Hooker 1865)
   Species of this large genus are native to the Balkan area of the Mediterranean and to southwestern Asia. They are well represented in Europe and Asia but are not common in the New World. The plants thrive on light, well-drained calcareous and barren limestone soils under low rainfall. They are not tolerant of wet soils. Species are somewhat resistant to frost and not easily winter-killed.
   Various species have been recommended and used over the years as cattle fodder, forage, and in soil erosion control.
   The type species is commonly known as esparcet or sainfoin. The former common name connotes "healthy hay" or "holy hay." The cultivation of sainfoin originated in the south of France in the 15th century, where it came to be highly valued as fodder in the 16th century (De Candolle 1908). It spread slowly into other areas, but its acceptance was dramatic because its cultivation led to the transformation of nearly valueless, dry, calcareous areas into productive lands. Sainfoin reached the United States about 200 years ago, but did not attract attention until it became recognized as a forage crop in the barren limestone areas of the West.
   Sainfoin is a long-lived, deep-rooted perennial whose fleshy roots reach a

diameter of 5 cm and depths of several meters. The stout, erect stems rise up to 1 m in height from the crown. The forage resembles vetch, is soft, succulent, relished by livestock, and withstands heavy grazing. In some areas it is used as a nutritious hay for racehorses. Sainfoin is competitive with grass mixtures, produces a high yield in pure stands, is easy to harvest, and seeds prolifically. Alfonsus (1929) considered sainfoin second only to clovers as a honey plant. It is perhaps the most important honey plant today in the chalky uplands of England.

NODULATION STUDIES

Early workers concluded that sainfoin plants and rhizobia showed a limited infective range (Maassen and Müller 1907; Simon 1907; Krüger 1914) and, accordingly, that this species merited separate group distinction. Broader, somewhat promiscuous host relationships were reported by Wilson (1939a), although rhizobial strains from sainfoin showed a more restricted infection range than did sainfoin as host. Isolates from about 12 diverse species symbiosed with sainfoin, yet sainfoin rhizobia were not reciprocally infective (Wilson 1944a). In all the infective patterns of sainfoin rhizobia and host, it is quite apparent that a more delicate balance prevails between the plant and the microsymbiont with respect to effectiveness than in most rhizobia:host associations. Response of 3 sainfoin varieties to selected homologous strains was unimpressive, although the plants were abundantly nodulated (Burton and Curley 1968). Similar results of plot trials were reported in Montana (Sims et al. 1968). Sainfoin rhizobia were present in 30 of 51 Holland soils that ranged from pH 4.0 to 7.8; in many of these, sainfoin had not been grown previously (Schreven 1972). Four strains from sainfoin and 6 strains from crownvetch, *Coronilla varia*, were interchangeably effective.

Absence of nodules on *O. petraea* (Bieb.) Fisch. in France (Clos 1896) was probably due to circumstantial conditions, since 2 other species reported as negative by Clos were nodulated in other geographical areas.

Nodulated species of *Onobrychis*	Area	Reported by
arenaria DC.	S. Africa	Grobbelaar & Clarke (1974)
biebersteinii Sirj.	S. Africa	N. Grobbelaar (pers. comm. 1965)
caput-galli (L.) Lam.	S. Africa	Grobbelaar & Clarke (1974)
chorissanica Bge.	Ariz., U.S.A.	Martin (1948)
crista-galli (L.) Lam.	Israel	Hely & Ofer (1972)
cyri Grossh.	S. Africa	N. Grobbelaar (pers. comm. 1965)
micrantha Schrenk	S. Africa	N. Grobbelaar (pers. comm. 1965)
squarrosa Viv.	Israel	M. Alexander (pers. comm. 1963)
transcaucasia Grossh.	U.S.S.R.	Roizin (1959)
viciifolia Scop. (= *sativa* Lam.; = *vulgaris* Hill.)	Widely reported	Many investigators

An ancient Greek plant name.

Type: *O. spinosa* L.

70–80 species. Herbs, annual or perennial, or dwarf shrubs, rarely small trees, some spiny, usually covered with glandular and simple hairs. Leaves pinnately 3-, rarely 1-foliolate; leaflets very rarely 2–5 pairs, usually toothed, terminal largest; stipels none; stipules adnate to petiole. Flowers solitary, or 2–3 on short, axillary peduncles, peduncles often bristlelike, or in spikes or racemes, attractive, rose, yellow, purple, or white; calyx tube short, campanulate or tubular; lobes equal; standard orbicular, short-clawed; wings obovate-oblong; keel incurved, more or less beaked; stamens monadelphous; alternate filaments dilated at apex; anthers alternately short and versatile, long and basifixed; ovary minutely stalked, 2–8-ovuled; style glabrous, inflexed; stigma terminal, subcapitate. Pod linear, oblong, ovate-rhomboid, sessile or minutely stalked, turgid or cylindric, continuous inside, dehiscent; seeds kidney-shaped, punctate, smooth or rugose. (De Candolle 1825b; Bentham and Hooker 1865)

Members of *Ononis* differ from other genera in Trifolieae by having monadelphous stamens. Hutchinson (1964) positioned this genus in Ononideae trib. nov.

Members of this genus are distributed throughout northern and central Europe, the Mediterranean area, western Asia, and North Africa.

O. repens L. is commonly known as restharrow.

Isoflavones are present in 28 *Ononis* species (Harborne et al. 1971). Plant parts of various members have been used in medicinal preparations, especially in home remedies for bladder ailments. The glucoside ononin, 4′methyldaidzin, $C_{22}H_{22}O_9$, from *O. spinosa* L., is a diuretic (Merck Index 1968).

NODULATION STUDIES

Swollen, club-shaped bacteroids were described in elongated nodules of *O. natrix* L., *O. repens*, and *O. viscosa* L. (Lechtova-Trnka 1931). One might expect a limited infective range of *Ononis* rhizobia, inasmuch as the genus is positioned in the tribe Trifolieae; however, rhizobia from *O. fructicosa* L., *O. natrix*, and *O. vaginalis* Vahl (Wilson 1944a,b) nodulated *Wisteria frutescens*, *Amorpha fruticosa* (Wilson 1944b), and *Astragalus nuttalianus* (Wilson and Chin 1947). A strain from *O. repens* formed nodules on *Dorycnium hirsutum*, but fixed little nitrogen (Brockwell and Neal-Smith 1966). These host-rhizobia-infection studies confirm, in principle, that the positioning of *Ononis* in the tribe Trifolieae is inappropriate.

Nodulated species of *Ononis*	Area	Reported by
alopecuroides L.	Kew Gardens, England	Pers. observ. (1936)
	S. Africa	N. Grobbelaar (pers. comm. 1971)
arvensis L.	France	Clos (1896)
	U.S.A.	Wilson (1939a)
(*cenisia* L.) = *cristata* Mill.	Italian Alps	Kharbush (1928)

Nodulated species of *Ononis* (cont.)	Area	Reported by
columnae All.	U.S.A.	Wilson (1939a)
fruticosa L.	U.S.A.	Wilson (1944b); Wilson & Chin (1947)
(*hircina* Jacq.) = *arvensis* L.	Germany	Lachmann (1891)
inermis Pall.	France	Naudin (1897c)
mitissima L.	S. Africa	Grobbelaar & Clarke (1974)
natrix L.	France	Lechtova-Trnka (1931)
	U.S.A.	Wilson (1944b); Wilson & Chin (1947)
	Libya	B. C. Park (pers. comm. 1960)
officinalis L.	S. Africa	Grobbelaar & Clarke (1974)
reclinata L.	Israel	Hely & Ofer (1972)
repens L.	Germany	Wydler (1860)
	France	Lechtova-Trnka (1931)
	Australia	Brockwell & Neal-Smith (1966)
rotundifolia L.	Italian Alps	Kharbush (1928)
serrata Forsk.	Israel	M. Alexander (pers. comm. 1963); Hely & Ofer (1972)
spinosa L.	Germany	Lachmann (1891)
	France	Wydler (1860); Naudin (1897c)
	Kew Gardens, England	Pers. observ. (1936)
subspicata Lag.	S. Africa	Grobbelaar & Clarke (1974)
vaginalis Vahl	U.S.A.	Wilson (1939a, 1944a,b)
viscosa L.	France	Lechtova-Trnka (1931)

Ophrestia H. M. Forbes   Papilionoideae: Phaseoleae, Glycininae

Name is an anagram of *Tephrosia*.

Type: *O. oblongifolia* (E. Mey.) H. M. Forbes

14 species. Herbs, stems ascending or procumbent, flexuous, branching from the base, often pubescent. Leaves subsessile, 1–5-foliolate; leaflets elliptic or lanceolate, obtuse, acute, with 5–7 veins prominent underneath; stipels none; stipules linear to linear-subulate. Flowers purple, few to many, on short

pedicels, in erect, axillary racemes on long peduncles, usually longer than the leaves; calyx campanulate, slightly oblique at the base, hairy, bilabiate; lobes acuminate, ovate-lanceolate, upper 2 broad, short-toothed, connate beyond the middle; standard oblong, with 2 ears at the juncture of a channeled claw, exterior hairy; wings sublanceolate, linear, long-clawed, eared, hairy; keel petals short, clawed, eared; stamens monadelphous; anthers uniform; ovary linear, pilose, 1–2-ovuled; style hooked at apex, very short, hairy at the base; stigma capitate. Pod hairy, oblong to lanceolate, 2-valved, valves twisting at dehiscence; seeds smooth, hilum short, aril prominent. (Forbes 1948; Lackey 1977a)

The occurrence of subsessile leaves, coriaceous leaflets, and the absence of stipels in addition to floral differences accounted for the segregation of species from *Tephrosia* and *Glycine* and the establishment of this genus (Forbes 1948). Kinships of *Ophrestia* with *Glycine* and *Teramnus*, tribe Phaseoleae, prompted Verdcourt (1970a) to conclude that *Paraglycine* and *Pseudoglycine* (Hermann 1962) were congeneric with *Ophrestia*. Lackey (1977a) proposed its segregation as the type genus of a new subtribe, Ophrestiinae.

Representatives of this genus occur in tropical South Africa, especially Natal and Transvaal, and Asia.

Nodulated species of *Ophrestia*	Area	Reported by
oblongifolia (E. Mey.) H. M. Forbes	S. Africa	Grobbelaar & Clarke (1974)
radicosa (A. Rich.) Verdc.		
var. *schliebenii* (Harms) Verdc.	Zimbabwe	W. P. L. Sandmann (pers. comm. 1970)
retusa H. M. Forbes	S. Africa	N. Grobbelaar (pers. comm. 1968)
unifoliolata (Bak. f.) Verdc.	Zimbabwe	Corby (1974)
(= *Glycine unifoliolata* Bak. f.)		
(= *Paraglycine unifoliolata* (Bak. f.) Hermann)		

Orbexilum Raf.   Papilionoideae: Galegeae, Psoraliinae

Type: *O. onobrychis* (Nutt.) Rydb.

8 species. Herbs, perennial, erect, up to about 1.5 m tall; stems herbaceous or woody at the base, with thickened roots or rootstock, rarely shrubby. Leaves scattered, pinnately 1- or 3-foliolate, rarely 5-foliolate, usually glandular-punctate. Flowers white or bluish purple, in long-stalked axillary spikes or racemes; calyx campanulate, not gibbous at the base, usually glandular-punctate; lobes 5, lowest lobe longest; standard broad-obovate or suborbicular, sides recurved or spreading, with a straight or slightly curved,

narrow claw, usually with 2 small ears at the base; wings about as long as the standard, crescent-shaped, rounded at the apex, narrow-clawed, lobed at the base; keel petals shorter than the wings, basal lobe less prominent, united at the apex and slightly with the adjacent wing at the base; stamens 10, diadelphous or weakly monadelphous; alternate anthers often smaller; ovary 1-ovuled; style bent or curved near the apex; stigma capitate. Pod rounded or ovate, somewhat flattened, exserted from the calyx, beak short, incurved, pericarp usually ribbed or wrinkled, thick, not coherent to the seed, indehiscent; seed 1, suborbicular to kidney-shaped. (Rydberg 1919a)

Orbexilum is closely related to, and often regarded as, a section of *Psoralea*.

Members of this genus are indigenous to eastern North America. They flourish in wooded areas and along river banks.

NODULATION STUDIES

Because all of the species were reported as *Psoralea* synonyms and the results of host-infection tests are conflicting (Wilson 1939a; Appleman and Sears 1943), it seems desirable to conduct further investigation, especially since there is evidence suggesting a high degree of specificity.

Nodulated species of *Orbexilum* [†]	Area	Reported by
onobrychis (Nutt.) Rydb. (= *P. onobrychis* Nutt.)	U.S.A.	Wilson (1939a); Appleman and Sears (1943)
pedunculatum (Mill.) Rydb. (= *P. pedunculata* (Mill.) Vail)	La., Ark., U.S.A.	Pers. observ. (U.W. herb. spec. 1970)
simplex (Nutt.) Rydb. (= *P. simplex* Nutt.)	Tex., Miss., U.S.A.	Pers. observ. (U.W. herb. spec. 1970)

[†] Reported as *Psoralea* species.

Ormocarpopsis R. Vig.   Papilionoideae: Hedysareae, Aeschynomeninae

The Greek suffix, *opsis*, denotes resemblance to *Ormocarpum*.

Type: *O. calcicola* R. Vig.

5 species. Shrubs, 3–4 m tall, deciduous. Leaves imparipinnate, 9–17-foliolate; leaflets alternate, small, short-petioluled, oblong, usually with base rounded, apex truncate or mucronate; stipules lanceolate or triangular, striate, persistent. Flowers yellow, axillary, 1–3, on short peduncle or subsessile; bracts and bracteoles short, scarious; calyx campanulate, asymmetric; petals longer than the calyx; keel slightly beaked; stamens joined, forming 2 bundles; anthers uniform; ovary small, very short-stalked, few-ovuled. Pod small, ovate or subspherical, sometimes somewhat compressed, seldom segmented, usually 1-seeded. (Viguier 1952)

Ormocarpopsis species are native to Madagascar. They thrive in wooded regions on rocky calcareous and sandy soils.

From Greek, *hormos*, "necklace," "chain," and *karpos*, "fruit," in reference to the beaded appearance of the pods.

Type: *O. verrucosum* Beauv.

30 species. Shrubs or small trees, up to about 6 m tall, deciduous, often glandular sticky-haired. Leaves imparipinnate, rarely 1-foliolate, usually alternate; leaflets oblong-elliptic, sometimes apiculate, acuminate, numerous and small or few and large; stipels absent; stipules striate, small, scarious, persistent. Flowers yellow, violet, or white, sometimes purple-veined, in short, axillary or terminal racemes, or 1–4, fascicled in leaf axils; bracts and bracteoles small, persistent; calyx broad-campanulate; lobes 5, ovate, longer than the tube, 2 upper lobes subconnate, shorter, broader, lowest lobe longest; standard orbicular, with a short, oblong, curved claw; wings obliquely ovate or obovate, clawed; keel broad, incurved, subacute or obtuse, subequal to the wings, clawed; stamens 10, variously disposed in groups, with 2 groups of 5 being the most common, always at least 1 slit on the dorsal side, never united into an undivided sheath; anthers uniform; ovary sessile or stalked, many-ovuled, often densely setose; style cylindric, slender, inflexed; stigma terminal. Pod linear, compressed, grooved longitudinally, surface often warty or bristly-haired, constricted into 2 or more, indehiscent, oblong segments, narrowed at each end; seeds light brown, elliptic, compressed. (Bentham and Hooker 1865; Baker 1929; Pellegrin 1948; Phillips 1951; Gillett 1966c)

Representatives of this genus are common in the tropics and subtropics of the Old World. They are present in humid, freshwater areas, seacoasts, and also on open, dry slopes, in thickets, and on savannas.

Several species serve limited use as green manure, as supports for pepper, and as shade for coffee and cacao.

NODULATION STUDIES

Nodules were observed on *O. kirkii* S. Moore in Zimbabwe (Corby 1974), and on *O. trichocarpum* (Taub.) Harms in India (Parker 1933), in South Africa (Grobbelaar et al. 1967), and in Zimbabwe (Corby 1974).

From Greek, *hormos*, "necklace," "chain," in reference to the native use of the seeds as beads.

Type: *O. coccinea* (Aubl.) Jacks.

100–120 species. Trees, evergreen, mostly tall, unarmed. Leaves imparipinnate or abruptly pinnate; leaflets 3–19, leathery, terminal leaflet largest; rachis and undersurface of leaves often rusty-pubescent; stipels minute or none; stipules small, linear. Flowers pink, violet, greenish white, some with contrasting color markings, in copious terminal panicles, seldom in axillary racemes; bracts and bracteoles small, deltoid or linear; calyx campanulate, bibracteolate at base; lobes 5, upper 2 subconnate, usually broader and incurved; petals 5, free, twice the length of the calyx; standard suborbicular;

wings oblique, oval-oblong; keel petals free, obtuse; stamens 10, free, alternate ones subequal, often 1 or more sterile; anthers versatile; ovary pubescent, subsessile, 2–6-ovuled; style filiform, involute at apex; stigma usually bilobed, introrse, lateral. Pod short-stalked within calyx remnants, orbicular or oblong, often beaked, turgid or compressed around the seeds, leathery to woody, continuous or septate within, 2-valved, usually dehiscent; seeds 1–6, elliptic-globose, hard, glossy, red to black, usually bicolored bright red with a black spot around the hilum. (Bentham and Hooker 1865; Merrill and Chen 1943; Rudd 1965)

About 50 species are indigenous to tropical America; the others are native to tropical Asia and Madagascar. They flourish in rain and riverine forests, on mountain slopes up to 2,000-m elevation, and in sea-level coastal areas; some South American species are common in dry savannas.

Ormosia wood is neither of choice quality nor commercially important. In general, it is coarse grained, unattractive, slow drying, and subject to attack by beetles and borers. The bark of some species has a strong acrid odor when freshly cut. The heartwood is scanty in relation to the sapwood, and varies from a salmon color, as in *O. krugii* Urb. of Puerto Rico, to a deep red, as in *O. hosei* Hemsl. & Wilson, the red-bean tree of China. The wood of the latter species is heavier than water and is prized for cabinetmaking. *Ormosia* wood serves local uses for shingles, small construction boxes, and fuel. The bright bicolored seeds of various species are used as ornaments.

The major *Ormosia* alkaloids, designated by the prefix "ormo," have a physiological action akin to that of morphine, $C_{17}H_{19}NO_3$.

NODULATION STUDIES

Büsgen's drawings of 1905 showed dichotomously branched nodules on *O. sumatrana* (Miq.) Prain. Whitish brown, round to elongated, 3–10 mm nodules were present on the field specimens examined by Norris (1969). Rudd's observation (1965) was of a Kew Herbarium specimen.

Nodulated species of *Ormosia*	Area	Reported by
coarctaca Jacks.	Guyana	Rudd (1965)
coccinea (Aubl.) Jacks.	Bartica, Guyana	Norris (1969)
coutinhoi Ducke	Bartica, Guyana	Norris (1969)
monosperma (Sw.) Urb.	Trinidad	DeSouza (1966)
(= *dasycarpa* Jacks.)	Sri Lanka	Wright (1903)
sumatrana (Miq.) Prain	Sumatra	Büsgen (1905)

Ormosiopsis Ducke   Papilionoideae: Sophoreae

The Greek suffix, *opsis*, denotes resemblance to *Ormosia*.

Type: *O. triphylla* Ducke

The generic distinctiveness of *Ormosiopsis* as set forth by Ducke (1925a, 1930) and recognized by Hutchinson (1964) and Willis (1966) was discounted by Rudd (1965). Segregation of the 3 *Ormosiopsis* species was based on the so-called terminal position of their stigmas; in *Ormosia* stigmas are usually in the lateral position.

These species are now present as new combinations in *Ormosia*.

Ornithopus L.   Papilionoideae: Hedysareae, Coronillinae
305 / 3773

From Greek, *ornis*, *ornithos*, "bird," and *pous*, "foot," in reference to the pod clusters resembling a bird's foot.

Type: *O. perpusillus* L.

15 species. Herbs, annual, stems much-branched, delicate, slender, spreading or erect, unarmed, some pubescent. Leaves imparipinnate; leaflets small, 3–18 pairs; stipels absent; stipules small, free, linear. Flowers minute, yellow, white, or pink in small heads or umbels on long, axillary peduncles, sometimes subtended by a floral leaf; bracts and bracteoles small; calyx tubular or campanulate; lobes 5, equal, or the upper 2 connate above; standard obovate or suborbicular; wings oblong; keel obtuse, nearly straight, shorter than the wings; stamens 10, diadelphous; alternate filaments usually dilated above; anthers uniform; ovary sessile, many-ovuled; style curved, not bearded; stigma capitate. Pod usually curved, cylindric or flattened, beaked, tomentose, with constrictions between the 3–8 segments, each segment reticulate, ovate or oblong, 1-seeded, indehiscent; seeds transversely oblong, ovate, or subglobose. (Bentham and Hooker 1865)

Members of this genus are distributed throughout the Iberian Peninsula, southern Europe along the Mediterranean, and tropical Africa. They thrive on sandy loam and slightly acidic soils under a maritime climate.

Ornithopus species, especially *O. compressus* L. and *O. sativus* Brot., are cultivated in various parts of the world for grazing, soil cover and improvement, and for hay and silage. These 2 species have become particularly popular in Western Australia for sandy soils where *Trifolium subterraneum* has not flourished (Parker and Oakley 1963, 1965). In many areas *O. sativus*, called serradella, is esteemed as a complement to white and red clovers and alfalfa, since it thrives under conditions not particularly favorable to these latter plants (Schofield 1950). Its role as a catch crop on light acid soils with pH 5.0–5.5, as a pioneer cover, and as a rich source of protein is well known. Serradella has shown promise as a winter legume for forage and soil improvement in sandy soils in the southeastern United States.

An indigenous South American species, *O. micranthus* (Benth.) Ar., is a cattle forage of minor importance in Argentina.

Nodulation Studies

Rhizobial strains from serradella nodules were acknowledged as strains of *Rhizobium lupini* about the turn of the century (Maassen and Müller 1907; Simon 1907). Accordingly, proposal of the taxon *Rhizobium ornithopi* nov. sp. for isolates from *O. isthmocarpus* Coss. (Hollande and Crémieux 1926), now considered a subspecies of *O. sativus*, was superfluous and unwarranted.

As one follows the spread of serradella from the Iberian Peninsula throughout central Europe (Schofield 1950), it becomes obvious that its growth and establishment were improved on sites where lupines had previously grown. However, many of the rhizobial strains isolated by Lange (1961) from native legumes in Western Australia nodulated *Lupinus digitatus* but did not nodulate *O. compressus* and *O. sativus*. A marked degree of strain selectivity was shown in field tests using these 2 species in combination with effective lupine rhizobia and homologous *Ornithopus* strains (Parker and Oakley 1963). Lime pelleting of the seeds reduced nodulation of both lupines and serradella (Parker and Oakley 1965).

The mutual symbiobility between plants and rhizobia of *Lupinus* and *Ornithopus* (Jensen 1964) and serradella and *Lotus uliginosus* (Jensen 1967) has been amply proved. Furthermore, Jensen and his co-workers have established that the symbiotic range of the serradella-lupine rhizobia is much broader than it was originally conceived to be. Mutual interchangeability, while not always effective, now includes species of at least 9 unrelated genera (Abdel-Ghaffar and Jensen 1966; Jensen and Hansen 1968).

Nuclear changes in rhizobia-filled nodule host cells of *O. perpusillus* L. were described in detail by Hocquette (1930). Differences in chromatographic and electrophoretic behavior of leghemoglobins from lupine and serradella nodules formed by the same strain of *R. lupini* led Dilworth (1969) to conclude that hemoglobins in leguminous nodules are determined genetically by the plant and not by the rhizobial strain.

Nodulated species of *Ornithopus*	Area	Reported by
compressus L.	France	Clos (1896)
	Israel	M. Alexander (pers. comm. 1963)
	S. Africa	Grobbelaar et al. (1967)
micranthus (Benth.) Ar.	Uruguay	Pers. observ. (U.W. herb. spec. 1971)
perpusillus L.	Widely reported	Many investigators
pinnatus (Mill.) Druce	S. Africa	Grobbelaar & Clarke (1974)
(= *ebracteatus* Brot.)	France	Clos (1896)
sativus Brot.	Widely reported	Many investigators
ssp. *isthmocarpus* (Coss.) Dostál	Morocco	Hollande & Crémieux (1926)

Ostryocarpus Hook. f.
361 / 3832

Papilionoideae: Dalbergieae,
Lonchocarpinae

From Greek, *ostryo*, derived from *ostrakon*, "shell," "saucer," and *karpos*, "fruit," in reference to the shell-shaped pods.

Type: *O. riparius* Hook. f.

4 species. Shrubs, climbing, tall, or lianas. Leaves long, imparipinnate; leaflets in few pairs, usually 1–2, obovate-elliptic; stipels absent; stipules minute or inconspicuous. Flowers whitish yellow, usually fragrant, in dense racemose panicles on short, axillary branchlets; bracts and bracteoles inconspicuous, caducous; calyx campanulate, almost truncate, sometimes rusty-

pubescent; teeth 5, short; standard broad, rhomboid, glabrous, short-clawed, without ears, reflexed; wings free, long, sickle-shaped; keel petals connate at back, incurved, obtuse; stamens diadelphous; anthers elliptic, dorsifixed, versatile; ovary short, sessile, few-ovuled; disk adnate to base of the calyx; style filiform, curved; stigma small, terminal. Pod broad-oval, flat, not winged, leathery, apiculate, indehiscent; seeds 1–3, transversely arranged, oblong or orbicular, flat. (Dunn 1911d; Pellegrin 1948; Hauman 1954b)

Representatives of this genus are native to tropical West Africa. One species occurs in the coastal area of Zaire, the other 3 species throughout West Africa (Dunn 1911d).

C. riparius Hook. f. occurs commonly in forest rivulets or semiaquatic habitats of tropical West Africa. The bark and the congealed reddish sap which exudes from the bark are used as a fish poison in Liberia. The bark is also the source of a strong fiber useful for making fish nets (Dalziel 1948).

Ostryoderris Dunn   Papilionoideae: Dalbergieae,
(366b)   Lonchocarpinae

Name implies relationship to *Derris* and *Ostryocarpus*.

Lectotype: *C. impressa* Dunn

6–7 species. Lianas, shrubby climbers, or small upright trees; branchlets thick; wounded tissue exudes a reddish resin. Leaves imparipinnate; leaflets usually 4–7 pairs, opposite or alternate; stipels, if present, and stipules very small, caducous. Flowers pink or white, many, small, solitary, in softly tomentose, thyrsoid, terminal panicles; bracts conspicuous at base of pedicels; bracteoles 2, subtending the calyx; calyx campanulate, tomentose; sepals 5, small, very short, upper 2 more or less joined; petals glabrous, clawed; standard slightly gibbous at the base; wings and keel petals entirely free; stamens 10, diadelphous, anthers dorsifixed, longitudinally dehiscent; ovary pubescent, few-ovuled; disk sometimes 10-lobed, joined to the calyx. Pod usually short-stalked, oblong-linear, flattened, leathery, winged along both sutures, nonseptate, indehiscent, valves coherent between the 1–3, lens-shaped seeds. (Dunn 1911d; Pellegrin 1948; Hauman 1954b)

The position 366b was assigned by Harms (1915a).

Members of this genus occur in tropical Africa, especially in Nigeria and Zaire. They inhabit savanna woodlands.

The leaves of *O. lucida* (Welw.) Bak. f., a liana in Zaire, yield a dye used by natives. The plant sap has fish-stupefying properties. *O. stuhlmannii* (Taub.) Dunn ex Bak. f., now regarded as *Xeroderris stuhlmannii* (Drummond 1972), is the largest tree species of the genus. Its cream-colored, red-streaked wood is used for dug-out canoes and construction purposes. The leaves are eaten by stock in East Africa. The copious, sticky, blood red exudate from the slash is used for tanning hides.

Nodulation Studies

Nodules were observed on *O. gabonica* (Baill.) Dunn, in Yangambi, Zaire (Bonnier 1957), and on *O. stuhlmannii* (see *Xeroderris*) in Zimbabwe (Corby 1974).

From Greek, *ous, otos,* "ear," and *pteron,* "wing"; the wing petals have an ear at the base.

Type: *O. burchellii* DC. (= *Vigna burchellii*)

2 species. Undershrubs, glabrous, trailing or suberect; branches long, cylindric, rooting. Leaves pinnately 3-foliolate; leaflets smooth, lanceolate, apiculate, mucronate; stipels paired, linear-blunt, persistent; stipules peltate, lanceolate, striate. Flowers 2–4, blue to purple, paired at the top of axillary peduncles; bracts subulate-lanceolate; bracteoles 2, subtending the calyx, closely nerved; calyx tube short, campanulate; lobes 4, upper 2 connate, lowest lobe longest; petals subequal in length; standard large, oblong-orbicular, short-clawed, with 2 callosities at the base; wings oblong, obtuse, eared at the base; keel petals free, slightly eared, clawed at the base, acuminate at the apex and united, angled on the back; stamens 10, monadelphous; vexillary filament partly joined to others into a staminal sheath split ventrally; anthers small, versatile; ovary linear, compressed, straight, smooth, 5–6-ovuled; style smooth, incurved, inner margin, grooved, apex slightly thickened, sparsely pubescent; stigma terminal, very oblique, thick, bilabiate, upper lip rounded, larger. Pod linear-lanceolate, turgid, nonseptate within but with spongy tissue between the seeds, 2-valved, valves curling at dehiscence; seeds black, kidney-shaped, hilum small. (De Candolle 1825b; Harms 1908a; Baker 1929).

Taubert (1894) listed *Otoptera* as one of several generic synonyms of *Vigna.* In repositioning genera in Phaseolinae, Harms (1908a) recognized its separate generic status by designating its position as 428, immediately akin to *Vigna* 427.

Representatives of this genus are native to Madagascar, South Africa, and tropical southern Africa.

O. burchellii DC. is highly regarded in northern Transvaal for its vigorous, dense, hedgelike growth on sandbanks. Subsequent to stabilization and grass cover its growth subsides; thus it serves in a pioneer manner for binding drifting sands. The pods are relished by insects and stock.

NODULATION STUDIES

Nodules were observed on this species in Zimbabwe by Sandmann (1970b) and Corby (1974). Sandmann described them as large, round, and "covered with a white furry coat."

Hooker (1879) attributes the origin of the name to Ojjain, a town in central India from which the seeds were sent in 1795 to William Roxburgh, 1751–1815, then at the Calcutta Botanic Gardens. According to Menninger (1962), the genus was named by Bentham for a river in northwest India where the tree is native.

Type: *O. oojeinensis* (Roxb.) Hochr. (= *Dalbergia ougeinensis* Roxb.)

1 species. Tree, small to medium; branches slender, terete, gray. Leaves pinnately 3-foliolate; leaflets large, often subtomentose beneath, deciduous; stipels present; stipules free, caducous. Flowers white or pale pink to lilac, small, on slender pedicels, in short, densely clustered, fascicled racemes in leaf axils and nodes on old wood; bracts small, scalelike; bracteoles minute, below the calyx, persistent; calyx campanulate; lobes small, obtuse, upper 2 subconnate into a broad, emarginate lip, lower lobe larger than the laterals; petals short-clawed, inserted in a fleshy disk at the base of the calyx tube; standard suborbicular; wings oblique, oblong, slightly joined to the keel; keel petals subequal, blunt, slightly incurved; stamens 10, diadelphous; anthers uniform; ovary sessile, many-ovuled; style filiform, incurved, awl-shaped; stigma minute, globose, capitate, terminal. Pod plano-compressed, linear, weakly dehiscent, joints 2–5, somewhat distinct, each segment with 1 kidney-shaped seed. (Bentham and Hooker 1865; Hooker 1879)

This monotypic genus is endemic in central and northern India. The plants grow along roadsides, up to 1,700-m altitude in foothills of the Himalayas.

The trees reach heights of 7–14 m, with trunk girths of 1.5 m. The trunks are often twisted, but provide hard, durable, reddish brown wood suitable for farm implements. The leaves are a source of fodder. The bark exudes a kinolike resin used locally as an astringent febrifuge and to alleviate diarrhea. The tree is host to the lac insect (Howes 1949); the bark contains a fish-poisoning substance. This species is the sole member of the tribe Hedysareae that contains isoflavones (Harborne et al. 1971)

NODULATION STUDIES

A rhizobial strain from this species was maintained at one time in the United States Department of Agriculture culture collection (L. T. Leonard personal communication 1940).

From Greek, *oxys*, "sharp," and *lobion*, "pod."

Type: *O. cordifolium* Andr.

25–30 species. Shrubs, bushy or diffuse, rarely undershrubs, some small, prostrate, others up to 3 m tall, with thick, woody rhizomes; branches usually pubescent. Leaves 1-foliolate, opposite, occasionally scattered, entire, sometimes recurved, some with sharp-pointed lobes; upper surfaces of some species bear small, glandular dots between the veins, undersurfaces silky-pubescent; stipules small or absent. Flowers yellow, orange, or purplish red, in few-flowered or dense axillary or terminal racemes, occasionally in fascicles or dense corymbs; bracts and bracteoles soon caducous; calyx lobes imbricate, subequal, upper 2 often broader and connate above; petals clawed; standard orbicular; wings oblong; keel straight to slightly incurved, subequal to wings; stamens free; ovary stalked or sessile, 4- to many-ovuled, villous; style incurved, filiform or thickened near the base; stigma minute, terminal. Pod ovoid or oblong, sessile or stalked, 1–3 cm long, mostly pubescent, turgid, continuous within or rarely subseptate between the seeds, valves leathery. (Bentham 1864; Bentham and Hooker 1865)

Natural distribution of *Oxylobium* species is limited to Australia. They occur in coastal areas, tablelands, highlands, and river areas.

NODULATION STUDIES

Nodules were recorded by Lange (1959) on field-excavated root systems of *O. atropurpureum* Turcz., *O. capitatum* Benth., *O. lanceolatum* (Vent.) Druce, *O. parviflorum* Benth., and *O. reticulatum* Meissn. in Western Australia. Single-strain isolates from nodules of the aforementioned species were judged to be members of the slow-growing, cowpea-type of rhizobia on the basis of cultural and unilateral host-specificity tests (Lange 1961). *Lupinus digitatus* was highly responsive to infection by native rhizobial strains of *Oxylobium*, but a reciprocal response was not obtained with 2 lupine strains on *O. lanceolatum* seedlings (Lange 1962).

Oxyrhynchus Brandegee
(419a)

Papilionoideae: Phaseoleae,
Phaseolinae

From Greek, *oxys*, "sharp," and *rhynchos*, "beak," and in reference to the beaked keel and pod.

Type: *O. volubilis* Brandegee

4 species. Herbs, twining, perennial. Leaves 3-foliolate; stipels present; stipules striate. Flowers in axillary, narrow, racemelike thyrses; bracts awl-shaped, striate; bracteoles ovate; calyx campanulate, bilabiate; lobes rounded, subequal; standard kidney-shaped, broader than long, deeply emarginate, with 2 reflexed ears at the base; wings free, as long as the keel; keel broad-falcate, with narrowed acute beak, the 2 petals partly united, minutely ciliate; stamens diadelphous; filaments glabrous, slightly enlarged at the base; ovary

linear, pubescent; style glabrous except near apex where bearded on each side with long hairs and at tip bearing similar hairs partly surrounding the stigma; stigma ellipsoid-obovoid, attached on the dorsal side just below the middle. Pod straight, cylindric, beaked, terete or compressed, thin-walled, on short pedicel, 2–3-seeded; seeds globose, each with a linear hilum extending over half the circumference of the seed, the hilum covered with a white caruncle; germination hypogeous. (Piper 1924b; Rudd 1967)

This genus, originally set forth by Brandegee (1912), was based on 1 species, *O. volubilis* Brandegee, and was positioned by Harms (1915a) in proximity to *Rhynchosia*. Later on, Harms called attention to its closer identity to *Dolichos*, *Vigna*, and *Dysolobium* (Piper 1924b).

Members of this genus are native to the southern United States, Mexico, and the West Indies.

The herbage of *O. alienus* Piper, an attractive vine worthy of culture as an ornamental, is probably palatable to cattle. The seeds of *O. volubilis*, known by the name "frijol monilla" in some areas of Mexico, are used as food.

NODULATION STUDIES

Nodules were not observed by Piper (1924b) on the thickish roots of *O. alienus*.

Oxystigma Harms
(40b) / 3488

Caesalpinioideae: Cynometreae

From Greek, *oxys*, "sharp"; the stigma is pointed.

Type: *O. mannii* (Baill.) Harms

5 species. Trees or shrubs. Leaves paripinnate or imparipinnate; leaflets few; petiolules not twisted, usually alternate, leathery, pellucid glandular-punctate; stipules small, caducous. Flowers very small, short-pedicelled in slender, spikelike, racemose panicles, axillary or terminal; bracts and bracteoles reduced, scalelike; calyx tube very short, nearly absent or obsolete; sepals 5, equal, petaloid, broad-imbricate; petals none; stamens 10 or fewer, exserted; filaments filiform; anthers small, dorsifixed, longitudinally dehiscent; ovary sessile, small, 2-ovuled; style filiform, awl-shaped; stigma pointed. Pod drupaceous, sessile, semiorbicular or ovate, obliquely apiculate, sometimes corrugated or flattened, winglike at the base, with 1 seed in the thickened distal portion, indehiscent. (Harms 1897; Léonard 1952)

Representatives of this genus are indigenous to West Africa. They are usually found in damp areas, on river banks, and in mangrove swamps.

The light, red wood is surprisingly soft, easy to cut and split, and somewhat resinous. It is used for canoes, household utensils, and furniture. *O. mannii* (Baill.) Harms is a source of an oil used in the preparation of shellac and in the porcelain industry.

481

From Greek, *oxys*, "sharp," and *tropis*, "keel," in reference to the sharp-pointed keel.

Type: *O. montana* (L.) DC.

300 species. Herbs, perennial, acaulescent, or shrublets with numerous tufts of short stems covered with scaly stipules arising from a thick rootstock. Leaves imparipinnate; leaflets entire, in numerous pairs; stems and leaves often silky-haired; stipels absent; stipules small, linear, free or adnate to the petiole. Flowers white, pale yellow, violet, or purple, in axillary spikes or racemes, or arising directly from the base of the stem or rootstock; bracts small, membranous; bracteoles minute or absent; calyx tubular or campanulate; teeth subequal; petals usually long-clawed; standard erect, ovate or oblong; wings oblong; keel with an erect or ascending sharp beak, equal to wings or slightly shorter; stamens 10, diadelphous; anthers uniform; ovary sessile or stalked; style filiform, straight or curved; stigma minute, terminal. Pod sessile or stalked, often 2-celled by the intrusion of the seed-bearing ventral suture, ovate or oblong, 2-valved, walls papery, turgid, often silky-haired; seeds kidney-shaped. (Bentham and Hooker 1865; Barneby 1952)

The generic status of *Oxytropis* is questioned by some botanists. Old World species of *Astragalus*, including *Oxytropis* forms, have a haploid chromosome number of 8 (Senn 1938b; Darlington and Wylie 1955). New World astragali have haploid numbers of 11 and 12, and rarely 13 and 14 (Vilkomerson 1943; Turner 1956). Accordingly, *Oxytropis* species with a chromosome number of 8 are related more closely to Old World astragali than the two groups of New World astragali are related to each other (Ledingham 1957). Other studies support the concept that Old World species of *Astragalus* and *Oxytropis* represent a phylogenetic line different from that of New World astragali (Ledingham 1960; Ledingham and Rever 1963).

Species of this large genus are distributed throughout north temperate, sub-Arctic, and Arctic regions; some species are circumpolar. Vasil'chenko (1965) postulated that the *Oxytropis* species originated no later than the Miocene epoch in the western mountains of central Asia and underwent extreme dispersal during the preglacial period. About 40 species in the United States are native to the western Rocky Mountain states and adjacent plains. Other species occur in the midwestern and northeastern states and northward into Canada. The preferred habitat is sandy or gravelly soil near glacial lakes, dry plains, or in meadows up to 4,000 m.

Nonpoisonous members of this genus are commonly called point vetches. The foliage is not particularly relished by cattle and horses, but is considered fair-quality forage for sheep. The young leaves of some of the edible species have a bitter taste. Species occurring within the Arctic Circle are browsed by reindeer (Roizin 1959).

The poisonous species share with some *Astragalus* species the common name "locoweed," but in contrast to the astragali that are respectively toxic only to cattle, or to sheep, or to horses, the poisonous *Oxytropis* forms are deleterious to all three. *O. lamberti* Pursh, which extends from western Canada into the western plains of the United States and into Texas, is one of the most dangerous. It is commonly called white loco because its white flowers look like snow in thickly seeded areas, rattle loco because its mechanically agitated pods make a noise similar to a rattlesnake, and also stemless loco because the foliage arises directly from a rhizome. It is a most persistent plant because of its deeply penetrating root system; also it is a prolific seeder, and the seeds have a long vitality. It can be eradicated only by physical removal of the plants from the soil.

Symptoms of locoism are varied. In general, cattle, horses, and sheep tend to stray from the herd, exhibit exasperating stubbornness, a twitching of the eyelids, a chomping of the jaws, irregularities in gait, dragging of the feet, and a lack of muscular control. Eating of some species is habit-forming. Purgative treatments followed by a return to normal feed are remedial; however, overingestion is often fatal.

NODULATION STUDIES

The 9 species examined in Alaska (Allen et al. 1964) were abundantly nodulated in areas ranging from the Matanuska Valley to Point Barrow and Umiat along the Arctic Coast. Most the the species were growing in the shallow 30–50-cm soil layer overlying permafrost. The nodules were 2–4 mm in length, plump, and often multilobed, with vivid scarlet red interiors indicative of leghemoglobin. The plants evinced nitrogen sufficiency. The ability of the symbiotic nitrogen process to function in a psychrophylic environment suggests a unique evolutionary adaptation.

The strains of rhizobia from the Alaskan species conformed culturally and physiologically with those from *O. chartacea* Fassett, an epibiotic endemic to two localities in northern Wisconsin (Bushnell and Sarles 1937). All strains were Gram-negative short rods with monotrichous flagella. Growth on yeast-extract mannitol mineral salts was sparse; alkaline reactions were produced on a variety of carbohydrate media, and the reaction in litmus milk was alkaline with serum zone formation.

Tests of the Wisconsin strains with plants representative of a broad spectrum of the so-called cross-inoculation groups evidenced no symbiotic associations (Bushnell and Sarles 1937). However, Wilson (1939a) judged the plants and rhizobia of *O. lamberti* to be quite promiscuous and nonspecific symbiotically. Serological studies showed similarities between rhizobia from *O. nigrescens* (Pall.) Fisch. and *Astragalus alpinus*, but not with alfalfa, clover, pea, and bean rhizobia (Kriss et al. 1941). This finding is pertinent, since several *Oxytropis* species were formerly classified as astragali. An inquiry into rhizobial host-infection relationships of these groups may provide a useful tool to strengthen the genetic lines of evidence.

Nodulated species of *Oxytropis*	Area	Reported by
arctica R. Br.	Alaska, U.S.A.	Allen et al. (1964)
campestris (L.) DC.	French Alps	Kharbush (1928)
carpatica Uechtr.	Czechoslovakia	I.B.P. Rhizobium Catalogue (1973)
chartacea Fassett	Wis., U.S.A.	Bushnell & Sarles (1937)
columbiana St. John	Wash., U.S.A.	Pers. observ. (U.W. herb spec. 1970)

483

Nodulated species of *Oxytropis* (cont.)	Area	Reported by
deflexa (Pall.) DC.	Alaska, U.S.A.	Allen et al. (1964)
(= *deflexa* var. *deflexa* Barneby)		
var. *sericea* T. & G.	Utah, U.S.A.	Pers. observ. (U.W. herb. spec. 1970)
foliolosa Hook.	Alaska, U.S.A.	Allen et al. (1964)
(= *deflexa* (Pall.) DC.		
var. *foliolosa* (Hook.) Barneby)		
glabra (Pall.) DC.	U.S.S.R.	Pers. observ. (U.W.
var. *elongata* (Lebd.) Popov		herb. spec. 1970)
gracilis (A. Nels.) Schum.	Alaska, U.S.A.	Allen et al. (1964)
(= *campestris* (L.) DC.		
var. *gracilis* Barneby)		
koyukukensis Pors.	Alaska, U.S.A.	Allen et al. (1964)
lapponica (Wahl.) Gay	S. Africa	N. Grobbelaar (pers. comm. 1965)
lamberti Pursh	N.Y., U.S.A.	Wilson (1939a)
leucantha Bge.	Alaska, U.S.A.	Allen et al. (1964)
(= *viscida* Nutt.		
var. *subsucculenta* (Hook.) Barneby)		
maydelliana Trautv.	Arctic U.S.S.R.	Kriss et al. (1941)
nigrescens (Pall.) Fisch.	Arctic U.S.S.R.	Kriss et al. (1941); Kriss (1947)
	Alaska, U.S.A.	Allen et al. (1964)
oreophila Gray	Utah, U.S.A.	Pers. observ. (U.W. herb. spec. 1970)
pilosa DC.	Estonia, U.S.S.R.	Pärsim (1966)
riparia Litw.	Wis., U.S.A.	J. C. Burton (pers. comm. 1962)
scammaniana Hult.	Alaska, U.S.A.	Allen et al. (1964)
sordida (Willd.) Pers.	Arctic U.S.S.R.	Roizin (1959)
varians (Rydb.) Hult.	Alaska, U.S.A.	Allen et al. (1964)
(= *campestris* (L.) DC.		
var. *varians* Barneby)		

Pachecoa Standl. & Steyerm.   Papilionoideae: Hedysareae, Stylosanthinae

Named for Don Mariano Pacheco Herrarte, Director General of Agriculture of Guatemala during the 1940s.

Type: *P. prismatica* (Sessé & Moc.) Standl. & Schub.

2–3 species. Shrubs, small, erect; stems stiff, with hairs short, eglandular, or longer, yellowish, viscid. Leaves imparipinnate; leaflets few, 3–9, alternate, entire, membranous, mucronate; stipels absent; stipules linear-subulate, striate-nerved. Flowers yellow orange, subsessile, on stiff, mostly 1–2-flowered axillary peduncles; bracts small, lanceolate, rigid; bracteoles awl-shaped, almost filiform, denticulate; calyx funnel-shaped; tube glabrous; lobes 5, deeply cleft, lanceolate, silky-pubescent, bristly, 4 subequal, lowest lobe falcate-subulate, slightly longer and narrower; standard orbicular, villous outside, clawed; wings oblique-obovate; keel incurved, obtuse; stamens monadelphous; stigma truncate, oblique, terminal, minute. Pod lomentaceous, erect, hard, linear-oblong, longitudinally nerved, sutures straight, shallowly constricted into subtetragonal segments, segments 1-seeded, densely pilose. (Standley and Steyermark 1943, 1946; Burkart 1957)

The undersides of old leaves have a dark red network between the veins caused by tannin in the large, star-shaped cells of the mesophyll. This unusual characteristic occurs also in *Chapmannia* and *Arthrocarpum* (Gillett 1966c).

Members of this genus occur in tropical Mexico, and Central and South America. They form thickets on damp or dry hillsides.

NODULATION STUDIES

Round nodules were depicted in drawings of the root system of *P. venezuelensis* Burk. in Argentina (Burkart 1957). This observation was confirmed on field-grown plants from Venezuelan savannas (Barrios and Gonzalez 1971).

Pachyelasma Harms (90b)	Caesalpinioideae: Dimorphandreae

From Greek, *pachys*, "thick," "stout," and *elasma*, "implement," in reference to the thick fruits.

Type: *P. tessmanii* (Harms) Harms

1 species. Tree, large, unarmed; branches glabrous. Leaves bipinnate, large; pinnae 2–5 pairs, opposite, rarely alternate; leaflets 9–20, alternate, oblong-oblanceolate, asymmetric, subcoriaceous; stipules lanceolate, caducous. Flowers small, dark red, on short pedicels, in long, multiflowered, spiciform racemes, axillary or terminal, solitary or fascicled; calyx tube shortly and broadly cupular; sepals 5, broad, subequal, short, edges ciliate, rounded, imbricate; petals 5, imbricate, ovate-oblong, margins fringed, 2–3 times as long as the sepals; stamens 10, similar, free; filaments filiform, equal, glabrous; anthers small, dorsally attached near the base with a somewhat broad connective; ovary short-stalked, lanceolate, glabrous; ovules 15–20, in 2 rows; style short, large, broad. Pod large, thick, stout, woody, linear-oblong or oblong-lanceolate, 4-angled, almost winged, compressed, straight or slightly curved, transversely divided, each locule containing 1 seed, indehiscent or tardily dehiscent, sutures thick; seeds 15–20, thick, transverse, oblong, laterally compressed. (Harms 1913, 1915a; Pellegrin 1948; Wilczek 1952)

This monotypic genus is endemic in West Africa.

This very large forest tree often reaches a height of 66 m, with a trunk diameter of 1.6 m; the tall cylindrical bole is commonly flanked by huge plank buttresses. The wood has an offensive odor. Natives of Nigeria and Cameroon use the crushed pods as a fish poison and for minor medicinal purposes.

*Pachyrhizus A. Rich. ex DC.   Papilionoideae: Phaseoleae,
(433) / 3908   Phaseolinae

From Greek, *pachys*, "thick," "stout," and *rhiza*, "root."

The original spelling of the generic name with a single *r* by De Candolle (1825b) was conserved (Rickett and Stafleu 1959–1961) over the earlier approved spelling with double *r* (Camp et al. 1947).

Type: *P. erosus* (L.) Urb.

6 species. Vines, annual or perennial, herbaceous or decumbent, erect or woody tall climbers; roots tuberous. Leaves pinnately 3-foliolate; leaflets often angulate, sinuately lobed or coarsely dentate, terminal leaflet broad, lateral leaflets asymmetric; stipels and stipules present. Flowers fascicled, large, purple, pink, or white, on long-stalked, axillary racemes, nodes somewhat thickened; bracts and 2 bracteoles small, setaceous, caducous; calyx campanulate; lobes equal, subequal to calyx tube, lanceolate, upper 2 united into a short, bidentate lip; standard broad-ovate, recurved, 2-eared at the base; wings falcate, oblong, as long as the blunt, incurved keel; stamens diadelphous; anthers uniform; ovary subsessile, many-ovuled; style with thickened inrolled apex, flattened, bearded along the inner side; stigma globose, lateral. Pod linear, bristly-haired, transversely depressed between the seeds, weakly septate inside; seeds 4–9, yellow, red, purple, or black, flattened, ovate, square or kidney-shaped. (De Candolle 1825b; Bentham and Hooker 1865; Burkart 1943; Purseglove 1968)

The original generic number 427 (Taubert 1894) was changed to 433 by Harms (1908a) because of the reorganization within the subtribe necessitated by various revisions and inclusions of new genera.

Representatives of this genus are widespread in the tropical and subtropical regions of the Old World and the New World. All species are probably indigenous to Central and South America (Clausen 1944). The plants are both wild and cultivated. They prefer loose sandy soil. Species occur in moist or wet thickets and open pine forests.

The common English name for all species is yam bean; the Spanish name, "jicama," is probably derived from the Aztec "xicama."

Yam beans were cultivated in Central America in pre-Columbian days and taken by the Spaniards into the Philippines by way of Mexico; from the Philippines they were introduced into Asia. *P. erosus* (L.) Urb. and *P. tuberosus* Spreng. are the most widely cultivated for their edible tubers. The small tubers of *P. ahipa* (Wedd.) Parodi, a nonclimbing species, are eaten infrequently in Bolivia and Argentina. The tubers of *P. panamensis* Clausen, *P. strigosus* Clausen, and *P. vernalis* Clausen are not eaten (Clausen 1944).

The young tender pods of *P. erosus* are boiled and eaten as a vegetable. The tubers are solitary or several, simple or lobed, and often turnip-shaped, with a brown skin covering a white, sweet, and agreeable-tasting flesh. Since the best tubers are obtained from nonflowering plants, growers in some areas remove the flower buds to increase and improve tuber production. The tubers are usually eaten raw, but may be cooked as a substitute for yams, *Dioscorea esculenta*. The palatable starch made from the tubers resembles arrowroot starch and is used in custards and puddings (Porterfield 1939). Keuchenius (1924) and Masefield (1956) regarded this species highly as a green manure in Java and Malaysia, respectively, although nodulation was sparse.

486

Under certain circumstances the tubers of all species may be poisonous (Clausen 1944). Natives in some areas believe those of wild species are always poisonous. The literature near the turn of the last century is replete with references to the uses of the plants as an insecticide and to the toxic properties of the seeds, leaves, and roots of *P. erosus* in relation to fish, dogs, and other vertebrates. Toxic principles in the seeds and to some extent, the leaves, have been identified as the glucoside pachyrrhizid, $C_{30}H_{18}O_8(OCH_3)_2$, pachyr-rhizine, $C_{19}H_{12}O_6$, various saponins, and related compounds (Baker and Lynn 1953; Shangraw and Lynn 1955; Simonitsch et al. 1957). The role of the yam bean as a source of rotenone and other insecticidal constituents has been disappointing (Norton and Hansberry 1945).

NODULATION STUDIES

P. erosus was reported nodulated in the Philippines (Bañados and Fernandez 1954).

Padbruggea Miq.   Papilionoideae: Galegeae, Tephrosiinae
(257d)

Named for Robert Padbrugge, Governor of the Island of Amboin and patron of Rumphius.

Type: *P. dasyphylla* Miq.

5 species. Robust lianas; young branches usually densely brown-hairy. Leaves imparipinnate; leaflets flat or revolute, base cordate to rounded, under surface brown-hairy; stipels present; stipules small, caducous. Flowers lax, panicled or corymbose, in leaf axils; bracteoles 2, awl-shaped, caducous, near top of pedicel; calyx campanulate; lobes 5, distinct, blunt; standard broadly rounded, violet with yellow green blotches, with 2 small, basal, inflexed auricles connected by a broad ligule, appressed brown-hairy on back; wings sometimes white with a velvet top, free from keel petals, not appendaged; keel petals shorter than wings, obtuse; stamens diadelphous; anthers uniform; ovary long-stalked, with dense brown hairs, 2-ovuled; style short, glabrous, curved; stigma terminal. Pod oblong, thick, subindehiscent, 1–2-seeded.

This genus has close kinships with, but is distinctive from, *Millettia* and *Wisteria* (Dunn 1911c; Harms 1915a; Backer and Bakhuizen van den Brink 1963). Hutchinson's concept (1964) of its synonymy with *Millettia* merits reconsideration.

Padbruggea species are distributed throughout Thailand, Malaysia, and Java. They occur in lowland woods, forest borders, and along watersides.

The seeds of *P. dasyphylla* Miq. are injurious if eaten raw, but are palatable and nourishing after roasting.

Named for Ch. F. Pahud, 1803–1873, Governor-General of the Dutch East Indies.

Type: *P. javanica* Miq.

3–4 species. Trees, up to 30 m tall, nonbuttressed, unarmed. Leaves paripinnate; leaflets entire, chartaceous, 3–6 pairs; stipules connate, caducous. Flowers in axils of bracts, borne in terminal or axillary, densely pubescent corymbs or racemes; bracts and bracteoles caducous; calyx tube long, pubescent; lobes 4, leathery, imbricate in bud; petal 1, perfect, wide, short-clawed, obovate-spathulate, others rudimentary or absent; stamens 7, perfect, free or united in a declinate sheath split above, pubescent at the base; staminodes 2–3, small; ovary stalked, stalk adnate to the calyx tube; style filiform, long; stigma terminal. Pod oblique oblong, asymmetric, broad, compressed, 2-valved, valves thickly woody, base rounded, septate between the seeds and lined with a thin, spongy pulp; seeds compressed, ovate or oblong, transverse, with an incomplete, yellow, orange, or red aril. (King et al. 1901; Backer and Bakhuizen van den Brink 1963)

Some botanists treat all *Pahudia* species as synonyms of *Afzelia* (Léonard 1950); however, in most Malesian floras *Pahudia* is the preferred name (de Wit 1941).

Species of this genus are native to the Philippines, Borneo, and Indonesia. They usually grow on shallow coastal soils and in forest borders.

The wood of *P. rhomboidea* Prain, known as tindalo, ranks among the finest of Philippine woods for cabinets, quality furniture, and musical instruments. The sapwood is creamy white; the heartwood is yellowish red, becoming very dark with age. The wood is valued for its durability, hardness, freedom from warping, and ability to take a high-gloss finish (Whitford 1911). A mediocre-quality balsam is marketed from the crushed bark of *P. galedupa* Backer.

NODULATION STUDIES

Plants of *P. rhomboidea* examined in Los Baños, Laguna, Philippines, lacked nodules (Bañados and Fernandez 1954; personal observation 1962).

The original spelling of the name was *Paloue*; *Palovea* is the alternate spelling (Cowan 1957).

Type: *P. guianensis* Aubl.

4 species. Trees, small to large, unarmed. Leaves 1-foliolate, short-petioled, entire, large, leathery, pinnately nerved; stipules minute, deciduous. Flowers showy, red, in short, axillary racemes or terminal spikes; bracts short, small, persistent; bracteoles usually colored, forming a 2-lobed sheath, per-

sistent, shorter than the calyx; calyx elongated-turbinate; lobes 4, slightly unequal, imbricate; petals 5, oblong, subsessile, unequal, 1 large, 4 small of which 2 are rudimentary; stamens 9, free, fertile; filaments long; anthers oblong or linear, sometimes pilose, versatile, longitudinally dehiscent; ovary stalked, stalk adnate to the calyx tube, many-ovuled; style long, slender; stigma subcapitate, small, terminal. Pod small, straight or scimitar-shaped, compressed, narrow, woody, 2-valved, upper suture thicker; seeds ovate, hard, compressed, nonarillate. (Bentham and Hooker 1865; Amshoff 1939; Cowan 1957)

Members of this genus occur in tropical South America and Guiana. They are usually present in woodlands and along streams.

Paloveopsis R. S. Cowan   Caesalpinioideae: Amherstieae

The Greek suffix, *opsis*, denotes resemblance to *Paloue*.

Type: *P. emarginata* R. S. Cowan

1 species. Tree, up to 11 m tall, with a trunk about 20 cm in diameter and gray bark; new branches pilose. Leaves 1-foliolate, subsessile, large, elliptic, leathery, acute toward the apex but blunt, emarginate at the tip; stipules glabrous, caducous. Flowers in short, axillary or terminal racemes; bracts small, caducous, minutely ciliate; bracteoles connate, obovate, apex ciliate; calyx stalked, small, funnel-shaped; lobes 4, glabrous, elliptic, equal, imbricate; petals 5, equal, linear-lanceolate, acute, minutely punctate, long-ciliate; stamens 9, 3 fertile with long filaments, unequally connate at the base, filaments of the 6 staminodia short, awl-shaped; anthers 3, oblong, glabrous; ovary short-stalked, oblong, smooth; style long, slender, smooth; stigma small, capitellate. Pod not described. (Cowan 1957)

This monotypic genus is endemic in Guyana.

Panurea Spruce ex Benth. & Hook. f.   Papilionoideae:
149 / 3609   Sophoreae

Named for Panuré, near the Uaupés River in northern Brazil, where the plant grows abundantly.

Type: *P. longifolia* Spruce ex Benth.

1 species. Tree, up to 10 m tall, much-branched. Leaves 1-foliolate, oblong, elliptic, leathery; petiolule very short; stipules, small, awl-shaped, caducous. Flowers small, whitish yellow, in short, axillary panicles; bracts and

bracteoles small; calyx turbinate; lobes short, broad, upper 2 joined into a bidentate lip; petals short-clawed; standard rounded; wings oblique-ovate; keel petals ovate, shorter than the wings, free; stamens free; anthers small, subglobose; ovary sessile, few-ovuled; style thick, short, hooked-inflexed; stigma terminal. Pod compressed, 2-valved, apex pointed, neither winged nor nerved; seeds compressed, obliquely positioned to the valves, black when dry. (Bentham and Hooker 1865)

This monotypic genus is endemic in northern Brazil. It is gregarious among scattered forest trees.

Papilionopsis Steenis — Papilionoideae: Galegeae, Tephrosiinae

Name alludes to the resemblance of this most unusual Indo-Australian genus to papilionoid members of the Leguminosae.

Type: *P. stylidioides* Steenis

1 species. Herb, small, 20–25 cm tall, perennial; stems glabrous. Leaves reduced to sessile, semiamplexicaulous, leaflike phyllodes, linear-triangular, parallel-veined, with a hooded apex tipped with small, scurfy, glandlike dots; main stem surrounded by a basal rosette of such phyllodes; stipels and stipules absent. Flowers small, bluish purple, few, on slender pedicels, in loose clusters at the apex of a very long peduncle covered with 3 types of hairs: some minute and crisp, some long, spreading unicellular setae, and those toward the apex long, multicellular with glandular tips; calyx campanulate, densely pubescent; lobes 4, equally cleft to the middle, anterior lobe bifid; petals clawed; standard distinctly oblique, obovate-orbicular, recurved; wings erect, base conspicuously eared; keel petals erect, spathulate, curved; stamens 10; filaments alternately long and short, united into a tube split above on the adaxial side; anthers equal, orbicular; ovary linear, 4-ovuled; style glabrous; stigma terminal. Pod not described, but as judged from the ovary, flattish, linear, beaked, and 4-seeded. (Steenis 1960b)

This monotypic genus is endemic in western Papua New Guinea. It occurs in the Central Mountains, Swart Valley, Kudabaka, at 1,600–2,000-m altitude.

Paralbizzia Kosterm. — Mimosoideae: Ingeae

The Greek prefix, *para*, denotes nearness or relationship to *Albizia*.

Type: *P. turgida* (Merr.) Kosterm.

This genus of questionable status is one of 10 genera established in 1954 by Kostermans in his revision of Indo-Malayan species in *Pithecellobium*. It consists of 3 species. Segregation was made on the basis of the pods being straight or slightly falcate, dehiscence occurring along both sutures, the valves not twisting after dehiscence, and the seeds being large.

Paramachaerium Ducke   Papilionoideae: Dalbergieae, Pterocarpinae

The Greek prefix, *para*, denotes resemblance to the genus *Machaerium*.

Type: *P. schomburgkii* (Benth.) Ducke

3 species. Trees, medium to tall, unarmed; trunk usually irregular. Leaves imparipinnate; leaflets several, alternate, oblong, acuminate, penninerved; stipules minute, early deciduous. Flowers violet, numerous, in terminal panicles; bracts persistent; bracteoles usually larger than the bracts; calyx campanulate, somewhat gibbous, obtuse at the base, bilabiate; standard clawed, orbicular, only slightly longer than the wings; keel petals joined; stamens 10, monadelphous; anthers versatile; disk glandular; ovary very short-stalked, several-ovuled; stigma capitate. Pod sessile, with seeds 1–3 at the base and a membranous, flat, apical wing about 3 times longer than broad. (Ducke 1935b; Dwyer 1965)

Representatives of this genus are distributed throughout Panama, Guyana, and Amazonas, Brazil. They flourish along streams and on land periodically inundated.

The trees usually exhibit poor form and the bark exudes a red sap. The wood lacks commercial potential (Record and Hess 1943; Brizicky 1960).

Paramacrolobium J. Léonard   Caesalpinioideae: Amherstieae

The Greek prefix, *para*, alludes to the derivation of the genus from *Macrolobium*.

Type: *P. coeruleum* (Taub.) J. Léonard (= *Macrolobium coeruleum*)

1 species. Tree, up to 40 m tall. Leaves paripinnate; leaflets 3–5 pairs, rarely 2 pairs, elliptic to oblong, base asymmetric, apex acuminate, subcoriaceous, glabrous, eglandular; petiolules twisted; rachis not grooved; stipels none; stipules intrapetiolar, united, oblong, semicylindric, biapiculate, persistent. Flowers in compact, corymblike panicles; bracteoles very long, glabrous; sepals 4, unequal, posterior sepal longer and bidentate; petals blue, 5, posterior petal very large, bilobed, the 2 laterals smaller, the 2 anterior petals filiform and very small; stamens 9, unequally united at the base, usually 3, rarely 5, fertile and exserted, usually 6, rarely 4, staminoidal, smaller; ovary linear, long-stalked, 1 side adnate to the calyx tube. Pod oblique-oblong, large, apiculate, glabrous, winged along 1 suture, finely warted but without longitudinal ribs, 2-valved, woody, nonseptate within, dehiscent; seeds 3–8, rectangular. (Léonard 1954c, 1957; Hutchinson and Dalziel 1958)

This monotypic genus is endemic in tropical Africa. It is more common in dryland forests than in forested, periodically inundated areas along rivers.

The hard, durable wood is used for railway crossties, general carpentry, and the manufacture of furniture and gongs.

NODULATION STUDIES

Rhizobial nodules are not known to occur on this species. However, Peyronel and Fassi (1960) reported an alternate mode of symbiosis with ectotrophic mycorrhiza for *P. coeruleum* (Taub.) J. Léonard and *P. fragrans* (Bak.) Ont.; the latter is more commonly regarded as *Macrolobium fragrans*.

The Greek prefix, *para*, denotes resemblance or nearness to *Piptadenia*.

Lectotype: *P. rigida* (Benth.) Brenan

2 species. Trees or shrubs, not prickly. Leaves bipinnate; pinnae opposite; leaflets small, in numerous pairs, seldom large, in few pairs; petiolar and jugal glands usually present. Flowers small, white, or green, mostly bisexual, in spicate racemes on axillary, solitary or fascicled peduncles; calyx campanulate, shortly 5-dentate; petals 5, often joined in the lower half; stamens 10, free, exserted; anthers crowned by a deciduous gland; ovary subsessile; ovules 3 or more. Pod usually stalked, broad, linear, flat, membranous or subcoriaceous, 2-valved, nonseptate, not pulpy; seeds compressed, distinctly winged, funicle filiform. (Brenan 1955, 1963)

Brenan (1963) proposed the name *Parapiptadenia* to differentiate 2 *Piptadenia* species described by Bentham (1842) from others included in *Piptadenia* (1840).

Members of this genus are indigenous to tropical eastern Brazil.

The lectotype *P. rigida* (Benth.) Brenan yields angico gum, a watersoluble, dark red resin similar to gum arabic. It is used in Brazil as mucilage and as a constituent of medicines.

NODULATION STUDIES

Nodules were observed on *Parapiptadenia rigida*, reported as *Piptadenia rigida*, in France by Lechtova-Trnka (1931), and in Argentina by Rothschild (1970).

Parasamanea Kosterm.   Mimosoideae: Ingeae

The Greek prefix, *para*, denotes resemblance or nearness to *Samanea*.

This genus as set forth by Kostermans (1954) consists of 1 Malaysian mimosoid species, *P. landakensis* (Kosterm.) Kosterm. (= *Pithecellobium landakense*. Segregation from *Pithecellobium* was based on the pod being woody, 1-seeded, and pulpy, and the valves not twisting after dehiscence. As yet, *Parasamanea* has not been accorded general acceptance (Hutchinson 1964).

Paratephrosia Domin   Papilionoideae: Galegeae,
(255b)   Tephrosiinae

The Greek prefix, *para*, denotes resemblance or nearness to *Tephrosia*.

Type: *P. lanata* Domin

1 species. Shrublet, densely silky-tomentose overall. Leaves dense on branches, pinnately 3-foliolate; stipules linear-subulate. Flowers fascicled in leaf axils; bracts linear-subulate; calyx tube very short, much shorter than the

lobes; lobes linear-subulate, elongated, free, subequal; petals shorter than the calyx, tomentose outside; standard transversely oblong-orbicular, entire, short-clawed; wings free; keel slightly incurved, obtuse; vexillary stamen free only at the base; anthers uniform; ovary 1-ovuled; style filiform, flattened, glabrous; stigma small, terminal. Pod sessile, longer than the calyx, obliquely semiovate, tomentose; seed suborbicular. (Domin 1912; Harms 1915a)

This monotypic genus is endemic in central Australia. The plants grow in arid regions.

Parenterolobium Kosterm.　　　　　Mimosoideae: Ingeae

The Greek prefix, *para*, denotes resemblance or nearness to *Enterolobium*.

Type: *P. rosulatum* Kosterm. (= *Pithecellobium rosulatum*)

1 species.

This genus was set forth by Kostermans (1954) to accommodate a single Asian species of *Pithecellobium*, distinctive for its flat, circinate, indehiscent pods that lack true septa and segment superficially. The genus lacks acceptance by conservative botanists.

This monotypic genus is endemic in Malaysia.

Parkia R. Br.　　　　　Mimosoideae: Parkieae
29 / 3469

Named for Dr. Mungo Park, 1771–1806, Scottish surgeon, naturalist, and explorer of the Niger River Basin, Africa. Park was drowned in the river during an attack by local inhabitants.

Type: *P. filicoidea* Welw. (= *P. africana* R. Br.)

50–60 species. Trees, unarmed, up to about 40 m tall with trunk diameters 1–1.5 m, often well-buttressed up to about 3 m from the base, crowns flat, spreading. Leaves bipinnate; pinnae pairs many; leaflets small, 10–70 pairs; petioles and rachis glanduliferous. Flowers 5-merous, minute, very numerous, borne in compact, globular or club-shaped heads, on long pendent peduncles, solitary and axillary, or several at apices of branches; basal flowers male or sterile, white or red; the upper perfect flowers yellow, pink, or red; bracteoles narrow, spathulate, subtending each flower in the head; calyx cylindric, tubular, subbilabiate; lobes 5, short, imbricate in bud; petals 5, narrowly linear-spathulate, free or connate at the base with the calyx and staminal tube; stamens 10, monadelphous, connate rather high up; filaments exserted, white or yellowish; anthers oblong, usually with a gland at the tip; ovary sessile or short-stalked, many-ovuled; style filiform; stigma terminal. Pod oblong or oblong-linear, pendulous, straight or curved, stalked, clustered, compressed, subcylindric, leathery or woody, often with a pulpy, fleshy interior, 2-valved, very tardily dehiscent or indehiscent; seeds ovoid to oblong, compressed, thick, transverse. (Bentham and Hooker 1865; Gilbert and Boutique 1952a; Brenan 1959)

Representatives of this genus are widespread in the tropics and subtropics of both hemispheres. About 20 species occur in the Amazon region of Brazil, 7 species in West and East Africa and Madagascar, and several in Asia. They thrive in rain forests, lowland river areas, and also in park savannas and drier forested regions up to 650-m altitude.

P. bicolor (Jacq. Benth., *P. biglobosa* (Jacq.) Benth., *P. clappertoniana* Keay, *P. filicoidea* Welw., *P. javanica* (Lam.) Merr., and *P. timoriana* (DC.) Merr. are among the most showy members of the genus. *Parkia* species are easily recognized by their malodorous, drumsticklike flower heads, 5–20 cm in diameter, that hang down below the foliage at the ends of long peduncles. The flowers of some species are pollinated by bats, which visit the flowers for the nectar contained in a ring near the junction of the flower head to the stalk. As the bats clasp the flower heads, their bodies become covered with pollen. That pollination does not take place when devices are fixed to the flower head to prevent bats, but not insects, from reaching the flowers is proof that bats are solely responsible for pollination (Lawson 1966).

Parkia species serve a multitude of uses in their native habitats (Watt and Breyer-Brandwijk 1962; Burkill 1966). Dalziel (1948) referred to *P. filicoidea*, commonly called African locust bean in West Africa, as "the most typical tree of the park savanna." Eggeling and Dale (1951) spoke of this species as "the only tree left unfelled in populous regions by reason of its great importance as food."

Since the synonym for *P. filicoidea*, *P. africana* R. Br., appears frequently in the literature, both names are used in this discussion to comply with that in the reported data.

The pods of *Parkia* species are good cattle fodder. The tough, membranous inner lining of the pods of *P. filicoidea* is used as cordage. Natives in various areas use the fresh pulp to make a refreshing diuretic drink. The pulp is also eaten as a concentrated food, similar to dates, or is collected, dried, sifted, and sold as a yellow meal that is eaten with rice and meats or in soups, and is made into cakes. The seeds of all species contain a high content of fat; those of *P. africana* yield about 24 percent of a semifluid, sweet oil. In Java the seeds of *P. speciosa* Hassk. are consumed raw or roasted in the pod. The roasted seeds of some species are used to prepare a coffeelike infusion. The bark yields 12–15 percent tannin. Decoctions are used as an astringent. The bark of *P. filicoidea* also yields a red brown coloring matter. The leaves of some species are used to make refreshing drinks and sauces, and as fodder for livestock, and, because of their high nitrogen and ash contents, they are used as green manure.

Parkia timber is of little commercial importance. The sapwood is a grayish yellow, the heartwood, a dull brown. The wood is coarse grained and textured, easy to work, and finishes smoothly; it varies in density from 256 to 545 kg/m^3. Primary uses have been reserved for house uprights, inner house construction, boxboards, bowls, implement handles, and fuel (Record and Hess 1943).

NODULATION STUDIES

Nodules were reported on *P. africana, P. biglandulosa* W. & A., and *P. javanica* in Hawaii (Allen and Allen 1936a) and on *P. speciosa* in Malaysia (Lim 1977). Six plantlets of *P. javanica* examined in the Philippines by Bañados and Fernandez (1954) lacked nodules. Also, nodules were absent on the brown roots of 2 specimens of *P. pendula* Benth. examined by Norris (1969) at Manaus, Amazonas, Brazil.

Five strains of rhizobia isolated from each of the aforementioned nodulated species in Hawaii were effective on host plantlets of *P. biglandulosa.* Reciprocal inoculation was obtained using seeds and strains of *Vigna sinensis* as an indicator plant of the cowpea miscellany (Allen and Allen 1936a). Two strains from *P. africana,* used as inocula for 20 members of 14 genera of the cowpea miscellany, produced nodules on all but 3 species, *Desmodium purpureum, Lespedeza sericea,* and *Phaseolus lunatus* (Allen and Allen 1939). Plant-host responses of *Cajanus cajan* and *Dolichos lablab* were highly effective.

Parkinsonia L.   Caesalpinioideae: Eucaesalpinieae
94 / 3551

Named for John Parkinson, 1567–1650, herbalist to King James I, and apothecary of London. Parkinson was author of *Paradisi in Sole Paradisus Terrestris,* 1629, which extolled the joys of a garden, and of *Theatrum Botanicum,* 1640, which dealt with botanical descriptions of known plants.

Type: *P. aculeata* L.

2–4 species. Shrubs, or small trees, up to 10 m tall; trunks usually short, up to about 37 cm in diameter; young bark green; branches slender, drooping, spinescent. Leaves bipinnate, alternate or fascicled, common petiole usually very short, with 1–2 pairs of pinnae; pinnae up to 40 cm long, flat, with many pairs of minute, sometimes scarcely developed, widely spaced, deciduous leaflets; stipules short, spinescent. Flowers yellow, in lax, short, axillary racemes; bracts small, caducous; bracteoles absent; calyx tube short-campanulate; lobes 5, subequal, membranous, narrowly imbricate; petals 5, spreading, longer than the calyx, subequal, uppermost petal within and broader; stamens 10, sometimes fewer, free; filaments flattened, villous at the base; anthers elliptic, uniform, longitudinally dehiscent; ovary short-stalked, free, multiovuled, glabrous or hairy; style filiform, twice folded in bud; stigma small, terminal. Pod linear, straight or twisted, pendent, indehiscent or weakly 2-valved, often constricted between the few, oblong, remote seeds. (Bentham and Hooker 1865; Amshoff 1939; Phillips 1951)

The genus is closely allied to *Cercidium* (Johnston 1924).

Members of this genus inhabit the dry tropics and subtropics of North and South America and the semidesert areas of Arizona, California, Mexico, and South Africa.

The type species, called Jerusalem thorn or paloverde, is planted as an ornamental and as a thorn fence or hedge for gardens and highways. Barrett

(1956) described it as a little tree resembling a miniature weeping willow at a distance. The pods are eaten by cattle.

Parkinsonia wood is hard and heavy, with a density of 750 kg / m³. It is a brittle wood, but is satisfactory for carvings because of its availability in small sizes, its fine texture, and its quality of finishing smoothly.

NODULATION STUDIES

Nodules were lacking on *P. aculeata* L. seedlings examined in New York (Wilson 1939c), in Arizona (Martin 1948), in Hawaii (personal observation 1962), and in South Africa (Grobbelaar et al. 1964), and on *P. africana* Sond. in South Africa (Grobbelaar et al. 1964).

Parochetus Buch.-Ham. ex D. Don Papilionoideae: Trifolieae
225 / 3686

From Greek, *para*, "near," and *ochetos*, "stream," in reference to the habitat of the plant.

Type: *P. communis* Buch.-Ham. ex D. Don

1 species. Herb, perennial, prostrate, creeping, rooting at the nodes. Leaves palmately 3-foliolate; leaflets obcordate or denticulate; stems and leaf undersurfaces finely pubescent; stipels absent; stipules free or briefly joined to the base of the petiole. Flowers blue or pale purple, rather large, 1–4, stalked, in umbels at the apex of axillary peduncles; flowers in the lower leaf axils very small and cleistagamous with pods ripening on or below the soil surface; bracts stipulelike at the base of pedicels; bracteoles absent; calyx campanulate, deeply cleft; lobes 5, acute, subequal, upper 2 connate above; petals free from staminal tube; standard cobalt blue, obovate, short-clawed, without ears or tubercles; wings pinkish, long-clawed, oblong-obovate; keel falcate, long-clawed, shorter than the wings, abruptly inflexed, subacute; stamens 10, diadelphous; filaments not dilated; anthers uniform; ovary sessile, many-ovuled; style long, glabrous, inflexed above; stigma small, terminal. Pod linear, base narrow, at length obliquely straight, turgid, acutely beaked, nonseptate within, 2-valved; seeds 8–20. (Bentham and Hooker 1865; Baker 1926; Backer and Bakhuizen van den Brink 1963)

This monotypic genus is endemic in Java, Asia, East Africa, and South Africa. It thrives in humid areas bordering streams and trenches and in moist grassy fields in mountainous regions up to 1,000-m altitude.

P. communis, Buch.-Ham. ex D. Don, blue oxalis or shamrock pea, is a useful ornamental for rock gardens and hanging baskets.

NODULATION STUDIES

Nodules were observed on this species in Sri Lanka (W. R. C. Paul personal communication 1951) and in South Africa (Grobbelaar and Clarke 1974).

Named for Charles Christopher Parry, 1823–1890, American botanist who collaborated with Asa Gray and collected plants during the United States–Mexico boundary survey, 1854–1858.

Type: *P. filifolia* T. & G.

2 species. Shrubs, small, low, up to 1 m tall, multibranched; branches whiplike, stems and leaves gland-dotted; foliage slightly aromatic. Leaves imparipinnate; leaflets linear-filiform to narrow-elliptic, in several–many pairs; stipels usually absent, if present, glandular, persistent; stipules minute, caducous. Flowers small, yellowish white, in short or long spikelike, dense, terminal racemes; bracts minute, deciduous; calyx turbinate, glandular or whitish-pubescent, usually externally gland-dotted; lobes 5, short, triangular, equal; petals none; stamens 10, free; filaments attached at the base of the calyx; anthers uniform; ovary 2-ovuled; style thick, exserted, apically hooked, slightly pilose; stigma glandlike, lateral. Pod small, oblique-obovate, short-beaked, narrowed at the base, conspicuously gland-dotted, indehiscent; seed oval, slightly flattened. (Rydberg 1919a)

Members of this genus are native to the western United States; they are especially common in New Mexico and from Arizona into Mexico. They grow on open, rolling, treeless areas up to 2,000-m altitude.

Both species are drought-resistant and prefer sandy, alkaline soils. Because of the sprouting habit of their horizontal roots, the plants are useful as sand-binders and in controlling wind erosion. Plantings of *P. filifolia* T. & G. and *P. rotundata* Woot. are highly recommended in some areas for this purpose. Indians used the stems to make brooms and baskets, the seeds as a toothache remedy, and the foliage as an insecticide; however, the last-named usage lacks commercial importance. The foliage is unpalatable to livestock.

NODULATION STUDIES

Strains of rhizobia from *P. filifolia* incited nodules on *Amorpha fruticosa*, *Crotalaria sagittalis*, *Wisteria frutescens* (Wilson 1944b), and *Astragalus nuttalianus* (Wilson and Chin 1947).

Passaea Adans.   Papilionoideae: Trifolieae

Type: *P. ornithopodioides* Scop.

1 species. Herb, annual, erect, up to 10 cm tall, simple or branched; stems with simple or glandular hairs. Leaves 3-foliolate; leaflets obovate-lanceolate, denticulate, terminal leaflet long-petioled, slightly glandular-pilose; stipules small, adnate to the petiole, subentire, glandular. Flowers yellow, glabrous, 1–2, on slender, glandular peduncles; calyx tube very short; lobes narrow-linear, filiform, glandular, 4–6 times longer than the calyx tube; standard rounded, apiculate, equal to the keel; wings oblong, obtuse, with a basal, reflexed auricle; keel straight, slightly beaked, longer than the wings; stamens

monadelphous; anthers very small, alternately ovate and dorsifixed, oblong and basifixed. Pod linear-compressed, torulose, septate between the 5–10 seeds, glandular-pilose; seeds ovoid, acutely tuberculate. (Original not seen; *fide* Hutchinson 1964)

This monotypic genus is endemic in the Mediterranean region.

Segregation from *Ononis* appears to have been based on septation between the 5–10 seeds and torulose pods.

Pearsonia Dümm.   Papilionoideae: Genisteae, Crotalariinae
(196a)

Named for Professor H. H. W. Pearson, 1870–1916, South African botanist.

Type: *P. sessilifolia* (Harv.) Dümm.

11 species. Herbs or small shrubs, perennial, branching simply from woody rootstocks, usually tawny silky-haired. Leaves sessile or short-petioled, 3-foliolate; leaflets linear-lanceolate; stipules usually absent. Flowers mostly yellow or white with purple, in lax, leaf-opposed or terminal racemes, rarely solitary on long peduncles; bracts and bracteoles small; calyx hairy, campanulate; lobes 5, lowermost lobe narrower than the others, 4 upper lobes somewhat joined in pairs; standard oblong-obovate, concave, not reflexed, short-clawed, usually villous; wings straight, oblong, long-clawed, not eared; keel narrow, shorter than the wings, straight, long-clawed, swollen; stamens mostly joined below, split above; anthers narrow-oblong, alternately basifixed and dorsifixed; ovary sessile, several- to many-ovuled, villous; style glabrous, straight or curved downward; stigma truncate. Pod much longer than the calyx, oblong-lanceolate, compressed, 2-valved, straight, beaked, hairy. (Dümmer 1912; Harms 1915a; Polhill 1974, 1976)

The concept of *Pearsonia* was revised and enlarged by Polhill (1974) to include, as synonyms, species of *Phaenohoffmannia*, *Gamwellia*, and *Edbakeria*, all quite similar to *Lotononis* species.

Representatives of this genus are common in Transvaal and Natal, South Africa, Zimbabwe, and Mozambique.

Nodulated species of *Pearsonia*	Area	Reported by
aristata (Schinz) Dümm.	S. Africa	Grobbelaar & Clarke (1972)
	Zimbabwe	Corby (1974)
atherstonei Dümm.	S. Africa	Grobbelaar & Clarke (1972)
cajanifolia (Harv.) Polh. ssp. *cryptantha* (Bak.) Polh. (= *Phaenohoffmannia cajanifolia* ssp. *cryptantha*)	S. Africa	N. Grobbelaar (pers. comm. 1973)
	Zimbabwe	Corby (1974)
grandifolia (Bolus) Polh. ssp. *latebracteolata* (Dümm.) Polh. (= *Phaenohoffmannia latebracteolata*)	Zimbabwe	Corby (1974)

Nodulated species of *Pearsonia* (cont.)	Area	Reported by
marginata Dümm.		
var. *marginata*	S. Africa	Grobbelaar & Clarke (1972)
metallifera Wild	Zimbabwe	Corby (1974)
mesopontica Polh.	Zimbabwe	Corby (pers. comm. 1973)
podalyriaefolia Dümm.	S. Africa	Grobbelaar & Clarke (1972)
propinqua Dümm.	S. Africa	Grobbelaar & Clarke (1975)
sessilifolia (Harv.) Dümm.		
(= *mucronata* Burtt Davy)	Zimbabwe	Corby (1974)
ssp. *filifolia* (Bolus) Polh.	S. Africa	Grobbelaar & Clarke (1975)
ssp. *marginata* (Schinz) Polh.	S. Africa	Grobbelaar & Clarke (1975)

Pediomelum Rydb.　　　　Papilionoideae: Galegeae, Psoraliinae

From Greek, *pedion*, "plain," "field," and *melon*, "apple"; literally, "prairie apple," a common name for the edible tubers.

Type: *P. esculentum* (Pursh) Rydb.

22 species. Herbs, perennial; roots round or fusiform, starchy, edible. Leaves more or less gland-dotted, palmately 3–7-foliolate, or terminal leaflet long-stalked. Flowers blue or purple, in axillary, peduncled, dense spikes or racemes; calyx deeply campanulate, lowest of the 5 lobes usually the longest; standard broad-oblanceolate or obovate, tapering into a curved, boat-shaped claw; wings oblong-oblanceolate, with a basal lobe; keel petals broader and shorter, united at the rounded apex, each adnate to base of the adjacent wing; stamens 10, diadelphous, vexillary stamen united with others only at the base or midway; anthers equal; ovary 1-ovuled; style abruptly bent above; stigma capitate. Pod oval, flattened, with a long, flat, sword-shaped beak as long as or longer than the body, pericarp thin, free from the seed, circularly dehiscent around the middle, the upper part and beak falling off, thus setting the seed free or carrying it along in falling. (Rydberg 1919a)

The genus has strong affinities with *Psoralea*, from which many of the species were segregated (Rydberg 1919a).

Members of this genus are indigenous to the United States. They occur in the midwestern states from Wisconsin to Texas and in sandy areas of the coastal plain from North Carolina to Florida. A few species occur in the Utah, Arizona, and California region on sandy soil up to 1,500 m.

The type species, a perennial herb of the Great Plains of the United States, is called prairie turnip or Indian potato for its farinaceous edible tubers. The French voyageurs gave it the names "pomme blanche" and "pomme de prairie." The Indians of the Great Plains called *P. hypogaeum* (Nutt.) Rydb. breadroot. The tubers of both species were once plentiful and were relished by bears, gophers, and meadow mice, as well as humans.

Nodulation Studies

P. esculentum (Pursh) Rydb., designated by its synonym *Psoralea esculenta*, was recorded nodulated in North Dakota by Bolley in 1893. Isolates from nodules of this species were monotrichously flagellated short rods that produced abundant, gummy, opaque growth on yeast-extract mannitol mineral salts and calcium glycerophosphate agars, and an acid reaction with no serum zone formation in litmus milk (Bushnell and Sarles 1937). Because these cultures lacked the ability to form nodules on species of leguminous plant representatives of all the "major" and "minor" cross-inoculation groups then recognized, this species was not assigned to any group (Bushnell and Sarles 1937). In later tests conducted by Wilson (1939a), *P. esculentum* was nodulated by diverse rhizobial isolates from *Glycine, Trifolium, Vigna*, and *Sesbania* species.

Nodules were also observed on the roots of *P. subacaulis* (T. & G.) Rydb. collected in Tennessee (personal observation, University of Wisconsin herbarium specimen 1971).

Peekelia Harms　　　　　　　　Papilionoideae: Phaseoleae, Phaseolinae

Named for Gerhard Peekel, 1876–1949, Roman Catholic missionary and plant collector in the Bismarck Archipelago.

Type: *P. papuana* (Pulle) Harms

1 species. Herb or shrublet, climbing. Leaves 3-foliolate. Flowers small, pedicelled 3–6, crowded in thickened fascicles on slender, elongated racemes; calyx obliquely cupular, upper half 5-cleft; upper lobes oblong, usually connate higher up, lateral lobes oblique-lanceolate; standard subreniform, apex emarginate, base cordate, short-clawed, with a callous fold and ear above the claw; wings long-clawed, obliquely obovate-rounded, subequal, undulate; keel petals clawed, curved almost at a right angle, apices acute and beaked, upper part laterally twisted; stamens 10, vexillary stamen connate with the other 9 in the middle, appendaged above the base; ovary linear, short-stalked, 3–4-ovuled; style slightly twisted laterally and bearded above, apex obliquely truncate; stigma small; annular disk at the base of the ovary, tubular, obliquely truncate. Pod oblong to suboval, valves papery, dehiscent; seeds 2–3, subglobose, shiny, dark brown. (Harms 1920)

Verdcourt (1978) merged the genus with *Oxyrhynchus*.

This monotypic genus is endemic in Papua New Guinea.

Pellegriniodendron J. Léonard   Caesalpinioideae:
                                                    Amherstieae

Named for Pietro Pellegrini, Italian botanist; he published with A. Aubréville
on African flora.

Type: *P. diphyllum* (Harms) J. Léonard (= *Macrolobium diphyllum*)

1 species. Tree. Leaves glabrous, subsessile, 2-foliolate; leaflets oblong to
obliquely oblong-lanceolate, unequal at the base, apex rounded, twisted and
briefly acuminate, multipinninerved; stipels very obscure, awl-shaped, subtending very short petiolules; stipules triangular, intrapetiolar, connate,
ribbed. Flowers in panicles, terminal; bracts striate, short, thick, early caducous; bracteoles 2, opposite, concave, obovate, valvate, sepaloid, small, persistent; sepals 5, subequal, narrow-lanceolate; petals 5, unequal, 1 large, long-
clawed, deeply bilobed, other 4 subequal, similar to the sepals; stamens 9, 3
large, fertile, exserted, 6 small; anthers dorsifixed, longitudinally dehiscent;
ovary compressed, short-stalked, stalk laterally adnate to the receptacle; style
pilose at the base. Pod compressed, oblique-oblong, valves glabrous, non-
ridged, curled on dehiscence, sutures not winged. (Léonard 1955, 1957)

This monotypic genus is endemic in tropical Africa.

Peltogyne Vog.   Caesalpinioideae: Amherstieae
50 / 3500

From Greek, *pelte*, "shield," and *gyne*, "woman," in reference to the form of
the stigma.

Type: *P. discolor* Vog.

27 species. Trees, medium to large, unarmed, usually with straight boles,
without buttresses; bark grayish black. Leaves pinnately 2-foliolate; leaflets
small to large, ovate-oblong, acuminate or blunt, waxy or leathery. Flowers
white, yellowish, or rose, small to medium, in short racemes aggregated in
dense, large, terminal, corymbose panicles; bracts and bracteoles small, triangular or orbicular, deciduous; calyx tube short, thick, campanulate; segments
4, imbricate; petals 5, sessile, oblong or obovate, unequal, uppermost petal
usually broader; stamens 10, free, glabrous; filaments long; anthers oblong,
uniform, longitudinally dehiscent; ovary short-stalked, stalk adnate to the
calyx tube, few-ovuled; style slender, short; stigma capitate-dilated, terminal,
lobed. Pods in clusters, short, oblique-orbicular, leathery, plano-compressed,
2-valved, sometimes narrowly winged along the upper suture, usually
1-seeded, without pulp, dehiscent or indehiscent; seed compressed, suborbicular. (Bentham and Hooker 1865; Amshoff 1939)

Peltogyne is a South American genus centered in the Amazon Basin,
Brazil, extending into Panama and Trinidad. Species occur along rivers and
in rain forests.

The hardwood timber is marketed in the American trade as amaranth,
purpleheart, or violet wood. *P. densiflora* Spruce, *P. catingae* Ducke, *P. excelsa*

Ducke, *P. gracilipes* Ducke, *P. lecointei* Ducke, *P. maranhensis* Huber, and *P. pubescens* Benth. are among the most prominent sources of wood (Record and Hess 1943).

The freshly cut heartwood is brown, but with exposure it becomes uniformly purple. The sapwood is off-white to gray brown. The wood has a high density, between 960 and 1040 kg/m^3, a medium to fine texture, and is valued for its strength and durability. The wood is used for furniture and specialty turnery items, and sliced into veneer for interior finishing and decorative inlay work.

NODULATION STUDIES

Nodules were not present on the brown roots of 9 seedlings of *P. venosa* Benth. examined by Norris (1969) in 3 sites in Bartica, Guyana.

Peltophoropsis Chiov.   Caesalpinioideae: Eucaesalpinieae

The Greek suffix, *opsis*, denotes resemblance to *Peltophorum*.

Type: *P. scioana* Chiov.

1 species. Shrub; branches with swollen nodes. Leaves bipinnate; pinnae pairs 2–4; leaflets small, elliptic, 3–6 pairs; stipules recurved, spiny. Flowers yellow, few, in loose racemes; bracts minute; calyx funnel-shaped; sepals 5, imbricate, subequal; petals 5, short-clawed, uppermost petal ovate, large, 4 others subequal, short, narrow; stamens 10, free, equal; anthers dorsifixed; ovary short-stalked, 2–3-ovuled; style thick; stigma small, subpeltate. Pod lanceolate, flattened, upper suture thickened, narrowed at each end, subwinged on each side, indehiscent; seeds mottled, 1–3, flattened, on a long funicle. (Roti-Michelozzi 1957)

Peltophorum species are unarmed and have 5 dissimilar petals, as compared with *P. scioana* Chiov., which has only spiny stipules and 4 subequal petals and 1 large uppermost petal.

This monotypic genus is endemic in Ethiopia and northern Kenya. The plants thrive on sandy soils over sedimentary rock, usually in areas having a mean annual rainfall of about 200 mm.

From Greek, *pelte*, "shield," and the verb *phoreo*, "bear," "carry," in reference to the shieldlike attachment of the stigma to the style.

Type: *P. dubium* (Spreng.) Taub.

15 species. Trees, tall, or shrubs, unarmed, deciduous. Leaves large, abruptly bipinnate; leaflets paired, small, numerous; stipules small, caducous. Flowers showy, yellow, in erect, paniculate racemes at the ends of branches; bracts narrow, lanceolate; bracteoles absent; calyx shortly turbinate-campanulate, often rusty-pubescent, globular before opening; lobes 5, obovate-orbicular, reflexed, subequal, imbricate, caducous; petals 5, orbicular or obovate, unequal, clawed, margins crimped or undulate; imbricate; stamens 10, free, declinate; filaments pilose at the base, sometimes glandular; anthers uniform, versatile, longitudinally dehiscent; ovary sessile or short-stalked, free, 2- or more-ovuled; style filiform, terete; stigma board-peltate. Pod oblong-lanceolate, flattened, slightly winged at the sutures, reddish or bronze when mature, indehiscent; seeds 1–2, rarely 3–4, plano-compressed, transverse. (Bentham and Hooker 1865; Baker 1930; Phillips 1951)

Members of this genus are common in the tropics and subtropics of both hemispheres. They occur along forested coastal areas, in high grassy woodlands, the dry bushveld of South Africa, and on well-drained soils on slopes up to 700 m.

The trees of this genus often reach 30 m or more in height, with umbrella-shaped crowns of dense green foliage. They are especially favored for their large spikes of sweet-scented, golden yellow flowers which bloom twice a year. They are planted as shade trees along roadsides and as park and garden ornamentals. In the Malay Peninsula, *P. dasyrrhachis* Kurz is cultivated on coffee and cacao plantations for shade.

The red brown wood of *P. vogelianum* Walp. in Argentina and Paraguay, the purplish wood of *P. adnatum* Griseb. in the West Indies, and the orange-colored wood of *P. linnaei* Benth. in Brazil are used locally for cabinets, furniture, and carpentry (Uphof 1968).

Infusions of various plant parts, especially of the astringent bark of *P. pterocarpum* Backer, are used for minor medicinal purposes in the East Indies. A yellow brown dye from the bark is used in the batik industry.

NODULATION STUDIES

The following species lacked nodules: *P. africanum* Sond. in South Africa (Grobbelaar et al. 1964), in Zimbabwe (Corby 1974), and in Singapore (Lim 1977); *P. pterocarpum* (= *P. inerme* (Roxb.) Naves) in Hawaii (Allen and Allen 1936b), in the Philippines (Bañados and Fernandez 1954), in Singapore (Lim 1977), and in Java (personal observation 1977).

From Greek, *pente*, "five," and *kleithron*, "bolt," in reference to the 5 imbricate calyx lobes and the 5 petals being joined together at the base.

Type: *P. macroloba* (Willd.) Ktze.

3 species. Trees, tall, unarmed, evergreen; stems, leaf rachis, and peduncles rusty-pubescent, eglandular. Leaves bipinnate; pinnae 10 or more pairs, leaflets numerous; stipels setaceous; stipules small, lanceolate. Flowers usually whitish yellow, bisexual or dioecious, in elongated spikes, often paniculate; calyx campanulate; lobes 5, short, broad, imbricate; petals 5, elliptic or oblong, valvate, united at the base or to the middle; stamens 5, fertile, alternate with petals and adnate to the base, staminodes 5, 10, or 15, filiform, opposite the petals; anthers tipped by a deciduous gland; ovary subsessile, many-ovuled; style filiform; stigma small, terminal. Pod large, long, broad, flattened, narrowed to the base, leathery to woody, brown, 2-valved, explosively dehiscent with valves curling backwards at maturity, with several, large, rhomboid seeds. (Bentham 1840; Bentham and Hooker 1865; Baker 1930; Gilbert and Boutique 1952a)

Representatives of this genus are found in tropical Africa and America. Two species occur in East Africa, *P. macrophylla* Benth., from Nigeria to Angola, and *P. eetveldeana* De Wild. & Th. Dur., in Gabon and the Congo Basin. *P. macroloba* (Willd.) Ktze. is indigenous to South America, especially in Amazonas, Brazil, and northward to the West Indies. All species occur in wet forest regions, along river banks, and in coastal savannas.

Seeds of all 3 species contain 30–36 percent nondrying, high-melting-point malodorous oil rich in glyceryl triesters of long-chain fatty acids (Hildritch and Williams 1964). The species are commonly called oil bean tree. *P. macrophylla* seeds are the source of owala butter or owala oil used in East Africa for soap, candles, lubricants, and medicinal lotions. The seeds are rich in protein, but not starch, and are consumed after roasting, or ground up as an ingredient for bread (Dalziel 1948). Seeds of *P. macroloba*, which contain the alkaloid paucine, $C_{27}H_{39}N_5O_5$ (Henry 1939), are used as an emetic and snakebite remedy in some primitive areas of South America.

The wood of all species is of good quality, but is not available in sufficient quantity to have economic importance in the timber trade. *P. eetveldeana* has white wood. The sapwood of *P. macrophylla* is pinkish white with heartwood striped or dark brown. The wood is hard, durable, resistant to termite, fungal, and insect attack, and has a density of 750–850 kg/m^3. The attractive finished wood is used for mortars and pestles, heavy construction and flooring, beams and house frames; the rough lumber is used for fuel on river boats in Brazil (Record and Hess 1943).

NODULATION STUDIES

Typical, slow-growing rhizobia were isolated from the brown black, woody-cortexed nodules of *P. macroloba* sent to the authors by DeSouza (1966) from Trinidad. *P. macrophylla*, examined by Bonnier (1957) in Yangambi, Zaire, lacked nodules.

Pentadynamis R. Br.
210a / 3671

From Greek, *pente*, "five," and *dynamis*, "power," probably in reference to the 5 large linear and 5 large ovoid anthers.

Type: *P. incana* R. Br.

1 species. Herbs or undershrubs, up to about 60 cm tall; stems and foliage hoary white silky-haired. Leaves 3-foliolate; leaflets sessile, linear, obtuse, terminal leaflet longest. Flowers yellow, many, in axillary racemes; calyx lobes acute, subequal; petals twice the length of the calyx; standard broad with callous ridges decurrent on the claw, without auricles; keel petals obtuse, as long as the wings; stamens 10, diadelphous; anthers alternately long and erect, short and versatile; ovary several-ovuled, pubescent; style incurved, bearded above on the inner side; stigma terminal. Unripe pod gray white pubescent, compressed, linear. (Bentham 1864; Bentham and Hooker 1865)

This monospecific genus was positioned near to, but distinctive from, *Crotalaria* on the basis of the nonbeaked keel, diadelphous stamens, and compressed pods (Bentham and Hooker 1865). Hutchinson (1964) considered it synonymous with *Crotalaria*.

This monotypic genus is endemic in South Australia. It occurs on sand hills.

Periandra Mart. ex Benth.
385 / 3859

From Greek, *peri*, "around," and *aner*, *andros*, "man"; the stamens are equal all around the style.

Type: *P. mediterranea* (Vell.) Taub.

6 species. Shrubs or herbs, erect or climbing. Leaves pinnately 3-foliolate or lower 1-foliolate; stipels present; stipules paired, striate, united above. Flowers blue or scarlet, solitary or many, on elongated peduncles axillary or irregularly racemose at ends of branches; bracts paired, persistent, stipulelike, upper ones connate; bracteoles generally large, striate, persistent, appressed to the calyx; calyx short-campanulate; lobes very short, upper 2 subconnate, lowermost lobe longer; standard broadly obovate or orbicular, short-clawed, incurved, folded; wings obliquely obovate or oblong; keel broad, incurved, scarcely shorter than the wings; stamens 10, vexillary stamen free or slightly connate with others; filaments filiform; anthers uniform; ovary subsessile, many-ovuled; style incurved, apex slightly clavate; stigma terminal, beardless. Pod subsessile, linear, acuminate, plano-compressed, each suture thickened, 2-valved; seeds compressed. (Bentham and Hooker 1865)

Members of this genus are native to Central America and Brazil.

Five of the 6 species occur in northern Brazil. The roots are used in the same manner as those of *Glycyrrhiza* species.

From the Greek verb *perikopto*, "cut all around," in reference to the deep cuts in the calyx.

Type: *P. mooniana* Thwait.

6 species. Trees, small or large, or tall shrubs. Leaves imparipinnate, usually alternate; leaflets progressively larger toward the apex, leathery; rachis grooved; stipels usually present. Flowers cream, white, or purple, in axillary racemes or terminal panicles; bracts and bracteoles minute, caducous; calyx large, campanulate, cleft to below the middle; lobes subequal or upper 2 shorter, subconnate, caducous; standard broad-orbicular, reflexed; wings and keel petals distinctly eared; wings falcate-obovate; keel incurved, blunt, petals free; stamens free, apically recurved; anthers versatile; ovary stalked, oblong, villous, encircled by a conspicuous disk, few-ovuled; style long, filiform, awl-shaped, apically involute; stigma terminal, introrse. Pod long-stalked, flat, linear-oblong, thinly woody, narrowly winged along both sutures but wing broader on lower margin, indehiscent; seeds 1–7, compressed, ovate-orbicular, dull red. (Bentham and Hooker 1865; Knaap-van Meeuwen 1962)

Pericopsis was conceived of as a monotypic genus indigenous to Sri Lanka. Recently Knaap-van Meeuwen (1962) reduced *Afrormosia* to *Pericopsis*, thereby incorporating 5 African species.

One species occurs in Sri Lanka, 5 in tropical Africa. They flourish in rain forests, fringe forests, and savannas.

The tree species are exploited for their fine wood. The wood of the Asian species, *P. mooniana* Thwait., is valued for furniture; the close-grained, yellow brown wood of *P. angolensis* (Bak.) van Meeuw. is used for general construction purposes and in the fashioning of curios. *P. elata* (Harms) van Meeuw. supplies kokrodua to the timber trade; it is a wood heavier and stronger than white oak and very resistant to decay. It is especially valued in the United States for boatbuilding, as an alternate for teak, and is also in high esteem for interior trims and decorative veneer (Kukachka 1970).

NODULATION STUDIES

Nodules were observed on *P. elata*, reported as *Afrormosia elata*, in Yangambi, Zaire (Bonnier 1957), and on *P. angolensis*, reported as *A. angolensis*, in Zimbabwe (Corby 1974).

Petaladenium Ducke   Papilionoideae: Sophoreae

From Greek, *petalon*, "petal," and *aden*, "gland," in reference to the glands on the margins of the wing petals.

Type: *P. urceoliferum* Ducke

1 species. Tree. Leaves imparipinnate; leaflets large; stipels absent. Flowers lilac, in lateral simple or slightly branched racemes; bracts (?); bracteoles large, subpersistent; calyx campanulate, slightly oblique; lobes 5, subequal,

valvate in bud; petals free, clawed at the base, caducous after flowering; standard suborbicular, broadly cordate at the base, apex inflexed, retuse in the middle; wings oblong, eared on each side at the base, narrowed at the apex, upper margin fringed with short-stalked glands; keel petals triangular-oblong, eared on each side at the base, keeled on the back; stamens 10, diadelphous; anthers dorsifixed, oblong-linear, connective mucronulate at the apex; ovary long-stalked, 4–5-ovuled; style filiform; stigma terminal, minute. Pod linear-oblong, very compressed, long-stalked, apex acuminate, 2-valved. (Ducke 1938b)

This monotypic genus is endemic in the upper Rio Negro area, Brazil.

Petalostemon Michx.   Papilionoideae: Galegeae, Psoraliinae
(247) / 3710

From Greek, *petalon*, "petal," and *stemon*, "stamen," in reference to the 4 petals joined to the top of the staminal tube.

Type: *P. candidum* (Willd.) Michx.

40 species. Herbs, mostly perennial, very rarely annual, erect, often branched at the base, deep-rooted; foliage more or less gland-dotted. Leaves imparipinnate; leaflets small, 3–many, entire, short-petioluled, of some species lemon-scented when crushed; stipels minute or absent; stipules minute, linear to awl-shaped, setaceous, persistent. Flowers white, rose, purple, or yellow, indistinctly papilionaceous, in densely flowered, terminal or leaf-opposed spikes; individual flowers usually subtended by 3–4 rows of large, broad-ovate, deciduous or persistent, membranous or setaceous bracts; bracteoles absent; calyx turbinate to campanulate, sessile; lobes 5, triangular-lanceolate or filiform-subulate, subequal, usually shorter than the calyx tube; all petals slender-clawed; standard free, long-clawed, inserted at the base of the calyx, usually cordate or truncate at the base; other 4 petals subsimilar, oblique-oblong to oval, short-clawed, acute at the base, attached to the top of the staminal tube, spreading; stamens 5, monadelphous, broad at the base, split above; filaments one-fifth to one-third free, alternating with the 4 petals; anthers often glandular-apiculate; ovary sessile, 2-ovuled; style filiform, awl-shaped; stigma small, terminal. Pod more or less compressed, oblique-obovate to semiorbicular or lunate, membranous, partially enclosed by the calyx, usually dehiscent; seeds 1–2, subreniform. (Bentham and Hooker 1865; Rydberg 1920)

Representatives of this genus are distributed throughout North America; they occur from Canada to Mexico and along the Atlantic coastal plain, from North Carolina to Florida. They thrive on sandy and rocky soil, dry prairies, river banks, and mesas up to 2,000-m altitude.

These plants serve as good ground cover, wild forage for sheep and goats, have erosion-control properties, and are good wild honey plants. The roots of slender prairie clover, *P. oligophyllum* (Torr.) Rydb., were eaten by the

Indians of New Mexico. The leaves of *P. purpureum* (Vent.) Rydb. served to make a tea. *P. candidum* (Willd.) Michx. is used as an emetic by the Hopi Indians.

NODULATION STUDIES

The absence of nodules in Arizona on *P. compactum* (Spreng.) Swezey and *P. flavescens* S. Wats. (Martin 1948) was presumably an environmental circumstance.

Six rhizobial strains isolated from nodules of *P. candidum, P. purpureum,* and *P. villosum* Nutt. were monotrichously flagellated short-to-medium rods that produced abundant, gummy, opaque growth on yeast-extract mannitol mineral salts and calcium glycerophate agars, and a very alkaline reaction with serum zone formation in litmus milk. Although rhizobia from these *Petalostemon* species nodulated *Amorpha fruticosa* in a few tests, no cross-symbiobility with any representatives of the wide spectrum of rhizobia-plant groups was obtained (Bushnell and Sarles 1937). Wilson (1939a), however, regarded *Petalostemon* species as quite versatile in cross-symbiotic associations, but he did not obtain nodulation with a strain from *Amorpha fruticosa*. Later he confirmed that *A. fruticosa, Wisteria frutescens* (Wilson 1944b), and also *Astragalus nuttalianus* (Wilson and Chin 1947) were nodulated by rhizobia from *P. purpureum*.

The close botanical kinship between *Amorpha* and *Petalostemon*, and the similarities in the cultural and physiological reactions of their rhizobial isolates, indicate that a mutual symbiotic association is not unexpected.

Nodulated species of *Petalostemon*	Area	Reported by
†*candidum* (Willd.) Michx.	Minn., Iowa, U.S.A.	Bolley (1893); Buchanan (1909)
microphyllum (T. & G.) Hell.	Tex., U.S.A.	Parks (1937)
multiflorus Nutt.	Kans., U.S.A.	M. D. Atkins (pers. comm. 1950)
oligophyllus (Torr.) Rydb.	N.Y., U.S.A.	Wilson (1939a)
†*purpureum* (Vent.) Rydb.	Wis., U.S.A.	Bushnell & Sarles (1937)
	N.Y., U.S.A.	Wilson (1939a, 1944b)
(= *violaceum* Michx.)	Minn., Iowa, U.S.A.	Bolley (1893); Buchanan (1909)
stanfieldii Small	Okla., U.S.A.	Pers. observ. (U.W. herb. spec. 1971)
†*villosum* Nutt.	Wis., U.S.A.	Bushnell & Sarles (1937)
	N.Y., U.S.A.	Wilson (1939a)

†Also reported as *Kuhnistera*, a rejected generic name.

From Greek, *petalon*, "petal," in reference to the petaloid style.

Type: *P. labicheoides* R. Br.

2 species. Shrubs, erect, up to about 2 m tall, almost glabrous; young shoots minutely silky-haired. Leaves imparipinnate; leaflets small, alternate, lanceolate; rachis sometimes spinescent; stipules narrow, early caducous. Flowers yellow, solitary, axillary; bracts and bracteoles small, caducous; calyx tube very short; lobes 5, subequal, scarcely joined at the base, overlapping, minutely hairy; petals 5, subequal, spreading, imbricate; stamens 5, 3 perfect; filaments very short; anthers linear, basifixed, opening by longitudinal slits; staminodes 2, small, with acuminate, imperfect anthers; ovary subsessile, free, several- to many-ovuled; style very distinctive, yellow, large, dilated, petaloid, gibbous above the ovary, 3-lobed, the 2 short lobes in front, the other lobe longer and terminating in a small stigma. Pod erect, compressed, ovate-oblong, oblique, 2-valved, dehiscent; seeds 4–6, ovate-oblong, compressed, with a small, fleshy aril. (Bentham 1864; Bentham and Hooker 1865)

Members of this genus are restricted to Queensland and New South Wales, Australia. They thrive in arid zones and rocky ranges.

The alkaloid tetrahydroharman, $C_{12}H_{14}N_2$, is found in leaves and stems of *P. labicheoides* R. Br. (Badger and Beecham 1951) and *P. labicheoides* var. *casseoides*; the latter also yields tryptamine $C_{10}H_{12}N_2$, *N*-methyltryptamine, and *N*-dimethyltryptamine (Johns et al. 1966). Tryptophane-derived alkaloids of the harman group are more commonly associated with seeds and roots of *Peganum harmala* (Zygophyllaceae). Harman derivatives cause a fall in blood pressure due to a weakening of the cardiac muscle (Henry 1939).

NODULATION STUDIES

All specimens of *P. labicheoides* examined in sandy soils of arid New South Wales and in potted rhizosphere soil lacked nodules (Beadle 1964).

Named for Dr. Robert Peter, 1805–1894, American botanist, active in floral research of Kentucky.

Type: *P. scoparia* Gray

4 species. Herbs or undershrubs, perennial; stems herbaceous, erect or decumbent, bushy, branched from near the somewhat woody base, terete, longitudinally striate; roots thick, tuberous. Leaves imparipinnate, 7–many-foliolate, petioled; leaflets small, many, narrow-oblong to elliptic; stipels absent; stipules paired as slender, stiff prickles. Flowers white to yellow or purplish tinged, in narrow racemes, terminal or leaf-opposed, not axillary; bracts narrow, awl-shaped; bracteoles absent; calyx cylindric-campanulate, gibbous above the base; lobes 5, deltoid to narrowly lanceolate-acuminate, unequal, upper 2 connate higher up; petals long-clawed; standard folded

lengthwise, sides reflexed; wings smaller than the standard, slightly eared; keel petals smaller and shorter than the wings, eared at the base; stamens diadelphous; anthers similar, oblong; ovary subsessile or short-stalked, laterally flattened, several-ovuled; style bent upward, becoming hard, spirally twisted and widened at the base; stigma terminal, slightly oblique, with an apical tuft of short hairs. Pod linear, narrow-oblong to nearly straight, subsessile, flattened, few-seeded, dehiscent along both sutures; seeds rounded, oblong, compressed. (Bentham and Hooker 1865; Rydberg 1923; Porter 1956)

Peteria species occur in Idaho, Wyoming, and the southwestern United States and northern Mexico. They thrive on desert washes, gravelly slopes of dry hills, and mesas or mountain valleys, up to 2,000-m altitude.

The tuberous roots of *P. glandulosa* (Gray) Rydb. (= *P. scoparia* var. *glandulosa* (Gray) S. Wats.), known as camote del monte, or mountain sweet potato, by natives of northern Mexico, are said to be edible.

**Petteria* C. Presl   Papilionoideae: Genisteae, Spartiinae
215 / 3676

Named for Franz Petter, 1798–1853, Austrian professor at Spalato; he wrote on the botany of Dalmatia.

Type: *P. ramentacea* (Sieb.) C. Presl

1 species. Shrub, unarmed. Leaves palmately 3-foliolate; stipules small, triangular, intrapetiolar. Flowers yellow, in dense, erect, terminal, leafless racemes; bracts membranous, inserted on the pedicel, caducous; bracteoles absent; calyx campanulate-tubular, bilabiate, upper lip 2-lobed, deeply cleft, each lobe broad-falcate, lower lip 3-toothed; standard orbicular; wings and keel oblong, almost straight, claws adnate to the staminal tube; keel pouched on each side; stamens monadelphous; anthers alternately short and versatile, long and basifixed; ovary sessile, many-ovuled; style minutely incurved, glabrous; stigma terminal, slightly oblique. Pod linear, oblong, straight or slightly curved, compressed, 2-valved, continuous inside; seeds usually 5–9. (Bentham and Hooker 1865)

This monotypic genus is endemic in the Balkans and Dalmatia.

NODULATION STUDIES

Nodules were observed on this species in France (Lechtova-Trnka 1931).

Phaenohoffmannia Ktze.   Papilionoideae: Genisteae,
                                             Crotalariinae

SEE *Pearsonia* Dümm.

From Greek, *phaselos*, Latin, *phaselus*, the name for a kind of bean.

Type: *P. vulgaris* L.

50–100 species. Herbs, annual or perennial, prostrate or erect, often climbing, rarely woody at the base. Leaves petioled, pinnately 3-foliolate; leaflet shapes variable, 3-lobed, asymmetric; stipels present; stipules striate, not extending below the point of attachment, persistent. Flowers few to many, pedicelled, in axillary racemes, generally fascicled 2–4 on thickened nodes near the apex of the rachis, colors various; bracteoles persistent, much larger than the persistent bracts; calyx campanulate, bilabiate, upper lip bidentate, lower lip tridentate; petals clawed; standard orbicular, auricled at the base, reflexed, somewhat twisted; wings often slightly longer than the standard, base adherent to the claws of the keel, and turning spirally with it; keel linear, beak coiled in 2–3 complete spiral turns; stamens 10, 1 free, often thickened, 9 connate below; ovary sessile or subsessile, 1- to many-ovuled; style without an apical appendage, thickened, curved and twisted within the beak of the keel, bearded; stigma elongated, oblique or lateral. Pod linear, oblong or falcate, subcylindric or compressed, 2-valved, dehiscent; seeds 1–many, oblong-reniform, hilum short, central. (Bentham and Hooker 1865; Hassler 1923; Phillips 1951; Wilczek 1954; Verdcourt 1971)

The original generic number 423 (Taubert 1894) was changed to 435 by Harms (1908a) because of the reorganization within the subtribe necessitated by various revisions and inclusions of new genera.

This genus is American in origin. The plants are both wild and domesticated. There are numerous cultivars. *Phaseolus* species have adapted to diverse soil types, rainfall levels, and temperatures.

Botanists have generally considered *Phaseolus* a conglomerate of closely related species that were distinguished with difficulty not only *inter se*, but also from species of other genera (Gillett 1966a; Verdcourt 1970d). Over the years two large groups of beans have entered into the composition of *Phaseolus*. In one category were the large-seeded forms of American and European origin, exemplified by *P. vulgaris* L., *P. coccineus* L., *P. lunatus* L., and their many cultivars. These phaseoli are indigenous to temperate regions that accommodate their neutral or long-day light requirements. Small-seeded Asian species of the so-called mung group constitute the second category; these tropical forms are characterized by having short-day photoperiod demands (Allard and Zaumeyer 1944). Prominent among these are the mat or moth bean, *P. aconitifolius* Jacq., adzuki bean, *P. angularis* (Willd.) Wight, green or golden gram, or mung, *P. aureus* Roxb., rice bean, *P. calcaratus* Roxb., and black gram, or urd bean, *P. mungo* L. Within the last decade basic taxonomic evidence has supported the transfer of this latter category to *Vigna* (Verdcourt 1968a, 1969; Ohwi and Ohashi 1969).

The common bean, *P. vulgaris*, was a food of the ancients. Evidence of its origin has been defined by the discovery of a well-preserved, uncharred pod valve in the Tehuacán Valley of Mexico that showed a carbon 14 date of 4975 B.C. Related evidence indicated that this specimen was a domesticate (Kaplan 1965). The controversy over the origin and history of the common bean was enlivened recently by the proposal of *P. aborigineus* Burk., an annual summer climber in northwest Argentina, as its wild ancestor (Burkart and Brücher 1953). Fascinating insights into the archaeology and ethnology of the common bean have been reviewed by MacNeish (1960), Towle (1961), and Heiser (1965).

The common bean is the best known and most widely cultivated of the phaseoli. Approximately 500 varieties are known. It is unsurpassed as a source of vegetable protein for man and animals, and also as a source of essential amino acids in complementing high carbohydrate consumption. The close relationship between beans and corn in the human diet is discussed and well documented by Williams (1952) and Kaplan (1965). Many species are also valuable as green manure, cover crops, and forage. The boiled or mashed tuberous roots of *P. adenanthus* G. F. W. Mey., *P. coccineus*, *P. diversifolius* Pers., *P. heterophyllus* Willd., *P. retusus* Benth. (= *P. metcalfei* Woot. & Standl.), and several other species are edible. The enormous fleshy roots of *P. retusus*, Metcalfe bean, may reach diameters of 10–15 cm and weigh as much as 13.5 kg (Smith 1898). This drought-resistant perennial was regarded in pioneering days as one of the most promising native forage plants for sheep and cattle of the western and southwestern United States. Wild hogs relish its root tubers and threaten its survival on open ranges.

At the turn of the century considerable quantities of prussic acid, HCN, were reported in colored seeds from wild forms of *P. lunatus* grown in Mauritius (Dunstan and Henry 1903, 1907–1908). The acid was traced to phaseolunatin, $C_{10}H_{17}NO_6$, which also occurs in flax and is better known today as linamarin. Under damp conditions, or when the seeds are moistened, masticated, and ingested, phaseolunatase, an enzyme also present in the seeds, releases the HCN from the glucoside. Linamarin is present in all parts of the plant, but occurs in largest amounts in the seeds. Its occurrence is highest in the small, plump, solid-colored or spotted seeds of wild tropical varieties. Its absence in white seeds of cultivated species is ascribed to the higher plant metabolism induced by improved nutrition under cultivation. Genetic breeding has undoubtedly played an important role also. The cyanogenetic principle in lima beans cultivated in the United States has always been below the level dangerous for consumption. Regulations governing imported beans are strict.

Poisoning from ingestion of lima beans containing lethal levels of HCN has been more common in animals than in man. Since lima beans rarely constitute an entire meal for humans, the hydrocyanic acid is diluted; also, the cyanogenetic glucoside and enzyme are destroyed by boiling. Animals, on the other hand, usually consume large quantities of the vines and raw seeds at one feeding. Lima bean ensilage is to be viewed with caution (Huffman et al. 1956). Ingestion by man of white seed only from cultivated varieties, discarding the water in which seed are boiled, and prolonged cooking of hard-coated seed in 50 percent vinegar-water are recommended ways to render lima beans nontoxic. Various aspects of this topic are summarized by Serrano (1923), Charlton (1926), Guérin (1930), Steyn (1934), Watt and Breyer-Brandwijk (1962), and Kingsbury (1964).

NODULATION STUDIES

The term *Bacillus radicicola–Phaseolus* was applied to bean rhizobia by Beijerinck (1888) to distinguish them from all other rhizobia. Later Schneider (1892) proposed the name *Rhizobium frankii* var. *majus* for rhizobia symbiotic with *P. vulgaris*, as distinctive from *R. frankii* var. *minus* for rhizobia symbiotic with *Pisum sativum*. Schneider's designations, however, were not valid taxonomically. *Rhizobium phaseoli* became the accepted designation in 1926. This was 1 of 10 *Rhizobium* species included by Dangeard in a new tribe Hyphoideae of the family Bacteriaceae. Despite Dangeard's differentiation of species on serological, host-infection, and cytological properties, many of his names were declared invalid because of his failure to observe rules of priority. The nomen *R. phaseoli*, however, was sustained.

Currently, *R. phaseoli* applies to peritrichously flagellated, often highly vacuolated, rod-shaped forms that are smaller than clover and pea rhizobia. Bacteroids are usually swollen rods with few branched forms. Growth on laboratory media is rapid. Colonies are raised, semitranslucent, mucilaginous, glistening, and white. Acid reactions are produced in carbohydrate media and in litmus milk with the formation of deep serum zones.

Strains of *R. phaseoli* are ubiquitous in soil, and, correspondingly, *P. vulgaris* is highly receptive to appropriate inocula. As many as 5,000 nodules were recorded on individual bean plants in greenhouse culture (Löhnis and Leonard 1926). However, most strains of *R. phaseoli* are functionally deficient in nitrogen fixation. The explanation for this lack of compatibility between bean plants and their microsymbiont is not clear. Malnutrition of the rhizobia resulting from large deposits of starch in the nodule host cells was suggested as one explanation by McCoy (1929).

Rhizobiologists have long been perplexed by the allocation of *Phaseolus* species to two different rhizobia:plant groups (Whiting and Hansen 1920). *P. vulgaris*, *P. angustifolius* Roxb., and *P. coccineus* are nodulated only by the fast-growing, peritrichously flagellated strains of *R. phaseoli*, whereas the small-seeded, tropical species of the Asian mung group are nodulated only by slow-growing, monotrichously flagellated rhizobia from members of the cowpea miscellany (Richmond 1926a). Lima bean, *P. lunatus*, a large-seeded species nodulated by cowpea-type rhizobia and not by *R. phaseoli*, is the sole exception (Richmond 1926a). The evidence seems conclusive that its origin was tropical America (Mackie 1943; Heiser 1965).

Inoculation experiments with *P. vulgaris* scion grafted onto *P. lunatus* rootstock, and vice versa, showed that infections occurred only when the root and the rhizobia were from homologous plant species. Seeds produced on the grafted plants were altered, since second-generation plants were infected by the rhizobia of both species; however, the nodules did not appear normal (Richmond 1926b).

The plant-inoculation results of Barua and Bhaduri (1967) emphasized further the demarcation between these two *Phaseolus* categories. Phasin, the phytohaemagglutinin present in seeds of *P. vulgaris*, but absent and minimal in *P. aureus* and soybean, respectively, was judged not to be a factor governing differences in host selectivity (Bhaduri et al. 1967).

The recent segregation and transfer of the Asian mung group into *Vigna* is a most acceptable explanation and taxonomic solution to a conundrum of many decades.

Nodulated species of *Phaseolus*	Area	Reported by
aborigineus Burk.	Argentina	Rothschild (1964, 1970)
†(*aconitifolius* Jacq.) = *Vigna aconitifolia* (Jacq.) Maréchal	Widely reported	Many investigators
acutifolius Gray	Widely reported	Many investigators
var. *latifolius* Freeman	Widely reported	Many investigators
adenanthus G. F. W. Mey.	Mexico	Pers. observ. (U.W. herb. spec. 1971)
	S. Africa	Grobbelaar & Clarke (1974)
†(*angularis* (Willd.) Wight) = *Vigna angularis* (Willd.) Ohwi & Ohashi	Widely reported	Many investigators
angustifolius Roxb.	Widely reported	Many investigators
‡(*atropurpureus* Moc. & Sessé) = *M. atropurpureum* Urb.	S. Africa	Grobbelaar et al. (1967)
†(*aureus* Roxb.) = *Vigna radiata* (L.) Wilczek	Widely reported	Many investigators
bracteatus Spreng.	Australia	McLeod (1962)
†(*calcaratus* Roxb.) = *Vigna umbellata* (Thunb.) Ohwi & Ohashi	Widely reported	Many investigators
candidus Vell.	Argentina	Rothschild (1964, 1970)
coccineus L. (= *multiflorus* Willd.)	Widely reported	Many investigators
erythroloma Mart. ex Benth.	Argentina	Rothschild (1964, 1970)
‡(*lathyroides* L.)	Hawaii, U.S.A.	Allen & Allen (1936a, 1939)
	Australia	Norris (1959b)
(= *semierectus* L.) = *M. lathyroides* (L.) Urb.	N.Y., U.S.A.	Wilson (1945a)
linearis HBK	Venezuela	Barrios & Gonzalez (1971)
lunatus L.	Widely reported	Many investigators
var. *macrocarpus* Benth. (= *limensis* Macf.)	Widely reported	Many investigators
†(*mungo* L.) = *Vigna mungo* (L.) Hepper	Widely reported	Many investigators
oleraceus [?]	Great Britain	Dawson (1900b)
pauciflorus Benth.	Ill., U.S.A.	Schneider (1893b)
peduncularis HBK	Venezuela	Barrios & Gonzalez (1971)
pilosus HBK	Venezuela	Barrios & Gonzalez (1971)

Nodulated species of *Phaseolus* (cont.)	Area	Reported by
polystachyus (L.) BSP	N.Y., U.S.A.	Wilson (1939b)
†(*radiatus* L.)	Widely reported	Many investigators
= *Vigna radiata* (L.) Wilczek		
sublobatus Roxb.	Java	Keuchenius (1924)
†(*trichocarpus* C. Wright)	Puerto Rico	Dubey et al. (1972)
= *Vigna longifolia* (Benth.) Verdc.		
†(*trilobatus* (L.) Schreb.)	Japan	Asai (1944)
= *Vigna trilobata* (L.) Verdc.		
†(*trilobus* Ait.)	India	Raju (1936)
= *Vigna trilobata* (L.) Verdc.		
trinervius Heyne ex Wall.	Trinidad	DeSouza (1963)
†(*viridissimus* Ten.)	U.S.D.A.	Kellerman & Robinson
= *Vigna mungo* (L.) Hepper		(1907)
vulgaris L.	Widely reported	Many investigators
var. *nanus* Aschers.	Ind., U.S.A.	Burrage (1900)

†Species heretofore reported nodulated as *Phaseolus* spp., now *Vigna* spp.

‡Segregated from a section of *Phaseolus* to form the genus *Macroptilium* (Urban 1928; León and Alain 1951).

Phylacium Benn.   Papilionoideae: Hedysareae, Desmodiinae
346 / 3816

From Greek, *phylax*, "guard," "sentinel."

Type: *P. bracteosum* Benn.

2–3 species. Herbs, twining, climbing. Leaves pinnately 3-foliolate; leaflets oblong, large, entire; stipels present; stipules narrow, caducous. Flowers white or pale violet, small, on short pedicels in fascicled, axillary racemes; bracts 1–3, large, concave, some much enlarged after flowering, persistent, enfolding the pod; calyx bibracteolate at the base; lobes acute, shorter than the tube, upper 2 connate into an entire lip; petals long-clawed; standard suborbicular, with a callus inside and with 2 basal, inflexed auricles; wings narrow, falcate, free, spurred at the base; keel blunt, incurved, much shorter than the narrow wings; vexillary stamen free or loosely adherent to others, otherwise connate with others into a sheath; anthers uniform; ovary subsessile, surrounded by an annular disk, 1-ovuled; style curved, thickened and glabrous above, apex awl-shaped; stigma capitate. Pod small, compressed, elliptic-semiorbicular, membranous, acuminate with style, reticulate, sessile within the withered calyx, indehiscent, 1-seeded. (Bentham and Hooker 1865)

Members of this genus are indigenous to Malaysia and Indochina. They occur in brushwood and light forested areas.

From Greek, *phyllon*, "leaf," and *karpos*, "fruit," in reference to the winged pod.

Type: *P. riedelii* Tul.

2 species. Trees, medium to large, up to about 25 m tall, with a trunk diameter of 1 m and a spreading crown, unarmed. Leaves paripinnate, deciduous; leaflets opposite, rarely alternate, pairs few to numerous, ovate-elliptic, slightly asymmetric, large, broad; stipules erect and conspicuous, resembling narrow, basally veined leaves with a tapered tip, or small and inconspicuous. Flowers purple or scarlet, fragrant, pedicelled, densely clustered on short racemes or fascicled at old leafless nodes during the dry season before new leaves appear; bracts and bracteoles early caducous; calyx tube short, cupular; lobes 4, free, leathery, elliptic, obtuse, ciliate, short, subequal, imbricate; petals 3, obovate, obtuse, imbricate, the innermost petal smaller; stamens 10, diadelphous; anthers uniform, ovate, versatile, introrse; ovary stalked, free, linear, glabrous, few-ovuled; style filiform, apically clavate; stigma small, globose, terminal. Pod flat, leathery, lanceolate, compressed, oblong, subfalcate, with a thin wing along the upper suture, indehiscent, 1–2 seeded; seeds compressed, kidney-shaped. (Tulasne 1843; Bentham and Hooker 1865; Standley and Steyermark 1946)

Representatives of this genus are found in northern Brazil, Panama, Guatemala, and Peru. They flourish on thinly forested hillsides and in ravines on dry, rocky soil.

Both species are exceptionally attractive as ornamentals. *P. septentrionalis* Donn. Smith, the monkey flower tree, has been introduced into many areas similar to its native habitat; however, its flowering performance has been somewhat erratic and disappointing. Blooming is at its height in Guatemala during the peak of the dry season (Standley and Steyermark 1946; Menninger 1962).

NODULATION STUDIES

This species was abundantly nodulated in Hawaii (personal observation 1938).

Phyllodium Desv.   Papilionoideae: Hedysareae, Desmodiinae

From Greek, *phyllodes*, "leaflike," in reference to the large foliaceous flower bracts.

Type: *P. elegans* (Lour.) Desv.

6–9 species. Herbs or small shrubs; branches slender, woody, some species gray-downy. Leaves pinnately 3-foliolate; leaflets oblong-ovate, leathery, terminal leaflet larger, lanceolate; stipels present; stipules striate, conspicuous. Flowers small, usually yellow, loosely clustered, umbellate-capitate, in long, axillary and terminal racemelike inflorescences; bracts conspicuous, 3-foliolate, the 2 lateral bracts large, leafy, almost concealing each umbel, the terminal bract setaceous; calyx narrowly campanulate; lobes 4, upper lobe entire or minutely 2-toothed, bibracteolate at the base; standard obovate, base narrow; wings spurred above the claw; stamens deciduous with the petals, 1 free midway, others united about two-thirds of their length; ovary few-ovuled; disk thin; style not thickened above. Pod usually 2-segmented, seldom 4, small, orbicular-truncate, often downy, compressed. (Schindler 1924)

This genus was formerly a section in *Desmodium*.

Members of this genus are common in tropical and subtropical eastern Asia and Taiwan. They thrive in grassy fields, thickets, and teak forests.

NODULATION STUDIES

P. pulchellum (L.) Desv. was reported nodulated as *Desmodium pulchellum* in Java (Keuchenius 1924) and in the Philippines (Bañados and Fernandez 1954).

Phyllota Benth.   Papilionoideae: Podalyrieae

175 / 3625

From Greek, *phyllon*, "leaf," and *ous, otos,* "ear"; the bracteoles of some species are leafy.

Type: *P. barbata* Benth.

9–10 species. Shrubs, small, heathlike; branches erect, pubescent. Leaves alternate or scattered, simple, entire, margins revolute; stipules minute or none. Flowers yellow or purplish red and yellow, 1–many, on short pedicels, axillary or terminal; bracts small or absent; bracteoles leaflike and persistent, adnate to the petiole below the calyx; calyx pubescent, upper lobes broader, sometimes united into an upper lip; petals clawed; standard suborbicular, longer than the lower petals; wings oblong; keel incurved; all stamens or outer 5 slightly joined to the base of the petals and sometimes forming with them a ring or short tube; ovary small, pubescent, sessile, 2-ovuled; style dilated or thickened at the base, awl-shaped above; stigma small, terminal. Pod ovate, 2-valved, turgid; seeds 1–2, kidney-shaped. (Bentham 1864; Bentham and Hooker 1865)

Distribution of *Phyllota* species is limited to Australia. They flourish in tablelands and coastal areas of Western Australia, South Australia, New South Wales, Queensland, and Victoria.

NODULATION STUDIES

Nodules have been reported on *P. pleurandroides* F. Muell. in South Australia (Harris 1953), on *P. phylicoides* Benth. in Queensland (Bowen 1956), and on *P. barbata* Benth. in Western Australia (Lange 1959).

Phylloxylon Baill.
(374) / 3846

Papilionoideae: Dalbergieae, Anomalae

From Greek, _phyllon_, "leaf," and _oxys_, "sharp," in reference to the flat, lance-like phylloclades which function in the absence of leaves.

Type: _P. decipiens_ Baill.

5 species. Shrubs or trees, branches stiff, flattened, margins notched, bearing small scales instead of leaves. Flowers purple, in small clusters or spikes at the notches of the phylloclades; bracts minute, triangular; calyx small, campanulate; teeth minute; petals equal length; standard ovoid, reflexed, clawed; wings narrower; keel petals straight, blunt; stamens 10, diadelphous; anthers small, globose; ovary sessile, few-ovuled; style short, incurved; stigma capitate. Pod leathery, swollen, tapering at both ends, 1–2-seeded, indehiscent.

Phylloxylon has priority over _Neobaronia_ (Harms 1900). Hutchinson (1964) positioned the genus in Geoffroeeae trib. nov.

Representatives of this genus are indigenous to Madagascar and Mauritius.

The bark of _P. ensifolius_ Baill. exudes a red resin used by the natives of Madagascar to stupefy fish. Two tree species, _P. xylophylloides_ Bak. and _P. xiphoclada_ Bak., yield a fine hard wood of considerable local value.

Physanthyllis Boiss.

Papilionoideae: Loteae

From Greek, _physa_, "bladder"; the calyx is bladderlike; the name alludes to resemblance to _Anthyllis_.

Type: _P. tetraphylla_ (L.) Boiss.

1 species. Herb, annual; branches procumbent, villous to hirsute. Leaves sessile, imparipinnate; leaflets 2–3 pairs, obovate-elliptic, entire, terminal leaflet largest; stipels and stipules absent. Flowers yellow, 1–7, in sessile, axillary fascicles; bracts and bracteoles absent; calyx inflated at anthesis, ovoid, finely silky-haired, narrowed and straight at the mouth; lobes 5, small, subequal; petals yellow, long-clawed; standard longest; keel tip often red; stamens diadelphous; filaments apically dilated; anthers uniform, suborbicular; ovary long-stalked, 2-ovuled; style glabrous; stigma terminal. Pod stalked, enclosed by the bladderlike calyx, constricted and septate between the 2 seeds, indehiscent. (Hutchinson 1964)

This genus, formerly a section of _Anthyllis_, was segregated on the basis of the inflated calyx and diadelphous stamens.

This monotypic genus is endemic in North Africa, from Morocco to Libya, in eastern Mediterranean areas, and along the coastal plains of Israel. The plants inhabit hot, dry, sandy areas.

Nodulation Studies

In 1960, B. C. Park forwarded Libyan specimens of this species to Kew Gardens, England, for identification; he later submitted evidence of its nodulation to the authors.

From Greek, *physa*, "bladder," in reference to the shape of the stigma.

Type: *P. venenosum* Balf.

3–5 species. Herbs or tall vines, woody at the base; branches glabrous. Leaves pinnately 3-foliolate; leaflets large, central leaflet broadest; stipels and stipules minute. Flowers medium, rose, greenish yellow, or white spotted with violet, in fascicled racemes on elongated, axillary peduncles or terminal branches; rachis thick, nodose; some species bloom before and again after leaf emergence; bracts small, caducous; calyx widely campanulate; teeth 5, deltoid, upper 2 more or less joined; standard ovate, enclosing wings and keel, edges crinkled; wings short, obovate-oblong, incurved, free; keel spurred, beak spirally twisted; stamens 10, diadelphous; anthers uniform; ovary short-stalked, few-ovuled; style thickened at base and curled inside the keel; stigma hooded, the concavity below it thickly bearded. Pod linear, cylindric, 2-valved, with woolly tissue between the seeds; seeds few, ellipsoid, thick, 1 edge nearly straight, the other grooved by the depressed hilum. (Bentham and Hooker 1865; Wilczek 1954)

The original generic number 422 (Taubert 1894) was changed to 437 by Harms (1908a) because of the reorganization within the subtribe necessitated by various revisions and inclusions of new genera.

Members of this genus are distributed throughout tropical West Africa; they are especially abundant near the mouths of the Niger and Old Calabar rivers. Species thrive in dry and humid forests that are periodically inundated by floods. The germination and growth of some species are benefited by burn-off of cleared forests and savannas.

In the middle of the 19th century attention was directed to the poisonous properties of *P. venenosum* Balf., the Calabar or ordeal bean, used by African natives in criminal trials by ordeal. The poisonous seeds are nearly tasteless and closely resemble those of *Mucuna* species, with the exceptions that they are more oblong than circular and have a grooved hilum (Dalziel 1948). Ingestion of the seeds results in paralysis of the central nervous system, respiratory failure, and usually heart failure. The pharmacological action of physostigmine, $C_{15}H_{21}N_3O_2$, the major alkaloid isolated to date, is the inhibition of acetylcholinesterase. This enzyme functions in the transmission of nerve impulses at nerve endings and in the conduction of impulses along nerve and muscle fibers. The reader is referred to the reviews of Coxworth (1965) and Robinson (1968) for further details.

Physostigmine is valuable in ophthalmic medicine to induce miosis and is used topically in glaucoma; its action is opposite to that of atropine. Physostigmine salicylate is useful in veterinary medicine to counteract rumen impaction in cattle and impaction colic in horses. The seeds of *P. venenosum* yield a heavy oil fatal to mice. The oblong, reddish brown seeds of *P. mesoponticum* Taub. are poisonous; an alkaloid giving the reactions of physostigmine has been identified (Verdcourt and Trump 1969).

Named for Charles Pickering, 1805–1878, botanist with the Wilkes Expedition to California in 1841.

Type: *P. montana* Nutt.

1 species. Shrub, 1–3 m tall, spiny, evergreen, rigid, densely branched, xerophytic. Leaves small, palmately 3-, rarely 1-foliolate, subsessile; leaflets elliptic, entire; stipules minute and caducous, or absent. Flowers about 2 cm long, rose purple, on short, solitary pedicels in short terminal racemes, pedicels minutely bibracteolate; calyx turbinate-campanulate; teeth 4, very short, subequal, broad; petals free, subequal; standard orbicular, sides reflexed, usually with a yellow or white spot at the base; wings oblong; keel petals free, oblong, narrower, straight; stamens 10, free, distinct; ovary short-stalked, many-ovuled; style incurved; stigma minute, terminal. Pod linear, flat, stalked, straight, somewhat constricted between the 6–10 seeds. (Bentham and Hooker 1865; Jepson 1936)

This monotypic genus is endemic in Baja California, Mexico, and southern California. It is commonly called chaparral pea. The plants grow on dry, rocky mountain slopes up to 1,600 m.

NODULATION STUDIES

This xerophytic species was observed abundantly nodulated in the California chaparral (Vlamis et al. 1961).

Pictetia DC.   Papilionoideae: Hedysareae, Aeschynomeninae
318 / 3786

Named for M.-A. Pictet, celebrated physician, colleague and friend of De Candolle.

Lectotype: *P. aculeata* (Vahl) Urb.

8 species. Shrubs, erect, or trees, small, deciduous; branches spiny or pubescent. Leaves imparipinnate; leaflets alternate, 3–25, small, rounded, tipped with a bristle; stipels none; stipules paired, spiny. Flowers yellow, in few-flowered racemes borne laterally or in axils of short twigs, or solitary; bracts and bracteoles early caducous; calyx campanulate, constricted at the base; lobes 5, unequal, 2 upper lobes short, obtuse, lower 3 longer, acuminate; petals and stamens inserted in the upper half of the calyx; standard rounded, reflexed, not eared; wings oblong; keel petals shorter than the wings, bent and connate near the apex; stamens diadelphous; ovary stalked, narrow, hairy, multiovuled; style smooth, slender; stigma small, terminal. Pod stalked, linear, flattened, warty or hairy, indehiscent, sometimes constricted between the 1–6 oblong joints. (De Candolle 1825b; Bentham and Hooker 1865; Harms 1908a)

Representatives of this genus are native to the West Indies, Mexico, and Cuba. They inhabit dry coastal thickets and pastures.

The strong, durable wood of the type species is suitable for small turnery items, but it is used mainly for fence posts, stakes, shingles, and firewood.

From Greek, *pilos*, "cap," in reference to the large, peltate stigma.

Type: *P. reticulatum* (DC.) Hochst.

3 species. Trees, evergreen, unarmed, small to medium, up to 10 m tall, erect, with spreading crown, or shrubby, scrambling, with trunk twisted, gnarled; branchlets smooth or rusty-pubescent. Leaves simple, broad, deeply bilobed at the base and the apex, typically *Bauhinia*-like, with 9–15 prominent main veins radiating from the base at the point of attachment with the long petiole; stipules narrow. Flowers white or pale yellow, on short stalks, in simple or panicled terminal or axillary racemes; male and female flowers usually borne on separate plants; bisexual flowers sometimes present; bracteoles subpersistent; calyx turbinate; sepals 5, triangular, free above, all coherent in the lower half; petals 5, spoon-shaped, crinkled, clawed; male flowers with 10 perfect stamens alternately long and short; ovary rudimentary; female flowers with 10 staminodes; ovary stalked, cylindric, pubescent; style thick; stigma large, peltate. Pod linear-elongated, broad, flat, narrowed at the base, apiculate, hard, indehiscent, surface checkered or cracked; seeds several, embedded in pulp. (De Wit 1956; Keay et al. 1964)

Although *Piliostigma* is conserved as a genus distinctive from *Bauhinia*, from which it was segregated (de Wit 1956), the position number accorded it by the *International Code of Botanical Nomenclature* (1966) is identical with that for *Bauhinia*.

Members of this genus occur in tropical Africa and Indo-Malaya. One species, *P. malabaricum* (Roxb.) Benth. var *acidum* de Wit, is confined to Asia, in the drier monsoon belt of India, Malaysia, and the Philippines. The 2 African species, *P. reticulatum* (DC.) Hochst. and *P. thonningii* (Schumach.) Milne-Redhead, inhabit dry and moist savannas, respectively. They are widespread and abundant in western and eastern territories from Senegal and Gambia to the Congo Basin, Ethiopia, Zimbabwe, Transvaal, and Mozambique.

In Africa, the *Piliostigma* species are often pioneer plants which gain dominance; they have considerable ecological and economic importance. Decoctions of plant parts are used by African natives for various medicinal purposes (Watt and Breyer-Brandwijk 1962). The unripe pods serve as a soap substitute; the ripe pods and leaves are good cattle fodder. The bark and branches yield tannin, the roots, a red brown dye, and the seeds, a blue black dye; however, these are of no commercial value. The reddish brown, tough, strong wood lacks market appeal.

Nodulation Studies

P. malabaricum var. *acidum*, reported as *Bauhinia malabarica*, was devoid of nodules in the Philippines (Bañados and Fernandez 1954). In Zimbabwe, no nodules were found on *P. thonningii* (Corby 1974).

From the Greek verb *pipto*, "fall," and *aden*, "gland," in reference to a deciduous gland on the anthers.

Lectotype: *P. latifolia* Benth.

11–15 species. Shrubs, erect or scandent, or trees, rarely lianas; stems
usually armed with prickles or with spiny stipules; trees usually about 30 m
tall, majestic, with trunk diameters up to 1 m, flat-topped and flat-branched.
Leaves bipinnate, large; pinnae opposite; leaflets small, numerous, broad;
petiole and rachis usually glandular. Flowers small, white or greenish, sessile,
in spicate or globose heads, on axillary peduncles, solitary or fascicled, sometimes in terminal panicles; bracts small, caducous; calyx campanulate; lobes 5,
short; petals usually connate to the middle, valvate; stamens 10, exserted, free
or nearly so; filaments and petals joined at the base to a small disk; anthers
small, roundish, tipped with a deciduous gland; ovary subsessile or stalked, 3-
to many-ovuled; style filiform. Pod linear, wide, flat, membranous or leathery,
without septa or pulp, 2-valved, valves straight or deeply constricted between
elliptic seeds, dehiscent along both sutures; seeds not winged, funicle filiform.
(Bentham 1840; Brenan 1955, 1963; Altschul 1964)

Piptadenia species are native to tropical America, Guiana, Venezuela, and
the West Indies.

Earlier estimates of 50–60 species have been reduced to 11–15 species.
The excluded species were transferred to *Newtonia, Piptadeniastrum* (Brenan
1955), *Parapiptadenia* (Brenan 1963), and to *Anadenanthera* (Altschul 1964).
Most of the species transferred to the first 2 genera are indigenous to tropical
Africa.

Comments here concerning *P. colubrina* (Vell.) Benth. and *P. peregrina*
(L.) Benth., *Anadenanthera* species, *P. rigida* Benth., now *Parapiptadenia rigida*,
and *P. africana* Hook. f., now *Piptadeniastrum africanum*, conform with the
literature prior to their segregation.

The economic value of this genus rests primarily on the excellence of the
timber of *P. excelsa* (Griseb.) Lillo, *P. macrocarpa* Benth., *P. paraguayensis*
(Benth.) Lindm., and *P. rigida*. The reddish brown wood of the last-named
species is called angico in the timber trade; it is hard, has a density of about
945 kg/m^3, and is easily mistaken for true mahogany, *Swietenia* sp. (Record
and Hess 1943). Its use is reserved for high-quality furniture and house
construction. Angico gum, obtained from *P. macrocarpa* and *P. rigida*, is intermediate in quality between Sudan and Senegal gum arabic (Howes 1949).

A snuff made from the pulverized seed of *P. macrocarpa* was used by the
aborigines of tropical South America as a hallucinative narcotic (Wurdack
1958). Inhalation of the snuff was followed by rigidity, staring, at times a
convulsion, and thereafter an excited, frenzied animation. The presence of
bufotenine and other indole bases derived from tryptophane (see Harborne
et al. 1971) are the probable causes of the hallucinogenic effects. Despite
documentation of the uses and effects of the snuff made from *P. macrocarpa*
by aborigines of the Caribbean and northern South America, the intoxicating
effects attributed to the snuff from *P. peregrina*, now renamed *Anadenanthera
peregrina* (L.) Speg., were not confirmed (Turner and Merlis 1959).

Nodulation Studies

No nodules were observed on *P. obliqua* Macbr. in Venezuela (Barrios and Gonzalez 1971). The presence of nodules on *P. rigida* (= *Parapiptadenia rigida*) in the gardens at Villa Thuret, Cap d'Antibes, France (Lechtova-Trnka 1931) was confirmed in 9 sites in Argentina by Rothschild (1970). The nodulation account for *P. peregrina* in Brazil (Campêlo and Döbereiner 1969) should now be attributed to *Anadenanthera peregrina*.

The red brown roots of *P. colubrina* Benth., an introduced tree in Singapore, did not bear nodules (Lim 1977). This species is now named *Anadenanthera colubrina*.

Piptadeniastrum Brenan   Mimosoideae: Piptadenieae

The Latin suffix, *astrum*, indicates inferiority or incomplete resemblance to *Piptadenia*.

Type: *P. africanum* (Hook. f.) Brenan (= *Piptadenia africana*)

1 species. Tree, unarmed, up to about 50 m tall, with a trunk diameter up to 1.8 m; base buttressed. Leaves bipinnate; pinnae alternate, up to 30 pairs; leaflets in many pairs, usually 30–50, small, linear, sessile, folding at sunset; rachis eglandular. Flowers malodorous, white to yellowish, on jointed pedicels, in dense, cylindric spikes in terminal or axillary aggregations; calyx campanulate; teeth 5, short, blunt; petals shortly united at the base; stamens 10, fertile, adnate to the corolla tube, united at the base; filaments long, curved, exserted; anthers dorsifixed, with apical gland shed during flowering; ovary subsessile; style somewhat curved; stigma truncate. Pod short-stalked, linear or curved, thin, smooth, flattened, 2-valved, dehiscent along 1 suture; seeds thin, flat, elliptic, surrounded by a broad, membranous wing and attached near the middle to the margin of the pod by a long funicle. (Brenan 1955, 1959)

This genus was set forth by Brenan in 1955 to distinguish the African species, *Piptadenia africana*, from the South American species. Separation was based on the dehiscence of the pod along 1 instead of both sutures and the absence of leaf glands.

This monotypic genus is endemic to West Africa. The plants occur from Senegal to Angola and the Congo Basin; they are also common in Liberia. They thrive in riverine and lower rain forests, and are light-demanding.

The bole of this majestic tree is straight, cylindrical, and seldom forked. The wood is of good quality, durable in fresh water, and somewhat termite-resistant. The sapwood is nearly white and less durable than the golden brown heartwood. The objectionable odor of the heartwood limits use of the wood for interior woodwork, but the timber has local value for heavy construction and flooring.

The Greek suffix, *opsis*, denotes a resemblance to *Piptadenia*.

Type: P. lomentifera Burk.

1 species. Shrub, arborescent, 3–7 m tall. Leaves relatively small, bipinnate; pinnae pairs 1; leaflets about 3 pairs, oblique-oblong; stipules short-spinescent. Flowers bisexual, uniform, sessile, in axillary, globose heads; calyx 5-toothed; lobes soft-pubescent, united; petals 5, soft-pubescent, free, valvate; stamens 10, free; anthers with apical gland; ovary stalked, several-ovuled; style filiform, bent; stigma terminal, inconspicuously concave. Pod stalked, linear, flattened, transversely jointed, segments subquadrate, 1-seeded, indehiscent; seeds flat. (Burkart 1944)

This monotypic genus is restricted to northern Paraguay.

Piptanthus Sweet   Papilionoideae: Podalyrieae
156 / 3616

From the Greek verb *pipto*, "fall," and *anthos*, "flower"; the petals, stamens, and calyx lobes soon drop; thus the flowers are short-lived.

Type: *P. nepalensis* Sweet

9 species. Shrubs, deciduous or half-evergreen. Leaves petioled, palmately 3-foliolate; leaflets entire, sessile, some with silky, gray-pubescence; stipules connate into 1, leaf-opposed. Flowers large, yellow, several together in short, terminal, upright racemes; bracts sheathlike, caducous; bracteoles absent; calyx campanulate; lobes 5, short, subsessile, equal, broadly deltoid or triangular; petals long-clawed; standard subequal to the wings, suborbicular, sides reflexed, emarginate, wings obovate-oblong; keel petals slightly longer than the wings, somewhat incurved, round-tipped, partly joined at the back; stamens free; filaments flattened; ovary stalked, linear, multiovuled; style filiform; stigma minute, terminal. Pod short-stalked, broad-linear, compressed, continuous within, 6–8 seeded; seeds kidney-shaped. (Bentham and Hooker 1865)

The genus is closely related to *Anagyris*.

Members of this genus are native to central Asia, China, and the Himalayas up to 2,600-m altitude. Several species have been introduced into the United States as ornamental shrubs.

P. nanus Popov yields the following quinolizidine alkaloids: sparteine, $C_{15}H_{26}N_2$, lupanine, $C_{15}H_{24}N_2O$, piptanthine (isopiptanthine), $C_{14}H_{24}N_2$, ormosanine (piptamine), $C_{20}H_{33}N_3$, and nanine, $C_{15}H_{24}N_2O_2$ (see Harborne et al. 1971). Piptamine and piptanthine have also been isolated from *P. mongolicus* Maxim., and cytisine from *P. nepalensis* Sweet. The occurrence of these so-called lupine alkaloids is chemotaxonomically significant because it demonstrates a closer phyletical relationship of *Piptanthus* species with Genisteae and Sophoreae than with Southern Hemisphere elements of the Podalyrieae (Mears and Mabry 1971; Turner 1971).

Nodulation Studies

Plantlets of *P. nepalensis* were promiscuously nodulated by rhizobial strains from diverse species of the cowpea miscellany and also from *Lens, Medicago, Trifolium,* and *Vicia* species. More than 50 percent of the plants tested were nodulated by strains from *Amorpha fruticosa, Lespedeza striata, Oxytropis lambertii, Spartium scoparium, Stizolobium deeringianum, Vicia villosa,* and *Wisteria chinensis* (Wilson 1939a).

---

*Piscidia L.   Papilionoideae: Dalbergieae, Lonchocarpinae
367 / 3839

From Latin, *piscis,* "fish," in reference to the fish-poisoning or stupefying properties of the plant parts.

Type: *P. piscipula* (L.) Sarg.

8 species. Shrubs, low, sprawling, or trees, up to 20 m tall; trunks to 20 cm in diameter, unarmed, deciduous. Leaves imparipinnate; leaflets 2–13 pairs, terminal leaflet often largest, leathery at maturity, elliptic to oblong, ovate or obovate; stipels absent; stipules paired, at base of petiole, oblique-ovate to kidney-shaped, early caducous. Flowers pink, cream, or white tinged with red, in few-flowered, showy, axillary or pseudoterminal panicles; bracts and bracteoles small, early caducous; calyx campanulate; lobes 5, short, broad, upper 2 partly joined; standard suborbicular, notched, outside often pubescent; wings usually longer than standard and keel; keel petals adherent to wings and connate below; stamens 10, monadelphous, vexillary stamen free at the base; anthers oblong, dorsifixed; ovary sessile, pubescent, many-ovuled; style filiform, incurved, glabrous above; stigma small, terminal. Pod on short stipe exserted beyond the persistent calyx, linear, flat, with 4 broad, longitudinal, membranous wings continuous or slightly constricted between the seeds, indehiscent; pods becoming brittle at maturity and breaking to release lustrous, reddish tan to dark brown, kidney-shaped seeds. (Bentham and Hooker 1865; Rudd 1969)

Piscidia species are distributed throughout Florida, Mexico, Central America, the West Indies, northern Peru, and Venezuela. They are common in dry woodlands and thickets; some species appear restricted to limestone areas.

In southern Florida, the type species, called Jamaican dogwood, is cultivated for its unusual pods and attractive flowers, which bloom before the new leaves appear. Natives throughout the West Indies and Central America value this species for the fish-stupefying properties of its bark, roots, and stems; they commonly call it the fish-fuddle tree. The pods also served as a source of arrow poison (Cheney 1931).

Rudd (1969) quotes the following fascinating account furnished by Sloane, a historian of Jamaica, in 1725: "The Bark of this Tree stamp'd and thrown into the standing Pool where Fish are, intoxicates them for some Time, they turning their Bellies up, and coming above Water, but if they are not presently caught, they come to themselves and recover . . . The Indians and Negro's make use of this Bark to take Fish . . . The Fish caught after this manner, are counted very wholesome and good food . . . This is a Providence of God to those barbarous People, being a natural Help for present Food and Sustenance."

The first chemical inquiry into an explanation for the analgesic properties of *P. piscipula* (L.) Sarg. resulted in the isolation of piscidia (Hart 1883), which was later shown to be a mixture of compounds (Freer and Clover 1901). Piscidic acid, $C_{11}H_{12}O_7$, was one of the crystalline substances isolated. Rotenone and ichthynone, $C_{23}H_{20}O_7$, a new compound toxic to goldfish in 1 ppm concentration, were identified in the root bark by Russell and Kaczka (1944). Auxence (1953) conducted the earliest pharmacognistic study of the plant parts of this species. Later chemical studies clarified the status of ichthynone as an isoflavone (Dyke et al. 1964; Schwarz et al. 1964) and identified the presence of various isoflavonoids and related compounds in root extracts of the Jamaican dogwood (Falshaw et al. 1966).

The wood has no commercial possibilities, but is of reasonably good quality. The sapwood is whitish; the heartwood is yellowish brown, becoming dark upon exposure. The wood has a medium to high luster, is without odor or taste, highly durable, hard, heavy, has a density of 800–900 kg/m^3, and finishes smoothly, but has a coarse texture. Green timber used as posts often roots. Other local uses of the wood are for heavy construction, railroad ties, wheel spokes, and fuel; it provides an excellent charcoal (Record and Hess 1943).

NODULATION STUDIES

Nodules 10–20 mm in length were common on the odoriferous roots of *P. piscipula* (= *P. erythrina* L.) growing in Hawaii (Allen and Allen 1936a). Rhizobia isolated from these plants effectively nodulated *Vigna sinensis* and vice versa. Colony growth on asparagus-potato and yeast-extract mannitol mineral salts agars was raised and mucilaginous, alkaline reactions were produced on all carbohydrate media and in litmus milk, and tyrosinase reactions were positive (Allen and Allen 1939). In cross-testing *Piscidia* rhizobia with 19 species of 14 different genera of the cowpea miscellany, only *Phaseolus lunatus* failed to nodulate; the majority of the crosses gave effective symbiosis.

Pisum L.   Papilionoideae: Vicieae
381 / 3855

From Greek, *pisos*, the name of a pea plant.
   Type: *P. sativum* L.
   6 species. Herbs, annual or perennial, glabrous, spreading or climbing by well-developed tendrils; roots white, taproots with many laterals. Leaves paripinnate; leaflets 1–3 pairs; rachis ending in a branched tendril or bristle; stipules large, sometimes larger than the leaflets, semicordate or semisagit-

tate. Flowers showy, purple, pink, or white, solitary or few and racemose on elongated, axillary peduncles; bracts minute, caducous; bracteoles absent; calyx obliquely campanulate, often dorsally gibbous; lobes subequal, large, upper 2 sometimes broader; standard broad-ovoid to almost orbicular, with short, broad claw; wings falcate-oblong, adnate to the keel; keel obtuse, incurved, shorter than the wings; stamens diadelphous, vexillary stamen free or connate at the middle with others; filaments dilated above; anthers uniform; ovary subsessile, many-ovuled; style bent abruptly upward, dilated at apex, bearded above; stigma subterminal. Pod compressed to cylindric, inflated, oblong-linear, obliquely acute, 2-valved; seeds subglobose. (Bentham and Hooker 1865)

Pisum species are native to the eastern Mediterranean and western Asia. The plants are both wild and cultivated. They thrive on well-drained soils of reasonable levels of fertility. A pH range between 5.5 and 7.0 is preferred. Cool temperature areas, 12–18°C, are optimum.

Pisum species have served man for ages. Wade (1937) credited Pickering's *Chronological History of Plants* (1879) with the statement that peas are the only culinary vegetable traced with certainty back to the Stone Age. It is said that before 1700 it was uncommon for peas to be eaten other than as dry, cooked seeds.

Genesis of the modern science of genetics began in 1856 with Mendel's epoch-making cross-breeding experiments with garden peas. Peas were also the first crop used in scientifically controlled experiments for the production of new varieties. Upon the advent of the canning industry in the middle 1800s, some of these varieties served, in turn, as experimental tools in the formulation of canning techniques and for technical improvements in canning machinery. In addition to their use as food plants for man, *Pisum* species serve as cattle forage, ground cover, and for making hay, silage, and green manure.

Unfavorable effects in animals following ingestion of pea-vine ensilage and field pea pasturage were summarized by Kingsbury (1964). Pisatin, $C_{17}H_{14}O_6$, an antifungal substance, was isolated and described from the endocarp of *P. sativum* L. pods inoculated with *Monilinia fructicola* (Cruickshank and Perrin 1961; Perrin and Bottomley 1962). This antibiotic is relatively weak and has lacked promise as a therapeutant, despite its broad biological spectrum (Cruickshank 1962). Of interest was the finding that fungi pathogenic to *P. sativum* were relatively insensitive to pisatin, whereas fungi nonpathogenic to *P. sativum* were generally highly sensitive.

Pisum species are differentiated with difficulty, and, similarly, their nomenclature is confused. *P. arvense* L. (= *P. sativum* ssp. *arvense* (L.) Poir.), field pea, is generally considered distinct from *P. sativum* L. (= *P. hortense* Aschers. & Graebn., *P. sativum* ssp. *hortense* Poir.), garden pea, but some botanists consider the differences purely artificial because of cross-fertility. *P. sativum* is not known in the wild state, nor is its origin defined, but the possibility of it having arisen by hybridization and back-crossing between *P. elatius* Bieb., a weed in Europe, and *P. arvense* is sometimes conjectured. Complicated aspects of this controversial topic are documented by Sutton (1911), Lamprecht (1956a,b), and Yarnell (1962).

NODULATION STUDIES

Throughout the years since their isolation (Nobbe et al. 1891), pea rhizobia have been used in every type of experimental study of symbiotic nitrogen fixation. Ability of these rhizobia to nodulate *Vicia* species, but not species of 5 other genera, initiated the cross-inoculation concept (Nobbe et al. 1895). Pea rhizobia were recognized by Beijerinck (1888) as 1 of the 7 varieties of *Bacillus radicicola*, as *Rhizobium frankii* var. *minus* by Schneider (1892), as *R. radicicola* by Hiltner and Störmer (1903), and in 1923 as *R. leguminosarum* in the first edition of Bergey's *Manual of Determinative Bacteriology*. This specific epithet currently applies to the rhizobia of species in 4 genera, *Lathyrus, Lens, Pisum,* and *Vicia* of the tribe Vicieae. Pea rhizobia are peritrichously flagellated, and form X-, Y-, and club-shaped bacteroids. Beijerinck's crude drawings of these unusual forms in his laboratory notebook, *Bacteria,* with the comment "Bacteroids of *Vicia Faba*; those of *Pisum sativum* almost identical. For *Trifolium pratense* small round vesicles," are of historical interest (Iterson et al. 1940). Growth of pea rhizobia on laboratory media is rapid. Colony growth has a tendency to spread, is mucilaginous, semitranslucent, and glistening.

Pisum nodules conform in morphology and histology to those of the Vicieae type (Spratt 1919); they are elongated, with well-defined apical meristems, are frequently branched, and occur in clusters.

Nodulated species of *Pisum*	Area	Reported by
arvense L.	Widely reported	Many investigators
var. *hibernicum*	Germany	Stapp (1923)
var. *vernale*	Germany	Stapp (1923)
hortense Aschers. & Graebn.	Canada	E. Rosylycky (pers. comm. 1969)
humile Boiss. & Nóe	Israel	Hely & Ofer (1972)
jomardi Schrank	Germany	Stapp (1923)
sativum L.	Widely reported	Many investigators

Pithecellobium Mart.   Mimosoideae: Ingeae
7 / 3441

From Greek, *pithekos,* "ape," "monkey," and *ellobion,* "earring," alluding to the descriptive name, "ape's earring," often given to the coiled pods. *Pithecollobium* and *Pithecolobium* occur in the literature as orthographic errors and are rejected spellings (Rickett and Stafleu 1959–1961; Mohlenbrock 1963a).

Type: *P. unguis-cati* (L.) Benth.

100–200 species. Shrubs or small trees, near evergreen, usually less than 10 m tall, some 20–30 m tall, unarmed or with spiny stipules or axillary spines. Leaves bipinnate; old leaves shed as new ones appear; pinnae pairs few; leaflets small, in many pairs, or large, in few pairs; petioles and rachis usually glandular. Flowers usually pink, white, or purplish, bisexual or polygamous, fragrant, borne in small, globose heads or cylindric spikes, on solitary or clustered peduncles in leaf axils or in terminal panicles; calyx campanulate or tubular, shortly 5-dentate; petals united below the middle forming a tubular, 5–6-lobed corolla; stamen numbers indefinite, mostly numerous, long-exserted, united at the base into a tube; anthers small; ovary sessile or stalked, multiovuled; style slender; stigma small, terminal. Pod variable in form, flat, straight, coiled or twisted, often constricted but not septate between the seeds, leathery, dehiscent or indehiscent; seeds orbicular or ovate, mostly dark brown or black, shiny or frosted, funicle filiform or expanded into a fleshy, white, pink, or red aril. (Bentham and Hooker 1865)

This genus is beset with taxonomic problems because of its size and diversity of synonyms. North American representatives of the genus were segregated into 7 genera mainly on the basis of pod characters (Britton and Rose 1928). Ten or more genera were recognized by Kostermans (1954) in his revision of Old World *Pithecellobium* species.

Members of this large genus occur in tropical and subtropical areas of the Old World and the New World. Major distribution is in tropical America and Asia. About 24 species are present in Mexico and Cuba. The plants flourish in alluvial areas, tidal forests, stream bottoms, on sandy loams and clay soils, on limestone cliffs, and in wet and dry forests. Some species are drought-resistant. Few species are cultivated.

P. arboreum (L.) Urb., a small tree known as coralillo in Mexico, *P. dulce* (Roxb.) Benth., the Manila tamarind, *P. unguiscati*, cat's claw, and *P. flexicaule* (Benth.) Coulter, a shrub or small tree sometimes called Texas ebony, are prominent members of the genus. Of these, *P. dulce* appears to be the best known because of its widespread naturalization. All species grow rapidly, are reasonably drought-resistant, and withstand heavy pruning and browsing by goats, but the spiny forms are usually avoided by cattle. They are generally satisfactory as hedge plants, highway and street shade, and for landscape gardening. Considerable use has been made of some species as nurse trees for cacao, coffee, and tea.

The barks of many species have high tannin contents; in some, the amount approximates 25 percent; fish-stupefying properties of the bark are somewhat common. The crushed pods of *P. parvifolium* Benth. yield a rich golden yellow dye, called algarovilla in the West Indies; a purple dye is extract-ed from the crushed pods of *P. jiringa* Prain ex King in Malaysia. The pulpy seeds of *P. dulce* are relished by monkeys and birds. In Mexico they are consumed raw or roasted, or used in a beverage resembling lemonade. The gum obtained from the trunk of this species is transparent and reddish brown and when dissolved in water is a good mucilage. Seeds of various species have been used as a coffee substitute and as ornaments. In Malaysia the leaves, green pods, and flowers of *P. lobatum* Benth. are eaten as vegetables. The flowers of most species are a source of a delicious honey. Leaf and bark decoctions serve varied medicinal purposes (Quisumbing 1951).

Unnamed alkaloids have been credited to a number of species (Willaman and Schubert 1961). *P. bigeminum* Mart. and *P. lobatum* yield pithecolobine, $C_{22}H_{46}N_4O_2$, an alkaloid that was isolated in impure form about the turn of the century from *P. saman* Benth., now renamed *Samanea saman* (Greshoff 1890). Purification and formulation of pithecolobine were reported by

Wiesner et al. (1952, 1953). *P. lobatum* is also the source of djenkolic acid, $C_7H_{14}N_2S_2O_4$, an uncommon sulphur-containing amino acid derived from cysteine (Veen and Hyman 1935).

Pithecellobium wood is used for general carpentry in many areas. The trees tend to be small and have poor timber form. The wood varies in structural details. The sapwood is yellowish white; the heartwood is reddish brown. The wood varies from soft to hard, and has a density of 575–960 kg/m³. It is strong and yet somewhat brittle, but it is moderately durable in soil and when exposed to weather. The wood of *P. racemiflorum* Donn. Smith is known as Surinam snakewood or zebrawood, because of its irregularly striped figuring; however, it should not be mistaken for true zebrawood, *Connarus guianensis* (Titmuss 1965). *P. arboreum* provides a reddish brown, streaked wood of excellent quality that takes a fine polish. The wood of *P. dulce*, known in the Philippines as kamachili, is malodorous when freshly cut. It is used for boxes, crates, wagon wheels, and fuel (Record and Hess 1943).

NODULATION STUDIES

Because several rhizobial cultures isolated from *P. dulce* failed to form nodules on seedlings of various plant groups, Carroll (1934a) considered this species highly selective. In other tests with this species (Wilson 1939a), rhizobia from *Baptisia australis*, *Crotalaria spectabilis*, and *Dalea alopecuroides* incurred nodulation.

The indigenous tree in Singapore, *P. jiringa*, was devoid of nodules (Lim 1977).

Nodulated species of *Pithecellobium*	Area	Reported by
adinocephalum Donn. Smith	Costa Rica	J. León (pers. comm. 1956)
caraboboënse Harms	Venezuela	Barrios & Gonzalez (1971)
cauliflorum Mart.	Belém, Pará, Brazil	Norris (1969)
collinum Sandw.	Bartica, Guyana	Norris (1969)
corymbosum (Rich.) Benth.	Venezuela	Barrios & Gonzalez (1971)
dulce (Roxb.) Benth.	Widely reported	Many investigators
ellipticum (Bl.) Hassk.	Java	Pers. observ. (1977)
lanceolatum Benth.	Venezuela	Barrios & Gonzalez (1971)
latifolium Benth.	Costa Rica	J. León (pers. comm. 1956)
(= *Zygia latifolia* (L.) Fawc. & Rend.)	Trinidad	DeSouza (1966)
montanum Benth.	Malaysia	Janse (1897)
tortum Mart.	Hawaii, U.S.A.	Pers. observ. (1940)
	Venezuela	Barrios & Gonzalez (1971)
trapezifolium Benth.	Trinidad	DeSouza (1966)
unguis-cati (L.) Benth.	Hawaii, U.S.A.	Pers. observ. (1940)
	Trinidad	DeSouza (1966)

Plagiocarpus Benth.
250 / 2713

Papilionoideae: Galegeae,
Brongniartiinae

From Greek, *plagios*, "oblique," and *karpos*, "fruit."

Type: *P. axillaris* Benth.

1 species. Shrublet with numerous stems, erect, densely hairy, from woody rootstock. Leaves sessile, mostly 3-foliolate, lower leaves simple; stipules minute. Flowers yellow, solitary, axillary, subtended by small linear bracts and paired bracteoles; calyx lobes 5, equal, triangular-lanceolate, longer than the tube; standard broad-ovate, heart-shaped at the base, short-clawed; wings free, narrow, conspicuously eared; keel narrow, incurved, blunt at apex; stamens all joined in a sheath split above; anthers subequal, large, alternately basifixed and dorsifixed; style slender, curved at the tip; stigma minute. Pod oblique-ovate, 1–2-seeded; seeds large, ellipsoid, flattened, with small hilum near 1 end surrounded by a collarlike, lipped aril. (Polhill 1976)

The original tribal placement of this genus in Galegeae *sensu lato* was changed to Lotononideae by Hutchinson (1964). Polhill (1976) assigned the genus to the tribe Bossiaeae, thereby recognizing a close affinity to Genisteae *sensu lato*.

This monotypic genus is endemic in tropical Australia.

Plagiolobium Sweet

Papilionoideae: Genisteae, Bossiaeinae

From Greek, *plagios*, "oblique," and *lobos*, "lobe," in reference to the sinuate leaf margins.

Type: *P. chorizemifolia* (DC.) Sweet

1 species. Shrub, small; rootstock woody; branches stout, villous or subglabrous, often rusty-tomentose. Leaves simple, ovate to lanceolate, margins sinuate, pungent-pointed, often undulate, leathery, reticulate, usually glabrous; stipules short, awl-shaped, spiny. Flowers dark blue, small, 2–5 together on short pedicels in leaf axils; bracteoles 2, awl-shaped, subtending the calyx; calyx silky-haired, bilabiate, upper lip bidentate, large, broad, obcordate, 3 lower lobes very small; standard orbicular, emarginate, short-clawed; wings spurred on 1 side, much longer than the keel; stamens diadelphous; anthers uniform, alternately basifixed and dorsifixed; ovary stalked, glabrous, 2-ovuled; style smooth, incurved; stigma terminal. Pod short-stalked, glabrous, oblong, about twice as long as broad, turgid, dehiscent, 2-seeded. (Bentham 1864)

Segregation of this species was based on the occurrence of diadelphous stamens, in contrast to all stamens in *Hovea* species being united into a sheath. Hutchinson (1964) recognizes the distinctiveness of *Plagiolobium*, but some Australian botanists do not (Burbidge 1963).

This monotypic genus is endemic in Western Australia.

NODULATION STUDIES

P. chorizemifolia (DC.) Sweet, reported as *Hovea chorizemifolia* by Lange (1959), was observed nodulated in Western Australia.

531

Plagiosiphon Harms　　　　　　　Caesalpinioideae: Cynometreae
(39a) / 3484

From Greek, *plagios*, "oblique," and *siphon*, "tube"; the calyx is obliquely cylindric.

Type: *P. discifer* Harms

5 species. Shrubs or trees, some with blunt spines on branches. Leaves paripinnate; leaflets sessile, emarginate, usually asymmetric, 1–numerous pairs; rachis grooved. Flowers several, in elongated racemes; bracteoles large, petaloid, reticulate, subtending the calyx, sometimes enveloping the flower but not valvate; bracts minute or absent; calyx tube elongated, obliquely funnel-shaped or cylindric, base inflated on 1 side; disk prominent, thick, at the base of the calyx; lobes 4, imbricate in bud, later reflexed, obtuse; petals 4–5, oblong-spathulate, equal or slightly longer than the calyx lobes, apex obtuse to subacute, narrowed to the base; stamens 8–10, free; filaments slender, hairy at the base; anthers small, oblong; ovary stipitate, oblique-oblong, densely pilose; ovules 2–6; style filiform, hairy below; stigma small. Pod stalked, compressed, oblong, rusty-pubescent, transversely nerved, valves inrolled after dehiscence. (Harms 1897; Léonard 1951, 1957)

Representatives of this genus are distributed throughout tropical East Africa. They occur along river banks or on rocky hills.

Plathymenia Benth.　　　　　　　Mimosoideae: Piptadenieae
26 / 3466

From Greek, *platys*, "broad," and *hymen*, "membrane," in reference to the pod.

Type: *P. foliolosa* Benth.

3–4 species. Shrubs or trees, unarmed. Leaves large, bipinnate; leaflets in many pairs, small, elliptic; petiolar and jugal glands usually present. Flowers sessile or minutely pedicelled, small, bisexual or polygamous, crowded in axillary, cylindric spikes or in panicles at the tips of branches; calyx campanulate; lobes 5, very short, equal; petals 5, equal, lanceolate, all joined at first, later free; stamens 10, free, equal, short-exserted; anthers topped with a large gland; ovary hairy, stalked, many-ovuled; style thin; stigma trim, concave, terminal. Pod stalked, broad, flat, thin, linear, straight, with a 2-valved, continuous exocarp, endocarp adhering to the seed, breaking away into 1-seeded, transverse joints; seeds transverse with a long funicle. (Bentham 1842; Bentham and Hooker 1865)

This genus is "intermediate as to fruit between *Entada* and *Piptadenia*, with the flowers of both genera, and the habit nearly of *Piptadenia*" (Bentham 1842).

Members of this genus are native to South America, especially from the lower Amazon Basin of Brazil to São Paulo. They are uncommon and usually occur on slopes. Some species do grow along river banks.

P. reticulata Benth. is a small tree in the northern limits of its Amazon range, but farther south it reaches a height of 40 m, with a cylindrical trunk 1 m in diameter and free of branches for 20–25 m (Record and Hess 1943). The heartwood is lustrous, pale to deep yellow, becoming a rich brown upon exposure. The sapwood is yellowish white. The grain of the wood is moderately fine, the planed surfaces are silky, the texture is fine and uniform. The wood is fairly resistant to decay, easy to work, and has a density of 560–640 kg/m³. It is so highly esteemed and so much in demand in Brazil that it features little in the export trade. The wood is known in the very limited trade as vinhático; it is not to be confused with another yellow wood, vinhático espinho, *Chloroleucos vinhatico* (Titmuss 1965). Major uses are for fine furniture, interior trim, parquet floors, general construction, and dug-out canoes.

Platycelyphium Harms   Papilionoideae: Sophoreae
(137a) / 3597a

From Greek, *platys*, "broad," and *kelyphos*, "pod."
    Type: *P. cyananthum* Harms
    1 species. Trees, young branches silky-haired, later glabrous. Leaves imparipinnate; leaflets opposite or subopposite, short-petioled, 2–3 pairs, broad-ovate, lower surface at first densely pubescent. Flowers few, blue, solitary, on long, slender pedicels along rachis of lax, lateral racemes from older branches; bracts and bracteoles none; calyx campanulate, cleft almost to the middle; lobes 5, equal, ovate-deltoid, hairy; petals short-clawed, glabrous, inserted at base of the calyx; standard suborbicular or broad-elliptic, emarginate; wings and keel petals obliquely oval-oblong, subequal, obtuse, free; stamens 10, free, equal, inserted with the petals at base of the calyx; anthers oblong, dorsifixed; ovary short-stalked, elongated, silky-haired, 1-ovuled; style narrow, pilose at the base, strongly curved downward near apex; stigma truncate. Pod short-stalked, flat, with calyx persistent, oblique-elliptic, narrowed at both ends, silky-pubescent, 1-seeded, indehiscent. (Harms 1905b, 1908a; Baker 1929)
    This monotypic genus is endemic in tropical East Africa.

Platycyamus Benth.   Papilionoideae: Phaseoleae, Phaseolinae
394

From Greek, *platys*, "broad," and *kyamos*, "bean."
    Type: *P. regnellii* Benth.
    2 species. Trees, usually large, handsome; trunks straight; crowns ample; foliage light green. Leaves imparipinnate; leaflets 3–11, large; stipels present; stipules caducous. Flowers large, attractive, violet purple to white, in densely pubescent, terminal panicles; flower solitary in each bract; bracts small, caducous; bracteoles minute, caducous; calyx lobes short, upper 2 connate into 1 emarginate lobe; standard suborbicular, narrowed at the base, not eared; wings subfalcate-oblong; keel petals subequal to the wings; all petals free; stamens diadelphous; anthers uniform; ovary sessile, few to many-ovuled; disk scalloped; style filiform, incurved, beardless; stigma small, ter-

minal. Pod large, broad, linear, compressed, usually tomentose, 2-valved, sutures prominent, upper suture narrow-winged; seeds 2–3, broadly kidney shaped, compressed. (Bentham and Hooker 1865)

The tribal position of *Platycyamus* is clearly unsatisfactory. Some botanists now exclude *Platycyamus* from Phaseoleae and refer it to Glycineae (Hutchinson 1964), Galegeae (Lackey 1977a), or Tephrosieae (R. M. Polhill unpublished data 1979).

Representatives of this genus are indigenous to tropical South America, Brazil, and Peru.

Both species, *P. regnellii* Benth., common in Minas Gerais, Brazil, and *P. ulei* Harms, the Peruvian species, are beautiful trees admired for their stature and papilionaceous flowers. The bark of the Brazilian species is said to be poisonous to animals. Various features of the wood render it unacceptable for fine furniture; its use is reserved for general construction and similar purposes.

Platylobium Sm.   Papilionoideae: Genisteae, Bossiaeinae
188 / 3648

From Greek, *platys*, "broad," and *lobion*, "pod," in reference to the shape of the fruit.

Type: *P. formosum* Sm.

3 species. Shrubs, up to about 1.5 m tall; branches slender, opposite, very leafy. Leaves opposite, rarely alternate, entire, cordate-ovate, revolute, acute, leathery, usually tipped with a minute spine; stipules small, in pairs, lanceolate, membranous, sometimes brownish. Flowers yellow, solitary, in opposite axils; bracts several, at base of pedicel, ovate, concave, hairy, rigid, dry, imbricate; bracteoles similar to but longer than bracts, subtending the calyx; calyx campanulate, obtuse, usually hairy, 2 upper lobes large, obovate, obtuse, free or shortly connate, lower 3 small, narrow, acute, spreading; all petals clawed; standard orbicular or kidney-shaped, deeply emarginate, spreading; wings oblong or obovate, shorter than the standard; keel petals often tipped with crimson, obtuse, adherent, shorter than the wings; stamens connate into a sheath split above; anthers uniform, orbicular, versatile; ovary sessile or stalked, usually villous or ciliate, many-ovuled; style awl-shaped, incurved, smooth; stigma terminal. Pod sessile or short-stalked, very flat, obtuse, somewhat scimitar-shaped, clothed with scattered hairs, continuous within, elastically 2-valved, with the valves rolling back but not separating along the upper suture; seeds usually about 8, black, compressed. (Smith 1794; Bentham 1864; Bentham and Hooker 1865)

This genus was assigned to the tribe Bossiaeae by Polhill (1976) in his restructuring of Genisteae *sensu lato*.

Representatives of this genus are indigenous to temperate eastern and southern Australia and Tasmania. They occur on coastal and tablelands, in forest and heath areas, and on light sandy soil.

The type species is a prostrate plant with large red and yellow pea flowers. It blooms throughout the year. The growth habit has erosion-control value.

NODULATION STUDIES

P. formosum Sm. and *P. obtusangulum* Hook. are nodulated. Rhizobial strains from the latter failed to nodulate selected species of *Medicago, Trifolium, Vicia, Pisum, Phaseolus,* and *Glycine* species (Ewart and Thomson 1912). Conversely, rhizobia from representative species of these genera and also from *Crotalaria, Desmodium, Lespedeza, Lupinus,* and *Vigna* species did not nodulate *P. obtusangulum* seedlings; however, nodulation was obtained with strains from *Albizia julibrissin, Apios tuberosa, Lotus corniculatus,* and *Stizolobium deeringianum* (Wilson 1939a). McKnight (1949a) proposed the inclusion of *Platylobium* in the cowpea cross-inoculation group on the basis of successful responses of cowpea to nodulation by rhizobia from *P. formosum.*

Platymiscium Vog.　　　　　　　　Papilionoideae: Dalbergieae,
358 / 3829　　　　　　　　　　　　Lonchocarpinae

From Greek, *platys,* "broad," and the Latin verb *misceo,* "mix."
　Type: *P. floribundum* Vog.
　30 species. Trees, small to medium, or shrubs, unarmed, deciduous. Leaves opposite or whorled, imparipinnate; leaflets 3–9, large, leathery; stipels none; stipules interpetiolar, lanceolate, leathery, caducous. Flowers many, yellow to bright orange, borne in clusters or semipendulous racemes in leaf axils or at the nodes of 1-year-old branches, usually before the new foliage appears; bracts and bracteoles small, caducous; calyx campanulate, rarely blunt, turbinate at the base; teeth 5, unequal, short; standard orbicular or ovate; wings free, oblique-oblong; keel oblong, straight or incurved, petals joined dorsally only near the apex; stamens 10, monadelphous, joined into a tube split above, rarely 1 free; anthers versatile; ovary long-stalked, 1-ovuled; style incurved, slender; stigma small, terminal. Pod stalked, papery or slightly leathery, oblong, flat, margins thin, smooth, 1-seeded, indehiscent; seeds large, flat, kidney-shaped. (Bentham and Hooker 1865; Amshoff 1939)

Members of this genus are common in tropical Mexico and Central and South America. They flourish in wet or moist forests and grassy savannas.

When the Spaniards landed on the northern edge of South America, the huge *Platymiscium* trees reminded them of oaks; therefore, they gave them the name "roble," or Spanish oak (Menninger 1962). These trees reach 30 m in height, with girths of 10 m. The hollow branches are often inhabited by fierce ants. *P. trinitatis* Benth. and *P. ulei* Harms in Brazil, *P. dimorphandrum* Donn. Smith in Guatemala, and *P. pinnatum* (Jacq.) Dug. (= *P. polystachyum* Benth.) in Panama and Venezuela are important lumber trees. All have rich, reddish brown heartwood with attractive purple, black, or red markings. The wood is hard, moderately resistant to decay and weathering, easy to work, and takes a high natural polish (Record and Hess 1943). It is suitable for durable construction, furniture, cabinets, veneer, and minor turnery objects, such as brush, tool, and knife handles, and billiard cues. *P. floribundum* Vog. and *P. pinnatum* have been introduced into various subtropical areas as garden and park ornamentals. The fragrant, orange yellow flowers of these small trees are violet-scented.

NODULATION STUDIES

Nodules have been observed on *P. stipulare* Benth. in Hawaii (personal observation 1938), on *P. trinitatis* Benth. in Belém, Brazil (Norris 1969), and on *P. pinnatum* in Venezuela (Barrios and Gonzalez 1971). Nodules on *P. stipulare* and *P. trinitatis* were round to elongated. The surfaces of old nodules were blackish brown because of a barklike outer cortical parenchyma.

Platypodium Vog.   Papilionoideae: Dalbergieae,
355 / 3826   Pterocarpinae

From Greek, *platys*, "broad," and *podion*, "foot"; the base of the pod is a broad wing.

Type: *P. elegans* Vog.

2–3 species. Trees, medium to large, up to 25 m tall, with trunk diameters in excess of 1 m., unarmed, evergreen. Leaves paripinnate or imparipinnate, large; leaflets 10–20, ovate, alternate, leathery; stipels absent; stipules linear, caducous. Flowers white or yellow, few, in lax, pendent, axillary racemes; bracts and bracteoles small, caducous; calyx campanulate, turbinate at the base; lobes 5, upper 2 broader, connate high up; standard large, orbicular, reflexed; wings oblique-oblong, shorter; keel petals similar, obtuse, dorsally connate; stamens diadelphous, vexillary and lowest stamens free, others connate into 2 4-anthered bundles; anthers versatile; ovary long-stalked, elongated, multiovuled; disk glandular; style filiform; stigma small, terminal. Pod stalked, pendent, samaroid, apex woody, base narrowed into an oblique-oblong, flat, membranous, venous wing, indehiscent; seed 1, seldom 2, oblong-reniform, borne near the top of the pod. (Bentham and Hooker 1865)

Platypodium species occur in tropical Central and South America, ranging from Panama to Brazil. They are gregarious in littoral forests of Panama and in interior plateaus of Brazil.

The wood is not exported, nor does it seem to have potential value.

Platysepalum Welw. ex Bak.   Papilionoideae: Galegeae,
261 / 3725   Tephrosiinae

From Greek, *platys*, "broad"; the 2 upper sepals are joined and are as large as the standard petal.

Type: *P. violaceum* Welw. ex Bak.

12 species. Small trees or shrubs, rarely lianas; young branches and

foliage often brownish-pubescent. Leaves imparipinnate; leaflets 2–8 pairs, opposite, oblong-elliptic, acuminate; stipels awl-shaped; stipules small, early caducous. Flowers red violet, blue violet, yellow, or white, some purple-blotched, many, on long, pilose rachis in erect, axillary or terminal panicles, rarely racemose; bracteoles 2, large, oblong, often persistent; calyx campanulate; tube short, bilabiate; lobes 5, upper 2 joined into a broad emarginate lip completely hooding the standard, lower 3 short, linear-lanceolate; standard glabrous, broad, apex deeply bilobed, short-clawed, base 2-pouched or 2-scaled; wings and keel blunt, equal, the latter long-clawed, incurved; stamens 10, the upper stamen free, providing access to nectaries rimming the ovary base; anthers oblong, versatile; ovary linear-lanceolate, sessile, velvety-haired, 4–7-ovuled; style smooth, curved; stigma minute, capitate. Pod linear, flat, woody, 2-valved, often velvety-tomentose, tardily dehiscent; seeds discoid. (Hauman 1954a; Gillett 1960b)

The genus is closely allied to *Millettia*.

Representatives of this genus are found in tropical West Africa, specifically Guinea, eastern Nigeria, and the upper regions of Zaire; they are uncommon in tropical East Africa. They grow in forested areas and along river banks.

The hard, durable wood of *P. violaceum* Welw. ex Bak. var. *vanhouttei* (De Wild.) Hauman resembles ebony. The trunk exudes a red resin. All of the species are ornamentally attractive.

Podalyria Willd. Papilionoideae: Podalyrieae
161 / 3621

Named for Podalirius, son of the Greco-Roman god of medicine, Aesculapius.

Type: *P. retzii* (J. F. Gmel.) Rick. & Stafl.

25 species. Shrubs, usually silky or silvery-pubescent. Leaves simple, linear-elliptic or suborbicular, often subcordate at the base, margins revolute, usually densely tomentose above and below; stipules awl-shaped, often caducous. Flowers lilac, pink, or white, 1–2, rarely 3–4, on axillary peduncles; bracteoles very small; calyx wide, campanulate, villous; tubes broader than long, shallow, intruse at the base, sometimes split on 1 side; lobes 5, subequal, lanceolate, usually acuminate, upper 2 usually joined higher up than the lower 3; standard suborbicular or oblong-obovate or bilobed, slightly longer than the wings, with a short, recurved claw; wings elliptic, oblong, oblique; keel petals shorter than the wings, broad-obovate, blunt, lower margin convex, upper margin straight, claw flat, straight; stamens 10, free or connate only at the base; filaments usually flattened and smooth at the base; ovary sessile, densely villous; ovules many and in 2 rows; style filiform; stigma minute, terminal, capitate. Pod turgid, ovoid or oblong, valves leathery. (Bentham and Hooker 1865; Phillips 1951)

Members of this genus are indigenous to South Africa. They are common in southwestern Cape Province and Natal. Most species thrive in wooded areas.

P. calyptrata (Retz.) Willd. is very sweet scented, and in full bloom has the appearance of a lilac bush. This species is cultivated sparingly in California. In South Africa it is called kuertje, or cancer bush. The silky foliage of *P. sericea* (Andr.) R. Br., satin bush, has ornamental appeal. A number of species are called rattle bush, because the seeds rattle in the ripe pods.

Three species, *P. buxifolia* Willd., *P. calyptrata*, and *P. sericea*, introduced into New Zealand for horticultural purposes, were examined by White (1944) for alkaloids. Sparteine and cytisine were absent, but optical forms and mixtures of lupanine, $C_{15}H_{24}N_2O$, none of which had formerly been found in members of the tribe Podalyrieae, were found in the foliage and seeds of each species. *P. buxifolia* yielded up to 1.64 percent of the *dl*-form with a trace of the *l*-form, *P. calyptrata*, 0.4 to 1.19 percent of the pure *l*-form, and *P. sericea*, 0.26 percent of the *d*-form, and also a *dl*-lupanine mixture. These quinolizidine (lupine) alkaloids have potential as insecticides.

NODULATION STUDIES

Nodules have been recorded on the following 6 species in South Africa (Grobbelaar and Clarke 1972, 1975): *P. biflora* (L.) Willd., *P. calyptrata*, *P. cuneifolia* Vent., *P. glauca* (Thunb.) DC., *P. myrtillifolia* Willd., and *P. sericea*.

Podocytisus Boiss. & Heldr.   Papilionoideae: Genisteae, Spartiinae

From Greek, *pous, podos*, "foot," and *kytisos*, the name for a kind of clover; the pod is stalked.

Type: *P. caramanicus* Boiss. & Heldr. (= *Laburnum caramanicum*)

1 species. Shrub, unarmed, erect, up to 2 m tall; branches virgate, terete when young, glaucous. Leaves petioled, 3-foliolate, scattered on new shoots; leaflets obovate, 1-nerved; stipules absent. Flowers yellow, in short, erect, terminal racemes; bracteoles 1–2, at apex of pedicels; calyx short-campanulate, bilabiate, upper lip ovate, very obtuse, shortly split, lower lip longer, minutely 3-toothed; standard orbicular, short-clawed; wings oblong, free, shorter than the keel, short-clawed; keel incurved, 2-toothed, as long as the standard; stamens monadelphous; anthers linear, alternately short and versatile, long and basifixed; ovary stalked, 6–9-ovuled; style rolled inward at a right angle to the ovary; stigma capitate, terminal. Pod stalked, thin, flat, papery, ovate-oblong, indehiscent, upper suture broadly winged; seeds 3–6, suborbicular, on a long, slender funicle, without strophiole. (Boissier 1849; Tutin et al. 1968)

Polhill (1976) included this genus in Genisteae *sensu stricto*.

This monotypic genus is endemic in Greece and Asia Minor.

Podopetalum F. Muell.   Papilionoideae: Sophoreae
134 / 3593

From Greek, *pous, podos,* "foot," and *petalon,* "petal"; the petals are long-clawed.

Type: *P. ormondii* F. Muell.

1 species. Small tree or shrub. Leaves opposite or subopposite, imparipinnate; leaflets large, lanceolate; stipels absent; stipules caducous. Flowers rose, racemose-paniculate; bracts minute, deltoid, persistent; bracteoles rudimentary; calyx shortly 5-toothed; teeth subequal in length, slightly imbricate in bud, deltoid, upper 2 close together; petals free, upper petal kidney-shaped, swollen in the middle, gradually narrowed into a long claw, other 4 longer, spathulate or orbicular, symmetric, long-clawed; stamens 10, free; anthers oblong, dorsifixed; disk 10-grooved, adnate to and half the length of the calyx tube; style filiform, inrolled; stigma terminal, very minute; ovary long-stalked, small, narrow, 6–7-ovuled. Pod stalked; seeds 1–4, oblong, red, hilum small.

This monotypic genus is endemic in Queensland, Australia.

Poecilanthe Benth.   Papilionoideae: Galegeae, Tephrosiinae
262 / 3727

From Greek, *poikilos,* "many-colored," and *anthos,* "flower"; the flowers are variegated.

Type: *P. falcata* (Vell.) Ducke (= *P. grandiflora* Benth.)

7 species. Trees, small to medium, up to about 15 m tall, unarmed. Leaves imparipinnate, 5–7-, rarely 1-foliolate, persistent, glabrous; leaflets alternate, lanceolate, sometimes very large, leathery; stipels minute or absent; stipules minute, inconspicuous. Flowers yellow, white, blue, red, or variegated, scattered or clustered on rachis in short, lateral or axillary racemes or panicles; bracteoles small; calyx turbinate-campanulate; lobes large, lanceolate, upper 2 united into 1 emarginate lobe; standard orbicular, not appendaged; wings falcate-oblong to obovate; keel incurved; keel petals glabrous, connate at the back; stamens 10, monadelphous, connate into a sheath split above; anthers alternately long and basifixed, short and dorsifixed; ovary subsessile or short-stalked, several-ovuled; style filiform, incurved, glabrous; stigma small, terminal. Pod disk-shaped, hard, subcoriaceous, 2-valved, tardily dehiscent, 1-seeded; seeds hard, shiny, obovate, compressed, with a basal hilum. (Bentham 1860; Burkart 1943)

Bentham (1860) and Burkart (1943) positioned this genus in the tribe Dalbergieae.

Members of this genus are common in tropical South America, especially in Brazil and Surinam. They are usually found on periodically inundated land along rivers.

Most of these species are little known and none have commercial importance. Because of their dense floral clusters, *P. falcata* (Vell.) Ducke and *P. parviflora* Benth. are planted as ornamentals in plazas and parks in Argentina; both of these species are small trees. The wood of *P. effusa* (Huber) Ducke, a small forest tree of Amazonas and Pará, Brazil, has good qualities of hardness, texture, and color, but it is apparently in short supply and has no special uses (Record and Hess 1943).

Poeppigia C. Presl
111 / 3569

Caesalpinioideae: Sclerolobieae

Named for Edward F. Poeppig, professor at the University of Leipzig; he collected in the Amazon area.

Type: *C. procera* C. Presl

1 species. Tree, tall, unarmed, usually with a straight, low-buttressed trunk. Leaves imparipinnate; leaflets numerous, small. Flowers yellow, small, in terminal, paniculate cymes; bracts and bracteoles narrow, membranous, caducous; calyx campanulate; lobes 5, subequal, slightly imbricate, more or less connate above the disk; petals 5, imbricate, subequal, oblong, uppermost petal within the others; stamens 10, free; filaments almost straight, glabrous; anthers ovate or oblong, versatile, longitudinally dehiscent; ovary on stalk obliquely inserted into the calyx tube, many-ovuled; style short, conic or incurved; stigma small, terminal. Pod elongated, membranous, flat, narrow-winged along the upper suture, indehiscent; seeds transverse, ovate, compressed. (Bentham and Hooker 1865; Standley and Steyermark 1946)

This monotypic genus is endemic in tropical South America and the West Indies. The plants are common in mixed forests in upland regions.

The wood lacks possibilities for commercial exploitation because of its scarcity and small size; however, it is of good quality and attractively veined. Its principal use is for cart axles, house supports, and semiheavy construction. The thin bark is rich in tannin (Record and Hess 1943).

Pogocybe Pierre
(90a) / 3546a

Caesalpinioideae: Eucaesalpinieae

From Greek, *pogon*, "beard"; the inner calyx is bearded.

Type: *P. entadoides* Pierre

1 species. Tree, small, unarmed; foliage glabrous. Leaves bipinnate; pinnae opposite in 4 pairs; leaflets alternate, 12–14 per pinna, elliptic, leathery, margins serrate-crenate. Flowers dioecious, short-stalked, arranged in a panicle of spiciform racemes; only male flowers described; calyx tube turbinate, bearded inside, shorter than the 5 calyx lobes; petals 5, short, alternating with the calyx lobes, slightly imbricate, hairy inside; stamens 10, inserted in 2 rows along the rim of the calyx tube, 5 aligned opposite the lobes, 5 shorter, opposite the petals; filaments joined into a tube, reflexed above; anthers all

turned inward, dorsifixed, pollen spherical. Pod not described. (Harms 1908a)

This genus bears rather close resemblance to *Wagatea* and was positioned accordingly (Harms 1908a); however, some authors consider it a synonym of *Gleditsia* (Hutchinson 1964).

This monotypic genus is endemic in Vietnam.

***Poiretia* Vent.
320 / 3789

Papilionoideae: Hedysareae,
Aeschynomeninae

Named for Jean Louis Marie Poiret, 1755–1834, French botanist.

Type: *P. scandens* Vent.

7–11 species. Herbs or shrublets, perennial, scandent, rarely suberect, gland-dotted. Leaves pinnate, leaflets 4, rarely 3, small; stipels inconspicuous; stipules sessile. Flowers small, yellow or rose, in axillary racemes or terminal panicles; bracts lanceolate; bracteoles small; calyx glandular-punctate; lobes 5, short-dentate, 2 upper lobes wider than others, all shorter than calyx tube; standard densely glandular, broad-orbicular, reflexed; wings falcate-oblong; keel strongly incurved, narrow-beaked, rarely blunt; stamens 10, monadelphous; anthers subreniform, alternate ones sometimes longer; ovary sessile, many-ovuled; style incurved; stigma terminal. Pod segmented, linear, pendulous, glandular-punctate, compressed, joints oblong or quadrate, membranous or subcoriaceous, reticulate or verrucose, indehiscent; seeds several, kidney-shaped. (Bentham and Hooker 1865; Burkart 1943; Standley and Steyermark 1946)

Poiretia species occur in the warm regions of North and South America, principally Brazil. They thrive in rocky thickets and open fields.

Uses of the plants are minor. *P. tetraphylla* (Poir.) Burk. is said to be toxic to cattle and probably to be a source of rotenone.

**Poissonia* Baill.
266 / 3731

Papilionoideae: Galegeae, Tephrosiinae

Named for Jules Poisson, 1833–1919, French botanist.

Type: *P. orbicularis* (Benth.) Hauman (= *P. solanacea* Baill.)

3 species. Undershrubs, densely grayish-pubescent. Leaves petioled, 1-foliolate; leaflets obovate or heart-shaped; large, pinnately nerved, jointed at the base; stipels none; stipules 2, linear-subulate. Flowers few, solitary, resupinate on a long, erect, axillary peduncle; calyx obconical-turbinate to campanulate; lobes 5, deeply cleft, long-subulate, subequal, upper 2 connate beyond the middle, imbricate; petals 5, short-clawed, smooth, violet; standard

suborbicular; wings oblique-obovate; keel incurved, blunt; stamens diadelphous; anthers uniform; ovary linear, short-stalked, multiovuled; style incurved, densely clothed in hairs below the capitate stigma. Pod short-stalked, with persistent calyx at the base, compressed, glabrous, depressed with oblique lines but nonseptate inside between the transverse, obovate, compressed seeds. (Hauman 1925)

Representatives of this species are distributed throughout the tropical Andes, from Peru to northern Argentina.

In Hauman's opinion (1925), this genus is more appropriately positioned in the subtribe Robiniinae.

Poitaea Vent.   Papilionoideae: Galegeae, Robiniinae
271 / 3736

Named for M. Poiteau, French botanist, author of *Flore Parisienne* (1808–1813).

Type: *P. galegioides* Vent.

5–6 species. Shrubs. Leaves imparipinnate, in 2 vertical rows; leaflets elliptic or oval, entire, membranous, usually numerous, 7–41; stipels absent; stipules bristly, awl-shaped with a broad base. Flowers rose purple to white, pendulous, on solitary pedicels in axillary racemes; bracts small; bracteoles absent; calyx cylindric-turbinate, acute at the base, somewhat jointed; lobes 5, very short, upper 2 adjacent; standard obovate, retuse, narrowed at the base, not reflexed; wings narrow, linear or linear-oblanceolate, longer or shorter than the standard; keel petals narrow, straight or slightly falcate, obtuse, longer than the standard, slightly united but tips free; stamens 10, diadelphous, 1 free, others united into a long sheath dilated above, often exserted; anthers uniform; ovary stalked, many-ovuled; style awl-shaped, straight or slightly incurved, glabrous; stigma minute, terminal. Pod flat, stalked, linear, 2-valved, with flattened, orbicular-obovate seeds. (Bentham and Hooker 1865; Rydberg 1924)

Members of this genus are native to the West Indies, Cuba, and the Dominican Republic.

Polystemonanthus Harms   Caesalpinioideae: Amherstieae
(70a) / 3525

From Greek, *polys*, "many," *stemon*, "stamen," and *anthos*, "flower."

Type: *P. dinklagei* Harms

1 species. Tree, small, with umbrella-shaped crown; branches and inflorescences rusty-tomentose. Leaves 30–50 cm long, paripinnate; leaflets 5–7 pairs, short-petioled, oblong or lanceolate, abruptly acuminate, leathery,

upper surface smooth, lower surface prominently veined, densely hairy. Flowers white or yellow brown, fragrant, large, on stout pedicels in terminal panicles; bracteoles large, thick, petaloid, terminating in a large glandular apicula, valvate, completely enveloping the flower bud, persistent; calyx tube thick, campanulate to funnel-shaped; lobes 4, broad-ovate, imbricate, densely silky-haired outside; petals 5, imbricate, subequal in length, outer 2 long-clawed, lateral 2 short-clawed, the innermost petal very narrow; stamens very numerous, more than 25, exserted; anthers dorsifixed, linear, deeply arrow-shaped; ovary long-stalked, stalk attached to 1 side of the calyx tube wall, surface silky; ovules 10–12; style long, smooth; stigma minute. Pod brown, short-stalked, flattened, apex rounded, apiculate, narrower toward the base, valves woody, sutures distinctly winged along the edges. (Harms 1897; Léonard 1957)

This monotypic genus is endemic in tropical West Africa. It grows along sandy beaches.

Pongamia Vent.
(364) / 3836

Papilionoideae: Dalbergieae,
Lonchocarpinae

From the Tamil name, *pongam*.

Type: *P. pinnata* (L.) Pierre (= *P. glabra* Vent.)

1 species. Tree up to about 25 m tall, fast-growing, glabrous, deciduous; branches drooping; trunk diameter up to about 60 cm; bark smooth, gray. Leaves imparipinnate, shiny; young leaves pinkish red, mature leaves glossy, deep green; leaflets 5–9, the terminal leaflet larger than the others; stipels none; stipules caducous. Flowers fragrant, white to lavender pink, paired along rachis in axillary, pendent, long racemes or panicles; bracts very caducous; bracteoles 2, small, subtending the calyx, often caducous; calyx campanulate or cup-shaped, truncate, short-dentate, lowermost lobe sometimes longer; standard suborbicular, broad, usually with 2 inflexed, basal ears, thinly silky-haired outside; wings oblique, long, somewhat adherent to the obtuse keel; keel petals coherent at apex; stamens monadelphous, vexillary stamen free at the base but joined with others into a closed tube; anthers dorsifixed, connective hairy; ovary subsessile to short-stalked, pubescent; ovules 2, rarely 3; style filiform, upper half incurved, glabrous; stigma small, terminal. Pod short-stalked, oblique-oblong, flat, smooth, thickly leathery to subwoody, not winged or thickened along the margins, indehiscent, 1-seeded; seed thick, kidney-shaped, hilum small. (Bentham and Hooker 1865; Backer and Bakhuizen van den Brink 1963)

This monotypic genus is endemic in tropical Asia. It is now widely dispersed throughout the tropics, Madagascar, Africa, and the Philippines. The plants occur mainly in coastal forests and along stream banks.

The pongam tree is cultivated for two purposes: (1) as an ornamental in gardens and along avenues and roadsides, for its fragrant *Wisteria*-like flowers, and (2) as a host plant for lac insects. It is appreciated as an ornamental throughout coastal India and all of Polynesia.

All parts of the tree serve a purpose. In the Philippines the bark is used for making strings and ropes. The bark also yields a black gum that is used to treat wounds caused by poisonous fish. In wet areas of the tropics the leaves serve as green manure and as fodder. Well-decomposed flowers are used by gardeners as compost for plants requiring rich nutrients. The black malodorous roots contain a potent fish-stupefying principle. In primitive areas of Malaysia and India root extracts are applied to abscesses; other plant parts, especially crushed seeds and leaves, are regarded as having antiseptic properties. The seeds contain pongam oil, a bitter, red brown, thick oil, 27–36 percent by weight, which is used as a liniment to treat scabies, herpes, and rheumatism and as an illuminating oil (Burkill 1966). The oil has a high content of triglycerides, and its disagreeable taste and odor are due to bitter flavonoid constituents, pongamiin and karanjin. The wood is yellowish white, coarse, hard, and beautifully grained, but is not durable. Use of the wood is limited to cabinetmaking, cart wheels, posts, and fuel.

NODULATION STUDIES

P. pinnata (L.) Pierre has been amply documented as a nodulated species, initially in Sri Lanka by Wright (1903), later in Hawaii by Allen and Allen (1936a, 1939) and in Australia by Norris (1959b). This species was tentatively placed in the cowpea miscellany (Allen and Allen 1936a) and later confirmed (Allen and Allen 1939). Two rhizobial strains from this species produced nodules on 18 species of 12 different genera in this category. Ineffective responses were obtained with both strains only on *Clitoria ternatea* and *Stizolobium utile* and by 1 strain on *Lespedeza stipulacea* and *Samanea saman*. The strains were culturally and physiologically typical of slow-growing rhizobia, a conclusion confirmed in principle by Norris' inclusion of strain CB 564 as a cowpea type in his survey of the role of calcium and magnesium in the nutrition of rhizobia.

Priestleya DC.   Papilionoideae: Genisteae, Lipariinae
183 / 3643

Named for Joseph Priestley, 1733–1894, English chemist; he discovered that green plants, in sunlight, emit oxygen (De Candolle 1825b).

Lectotype: *P. myrtifolia* DC.

15 species. Shrubs, usually silky-haired, much-branched, less than 1 m tall. Leaves simple, entire, linear-lanceolate, usually turning black when dry; stipules absent. Flowers yellow, in terminal heads or racemes, seldom axillary; bracts large, ovate-elliptic or inner ones bristly; bracteoles setaceous, caducous; calyx villous, campanulate, sometimes intruse at the base; lobes 5, ovate or oblong-ovate, obtuse or acute, longer than the tube, subequal, lowermost

lobe longest; standard suborbicular or obovate, short-clawed, usually reflexed; wings subfalcate, eared, short-clawed; keel short, incurved, beaked or blunt; stamens diadelphous; filaments linear-oblong; anthers subuniform; ovary sessile, 2- to several-ovuled; style awl-shaped or cylindric, as long as the keel, often pilose, sometimes apically spurred; stigma small, terminal. Pod linear-oblong, some beaked, villous, compressed or slightly swollen, 2-valved, nonseptate; seeds few. (De Candolle 1825b,c; Bentham and Hooker 1865; Harvey 1894; Phillips 1951)

Polhill (1976) assigned this genus to the tribe Liparieae in his revision of Genisteae *sensu lato*.

Members of this genus are native to South Africa; 5 species are restricted to the Cape Peninsula. They flourish on mountain slopes and gravelly hill sites.

NODULATION STUDIES

Nodules have been observed on the following species growing in their native South African habitats: *P. calycina* L. Bolus, *P. elliptica* DC., *P. sericea* E. Mey., *P. tomentosa* (L.) Druce, and *P. vestita* DC. (Grobbelaar and Clarke 1972).

Prioria Griseb.   Caesalpinioideae: Cynometreae
41 / 3489

Named for Richard Chandler Alexander Prior, 1809–1902, British botanist.

Type: *P. copaifera* Griseb.

1 species. Tree, tall, unarmed. Leaves paripinnate; leaflets usually 2 pairs, long, elliptic, gland-dotted, leathery, glabrous, dull olive green; stipules scalelike, caducous. Flowers small, cream or yellowish tan, fragrant, sessile, in interrupted spikes paniculate at apex of branches; bracts inconspicuous; bracteoles 2-lobed, coherently cup-shaped; calyx tube short; lobes 5, orbicular, subpetaloid, short, imbricate; petals absent; stamens 10, free, subequal; anthers thick, uniform, apiculate, connective thick; ovary sessile or short-stalked, free, 2-ovuled; style short, awl-shaped; stigma minute, terminal. Pod obliquely obovate-orbicular, 7–10 cm long, flat, leathery to woody, 2-valved; seed 1, flat, large, pendulous at dehiscence. (Bentham and Hooker 1865; Fawcett and Rendle 1920)

This monotypic genus is endemic in west central Mexico, Central America, Colombia, and Jamaica. The plants grow on rich loamy soils along coastal plains and in mucky river areas that are inundated periodically; they are usually found less than 33 m above sea level.

These trees reach heights of 40 m, with straight symmetrical boles clear of branches 20 m from the base; trunk diameters measuring from 0.5 to 1 m are usual. The small core of heartwood is black and surrounded by a broad zone of pinkish to red sapwood. Improved methods of wood processing have resulted in an increased demand for this wood, known as cativo in the timber trade. The texture of cativo resembles that of mahogany (*Swietenia macrophylla*), but it is not as durable. Its uses are reserved for furniture, cabinet parts, crates, and as a veneer. The pulp blends satisfactorily with other pulps for paper making. The resinous gum exudate from the freshly cut wood is used medicinally in some areas. In 1958 an estimated 4.3 million meters of cativo logs and 2.3 million meters of sawn cativo lumber were imported from Costa Rica and Colombia by the United States (Cooper 1928; Barbour 1952; Kukachka and Kryn 1958). The large seeds are edible, and near the turn of the last century they were being sold in the markets of Panama (Taubert 1894).

Priotropis W. & A.   Papilionoideae: Genisteae, Crotalariinae
210 / 3670

From Latin, *prior*, "superior," and Greek, *tropis*, "keel," in reference to the prominently beaked keel.

Type: *P. cytisoides* W. & A.

2 species. Shrubs or undershrubs, with slender, glabrous branches. Leaves long-petioled, palmately 3-foliolate; leaflets smooth, membranous, oblong, both ends narrowed; stipules none. Flowers yellow, 12–20, in short-stalked, densely flowered, leaf-opposed and terminal racemes; calyx campanulate, finely silky; lobes lanceolate, subequal, free; standard suborbicular, with 2 callosities above the short claw; wings obovate-oblong, shorter than the standard; keel broad, with a long, distinct, ascending beak; stamens monadelphous, all connate into a sheath split above; anthers alternately small and versatile, long and basifixed; ovary stalked, many-ovuled; style long, strongly incurved, inner side barbate; stigma terminal. Pod stalked, oblong, flattened, not septate, 2-valved, 5–6-seeded. (Bentham and Hooker 1865)

The vegetative and floral characters of members of this genus agree entirely with those of *Crotalaria* species (Polhill 1968b).

Both species occur in Nepal, and tropical eastern Himalayan areas, up to 2,000-m altitude.

Prosopidastrum Burk.   Mimosoideae: Adenanthereae

The Latin suffix, *astrum*, indicates inferiority or incomplete resemblance to *Prosopis*.

Type: *P. mexicanum* (Dressl.) Burk. (= *Prosopis globosa* var. *mexicana*)

2 species. Shrubs, xerophytic, glabrous or subpubescent, spreading, de-

ciduous; young branches green and angular, prominently grooved, subspinescently tipped. Leaves small, bipinnate; pinnae 1 pair; leaflets oblong, small, 2–4 pairs; stipules short, dry, spinescent, recurved, persistent, awl-shaped. Flowers 5-parted, small, greenish or yellow, short-pedicelled in solitary, axillary, small, globose heads; calyx campanulate; lobes short; petals oblong-lanceolate, partly united; stamens 10, free, short-exserted; anthers elliptic, introrse, tipped with a globose gland; ovary oval, pubescent; ovules 12–20; style filiform; stigma small, concave. Pod lomentaceous, linear, flattened, dry, pericarp thin, subcoriaceous, not differentiated into layers, without fleshy mesocarp or firm endocarp as in *Prosopis*, dividing into 1-seeded square segments, the 2 valves weakly segmented transversely, dehiscent; seeds oval, hard, with endosperm thinly mucilaginous. (Burkart 1964b)

The genus was established by the segregation of species from section *Lomentaria, Prosopis*.

The species are geographically disjunctive; *P. mexicanum* (Dressl.) Burk. occurs near the coast and in arroyas of San Fernando in Baja California, Mexico; *P. globosa* (Gillies ex H. & A.) Burk. occurs in southern and western Argentina in semidesert regions.

Prosopis L.	Mimosoideae: Adenanthereae
19 / 3454	

From an ancient Greek name for burdock, an entirely different prickly plant.

Type: *P. cineraria* (L.) Druce (= *P. spicigera* L.)

45 species. Trees or shrubs, unarmed or armed with prickles or straight, stout, persistent, stipular spines. Leaves bipinnate, deciduous; pinnae 2–4 pairs, leaflets few- or many-paired; petioles tipped with a minute gland; stipules small, linear, membranous or spinescent, caducous, or absent. Flowers greenish white, small, subsessile, in axillary, peduncled, cylindric spikes, rarely in globose heads; calyx campanulate; lobes 5, short, equal; petals 5, at first connate below the middle, ultimately free, glabrous or tomentose on inner surface near apex, sometimes puberulent on outer surface; stamens 10, free, short-exserted, inserted with petals on the margin of a minute disk adnate to the calyx tube, those opposite calyx lobes longer; filaments filiform; anthers oblong, versatile, dehiscent by marginal pores, connectives tipped with a minute, deciduous gland; ovary sessile or stalked, villose, multiovuled; style filiform; stigma terminal, minute. Pod linear, thick, compressed or subterete, straight or falcate, exocarp thin, leathery to woody, pale yellow, mesocarp thick, spongy, consisting of a sweet pulp surrounding each seed, endocarp papery and continuous with septa between the seeds or more or less surrounding each seed, indehiscent; seeds obliquely and separately enclosed, oblong, compressed, light brown. (Bentham and Hooker 1865; Standley 1920–1926; Burkart 1940; Standley and Steyermark 1943; Gilbert and Boutique 1952a; Brenan 1959)

Species delineation is confused by synonymy and variable forms. Some botanists consider *P. juliflora* (Sw.) DC. the North American form of the complex (Standley 1920–1926) and regard its 2 varieties, *velutina* Woot. and *glandulosa* Torr., as distinct species. *P. chilensis* (Mol.) Stuntz, the variable species of South America, extends southward to Chile and Argentina (Burkart 1940). All of these are commonly called algarroba and mesquite.

Prosopis species are widespread in subtropical North and South America, Asia, and Africa, but uncommon in Europe and Australia. Species occur on all soil types, on dry ranges, grassland, and overgrazed areas and river valleys. Root systems tend to be extensive. The plants are dominant or codominant in many arid ecosystems receiving less than 8 cm rainfall a year.

All species are regarded favorably in their communities. Burkart (1952) cites the comment of Casteñada Vega that "los algarrobos son los reyes de los montes en la República Argentina." This remark is often amplified to consideration of algarroba as one of the king crops of the world. *Prosopis* species are used as windbreaks, living fences, honey plants, soil-binders, and for cattle shade and forage. Rock (1920) appraised *P. juliflora* as "the most valuable of all the introduced trees in the Hawaiian Islands."

P. stephaniana (Willd.) Spreng., a dominant, wild species in Iraq, occurs in desert areas, along river banks, and in the mountainous foothills. Prebiblical records of the Sumerians designated this species "eri-tilla," meaning "plant of the city of life" (Khudairi 1957). Mesopotamian soil, under cultivation for 5,000 years, owes its fertility to the planting of this species during fallow years. The roots commonly penetrate to soil depths of 2 m; root branches often exceed 5 m.

The wood is highly esteemed. In some desert areas it is among the most valuable. The sapwood is a light to dull yellow; the heartwood varies from a yellowish to dark reddish, purplish brown. The wood is compact, close grained, heavy, takes a good polish, and is resistant to borers, termites, and general decay; it has a density of 800–1090 kg/m^3 (Record and Hess 1943). The wood of *P. africana* (Guill. & Perr.) Taub., known as ironwood in Gambia, is hard enough to blunt an axe; that of *P. chilensis* is credited with accoustical properties. *Prosopis* wood is used for implement handles, turnery articles, walking sticks, street-paving blocks, fence posts, parquet floors, cabinets, house construction, and as heavy construction timber. It also yields good charcoal and is a good fuel. In earlier days, cattlemen in the western United States sought the wood for branding-iron fires because of its ability to hold heat.

The bark of *P. africana* contains 14–16 percent tannin (Dalziel 1948). The roots of *P. juliflora* yield 6–7 percent tannin. Mesquite bark is also a source of native coloring matter used to impart a reddish brown tint to leather and cloth.

The pods of *P. juliflora* have economic value. Prior to 1948 about 500,000 bags of pods from this species, known as kiawe, were collected annually in Hawaii for fodder (Neal 1948). In the southwestern United States the Pima Indians ground the pods into a meal, pinole, that was used to make a nutritious bread. The pods were also collected and sold to pioneer settlers for cattle food.

The pods are a rich source of grape sugar, which contributes to their palatability as cattle forage and in making alcoholic beverages. A refreshment called mesquitatole, native beers, and mescal and mesquite wines are popular concoctions in Mexico and Central America. Cakes made in Mexico from the meal are known as mesquitamales.

The fragrant, yellowish green flowers of all species are attractants for bees and a rich source of clear, delicious honey. *P. juliflora* is known in some areas as honey mesquite. In the 1940s nearly 181 metric tons of this honey were produced in Hawaii annually and shipped to the mainland United States (Neal 1948). The potential of this species as a commercial honey crop in the southwestern United States merits recognition.

Mesquite gum, a reddish amber exudate often conspicuous in 5-cm-long ovoid tears on trunks and branches, resembles gum arabic, dissolves readily in water, has emulsifying properties, and is an excellent mucilage. It is a potential source of L-arabinose and D-glucuronic acid (Mantell 1947). It can be used in making gumdrop candy (Howes 1949).

Macerated leaves of *P. africana* are ingested by natives in tropical Africa as an aphrodisiac and to insure male fertility (Uphof 1968).

The leaves of *P. ruscifolia* Griseb. yield vinaline, an alkaloid having antibacterial and antifungal properties (Cercós 1951). Intravenous injections of 0.25 mg/kg and 0.01–0.1 mg/15 were lethal for rabbits and rats, respectively; oral doses of 50 mg/kg of weight of rabbits were nonlethal. Pharmacological actions of prosopine and prosopinine, alkaloids isolated from *P. africana* by Ratle et al. (1966), are yet to be defined.

Nodulation Studies

The lack of nodules on 3 of the reported species is believed to be due to unfavorable seasonal conditions at the time of the examinations or to insufficient numbers of observations.

Root nodules on *Prosopis* species have apical meristems; hence their mature shape is cylindrical. A prominent 3–4 layer band of tannin-filled cells, conspicuous histologically in the nodule cortex, encircles the bacteroid zone (personal observation). This rigid layer contributes to durability and may account for the occurrence of perennial nodules on *P. stephaniana* in Iraq (Khudairi 1969).

Rhizobia from nodules of *Prosopis* species are slow growing on laboratory media and produce alkaline reactions in litmus milk and on carbohydrate substrates (personal observation). The species are promiscuously susceptible to versatile rhizobia (Wilson 1939a; Burton personal communication 1971). Reciprocal infection tests using species from mimosoid genera *Mimosa*, *Desmanthus*, *Leucaena*, and *Prosopis* evidenced a greater interchangeability and effectiveness *inter se* than with species in the 2 other subfamilies (personal observation).

549

Nodulated species of *Prosopis*	Area	Reported by
chilensis (Mol.) Stuntz	Sri Lanka	W. R. C. Paul (pers. comm. 1951)
dulcis Kunth	France	Lechtova-Trnka (1931)
juliflora (Sw.) DC.	Hawaii, U.S.A.	Pers. observ.
	N.Y., U.S.A.	Wilson (1939a)
var. *glandulosa* Cockerell	Wis., U.S.A.	J. C. Burton (pers. comm. 1971)
stephaniana (Willd.) Spreng.	Iraq	Khudairi (1957)
vidaliana [?]	U.S.D.A.	L. T. Leonard (pers. comm. 1938)

Nonnodulated species of *Prosopis*	Area	Reported by
nandubey Lorentz	France	Lechtova-Trnka (1931)
(*pubescens* Benth.)	Ariz., U.S.A.	Martin (1948)
= *Strombocarpa odorata* (Torr. & Frém.) Gray		
velutina Woot.	Ariz., U.S.A.	Martin (1948)

Pseudarthria W. & A.
338 / 3808

Papilionoideae: Hedysareae,
Desmodiinae

From Greek, *pseudo-*, "false," "resembling," and *arthron*, "joint"; the depressions between the seeds give the pods the appearance of being jointed.
Type: *P. viscida* (L.) W. & A.
12 species. Herbs or shrublets, perennial, small, up to about 1 m tall, diffuse or erect; foliage and slender stems silky-gray or sticky-haired. Leaves petioled, pinnately 3-foliolate; leaflets elliptic; stipels membranous or subcoriaceous; stipules free, lanceolate-subulate, striate, scarious. Flowers small, red, purple, pink, occasionally white, fascicled or paired on rachis of terminal or axillary panicled racemes; bracts narrow; bracteoles very caducous or absent; calyx campanulate, more or less bilabiate; lobes 5, linear, subequal, 2 upper lobes almost entirely joined higher up, lower lip trilobed; standard obovate or suborbicular; wings oblique-oblong, equaling the keel, free, eared, short-clawed; keel oblong, blunt, eared, clawed; stamens diadelphous; anthers uniform; ovary subsessile or stalked, several- to many-ovuled; style filiform, inflexed, awl-shaped; stigma small, capitate, terminal. Pod linear or oblong, flat, broad, hooked-hairy, the membranous sutures straight or shallowly incised but not jointed, 2-valved, valves reticulate, continuous within, dehiscent; seeds usually 6, compressed, subreniform. (Baker 1929; Phillips 1951; Léonard 1954a; van Meeuwen et al. 1961)

These drought-resistant species bear marked similarities to *Desmodium* species, from which they differ primarily in not having jointed pods.

Members of this genus are native to tropical Asia and Africa. They grow in dry thickets and grasslands, along roadsides and forest edges.

The powdered roots of *P. alba* A. Chev. form an ingredient of an aphrodisiac and an antidysenteric medicine in the Ivory Coast area (Kerharo and Bouquet 1950). Natives of Zaire apply hot boiled leaves of *P. confertiflora* Bak. as a pain remedy for rheumatism.

NODULATION STUDIES

Two species are reported nodulated. The early report by Wright (1903) for *P. viscida* (L.) W. & A. in Sri Lanka was confirmed by Keuchenius (1924) in Java. Seeds and rhizosphere soil from plants of *P. hookeri* W. & A., sent by H. D. L. Corby from Zimbabwe to the authors in 1956, produced well-nodulated seedlings in the greenhouse at the University of Wisconsin. A purplish black pigmentation, often with a tinge of red, was observed in the bacteroid area of nodules of *P. hookeri* in Zimbabwe (Sandmann 1970b).

Pseudeminia Verdc.   Papilionoideae: Phaseoleae, Phaseolinae

Name means "false *Eminia*."

Type: *P. comosa* (Bak.) Verdc.

4 species. Herbs, perennial, creeping, twining or erect, with tuberous roots. Leaves pinnately 3-foliolate, petioled; leaflets papery, pinnately veined; stipels filiform or linear-lanceolate, persistent; stipules lanceolate, veined, persistent. Flowers purplish, blue, mixed with green, pedicelled in short, axillary racemes; bracts conspicuous, linear-lanceolate, veined, persistent; bracteoles 2, filiform, bristly, persistent; calyx campanulate; lobes triangular lanceolate, subequal, densely bristly-hairy, 2 upper lobes connate at the base or to the middle; petals smooth, clawed; standard elliptic with basal ears; wings scarcely elliptic, spurred at the base, folded, slightly adherent above to keel; keel elliptic-oblong, obtuse, beaked, not incurved; stamens diadelphous, 1 free, geniculate; anthers uniform; ovary spindle-shaped, hairy, 2-ovuled; disk slightly tubular, scarcely undulate; style filiform, hairy at the base, upper part filiform, glabrous, rectangularly reflexed; stigma terminal, minute. Pod oblong, compressed, densely setose, barely septate, dehiscent; seeds irregular, oblong-reniform. (Verdcourt 1970c)

From studies of the pollen grains, Verdcourt (1970c) concluded that *Pseudeminia* was not closely related to any other African Phaseoleae. The kinship appears closer to the subtribe Glycininae than to Phaseolinae (Lackey 1977a).

Representatives of this genus are widespread in tropical East Africa. They thrive on sandy soils, along roadsides, in woodlands, and in mixed dry scrub.

These species are suitable ground cover because of their creeping habit. *P. comosa* (Bak.) Verdc. is cultivated with cassava and sisal. The tuberous root contains saponins used in native soaps.

NODULATION STUDIES

Corby (1974) observed nodules on *P. comosa* in Zimbabwe.

Pseudoberlinia Duvign. Caesalpinioideae: Amherstieae

Name means "false *Berlinia*."
  Type: *P. paniculata* (Benth.) Duvign.
  3–4 species. Trees or shrubs, up to 12–15 m tall, with trunk diameters up to 80 cm. Leaves paripinnate, 3–5-paired; leaflets petioluled, very asymmetric at the base; stipules generally caducous, sometimes large. Flowers small to medium, usually white, fragrant, in terminal or subterminal panicles; bracts medium or small, bracteoles forming an involucre beneath the flower, valvate, suborbicular or oval; sepals 5, free, oval or triangular; petals 5, middle petal clawed at the base, lateral petals narrower but of same length; stamens 10, diadelphous, 1 free, 9 united at the base, slightly exserted; ovary hairy, few-ovuled; stigma capitate. Pod flat, woody. (Hauman 1952)
  Distinctiveness of this genus from *Julbernardia* is a matter of opinion among botanists; recognition is given here to the opinion of Hauman (1952).
  Representatives of this genus are widespread in tropical Africa. They inhabit dry, forest areas.
  The wood is of little value. The fibrous bark of *P. baumii* (Harms) Duvign. is used in making small boats, nets, and baskets.

Pseudoentada Britt. & Rose Mimosoideae: Piptadenieae

Name means "false *Entada*"; the species was segregated from *Entada* because of its prickly character.
  Type: *P. patens* (H. & A.) Britt. & Rose
  1 species. Vine, high climbing, armed with numerous, short, reflexed prickles. Leaves bipinnate; pinnae 2–4 pairs; leaflets 3–6 pairs, orbicular to obovate, obtuse or rounded, glabrous above, sometimes hairy along midvein below; petiole and rachis prickly, glandular between pinnae and leaflet pairs. Flowers yellowish white, in slender, terminal, spiked panicles; calyx minute; teeth 5; petals 5, united near the base; stamens 10, exserted. Pod flat, valves thin, separating from the margin, transversely jointed into 1-seeded segments. (Rydberg 1923).
  This monotypic genus is endemic in Nicaragua.

Pseudoeriosema Hauman Papilionoideae: Phaseoleae, Glycininae

Name means "false *Eriosema*."
  Type: *P. andongense* (Welw.) Hauman
  8 species. Herbs, erect, up to about 45 cm tall, or creeping, or undershrubs, covered with long, white hairs. Leaves 1–3-foliolate; leaflets large, with a network of prominent veins on lower surfaces. Flowers white, pale

pink, mauve, or yellow, very small, in terminal or axillary, rather dense sub-capitate racemes; bracts larger than bracteoles, both stiff, linear, persistent; calyx teeth 5, upper 2 joined, acuminate, hairy; petals long-clawed, subequal, outer surfaces silky; stamens 10, diadelphous; ovary villous, 2-ovuled; style short, curved; stigma smooth, capitate. Pod oblong-oval, long-hairy, 2-valved, dehiscent; seeds 2, with conspicuous arilloid strophiole around the hilum. (Hauman 1954c; Verdcourt 1970a)

The majority of the species have *Glycine* synonyms.

Members of this genus are native to northeastern and central Africa; 5 species occur in Zaire. They flourish along coastal areas, in cleared forests, and on swampy grasslands.

The roots of *P. homblei* (De Wild.) Hauman var. *latistipulatum* Hauman are consumed as a vegetable.

Pseudoglycine Hermann   Papilionoideae: Phaseoleae, Glycininae

This genus, originally segregated from *Glycine* by Hermann (1962), is now considered congeneric with *Ophrestia* (q.v.).

Pseudolotus Rech. f.   Papilionoideae: Loteae

Name means "false *Lotus*."

Type: *P. makranicus* Rech. f. (= *Lotus makranicus*)

1 species. Herb, perennial, with slender rhizome; stems herbacous, densely leafy, base woody. Leaves palmately 3–5-foliolate, on short petioles; leaflets obovate, small, entire, villous; petioles persistent, later hardened; stipules very small, glanduliferous, purple. Flowers 1–2, together, axillary; bracts large, foliaceous, similar to upper leaves; calyx short-campanulate; lobes 5, equal, awl-shaped, slightly longer than the tube; petals short-clawed; standard orbicular; wings about as long as the keel petals; keel petals not beaked, shorter than standard; stamens all connate into a tube unequally split adaxially, apex not dilated; free parts of filaments not thickened; anthers equal, ellipsoid; ovary linear, about 3-ovuled; style glabrous. Pod sessile, linear, compressed, membranous, constricted between the 3 seeds. (Rechinger 1957)

This genus closely resembles *Lotus*, but differs in the persistence of petioles, the glandular stipules, and the stamens entirely connate.

This monotypic genus is endemic in Afghanistan.

Name means "false *Macrolobium*."

Type: *P. mengei* (De Wild.) Hauman

1 species. Tree, medium-sized, branches slender, glabrous. Leaves paripinnate; leaflets medium, opposite or subopposite, 2–3 pairs, papery, symmetric at the base, with translucent dots; stipules caducous. Flowers small to medium, polygamous, in terminal and axillary, paniculate inflorescences; bracteoles 2, forming an involucre in bud, opposite, valvate, free, somewhat thick; calyx very shortly tubular; lobes 5, imbricate, 3 free, subcordate, tapered to a point, the other 2 united almost to the apex; petals 5, 3 well-developed, equal in length, clawed, middle petal large, suborbicular, other 2 reduced to scales; stamens 10, rarely 13, free, all fertile, exserted; ovary stalked, stalk adnate to the side of the receptacle, hairy, few-ovuled; style exserted; stigma capitate. Pod flat, oblong, ends acuminate, transversely septate between the seeds, 2-valved, valves inrolled upon dehiscence. (Hauman 1952; Léonard 1957)

This monotypic genus is endemic in equatorial Africa. It occurs in dense primitive and secondary forests.

The red, durable wood is used for gongs; the crushed flowers have fish-paralyzing properties.

Pseudoprosopis Harms   Mimosoideae: Adenanthereae
(24a) / 3462a

Name means "false *Prosopis*"; the species bear a close resemblance to *Prosopis*, except for the dehiscent pods.

Type: *P. fischeri* (Taub.) Harms

4 species. Shrubs, lianas or small trees, unarmed; branches often scrambling, zigzag. Leaves bipinnate; pinnae 1–6 pairs; leaflets few to many pairs, 2–15, small, oblong, rounded or emarginate at the apex; petiole and rachis without glands. Flowers fragrant, white or cream, short-pedicelled, in dense spikes or catkinlike racemes, on short, thick, axillary or terminal peduncles; calyx shortly cupular, oblique; lobes 5, joined, short-toothed, unequal, uppermost lobe longest; petals 5, lanceolate, free or almost so, valvate in bud, inflexed at the apex; stamens 10, all fertile; filaments long, exserted, free or united at the base; anthers with large, caducous, apical gland; ovary oblong, short-stalked, usually densely villous, multiovuled; style filiform, smooth. Pod narrow, compressed, linear or curved, woody, erect, the 2 valves recurved after dehiscing downward from the apex; seeds 5–10, obliquely placed in the pod, each in a depression, flat, subquadrate-orbicular or rhomboid, brown, shiny, with a central areola. (Harms 1908a, 1913; Gilbert and Boutique 1952a; Brenan 1959)

Representatives of this genus are distributed throughout tropical Africa. They flourish in evergreen and deciduous thickets, usually in sandy areas.

Name means "false *Samanea*."

Type: *P. guachapele* (HBK) Harms

1 species. Tree, medium to large, unarmed; bark rough; branches, petioles, and inflorescences tomentose-hairy. Leaves bipinnate, large; pinnae in few pairs; leaflets obovate-oblong, 4–6 pairs. Flowers yellowish white, on long pedicels, umbellate, the central flowers nectariferous, sterile, very large; calyx tubular-cupular or narrowly campanulate; lobes small, short; petals united into a tube exserted beyond the calyx, apex 5–6-lobed; stamens 15–20; filaments united at the base into a short tube, long-exserted; ovary short-stalked, many-ovuled. Pod narrow, linear, flat, transversely furrowed, apex long-beaked, valves thin, ventral margin slightly thickened, usually undulate, tardily dehiscent, dorsal margin thicker, straight, indehiscent; seeds 12–15, elliptic, compressed, white or yellowish, with a slender funicle. (Harms 1930)

This monotypic genus is endemic in Central and South America, from Panama to Venezuela and Ecuador.

The trees provide excellent shade in parks and gardens. The wood has a golden luster, moderate weight and hardness, a medium-coarse texture, moderate resistance to decay, and finishes smoothly and attractively. Its primary use for furniture is limited by its scarcity (Record and Hess 1943).

Pseudovigna Verdc.   Papilionoideae: Phaseoleae,
                                       Phaseolinae

Name means "false *Vigna*."

Type: *P. argentea* (Willd.) Verdc. (= *Dolichos argenteum*)

1 species. Herb, perennial, prostrate or climbing; stems and pods covered with brown, bristly hairs. Leaves prominently nerved, pinnately 3-foliolate; leaflets papery, silvery-pilose underneath; stipels filiform, persistent; stipules ovate-lanceolate, nervose, persistent. Flowers white, pink, mauve, 2–8, pedicelled, in small heads on long, axillary peduncles; bracts ovate-lanceolate, subpersistent, nervose; bracteoles 2, linear-lanceolate; calyx campanulate; lobes unequal, long, pointed, upper 2 joined except near apex, covered with brown hairs; all petals clawed, with basal ears; standard obovate-oblong, eared at the base; wings oblong; keel petals broad, blunt, somewhat connate; stamens diadelphous; alternate filaments longer; anthers subuniform; ovary subsessile, oblong, hairy at the base; style glabrous, base slightly flattened, curved in the middle, apex cylindric; stigma terminal, apex flat, ringed with a fringe of hairs. Pod oblong, compressed, septate, densely covered with stiff, brown hairs, dehiscent; seeds 1–3, black, oblong, kidney-shaped. (Verdcourt 1970c)

Lackey (1977a) transferred *Pseudovigna* from the subtribe Phaseolinae to Glycininae. This change awaits general acceptance.

This monotypic genus is endemic in tropical Africa, specifically Kenya, Tanzania, and Mozambique. It grows along coastal banks and roadsides, and in grasslands.

NODULATION STUDIES

Nodules were observed on this species in Zimbabwe (H. D. L. Corby personal communication 1972).

Psophocarpus DC.   Papilionoideae: Phaseoleae, Phaseolinae
(431) / 3914

From Greek, *psophos*, "noise," and *karpos*, "fruit," in reference to the explosive noise of the ripe seed pods upon dehiscence.

Type: *P. tetragonolobus* (L.) DC.

8–10 species. Herbs, twining, tall perennial; roots numerous, thick, main root especially tuberous. Leaves pinnately 3-foliolate, rarely 1-foliolate, ovate-rhomboid, long-petioled; leaflets short-petioluled, ovate, somewhat pubescent beneath; stipels small, 2-parted, glabrous; stipules usually lanceolate, membranous. Flowers rather large, white, blue, or violet, sometimes splashed with green, in fascicled racemes near the tips of elongated, axillary peduncles on swollen nodes; bracts small, caducous; bracteoles 2, large, ovate, persistent, below the calyx; calyx tubular-campanulate, bilabiate; 2 upper lobes joined into 1 emarginate or bidentate lip, lower lip with 3 lobes, free, lowermost lobe longest; petals exserted, equal; standard suborbicular, longer than other petals, glabrous, with 2 basal ears; wings longer than the keel, spurred, oblique-obovate; keel blunt, incurved at the apex, beaked and obtuse; stamens 10, diadelphous, vexillary stamen free at the base but joined with the others at the middle; anthers uniform, alternately dorsifixed and subbasifixed; ovary subsessile to short-stalked, 3- to many-ovuled; disk cylindric, more or less lobed; style thickened above the ovary, incurved, bearded lengthwise; stigma densely hairy-tufted. Pod quadrangular, elongated, 4-winged lengthwise, subcoriaceous, 2-valved, septate within, filled between the seeds, dehiscent; seeds transversely ellipsoid to oblong, hilum lateral, oblong. (Wilczek 1954; Backer and Bakhuizen van den Brink 1963)

The original generic number 429 (Taubert 1894) was changed to 431 by Harms (1908a) because of the reorganization within the subtribe necessitated by various revisions and inclusions of new genera.

Members of this genus are native to tropical Africa; their origin is presumed to be Madagascar. They are well distributed in southeastern Asia, Indonesia, and Melanesia. The plants are both wild and cultivated and prefer loam soils. Some species are best adapted to hot wet climates, others have drought-resistant properties.

All species have merit as green manure, ground cover, fodder, and as soil-restorative plants; however, attention has focused largely on *P. lukafensis* (De Wild.) Wilczek, *P. palustris* Desv., and *P. tetragonolobus* (L.) DC. The last-named species, known throughout tropical Africa as Goa bean, asparagus pea, winged bean, and four-angled bean, is cultivated as a garden plant in southern Asia, where the young pods are eaten in the manner of French beans. The young sprouts, leaves, and flowers also serve as vegetable ingredients in soups. The ripe roasted seeds are a delicacy in various tropical areas. The lateral roots of plants older than 1 year become enlarged storage organs of starch and carbohydrate materials and are edible after prolonged cooking.

NODULATION STUDIES

Thompstone and Sawyer (1914) wrote of the extraordinary development of nodules on *P. tetragonolobus* in Burma, where it was used as a rotation green manure in the cultivation of sugar cane.

Masefield's studies (1952, 1957, 1961) on nodulation also emphasized the exceptional capacity of this species to form large and numerous nodules. A sampling of plants grown at the Kuala Lumpur Agricultural Station, Malaysia, on a soil with a 17.7 cm water table, showed a mean fresh weight of 21.65 g of nodules per plant, or an equivalent of 600 lbs of nodules per acre, or 673 kg/ha (Masefield 1957). The mean number of nodules per plant was 441. The largest nodule weighed 0.53 g and others exceeded 0.4 g. This impressive yield of nodules, which was higher than that observed by Masefield in previous surveys of tropical and temperate crops, was confirmed, in principle, in a later study conducted under tropical greenhouse conditions in Great Britain (Masefield 1961). Nodule weight from 5 109-day-old plants grown in earthenware jars at temperatures between 30° and 37.8°C was 23.12 g per plant. The largest single nodule weighed 0.6 g, with an extreme diameter of 1.2 cm. Nitrogen-fixation ability was evidenced by red pigmentation of the inner nodule tissue. The crude protein percentage of the shoot, based on dry matter, was 25.62 percent. Masefield's conclusion (1961) that "weight of nodules per plant is the highest ever recorded for any annual or perennial legume in the first few months of growth," appears justified.

P. palustris, a species introduced into Malaysia from Sumatra, was not observed nodulated in the field (Masefield 1957), but successful nodulation was obtained in the Oxford greenhouse tests by using as an inoculum a mixture of dried ground nodules from *P. tetragonolobus* to which was added some Zambian soil in which plants of *P. lukafuensis* had grown. After 107 days the plants averaged 35 nodules, with a fresh weight of 2.03 g of nodules per plant. This plant species weighed about one-third that of *P. tetragonolobus*. The largest nodule weighed 0.23 g and was 1 cm in diameter. Comparatively, the fresh weight of nodules per plant of *P. palustris* was about one-eleventh, the number of nodules about one-eighteenth, and the crude protein percentage about one-twenty-second of the values for *P. tetragonolobus*. To what extent these values would have been improved by use of a highly effective strain from *P. palustris* is unknown.

Cultural and physiological reactions of isolates from *P. tetragonolobus* conformed with the slow-growing rhizobial type (Palacios and Bari 1936a). The colonies were glistening, opaque, up to 3 mm in diameter, smooth, entire, and nonmucilaginous. Alkaline reactions were produced in litmus milk and on nitrate media containing various individual carbohydrates. Profuse growth was obtained on mannose and mannitol media.

Reciprocal cross-inoculation experiments conducted by Palacios and Bari

(1936a) between rhizobia and plants of *P. tetragonolobus, Cajanus indicus,* and *Dolichos biflorus* were positive. An effective response of *P. tetragonolobus* in field tests to a nonspecific cowpea-type inoculum was reported by DeSouza (1963).

Nodulated species of *Psophocarpus*	Area	Reported by
lancifolius Harms	Zimbabwe	Corby (1974)
lukafuensis (De Wild.) Wilczek	Zambia	Masefield (1961)
palustris Desv.	Java	Keuchenius (1924)
	Sri Lanka	W. R. C. Paul (pers. comm. 1951)
	Zimbabwe	Sandmann (1970a)
scandens (Endl.) Verdc.	S. Africa	Grobbelaar & Clarke (1974)
tetragonolobus (L.) DC.	Burma	Thompstone & Sawyer (1914)
	Java	Keuchenius (1924)
	India	Palacios & Bari (1936a)
	Nigeria; Malaysia	Masefield (1952, 1957)
	Philippines	Bañados & Fernandez (1954)
	Trinidad	DeSouza (1963)

Psoralea L.   Papilionoideae: Galegeae, Psoraliinae
240 / 3703

From Greek, *psoraleos,* "scurfy," "scabby," in reference to the resinous black or pellucid dots throughout the foliage.

Type: *P. pinnata* L.

120 species. Perennial herbs, shrubs, or shrublets, erect, often aromatic and covered throughout with numerous glands, some hairy, some with rhizomes or thick roots. Leaves imparipinnate, palmately 3- to many-foliolate, rarely 1-foliolate; leaflets entire or dentate, linear, lanceolate, or obovate, rarely heart-shaped at the base, gland-dotted; stipules free, lanceolate to linear, wholly or partly adnate to the petiole, usually persistent. Flowers blue, purple, rose, or white, rarely yellow, capitate or spicate on long, axillary peduncles, rarely fascicled or solitary; bracts 2–3, sometimes ovate, acuminate, membranous; bracteoles absent; calyx campanulate, sometimes gibbous or oblique, usually gland-dotted, glabrous, sometimes bilabiate; lobes 5, ovate, subequal or lowermost lobe longest, upper 2 usually connate; petals 5; standard orbicular or ovate, usually short-clawed, sides usually reflexed; wings oblong, subfalcate, eared, clawed, longer than the keel petals; keel petals incurved, obtuse, subcoherent in the middle, sometimes eared and clawed; stamens diadelphous, vexillary stamen usually free, sometimes partly united with the others; anthers equal, uniform, in some species alternate ones smaller; ovary subsessile, usually 1-ovuled, glabrous; style slender, generally swollen at the base, incurved above, persistent; stigma capitate, terminal. Pod

short, ovate, usually flattened or turgid, sometimes conspicuously wrinkled, usually concealed within the calyx, indehiscent, 1-seeded; seed without an appendage, funicle very short. (Bentham and Hooker 1865; Forbes 1930; Phillips 1951)

Rydberg (1919a) divided the North American species of *Psoralea* into 5 genera and created 3 additional genera for 3 introduced species. Few American botanists have accepted Rydberg's segregate genera; rather, most botanists continue to treat them as infrageneric taxa within the earlier concept of the genus. However, Hutchinson (1964) included them within Psoralieae trib. nov. In order to satisfy both schools of thought, the segregate genera are herein listed individually, and the pertinent revised nomenclature for the nodulated *Psoralea* species is hereinafter indicated.

Perhaps 100 species of this large genus are widely distributed throughout the warmer parts of the globe. They are most abundant in South Africa and North America. About 50 species occur in the United States, primarily in the southern and southwestern areas, 10 species in subtropical South America, and 15 species in Queensland and New South Wales, Australia. They inhabit sandy prairies, river banks, and coastal areas.

The scurfy, gland-dotted foliage characteristic of *Psoralea* species is denoted by the common names given various American species, such as hoary scurfpea, *P. canescens* Michx., slender scurfpea, *P. gracilis* Chapm., slim-flower scurfpea, *P. tenuiflora* Pursh.

P. esculenta Pursh, Indian breadroot, is perhaps the best-known species in the United States because of its popularity among the early French voyageurs, who gave it the names "pomme blanche" and "pomme de prairie." The edible tuberous root, often weighing as much as 450 g, was an important food of the American Indians. The tuberous roots of *P. hypogaea* Nutt., little scurfpea, or little breadfruit, a purple-flowered herb of the dry plains from Nebraska to Texas, was also used as food by the early settlers (Medsger 1939). The starchy tubers of *P. castorea* S. Wats. and *P. mephitica* S. Wats. were consumed raw, cooked, or ground into flour for mush or bread by the Indians of Utah and Arizona. The roots of the latter species furnished a yellow dye used by a California Indian tribe. The inner bark of *P. macristachya* DC. was a source of fiber to make a coarse thread. Aromatic teas are prepared in South Africa from leaves of *P. decumbens* Ait. and in Chile from *P. glandulosa* L.; the latter is said to relieve stomach distress and to have vermifuge properties. Root infusions of *P. pinnata* L. in South Africa are reputedly imbibed to treat hysteria. This bright blue-flowered species is an excellent border plant introduced overseas as an ornamental. *P. tenuiflora* is a good honey plant. *P. patens* Lindl., in Australia, is salt-tolerant. All species have soil-binding properties and are good wildlife food, but they have little browse value. Their cultivation has been recommended but is seldom practiced. The aromatic pods of *P. corylifolia* L. have potential in perfume manufacture.

NODULATION STUDIES

Bushnell and Sarles (1937) described *P. esculenta* rhizobia as monotrichous short rods that produced gummy opaque colonies on yeast-extract mannitol mineral salts and calcium glycerophosphate agars, and an acid reaction in litmus milk. Plantlets of this species were nodulated by rhizobia from *Caragana frutescens, Glycine max, Oxytropis lambertii, Phaseolus vulgaris, Stizolobium deeringianum,* and *Vicia villosa* in tests conducted by Wilson (1939a). Reciprocal cross-inoculations between *Vigna sinensis* and *P. corylifolia* in western India led Y. S Kulkarni (personal communication 1975) to conclude that *Psoralea* rhizobia are part of the cowpea rhizobial conglomerate.

Nodulated species of *Psoralea*	Area	Reported by
† (*acaulis* Stev.)	France	Laurent (1891)
= *Asphalthium acaulis* (Stev.) Ktze.	N.Y., U.S.A.	Wilson (1939a)
	S. Africa	Grobbelaar et al. (1967)
adscendens F. Muell.	Australia	N. C. W. Beadle (pers. comm. 1962)
† (*americana* L.)	S. Africa	Grobbelaar et al. (1967)
= *Cullen americanum* (L.) Rydb.		
aphylla L.	S. Africa	Grobbelaar & Clarke (1972)
† (*argophylla* Pursh)	U.S.A.	Bolley (1893); Alway & Pinckney (1909); Warren (1909); Pers. observ. (U.W. herb. spec. 1970)
= *Psoralidium argophyllum* (Pursh) Rydb.		
asarina (Berger) Salt.	S. Africa	Grobbelaar & Clarke (1972)
†(*bituminosa* L.)	France	Naudin (1897d)
= *Asphalthium bituminosum* (L.) Ktze.		
candicans E. & Z.	S. Africa	Grobbelaar et al. (1967)
capitata L. f.	S. Africa	Grobbelaar & Clarke (1972)
cordata Thunb.	S. Africa	Grobbelaar & Clarke (1972)
†(*corylifolia* L.)	Sri Lanka	W. R. C. Paul (pers. comm. 1951)
= *Cullen corylifolia* (L.) Medik.	India	Y. S. Kulkarni (pers. comm. 1972)
decumbens Ait.	S. Africa	Grobbelaar & Clarke (1972)
eriantha Benth.	Australia	Norris (1959b); Beadle (1964)
†(*esculenta* Pursh)	U.S.A.	Bolley (1983); Bushnell & Sarles (1937); Wilson (1939a); Pers. observ. (U.W. herb. spec. 1970)
= *Pediomelum esculentum* (Pursh) Rydb.		
†(*floribunda* Nutt.)	Kans., U.S.A.	Warren (1910)
= *Psoralidium floribundum* (Nutt.) Rydb.		

Nodulated species of *Psoralea* (cont.)	Area	Reported by
foliosa Oliv.	Zimbabwe	Corby (1974)
fruticans (L.) Bruce	S. Africa	Grobbelaar & Clarke (1972)
imbricata (L. f.) Thunb.	S. Africa	Grobbelaar & Clarke (1972)
laxa Salt.	S. Africa	Grobbelaar & Clarke (1972)
obtusifolia DC.	S. Africa	Mostert (1955)
	Zimbabwe	Corby (1974)
officinalis [?]	Kew Gardens, England	Pers. observ. (1936)
oligophylla E. & Z.	S. Africa	Grobbelaar & Clarke (1972)
†(*onobrychis* Nutt.) = *Orbexilum onobrychis* (Nutt.) Rydb.	U.S.A.	Wilson (1939a); Appleman & Sears (1943)
†(*orbicularis* Lindl.) = *Hoita orbicularis* (Lindl.) Rydb.	Calif., U.S.A.	Pers. observ. (U.W. herb. spec. 1970)
patens Lindl.	Australia	Harris (1953); Beadle (1964)
var. *cinera*	Australia	Beadle (pers. comm. 1962)
†(*pedunculata* (Mill.) Vail) = *Orbexilum pedunculatum* (Mill.) Rydb.	La., U.S.A.	Pers. observ. (U.W. herb. spec. 1970)
pinnata L.	S. Africa	Grobbelaar et al. (1967)
psoralioides (Walt.) Cory	Va., U.S.A.	Erdman (1948b); Pers. observ. (U.W. herb. spec. 1970)
var. *eglandulosa* (Ell.) Freeman	Ky., U.S.A.	Pers. observ. (U.W. herb. spec. 1970)
restioides E. & Z.	S. Africa	Grobbelaar & Clarke (1972)
rotundifolia L.	S. Africa	Grobbelaar & Clarke (1972)
†(*simplex* Nutt.) = *Orbexilum simplex* (Nutt.) Rydb.	Tex., U.S.A.	Pers. observ. (U.W. herb. spec. 1970)
tenax Lindl.	Australia	Bowen (1956); Beadle (1964)
†(*tenuiflora* Pursh) = *Psoralidium tenuiflorum* (Pursh) Rydb.	N. Dak., U.S.A.	Bolley (1893)
tomentosa Thunb.	S. Africa	Grobbelaar & Clarke (1972)
wilmsii Harms	S. Africa	Grobbelaar & Clarke (1972)

†Reported as *Psoralea* species; also recorded under the alternate generic name.

Name is the diminutive of *Psoralea*.

Type: *P. tenuiflorum* (Pursh) Rydb.

14 species. Herbs, perennial, with rootstocks. Leaves gland-dotted, palmately 3–5-foliolate. Flowers blue or purple, sometimes whitish with keel purple-tipped, in long-peduncled, interrupted spikes or racemes bearing 1–4 flowers at each node; calyx campanulate; tube short; lobes 5, lowest lobe longest; standard rounded-obovate to orbicular, usually with narrow, basal lobes and a short, straight, narrow claw; blades of the wings and keel-petals united below, obliquely oblanceolate, oblong, or obovate, rounded at the apex, each with a small, basal lobe, those of the wings twice as long as their claws and as long as the blades of the keel petals; stamens 10, diadelphous; style bent above; stigma capitate. Pod lunate, compressed, gland-dotted but not wrinkled, pericarp leathery, beak short, straight; seeds brown, kidney-shaped. (Rydberg 1919a)

Members of this genus are widespread in the midwestern United States, ranging from North Dakota to Texas. They grow on sandy plains and prairies.

The species often form dense clumps and are effective sand-binders for erosion control. *P. tenuiflorum* (Pursh) Rydb. is a good honey plant.

Nodulation Studies

Three species were reported nodulated as members of *Psoralea* prior to their segregation: *P. argophyllum* (Pursh) Rydb. (Bolley 1893; Warren 1909; personal observation University of Wisconsin herbarium specimen 1971), *P. floribundum* (Nutt.) Rydb. (Warren 1910), and *P. tenuiflorum* (Bolley 1893; Martin 1948).

From Greek, *psora*, "scab," and *batos*, "bramble," in reference to the glandular pustules and spiny stipules.

Type: *P. benthamii* (Brandegee) Rydb.

2 species. Shrubs, low, usually less than 1 m tall; branches white-tomentose, with orange to brown pustular glands. Leaves pinnate; leaflets orbicular; stipules awl-shaped, spinescent. Flowers ascending, spicate; bracts narrow, tipped with a gland; calyx deeply campanulate, usually white-tomentose and with scattered glands; lobes oblong; petals yellow, fleshy, distinct, inserted at base of the staminal tube; standard broader than other petals; stamens 9, monadelphous, alternately long and short; ovary 2-ovuled; stigma capitate, large. Pod ovoid or ellipsoid, exserted, turgid, tomentose and glandular, sutures not prominent, beak slender; seeds 1–2. (Rydberg 1919a)

Psorabatus is closely related to *Dalea*; both species have *Dalea* synonyms.

Both species are native to the islands off the west coast of Baja California, Mexico.

From Greek, *psora*, "scab," and *dendron*, "tree"; hence "scab tree," because of the gland-dotted foliage.

Type: *P. johnsoni* (S. Wats.) Rydb.

12 species. Shrubs, or small trees; stems smooth, gray or straw-colored with age; branches often spinescent, pubescent, gland-dotted when young; glands, particularly on the peduncles, often awl-shaped or conical. Leaves simple or pinnate, glandular-punctate. Flowers pedicelled, in loose or crowded racemes; rachis sometimes spinescent, subtended by 2 small, awl-shaped bractlets, and a caducous bract at the base of the short pedicels; calyx turbinate, glandular or hairy; petals subequal in length or the keel rarely longer, all inserted at the base of the staminal tube; standard usually deeply emarginate apically and 2-lobed basally; wings and keel petals alike, the latter partly joined along lower edge, oval, with a conspicuous rounded basal lobe; stamens monadelphous, 9 or 10; ovary 2–6-ovuled. Pod flattened, longer than the calyx, long-beaked, oblique, sutures prominent, conspicuously gland-dotted. (Rydberg 1919a)

The genus was established to accommodate certain species formerly in *Dalea* (Rydberg 1919a).

Representatives of this genus are indigenous to North America. They are common in the southwestern United States and Baja California and northwestern Mexico. They flourish in sandy and rocky gullies up to 1,500-m altitude.

P. spinosum (Gray) Rydb. (= *Dalea spinosa*), a tree species up to 7 m tall, has showy deep blue flowers and attractive wood. The brownish gray heartwood is conspicuously variegated with brown stripes. The wood is hard, easy to work, finishes well, and is considered desirable for decorative articles, but is available only in small quantities (Record and Hess 1943).

Nodulation Studies

A specimen of *P. spinosum* in the University of Wisconsin Herbarium, collected in California, bore numerous effective-type nodules (personal observation).

Psorothamnus Rydb.   Papilionoideae: Galegeae, Psoraliinae

From Greek, *psora*, "scab," and *thamnos*, "bush," in reference to the gland-dotted foliage.

Type: *P. emoryi* (Gray) Rydb.

8 species. Shrubs, intricately branched, glandular-pustulate; glands red or orange, turning brown or black. Leaves deciduous, simple or pinnate. Flowers sessile, blue or purple, in dense, short, globose spikes; bracts lanceolate, caducous; bracteoles none or represented by glands; calyx turbinate, usually pubescent, strongly 10-ribbed, with a row of glands in each furrow; petals subequal, inserted at the base of the staminal tube, often with a gland

near apex, except the keel petals; standard suborbicular or round-cordate, other petals with a rounded basal lobe, keel petals much broader; stamens 9 or 10, monadelphous, united high up into a tube. Pod oblique-obovoid, turgid, hairy above, smooth at the base, slightly longer than the calyx tube, but shorter than the lobes, sutures not prominent. (Rydberg 1919a)

Members of this genus are native to North America; they are common in Utah, the Mojave Desert, and Baja California and northwestern Mexico. They thrive on sandy mesas and arroyas below 300-m altitude.

Indian tribes of the southwestern United States wove baskets from the slender stems of *P. emoryi* (Gray) Rydb. (= *Dalea emoryi*). Tea brewed from leaves of *P. polyadenius* Rydb. (= *Dalea polyadenia*) was a cold and cough remedy of the Nevada Indians.

Nodulation Studies

According to Wilson (1939b), a rhizobial strain from *P. scoparius* (Gray) Rydb., reported as *Parosela scoparia*, nodulated *Crotalaria grantiana* but not *C. verrucosa*.

*Pterocarpus Jacq.   Papilionoideae: Dalbergieae,
357 / 3828   Pterocarpinae

From Greek, *pteron*, "wing," and *karpos*, "fruit," in reference to the winged pod.

Type: *P. officinalis* Jacq.

60–70 species. Trees, usually huge, deciduous or evergreen, handsome, usually large crowned, unarmed. Leaves imparipinnate, rarely unifoliate; leaflets mostly alternate, few to many; stipels absent; stipules usually small, sometimes large and foliaceous, very caducous. Flowers yellow or orange, rarely whitish mixed with violet, usually large, showy, fragrant, in simple, lax, axillary or terminal, panicled racemes; bracts and bracteoles 2, small, caducous; calyx campanulate-tubular, especially turbinate at the base, somewhat leathery; teeth 5, short, upper 2 more or less connate; petals glabrous; standard orbicular or broad-ovate, neither eared nor gibbous; wings obliquely obovate or oblong; keel petals free or briefly joined at the apex, equal to or shorter than the wings; stamens usually monadelphous, connate into a sheath split above or the sheath equally split into 2 bundles, or vexillary stamen free; anthers versatile; ovary sessile or stalked, few-ovuled; disk absent; style short, filiform, incurved; stigma small, terminal. Pod samaroid, often large, muchcompressed, stalked within the calyx, orbicular or ovate, rarely oval-oblong, oblique or falcate with a lateral or rarely terminal style, winged all around, septate within, indehiscent; seeds 1–3, oblong-subreniform, separated by hard septa. (Bentham and Hooker 1865; Amshoff 1939; Standley and Steyermark 1946; Hauman 1954b)

Species of this pantropical genus are widespread in the tropics of the Old World and the New World. They are well represented in the Malayan Archipelago and central Africa. Most species are hygrophilic; they thrive in rain forests and along inundated river banks, swamps, and lagoons; some species inhabit dry, upland soils.

Some of the most handsome large-crowned deciduous trees of the leguminous family are members of this genus. They grow rapidly, reach a height of 25 m or more, with straight trunks 60 cm or more in diameter, and are crowned with graceful, feathery, compound leaves.

The flowers are profuse, showy, and fragrant. *P. podocarpus* Blake, a medium-sized tree, is cultivated as an ornamental in Venezuela. Its numerous, small, copper yellow, butterfly-shaped flowers bloom while the tree is bare of leaves. *P. echinatus* Pers., *P. rotundifolius* (Sond.) Druce, *P. sericeus* Benth., and *P. stevensonii* Burtt Davy are magnificent shade trees that merit inclusion in tropical gardens. The first 3 species are propagated in Zimbabwe for their exceptional beauty; the last-named species is a favorite ornamental in the Philippines. The persistent, pendulant, winged, indehiscent pods add to the beauty of various species.

A blood red, kino-type gum is obtained from most species by tapping the trunks; thus the word "blood" features in many common names. The first supplies of this gum for European medicines were obtained from *P. erinaceous* Poir. as early as 1757 (Burkill 1966). At the turn of the last century kino gum was in great demand as an astringent in treating diarrhea and dysentery. It was also used to a limited extent in the preparation of some wines. The gum from *P. marsupium* Roxb., known as Malabar kino, became the official kino of the British Pharmacopoeia. It consists of 70–80 percent kino-tannic acid (Merck 1968). Santalin, a blood red ether-soluble and alcohol-soluble dye from the heartwood of *P. santaloides* L. f., was important in pharmacy and in the leather, wool, and graphic arts industries prior to the use of synthetic dyes.

Pterocarpus timber of the Asian and African species ranks very high among the world's most handsome high-quality woods. The wood of the American species lacks commercial value because the heartwood is poorly developed and the sapwood is soft and perishable. The timber of *P. indicus* Willd. is marketed as narra, amboyna, or Philippine mahogany throughout the Malay Peninsula. This beautiful, termite-resistant wood is known as Burmese rosewood and is prized for its roselike fragrance. Prickly narra, *P. echinatus*, and Blanco's narra, *P. blancoi* Merr., are almost indistinguishable from true narra. The wood of *P. santalinus* L. f. is marketed as red sandalwood, and that of *P. angolensis* DC., muninga or bloodwood, as a substitute for Indian teak. In equatorial Africa, *P. soyauxii* Taub. is commercially important in the timber trade as African padauk, a red wood exported as veneer (Kukachka 1970). *P. marsupium* is, next to teak and rosewood (*Dalbergia*), the most widely cultivated tree in India.

NODULATION STUDIES

Rhizobia isolated from nodules of *P. echinatus*, *P. indicus*, *P. marsupium*, and *P. vidalianus* Rolfe were effective on *P. indicus* (Allen and Allen 1936a). Five strains from each species nodulated plantlets of *Vigna sinensis* but not plants representative of other cross-inoculation groups. Reciprocally, rhizobial strains from *V. sinensis* nodulated *P. indicus* effectively.

Nodulated species of *Pterocarpus*	Area	Reported by
angolensis DC.	Zimbabwe	Corby (1974)
antunesii (Taub.) Harms	Zimbabwe	Corby (1974)
brenanii (L.) Barb. & Torre	Zimbabwe	Corby (1974)
echinatus Pers.	Malaysia	Wright (1903)
	Hawaii, U.S.A.	Allen & Allen (1936a)
	Singapore	Lim (1977)
indicus Willd.	Malaysia	Wright (1903)
	Hawaii, U.S.A.	Allen & Allen (1936a)
	Philippines	Bañados & Fernandez (1954)
marsupium Roxb.	Hawaii, U.S.A.	Allen & Allen (1936a)
officinalis Jacq.	Trinidad	DeSouza (1966)
podocarpus Blake	Venezuela	Barrios & Gonzalez (1971)
rotundifolius (Sond.) Druce	S. Africa	Grobbelaar et al. (1967)
ssp. *polyanthus* (Harms) Mend. & Sousa		
var. *martinii* (Dunkl.) Mend. & Sousa	Zimbabwe	Corby (1974)
var. *polyanthus*	Zimbabwe	Corby (1974)
ssp. *rotundifolius*	Zimbabwe	Corby (1974)
soyauxii Taub.	Zaire	Bonnier (1957)
vidalianus Rolfe	Hawaii, U.S.A.	Allen & Allen (1936a)

Pterodon Vog.
372 / 3844

Papilionoideae: Dalbergieae, Geoffraeinae

From Greek, *pteron*, "wing," and *odous, odontos*, "tooth."

Type: *P. emarginatus* Vog.

4 species. Trees, medium to tall, unarmed. Leaves paripinnate; leaflets opposite or alternate, glandular-punctate; stipels small, caducous. Flowers rose, pale lilac, or white, showy, in terminal panicles leafy at the base; bracts very caducous; bracteoles membranous, similar to the calyx lobes, early caducous; calyx tube very short, bilabiate, 2 upper lobes large, winglike, petaloid, lower lip very small, with 3 short teeth; standard broadly ovate or orbicular, emarginate; wings obovate or oblong, falcate, shortly 2-parted, free; keel petals similar to the wings but slightly smaller and entire, slightly coherent at the back; stamens all joined into a sheath split above; anthers versatile, uniform, dehiscent by longitudinal slits; ovary stalked or subsessile, 1-ovuled; style slender, incurved; stigma terminal, capitate. Pod drupaceous, elongated-ovate, flattened, oblique, indehiscent; seed 1, contained within a thin, oily sarcocarp that separates at maturity from the woody, winged exocarp containing a pungent, balsam oil. (Bentham and Hooker 1865)

Pterodon species occur in Brazil and Bolivia. They prefer dry regions.

The timber is well known locally as being suitable for cart wheels and other items requiring durability. The hard, heavy, coarsely textured wood of *P. pubescens* Benth., a sparsely foliaged tree, is the best known.

From Greek, *pteron*, "wing," and *gyne*, "woman," in reference to the winged ovary.

Type: *P. nitens* Tul.

1 species. Tree, 5–12 m tall, unarmed, evergreen. Leaves abruptly pinnate, large; leaflets generally alternate, small, elliptic or oval, subsymmetric, shiny, subcoriaceous; stipules minute to inconspicuous. Flowers yellow, small, pedicelled, in short, loose, catkinlike, axillary racemes; bracts small, scalelike; bracteoles absent; calyx tube short; lobes 5, subpetaloid, subequal, free, overlapping; petals 5, subequal, imbricate; stamens 10, free, subequal or alternately shorter; anthers uniform, versatile, short, longitudinally dehiscent; ovary short-stalked, compressed, downy, 1-ovuled, winged along the upper side; style short; stigma blunt, terminal. Pod samaroid, leathery, planocompressed, reddish brown, shiny, with 1 pendulous, flattened, oblong-obovate seed in a reticulate segment or chamber at the base of the thin, rigid, incurved, parallel-veined wing, indehiscent. (Tulasne 1843; Bentham and Hooker 1865)

This monotypic genus is endemic in South America, specifically northern Argentina, Bolivia, Paraguay, Brazil, and the lower slopes of the Andes.

In some areas the trees reach a height of 33 m or more, with trunk diameters in excess of 1 m. The wood, known as tipa, tipa colorada, or viraro, is hard, heavy, and strong with a density of 800–960 kg/m^3. The freshly cut heartwood is pinkish brown but darkens to a mahogany color upon exposure. It is an excellent wood for cabinetwork, furniture, interior finishing, railway crossties, and cooperage (Record and Hess 1943).

Nodulation Studies

Plants of this species grown in rhizosphere soil of abundantly nodulated *Acacia visco* in Argentina lacked nodules (Rothschild 1970).

From Greek, *pteron*, "wing," and *lobion*, "pod."

Type: *P. stellatum* (Forsk.) Brenan (= *P. exosum* (J. F. Gmel.) Bak. f.)

10–12 species. Lianas, shrubs, tall, high-climbing or small trees, armed with recurved spines. Leaves bipinnate; leaflets small, numerous; stipels and stipules inconspicuous. Flowers white or yellow, small, numerous, in lax, paniculate, terminal racemes; bracts caducous; bracteoles absent; calyx shallowly cup-shaped; tube short; lobes 5, imbricate, lowermost concave, longer than the tube; petals 5, obovate, spreading, slightly unequal, imbricate, uppermost petal innermost; stamens 10, free, declinate or subequal, longer than the petals; filaments villous at the base or glabrous; anthers uniform, longitudinally dehiscent; ovary sessile, free, villous, 1-ovuled; style short or elongated and apically clavate; stigma terminal, truncate or concave. Pod samaroid, subsessile, compressed, oblong, 1-seeded, oblique-ovate at the base, membranously winged at the apex, indehiscent. (Bentham and Hooker 1865; Baker 1930; Phillips 1951; Roti-Michelozzi 1957)

Representatives of this genus are indigenous to tropical and subtropical Africa and Asia. They grow in wooded savannas and sclerophyll forests up to 1,500-m altitude.

In some sections of East Africa, the macerated leaves of *P. stellatum* (Forsk.) Brenan are mixed with dissolved iron rust to impart a black color to leather (Uphof 1968).

Nodulation Studies

Roots of *P. stellatum* examined by Grobbelaar et al. (1967) in South Africa and by Corby (1974) in Zimbabwe lacked nodules.

Pteroloma DC.   Papilionoideae: Hedysareae, Desmodiinae

From Greek, *pteron*, "wing," and *loma*, "border," "margin," in reference to the winged petioles.

Type: *P. triquetrum* (DC.) Benth.

5 species. Herbs or shrublets. Leaves 1-foliolate; petiole winged by large, broad-lanceolate, striate stipels. Flowers small, on slender pedicels, fascicled 3–7 in lax racemes; calyx turbinate-campanulate, subtended by 2 bracteoles; calyx lobes 4, the upper lobe broad, minutely bidentate; standard orbicular, with a small ear on each basal margin and a triangular callosity above the claw; wings oblong, spurred or eared; keel bowed, upcurved, narrowed to a blunt beak; stamens diadelphous, vexillary stamen free at the base, upcurved and briefly joined with the others but again free above the middle; anthers uniform; ovary sessile; ovules several; style glabrous above; stigma capitate. Pod flat, straight, exserted, segments 5–8, 1-seeded, quadrate. (Schindler 1924; Backer and Bakhuizen van den Brink 1963)

The species in this genus originally constituted section *Pteroloma* of *Desmodium*. Schindler (1924) elevated this section to generic rank in his revision of *Desmodium*.

Pteroloma species are indigenous to tropical Southeast Asia, India, and Papua New Guinea. They inhabit grassy wilds, thickets, and secondary forests.

The type species is a satisfactory green-manuring crop in the Far East; in most agricultural accounts it is designated as *Desmodium triquetrum*. It produces good foliage in both sunlight and partial shade. Leaf extracts reputedly have medicinal properties; dried leaves contain 7–8 percent tannin.

Nodulation Studies

P. auriculatum (DC.) Schindl. and *P. triquetrum* (DC.) Benth., reported by Keuchenius (1924) as *Desmodium auriculatum* and *D. triquetrum*, respectively, were nodulated in Java.

From Greek, *ptyche*, "fold," and *lobion*, "pod."
> Type: *P. biflorum* (E. Mey.) Brummitt
> 3 species.
> *Sylitra* and *Ptycholobium*, both closely allied to *Tephrosia*, were treated as separate genera by Harms (1915b). Brummitt (1965) proposed the conservation of *Ptycholobium* Harms over *Sylitra*, with *P. biflorum* (E. Mey.) Brummitt (= *S. biflora*) as the type species.

NODULATION STUDIES

Nodules have been observed on the roots of *P. biflorum* ssp. *biflorum* in South Africa (N. Grobbelaar personal communication 1974) and on *P. contortum* (N. E. Br.) Brummitt and *P. plicatum* (Oliv.) Harms in Zimbabwe (Corby 1974).

Ptychosema Benth.   Papilionoideae: Galegeae, Tephrosiinae
253 / 3716

From Greek, *ptyche*, "fold," and *semeia*, "standard."
> Type: *P. pusillum* Benth.
> 2 species. Herbs, small, finely stemmed, usually ascending, up to about 20 cm tall, diffuse, subglabrous, perennial; roots generally spreading underground, giving rise to leafy branches above. Leaves imparipinnate; leaflets small, narrow-obovate, entire, prominently veined below; stipels absent; stipules small, straight, narrow, sparsely haired. Flowers large, red brown, yellow, or pink purplish, 1–few, in erect, terminal racemes on slender peduncles; bracts 2, at the base of the pedicels; bracteoles 2, near the middle of the pedicels; calyx turbinate, weakly bilabiate; 2 upper lobes slightly broader and connate higher up into a truncate, emarginate lip; standard suborbicular, erect, notched; wings sickle-shaped, obtuse, short-clawed, free; keel blunt, shorter than the wings, short-clawed, not beaked; stamens 10, all united into a sheath split above; anthers uniform, subequal, dorsifixed; ovary short-stalked, glabrous, several-ovuled; style short, inflexed; stigma oblique, terminal. Pod rounded-oblong, flattened, on a slender stalk; seeds elliptic-oblong, with a rim-aril. (Bentham 1864; Bentham and Hooker 1865; Lee 1973)
> Polhill (1976) assigned this genus to the tribe Bossiaeae in his reorganization of Genisteae *sensu lato*.
> Both species are quite rare; they are found in the tablelands, sand hills, and scrub areas of western and central Australia.

Named for M. N. Puerari, 1765–1845, Danish botanist.

Type: *P. tuberosa* (Roxb.) Benth.

25 species. Vines, twining, trailing or high-climbing, herbaceous or woody, perennial, some with large, tuberous rootstocks. Leaves pinnately 3-foliolate; leaflets ovate-rhombic, large, entire or sinuately 2–3-lobed; stipels present; stipules ovate-lanceolate. Flowers blue, pink, or purple, 1–several, pedicelled, in long, dense, terminal or axillary racemes or panicles; rachis usually nodose; bracts caducous, small or narrow; bracteoles small, subtending the calyx, caducous or subpersistent; calyx campanulate; lobes 5, upper 2 completely or partly joined into an entire or a 2-toothed lobe, longer than the lateral lobes, lowest lobe longest; standard obovate, with 2 inflexed ears, short-clawed, sometimes blotched with yellow inside; wings narrow, oblong or obovate, falcate, eared, often adherent to the middle of the keel; keel petals subequal to the wings, straight or slightly curved, short-clawed, eared, sometimes beaked; stamens 10, usually the vexillary stamen free at the base, connate in the middle with the others, sometimes diadelphous with the vexillary stamen completely free; anthers uniform; ovary subsessile or short-stalked, many-ovuled; style filiform, beardless, inflexed above; stigma small, capitellate. Pod linear, compressed or cylindric, membranous or subcoriaceous, continuous or septate between the seeds, some pubescent, 2-valved, dehiscent, many-seeded; seeds compressed, suborbicular or transversely oblong. (De Candolle 1825b,c; Bentham and Hooker 1865; Amshoff 1939)

The transfer of this genus to the subtribe Glycininae is under consideration (Lackey 1977a).

Members of this genus are native to tropical Southeast Asia, Malaysia, Polynesia, and Japan. They occur along roadsides, in brushwood, light forests, and thickets. Cultivated species are commonly called kudzu vine.

The members of this genus are fast growing and deep rooted, and provide dense foliage (Chevalier 1951); all serve as excellent soil cover, are suitable for erosion- and weed-control, and provide good hay and pasture.

Cattle require time to accept *Pueraria* as palatable. Thunberg kudzu, *P. triloba* (Lour.) Makino, has foliage with nutritive value only slightly less than that of alfalfa. Some clay soils in Florida have yielded 3 cuttings of hay in a season (Piper 1924a). This species was introduced into the United States in the late 1800s from Japan, where it was valued for its edible tubers, which are boiled as a vegetable, or manufactured into Japanese arrowroot, a cooking starch. The stems of *P. triloba* are retted for a fiber, called kudzi in Japan and ko in China, used for cordage and in the manufacture of grass cloth (Burkill 1966). *P. phaseoloides* (Roxb.) Benth. is also useful as a fiber plant and as a green manure on rubber and coffee plantations in Malaysia. In Ghana it is considered excellent continuous cover for cacao because it is shade tolerant (Jordan and Opoku 1966).

NODULATION STUDIES

Synonymy has complicated the nodulation reports of *P. triloba* (= *P. hirsuta* Schneid., *P. thunbergiana* Benth., *P. lobata* (Willd.) Ohwi), and of *P. phaseoloides* (= *P. javanica* Benth.). Numerous investigators have observed these species copiously nodulated in diverse geographical areas (Keuchenius 1924; Bañados and Fernandez 1954; Bonnier 1957; Rothschild 1963). Rothschild's (1963) morphological and histological study of *P. triloba* nodules showed no unusual features.

Pueraria rhizobia are monotrichously flagellated, slow growing, and produce alkaline reactions in carbohydrate media, and in litmus milk without serum zone formation. Host-infection studies identify *Pueraria* species as members of the cowpea miscellany (Löhnis and Leonard 1926; Posadas 1941).

Pultenaea Sm.   Papilionoideae: Podalyrieae
176 / 3636

Named for Dr. Richard Pulteney, 1730–1801, English botanist.

Type: *P. stipularis* Sm.

100 species. Shrubs, erect or decumbent, rigid, tall, some straggling; young branches commonly pubescent. Leaves simple, mostly alternate, rarely opposite or in whorls of 3; stipules linear-lanceolate or awl-shaped, brown, scarious, setaceous, often somewhat connate within the leaf axil. Flowers orange yellow, solitary or axillary or densely crowded in terminal heads or clusters; bracts imbricate, scarious, brown, surrounding each flower; bracteoles adnate to calyx tube or subtending the calyx, persistent; calyx lobes 5, slightly bilabiate, upper 2 somewhat united into an upper lip and usually larger and longer than the lower 3, rarely all equal; petals long-clawed; standard suborbicular, equal to or longer than the lower petals; wings oblong; keel often purple, incurved; stamens 10, free; ovary sessile, villous, occasionally glabrous, rarely short-stalked, 2-ovuled; style awl-shaped, filiform or dilated at the base; stigma small, terminal. Pod ovate, often beaked, small, compressed or turgid, 2-valved, dehiscent; seeds 1–2, kidney-shaped. (Bentham 1864; Bentham and Hooker 1865)

Members of this genus present considerable diversity in foliage, inflorescence, calyx, and other characters (Bentham 1864).

Representatives of this large genus are limited in distribution to Australia. They are widespread in Queensland, Victoria, South Australia, and Tasmania; perhaps more than 60 species occur in New South Wales. They flourish in coastal areas, tablelands, stony slopes, sandy hills, marshes, and moist grassy places. Species are often found at altitudes exceeding 1,500 m in the Australian Alps.

Most species are wild plants. Several showy species, exemplified by *P. spinosa* Williams., with brilliant deep brick red flowers, *P. boormani* Williams., with red and yellow flowers, and the yellow-flowered *P. cinerascens* Maid. & Betche, with ashy silver leaves, thrive in dry areas and are cultivated as ornamental shrubs.

NODULATION STUDIES

Strong's (1938) hypothesis that rhizobia of the indigenous *Pultenaea* species, on deep sandy virgin soils of southern and southeastern Australia, could symbiose naturally with an introduced flora of pasture medics and clovers was negated by the findings of Lange (1961). Rhizobial strains of 3 *Pultenaea* species had a natural affinity only with cowpea, soybean, and lupine, not with peas, clover, vetch, or medic. The addition of *P. myrtoides* A. Cunn. and *P. villosa* Willd. to the cowpea cross-inoculation complex had previously been recommended by McKnight (1949a). Norris (1958a) concurred that *Vigna sinensis* and *Phaseolus lathyroides*, as marker plants, were well nodulated by *P. paleacea* Willd. and *P. villosa*.

Nodulated species of *Pultenaea*	Area	Reported by
acerosa R. Br.	Adelaide (C.S.I.R.O.)	Harris (1953)
canaliculata F. Muell.	Adelaide (C.S.I.R.O.)	Harris (1953)
daphnoides Wendl.	Adelaide (C.S.I.R.O.)	Harris (1953)
densifolia F. Muell.	Adelaide (C.S.I.R.O.)	Harris (1953)
elliptica Sm.	N.S. Wales	Hannon (1949)
ericifolia Benth.	W. Australia	Lange (1959)
involucrata Benth.	Adelaide (C.S.I.R.O.)	Harris (1953)
largiflorens F. Muell. var. *latifolia* Williams.	Adelaide (C.S.I.R.O.)	Harris (1953)
laxiflora Benth. var. *pilosa* Williams.	Adelaide (C.S.I.R.O.)	Harris (1953)
microphylla Sieb.	Queensland	Bowen (1956)
myrtoides A. Cunn.	Queensland	McKnight (1949a)
ochreata Meissn.	W. Australia	Lange (1959)
paleacea Willd.	Queensland	Norris (1958a)
prostrata Benth.	Adelaide (C.S.I.R.O.)	Harris (1953)
reticulata (Sm.) Benth.	W. Australia	Lange (1959)
retusa Sm.	Queensland	Bowen (1956)
scabra R. Br.	Adelaide (C.S.I.R.O.)	Harris (1953)
stipularis Sm.	N.S. Wales	Hannon (1949)
strobilifera Meissn.	W. Australia	Lange (1959)
tenuifolia R. Br.	Adelaide (C.S.I.R.O.)	Harris (1953)
teretifolia Williams. var. *brachyphylla*	Kangaroo Island	Harris (1953)
verruculosa Turcz.	W. Australia	Lange (1959)
villosa Willd.	Queensland	McKnight (1949a)

Pycnospora R. Br. ex W. & A.
339 / 3809

From Greek, *pyknos*, "dense," and *spora*, "seed"; the pods are many-seeded.

Type: *P. lutescens* (Poir.) Schindl.

1 species. Herbs or shrublets, perennial, prostrate or ascending, spreading; stems tufted. Leaves pinnately 3-foliolate; leaflets obovate, entire; stipels present; stipules linear, free, membranous, striate. Flowers small, purplish, paired along rachis of slender, terminal racemes, rarely in panicles; bracts membranous, caducous; bracteoles, absent; calyx cup-shaped, deeply cleft, divided to or below the middle; teeth 5, long, subequal, 2 upper teeth united higher up; standard suborbicular, narrowed at the base into a claw; wings oblique-oblong, joined to the keel; keel petals incurved, obtuse, each side slightly appendaged; stamens 10, diadelphous, vexillary stamen free above but united with the other 9 below; anthers uniform; ovary sessile, many-ovuled; disk present; style inflexed, not barbate, awl-shaped; stigma minute, terminal, capitate. Pod small, oblong, turgid, 2-valved, pubescent, continuous within, transversely veined, not jointed, style, stamens, and calyx persistent on the pod, dehiscent; seeds 8–10. (Bentham and Hooker 1865; Baker 1929; Léonard 1954a; Backer and Bakhuizen van den Brink 1963)

This species combines the growth habit of *Desmodium* and the fruiting characters of *Crotalaria*.

This monotypic genus is endemic in tropical and subtropical Indo-Malaya, Southeast Asia, northeastern Australia, and Africa. The plants thrive on grassland savannas, abandoned fields, and deteriorated pasture lands.

Pynaertiodendron De Wild.   Caesalpinioideae: Amherstieae

Named for Pynaert, collector of tropical plants in Zaire.

Type: *P. congolanum* De Wild.

2 species. Trees, 10–12 m tall, unarmed. Leaves simple, entire, ovate, symmetric, not pellucid-punctate; petioles stout, roughly grooved or wrinkled. Flower small, many, in short axillary or terminal racemes, with thin, dry bracts crowded at the base; bracteoles 2, opposite, concave, valvate, petaloid, covered with a white pubescence, persistent, completely enveloping the flower bud; calyx tube very short, cuplike; lobes 4, minute, scalelike; petal 1, rarely 2, short-clawed, broad-elliptic, enveloping the ovary in bud; stamens 3, free, fertile; filaments short; anthers dorsifixed, longitudinally dehiscent; ovary short-stalked; disk present; style slender. Pod not described. (Léonard 1952)

These species differ from those of *Cryptosepalum* primarily in having simple leaves. Some botanists regard the 2 genera synonymously and merge all species into *Cryptosepalum* (Léonard 1957; Hutchinson 1964).

Both species are indigenous to tropical Africa.

Named for Carl Gottlieb Rafn, 1769–1808, Danish botanist.

Type: *R. amplexicaulis* (L.) Thunb.

32 species. Shrubs or shrublets, glabrous, often gray green, sometimes singly branched from an underground rootstock. Leaves simple, ovate, lanceolate, rarely slender or linear, entire, 1-nerved or reticulately venose; stipules absent. Flowers yellow, in short, terminal racemes or solitary in axils of leaflike bracts; bracteoles foliaceous or absent; calyx campanulate; lobes 5, lanceolate, acuminate, lowest lobe narrower and smaller than others; standard ovate or suborbicular, sometimes bilobed; wings falcate-oblong, clawed; keel incurved, beaked or truncate, sometimes clawed and gibbous; stamens monadelphous, all connate into a sheath split above; anthers alternately short and versatile, long and basifixed; ovary sessile or stalked, glabrous, 2- to many-ovuled; style semiterete, incurved; stigma small, capitate. Pod compressed, linear or lanceolate, bordered with a narrow wing on the upper suture, 2-valved, valves leathery, nonseptate. (Bentham and Hooker 1865; Harvey 1894; Phillips 1951; Dahlgren 1963b)

Polhill (1976) assigned this genus to the tribe Crotalarieae in his reorganization of Genisteae *sensu lato.*

The genus is similar morphologically to *Wiborgia* and *Lebeckia* (Dahlgren 1963b).

Members of this genus are native to the southwestern region of the Cape Province of South Africa. The plants are usually found on sandy flats, hill slopes, and plateaus.

Rafnia species are useful for soil-building and as ground cover. In early Cape colonial days, decoctions of plant parts of *R. amplexicaulis* (L.) Thunb. were used as a demulcent to treat catarrh and coughs. The root has a licorice flavor. A tea brewed from the leaves of *R. perfoliata* E. Mey. has diuretic properties.

NODULATION STUDIES

The presence of nodules on the following species has been reported from South Africa by Grobbelaar and Clarke (1972): *R. amplexicaulis, R. angulata* Thunb., *R. crassifolia* Harv., *R. cuneifolia* Thunb., *R. elliptica* Thunb., *R. opposita* Thunb., *R. ovata* E. Mey., *R. perfoliata,* and *R. triflora* Thunb.

Raimondianthus Harms   Papilionoideae: Hedysareae,
                                          Aeschynomeninae

Named for Raimondi; he collected the specimens in Peru.

Type: *R. platycarpus* Harms

1 species. Shrublet, small, scandent; stems densely velvety-haired. Leaves imparipinnate; leaflets oblong or ovate, 2–3 pairs, petioluled, lower surfaces densely pilose; stipules lanceolate. Flowers on slender pedicels, in axillary or terminal divergent panicles; bracts small, deltoid; bracteoles 2, subtending the calyx; calyx obliquely cupular to funnel-shaped; lobes very short, upper pair slightly connate; petals long-exserted; standard broad-obovate or truncate

near the broad claw; wings oblanceolate-oblong, bluntly appendaged and pilose at the base towards the claw; keel petals narrow-clawed, obliquely oblong-oblanceolate, blunt; anthers small; ovary sessile, 6–8-ovuled; style slender, base densely pubescent; stigma minute. Pod sessile, lanceolate, straight or rarely curved, compressed, flat, papery, subglabrous, indehiscent (?); oblique veins converge at a median line of the longitudinal axis across the humps of the underlying 6–8 seeds; along this line the pod has a rather broad flat winged margin. (Harms 1928b)

The genus closely resembles *Isodesmia*.

This monotypic genus is endemic in Peru.

Ramirezella Rose   Papilionoideae: Phaseoleae, Phaseolinae

Named for Dr. José Ramirez, of the Instituto Medico Nacional de México, expert in the natural history of Mexico.

Type: *R. strobilophora* (Robins.) Rose

9 species. Vines, tall, twining, perennial; stems herbaceous or woody, pubescent or glabrous. Leaves pinnately 3-foliolate; leaflets ovate, acuminate; stipels and stipules present. Flowers purple or white, large, showy, crowded in axillary racemes; bracts large, striate, overlapping in bud; bracteoles at base of calyx, small, ovate; calyx campanulate, small, colored; teeth 5, short, equal, obtuse or rounded; standard rounded, with scalelike appendages at the base; wings eared on upper side; keel elongated, erect at base, bent at right angle at middle, incurved at the apex, beaked; stamens 10, diadelphous; style bearded near the tip; stigma oblique. Pod straight, oblong, turgid, dehiscent; seeds orbicular, embedded in white, spongy tissue. (Rose 1903)

The genus was structured with *Vigna strobilophora* as the type species. Morton (1944) regarded the distinctions that set *Ramirezella* apart from *Vigna* as arbitrary, since many species of *Phaseolus* show the same characters. He proposed *Phaseolus strobilophora* as the new name combination; however, both Piper (1926) and Hutchinson (1964) accepted *Ramirezella* as valid. Using the type species in host-infection studies with rhizobia of the bean and cowpea type would be a practical symbiotaxonomic approach to confirm whether the host kinship is closer to *Phaseolus* or *Vigna*.

Representatives of this genus are indigenous to Mexico and Central America.

These twining vines commonly ascend to the tops of trees in their native habitats. The purple and white floral racemes, especially those of *R. oranta* Piper, rival those of *Wisteria* species in beauty.

NODULATION STUDIES

Nodulation reports are lacking for *Ramirezella* species *per se*, but nodules were observed on plants introduced into a botanical garden in France from Texas as *Vigna strobilophora* (Naudin 1897a).

Ramorinoa Speg.

Papilionoideae: Dalbergieae,
Lonchocarpinae

Type: *R. girolae* Speg.

1 species. Tree, small; branches rigid, cylindric, spine-tipped. Leaves modified to very small scales. Flowers, about 1 cm long, on short pedicels, in simple or branched, erect, lateral racemes; rachis slender, nodose; bracts small, caducous; calyx campanulate, pubescent; lobes 5, triangular, some shorter than the calyx tube, the higher ones slightly united; corolla orange, smooth, membranous; petals subequal; standard suborbicular; stamens 10, diadelphous, 1 free, others united to the middle; filaments threadlike; anthers uniform, dorsifixed, versatile, elliptic; ovary sessile, pubescent, oval, 2–6-ovuled; style smooth, curved at a right angle; stigma terminal, small, globose. Pod indehiscent, dry, reticulate, obovoid or elliptic, somewhat flat, smooth, with slight transverse depressions between the seeds, narrow-winged all around; seeds 1–4, transverse, oval. (Burkart 1943)

This monotypic genus is found exclusively in western Argentina. It occurs in sandy, dry areas.

The wood of this xerophytic species, known as chica, serves various local uses. The seeds are used as a substitute for coffee (Burkart 1943).

Recordoxylon Ducke

Caesalpinioideae: Sclerolobieae

Named for Professor S. J. Record, School of Forestry, Yale University; his observation of the wood structure led Ducke to revise his original classification.

Type: *R. amazonicum* Ducke

2 species. Trees, large, unarmed. Leaves glabrous, imparipinnate; leaflets 5–7, mostly opposite, medium to large. Flowers golden yellow, in conspicuous, reddish-tomentose, terminal panicles of few-flowered racemes; calyx densely reddish silky-haired; lobes 5, imbricate, caducous; petals 5, large, subequal, short-clawed; stamens 10; filaments free, glabrous; ovary short-stalked, silky-haired. Pod small, flat, linear-oblong, glabrous, very narrow-winged along the upper suture, valves thin, leathery, tardily and nonelastically dehiscent, not filled between the seeds; seeds few, brown, flattened but not winged, subkeeled on 1 side. (Ducke 1934b; Record and Hess 1943)

Circumstances leading to the taxonomic reconsideration of *Melanoxylon amazonicum* and the creation of this genus are reviewed in full by Ducke (1934b).

Both species are found in Amazonas, Brazil.

In the flowering stage *R. amazonicum* Ducke closely resembles *Melanoxylon* except that the pods are different; the wood resembles that of *Bowdichia* species. The other species, *R. stenopetalum* Ducke, discovered in the nonflooded forest area near São Paulo de Olivença in western Amazonas, was viewed by Record and Hess (1943) as a "probable geographical variety of *R. amazonicum*."

Named for Esprit Requien of Avignon, 1788–1851, French botanist.

Type: *R. obcordata* (Lam.) DC.

3 species. Shrublets not exceeding 1 m in height; stems subwoody, sparsely branched, densely tomentose. Leaves 1-foliolate, obcordate with sharp apical point, hooked or straight; stipules awl-shaped, free. Flowers axillary, small, subsessile, solitary or few-clustered; calyx campanulate; teeth 5, all acute, lowest tooth longest; standard obovate; keel petals free, blunt; stamens monadelphous, the tube split above; ovary sessile, 1-ovuled; style short, incurved. Pod oval, compressed, membranous, usually pubescent, 1-seeded, apex acute, hooked. (De Candolle 1825b; Harvey 1894)

This genus was formerly a section in *Tephrosia* (Taubert 1894).

Members of this genus are native to South Africa and tropical Africa.

NODULATION STUDIES

Corby (1974) observed nodules on *R. pseudosphaerosperma* (Schinz) Brummitt and *R. sphaerosperma* DC. in Zimbabwe.

Retama Raf. Papilionoideae: Genisteae, Spartiinae
/ 3675a

Name was derived from the Arabic, *ratam*, the name of a desert shrub mentioned in Biblical times. Disagreement over the proper name for this small genus of little economic value lasted many years. As Heywood (1968) stated: "It is unfortunate that the generic name *Retama* Boiss. (1839) is antedated by *Lygos* Adanson (1763). This is especially so since *Retama* is a name familiar in horticulture as well as being a common name in Spain and other Mediterranean countries for species belonging to this genus." This nomenclatural problem, however, has recently been resolved by the conservation of *Retama* Raf. over *Lygos* Adans. (*International Code of Botanic Nomenclature* 1972).

Lectotype: *R. monosperma* (L.) Boiss.

3 species. Shrubs, erect, up to about 3 m tall, unarmed, much-branched. Leaves simple, linear, silky-pubescent, early deciduous. Flowers large, yellow or white, in dense or lax axillary racemes; calyx urn-shaped, campanulate or turbinate, bilabiate, the upper lip with 2 broad, triangular teeth; standard ovate or rhombic, somewhat hairy; wings short, oblong; keel blunt; stamens monadelphous; style filiform, incurved. Pod small, globose, short-beaked, indehiscent, tardily or incompletely dehiscent along the lower suture, 1–2-seeded. (As *Lygos* Adans., Tutin et al. 1968)

Retama species are native to Mediterranean regions, North Africa, southern Spain, Portugal, and Israel. The plants usually grow on maritime sands, in dry places, and on sandy desert soils.

Various *Baptisia*-type lupine alkaloids have been ascribed to *Retama* species (Turner 1971). These compounds undoubtedly account for the reputed use of plant parts as a vermifuge and purgative. Sheep relish the pods. The wood serves in desert areas as a source of charcoal.

NODULATION STUDIES

Nodules were observed on *R. raetam* (Forsk.) Webb. & Berth. in Libya (B. C. Park personal communication 1960) and on *R. monosperma* (L.) Boiss. (reported as *Lygos monosperma* (L.) Heyw.) and *R. sphaerocarpa* (L.) Boiss. (reported as *L. sphaerocarpa* (L.) Heyw.) in South Africa (Grobbelaar and Clarke 1974).

*Rhodopis Urb.   Papilionoideae: Phaseoleae, Erythrininae
(395a) / 3871

From Greek, *rhodon,* "rose," and *ops, opos,* "countenance," "face," in reference to the rosy-cheeked appearance of the flowers.

Type: *R. planisiliqua* (L.) Urb.

2 species. Shrubs, climbing; young branches retrorsely pilose, later glabrous. Leaves narrow, 1-foliolate; petiole jointed at the base and apex; stipels present; stipules small, awl-shaped. Flowers showy, rose, many, in fascicles in long racemes; bracts awl-shaped; calyx tubular; lobes 5, upper 2 united into 1 ovate, short-acuminate lobe, 2 lateral lobes much smaller, lowermost lobe subequal, lanceolate; standard oblong, erect, eared at the base; wings linear-oblong, eared, long-clawed; keel petals joined at the back, eared, longer than the wings; claws of wings and keel adnate to base of staminal tube; stamens 10, diadelphous, vexillary stamen free only at the base, then united with the others; anthers uniform, linear-oblong; ovary short-stalked, surrounded at the base by a short conical-tubular disk, many-ovuled; style not bearded; stigma terminal, small. Pod short-stalked, flattened, straight or slightly curved, short-beaked, 2-valved, woody, dorsal suture not thickened; seeds many, horizontal, separated by thin membranes. (Urban 1900; Harms 1908a)

Both species are found in the West Indies, especially in Haiti and the Dominican Republic.

*Rhynchosia Lour.   Papilionoideae: Phaseoleae, Cajaninae
419 / 3897

From Greek, *rhynchos,* "beak," in reference to the conspicuously beaked keel; the pods also have a short beak.

Type: *R. volubilis* Lour.

200–300 species. Herbs, stems ridged or angular, trailing or weakly twining, perennial, or woody undershrubs, sometimes with a thick, woody rootstock. Leaves usually pinnately 3-foliolate, rarely palmately 1–3-foliolate; leaflets variously shaped, usually rhomboid and somewhat acuminate, or heart-shaped, with resinous dots on the underside; stipels minute or absent;

stipules ovate or lanceolate, sometimes fugacious. Flowers large or small, pedicelled, yellow streaked with red or purple, white, or greenish, in many-flowered axillary racemes, sometimes paniculate, rarely solitary or in dense, oblong, subsessile clusters; bracts often leafy, caducous; bracteoles absent; calyx campanulate, short, prominently nerved; teeth 5, linear or ovate, long-acuminate, usually longer than the calyx tube; 2 upper lobes somewhat connate, lower lobe longer; standard obovate or orbicular, spreading or reflexed with a short, deeply channeled claw, eared at the base, gland-dotted on the back; wings oblong or oblong-linear, with a straight or curved claw, narrow, usually shorter than the keel petals; keel petals obtuse, partly joined, usually glabrous, apically incurved; stamens diadelphous, vexillary stamen often kneed; anthers uniform; ovary subsessile, lanceolate or ovate, 2-ovuled, usually densely hairy; disk small, basal, cupular, hairy, sometimes glandular; style glabrous, incurved and thickened above; stigma terminal, capitate. Pod compressed, globose to falcate, usually velvety and glandular, rarely septate, 2-valved, dehiscent; seeds 2, blue black or brown, compressed-globose or subreniform. (Bentham and Hooker 1865; Amshoff 1939; Pellegrin 1948; Hauman 1954c; Grear 1978)

Representatives of this large genus inhabit warm latitudes of both hemispheres. Of the approximately 110 species found in tropical Africa, about 42 occur in the Congo Basin, 80 in South Africa. There are 10 species in northern Argentina, 15 in the southern United States, and 6 in Australia. *Rhynchosia* species are common in brushwood and forest clearings, jungle borders, and in dry, sandy and silty clay soils of open savannas and mesas. Some species grow up to 1,800-m altitude.

Several species, commonly called rosary bean, have attractive red, blue, black, mottled, or bicolored seeds that are used for necklaces, bracelets, or rosaries. The steel blue seeds of *R. calycina* Guill. & Perr. reputedly were used along the Gold Coast as standards to weigh gold dust (Dalziel 1948). Seeds of *R. pyramidalis* (Lam.) Urb. and *R. longeracemosa* Mart. & Gal. have narcotic properties utilized by certain South Mexican Indian tribes (Uphof 1968). Poisonous properties once imputed to the seeds of *R. phaseoloides* DC. and *R. minima* (L.) DC. have been disproved. These two species are considered important palatable pasture plants for cattle and horses in South America and tropical Africa, respectively (Whyte et al. 1953). The roots of *R. albiflora* (Sims) Alst. are used in Zambia to sweeten porridge. The seeds of *R. corylifolia* Mart. yield an alcohol-soluble yellow dye. The leaves of *R. manobotrya* Harms, a shrub, are boiled as a vegetable by natives of the Congo Basin. All species serve as ground cover and in erosion control. *R. texana* T. & G. has particular ground-cover potential and responds well to cultivation. *R. buettueri* Harms, known for its beautiful long reddish purple floral racemes, has ornamental potential for cultivation outside of its native southern Nigeria.

NODULATION STUDIES

Rhizobia isolated from nodules of *R. erecta* (Walt.) DC. (= *Dolicholus erectus*) were monotrichously flagellated (Shunk 1921). Cultural and biochemical data are lacking. The ability of strains from *R. minima* to nodulate *Amorpha*, *Crotalaria*, and *Wisteria* species (Wilson 1944b), nodulation of *R. minima* by 9 of

12 strains of *Rhizobium japonicum* (Wilson 1945a), and the effective response of cowpea nodulated by strains from *Rhynchosia goetzei* Harms (Sandmann 1970a) prompt the inclusion of *Rhynchosia* species in the cowpea miscellany.

Nodulated species of *Rhynchosia*	Area	Reported by
adenodes E. & Z.	S. Africa	Grobbelaar & Clarke (1972)
albissima Gandoger	Zimbabwe	Corby (1974)
angulosa Schinz	S. Africa	Grobbelaar & Clarke (1972)
buchananii Harms	Zimbabwe	Corby (1974)
capensis (Burm.) Schinz	S. Africa	Grobbelaar & Clarke (1972)
caribaea (Jacq.) DC.	S. Africa	Grobbelaar et al. (1964)
var. *picta* (E. Mey.) Bak. f.	S. Africa	Grobbelaar & Clarke (1972)
ciliata Druce	S. Africa	Grobbelaar & Clarke (1975)
clivorum S. Moore	S. Africa	Grobbelaar & Clarke (1972)
	Zimbabwe	Corby (1974)
confusa Burtt Davy	S. Africa	Grobbelaar et al. (1967)
cooperi (Harv. ex Bak. f.) Burtt Davy	S. Africa	Grobbelaar & Clarke (1975)
densiflora DC.	S. Africa	Grobbelaar & Clarke (1972)
ssp. *chrysadenia* (Taub.) Verdc.	Zimbabwe	Corby (1974)
divaricata Bak.	Zimbabwe	Corby (1974)
effusa Druce	S. Africa	N. Grobbelaar (pers. comm. 1963)
elegantissima Schinz	S. Africa	Grobbelaar et al. (1967)
erecta (Walt.) DC.	S. Africa	Grobbelaar & Clarke (1972)
(= *Dolicholus erectus*)	N.C., U.S.A.	Shunk (1921)
fleckii Schinz	Zimbabwe	Corby (1974)
goetzei Harms	Zimbabwe	Corby (1974)
hirta (Andr.) Meikle & Verdc.	Zimbabwe	Corby (1974)
holosericea Schinz	Zimbabwe	Corby (1974)
holstii Harms	Zimbabwe	Corby (1974)
hosei Griseb.	Puerto Rico	Dubey et al. (1972)
insignis (O. Hoffm.) R. E. Fries	Zimbabwe	Corby (1974)
jacottetii Schinz	S. Africa	Grobbelaar et al. (1967)
komatiensis Harms	S. Africa	Grobbelaar & Clarke (1972)
memnonia DC.	Australia	D. Norris (pers. comm. 1962)
minima (L.) DC.	Widely reported	Many investigators
var. *minima*	Zimbabwe	Corby (1974)
var. *prostrata* (Harv.) Meikle	Zimbabwe	Corby (1974)
	S. Africa	Grobbelaar & Clarke (1972)
mollissima (Ell.) Shuttlew. ex W. Wats.	Java	Keuchenius (1924)
monophylla Schltr.	S. Africa	Grobbelaar & Clarke (1972)

Nodulated species of *Rhynchosia* (cont.)	Area	Reported by
nervosa Benth.		
var. *petiola*	S. Africa	Grobbelaar & Clarke (1972)
nitens Benth.	S. Africa	Grobbelaar et al. (1967)
nyasica Bak.	Zimbabwe	Corby (1974)
pentheri Schltr.		
var. *hutchinsoniana* Burtt Davy	S. Africa	Grobbelaar & Clarke (1975)
var. *pentheri*	S. Africa	Grobbelaar & Clarke (1972)
phaseoloides DC.	Java	Keuchenius (1924)
procurrens (Hiern) Schum.		
ssp. *floribunda* (Bak.) Verdc.	Zimbabwe	Corby (1974)
pyramidalis (Lam.) Urb.	S. Africa	N. Grobbelaar (pers. comm. 1965)
	Venezuela	Barrios & Gonzalez (1971)
resinosa (Hochst. ex A. Rich.) Bak.	Zimbabwe	Corby (1974)
reticulata (Sw.) DC.	Puerto Rico	Dubey et al. (1972)
rufescens (Willd.) DC.	Java	Keuchenius (1924)
senna H. & A.	Argentina	Rothschild (1970)
sordida Schinz	S. Africa	Grobbelaar et al. (1967)
	Zimbabwe	Corby (1974)
stipata Meikle	Zimbabwe	Corby (1974)
sublobata (Schumach.) Meikle	Zimbabwe	Corby (1974)
	S. Africa	Grobbelaar & Clarke (1975)
swynnertonii Bak. f.	Zimbabwe	Corby (1974)
texana T. & G.	Ariz., U.S.A.	Martin (1948)
thorncroftii (Bak. f.) Burtt Davy	S. Africa	Grobbelaar & Clarke (1972)
totta DC.	S. Africa	Grobbelaar et al. (1967)
var. *elongatifolia*	Zimbabwe	Corby (1974)
var. *totta*	Zimbabwe	Corby (1974)
var. *venulosa* (Hiern) Verdc.	Zimbabwe	Corby (1974)
tricuspidata Bak. f.	Zimbabwe	Corby (1974)
venulosa (Hiern) Schum.	S. Africa	Grobbelaar et al. (1967)
volubilis Lour.	Japan	Asai (1944)

Rhynchotropis Harms
(239a) / 3701

Papilionoideae: Galegeae, Indigoferinae

From Greek, *rhynchos*, "beak," and *tropis*, "keel."

Type: *R. poggei* (Taub.) Harms

2 species. Herbs, erect, perennial; stems numerous, slender, angular, appressed-pubescent, arising from a fleshy or woody rootstock. Leaves simple, sessile, linear or narrow-lanceolate, entire; stipules persistent, awl-

shaped. Flowers in lax, elongated, axillary and paniculate racemes; bracts awl-shaped; bracteoles absent; calyx campanulate; lobes 5, subulate-lanceolate; standard suborbicular, shortly acuminate; wings shortly acuminate; keel long, beaked, petals joined at the back; stamens 10, diadelphous, vexillary stamen coherent with others at the base, alternately longer and shorter; anthers hairy-tufted at the base and apex; ovary linear, 4–6-ovuled; style dilated at the base, narrowed at the apex; stigma terminal, oblique. Pod short-stalked, linear or oblanceolate-linear, beaked, subcylindric, appressed-pubescent, with thin septa between the 2–4 seeds. (Harms 1908a)

This genus is positioned between, and easily confused with, *Cyamopsis* and *Indigofera*.

Both species are native to tropical West Africa. They usually occur in dry, lightly forested areas.

Rhytidomene Rydb.   Papilionoideae: Galegeae, Psoraliinae

From Greek, *rhytidos*, "wrinkled," and *mene*, "moon," in reference to the cross-wrinkled, crescent-shaped pod.

Type: *R. lupinellus* (Michx.) Rydb.

1 species. Herb, perennial; stems clustered, erect, slightly glandular, arising from an elongated, creeping rhizome. Leaves scattered, palmately 5–7-foliolate; leaflets linear-filiform, gland-dotted; stipules small, setaceous. Flowers blue, small, few on short pedicels in long-peduncled, lax axillary racemes; bracts minute, ovate, toothed; calyx accrescent, campanulate; tube short, sparingly glandular-punctate below, not gibbous at the base; lobes 5, lowest lobe longest, awl-shaped; standard suborbicular, claw short, doubly bent; wings narrowly oblique-obovate, apex rounded, basal lobe small; keel petals shorter, united at the apex, slightly lobed at the base, each petal adnate at the base to the adjacent wing; stamens 10, vexillary stamen entirely free or united above to the others; ovary 1-ovuled; style upcurved. Pod about 1 cm long, exserted from the calyx, compressed, crescent-shaped, obliquely cross-wrinkled, gland-dotted, pericarp leathery, short-beaked, indehiscent, 1-seeded; seeds brown, kidney-shaped. (Rydberg 1919a)

This monotypic genus is endemic in warm temperate North America, on coastal plains from North Carolina to central Florida.

Riedeliella Harms   Papilionoideae: Sophoreae
(124a) / 3582a

Named for Ludwig Riedel, botanist with the Langsdorf Expedition; his collection from southern Brazil was the basis for the generic description.

Type: *R. graciliflora* Harms

2 species. Shrubs, erect or scandent, up to about 2 m tall; branches

glabrous or densely covered with white hairs and occasionally with yellow, broad-based setae. Leaves imparipinnate, 5–9-foliolate; leaflets opposite or subopposite, subcoriaceous, broad-lanceolate to ovate, rounded at the base, apex obtuse to acute, glabrous, villous or strigose; rachis smooth or hairy; stipels and stipules absent. Flowers yellow, sessile or stalked, racemose or paniculate; bracts lanceolate, subpersistent; bracteoles 2, smaller; calyx campanulate; teeth 5, subequal, very short, deltoid; petals 5, subequal, nearly free, inserted at the base of calyx tube, clawed; stamens 10, united near the base, filaments equal, long, slender, exserted; anthers subquadrate-ovate; ovary elongated, oblong, very short-stalked, glabrous or sericeous, 4–6-ovuled; style slender. Pod semilunar, compressed, broad-winged; seeds few, compressed, curved, purplish. (Harms 1908a; Mohlenbrock 1962b)

Riedeliella was among 8 genera segregated from the tribe Sophoreae to form Cadieae trib. nov. (Hutchinson 1964). It is closely related to *Sweetia* (Harms 1908a; Mohlenbrock 1962b).

Both species occur in Brazil and Paraguay.

Robinia L.   Papilionoideae: Galegeae, Robiniinae
268 / 3733

Named for Jean Robin, 1550–1629, and his son, Vespasian, 1579–1662, Royal Gardeners in Paris during the reign of Henry IV. According to one historical record, seeds of *R. pseudoacacia*, the black locust, were sent to Jean from Canada and cultivated by him on a large scale about 1601. Another record cites the planting of the seed by Vespasian about 1635, who was at that time arborist to Louis XIII in the Jardin des Plantes, Paris (Power and Cambier 1890).

Type: *R. pseudoacacia* L.

20 species. Shrubs or trees, small, medium, or large, deciduous; trunks usually straight, long, slender; branches slender, terete, angled, and zigzag; bark furrowed; stipular spines on twigs and young branchlets. Leaves imparipinnate, petioled; leaflets 6–20 pairs; small, oblong, penniveined, petioluled; stipels present; stipules setaceous, spinescent at maturity, persistent. Flowers white, pink, or lavender, on long, slender pedicels, in short, pendent, lax, axillary racemes; bracts small, acuminate, membranous, very caducous; bracteoles early caducous or absent; calyx campanulate, slightly bilabiate; lobes 5, 2 upper lobes shorter than others and coherent in part, lower 3 longer, deeply cleft, subequal; petals short-clawed; standard large, rounded, reflexed, scarcely longer than the oblong-falcate, curved, free wings; keel petals incurved, obtuse, partly united below, eared at base; stamens 10, diadelphous, inserted with the petals on a tubular disk, upper stamen free or connate in the middle with the staminal tube; anthers ovoid, uniform or alternate ones smaller; ovary linear-oblong, stalked, inserted at the base of the calyx, multiovuled; ovules suspended in 2 ranks; style awl-shaped, inflexed, bearded on the inner side above; stigma small, terminal. Pod small, short-stalked, linear, flat, glabrous or hairy, some narrow-winged

along the upper suture, continuous inside, 2-valved, valves thin and membranous, dehiscent, several- to many-seeded; seeds hard, oblong-oblique. (Bentham and Hooker 1865; Rehder 1940; Sargent 1949)

Members of this genus are native to temperate regions of North and Central America, but have become widely distributed in many other areas through introduction. They occur in open and forested areas, in gullies, on hillsides, and in mountain areas up to about 2,000-m altitude. Species grow well on poor sites and thrive on well-drained, fertile soils.

Black locust, *R. pseudoacacia* L., is native to the eastern United States, with its best development in the Appalachian area. Best growth occurs on a well-drained, calcareous soil on sites lacking a pronounced subsoil. It is intolerant of shade and is seldom found in dense forests except where it is dominant. Ordinarily, it is a medium-sized deciduous tree of about 20 m in height and has a fluted or twisted trunk (Rendle 1969), but it may reach a height of 30 m, with a trunk diameter slightly exceeding 1 m. The seeds average about 52,800 per kg and 68 percent germination; they remain viable for about 10 years in cool storage. This species was probably the first native American tree to be planted in England (Medsger 1939).

Various early investigators of rhizobia:legume symbiosis commented on the nitrogen-fixing value of the black locust (Frank 1890a; Nobbe et al. 1891; Nobbe et al. 1895; Nobbe and Hiltner 1899; Nobbe et al. 1908; Mattoon 1930). Ferguson's early report (1922) of the beneficial effect of black locust in growth association with catalpa has been confirmed in reports of its association with white ash, tulip poplar, black oak, and chestnut oak (Chapman 1934). Heights and diameters of these other trees were inversely proportional to increased distances from *Robinia* sites. Reports of this nature have contributed substantially to the practice of using black locust in mixed plantings. Aside from the benefit from symbiotic nitrogen fixation (McIntyre and Jeffries 1932), considerable calcium, magnesium, and potassium are returned to the soil from black locust leaf litter (Garman and Merkle 1938). Because of its ability to survive on acid soil banks, this species is planted extensively on land strip-mined for coal (McAlister 1971).

The wood resembles golden oak (Titmuss 1965). Freshly cut locust heartwood varies from greenish yellow to dark brown, but upon exposure it darkens to an attractive golden, or dark russet brown. The sapwood is creamy white. The wood has a high luster, is odorless, yet sometimes "beany"; the texture is coarse, close grained, and usually straight. It is valued for its strength, durability, flexibility, and resistance to shrinkage. Everett (1969) cited the black locust as the most important lumber tree introduced into Romania and Hungary. It is one of the hardest of American woods and very difficult to work; at 15 percent moisture its density is 640–800 kg/m³. The wood's resistance to rot is attributed to the occurrence of about 4 percent taxifolin, an isomer of dihydroquercetin, or dihydrorobinetin, a growth inhibitor of wood-destroying fungi (Freudenberg and Hartmann 1953).

The wood is used for fence posts, wagon-wheel hubs, telephone and telegraph glass insulator pins, and tree nails. One of its traditional uses has been for wooden pins in shipbuilding. About 2.8 million board feet (7,000 m³) were used in all manufacturing industries in the United States in 1960

(McAlister 1971). Of this amount, about 70 percent was used in lumber and wood products and the remainder in household furniture.

Chemical substances associated with the poisoning of horses, cattle, sheep, and man have been isolated from black locust bark, flowers, leaves, young shoots, and seed (Watt and Breyer-Brandwijk 1962; Kingsbury 1964). Severity of poisoning ranges from lassitude and weakness to death within several days; the effects vary with the animal concerned. Robin, a heat-labile phytotoxin (Power and Cambier 1890; Power 1901), and robitin, a glucoside (Tasaki and Tanaka 1918), are the most commonly mentioned toxic compounds. The highly toxic action of robin is similar to that of abrin, the toxalbumin of jequirity seeds (see *Abrus*) and ricin, the toxic protein of castor bean. Robinin, $C_{33}H_{40}O_{19}$, an aromatic glucoside is found in the white flowers (Sando 1932).

The wood of *Pseudoacacia* var. *rectissima* Raber, Long Island yellow locust, or shipmast locust, has a straighter bole and greater resistance to decay and wood borers than that of the type (Raber 1936; McAlister 1971). Durability of the wood in soil as posts and surveyor stakes exceeds that of the common locust by 50–100 percent. Supply of this variety is limited, since it seldom bears pods and therefore must be propagated vegetatively from sprouts or root cuttings.

The other major *Robinia* species are distinctive from black locust and of lesser importance. *R. neomexicana* Gray, a shade-tolerant, spiny shrub, typifies the smaller species of the genus. Its tendency to establish dense thickets in limestone soil on dry slopes has value in erosion control and the revegetation of gullies. It is a good browse for goats and deer and provides good cover for wildlife and acceptable forage for cattle, horses, and sheep. The hard, close-grained, yellow-brownish wood is used for small turnery objects, fuel, and charcoal. *R. hispida* L. and *R. viscosa* Vent. are popular as showy ornamentals in many parks of Argentina. The latter species is known as clammy locust in the United States.

Robinia species have fragrant flowers and are a good honey source. In the Danube Basin this so-called acacia honey commands a higher price than other kinds of honey (Farcas 1940). Indian tribes of the United States use the roots and barks of *Robinia* species for various medicinal purposes.

Nodulation Studies

It appears that black locust was the first leguminous tree studied relative to symbiont typification, rhizobia:plant group status, and symbiotic nitrogen fixation. In 1887 Tschirch described the bacteroids in *Robinia* nodules as curved rods with swollen ends. Isolation of *Robinia* rhizobia by Nobbe et al. followed in 1891. Until Maassen and Müller's host-infection studies in 1907, the morphology of these organisms played a role in fashioning the taxonomic structure of the genus *Rhizobium* (Schneider 1892; Hiltner 1900; Hiltner and Störmer 1903); however, with the advent of emphasis on host relationships as a taxonomic criterion, species status of *Robinia* rhizobia was nullified.

Robinia rhizobia produce abundant mucilaginous growth on laboratory media and an alkaline reaction with serum zone formation in litmus milk. Type of flagellation awaits clarification; peritrichous flagellation was reported by Shunk (1921) and Ishizawa (1953); monotrichous flagellation was observed by Bushnell and Sarles (1937). Serological affinities between *Robinia* rhizobia and those of clover, beans, cowpea, soybean, and *Lespedeza* species were negative (Carroll 1934b). Positive kinships claimed by Fehér and Bokor (1926) between rhizobia of *Robinia, Laburnum, Amorpha,* and *Gleditsia* are questionable, since *Gleditsia triacanthos* is a nonnodulating species.

Host-infection studies using *Robinia* rhizobia were extensive during the 1890–1891 period. In 1891 Nobbe et al. called attention to increases of 22-fold and 105-fold in dry weight and nitrogen content, respectively, of inoculated versus noninoculated black locust plants. A later report emphasized the value of rhizobia in promoting growth of black locust on soils poor in nitrogen (Nobbe et al. 1895). Throughout the 1890s there was a growing awareness that rhizobia differed culturally and physiologically and in their ability to infect and benefit various other host plants. Thus, it is quite understandable that in 1896 United States Patent 570,813 was granted to Nobbe and Hiltner on the claim that the most effective nodules were produced only on that species from whose root nodules the rhizobia *per se* were obtained. With less closely allied species there was less effective nodulation, and complete incompatibility ensued with taxonomically unrelated species. A high degree of efficiency of *Robinia* rhizobia with *Robinia* plants, a much lesser efficiency with *Colutea* species, and absolute uselessness of *Robinia* rhizobia with peas were cited as examples. Nonpromiscuity of *Robinia* species to nodulation by heterologous strains and, similarly, noninfectiveness of *Robinia* rhizobia for other species have been repeatedly shown over the years (Maassen and Müller 1907; Burrill and Hansen 1917; Conklin 1936; Raju 1936; Thorne and Walker 1936; Bushnell and Sarles 1937; Ishizawa 1954). Wilson's data (1939a) showed certain exceptions, but were not convincing because of the atypical, collar-type or crotch-type of nodule produced. In another study (Gregory and Allen 1953), a rhizobial strain from *Caragana arborescens* produced nodules on black locust, but the response was not reciprocal.

Robinia nodules are round and creamy white when young but become elongated as a result of apical meristematic activity. Old nodules have a protective, barklike basal exterior dark with tannin deposit. Perennial nodules show constrictions between successive periods of dormancy and renewed growth. Spratt (1919) referred to *Robinia* nodules as the Mimosoideae type. Infection threads tend to be prevalent and conspicuous in the host cells (Lechtova-Trnka 1931). Invaded host cells show marked hypertrophy following rhizobial invasion.

Nodulated species of *Robinia*	Area	Reported by
hispida L.	Wis., U.S.A.	J. C. Burton (pers. comm. 1970)
kelseyi Hutch.	N.Y., U.S.A.	Wilson (1939a)
pseudoacacia L.	Widely reported	Many investigators
var. *rectissima* Raber	N.Y., U.S.A.	Wilson (1944b)
var. *umbraculifera* DC.	Japan	Pers. observ. (1962)
viscosa Vent.	Widely reported	Many investigators

Named for Walter Robyns, Belgian botanist.

Type: *R. vanderystii* Wilczek

1 species. Herb, annual, prostrate or ascending. Leaves petioled, palmately 3-foliolate, or the uppermost simple; stipels absent; stipules 2, free, narrow, elliptic or lanceolate. Flowers subsessile, in subcapitate 5–9-flowered, short-peduncled, axillary or terminal racemes, shorter than the leaves; bracts linear-lanceolate, persistent; bracteoles none; calyx silky-haired, about as long as the corolla; lobes 5, subulate-lanceolate, about as long as the tube, valvate; standard spathulate, clawed, not appendaged, slightly longer than the wings and keel; wings narrow, somewhat lanceolate; keel petals similar to the wings, erect, not beaked, free; fertile stamens 5, 3 anterior ones contiguous, all joined at the base with 4 staminodes; anthers suborbicular, subbasifixed; ovary sessile, 6–10-ovuled, outside silky-tomentose; style erect, straight, glabrous, persistent; stigma terminal. Pod compressed, dehiscent along 1 suture, linear-lanceolate, nonseptate, apex acute; seeds subreniform, funicle short, filiform. (Wilczek 1953)

Polhill (1976) assigned this genus to the tribe Crotalarieae, which he segregated from Genisteae *sensu lato.*

This monotypic genus is endemic in tropical Africa, specifically in Zaire and Zambia.

*Rothia Pers.   Papilionoideae: Genisteae, Crotalariinae

199 / 3659

Named for Albrecht Wilhelm Roth, botanist and physician of Bremen, Germany, 1757–1834.

Type: *R. trifoliata* (Roth) Pers.

2 species. Herbs, annual, diffuse or prostrate, up to about 15 cm tall, copiously branched, usually soft-hairy overall. Leaves palmately 3-foliolate, petioled; leaflets oblanceolate-oblong, acute or obtuse, subsessile; stipules free, minute, persistent. Flowers yellow or violet, very small, solitary or 2–4, in short, leaf-opposed racemes; bracts and bracteoles setaceous; calyx short, turbinate, deeply cleft, usually silky-haired; lobes 5, narrow, subequal, as long as the calyx tube, 2 upper lobes broader and arched; petals briefly exserted, narrow, nearly straight, distinctly clawed; standard ovate-oblong; wings and keel petals narrow, the latter slightly coherent; stamens monadelphous, united into a sheath split above; anthers small, uniform; ovary sessile, many-ovuled, usually 20 or more; style straight, glabrous; stigma terminal. Pod linear, slender, nearly straight, flattened, acute, nonseptate, dehiscent along the upper suture. (Bentham 1864; Bentham and Hooker 1865; Hooker 1879)

Polhill (1976) assigned this genus to the tribe Crotalarieae, which he segregated from Genisteae *sensu lato.*

Both species occur in tropical northern Australia, eastern India, and Africa. They flourish in sandy habitats.

NODULATION STUDIES

Nodules were observed on *R. hirsuta* (Guill. & Perr.) Bak., growing in native habitats of South Africa (Grobbelaar et al. 1964) and Zimbabwe (Corby 1974).

Rudolphia Willd.   Papilionoideae: Phaseoleae, Erythrininae
396 / 3872

Named for Karl Asmund Rudolphi, 1771–1832, German botanist.

Type: *R. volubilis* Willd.

1 species. Herb, twining. Leaves 1-foliolate, heart-shaped, acuminate; stipels present; stipules small, caducous. Flowers purple, red, or flesh-colored in elongated, axillary racemes or fascicled on the rachis; bracts and bracteoles small, caducous; calyx tubular, red or purple; 2 upper lobes rounded and joined almost to the apex, the 2 lateral lobes much smaller, the lowest lobe long, acuminate, bent; standard emarginate, rose, large, oblong, narrow-clawed, erect, folded; wings and keel petals free, narrow, and much shorter; stamens diadelphous; anthers uniform, ovate; ovary short-stalked, multiovuled; style long, straight, dilated in the middle; stigma terminal, subcapitate. Pod sessile, linear, flat, short-beaked with the persistent style, 2-valved, pulpy inside, valves tortuous at dehiscence; seeds obovate, flat. (Urban 1900; Perkins 1907)

Species segregated later from *Rudolphia* constitute *Rhodopis* (Urban 1900; Harms 1908a) and *Neorudolphia* (Britton and Wilson 1924), 2 other closely related genera. Some botanists consider these designations synonymous (Hutchinson 1964).

This monotypic genus is endemic in the West Indies, specifically Cuba and Puerto Rico. It is common in primeval forests at high altitudes and in wet thickets.

Sabinea DC.   Papilionoideae: Galegeae, Robiniinae
277 / 3742

Named for Joseph Sabine, Esq., once Secretary of the Society of Horticulture of London.

Type: *S. florida* (Vahl) DC.

4 species. Trees or shrubs, up to 6 m tall, unarmed. Leaves paripinnate; leaflets deciduous, entire; stipels minute when present; stipules short. Flowers purplish or red, in fascicles or short racemes in old leaf axils, appearing with or before the new leaves; bracts small; bracteoles absent; calyx campanulate or turbinate, truncate, open, as broad as tall; lobes 5, minute; standard broad, suborbicular, spreading or reflexed, short-clawed; wings falcate-oblong,

short-clawed, free, with a basal ear; keel petals nearly semicircular, incurved, rounded at apex, obtuse, clawed, equal to or longer than the wings; stamens 10, diadelphous; filaments subequal or the lower 5 longer and more connate; anthers uniform; ovary stalked, many-ovuled; style incurved, glabrous; stigma minute, terminal. Pod stalked, flat, linear, glabrous, acute at each end, continuous within; seeds compressed, ovate. (De Candolle 1825c; Bentham and Hooker 1865)

Members of this genus occur in the West Indies, Panama, and the Dominican Republic.

Each species is an attractive ornamental and merits consideration for avenue and garden plantings. In many areas of the West Indies these trees are among the most showy of the native plants and are unsurpassed for their bright color and feathery foliage.

Sakoanala R. Vig.   Papilionoideae: Sophoreae

Type: *S. madagascariensis* R. Vig.

2 species. Trees, 10–15 m tall. Leaves imparipinnate, deciduous; leaflets 11–13, alternate or subopposite, large, elliptic, sometimes ovate, slightly asymmetric, rounded at the base, glabrous, apex subacuminate; stipels absent. Flowers beautiful, violet blue, in large, terminal panicles, appearing before the leaves; bracteoles very small; calyx campanulate; lobes small, semiorbicular; standard orbicular, emarginate, clawed; wings oblong, shorter than the standard, not eared; keel petals oblong; stamens 10, smooth, free; anthers dorsifixed; ovary stalked, smooth, many-ovuled; style curved, glabrous. Pod very flat, papery, narrow-winged, rounded at each end, central region densely reticulate; seeds several, transverse, small. (Viguier 1952)

Both species are indigenous to Madagascar. They occur in forests along the eastern shore and in the mountainous region of the northwest.

Salweenia Bak. f.   Papilionoideae: Galegeae, Robiniinae

Name refers to the Salween Gorge, Chando District, Tibet, the type locality.

Type: *S. wardii* Bak. f.

1 species. Shrub, small, erect, up to about 1 m tall. Leaves imparipinnate; leaflets entire, narrow, 4–5 pairs; stipels none; stipules minute, leafy. Flowers short-pedicelled, in terminal fascicles; bracts and bracteoles minute; calyx campanulate; teeth 5, short, deltoid, upper 2 partly connate; standard orbicular-obovate, emarginate, shorter than the wings and keel, short-clawed; wings and keel equal in length, long-clawed; stamens 10, diadelphous; an-

thers dorsifixed, uniform; ovary stalked, few-ovuled. Pod stalked, flat, linear, slightly constricted, submembranous, 2-valved, dehiscent, 4–5-seeded. (Baker 1935a)

The genus is closely allied to *Caragana*, but its nonspinous stipules and submembranous, stipitate pods are distinguishing features.

This monotypic genus is endemic in southern Tibet. The plants grow at altitudes between 3,000 and 3,500 m on rocky, dry slopes and along stream banks.

Samanea (Benth.) Merr.   Mimosoideae: Ingeae

From Spanish, *saman*, derived from French Caribbean vernacular, *zamang*, "rain tree."

Type: *S. saman* (Jacq.) Merr.

20 species. Trees, large, with evergreen, wide-spreading crowns, or shrubs, unarmed. Leaves abruptly bipinnate; pinnae 4–6 pairs, opposite, with glands interplaced or on petiole base; leaflets paired, few or many, oblong, linear; stipules lanceolate, small, deciduous. Flowers 5-merous, pink or yellowish, in globose heads or umbels; peduncles solitary, paired or in loose fascicles in upper leaf axils; calyx tubular or campanulate, shortly 5-lobed; petals connate to the middle, valvate; stamens many, usually pink, united at the base into a tube, long-exserted; anthers minute, not glandular; ovary sessile, many-ovuled; style filiform; stigma minute, capitate. Pod stalked, compressed, curved or somewhat straight, leathery or fleshy, septate between the seeds, mesocarp pulpy, sweetish, sutures thick, persistent, indehiscent; seeds numerous, transverse, flattish, oblong-ovate, shiny. (Merrill 1916)

Samanea species are indigenous to tropical Central America, the West Indies, and northern South America. Members of this genus have been widely introduced and naturalized in many parts of the Old World tropics.

The best-known species, *S. saman* (Jacq.) Merr., the monkey pod, rain tree, or cow tamarind, is frequently discussed under its synonyms, *Pithecellobium saman* and *Enterolobium saman*. This tree species usually reaches a height of 30 m, but specimens with enormous dimensions are not uncommon. An old saman, west of Government House in Trinidad, was recorded to be about 50 m tall, with a trunk diameter over 2.6 m and a branch spread of about 60 m (Pertchik and Pertchik 1959). The crown of a magnificent tree observed in 1962 by the authors in Suva, Fiji, covered over 0.2 ha (about one-half acre). The tree house in the 1960 Walt Disney movie *Swiss Family Robinson*, filmed on the Island of Tobago, was built in the 80-m-diameter branched canopy of a gigantic 60-m tall saman tree. Underneath a famed tree near Maracay, Venezuela, it is said that Simon Bolivar encamped his entire liberation army (Pertchik and Pertchik 1959); accordingly, in this area the tree is known by the common name "saman de guerra." This specimen was described in 1799 by Humboldt during his exploration of the Orinoco region.

Folding of the leaflets of *S. saman* at night and on cloudy days is one explanation for the popular name "rain tree," since the leaves then offer no protection from rain. Another explanation stems from the fact that the species is host to sap-sucking cicada insects that feed on the foliage and eject limpid streams of honey-dew juices which dampen those who sit or sleep beneath the tree.

S. saman grows rapidly, is hardy, indifferent to soil types, and thrives from sea level to 1,000 m in both wet and dry climates, but it is seldom found in forest stands. Young specimens transplant easily. According to Pertchik and Pertchik (1959), "when Jamaica imported its beef cattle from Venezuela, it was the custom to accompany the animals during the voyage with bags of the tree's pods for feed." Since fertility of the seed is unaffected by the digestive processes in the rumen and, moreover, seed germination is thus enhanced, dissemination of the rain tree in the West Indies followed the growth of the cattle industry. In Cuba and other areas it is a desirable pasture tree commonly called cow tamarind. The sweet pods are also relished by goats and hogs. Use of the pods for human consumption, except to prepare a sweet native beverage resembling lemonade, is negligible.

This species is favored in botanical and private gardens and along roadsides for shade, and is cultivated agriculturally as shade for tea, coffee, teak, and livestock. Because of its shallow, radiating root system and topheavy crown, it is undesirable near buildings.

The wood of *S. saman* is lightweight, with a density of 720–880 kg/m^3, has a golden brown color with dark streaks, takes a fine finish, and is resistant to dry-wood termites. The timber is not harvested commercially because of the scarcity of timber stands. Its use is reserved for bowls, trays, carvings, furniture, and paneling. In Central America, cross sections of the trunk are used for two-wheeled carts. Similar uses are made of the wood of *S. saminiqua* Pitt. and *S. polycephala* (Benth.) Pitt. in Venezuela, and of *S. pedicellaris* (DC.) Killip, from the lower Amazon area southward to Rio de Janeiro.

Improved growth, yield, protein content, and nutritive quality were recorded for *Axonopus compressus*, a pasture grass, grown under *S. saman* in Malaysia (Jagoe 1949). The benefit by association was presumptively attributed to nitrogen made available in the soil by excretion or decomposition of the leguminous nodules. *Samanea* bark lacks tannin but does yield an inferior gum. The bark and seeds yield a minor saponinlike alkaloid, pithecolobine, $C_{22}H_{46}N_4O_2$ (Wiesner et al. 1952; Wiesner et al. 1953).

NODULATION STUDIES

Nodules were recorded on *S. saman* in Hawaii (Allen and Allen 1936a), in Malaysia (Jagoe 1949), and in the Philippines (Bañados and Fernandez 1954).

Colonies of *Samanea* rhizobia are raised and mucilaginous, and growth is moderately fast. Alkaline reactions are produced in litmus milk.

Reciprocal cross-inoculation of *S. saman* and *V. unguiculata* (= *V. sinensis*), by rhizobial strains from each other's host, was shown by the authors in 1936(a). In later tests (Allen and Allen 1939), *S. saman* was nodulated by each of 51 rhizobial strains from 26 plant species in the cowpea miscellany. *S. saman* was benefited by 28 strains. Strains from *S. saman* failed to nodulate only 2 of

20 species tested, namely *Cytisus scoparius* and *Phaseolus lunatus*. Only *Canavalia enviformis* and *Clitoria ternatea* showed ineffective reponses to nodulation by the *Samanea* strains. In later tests, 2 *Samanea* strains markedly benefited the growth of peanut plants (Allen and Allen 1940a). This mimosaceous species showed positive affinities with the so-called cowpea miscellany in all tests.

Saraca L.   Caesalpinioideae: Amherstieae
47 / 3497

Name is probably a corruption of the vernacular Indian name *asok*, or *asoka*, "without grief."

Type: *S. indica* L.

25 species. Shrubs or trees, medium-sized, usually 7–10 m tall, but occasionally up to about 20 m, evergreen, unarmed. Leaves paripinnate, large; new leaf shoots developing after dry spells are at first pink to purplish and hang laxly as tassels, later stiffening and becoming dark green; leaflets few, usually up to 7 pairs, leathery, subsessile, ovate-lanceolate, mostly acuminate, uppermost leaflet largest, occasionally with a pair of tuberclelike glands at the margin of the leaf base and at the tip; stipules connate, small, caducous. Flowers fragrant, especially at night, yellow, rose, or red, with dark-colored center or eyespot, mostly bisexual, pedicelled in upturned, rounded, paniculate corymbs on twigs, the older wood of branches, and on tree trunks; bracts small, caducous; bracteoles 2, often showy, yellow to red, somewhat persistent, shorter than the calyx tube; calyx tube long, cylindric; lobes 4, petaloid, often showy, orange, ovate, subequal, imbricate in bud; petals absent; stamens 3–10, often partly abortive; filaments free, yellow or red, smooth, long-exserted; fertile anthers purplish, small, oval, versatile, dorsifixed, dehiscent by slits; ovary oblong, hairy, many-ovuled, stalk partly adnate to the calyx tube; style filiform; stigma terminal. Pod oblong or linear, compressed or almost turgid, leathery to woody, dark red or purple black, 2-valved; seeds 1–8, often laterally compressed, ovoid or angled. (Bentham and Hooker 1865; Zuijderhoudt 1967)

The recent revision of this genus by Zuijderhoudt (1967) reduced the number of species to 8 and designated *S. asoca* (Roxb.) De Wilde the type species. The International Association for Plant Taxonomy, however, has not passed a ruling on this revision.

Representatives of this genus are common in tropical Asia and Indo-Malaya. The species thrive on cool, moist to wet, well-drained soil, especially along shaded stream banks and in rocky and jungle areas up to 1,300-m elevation. The rocky mountain and hill streams that tunnel through Malaysian forests are commonly called saraca streams because their banks are predominantly flanked with these trees.

According to legend, Gautama Buddha was born under the "sorrowless tree," *S. indica* L. It is regarded as a sacred tree, and frequently planted in gardens of Buddhist and Hindu temples.

Saraca species are among the most beautiful tropical trees that thrive under shade. *S. declinata* Miq., *S. indica*, *S. palembanica* (Miq.) Bak., *S. thaipingensis* Cantley, and *S. trianda* Bak. are highly prized as ornamentals, and have been introduced for that purpose into Florida. Several species are used successfully as shade on coffee plantations. Decoctions of the bark are used as an astringent and uterine tonic in Asia. *Saraca* wood is soft and reddish in color. The supply is limited and of no commercial importance.

NODULATION STUDIES

Four species, *S. declinata*, *S. indica*, and *S. thaipingensis* in Hawaii (Allen and Allen 1936b) and *S. asoca* in western India (V. Schinde personal communication 1977), lacked nodules. Seedlings of the first-named species examined in the Philippines (personal observation 1962) confirmed the negative findings of Bañados and Fernandez (1954) for that area. These observations contradict Corner's (1952) statement that in Malaysian forests saracas grow "mostly at the very edge of the stream and their dark fibrous roots, beset with nodules, trail out in bundles in the running water." Corner's comments are accepted, however, as evidence that some species of *Saraca* bear nodules in their native habitat. Indeed, Lim (1977) recorded the presence of sparse nodules on *S. declinata* in Singapore.

Sarothamnus Wimm.   Papilionoideae: Genisteae, Cytisinae
/ 3682a

From Greek, *saron*, "broom," and *thamnos*, "shrub."

Type: *S. scoparius* (L.) W. D. J. Koch

1 species. Shrub, usually 0.5–2 m tall; branches slender, wandlike, erect, 5-angled. Leaves short-petioled, 3-foliolate, or 1-foliolate and subsessile on young twigs; leaflets small, obovate to lanceolate, underside silky-haired, deciduous, or later reduced to scales. Flowers yellow, 1–2, pedicelled, in lower leaf axils; bracteoles scalelike, caducous; calyx obscurely bilabiate, lips ovoid, uppermost lip 2-toothed, lower lip 3-toothed; petals unequal; standard ovate, reflexed; wings short; keel petals joined, incurved, blunt, pubescent on the back; stamens 10, all connate into a tube, free above; filaments alternately longer and shorter; ovary short-stalked, villous; style smooth, very long, spirally incurved; stigma capitate, terminal. Pod oblong, strongly compressed, slightly beaked, with yellowish or white hairs along the sutures, valves smooth, black at maturity, twisting during explosive dehiscence; seeds numerous, blackish brown, small, ellipsoid. (Gams 1924; Komarov 1971)

This monotypic genus is endemic in Europe, from southern Sweden to western Ukraine. The plants grow in conifer and mixed woodlands and are gregarious on silica soils and sandy heaths.

This species is a useful sand-binder and serves as browse for sheep and goats. The twigs are a fiber source used as a substitute for jute and in the manufacture of brooms. The bark yields a yellow and brown dye used to color paper and textiles. In France the buds and leaves are sometimes eaten in salads; in western Germany the yound pods are cooked in the same manner as green beans. Young buds and seeds have served as a coffee substitute and sometimes are marinated in vinegar, salt, or alcohol as a relish. Flowering tops of the plants have been used as a cardiac tonic and a diuretic in folk medicine. Sarothamnine, $C_{30}H_{50}N_4$, a bisparteine-type alkaloid, has been isolated from branch tips (Delaby et al. 1949).

NODULATION STUDIES

S. scoparius (L.) W. D. J. Koch, commonly called broom, is known by many synonyms, such as *S. vulgaris* Wimm., *Genista scoparia*, *Cytisus scoparius*, and *Spartium scoparium*. Reports of nodulation are widespread under each of these names. Drawings by Gams (1924) depicted the nodules as elongated, lobed or digitate, containing swollen bacteroids. Histological studies by Lechtova-Trnka (1931) revealed nothing unusual in the infection strands and cell hypertrophy of the invaded cells.

Rhizobial isolates showed responses restricted to the homologous host in early cross-infection trials by Maassen and Müller (1907) and were thus regarded as a separate group. More recently, broad spectrum testing of *S. scoparius* has shown a natural alliance with members of the promiscuous cowpea-lupine-soybean cross-inoculation grouping (Allen and Allen 1939; Wilson 1939a).

Sauvallella Rydb.   Papilionoideae: Galegeae, Robiniinae

Named for Francisco A. Suavalle, 1807–1879, Cuban botanist; he published a revised catalogue of plants collected in Cuba by Charles Wright.

Type: *S. immarginata* (Ch. Wright) Rydb.

1 species. Shrub. Leaves imparipinnate; leaflets subopposite, several pairs; stipels minute, lanceolate; stipules free, awl-shaped. Flowers in axillary racemes; bracts small, bracteoles absent; calyx short-turbinate, broader than long; teeth minute, lower 3 triangular-subulate, upper 2 close together, blunt; standard suborbicular, short-clawed; wings oblique-obovate, short-clawed, with a prominent basal ear; keel petals crescent-shaped, united to the apex, longer than the other petals; stamens diadelphous; ovary glabrous, many-ovuled; style glabrous, slightly incurved; stigma terminal. Pod flat, linear, stalked, many-seeded. (Rydberg 1924)

The genus is intermediate between *Gliricidia* and *Poitaea*. According to León and Alain (1951), this species is identical with *Cajalbania immarginata*.

This monotypic genus is endemic in Cuba. The plants grow in pine forests.

Named for G. Scheffler; he collected the specimens in Tanzania, Rwanda, and Burundi, formerly German East Africa, about 1900.

Type: *S. usambarense* Harms

6 species. Trees, large, up to about 30 m tall, with trunks 30–40 cm in diameter, or shrubs. Leaves imparipinnate; leaflets alternate, under surfaces gland-dotted; stipels absent; stipules small, caducous. Flowers white, in axillary and terminal racemes or panicles; bracts and bracteoles minute, caducous; calyx campanulate, glandular; teeth very short, covered with rusty red, soft hairs; standard longer than the calyx and other petals, suborbicular to obovate, with short, broad claw, rusty velvety-haired; wings narrow, obtuse, slightly hairy, sparsely glandular; keel obtuse, outside beset with small, globose glands; stamens diadelphous; ovary long-stalked, 3–4-ovuled, densely glandular and more or less rusty tomentose; style short, awl-shaped; stigma small, terminal. Pod stalked, obliquely semiobovate, oblanceolate, or subfalcate, asymmetric, narrowed toward the base, curved, briefly apiculate, woody, tomentellous and glandular outside, tardily dehiscent, valves thick, inside hairy, 1–2-seeded. (Harms 1908a; Pellegrin 1948; Hauman 1954a)

Members of this genus are widespread in tropical Africa.

The wood is of good quality and is used, when available, in the manufacture of musical instruments.

From the Greek verb *schizo*, "divide," and *lobion*, "pod"; the inner and outer layers of the pod separate at maturity.

Type: *S. parahybum* (Vell.) Blake (= *S. excelsum* Vog.)

4–5 species. Trees, tall, rapid-growing, unarmed, deciduous, usually with cylindric boles, high buttresses, and wide-spreading, open crowns. Leaves bipinnate, large; pinnae 15–20 pairs, fernlike; leaflets small, elliptic, 10–20 pairs, stipules absent. Flowers golden yellow, large, showy, profuse, in axillary, semierect racemes or terminal panicles; bracts minute; bracteoles absent; calyx tube obliquely turbinate; lobes 5, overlapping, reflexed at flowering; petals 5, clawed, subequal, overlapping, uppermost petal innermost; stamens 10, free, subdeclinate; filaments villous, basally rough; anthers uniform, longitudinally dehiscent; ovary subsessile, affixed to 1 side of calyx tube, many-ovuled; style filiform; stigma minute, terminal. Pod flat, spoon-shaped, exocarp firm, leathery, separating from the thin, membranous, winglike endocarp layer at maturity, 2-valved, tardily dehiscent, 1-seeded; seed large, oblong, compressed, located near apex. (Bentham and Hooker 1865; Britton and Rose 1930; Backer and Bakhuizen van den Brink 1963)

Representatives of this genus range from Brazil to southern Mexico. They flourish on well-drained soils in moist, mixed and secondary forests on plains or hillsides.

The type species, named for the Parahyba River in Brazil, is listed as one of the most magnificent fast-growing flowering trees of the tropics. Three-year-old plants are often 7–8 m tall; forest specimens reach heights of 30–35 m. The trunk is straight and free of branches except for the crowning cluster of leaves, which resemble tree-fern fronds. The trees are leafless at the time of flowering, and so are especially beautiful when covered by masses of large, showy, yellow gold blossoms. The plants are readily propagated from seeds and are cultivated as ornamentals in Sri Lanka, Java, Florida, and various Pacific Islands.

The timber is rarely utilized, possibly because it has a fecal odor when freshly cut and is not resistant to decay or insect attack. The soft, spongy, light-colored young wood has potential as a source of paper pulp, but thus far the trees are not grown commercially.

Nodulation Studies

Nodules were absent on seedlings and mature trees of the type species in Hawaii (personal observation 1940) and in the Philippines (Bañados and Fernandez 1954; personal observation 1962).

Schizoscyphus Schum. ex Taub.   Caesalpinioideae:
38 / 3482   Cynometreae

From the Greek verb *schizo*, "divide," and *skyphos*, "cup," in reference to the deep split along the ventral side of the calyx tube.

Type: *S. roseus* (Schum.) Taub.

1 species. Tree, medium-sized, up to 9 m tall; branches slender, deeply 4-grooved. Leaves subsessile, paripinnate; leaflets 6–8 pairs, sessile, oblong-lanceolate, asymmetric, covered with deciduous scales prior to unfolding. Flowers showy, rose, many, on short, rusty-haired peduncles, in terminal racemes; bracts caducous; calyx obliquely and narrowly turbinate, striate, unilaterally split in front; lobes 4, reflexed, oblong, membranous; petals 3, very small, lanceolate, acute, equal; stamens more than 30, connate beyond the calyx tube into a sheath split ventrally like the calyx; filaments long-exserted; anthers apiculate, versatile; ovary sessile, 1-ovuled; style elongated, curved; stigma funnel-shaped. Pod not described.

This monotypic genus is endemic in Papua New Guinea. The plants grow in upland forests.

Named for Richard van der Schot, Chief Gardener at the Schönbrunn Imperial Garden, Vienna, and traveling companion of Jacquin.

Type: *S. afra* (L.) Thunb.

20 species. Shrubs or small trees, much-branched, unarmed, deciduous before flowering. Leaves paripinnate; leaflets leathery, 3–18 pairs, lowermost usually smaller than upper pairs; stipules small, caducous. Flowers bright red, pink, or flesh-colored, many, in short, dense, showy panicles, often arising from old wood; bracts and bracteoles small, ovate-oblong, membranous, caducous; calyx turbinate or cylindric; lobes 4, overlapping, not reflexed at flowering, sometimes pink, caducous; petals 5, ovate or oblong, subequal, longer than and inserted in mouth of calyx tube, subsessile, imbricate, uppermost petal innermost, sometimes small and scalelike; stamens 10, free or joined shortly at the base, inserted with petals; anthers uniform, longitudinally dehiscent; ovary stalked, stalk attached to 1 side of calyx tube, several- to many-ovuled; style longer than stamens; stigma small, terminal. Pod flat, linear, oblong or falcate, leathery or woody, upper suture often thickened or winged, subdehiscent, 1- to few-seeded; seeds globose, compressed, with or without a yellow, fleshy, cupular aril. (Bentham and Hooker 1865; Baker 1930; Pellegrin 1948; Léonard 1952, 1957)

Species differentiation within this genus is fraught with difficulty. Of 15 or more specific names recognized in southern Africa, Codd (1956) concluded that only 4 species were validly distinctive: *S. afra* (L.) Thunb., *S. brachypetala* Sond., *S. capitula* Bolle, and *S. latifolia* Jacq.

Members of this genus are widespread in tropical Africa; about 10 species are indigenous to South Africa. They thrive in river bottom areas, high forests, bushveld, savannas, and dry scrub.

These species are mostly small- to medium-sized, drought-resistant, slow-growing trees found in dry areas, but some are cultivated for their ornamental beauty. The handsome clusters of flowers filled with nectar are attractants for bees and small birds. Trees laden with pods that range in color from green to rust to pink are as striking a sight as ones in full bloom. The first European settlers in South Africa applied the name "boerboon," "farmer's bean," or "Hottentot's bean" to *Schotia* species because the large seeds were roasted and eaten in times of food scarcity. They are used as an ingredient for porridge or ground into a meal for a native bread.

The economic value of *Schotia* species is negligible. The wood, which resembles walnut, is hard, durable, and tough, but only a limited supply is available. It is used for furniture and small objects. The wood dust is an eye irritant. The bark yields appreciable tannin, that of *S. afra*, about 16 percent. The foliage is browsed by livestock. Natives have used decoctions of plant parts for various medicinal purposes. The seed arils are eaten by birds.

NODULATION STUDIES

Nodules were lacking on specimens of *S. brachypetala* in Hawaii (Allen and Allen 1936b), South Africa (Grobbelaar et al. 1964), and Zimbabwe (Corby 1974), on *S. capitata* in South Africa (Grobbelaar and Clarke 1972) and Zimbabwe (Corby 1974), and on *S. latifolia* in Hawaii (Allen and Allen 1936b) and South Africa (Grobbelaar and Clarke 1972).

The Latin suffix, *astrum*, indicates inferiority or incomplete resemblance to *Schrankia*.

Type: *S. insigne* Hassl.

1 species. Shrub or small tree, unarmed. Leaves bipinnate; pinnae pairs few; leaflet pairs many; stipels present. Flowers sessile, in heads disposed along cylindric, foliaceous spikes; calyx campanulate, 4-lobed, valvate; petals 4, equal, valvate to the middle; stamens 8, free, very long, exserted; anthers small, without glands; ovary short-stalked, multiovuled; style slender; stigma small, terminal. Pod stalked, linear, about 15 cm long, subquadrangular, somewhat torulose, beaked, valves entire, separating from the septate, jointed, dilated sutures, elastically dehiscent; seeds oblong. (Hassler 1919)

This monotypic genus is endemic in northern Paraguay. The plants inhabit arid mountain forests.

Schrankia Willd.   Mimosoideae: Eumimoseae

13 / 3448

Named for Franz von Paula von Schrank, 1747–1835, Professor of Botany, Munich.

Type: *S. quadrivalvis* (L.) Merr. (= *S. aculeata* Willd.)

20 species. Subshrubs or herbs, perennial, deciduous, semierect to prostrate; stems weak, not woody, often angular, with short, recurved prickles. Leaves bipinnate; pinnae 1–8 pairs; leaflets usually sessile, small, 4-numerous, glabrous, inequilateral, often sensitive; petiole not glandular; rachis bristly; stipules setaceous. Flowers small, perfect or polygamous, rose, purplish, or pinkish white, 4-or 5-parted, sessile, in globose heads or short cylindric spikes on prickly, axillary, solitary or fascicled peduncles; calyx minute, inconspicuously toothed; petals united into a funnel-form corolla, deeply lobed above; stamens as many or twice as many as the petals, free, exserted; filaments filiform in bisexual flowers, flattened in the male flowers; anthers small, eglandular; ovary subsessile, multiovuled; style filiform; stigma terminal, obtuse. Pod linear, narrow, flattened or tetragonal, acute or acuminate, densely prickly, nonseptate, valves separating from the wide, thickened, persistent margins; seeds longitudinal, oblong, quadrangular. (Bentham and Hooker 1865; Isely 1973)

Schrankia species are well distributed from the southern United States to northern Argentina. They flourish on dry prairies and in open woods on sandy soil.

Several species are commonly called sensitive briar; however, the leaves are not as sensitive as those of *Mimosa* species.

Nodulation Studies

Nodules have been observed on *S. leptocarpa* DC. in Venezuela (Barrios and Gonzalez 1971), and on *S. uncinata* Willd. in France (Lechtova-Trnka 1931) and in the United States (Leonard 1925; Wilson 1944a,b; Wilson and Chin 1947). Lechtova-Trnka (1931) described typical nodule development and dissemination of rhizobia by infection threads. A wide range of host infectiveness was shown by rhizobia from *S. uncinata* (Wilson 1944a,b; Wilson and Chin 1947).

Sclerolobium Vog.   Caesalpinioideae: Sclerolobieae
113 / 3571

From Greek, *skleros*, "hard," and *lobion*, "pod."
    Type: *S. denudatum* Vog.
    25–30 species. Trees, small to medium, rarely large, unarmed, twigs of some species hollow and inhabited by ants. Leaves paripinnate, rarely imparipinnate, 1–3-foliolate; leaflets oblong, acuminate, leathery; stipules leafy, deeply cleft or minute, deciduous or persistent. Flowers small, yellow or white, in dense racemes or spicate terminal panicles; bracts minute or conspicuously awl-shaped in bud; bracteoles awl-shaped, often absent; calyx tube short, campanulate; lobes 5, subequal, imbricate; petals 5, small, linear or ovoid, subequal, often pilose, imbricate; stamens 10, free; filaments hairy at the base, plicate in bud; anthers versatile, longitudinally dehiscent; ovary short-stalked; ovules few; style slender, smooth; stigma small, terminal. Pod short-stalked, compressed, ovate, epicarp loose, detachable, mesocarp thinly fibrous, endocarp thin and hard, indehiscent, 1–2-seeded; seeds large, compressed, orbicular-reniform. (Bentham and Hooker 1865; Amshoff 1939; Dwyer 1957b)
    Representatives of this genus are indigenous to tropical northern South America.
    S. chrysolobium Poepp. & Endl. is one of the tallest and best-known trees in Venezuela, reaching heights up to 30 m. The wood is readily subject to decay, rather coarse, varies in density between 400 and 700 kg/m³, and is used for small carpentry items and as a fuel (Record and Hess 1943).

NODULATION STUDIES

Numerous round nodules, 3–4 mm in diameter, were observed on the roots of *S. micropetalum* Ducke in Bartica, Guyana (Norris 1969) and on *S. aureum* (Tul.) Benth. var. *grandiflorum* Dwyer in Venezuela (Barrios and Gonzalez 1971).

Scorodophloeus Harms   Caesalpinioideae: Cynometreae
(39b) / 3485

From Greek, *skorodon*, "garlic," and *phloios*, "bark."
    Type: *S. zenkeri* Harms
    2 species. Trees, large, tall, evergreen, unarmed. Leaves usually imparipinnate; leaflets small, unequal, 3–20, alternate, sometimes subopposite

to opposite, subsessile, entire, not pellucid-punctate; stipules free, linear-lanceolate, very caducous. Flowers arranged spirally in dense, terminal or axillary racemes; bracteoles linear-lancolate, paired below the middle of the pedicels, not forming an involucre around the bud; calyx narrowly subcylindric-turbinate, elongated; lobes 4, ovate, obtuse, imbricate, later reflexed; petals 5, obovate, margins fringed, free, imbricate, subequal, longer than the lobes, narrowed toward the base, short-clawed, blunt; stamens 10, free, exserted; filaments slender, free, glabrous, alternately long and short; anthers dorsifixed, longitudinally dehiscent; ovary compressed, stalked, stalk adnate to the calyx tube, glabrous with base puberulent, 2-ovuled; disk absent; style elongated, slender; stigma small, capitellate. Pod stalked, compressed, asymmetrically oblong, apiculate, leathery or woody, transversely veined, 2-valved, curled on dehiscence; seeds large, compressed, suborbicular. (Harms 1908a; Pellegrin 1948; Léonard 1951, 1952, 1957; Brenan 1967)

Both species occur in tropical Africa. They flourish along river banks and in forests.

S. zenkeri Harms, a large tree often 30–40 m tall, with a trunk diameter of 60–100 cm, yields a very hard, durable wood which emits a strong garlic odor. For this reason the trees and the wood are especially objectionable after a rain. The bark and leaves are used by natives as a condiment, or substitute for garlic.

NODULATION STUDIES

Nodules were absent on this species in Yangambi, Zaire (Bonnier 1957).

Scorpiurus L. Papilionoideae: Hedysareae, Coronillinae
303 / 3771

From Greek, *skorpios*, "scorpion" and *oura*, "tail," in reference to the appearance of the pods.

Type: *S. sulcatus* L.

7 species. Herbs, acaulescent or decumbent, annual. Leaves simple, lanceolate, entire, 3–5, parallel-veined, narrowed into the petiole; stipules long, linear, partly adnate to the petiole. Flowers yellow or purplish, small, nodding, solitary or umbellate, 3–6, on axillary peduncles; bracts minute; bracteoles absent; calyx campanulate; teeth 5, upper 2 connate above; petals long-clawed; standard orbicular; wings obliquely obovate or oblong; keel incurved, sharp-beaked; stamens 10, diadelphous; alternate filaments dilated above; anthers uniform; ovary sessile, cylindric, multiovuled; style inflexed, dilated in the middle; stigma terminal, capitate. Pod lomentaceous, subterete, circinately inrolled, deeply grooved, ribs often warty or glandular, the 1-seeded joints scarcely separating and subcontinuous, indehiscent; seeds ovoid-globose, cotyledons twisted, folded, and elongated. (Bentham and Hooker 1865)

Species delineation within this genus has presented problems because of the varietal and continuous range of intermediate forms (Heyn and Raviv 1966).

Members of this genus range from the southwestern Mediterranean area into south central Europe.

S. vermiculatus L. is cultivated on a small scale in Argentina, where plant parts are used as spicy supplements in salads. Several species have possibilities as a pasture plant for sheep.

Nodulated species of *Scorpiurus*	Area	Reported by
muricatus L.	Israel	Hely & Ofer (1972)
var. *subvillosus* (L.) Fiori	Yugoslavia	Pers. observ. (U.W. herb. spec. 1970)
(= *subvillosus* L.)	France	Naudin (1897a)
vermiculatus L.	Germany	Wydler (1860)
	France	Clos (1896)

Scottia R. Br. ex Ait. Papilionoideae: Genisteae, Bossiaeinae

Named for Robert Scott, 1757–1808, physician and Professor of Botany in Dublin.

Type: *S. dentata* R. Br. ex DC.

6 species. Shrubs, erect, glabrous, up to about 3 m tall. Leaves opposite, lanceolate-orbicular, entire, very small, sometimes triangular, sinuate or dentate; stipules minute. Flowers yellow orange, solitary, pedicelled, axillary; bracts broad, rigid; bracteoles long, caducous; calyx persistent; lobes 5, mostly unequal in length, upper 2 broad, obtuse, large, others small; standard 2–3 times longer than the calyx; stamens all connate into a sheath split above; anthers uniform, versatile, dorsifixed; ovary long-stalked, 2–8-ovuled. Pod linear, glabrous, long-stalked, exserted, valves leathery, margins thick, completely dehiscent. (Bentham 1864)

These 6 species formerly constituted series *Oppositifoliae* in *Bossiaea* (Bentham 1864; Burbidge 1963). Generic status was based on the occurrence of opposite leaves and long-stalked pods in contrast to alternate leaves and sessile or short-stalked pods in *Bossiaea* species.

Scottia species are indigenous to Western Australia.

NODULATION STUDIES

S. aquifolium R. Br. ex DC. and *S. dentata* R. Br. ex DC. were reported nodulated as *Bossiaea aquifolium* and *B. dentata*, respectively. Rhizobial isolates from these species produced good nodulation on cowpeas and to a lesser extent on *Lupinus* species, *Glycine hispida*, and *Phaseolus lathyroides*, now *Macroptilium lathyroides*; nodules were not produced on clover and medic (Lange 1961). Inclusion of the *Scottia* species within the cowpea miscellany appears justified.

From Latin, *securis*, "ax," and the verb *gero*, "carry."
    Type: *S. securidaca* (L.) Dalla Torre & Sarntheim.
    1 species. Herb, annual, glabrous; stems several, hollow, from rootstock. Leaves imparipinnate, 5–7 pairs, crowded in the upper part of the rachis, entire, oblong-obovate, glaucescent underneath; stipules small, membranous, oblong-linear. Flowers nodding, yellow, 4–8, in umbellate racemes on long axillary or terminal peduncles; bracts minute, deflexed; bracteoles none; calyx short-campanulate, slightly bilabiate; lobes subequal, triangular-lanceolate, upper 2 broader and connate higher up than the lower 3; petals free from the staminal tube; standard suborbicular; wings oblique-oblong; keel incurved, subbeaked; stamens 10, diadelphous; alternate filaments slightly dilated upwards; anthers uniform; ovary sessile, many-ovuled; style incurved, glabrous; stigma capitate. Pod linear, falcate, terminating in a hooked beak due to the persistent style, plano-compressed, tardily 2-valved to indehiscent, valves papillose, constricted between brown, 4-angled, flattened seeds. (De Candolle 1825b; Bentham and Hooker 1865)
    This monotypic genus is endemic in the Mediterranean region, from Spain to the Caucasus. The plants grow in deciduous woods as scrub and weeds. They are usually abundant along roadsides and on cultivated fields.
    All plant parts have a bitter, unpleasant taste.

NODULATION STUDIES

    This species was nodulated in the hothouse at Kew Gardens, England (personal observation 1936), in Australia (Brockwell et al. 1966), and in South Africa (Grobbelaar and Clarke 1974). Rhizobial isolates were ineffective in association with several *Lotus* species (Brockwell et al. 1966).

Sellocharis Taub.   Papilionoideae: Genisteae, Crotalariinae
207 / 3667

Named for Friederich Sellow, 1789–1831, German botanist; he traveled and collected extensively in Brazil.
    Type: *S. paradoxa* Taub.
    1 species. Shrublet, erect, young branches yellow silky-haired, later smooth. Leaves sessile, sublinear-lanceolate, mucronate, usually 6 in pairs at right angles, giving the foliage a whorled appearance; stipels and stipules none. Flowers solitary or paired in axils of verticillate leaves; calyx subcampanulate, bilabiate, rusty-haired; teeth 5, upper 2 triangular, lower 3 longer and connate almost to the apex; standard orbicular, deeply emarginate, short-clawed; wings equal to the standard, oblong, short-clawed; keel petals about half as long as the standard, long- and slender-clawed, inconspicuously eared; stamens connate into a sheath split above; anthers globose, uniform; ovary subsessile, linear, compressed, multiovuled; style short, straight, awl-shaped; stigma capitate, thick. Pod subsessile, linear, flattened, tawny-

pubescent, upper suture thick, ending with a persistent style, 2-valved; seeds many.

This monotypic genus is endemic in southeastern Brazil.

Polhill (1976) assigned this genus to the tribe Crotalarieae in his reorganization of Genisteae *sensu lato*.

Serialbizzia Kosterm.   Mimosoideae: Ingeae

Type: *S. acle* (Blanco) Kosterm.

The 2 species of this Malaysian genus were segregated by Kostermans (1954) from the *Pithecellobium* complex because of their distinctive, straight, strap-shaped, indehiscent pods, which crack or break irregularly. Acceptance of the genus is in abeyance, awaiting a comprehensive revision of *Pithecellobium* on a world basis (Hutchinson 1964). Bentham's concept (1875) of *Pithecellobium* continues to have wide adherence.

Members of this proposed genus occur in Indochina, western Malaysia, and Celebes.

Serianthes Benth.   Mimosoideae: Ingeae
5 / 3439

From Greek, *serikos*, "silky," and *anthos*, "flower," in reference to the dense silky hairs on the outer surface of the petals.

Type: *S. dilmyi* Fosberg (= *S. grandiflora* (Wall.) Benth.)

13 species. Trees, small to large, rarely shrubs, unarmed; twigs warty, with crowded leaf scars; young twigs rusty-tomentose. Leaves large, bipinnate; pinnae pairs many; leaflets small, sessile, asymmetrically oblong, many pairs; main rachis with elevated, disklike glands; stipules obscure. Flowers white or yellow, subsessile on thick pedicels, few, large, in axillary corymbose panicles; calyx broad, campanulate, actinomorphic; lobes 5, short, thick; petals 5, equal, free, outer surface silky-haired; stamens very numerous, 200–500; filaments often red or yellowish white, much longer than the petals, free above, but all joined at the base with the ovary into a tube; anthers minute; ovary sessile, glabrous, multiovuled; style filiform. Pod sessile, oblong, thick, woody, variously cross-veined, usually covered with dense, brown hairs, margins often thickened, indehiscent; seeds shiny, oblong, arranged transversely in the pod between transverse septa. (Bentham and Hooker 1865; Fosberg 1960)

The genus is closely allied to *Wallaceodendron* and *Albizia*.

Representatives of this genus are distributed almost exclusively on South

Pacific islands and the Malay Peninsula (Fosberg 1960). They flourish in littoral zones on soil derived from calcareous or serpentine rocks.

The yellow-tinged wood of *S. myriadenia* (Bert. ex Guill.) Planch. is highly valued for canoes and paddles by South Pacific natives. The red seeds are used in leis and are edible when roasted. *S. nelsonii* Merr., restricted in habitat to the islands of Rota and Guam, presumably became extinct on the latter as a result of the devastating typhoon of 1962 (Stone 1971).

Sesbania Scop.   Papilionoideae: Galegeae, Robiniinae
281 / 3747

Sesbania is Latinized version of the French, *sesban*, from Arabic, *saisaban*, or from Persian, *sisaban*.

Type: *S. sesban* (L.) Merr.

70 species. Herbs, shrubs, annual or perennial, or small to medium, slender trees, fast-growing, short-lived, ever-green or deciduous. Leaves paripinnate, long, narrow; leaflets in many pairs, rounded or oblong, usually asymmetric at the base, often glaucous; stipels minute or absent; stipules small, foliaceous, awl-shaped, caducous or somewhat persistent. Flowers attractive, yellow, red, purplish, variegated or streaked, seldom white, large or small, on slender pedicels, solitary or paired in short, lax, axillary racemes, usually unpleasantly scented; bracts and bracteoles setaceous, caducous or rarely persistent; calyx tube broad, campanulate, sometimes truncate; lobes 5, short-dentate, subequal, usually subulate-triangular; all petals long-clawed; standard orbicular or ovate, spreading or reflexed, with 2 free or adnate appendages on the claw; wings falcate-oblong or linear, sometimes eared; keel incurved, boat-shaped or plano-compressed, obtuse, sometimes acuminate; stamens diadelphous, vexillary stamen bent; anthers dorsifixed, uniform or alternate 5 slightly longer; ovary sessile or stalked, many-ovuled, sometimes villous; style glabrous, incurved; stigma small, capitate. Pod linear, often 10 cm or more long, cylindric or compressed, rarely oblong, sometimes 4-angled or 4-winged, pendulous, 2-valved, subindehiscent or dehiscent, transversely septate between numerous seeds; seeds oblong or subquadrate. (Bentham and Hooker 1865; Phillips and Hutchinson 1921; Cronquist 1954a; Gillett 1963b)

Sesbania species are widespread in the warmer latitudes of both hemispheres. They are common along stream and swamp banks and in moist and inundated bottomlands. The plants grow both wild and cultivated. A comprehensive account of African species was published by Gillett (1963b).

S. grandiflora Poir., a much-cultivated tree for roadsides and gardens in the tropics, typifies the ornamental species of this genus. This rapid-growing tree reaches a height of about 10 m, with a trunk diameter of about 0.3 m. The beauty of the tree lies in its graceful habit, dark green foliage, and large, butterfly-shaped, pink to magenta red flowers that are borne in short, pendulous clusters. The upper petals are usually darker than the lower ones.

Single- and double-flowered varieties are known; the white-flowered variety is uncommon and usually in demand. However, this tree has 3 undesirable features: (1) it is short-lived; (2) it tends to be shallow-rooted and subject to uprooting by high winds; and (3) it is a prolific seeder that causes objectionable pod litter on lawns.

All *Sesbania* species have soil-improvement properties; however, their use as ground cover and for soil erosion has been limited. *S. exaltata* (Raf.) Cory and *S. macrocarpa* Muhl. were once used extensively for these purposes in citrus groves. *S. tomentosa* H. & A., known as ohai in Polynesia, is particularly effective in preventing erosion in many coastal areas. *S. aegyptiaca* Poir., *S. cinerascens* Welw. ex Bak., *S. grandiflora*, *S. punctata* DC., *S. roxburghii* Merr., and *S. sesban* (L.) Merr. are noteworthy for their use as windbreaks, live fences, supports for peppers, and as shade on coffee and tea plantations.

Various species yield fiber suitable for ropemaking and fish nets. A Japanese patent specifies the stems of *S. aegyptiaca* as a substitute for hemp (Matsuoka 1920). Dundee fiber, or dhaincha, is made from the stems of *S. cannabina* Roxb. and *S. aculeata* Poir. in India as a substitute for jute and hemp. The long stems of *S. exaltata* also have fiber quality; the pith is used in making a paper similar to rice paper.

The flowers of *S. grandiflora* are eaten as a vegetable, or salad, and used in curries. The leaves are reportedly a source of ascorbic acid. The clear, garnet red, astringent katurai gum obtained from the trunk turns black upon exposure to air. The gum qualities are similar to those of gum arabic. Bark decoctions are astringent. Leaf juices have diuretic and laxative properties. The seed oil of *A. aegyptiaca* is a rich source of oleic, linoleic, and stearic fatty acids (Farooq et al. 1954). The seeds of *S. sesban* are used medicinally in many areas; the leaves contain a saponin.

S. sesban and *S. speciosa* Taub. are excellent forage and fodder plants in India and Taiwan. The former is valuable in saline areas on land subject to flooding, and is cultivated profitably on irrigated pastures. *S. cinerascens* and *S. sesban* var. *zambesiaca* are invaluable cattle graze in the flood plains of the Chambeshi River in Zambia and Malawi. However, in North America, especially in southern Texas and Florida, symptoms of poisoning among cattle, sheep, goats, and poultry have been ascribed to the consumption of leaves and seeds of *S. drummondii* (Rydb.) Cory, *S. exaltata*, *S. punicea* (Cav.) DC., and *S. vesicarium* (Jacq.) Ell. (Marsh and Clawson 1920; Muenscher 1939; Kingsbury 1964). The toxic substances were not identified but are presumed to be saponins.

Sesbania wood is soft and lightweight, coarse in texture, saws woolly, and has a low durability. It is used mainly for making toys, small objects of art, and gunpowder charcoal.

Nodulation Studies

Young, round nodules on *Sesbania* species have hemispherical meristems which become interrupted and apical in old, multilobed nodules (Harris et al. 1949). Nodules on 12-month-old plants form a compact, clustered mass along the taproot, yet each nodule remains distinctive and may be severed readily from the mass without disruption of adjacent ones.

Five explanations have been offered for the longevity of nodules on this species: (1) continuously active meristematic areas; (2) functional longevity of the infection threads; (3) an elaborately developed vascular system; (4) the prevalence of noninvaded cortical sclerenchyma; and (5) a gradual disintegration of the bacteroid area.

During their study, Harris et al. (1949) frequently observed 12–18 rootlets emerging from large, multilobed nodules on the 8–12-month-old plants. Young rootlets bore abundant root hairs that showed typical curling and, upon microscopic examination, revealed rhizobial infection threads. Histological sections showed rootlet initials arising within the endodermis of vascular bundles and digesting their way outwardly at right angles to the vascular strand from which they were derived. Root origin within these nodules was postulated to have resulted from a stimulation, or irritation, of the host cells caused by the accumulation of by-products within the bacteroid area. It was speculated that four factors account for the improbable occurrence of rootlets in nodules of herbaceous leguminous species: (1) the fragile vascular junction of the nodules with the root; (2) the lesser development and probable inadequacy of the nodule vascular system; (3) the lack of a sclerenchyma or protective layer in the nodule cortex; and (4) a shorter growing season.

Sesbania seedlings are highly receptive to infection by their homologous rhizobia (Harris et al. 1949); however, different *Sesbania* species have restricted susceptibility profiles and *Sesbania* rhizobia have a rather restricted host range. Raju (1936, 1938a), Toxopeus (1936), and Briscoe and Andrews (1938) concluded that *Sesbania* species and their microsymbionts constituted a separate rhizobia:plant group. A broader insight into this topic was provided by Johnson and Allen (1952a). Thirty-nine strains of rhizobia from 6 *Sesbania* species ineffectively nodulated beans and cowpea plants, but nodulation was not reciprocal. All the *Sesbania* strains nodulated all of the *Sesbania* species, but the degrees of effectiveness varied markedly. Rhizobia from alfalfa, clover, peas, soybeans, and lupines failed to nodulate *Sesbania* species and vice versa. Some reciprocal affinity between cowpea, soybean, and *Sesbania* species was achieved by Rangaswami and Oblisami (1962), but in all other plant tests the results agreed with those of previous workers.

Nodulated species of *Sesbania*	Area	Reported by
aculeata Poir.	Widely reported	Many investigators
aegyptiaca Poir.	Widely reported	Many investigators
bispinosa (Jacq.) W. F. Wight	Australia	Bowen (1956)
var. *bispinosa*	S. Africa	Grobbelaar et al. (1967)
	Zimbabwe	Corby (1974)
brevipeduncula Gillett	Zimbabwe	Corby (1974)
cannabina Roxb.	India	Mann (1906)
cavanillesii S. Wats.	Tex., U.S.A.	Pers. observ. (1938)
cinerascens Welw. ex Bak.	Sri Lanka	W. R. C. Paul (pers. comm. 1951)
	S. Africa	Grobbelaar et al. (1964)
coerulescens Harms	Zimbabwe	Corby (1974)
	S. Africa	Grobbelaar & Clarke (1975)

Nodulated species of *Sesbania* (cont.)	Area	Reported by
drummondii (Rydb.) Cory	N.Y., U.S.A.	Wilson (1939a)
exaltata (Raf.) Cory	Widely reported	Many investigators
grandiflora Poir.	Widely reported	Many investigators
greenwayi Gillett	Zimbabwe	H. D. L. Corby (pers. comm. 1974)
longifolia DC.	Hawaii, U.S.A.	Pers. observ. (1940)
macrantha Welw. ex Phill. & Hutch.		
var. *levis*	S. Africa	Grobbelaar et al. (1967)
	Zimbabwe	Corby (1974)
var. *macrantha*	Zimbabwe	Corby (1974)
macrocarpa Muhl.	Widely reported	Many investigators.
microphylla Harms	Taiwan	Pers. observ. (1962)
	Zimbabwe	Corby (1974)
mossambicensis Klotz.	S. Africa	Mostert (1955)
ssp. *mossambicensis*	Zimbabwe	Corby (1974)
rogersii Phill. & Hutch.	Zimbabwe	Corby (1974)
rostrata Brem. & Oberm.	Zimbabwe	Corby (1974)
roxburghii Merr.	Hawaii, U.S.A.	Pers. observ. (1940)
sericea (Willd.) Link	West Indies	DeSouza (1966)
sesban (L.) Merr.	Hawaii, U.S.A.	Pers. observ. (1934)
ssp. *sesban* var. *muricata* Baq.	Pakistan	A. Mahmood (pers. comm. 1977)
var. *nubica* Chiov.	S. Africa	Grobbelaar et al. (1967)
	Zimbabwe	Corby (1974)
var. *zambesiaca* Gillett	Zimbabwe	Corby (1974)
sonorae Rydb.	Mexico	Pers. observ. (1948)
speciosa Taub.	India	Raju (1938a); Ranga-swami & Oblisami (1962)
	Philippines	Pers. observ. (1962)
sphaerosperma Welw.	S. Africa	Grobbelaar & Clarke (1975)
tomentosa H. & A.	Hawaii, U.S.A.	Pers. observ. (1940)
	Wis., U.S.A.	Johnson & Allen (1952a,b)
transvaalensis Gillett	S. Africa	Grobbelaar & Clarke (1975)

Sewerzowia Regel & Schmalh.
296 / 3764

Papilionoideae: Galegeae, Astragalinae

Named for Sewerzow, Russian traveler-botanist.

Type: *S. turkestanica* Regel & Schmalh.

1 species. Herb, annual, erect. Leaves imparipinnate; stipels absent; stipules free, awl-shaped. Flowers small, few to several, in axillary racemes; calyx tubular; teeth awl-shaped; petals long-clawed; standard erect, truncate at apex; wings oblong; keel straight, blunt, shorter than the wings; stamens 10, diadelphous; ovary sessile, many-ovuled; style short, thick, stigma capi-

tate, terminal. Pod sessile, elliptic, 3-edged, dorsal surface flat, ventral side keeled; pod divided inside into 2 compartments by the intrusion of a double septum from the dorsal suture, valves boat-shaped, rimmed with thorny teeth; seeds flat, ovate-reniform.

This monotypic genus is endemic in Turkestan.

Shuteria W. & A.　　　　　　Papilionoideae: Phaseoleae, Glycininae
389 / 3863

Named for James Shuter, Naturalist of the Madras Establishment.

Type: *S. vestita* W. & A.

5 species. Herbs, slender, climbing or creeping, glabrous or densely hairy. Leaves pinnately 3-foliolate; stipels present; stipules striate. Flowers small, white or pink, with red or violet veins, medium-sized, solitary, paired or fascicled along the rachis of axillary racemes; bracts and bracteoles paired at calyx base, acute, striate, persistent; calyx tube gibbous; lobes long, upper 2 joined; standard obovate, suberect, short-clawed; keel shorter than and adherent to the narrow wings; stamens diadelphous, anthers small, uniform; ovary subsessile, many-ovuled; style incurved, not bearded; stigma terminal, small. Pod sessile within the calyx, flat, linear, 2-valved, indistinctly septate between the 4 or more seeds. (Bentham and Hooker 1865; Hooker 1879; Backer and Bakhuizen van den Brink 1963)

Shuteria species are widespread in tropical and subtropical Asia, especially Malaysia, the Philippines, and India. They flourish in brushwood forest and forest borders.

In Java *S. vestita* W. & A. has proved satisfactory as a green-manuring crop on coffee plantations and on tea and cinchona estates at altitudes up to 2,200 m.

NODULATION STUDIES

Keuchenius (1924) reported nodules on *S. vestita* in Java.

Sindora* Miq.　　　　　　Caesalpinioideae: Cynometreae

From *sindor* or *sendok*, the Malay name for these trees.

Type: *S. sumatrana* Miq.

18–20 species. Trees, large, erect, deciduous, generally unarmed, some prickly, balsamiferous. Leaves paripinnate; leaflets mostly opposite in Asian species, alternate in African species, 2–10 pairs, leathery; stipules leafy, caducous, sometimes with scattered, translucid dots when young. Flowers small, on short pedicels, in terminal panicles with racemiform branches; bracts ovate; bracteoles 2, linear, caducous; calyx tube short; lobes 4, valvate, subequal, slightly imbricate, with soft bristles outside, uppermost segment largest; petal 1, large, located inside the upper calyx lobe, sessile, oblong, margins incurved, hairy on the back, others rudimentary or absent; stamens 9–10, hairy at the base, shortly and obliquely united, declinate, upper stamen absent or shorter and reduced to a staminode, the next 2 longest, fertile, 7 others imperfect or sterile; anthers dorsifixed, longitudinally dehiscent; ovary free, broad, hairy, 2–6-ovuled, stalk short; disk thick in Asian species, absent in African species; style slender; stigma small, terminal. Pod flat, oval or orbicular, glabrous, often prickly, margined and transversely nerved, 2-valved or indehiscent; seeds 1–3, black, shiny, with a large fleshy aril. (Bentham and Hooker 1865; Baker 1930; Pellegrin 1948; Léonard 1957; Backer and Bakhuizen van den Brink 1963)

Various differences are recognized between the Asian and the African species.

Members of this genus are native to Indochina, Malaysia, Sumatra, and the Philippines; 1 species is endemic in tropical Africa. They inhabit rocky hill forests, seashore areas, and limestone ridges.

S. coriacea Prain, common in the lowland forests of the Malay Peninsula, often reaches a height of 50 m, with a trunk clear of branches 20 or more meters from the ground. *S. javanica* (K. & V.) Backer, *S. leiocarpa* Backer, *S. siamensis* Teysm., *S. supa* (Blanco) Merr., and *S. velutina* Bak. are among the most prominent species. *S. klaineana* Pierre is the only African species.

Sindora wood varies considerably in quality. In Asia the best wood is supplied by *S. intermedia* Bak. and *S. siamensis*, gigantic trees over 35 m tall. It is strong and durable but not available in large quantities. Supá, the wood of *S. supa*, in the Philippines, is highly prized for interior house trimming, naval construction, furniture, and cabinetmaking. The sapwood is cream-colored or pinkish. The heartwood is yellow when freshly cut but upon aging and exposure it becomes reddish brown. It has a density of about 700 kg/m^3, is hard to work, but is highly appreciated for its durability. The wood of *S. inermis* Merr., a species mostly confined to the Cotabato and Davao areas, Philippines, resembles that of *S. supa*. The woods of both species have a very pleasant, aromatic fragrance. In contrast, the wood of *S. leiocarpa*, a tree in Sumatra, is reddish gray, soft, and decays readily under moist to wet conditions.

All *Sindora* species yield oil, but the kayu-gálu oil of *S. inermis* (Brown 1921) and supá oil of *S. supa* are the most commercially valuable. All of these oils have similar properties and uses and are easily obtained by making V-cuts in the trunks. A freshly cut tree may yield 10 liters of supá oil. The oil is a mixture of sesquiterpenes that are nondrying, limpid, light yellow in color, homogeneous, and pleasantly fragrant (Clover 1906). It serves diverse uses, in perfumes, paints, varnishes, transparent papers, for illumination, and to some extent medicinally in treating skin diseases (Quisumbing 1951).

NODULATION STUDIES

Plantlets of *S. supa* examined by Bañados and Fernandez (1954) and by the authors (personal observation 1962) in the Laguna and Calamba areas of the Philippines lacked nodules.

The Greek suffix, *opsis*, denotes resemblance to *Sindora*.

Type: *S. le-testui* (Pellegr.) J. Léonard

1 species. Tree. Leaves paripinnate; leaflets numerous, alternate, acuminate, conspicuously pellucid-punctate, each with a marginal, unilateral gland near the base, with the principal nerve terminating in a small porelike gland visible underneath; stipules caducous. Flowers on short pedicels, in 2 rows along very short rachis of axillary racemes; bracts small; bracteoles 2, small, not enveloping the bud, caducous; lobes 4, subvalvate; petal 1, elliptic, clawed, glabrous, not fleshy, caducous; stamens 10, vexillary stamen subfree, others united at the base; filaments alternately long and short; anthers similar, equal in length; ovary short-stalked, 5–7-ovuled; disk absent. Pod oblong-lanceolate, acuminate, 7–10 cm long, narrowed at the base, very compressed, 2-valved, leathery, mesocarp reticulate, dehiscent; seeds 1–2, oblong, thick, not arillate. (Léonard 1957)

This monotypic genus is endemic in tropical Africa.

Sinodolichos Verdc.   Papilionoideae: Phaseoleae, Phaseolinae

Name means "China *Dolichos*."

Type: *S. lagopus* (Dunn) Verdc.

2 species. Herbs, perennial, twining, semiwoody. Leaves pinnately 3-foliolate, petioled; leaflets ovate or rhomboid, papery to subcoriaceous, acuminate, prominently pinnately veined; stipels linear, persistent; stipules deltoid, subpersistent, sinewy. Flowers purple, on short pedicels, many or few, in short, axillary racemes; bracts ovate, caducous, sinewy; bracteoles 2, conspicuous, ovate, subpersistent; calyx short, bristly-haired; tube campanulate; lobes 4, linear-lanceolate, long, narrow, 2 upper lobes connate into a narrow short bidentate lip; petals glabrous, clawed, becoming black when dry; standard rounded or elliptic-oblong, eared at the base, not appendaged; wings hardly elliptic-oblong, folded, spurred; keel oblong-elliptic, erect, somewhat incurved or beaked, base obtuse; stamens diadelphous, vexillary stamen appendaged at the base; filaments unequal; anthers subuniform; ovary linear, subsessile, about 10-ovuled; disk annular, slightly lobate; style erect or incurved above, thickened, subuniformly cylindric; stigma funnel-shaped, capitate, penicillate. Pod linear-oblong, flattened, densely rusty bristly-haired, hardly septate, dehiscent; seeds 3–10, oblong, somewhat arillate, reticulate, granulose-rugose, hilum elliptic. (Dunn 1903; Verdcourt 1970c)

Verdcourt (1970c) and Lackey (1977a) classified *Sindolichos* within the subtribe Glycininae.

Both species occur in China and Burma. They thrive in roadside thickets on sandy soil.

Named for Smirnov; he first found the plant in the Kyzyl Kum desert, Turkestan, central Soviet Union.

Type: *S. turkestana* Bge.

1 species. Shrub, up to about 1 m tall, with strong, erect stems and many long, slender, white-pubescent branches. Leaves 1–3-foliolate, obovate, jointed at the base, deciduous; stipules absent. Flowers lilac or white, few, in short, lax, axillary racemes; calyx white-tomentose, somewhat 2-lipped; lobes 5, triangular-lanceolate; standard subreniform-orbicular, emarginate, not callous at the base; wings subfalcate; keel blunt, incurved; stamens diadelphous; ovary short-stalked, many-ovuled; style inflexed, cylindric-slender, thickly bearded along the back; stigma capitate. Pod large, ovaloid, bladder-like, deeply grooved and cross-nerved along the membranous valve walls; seeds several, kidney-shaped, flattened, smooth. (Komarov 1971)

This monotypic genus is endemic in Turkestan. The plants grow in the sands of the desert.

Smirnovine-type alkaloids, i.e., smirnovine, smirnovinine, and sphaerophysine, having the same general formula, $C_{10}H_{22}N_4$, occur in the roots and leaves of *Astragalus tibetanus*, *S. turkestana*, *Eremosparton aphyllum*, *E. flaccidum*, *Sphaerophysa salsula*, and *Galega officinalis* (see Harborne et al. 1971). Medicinally, these compounds have ganglion-blocking properties. The limited distribution of these smirnovine alkaloids attests to the relatedness of these genera within the tribe Galegeae (Mears and Mabry 1971).

Named for Sir James Smith, 1759–1828, founder and first president of the Linnean Society of London. In 1784 he purchased the unpublished manuscripts and herbarium collection of Linnaeus for a thousand English guineas.

Type: *S. sensitiva* Ait.

30 species. Herbs, annual or perennial, or shrublets up to about 1 m tall, base woody, sometimes villous, not glandular. Leaves usually paripinnate; leaflets small, 6–12 pairs, asymmetric at the base, oblong or linear, petioluled, sometimes ciliate; stipels none; stipules persistent, membranous, with a 2-eared spur below the point of attachment. Flowers yellow, orange, blue, lilac, axillary, fascicled or in stalked, generally dense, small racemes or umbel-like scorpioid cymes; bracts small, entire, thin, brownish, caducous; bracteoles persistent, scarious, attached below the calyx; calyx dry, membranous, deeply bilabiate, upper lip entire or emarginate, lower lip 3-lobed or entire, membranous, dry, often bristly-haired; standard obovate, suborbicular or spathulate, usually sessile, sometimes adnate to the staminal tube; wings obliquely oblong-linear, broad, rarely obovate, with a short linear claw; keel petals elliptic, incurved, united at the back, free above; stamens 10, joined in a sheath slit above and later below forming 2 phalanges of 5; filaments alternately long and short; anthers uniform; ovary linear, short-stalked; ovules 2–9; disk cupu-

lar; style smooth, linear, incurved; stigma minute, terminal. Pod stalked, consisting of 2–9 flattened segments folded within the persistent, accrescent calyx, segments semiorbicular, smooth or knobby, indehiscent but at length separate; seeds kidney-shaped. (Bentham and Hooker 1865; Baker 1929; Phillips 1951; Dewit and Duvigneaud 1954a,b)

Smithia and *Kotschya* are closely related. The latter taxon was formerly regarded as a subgenus of *Smithia*; hence most *Kotschya* species bear *Smithia* synonyms. The various characteristics separating these 2 genera were outlined by Dewit and Duvigneaud (1954a); the occurrence of spurred stipules in *Smithia* species in contrast to nonspurred stipules in *Kotschya* species was a primary distinction.

Smithia species are paleotropical, especially in Asia and Madagascar; 2 species occur in equatorial Zaire. The plants inhabit sunny, humid, sandy, and grassy localities.

In Sri Lanka, *S. blanda* Wall. is good forage above 1,500 m (W. R. C. Paul personal communication 1951). The leaves of *S. sensitiva* Ait. contain saponins and are used as a crude soap substitute in India. In Indonesia the leaves are reportedly palatable to cattle, consumed by humans as pot herbs, and also administered medicinally for certain urinary difficulties (Heyne 1927).

Nodulated species of *Smithia*	Area	Reported by
begimina Dalz.	W. India	V. Schinde (pers. comm. 1977)
blanda Wall.	Sri Lanka	W. R. C. Paul (pers. comm. 1951)
capitata Dalz.	W. India	Bhelke (1972)
conferta Sm.	W. India	V. Schinde (pers. comm. 1977)
erubescens (E. Mey.) Bak. f.	S. Africa	Grobbelaar & Clarke (1975)
eylesii S. Moore	Zimbabwe	H. D. L. Corby (pers. comm. 1956)
purpurea HBK	W. India	Bhelke (1972)
pycnantha Benth.	W. India	Bhelke (1972)
sensitiva Ait.	W. India	V. Schinde (pers. comm. 1977)
setulosa Dalz.	W. India	V. Schinde (pers. comm. 1977)

Soemmeringia Mart.

326 / 3795

Papilionoideae: Hedysareae, Aeschynomeninae

Named for Samuel Thomas von Soemmering, 1755–1830, German naturalist and anatomist.

Type: *S. semperflorens* Mart.

1 species. Annual herb. Leaves imparipinnate; leaflets many, usually ser-

rate, narrowly obovate-oblanceolate, with a median nerve and few ascending lateral nerves; rachis sharply pointed; stipels none; stipules lanceolate. Flowers yellow, 1–2, pedicelled, axillary; bracts small, awl-shaped; bracteoles lanceolate, striate, persistent; calyx deeply bilabiate, upper lip bidentate, lower tridentate; petals papery, persistent, reticulate-veined; standard subsessile, orbicular-reniform; wings small; keel petals blunt, incurved, longer than the wings but shorter than the standard; stamens in 2 lateral bundles; filaments split below, later split above; anthers uniform; ovary stalked, many-ovuled; style incurved; stigma small, terminal. Pod stalked, shorter than the standard, upper margin straight, lower margin weakly convex, reticulate, undulate between the seeds. (Bentham and Hooker 1865)

This monotypic genus is endemic in northeastern Brazil.

Sophora L.

142 / 3602

Papilionoideae: Sophoreae

From Arabic, *sufayra*, the name of a *Sophora* tree.

Type: *S. tomentosa* L.

50–80 species. Trees, shrubs, small to medium, deciduous or evergreen, rarely perennial herbs, unarmed. Leaves imparipinnate; leaflets entire, numerous and small, or few and large, sometimes densely tomentose; stipels bristly or absent; stipules minute, caducous. Flowers usually yellow, white, occasionally blue violet, borne in simple, terminal racemes or leafy panicles; bracts and bracteoles often absent, otherwise both linear, minute, caducous; calyx tubular-campanulate, obliquely truncate, sometimes tomentose; teeth 5, very short, subequal or upper 2 subconnate and larger; standard broad, obovate, erect or reflexed, usually shorter than the keel; wings oblique-oblong, free; keel petals overlapping or joined dorsally; stamens 10, free, rarely joined at the base into a ring; filaments linear; anthers versatile, sometimes with a small, blunt, apical gland; ovary short-stalked, several- to many-ovuled; style cylindric, incurved; stigma small, terminal, rounded. Pod stalked, cylindric or moniliform, sometimes winged, fleshy, woody, or leathery, compressed between the seeds, indehiscent or tardily dehiscent; seeds oblong or oval, often yellow, somewhat compressed, cotyledons thick, fleshy. (Bentham and Hooker 1865; Burkart 1952; Rudd 1968b, 1970a)

Members of this genus are distributed throughout the warm regions of the Old World and the New World. They are adapted to sandy soil, coastal dune and dry areas, primeval mixed forests, and calcareous soils.

The medium-sized, graceful trees of this genus are especially noteworthy for the beauty of their floral racemes. *S. flavescens* Ait., *S. linearifolia* Griseb., *S. macrocarpa* Sm., *S. rhynchocarpa* Griseb., *S. viciifolia* Hance, and *S. violacea* Thw. are cultivated ornamentals in many parts of the world. The gorgeous, sulphur yellow blossoms of *S. tetraptera* J. Mill., four-wing sophora, or kowhai, the national flower of New Zealand, resemble European *Laburnum*. *S.*

chrysophylla (Salisb.) Seem. is the handsome evergreen, mamani tree, whose leaves are browsed by wild cattle and horses on the upper slopes of Hawaiian volcanoes. *S. japonica* L., the Japanese pagoda tree, is prevalent in temple gardens. Two species, *S. affinis* T. & G. and *S. secundiflora* (Ort.) Lag. ex DC., are native to the southwestern United States. The latter, known locally as Texas mountain laurel or mescal bean, is commonly cultivated for its fragrant, handsome, violet blue flowers.

Native uses of plant parts have been diverse. At one time a yellow dye from pods of *S. japonica* was valued highly in the batik and silk industries of the Orient. The leaves and pods were also used as an opium adulterant. Numerous medicinal astringent properties have been attributed to bark and root decoctions of several Asian species. Seeds, leaves, and flower buds yield a variety of flavonoid or isoflavonoid glucosides, quercetin, alkaloids, and related compounds (Watt and Breyer-Brandwijk 1962). Many of these compounds were given distinctive names, such as sophoricoside, sophorin, and others; however, several are now considered synonymous, or secondary to other terms. Anagyrine, cytisine, methylcytisine, matrine, and sparteine are among the most poisonous and abundant alkaloids present (Harborne et al. 1971). Yields of 3 percent crude alkaloids in plant parts are not uncommon.

The green leaves and red seeds of the mescal bean have accounted for losses of cattle, sheep, and goats. Toxicity of the leaves varies considerably; mature autumn and winter leaves were much more toxic for range sheep than young leaves during spring months (Boughton and Hardy 1935). The seeds are not particularly harmful unless the hard seed coats are broken. Poisoning of goats occurs more commonly from the ingestion of a large amount of leaves at one time than from continued nibbling over a period of days. Cattle are extremely susceptible and die within a few hours after eating relatively small amounts of the leaves. Mescal beans were used by the Plains Indian tribes of the United States as an hallucinogen in the ceremonial Red Bean Dance. *S. sericea* Nutt., one of the few herbaceous species, is called a locoweed in some southwestern states because of the crazed effect it has on horses. Matrine is the principal alkaloid constituent of the dried roots of *S. flavescens*, the Chinese drug kuh-seng, used to treat dysentery.

The wood of *S. japonica* has properties similar to chestnut wood; it is valued in the Orient for cabinetwork. Other *Sophora* wood lacks commercial importance because of mediocre quality and limited quantity; however, it is durable and suitable for fence posts, wheel shafts, and small turnery objects.

NODULATION STUDIES

Cultural, biochemical, and host-infection studies of *Sophora* rhizobia are sparse. Strains isolated from indigenous New Zealand species were slow growing and produced acid (R. M. Greenwood personal communication 1975). Ishizawa's conclusion (1953, 1954) that minor differences shown by strains from *S. angustifolia* S. & Z. merited distinction from those of *S. japonica*, which was designated by him as *Styphnolobium japonicum*, seems unwarranted. Both displayed affiliation with the cowpea miscellany.

Nodules on mature tree specimens of *S. japonica* and *S. tomentosa* L. in Hawaii and *S. secundiflora* in Texas suggested perennial growth (personal observation 1936, 1939). Proximal regions of nodules were thick, brown, and woody, in contrast to the flesh-colored, forked, apical regions of the same nodule. No unusual histological features of *S. moorcroftiana* Benth. nodules were noted by Lechtova-Trnka (1931).

Presumably, the absence of nodules on *S. arizonica* S. Wats., recorded by Martin (1948), was a reflection of the arid seasonal circumstances in Arizona.

Nodulated species of *Sophora*	Area	Reported by
angustifolia S. & Z.	Japan	Asai (1944); Ishizawa (1953)
chrysophylla (Salisb.) Seem.	N. Zealand	R. M. Greenwood (pers. comm. 1975)
davidii (French) Ravol.	S. Africa	Grobbelaar & Clarke (1974)
flavescens Ait.	S. Africa	N. Grobbelaar (pers. comm. 1969)
formosa Kearn. & Peeb.	U.S.A.	Burton et al. (1974)
gypsophila var. *guadalupensis* Turn. & Powell	U.S.A.	J. C. Burton (pers. comm. 1974)
howinsula (Oliv.) P. S. Green	N. Zealand	R. M. Greenwood (pers. comm. 1975)
inhambanensis Klotz.	S. Africa	Grobbelaar & Clarke (1972)
japonica L.	Hawaii, U.S.A.	Pers. observ. (1940)
(= *Styphnolobium japonicum* Schott)	Japan	Ishizawa (1953)
(*microphylla* Ait.)	N. Zealand	R. M. Greenwood (pers. comm. 1975)
= *tetraptera* J. Mill.		
moorcroftiana Benth.	France	Lechtova-Trnka (1931)
prostrata Buchan.	N. Zealand	R. M. Greenwood (pers. comm. 1975)
secundiflora (Ort.) Lag. ex DC.	Tex., U.S.A.	J. C. Burton (pers. comm. 1973)
tetraptera J. Mill.	N. Zealand	Pers. observ. (1962); R. M. Greenwood (pers. comm. 1975)
tomentosa L.	Hawaii, U.S.A.	Pers. observ. (1940)
velutina Lindl. ssp. *zimbabweensis*	Zimbabwe	Corby (1974)
viciifolia Hance	N. Zealand	R. M. Greenwood (pers. comm. 1975)

Sopropis Britt. & Rose   Mimosoideae: Adenanthereae

Name is an anagram of *Prosopis*.

Type: *S. palmeri* (S. Wats.) Britt. & Rose

1 species. Shrub or small tree, branches loosely arranged, densely and finely pubescent, armed with straight, needlelike, stipular spines. Leaves

bipinnate; pinnae pair 1; leaflets alternate, oblong, small. Flowers fragrant, small, bisexual, sessile, in long, densely flowered, short-peduncled, axillary spikes; calyx campanulate; teeth 5; petals nearly united, outside softly haired; stamens 10, free, distinct, long-exserted; anthers tipped with a large, deciduous gland; ovary stalked, many-ovuled, tomentose; style filiform, loosely villous; stigma small, terminal. Pod linear, compressed, septate but not constricted between flattened seeds, mesocarp fibrous. (Rydberg 1928)

This monotypic genus is endemic in Baja California, Mexico.

Spartidium Pomel   Papilionoideae: Crotalarieae

From Greek, *spartos*, the name of a broom plant.

Type: *S. saharae* (Coss.) Pomel

1 species. Shrub, virgate, largely leafless. Leaves only on young shoots, simple, narrow-lanceolate. Flowers yellow, in lax terminal racemes; bracts and bracteoles setaceous, caducous; calyx campanulate, short-lobed, upper cleft deepest, lowest lobe narrowest; standard ovate, pubescent; keel half elliptic, long-clawed; stamens in sheath, open on upper side; anthers 4 basifixed longer, alternating with 6 shorter, dorsifixed; ovary stalked; style slender; stigma small, terminal. Pod membranous, short-stalked, flat, narrow-oblong, several-seeded. (Polhill 1976)

This monotypic genus is endemic in North Africa, ranging from Libya to Algeria. The plants grow in sand dunes. They are similar in appearance to *Lebeckia* and *Retama*.

Spartium L.   Papilionoideae: Genisteae, Spartiinae
213 / 3674

From Greek, *spartos*, the name of a broom plant.

Type: *S. junceum* L.

1 species. Shrub, erect, 3–4 m tall, unarmed; branches slender, rushlike, glaucous green, leafless or nearly so. Leaves 1-foliolate, sessile, entire, small, sparse, usually only on young stems; stipules none. Flowers large, yellow, sweet-scented, showy, 5–20, in loose, terminal racemes; bracts and bracteoles small, caducous; calyx spathelike, membranous, dry, split above dorsally, hence irregularly unilabiate, rarely bilabiate; teeth 5, short, subequal; standard free, large, obovate, recurved; wings obovate, short; keel incurved, pointed, longer than the wings, hairy along the lower margin, claws of the keel petals adnate to the staminal tube; stamens 10, monadelphous; anthers alternately short and versatile, long and basifixed; ovary sessile, multiovuled; style incurved, smooth; stigma oblong. Pod oblong-linear, smooth, flat,

2-valved, subseptate inside between the many seeds, dehiscent. (Rehder 1940; Rothmaler 1949)

This monotypic genus is endemic in regions bordering the Mediterranean, southwestern Europe, and the Canary Islands. It has been introduced into many areas.

Polhill (1976) included this genus in Genisteae *sensu stricto*.

S. junceum L., Spanish or weaver's broom, is a popular ornamental in many parts of the world, despite its susceptibility to green fly pests. It is an excellent source of wildlife food in its native habitats. A fiber with textile possibilities comparable to those of jute is made from its branches.

Reports of the presence of sparteine in *S. junceum* in Europe (Sanna and Chessa 1927; Jarétzky and Axer 1934) were not confirmed in New Zealand by White (1943). This discrepancy was attributed to environmental differences. However, White found considerable amounts of cytisine throughout the plant to account for its toxicity. Since White's study, lesser amounts of other quinolizidine alkaloids, i.e., methyl cytisine, anagyrine, and thermopsine, have been identified in this species (Mears and Mabry 1971).

NODULATION STUDIES

Nodulation of this species is well documented in various European and American sources. Wilson's results (1939a) showed that *S. junceum* was susceptible to nodulation by rhizobia from diverse leguminous species of the cowpea miscellany.

Spathionema Taub.   Papilionoideae: Phaseoleae, Phaseolinae
(430) / 3913

From Greek, *spathe*, "spatula," and *nema*, "thread"; the filamentous vexillary stamen is spathulate at the apex.

Type: *S. kilimandscharicum* Taub.

1 species. Shrublet, climbing; stems slender, glabrous, whiplike. Leaves 3-foliolate; leaflets broad-ovate, shortly acuminate; stipels and stipules present. Flowers blue violet, large, showy, in lateral, few-flowered racemes along nodose rachis, in bloom before leaves appear; calyx subcampanulate, inside villous; lobes 5, upper 2 connate into 1 emarginate lobe, lateral 2 broad, blunt, lowest lobe triangular-acute; standard suborbicular, 2-eared, short-clawed; wings obliquely obovate-oblong, slightly longer than the standard, adherent to the keel; keel petals falcate, whitish blue, shorter than the wings, clawed, joined midway along the back; stamens 10, vexillary stamen spathulate at the tip, free except at the base, other 9 longer, connate into a sheath; anthers oval, dorsifixed, dehiscent by means of a longitudinal fissure; ovary short-stalked, 2–3-ovuled, linear; style slim at the base, thickened and bearded below the cup-shaped, terminal stigma. Pod short-oblong, flat, 2-valved; seeds kidney-shaped, flattened, dark brown, shiny. (Harms 1897, 1908a; Verdcourt 1970d)

This genus was first positioned as 428a (Harms 1897) and later changed to 430 (Harms 1908a). The genus is closely allied with *Vigna* (Verdcourt 1970d).

This monotypic genus is endemic in tropical East Africa. The plants occur along a crater rim on Mount Kilimanjaro.

Spatholobus Hassk. Papilionoideae: Phaseoleae, Galactiinae
402 / 3878

From Greek, *spathe*, "spatula," and *lobos*, "pod," in reference to the shape of the pod.
Type: *S. littoralis* Hassk.
15 species. Lianas, robust, woody, some climbing 50 m or more, often tomentose. Leaves pinnately 3-foliolate, undersurfaces usually white-, yellow-, or brown-hairy; stipels and stipules caducous. Flowers small, purple, pink, or white, in large, many-flowered, terminal panicles and in the highest leaf axils; bracteoles small, narrow, briefly dentate; calyx campanulate; lobes lanceolate or oblong-deltoid, 2 upper lobes nearly completely joined into 1 entire or emarginate lobe; standard obovate or suborbicular, obtuse, without basal ears; wings oblique-oblong, free; keel petals obtuse, shorter than the wings, nearly straight; stamens 10, diadelphous, all perfect or alternate ones reduced to staminodes; anthers minute, uniform; ovary sessile or short-stalked, 2-ovuled; style incurved, glabrous; stigma small, terminal, capitate. Pod sub-sessile or stalked, linear or curved, base flat, narrow and empty, apex broad, 1–2-seeded, tardily dehiscent only at the top. (Bentham and Hooker 1865; Backer and Bakhuizen van den Brink 1963)

Lackey (1977a) placed the genus tentatively in the tribe Dalbergieae.

Representatives of this genus are native to Southeast Asia, Malaysia, and the Philippines. They thrive in secondary forests and brushwood areas.

The bark exudes an astringent red brown sap, which forms a kino when dry. It has local medicinal uses for coughs and colic. The stout stems serve as rough cordage (Burkill 1966). The seeds of *S. roxburghii* Benth. have a high content of triglyceride oils (Namboodiripad 1966).

Sphaerolobium Sm. Papilionoideae: Podalyrieae
171 / 3631

From Greek, *sphaira*, "ball," and *lobion*, "pod"; the pod is globular.
Type: *S. vimineum* Sm.
12 species. Shrubs or undershrubs, low, glabrous; stems cylindric or winged, erect, slender, often leafless, wiry, reedlike and with perpendicular ridges. Leaves, when present, narrow, entire, alternate, irregularly opposite, or whorled. Flowers yellow or red, numerous, in terminal or lateral racemes, sometimes in clusters; bracts and bracteoles caducous; calyx lobes imbricate,

upper 2 larger, falcate, united into a lip; petals short-clawed; standard broad, orbicular, emarginate; wings shorter than the standard, oblong, often falcate; keel obtuse or straight; stamens free; ovary stalked, 2-ovuled; style incurved, usually awl-shaped or dilated at base, usually ringed with hairs below the stigma. Pods small, short-stipitate, subglobose or compressed, about 3 mm in diameter, 1–2-seeded. (Bentham 1864; Bentham and Hooker 1865)

Members of this genus are indigenous to Western Australia; 1 species occurs in New South Wales. All species are absent or extremely rare in Northern Territory. They inhabit mountain ranges, heathy swamp areas, and coastal and inland tablelands.

Nodulation Studies

Nodulation was observed by Lange (1959) on excavated root systems of *S. alatum* Benth., *S. grandiflorum* (R. Br.) Benth., and *S. medium* R. Br., all indigenous to the South-West Province of Western Australia. Single strains from each species were culturally similar to strains of slow-growing rhizobia. Each strain nodulated *Vigna sinensis* and some, but not all, plants of *Phaseolus lathyroides* and *P. vulgaris*; however, with few exceptions, they lacked ability to nodulate *Lupinus* and *Glycine* species (Lange 1961).

Sphaerophysa DC.   Papilionoideae: Galegeae, Coluteinae

From Greek, *sphaira*, "ball," and *physa*, "bladder"; the pod is inflated and globose.

Type: *S. salsula* (Pall.) DC.

3 species. Herbs, perennial, or shrublets. Leaves imparipinnate, elliptic to oblong-oval, obtuse; leaflets 3–many, entire; stipels absent; stipules small. Flowers numerous, in axillary racemes; calyx campanulate; teeth 5, upper 2 close together; standard suborbicular, sides reflexed; wings falcate-oblong; keel incurved, obtuse; stamens 10, diadelphous; anthers uniform; ovary stipitate, many-ovuled; style incurved, not hooked at the tip, bearded along the upper part on the adaxial side; stigma terminal, capitate, oblique. Pod long-stalked, glabrous or with scattered hairs, inflated, membranous or leathery; seed-bearing suture indented, tardily dehiscent, containing kidney-shaped, brown, smooth, dull-surfaced seeds. (De Candolle 1825a,b; Bentham and Hooker 1865)

Representatives of this genus are distributed from the Caucasus Mountains to northern China. The species are generally considered halophytes, thus occurring in saline, sandy or clay riverside soils and related areas.

Spherophysine, $C_{10}H_{22}N_4$, a ganglion-blocking alkaloid having the same general formula as smirnovine and smirnovinine (Matyukhina and Ryabinin 1964), is found in plant parts of *S. salsula* (Pall.) DC. (Rubinshtein and Menshikov 1944). Its occurrence also in species of *Astragalus, Galega, Eremosparton*, and *Smirnowia* attests to the kinships of these genera and their proper inclusion in the same tribe, Galegeae (Mears and Mabry 1971).

From Greek, *sphen*, "wedge," in reference to the shape of the distinctive cuneate style.

Type: *S. marginata* E. Mey.

18 species. Shrubs, low, prostrate or climbing, often with herbaceous branches from large, underground rootstocks. Leaves pinnately 3-foliolate; leaflets elliptic or lanceolate, lateral ones sometimes lobed or asymmetric; stipels and stipules small, persistent. Flowers yellow, violet, or red, large, few, usually congested at the apices of long, axillary peduncles; bracts small, striate, or none; bracteoles 2, at base of the calyx; calyx campanulate, glabrous; lobes 5, ovate, blunt, short, broad, rounded, lowest lobe longer; standard suborbicular, with 2 inflexed ears; wings narrow, clawed, longer than the blunt keel; stamens 10, diadelphous; 5 anthers dorsifixed, 5 subbasifixed; ovary linear, subsessile, many-ovuled, surrounded by a cupular disk at the base; style stiff, twisted, channeled below, wide, cuneiform, hairy at the apex; stigma terminal. Pod narrow, linear, straight, leathery, sutures prominent, 2-valved, spirally twisted after dehiscence. Seeds black, many, ellipsoid or cylindric, minutely papillose. (Harms 1899, 1908a; Baker 1929; Wilczek 1954)

Members of this genus are closely allied to and were formerly grouped with *Vigna*; hence most of them bear *Vigna* synonyms (Harms 1899, 1911). Gillett (1966a) considers them most closely related to *Lablab*.

Sphenostylis species are distributed throughout tropical Africa; 1 species is endemic in India. They inhabit savannas and forest clearings. Species grow both wild and cultivated.

Several species, notably *S. stenocarpa* (Hochst. ex A. Rich.) Harms, are cultivated for their seeds, which are edible after soaking and boiling, and their yamlike tubers, which taste like potatoes (Busson 1965). According to Dalziel (1943), a ground-up bean meal is said to restore sobriety to intoxicated persons. The yellow flowers of *S. schweinfurthii* Harms are not only ornamentally attractive and worthy of horticultural exploitation but are cooked and relished as a vegetable.

Nodulated species of *Sphenostylis*	Area	Reported by
angustifolia Sond.	S. Africa	Grobbelaar et al. (1964)
marginata E. Mey.	S. Africa	Grobbelaar et al. (1967)
ssp. *erecta* (Bak. f.) Verdc.	Zimbabwe	Corby (1974)
ssp. *obtusifolia* (Harms) Verdc.	Zimbabwe	Sandmann (1970b); Corby (1974)
stenocarpa (Hochst. ex A. Rich.) Harms	Zimbabwe	Corby (1974)

Sphinctospermum Rose   Papilionoideae: Galegeae, Tephrosiinae
(255a)

From Greek, *sphinktos*, "constricted," and *sperma*, "seed"; the seeds are constricted at the middle.
     Type: *S. constrictum* (S. Wats.) Rose
     1 species. Herb, annual, small, herbage sparsely strigose; stems erect, slender. Leaves simple, linear, nerves ascending, lateral; stipules awl-shaped. Flowers small, inconspicuous, yellow, purple-spotted, single, rarely paired, on short pedicels, axillary; calyx teeth 5, acuminate, upper 2 more united than the others; petals subequal in length; standard orbicular, emarginate, short-clawed; wings oblong; stamens 10, 1 free to the base, others united halfway up; style slender, hairy near apex; ovary sessile, many-ovuled. Pod linear, 2-valved, septate between the seeds; seeds 6–10, oblong, 4-angled, constricted in the middle suggesting miniature vertebrae, dull brown, minutely roughened. (Rose 1906; Harms 1915a)
     This monotypic genus is endemic in Baja California, Mexico, and Arizona. The plants grow in open, sandy areas up to about 1,200-m altitude.

Spirotropis Tul.   Papilionoideae: Sophoreae
132 / 3590

From Greek, *speira*, "anything coiled or wound," and *tropis*, "keel"; the keel is conspicuously convolute.
     Type: *S. longifolia* (DC.) Baill.
     1 species. Tree, unarmed. Leaves imparipinnate; leaflets large, few pairs, opposite or subalternate, leathery, undersides velvety; stipules spathulate-oblanceolate, foliaceous. Flowers purple red, in terminal paniculate racemes; bracts and bracteoles minute; calyx tubular, bilabiate, upper lip bidentate, lower tridentate, at length split to the base into 2 reflexed segments; petals very short-clawed; standard elliptic, broader than the other petals; wings linear-oblong, straight, shorter than the standard; keel petals same shape and length as the wings, convex, eventually coiled longitudinally; stamens free, anthers basifixed; ovary subsessile, few-ovuled; style filiform; stigma small, terminal. Pod flat, oblong, both ends narrowed, not winged, reticulate, indehiscent. (Bentham and Hooker 1865)
     This monotypic genus is endemic in French Guiana and Surinam.

Stachyothyrsus Harms Caesalpinioideae: Eucaesalpinieae
(90a) / 3547

From Greek, _stachys_, "spike," and _thyrsos_, "panicle," in reference to the inflorescence.

Type: _S. staudtii_ Harms

2 species. Trees, up to about 25 m tall, evergreen. Leaves bipinnate; pinnae opposite, 2 pairs; leaflets 3–4 pairs, opposite or subopposite, petioluled, oblong, acuminate, leathery, glabrous. Flowers many, in large, terminal, pyramidal panicles of 10–20 elongated spikes; bracteoles none; calyx tube short, broadly cupular; lobes 5, oblong, semiorbicular; petals 5, white, equal, oblong, imbricate in bud, 2–3 times longer than the calyx; stamens 10, unequal, the 5 longer ones alternate with the petals, with filaments club-shaped and broader toward the apex, the 5 shorter ones opposite the petals, with slender filaments; anthers ovoid, short-apiculate, subbasifixed, dehiscent by longitudinal slits; ovary obliquely and shortly oblong; ovules 2–3, arranged vertically; style short, thick; stigma terminal. Pod flat, oblong-oblanceolate, acuminate, thinly woody, reticulate, narrowed toward the base, 2–3-seeded. (Harms 1897)

Léonard and Voorhoeve concluded that the monotypic genus _Kaoue_ Pelleg. is congeneric with _Stachyothyrsus_; in consequence, _K. stapfiana_ (A. Chev.) Pelleg. was merged into this genus as _S. stapfiana_ (A. Chev.) J. Léonard & Voorhoeve (see Voorhoeve 1965).

Both species are native to tropical Africa, Cameroon, and Liberia. They are abundant in high and secondary forests.

The timber has little value other than for fuel.

Stahlia Bello Caesalpinioideae: Cynometreae
36 / 3480

Named for Agustín Stahl, 1842–1917, physician and botanist of Bayamon, Puerto Rico; author of _Estudios sobre la Flora de Puerto Rico_, 1883–1888.

Type: _S. maritima_ Bello

1 species. Tree, unarmed, evergreen, usually small but may reach a height of 20 m, with a trunk diameter between 0.3 and 1 m; young bark smooth, gray, becoming thick and furrowed with age. Leaves paripinnate; leaflets usually 4–6 pairs, asymmetric, thin, ovate-acuminate, undersurfaces with prominent, scattered, black, glandular dots; petiolule red. Flowers medium-sized, cream or pale yellow tinged with pink, in axillary or terminal racemes; bracts connate at the base, membranous, caducous, not imbricate; bracteoles absent; calyx funnel-shaped; lobes 5, large, subequal, free, obtuse, overlapping; petals 5, subequal, imbricate, dorsally verrucose; stamens 10, free; filaments linear-subulate, woolly; anthers uniform; ovary sessile, 2-ovuled; style slender, truncate, curved. Pod red, oval, smooth, fleshy, somewhat compressed, leathery, 1–2-seeded, indehiscent or tardily dehiscent. (Urban 1899; Perkins 1907; Little and Wadsworth 1964)

This monotypic genus is endemic in the West Indies and Puerto Rico. The plants flourish in coastal areas, in or near mangrove swamps, and in marshy deltas.

The wood, known as cóbana negra, takes a lustrous finish, resembles mahogany, and was once popular for furniture. Its hardness, density of about 1280 kg/m³, and resistance to decay contribute to its value for construction purposes. Demands for the wood have accounted for its present scarcity (Record and Hess 1943; Little and Wadsworth 1964). The pods have a pleasant odor of ripe apples, but the flesh is tasteless.

Stauracanthus Link   Papilionoideae: Genisteae, Cytisinae

From Greek, *stauros*, "cross," and *akantha*, "spine," in reference to the spines on the angular branches and branchlets.

Type: *S. genistoides* (Brot.) Samp.

1 species. Shrub, tall; branches and spines opposite, clothed with silvery, silky hairs. Leaves reduced to very small, scalelike, blunt, subopposite phyllodes, not spine-tipped. Flowers densely villous, in short lateral or terminal, spineless racemes, or in small panicles on shoots; calyx clothed with golden brown hairs, 2-lipped, lower lip tridentate, long, curved, upper lip bidentate to the middle, shorter; petals subequal to and somewhat included by the calyx; standard rather silky-pubescent outside; stamens monadelphous; anthers alternately basifixed and dorsifixed. Pod very linear-oblong, longer than and conspicuously exserted from the calyx, dehiscent, 5–6-seeded. (Rothmaler 1941)

A comprehensive survey of leaf flavonoids in 23 genera in Genisteae (Harborne 1971) supported Rothmaler's (1941) circumspection of the tribe and the inclusion therein of *Stauracanthus*.

This monotypic genus is endemic in southwestern Spain, Portugal, and northwestern Africa.

Steinbachiella Harms   Papilionoideae: Dalbergieae, Pterocarpinae

Named for Steinbach; he collected the specimens in Bolivia.

Type: *S. leptoclada* Harms

1 species. Tree, up to 10 m tall; young branches slender, pubescent. Leaves imparipinnate; leaflets 11–13, alternate, oblanceolate or narrow-oblong; stipules lanceolate, acute, striate, caducous. Flowers yellow, 1–4, on slender, apically jointed pedicels, in short racemes below the axils of new leaves; bracteoles 2, obovate-oblong, caducous, subtending the calyx; calyx oblique, cupular to funnel-shaped; lobes 5, lowermost lanceolate, 2 laterals broad-lanceolate, upper 2 slightly connate, oblique-deltoid; petals short-clawed, subequal; standard suborbicular; wings oblique, broad-obovate; keel

623

petals obovate, falcate, blunt; stamens 10, 1 free except at the middle, 9 joined into a tube; anthers small, slightly unequal; ovary long-stalked, lanceolate, 1–2-ovuled; style filiform, smooth, acute; stigma minute. (Harms 1928a)

This monotypic genus is endemic in Bolivia. The plants thrive in the pampas.

Stemonocoleus Harms   Caesalpinioideae: Cynometreae
(43a) / 3491a

From Greek, *stemon*, "stamen," and *koleon*, "sheath," in reference to the insertion of the filaments on a staminal disk.

Type: *S. micranthus* Harms

1 species. Tree, glabrous, large, up to 30 m tall; bole straight, buttressed; crown spreading. Leaves pinnate; leaflets 4–10, alternate, on short petioles, oblong or ovate, rounded or notched at the apex, base obliquely acute or obtuse, pellucid-punctate, lateral nerves prominent above and below. Flowers very small, in 2 rows, in very short pedicels, crowded toward the ends of terminal, panicled racemes; bracts broad, boat-shaped and ovate at the base, enclosing the bud; bracteoles 2, small, narrow-oblong to lanceolate, slightly keeled on the back and fringed toward the apex; calyx tube short, funnel-shaped, thickened at the base with staminal disk protruding and sheathlike with 1 side open; lobes 4–5, slightly unequal, imbricate; petals none; stamens 4, inserted below the rim of a unilateral disk; staminodes none; anthers dorsifixed, turned inward; ovary subsessile or very short-stalked, glabrous, 2–3-ovuled; style elongated, coiled inward; stigma minute, capitellate. Pod samaroid, papery, large, cuneate and twisted basally, rounded at apex, indehiscent; seeds 1–2, flat, suboblong, located centrally. (Harms 1905b, 1908a; Léonard 1957)

This monotypic genus is endemic in equatorial West Africa, notably in Gabon, Ivory Coast, and southern Nigeria.

Stenodrepanum Harms   Caesalpinioideae: Eucaesalpinieae

From Greek, *stenos*, "narrow," and *drepanon*, "sickle," in reference to the pod.

Type: *S. bergii* Harms

1 species. Shrub, erect. Leaves bipinnate; leaflets very small, 5–9 pairs, elliptic-oblong, obtuse, slightly asymmetric at the base, median nerve slightly visible, margins scarcely crenate, somewhat glandular; stipules membranous, oval, free, glabrous. Flowers yellow, many, racemose; calyx briefly cup-shaped; lobes 5, glabrous, subequal, 1 slightly broader or oblong, obtuse; petals 5, subequal, slightly longer than the lobes, 1 broader, obovate, with a

broad blade, the others oblong-oblanceolate, narrowed into a claw, slightly glandular-warty; stamens 10; filaments hirsute, slightly glandular; ovary short-stalked, densely glandular-verruculose; stigma small, obliquely capitellate, about 8-ovuled. Pod narrowly linear-lanceolate, more or less falcate, subrostrate, obliquely striate lengthwise, faintly constricted between the 5–9, flattened, narrow, shiny, brownish black seeds. (Harms 1921b)

This monotypic genus is endemic in central Argentina.

In Harms's opinion the genus seemed more closely allied to *Hoffmanseggia* than to *Caesalpinia*, and the pods most similar to those of *Parkinsonia*.

Storckiella Seem.   Caesalpinioideae: Cassieae
82 / 3538

Named for Jacob Storck, assistant to Dr. Berthold Seeman during his Fiji explorations, 1860–1861.

Type: *S. vitiensis* Seem.

3 species. Trees, tall, unarmed, up to about 25 m tall; bark scaly, fissured. Leaves imparipinnate; leaflets alternate, leathery; stipules minute, caducous. Flowers golden yellow, in terminal panicles; bracts and bracteoles very caducous; calyx tube very short, turbinate; lobes 3–5, slightly unequal, imbricate; petals 3–5, oblong, imbricate, upper inner petal often incompletely developed; stamens 10–12; filaments glabrous, filiform, free; anthers linear, basifixed, dehiscent by apical pores or by short slits; ovary subsessile, multiovuled; style awl-shaped; stigma terminal, blunt. Pod oblong or subfalcate, flattened, leathery, upper suture broad-winged, 2-valved; seeds brown, transverse, suborbicular, flattened. (Bentham and Hooker 1865)

Species of this genus are indigenous to New Caledonia and Fiji.

The trees closely resemble *Cassia* shower trees and are exceptionally attractive in their native habitats for their graceful stature and masses of golden yellow flowers.

Stracheya Benth.   Papilionoideae: Hedysareae, Euhedysarinae
308 / 3776

Named for General Sir Richard Strachey, F.R.S., 1817–1908; he collected plants in the Central Himalayas.

Type: *S. tibetica* Benth.

1 species. Shrublet, low, nearly stemless, densely tufted from a woody rootstock. Leaves petioled, imparipinnate; leaflets 11–15, entire, oblong, blunt, faintly hairy; stipels absent; stipules scarious, villous. Flowers reddish, quite large, 1–4, on short pedicels on very short axillary peduncles; bracts lanceolate; bracteoles smaller, both persistent; calyx tube turbinate; teeth 5,

upper 2 longer than others and somewhat connate; standard broad-obovate, narrowed to the base; wings shorter than the standard, falcate-oblong; keel as long as the standard, blunt, incurved; stamens diadelphous, vexillary stamen entirely free, or free only at the base, then connate in the middle with the others; anthers uniform; ovary linear, subsessile, 4- or more-ovuled; style long, filiform, abruptly incurved above; stigma minute, terminal. Pod not stalked, linear, straight, flattened, rigid, exserted from the calyx, sutures continuous, bristly-toothed, not visibly jointed, joints scarcely separating, indehiscent; seeds kidney-shaped. (Bentham and Hooker 1865; Hooker 1879)

This monotypic genus is endemic in Tibet. The plants grow in the Himalayan alpine region between 4,300- and 5,750-m altitude.

Streblorrhiza Endl.   Papilionoideae: Galegeae, Robiniinae
284 / 3752

From Greek, *streblos*, "twisted," and *rhiza*, "root," in reference to the shape of the roots.

Type: *S. speciosa* Endl.

1 species. Shrub, tall, scandent, glabrous. Leaves imparipinnate; leaflets few, rather large; stipels absent; stipules small. Flowers flesh-colored, showy, large, in axillary racemes; bracts small; bracteoles minute; calyx quite broad; lobes 5, lower 3 short, upper 2 very short; standard ovate, erect, subsessile, narrowed at the base; wings short, narrow, small; keel incurved, acute, subequal to the standard; stamens diadelphous; anthers uniform; ovary stalked, multiovuled; style slender, incurved, glabrous; stigma terminal. Pod compressed, broad-oblong, obliquely subfalcate, 2-valved, valves leathery, continuous within and pubescent; seeds kidney-shaped, funicle filiform. (Bentham and Hooker 1865)

This monotypic genus is endemic in Norfolk Island.

Willis (1966) suggested that this species is extinct.

Strombocarpa Engelm. & Gray   Mimosoideae: Adenanthereae

From Greek, *strombos*, "something spun round," "a top," and *karpos*, "fruit," in reference to the shape of the pods.

Type: *S. strombulifera* (Willd.) Gray

6 species. Shrubs or trees, up to 30 m tall, deciduous. Leaves bipinnate; pinnae pairs few, usually with small, round, sessile glands between the pinnae; leaflets small, 5–9 pairs; stipules spiny, adnate to petiole. Flowers sessile, many, yellow to brown or greenish white, in spikes or globular heads in leaf axils; calyx tube short, campanulate, thin, membranous; teeth 5, very short;

petals 5, regular, valvate; stamens 10, free, short; anthers with terminal, deciduous gland; ovary small, many-ovuled; style slender, exserted. Pod tightly coiled into a cylindric spiral, 2.5–5 cm long, with up to 30 spiral turns, each about 2 mm in diameter, not stipitate, not beaked, and not dehiscent. (Britton and Rose 1928)

Strombocarpa was formerly a section of *Prosopis*. Members of the former are commonly called screw bean, screw pod, or tornillo; those of the latter, mesquite. The occurrence in *Strombocarpa* of spiny stipules instead of spiny shoots and the presence of tightly coiled pods instead of linear ones are among the most constant differences between the 2 taxa (Benson 1941).

Representatives of this genus are indigenous to temperate southwestern United States, Mexico, and South America. They thrive in bottomlands along desert streams, near waterholes, and in ravines up to 800-m altitude.

The economic importance of *Strombocarpa* species coincides with that of *Prosopis* species.

NODULATION STUDIES

S. odorata (Torr. & Frém.) Gray (reported as *Prosopis pubescens*) lacked nodules in arid areas of Arizona and in greenhouse seedling tests (Martin 1948). Inasmuch as related *Prosopis* species are generally nodulated, this observation appears inconclusive.

Strongylodon Vog.   Papilionoideae: Phaseoleae, Erythrininae
397 / 3873

From Greek, *strongylos*, "round," and *odous, odontos*, "tooth," in reference to the round-toothed calyx.

Type: *S. ruber* Vog. (= *S. lucidus* (Forst. f.) Seem.)

20 species. Shrubs or shrublets, winding, or woody vines. Leaves pinnately 3-foliolate; stipels usually present; stipules small. Flowers large, papilionaceous, red, blue, violet, or greenish, fascicled in showy, long, pendulous, axillary racemes; rachis nodose; bracts small; bracteoles orbicular, small, caducous; calyx campanulate; teeth broad, round, obtuse, upper 2 slightly united; petals unequal; standard ovate-oblong, acute, recurved, with 2 appendages above the claw; wings obtuse, shorter than the standard, adherent to the keel petals; keel much recurved, beaked, as long as the standard; stamens 10, diadelphous; anthers uniform; ovary stalked, 1- to few-ovuled; basal disk dentate; style filiform, not barbate; stigma small, terminal, capitate. Pod stalked, beaked, ovate-oblong, 2-valved, valves convex, leathery, tardily dehiscent; seeds 1–few, thick, large, orbicular, stony, half-surrounded by a linear hilum. (Bentham and Hooker 1865; Steiner 1959)

Members of this genus are distributed throughout Oceania, from Hawaii, Polynesia, and the Indo-Malaya Archipelago to Madagascar. They flourish in damp ravines and in forests at low to medium altitudes.

These plants are most notable for their indigo blue to brilliant scarlet, crescent-shaped flowers that hang in large, pendent racemes. Nine of the 10 species occurring in the Philippines are endemic. The flowers of few species, if any, surpass in beauty the blue green blossoms of *S. macrobotrys* Gray, the Philippine jadevine. The individual flowers, 6–7.5 cm long, hang in racemes 1 m long. The indehiscent pod contains 3–10 large, stony seeds that lose their vitality within 2 weeks (Steiner 1959). The type species is a liana commonly found at 150–800-m elevations, where it festoons the tallest trees in rainforest areas. Its brilliant scarlet flowers on reddish pedicels are commonly fascicled in twos or threes in racemes 30–50 cm long. Hawaiians call this woody vine "nukuiwi" for the red bird "iwi" with the curved beak "nuku."

NODULATION STUDIES

Nodules have been reported on *S. macrobotrys* in the Philippines (Bañados and Fernandez 1954; personal observation 1962) and in Trinidad (DeSouza 1963), on *S. ruber* Vog. in Hawaii (personal observation 1962), and on the latter species as *S. lucidus* (Forst. f.) Seem. in the Bogor Botanic Garden, Java (personal observation 1977).

Strophostyles Ell.   Papilionoideae: Phaseoleae, Phaseolinae

From Greek, *strophe*, "a twisting," "a turning," in reference to the shape of the style.

Type: *S. helvola* (L.) Ell.

3–4 species. Vines, annual or perennial, herbaceous; stems twining or trailing, retrorsely hairy. Leaves long-petioled, pinnately 3-foliolate; leaflets lanceolate or ovately 3-lobed, terminal leaflet petioluled; stipels present; stipules striate, small, persistent. Flowers pink, purple, or cream-colored, few to several, closely clustered, subsessile, in long peduncled, axillary racemes; flowers subtended by a small bract and 2 bracteoles; calyx campanulate, more or less bilabiate; 2 upper lobes almost completely connate, lowermost lobe longest; standard orbicular to ovate; wings oblong, shorter than the keel; keel strongly incurved, not spirally twisted, beaked; stamens 10, diadelphous; ovary subsessile; style strongly incurved, bearded along the inner side. Pod linear, flat or cylindric, valves twisting spirally and laterally after dehiscence; seeds several to many, quadrate or oblong, woolly. (Wilbur 1963; Steyermark 1963)

The genus is closely related to and often united with *Phaseolus*.

Members of this genus are indigenous to the eastern and midwestern United States. They flourish in dry, sandy areas.

These plants are commonly called mealybean, fuzzybean, or wildbean. All members of the genus have erosion-control value. The seeds and herbage are good wildlife food.

NODULATION STUDIES

The inability of *Strophostyles* rhizobia to nodulate species of other genera and vice versa prompted Burrill and Hansen (1917) and Bushnell and Sarles (1937) to designate a monospecific *Strophostyles* plant-infection group. This conclusion merits confirmation in view of the taxonomic complexities and close interrelationships now recognized among members of *Strophostyles*, *Phaseolus*, and *Vigna*.

Nodulated species of *Strophostyles*	Area	Reported by
helvola (L.) Ell.	Ill., U.S.A.	Burrill & Hansen (1917)
	Wis., U.S.A.	Bushnell & Sarles (1937)
	Ind., U.S.A.; Canada	Pers. observ. (U.W. herb. spec. 1972)
var. *missouriensis* (S. Wats.) Britt.	Mo., U.S.A.	Steyermark (1963)
leiosperma (T. & G.) Piper	Wis., U.S.A.	Bushnell & Sarles (1937)
	Minn., N. Dak., U.S.A.	U.S.D.A. *Rhizobium* collection (1938)
	Kans., N. Dak., U.S.A.	M. D. Atkins (pers. comm. 1950)
(= *pauciflora* (Benth.) S. Wats.)	Ark., Ill., Tex., U.S.A.	Pers. observ. (U.W. herb. spec. 1972)
umbellata (Muhl.) Britt.	Penn., U.S.A.	Pers. observ. (U.W. herb. spec. 1972)

Stryphnodendron Mart.   Mimosoideae: Adenanthereae
23 / 3461

From Greek, *stryphnos*, "astringent," and *dendron*, "tree."

Type: *C. barbatimam* Mart.

15 species. Trees, unarmed, usually small, with thick twigs and thin foliage; some Amazon species are very large. Leaves bipinnate; pinnae numerous; leaflets small, broad, mostly alternate, in numerous pairs; petiole or rachis glanduliferous. Flowers 5-parted, small, white to purplish, hermaphroditic or subpolygamous, profuse, borne in short-stalked, axillary, cylindric spikes; calyx campanulate, short-toothed; petals joined below, free above the middle, valvate; stamens 10, free, slightly exserted; anthers tipped with a deciduous apical gland; ovary short-stalked, many-ovuled; style slender; stigma terminal. Pod linear, narrow, compressed, indehiscent, septate, septa continuous with the endocarp, mesocarp subpulpy, fleshy; seeds several, transverse, funicle filiform. (Bentham and Hooker 1865; Britton and Rose 1928)

Representatives of this genus are native to tropical South America, Brazil, Costa Rica, and Guiana.

629

The bark of *S. barbatimam* Mart., known as barbatimão in Brazil, yields up to 40 percent tannin; decoctions of the bark and leaves serve native uses as tonics, antidiarrhetics, and hemostatic astringents. The unattractive wood of these twisted and gnarled trees is used primarily for general purposes, but it does play a role in the making of small furniture and turnery articles because of its ability to take a high polish (Record and Hess 1943).

NODULATION STUDIES

In Zimbabwe, Corby (personal communication 1977) observed nodules on roots of *S. adstringens* Cov. and *S. barbatimam* grown from imported seeds.

Stuhlmannia Taub.   Caesalpinioideae: Eucaesalpinieae
(102c)

Named for Franz Ludwig Stuhlmann, 1863–1928, Director of the Biologische Landwirtschaftliche Institut, Hamburg; he collected extensively in Sri Lanka and Java.

Type: *S. moavi* Taub.

1 species. Tree, with very rough bark. Leaves imparipinnate; leaflets subopposite, subsessile, 3–6 pairs, large, asymmetric at the base; stipels and stipules absent. Flowers yellow, often paired, many, in terminal, sometimes axillary, racemes; bracts linear-oblong, caducous, minute; bracteoles absent; calyx shortly cupular; lobes 5, narrow, sublinear, subacute, free nearly to the base, valvate; petals 5, clawed, subequal, obovate-spathulate; stamens 10; filaments thickened below, hairy and united at the base, alternately long and short; anthers small, oblong, subbasifixed, longitudinally dehiscent; ovary short-stalked, free, oblong, glandular, 2-ovuled; style glandular below, ciliate around the small, terminal stigma. (Harms 1897; Baker 1930)

Harms (1897) originally positioned this genus in Cassieae as 81a. Following his recognition of its closer taxonomic kinship with members of the tribe Eucaesalpinieae, Harms (1915a) changed *Stuhlmannia* to 102c.

This monotypic genus is endemic in tropical East Africa.

Stylosanthes Sw.   Papilionoideae: Hedysareae, Stylosanthinae
332 / 3802

From Greek, *stylos*, "pillar," and *anthos*, "flower," in reference to the stalklike calyx tube.

Lectotype: *S. procumbens* Sw.

50 species. Herbs, usually low, perennial, rarely subshrubs, some with bristly- or viscid-haired stems. Leaves pinnately 3-foliolate; pinnae sometimes elliptic and pungent; stipels absent; stipules sheathlike, with conspicuous awl-shaped tips, fused to the base or lower part of the petiole. Flowers yellow orange, rarely white, often streaked with purple, on short pedicels, in dense,

terminal or axillary, leafy-bracted spikes or heads, in pairs or solitary; apetal-ous flowers uncommon and inconspicuous; bracts thinly membranous, glumelike; bracteoles 2–3, hyaline; calyx tube short, campanulate, slender, columnar; lobes 5, membranous, very unequal, upper 4 connate, lowest lobe longest, free, narrow, distinct; petals short-clawed, inserted with the stamens at the apex of the tube; standard orbicular-obovate, slightly emarginate, ses-sile; wings oblong, free; keel petals incurved, eared at the base, about as long as the wings, often beaked; stamens 10, monadelphous, tube later slit; anthers alternately long and subbasifixed, short and versatile; ovary subsessile, 2–3-ovuled; style very long, filiform, becoming broken at the middle or near the base after flowering, lower part persistent, recurved; stigma minute, terminal. Pod sessile, compressed, 1–2-jointed, segments reticulate, tuberculate or pubescent, hooked at the apex with the persistent style base, beak usually curved or hooked, dehiscent at apex or indehiscent; seeds compressed, ovoid, smooth, shiny. (Bentham and Hooker 1865; Baker 1929; Amshoff 1939; Phillips 1951; Mohlenbrock 1957)

The genus is taxonomically complex (Mohlenbrock 1957, 1962a).

Members of this genus are native to tropical Asia, Africa, and America, and warm eastern and southern regions of the United States. They thrive in limestone and gravelly waste places, dry sandy woodlands, dry rice fields, and in open pastures.

Because many members of this genus exhibit vigorous growth, deep-rooting habit, and drought resistance, they are valued as green manures, cover plants, and for soil conservation. In warm regions where lucerne cannot be grown, *S. mucronata* Willd., called wild lucerne in Australia, *S. gracilis* Taub., *S. guyanensis* Sw., Brazilian lucerne, and *S. sundaica* Taub., Townsville clover, have become important substitutes. These species are similar in ap-pearance to lucerne, but have a coarser foliage, greater vigor, and a tendency to root at the nodes.

Brazilian lucerne has been especially valued as ground cover on Malay-sian tea plantations because of its prostrate spreading growth, ability to smother weeds, profusely nodulated root system, role in humus formation, and beneficial effect on the earthworm population (Heaton 1954; Vivian 1959; Chandapillai 1972). MacTaggart (1937) attributed the founding of the milk industry in northern Australia to the abundance of *S. sundaica*. In South America, *S. erecta* Beauv. is propagated as graze and fodder for horses and beef cattle.

NODULATION STUDIES

Comprehensive host-infection tests have shown that *S. guyanensis* is highly susceptible to nodulation by strains from diverse species of the cowpea miscellany and, correspondingly, that *S. guyanensis* rhizobia have a promiscu-ous host range (Allen and Allen 1939). These findings undoubtedly explain the high incidence of nodulation of this species by indigenous rhizobia in tropical soils (MacTaggart 1937; Heaton 1954; Vivian 1959; Chandapillai 1972). In screening tests of 21 accessions of 7 *Stylosanthes* species in combina-tion with 25 rhizobial strains from *Stylosanthes, Arachis, Zornia,* and *Dolichos* species, *S. erecta* was effectively nodulated by 90 percent of the strains, whereas *S. hamata* (L.) Taub. was nodulated only by 6 strains, and none of the associations was effective (Mannetje 1969). The nodulation patterns of 6 accessions indicated highly specific *Rhizobium* requirements. Mannetje rightly emphasized the need to select specific strains for agronomic use where high yields and percentages of fixed nitrogen are essential.

Nodulated species of *Stylosanthes*	Area	Reported by
biflora (L.) BSP	N.C., U.S.A.	Shunk (1921)
	Md., U.S.A.	Pers. observ. (U.W. herb. spec. 1970)
bojeri Vog.	Australia	Norris (1959b)
	S. Africa	N. Grobbelaar (pers. comm. 1961); B. Strijdom (pers. comm. 1968)
capitata Vog.	Venezuela	Barrios & Gonzalez (1971)
erecta Beauv.	Australia	Mannetje (1969)
fruticosa (Retz.) Alst.	Sri Lanka	W. R. C. Paul (pers. comm. 1951)
	S. Africa	Grobbelaar & Clarke (1972)
	Zimbabwe	Corby (1974)
gracilis Taub.	Sri Lanka	W. R. C. Paul (pers. comm. 1951); Heaton (1954)
	Zaire	Bonnier (1957)
	Australia	Norris (1959b)
	S. Africa	N. Grobbelaar (pers. comm. 1962)
guyanensis Sw.	Hawaii, U.S.A.	Allen & Allen (1939)
	S. Africa	N. Grobbelaar (pers. comm. 1963)
	Zimbabwe	Sandmann (1970a)
ssp. *guyanensis*	Venezuela	Barrios & Gonzalez (1971)
hamata (L.) Taub.	West Indies; Kenya	Pers. observ. (U.W. herb. spec. 1970)
	Venezuela	Barrios & Gonzalez (1971)
humilis HBK	Zimbabwe	Sandmann (1970a)
leiocarpa Vog.	S. Africa	N. Grobbelaar (pers. comm. 1963)
mexicana Taub.	Venezuela	Barrios & Gonzalez (1971)
montevidensis Vog.	S. Africa	Grobbelaar et al. (1967)
mucronata Willd.	Australia	Mannetje (1969)
riparia Kearn.	Va., U.S.A.	Pers. observ. (U.W. herb. spec. 1970)
sundaica Taub.	Sri Lanka	W. R. C. Paul (pers. comm. 1951)
	Australia	Norris (1959b)
tuberculata Blake	Venezuela	Barrios & Gonzalez (1971)
viscosa Sw.	Australia	Mannetje (1969)

Named for James Sutherland, Superintendent of the Botanic Gardens and botanist of Edinburgh in the late 17th century; he published a catalogue of plants in the Botanic Gardens in 1683.

Type: *S. frutescens* (L.) R. Br.

6 species. Shrubs, up to 2 m tall, hoary-pubescent. Leaves imparipinnate; leaflets linear-oblong, many pairs, pubescent on 1 or both surfaces; stipules minute, lanceolate, withering. Flowers large, bright red or scarlet, few, in short, axillary racemes; bracts and bracteoles small; calyx campanulate; lobes 5, subequal; standard erect, sides reflexed, shorter than the wings and keel petals; wings small, clawed, oblong, obtuse or acute, eared; keel lanceolate, incurved, erect, claws connate except at the base; stamens 10, diadelphous; anthers uniform; ovary stalked, many-ovuled; style bearded along the back and in front just below the small, capitate, terminal stigma. Pod oblong-ellipsoid, papery, semitranslucent, inflated, subindehiscent; seeds flat, kidney-shaped, funicle filiform. (Bentham and Hooker 1865; Baker 1929; Phillips 1951)

Representatives of this genus are native to Namibia and South Africa. They grow on dry slopes, hills, flats, and sand dunes.

These shrubs are cultivated in gardens for their showy flowers and decorative, bladderlike pods; such common names as gansies, goslings, turkey bells and balloon pea, refer to the inflated pods. The vernacular name, "kankerbos" or "cancer bush," for *S. frutescens* (L.) R. Br., alludes to the native use of leaf infusions for the treatment of intestinal and uterine ailments, including cancer (Watt and Breyer-Brandwijk 1962).

NODULATION STUDIES

In 1896 Clos observed nodules on young roots of an exotic specimen of *S. frutescens* introduced into France. Grobbelaar et al. (1967) also reported nodules on *S. frutescens* growing in native habitats in South Africa. Rhizobial strains from diverse members of the cowpea miscellany produced nodules on this species in experimental tests conducted by Wilson (1939a). An annual report from the University of Pretoria (1954) cited nodulation of *Psoralea* and *Vigna* species by rhizobia isolated from *S. macrophylla* [?].

Swainsona Salisb.　　　　　Papilionoideae: Galegeae, Coluteinae

289 / 3757

Named for Isaac Swainson, 1746–1806; he maintained a private botanical garden at Twickenham, near London, about 1789.

Type: *S. galegifolia* (Andr.) R. Br. (= *S. coronillifolia* Salisb.)

50–60 species. Herbs, perennial, rarely annual or small undershrubs, glabrous or pubescent with appressed hairs. Leaves imparipinnate; leaflets numerous, lanceolate to oblong, small, entire; stipels absent; stipules herbaceous, oblique, with broad base, sometimes setaceous. Flowers ornamentally attractive, purple, red, pink, rarely white, yellow, or mixed, in loosely spaced, erect, axillary racemes; bracts small, membranous, narrow; bracteoles small, usually affixed to the pedicels just below the calyx or appressed to the calyx tube; calyx campanulate; lobes 5, subequal or upper 2 shorter, pubescent inside; standard orbicular-reniform, spreading or reflexed, often with a patch of yellow, green, or white and 2 short callosities above the short claw; wings short, oblong or twisted, free, usually shorter than the keel; keel petals broad, pouched, incurved or spirally coiled; stamens 10, diadelphous; anthers uniform; ovary sessile or stalked, many-ovuled; style incurved, blunt or curled back at the apex, longitudinally bearded on the inner edge; stigma small, terminal. Pod bladderlike or oblong, turgid or inflated, leathery or membranous, more or less 2-celled by the intrusion of the upper suture, 2-valved or subindehiscent; seeds small, kidney-shaped. (Bentham and Hooker 1865; Lee 1948; Allan 1961)

Swainsona species are limited in distribution to Australia. They are widespread in New South Wales, Western Australia, and southern central Australia in regions of less than 50 cm rainfall (Lee 1948). They flourish on sandy soil, saltbush plains, rocky ridges, and clay soils. One species, *S. novazelandiae* Hook. f., occurs in the high montane and subalpine areas of New Zealand.

Reputedly, positive or suspected livestock poisoning has resulted from ingestion of the foliage of *S. greyana* Lindl., *S. lessertiifolia* DC., *S. luteola* F. Muell., *S. microphylla* Gray, *S. oroboides* F. Muell., and *S. procumbens* F. Muell.; however, *S. stipularis* F. Muell. is regarded as good fodder in some areas. Symptoms of *Swainsona* poisoning resemble those of selenium, i.e., a form of insanity evidenced by erratic walking, defective vision, staggering, trembling, a propensity to climb, and a tendency of the animals to jump over small twigs as though they were almost half a meter high (Hurst 1942; Webb 1948). Animals exhibiting these symptoms are referred to as being "pea-struck." *S. greyana* ssp. *cadelli* is suspected of being fatal to bees (Hurst 1942). Toxic properties of *Swainsona* species have not been defined.

NODULATION STUDIES

Perhaps the earliest record of nodules on *S. galegifolia* (Andr.) R. Br. was Munson's (1899) account of their use as crushed nodule inoculum for pea, bean, vetch, lupine, and soybean culture in greenhouse tests. The results were poor and, in fact, questionable because of contamination in the control pots. Munson himself concluded that these experiments at the Maine Agricultural Experiment Station "do not justify the recommendation of germ cultures for leguminous crops." In later tests Ewart and Thomson (1912) also used crushed nodules of this species as inocula for potted seedlings of beans, peas, medic, vetch, clover, and soybeans; nodules were not produced. Wilson's data (1939a) support the inclusion of this species in the cowpea miscellany. Host-infection studies were not reported by Lange (1959) and Beadle (1964). Beadle's earlier comment (1959) concerning the role of *Swainsona* species in arid and semiarid plant communities of western New South Wales and Queensland merits citation here: ". . . the herbaceous legumes which are abundant in the saltbush and bluebush communities undoubtedly add large quantities of nitrogen to the soil. For example it is estimated that the purple pea *Swainsona swainsonioides*, which carpets the ground under certain conditions, may fix nitrogen at the rate of 250 lbs. nitrogen per acre per annum, which is a highly significant figure. Such an addition, however, is likely to occur only about once every ten years."

Wide, tortuous infection threads containing rhizobia 4–6 abreast and abundant starch granules in invaded cells were depicted in *S. galegifolia* nodules by Lechtova-Trnka (1931).

Nodulated species of *Swainsona*	Area	Reported by
burkittii F. Muell. ex Benth.	N.S. Wales, Australia	Beadle (1964)
campylantha F. Muell.	N.S. Wales, Australia	Beadle (1964)
canescens F. Muell.	N.S. Wales, Australia	Pers. observ. (1962)
	N. Zealand	R. M. Greenwood (pers. comm. 1975)
fissimontana J. M. Black	N.S. Wales, Australia	Beadle (1964)
flavicarinata J. M. Black	N.S. Wales, Australia	Beadle (1964)
galegifolia (Andr.) R. Br. (= *coronillifolia* Salisb.)	Widely reported	Many investigators
greyana Lindl.	N.S. Wales, Australia	Beadle (1964)
lessertiifolia DC.	N.S. Wales, Australia	Beadle (1964)
microphylla Gray	N.S. Wales, Australia	Beadle (1964)
novazelandiae Hook. f.	N. Zealand	R. M. Greenwood (pers. comm. 1975)
†*occidentalis* F. Muell.	W. Australia	Lange (1959)
oroboides F. Muell. ssp. *oroboides* A. Lee	N.S. Wales, Australia	Beadle (1964)
phacoides Benth. ssp. *phacoides* A. Lee	N.S. Wales, Australia	Beadle (1964)
procumbens F. Muell.	N.S. Wales, Australia	Beadle (1964)
rigida J. M. Black	N.S. Wales, Australia	Beadle (1964)
stipularis F. Muell.	N.S. Wales, Australia	Beadle (1964)
swainsonioides (Benth.) A. Lee	N.S. Wales, Australia	Beadle (1964)

†Erroneously reported as *S. Incei* Price.

Named for Olaf Swartz, 1760–1818, Swedish botanist.

Type: *S. guianensis* (Aubl.) Urb. (= *S. alata* Willd.)

130 species. Trees, up to about 30 m tall, or small shrubs, seldom lianas, unarmed. Leaves imparipinnate 3–31-foliolate, rarely 1-foliolate; leaflets ovate to oblong, often minutely haired, 1–15 pairs; petiole and rachis often winged or broad-margined; stipules variously shaped, minute or leafy, usually caducous. Flowers white to orange yellow, pedicelled, in simple or panicled racemes, usually on stems or older branches, rarely in leaf axils or solitary; bracts small, usually triangular, caducous; bracteoles very small and caducous, or not present; calyx globose or ovoid in bud due to fusion of the segments, but rupturing at anthesis into 3–5 irregular segments; petal 1, broad, large, white, yellow, or violet, or with 2 laterals; stamens mostly dimorphic, very numerous, usually more than 30, free or slightly connate at the base, declinate and incurved; anthers dorsifixed, larger on the longer filaments, longitudinally dehiscent; ovary stalked, linear, slightly incurved, multiovuled; style narrow, short or long; stigma small, terminal, rarely capitate. Pod oval or elongated, moniliform, flattened or rounded, leathery or fleshy, surface smooth or rough, 2-valved, tardily dehiscent or indehiscent, 1- to several-seeded; seeds ovoid-reniform, with or without a white, yellow, or red aril. (Bentham and Hooker 1865; Gilbert and Boutique 1952b; Brenan 1967; Cowan 1968)

Positioning of this genus has provoked differences of opinion over many years. In 1951 Corner elevated the 9 genera of the tribe to subfamily status, Swartzioideae, with its placement intermediate between subfamilies Caesalpinioideae and Papilionoideae. Cowan's monograph (1968) deals with the *Swartzia* assemblage as a genus in the tribe Swartzieae, Caesalpinioideae.

Species of this large genus are distributed throughout tropical Mexico, the West Indies, and Central and South America; the center of distribution is in northern South America. Plants thrive on rocky beaches, dry sandy plains, in wet ravines, and in inundated forest areas.

Several of the tall, wide-spreading trees of the South American group are exceptionally beautiful when in flower (Menninger 1962). The fresh, odoriferous fruits of *S. fistuloides* Harms and *S. madagascariensis* Desv., the 2 African species, are used to stupefy fish. At one time the possibility of using the fruit of the former species to kill the bilharzial snail was considered (Coates Palgrave 1957). The pericarp and fruit yield a catechu tannin, an amorphous, powerfully hemolytic saponoside, and a yellow flavone pigment to which the name "swartziol" was tentatively applied. This compound was later judged identical with kaempferol, a tetrahydroxyflavone, $C_{15}H_{10}O_6$, (Paris and Bézanger-Beauquesne 1956). Rotenone, alkaloids, and anthraquinones were not detected. The saponins of this species have been studied in detail (Sandberg et al. 1958). The black, shiny pods of *S. madagascariensis* are valued as cattle feed in Zimbabwe (Walters 1924).

Swartzia wood is of little commercial importance. The brown to black heartwood is slow in forming. The wood is hard, tends to be heavier than water, resistant to decay, and difficult to work. Its use is limited to heavy construction purposes.

NODULATION STUDIES

Nodulation of *Swartzia* species has phylogenetic significance because of their more or less "bridge position" between the subfamily Caesalpinioideae, whose members tend to lack nodules, and the Papilionoideae, wherein nodulation is common. The first recordings of nodulation on *Swartzia* species are credited to DeSouza (1966), who commented: "On *Swartzia trinitensis* Urb., nodulation was very well developed. The tree, about ten feet tall, was growing in a typical tropical rain forest environment at Blanchisseuse. Nodulation was quite extensive on the lateral and secondary lateral roots; the nodules were branched and borne in clusters and were of the hard, 'woody' type. Internal pigmentation was observed to be dark brown. The mode of infection by the bacteria leading to nodule initiation may not have been through root hairs as these were quite scarce." Nodules on the seedlings examined in Guyana by Norris (1969) varied from few to numerous, round to elongated; those on *S. oblanceolata* Sandw. were of appreciable size.

Cultural and biochemical characteristics of the rhizobia isolated by the authors from nodules provided by DeSouza resembled those of slow-growing cowpea-type rhizobia. Lack of germination of *Swartzia* seed prevented host-infection tests; however, 4 strains from *S. pinnata* Willd. and 2 strains from *S. trinitensis* nodulated *Vigna unguiculata* and *Crotalaria striata* in Leonard assembly tests (personal observation 1965).

Nodulated species of *Swartzia*	Area	Reported by
benthamiana Miq.	Bartica, Guyana	Norris (1969)
leiocalycina Benth.	Bartica, Guyana	Norris (1969)
madagascariensis Desv.	Zimbabwe	Corby (1974)
oblanceolata Sandw.	Bartica, Guyana	Norris (1969)
pinnata Willd.	Trinidad	DeSouza (1966)
simplex (Sw.) Spreng.	Trinidad	DeSouza (1966)
trinitensis Urb.	Trinidad	DeSouza (1966)

Sweetia Spreng.　　　　　　　　　　Papilionoideae: Sophoreae

124 / 3582

Named for Robert Sweet, 1783–1835, English botanist.

Type: *S. fruticosa* Spreng.

20 species. Trees, small to large, up to about 40 m tall, unarmed; trunks and branches smooth with conspicuous lenticels. Leaves imparipinnate; leaflets 3–21, usually large, petioluled, asymmetric at the base, lanceolate to ovate, pairs variable, glossy, leathery; stipels minute or none; stipules small, caducous. Flowers small, usually fragrant, yellowish white or greenish, in small, terminal racemes; bracts and bracteoles narrow, minute, caducous; calyx turbinate-campanulate, entire; lobes subequal, valvate; petals 5, equal, free, spreading, erect, upper petal outermost and often broader; stamens 10, occasionally 5, free, subequal, longer than the petals; filaments inflexed; an-

thers uniform, ovate; ovary sessile or short-stalked, 2–4-ovuled; style filiform; stigma small, terminal or truncate. Pod oblong-linear or lanceolate, flat, leathery, plano-compressed, some with obscure, narrow wing near apex, indehiscent; seeds 2–4, ovate or orbicular, compressed, cotyledons thick. (Bentham and Hooker 1865; Amshoff 1939; Mohlenbrock 1963c)

Early botanists aligned this genus with Caesalpinioideae because of the absence of a typical papilionaceous flower.

Distribution of this genus is confined to tropical and subtropical regions in the Western Hemisphere, with Brazil the center of distribution; it is widespread from southern Mexico to northern Argentina (Mohlenbrock 1963c).

S. elegans Benth. and *S. nitens* (Vog.) Benth., of South America, and *S. panamensis* Benth., of North America, are among the best-known members of this genus. *Sweetia* wood does not feature in the export timber trade, but it is used locally for heavy construction purposes.

In earlier days an extract of the bark of *S. panamensis* was used medicinally by natives in the treatment of malaria, scrofula, and syphilis. In 1964 Fitzgerald et al. isolated 6 alkaloids from extracts of which sweetinine, $C_{20}H_{33}N_3$, was considered a newly discovered one, but it has marked similarity to and may be identical with ormosanine from *Ormosia* species.

NODULATION STUDIES

Norris (1969) observed brown, elongated nodules on 7 seedlings of *S. praeclara* Sandw. in Bartica, Guyana.

Sylitra E. Mey.   Papilionoideae: Galegeae,
254 / 3717   Tephrosiinae

Type: *S. biflora* E. Mey.

2–3 species. Undershrubs, slender, perennial, thinly covered with grayish hairs; taproots usually very long. Leaves usually 1-, sometimes palmately 3-, rarely 5-foliolate; leaflets entire, linear or lanceolate, obliquely veined, apiculate, with gray appressed hairs; petioles short, jointed at the apex; stipules small. Flowers very small, paired, in leaf axils; bracts and bracteoles inconspicuous or absent; calyx turbinate-campanulate, sub-2-lipped, narrow, hairy; lobes 5, subequal, ovate-lanceolate, shorter than the tube; standard obovate-spathulate, pubescent, narrowed into a claw; wings falcate-oblong, equal to the standard in length, adnate to the keel, eared, with a linear claw; keel blunt, shorter than the wings, with a long-linear claw; stamens monadelphous, vexillary stamen free at the base only, then connate with others into a tube slit along the top; anthers uniform; ovary sessile, many-ovuled; style inflexed, glabrous; stigma terminal, capitate. Pod oblong, flat, membranous, densely hairy, contorted or undulate, almost dehiscent; seeds 4–6, suborbicular. (Baker 1926; Phillips 1951)

Members of this genus have the habit of *Tephrosia* and the pod character of *Lessertia*. Brummitt (1965) demonstrated that *Sylitra* is an orthographic variant to be regarded as an illegitimate homonym, and proposed its merger with *Ptycholobium* because of its close relationship. There has been a tendency recently, however, to merge both *Ptycholobium* and *Sylitra* into *Tephrosia*.

Sylitra species occur in Namibia and South Africa.

Sympetalandra Stapf   Caesalpinioideae: Dimorphandreae
(34a) / 3474a

From Greek, *sym-*, "joined together," *petalon*, "petal," and *aner*, *andros*, "man."

Type: *S. borneensis* Stapf

1 species. Tree, small, smooth. Leaves paripinnate; leaflet pairs 2, gland-dotted, leathery; rachis terminates in a spicule. Flowers small, short-stalked in dense, axillary, spikelike racemes or panicles near apex of twigs; bracts inconspicuous, bracteoles lacking; calyx campanulate, broad; lobes 5, short, imbricate in bud; petals 5, equal, oblong, imbricate, uppermost petal innermost, all joined into a short tube at base; stamens 10, all fertile, free; filaments smooth, alternately short and long, the latter exserted; anthers uniform, basifixed, tipped with a deciduous gland, opening by longitudinal slits; ovary stalked, stipe free; ovules 2; style length equal to long stamens; stigma terminal, punctiform. Pod not described.

Harms (1908a) positioned this genus between *Dimorphandra* and *Burkea*. This monotypic genus is endemic in Malaysia and Borneo.

Tachigalia Aubl.   Caesalpinioideae: Amherstieae
52 / 3502

Name is derived from *tachigali*, a word used by the Indians of Guiana for trees inhabited by *tachi*, stinging ants of the genus *Pseudomyrma*.

Type: *T. paniculata* Aubl.

22 species. Trees, small or large. Leaves paripinnate; leaflets 3- to 15-paired, petioluled, papery to leathery, asymmetric, apex acuminate or mucronate, base blunt or heart-shaped, closely crowded, upper pairs usually larger; petiole and basal part of the rachis often triangular, hollow, and myrmecophilous; stipules paired, segmented, foliaceous. Flowers yellow, white, or orange, in elongated, dense, panicled, axillary or terminal racemes; bracteoles awl-shaped, pubescent, soon caducous; calyx campanulate, urceolate or triangular, often pubescent on the outside; lobes 5, imbricate, reflexed at anthesis, unequal, lower 2 outermost, smaller and thicker, margins ciliate; petals 5, subequal, obovate, imbricate, petaloid to fleshy, uppermost one less oblique; stamens 10, unequal; filaments pubescent at the base, uniform, or 7 longer, more slender and awl-shaped, other 3 much-reduced, thicker, shorter; anthers ovate, versatile, uniform, glabrous, longitudinally dehiscent;

ovary stalk joined to the middle or apex of the calyx tube; ovary narrow-oblong, densely pubescent, 4–15-ovuled; style filiform, pubescent below; stigma terminal, small, capitate, rarely 2-parted. Pod short-stalked, oblong, compressed, thin, membranous, valves leathery, separating at maturity; seed 1, ovate, compressed, surrounded by a thin, broad wing. (Bentham and Hooker 1865; Amshoff 1939; Dwyer 1954)

Members of this genus are widespread in tropical Central and South America.

The myrmecophytic species of this genus are commonly called tachy. Excellent accounts of the colonization of the hollow foliar axils of *T. paniculata* Aubl. as nesting chambers for ants and other insects are given by Bailey (1923) and Dwyer (1954). The ferocity of the stinging ants inhabiting these species may account for the paucity of herbarium collections (Wheeler 1921).

NODULATION STUDIES

Brown roots of 2 unidentified *Tachigalia* species examined in Manaus, Amazonas, Brazil, by Norris (1969) lacked nodules.

Talbotiella Bak. f.   Caesalpinioideae: Cynometreae

Named for P. Amaury Talbot of the Nigerian Administrative Service and Mrs. Talbot; they sent their botanical collection from southern Nigeria to the National Herbarium, London.

Type: *T. eketensis* Bak. f.

3 species. Shrubs, bushy, or trees, up to 20 m tall; trunk thick; foliage heavy. Leaves paripinnate; leaflets opposite, 15–20 pairs, mauve pink at emergence, truncately eared on the lower side, obtuse at apex, green at maturity, not pellucid-punctate; rachis grooved, slightly winged. Flowers white, lax, in more than 2 rows, on long, slender, pilose pedicels, in slender, axillary and terminal, softly pilose racemes; scales at the base of young shoots and peduncles, imbricate, dry, brown, persistent; bracts pink; bracteoles oblong-linear, persistent, colored, petaloid; calyx briefly funnel-shaped; lobes 4, white, glabrous, broadly overlapping; petals lacking; stamens 8–10, free; anthers orange; ovary stalked, stalk joined to the side of the receptacle, hairy, 2-ovuled; disk absent. Pod flat, stalked, young immature pods prominently transversely nerved, nerves less prominent on mature pods, 2-valved, valves leathery to subwoody and inrolled after dehiscence; seeds suborbicular, not arillate. (Baker 1914; Léonard 1951, 1957)

This genus is closely allied to *Hymenostegia*.

Representatives of this genus are indigenous to tropical West Africa. They usually abound in loose, drifting sands of damp, sandy coastal areas and in freshwater swamp forests; hence they often have stilt roots.

The wood of *T. gentii* Hutch. & Greenway in Ghana serves local construction purposes.

From Arabic, *tamr*, "dried date," and *hindi*, "Indian."

Type: *T. indica* L.

1 species. Tree, semievergreen, multibranched, unarmed. Leaves paripinnate; leaflets 10–20 pairs, oblong, entire or emarginate, asymmetric, not pellucid-punctate; stipules minute, caducous. Flowers small, yellow streaked with pink, in few-flowered, drooping racemes or panicles, axillary or terminal; bracts ovate-oblong, colored, concave, caducous; bracteoles 2, opposite, concave, sepaloid, red, caducous, enclosing flower in bud; calyx tube narrow, turbinate, deeply cleft; lobes 4, imbricate, reddish outside, yellow inside, subequal and submembranous; petals 5, upper 3 narrow, oblong-obovate, subequal, slight-clawed, imbricate, uppermost petal innermost, narrower, lower 2 small, scalelike; stamens united into a sheath split above, only 3 fertile, alternating with 4 staminodes; filaments short; anthers dorsifixed, oblong, longitudinally dehiscent; ovary flattened, stalked, stalk adnate to the calyx tube, 8–10- to many-ovuled; style elongated, thick; stigma terminal, blunt, subcapitate. Pod oblong to linear-oblong, usually curved, thick, subcylindric to subcompressed, indehiscent, exocarp brittle, constricted or septate between the seeds, mesocarp pulpy-fibrous, yellow, acid; seeds 1–10, obovate-orbicular, compressed, embedded in a pulp and joined by tough fibers. (Standley and Steyermark 1946; Léonard 1952)

This monotypic genus is probably native to tropical Africa. It is now cultivated pantropically, especially in India. It requires well-drained soils and is well adapted to dry savannas and poor soils in lowland areas. The plants are often found growing wild on stream and river banks and in proximity to termite mounds. They do not thrive or set fruit in regions of high rainfall or wet monsoons.

The tamarind is a handsome, slow-growing tree that reaches an average height of 15 m, has a stout, slightly buttressed trunk, and a rounded, dense crown. It is ornamentally attractive and useful for shade in gardens, parks, and avenues. Ages of 150 and 200 years have been recorded for specimens in Hawaii (Neal 1948) and India (Cowen 1965), respectively.

The tamarind is best known for the thick, sticky, brown, pleasant-tasting pulp that surrounds the seeds. This tart syrupy pulp constitutes about 40 percent of the indehiscent pod. The approximate composition of the pulp, in percent, is as follows: water, 20.6; protein, 3.1; fat, 0.4; carbohydrates, mostly sugars, 70.8; fiber, 3.0; and ash, 2.1 (Purseglove 1968).

A high content of tartaric acid accounts for the taste of the pulp; acetic, citric, malic, and succinic acids are also present. The pulp is used as a confectionary eaten with or without salt, as an ingredient in curries, chutneys, preserves, pickles, sherbets, and beverages, as a base for fermented drinks, and as a gentle laxative (Quisumbing 1951).

All plant parts serve a purpose (Watt and Breyer-Brandwijk 1962). After the removal of the outer covering, the seeds may be eaten raw, boiled, or fried. They are also a source of starch and may be ground into a flour. Young seeds yield 10–15 percent by weight of a high-quality, amber-colored, sweet-tasting oil; a yield of about 4 percent is obtained from old, dried seeds. At one time the oil was in demand in the preparation of varnishes and paints, in the finishing of various Indian cloths, and as an illuminant. A yellow dye is available from the leaves. The bark yields considerable tannin; an ink is made from the burned bark. Native medicinal preparations from plant parts serve varied purposes.

Tamarind wood is hard, has a density of 900 kg/m₃, and takes a fine polish but is very difficult to work. Its use is reserved for small carpentry items, for fuel, and for making charcoal.

NODULATION STUDIES

The nonnodulating character of this species is documented by reports from Germany (Morck 1891), Hawaii (Allen and Allen 1936b), the Philippines (Bañados and Fernandez 1954; personal observation 1962). Venezuela (Barrios and Gonzalez 1971), Zimbabwe (Corby 1974), and Singapore (Lim 1977).

Taralea Aubl.   Papilionoideae: Galegeae, Tephrosiinae

From *tarala*, the vernacular name for the plant in Guiana.

Type: *T. oppositifolia* Aubl.

8 species. Trees or shrubs. Leaves alternate or opposite, imparipinnate, not glandular or pellucid-punctate; leaflets leathery, alternate or opposite; rachis tip often projected beyond the leaflets; stipels none; stipules small, caducous. Flowers red or violet, in terminal panicles; bracts and bracteoles small, caducous; calyx bilabiate; 2 large, petaloid lobes form the upper lip, lower lip very short, 3-toothed; petals clawed; standard rounded, reflexed; wings oblong, apex notched; keel narrow, not curved; stamens 10, all joined; anthers ovoid, versatile; ovary short-stalked, 1-ovuled; style slender, smooth; stigma small, terminal. Pod flat, rounded, 2-valved, elastically dehiscent; seeds not odorous. (Bentham and Hooker 1865; Ducke 1934c, 1938a; Amshoff 1939)

Rejection of *Taralea* as a taxonomic entity, and its inclusion in *Dipteryx* was deplored by Ducke (1934c, 1940a), who urged its reestablishment as a distinct genus; Hutchinson (1964) accorded *Taralea* separate generic status.

Members of this genus are common in tropical South America, especially Venezuela, Brazil, and Guiana. Three species occur only along rivers and in inundated plains and marshes; 2 species inhabit forested mountains.

The seeds of *T. oppositifolia* Aubl. contain an unscented oil that is extracted for industrial purposes. The very hard and heavy wood is suitable for construction, but is not attractive for finished products (Record and Hess 1943).

Named for Tavernier, French voyageur; his journeys were responsible for the introduction of many new products from the Orient.

Type: *T. nummularia* DC.

7 species. Shrublets, copiously branched; branches slender, canescent, rigid. Leaves few, pinnately 3-foliolate, seldom simple; leaflets obovate or orbicular; stipels absent; stipules papery. Flowers few, rose, red, or white, in axillary racemes; bracts minute; bracteoles small; calyx short, campanulate; lobes 5, subequal, distinct, setaceous; petals papery, persistent; standard obovate, slightly clawed; wings small, shorter than standard; keel as long as the standard, obliquely truncate at the apex; stamens 10, vexillary stamen free at the base and connate in the middle with others or later entirely free; anthers uniform; ovary stalked, 1–4-ovuled; style filiform, long, incurved; stigma small, terminal. Pod flattened, consisting of 1–4, rounded, 1-seeded, indehiscent segments; seeds kidney-shaped. (De Candolle 1825b; Bentham and Hooker 1865)

Representatives of this genus are scattered from the Upper Nile region into India. They flourish in desert areas and are grazed by camels.

NODULATION STUDIES

Plants of *T. cunnifolia* Arn. growing wild in hilly areas of Maharashta State, western India, were well nodulated (Y. S. Kulkarni personal communication 1978).

Teline Medik.   Papilionoideae: Genisteae, Cytisinae

From Greek, *teline*, a plant name used by Dioscorides.

Type: *T. linifolium* (L.) Webb and Berth.

2 species. Shrubs, erect, unarmed, usually much-branched, pubescent. Leaves 3-foliolate, subsessile or petioled; leaflets obovate or linear-lanceolate, hairy on 1 surface. Flowers yellow, in subcorymbose, axillary or terminal racemes, lower ones axillary; calyx tubular-campanulate, bilabiate, upper lip with 2 elongated teeth, lower lip minutely 3-toothed; standard broad-ovate, equal to or longer than the wings and keel; keel petals rounded at the apex, pubescent overall; stamens 10, monadelphous. Pod tapered toward the base, pubescent, compressed, dehiscent, 2–6-seeded. (Tutin et al. 1968; Komarov 1971)

Both species occur in Mediterranean regions, the Azores, and the Canary Islands. They are found in rocky scrub and woodland areas. Throughout the Soviet Union the name for these plants is lozhnodrok, or false broom.

NODULATION STUDIES

Both species, *T. linifolius* (L.) Webb & Berth. (= *Cytisus linifolius* = *Genista linifolia*) and *T. monspessulana* (L.) K. Koch (= *Cytisus monspessulanus*) were nodulated in South Africa (Grobbelaar and Clarke 1974) in tests with imported seeds.

Named for John Templeton, 1766–1825, Irish botanist.

Type: *T. retusa* (Vent.) R. Br.

10 species. Shrubs, rarely undershrubs, with woody rootstocks, glabrous, sometimes leafless; branches angular or grooved. Leaves, when present, simple, entire, often reduced to small scales; stipules minute or spinescent. Flowers red, yellow, or mixed with purple, solitary or paired in axils; bracts minute, imbricate, sometimes dry and rigid, at the base of pedicels; bracteoles persistent at or above middle of the pedicel; calyx lobes 5, upper 2 united into a lip, ciliolate, seldom free, lowest lobe longest; standard suborbicular or ovate, recurved; wings narrow, usually shorter than the standard; keel petals slightly connate, as long as the standard or slightly shorter; stamens united except at the top; anthers alternately long, erect, and basifixed, short, versatile, and dorsifixed; ovary sessile or stalked, several-ovuled; style recurved, glabrous; stigma small, terminal. Pod sessile or stalked, compressed, oblong or linear, usually oblique, leathery, 2-valved, dehiscent. (Bentham 1864; Bentham and Hooker 1865)

This genus was assigned to the tribe Bossiaeae by Polhill (1976) in his reorganization of Genisteae *sensu lato*.

Members of this genus are limited in distribution to Australia. They thrive on clay, sandy, saline, and coastal limestone soils. Some species are tolerant of salt spray.

These occasionally leafless shrubs commonly reach heights of 1.5–2 m and are valued as ornamentals for their display of attractive flowers. The presence of sparteine (Fitzgerald 1964) and cytisine (White 1951) in the foliage of *T. egena* (F. Muell.) Benth. and *T. retusa* (Vent.) R. Br., respectively, accounts for the poisonous properties of these species.

NODULATION STUDIES

Two rhizobial strains from *T. retusa* were listed in the Type Collection of the Commonwealth Scientific and Industrial Research Organization, Division of Soils, Adelaide, in 1953 (Harris). Nodulation of *T. sulcata* (Meissn.) Benth. and *T. retusa* was reported by Lange in 1959 and 1961. Nodules were not observed by Beadle (1964) on plants of *T. egena* grown in potted soil to the 3-leaf stage or on plants in arid-zone habitats.

A single rhizobial strain from *T. retusa* showed cultural characteristics akin to those of the slow-growing cowpea, lupine, and soybean type (Lange 1961). *Glycine hispida* was not nodulated by this strain, and not all *Lupinus* species were nodulated, but *Phaseolus lathyroides* and *Vigna sinensis* were receptive. In reciprocal tests, 2 strains of lupine rhizobia failed to nodulate *T. retusa* seedlings (Lange 1962).

From Greek, *tephros*, "ash-colored," in reference to the gray pubescence on the leaves.

Type: *T. villosa* (L.) Pers.

400 species. Shrubs, shrublets or herbs, annual and perennial; stems erect, decumbent, or prostrate, usually arising from a woody crown; roots woody. Leaves imparipinnate; leaflets numerous, seldom 1–3, petioluled, glabrous to pubescent above, usually silky-pubescent beneath, lateral nerves conspicuously parallel; stipels none; stipules slender, variable, either bristly, striate, or foliaceous. Flowers red, purple, or white, small or large, in leaf-opposed, axillary or terminal racemes; bracts persistent or caducous; bracteoles none; calyx campanulate; lobes 5, subequal or the lowest lobe longest, varying from deltoid to awl-shaped, rarely upper 2 connate partly or completely; petals clawed; standard suborbicular, without basal ears, silky on back; wings obliquely obovate or oblong, slightly adherent to the blunt, incurved, nonbeaked keel; stamens 10, vexillary stamen free at the base, somewhat connate with others at the middle, upwards quite free; anthers uniform; ovary sessile, rarely substalked, many-ovuled, rarely 1-ovuled; style incurved or inflexed, sometimes dorsally flattened, glabrous or variously bearded; stigma terminal. Pod sessile, exserted from the dried calyx, linear, rarely ovate or oblong, compressed, obliquely narrowed and beaked at the apex, nonseptate within or thinly septate, 2-valved, dehiscent, many-seeded. (Bentham and Hooker 1865; Amshoff 1939; Pellegrin 1948; Cronquist 1954a; Hutchinson and Dalziel 1958)

Representatives of this large genus are distributed throughout warm regions of both hemispheres. Seventy species occur in South Africa, 50 in equatorial Africa. About 30 species are native to the United States, mostly in the Southwest. They are well adapted to sandy soils. The plants are generally found wild in grassy fields, thickets, and along roadsides.

Interest in *Tephrosia* species as a source of contact insecticides was aroused by their native use as fish and arrow poisons and the similarity in their physiological action to that of *Derris* and *Lonchocarpus* species (Little 1931). Early studies by Hanriot (1907a,b) identified 3 toxic principles in the leaves of *T. vogelii* Hook. f. One of these, tephrosin, $C_{31}H_{26}O_{10}$, killed fish in 1 hour at a concentration of 1:25 ppm; the 2 other substances were only slightly poisonous to fish. Confirmation of Hanriot's finding and the correction of his formula and melting point of tephrosin were followed by the isolation of toxicarol, $C_{23}H_{22}O_7$, deguelin, an isomer of rotenone (Clark 1931a,b), and rotenone, $C_{23}H_{22}O_6$ (Clark 1933) from *Tephrosia* species.

At one time cultivation of *Tephrosia* species for insecticidal purposes held great promise because they thrive in areas and under conditions where *Derris* and *Lonchocarpus* species do not; but they proved unimportant commercially because of low and variable yield of insecticidal compounds (Roark 1937; Jones 1942; Irvine and Freyre 1959a,b). The following species were basic as sources of rotenoid compounds: *T. grandiflora* Pers., *T. macropoda* Harv., *T. toxicaria* (Sw.) Pers., *T. virginiana* (L.) Pers., and *T. vogelii.*

Although not all species are toxic and animal susceptibility to obnoxious species is variable, *Tephrosia* species have only limited forage value. The foliage is coarse and unpalatable. *Tephrosia* roots are long, fibrous, thick, yellowish in color, and have a characteristic odor which is attributed to tephrosal, a volatile oil. *T. candida* (Roxb.) DC. exemplifies the *Tephrosia* species found satisfactory for green-manuring, soil cover, soil improvement, and reclamation, but the thick, deep roots are removed with difficulty. The leaf litter tends to be abundant and high in nitrogen content, nodulation is copious, and plant growth is prolific. Among 16 varietal introductions of *T. vogelii* in Puerto Rico, the leaf nitrogen content ranged between 4.18 and 4.83 percent (Irvine and Freyre 1959a). Powdered seeds of *T. vogelii* have been used successfully to eliminate European carp from inland fisheries before restocking with edible and game fish (Blommaert 1950). All carp were killed within 5 hours by a treatment of 10 ppm. The toxic effect was completely dissipated within 144 hours following dispersal of the powdered seed in water. Perennial, shrubby growth, dense foliage, and good root anchorage have prompted the use of *Tephrosia* species in Africa and Asia for hedges, windbreaks, as shade on tea and coffee plantations, and to prevent soil erosion.

NODULATION STUDIES

Reciprocal and cross-inoculation between plants and rhizobia of 3 *Tephrosia* species and *Vignu sinensis* were reported by the authors in 1936(a). In later tests (Allen and Allen 1939), diverse members of the cowpea miscellany, with the exception of *Phaseolus lunatus*, were nodulated by strains from *T. candida* and *T. noctiflora* Boj. ex Bak. nodules. Wilson (1939a) reported promiscuous susceptibilities of 4 *Tephrosia* species to rhizobia from various rhizobia:plant groups. Consideration of *T. virginiana* var. *holosericea* as "unassigned to any cross-inoculation group" is viewed as an anomaly (Bushnell and Sarles 1937).

Nodulated species of *Tephrosia*	Area	Reported by
acaciaefolia Welw. ex Bak.	S. Africa	Grobbelaar et al. (1964)
adunca Benth.	Venezuela	Barrios & Gonzalez (1971)
aemula Harv.	S. Africa	Grobbelaar et al. (1967)
aequilata Bak.		
ssp. *australis* Brummitt	Zimbabwe	Corby (1974)
bidwilli Benth.	N.S. Wales, Australia	Beadle (1964)
burchellii Burtt Davy	S. Africa	Grobbelaar et al. (1967)
candida (Roxb.) DC.	Widely reported	Many investigators
capensis (Jacq.) Pers.	S. Africa	Grobbelaar et al. (1967)
var. *acutifolia* E. Mey.	S. Africa	Grobbelaar & Clarke (1972)
var. *hirsuta* Harv.	S. Africa	Grobbelaar & Clarke (1975)
cathartica (Sessé & Moc.) Urb.	Puerto Rico	Dubey et al. (1972)
cephalantha Welw. ex Bak.		
var. *decumbens* Welw. ex Bak.	Zimbabwe	Corby (1974)
chimanimaniana Brummitt	Zimbabwe	Corby (1974)
cinerea (L.) Pers.	Venezuela	Barrios & Gonzalez (1971)

Nodulated species of *Tephrosia* (cont.)	Area	Reported by
cordata Hutch.	S. Africa	Grobbelaar & Clarke (1975)
contorta N. E. Br.	S. Africa	Grobbelaar & Clarke (1975)
coronilloides Welw. ex Bak.	Zimbabwe	Corby (1974)
dasyphylla Welw. ex Bak.		
ssp. *dasyphylla*	Zimbabwe	Corby (1974)
decora Welw. ex Bak.	Zimbabwe	Corby (1974)
dichotoma Desv.	Philippines	Bañados & Fernandez (1954)
diffusa Harv.	S. Africa	Grobbelaar et al. (1967)
dregeana E. Mey.	S. Africa	N. Grobbelaar (pers. comm. 1974)
elata Deflers		
ssp. *elata*	Zimbabwe	Corby (1974)
ssp. *heckmanniana* (Harms) Brummitt		
var. *heckmanniana*	Zimbabwe	Corby (1974)
elongata E. Mey.		
var. *elongata*	S. Africa	Grobbelaar & Clarke (1975)
var. *lasiocaulos*	Zimbabwe	Corby (1974)
var. *pubescens*	S. Africa	Grobbelaar & Clarke (1975)
euchroa Verdoorn	Zimbabwe	Corby (1974)
festina Brummitt	Zimbabwe	Corby (1974)
forbesii Bak.		
ssp. *interior* Brummitt	Zimbabwe	Corby (1974)
glomeruliflora Schinz	S. Africa	N. Grobbelaar (pers. comm. 1974)
grandibracteata Merx.	Zimbabwe	Corby (1974)
grandiflora Pers.	U.S.A.	Wilson (1939a)
hookeriana W. & A.	Sri Lanka	Holland (1924)
incarnata Brummitt	Zimbabwe	Corby (1974)
linearis (Willd.) Pers.		
var. *discolor* (E. Mey.) Brummitt	S. Africa	Grobbelaar & Clarke (1972)
	Zimbabwe	Corby (1974)
var. *linearis*	Zimbabwe	Corby (1974)
longipes Meissn.		
ssp. *longipes*		
var. *longipes*	Zimbabwe	Corby (1974)
var. *lurida* (Sond.) Gillett	S. Africa	Grobbelaar et al. (1964)
ssp. *swynnertonii*	Zimbabwe	Corby (1974)
lupinifolia DC.	S. Africa	Grobbelaar & Clarke (1975)
lurida Sond.		
var. *drummondii* Brummitt	Zimbabwe	Corby (1974)
var. *lurida*	Zimbabwe	Corby (1974)
macropoda Harv.	S. Africa	Grobbelaar et al. (1967)
maxima Pers.	Java	Keuchenius (1924)
micrantha Gillett	Zimbabwe	Corby (1974)
montana Brummitt	Zimbabwe	Corby (1974)

Nodulated species of *Tephrosia* (cont.)	Area	Reported by
multijuga R. G. N. Young	S. Africa	Grobbelaar et al. (1964)
noctiflora Boj. ex Bak.	Widely reported	Many investigators
paniculata Welw. ex Bak.		
var. *paniculata*	Zimbabwe	Corby (1974)
paradoxa Brummitt	Zimbabwe	Corby (1974)
plicata Oliv.	S. Africa	Grobbelaar et al. (1967)
polystachya E. Mey.	S. Africa	Grobbelaar et al. (1967)
var. *hirta* Harv.	S. Africa	Grobbelaar et al. (1967)
var. *latifolia* Harv.	S. Africa	Grobbelaar & Clarke (1972)
polystachyoides Bak. f.	S. Africa	Grobbelaar et al. (1967)
praecana Brummitt	Zimbabwe	Corby (1974)
pseudocapitata H. M. Forbes	S. Africa	Grobbelaar & Clarke (1975)
pumila (Lam.) Pers.		
var. *pumila*	Zimbabwe	Corby (1974)
purpurea (L.) Pers.	Widely reported	Many investigators
ssp. *altissima* Brummitt	Zimbabwe	Corby (1974)
ssp. *leptostachya* (DC.) Brummitt var. *delagoensis* (H. M. Forbes) Brummitt		
var. *leptostachya*	Zimbabwe	Corby (1974)
var. *pubescens* (Bak.)	Zimbabwe	Corby (1974)
Brummitt	Zimbabwe	Corby (1974)
radicans Welw. ex Bak.	S. Africa	Grobbelaar et al. (1964)
	Zimbabwe	Corby (1974)
reptans Bak.	S. Africa	Grobbelaar et al. (1964)
var. *reptans*	Zimbabwe	Corby (1974)
rhodesica Bak. f.		
var. *evansii* (Hutch. & Burtt Davy) Brummitt	Zimbabwe	Corby (1974)
var. *polystachyoides* (Bak. f.) Brummitt	Zimbabwe	Corby (1974)
var. *rhodesica*	Zimbabwe	Corby (1974)
rupicola Gillett	Zimbabwe	Corby (1974)
semiglabra Sond.	S. Africa	Grobbelaar et al. (1964)
sessiliflora (Poir.) Hassl.	Venezuela	Barrios & Gonzalez (1971)
shiluwanensis Schinz	S. Africa	Grobbelaar et al. (1967)
sparsiflora H. M. Forbes	S. Africa	Grobbelaar & Clarke (1972)
spathacea Hutch. & Burtt Davy	S. Africa	Grobbelaar et al. (1967)
sphaerosphora F. Muell.	S. Africa	B. W. Strijdom (pers. comm. 1967)
spicata (Walt.) T. & G. (= *Cracca spicata* Walt.)	N.C., U.S.A.	Shunk (1921)
stormsii De Wild.		
var. *stormsii*	Zimbabwe	Corby (1974)
subtriflora Backer	Pakistan	A. Mahmood (pers. comm. 1977)
subulata Hutch. & Burtt Davy	S. Africa	Grobbelaar et al. (1967)

Nodulated species of *Tephrosia* (cont.)	Area	Reported by
tenella Gray	Venezuela	Barrios & Gonzalez (1971)
tenuis Wall.	India	Satyanarayan & Gaur (1965)
toxicaria (Sw.) Pers.	Sri Lanka	W. R. C. Paul (pers. comm. 1951)
tzaneenensis H. M. Forbes	S. Africa	Grobbelaar et al. (1964)
uniflora Pers.		
ssp. *uniflora*	Zimbabwe	Corby (1974)
vestita Vog.	Philippines	Pers. observ. (1962)
villosa (L.) Pers.	Sri Lanka	W. R. C. Paul (pers. comm. 1951)
ssp. *ehrenbergiana* (Schweinf.) Brummitt		
var. *ehrenbergiana*	Zimbabwe	Corby (1974)
virginiana (L.) Pers.	N.Y., U.S.A.	Wilson (1939a)
(= *Cracca virginiana* L.)	N.C., U.S.A.	Shunk (1921)
var. *holosericea* (Nutt.) T. & G.	Wis., U.S.A.	Bushnell & Sarles (1937)
vogelii Hook. f.	Widely reported	Many investigators
zombensis Bak.	S. Africa	N. Grobbelaar (pers. comm. 1968)
zoutpansbergensis Brem.	Zimbabwe	Corby (1974)

Teramnus (P. Br.) Sw.   Papilionoideae: Phaseoleae, Glycininae
391 / 3866

From Greek, *teramnon*, "room," "chamber," in reference to the seeds being in separate compartments.

Type: *T. volubilis* Sw.

8 species. Herbs, low, slender, trailing or twining, perennial, variably pubescent. Leaves pinnately 3-foliolate; leaflets ovate to lanceolate; stipels present; stipules small. Flowers purplish, minute, on slender, axillary peduncles in pairs or fascicles or in interrupted racemes longer than the leaves; bracts small; bracteoles 2, linear or lanceolate, subtending the calyx; calyx tubular-campanulate, sometimes densely hirsute; lobes 5, acuminate, upper 2 usually joined higher up; standard obovate, narrowed at the base, long-clawed, not callous; wings narrow, oblique, oblong, joined to the shorter, straight, blunt keel; stamens monadelphous, 5 longer with fertile anthers alternating with 5 staminodes; ovary sessile, many-ovuled; style short, thick, beardless; stigma capitate. Pod linear, 2-valved, septate inside between seeds and tipped with a persistent, hooked style, spirally twisted after dehiscence; seeds up to 8, hilum lateral, short, without aril. (Bentham and Hooker 1865; Baker 1929; Pellegrin 1948; Phillips 1951; Hauman 1954c)

For many years the genus was confounded with *Glycine*.

Representatives of this genus inhabit the tropics and subtropics of both

649

hemispheres. They are mostly native to Africa and Asia; only 2 species occur in America. *Teramnus* species flourish in thickets, open rocky, dry, and waste areas, lava plains, and forest clearings. The plants are drought-resistant but susceptible to frosts.

The species have promise as pasture plants. *T. labialis* (L. f.) Spreng. and *T. uncinatus* (L.) Sw. are palatable leafy plants with forage value in fallow fields for cattle and horses in Brazil and Central America. All species are free rooting and provide dense ground cover in acceptable areas. Attempts to establish *T. labialis* as green manure in Malaysia were unsuccessful; in Kenya it was not palatable to sheep (Whyte et al. 1953).

NODULATION STUDIES

The nodules on *T. uncinatus* resemble those of *Vigna sinensis* in size and root location.

Teramnus rhizobia produce slow, scanty growth and alkaline reactions on carbohydrate media and in bromcresol purple milk (Galli 1959).

Extensive reciprocal cross-inoculation tests between rhizobial strains and plants of *T. uncinatus* with those of *Indigofera, Desmodium, Centrosema, Stylosanthes, Tephrosia*, and *Vigna* species evidenced effective nodulation (Galli 1958). No nodules were induced in reciprocal tests using rhizobia and plants of *Medicago, Trifolium, Vicia, Lupinus, Soya, Phaseolus*, and *Leucaena* species.

Nodulated species of *Teramnus*	Area	Reported by
labialis (L. f.) Spreng.	Java	Keuchenius (1924)
	Kenya	Bumpus (1957)
	W. India	Y. S. Kulkarni (pers. comm. 1972)
ssp. *labialis*		
var. *abyssinicus* (A. Rich.) Verdc.	Zimbabwe	Corby (1974)
var. *arabicus* Verdc.	Zimbabwe	Corby (1974)
repens (Taub.) Bak. f.	Kenya	Whyte et al. (1953)
uncinatus (L.) Sw.	Brazil	Galli (1958)
ssp. *ringoetii* (De Wild.) Verd.	Zimbabwe	Corby (1974)
volubilis Sw.	Venezuela	Barrios & Gonzalez (1971)

Terua Standl. & Hermann   Papilionoideae: Galegeae

Named for Terua Williams (Mrs. Louis O. Williams) of the Escuela Agricola Panamericana, Tegucigalpa, Honduras.

Type: *T. vallicola* Standl. & Hermann

1 species. Shrub or small tree, up to 6 m tall; branches alternate, dotted with yellow lenticels. Leaves imparipinnate, glabrous; leaflets 5, entire, irregu-

larly translucent-punctate; stipels absent; stipules present. Flowers red purple, few, in axillary panicles; bracts and bracteoles ovate; calyx urceolate in bud, broadly cup-shaped at maturity; wings strongly adherent to the apical half of the keel; stamens monadelphous, vexillary stamen free at base, then united with other 9; anthers versatile, lanceolate; style curved; stigma minute, terminal, glabrous. Pod without resin ducts, compressed, stalked, straight, both sutures thickened, elastically dehiscent along the ventral suture, 1–4-seeded; seeds reddish brown, subreniform, compressed, hilum oval, surrounded by a membranous collar. (Hermann 1949)

Hermann (1949) construed the genus as most closely related to *Willardia*, in Galegeae, and regarded it as a more markedly transitional type between Galegeae and Dalbergieae than any genus so far known.

This monotypic genus is endemic in Honduras. The plants thrive along river banks, on grassy slopes, and in dense moist forests at altitudes of 800–1,000 m.

Tessmannia Harms   Caesalpinioideae: Amherstieae
(56a)

Named for G. Tessmann, botanical collector in Equatorial Guinea at the beginning of the 20th century.

Type: *T. africana* Harms

10–12 species. Trees large, up to 50 m in height, with tall, straight, cylindric boles and lower trunk diameters of 1–1.5 m, often buttressed. Leaves usually imparipinnate; leaflets 7–44, alternate, opposite, or subopposite, subsessile, generally emarginate, pellucid-punctate. Flowers small to medium, in axillary or terminal racemes or panicles; bracteoles small, very caducous; calyx with short receptacle; lobes 4, subvalvate in bud, united at the base, entire, thick, usually silky-haired inside and warty outside, uppermost lobe oblong and twice as broad as other 3; petals 5, rose or white, free, imbricate, longer than the lobes, subequal or uppermost petal larger, linear-lanceolate, clawed, narrower than the others, margins crisped, other 4 petals similar, oblanceolate, clawed, downy inside along the median groove; stamens 10, 1 free, 9 joined into a hairy sheath; filaments bent in bud, hairy or villous below, alternately long and short; anthers oblong, dorsifixed; ovary long-stalked, stalk hairy, several-ovuled, usually 7–8; style long, coiled or curved, smooth; stigma capitellate. Pod suborbicular or ovate, flattened, woody, warty or smooth, tardily dehiscent; seeds usually oblong, black, shiny. (Harms 1910; Pellegrin 1948; Léonard 1949, 1952)

The genus is closely related to the African genus *Baikiaea* and the Asian genus *Sindora* (Harms 1915a).

The freshly cut wood exudes an abundant, sticky, greenish black sap, which has the properties of copals; when properly aged, the wood is valued for fine furniture, carpentry, and railroad ties.

Tetraberlinia (Harms) Hauman
Caesalpinioideae:
Amherstieae

The prefix, _tetra-_, "four," refers to the presence of 4 sepals, in contrast to the 5 sepals of _Berlinia_ species.

Type: _T. bifoliolata_ (Harms) Hauman (= _Berlinia bifoliolata_)

1 species. Tree, evergreen, large, up to 42 m tall; bole straight, cylindric, often free of branches up to 20 m, flanked by low, thick root swellings or spurs, crown small, rounded. Leaves paripinnate; leaflets sessile, often large, papery, somewhat asymmetric; stipules foliaceous, early caducous. Flowers yellow, sessile in bud, short-pedicelled in flower, medium-sized, solitary in terminal or axillary panicles; bracts concave, golden silky-haired outside, smooth inside; bracteoles 2, opposite, valvate, somewhat connate with calyx at the base; calyx short, cup-shaped, hairy, grooved or ridged inside; lobes 4, linear-lanceolate, 2 connate near the top; petals 5, upper petal clawed, apically truncate, other 4 linear, somewhat longer than the calyx lobes, glabrous; stamens 10, pinkish, 1 free in the axil of the upper petal, 9 shortly connate at their base, filiform, exserted; anthers basifixed; ovary densely brown-pilose, 4-ovuled; style long; stigma terminal, capitate. Pod oblong, curved, acuminate, dehiscent, valves twisting. (Hauman 1952; Voorhoeve 1965)

Some botanists prefer the retention of _Tetraberlinia_ as a section of the complicated _Berlinia_ complex; its distinctiveness as a genus is a matter of opinion.

This monotypic genus is endemic in tropical West Africa. The plants grow in shady forested areas.

*Tetragonolobus Scop.
Papilionoideae: Loteae

/ 3699

From Greek, _tetra-_, "four," _gonia_, "angle," and _lobos_, "pod"; the pod is 4-cornered in cross section.

Type: _T. maritimus_ (L.) Roth (= _Lotus siliquosus_)

15 species. Herbs, annual or perennial, resembling _Lotus_ except leaves 3-foliolate; stipules foliaceous; flowers solitary or paired, with calyx teeth equal; and pod tetragonal, square in cross section, usually with angles winged.

Species in this genus formerly constituted a section of _Lotus_.

Members of this genus are distributed throughout the Mediterranean region, and central and southern Europe extending into Ukraine and Asia Minor. Species flourish in meadows and gullies.

T. palaestinus Boiss., square pod or winged pea, showed considerable promise as a green manure, winter cover, and hay crop in Israel, where it was copiously nodulated in the wild state (Itzhaky 1954).

NODULATION STUDIES

Simon (1914) placed _T. purpureus_ Moench in the _Lotus_ cross-inoculation group on the basis of compatible relationships with _Anthyllis vulneraria_ and _Lotus uliginosus_.

Nodulated species of *Tetragonolobus*	Area	Reported by
conjugatus (L.) Link	France	Clos (1896)
maritimus (L.) Roth		
(= *siliquosus* Roth)	Morocco	Hollande & Crémieux (1926)
(= *Lotus siliquosus* L.)	France	Dangeard (1926)
palaestinus Boiss.	Israel	Itzhaky (1954)
purpureus Moench	France	Naudin (1897d)
	Germany	Simon (1914)
(= *Lotus tetragonolobus* L.)	U.S.A.	Wilson (1939a)
requienii (Mauri ex Sanguinetti) Sanguinetti	S. Africa	Grobbelaar & Clarke (1974)

Tetrapleura Benth.   Mimosoideae: Adenanthereae

21 / 3457

From Greek, *tetra-*, "four," and *pleura*, "rib," in reference to the 4-ribbed pod.

Type: *T. tetraptera* (Schum. & Thonn.) Taub.

2 species. Trees, large, often 25 m tall, with trunk girths 3 m or more, unarmed, deciduous; bole usually buttressed. Leaves bipinnate, opposite; pinnae 5–9 pairs; leaflets 12–24, alternate, small, elliptic with asymmetric base and notched apex, glabrous or minutely pubescent; rachis eglandular. Flowers small, pink or creamy yellow, slender-stalked, densely crowded in solitary or paired, axillary, spicate racemes; bracts quite persistent; calyx shallow, campanulate; teeth short, valvate in bud; petals linear-lanceolate, free; stamens 10, free, exserted, uniform; anthers with an apical gland; ovary sessile, cylindric, many-ovuled; style filiform, exserted; stigma indistinct. Pod pendent on a stout stalk, oblong, thick, 15–25 cm long by 5 cm wide, with 4 winglike, longitudinal ribs, sutures acute, indehiscent, septate between the seeds, septa continuous with the endocarp; seeds small, hard, black, ovate, transverse, compressed. (Bentham 1842; Baker 1930; Pellegrin 1948; Brenan 1959)

Both species are indigenous to tropical Africa. They thrive in lowland rain-forest areas.

The dark, red brown pods of *T. tetraptera* (Schum. & Thonn.) Taub. are distinctive. Two of the longitudinal ridges are hard, woody, and sutural. The other two are winglike and filled with a soft, sugary, edible pulp with a caramellike odor. A small amount of saponin in the pulp accounts for its use as a pomade and for laundry purposes. The fruits also serve minor medicinal uses and have fish-poisoning properties (Dalziel 1948; Keay et al. 1964).

The wood is of medium durability. The sapwood is white; the light reddish tinge of the heartwood becomes a deep yellow brown upon aging and exposure. Limited use is made of the wood for small furniture items, doors, and window frames (Eggeling and Dale 1951; Keay et al. 1964).

From Greek, *tetrapteros*, "4-winged," and *karpos*, "fruit."

Type: *T. geayi* Humbert

1 species. Tree, small, shrubby, 5–6 m tall, unarmed, xerophytic. Leaves bipinnate; leaflets mostly alternate, apical ones paired; stipules caducous. Flowers unisexual, dioecious, small, pale green, in axillary racemes, pedicels jointed near the base; bracts and bracteoles very small; calyx campanulate, concave; lobes 4, equal, 1 posterior; petals 4, equal, imbricate; male flowers with stamens 4, free, equal; anthers introrse, dorsifixed, longitudinally dehiscent; ovary rudimentary; female flowers with 4 staminodes; ovary stalked, free, 4-plicate, 1-ovuled; style short; stigma capitate-reniform. Pod samaroid, indehiscent, dry, membranous, obovate, dorsiventrally flattened, 4-winged, outer wings broader; seed 1, ovoid, cotyledons flat. (Humbert 1939)

This monotypic genus is endemic in Madagascar. The plants grow in dry regions.

Teyleria Backer   Papilionoideae: Phaseoleae, Glycininae

Named for Pieter Teyler van der Hulst, 1702–1778, silk manufacturer at Haarlem, The Netherlands; his fortune made possible the Teyler's Foundation.

Type: *T. koordersii* (Backer) Backer

1 species. Herb, twining, perennial; stems angular, densely retrorsely hairy at the angles, grooved above. Leaves pinnately 3-foliolate, veins prominent; leaflets ovate-oblong, appressed pubescent; stipels filiform-subulate, small; stipules small, basifixed. Flowers small, white, in axillary, often irregularly branched racemes, in groups of 3 at nodes along the rachis; bracts small, persistent; bracteoles 2, erect, narrow, subtending the calyx, persistent; calyx campanulate; teeth 5, narrow, acute, longer than the calyx tube, lowest tooth longest, upper 2 joined into a 2-lobed lip, margins bristly; standard obovate, not eared or reflexed; wings narrow, adherent to the keel; keel petals violet-tipped, shorter than the wings, long-clawed, obtuse, adherent in part; stamens 10, diadelphous; anthers oval, small, dorsifixed; ovary sessile, 6–7-ovuled; style short, curved, glabrous above; stigma small, capitate. Pod sessile within the persistent calyx, borne in clusters, compressed, linear, flat, septate between the seeds, beaked, 2-valved; seeds reddish brown, squarish, compressed. (Backer 1939)

Teyleria is closely related to *Glycine* (Backer 1939; Hermann 1962).

This monotypic genus is endemic in southern China and Java. The plants flourish along sandy slopes and in brushwood, teakwood, and mixed forests.

From Greek, *thermos*, "lupine," and *opsis*, "resemblance," in reference to floral similarities.

Type: *T. lanceolata* R. Br.

15–20 species. Herbs, perennial, stout, usually less than 1 m tall, more or less pubescent throughout; rootstocks rhizomatous, woody, often creeping; stems annual, clustered, erect, branching from the base. Leaves palmately 3-foliolate; leaflets linear, elliptic, oblanceolate, or obovate; stipels absent; stipules free, conspicuous, ovate-cordate to lanceolate-acuminate, often foliaceous, usually persistent. Flowers large, showy, bright yellow or rarely purple, pedicelled, in erect, lax, or compact, long-stalked, terminal or leaf-opposed racemes; bracts conspicuous, leafy, simple or connate with 1–2 lateral stipules, semipersistent; bracteoles absent; calyx narrow-campanulate, bilabiate; lobes subequal, separate or upper 2 connate, truncate to emarginate, shorter than the lower, 3-dentate, deltoid-lanceolate lip; standard suborbicular, shorter than the wings, sides reflexed; wings oblong; keel petals nearly straight, subequal, partly joined above at the back, equal to or slightly longer than the wings; stamens 10, free, incurved; ovary sessile or nearly so, many-ovuled; style incurved; stigma minute, terminal. Pod sessile or short-stalked in the calyx, usually linear, sometimes lomentaceous, elliptic or ovate-inflated, straight or slightly curved, valves slightly leathery, constricted and septate between the 12–18 seeds, dehiscent. (Bentham and Hooker 1865; Larisey 1940b)

The genus has its closest affinity to *Baptisia* (Larisey 1940b), but the leaves do not blacken upon drying. The species are commonly called golden peas.

Representatives of this genus are widespread in temperate North America, especially the southeastern and western United States, and in Siberia, the Himalayas, China, and Japan. They inhabit semidry areas, foothills, sandy washes, and stream banks. The plants are often found up to 3,600-m altitude.

The drought-resistant feature of *Thermopsis* species, their deep-rooting habit, and their spreading by underground rootstocks to form broad patches are effective in erosion control. *T. montana* Nutt. is an excellent soil-binder in the yellow pine belt of the western United States. *T. caroliniana* M. A. Curt. is commonly used as a bright perennial ornamental. *T. rhombifolia* (Nutt.) Richards., false lupine, a common herb of the dry plains of Canada southward and eastward into Nebraska and Colorado, is reputedly poisonous to cattle. *T. lanceolata* R. Br. is used as an expectorant herb in the Ukraine.

Thermopsis species lack forage qualities because they contain quinolizidine alkaloids, such as anagyrine, cytisine, homothermopsine, *N*-methylcytisine, sparteine, and thermopsine (Harborne et al. 1971). These unique alkaloids occur primarily in members of the tribes Sophoreae, Podalyrieae, and Genisteae.

NODULATION STUDIES

Oval to branched nodules were observed on *T. lupinoides* R. Br. in Germany (Wendel 1918), on *T. fabacea* DC. in Kew Gardens, England (personal observation 1936), and on *T. caroliniana* and *T. fabacea* growing in the Polar-Alpine Botanical Gardens of the Kolsky Branch of Academy of Sciences, Kirovsk, Murmansk Region, Soviet Union (Roizin 1959).

T. *caroliniana* was nodulated by strains from about 21 species of the cowpea miscellany as well as by strains from *Medicago, Trifolium, Vicia,* and *Lotus* species (Wilson 1939a). A strain from *T. caroliniana* was reciprocally versatile. In contrast, *T. fabacea* was nodulated only by strains from *Albizia julibrisin, Apios tuberosa,* and *Desmodium canadense.* The inconsistency of these results indicates the need for further study.

Nodulated species of *Thermopsis*	Area	Reported by
caroliniana M. A. Curt.	N.Y., U.S.A.	Wilson (1939a)
	Arctic U.S.S.R.	Roizin (1959)
fabacea DC.	Kew Gardens, England	Pers. observ. (1936)
	Arctic U.S.S.R.	Roizin (1959)
lanceolata R. Br.		
(= *lupinoides* R. Br.)	Germany	Wendel (1918)
montana Nutt.	Ariz., U.S.A.	Martin (1948)
	S. Africa	N. Grobbelaar (pers. comm. 1970)
rhombifolia (Nutt.) Richards.	Canada	Pers. observ. (U.W. herb. spec. 1970)

Thornbera Rydb.   Papilionoideae: Galegeae, Psoraliinae

Named for Professor John James Thornber, University of Arizona.

Type: *T. albiflora* (Gray) Rydb.

12–14 species. Herbs, mostly perennial, branches and leaves gland-dotted. Leaves imparipinnate; stipels and stipules present. Flowers usually purple or white, in dense spikes; calyx campanulate, 10-ribbed, 5-lobed; petals free; standard heart-shaped or ovate, long-clawed, inserted on the staminal sheath; wing and keel petals subsessile, similar, oblong or oblanceolate to obovate, not lobed at the base, short-clawed, inserted at the mouth of the staminal tube; stamens 9–10; filaments all joined below, free above; ovary sessile, 2-ovuled; style filiform; stigma minute. Pod small, oblique-obovate, slightly compressed, membranous, enclosed by the calyx, 1-seeded, indehiscent. (Rydberg 1919b)

The members of this genus are closely akin to *Petalostemon* and *Dalea* species.

Thornbera species are scattered throughout New Mexico, Arizona, and northern Mexico. They grow on mountains at 1,000–2,300-m altitude.

NODULATION STUDIES

Martin (1948) observed nodules on *T. ordiae* (Gray) Rydb. in Arizona, but not on *T. albiflora* (Gray) Rydb. (both reported as *Dalea* ssp.).

From Greek, *thylakos*, "pouch," and *anthos*, "flower"; thick concave bracteoles form a persistent involucre below the flower.

Type: *T. ferrugineus* Tul.

1 species. Shrub, unarmed. Leaves paripinnate; leaflets few, subcoriaceous; stipules (?). Flowers in short, terminal panicles; bracts thick, shell-shaped, leathery, caducous; bracteoles thick, concave, enclosing the young bud, forming a 2-lobed, persistent involucre below the flower; calyx tube very short; lobes 5, ovate, rounded, petaloid, ciliate, imbricate; petals 5, small, narrow, obovate, imbricate, uppermost one innermost; stamens 10; filaments shortly connate at the base, inflexed at the apex; anthers short, uniform, longitudinally dehiscent; ovary subsessile, free, few-ovuled; style elongated, inrolled in bud; stigma peltate. Pod not described. (Bentham and Hooker 1865)

This monotypic genus is endemic in the Amazon region of Brazil.

Tipuana Benth.   Papilionoideae: Dalbergieae, Pterocarpinae
354 / 3825

Name refers to the Tipuani Valley, Bolivia, where these trees are abundant.

Type: *T. tipu* (Benth.) Ktze. (= *T. speciosa*)

1 species. Tree, fast-growing, medium to large, unarmed, deciduous. Leaves imparipinnate, mostly alternate; leaflets alternate, elliptic, obovate-oblong, emarginate, few to many, subcoriaceous; stipels none; stipules minute, caducous. Flowers bright yellow, appearing at leaf renewal, in terminal, lax panicles; bracts and bracteoles small, very caducous after flowering; calyx elongated, tubular at base, campanulate above; lobes 5, short, broad, upper 2 large, subobtuse, connate higher up, lower 3 triangular, subequal; petals clawed, very unequal in size; standard very broad, reflexed, not appendaged; wings oblong, slightly longer than standard; keel petals oblong, blunt, short, smaller than wings, dorsally coherent; stamens 10, diadelphous; anthers elliptic, dorsifixed; ovary short-stalked, silky-pubescent, without disk, 1 to few-ovuled; style short, incurved; stigma terminal, glabrous. Pod samaroid, about 8 cm long, persistent, leathery, winged at the top, with 1–3, elongated seeds at base, indehiscent. (Ducke 1930; Burkart 1943)

This monotypic genus is endemic in Bolivia, Brazil, and the warm regions of Argentina.

T. tipu (Benth.) Ktze., pride of Bolivia, is a fast-growing subevergreen, planted for shade and ornament in parks and along avenues in many South American cities. It has been introduced successfully into Algeria, southern France, Florida, and California. Ordinarily the tree is not more than 10 m tall, but in native forested areas it reaches a height of 35 m, with a bole approximately 1.75 m in diameter above the buttressed base. In the timber trade the wood is called tipa.

Uses of the tree are varied in its native habitats. The leaves and unripe fruits have saponin and peroxidase properties (Burkart 1943). A red resin exuded from the bark may be used for tanning and dyeing. The pale yellow to brownish wood is highly valued for cabinetwork, fine carpentry, carved collectors' items, and high-grade furniture. It has a density of 640–750 kg/m³, finishes smoothly, and takes a fine polish (Record and Hess 1943).

NODULATION STUDIES

Nodules were present on this species in Hawaii (personal observation 1940), in South Africa (Grobbelaar et al. 1964), and in Argentina (Rothschild 1970).

Trachylobium Hayne   Caesalpinioideae: Amherstieae
48 / 3498

From Greek, *trachys*, "rough," and *lobion*, "pod," in reference to the rough and resinously warty surface of the pods.

Type: *T. verrucosum* (Gaertn.) Oliv.

1 species. Tree, medium-sized, up to about 25 m tall, unarmed, resiniferous. Leaves 2-foliolate, on short petioles; leaflets leathery, short-stalked, asymmetric, sprinkled with translucid-punctate dots; stipules very caducous. Flowers white, in 2 rows, in densely and finely hairy, lax, terminal, corymbose panicles; bracts and bracteoles oval or orbicular, concave, caducous before flowering; calyx short, narrow-turbinate; lobes 4, imbricate in bud; petals 5, upper 3 suborbicular, large, subequal, clawed, lower 2 minute, scalelike, sometimes all clawed, imbricate, uppermost petal innermost; stamens 10, free or slightly connate at the base; filaments glabrous above, villous at the base; anthers oblong, uniform, longitudinally dehiscent; ovary short-stalked, stalk adnate to the calyx tube, 4–5-ovuled; disk hirsute; style filiform; stigma small, terminal. Pod drupaceous, thick, leathery, rough, resinously warty, indehiscent, containing thick seeds without arils. (Bentham and Hooker 1865; Léonard 1957)

This monotypic genus is endemic in tropical East Africa and Madagascar.

The resin obtained from *T. verrucosum* (Gaertn.) Oliv. is marketed under a variety of names, but it is best known as East African or Zanzibar copal. It is obtained as an exudate from living trees by tapping, or in a semifossilized form from soil beneath the trees, or in the fully fossilized state as yellowish red, disklike plates, in areas where the trees formerly grew (Howes 1949). The fully aged product is the best quality for high-grade varnishes.

NODULATION STUDIES

Wright (1903) reported nodules on this species growing in Sri Lanka; however, because nodulation is a relatively rare occurrence in Amherstieae, this report merits confirmation.

From Latin, *tres*, "three," and *folium*, "leaf."

   Type: *T. pratense* L.

   250–300 species. Herbs, annual, biennial or perennial, erect, ascending or creeping. Leaves palmately 3-foliolate, rarely 5- or 7-foliolate; leaflets obovate, margins usually toothed; stipels none; stipules large, conspicuous, adnate to the base of the petiole. Flowers mostly purple, red, or white, rarely yellow, sessile or pedicelled, in axillary or terminal, sessile or peduncled spikes, clustered heads, or umbels, rarely solitary; bracts absent or small, membranous, persistent or caducous, sometimes the outer bracts connate into a dentate or lobed whorl; bracteoles absent; calyx campanulate or cylindric; lobes 5, narrow-linear to triangular, subequal or lower lobes longer, sometimes upper 2 more or less joined; petals persistent, clawed, all or lower 4 somewhat fused to the staminal tube; standard oblong or ovate, sessile, longer than the wing and keel petals; wings oblong, eared, narrow; keel petals obtuse, oblong, clawed, shorter than the wings; stamens 10, vexillary stamen free or rarely connate in the middle with the others; all or only alternate filaments apically dilated; anthers uniform; ovary sessile or subsessile, few-ovuled; style slender, apically incurved, usually glabrous; stigma hooked or capitate. Pod small, oblong or ovate, compressed or subcylindric, membranous, often enclosed by the persistent calyx, indehiscent or scarcely dehiscent; seeds usually 1–2.(Bentham and Hooker 1865; Tutin et al. 1968; Gillett and Cochrane 1973)

   Representatives of this large genus are distributed throughout Europe, western Asia, North Africa, and western North America (Isely 1951). About 85 species are native to the United States (Pieters and Hollowell 1937); about 20 species are endemic in South America. They are widespread throughout temperate and subtropical regions of the Northern Hemisphere. No species is native to Australia. The plants are annuals and perennials, wild and cultivated. There are numerous cultivars, subspecies, and varieties. Many introductions are naturalized along roadsides, old fields, and waste places.

   Trifolium species are important throughout the world as honey plants, pasture forage, hay, silage, and for soil improvement and green manuring. Within this large genus are species suitable for all seasons, types of soil, levels of fertility, and almost all climatic conditions.

   T. pratense L. is perhaps the most widely grown of the true clovers. Its introduction into European agriculture had a more pronounced effect on civilization than did the potato, or any other introduced forage plant (Piper 1924a). Red clover is second to alfalfa in importance among leguminous forage crops in the United States.

   Red clover had its origin within those countries that border the Mediterranean and Red Sea, particularly in Asia Minor and southeastern Europe (Fergus and Hollowell 1960). The migration and distribution of red clover throughout Europe was described by Merkenschlager (1934). The introduction date of this species into the United States is not definitely known, but it may have been as early as 1625 by way of a ship from the Netherlands that brought livestock and "all sorts of seed." Red clover was being cultivated at that time as livestock feed in Flanders, whence it entered Great Britain. The earliest definite reference to clover in the Colonies was about 1663; April 10, 1729, was the date of the first newspaper advertisement of clover seed for sale (Fergus and Hollowell 1960).

Clovers have been the subject of extensive chemical investigations because of their association with animal health (Watt and Breyer-Brandwijk 1962). *T. hybridum* L., *T. pratense*, and *T. repens* L. are among those considered the cause of a dermatitis associated with photosensitization. Symptoms of trefoil dermatitis, or trifoliosis, following ingestion of foliage and coincident exposure to sunlight are nervous and digestive disturbances and swelling and itching of the skin, often followed by exudation of a fluid that forms crusts and gives rise to bleeding. Although the glucosides isoquercitrin, $C_{21}H_{20}O_{12}$, kaempferol, $C_{15}H_{10}O_6$, and other substances have been implicated, the cause of the dermatitis has not yet been isolated.

Estrogenic properties of various clovers are attributed to genistein, $C_{15}H_{10}O_5$, formononetin, $C_{16}H_{12}O_4$ (Bradbury and White 1951), and a coumarin derivative, coumestrol, $C_{15}H_8O_5$ (Bickoff et al. 1958a,b; Lyman et al. 1959).

NODULATION STUDIES

Clover rhizobia were among the first root-nodule bacteria to have been studied. Upon recognizing that not all root-nodule bacteria were identical, Beijerinck (1888) considered them a variety, *trifoliorum*, of *Bacillus radicicola*. Following Frank's proposal of the generic name *Rhizobium* (1890b), they became known as *R. mutabile* (Schneider 1892) because of their bacteroidal shapes. The taxon *R. trifolii* by which these organisms are known today, was proposed by Dangeard in 1926.

Clover rhizobia produce rapid growth on laboratory media and mucilaginous, transparent colonies that become white with age. An acid reaction and a pronounced serum zone are produced in litmus milk. Rod-shaped forms of the organisms are peritrichously flagellated; bacteroidal forms are pear-shaped, swollen, vacuolated, and rarely X- or Y-shaped.

The literature concerning strain variation and host-specificity reactions of clover rhizobia is voluminous. To date, *T. ambiguum* Bieb., Kura clover, has been the most distinctively nonresponsive species to nitrogen fixation (Parker and Allen 1952); however, effective symbiosis was reported by Erdman and Means (1956). This species attracted widespread interest as a honey plant because of the high sugar content of the nectar.

In general, host affinities of *R. trifolii* strains are limited to *Trifolium* species; similarly, clover species constitute a highly specialized symbiotic plant group. Symbiotic affinites within the genus closely parallel and reflect phylogenetic relationships. Rhizobial strains from endemic European clovers lack symbiobility with most native African species and vice versa (Norris 1959a; Norris and Mannetje 1964). Recognizing such specific selectivity as a useful taxonomic tool, Norris (1965) coined the term "symbiotaxonomy," in analogy to cytotaxonomy and chemotaxonomy.

Nodulated species of *Trifolium*	Area	Reported by
affine C. Presl	Wis., U.S.A.	Burton et al. (1974)
africanum Ser.	Widely reported	Many investigators
var. *lydenburgense* Gillett	S. Africa	U. of Pretoria (1958)
agrarium L.	Widely reported	Many investigators
(= *strepens* Crantz)		
albopurpureum T. & G.	Calif., U.S.A.	Holland (1966)
(= *macraei* H. & A.)		
alexandrinum L.	Widely reported	Many investigators
alpestre L.	Germany	Morck (1891)
	Italy	Kharbush (1928)
alpinum L.	Italy	Kharbush (1928)
	France	Lechtova-Trnka (1931); Pers. observ. (U.W. herb. spec. 1971)
ambiguum Bieb.	Widely reported	Many investigators
amphianthum T. & G.	U.S.D.A.	L. T. Leonard (pers. comm. 1930)
angustifolium L.	France	Clos (1896)
argentifolium [?]	France	Clos (1896)
arvense L.	Widely reported	Many investigators
baccarinii Chiov.	Australia	Norris & Mannetje (1964)
	S. Africa	Saubert & Scheffler (1967a,b)
badium Schreb.	French Alps	Kharbush (1928)
beckwithii Brew.	U.S.D.A.	L. T. Leonard (pers. comm. 1930); J. C. Burton (pers. comm. 1971)
bejariense Moric.	U.S.D.A.	L. T. Leonard (pers. comm. 1930)
berytheum Boiss. & Blanche	Wis., U.S.A.	Burton et al. (1974)
bifidum Gray	Calif., U.S.A.	Holland (1966)
bocconei Savi	France	Clos (1896)
	Wis., U.S.A.	Burton et al. (1974)
boissieri Guss. ex Boiss.	Wis., U.S.A.	Burton et al. (1974)
brandegei S. Wats.	Ariz., U.S.A.	Martin (1948)
burchellianum Ser.	S. Africa	U. of Pretoria (1955–1956); Norris & Mannetje (1964)
var. *johnstonii* (Oliv.) Gillett	Kenya	Norris & Mannetje (1964)
campestre Schreb.	France	Clos (1896)
	Libya	B. C. Park (pers. comm. 1960)
	Israel	Hely & Ofer (1972)
carolinianum Michx.	U.S.A.	Ball (1909); Pieters (1927); Carroll (1934a)

Nodulated species of *Trifolium* (cont.)	Area	Reported by
cernum Brot.	Australia	Jensen (1942)
	S. Africa	Grobbelaar et al. (1967)
cheranganiense Gillett	Kenya	Bumpus (1957); Norris & Mannetje (1964)
cherleri L.	S. Africa	Grobbelaar et al. (1967)
ciliolatum Benth.	Calif., U.S.A.	Holland (1966)
clusii Gren. & Godr.	Israel	Hely & Ofer (1972)
clypeatum L.	Wis., U.S.A.	Burton et al. (1974)
compactum Post	Wis., U.S.A.	Burton et al. (1974)
cryptopodium Steud. ex A. Rich.	Centr. Africa	D. O. Norris (pers. comm. 1959)
dalmaticum Vis.	Yugoslavia	V. J. Radulovíc (pers. comm. 1968)
dasyurum C. Presl	Wis., U.S.A.	Burton et al. (1974)
depauperatum Desv.	Calif., U.S.A.	Holland (1966)
desvauxii Boiss. & Bl.	Australia	M. S. Chowdhary (pers. comm. 1965)
dichotomum H. & A.	Calif., U.S.A.	Holland (1966)
diffusum F. E. Ehrh.	S. Africa	Grobbelaar et al. (1967)
dubium Sibth. (= *minus* Relhan)	Widely reported	Many investigators
elegans Savi	Italy	Pichi (1888)
expansum W. & K.	U.S.S.R.	Roizin (1959)
filiforme L.	Germany	Lachmann (1891); Stapp (1923)
flavulum Greene	Calif., U.S.A.	Holland (1966)
formosum D'Urv.	S. Africa	Grobbelaar et al. (1967)
fragiferum L. (= *involucratum* Dulac)	Widely reported	Many investigators
fucatum Lindl.	U.S.D.A.	Erdman (1948b)
glanduliferum Boiss.	Israel	M. Alexander (pers. comm. 1963)
globosum L.	Wis., U.S.A.	Burton et al. (1974)
glomeratum L.	Widely reported	Many investigators
heterodon Gray	Germany	Schneider (1905)
hirtum All.	U.S.D.A.	Erdman (1948b)
hybridum L.	Widely reported	Many investigators
incarnatum L.	Widely reported	Many investigators
involucratum Ort. (= *willdenovii* Spreng.)	Wis., U.S.A.	Wipf (1939)
(= *wormskjoldii* Lehm.)	U.S.D.A.	Erdman (1947)
isodon Murb.	Wis., U.S.A.	Burton et al. (1974)
isthmocarpum Brot.	Brazil	Norris & Mannetje (1964)
johnstonii Oliv.	N.Y., U.S.A.	Wilson (1939a)
	Kenya	Bumpus (1957)
kingii S. Wats.	Utah, U.S.A.	Pers. observ. (U.W. herb. spec. 1971)
(= *haydenii* Porter)	Czechoslovakia	E. Hamatová (pers. comm. 1970)
var. *productum* Jepson	U.S.D.A.	L. T. Leonard (pers. comm. 1938)

Nodulated species of *Trifolium* (cont.)	Area	Reported by
lappaceum L.	Ariz., U.S.A.	Martin (1948)
	U.S.D.A.	Erdman (1948b)
leucanthum Bieb.	Wis., U.S.A.	Burton et al. (1974)
ligusticum Balb. ex Lois.	S. Africa	Grobbelaar & Clarke (1974)
lupinaster L.	U.S.S.R.	Roizin (1959)
var. *albiflorum*	U.S.S.R.	Pers. observ. (U.W. herb. spec. 1971)
maritimum Huds.	Yugoslavia	V. J. Radulovíc (pers. comm. 1968)
masaiense Gillett	Tanzania	Norris & Mannetje (1964)
medium L.	Widely reported	Many investigators
meduseum Blanche ex Boiss.	Wis., U.S.A.	Burton et al. (1974)
melanthum H. & A.	Calif., U.S.A.	Holland (1966)
meneghinianum Clem.	Wis., U.S.A.	Burton et al. (1974)
michelianum Savi	U.S.D.A.	Erdman (1948b)
microcephalum Pursh	Calif., U.S.A.	Holland (1966)
microdon H. & A.	Calif., U.S.A.	Holland (1966)
montanum L.	Italian Alps	Kharbush (1928)
	Arctic U.S.S.R.	Roizin (1959)
mutabile Portenschl.	Wis., U.S.A.	Burton et al. (1974)
nigrescens Viv.	U.S.D.A.	Erdman (1948b)
occidentale Coombe	Wis., U.S.A.	Burton et al. (1974)
ochroleucon Huds.	Turkey	Erdman & Means (1956)
ornithopodioides L.	Great Britain	Nutman (1956)
palaestinum Boiss.	S. Africa	N. Grobbelaar (pers. comm. 1961)
	Israel	M. Alexander (pers. comm. 1963)
pallidum Waldst. & Kit.	Wis., U.S.A.	Burton et al. (1974)
pannonicum Jacq.	Widely reported	Many investigators
parviflorum F. E. Ehrh.	U.S.D.A.	Pieters (1927)
pauciflorum Nutt. (= *oliganthum* Steud.)	Calif., U.S.A.	Holland (1966)
petrisavii Clem.	Wis., U.S.A.	Burton et al. (1974)
philisticum Zoh.	Wis., U.S.A.	Burton et al. (1974)
physodes Stev. ex Bieb.	Wis., U.S.A.	Burton et al. (1974)
polymorphum Poir.	Argentina	Schiel (1945)
	Uruguay	C. Batthynay (pers. comm. 1961)
polystachyum Fresen.	S. Africa	N. Grobbelaar (pers. comm. 1963)
pratense L.	Widely reported	Many investigators
var. *nivale*	Italian Alps	Malan (1938)
var. *perenne* Host	Widely reported	Many investigators
var. *sativum* Schreb. (= *sativum* Crome)	Widely reported	Many investigators
procumbens L.	Widely reported	Many investigators
pseudostriatum Bak. f.	Uganda	Norris & Mannetje (1964)

Nodulated species of *Trifolium* (cont.)	Area	Reported by
purpureum Loisel.	Libya	B. C. Park (pers. comm. 1960)
	Israel	M. Alexander (pers. comm. 1963); Hely & Ofer (1972)
reflexum L.	Ind., U.S.A.	Burrage (1900)
	Fla., U.S.A.	Carroll (1934a)
repens L.	Widely reported	Many investigators
var. *latum*	Ala., U.S.A.	Duggar (1935)
var. *pallescens*	Italian Alps	Malan (1938)
resupinatum L. (= *suaveolens* Willd.)	Widely reported	Many investigators
rubens L.	Germany	Lachmann (1891); Pers. observ. (1936)
ruepellianum Fresen.	Kenya	Norris (1956); Norris & Mannetje (1964)
rydbergii Greene	Mont., U.S.A.	Pers. observ. (1968)
scabrum L.	France	Clos (1896)
semipilosum Fresen.	Australia	Norris (1956)
	Kenya	Bumpus (1957)
var. *glabrescens*	Kenya	Strange (1958); Norris & Mannetje (1964)
var. *kilimanjaricum*	Kenya	Norris & Mannetje (1964)
var. *semipilosum*	Kenya	Norris & Mannetje (1964)
somalense Taub. ex Harms	S. Africa	B. W. Strijdom (pers. comm. 1967)
	Zimbabwe	W. P. L. Sandmann (pers. comm. 1970)
spadiceum L.	U.S.D.A.	Erdman & Means (1956)
	U.S.S.R.	Pärsim (1967)
	Scandinavia	Pers. observ. (U.W. herb. spec. 1971)
spinosum [?]	S. Africa	U. of Pretoria (1958)
spumosum L.	S. Africa	Saubert & Scheffler (1967a,b)
squamosum L.	S. Africa	Grobbelaar & Clarke (1974)
squarrosum L.	U.S.S.R.	Lopatina (1931)
stellatum L.	S. Africa	Grobbelaar et al. (1967)
	Italy	Pers. observ. (U.W. herb. spec. 1971)
	Israel	Hely & Ofer (1972)
stenophyllum Boiss.	Wis., U.S.A.	Burton et al. (1974)

Nodulated species of *Trifolium* (cont.)	Area	Reported by
steudneri Schweinf.	Australia	Norris (1956)
	Kenya	Bumpus (1957)
	S. Africa	N. Grobbelaar (pers. comm. 1963)
	Zimbabwe	W. P. L. Sandmann (pers. comm. 1970)
striatum L.		
var. *brevidens*	S. Africa	N. Grobbelaar (pers. comm. 1963)
strictum L.	S. Africa	Grobbelaar et al. (1967)
subterraneum L.	Widely reported	Many investigators
ssp. *brachycalycium* Katz. & Morley	Israel	Hely & Ofer (1972)
ssp. *subterraneum* Katz. & Morley	Australia	Gibson & Brockwell (1968)
ssp. *yanninicum* Katz. & Morley	Australia	Gibson & Brockwell (1968)
supinum Savi	Wis., U.S.A.	J. C. Burton (pers. comm. 1971)
tembense Fresen.	Australia	Norris (1956)
	Kenya	Bumpus (1957)
thalii Vill.	Italy	Malan (1938)
tomentosum L.	Australia	Vincent (1945)
	Libya	B. C. Park (pers. comm. 1960)
	Israel	M. Alexander (pers. comm. 1963)
var. *curvisepalum* (Täckh.) Thièb.	Israel	Hely & Ofer (1972)
tridentatum Lindl.	Calif., U.S.A.	Holland (1966)
uniflorum L.	Libya	B. C. Park (pers. comm. 1960)
usambarense Taub.	Kenya	Bumpus (1957)
	Tanzania	Norris & Mannetje (1964)
	Zimbabwe	W. P. L. Sandmann (pers. comm. 1970)
vavilovii Gig.	Wis., U.S.A.	Burton et al. (1974)
vernum Phil.	Wis., U.S.A.	Burton et al. (1974)
vesiculosum Savi	Ga., U.S.A.	Beaty et al. (1963)
violaceum David	France	Demolon & Dunez (1938)
(*willdenovii* Spreng.) = *involucratum* Ort.	U.S.D.A.	Erdman (1947)
(*wormskjoldii* Lehm.) = *involucratum* Ort.	U.S.D.A.	Erdman (1947)
xerocephalum Fenzl	Israel	M. Alexander (pers. comm. 1963)

From Latin, *trigonus*, "3-angled," with diminutive suffix, in reference to the small, triangular appearance of the flower.

Type: *T. foenum-graecum* L.

70–75 species. Herbs, annual, erect, often strongly scented. Leaves pinnately 3-foliolate; leaflets toothed; stipules adnate to the petiole. Flowers yellow, blue, or white, in heads, umbels, or short, dense racemes in leaf axils, rarely solitary; bracts minute; bracteoles none; calyx tube short, campanulate; lobes ovate, acuminate, subequal, about as long as the calyx tube; petals free from the staminal tube, caducous; standard obovate or oblong, sessile or narrowed into a broad claw; wings oblong, eared, clawed; keel petals oblong, shorter than the wings, obtuse, clawed; vexillary stamen free or connate to the middle with others; filaments not dilated; anthers uniform; ovary sessile or short-stalked, few- to many-ovuled; style filiform or thick, glabrous; stigma terminal. Pod variable, oblong or oblong-linear, compressed or terete, thick, beaked, nonseptate, indehiscent or opening along only 1 suture, 1- to many-seeded. (Bentham and Hooker 1865; Tutin et al. 1968)

Members of this genus are native to the Mediterranean area. They are widely distributed from the Canary Islands to the Near East and from North Africa into southwestern Asia and India. Isolated representatives appear in Australia and South Africa. Species thrive in loam soil. Growth demands are minimal, but all species grow best in well-drained soils in areas receiving 50–150 cm of rainfall.

The type species, commonly known as fenugreek, Greek clover, or Greek hay, was widely cultivated during ancient times throughout the Mediterranean area and southern Asia. Young plants of this cloverlike annual were relished as salad by the Egyptians in Biblical times (Moldenke and Moldenke 1952).

Crushed fenugreek seeds have been used as an ingredient of poultices for burns, as a cosmetic, in hair oils, and in cough, diuretic, and laxative tonics. The seeds are mucilaginous when mixed with water; they may be consumed raw or boiled. Egyptian and Indian women eat them to promote lactation. The powdered seeds are used locally as a source of yellow dye and as a spice, and commercially as an ingredient of curry powder for food flavoring. The foliage was one of the components of the celebrated Egyptian incense kuphi, a holy smoke used in fumigation and embalming rites (Rosegarten 1969). Chopped foliage of *T. caerulea* (L.) Ser. is used to flavor green cheese and herb bread in Switzerland.

All species of the genus have soil-binding and soil-improvement properties. The protein content of the foliage often exceeds 20 percent as a result of prolific nodulation under natural conditions. *T. foenum-graecum* L., *T. hamosa* L., *T. laciniata* Desf., *T. marginata* Hochst. & Steud., and *T. occulta* Del. serve as forage in the eastern Mediterranean area. *T. arabica* Del. and *T. stellata* Forsk. are foraged by animals in the desert areas of the Sahara, Palestine, and the Dead Sea. *T. suavissima* Lindl. is well adapted to wet, swampy habitats. Despite the cloverlike nature of fenugreek and its freedom from insect enemies and diseases, it has had only limited use in the United States as a winter green manure in orchards of the West Coast.

The alkaloid trigonelline, $C_7H_7NO_2$, the *N*-methylbetaine of nicotinic acid, has been isolated from plant parts, mainly seeds, of *T. caerulea, T. cretica* Boiss., *T. foenum-graecum, T. lilacina* Boiss., *T. radiata* Boiss., and *T. spinosa* L. (Willaman and Schubert 1961). This alkaloid exerts no pharmacological action; it is found also in plant parts of other species. *Trigonella* seeds also yield choline, a semicrystalline white saponin, a lactation-stimulating oil, and various gums; the gums may approximate 23 percent of the seed content (Casares López et al. 1950; Watt and Breyer-Brandwijk 1962).

Several decades ago attention was drawn to *T. foenum-graecum* seeds as a source of diosgenin, $C_{27}H_{42}O_3$, a medicinally important steroidal sapogenin (Soliman and Mustafa 1943; Fazli and Hardman 1968; Rosengarten 1969); however, its importance diminished as other plant sources were reported (Marker et al. 1943, 1947), and especially when diosgenin was synthesized by Mazur et al. in 1960. Prior to its synthesis the major source of diosgenin was wild yams, i.e., the tubers of *Dioscorea* species obtained from Mexico and Central America.

NODULATION STUDIES

Fuchs pictured nodules on *foenograecum* in the first edition of his *De Historia Stirpium Commentarii Insignes*, published in 1542. Thus, this species ranks with *Aphacia (= Lathyrus), Vicia,* and *Faba* in being among the first genera in which nodules were recorded.

Rhizobia from nodules of *T. foenum-graecum* are akin culturally and biochemically to alfalfa rhizobia and are of the species *Rhizobium meliloti*. Bacteroids were described by Morck (1891) as rods, sometimes swollen or forked at one end. Simon and Krüger in 1914, working independently, and Burrill and Hanson in 1917 demonstrated host-plant relationships with *Medicago* species; their findings have not been challenged.

The taxonomic transfer of *Trigonella ornithopodioides* (L.) DC. to the genus *Trifolium* appears justified in the light of rhizobial kinships (Nutman 1956).

Nodulated species of *Trigonella*	Area	Reported by
anguina Del.	Zimbabwe	W. P. L. Sandmann (pers. comm. 1970)
arabica Del.	Israel	Hely & Ofer (1972)
aurantiaca Boiss.	Iraq	Pers. observ. (U.W. herb. spec. 1970)
balansae Boiss. & Reut.	R.I., Calif., U.S.A.	Pers. observ. (U.W. herb. spec. 1970)
berthyea Boiss. & Blanche	Israel	M. Alexander (pers. comm. 1963)
caerulea (L.) Ser.	Widely reported	Many investigators
calliceras Fisch. ex Bieb.	S. Africa	Grobbelaar et al. (1967)
coelesyriaca Boiss.	Israel	M. Alexander (pers. comm. 1963)
cylindracea Desv.	Israel	M. Alexander (pers. comm. 1963)
foenum-graecum L.	Widely reported	Many investigators
geminiflora Bge.	U.S.S.R.	Pers. observ. (U.W. herb. spec. 1970)

Nodulated species of *Trigonella* (cont.)	Area	Reported by
gladiata Stev. ex Bieb.	S. Africa	Grobbelaar et al. (1967)
hamosa L.	S. Africa	Mostert (1955)
hybrida Pourr.	France	Vuillemin (1888)
monantha C. A. Mey.	Pakistan	A. Mahmood (pers. comm. 1977)
monspeliaca L.	S. Africa	Grobbelaar et al. (1967)
occulta Del.	India	Narayana & Gothwal (1964)
(*ornithopodiodes* (L.) DC.) = *Trifolium ornithopodiodes* L.	Australia	Harris (1953)
stellata Forsk.	Libya	B. C. Park (pers. comm. 1960)
	Israel	Hely & Ofer (1972)
suavissima Lindl.	Australia	Bowen (1956); Norris (1959b); Beadle (1964)

Triplisomeris Aubrév. & Pellegr.

Caesalpinioideae:
Amherstieae

From Latin, *triplus*, "three," and *isomerus*, "with equal parts," in reference to 3 of the petals.

Type: *T. explicans* (Baill.) Aubrév. & Pellegr.

3 species. Scandent shrubs or small trees, unarmed, evergreen. Leaves paripinnate; leaflets few-paired, symmetric, obovate-elliptical, without translucent gland dots; stipules not intrapetiolar. Flowers small, many, in very long pendent racemes; bracteoles 2, well-developed; calyx tube cup-shaped; petals 5, upper 3 large, equal, lower 2 rudimentary; fertile stamens 3; anthers open by slits; staminodes usually 6; ovary short-stalked. Pod compressed, woody, 2-valved, not noticeably ridged or wrinkled, long persistent on the tree, valves twisting after dehiscence; seeds subquadrangular, thick, brown, waxy. (Aubréville and Pellegrin 1958)

Species of this genus were originally included in *Macrolobium*, which is now reserved entirely for the distinctive American elements (Cowan 1953). Some of the excluded African species were transferred to *Anthonotha*, which in turn was partly fragmented by Aubréville and Pellegrin (1958) to establish 2 new genera, *Isomacrolobium* and *Triplisomeris*. These new taxa are regarded as synonyms of *Macrolobium* by Hutchinson (1964).

Members of the genus are native to Gabon and Ivory Coast. They grow in forested coastal areas.

Tylosema (Schweinf.) Torre & Hillc.

From Greek, *tylos*, "knob," and *sema*, "sign," "mark"; 1 petal has 2 basal callosities.

Type: *T. fassoglensis* (Kotschy) Torre & Hillc.

3–4 species. Vines herbaceous or woody, creeping or high climbing, usually with forked tendrils and a large underground tuber; shoots, leaflets, and inflorescences at first brownish, silky-haired, later glabrous. Leaves broad, heart-shaped at the base, emarginate to subbilobed at the apex; stipules 2, lanceolate, persistent. Flowers yellow, profuse in axillary panicles 20–30 cm long, near the tips of branches; bracteoles filiform, persistent; calyx short-campanulate, rusty pubescent; lobes 5, free, oblong-lanceolate, acute, upper 2 joined; petals unequal, upper petal reflexed, very small, with a prominent, erect, bifid callus at the base; stamens 2, fertile; staminodes 8, shorter, unequal, deformed; ovary flat, 2-ovuled, hairy along dorsal margin; style slender, recurved; stigma pointed. Pod oblong, asymmetric, apiculate, woody, smooth; seeds 2, large, elliptic, black. (Brenan 1967)

Tylosema and *Bauhinia* are considered congeneric by some botanists (Wilczek 1952; Hutchinson 1964).

Tylosema species are scattered throughout tropical Africa, Angola, Ethiopia, Kenya, and Zimbabwe. They thrive in wooded savannas.

The seeds and tubers are consumed by natives in equatorial Africa.

NODULATION STUDIES

No nodules were present on *T. fassoglensis* (Kotschy) Torre & Hillc. examined in Zimbabwe (Corby 1974).

Uittienia Steenis

Caesalpinioideae: Cassieae

Named for Hendirk Uittien, 1898–1944, outstanding authority on the botany of Suriname, Curator of the Herbarium, State University, Utrecht, The Netherlands, and later a Professor of Botany in the Colonial College of Agriculture, Deventer, The Netherlands.

Type: *U. modesta* Steenis

1 species. Tree, 14–30 m tall. Leaves 1-foliolate, ovate-lanceolate, midrib elevated; petioles thickened at both ends, jointed; stipules minute, linear, caducous. Flowers on long pedicels, in short, axillary, corymbiform thyrses; calyx tube absent; calyx lobes 5, subtriangular, nerved, free, caducous, narrowly imbricate in bud; petals 5; stamens 5, 2 abortive, filiform or small and sterile, 3 fertile, inserted on the rim of a hairy, distinct, cushion-shaped disk; filaments long, subterete, free; anthers basifixed, symmetric, heart-shaped, longitudinally dehiscent; ovary cylindric, sometimes distinctly oblique, 1–2-ovuled, stalk thick, flat; style filiform; fruit not described by Steenis (1948). Pod subobliquely globular, very large, woody, indehiscent, 1-seeded; seed subglobular, surrounded by a thick, soft, pulpy endocarp. (See Hutchinson 1964)

A later description of the fruit was relayed to Hutchinson (1964) by Steenis. Staminodes were not observed by Steyaert (1953); for this and other reasons Steyaert considered *Uittienia* a subgenus of *Dialium*.

This monotypic genus is endemic in Borneo. The plants grow in primary forests usually below 100-m altitude.

Uleanthus Harms   Papilionoideae: Sophoreae
(133a) / 3591a

Named for Ernest Heinrich George Ule, 1854–1915, Brazilian botanist; he discovered the plants in the Amazonas region.

Type: *U. erythrinoides* Harms

1 species. Tree, small. Leaves imparipinnate; leaflets 1–2 pairs, on short petiolules, oblong or elliptic, acuminate. Flowers handsome, in lax, few-flowered, axillary racemes or on trunks or branches; calyx red, funnel-shaped, with thickened base; lobes 4–5, unequal, valvate in bud, upper 2 larger and broader; petals blue or rose purple; standard large, rounded, clawed, other petals very small, linear- lanceolate; keel petals free; stamens 10, free, smooth; anthers small, oblong, versatile; ovary stalked, small, narrow, hairy, 5–8-ovuled; style filiform, hairy at base, smooth above; stigma minute. Pod on slender, short stalk, oblique, 2–3 cm wide for three-fourths of its length, ending in a needle point, sparsely tomentose, 2-valved, valves thin, woody, coiling upon dehiscence, 3–5-seeded. (Harms 1908a)

Uleanthus has close affinity with *Bowdichia*, but the pods and seeds closely resemble those of *Alexa* (Ducke 1922).

This monotypic genus is endemic above the cataracts on the Madeira and Tapjoz rivers, tributaries of the Amazon. The plants thrive in upland, high forests.

The commercial possibilities of this low-lustered, hard, fine-textured wood have not been fully explored (Record and Hess 1943).

Ulex L.   Papilionoideae: Genisteae, Cytisinae
220 / 3681

A Latin name used by Pliny for a spiny shrub.

Type: *U. europaeus* L.

20 species. Shrubs, dense, main stems erect, 1 m tall, many terminal and lateral branches usually reduced to green spines. Leaves of seedlings 3-foliolate, usually alternate, but reduced on mature plants to scales or narrow, spinelike phyllodes; stipels and stipules absent. Flowers yellow, solitary or in small clusters in axils of spines or in short racemes or umbels; bracts small; calyx bilabiate to the base, upper lip bidentate, lower lip tridentate,

membranous, persistent, subtended by 2 small, broad, basal bracteoles; petals free, clawed, subequal; standard ovate; wings narrow, obtuse; keel petals obtuse, lower margin pilose; stamens monadelphous, anthers alternately long and basifixed, short and dorsifixed; ovary sessile, multiovuled; style incurved, glabrous; stigma subcapitate, terminal. Pod ovoid or linear-oblong, compressed or turgid, densely hairy, scarcely exserted from the persistent calyx, nonseptate, 2-valved, valves twisting, explosively dehiscent; seeds 1–6. (Bentham and Hooker 1865; Rothmaler 1941)

Polhill (1976) included this genus in Genisteae *sensu stricto*.

Representatives of this genus are native to Great Britain, western Europe, the Iberian Peninsula, and North Africa. Species have been widely introduced and naturalized elsewhere. The plants inhabit sandy waysides, well-drained acid heath soils, and rocky coastal headlands.

All of the species are commonly called gorse, furze, or whin. *U. europaeus*, *U. gallii* Planch., and *U. minor* Roth. cover vast heathland areas in Great Britain and Ireland. In Europe *U. europaeus* is sometimes cultivated for cattle bedding, hedges to enclose livestock, and for fodder; however, it is not widely used as a fodder because of its spiny nature and alkaloid content. It was introduced into Hawaii and various South Pacific areas for use as a "living fence." It is not uncommon for plants to escape and become a nuisance difficult to eradicate. *U. europaeus* occurs along the coastal strips of the North Atlantic and North Pacific United States, where it serves as a sandy soil-binder. In addition to its tendency to form an impenetrable barrier under good growth conditions, gorse is highly inflammable and burns with explosive force. The town of Bandon, Oregon, owed its destruction by fire in 1936 to gorse that had been featured as an ornamental by the early inhabitants (Holbrook 1936). Sparks from a nearby forest fire ignited the gorse along the streets and in fields adjoining the town. The town of 1,800 people was destroyed in less than 5 hours.

Early studies concerning alkaloids in *Ulex* species are documented by Clemo and Raper (1935) and White (1943). Variations in alkaloid occurrence and concentration have been attributed to seasonal growth and seed-ripening differences. Recent studies confirm the presence of anagyrine and cytisine in plant parts of 3 species. An earlier study by Jaretzky and Axer (1934) showed the absence of sparteine in *U. africanus* Webb, *U. aphyllus* Link, *U. parviflorus* Pourr., and *U. spartioides* Webb.

NODULATION STUDIES

Wydler alluded to the occurrence of nodules on *U. europaeus* as early as 1860. On the basis of limited host-infection tests, Simon (1914) concluded that this species was nodulated only by its homologous rhizobia; however, Pieters (1927) listed it in the cowpea miscellany. Wilson (1939a) showed that it was nodulated by rhizobia from a cowpea affiliate, *Stizolobium deeringianum*. Further inquiries into the nodulation profile of *Ulex* species are warranted.

In a study of nodule longevity cycles, Pate (1958a) observed that the expanding root systems of *U. europaeus* and *U. gallii* were progressively nodulated in normal seasons in Northern Ireland. Nodules at least 20 months old were observed on plants 3 or more years old. Tissue development in tagged nodules was similar to that of perennial nodules on *Sesbania grandiflora* (Harris et al. 1949) and *Caragana arborescens* (Allen et al. 1955). Nodules on *U. europaeus* did not persist into the third season. Pate's observations agreed with those of Spratt (1919) relative to lobed nodules on *Sophora* and *Acacia* and with those of Jimbo (1927) concerning *Wisteria* nodules. Growth substances in *U. europaeus* nodules were described in a later paper (Pate 1958b).

Umtiza Sim   Caesalpinioideae: Amherstieae

Type: *U. listeriana* Sim

1 species. Tree, evergreen; branches and twigs often modified into stout spines, with or without leaves. Leaves paripinnate; leaflets glossy, sessile, oblong or oblong-linear; stipels and stipules none. Flowers small, regular, in short panicles, at the apex of short, lateral branches; calyx campanulate; sepals 5, free, shorter than the tube; petals 5, free, imbricate, longer than the sepals, inserted in the mouth of the calyx tube with the 10, free, equal stamens; anthers kidney-shaped, versatile; ovary sessile, 2-ovuled, style exserted, thick, cylindric, longer than the stamens; stigma truncate. Pod thin, flattened, pointed, membranous, dehiscent, 1-seeded. (Phillips 1951)

The positioning of *Umtiza* in proximity to *Schotia* by Phillips (1951) was not completely acceptable to Léonard (1957).

This monotypic genus is common in the Cape of Good Hope, South Africa. The plants grow in forests near East London and King William's Town. According to Riley (1963), *Umtiza* is the only genus of the subfamily Caesalpinioideae endemic in South Africa.

Uraria Desv.   Papilionoideae: Hedysareae, Desmodiinae
341 / 3811

From Greek, *oura*, "tail," in reference to the long, taillike terminal racemes.

Type: *U. picta* (Jacq.) Desv. ex DC.

35 species. Herbs or erect undershrubs, woody, stems somewhat pubescent. Leaves imparipinnate, 1–9-foliolate; leaflets large, prominently veined; stipels acuminate; stipules large, ovate-lanceolate, acutely pointed, persistent. Flowers pink, purple, or yellow, small, paired along rachis in dense, simple, spikelike, terminal racemes up to 30 cm long, pedicels in pairs, hooked at the tips causing inversion of flowers; bracts ovate or lanceolate-acuminate, persistent or caducous; bracteoles none; calyx jointed with the pedicel; lobes pointed, spreading, the 2 upper ones shortest, becoming lowest by resupination of the flower, the other 3 much longer; standard subsessile, obovate or orbicular, narrowed into a claw, glabrous, without basal ears or tubercles;

wings falcate-oblong, adhering to the longer, slightly incurved, blunt keel; stamens 10, diadelphous; anthers uniform; ovary sessile or short-stalked, 2- to many-ovuled; style glabrous, inflexed above; stigma small, capitate. Pod subsessile, enclosed in the nonenlarged calyx, 2–7 jointed, each segment ovate, flattened and nearly separated, but connected by the sutures, indehiscent; seeds brown, lenticular or subreniform, flattened. (Bentham and Hooker 1865; Schindler 1917; Baker 1929; Pellegrin 1948; Hutchinson and Dalziel 1958)

Uraria Desv. is distinguished from *Urariopsis* by the pod joints being connected by their sutures, the flowers being in dense, simple racemes, and the 3 lowermost calyx lobes being more than 3 times longer than the 2 uppermost ones (Backer and Bakhuizen van den Brink 1963).

Uraria is a paleotropical genus. Species are well distributed in tropical Asia and Africa. *U. cylindracea* Benth., *U. lagopoides* DC., and *U. picta* (Jacq.) Desv. ex DC. occur in Queensland and northern Australia. The plants prefer open sandy soil, grasslands, and deforested areas, up to 3,000-m altitude.

All species are good ground cover in dry wasteland areas and are grown for this purpose in village groves, along roadsides, and in teak forests. *U. aequilobata* Hosokawa and *U. macrostachys* Wall. are especially used for this purpose in Taiwan. *U. lagopoides*, a prostrate herb with flowers that unfold in the evening, has shown promise as a green manure in Java and India. The pulverized leaves of *U. picta* are used medicinally in southern Nigeria as a remedy for gonorrhea. In some parts of West Africa the plant is credited with having magical charms against sharp weapons. In India it is used as an antidote against the bite of certain vipers. Crushed leaves of *U. crinita* Desv. are applied externally in Malaysia for delousing purposes (Burkill 1966).

NODULATION STUDIES

Bañados and Fernandes (1954) observed nodules on 18 of 20 plants of *U. lagopodioides* (L.) Desv. ex DC. examined in the Philippines. Rhizobial isolates from this species tested in Papua New Guinea by Trinick (1968) cross-inoculated ineffectively with *Phaseolus lathyroides*, but not at all with *Leucaena leucocephala*. In the Bogor Botanic Garden, Java, nursery plants of *U. spinosa* Desv. were well nodulated (personal observation 1977).

Urariopsis Schindl.   Papilionoideae: Hedysareae,
                                                    Desmodiinae

The Greek suffix, *opsis*, denotes resemblance to *Uraria*.

Type: *U. cordifolia* (Wall.) Schindl.

1 species. Shrub, small, erect. Leaves 1-foliolate, ovate with heart-shaped base, densely ciliate, both surfaces hairy, lower surface strongly veined; petiole with 2 apical stipels; stipules long-acuminate, base broadly heart-shaped. Flowers pink, paired in paniculately branched, terminal and axillary racemes; bracts large, acuminate, caducous; bracteoles none; calyx campanu-

late, jointed with the persistent, hooked pedicel, deeply 5-parted; lobes sub-equal, 2 upper ones connate; petals small; standard obovate, subsessile, basal ears or tubercles absent; wings adherent to the longer keel; stamens diadelphous; anthers uniform; ovary short-stalked; style bent at a right angle, upper part glabrous; stigma terminal. Pod 2–5-jointed, submoniliform, seg-ments flat, peltately connected by short stalks, indehiscent. (Schindler 1917; Backer and Bakhuizen van den Brink 1963)

This monotypic genus is endemic in Southeast Asia and western Malaysia. The plants grow in low grassy localities in monsoon regions.

Uribea Dugand & Romero   Papilionoideae: Sophoreae

Named for Lorenzo Uribe, S.J., Director of the Instituto de Ciencias Naturales, Universidad Nacional, Bogotá, Colombia, and his father, Joaquin Antonio Uribe, 1858–1935, teacher, naturalist, and author of *Florae Sonsonen-sis* and *Florae Antioquiensis*.

Type: *U. tamarindoides* Dugand & Romero

1 species. Tree, 15–20 m tall. Leaves imparipinnate, deciduous; leaflets alternate, lanceolate or elliptic, pinnately veined; stipels and stipules present. Flowers small, rose or violet, in short, axillary, solitary racemes; bracts and bracteoles minute; calyx lobes 5, 2 uppermost joined except near the apex, others ovate, subequal; standard slightly broader than other petals, obovate-elliptic, margins inflexed, broad-clawed; wings and keel petals quite similar, clawed; stamens 10; filaments free or united briefly at the base, persistent, arising from a thin disk at the base of the ovary; anthers small, subbasifixed, acute; ovary stalked, few-ovuled; style awl-shaped, glabrous; stigma minute. Pod slightly compressed, short-stalked, linear-oblong or elliptic, indehiscent; seeds compressed, rounded. (Dugand 1962a)

This monotypic genus is endemic in Costa Rica, Colombia, and Panama.

The heavy durable wood is relatively unavailable and has limited use (Dugand 1962b).

Vandasia Domin   Papilionoideae: Phaseoleae, Glycininae

Named for Professor Karel Vandas, 1861–1923, Czechoslovakian botanist, professor at the Technische Hochschule, Brno.

Type: *V. retusa* (Benth.) Domin

1 species. Shrub, attractive, high-climbing; shoots and inflorescence silky-pubescent; branches usually angular. Leaves 3-foliolate, large; leaflets

obcordate or obovate-truncate and emarginate, venation peculiarly transverse, finely reticulate; stipels present; stipules ovate-lanceolate, striate, reflexed. Flowers handsome, showy, large, numerous, violet with yellow spots, fascicled and racemose in terminal panicles; bracts minute, caducous; bracteoles absent; calyx hoary-pubescent; lobes 5, short, obtuse, upper 2 connate; standard broad-orbicular, emarginate, not eared; wings adherent to the keel; keel subequal to the wings, much incurved, blunt; stamens 10, diadelphous; anthers uniform; ovary subsessile; ovules 8–10; style thickened, incurved at the base, later straight and slender; stigma small, terminal. Pod broad, linear, flattened, silky-villous but somewhat turgid, without pithy partitions inside, 2-valved; seeds large, oblong. (Domin 1926)

Segregation of this species from *Hardenbergia* was made on the basis of the peculiarly obcordate wings and the length of the wings exceeding that of the keel.

This monotypic genus is endemic in Queensland, Australia, and Papua New Guinea.

Vatairea Aubl.   Papilionoideae: Dalbergieae, Pterocarpinae

Type: *V. guianensis* Aubl.

7 species. Trees, large, up to 40 m tall, with trunk diameters about 2 m and supported by narrow buttresses, unarmed; usually leafless during flowering and fruiting seasons. Leaves imparipinnate, crowded at the end of branches; leaflets several, usually 5–17, alternate, mostly large, leathery, pinnately nerved; stipels absent; stipules small, caducous. Flowers small, white to purplish, in large, terminal panicles; bracts and bracteoles minute; calyx turbinate-campanulate, thick, narrowed at the base; lobes 5, short, equal; petals clawed, subequal; standard orbicular or ovate; wings narrow-ovate; keel petals similar but briefly connate dorsally; stamens monadelphous, connate into a split sheath; anthers versatile or subbasifixed; ovary subsessile or short-stalked, 1–3-ovuled, but usually only 1 ovule perfect; style short. Pod samaroid, woody, with a large, terminal, transverse wing at the top and a persistent style, rarely round and corky with a rudimentary wing, indehiscent, 1-seeded. (Ducke 1922, 1930)

The genus is closely aligned to *Pterocarpus* and *Andira*.

Members of this genus are widespread in Brazil, Guiana, and the Atlantic coastal regions of tropical Central America and Mexico. They thrive on savannas, hillsides, and alluvial plains.

The wood of these species often has an unpleasant odor when freshly cut. It lacks commercial possibilities except for local use. It is of high quality, being resistant to decay and insect attacks, and being moderately hard and heavy (Record and Hess 1943). Various native medicinal properties are attributed to the sap and bark decoctions, especially for treatment of eczema.

675

Vataireopsis Ducke   Papilionoideae: Dalbergieae,
                                          Pterocarpinae

The Greek suffix, _opsis_, denotes the close resemblance of this genus to _Vatairea_.

Type: _V. speciosa_ Ducke

3 species. Trees, medium to tall, unarmed, defoliated during flowering and fruiting. Leaves large, imparipinnate, congested at the tops of branches; leaflets linear-oblong, alternate, 30–50; stipels minute; stipules awl-shaped, caducous. Flowers blue violet, numerous, in large, terminal panicles; bracts and bracteoles small, caducous; calyx turbinate-campanulate, incurved in bud; teeth 5, unequal, short, upper 2 connate; petals 5, subequal, smooth, clawed; standard broad-orbicular, not appendaged at the base; wings and keel narrower, free; stamens 10, monadelphous, joined only in lower third of length; anthers small, ovate, versatile, longitudinally dehiscent; ovary stalked, 1–2-ovuled; style short, persistent. Pod samaroid, compressed, with 1 large seed at the base, 2 small lateral wings, the upper part broad-winged and with a cuspidate tip. (Ducke 1932, 1933)

Representatives of this genus are indigenous to Amazonas, Brazil. They flourish in upland forests.

The coarse-textured, yellow wood of _V. araroba_ (Aguiar) Ducke serves limited use for general construction and carpentry. Its bark yields araroba, or Goa powder, a chrysarobin compound used to treat psoriasis and other chronic skin diseases.

Vatovaea Chiov.   Papilionoideae: Phaseoleae, Phaseolinae

Named for A. Vàtova; he collected this plant in Ethiopia in 1937.

Type: _V. pseudolablab_ (Harms) Gillett (= _V. biloba_ Chiov.)

1 species. Woody climber. Leaves sparse, pinnately 3-foliolate, 2 lateral leaflets smaller, asymmetric; stipels and stipules truncate at the base. Flowers 3–5, at callous nodes of pseudoracemes on rachis longer than the peduncle; calyx tube narrow at the base, hemispherical above; 2 upper calyx lobes lanceolate, somewhat connate, lower lobes triangular with ciliate margins; standard deep rose purple, suborbicular, reflexed, with a pair of large, stiff, erect appendages at the base; wings pale rose, oblong, with narrow pointed spur at the back; keel greenish, not spurred, apex rounded, curved upward at a right angle to the base; stamens monadelphous; filaments all joined into a sheath split above; ovary oblong, 6–12-ovuled, hairy at the base; style curved through 180°, bearded inside along the upper half, with a reflexed appendage above the stigma; stigma lateral, bent downward, hairy-tufted on the inner margin. Pod broad, flat, valves twisted after dehiscence; seeds small, hilum elliptic. (Chiovenda 1951; Gillett 1966a)

The genus has close affinity with _Physostigma_.

This monotypic genus is endemic in East Africa, in Ethiopia, Sudan, Kenya, and Tanzania. The plants occur in low rainfall areas up to 1,350-m altitude.

676

From Latin, *vermis*, "worm," and *frux*, "fruit"; the pod shape resembles that of a small grubworm.

Type: *V. abyssinica* (A. Rich.) Gillett (= *Helminthocarpon abyssinica*)

1 species. Herb, slender, silky-haired, decumbent. Leaves imparipinnate; leaflets many, entire; stipules minute. Flowers small, red yellow, in axillary, 4–6-flowered, short-stalked umbels; bracts obsolete; bracteoles none; calyx short, tubular, with 2 upper teeth broader than the others; petals free, long-clawed; standard suborbicular; wings obovate-oblong, partly adnate to the blunt, slightly upcurved keel; stamens united to form a tube or vexillary stamen almost free; alternate filaments broadened at the tips; anthers uniform; ovary sessile, 2-ovuled; style incurved; stigma terminal. Pod linear, weakly 4-sided, circinate-incurved, leathery, transversely veined, not jointed, indehiscent, subseptate within.

The name *Vermifrux* has been proposed to replace *Helminthocarpon* A. Rich (Gillett 1966b), a later homonym for *Helminthocarpon* Fée.

This monotypic genus is endemic in Ethiopia, Eritrea, Sudan, and Yemen.

Vicia L.   Papilionoideae: Vicieae
378 / 3852

The Latin name of a vetch plant.

Type: *V. sativa* L.

150 species. Herbaceous vines, annual, biennial or perennial, climbing or trailing, mostly tendril-climbing, rarely low or suberect. Leaves paripinnate; leaflets small, entire or dentate at apex, linear or oblong, petioluled, usually in numerous pairs; rachis usually ending in a simple or branched tendril or recurved bristle; stipels absent; stipules small, semisagittate. Flowers blue, violet, yellow, or white, sessile or long-stalked, solitary or fascicled in leaf axils or in elongated, axillary racemes; bracts minute, caducous; bracteoles absent; calyx campanulate to somewhat turbinate; tube oblique and obtuse at the base; lobes 5, well-developed, subequal or upper 2 short and lowest lobe longer; standard obovate or oblong, narrowed at the base, broad-clawed, emarginate; wings oblique-oblong, joined to the middle of the keel; keel petals shorter than the wings, falcate-oblong or broad; stamens monadelphous or diadelphous, vexillary stamen free or slightly joined with others into a sheath with an oblique mouth; filaments filiform; anthers uniform; ovary subsessile or stalked, many-ovuled, rarely 2-ovuled; style inflexed, filiform, compressed dorsally, pubescent all around or with a tuft of hairs at the apex; stigma terminal. Pod usually compressed, oblong or linear, flat or swollen, nonseptate between the seeds, 2-valved, dehiscent; seeds 2–many, globular, rarely compressed, cotyledons thick. (Bentham and Hooker 1865; Boutique 1954)

Representatives of this large genus are widely distributed in temperate areas. Many species are native to Europe and adjacent Asian regions, most are now common to all continents except Australia. Of the 35 species found in the United States, perhaps 25 are native (Schoth and McKee 1962). The plants are both wild and cultivated. Species thrive on diverse soil types, with a preference for well-drained, sandy prairies, pinelands, and upland and loamy soils. They occur in waste areas, along roadsides, in thickets, along river banks, in open woods, and along swamp borders.

Vicia species are among the most important leguminous plants grown in both the Old World and the New World. They are used extensively as cover crops in orchards and citrus groves, for erosion control along roadsides, as winter annuals, for soil improvement, green manure, livestock forage, hay and silage, and as wildlife food and shelter. Seeds of *V. calcarata* Desf., *V. ervilia* (L.) Willd., and *V. faba* L. are eaten boiled, roasted, or in soups, and are also made into flour. The last-named species is one of the oldest cultivated plants.

The foliage of all vetches is edible and palatable to livestock. However, poisoning of farm animals has been associated with the occurrence of vicine, $C_{10}H_{16}N_4O_7$, convicine, $C_{10}H_{15}N_3O_8$, and similar amino glucosides in the seeds of *V. faba* and *V. sativa* L. (Steyn 1934; Webb 1948). *V. faba* has also been associated with favism or fabism, an acute hemolytic anemia of man and animals. This unusual malady follows the ingestion of *V. faba* seeds, or the inhalation of pollen, by individuals whose erythrocytes are deficient in glucose-6-phosphate dehydrogenase (Watt and Breyer-Brandwijk 1962; Kosower and Kosower 1967). The exact role of *V. faba* seeds in the complicated chain reaction resulting in favism is unknown. The occurrence of favism is rare and is apparently linked with genetic factors in the individuals affected.

Two amino acids, dopa, dioxyphenylalanine, and L-dopa, β-(3,4-dihydroxyphenyl)-L-alanine, are present in high concentrations in *Vicia faba* (Guggenheim 1913; Sealock 1955). The recent success of L-dopa in the treatment of Parkinson's disease prompted Natelson (1969) to suggest that a high intake of *V. faba* beans may be of therapeutic value and much less costly than the purified amino acid extract.

NODULATION STUDIES

Vicia species were widely used in early nodular studies. The earliest known drawings of *Vicia* nodules were those included in Fuchs's *De Historia Stirpium Commentarii Insignes*, published in 1542. However, no mention of nodular growths was made in the text.

Malpighi's *Anatome Plantarum*, a booklet published in 1679, featured a plate again depicting *V. faba* with a nodulated root system. Malpighi believed the protuberances to be galls caused by worms or insect larvae. Prior to this publication, such swellings were considered normal outgrowths.

Several contributions were made to the field of rhizobiology by Eriksson, who published a study of *V. faba* nodules in 1873. This was the earliest account of X- and Y-shaped bacteroids within mature nodule cells. Their significance in nitrogen fixation was not recognized until much later. Hyphallike strands, which Eriksson interpreted as fungi and the cause of nodule formation, were seen in young nodule cells. These hyphae were undoubtedly infection threads. In the field of histology Eriksson showed that division and differentiation of root cortical cells gave rise to nodule tissues.

In a comprehensive study published in 1887, Ward presented evidence that nodule formation was a result of infection. Using *V. faba* as experimental test plants, Ward found that plants grown in burnt soil were devoid of nodules. However, nodules always formed on plants grown in sterile soil after it had been watered with washings of common garden soil. Also, nodulation invariably resulted when pieces of old tubercles were placed among the roots. Furthermore, Ward recognized *Vicia* root hairs as the site of infection, associated the entry of the infective agent with a refractive hyaline spot at the tip or side of the hair, and described the deformation of the hair, or the shepherd's crook as it is known today, and the progression of the infective agent down the hair into the root cortex. Presumably, this progressive filament was an infection thread. The excellence of Ward's paper is marred only by his misconception of a fungus as the causal agent.

On May 25, 1887, Beijerinck confirmed Eriksson's observation of X- and Y-shaped bacteroids of *V. faba* nodules as evinced by crude pencil drawings and a handwritten entry in his notebook (Iterson et al. 1940). On the same page Beijerinck recorded on May 31 that on May 26 "a small quantity of a ground-up nodule of *V. faba* was sown on a solid culture medium made by adding gelatine to a decoction of roots of the same plant." This entry "was the beginning of an enormous amount of experimental work leading to the isolation of *Bacillus radicicola* and to the experimental proof that this bacterium — or closely related varieties and species — is responsible for the formation of the nodules on the roots of *Leguminosae* in general" (Iterson et al. 1940).

Six months later at a meeting of the Koninklijke Akademie van Wetenschappen, Amsterdam, Beijerinck announced the pure culture isolation of the root-nodule-forming organism, described its principal characteristics, and proposed the name *Bacillus radicicola*.

In his paper of 1888, Beijerinck made two more original contributions to the field of rhizobial studies. One was the observation of swarmers and bacteroid shapes of rhizobia from *V. faba* and *V. hirsuta* (L.) S. F Gray grown on a *V. faba* stem-extract gelatin medium. These unusual morphological forms had been observed previously only in nodule smears or histological sections. The other was Beijerinck's terminology of the root nodule bacteria from *V. faba* as var. *Fabae*, 1 of 7 varieties of *B. radicicola*. This was the precursor of species differentiation of the root nodule bacteria.

Beijerinck continued his groundbreaking work when he described star-shaped bacteroids of *V. faba* rhizobia in cultures grown on a *V. faba* stem-extract, peptone, cane sugar medium (Beijerinck 1890). These abnormal formations were interpreted to be indicative of a kinship of *B. radicicola* with actinomycetes. Today these star-shaped forms are construed as a taxonomic characteristic linking rhizobia and agrobacteria. Also, that paper presented for the first time direct experimental proof of the nodulating ability of a pure culture of *B. radicicola*. Aseptically cultivated seedlings of *V. faba* served as the test plant. Differences between root nodule bacteria from *Ornithopus sativa* and *V. faba* were recognized to be more striking than were formerly suspected. In conclusion, Beijerinck used the species designation of *Bacillus fabae* in contradistinction to *Bacillus ornithopi*.

In 1895 Nobbe et al. reported that rhizobia from *V. sativa*, *V. faba*, and *Pisum sativum* were mutually interchangeable, but rhizobia from garden pea, *Pisum sativum*, did not nodulate *Anthyllis*, *Medicago*, *Trifolium*, *Robinia*, or *Ornithopus* species. This was the genesis of the cross-inoculation concept, and particularly of the pea cross-inoculation group.

Duggar's field trial experiments of 1897 in Alabama with hairy vetch, *V. villosa* Roth, pioneered rhizobia-seed inoculation practices in the United States. Two plots were sown with seed that had been dipped in an aqueous suspension of soil from a garden where vetch had been grown; the seed sown on 2 other plots was untreated. The contrasting yields were impressive. The green forage yield of the treated plants was over 10 times that of the untreated ones; the increase in hay was 995 percent.

In 1937 Mothes and Pietz described chemically the red pigment in effective nodules of *V. faba*. This pigment, now known as leghemoglobin, had been accepted passively for many decades as a macroscopic difference between effective and ineffective nodules of various leguminous species. The extraction and characterization of this hemoprotein stimulated a revival of biochemical studies of symbiotic nitrogen fixation.

Vicia rhizobia, as members of the species *Rhizobium leguminosarum*, form nodules on the roots of *Lathyrus*, *Pisum*, and *Lens* species, and vice versa. Growth on mannitol agar is rapid, with a tendency to spread, colonies are raised, semitranslucent, and white, and consistency is slimy to viscous. Acid reactions are produced in carbohydrate media, and in litmus milk with the formation of a conspicuous serum zone. *Vicia*-type nodules have apical meristems, are elongated, oval to multibranched, and may occur in coralloid clusters.

Nodulated species of *Vicia*	Area	Reported by
acutifolia Ell.	Fla., U.S.A.	Carroll (1934a)
alba Moench	N.C., U.S.A.	Shunk (1921)
americana Muhl.	Widely reported	Many investigators
var. *linearis* S. Wats.	N.Dak., U.S.A.	Bolley (1893)
	Kans., Neb., U.S.A.	M. D. Atkins (pers. comm. 1950)
amoena Fisch.	Ariz., U.S.A.	Martin (1948)
amphicarpa [?]	Wis., U.S.A.	Burton et al. (1974)
angustifolia L.	Widely reported	Many investigators
var. *segetalis* (Thuill.) K. Koch	Italy	Pichi (1888)
articulata Hornem.	Australia	Riceman & Powrie (1952)
	S. Africa	Grobbelaar & Clarke (1974)
astragalus S. F. Gray	S. Africa	Grobbelaar & Clarke (1974)
baborensis Bat. & Trab.	S. Africa	N. Grobbelaar (pers. comm. 1963)

Nodulated species of *Vicia* (cont.)	Area	Reported by
benghalensis (= *atropurpurea* Desf.)	Widely reported	Many investigators
biennis L.	S. Africa	Grobbelaar & Clarke (1974)
	Wis., U.S.A.	Burton et al. (1974)
bithynica L. (*calcarata* Desf.)	Europe	Treviranus (1853)
	Australia	Riceman & Powrie (1952)
= *monantha* Retz.	Libya	B. C. Park (pers. comm. 1960)
caroliniana Walt.	Wis., U.S.A.	Bushnell & Sarles (1937)
	N.C., U.S.A.	Shunk (1921)
cassubica L.	U.S.S.R.	Pärsim (1967)
cordata Wulf.	Arctic U.S.S.R.	Roizin (1959); Pers. observ. (U.W. herb. spec. 1971)
cracca L.	Widely reported	Many investigators
var. *japonica* Miq.	Japan	Asai (1944)
dasycarpa Ten.	Widely reported	Many investigators
disperma DC.	Widely reported	Many investigators
dumetorum L.	S. Africa	Grobbelaar et al. (1967)
ervilia (L.) Willd.	Ala., U.S.A.	Duggar (1935)
	Arctic U.S.S.R.	Roizin (1959)
exigua Nutt.	Ariz., U.S.A.	Martin (1948)
faba L.	Widely reported	Many investigators
var. *equina*	Australia	Riceman & Powrie (1952)
ferruginea Boiss.	S. Africa	N. Grobbelaar (pers. comm. 1963)
floridana S. Wats.	Fla., U.S.A.	Carroll (1934a)
	N.Y., U.S.A.	Wilson (1939a)
fulgens Batt.	U.S.S.R.	Lopatina (1931)
graminea Sm.	S. Africa	N. Grobbelaar (pers. comm. 1963)
	Argentina	Rothschild (1970)
grandiflora Scop.	S. Africa	N. Grobbelaar (pers. comm. 1963)
	Wis., U.S.A.	Burton et al. (1974)
hirsuta (L.) S. F. Gray	Widely reported	Many investigators
hybrida L.	Wis., U.S.A.	Burton et al. (1974)
hyrcanica Fisch. & Mey.	Yugoslavia	Sarić (1964)
lathyroides L.	Germany	Treviranus (1853)
leavenworthii T. & G.	Tex., U.S.A.	Pers. observ. (U.W. herb. spec. 1971)
ludoviciana Nutt.	U.S.A.	Leonard & Dodson (1933)
	Arctic U.S.S.R.	Roizin (1959)
lutea L.	Widely reported	Many investigators
ssp. *vestita* (Boiss.) Rouy	S. Africa	Grobbelaar & Clarke (1974)
macrocarpa Bert.	Portugal	D'Oliveira & Loureiro (1941)

Nodulated species of *Vicia* (cont.)	Area	Reported by
melanops Sibth. & Sm.	S. Africa	U. of Pretoria (1957)
micrantha Nutt.	Wis., U.S.A.	Bushnell & Sarles (1937)
	Tex., U.S.A.	Pers. observ. (U.W. herb. spec. 1971)
monantha Retz.	U.S.A.	McKee et al. (1931); Duggar (1935)
(= *calcarata* Desf.)	Arctic U.S.S.R.	Roizin (1959)
narbonensis L.	Germany	Morck (1891)
	Israel	Hely & Ofer (1972)
nipponica Matsum.	Japan	Ishizawa (1953)
noëana Boiss.	Ariz., U.S.A.	Martin (1948)
obscura Vog.	Brazil	Galli (1958)
palaestina Boiss.	Israel	Hely & Ofer (1972)
pannonica Crantz	Widely reported	Many investigators
peregrina L.	Yugoslavia	Sarić (1963)
	Israel	Hely & Ofer (1972)
pisiformis L.	U.S.S.R.	Rasumowskaja (1932)
pubescens (DC.) Link	S. Africa	Grobbelaar & Clarke (1974)
sativa L.	Widely reported	Many investigators
var. *angustifolia* Ser.	S. Africa	Grobbelaar & Clarke (1974)
sepium L.	Widely reported	Many investigators
sicula Guss.	U.S.A.	Lipman & Fowler (1915)
silvatica L.	Arctic U.S.S.R.	Roizin (1959)
	Poland	Zimny (1962)
tenuifolia Roth	Wis., U.S.A.	Bushnell & Sarles (1937)
tenuissima (Bieb.) Schinz & Thell.	S. Africa	Grobbelaar & Clarke (1974)
tetrasperma (L.) Schreb.	Japan	Asai (1944)
	U.S.A.	Wilson (1944b)
(= *Ervum tetraspermum* L.)	Poland	Zimny (1962)
unijuga A. Br.	Japan	Asai (1944)
vexillata Benth.	Japan	Asai (1944)
villosa Roth	Widely reported	Many investigators
var. *gore*	U.S.A.	Wilson (1939a)
ssp. *varia* (Host) Corb.	S. Africa	Grobbelaar & Clarke (1974)

Vigna Savi   Papilionoideae: Phaseoleae, Phaseolinae
(427) / 3905

Named for Dominico Vigna, Professor of Botany at Pisa during the first half of the 17th century.

Type: *V. luteola* (Jacq.) Benth. (= *V. repens* (L.) Ktze.)

100–150 species. Herbs, annual or perennial, prostrate or twining, rarely erect. Leaves pinnately 3-foliolate; stipels and stipules present. Flowers yellow, white, pink, violet, or blotched, fasicled-racemose on elongated rachis near the apex of axillary, often villous, peduncles; bracts and bracteoles small, caducous or persistent; calyx campanulate, sometimes bristly; lobes lanceolate, upper 2 often connate; standard orbicular or broad-ovate, with inflexed, basal ears; wings shorter than the standard, falcate, eared, with a linear claw; keel subequal to the wings, with an incurved but not a spirally coiled beak; stamens diadelphous; anthers uniform; ovary sessile, several- to many-ovuled, often surrounded by a villous, cupular disk; style linear, often thickened above, bearded lengthwise along the inner side, usually ending beyond the stigma in a distinct beak; stigma oblique, subterminal or introrsely lateral, seldom terminal. Pod linear, cylindric, straight or curved, 2-valved, filled with spongy tissue between small, somewhat square or kidney-shaped seeds. (Bentham and Hooker 1865; Phillips 1951; Wilczek 1954; Hutchinson and Dalziel 1958)

The original generic number 426 (Taubert 1894) was changed to 427 by Harms (1908a) because of the reorganization within the subtribe necessitated by various revisions and inclusions of new genera.

The genus *Vigna* lacks taxonomic simplicity. Old World species were sometimes included in the genus *Dolichos* (Verdcourt 1968a). Similarites of *Vigna* and *Phaseolus* species have been a source of confusion for decades (Verdcourt 1970d). Certain taxonomic and nomenclatural problems have their origin in historical ambiguities. The ability of major members to cross-fertilize readily and produce fertile hybrids causes other taxonomic complications. The morphological differences between many species of these 2 closely related genera are quite insignificant and as Gillett (1966a) noted, no botanist has yet drawn a satisfactory line of distinction between them. Within the past decade, various small-seeded Asian *Phaseolus* species of the mung group have been transferred to the genus *Vigna* (Ohwi and Ohashi 1969; Verdcourt 1969).

Members of this large genus are native to tropical and subtropical regions of both hemispheres. The plants grow both wild and cultivated, and are adapted to a wide variety of soil types. The species prefer well-drained areas and often are chosen for low-fertility sites.

Vigna species serve as pot herbs, vegetables, hay, silage, pasture, green manure, soil cover, and soil-improvers. The fresh or dried seeds have been a popular and important source of protein in man's diet for centuries. The cowpea, or black-eyed pea, *V. unguiculata* (L.) Walp. (= *V. sinensis* (L.) Endl. ex Hassk.) has long been heralded as the most economically important member of the genus. It was the "phaseolus" of Pliny, Columella, and other Roman writers (Piper 1924a). Before the discovery of America it was the bean most commonly cultivated in the Old World for human food. In the last decade, the prestige of this genus as a reservoir of vegetable protein has been augmented immeasurably by the incorporation into *Vigna* of such important small-seeded Asian pulses as gram, mung, rice, adzuki, tepary, urd, and mat or moth beans, all formerly *Phaseolus* species (see *Phaseolus*). Good harvests from these crops often mean the difference between protein sufficiency and deficiency in India and eastern Asia. Several African species have large, spindle-shaped, edible root tubers. Those of *V. vexillata* (L.) A. Rich. in Ethiopia, *V. pseudotriloba* Harms in Namibia, and *V. stenophylla* Harms = *V.*

stenophylla (Harv.) Burtt Davy in Cameroon (Harms 1911) are supplemental in native diets as a kind of yam or potato substitute.

Three cultivated forms of cowpea, the catjany or Hindu cowpea, the asparagus or yard-long bean, and the black-eyed cowpea, have been prominent in agricultural history. Differences between these forms are primarily in shape and length of their pods and seed characteristics. These forms were previously regarded as separate species, namely, *V. catjang* Walp., *V. sesquipedalis* (L.) Fruhw., and *V. sinensis*, respectively; however, all of them cross-fertilize readily, and numerous intergradations are recognized. Currently, *V. sesquipedalis* and *V. sinensis* are considered synonymous with *V. unguiculata*, the wild prototype, indigenous to tropical Africa. Purseglove's suggestion (1968), that the cultivated cowpea be designated *V. sinensis* and that the name *V. unguiculata* be reserved only for the wild form has considerable merit.

The history of the cowpea and its dissemination has many versions; however, it is generally agreed that *V. unguiculata* originated in tropical Africa and underwent cultivation at a very early time following its dispersal. It presumably entered the United States by way of the West Indies in the 18th century. Carrier (1923) believed that it might have been introduced accidentally by some European fisherman or exploring expedition. Wight (1907) credited Roman's *Natural History of East and West Florida*, published in 1775, with the first unmistakable mention of the white black-eyed pea in the United States. *V. unguiculata* was introduced into Jamaica between 1672 and 1687 and reached the Potomac between 1790 and 1795. Washington reputedly wrote of the cowpea in a letter dated 1797. The first use of the term "cowpea" is attributed to Thomas Jefferson.

Nodulation Studies

Wild and cultivated cowpea plants are commonly found naturally nodulated in diverse soils and geographical areas; experimentally, it is one of the plants most susceptible to inoculation and, in turn, it is one of the most responsive to nitrogen fixation; an ineffective host response is exceptional. General acceptance of the cowpea, as an indicator or test plant in the screening of rhizobia to determine their plant group affiliations, resulted from their successful use in cross-inoculation tests by Garman and Didlake (1914).

Today the cowpea group, or miscellany, is a large, versatile collection of tropical and subtropical species within which slow-growing nodule-forming bacteria, known as cowpea or cowpea-type rhizobia, are mutually interchangeable.

Morphological, cytological, and phytochemical evidence strongly supports the segregation and transfer of the tropical Asian mung group from *Phaseolus* to *Vigna*. An explanation for the inability of these small-seeded, tropical species to symbiose with strains of *Rhizobium phaseoli* from nodules of the large-seeded European *Phaseolus* species has long been an enigma (Richmond 1926a; Raju 1936; Barua and Bhaduri 1967). Designation of these tropical forms as *Vigna* species is compatible with their identity as members of the cowpea miscellany and is highly consistent with symbiotaxonomic evidence.

Cowpea rhizobia closely resemble those of lupine and soybean as slow-growing forms, fastidious in their carbon and nitrogen requirements, producing an alkaline reaction in litmus milk without serum zone formation, and in overlapping infection affinities with *Glycine* and *Lupinus* species. Fusion of cowpea-type rhizobia into *Rhizobium japonicum* was proposed by Walker and Brown (1935). Data from computer techniques provided the background for Graham's reasoning (1964) that all slow-growing rhizobia, i.e., the *R. lupini, R. japonicum*, and cowpea complex, merited the generic distinction *Phytomyxa*. Later results by DeLey and Rassel (1965) were in general agreement that the slow-growing, subpolar flagellated rhizobia with high guanine + cytosine content, 62.8–65.5 percent, constitute 1 genetic group. Despite the pros and cons of these various proposals, cowpea rhizobia still await species designation.

Norris' evolutionary concept of rhizobia:plant symbiosis (1958a, 1965) has relevance to the cowpea miscellany. The main points advanced by Norris were (1) rhizobia symbiosis arose in acid tropical soils under wet conditions and low availability of essential nutrients; (2) ancient tropical leguminous plants were symbiotically promiscuous; (3) promiscuously infective cowpea-type rhizobia were common to all tropical species and, therefore, were the ancestral symbiont; and (4) evolution and dissemination of leguminous plants from tropical to temperate environments resulted in symbiotic specialization. As a consequence, this specialization is reflected by rhizobial strain selection and host specificities culminating in rhizobia:plant groupings. Parker (1968), on the other hand, argued that legumes evolved on calcareous soils in a subhumid environment, and that ancestral rhizobia were peritrichously flagellated organisms similar to *R. meliloti* strains. Most of the evidence, however, does not support the latter position. The 3 entirely temperate tribes, Loteae, Trifolieae, and Vicieae, all with calcicolic habit, are regarded unanimously by botanists as being most recently evolved. Each has a narrow, highly specialized spectrum of symbiotic associations. Altogether they comprise only about 1,000 species in 20 genera, less than 5 percent of the total number of leguminous species.

Nodulated species of *Vigna*	Area	Reported by
†*acontifolia* (Jacq.) Maréchal (= *Phaseolus acontifolius* Jacq.)	Widely reported	Many investigators
†*angularis* (Willd.) Ohwi & Ohashi (= *Phaseolus angularis*)	Widely reported	Many investigators
angustifoliolata Verdc.	S. Africa	Grobbelaar & Clarke (1975)
capensis Walp.	S. Africa	N. Grobbelaar (pers. comm. 1961)
coerula Bak.	S. Africa	N. Grobbelaar (pers. comm. 1966)
davyi Bolus	S. Africa	Grobbelaar & Clarke (1975)
decipiens Harv.	S. Africa	Grobbelaar & Clarke (1975)
desmodioides Wilczek	Zaire	Bonnier (1960)
frutescens A. Rich.	Zimbabwe	Corby (1974)
gazensis Bak. f.	Zimbabwe	Corby (1974)
gracilis Hook. f.	Kenya	Whyte et al. (1953)

Nodulated species of *Vigna* (cont.)	Area	Reported by
hispida E. Mey.	S. Africa	Grobbelaar et al. (1964)
hosei (Craib) Backer	Taiwan	Pers. observ. (1962)
(= *oligosperma* Backer)	Sri Lanka	Holland (1924)
juncea Milne-Redhead		
var. *major* Milne-Redhead	Zimbabwe	Corby (1974)
khandalensis (Sant.) Rag. & Wad.	W. India	V. Schinde (pers. comm. 1977)
lanceolata Benth.	Queensland, Australia	Bowen (1956)
	N.S. Wales, Australia	Beadle (1964)
†*longifolia* (Benth.) Verdc.		
(= *Phaseolus trichocarpus* C. Wright)	Puerto Rico	Dubey et al. (1972)
luteola (Jacq.) Benth.	Trinidad	DeSouza (1966)
	Zimbabwe	Corby (1974)
macrorhyncha (Harms) Milne-Redhead	Zimbabwe	Corby (1974)
marina (Burm.) Merr.	Java	Keuchenius (1924)
	Hawaii, U.S.A.	Allen & Allen (1936a)
	India	W. R. C. Paul (pers. comm. 1951)
monophylla Taub.	Zimbabwe	Corby (1974)
†*mungo* (L.) Hepper	Widely reported	Many investigators
(= *Phaseolus mungo* L.)		
(= *P. viridissimus* Ten.)		
nervosa Markött.	Zimbabwe	Corby (1974)
(= *galpinii* Burtt Davy)	S. Africa	N. Grobbelaar (pers. comm. 1965)
nuda N. E. Br.	Zimbabwe	Corby (1974)
oblongifolia A. Rich.		
var. *oblongifolia*	Zimbabwe	Corby (1974)
var. *parviflora* (Bak.) Verdc.	Zimbabwe	Corby (1974)
owahuensis Vog.	Hawaii, U.S.A.	Allen & Allen (1936a)
pilosa (Willd.) Bak.	Java	Keuchenius (1924)
pygmaea R. E. Fries	Zimbabwe	Corby (1974)
racemosa Hutch. & Dalz.	W. Africa	Dalziel (1948)
†*radiata* (L.) Wilczek	Widely reported	Many investigators
(= *Phaseolus radiatus* L.)		
(= *P. aureus* Roxb.)		
repens Bak.	U.S.A.	Wilson (1944b)
reticulata Hook. f.	Zimbabwe	Corby (1974)
retusa Walp.	U.S.A.	Wilson (1939c)
sandwicensis Gray	Hawaii, U.S.A.	Allen & Allen (1936a)
stenoloba (Harv.) Burtt Davy	S. Africa	N. Grobbelaar (pers. comm. 1968)
stenophylla (Harv.) Burtt Davy	S. Africa	Grobbelaar et al. (1967)
var. *lata*	S. Africa	N. Grobbelaar (pers. comm. 1971)
(*strobilophora* Robins.)	France	Naudin (1897a)
= *Ramirezella strobilophora* Rose		
†*trilobata* (L.) Verdc.		
(= *Phaseolus trilobatus* (L.) Schreb.	Japan	Asai (1944)

Nodulated species of *Vigna* (cont.)	Area	Reported by
(=*P. trilobus* Ait.)	India	Raju (1936)
†*umbellata* (Thunb.) Ohwi & Ohashi (= *Phaseolus calcaratus* Roxb.)	Widely reported	Many investigators
unguiculata (L.) Walp. (= *sinensis* (L.) Endl. ex Hassk.) ssp. *cylindrica* (L.) van Eseltine	Widely reported	Many investigators
(= *catjang* Walp.)	India	Joshi (1920); Raju (1936)
(= *cylindrica* (L.) Skeels)	S. Africa	N. Grobbelaar (pers. comm. 1966)
ssp. *dekindtiana* (Harms) Verdc.	Zimbabwe	Corby (1974)
	S. Africa	Grobbelaar & Clarke (1975)
ssp. *sesquipedalis* (L.) Verdc. (= *sesquipedalis* (L.) Fruhw.)	Widely reported	Many investigators
vexillata (L.) A. Rich.	U.S.A.	Wilson (1939a)
	Australia	McLeod (1962)
	S. Africa	B. W. Strijdom (pers. comm. 1970)
var. *angustifolia* (Schumach.) Bak.	Zimbabwe	Corby (1974)
var. *vexillata*	Zimbabwe	Corby (1974)
wilmsii Burtt Davy	S. Africa	Grobbelaar et al. (1964)

†Species heretofore reported nodulated as *Phaseolus* spp., now classified as *Vigna* spp.

Viminaria Sm.
172 / 3632

Papilionoideae: Podalyrieae

From Latin, *vimen*, "twig," in reference to the leafless, long, twiglike branches.

Type: *V. juncea* (Schrad.) Hoffmgg. (= *V. denudata* Sm.)

1 species. Shrubs, glabrous, branches erect, whip- or rushlike, up to 3–7 m tall. Leaves rare, alternate if present, usually reduced to filiform petioles; stipules small, scarious. Flowers small, attractive, in terminal racemes often 20 cm long; bracteoles absent; calyx lobes 5, almost equal, short, free; petals long-clawed; standard orange yellow, orbicular, usually twice as long as the oblong wings; keel blunt, curved, red, about equal to wings; stamens free; ovary subsessile, 2-ovuled; style filiform; stigma small, terminal. Pod small, black, sessile, ovoid-oblong, obliquely beaked with remnant of style, indehiscent, usually containing 1 red, strophiolate seed. (Bentham 1864; Bentham and Hooker 1865)

This monotypic genus is endemic in temperate Australia. The plants thrive in the coastal areas and sandy soils of New South Wales, Queensland, Victoria, and South and Western Australia.

NODULATION STUDIES

A rhizobial strain listed as *V. denudata* Sm. was maintained in the type culture collection by the Commonwealth Scientific and Industrial Research Organization in Adelaide in 1953 (Harris 1953). A pure culture isolated from this species in southwestern Australia by Lange (1961) had cultural characteristics akin to the slow-growing group of rhizobia. In greenhouse tests it produced nodules on *Glycine hispida*, *Phaseolus lathyroides*, and *Vigna sinensis*, and displayed selectivity on *Lupinus* species by its infection of *L. albus* and *L. digitatus*, but not of *L. pilosus*, *L. subcarnosus*, or *L. villosus*. In a later study, *Viminaria* seedlings were not nodulated by rhizobial strains from *Lupinus digitatus* and *L. angustifolius* (Lange 1962); therefore, the infection was non-reciprocal.

Electron micrographs of the bacteroid zone of effective nodules of *V. juncea* (Schrad.) Hoffmgg. by Dart and Mercer (1966) revealed fine structural organization of the host cells similar to that established for the cowpea-soybean miscellany. The rhizobia retained their rod form; as many as 8 elongated rods were enclosed within each membrane envelope.

Virgilia Poir. Papilionoideae: Sophoreae
148 / 3608

Named for the poet Virgil, 70–19 B.C.

Type: *V. capensis* (L.) Poir. (= *V. oroboides* (Berg.) Salt.)

2 species. Trees, fast-growing, evergreen, medium-sized; twigs densely pubescent. Leaves imparipinnate; leaflets small, linear-lanceolate, 3–10 pairs, leathery; stipels absent; stipules narrow, linear, acuminate, caducous. Flowers pink to rose-mauve, showy, many, in short, terminal racemes; bracts broad, very caducous; bracteoles none; calyx widely campanulate, intruse at the base, silky-haired, bilabiate, upper lip bidentate, longer than the tube, lower lip tridentate, longer than the tube; all petals long-clawed; standard strongly recurved, orbicular; wings oblong, falcate; keel incurved, beaked, shorter than the wings, petals joined dorsally; stamens 10, free, glandular tissue at their base exudes aromatic nectar; filaments linear, narrow above, villous; anthers small, linear, versatile; ovary sessile or short-stalked, densely villous, few-ovuled; style curved, smooth, incurved, awl-shaped; stigma small, terminal. Pod linear-oblong, plano-compressed, leathery, densely tomentose, with thick margins, 2-valved, straw-colored at maturity; seeds black, ovoid-flattened. (Bentham and Hooker 1865)

Both species are indigenous to South Africa and are confined to a narrow coastal strip from van Staaden's Pass, near Port Elizabeth, to the Cape Peninsula (J.F.V. Phillips 1926; Palmer and Pitman 1961). The plants flourish in humid forest margins and on hillsides up to about 1,500-m altitude, on acid sandstone soil and clay loams.

V. divaricata Adamson is the second species; however, some botanists doubt that it is distinct from *V. capensis* L. (Poir), since the differences are primarily in flowering habit. *V. divaricata*, of the Keurbooms River area, bears a profusion of deep pink, pealike blossoms in the spring and is luxuriant in both foliage and flowers, whereas the flowers of *V. capensis*, native to the Cape Peninsula, are mauve pink, and borne intermittently from early summer to autumn, and both foliage and flowers are less abundant (Palmer and Pitman, 1961). Both species are fast growing and bear the Afrikaans common name "keurboom," meaning "choice tree." The keurboom reaches a height of 12–16 m, with a clear bole up to 5–8 m and a girth of 70–150 cm. The fragrant flowers are commonly pollinated by sunbirds. The seeds reputedly retain vitality up to 30 years, germinate profusely in burned-over areas, and are disseminated by streams, heavy rains, and wild animals in browsed foliage. Young seedlings transplant readily, but the shallow, surface-spreading roots provide poor anchorage. *V. divaricata* is usually preferred for ornamental purposes because it has a bush-type growth, low branches, a height seldom over 9 m, and also a showy profusion of sweetly scented flowers.

Keurboom bark emits a transparent gum which has chemical similarities to peach and cherry gums (Stephen 1957). Bush women once used the gum as a starch substitute. The foliage and seeds contain the following quinolizidine alkaloids: cytisine, $C_{11}H_{14}N_2O$, virgiline, $C_{15}H_{24}N_2O_2$, lupanine, $C_{15}H_{24}N_2O$, and oroboidine, $C_{20}H_{27}N_3O_3$ (see Harborne et al. 1971).

Little use is made of the light, soft wood other than for yokes, wagon planks, and fuel.

NODULATION STUDIES

In 1926 J. F. V. Phillips wrote of the isolation of rhizobia from nodules of *V. capensis* on wood-ashes-maltose agar, their dissimilarity from the microsymbiont of *Podocarpus thunbergii* and *P. elongata*, and their ability to improve the growth of host seedlings. Wilson (1939a) reported nodulation of this species by rhizobia from *Lespedeza sericea*, *Lupinus perennis*, *Sesbania macrocarpa*, and *Swainsona coronillaefolia*, and in a later report (Wilson 1944b) cited nodulation of *Amorpha fruticosa* and *Wisteria frutescens*, but not *Crotalaria sagittalis*, by isolates from *V. capensis*.

Voandzeia Thou.   Papilionoideae: Phaseoleae, Phaseolinae
(429) / 3903

Name derived from Malagasy, *voandzou*, meaning a seed or grain that satisfies well.

Type: *V. subterranea* Thou.

1 species. Herb, annual, creeping, prostrate, much-branched; stems bunched or trailing, rooting at the nodes. Leaves pinnately 3-foliolate, long-petioled; leaflets lanceolate-elliptic, narrowed at the base, entire, glabrous; terminal leaflet large, subtended by 2 small, persistent stipels, lateral leaflets by 1; petiole erect, grooved, thickened at the base; petiolules short, pulvinules hairy; stipules striate, persistent. Flowers small, 1–3, yellowish, axillary, on short, thick, hairy peduncles arising from stem nodes at the ground level and having a swollen glandular apex that pushes into the soil; bracts small; bracteoles 2, small, hairy, membranous, more or less persistent; calyx campanulate, glabrous, bilabiate; lobes 5, broad, upper 2 connate into 1 emarginate lip, lower lip 3-lobed; standard notched, orbicular, with 2 small, inflexed, basal ears; wings oblanceolate, falcate; keel petals subequal to the wings, slightly incurved, obtuse; stamens diadelphous; anthers uniform; ovary small, subses-

sile, few-ovuled, usually 2, rarely 3 ovules; style incurved, long-bearded above; stigma introrsely lateral, oblong. Pod irregularly subglobose, semiorbicular to semioval, 2-valved, 1–2-seeded, brown, hard and wrinkled when mature, ripening underground by the geotropic lengthening of the peduncle, indehiscent; seeds subglobose, smooth, cream, brown, or black, mottled or blotched, without aril. (Bentham and Hooker 1865; Wilczek 1954; Hutchinson and Dalziel 1958; Backer and Bakhuizen van den Brink 1963; Doku and Karikari 1971)

The occurrence of normal bisexual flowers and also of apetalous fertile female ones by this species is questioned by Backer and Bakhuizen van den Brink (1963), who interpret the latter as being old, normal flowers that have lost corolla and stamens.

The original generic number 425 (Taubert 1894) was changed to 429 by Harms (1908a) because of the reorganization within the subtribe necessitated by various revisions and inclusions of new genera.

This monotypic genus is endemic in West Africa. The plants are uncommon in the wild state. They are now widely distributed throughout tropical and subtropical Africa and have been introduced and cultivated in Malaysia, Java, and South America. *Voandzeia* plants are best suited for poor, well-drained, sandy soils with a pH range of 5–6.5. They require high temperature and ample sunshine for maximum growth.

The widespread cultivation of this indigenous pulse on the African continent accounts for its many common names, such as Bambara bean, Congo goober, voandzou, Madagascar peanut, juga bean, and earth pea. It probably had its origin in the vicinity of Bambara, a city in Chad; however, it occurs wild in various other areas in Africa (Rassel 1960; Doku and Karikari 1971). Cultivation of this pest- and disease-resistant legume in Africa preceded the introduction of the peanut. Its dissemination into other areas of the world paralleled the great slave-trade period. Today Bambara groundnut ranks behind peanut and cowpea as the third most important leguminous crop in Africa in production and consumption.

The Bambara groundnut closely resembles the peanut in growth habit, but it produces less foliage as green manure. The seeds also have a lower content of calcium, phosphoric acid, and oil; however, the potash content is higher. Its agronomic value is one of either an intercycle or solo crop. Many ecotypes and cultivars have been developed as a result of hybridization and natural and artificial selection.

The nuts have a high nutritive value, and there is no record of toxicity resulting from their consumption (Stanton 1966). They are eaten raw, grilled, boiled, or roasted; they are also pounded into a flour or meal that is used in soups or cakes. The leafy foliage is also palatable to stock.

NODULATION STUDIES

Nodulation of *V. subterranea* Thou. has been recorded in various geographical areas: France (Naudin 1897a), the United States (Sornay 1916), Trinidad (DeSouza 1963), Africa (Stanton 1966), and Zimbabwe (Sandmann 1970b). The assignment of this species to the cowpea miscellany relative to rhizobial association is justified, inasmuch as *Voandzeia* is closely related to *Vigna* (Harms 1908a).

From Galibi, *wákapu*, the Caribbean Indian name for angelim.

Type: *V. americana* Aubl.

4 species. Trees, medium to tall, slender, unarmed; young branches rusty-haired. Leaves imparipinnate; leaflets ovate-elongated, 2–4 pairs, acuminate, smooth, large, leathery; petiole glandular; stipules caducous. Flowers small, yellow, in erect, terminal, many-flowered, panicled racemes; calyx half-cylindric or reversed cone-shaped; lobes 5, equal, imbricate; petals 5, equal, imbricate, sparsely hairy, longer than the calyx lobes, subspathulate; stamens 10; filaments free, base dilated; anthers subsagittate, linear, with 2 longitudinally dehiscent compartments; ovary short-stalked, 1-ovuled; style short, slightly incurved; stigma small. Pod asymmetrically obovate, long-narrowed at the base, pericarp thick, warty or longitudinally grooved, dehiscent along only 1 margin, 1-seeded; seed ovate, smooth. (Harms 1915a; Amshoff 1939)

This genus has a taxonomic history closely interwoven with *Andira* (Taubert 1894; Harms 1915a; Camp et al. 1947); however, the genera are clearly distinctive on the basis of pod characteristics.

Representatives of this genus are common in South America, especially in Guiana and northern Brazil. They occur in forests on noninundated land, often on sandy-clay soils.

The wood of all species is similar; it is scarce and expensive, but in demand for fine cabinetwork, parquet flooring, and furniture. The wood of *V. americana* Aubl. is known as partridgewood in the United States timber trade and as bruinhart in Surinam. It is hard, heavy, coarse grained, and resistant to decay, and to termite and insect attack. The heartwood is an attractive reddish brown, with distinctive brown vessel markings (Record and Hess 1943). This species reaches its best development in Pará, Brazil, where it is a valued commercial wood (Kukachka 1970).

Wagatea Dalz.   Caesalpinioideae: Eucaesalpinieae
90 / 3546

Named for Jakob Waga, 1800–1872, Polish botanist.

Type: *W. spicata* Dalz.

1 species. Liana, robust, tall, climbing, or low shrub; branchlets and main rachis of leaves and inflorescences prickly. Leaves abruptly bipinnate; pinnae 4–6 pairs; leaflets oblong, 5–7 pairs, upper surface glabrous or thinly pubescent, lower side pale green, densely pubescent; stipules minute. Flowers borne on thick rachis in long, densely flowered, spikelike racemes or branched panicles; bracts awl-shaped, caducous; bracteoles absent; calyx scarlet, cupular-campanulate; tube grooved longitudinally at the base; lobes 5, hardly longer than the calyx tube, rounded, imbricate in bud, the lowest lobe outermost, longest, very concave; petals 5, orange, erect, oblong-spathulate, rounded, imbricate, uppermost petal inside and broader; stamens 10, free, slightly declinate, not exserted, base thickened and densely pubescent; anthers uniform, longitudinally dehiscent; ovary sessile, 4–6-ovuled; style

somewhat clavate at the apex; stigma oblique, bilobed, concave. Pod oblong-linear, acute, leathery, depressed transversely between the seeds, sutures thick, not winged, indehiscent; seeds 1–4, obovate-oblong. (Bentham and Hooker 1865; Backer and Bakhuizen van den Brink 1963)

This monotypic genus is endemic in southwestern India. It has been introduced into Java, where it is cultivated as an ornamental.

Wallaceodendron Koord.　　　　　　　Mimosoideae: Ingeae
(7a) / 3442

Named for Alfred Russel Wallace, 1823–1913, English naturalist and explorer.

Type: *W. celebicum* Koord.

1 species. Tree, unarmed, up to 40 m tall, erect, buttressed at the base. Leaves bipinnate; pinnae pairs few, each with 3–5 pairs of large leaflets; petiolar glands none; jugal glands large. Flowers large, in few-flowered, erect, axillary, paired racemes; bracts caducous; bracteoles none; calyx narrow-campanulate; lobes 5, minute, valvate; petals valvate in bud, connate when young, later 3–5, free, joined to base of the staminal tube; stamens many, exserted, all joined into a tube at the base; anthers minute, versatile, without glands; ovary sessile, many-ovuled; style filiform; stigma small. Pod oblong, compressed, thick, leathery, with sutures thickened, continuous, persistent, pericarp nonseptate, 2-valved, endocarp distinct and free from pericarp, transversely segmented into 1-seeded joints; each oblong, compressed seed surrounded by an individual envelope, but not pulpy. (Harms 1900)

This monotypic genus is endemic in Celebes and the Philippines. The plants grow chiefly at low altitudes near seashores.

The timber is known in the Philippines as banúyo, or ironwood, and is sometimes marketed as Derham mahogany in the United States. The rather heavy, hard, red brown wood is used for ship cabins, interior home finishing, furniture, and musical instruments, but it is generally in short supply.

NODULATION STUDIES

Two large trees of *W. celebicum* Koord. examined in Hawaii were copiously nodulated (personal observation 1940).

Walpersia Harv.　　　　　　Papilionoideae: Genisteae, Lipariinae
187 / 3647

Named for W. G. Walpers, 1816–1853, German botanist and author of *Repertorium Botanices Systematicae*.

Type: *W. burtonioides* Harv.

1 species. Shrubs, small, up to 1 m tall; branches many, erect, villous.

Leaves spirally inserted, petioled, simple, linear, acute, margins reflexed, midrib prominent underneath; stipules absent. Flowers yellow, on slender, hairy pedicels in upper leaf axils; calyx campanulate, subtended by 2 leafy bracteoles; lobes 5, upper 2 broader than lower 3; petal claws adnate at the base to the short staminal tube; standard ovate, with a small callosity near the claw; wings oblong, eared; keel incurved, bluntly spurred or pouched at each side; stamens monadelphous, all connate into a short sheath split above; anthers alternately short and versatile, long and basifixed; ovary sessile, silky, 2-ovuled; style slender, long. Pod not known. (Bentham and Hooker 1865; Harvey 1894; Phillips 1951)

Dahlgren's (1965b) later examination of the original specimen sheet of this species in Sonder's Herbarium in Stockholm has raised considerable doubt about the validity of this genus.

This monotypic genus is endemic in South Africa. The plants occur primarily in the southwestern area of the Cape Province.

Weberbauerella Ulbrich   Papilionoideae: Hedysareae,
(323a)   Aeschynomeninae

Named for August Weberbauer, 1871–1948, Peruvian botanist.

Type: *W. brongniartioides* Ulbrich

2 species. Shrublets with dark brown, subterranean, long, ovate, tuberous rhizomes; branches, leaflets, calyx, and petals gland-dotted. Leaves imparipinnate; leaflets small, oval to elliptic, 17–20 pairs, much smaller toward the apex of the rachis; stipels none. Flowers yellow, streaked with brown, medium-large, several, in axillary racemes; bracts short, lanceolate; calyx obliquely campanulate, bilabiate, upper lip formed by 2 broad connate lobes, lower lip 3-lobed, lobes acutely lanceolate; standard short-clawed, nearly round, but creased at the center with margins folded inward; keel petals with triangular ears, short-clawed, joined near the base; stamens 10, monadelphous; anthers equal; ovary stalked, linear, hairy, transversely septate, several-ovuled; style slender; stigma punctiform. Pod not seen. (Ulbrich 1906)

The genus is positioned between *Ormocarpum* and *Aeschynomene* (Harms 1915a). *W. raimondiana* Ferreyra is the other species (Ferreyra 1951).

Both species are native to Peru. They flourish on sandy soil.

Wenderothia Schlechtd.   Papilionoideae: Phaseoleae,
                                                            Diocleinae

Named for Georg Wilhelm Franz Wenderoth, 1774–1861, German botanist.

Type: *W. discolor* Schlechtd.

12 species. Herbs, vines or shrublets, trailing or climbing, perennial, stems woody. Leaves pinnately 3-foliolate; leaflets papery to leathery, entire,

ovate, acuminate, usually pubescent; petioles grooved above; petiolules fleshy; stipules caducous, thin, not striate. Flowers numerous, in axillary, racemelike thyrses, each node along the rachis bearing 2–3 flowers; bracts minute, caducous; bracteoles orbicular, caducous; calyx leathery, tubular-campanulate, bilabiate, upper lip large, entire, lower lip small, tridentate; standard large, reflexed, emarginate, not eared, with 2 callosities near the base; wings free, narrow, eared; keel falcate, beaked, sometimes with tip coiled; stamens diadelphous, vexillary stamen more or less free; anthers uniform; ovary several-ovuled; style smooth, stigma terminal, capitate. Pod stalked, often pubescent, oblong or linear, straight or curved, compressed, each valve with 1–3 extra longitudinal ribs in addition to sutural ribs; seeds several, oblong, compressed or lens-shaped, with short hilum. (Piper 1925; Sauer 1964)

Piper (1925) regarded *Wenderothia* as a genus distinct from *Canavalia*; however, Sauer (1964) retained this group as a subgenus of *Canavalia*, an opinion now widely accepted.

Members of this genus occur in the tropics and subtropics of the New World.

None of the species have economic value, but several have ornamental possibilities.

Whitfordiodendron Elmer
(257c)

Papilionoideae: Galegeae,
Tephrosiinae

Named for Dr. H. N. Whitford, Chief of the Division of Investigation, Bureau of Forestry, the Philippines.

Type: *W. scandens* Elmer

9 species. Lianas, large, woody; rarely trees. Leaves imparipinnate; leaflets leathery, oblong-ovate, tips acuminate, recurved, basal pair usually smaller; stipels absent; stipules small, acute. Flowers red or purple, some with yellow markings, crowded or bunched in long, panicled racemes at the tips of branches; bracts and bracteoles broad, acuminate or narrow, spinelike; calyx short, 5-toothed; 2 upper lobes somewhat joined; petals silky-haired along the outside median line; standard orbicular, sometimes notched, short-clawed, reflexed; wings and keel petals similar, oblong, semisagittate; stamens 10, diadelphous; anthers uniform; ovary stalked, hairy, 1–3-ovuled; style incurved; stigma small, terminal. Pod large, subsessile, ovate, short-beaked, slightly compressed, 1-seeded, indehiscent; seed large. (Elmer 1910; Harms 1915a)

Some botanists regard *Whitfordiodendron* as one of the synonyms of *Millettia* (Hutchinson 1964), others consider it distinctive for its 1-seeded, indehiscent pods.

Representatives of this genus are indigenous to Burma, Malaysia, the Philippines, and Taiwan.

Seeds of *W. erianthum* Dunn in Malaysia (Burkill 1966) and *W. scandens*

Elmer in the Philippines (Elmer 1910) are boiled and eaten as a vegetable. The heavy wood of *W. pubescens* Burkill is used for house beams and rafters in Burma.

Wiborgia Thunb.

201 / 3661

Papilionoideae: Genisteae, Crotalariinae

Named for Eric Wiborg, Danish botanist.

Lectotype: *W. obcordata* Thunb.

10 species. Shrubs, rigid, slender, often many-branched, spinescent, glabrous or silky-pubescent. Leaves palmately 3-foliolate; leaflets linear, oblong, or obovate, sometimes mucronate, turning black when dry; stipules small or absent. Flowers yellow or cream, usually in dense, many-flowered, terminal or unilateral racemes; bracts and bracteoles small; calyx obliquely campanulate, subbilabiate, glabrous; lobes 5, shorter than calyx tube, subequal; petal claws long, slender; standard oblong or ovate, apart from other petals, sharply reflexed; wings shorter than standard and keel, oblong; keel beaked or incurved, longer than standard; stamens monadelphous, all joined briefly at the base; anthers unequal, alternately short and versatile, long and basifixed; ovary stalked, few-ovuled; stigma terminal, minutely capitate. Pod stalked, ovate, flat, winged along the upper suture, lower margin usually sharp and thin, valves sometimes ridged, indehiscent; seeds with filiform funicle. (Bentham and Hooker 1865; Phillips 1951; Dahlgren 1963b)

The genus is closely related to *Aspalathus* (Dahlgren 1963b).

Polhill (1976) assigned this genus to the tribe Crotalarieae in his reorganization of Genisteae *sensu lato*.

Representatives of this genus are indigenous to Cape Province, South Africa. They thrive on sandy flats and dry, rocky areas.

NODULATION STUDIES

W. armata Harv. and *W. obcordata* Thunb. are nodulated in their native habits in South Africa (Grobbelaar and Clarke 1972).

Willardia Rose

272 / 3737

Papilionoideae: Galegeae, Robiniinae

Named for Alexander Willard, United States Consul at Guaymas, Mexico, for 25 years.

Type: *W. mexicana* (S. Wats.) Rose

6 species. Shrubs or trees, small, unarmed. Leaves imparipinnate; leaflets numerous, entire; stipels none; stipules rudimentary. Flowers showy, lilac, in axillary racemes; calyx truncate, margin trim; teeth small, equal; petals sub-

equal; standard orbicular, spreading; wings falcate; keel slightly incurved; stamens 10, the vexillary stamen connate above with others into a tube and free only at the base; anthers uniform; ovary subsessile, several-ovuled; style inflexed, glabrous or sparsely hairy at the base; stigma capitate, minute. Pod linear-oblong, flat, nonseptate within, 2-valved, dehiscent; seeds kidney-shaped, flat. (Rose 1891; Hermann 1947b)

Members of this genus are often mistaken for *Lonchocarpus* species but are distinguished by having dehiscent pods.

Willardia species are distributed throughout Mexico and Central America.

W. mexicana (S. Wats.) Rose reaches a height of about 12 m in western Mexico. The wood, known as palo piojo or nesco by the mountain dwellers, is useful for fuel, props, and mine supports.

__Wisteria__ Nutt.   Papilionoideae: Galegeae, Tephrosiinae
(258) / 3722

Named for Caspar Wistar, 1761–1818, Professor of Anatomy, University of Pennsylvania, one of the early owners of the present site of Vernon Park, Philadelphia. The orthographic error *Wisteria* is conserved.

Type: *W. frutescens* (L.) Poir.

10 species. Shrubs, deciduous, twining, or tall-climbing, woody vines. Leaves imparipinnate; leaflets petioluled, alternate, oblong-elliptic, entire, 3–10 pairs, the apical leaf usually largest; stipels present; stipules small, caducous. Flowers blue, purplish, pink, or white, showy, large, pedicelled, in long, pendent, terminal racemes; bracts caducous; bracteoles none; calyx campanulate, upper lip bilobed, united almost to apex, lower lip shortly trilobed; lobes 5, lower 3 narrow and distinct but often longer than those above; petals subequal; standard large, roundish, reflexed, usually with 2 hornlike callosities at the base; wings oblong-falcate, doubly eared at the base; keel petals upwardly incurved, scythe-shaped, free, obtuse, eared at the base, coherent at the apex; stamens 10, diadelphous, 1 free or connate in the middle with the others; filaments not dilated; anthers uniform; ovary stalked, stalk surrounded by a collarlike, intrastaminal, glandular disk, many-ovuled; style inflexed, glabrous; stigma small, capitate, terminal. Pod stalked, elongated, torulose, slightly constricted between the seeds but not septate within, smooth or pubescent, 2-valved, valves leathery, tardily dehiscent; seeds several, large, kidney-shaped. (Bentham and Hooker 1865)

The presence of the intrastaminal disk, an anatomical feature unknown in Galegeae, and a single series instead of 2 series of vascular strands serving the perianth, stamens, and disk, have been offered as evidence that *Wisteria* merits positioning in Phaseoleae (Moore 1936).

Members of this genus are native to mild, temperate China, Japan, and eastern North America. They are usually found on low, moist, fertile alluvial soils, in woods, and along river banks.

The profuse, fragrant flowers of all *Wisteria* species have remarkable beauty. *W. floribunda* (Willd.) DC., Japanese wisteria, *W. frutescens* (L.) Poir., American wisteria, *W. macrostachya* Nutt., Kentucky wisteria, and *W. sinensis* (Sims) Sweet, Chinese wisteria, are widely cultivated as ornaments for homesites and gardens. The woody vines often exceed 10 m in height and require substantial supporting frames.

The bark of *W. brachybotrys* S. & Z. in Japan yields a fiber used locally for cloth and thread (Uphof 1968). Various bits of evidence allude to the toxic nature of *Wisteria* bark, seeds, and pods (Watt and Breyer-Brandwijk 1962; Kingsbury 1964).

Nodulation Studies

Cultural and biochemical characterizations of *Wisteria* rhizobia are not well defined. Serologically different strains were isolated by Jimbo (1930) from different nodules of the same plant. Carroll's conclusion from host-infection (1934a) and agglutination tests (1934b) that *W. chinensis* DC. was highly selective for rhizobia was not confirmed by Wilson (1939a, 1944b, 1945a) and Wilson and Chin (1947).

Perennial, apical-type, 4–6-year-old, moniliform nodules on *W. sinensis* were described by Molisch (1926) and Jimbo (1927). Each annual growth portion was larger than the previous one, easily distinguished by transverse indentations, and lighter in color. Presumably, rigidity and longevity of these nodules were attributable to a layer of stone cells in the nodule cortex and a continuous endodermis surrounding the vascular system. Hollow, cortical-shelled nodules were observed.

Nodulated species of *Wisteria*	Area	Reported by
brachybotrys S. & Z.	Japan	Jimbo (1930); Asai (1944)
chinensis DC.	Widely reported	Many investigators
var. *grandiflora*	France	Lechtova-Trnka (1931)
floribunda (Willd.) DC.	Japan	Jimbo (1930); Asai (1944)
var. *alba* DC.	N.Y., U.S.A.	Wilson & Chin (1947)
frutescens (L.) Poir.	N.Y., U.S.A.	Wilson (1945a)
sinensis (Sims) Sweet	Japan	Jimbo (1927)
venusta Rehder & Wilson	Japan	Jimbo (1930)

Xanthocercis Baill.   Papilionoideae: Sophoreae
(122a) / 3580a

From Greek, *xanthos*, "yellow," and *Cercis*.

Type: *X. madagascariensis* Baill.

2 species. Trees, small to large, 16 m or more tall, with trunks about 1.75 m in diameter, evergreen. Leaves imparipinnate; leaflets 5–15, alternate, oblong, shiny. Flowers white, cream, yellow, or mauve, pedicelled, often silky-villous, in lax, terminal or axillary panicles; bracts and bracteoles small; calyx

cup-shaped, thickened and disklike at the base, truncate or short-toothed, grayish-downy; petals 5, clawed, free, subequal, 4 lanceolate, with a silky-haired median line; standard broad, oblong, eared at the base; stamens 10, diadelphous; ovary long-stalked, linear, 10–20-ovuled; style very short. Pod stalked, drupaceous, spherical or oval, yellowish to brownish, fleshy, indehiscent; seeds 1–3, kidney-shaped, black, shiny, thick. (Dumaz-le-Grand 1952)

Taubert (1894) originally positioned *Xanthocercis* as 375 in Dalbergieae, subtribe Anomalae. Its current position as 122a (Harms 1908a) in Sophoreae is in keeping with its congeneric status with and priority over *Pseudocadia* (Dumaz-le-Grand 1952).

Both species inhabit tropical East Africa and Madagascar. They flourish in hot dry areas in bushveld or wooded flats and along river banks in deep alluvial sandy soil.

The hard, heavy, whitish gray wood serves various local uses. Both species are protected by limited conservation regulations. The fruit has an agreeable astringent flavor and is relished by birds, monkeys, and antelope, but is rarely consumed by man except in times of famine.

NODULATION STUDIES

Nodules were observed on *X. zambesiaca* (Bak.) Dumaz-le-Grand in Zimbabwe by Corby (1974).

Xerocladia Harv.   Mimosoideae: Adenanthereae
18 / 3453

From Greek, *xeros*, "dry," and *klados*, "branch."

Type: *X. viridiramis* (Burch.) Taub. (= *X. zeyheri* Harv.)

1 species. Shrub or undershrub, small, rigid, much-branched, armed with short, paired, spiny, recurved stipules. Leaves bipinnate, few, small; pinnae 1–2 pairs; leaflets small, few pairs. Flowers small, subsessile, in axillary, globose heads; calyx campanulate; lobes 5, longer than the tube, oblong, blunt, woolly-edged, nearly entirely free; petals 5, oblong, valvate, free almost to the base; stamens 10, free, short-exserted; filaments free, glabrous, alternately shorter; anthers tipped with minute, deciduous gland; ovary sessile or short-stalked, villous; ovule 1, rarely 2; style filiform, thickened toward apex; stigma small, truncate. Pod short-stalked, broad, falcate-ovoid, flattened, laterally apiculate, slightly winged along the lower suture, indehiscent; seeds usually 1, ovate, compressed. (Bentham and Hooker 1865; Harvey 1894; Baker 1930; Phillips 1951)

This monotypic genus is endemic in temperate Namibia and Namaqualand.

NODULATION STUDIES

Grobbelaar (personal communication 1974) observed nodules on this species in South Africa.

Xeroderris Roberty

Papilionoideae: Dalbergieae,
Lonchocarpinae

From Greek, *xeros*, "dry," and *Derris*, a closely related genus.

Lectotype: *X. stuhlmannii* (Taub.) Mend. & Sousa

2 species. Trees, 10–15 m tall, deciduous; bark fissured, flaking off in thick, irregular patches. Leaves imparipinnate, tufted at apex of branches, with short, rusty hairs when young; leaflets 4–7 pairs, ovate-elliptic, terminal leaflet largest; stipules inconspicuous. Flowers white, in slender rusty-haired racemes clustered on terminal shoots prior to leaf renewal; calyx broadly cup-shaped; teeth 5, broad, rusty-haired; standard short-clawed, slightly bulged; stamens 10, diadelphous. Pod flat, thin, tapered at both ends, both margins narrow-winged, usually 1-seeded, dark velvety brown, brittle when ripe. (Roberty 1954)

Hutchinson (1964) considered this genus synonymous with *Ostryoderris*, but more recently Drummond (1972) has regarded it as distinctive.

Both species occur in tropical Africa, from Senegal to Zaire and East Africa. They thrive in hot, dry savanna woodlands.

The soft, red-streaked, cream-colored wood of *X. stuhlmannii* (Taub.) Mend. & Sousa is used for dug-out canoes and small construction. The copious red exudate from the slash serves in the tanning of hides. The leaves are eaten by stock in East Africa.

NODULATION STUDIES

Corby (1974) reported nodules on *X. stuhlmannii* growing in Zimbabwe.

Xylia Benth.

Mimosoideae: Piptadenieae

24 / 3461

From Greek, *xylon*, "wood"; the trees of this genus have hard wood.

Type: *X. xylocarpa* (Roxb.) Taub.

13–15 species. Trees, small to large, unarmed. Leaves bipinnate; pinnae 1 pair; leaflets large and few-paired, or small and numerous pairs; petiole with a gland near apex and a small gland at the junction of the terminal leaflets; stipules small, linear, caducous. Flowers pale green, white, or yellow, small, sessile or short-pedicelled, in long-peduncled globose heads, axillary, racemose or subfascicled at apices of branches; calyx tubular-campanulate; lobes 5, valvate in bud; petals 5, slightly joined at the base; stamens 10, sometimes more, exserted; filaments free; anthers ovate, usually with a caducous apical gland at an early stage; ovary sessile, pubescent, several- to many-ovuled; style filiform; stigma terminal, very small. Pod sessile, oblong-falcate, large, thick, flat, woody, septate within, elastically dehiscent from apex with both valves recurving; seeds transverse, obovate, compressed, brown, glossy, unwinged, in depressions in the upper part of the valve. (Bentham 1842; Bentham and Hooker 1865; Pellegrin 1948; Gilbert and Boutique 1952a; Brenan 1959)

Xylia species are paleotropical and flourish in equatorial Africa, Madagascar, southern India, and Burma. They grow in rain forests and along streams; best growth occurs on alluvial soil.

X. evansii Hutch., *X. ghesquierei* Robyns, and *X. xylocarpa* (Roxb.) Taub. reach heights of about 30 m, with trunk diameters up to 0.75 m, and yield valuable hard wood; *X. xylocarpa* (= *X. dolabriformis* Benth.) is known as pyinkado or pyingado in the timber trade. In general, *Xylia* wood is reddish brown, durable, difficult to work, and has a density of 880–960 kg/m^3; the planed surfaces have an oily appearance. It is valued for pilings, bridge supports, and related heavy-duty purposes. The bark of all species is a rich source of tannin. The seeds yield a yellow, nondrying oil (Titmuss 1965; Burkill 1966).

NODULATION STUDIES

Specimens of *X. torreana* Brenan examined in Zimbabwe (Corby 1974) and *X. dolabriformis* in western India (V. Schinde personal communication 1977) lacked nodules.

Yucaratonia Burk.   Papilionoideae: Galegeae, Robiniinae

Named derived from *yuca ratón*, the colloquial name in Ecuador for a plant eaten by rodents.

Type: *Y. brenningii* (Harms) Burk. (= *Sesbania brenningii*)

1 species. Tree or small shrub, 2–6 m tall, deciduous; trunk 10–18 cm in diameter; bark gray, with longitudinal fissures. Leaves large, imparipinnate; young leaflets silky-pubescent, later smooth, elliptic, usually alternate, margins entire, pinnately nerved, many, oblong with apex notched, petioluled; rachis elongated, much-grooved; stipels none; stipules free, awl-shaped, caducous. Flowers white, large, single or several on slender pedicels in axillary, lax, pendent racemes; bracts inconspicuous, lanceolate, pubescent, caducous; bracteoles absent; calyx membranous, campanulate-truncate; teeth 5, small, pubescent; standard obovate, strongly reflexed with upper margin cupped inward; keel and wings equal in length; stamens 10, diadelphous, vexillary stamen completely free; anthers elliptic, uniform; ovary linear, stalked, with an encircling disk at the base, multiovuled; style short, smooth, incurved; stigma terminal, capitate. Pod elongated, 15–27 cm long, linear-compressed, apex shortly acute, leathery, 2-valved, elastically dehiscent, many-seeded; seeds triangular-oval, transversely positioned, coffee-colored, noncarunculate, hilum apical. (Burkart 1969)

This genus occupies a position between *Sesbania* and *Gliricidia*.

This monotypic genus is endemic in Ecuador and northern Argentina.

Named for Professor H. C. Zen, Executive Secretary of the China Foundation for the Promotion of Education and Culture; he allocated annual grants to the Botanical Institute of Sun Yatsen University during its difficult early years.

Type: *Z. insignis* Chun

1 species. Tree, tall, unarmed. Leaves imparipinnate, deciduous; leaflets alternate, 9–13 pairs, entire; stipels and stipules absent. Flowers red, bisexual, on short pedicels in terminal, long-peduncled, cymose-paniculate inflorescences; bracteoles present; sepals 5, unequal, 1 much smaller, free, imbricate; petals 5, subequal, upper 2 lateral petals strongly involute, each enveloping a pair of free, fertile, geniculate stamens, the 5th stamen usually reduced to an inconspicuous, short, filiform staminode, all inserted on the margin of a small, sinuate-lobate disk; fertile anthers uniform, longitudinally dehiscent, linear-oblong, basifixed; ovary short-stalked, elongated, silky-pubescent, few-ovuled; style short, inflexed; stigma subterminal. Pod reddish brown, compressed, membranous, elliptic-oblong, falcate, apex acuminate, reticulate, upper suture broad-winged, indehiscent; seeds few, orbicular, located near the middle of the valves, compressed, dark brown, smooth, shiny, on long, slender funicles. (Chun 1946)

This monotypic genus is endemic in southern China. The plants grow on rocky hills in forested ravines.

Z. insignis Chun yields a high-quality wood suitable for cabinetmaking, but it is not marketed because of the small supply. This shortage is a consequence of the irreparable loss of the priceless living plant collection in experimental gardens during the invasion of southern China in World War II (Chun 1946).

Zenkerella Taub.   Caesalpinioideae: Cynometreae

(39) / 3483

Named for Jonathan Karl Zenker, 1799–1837, German botanist.

Type: *Z. citrina* Taub.

6 species. Trees and shrubs, small, deciduous. Leaves simple, entire, broad-elliptic, acuminate, not pellucid-punctate; petioles transversely creased; stipules present. Flowers white or greenish yellow, few, disposed in 2 rows, in very short, axillary fascicled racemes; bracteoles not petaloid, caducous; calyx tube elongated, cupular-turbinate or funnel-shaped; lobes 4–5, free, imbricate, tips reflexed at anthesis; petals 5, free, subequal, elliptic or spathulate; stamens 10, free or joined only at the base; ovary smooth or pubescent, long-stalked, stipe adnate to the calyx tube wall, 2–3-ovuled; disk absent. Pod small, stalked, flattened, semiorbicular, beaked, 2-valved, leathery, sutures not winged, dehiscent, 1–3-seeded. (Léonard 1951, 1957)

This genus is regarded distinctive from, but closely akin to, *Cynometra* (Léonard 1957).

Members of this genus are distributed throughout tropical East and West Africa, ranging from Nigeria to Gabon and Tanzania.

Type: *Z. falcata* Maxim. & Nees

8–10 species. Shrubs, large or medium, to tall trees with straight trunks, unarmed. Leaves 1-foliolate, leathery, short-petioled, margins toothed at remote intervals; stipules rigid, persistent. Flowers small, yellow or rose, in terminal, panicled racemes; bracts small; bracteoles minute or absent; calyx tube very short, disklike, upper or expanded part unsegmented and acuminate in bud, cleft and reflexed at anthesis or caducous; petals 5, imbricate; standard broader than other petals and outermost; stamens 9–13, usually 10, subhypogynous; filaments very short; anthers uniform, linear, acuminate, basifixed, dehiscent by longitudinal slits; ovary subsessile or stalked, many-ovuled; style short, awl-shaped; stigma small, obliquely terminal. Pod green, ovoid, thick, 2-valved, with 1–2 flattened, ovoid seeds. (Bentham and Hooker 1865; Britton and Rose 1930; Standley and Steyermark 1946)

Representatives of this genus are native to Brazil and Central America. They thrive in wet coastal forests.

The bark is a source of tannin, the pods serve various medicinal purposes, and the sweet-scented flowers attract bees. *Zollernia* wood has a waxy appearance, a mild, fragrant odor, and finishes smoothly with a fine polish. The scarce wood is highly prized for tool handles, billiard cue butts, and other small articles of turnery (Record and Hess 1943).

Zornia J. F. Gmel.   Papilionoideae: Hedysareae,

334 / 3804   Stylosanthinae

Named for Johann Zorn, 1739–1799, apothecary and botanical author, native of Bavaria, Germany.

Type: *Z. bracteata* J. F. Gmel.

75 species. Herbs, mostly perennial, few annual, low, some shrubby, up to 1–2 m tall, others prostrate, spreading; stems wiry, diffusely branched, woody at the base. Leaves palmately 2–4-, rarely 3-foliolate; leaflets elliptic, lanceolate or linear, often marked with transparent black dots, lower leaflets on some species differ in shape from upper ones; petioles arise between paired, subfoliaceous, generally persistent and punctate stipules; stipels absent. Flowers yellow or yellow orange, frequently red- or purple-streaked, subsessile, solitary or numerous, in interrupted spikelike racemes, terminal and axillary; bracts usually paired, spurred, ovate or narrow-elliptic, conspicuously large, persistent, longitudinally nerved, enclosing the sessile flowers; bracteoles none; calyx membranous, very short; lobes 5, ciliate, upper 2 connate, lower lobe oblong or lanceolate, subequal to the upper lip, lateral 2 much smaller than other 3; standard suborbicular, clawed; wings usually clawed, obliquely obovate or oblong, shorter and narrower than the standard;

keel incurved, subrostrate, clawed; stamens 10, monadelphous; lower half of filaments all joined into a tube; anthers alternately small and short, large and long, the latter subbasifixed; ovary sessile, linear, several- to many-ovuled; style filiform; stigma small, terminal. Pod lomentaceous, flattened, upper suture straight, lower suture deeply indented, segments prominently nerved, often bristly or glandular, indehiscent; 1-seeded; seeds suborbicular or kidney-shaped. (Bentham and Hooker 1865; Baker 1929; Amshoff 1939; Phillips 1951; Léonard 1954a; Milne-Redhead 1954; Mohlenbrock 1961)

Estimates of the size of the genus vary. Mohlenbrock (1961) recognized only 75 species of the more than 100 species, subspecies, and varieties described before 1953.

Zornia species are widespread in tropical and subtropical regions throughout the world. About 30 species in the Western Hemisphere are distributed from the southwestern United States into Buenos Aires, Argentina. Australia, Mexico, Central America, Africa, and Brazil each have 10–15 native species. The habitats are somewhat varied, but the plants usually grow in semidry areas, grassy savannas, along roadsides, most often in sandy soils. Species in the United States are found in the coastal prairie areas, ranging from Virginia to Florida and westward into Arizona.

Z. diphylla Pers., *Z. gracilis* DC., and *Z. latifolia* Sm. are regarded as good natural forage in Argentina, where they tend to be abundant, especially in overgrazed pasture lands. *Z. diphylla* has shown promise as a fodder crop. Its foliage is reputedly stored as hay for horses in the French Sudan, Nigeria, and Zimbabwe, but Hurst (1942) reported that horses that had eaten the foliage of this species in Australia suffered the impaired use of limbs and damage to eyesight. In some parts of Africa and Central America the plant parts are used medicinally to cure dysentery, and as a febrifuge and diuretic. This species apparently contains saponins that produce lather in water and is occasionally referred to as poor man's soap.

NODULATION STUDIES

The keen perception of Bentham (1859–1870) is shown by his depiction of nodules on *Z. diphylla* in Martius' *Flora Brasiliensis* 15, part 1, plate 21. The nodulated adventitious roots of this species radiated 1 m or more immediately below the soil surface in white sand forest clearings in Guyana (Whitton 1962). Each of the 4 species that Barrios and Gonzalez (1971) observed in Venezuela in sandy savannas not subject to inundation in the rainy season bore round nodules. The area had dominant grass formation of *Trachypogon* species. The observations were made during the rainy season.

Vyas and Prasad (1959) reported symbiotic compatibility between rhizobia from *Z. diphylla* and plants of *Pisum sativum*, but this result is viewed with skepticism. The results of plant experiments by Mannetje (1969) and Sandmann (1970a) showing symbiobility between *Zornia* rhizobia and *Stylosanthes* species seem more conclusive and plausible in view of the close tribal kinships of these 2 genera.

Nodulated species of *Zornia*	Area	Reported by
capensis Pers.		
ssp. *capensis*	Zimbabwe	Corby (1974)
curvata Mohl.	Venezuela	Barrios & Gonzalez (1971)
dictiocarpa DC.	Australia	Mannetje (1969)
diphylla Pers.	Brazil	Bentham (1859–1870)
	Australia	Bowen (1956)
	Guyana	Whitton (1962)
gibbosa Span.	W. India	V. Schinde (pers. comm. 1977)
glochidiata Reich. ex DC.	S. Africa	Grobbelaar et al. (1964)
	Zimbabwe	Corby (1974)
herbacea Pitt.	Venezuela	Barrios & Gonzalez (1971)
latifolia Sm.		
var. *latifolia*	Venezuela	Barrios & Gonzalez (1971)
linearis J. F. Gmel.	S. Africa	Grobbelaar et al. (1964)
marajoara Huber	Venezuela	Barrios & Gonzalez (1971)
milneana Mohl.	S. Africa	Grobbelaar et al. (1964)
	Zimbabwe	Corby (1974)
pratensis Milne-Redhead		
ssp. *pratensis*	Zimbabwe	Corby (1974)

Zuccagnia Cav.   Caesalpinioideae: Eucaesalpinioideae
101 / 3558

Named for Attilio Zuccagni, 1754–1807, Italian botanist.

Type: *Z. punctata* Cav.

1 species. Shrub or small tree, about 5 m tall, with a trunk diameter about 20 cm; branches many, glutinous, gray to blackish. Leaves paripinnate, short; leaflets small, alternate or subopposite, 5–13 pairs, elliptic-oblong, acuminate, leathery, conspicuously glandular-punctate; stipules minute. Flowers yellow, in lax terminal racemes; bracts small, caducous; bracteoles none; calyx turbinate; lobes 5, subequal, lowermost lobe outside; petals 5, obovate-orbicular, free, subequal, uppermost petal inside and broader, imbricate; stamens 10, free; filaments pilose at the base; anthers ovoid, uniform; ovary short-stalked, 1-ovuled; style slender; stigma concave. Pod ovoid, short, about 1 cm long, compressed, leathery, red bristly-haired, 2-valved, dehiscent, 1-seeded. (Bentham and Hooker 1865; Burkart 1952)

This monotypic genus is endemic in central Argentina and Chile. Its xerophytic plants grow in brushlands and thickets up to 2,700-m altitude.

This plant is used for crude roofing material, supports for vines, and as fuel. Decoctions of plant parts have served various minor medicinal purposes. Alcoholic extraction of the sticky substance in the leaves yields a yellow dye for coloring wool (Burkart 1952).

APPENDICES

Summary of Nodulation Data

Mimosoideae

Genus	Number of species in genus	Nodulation data		
		+	−	Total
Acacia Mill.	800–900	206	11	217
Adenanthera L.	12		3	3
Affonsea A. St.-Hil.	7–8			
Albizia Durazz.	150	32		32
Amblygonocarpus Harms	1		1	1
Anadenanthera Speg.	2	1		1
Archidendron F. Muell.	30–33			
Arthrosamanea Britt. & Rose	10			
Aubrevillea Pellegr.	2			
*Calliandra Benth.	150	10	3	13
Calpocalyx Harms	6–11			
Carthormion Hassk.	15	1		1
Cedrelinga Ducke	1			
Cylicodiscus Harms	2			
Cylindrokelupha Kosterm.	7			
Delaportea Thorel ex Gagnep.	1			
Desmanthus Willd.	40	5		5
Dichrostachys W. & A.	20	9		9
Dinizia Ducke	1			
Elephantorrhiza Benth.	8–10	4		4
*Entada Adans.	30–40	5		5
Enterolobium Mart.	8–10	2		2
Fillaeopsis Harms	1			
Gagnebina Neck.	2			
Goldmania Rose ex M. Micheli	2			
Hansemannia Schum.	4			
Indopiptadenia Brenan	1			
Inga Scop.	150–300	13	4	17

* = Conserved generic name.

Summary of Nodulation Data

Genus	Number of species in genus	Nodulation data		
		+	−	Total
Leucaena Benth.	50	3		3
Lysiloma Benth.	30–50	1		1
Macrosamanea Britt. & Rose	8			
Marmoroxylon Killip ex Record	1			
Mimosa L.	600	22	3	25
Mimozyganthus Burk.	1			
Monoschisma Brenan	2			
Morolobium Kosterm.	1			
Neptunia Lour.	10–15	4		4
Newtonia Baill.	15		2	2
Paralbizzia Kosterm.	3			
Parapiptadenia Brenan	2	1		1
Parasamanea Kosterm.	1			
Parenterolobium Kosterm.	1			
Parkia R. Br.	50–60	4	1	5
Pentaclethra Benth.	3	1		1
Piptadenia Benth.	11–15	1	2	3
Piptadeniastrum Brenan	1			
Piptadeniopsis Burk.	1			
Pithecellobium Mart.	100–200	13	1	14
Plathymenia Benth.	3–4			
Prosopidastrum Burk.	2			
Prosopis L.	45	6	3	9
Pseudoentada Britt. & Rose	1			
Pseudoprosopis Harms	4			
Pseudosamanea Harms	1			
Samanea (Benth.) Merr.	20	1		1
Schranckiastrum Hassl.	1			
Schrankia Willd.	20	2		2
Serialbizzia Kosterm.	2			
Serianthes Benth.	13			
Sopropis Britt. & Rose	1			
Strombocarpa Engelm. & Gray	6		1	1
Stryphnodendron Mart.	15	2		2
Tetrapleura Benth.	2			
Wallaceodendron Koord.	1	1		1
Xerocladia Harv.	1	1		1
Xylia Benth.	13–15		2	2

Subfamily Mimosoideae	Estimated number of species	Species reported		
		+	−	Total
66 genera				
31 genera examined	2,506–2,920	351	37	388
18 with all species +				
5 with all species −				
8 with both + and − species				

Caesalpinioideae

Genus	Number of species in genus	Nodulation data		
		+	−	Total
Acrocarpus Wight ex Arn.	3			
Afzelia Sm.	30		2	2
Aldina Endl.	12			
Amherstia Wall.	1	1	1	1[a]
Amphimas Pierre ex Harms	4			
Androcalymma Dwyer	1			
Anthonotha Beauv.	15			
Apaloxylon Drake del Castillo	2			
Aphanocalyx Oliv.	3			
Aprevalia Baill.	1			
Apuleia Mart.	2			
Arcoa Urb.	1			
Augouardia Pellegr.	1			
Baikiaea Benth.	10		2	2
Baphiopsis Benth. ex Bak.	2			
Batesia Spruce ex Benth.	1			
Bathiaea Drake del Castillo	1			
Baudouinia Baill.	2–6			
Bauhinia L.	550–575	1	26	27
Berlinia Soland. ex Hook. f.	15		1	1
Brachystegia Benth.	70		8	8
Brandzeia Baill.	1			
Brenierea Humbert	1			
Brownea Jacq.	30	1	5	6
Browneopsis Huber	3			
Burkea Benth.	1		1	1
Bussea Harms	4–6			
Caesalpinia L.	200		14	14
Campsiandra Benth.	3			
Cassia L.	600	44	55	99
Cenostigma Tul.	6			

* = Conserved generic name.

710

Genus	Number of species in genus	Nodulation data		
		+	−	Total
Ceratonia L.	1		1	1
Cercidium Tul.	8–10		2	2
Cercis L.	5–7	1	2	3
Chidlowia Hoyle	1			
Colophospermum Kirk ex J. Léonard	1		1	1
Colvillea Boj. ex Hook.	1	1		1
Conzattia Rose	3			
*Copaifera L.	25	1	1	1[a]
Cordeauxia Hemsl.	1	1		1
Cordyla Lour.	5		1	1
*Crudia Schreb.	50–55		1	1
Cryptosepalum Benth.	12–15		1	1
Cyathostegia (Benth.) Schery	2			
Cymbosepalum Bak.	1			
Cynometra L.	60–70		3	3
Daniellia Benn.	12			
Dansera Steenis	1			
Delonix Raf.	3		2	2
Denisophytum R. Vig.	1			
Detarium Juss.	3–4			
Dewindtia De Wild.	1			
Dialum L.	70	1	2	3
Dicorynia Benth.	2			
Dicymbe Spruce ex Benth. & Hook. f.	7–10	2		2
Didelotia Baill.	8			
Dimorphandra Schott	25	1		1
Diptychandra Tul.	3			
Distemonanthus Benth.	1			
Duparquetia Baill.	1			
Eligmocarpus R. Cap.	1			
Elizabetha Schomb. ex Benth.	10			
Endertia Steenis & de Wit	1			
Englerodendron Harms	1			
Eperua Aubl.	11	1	3	3[a]
Erythrophleum Afzel.	15–17	2		2
Eurypetalum Harms	3–4			
Exostyles Schott ex Spreng.	2			
Gilbertiodendron J. Léonard	26	1		1
Gillettiodendron Verm.	5–7			
Gleditsia L.	12		5	5

Summary of Nodulation Data

Genus	Number of species in genus	Nodulation data +	−	Total
Goniorrhachis Taub.	1			
Gossweilerodendron Harms	1	1		1
Griffonia Baill.	3–4			
Guibourtia Benn. emend J. Léonard	15–18		2	2
Gymnocladus Lam.	3		1	1
Haematoxylon L.	3		1	1
Hardwickia Roxb.	2	1		1
Heterostemon Desf.	7–8			
Hoffmanseggia Cav.	45		4	4
Holocalyx M. Micheli	2		1	1
Humboldtia Vahl	5–8			
Hylodendron Taub.	1			
Hymenaea L.	25–30	1	1	1 [a]
Hymenostegia (Benth.) Harms	20			
Intsia Thou.	7–9	1	2	3
Isoberlinia Craib & Stapf	7			
Isomacrolobium Aubrév. & Pellegr.	10			
Jacqueshuberia Ducke	2			
Julbernardia Pellegr.	9–11		1	1
Kalappia Kosterm.	1			
Kingiodendron Harms	4			
Koompassia Maing.	2		1	1
Labichea Gaud. ex DC.	8			
Lebruniodendron J. Léonard	1			
Lecointea Ducke	3			
Lemuropisum H. Perrier	1			
Leonardoxa Aubrév.	3			
Leucostegane Prain	2			
Librevillea Hoyle	1			
Loesenera Harms	4			
Lophocarpinia Burk.	1			
Lysidice Hance	1	1		1
Macroberlinia (Harms) Hauman	1			
Macrolobium Schreb.	50		1	1
Maniltoa Scheff.	15	1	1	2
Martiodendron Gleason	4			
Melanoxylon Schott	2			
Mendoravia R. Cap.	1			

Genus	Number of species in genus	Nodulation data +	−	Total
Mezoneurum Desf.	35		2	2
Michelsonia Hauman	2			
Microberlinia A. Chev.	2			
Mildbraediodendron Harms	1			
Moldenhauera Schrad.	5			
Monopetalanthus Harms	8–10			
Mora Schomb. ex Benth.	10	1	2	2 [a]
Neochevalierodendron J. Léonard	1			
Oddoniodendron De Wild.	1			
Oxystigma Harms	5			
Pachyelasma Harms	1			
Pahudia Miq.	3–4		1	1
Paloue Aubl.	4			
Paloveopsis R. S. Cowan	1			
Paramacrolobium J. Léonard	1			
Parkinsonia L.	2–4		2	2
Pellegriniodendron J. Léonard	1			
Peltogyne Vog.	27		1	1
Peltophoropsis Chiov.	1			
Peltophorum (Vog.) Benth.	15		2	2
Petalostylis R. Br.	2		1	1
Phyllocarpus Riedel ex Tul.	2	1		1
Piliostigma Hochst.	3		2	2
Plagiosiphon Harms	5			
Poeppigia C. Presl	1			
Pogocybe Pierre	1			
Polystemonanthus Harms	1			
Prioria Griseb.	1			
Pseudoberlinia Duvign.	3–4			
Pseudomacrolobium Hauman	1			
Pterogyne Tul.	1		1	1
Pterolobium R. Br. ex W. & A.	10–12		1	1
Pynaertiodendron De Wild.	2			
Recordoxylon Ducke	2			
Saraca L.	25	1	4	4 [a]
Schizolobium Vog.	4–5		1	1
Schizoscyphus Shum. ex Taub.	1			
Schotia Jacq.	20		3	3
Sclerolobium Vog.	25–30	2		2

Summary of Nodulation Data

Genus	Number of species in genus	Nodulation data +	−	Total
Scorodophloeus Harms	2		1	1
Sindora Miq.	18–20		1	1
Sindoropsis J. Léonard	1			
Stachyothyrus Harms	2			
Stahlia Bello	1			
Stemonocoleus Harms	1			
Stenodrepanum Harms	1			
Storckiella Seem.	3			
Stuhlmannia Taub.	1			
*Swartzia Schreb.	130	7		7
Sympetalandra Stapf	1			
Tachigalia Aubl.	22		2	2
Talbotiella Bak. f.	3			
Tamarindus L.	1		1	1
Tessmannia Harms	10–12			
Tetraberlinia (Harms) Hauman	1			
Tetrapterocarpon Humbert	1			
Thylacanthus Tul.	1			
Trachylobium Hayne	1	1		1
Triplisomeris Aubrév. & Pellegr.	3			
Tylosema (Schweinf.) Torre & Hillc.	3–4		1	1
Uittienia Steenis	1			
Umtiza Sim	1			
Vouacapoua Aubl.	4			
Wagatea Dalz.	1			
Zenia Chun	1			
Zenkerella Taub.	6			
Zollernia Nees	8–10			
*Zuccagnia Cav.	1			

[a] 1 species reported +/−.

Subfamily Caesalpinioideae	Estimated number of species	Species reported +	−	(+/−)	Total
177 genera					
65 genera examined	2,716–2,816	72	180	6	258
13 with all species +					
39 with all species −					
13 with both + and − species					

Papilionoideae

Genus	Number of species in genus	Nodulation data		
		+	−	Total
Abrus Adans.	12–15	6		6
Adenocarpus DC.	15–20	3		3
Adenodolichos Harms	15	1		1
Adesmia DC.	230–250	7		7
Aenictophyton A. Lee	1			
Aeschynomene L.	150–250	44		44
Afgekia Craib	1			
Afrormosia Harms	6	2		2
Airyantha Brummitt	2			
Alepidocalyx Piper	3			
Alexa Moq.	9		1	1
Alhagi Tourn. ex Adans.	5	1		1
Alistilus N. E. Br.	1	1		1
Alysicarpus Desv.	30	14		14
Amburana Schwacke & Taub.	2			
Amicia HBK	8			
Ammodendron Fisch. ex DC.	8			
Ammothamnus Bge.	4			
Amorpha L.	20–25	8		8
Amphicarpaea Ell.	3–4	4		4
Amphithalea E. & Z.	15	1		1
Anagyris L.	2	1		1
Anarthrophyllum Benth.	15			
Andira Juss.	35	1	1	2
Angylocalyx Taub.	10–14			
Antheroporum Gagnep.	2			
Anthyllis L.	30–50	6		6
Antopetitia A. Rich.	1	1		1
Aotus Sm.	15	4		4
Apios Fabr.	10	1		1
Apoplanesia C. Presl	1			
Apurimacia Harms	4			
Arachis L.	9–19	10		10
Argyrolobium E. & Z.	70	24		24

* = Conserved generic name.

715

Summary of Nodulation Data

Genus	Number of species in genus	Nodulation data +	Nodulation data −	Nodulation data Total
Arillaria Kurz	1			
Arthrocarpum Balf. f.	2			
Arthroclianthus Baill.	20			
Artrolobium Desv.	5	1		1
Aspalathus L.	245	60		60
Asphalthium Medik.	2	2		2
Astragalus L.	1,500–2,000	94	8	102
Ateleia (Moc. & Sessé ex DC.) Dietr.	17			
Atylosia W. & A.	36	3		3
Austrodolichos Verdc.	1			
Bakerophyton Hutch.	3	1		1
Balisaea Taub.	1			
Baphia Lodd.	60–100	2		2
Baphiastrum Harms	12			
Baptisia Vent.	35–50	6		6
Barbiera DC.	1			
Barklya F. Muell.	1			
Baueropsis Hutch.	1			
Baukea Vatke	1			
Behaimia Griseb.	1			
Belairia A. Rich.	5			
Bembicidium Rydb.	1			
Benedictella Maire	1			
Bergeronia M. Micheli	1			
Biserrula L.	1	1		1
Bolusanthus Harms	1	1		1
Bolusia Benth.	4	1		1
Borbonia L.	12–13	1		1
Bossiaea Vent.	40	12		12
Bowdichia HBK	2		1	1
Bowringia Champ. ex Benth.	2			
Brachysema R. Br.	16	5		5
Brongniartia HBK	60			
Brya P. Br.	8	1		1
Bryaspis Duvign.	1			
Buchenroedera E. & Z.	23	2		2
Burkillia Ridley	1			
Burtonia R. Br.	12	1		1
Butea Roxb. ex Willd.	3		1	1
Cadia Forsk.	5			
Cajalbania Urb.	1			
Cajanus DC.	2	1		1

Genus	Number of species in genus	Nodulation data +	Nodulation data −	Nodulation data Total
Calophaca Fisch.	5			
Calopogonium Desv.	12	3		3
Calpurnia E. Mey.	6	3		3
Calycotome Link	5	3		3
Camoensia Welw. ex Benth.	2		1	1
Camptosema H. & A.	12–15			
Campylotropis Bge.	65			
Canavalia DC.	50	8		8
Caragana Lam.	80	12		12
Carmichaelia R. Br.	41	18		18
Carrissoa Bak. f.	1			
Cascaronia Griseb.	1			
Castanospermum A. Cunn.	2		1	1
Catenaria Benth.	1	1		1
Caulocarpus Bak. f.	1			
Centrolobium Mart. ex Benth.	5			
Centrosema Benth.	50	8		8
Chadsia Boj.	18			
Chaetocalyx DC.	11		2	2
Chapmannia T. & G.	1			
Chesneya Lindl.	18			
Chloryllis E. Mey.	1			
Chordospartium Cheesem.	1	1		1
Chorizema Labill.	16	7		7
Christia Moench	12	1		1
Chrysoscias E. Mey.	6			
Cicer L.	14	2		2
Cladrastis Raf.	5		3	3
Clathrotropis Harms	4	2		2
Cleobulia Mart. ex Benth.	3			
Clianthus Soland. ex Lindl.	8	4		4
Climacorachis Hemsl. & Rose	2			
Clitoria L.	70	8		8
Clitoriopsis Wilczek	1			
Cochlianthus Benth.	2			
Codariocalyx Hassk.	2	2		2
Coelidium Vog. ex Walp.	15			
Collaea DC.	3			
Cologania Kunth	15	2		2
Colutea L.	27	3		3
Condylostylis Piper	2			
Corallospartium Armstr.	1	1		1
Corethrodendron Fisch. & Basiner	1			
Cornicina Boiss.	1			

Summary of Nodulation Data

Genus	Number of species in genus	Nodulation data +	−	Total
Coronilla L.	55	12		12
Coroya Pierre	1			
Corynella DC.	6			
Coursetia DC.	25	1		1
Cracca Benth.	10	2	1	3
Craibia Harms & Dunn	10	1		1
Cranocarpus Benth.	2			
Craspedolobium Harms	1			
Cratylia Mart. ex Benth.	5–8	1		1
Crotalaria L.	550	145		145
Cruddasia Prain	1			
Cullen Medik.	2	2		2
Cupulanthus Hutch.	1			
Cyamopsis DC.	3–4	3		3
Cyclocarpa Afzel. ex Urb.	1	1		1
Cyclolobium Benth.	6			
Cyclopia Vent.	15–20	4		4
Cylista Ait.	7			
Cymbosema Benth.	1			
Cytisopsis Jaub. & Spach	1			
Cytisus L.	50	22		22
Dahlstedtia Malme	2			
Dalbergia L. f.	100–300	15	1	16
Dalbergiella Bak. f.	3	1		1
Dalea L.	150	8	6	14
Dalhousiea R. Grah.	3		1	1
Daviesia Sm.	60	13		13
Decorsea R. Vig.	4	1		1
Dendrolobium Benth.	12–17			
Derris Lour.	70–80	5	1	6
Desmodium Desv.	350–450	76		76
Dewevrea M. Micheli	2	1		1
Dicerma DC.	3	1		1
Dichilus DC.	5	3		3
Dicraeopetalum Harms	1			
Didymopelta Regel & Schmalh.	1			
Dillwynia Sm.	15	9		9
Dioclea HBK	30–50	2		2
Diphyllarium Gagnep.	1			
Diphysa Jacq.	18			
Diplotropis Benth.	14		1	1
Dipogon Liebm.	1	1		1
Dipteryx Schreb.	10		1	1

Genus	Number of species in genus	Nodulation data		
		+	−	Total
Discolobium Benth.	8			
Dolichopsis Hassl.	1	1		1
Dolichos L.	60	18		18
Dorycnium Vill.	12	5		5
Dorycnopsis Boiss.	1			
Drepanocarpus G. F. W. Mey.	15	1		1
Droogmansia De Wild.	30	1		1
Dumasia DC.	10	2		2
Dunbaria W. & A.	15	2		2
Dussia Krug & Urb.	10			
Dysolobium Prain	4			
Ebenus L.	20			
Echinosophora Nakai	1			
Echinospartum (Spach) Rothm.	4			
Edbakeria R. Vig.	1			
Eleiotis DC.	1			
Eminia Taub.	5	1		1
Endomallus Gagnep.	2			
Eremosparton Fisch. & Mey.	5			
Erichsenia Hemsl.	1		1	1
Erinacea Adans.	1			
Eriosema DC. ex Desv.	140	29		29
Erythrina L.	108	27		27
Etaballia Benth.	1			
Euchilopsis F. Muell.	1			
Euchlora E. & Z.	1			
Euchresta Benn.	4–5			
Eutaxia R. Br.	8	5		5
Eversmannia Bge.	2			
*Eysenhardtia Kunth	15	1	1	2
Factorovskya Eig	1			
Fagelia Neck.	1	1		1
Ferreirea Allem.	2			
Fiebrigiella Harms	1			
Fissicalyx Benth.	1			
Flemingia Roxb. ex Ait.	40	5		5
Fordia Hemsl.	12			
Galactia P. Br.	50	6		6
Galega L.	6–8	7		7
Gamwellia Bak. f.	1			
Gastrolobium R. Br.	44	8		8

Summary of Nodulation Data

Genus	Number of species in genus	Nodulation data		
		+	−	Total
Geissaspis W. & A.	3	2		2
Genista L.	80	11		11
Genistidium I. M. Johnston	1			
Geoffroea Jacq.	6			
Gliricidia HBK	6–9	1		1
Glottidium Desv.	5	1		1
Glycine Willd.	10	6		6
Glycyrrhiza L.	30	4		4
Gompholobium Sm.	24	9		9
Gonocytisus Spach	3			
Goodia Salisb.	2	2		2
Gourliea Gillies ex Hook.	1	1		1
Gueldenstaedtia Fisch.	10			
Halimodendron Fisch. ex DC.	1	1		1
Hallia Thunb.	9			
Hammatolobium Fenzl	3			
Hanslia Schindl.	1			
Haplormosia Harms	2			
Hardenbergia Benth.	2	2		2
Harpalyce Moc. & Sessé ex DC.	25			
Haydonia Wilczek	2	1		1
Hebestigma Urb.	1			
Hedysarum L.	70–100	13		13
Hegnera Schindl.	1			
Herpyza Ch. Wright	1			
Hesperolaburnum Maire	1			
Hesperothamnus Brandegee	6			
Heylandia DC.	1	1		1
Hippocrepis L.	10–12	4		4
Hoita Rydb.	11	1		1
Holtzea Schindl.	1			
Hosackia Dougl. ex Benth.	50	5		5
Hovea R. Br.	10–12	4		4
Humularia Duvign.	40			
Hybosema Harms	1			
Hymenocarpos Savi	1	1		1
Hymenolobium Benth.	8–12	1		1
Hypocalyptus Thunb.	3	1		1
Indigofera L.	800	191	3	194
Inocarpus Forst.	2		1	1
Isodesmia Gardn.	2			
Isotropis Benth.	10	2		2

Genus	Number of species in genus	Nodulation data		
		+	−	Total
Jacksonia R. Br.	40	8		8
Jansonia Kipp.	1			
Kennedia Vent.	15	11		11
Kerstania Rech. f.	1			
Kerstingiella Harms	1	1		1
Kostyczewa Korsh.	2			
Kotschya Endl.	30	6		6
Kummerowia Schindl.	2	2		2
Kunstleria Prain	9			
Lablab Adans.	1	1		1
Laburnum Fabr.	4	4		4
Lamprolobium Benth.	2			
Lathriogyne E. & Z.	1			
Lathyrus L.	130	45		45
Latrobea Meissn.	6	1		1
Lebeckia Thunb.	46	7		7
Lennea Klotz.	6			
Lens Mill.	5	2		2
Leptoderris Dunn	20			
Leptodesmia Benth.	5			
Lespedeza Michx.	140	30		30
Lessertia DC.	60	17		17
Leucomphalos Benth.	1			
Leycephyllum Piper	1			
Liparia L.	4			
Listia E. Mey.	1	1		1
Loddigesia Sims	1	1		1
Lonchocarpus HBK	175	11	1	12
Lotononis E. & Z.	110–120	31		31
Lotus L.	100–120	58		58
Luetzelburgia Harms	5			
Lupiniphyllum Gillett	1	1		1
Lupinus L.	150	56		56
Luzonia Elmer	1			
Lyauteya Maire	1			
Maackia Rupr. & Maxim.	10–12	1	1	2
Machaerium Pers.	150	3		3
Macropsychanthus Harms	3			
Macroptilium Urb.	15	2		2
Macrotyloma (W. & A.) Verdc.	25	11		11
Margaritolobium Harms	1			

Summary of Nodulation Data

Genus	Number of species in genus	Nodulation data		
		+	−	Total
Mastersia Benth.	2	1		1
Mecopus Benn.	1			
Medicago L.	50–100	40		40
Meladenia Turcz.	3			
Melilotus Mill.	20	16		16
Melliniella Harms	1			
Melolobium E. & Z.	30	5		5
Microcharis Benth.	5			
Millettia W. & A.	180	6		6
Minkelersia Mart. & Gal.	4			
Mirbelia Sm.	20–25	6		6
Monarthrocarpus Merr.	2			
Monoplegma Piper	1			
Monopteryx Spruce ex Benth.	3			
Mucuna Adans.	150	16		16
Muellera L. f.	2			
Muelleranthus Hutch.	3			
Mundulea Benth.	15–20	1		1
Myrocarpus Allem.	4	1		1
Myrospermum Jacq.	1			
Myroxylon L. f.	6	1		1
Nemcia Domin	12	3		3
Neocollettia Hemsl.	1			
Neocracca Ktze.	1			
Neodielsia Harms	1			
Neodunnia R. Vig.	4			
Neoharmsia R. Vig.	1			
Neorautanenia Schinz	3–8	4		4
Nepa Webb	1			
Nephrodesmus Schindl.	7			
Nephromeria (Benth.) Schindl.	10	1		1
Nesphostylis Verdc.	1			
Nissolia Jacq.	10–12			
Nogra Merr.	3			
Notodon Urb.	4			
Notospartium Hook. f.	3	1		1
Oleiocarpon Dwyer	1			
Olneya Gray	1	1		1
Onobrychis Mill.	100	10		10
Ononis L.	70–80	18		18
Ophrestia H. M. Forbes	14	4		4
Orbexilum Raf.	8	3		3

722

Genus	Number of species in genus	Nodulation data		
		+	−	Total
Ormocarposis R. Vig.	5			
Ormocarpum Beauv.	30	2		2
Ormosia Jacks.	100–120	5		5
Ormosiopsis Ducke	3			
Ornithopus L.	15	6		6
Ostryocarpus Hook. f.	4			
Ostryoderris Dunn	6–7	2		2
Otoptera DC.	2	1		1
Ougeinia Benth.	1	1		1
Oxylobium Andr.	25–30	5		5
Oxyrhynchus Brandegee	4		1	1
Oxytropis DC.	300	22		22
Pachecoa Standl. & Steyerm.	2–3	1		1
Pachyrrhizus A. Rich. ex DC.	6	1		1
Padbruggea Miq.	5			
Panurea Spruce ex Benth. & Hook. f.	1			
Papilionopsis Steenis	1			
Paramachaerium Ducke	3			
Paratephrosia Domin	1			
Parochetus Buch.-Ham. ex D. Don	1	1		1
Parryella T. & G.	2	1		1
Passaea Adans.	1			
Pearsonia Dümm.	11	11		11
Pediomelum Rydb.	22	2		2
Peekelia Harms	1			
Pentadynamis R. Br.	1			
Periandra Mart. ex Benth.	6			
Pericopsis Thwait.	6	2		2
Petaladenium Ducke	1			
Petalostemon Michx.	40	7	2	9
Petalostylis R. Br.	2		1	1
Peteria Gray	4			
Petteria C. Presl	1	1		1
Phaseolus L.	50–100	21		21
Phylacium Benn.	2–3			
Phyllodium Desv.	6–9	1		1
Phyllota Benth.	9–10	3		3
Phylloxylon Baill.	5			
Physanthyllis Boiss.	1	1		1
Physostigma Balf.	3–5			
Pickeringia Nutt. ex T. & G.	1	1		1
Pictetia DC.	8			

Summary of Nodulation Data

Genus	Number of species in genus	Nodulation data +	−	Total
Piptanthus Sweet	9	1		1
Piscidia L.	8	1		1
Pisum L.	6	6		6
Plagiocarpus Benth.	1			
Plagiolobium Sweet	1	1		1
Platycelyphium Harms	1			
Platycyamus Benth.	2			
Platylobium Sm.	3	2		2
Platymiscium Vog.	30	3		3
Platypodium Vog.	2–3			
Platysepalum Welw. ex Bak.	12			
Podalyria Willd.	25	6		6
Podocytisus Boiss. & Heldr.	1			
Podopetalum F. Muell.	1			
Poecilanthe Benth.	7			
Poiretia Vent.	7–11			
Poissonia Baill.	3			
Poitaea Vent.	5–6			
Pongamia Vent.	1	1		1
Priestleya DC.	15	5		5
Priotropis W. & A.	2			
Pseudarthria W. & A.	12	2		2
Pseudeminia Verdc.	4	1		1
Pseudoeriosema Hauman	8			
Pseudolotus Rech. f.	1			
Pseudovigna Verdc.	1	1		1
Psophocarpus DC.	8–10	5		5
Psoralea L.	120	25		25
Psoralidium Rydb.	14	3		3
Psorobatus Rydb.	2			
Psorodendron Rydb.	12	1		1
Psorothamnus Rydb.	8	1		1
Pterocarpus Jacq.	60–70	14		14
Pterodon Vog.	4			
Pteroloma DC.	5	2		2
Ptycholobium Harms	3	3		3
Ptychosema Benth.	2			
Pueraria DC.	25	2		2
Pultenaea Sm.	100	23		23
Pycnospora R. Br. ex W. & A.	1			
Rafnia Thunb.	32	9		9
Raimondianthus Harms	1			
Ramirezella Rose	9	1		1

724

Genus	Number of species in genus	Nodulation data		
		+	−	Total
Ramorinoa Speg.	1			
Requienia DC.	3	2		2
Retama Raf.	3	3		3
Rhodopis Urb.	2			
Rhynchosia Lour.	200–300	57		57
Rhynchotropis Harms	2			
Rhytidomene Rydb.	1			
Riedeliella Harms	2			
Robinia L.	20	6		6
Robynsiophyton Wilczek	1			
Rothia Pers.	2	1		1
Rudolphia Willd.	1			
Sabinea DC.	4			
Sakoanala R. Vig.	2			
Salweenia Bak. f.	1			
Sarothamnus Wimm.	1	1		1
Sauvallella Rydb.	1			
Schefflerodendron Harms	6			
Scorpiurus L.	7	3		3
Scottia R. Br. ex Ait.	6	2		2
Securigera DC.	1	1		1
Sellocharis Taub.	1			
Sesbania Scop.	70	34		34
Sewerzowia Regel & Schmalh.	1			
Shuteria W. & A.	5	1		1
Sinodoliches Verdc.	2			
Smirnowia Bge.	1			
Smithia Ait.	30	10		10
Soemmeringia Mart.	1			
Sophora L.	50–80	16	1	17
Spartidium Pomel	1			
Spartium L.	1	1		1
Spathionema Taub.	1			
Spatholobus Hassk.	15			
Sphaerolobium Sm.	12	3		3
Sphaerophysa DC.	3			
Sphenostylis E. Mey.	18	5		5
Sphinctospermum Rose	1			
Spirotropis Tul.	1			
Stauracanthus Link	1			
Steinbachiella Harms	1			
Stracheya Benth.	1			
Streblorrhiza Endl.	1			

Summary of Nodulation Data

Genus	Number of species in genus	Nodulation data +	−	Total
Strongylodon Vog.	20	2		2
Strophostyles Ell.	4	4		4
Stylosanthes Sw.	50	18		18
Sutherlandia R. Br.	6	2		2
Swainsona Salisb.	50–60	17		17
Sweetia Spreng.	20	1		1
Sylitra E. Mey.	2–3			
Taralea Aubl.	8			
Taverniera DC.	7	1		1
Teline Medik.	2	2		2
Templetonia R. Br.	10	2	1	3
Tephrosia Pers.	400	95		95
Teramnus (P. Br.) Sw.	8	7		7
Terua Standl. & Hermann	1			
Tetragonolobus Scop.	15	5		5
Teyleria Backer	1			
Thermopsis R. Br.	15–20	5		5
Thornbera Rydb.	12–14	1	1	2
Tipuana Benth.	1	1		1
Trifolium L.	250–300	141		141
Trigonella L.	70–75	19		19
Uleanthus Harms	1			
Ulex L.	20	2		2
Uraria Desv.	35	2		2
Urariopsis Schindl.	1			
Uribea Dugand & Romero	1			
Vandasia Domin	1			
Vatairea Aubl.	7			
Vataireopsis Ducke	3			
Vatovaea Chiov.	1			
Vermifrux Gillett	1			
Vicia L.	150	65		65
Vigna Savi	100–150	46		46
Viminaria Sm.	1	1		1
Virgilia Poir.	2	1		1
Voandzeia Thou.	1	1		1
Walpersia Harv.	1			
Weberbauerella Ulbr.	2			
Wenderothia Schlechtd.	12			
Whitfordiodendron Elmer	9			
Wiborgia Thunb.	10	2		2

Genus	Number of species in genus	Nodulation data		
		+	−	Total
Willardia Rose	6			
Wisteria Nutt.	10	8		8
Xanthocercis Baill.	2	1		1
Xeroderris Roberty	2	1		1
Yucaratonia Burk.	1			
Zornia J. F. Gmel.	75	12		12

Subfamily Papilionoideae	Estimated number of species	Species reported		
505 genera		+	−	Total
269 genera examined	12,215–	2,416	46	2,462
241 with all species +	13,792			
14 with all species −				
14 with both + and − species				

REFERENCE
MATERIAL

References

Abdel-Ghaffar, A. S., and Jensen, H. L. 1966. The rhizobia of *Lupinus densiflorus* Benth., with some remarks on the classification of root-nodule bacteria. *Arch. Mikrobiol.* 54:393–405.

Acree, F., Jr., Jacobson, M., and Haller, H. L. 1944. *Amorpha fruticosa* contains no rotenone. *Science* 99 (2562):99–100.

Adams, A. B., and Riches, J. H. 1930. Root nodules on lupines. *J. Dep. Agric. West. Aust.* 2d ser. 7:556.

Aggarawal, K. L. 1934. Soil flora in Deodar Forests and its importance. *Indian For.* 60 (9):602–607.

Aitken, J. B. 1930. The Wallabas of British Guiana. *Trop. Woods* 23:1–5.

Akkermans, A. D. L., Abdulkadir, S., and Trinick, M. J. 1978. N-fixing root nodules in Ulmaceae: *Parasponia* or (and) *Trema* spp.? *Plant Soil* 49:711–715.

Alfonsus, L. 1929. Sainfoin next to the clovers as a honey plant. *Am. Bee J.* 69 (3):113–114.

Ali, S. I. 1967. Proposal for the conservation of the generic name 3896 *Cylista* Wight & Arnott (1834) (LEGUMINOSAE) vs. *Cylista* Ait. (1789). *Taxon* 16:247.

Ali, S. I. 1968. *Paracalyx* Ali, a new papilionaceous genus. *Univ. Stud.* 5 (3):93–97.

Allan, H. H. 1961. Papilionaceae. *In Flora of New Zealand*, 1:367–397. Wellington, N.Z.: R. E. Owen, Government Printer.

Allard, H. A., and Zaumeyer, W. J. 1944. Responses of beans (*Phaseolus*) and other legumes to length of day. *U. S. Dep. Agric. Tech. Bull.* 867. 24 pp.

Allen, E. K., and Allen, O. N. 1933. Attempts to demonstrate nitrogen-fixing bacteria within the tissues of *Cassia tora*. *Am. J. Bot.* 20:79–84.

Allen, E. K., and Allen, O. N. 1950. Biochemical and symbiotic properties of the rhizobia. *Bact. Rev.* 14 (4):273–330.

Allen, E. K., and Allen, O. N. 1958. Biological aspects of symbiotic nitrogen fixation. *Handb. Pflphysiol.* 8:48–118.

Allen, E. K., and Allen, O. N. 1961. The scope of nodulation. *Recent Adv. Bot.* 1:585–588.

Allen, E. K., and Allen, O. N. 1976. The nodulation profile of the genus *Cassia*. *In* Nutman, P. S., ed. *Symbiotic Nitrogen Fixation in Plants. Int. Biol. Progr.* 7:113–122.

Allen, E. K., Allen, O. N., and Klebesadel, L. J. 1964. An insight into symbiotic nitrogen-fixing plant associations in Alaska. *In* Dalhgren, G., ed. *Science in Alaska. Proc. 14th Alaskan Sci. Conf., Anchorage, Alaska. 1963.* 54–63.

Allen, E. K., Allen, O. N., and Newman, A. S. 1953. Pseudonodulation of leguminous plants induced by 2-bromo-3, 5-dichlorobenzoic acid. *Am. J. Bot.* 40:429–435.

Allen, E. K., Gregory, K. F., and Allen, O. N. 1955. Morphological development of nodules on *Caragana arborescens* Lam. *Can. J. Bot.* 33:139–148.

Allen, O. N., and Allen, E. K. 1936a. Root nodule bacteria of some tropical leguminous plants: I. Cross-inoculation studies with *Vigna sinensis* L. *Soil Sci.* 42 (1):61–77.

Allen, O. N., and Allen, E. K. 1936b. Plants in the sub-family Caesalpinioideae observed to be lacking nodules. *Soil Sci.* 42:87–91.

Allen, O. N., and Allen, E. K. 1939. Root nodule bacteria of some tropical leguminous plants: II. Cross-inoculation tests within the cowpea group. *Soil Sci.* 47 (1):63–76.

Allen, O. N., and Allen, E. K. 1940a. Response of the peanut plant to inoculation with rhizobia, with special reference to morphological development of the nodules. *Bot. Gaz.* 102 (1):121–142.

Allen, O. N., and Allen, E. K. 1940b. False nodulation in certain leguminous species. *Proc. Hawaiian Acad. Sci. Spec. Publ. Bernice P. Bishop Mus.* 35:15–16.

Allen, O. N., and Allen, E. K. 1947. A survey of nodulation among leguminous plants. *Soil Sci. Soc. Am. Proc.* 12:203–208.

Allen, O. N., and Allen, E. K. 1954. Morphogenesis of the leguminous root nodule. *In Abnormal and Pathological Plant Growth. Brookhaven Symp. Biol.* 6:209–234.

Allen, O. N., and Hamatová, E., comps. 1973. IBP World Catalogue of *Rhizobium* Collections. Skinner, F.A., ed. London: International Biological Program. 282 pp.

Alpin, T. E. H. 1967. Poisonous plants of Western Australia. *J. Dep. Agric. West Aust.* ser. 4. 8 (2):42–52.

Alpin, T. E. H. 1971. Poisonous plants of Western Australia. *J. Dep. Agric. West Aust.* ser. 4. 12 (1):12–18.

Altamirano, F. 1905. Algunas palabras acerca de una planta que se dice puede utilizarse para predecir los fenómenos meteorológicos. *An. Inst. Méd. Nac. Mexico* 7:232–247.

Altschul, S. von R. 1964. A taxonomic study of the genus *Anadenanthera.* *Contrib. Gray Herb.* 193:1–65.

Alway, F. J., and Pinckney, R. M. 1909. On the relation of native legumes to the soil nitrogen of Nebraska prairies. *J. Ind. Eng. Chem.* 1:771–772.

Amshoff, G. H. 1939. Papilionaceae. *In* Pulle, A., ed. *Flora of Suriname (Netherlands Guyana)*, 2 Part 2:1–257. Amsterdam: Kolonial Institut.

Anchel, M. 1949. Identification of the antibiotic substance from *Cassia reticulata* as 4,5-dihydroxyanthraquinone-2-carboxylic acid. *J. Biol. Chem.* 177:169–177.

Anslow, R. C. 1957. Investigation into the potential productivity of "*Acacia*" (*Leucaena glauca*) in Mauritius. *Rev. Agric. Sucr. ile Maurice* 36:39–49. *Chem. Abst.* 51:14903 (1957).

Appleman, M. D., and Sears, O. H. 1943. Further evidence of interchangeability among the groups of *Rhizobium leguminosarum. Soil Sci. Soc. Am. Proc.* 7:263–264.

Aquino, D. I., and Madamba, A. L. 1939. A study of root nodule bacteria of certain leguminous plants. *Philipp. Agric.* 28:120–132.

Arora, N. 1954. Morphological development of the root and stem nodules of *Aeschynemone indica* L. *Phytomorphology* 4:211–216.

Arora, N. 1956a. Histology of the root nodules of *Cicer arietinum* L. *Phytomorphology* 6 (3,4):367–378.

Arora, N. 1956b. Morphological development of root nodules on *Crotalaria juncea. Proc. 43rd Indian Sci. Congr. Agra, 1956.* 244.

Arora, N. 1956c. Morphological study of the root nodules on *Cajanus indicus. Proc. 43rd Indian Sci. Congr. Agra, 1956.* 244–245.

Asai, T. 1944. Über die Mykorrhizenbildung der Leguminosen-Pflanzen. *Jap. J. Bot.* 13:463–485.

Asō, K., and Ohkawara, S. 1928. Studies on the nodule bacteria of Genge. *Proc. Pap. 1st Int. Congr. Soil Sci. Comm.* 3:183–184.

Aubréville, A. 1968. *Leonardoxa* Aubréville, genre nouveau de Césalpinioidées Guinéo-Congolaises. *Adansonia* ser. 2. 8 (2):177–179.

Aubréville, A., and Pellegrin, F. 1958. De quelques Césalpiniées africaines. *Bull. Soc. Bot. Fr.* 104:495–498.

Auxence, E. G. 1953. A pharmacognostic study of *Piscidia erythrina. Econ. Bot.* 7:270–284.

Avasthi, B. K., and Tewari, J. D. 1955a. A preliminary phytochemical investigation of *Desmodium gangeticum. J. Am. Pharm. Assoc.* 44:625–627.

Avasthi, B. K., and Tewari, J. D. 1955b. A preliminary phytochemical investigation of *Desmodium gangeticum.* II. Chemical constitution of the lactone. *J. Am. Pharm. Assoc.* 44:628–629.

Axelrod, D. I. 1958. Evolution of the Madro-Tertiary geoflora. *Bot. Rev.* 24:431–509.

Backer, C. A. 1939. *Teyleria* Backer. *Jard. Bot. Buitenzorg Bull.* ser. 3. 16:107.

Backer, C. A., and Bakhuizen van den Brink, R. C., Jr. 1963. *Flora of Java,* 1. 648 pp. Groningen, The Netherlands: N. V. P. Noordhoff.

Badger, G. M., and Beecham, A. F. 1951. Isolation of tetrahydroharman from *Petalostyles labicheoides. Nature* 168 (4273):517–518.

Baeyer, A. 1878. Synthese des Indigoblaus. *Berichte* 11:1296–1297.

Baeyer, A. 1879. Ueber die Einwirkung des Fünffachchlorphosphors auf Isatin und auf verwandte Substanzen. *Berichte* 12:456–461.

Bailey, C. J., and Boulter, D. 1971. Urease, a typical seed protein of the Leguminosae. *In* Harborne, J. B., Boulter, D., and Turner, B. L., eds. *Chemotaxonomy of the Leguminosae,* 485–502. New York: Academic Press.

Bailey, I. W. 1923. Notes on neotropical ant-plants. II. *Tachigalia paniculata* Aubl. *Bot. Gaz.* 75:27–41.

Baillon, H. 1866. *Baudouinia, nov. gen. Adansonia* 6:193–194.

Baillon, H. 1869. Description du nouveau genre *Brandzeia. Adansonia* 9:215–218.

Baker, B. Y., and Lynn, E. V. 1953. Examination of the seed of *Pachyrrhizus erosus. J. Am. Pharm. Assoc.* (S.E.) 42:117–118.

Baker, E. G. 1914. Plants from the Eket District, S. Nigeria. Leguminosae. *J. Bot.* 52:2–3.

Baker, E. G. 1926–1930. *The Leguminosae of Tropical Africa.* Part I. 1926.

Papilionaceae, 1–215. Ghent: Erasmus Press. Part II. 1929. Papilionaceae, 216–607. Ostend: Unitas Press. Part III. 1930. Caesalpinieae; Mimoseae, 608–653. Ostend: Unitas Press.

Baker, E. G. 1932. Lista das leguminosae Africanas, colhidas em Angola por Carrisso e Mendonça (iter Angolanum 1927), e Mario de Castro, e em Mocambique por Gomes e Souza e Pomba Cuerra. *Bol. Soc. Broteriana.* ser. 2. 8:102–115.

Baker, E. G. 1935a. New genus of Leguminosae from Tibet. *J. Bot.* 73:134–136.

Baker, E. G. 1935b. New genus of Leguminosae from Northern Rhodesia. *J. Bot.* 73:160–162.

Baldridge, J. D. 1957. The lespedezas. Part II. Culture and utilization. *In* Norman, A. G., ed. *Advances in Agronomy*, 9:113–157. New York: Academic Press.

Balfour, I. B. 1882. *Arthrocarpum*, Balf. fil. *Proc. R. Soc. Edinb.* 11:510–511.

Balfour, I. B. 1888. *Arthrocarpum*. *In* Balfour, I. B., and others. *Botany of Socotra. Trans. R. Soc. Edinb.* 31:80–81.

Ball, O. M. 1909. A contribution to the life history of *Bacillus* (Ps.) *radicicola* Beij. *Centralbl. Bakt.* ser. 2. 23:47–59.

Bamber, M. K., and Holmes, J. A. 1911. Green manuring. *Circ. Agric. J. R. Bot. Gard. Ceylon* 5 (17):217–230.

Bañados, L. L., and Fernandez, W. L. 1954. Nodulation among the Leguminosae. *Philipp. Agric.* 37:529–533.

Barbour, W. R. 1952. Cativo. *J. For.* 50 (2):96–99.

Barclay, P. C. 1957. Improvement of *Lotus uliginosus. Proc. 19th Conf. N. Z. Grassl. Assoc.* 75–81.

Barneby, R. C. 1952. A revision of the North American species of *Oxytropis* DC. *Proc. Calif. Acad. Sci.* ser. 4. 27 (7):177–309.

Barneby, R. C. 1964. *Atlas of North American Astragalus.* Part I. The Phacoid and Homaloboid Astragali. Part II. The Cercidothrix, Hypoglottis, Piptoloboid, Trimeniaeus and Orophaca Astragali. *Mem. N. Y. Bot. Gard.* 13. 1188 pp.

Barrett, M. F. 1956. *Common Exotic Trees of South Florida (Dicotyledons)*. Gainesville: University of Florida Press. 414 pp.

Barrios, S., and Gonzalez, V. 1971. Rhizobial symbiosis on Venezuelan savannas. *Plant Soil* 34 (3):707–719.

Barua, M., and Bhaduri, P. N. 1967. Rhizobial relationship of the genus *Phaseolus*. I. Cross-inoculation performance of different species to *R. phaseoli, R. japonicum* and the cowpea organism. *Can. J. Microbiol.* 13:910–913.

Beadle, N. C. W. 1959. Some aspects of ecological research in semi-arid Australia. *In* Keast, A., Crocker, R. L., and Christian, C. S., eds. *Biogeography and Ecology in Australia. Monogr. Biol.* 8:452–460.

Beadle, N. C. W. 1964. Nitrogen economy in arid and semi-arid plant communities. Part III. The symbiotic nitrogen-fixing organisms. *Proc. Linn. Soc. N. S. W.* 89 (2):273–286.

Beath, O. A., Gilbert, C. S., and Eppson, H. F. 1939. The use of indicator plants in locating seleniferous areas in western United States. I. General. *Am. J. Bot.* 26:257–269.

Beath, O. A., Gilbert, C. S., and Eppson, H. F. 1940. The use of indicator plants in locating seleniferous areas in western United States. III. Further studies. *Am. J. Bot.* 27:564–573.

Beaty, E. R., Powell, J. D., and McCreery, R. A. 1963. AMCLO: A high yielding winter clover. *Ga. Agric. Exp. Stn. Circ.* 35:5–10.

Becker, R. B., Neal, W. M., Arnold, P. T. D., and Shealy, A. L. 1935. A study of the palatability and possible toxicity of 11 species of *Crotalaria*, especially of *C. spectabilis* Roth. *J. Agric. Res.* 50 (11):911–922.

Beeley, F. 1938. Nodule bacteria and leguminous cover plants. *J. Rubber Res. Inst. Malaya.* 8 (2):149–162.

Beijerinck, M. W. 1888. Die Bacterien der Papilionaceen-Knöllchen. *Bot. Ztg.* 46:725–735; 741–750; 757–771; 781–790; 797–804.

Beijerinck, M. W. 1890. Künstliche Infektion von *Vicia faba* mit *Bacillus radicicola*: Ernährungsbedingungen dieser Bakterien. *Bot. Ztg.* 48:837–843.

Belikov, A. S., Ban'kovskiĭ, A. I., and Tsarev, M. V. 1954. Alkaloid from *Gleditschia triacanthos* [in Russian]. *Zh. Obsch. Khim.* 24:919–922. *Chem. Abstr.* 48:11727b (1954).

Bell, E. A. 1958. Canavanine and related compounds in Leguminosae. *Biochem. J.* 70:617–619.

Bell, E. A. 1971. Comparative biochemistry of non-protein amino acids. *In* Harborne, J. B., Boulter, D., and Turner, B. L., eds. *Chemotaxonomy of the Leguminosae*, 179–206. New York: Academic Press.

Benjamin, M. S. 1915. A note on the occurrence of urease in legume nodules and other plant parts. *J. R. Soc. N. S. W.* 49:78–80.

Bennetts, H. W. 1935. An investigation of plants poisonous to stock in Western Australia. *J. Dep. Agric. West. Aust.* ser. 2. 12:431–441.

Benson, L. 1941. The mesquites and screwbeans of the United States. *Am. J. Bot.* 28:748–754.

Bentham, G. 1839. Description of the Mora Tree, [as given] by Mr. Robert H. Schomburgk. *Trans. Linn. Soc. Lond.* 18:207–211.

Bentham, G. 1840. Contributions towards a Flora of South America: enumeration of plants collected by Mr. Schomburgk in British Guiana. *J. Bot.* 2:38–163.

Bentham, G. 1842. Notes on Mimoseae, with a short synopsis of species. *J. Bot.* 4:323–418.

Bentham, G. 1859–1870. Leguminosae. *In* Martius, K. F. P. von, ed. *Flora Brasiliensis*, 15. Parts 1–3. Leipzig: Monachius.

Bentham, G. 1860. A synopsis of the Dalbergieae, a tribe of the Leguminosae. *J. Proc. Linn. Soc. Bot.* suppl. 4:1–128.

Bentham, G. 1864 (reprint 1967). *Flora Australiensis: A Description of the Plants of the Australian Territory*, 2. 521 pp. Ashford, Eng.: L. Reeve.

Bentham, G. 1871. Revision of the genus *Cassia. Trans. Linn. Soc. Lond.* 27: 503–591.

Bentham, G. 1875. Revision of the suborder Mimoseae. *Trans. Linn. Soc. Lond.* 30 (3):355–668.

References

Bentham, G., and Hooker, J. D. 1865. *Genera Plantarum*, 1. 1040 pp. London: Reeve.

Bergey, D. H. 1923. *Manual of Determinative Bacteriology*. 1st ed. Baltimore: Williams and Wilkins. 442 pp.

Bergey, D. H. 1925. *Manual of Determinative Bacteriology*. 2d ed. Baltimore: Williams and Wilkins. 462 pp.

Bhaduri, P. N., Das, S. N., and Lahiri, K. K. 1967. Growth response of nodule bacteria to phytohemoagglutinin. *Experientia* 23:784–785.

Bhelke, V. 1972. Some new records of nodulated wild leguminous plants. *Curr. Sci.* 41 (12):467.

Bhide, V. P. 1956. Cross-inoculation studies with some rhizobia of the cowpea group. *Indian Phytopathol.* 9:198–201.

Bickoff, E. M., Lyman, R. L., Livingston, A. L., and Booth, A. N. 1958a. Characterization of coumestrol, a naturally occurring plant estrogen. *J. Am. Chem. Soc.* 80:3969–3971.

Bickoff, E. M., Booth, A. N., Lyman, R. L., Livingston, A. L., Thompson, C. R., and Kohler, G. O. 1958b. Isolation of a new estrogen from Ladino clover. *J. Agric. Food Chem.* 6 (7):536–537.

Birdsong, B. A., Alston, R., and Turner, B. L. 1960. Distribution of canavanine in the family Leguminosae as related to phyletic groupings. *Can. J. Bot.* 38:499–505.

Black, J. M. 1948. *Flora of South Australia*. Part II. Casuarinaceae-Euphorbiaceae, 255–521. 2d ed. Adelaide: K. M. Stevenson, Government Printer.

Blohm, H. 1962. *Poisonous Plants of Venezuela*. Cambridge: Harvard University Press. 136 pp.

Blommaert, K. L. J. 1950. The plant *Tephrosia vogelii* Hooker as a fresh water fish poison. *Trans. R. Soc. S. Afr.* 32:247–264.

Bock, H. H. 1556. *Kreuter Büch*. Strassburg: Josiam Rihel. 424 pp.

Boekelheide, V. 1960. The *Erythrina* alkaloids. *In* Manske, R. H. F., ed. *The Alkaloids, Chemistry and Physiology*, 7:201–227. New York: Academic Press.

Boissier, E. 1849. *Podocytisus* Boiss. et Heldr. *Diagn. Plant. Orient. Nov.* ser. 1. 9 (7):7–8.

Bojer, W. 1843. Descriptiones plantarum rariorum quas in insulis Africae australis detexit. *Ann. Sci. Nat. Bot.* sér. 2. 20:95–106.

Bolley, H. L. 1893. Notes on root tubercles (Wurzelknollchen) of indigenous and exotic legumes in virgin soil of northwest. *Agric. Sci.* 7:58–66.

Bolton, J. L. 1962. *Alfalfa — Botany, Cultivation, and Utilization*. New York: Interscience Publishers. 473 pp.

Bolton, J. L., Goplen, B. P., and Baenziger, H. 1972. World distribution and historical developments. *In* Hanson, C. H., ed. *Alfalfa Science and Technology*, 1–34. Agron. Ser. 15. Madison, Wis.: American Society of Agronomy.

Bonnier, C. 1957. Symbiose *Rhizobium*-légumineuses en région équatoriale. *Inst. Nat. Étude Agron. Congo Belge (I.N.É.A.C.)* sér. Sci. 72. 67 pp.

Bonnier, C. 1958. Discussion: Chap. 13, Some factors affecting nodulation in the tropics, by G. B. Masefield. *In* Hallsworth, E. G., ed. *Nutrition of the Legumes*, 212. New York: Academic Press.

Bonnier, C. 1960. Symbiose *Rhizobium*-légumineuses: aspects particuliers aux régions tropicales. *Ann. Inst. Pasteur* 98:537–556.

Boughton, I. B., and Hardy, W. T. 1935. Mescalbean (*Sophora secundiflora*) poisonous for livestock. *Tex. Agric. Exp. Stn. Bull.* 519. 18 pp.

Boulter, D., Thurman, A. D., and Derbyshire, E. 1967. A disc electrophoretic study of globulin proteins of legume seeds with reference to their systematics. *New Phytol.* 66:27–36.

Boutique, R. 1954. Vicieae. *In* Robyns, W. *Flore du Congo Belge et du Ruanda-Urundi*, 6:76–86. Brussels: Publ. Inst. Nat. Étude Agron. Congo Belge (I.N.É.A.C.).

Bowden, K., Brown, B. G., and Batty, J. E. 1954. 5-hydroxytryptamine: its occurrence in cowhage. *Nature* 174 (4437):925–926.

Bowen, G. D. 1956. Nodulation of legumes indigenous to Queensland. *Queensl. J. Agric. Sci.* 13:47–60.

Bowen, G. D. 1959a. Field studies on nodulation and growth of *Centrosema pubescens* Benth. *Queensl. J. Agric. Sci.* 16 (4):253–265.

Bowen, G. D. 1959b. Specificity and nitrogen fixation in the *Rhizobium* symbiosis of *Centrosema pubescens* Benth. *Queensl. J. Agric. Sci.* 16:267–281.

Bowen, G. D., and Kennedy, M. M. 1961. Heritable variation in nodulation of *Centrosema pubescens* Benth. *Queensl. J. Agric. Sci.* 18 (2):161–170.

Bradbury, R. B., and White, D. E. 1951. The chemistry of subterranean clover. Part I. Isolation of formononetin and genistein. *J. Chem. Soc. Lond.* 1951:3447–3449.

Brandegee, T. S. 1912. *Oxyrhynchus gen. nov. Leguminosarum. In Plantae Mexicanae Purpusianae*, IV. *Univ. Calif. Publ. Bot.* 4 (15):269–281.

Brandegee, T. S. 1919. *Hesperothamnus*, gen. nov. *Leguminosarum. In Plantae Mexicanae Purpusianae. Univ. Calif. Publ. Bot.* 6:497–503.

Braz, F. R., and Gottlieb, O. R. 1968. Chemistry of Brazilian Leguminosae. XVIII. Oxyayanine A derivatives, hexaoxygenated flavones from *Apuleia leiocarpa* [in Portuguese]. *Ann. Acad. Bras. Cienc.* 40 (2):151–153. *Chem. Abstr.* 70:35022f (1969).

Braz, R., Jr., Eyton, W. B., Gottlieb, O. R., and Magalhaes, M. T. 1968. Chemistry of Brazilian Leguminosae. XII. Apuleins: hepta-oxygenated flavones from *Apuleia leiocarpa* [in Portuguese]. *Ann. Acad. Bras. Cienc.* 40 (1):23–27. *Chem. Abstr.* 70:822q. (1969).

Brenan, J. P. M. 1955. Notes on Mimosoideae I. *Kew Bull.* 1955 (2):161–192.

Brenan, J. P. M. 1959. *Flora of Tropical East Africa*: Leguminosae, Subfamily Mimosoideae. London: Crown Agents for Overseas Governments and Administrations. 173 pp.

Brenan, J. P. M. 1960. Nomen conservandum propositum. *Taxon* 9:193–194.

Brenan, J. P. M. 1963. Notes on Mimosoideae: VIII. *Kew Bull.* 17 (2):227–228.

Brenan, J. P. M. 1967. *Flora of Tropical East Africa*: Leguminosae, Subfamily Caesalpinioideae. London: Crown Agents for Overseas Governments and Administrations. 230 pp.

Brenan, J. P. M., ed. 1970. *Flora Zambesiaca*. Leguminosae (Mimosoideae), 3: Part 1. 153 pp. London: Crown Agents for Overseas Governments and Administrations.

References

Brenan, J. P. M., and Brummitt, R. K. 1965. New and little known species from the Flora Zambesiaca area: 19. Leguminosae-Mimosoideae. *Boc. Soc. Brot.* sér. 2. 39:189–205.

Brett, C. H. 1946. Insecticidal properties of the indigobush (*Amorpha fruticosa*). *J. Agric. Res.* 73 (3):81–96.

Brewbaker, J. L., and Hylin, J. W. 1965. Variations in mimosine content among *Leucaena* species and related Mimosaceae. *Crop Sci.* 5 (4):348–349.

Brink, R. A., and Roberts, W. L. 1937. The coumarin content of *Melilotus dentata. Science* 86:41–42.

Briscoe, C. F., and Andrews, W. B. 1938. Inoculation of sesban. *J. Am. Soc. Agron.* 30 (2):135–138.

Britton, N. L. 1920. Cuban plants new to science. *Mem. Torrey Bot. Club* 16 (2):57–118.

Britton, N. L., and Killip, E. P. 1936. Mimosaceae and Caesalpiniaceae of Colombia, *Ann. N. Y. Acad. Sci.* 35:101–208.

Britton, N. L., and Rose, J. N. 1928. Mimosaceae. *North Am. Flora* 23:1–194.

Britton, N. L., and Rose, J. N. 1930. Caesalpiniaceae. *North Am. Flora* 23:201–349.

Britton, N. L., and Wilson, P. 1924. *Scientific survey of Porto Rico and the Virgin Islands,* 5: Part 3, *Neorudolphia,* 426–427. New York: New York Academy of Sciences.

Brizicky, G. K. 1960. A new species of *Paramachaerium* from Panama. *Trop. Woods* 112:58–64.

Brockwell, J., and Hely, F. W. 1966. Symbiotic characteristics of *Rhizobium meliloti*: an appraisal of the systematic treatment of nodulation and nitrogen fixation interactions between hosts and rhizobia of diverse origins. *Aust. J. Agric. Res.* 17:885–899.

Brockwell, J., and Neal-Smith, C. A. 1966. Effective nodulation of hairy canary clover. *Dorycnium hirsutum* (L.) Ser. in DC. *Aust. C.S.I.R.O. Div. Plant. Ind. Field. Stn. Rec.* 5 (1):9–15.

Brockwell, J., Hely, F. W., and Neal-Smith, C. A. 1966. Some symbiotic characteristics of rhizobia responsible for spontaneous, effective field nodulation of *Lotus hispidus. Aust. J. Exp. Agric. Ani. Husb.* 6:365–370.

Brown, N. E. 1921. New plants from tropical and South Africa collected by Archdeacon F. A. Rogers. *Kew Bull. Misc. Inf.* 8:294–295.

Brown, W. H., ed. 1921. Minor Products of Philippine Forests. *Dep. Agric. Nat. Res. Bur. For. Bull.* 22. 410 pp.

Brown, W. L. 1960. Ants, acacias and browsing mammals. *Ecology* 41 (3):587–592.

Brummitt, R. K. 1965. Nomina conservanda proposita II, 1965. *Regnum Veg.* 40:23–24.

Brummitt, R. K. 1968a. A new genus of the tribe Sophoreae (Leguminosae) from western Africa and Borneo. *Kew Bull.* 22 (3):375–386.

Brummitt, R. K. 1968b. The genus *Baphia* (Leguminosae) in east and northeast tropical Africa. *Kew Bull.* 22 (3):513–536.

Brummitt, R. K. 1970. Notes on two south-east Asian species of Leguminosae, *Cathormion umbellatum* and *Pericopsis mooniana. Kew Bull.* 24:231–234.

Brummitt, R. K., and Ross, J. H. 1974. Proposal to conserve the generic name

3557 *Hoffmannseggia* Cav. nom. illegit. over *Larrea Ortega* (Leguminosae-Caesalpinioideae). *Taxon* 23:433–435.

Brunchorst, J. 1885. Ueber die Knöllchen an den Leguminosenwurzeln. *Ber. Dtsch. Bot. Ges.* 3:241–257.

Bryant, L. H. 1946. The exploitation of New South Wales tannin-bearing species. *In Symp. on Native Tanning Materials of Australia, Sydney, B.L.M.R.A.* 7–9.

Buchanan, R. E. 1909a. The gum produced by *Bacillus radicicola. Centralbl. Bakt.* ser. 2. 22:371–396.

Buchanan, R. E. 1909b. The bacteroids of *Bacillus radicicola. Centralbl. Bakt.* ser. 2. 23:59–91.

Buchanan, R. E. 1926. What names should be used for the organisms producing nodules on the roots of leguminous plants? *Proc. Iowa Acad. Sci.* 33:81–90.

Buckhout, W. A. 1889. Experiments on the production of root tubercles. *Annu. Rep. Penn. State Coll.* 1889 Part 2:177–181.

Buendia Lazaro, F. 1966. *Semillas y Plantulas de Leguminosas Pratenses Españolas.* Madrid: Ministerio de Agricultura. 248 pp.

Bumpus, E. D. 1957. Legume nodulation in Kenya. I. Exploratory field experiments. *East Afr. Agric. J.* 23 (2):91–99.

Bunting, A. H. 1955. A classification of cultivated groundnuts. *Emp. J. Exp. Agric.* 23:158–170.

Burbidge, N. T. 1963. *Dictionary of Australian Plant Genera.* Sydney: Angus and Robertson, Halstead Press. 345 pp.

Burger, M. M., and Noonan, K. D. 1970. Restoration of normal growth by covering of agglutinin sites on tumour cell surface. *Nature* 228:512–515.

Burkart, A. 1939a. Leguminosas-Hedisareas de la República Argentina. *Darwiniana* 3 (2):261–283.

Burkart, A. 1939b. Descripción de *Mimozyganthus* nuevo género de Leguminosas y sinopsis preliminar de los géneros argentinos de Mimosóideae. *Darwiniana* 3 (3):445–469.

Burkart, A. 1940. Materiales para una monographía del genero *Prosopis* (Leguminosae). *Darwiniana* 4:57–128.

Burkart, A. 1943. *Las Leguminosas Argentinas — Silvestres y Cultivadas.* Buenos Aires: Acme Agency. 590 pp.

Burkart, A. 1944. Tres nuevas leguminosas del Paraguay. *Darwiniana* 6 (3):477–482.

Burkart, A. 1949a. La posicion sistematica del "chanar" y las especies del genero *Geoffroea* (Leguminosae-Dalbergieae). *Darwiniana* 9 (1):9–23.

Burkart, A. 1949b. Leguminosas nuevas o criticas. III. *Darwiniana* 9:63–96.

Burkart, A. 1952. *Las Leguminosas Argentinas — Silvestres y Cultivadas.* 2d ed. Buenos Aires: Acme Agency. 569 pp.

Burkart, A. 1957. Leguminosas nuevas o criticas, V. *Darwiniana* 11:256–271.

Burkart, A. 1964a. Contribución al estudio del género *Adesmia* (Leguminosae), V. *Darwiniana* 13:9–66.

Burkart, A. 1964b. Leguminosas nuevas o criticas, VI. *Darwiniana* 13:428–448.

Burkart, A. 1967. Sinopsis del género sudamericano de leguminosas *Adesmia*

DC. (Contributión al estudio del género *Adesmia*, VII). *Darwiniana* 14:463–568.

Burkart, A. 1969. Leguminosas nuevas o criticas VII. *Darwiniana* 15:501–549.

Burkart, A. 1970. Las Leguminosas — Faseólias Argentinas. Los generos *Mucuna, Dioclea* y *Camptosema. Darwiniana* 16:175–218.

Burkart, A. 1971. El género *Galactia* (Legum.-Phaseoleae) en Sudamérica, con especial referencia a la Argentina y paises vecinos. *Darwiniana* 16:663–796 + 6 plates.

Burkart, A., and Brücher, H. 1953. *Phaseolus aborigineus* Burkart, die mutmassliche andine Stammform der Kulturbohne [in German, English summary]. *Züchter* 23 (3):65–72.

Burkart, A., and Vilchez, O. 1971. Valoración e ilustración del género *Fiebrigiella* Harms (Leguminosae-Hedysareae). *Darwiniana* 16:659–662.

Burkill, I. H. 1966. *A Dictionary of the Economic Products of the Malay Peninsula.* 1 (A–H). 1–1240; 2 (I–Z). 1241–2444. 2d ed. London: Crown Agents for the Colonies.

Burrage, S. 1900. Description of certain bacteria obtained from nodules of various leguminous plants. *Indiana Acad. Sci. Proc.* 16:157–161.

Burrill, T. J., and Hansen, R. 1917. Is symbiosis possible between legume bacteria and non-legume plants? *Ill. Agric. Exp. Stn. Bull.* 202:115–181.

Burrows, B. F., Finch, N., Ollis, W. D., and Sutherland, I. O. 1959. Mundulone. *Chem. Soc. Lond. Proc.* 1959:150–152.

Burton, J. C. 1952. Host specificities among certain plants in the cowpea cross-inoculation group. *Soil Sci. Soc. Am. Proc.* 16:356–358.

Burton, J. C. 1972. Nodulation and symbiotic nitrogen fixation. *In* Hanson, C. H., ed. *Alfalfa Science and Technology.* Agron. Ser. 15:229–246. Madison, Wis.: American Society of Agronomy.

Burton, J. C., and Curley, R. L. 1968. Nodulation and nitrogen fixation in sainfoin (*Onobrychis sativa*, Lam.) as influenced by strains of rhizobia. *In* Cooper, C. S., and Carleton, A. E., eds. *Sainfoin Symposium. Montana Agric. Exp. Stn. Bull.* 627. 109 pp.

Burton, J. C., and Erdman, L. W. 1940. A division of the alfalfa cross-inoculation group correlating efficiency in nitrogen fixation with source of *Rhizobium meliloti. J. Am. Soc. Agron.* 32:439–450.

Burton, J. C., and Wilson, P. W. 1939. Host plant specificity among the *Medicago* in association with root-nodule bacteria. *Soil Sci.* 47:293–303.

Burton, J. C., Curley, R. L., and Martinez, C. J. 1974. Rhizobia inoculants for various leguminous species. *Nitragin Inf. Bull.* 101:1–3.

Büsgen, M. 1905. Studien über die Wurzelsysteme einiger dikotyler Holzpflanzen. *Flora* 95:58–94.

Bushnell, O. A., and Sarles, W. B. 1937. Studies on the root-nodule bacteria of wild leguminous plants in Wisconsin. *Soil Sci.* 44 (6):409–423.

Busson, F. 1965. *Plantes Alimentaires de l'Ouest Africain.* Marseille: Imprimerie Leconte, 568 pp.

Byl, P. A. van der. 1914. The anatomy of *Acacia mollissima* (Willd.) with special reference to the distribution of tannin. *Union S. Afr. Dep. Agric. Div. Bot. Sci. Bull.* 3:32 pp.

Camp, B. J., and Norvell, M. J. 1966. The phenylethylamine alkaloids of native range plants. *Econ. Bot.* 20:274–278.

Camp, W. H., Rickett, H. W., and Weatherby, C. A., comps. and eds. 1947. International rules of botanical nomenclature. *Brittonia* 6 (1):1–120.

Campbell, H. A., and Link, K. P. 1941. Studies on the hemorrhagic sweetclover disease. IV. The isolation and crystallization of the hemorrhagic agent. *J. Biol. Chem.* 138:21–33.

Campbell, J. S., and Gooding, H. J. 1962. Recent developments in the production of food crops in Trinidad. *Trop. Agric. (Trinidad)* 39:261–270.

Campêlo, A. B., and Campêlo, C. R. 1970. Eficiência da inoculação cruzada entre espécies da subfamília Mimosoideae. *Pesqui. Agropecu. Bras.* 5:333–337.

Campêlo, A. B., and Döbereiner, J. 1969. Estudo sôbre inoculacão cruzada de algumas leguminosas florestais. *Pesqui. Agropecu. Bras.* 4:67–72.

Cannon, H. L. 1957. Description of indicator plants and methods of botanical prospecting for uranium deposits on the Colorado Plateau. *U. S. Geol. Surv. Bull.* 1030-M:399–516.

Cannon, H. L. 1960a. Botanical prospecting for ore deposits. *Science* 132 (3427):591–598.

Cannon, H. L. 1960b. The development of botanical methods of prospecting for uranium on the Colorado Plateau. *U. S. Geol. Surv. Bull.* 1085-A:1–50.

Cannon, H. L. 1964. Geochemistry of rocks and related soils and vegetation in the Yellow Cat area, Grand County, Utah. *U. S. Geol. Surv. Bull.* 1176. 127 pp.

Cannon, H. L. 1971. The use of plant indicators in ground water surveys, geologic mapping, and mineral prospecting. *Taxon* 20 (2–3):227–256.

Cappelletti, C. 1928. I tubercoli radicali delle leguminose considerati nei loro rapporti immunitari e morfologici. *Ann. Bot. (Rome)* 17 (5):1–87.

Capuron, R. 1968a. *Eligmocarpus* R. Capuron, gen. nov. *Adansonia* 8:205–208.

Capuron, R. 1968b. Les Swartziees de Madagascar. *Adansonia* 8:217–222.

Carmin, J. 1950. Plants in Israel, their biology, diseases and cryptogamic inhabitants. 3. *Alhagi maurorum* Medik. *Bull. Indep. Biol. Lab. (Kefar-Malal)* 6 (2) (48). 6 pp.

Carr, W. R. 1957. Notes on some Southern Rhodesian indigenous fruits, with particular reference to their ascorbic acid content. *Food Res.* 22:590–596.

Carrier, L. 1923. *The Beginnings of Agriculture in America.* 1st ed. New York: McGraw-Hill. 323 pp.

Carroll, W. R. 1934a. A study of *Rhizobium* species in relation to nodule formation on the roots of Florida legumes. I. *Soil Sci.* 37:117–135.

Carroll, W. R. 1934b. A study of *Rhizobium* species in relation to nodule formation on the roots of Florida legumes: II. *Soil Sci.* 37:227–241.

Casares López, R., Garcia Olmedo, R., and Peralta, T. 1950. A study of the composition of the seed of *Trigonella foenum-graecum*. *An. Bromatol.* 2:353–360. *Chem. Abstr.* 9140 (1951).

Cass Smith, W. P. 1941. Conserve nitrogen fertilisers by growing inoculated legumes. *J. Dep. Agric. West. Aust.* ser. 2. 18:87–89.

Cercós, A. P. 1951. Antibacterial activity of vinaline, alkaloid of *Prosopis rus-*

cifolia [in Spanish, English summary]. *Rev. Argent. Agron.* 18:200–209. *Chem. Abstr.* 46:11311e (1952).

Chakravarty, H. L. 1969. Flower-structure of Caesalpiniaceae (*Dialium*: an African apocarpus relict). *J. Indian Bot. Soc.* 48:191–193.

Chandapillai, M. M. 1972. Studies on the nodulation of *Stylosanthes guyanensis* Aubl. I. Effect of added organic matter in four types of Malaysian soil. *Trop. Agric. (Trinidad)* 49 (3):205–213.

Chapman, A. G. 1934. The effect of Black Locust on associated species with special reference to forest trees. *Am. Soil Surv. Assoc. Bull.* 15:39–41.

Charlton, J. 1926. The selection of Burma beans (*Phaseolus lunatus*) for low prussic acid content. *Mem. Dep. Agric. India*, chem. ser. 9:1–36.

Chen, H. K., and Shu, M. K. 1944. Note on the root-nodule bacteria of *Astragalus sinicus* L. *Soil Sci.* 58:291–293.

Cheney, R. H. 1931. Geographic and taxonomic distribution of American plant arrow poisons. *Am. J. Bot.* 18 (2):136–145.

Chestnut, V. K. 1899. Preliminary catalogue of plants poisonous to stock. *Annu. Rep. U. S. Dep. Agric. Bur. Anim. Ind.* 15 (1898):387–420.

Chestnut, V. K., and Wilcox, E. V. 1901. Stock-poisoning plants in Montana. A preliminary report. *U. S. Dep. Agric. Div. Bot. Bull.* 26. 150 pp.

Chevalier, A. 1933. Monographie de l'Arachide. Part I. *Rev. Bot. Appl. Agric. Trop.* 13:689–789.

Chevalier, A. 1946. Sur diverses Légumineuses Caesalpiniées à feuilles multi et parvifoliolées vivant dans les forêts de l'Afrique tropicale et donnant des bois recherchés. *Rev. Int. Bot. Appl. Agric. Trop.* 26 (289–290):585–621.

Chevalier, A. 1951. Le kudzu (*Pueraria hirsuta*) et quelques autres légumineuses anti-érosives à cultiver dans les pays tropicaux. *Rev. Bot. Appl. Agric. Trop.* 31:159–172.

Chevalier, A., and Sillans, R. 1952. Sur quatre *Droogmansia* de l'Afrique tropicale au N W de l'equateur. *Rev. Int. Bot. Appl. Agric. Trop.* 32:44–53.

Chiovenda, E. 1951. Plantae novae vel minus notae ex Aethiopia. *Webbia* 8:229–240.

Chrtkova-Zertova, A. 1966. Bemerkungen zur Taxonomie von *Lotus uliginosus* Schkuhr und L. *pedunculatus* Cav. *Folia Geobot. Phytotaxon.* 1:78–87.

Chun, W. Y. 1946. A new genus in the Chinese Flora. *Sunyatsenia* 6 (3–4):195–198.

Clark, E. P. 1930. Some constituents of *Derris* and "cube" roots other than rotenone. *Science* 71:396.

Clark, E. P. 1931a. Deguelin. I. The preparation, purification and properties of deguelin, a constituent of certain tropical fish-poisoning plants. *J. Am. Chem. Soc.* 53(1).313–317.

Clark, E. P. 1931b. *Tephrosia*. I. The composition of tephrosin and its relation to deguelin. *J. Am. Chem. Soc.* 53:729–732.

Clark, E. P. 1933. The occurrence of rotenone and related compounds in the roots of *Cracca virginiana*. *Science* n.s. 77 (1955):311–312.

Clark, E. P. 1943. The occurrence of rotenone and related substances in the seeds of the Berebeara Tree. A procedure for the separation of Deguelin and Tephrosin. *J. Am. Chem. Soc.* 65:27–29.

Clausen, R. T. 1944. A botanical study of the yam beans (*Pachyrrhizus*). *Cornell Univ. Agric. Exp. Stn. Mem.* 264. 38 pp.

Clemo, G. R., and Raper, R. 1935. The alkaloids of *Ulex europaeus*. Part I. *J. Chem. Soc.* 1935:10–11.

Cloonan, M. J. 1963. Black nodules on *Dolichos*. *Aust. J. Sci.* 26 (4):121.

Clos, D. 1893. Revision des tubercules des plantes et des tuberculoides des Légumineuses. *Acad. Sci. Inscript. Belles-Lett. (Toulouse) Mem.* ser. 9. 5:381–405.

Clos, D. 1896. Caractères extérieurs et modes de répartition des petits tubercules ou tuberculoïdes des légumineuses. *C. R. Acad. Sci.* 123:407–410.

Clough, G. W. 1925. Lathyrism. *Vet. Rec.* n.s. 5:839–840.

Clover, A. M. 1906. Philippine wood oils. *Philipp. J. Sci.* sec. A. 1:191–202.

Coates Palgrave, O. H. 1957. *Trees of Central Africa*. Salisbury: Robert Mac-Lehose for the National Publications Trust. 466 pp.

Codd, L. E. 1956. The *Schotia* species of Southern Africa. *Bothalia* 6 (3):513–533.

Coit, J. E. 1951. Carob or St. John's Bread. *Econ. Bot.* 5:82–96.

Collett, H., and Hemsley, W. B. 1890. On a collection of plants from Upper Burma and the Shan States. *J. Linn. Soc. Bot.* 28:1–150.

Conklin, M. E. 1936. Studies of the root nodule organisms of certain wild legumes. *Soil Sci.* 41:167–185.

Conn, H. J. 1948. *The History of Staining*. 2d ed. Geneva, N. Y.: Biotech Publications. 143 pp.

Conn, H. J. 1953. *Biological Stains*. 6th ed. Geneva, N. Y.: Biotech Publications. 367 pp.

Cook, O. F. 1919. *Olneya* beans. *J. Hered.* 10 (7):321–331.

Cook, O. F., and Collins, G. N. 1903. Economic Plants of Porto Rico. *Contrib. U.S. Natl. Herb.* 8 (2):57–269.

Cooper, G. P. 1928. Some interesting trees of Western Panama. *Trop. Woods* 14:1–8.

Corby, H. D. L. 1971. The shape of leguminous nodules and colour of leguminous roots. *Plant Soil* spec. vol. 1971:305–314.

Corby, H. D. L. 1974. Systematic implications of nodulation among Rhodesian legumes. *Kirkia* 9 (2):301–329.

Corner, E. J. H. 1951. The leguminous seed. *Phytomorphology* 1:117–150.

Corner, E. J. H. 1952. *Wayside Trees of Malaya*, 1. 772 pp. 2d ed. Singapore: Government Printing Office.

Couch, J. F. 1926. Relative toxicity of the lupine alkaloids. *J. Agric. Res.* 32 (1):51–67.

Cowan, R. S. 1953. A taxonomic revision of the genus *Macrolobium* (Leguminosae-Caesalpinioideae). *Mem. N. Y. Bot. Gard.* 8 (4):257–342.

Cowan, R. S. 1957. Tropical American Leguminosae — III. *Brittonia* 8:251–253.

Cowan, R. S. 1968. *Swartzia* (Leguminosae, Caesalpinioideae, Swartzieae). *Flora Neotropica*. Monograph No. 1. New York: Hafner. 228 pp.

References

Cowan, R. S. 1975. A monograph of the genus *Eperua* (Leguminosae: Caesalpinioideae). *Smithson. Contrib. Bot.* 28. 45 pp. + 13 figs. 2 tables.

Cowen, D. V. 1965. *Flowering Trees and Shrubs in India.* 4th ed. rev. and enlar. Bombay: Thacker. 159 pp.

Cox, D. H., Harris, D. L., and Richard, T. A. 1958. Chemical identification of *Crotalaria* poisoning in horses. *J. Am. Vet. Med. Assoc.* 133:425–426.

Coxworth, E. 1965. Alkaloids of the Calabar Bean. *In* Manske, R. H. F., ed. *The Alkaloids, Chemistry and Physiology,* 8:27–45. New York: Academic Press.

Craib, W. G. 1927. *Afgekia* Craib, gen. nov. *Kew Bull.* 1927:376–378.

Craig, L. E. 1955. Curare-like effects. *In* Manske, R. H. F., ed. *The Alkaloids, Chemistry and Physiology,* 5:265–293. New York: Academic Press.

Cram, W. H. 1952. Parent-seedling characteristics and relationships in *Caragana arborescens,* Lam. *Sci. Agric.* 32:380–402.

Crampton, B. 1946. Hairy canary clover. *Calif. Agric.* 18 (3):12–13.

Cranmer, M. F., and Mabry, T. J. 1966. The lupine alkaloids of the genus *Baptisia* (Leguminosae). *Phytochemistry* 5:1133–1138.

Cranmer, M. F., and Turner, B. L. 1967. Systematic significance of lupine alkaloids with particular reference to *Baptisia* (Leguminosae). *Evolution* 21 (3):508–517.

Crocker, R. L., and Wood, J. G. 1947. Some historical influences on the development of the South Australian vegetation communities and their bearings on concepts and classification in ecology. *Trans. R. Soc. S. Aust.* 71:91–136.

Crombie, L., and Whiting, D. A. 1963. The extractives of *Neorautanenia pseudopachyrrhiza*; the isolation and structure of a new rotenoid and two new isoflavanones. *J. Chem. Soc. Lond.* 1963:1569–1579.

Cronquist, A. 1954a. Galegeae. *In* Robyns, W. *Flore du Congo Belge et du Ruanda-Urundi,* 5:72–175. Brussels: Publ. Inst. Nat. Étude Agron. Congo Belge (I.N.É.A.C.).

Cronquist, A. 1954b. Genre *Dalbergia. In* Robyns, W. *Flore du Congo Belge et du Ruanda-Urundi,* 6:52–75. Brussels: Publ. Inst. Nat. Étude Agron. Congo Belge (I.N.É.A.C.).

Cruickshank, I. A. M. 1962. Studies on phytoalexins. IV. The antimicrobial spectrum of pisatin. *Aust. J. Biol. Sci.* 15:147–159.

Cruickshank, I. A. M., and Perrin, D. R. 1961. Studies on phytoalexins. III. The isolation, assay, and general properties of a phytoalexin from *Pisum sativum* L. *Aust. J. Biol. Sci.* 14:336–348.

Dadarwal, K. R., Singh, C. S., and Subbarao, N. S. 1974. Nodulation and serological studies of rhizobia from six species of *Arachis. Plant Soil* 40:535–544.

Dahlgren, R. 1960. Revision of the genus *Aspalathus,* 1. The species with flat leaflets. *Opera Bot.* 4:1–393.

Dahlgren, R. 1961. Revision of the genus *Aspalathus,* 2. The species with ericoid and pinoid leaflets, 1; 2. *Opera Bot.* 6 (2):9–69; 75–120.

Dahlgren, R. 1963a. Revision of the genus *Aspalathus,* 2. The species with ericoid and pinoid leaflets, 3. The *Aspalathus ciliaris* group and some related species. *Opera Bot.* 8 (1):1–183.

Dahlgren, R. 1963b. Studies on *Aspalathus* and some related genera in South Africa. *Opera Bot.* 9 (1). 301 pp.

Dahlgren, R. 1963c. The genus *Borbonia* L. incorporated in *Aspalathus* L. *Bot. Not.* 116 (2):185–192.

Dahlgren, R. 1965a. Revision of the genus *Aspalathus*, 2. The species with ericoid and pinoid leaflets, 4. *Opera Bot.* 10 (1):1–231.

Dahlgren, R. 1965b. The riddle of *Walpersia* Harv. *Bot. Not.* 118 Fasc. 1:97–103.

Dahlgren, R. 1968. Revision of the genus *Aspalathus*. Part III. The species with flat and simple leaves. *Opera Bot.* 22:1–126.

Dalechamps, J. 1587. *Historia Generalis Plantarum*, 1:487–488. Lyons: G. Rovillium.

Dalla Torre, C. G., and Harms, H. 1900–1907 (reprint 1958). *Genera Siphonogamarum ad Systema Englerianum Conscripta*. Leipzig: Engelmann. 921 pp. (Reprint. Lehre, Germany: J. Cramer. 568 pp.)

Daly, R. F., and Egment, A. C. 1966. Statistical supplement to "A look ahead for food and agriculture." *Agric. Econ. Res.* 18 (1):1–9.

Dalziel, J. M. 1948. *The Useful Plants of West Tropical Africa*. London: Crown Agents for the Colonies. 612 pp.

Damodaran, M., and Sivaramakrishnam, P. M. 1937. New sources of urease for determination of urea. *Biochem. J.* 31:1041–1046.

Dangeard, P. A. 1926. Recherches sur les tubercules radicaux des Légumineuses. *Botaniste* ser. 16. 270 pp.

Daniel, T. M., and Wisnieski, J. J. 1970. The reaction of Concanavalin-A with mycobacterial culture filtrates. *Am. Rev. Respir. Dis.* 101:762–764.

Dann, O., and Hofmann, H. 1963. Chromane. XV. Synthese von (±)-Brasilin. *Ann. Chem.* 667:116–125.

Darlington, C. D., and Wylie, A. P. 1955. *Chromosome Atlas of Flowering Plants*. 2d ed. London: Allen and Unwin. 519 pp.

Dart, P. J., and Mercer, F. V. 1966. Fine structure of bacteroids in root nodules of *Vigna sinensis*, *Acacia longifolia*, *Viminaria juncea* and *Lupinus angustifolius*. *J. Bact.* 91:1314–1319.

Daubeny, C. 1857. *Lectures on Roman Husbandry*. Oxford: James Wright, printer to the University. 328 pp.

Davidson, W. M. 1930. The relative value as contact insecticides of some constituents of *Derris*. *J. Econ. Entomol.* 23:877–879.

Davis, P. H., and Plitmann, U. 1970. *Lens*. In Davis, P. H., ed. *Flora of Turkey*, 3:325–328. Edinburgh: Edinburgh University Press.

Davis, T. A. W., and Richards, P. W. 1934. The vegetation of Moraballi Creek, British Guiana: an ecological study of a limited area of tropical rain forest. Part 2. *J. Ecol.* 22:106–155.

Dawson, M. 1900a. Nitragin and the nodules of leguminous plants. *Philos. Trans. R. Soc. Lond.* ser. B. 192:1–28.

Dawson, M. 1900b. Further observations on the nature and functions of the nodules of leguminous plants. *Philos. Trans. R. Soc. Lond.* ser. B.193 (1):51–67.

De Candolle, A. L. 1908. *Origin of Cultivated Plants*. Int. Sci. Ser. 48:468 pp. New York: D. Appleton.

References

De Candolle, A. P. 1825a. *Prodromus Systematis Naturalis Regni Vegetabilis*, 2. 644 pp. Paris: Treuttel and Würtz.

De Candolle, A. P. 1825b (reprint 1966). *Mémoires sur la Famille des Légumineuses.* Lehre, Germany: J. Cramer. 525 pp.

De Candolle, A. P. 1825c. Notice sur quelques genres et espèces nouvelles de légumineuses. *Ann. Sci. Nat.* 4:90–103.

DeLey, J., and Rassel, A. 1965. DNA base composition, flagellation and taxonomy of the genus *Rhizobium. J. Gen. Microbiol.* 41:85–91.

Demolon, A., and Dunez, A. 1938. Observations agronomiques sur la symbiose bactérienne des légumineuses. *C. R. Acad. Sci.* 206:701–703.

DeSouza, D. I. A. 1963. A study of the nodulation of some indigenous and introduced legumes in Trinidad. Diploma of Tropical Agriculture Thesis, The Imperial College of Tropical Agriculture, Univ. West Indies, Trinidad. 56 pp.

DeSouza, D. I. A. 1966. Nodulation of indigenous Trinidad legumes. *Trop. Agric. (Trinidad)* 43:265–267.

Desvaux, A. 1826. Observations sur la famille des Légumineuses. *Ann. Sci. Nat.* 9:404–431.

Deutsch, G., Döbereiner, J., and Tokarnia, C. H. 1965. Fotossensibilidade hepatogenica em bovinos na intoxicacao pela fava de *Enterolobium gummiferum* (Mart.) MacBr. *Congr. Int. Pastagens, São Paulo, 1965.* 2:1279–1282.

de Wit, H. C. D. 1941. Notes on the genera *Intsia* and *Pahudia* (Legum.) *Bull. Bot. Gard. Buitenzorg,* ser. 3. 17:139–154.

de Wit, H. C. D. 1947a. Revision of the genus *Koompassia* Maingay ex Benth. (Legum.) *Bull. Bot. Gard. Buitenzorg,* ser. 3. 17:309–312.

de Wit, H. C. D. 1947b. *Endertia* van Steenis et de Wit. A new leguminous genus from Borneo (Legum.-Caes.) *Bull. Bot. Gard. Buitenzorg,* ser. 3. 17 (3):323–327.

de Wit, H. C. D. 1952. A revision of the genus *Archidendron* F. Muell. (Mimosaceae). *Reinwardtia* 2 (1):69–96.

de Wit, H. C. D. 1956. Revision of Malaysian *Bauhinieae. Reinwardtia* 3:381–541.

de Wit, H. C. D. 1961. Typification and correct names of *Acacia villosa* Willd. and *Leucaena glauca* (L.) Benth. *Taxon* 10:50–54.

Dewit, J., and Duvigneaud, P. 1954a. Les *Smithia* Leguminosae du Congo méridional *Bull. Soc. R. Bot. Belg.* 86:207–214.

Dewit, J., and Duvigneaud, P. 1954b. *Kotchya* Endl. et *Smithia. In* Robyns, W. *Flore du Congo Belge et du Ruanda-Urundi,* 5. 377 pp. Brussels: Publ. Inst. Nat. Étude Agron. Congo Belge (I.N.É.A.C.).

Diatloff, A., and Diatloff, G. 1977. *Chaetocalyx* — a non-nodulating papilionaceous legume. *Trop. Agric. (Trinidad)* 54 (2):143–147.

Dijkman, M. J. 1950. *Leucaena* — A promising soil-erosion-control plant. *Econ. Bot.* 4:337–349.

Dilworth, M. J. 1969. The plant as the genetic determinant of leghemoglobin production in legume root nodules. *Biochim. Biophys. Acta* 184 (2):432–441.

746

Döbereiner, J. 1971. *Rhizobium* research in Brazil. *Proc. 4th Aust. Legume Nodulation Conf. Div. Plant Ind. C.S.I.R.O., Canberra.* Paper 18.

Döbereiner, J., Scott, D. B., Burris, R. H., and Hollaender, A., eds. 1978. *Limitations and Potentials of Biological N_2 Fixation in the Tropics.* Basic Life Sci. 10. 398 pp. New York: Plenum Press.

Doku, E. V., and Karikari, S. K. 1971. Bambarra groundnut. *Econ. Bot.* 25 (3):255–262.

D'Oliveira, M. de L., and Loureiro, S. M. de. 1941. Observações sôbre alguns isolamentos de *Rhizobium* e sua associação com espécies de Leguminosas. *Actas I Congr. Nac. Ciênc. Nat., Bol. Soc. Portug. Ciênc. Nat.* 13:366–371.

Domin, K. 1912. Fifth contribution to the flora of Australia. *Feddes Repert. Specierum Nov. Reg. Veg.* 11:261–264.

Domin, K. 1923. *Nemcia*, a new genus of the Leguminosae. *Preslia* 2:26–31.

Domin, K. 1926. Beiträge zur Flora und Pflanzengeographie Australiens. *Bibl. Bot.*,22:774–775.

Drummond, R. B. 1972. A list of Rhodesian legumes. *Kirkia* 8 (2): 209–229.

Dubey, H. D., Woodbury, R., and Rodríguez, R. L. 1972. New records of tropical legume nodulation. *Bot. Gaz.* 133 (1):35–38.

Ducke, A. 1922. Plantes nouvelles ou peu connues de la région amazonienne. Part 2. *Arch. Jard. Bot. Rio de Janeiro* 3:3–281.

Ducke, A. 1925a. Plantes nouvelles ou peu connues de la région amazonienne. Part 3. *Arch. Jard. Bot. Rio de Janeiro* 4:1–210.

Ducke, A. 1925b. As leguminosas do estado do Pará. *Arch. Jard. Bot. Rio de Janeiro* 4:211–341.

Ducke, A. 1930. Plantes nouvelles ou peu connues de la région amazonienne. Part 4. *Arch. Jard. Bot. Rio de Janeiro* 5:101–187.

Ducke, A. 1932. *Vataireopsis* Ducke nov. gen. Neue Arten aus der Hylaea Brasiliens. *Notizbl. Bot. Gart. Mus. Berlin-Dahlem* 11 (105):471–483.

Ducke, A. 1933. Plantes nouvelles au peu connues de la région amazonienne. Part 5. *Arch. Jard. Bot. Rio de Janeiro* 6:1–106.

Ducke, A. 1934a. Revision of the species of the genus *Elizabetha* Schomb. *Trop. Woods* 37:18–27.

Ducke, A. 1934b. *Recordoxylon:* a new genus of Leguminosae-Caesalpinioideae. *Trop. Woods* 39:16–18.

Ducke, A. 1934c. Les genres *Coumarouna* Aubl. et *Taralea* Aubl. *Rev. Bot. Appl. Agric. Trop.* 14:400–407.

Ducke, A. 1935a. New species of the genus *Dimorphandra* Schott section *Pocillum* Tul. *J. Wash. Acad. Sci.* 25 (4):193–198.

Ducke, A. 1935b. Note on the genus *Paramachaerium. Trop. Woods* 41:6–7.

Ducke, A. 1938a. Die Gattungen *Coumarouna* Aubl. und *Taralea* Aubl. *Notizbl. Bot. Gart. Mus. Berlin-Dahlem* 14 (121):120–127.

Ducke, A. 1938b. Plantes nouvelles ou peu connues de la region Amazonienne (X série). *Arch. Inst. Biol. Veg. Rio de Janeiro* 4 (1):6–64.

Ducke, A. 1940a. Revision of the species of the genus *Coumarouna* Aubl. or *Dipteryx* Schreb. *Trop. Woods* 61:1–10.

Ducke, A. 1940b. Notes on the Wallaba trees (*Eperua* Aubl.). *Trop. Woods* 62:21–28.

References

Ducke, A. 1940c. Additions to "Revision of the species of the genus *Elizabetha* Schomb." *Trop. Woods* 62:32–33.

Dugand, A. 1962a. Acerca de un nuevo genero de leguminosas (Lotoideae-Sophoreae). *Mutisia* 27:1–12.

Dugand, A. 1962b. La madera de *Uribea tamarindoides* (Leguminosae-Lotoideae-Sophoreae). *Mutisia* 27:13–16.

Duggar, J. F. 1897. Soil inoculation for leguminous plants. *Ala. Agr. Exp. Stn. Bull.* 87:459–488.

Duggar, J. F. 1934. Root nodule formation as affected by planting of shelled or unshelled seeds of bur clovers, black medic, hubam, and crimson and subterranean clovers. *J. Am. Soc. Agron.* 26:919–923.

Duggar, J. F. 1935. Relative promptness of nodule formation among vetches, vetchlings, winter peas, clovers, melilots, and medics. *J. Am. Soc. Agron.* 27:542–545.

Dumaz-le-Grand, N. 1952. Contribution à l'étude des légumineuses de Madagascar IV. Les genres *Xanthocercis* H. Bn. et *Pseudocadia* Harms. *Bull. Soc. Bot. Fr.* 99:313–315.

Dümmer, R. A. 1912. *Pearsonia*, a new genus of Leguminosae. *J. Bot.* 50:353–358.

Dümmer, R. A. 1913. A synopsis of the species *Lotononis* Eckl. & Zeyh. and *Pleiospora* Harv. *Trans. R. Soc. S. Afr.* 3:275–336.

Dunn, S. T. 1903. *Dolichos lagopus* Dunn. Descriptions of new Chinese plants. *J. Linn. Soc. Bot.* 35:483–518.

Dunn, S. T. 1910. *Leptoderris* Dunn. *Kew Bull.* 1910:386–391.

Dunn, S. T. 1911a. *Craibia*, a new genus of Leguminosae. *J. Bot.* 49:106–109.

Dunn, S. T. 1911b. Some additions to the leguminous genus *Fordia*. *Kew Bull. Misc. Inf.* 1911:62–64.

Dunn, S. T. 1911c. *Adinobotrys* and *Padbruggea*. *Kew Bull. Misc. Inf.* 1911:193–198.

Dunn, S. T. 1911d. *Ostryocarpus* and a new allied genus *Ostryoderris*. *Kew Bull. Misc. Inf.* 1911:362–364.

Dunn, S. T. 1912. A revision of the genus *Millettia* Wight et Arn. *J. Linn. Soc. Lond. Bot.* 41:123–243.

Dunstan, W. R., and Henry, T. A. 1903. Cyanogenesis in plants. Part III. On phaseolunatin, the cyanogenetic glucoside of *Phaseolus lunatus*. *Proc. R. Soc. Lond.* 72:285–294.

Dunstan, W. R., and Henry, T. A. 1907–1908. The poisonous properties of the beans of *Phaseolus lunatus*. *J. Board Agric. Gr. Br.* 14:722–731.

Dupuy, H. P., and Lee, J. G. 1954. The isolation of a material capable of producing experimental lathyrism. *J. Am. Pharm. Assoc.* 43:61–62.

Dussy, J., and Sannié, C. 1947. Sur un rhamnoside nouveau extrait des feuilles d'*Erythrophleum guineense*. *C. R. Acad. Sci.* 225:693–695.

Dutta, N. L. 1954. The sugar constituents of the saponins from *Entada scandens*. *J. Sci. Ind. Res. (India)* 13B:672.

Dutta, N. L. 1955. Chemical examination of *Mundulea suberosa* Benth. *J. Sci. Ind. Res. (India)* 14B:424–425.

Dutta, N. L. 1959. Chemical investigation of *Mundulea suberosa*. II. Constitution of munetone, the principal crystalline product of the root bark. *J. Indian Chem. Soc.* 36:165–170.

Duuren, B. L. van. 1961. Chemistry of edulin, neorautone, and related compounds from *Neorautanenia edulis* C. A. Sm. *J. Org. Chem.* 26:5013–5020.

Duuren, B. L. van, and Groenewoud, P. W. G. 1950a. South African fish poisonous plants. I. Preliminary investigations of the tuber of *Neorautanenia edulis. J. S. Afr. Chem. Inst.* n.s., 3:29–34.

Duuren, B. L. van, and Groenewoud, P. W. G. 1950b. South African fish poisonous plants. II. Determination of the fish toxic properties of the organic compounds isolated from the tuber. *J. S. Afr. Chem. Inst.* n.s. 3:35–40.

Duvigneaud, P. 1954a. Le genre *"Geissaspis"* dans le Congo méridional et les pays limitrophes. *Bull. Soc. R. Bot. Belg.* 86:145–205.

Duvigneaud, P. 1954b. Genres *Humularia. In* Robyns, W. *Flore du Congo Belge et du Ruanda-Urundi*, 5:300–330. Brussels: Publ. Inst. Nat. Étude Agron. Congo Belge (I.N.É.A.C.).

Duvigneaud, P. 1959. Étude sur la végétation du Katanga et de ses sols métallifères. 2. Plantes "cobaltophytes" dans le Haut-Katanga. *Bull. Soc. R. Bot. Belg.* 91:111–134.

Duvigneaud, P., and Brenan, J. P. M. 1966. The genus *Cryptosepalum* Benth. (Leguminosae) in the areas of the "Flora of Tropical East Africa" and the "Flora Zambesiaca." *Kew Bull.* 20 (1):1–23.

Duvigneaud, P., and Timperman, J. 1959. Études sur la végétation du Katanga et de ses sols métallifères. 3. Études sur le genre *Crotalaria. Bull. Soc. R. Bot. Belg.* 91:135–162.

Dwyer, J. D. 1954. The tropical American genus *Tachigalia* Aubl. (Caesalpiniaceae). *Ann. Mo. Bot. Gard.* 41:223–260.

Dwyer, J. D. 1957a. *Androcalymma*, a new genus of the tribe Cassieae (Caesalpiniaceae). *Ann. Mo. Bot. Gard.* 44:295–297.

Dwyer, J. D. 1957b. The tropical American genus *Sclerolobium* Vogel (Caesalpiniaceae). *Lloydia* 20 (2):67–118.

Dwyer, J. D. 1965. Flora of Panama. Part V. Fascicle 4. Family 83. Leguminosae, subfamily Papilionoideae (in part). *Ann. Mo. Bot. Gard.* 52 (1):1–54.

Dyke, S. F., Ollis, W. D., Sainsbury, M., and Schwarz, J. S. P. 1964. The extractives of *Piscidia erythrina* L. II. Synthetical evidence concerning the structure of ichthynone. *Tetrahedron* 20:1331–1338.

Eckhardt, M. M., Baldwin, I. L., and Fred, E. B. 1931. Studies of the root-nodule organism of *Lupinus. J. Bacteriol.* 21 (4):273–285.

Edwards, S. F., and Barlow, B. 1909. Legume bacteria. Further studies of the nitrogen accumulation in the Leguminosae. *Ont. Agric. Coll. Bull.* 169:1–32.

Eggeling, W. J., and Dale, I. R. 1951. *The Indigenous Trees of the Uganda Protectorate.* 2d ed. Entebbe, Uganda: Government Printer. 491 pp.

Eig, A. 1927. A second contribution to the knowledge of the flora of Palestine. *Zionist Organ. Inst. Agr. Nat. Hist. Bull.* 6:11–19.

Elmer, A. D. E. 1907. Some new Leguminosae. *Leafl. Philipp. Bot.* 1:220–232.

References

Elmer, A. D. E. 1910. A new genus and new species of Leguminosae. *Leafl. Philipp. Bot.* 2:689;743.

Erdman, L. W. 1947. Strain variation and host specificity of *Rhizobium trifolii* on different species of *Trifolium*. *Soil Sci. Soc. Am. Proc.* 11:255–259.

Erdman, L. W. 1948a. Strains of *Rhizobium* effective on Guar, *Cyamopsis tetragonoloba*. *J. Am. Soc. Agron.* 40 (4):364–369.

Erdman, L. W. 1948b. Legume inoculation — What it is. What it does. *U. S. Dep. Agric. Farmers Bull.* 2003. 20 pp.

Erdman, L. W. 1950. The effectivity of different strains of *Rhizobium* on annual and perennial lespedezas. *Soil Sci. Soc. Am. Proc.* 15:173–176.

Erdman, L. W., and Means, U. M. 1950. Strains of *Rhizobium* effective on the trefoils, *Lotus corniculatus* and *Lotus uliginosus*. *Soil Sci. Soc. Am. Proc.* 14:170–175.

Erdman, L. W., and Means, U. M. 1956. Strains of rhizobia effective on *Trifolium ambiguum*. *Agron. J.* 48:341–343.

Erdman, L. W., and Walker, R. H. 1927. Occurrence of the various groups of legume bacteria in Iowa soils. *Iowa Acad. Sci. Proc.* 34:53–57.

Eriksson, J. 1874. Studier öfver leguminosernas rotknölar. *Acta Univ. Lund. Lunds Univ. Års-Shrift. Part II. Afdel. Math. Nat.* 10 (8):1–30.

Esser, J. A. 1947. An old crop with a new future. *Progr. Res.* 1 (3):7–9. (Reprinted in *Chem. Dig.* 6 (15):229, 1947; also see *Econ. Bot.* 2:223, 1948).

Everett, T. H. 1969. *Living Trees of the World*. New York: Doubleday. 315 pp.

Ewart, A. J., and Thomson, N. 1912. On the cross-inoculation of the root tubercle bacteria upon the native and the cultivated Leguminosae. *R. Soc. Victoria Proc.* n.s. 25 (2):193–200.

Fadiman, J. 1965. *Genista canariensis:* a minor psychedelic. *Econ. Bot.* 19:383.

Falshaw, C. P., Ollis, W. D., Moore, J. A., and Magnus, K. 1966. The extractives of *Piscidia erythrina* L.III. The constitutions of lisetin, piscidone and piscerythrone. *Tetrahedron* Suppl. 7:333–348.

Farcas, A. 1940. Acacia (*Robinia pseudoacacia*) as a melliferous plant. *Bee World* 21:47–48.

Farmer, R. H., and Campbell, W. G. 1950. Isolation of a saponin from the heart-wood of the Mora tree and of a related species, morabukea. *Nature* 165 (4189):237.

Farooq, M. O., and Siddiqui, M. S. 1954. Chemical investigation of the seed oil of *Leucaena glauca*. *J. Am. Oil Chem. Soc.* 31:8–9. *Chem. Abstr.* 48:3047 (1954).

Farooq, M. O., Ahmad, M. S., and Malik, M. A. 1954. Chemical investigation of seed oil of *Sesbania aegyptica*. *J. Sci. Food Agric.* 5:498–500.

Fassi, B., and Fontana, A. 1961. Le micorrize ectotrofiche di *Julbernardia seretii* Cesalpiniacea del Congo. *Allionia* 7:131–157. *Biol. Abstr.* 24810 (1963).

Fassi, B., and Fontana, A. 1962a. Micorrize ectotrofiche di *Brachystegia laurentii* e di alcune altre Cesalpiniacee minori del Congo. *Alliona* 8:121–131.

Fassi, B., and Fontana, A. 1962b. Micorrize ectotrofiche nella cesalpiniacea *Julbernardia seretii*. *Nuovo G. Bot. Ital.* 69:173–174.

Fawcett, W., and Rendle, A. B. 1920. *Flora of Jamaica. Dicotyledons. Families Leguminosae to Callitrichaceae*, 4. 369 pp. London: Longmans, Green.

Fazli, F. R. Y., and Hardman, R. 1968. The spice, fenugreek (*Trigonella foenumgraecum* L.): its commercial varieties of seed as a source of diosgenin. *Trop. Sci.* 10 (2):66–78.

Fearing, O. S. 1959. A cytotaxonomic study of the genus *Cologania* and its relationship to *Amphicarpaea* (Leguminosae-Papilionoideae). Ph.D. Thesis. University of Texas-Austin. 140 pp.

Federov, S. I. 1940. Grafting and the nature of nodule bacteria. *Jarovizacija* 1 (28):88–89. *Herbage Abstr.* 11 (139):28 (Feb. 1941).

Fehér, D., and Bokor, R. 1926. Untersuchungen über die bakterielle Wurzel-symbiose einiger Leguminosenhölzer. *Planta* 2 (4/5):406–413.

Fergus, E. N., and Hollowell, E. A. 1960. Red clover. *Adv. Agron.* 12:365–436.

Ferguson, J. A. 1922. Influences of locust on the growth of catalpa. *J. For.* 20:318–319 (Notes).

Ferreyra, R. 1951. Una nueva leguminosae del Peru. *Publ. Mus. Hist. Nat. "Javier Prado," Univ. Nac. Mayor San Marcos.* ser. B, Bot. 3:1–5.

Ferris, E. B. 1922. Peanuts. *Miss. Agric. Exp. Stn. Bull.* 208. 14 pp.

Fish, M. S., Johnson, N. M., and Hornung, E. C. 1955. *Piptadenia* alkaloids. Indole bases of *P. peregrina* (L.) Benth. and related species. *J. Am. Chem. Soc.* 77:5892–5895.

Fitzgerald, J. S. 1964. Alkaloids of the Australian Leguminosae. II. Occurrence of sparteine in *Templetonia egena. Aust. J. Chem.* 17 (1):159.

Fitzgerald, J., La Pidus, J. B., and Beal, J. L. 1964. Sweetinine, an alkaloid from *Sweetia panamensis. Lloydia* 27 (2):107–110.

Folkers, K., and Major, R. T. 1937. Isolation of erythroidine, an alkaloid of curare action, from *Erythrina americana* Mill. *J. Am. Chem. Soc.* 59:1580–1581.

Folkers, K., and Unna, K. 1938. *Erythrina* alkaloids. II. A review, and new data on the alkaloids of species of the genus *Erythrina. J. Am. Pharm. Assoc.* 27:693–699.

Folkers, K., and Unna, K. 1939. *Erythrina* alkaloids. V. Comparative curare-like potencies of species of the genus *Erythrina. J. Am. Pharm. Assoc.* 28 (12):1019–1028.

Forbes, H. M. L. 1930. The genus *Psoralea* Linn. *Bothalia* 3:116–136.

Forbes, H. M. L. 1948. I. The revision of South African species of the genus *Tephrosia* Pers. II. The segregation therefrom of the genus *Ophrestia* Forbes. *Bothalia* 4:951–1001; 1003–1008.

Forde, M. B. 1974. The *Lotus pedunculatus-uliginosus* problem. *Lotus Newsl.* 5:3–7.

Fosberg, F. R. 1960. *Serianthes* Benth. (Leguminosae-Mimosoideae-Ingeae). *Reinwardtia* 5:293–317.

Fosberg, F. R. 1965. Revision of *Albizia* Sect. *Pachysperma* (Leguminosae-Mimosoideae). *Reinwardtia* 7 (1):71–90.

Fowler, G. J., and Srinivasian, M. 1921. The biochemistry of the indigenous indigo dye vat. *J. Indian Inst. Sci.* 4:205–221.

Frank, B. 1879. Ueber die Parasiten in den Wurzelanschwellungen der Papilionaceen. *Bot. Ztg.* 37:377–388; 393–400.

Frank, B. 1890a. Ueber Assimilation von Stickstaff aus der Luft durch *Robinia Pseudacacia. Ber. Dtsch. Bot. Ges.* 8:292–294.

References

Frank, B. 1890b. Ueber die Pilzsymbiose der Leguminosen. *Landwirtsch. Jahrb.* 19:523–640.

Fraps, G. S., and Carlyle, E. C. 1936. Locoine, the poisonous principle of loco weed, *Astragalus earlei. Tex. Agric. Exp. Stn. Bull.* 537. 18 pp.

Fred, E. B., and Davenport, A. 1918. Influence of reaction on nitrogen-assimilating bacteria. *J. Agric. Res.* 14:317–336.

Fred, E. B., Baldwin, I. L., and McCoy, E. 1932. *Root Nodule Bacteria and Leguminous Plants. Univ. Wisconsin Stud. Sci.* 5. 343 pp.

Freeman, G. F. 1918. The purple hyacinth bean. *Bot. Gaz.* 66:512–523.

Freer, P. C., and Clover, A. M. 1901. On the constituents of Jamaica dogwood. *Am. Chem. J.* 25:390–413.

Freise, F. W. 1936. Eine bisher unbekannte Physostigmin-Droge. *Pharm. Zentralhalle* 77:378–379.

Freudenberg, K., and Hartmann, L. 1953. Constituents from *Robinia pseudoacacia. Naturwissenschaften* 40:413. *Chem. Abstr.* 48:7128 (1954).

Friesner, G. M. 1926. Bacteria in the roots of *Gleditsia triacanthos* L. *Indiana Acad. Sci. Proc.* 34:215–222.

Fuchsius, L. 1542. *De Historia Stirpium Commentarii Insignes.* Basil: In Officina Isingriniana. 896 pp.

Gagnepain, F. 1911. *Delaportea* Thorel mss., *gen nov. In Mimosées Nouvelles. Lecomte Not. Syst.* 2 Article 24:113–120.

Gagnepain, F. 1915. Papilionacées nouvelles ou critiques. *Lecomte Not. Syst.* 3:180–206.

Galli, F. 1958. Inoculações cruzadas com bactérias dos nódulos de leguminosas tropicais. *Rev. Agric.* 33:139–150.

Galli, F. 1959. Caracteres culturais das bactérias dos nódulos de algumas leguminosas tropicais. *An. Esc. Super. Agric. "Luiz de Queiroz" Univ. São Paulo* 16:113–122.

Gallois, N., and Hardy, E. 1876. Sur l'*Erythrophleum guineense* et l' *Erythrophleum couminga. Bull. Soc. Chim. Paris* 26:39–42.

Gams, H. 1924. Leguminosae. *In* Hegi, G. *Illustrierte Flora von Mittel-Europa*, 4 (3):1113–1748. Munich: J. F. Lehmanns Verlag.

Gardner, C. A. 1941. Contributiones Florae Australiae Occidentalis XI. *J. R. Soc. West. Aust.* 27:165–209.

Garman, H., and Didlake, M. 1914. Six different species of nodule bacteria. *Ky. Agric. Exp. Stn. Bull.* 184:341–363.

Garman, W. H., and Merkle, F. G. 1938. Effect of locust trees upon the available mineral nutrients of the soil. *J. Am. Soc. Agron.* 30:122–124.

Garratt, G. A. 1922. Poisonous woods. *J. For.* 20:479–487.

Gavigan, J. C., and Curran, P. L. 1962. Experiments with the genera *Lotus* and *Anthyllis* and their associated rhizobia. *Sci. Proc. R. Dublin Soc.* ser. B. 1 (5):37–46.

Geoffroy, E. 1892. Sur le *Robinia nicou* et son principe actif. *J. Pharm. Chim.* 26:454–455.

Geoffroy, E. 1895. Contribution à l'étude du *Robinia nicou* au point de vue botanique, chimique et physiologique. *Ann. Inst. Colon. Marseille* 2:1–86.

Georgi, C. D. V., and Curtler, E. A. 1929. The periodic harvesting of tuba root (*Derris elliptica*, Benth.). *Malay. Agric. J.* 17:325–334.

Georgi, C. D. V., and Teik, G. L. 1932. The rotenone content of Malayan tuba root. *Malay. Agric. J.* 20:498–507.

Gerrans, G. C., and Harley-Mason, J. 1964. The alkaloids of *Virgilia oroboides. J. Chem. Soc. Lond.* 1964:2202–2206.

Gershon, D. 1961. Genetic studies of effective nodulation in *Lotus* spp. *Can. J. Microbiol.* 7:961–963.

Ghatak, N. 1934. Chemical examination of the kernels of the seeds of *Caesalpinia bonducella. Proc. Acad. Sci.* (United Provinces Agra Oudh, India) 4:141–146.

Ghosal, S., and Banerjee, P. K. 1969. Alkaloids of the roots of *Desmodium gangeticum. Aust. J. Chem.* 22 (9):2029–2031.

Ghosal, S., and Mukherjee, B. 1965. Occurrence of 5-methoxy-*N*, *N*-dimethyltryptamine oxide and other tryptamines in *Desmodium pulchellum* Benth. ex Baker. *Chem. Ind. Lond.* 1965 (19):793–794.

Ghosal, S., and Mukherjee, B. 1966. Indole-3-alkylamine bases of *Desmodium pulchellum. J. Org. Chem.* 31 (7):2284–2288.

Ghose, T. P., and Krishna, S. 1937. Occurrence of rotenone in *Millettia pachycarpa. Curr. Sci.* 6:57.

Ghosh, A. K. 1944. Rise and decay of the indigo industry in India. *Sci. Ind.* 9 (11):487–493; 9 (12):537–542.

Gibbons, R. W., Bunting, A. H., and Smartt, J. 1972. The classification of varieties of groundnut (*Arachis hypogaea* L.) *Euphytica* 21:78–85.

Gibson, A. H. 1958. The routes of infection in hosts of the cowpea cross-inoculation group. *Rhizobium Newsl.* 3 (2):6–7.

Gibson, A. H., and Brockwell, J. 1968. Symbiotic characteristics of subspecies of *Trifolium subterraneum* L. *Aust. J. Agric. Res.* 19:891–905.

Gilbert, G., and Boutique, R. 1952a. Mimosaceae. *In* Robyns, W. *Flore du Conge Belge et du Ruanda-Urundi,* 3:137–233. Brussels: Publ. Inst. Nat. Étude Agron. Congo Belge (I.N.É.A.C.).

Gilbert, G., and Boutique, R. 1952b. Swartzieae. *In* Robyns, W. *Flore du Congo Belge et du Ruanda-Urundi,* 3:550–554. Brussels: Publ. Inst. Nat. Étude Agron. Congo Belge (I.N.É.A.C.).

Gillett, J. B. 1960a. The genus *Craibia. Kew Bull.* 14:189–197.

Gillett, J. B. 1960b. A key to the species of *Platysepalum* Baker, with notes. *Kew Bull.* 14:464–467.

Gillett, J. B. 1963a. *Galega* L. (Leguminosae) in tropical Africa. *Kew Bull.* 17 (1):81–85.

Gillett, J. B. 1963b. *Sesbania* in Africa (excluding Madagascar) and southern Arabia. *Kew Bull.* 17 (1):91–159.

Gillett, J. B. 1964. *Biserrula* L. (Leguminosae) in tropical Africa and the number of fertile stamens in this genus. *Kew Bull.* 17 (3):503–506.

Gillett, J. B. 1966a. Notes on Leguminosae (Phaseoleae). *Kew Bull.* 20 (1):103–111.

Gillett, J. B. 1966b. *Vermifrux* Gillett (Leguminosae-Papilionoideae), a new name for *Helminthocarpon* A. Rich. *Kew Bull.* 20 (2):245.

Gillett, J. B. 1966c. The species of *Ormocarpum* Beauv. and *Arthrocarpum* Balf. f. (Leguminosae) in south-western Asia and Africa (excluding Madagascar). *Kew Bull.* 20 (2):323–355.

References

Gillett, J. M., and Cochrane, T. S. 1973. Preliminary reports on the flora of Wisconsin. No. 63. The genus *Trifolium* — the clovers. *Trans. Wis. Acad. Sci. Arts Lett.* 61:59–74.

Gillis, W. T., and Stearn, W. T. 1974. Typification of the names of the species of *Leucaena* and *Lysiloma* in the Bahamas. *Taxon* 23 (1):185–191.

Gleason, H. A. 1935. Some necessary nomenclatural changes. *Phytologia* 1 (3):141–142.

Glicksman, M., and Schachat, R. E. 1959. Gum arabic. *In* Whistler, R. L., and BeMiller, J. N., eds. *Industrial Gums — Polysaccharides and Their Derivatives*, 231–291. New York: Academic Press.

Godbole, S. N., Paranjpe, D. R., and Shrikhande, J. G. 1929. Some constituents of *Caesalpinia bonducella* nut (Flem). Part I. Bonducella nut oil. *J. Indian Chem. Soc.* 6:295–302.

Godfrin, M. J. 1884. Recherches sur l'anatomie comparée des cotylédons et de l'albumen. *Ann. Sci. Nat.* ser. 6. 19:5–158.

Goldstein, A. M., and Alter, E. N. 1959. Guar Gum. *In* Whistler, R. L., and BeMiller, J. N., eds. *Industrial Gums — Polysaccharides and Their Derivatives*, 321–341. New York: Academic Press.

Gonzalez, M. L., Basurco, J. C. P., and Schiel, E. 1969. Inoculación de *Vicia benghalensis* en area representiva de los suelos francos del Valle de Tulum (Provincia de San Juan). *Rev. Invest. Agropecu.* ser. 2 Biol. Prod. Veg. 6 (15):243–253.

Goodchild, D. J., and Bergersen, F. J. 1966. Electron microscopy of the infection and subsequent development of soybean nodule cells. *J. Bacteriol.* 92:204–213.

Goodding, L. N. 1939. Native legumes in Region 8. *U. S. Dep. Agric. Reg. Bull.* 55. Plant Study Ser. 1. 47 pp.

Gooding, E. G. B., Loveless, A. R., and Proctor, G. R. 1965. *Flora of Barbados. Overseas Res. Publ.* 7:486 pp. London: Her Majesty's Stationery Office.

Goor, G. A. W. van de. 1954. The value of some leguminous plants as green manure in comparison with *Crotalaria juncea. Neth. J. Agric. Sci.* 20 (1):37–43.

Goosen, A. 1963. The alkaloids of the Leguminosae. I. The structure of calpurnine from *Calpurnia subdecandra. J. Chem. Soc. Lond.* 1963:3067–3068.

Gordienko, I. I. 1960. Biogeocoenotic relationships between the nodule bacteria and Leguminosae, Dnieper broom (*Cytisus borysthenicus*) and white acacia (*Robinia pseudoacacia*) on Olesh sands [in Russian, English summary]. *Pochvovedenie* 8:71–74.

Goto, M., Noguchi, T., and Watanabe, T. 1958. Useful components in natural sources. XVII. Uterus contracting ingredients in plants. (2) Uterus contracting ingredients in *Lespedeza bicolor* var. *japonica* [in Japanese]. *Yakugaku Zasshi* 78:464–467. *Chem. Abstr.* 52:14082 (1958).

Graham, P. H. 1963. A study on the application of serodiagnostic and computative methods to the taxonomy of the root-nodule bacteria of legumes. Ph.D. thesis, University of Western Australia.

Graham, P. H. 1964. The application of computer techniques to the taxonomy of the root-nodule bacteria of legumes. *J. Gen. Microbiol.* 35:511–517.

Granick, S. 1937. Urease distribution in *Canavalia ensiformis*. *Plant Physiol.* 12:601–623.

Gray Herbarium Index, Harvard University. 1968. Boston: G.K. Hall. Vols. 1–10.

Grear, J. W., Jr. 1970. A revision of the American species of *Eriosema* (Leguminosae-Lotoideae). *Mem. N. Y. Bot. Gard.* 20 (3):1–98.

Grear, J. W. 1978. A revision of the New World species of *Rhynchosia* (Leguminosae-Faboideae). *Mem. N. Y. Bot. Gard.* 31:1–168.

Greenway, P. J. 1936. *Mundulea* fish poison. *Kew Bull.* 4:245–250.

Greenway, P. J. 1941. Dyeing and tanning products in East Africa. *Bull. Imp. Inst.* 39:222–245.

Greenway, P. J. 1947. Yeheb. *East Afr. Agric. J.* 12:216–219.

Gregory, K. F., and Allen, O. N. 1953. Physiological variations and host plant specificities of rhizobia isolated from *Caragana arborescens* L. *Can. J. Bot.* 31:730–738.

Greshoff, M. 1890. Mittheilungen aus dem chemisch-pharmakologischer Laboratorium des Botanischen Gartens zu Buitenzorg (Java). *Berichte* 23 (2):3537–3550.

Griffiths, F. P. 1949. Production and utilization of alfalfa. *Econ. Bot.* 3:170–183.

Griffiths, L. A. 1962. On the co-occurrence of coumarin, o-coumaric acid, and melilotic acid in *Gliricidia sepium* and *Dipteryx odorata*. *J. Exp. Bot.* 13 (38):169–175.

Grobbelaar, N., and Clarke, B. 1972. A qualitative study of the nodulating ability of legume species: List 2. *J. S. Afr. Bot.* 38 (4):241–247.

Grobbelaar, N., and Clarke, B. 1974. A qualitative study of the nodulating ability of legume species: List 4. *Agroplantae* 6:59–64.

Grobbelaar, N., and Clarke, B. 1975. A qualitative study of the nodulating ability of legume species: List 3. *J. S. Afr. Bot.* 41 (1):29–36.

Grobbelaar, N., Beijma, M. C. van, and Saubert, S. 1964. Additions to the list of nodule-bearing legume species. *S. Afr. J. Agric. Sci.* 7:265–270.

Grobbelaar, N., Beijma, M. C. van, and Todd, C. M. 1967. A qualitative study of the nodulating ability of legume species: List 1. *Publ. Univ. Pretoria* n.s. 38. 9 pp.

Guérin, P. 1930. L'acide cyanhydrique chez les Vesces. Sa répartition dans les divers organes des Légumineuses-Papilionacées à glucoside cyanogénétique. *C. R. Acad. Sci.* 190:512–514.

Guggenheim, M. 1913. Dioxyphenylalanin, eine neue Aminosäure aus *Vicia Faba*. *Z. Physiol. Chem.* 88:276–284.

Gunawardena, D. C. 1968. *Genera et Species Plantarum Zeylaniae*. Colombo: Lake House Investments. 268 pp.

Habish, H. A. 1970. Effect of certain soil conditions on nodulation of *Acacia* spp. *Plant Soil* 33:1–6.

References

Habish, H. A., and Khairi, S. M. 1968. Nodulation of legumes in the Sudan: cross-inoculation groups and the associated *Rhizobium* strains. *Exp. Agric.* 4:227–234.

Habish, H. A., and Khairi, S. M. 1970. Nodulation of legumes in the Sudan. *Rhizobium* strains and cross-inoculation of *Acacia* spp. *Exp. Agric.* 6 (2):171–176.

Hagerup, O. 1928. En hygrofil Baelgplante (*Aeschynemone aspera* L.) med Bakterieknolde paa Staengelen. *Dan. Bot. Ark.* 5 (14):1–9.

Hance, H. F. 1867. *Lysidice. J. Bot. Br. Foreign* 5:298–299.

Hannon, N. J. 1949. The nitrogen economy of the Hawkesbury sandstone soils around Sydney. The role of the native legumes. Honours Thesis, University of Sydney.

Hanriot, M. 1907a. Sur les substances actives du *Tephrosia vogelii*. *C. R. Acad. Sci.* 144:150–152.

Hanriot, M. 1907b. Sur la toxicité des principes définis du *Tephrosia vogelii* (Légumineuses). *C. R. Acad. Sci.* 144:498–500.

Hansberry, R., and Lee, C. 1943. The yam bean, *Pachyrrhizus erosus* Urban, as a possible insecticide. *J. Econ. Entomol.* 36:351–352.

Hanson, C. H., ed. 1972. *Alfalfa Science and Technology.* Agron. Ser. 15. 812 pp. Madison, Wis.: American Society of Agronomy.

Hanson, C. H., Tysdal, H. M., and Davis, R. L. 1962. Alfalfa. *In* Hughes, H. D., Heath, M. E., and Metcalfe, D. S., eds. *Forages*, 127–138. Ames, Iowa: Iowa State University Press.

Harborne, J. B., Boulter, D., and Turner, B. L., eds. 1971. *Chemotaxonomy of the Leguminosae.* New York: Academic Press. 612 pp.

Harms, H. 1897. Leguminosae. *In* Engler, A., and Prantl, K. *Die Natürlichen Pflanzenfamilien.* Supplement I of Addendum I to Part III, ser. 3:190–204. Leipzig: Engelmann.

Harms, H. 1899. Leguminosae africanae II. *Bot. Jahrb.* 26:308–310.

Harms, H. 1900. Leguminosae. *In* Engler, A., and Prantl, K. *Die Natürlichen Pflanzenfamilien.* Supplement I of Addendum II to Part III, ser. 3:30–34. Leipzig: Engelmann.

Harms, H. 1905a. *Neodielsia* Harms n. gen. pp. 68–69. *In* Diels, L. *Beiträge zur Flora des Tsin ling shan und andere Zusätze zur Flora von Central-China. Bot. Jahrb.* 36 (82):1–134.

Harms, H. 1905b. Zwei neue Gattungen der Leguminosae aus dem tropischen Afrika. *Bot. Jahrb.* 38:74–79.

Harms, H. 1908a. Leguminosae. *In* Engler, A., and Prantl, K. *Die Natürlichen Pflanzenfamilien.* Supplement II of Addendum III to Part III, ser. 3:145–177. Leipzig: Engelmann.

Harms, H. 1908b. Über Geokarpie bei einer afrikanischer Leguminosae. *Ber. Dtsch. Bot. Ges.* 26a:225–231.

Harms, H. 1908c. Zur nomenclatur des Perubalsambaumes. *Notizbl. Bot. Gart. Berlin* 5:85–98.

Harms, H. 1908d. Leguminosae andinae. *Bot. Jahrb.* 42:95–96.

Harms, H. 1909. Über Kleistogamie bei der Gattung *Argyrolobium. Ber. Dtsch. Bot. Ges.* 27:85–97.

Harms, H. 1910. Leguminosae africanae. V. *Bot. Jahrb.* 45:293–316.

Harms, H. 1911. Über einige Leguminosen des tropischen Afrika mit essbaren Knollen. *Notizbl. Königl. Bot. Gart. Mus. Berlin-Dahlem* 5 (48): 199–211.

Harms, H. 1913. Leguminosae africanae, VI. *Bot. Jahrb.* 49:419–454.

Harms, H. 1914. Leguminosae africanae. VII. 1. Eine neue Gattung der Leguminosen aus dem trop. Afrika. *Bot. Jahrb.* 51:359–368.

Harms, H. 1915a. Leguminosae. *In* Engler, A., and Prantl, K. *Die Natürlichen Pflanzenfamilien.* Supplement III of Addendum IV to Part III, ser. 3:119–151. Leipzig: Engelmann.

Harms, H. 1915b. Leguminosae. *In* Engler, A., and Drude, O. *Die Vegetation der Erde. IX. Die Pflanzenwelt Afrikas*, 3 (1):327–698. Leipzig: Engelmann.

Harms, H. 1917a. Weitere Beobachtung über Kleistogamie bei afrikanischen Arten der Gattung *Argyrolobium. Ber. Dtsch. Bot. Ges.* 35:175–186.

Harms, H. 1917b. Eine neue Gattung der Leguminosae aus dem tropischen Afrika, *Haplormosia* Harms. *Feddes Repert. Specierum Nov. Regni Veg.* 15:22–24.

Harms, H. 1920. Eine neue Gattung der Leguminosen-Papilionatae aus Papuasien. *Notizbl. Bot. Gart. Mus. Berlin-Dahlem* 7 (68):26–27.

Harms, H. 1921a. Einige Leguminosen aus China. *Feddes Repert. Specierum Nov. Regni. Veg.* 17:133–137.

Harms, H. 1921b. Eine neue Gattung der Leguminosae-Caesalpinioideae aus Argentina. *Notizbl. Bot. Gart. Mus. Berlin-Dahlem* 7 (70):500–501.

Harms, H. 1922. Über *Luetzelburgia* eine neue Gattung der Leguminosen aus Brasilien. *Ber. Dtsch. Bot. Ges.* 40:177–179.

Harms, H. 1923a. Leguminosae americanae novae. IV. *Feddes Repert. Specierum Nov. Regni Veg.* 19:9–18.

Harms, H. 1923b. Leguminosae americanae novae. V.. *Feddes Repert. Specierum Nov. Regni Veg.* 19:61–70.

Harms, H. 1928a. Planta Steinbachianae III. Leguminosae. *Notizbl. Bot. Gart. Mus. Berlin-Dahlem* 10:345–346.

Harms, H. 1928b. Eine neue Gattung der Leguminosae-Papilionatae aus Peru. *Notizbl. Bot. Gart. Mus. Berlin-Dahlem* 10 (94):387–388.

Harms, H. 1930. Zur Kenntnis von *Lysiloma guachapele* (H. B. K.) Benth. *Notizbl. Bot. Gart. Mus. Berlin-Dahlem* 11:52–56.

Harper, S. H. 1940. The active principles of leguminous fish-poison plants. Part IV. The isolation of malaccol from *Derris malaccensis. J. Chem. Soc. Lond.* 1940:309–314.

Harris, D. R. 1967. New light on plant domestication and the origins of agriculture: a review. *Geogr. Rev.* 57:90–107.

Harris, J. O., Allen, E. K., and Allen, O. N. 1949. Morphological development of nodules on *Sesbania grandiflora* Poir. with reference to the origin of nodule rootlets. *Am. J. Bot.* 36 (9):651–661.

Harris, J. R. 1953. List of species of bacterial genus *Rhizobium* maintained in the type collection of C.S.I.R.O. Adelaide: Division of Soils. Tech. Memo. 10/53. 6 pp.

Harrison, F. 1913. *Roman Farm Management.* New York: Macmillan. 365 pp.

References

Harrison, F. C., and Barlow, B. 1906. Cooperative experiments with nodule-forming bacteria. *Ontario Dep. Agric. Bull.* 148:1–19.

Hart, E. 1883. Piscidia, the active principle of Jamaica dogwood (*Piscidia Erythrina*). *Am. Chem. J.* 5:39–40.

Harvey, W. H. 1894. Leguminosae. *In* Harvey, W. H., and Sonder, O. W. *Flora Capensis*, 2:1–285. Ashford, Eng.: L. Reeve.

Hassler, E. 1919. Ex herbario Hassleriano: novitates paraguarienses. XXIII. Leguminosae VIII. *Feddes Repert. Specierum Nov. Reg. Veg.* 16:151–166.

Hassler, E. 1923. Revisio specierum austro-americanarum generis *Phaseoli* L. *Candollea* 1:417–472.

Hauke-Pacewiczowa, T. 1952. Selection of *Rhizobium* suitable for inoculation of *Galega officinalis* [in Polish]. *Acta Microbiol. Pol.* 1 (1):37–39.

Hauman, L. 1925. The genus *Poissonia* Baillon. *Kew Bull. Misc. Inf.* 6:276–279.

Hauman, L. 1952. Genres *Berlinia* et voisins. *In* Robyns, W. *Flore du Congo Belge et du Ruanda-Urundi*, 3:376–407. Brussels: Publ. Inst. Nat. Étude Agron. Congo Belge (I.N.É.A.C.).

Hauman, L. 1954a. Galegeae. *In* Robyns, W. *Flore du Congo Belge et du Ruanda-Urundi*, 5:4–72. Brussels: Publ. Inst. Nat. Étud. Agron. Congo Belge (I.N.É.A.C.).

Hauman, L. 1954b. Dalbergieae. *In* Robyns, W. *Flore du Congo Belge et du Ruanda-Urundi*, 6:4–52. Brussels: Publ. Inst. Nat. Étude Agron. Congo Belge (I.N.É.A.C.).

Hauman, L. 1954c. Phaseoleae. *In* Robyns, W. *Flore du Congo Belge et du Ruanda-Urundi*, 6:148–259. Brussels: Publ. Inst. Nat. Étude Agron. Congo Belge (I.N.É.A.C.).

Hauptmann, H., and Nazario, L. L. 1950. Some constituents of the leaves of *Cassia alata* L. *J. Am. Chem. Soc.* 72 (1):1492–1495.

Heath, M. E., Metcalfe, D. S., and Barnes, R. F., eds. 1973. *Forages*. 3d ed. Ames, Iowa: Iowa State University Press. 755 pp.

Heaton, G. L. 1954. Notes on *Stylosanthes gracilis* (Brazilian clover) as a ground cover for tea. *Tea Q.* 25 Part 1:8–9.

Heckel, E., and Schlagdenhauffen, F. 1886. Des graines de bonduc et de leur principe actif fébrifuge. *C. R. Acad. Sci.* 103:88–91.

Heiser, C. B. 1965. Cultivated plants and cultural diffusion in nuclear America. *Am. Anthropol.* 67:930–949.

Heller, C. A. 1962. Wild, edible and poisonous plants of Alaska. *Univ. Alaska Ext. Bull.* F-40. 87 pp.

Hellriegel, H., and Wilfarth, H. 1888. Untersuchungen über die Stickstoffnahrung der Gramineen und Leguminosen. Addendum. *Z. Ver. Rübenzucker-Ind. Dtsch. Reichs.* 234 pp.

Helten, W. M. van. 1915. Resultaten verkregen in den cultuurtuin met eenige nieuwe groenbemesters. *Dep. Landb., Nivj. Handel. Meded. Cult.* 2:28–32.

Hely, F. W., and Brockwell, J. 1964. Frequencies of annual species of *Medicago* on brown acid soils of the Macquarie region of New South Wales. *Aust. J. Agric. Res.* 15:50–60.

Hely, F. W., and Ofer, I. 1972. Nodulation and frequencies of wild leguminous species in the northern Negev region of Israel. *Aust. J. Agric. Res.* 23:267–84.

Hemsley, W. B. 1907. Diagnoses Africanae XIX. *Kew Bull. Misc. Inf.* 1907:361.

Hendry, G. W. 1923. Alfalfa in history. *J. Am. Soc. Agron.* 15:171–176.

Henke, L. A. 1945. Protein sources and supplements for dairy cows in Hawaii. *Hawaii Agric. Exp. Stn. Bull.* 95. 21 pp.

Henke, L. A. 1958. Value of *Leucaena glauca* as a feed for cattle. *8th Pacific Sci. Congr. Proc.* ser. B. 4:591–600.

Henry, T. A. 1939. *The Plant Alkaloids.* 3d ed. Philadelphia: P. Blakiston's. 689 pp.

Henson, P. R. 1957. The lespedezas. *Adv. Agron.* 9:113–157.

Henson, P. R., Baldridge, J. D., and Helm, C. A.. 1962. The lespedezas. *In* Hughes, H. D., Heath, M. E., and Metcalfe, D. S., eds. *Forages*, 169–179. 2d ed. Ames, Iowa: Iowa State University Press.

Hermann, F. J. 1947a. The Amazonian varieties of *Lonchocarpus nicou*, a rotenone-yielding plant. *J. Wash. Acad. Sci.* 37 (4):111–113.

Hermann, F. J. 1947b. Studies in *Lonchocarpus* and related genera, I: a synopsis of *Willardia*. *J. Wash. Acad. Sci.* 37 (12):427–430.

Hermann, F. J. 1949. Studies in *Lonchocarpus* and related genera. V: new species from Middle America and the *Lonchocarpus quatemalensis* complex. *J. Wash. Acad. Sci.* 39 (9):306–313.

Hermann, F. J. 1954. A synopsis of the genus *Arachis*. *U. S. Dep. Agric. Agric. Monogr.* 19. 26 pp.

Hermann, F. J. 1962. A revision of the genus *Glycine* and its immediate allies. *U. S. Dep. Agric. Tech. Bull.* 1268. 82 pp.

Heumann, K. 1890a. Neue Synthesen des Indigos und verwandter Farbstoffe. *Berichte* 23:3043–3045.

Heumann, K. 1890b. Neue Synthesen des Indigos und verwandter Farbstoffe. *Berichte* 23:3431–3435.

Heyn, C. C., and Raviv, V. 1966. Experimental taxonomic studies in the genus *Scorpiurus* (Papilionaceae). *Bull. Torrey Bot. Club* 93 (4):259–267.

Heyne, K. 1927. *Nuttige Planten van Nederlandsch Indie.* Legumes: Mimosoideae and Caesalpinioideae, 1:699–732. Legumes: Caesalpinioideae and Papilionoideae 2:733–850. Buitenzorg: Dep. Landbouw, Nijv. Handel Nederlandsch-Indie.

Heywood, V. H., ed. 1968. Flora Europaea. Notulae systematicae ad Floram Europaeam spectantes. *Feddes Repert. Specierum Nov. Reg. Veg.* 79:1–68.

Higbee, E. C. 1948. *Lonchocarpus, Derris*, and *Pyrethrum* cultivation and sources of supply. *U. S. Dep. Agric. Misc. Publ.* 650. 36 pp.

Hildritch, T. P., and Williams, P. N. 1964. *The Chemical Constitution of Natural Fats.* 4th ed. New York: John Wiley. 745 pp.

Hiltner, L. 1900. Ueber die Bakteroiden der Leguminosen-knöllchen und ihre willkürliche Erzeugung ausserhalb der Wirtzpflanzen. *Centralbl. Bakt.* ser. 2. 6:273–281.

Hiltner, L., and Störmer, K. 1903. Neue Untersuchungen über die Wurzelknöllchen der Leguminosen und deren Erreger. *Arb. K. Gesundheitsamt. Biol.* 3:151–307.

Hirst, E. L. 1942. Recent progress in the chemistry of the pectic materials and plant gums. *J. Chem. Soc.* 1942:70–78.

References

Hocquette, M. 1930. Évolution du noyau dans les cellules bactérifères des nodosités d'*Ornithopus perpusillus* pendant les phénomènes d'infection et de digestion intracellulaire. *C. R. Acad. Sci.* 191:1363–1365.

Hoehne, F. C. 1939. *Plantas e Substancias Vegetais Tóxicas e Medicinais.* São Paulo: "Graphicars." 355 pp.

Holbrook, S. H. 1936. The tragedy of Bandon . . . *Am. For.* 42:494–497.

Hollaender, A., ed. 1977. *Genetic Engineering for Nitrogen Fixation.* Basic Life. Sci. 9. 538 pp. New York: Plenum Press.

Holland, T. H. 1924. Some green manures and cover plants. *Dept. Agric. Ceylon Leafl.* 30. 6 pp.

Holland, T. H. 1966. Serological characteristics of certain root-nodule bacteria of legumes. *Antonie van Leeuwenhoek J. Microbiol. Serol.* 32:410–418.

Hollande, A. C., and Crémieux, G. 1926. Remarques au sujet d'un microbe radicicole (*Rhizobium ornithopi* n. sp.) isolé des nodosités des racines d'une Légumineuse du Maroc (*Ornithopus isthmocarpus* Coss.) *C. R. Soc. Biol.* 95 (35):1316–1317.

Holm, T. 1924. *Apios tuberosa* Moench. *Am. Mid. Nat.* 9 (3):5–23.

Holman, H. J. 1940. *A Survey of Insecticide Materials of Vegetable Origin.* London: Plant and Animal Products Department, Imperial Institute. 755 pp.

Hooker, J. D. 1879. *The Flora of British India*, Sabiaceae to Cornaceae. 2. Ashford, Eng.: L. Reeve. 792 pp.

Horne, D. B. 1966. The distribution of flavonoids in *Baptisia nuttalliana* and *B. lanceolata* and their taxonomic implications. *Diss. Abstr.* 26:6987; 6988.

Horrell, C. R. 1958. Herbage plants at Serere Experiment Station, Uganda 1954–1957. II: Legumes. *East Afr. Agric. J.* 24 (2):133–138.

Hosaka, E. Y., and Ripperton, J. C. 1944. Legumes of the Hawaiian Ranges. *Hawaii Agric. Exp. Stn. Bull.* 93. 80 pp.

Hosoda, K. 1928. The relation of hydrogen-ion concentration of media to the growth of *B. radicicola*. 1. The optimum and final hydrogen-ion concentration of *B. radicicola. Trans. Tottori Soc. Agric. Sci.* 1:89–104.

Houseman, P. A. 1944. *Licorice: Putting a Weed to Work.* London: Royal Institute of Chemistry of Great Britain and Ireland Monograph. 15 pp. 2 plates.

Howard, A., Howard, G. L. C., and Kahn, A. R. 1915. Some varieties of Indian gram (*Cicer arietinum* L.). *Mem. Dep. Agric. India* Bot. ser. 7 (6):213–285.

Howes, F. N. 1949. *Vegetable Gums and Resins.* Waltham, Mass.: Chronica Botanica. 188 pp.

Hoyle, A. C. 1932. *Chidlowia*, a new tree genus of Caesalpiniaceae from West Tropical Africa. *Kew Bull.* 1932:101–103.

Hoyle, A. C. 1952. *Brachystegia. In* Robyns, W. *Flore du Congo Belge et du Ruanda-Urundi*, 3:446–482. Brussels: Publ. Inst. Étude Agron. Congo Belge (I.N.É.A.C.).

Huffman, W. T., Moran, E. A., and Binns, W. 1956. Poisonous plants. *U.S. Dep. Agric. Yearbook* 1956:118–130.

Hughes, H. D. 1962. Birdsfoot trefoil. *In* Hughes, H. D., Heath, M. E., and Metcalfe, D. S., eds. *Forages*, 187–204. Ames, Iowa: Iowa State University Press.

Humbert, H. 1939. Un type aberrant de Légumineuses-Césalpiniées de Madagascar. *C. R. Acad. Sci.* 208:372–375.

Humbert, H. 1959. *Brenierea*, genre nouveau remarquable de Légumineuses-Césalpiniées du Sud de Madagascar. *C. R. Acad. Sci.* 249:1597–1600.

Hurst, E. 1942. *The Poison Plants of New South Wales.* Sydney: Poison Plants Committee, N.S. Wales. 498 pp.

Hutchinson, J. 1964. *The Genera of Flowering Plants.* Dicotyledones, 1. 516 pp. Oxford: Clarendon Press.

Hutchinson, J., and Dalziel, J. M. 1958. *Flora of West Tropical Africa.* 2d ed. rev. Keay, R. W. J. I Part 2. 828 pp. London: Crown Agents for Overseas Governments and Administrations.

Hutton, E. M., and Bonner, I. A. 1960. Dry matter and protein yields in four strains of *Leucaena glauca* Benth. *J. Aust. Inst. Agric. Sci.* 26:276–277.

Hutton, E. M., and Gray, S. G. 1959. Problems in adapting *Leucaena glauca* as a forage for the Australian tropics. *Emp. J. Exp. Agric.* 27:187–196.

Hymowitz, T. 1970. On the domestication of the soybean. *Econ. Bot.* 24 (4):408–421.

Hymowitz, T. 1972. The trans-domestication concept as applied to guar. *Econ. Bot.* 26 (1):49–60.

Inouye, C., and Maeda, K. 1952. Studies on the root nodules of peanut, I [in Japanese, English summary]. *Res. Rep. Kôchi Univ.* 1 (31). 6 pp.

Inouye, C., Yamasaki, T., and Maeda, K. 1953. Morphological studies on the root nodules of Chinese milk vetch [in Japanese, English summary]. *Res. Rep. Kôchi Univ.* 2 (35). 6 pp.

International Code of Botanical Nomenclature. 1966. Lanjouw, J., and others, eds. *Regnum Veg.* 46. 402 pp.

International Code of Botanical Nomenclature. 1972. Stafleu, F. A., and others, eds. *Regnum Veg.* 82. 426 pp.

International Institute of Agriculture. 1936. *Use of Leguminous Plants in Tropical Countries as Green Manure, as Cover and as Shade.* Rome: The Institute. 262 pp.

Irvine, J. E., and Freyre, R. H. 1959a. The occurrence of rotenoids in some species of *Tephrosia. J. Agric. Food Chem.* 7:106–107.

Irvine, J. E., and Freyre, R. H. 1959b. Varietal differences in the rotenoid content of *Tephrosia vogelii. Agron. J.* 51:664–665.

Irwin, H. S. 1964. Monographic studies in *Cassia* (Leguminosae-Caesalpinioideae). I. Section Xerocalyx. *N. Y. Bot. Gard. Mem.* 12 (1):1–114.

Irwin, H. S., and Barneby, R. C. 1976. Notes on the generic status of *Chamaecrista* Moench (Leguminosae: Caesalpinioideae). *Brittonia* 28 (1):28–36.

Irwin, H. S., and Turner, B. L. 1960. Chromosomal relationships and taxonomic considerations in the genus *Cassia. Am. J. Bot.* 47:309–318.

Isely, D. 1951. The Leguminosae of the north-central United States: I. Loteae and Trifoleae. *Iowa State Coll. J. Sci.* 25 (3):439–482.

Isely, D. 1954. Keys to sweet clovers (*Melilotus*). *Proc. Iowa Acad. Sci.* 61:119–131.

References

Isely, D. 1955. The Leguminosae of the north-central United States. II. Hedysareae. *Iowa State Coll. J. Sci.* 30 (1):33–118.

Isely, D. 1973. Leguminosae of the United States: I. Subfamily Mimosoideae. *N. Y. Bot. Gard. Mem.* 25 (1):1–152.

Ishizawa, S. 1953. Studies on the root-nodule bacteria of leguminous plants. I [in Japanese, English summary]. *J. Sci. Soil Manure (Japan)* 23:125–130; 189–195; 241–244.

Ishizawa, S. 1954. Studies on the root-nodule bacteria of leguminous plants. II. Part 1. a. Cross-inoculation test [in Japanese, English summary]. *J. Sci. Soil Manure (Japan)* 24:297–302.

Ishizawa, S. 1972. Root-nodule bacteria of tropical legumes. *JARQ: Jap. Agric. Res. Q.* 6 (4):199–211.

Ishizawa, S., and Toyoda, H. 1955. Comparative study on effective and ineffective nodules of leguminous plants. *Soil Plant Food* 1:47–48.

Itano, A., and Matsuura, A. 1934. Studies on the nodule bacteria of *Astragalus sinicus* (Genge) III. *Ber. Ōhara Inst. Landw. Forsch.* 6:341–369.

Itano, A., and Matsuura, A. 1936. Studies on nodule bacteria V. Influence of plant extract as accessory substance on the growth of nodule bacteria. *Ber. Ohara Inst. Landw. Forsch.* 7 (2):185–214.

Iterson, G. van, Jr., Den Dooren de Jong, L. E., and Kluyver, A. J. 1940. *Martinus Willem Beijerinck, His Life and His Work.* The Hague: Martinus Nijhoff. 192 pp.

Itzhaky, D. 1954. Some trials in the cultivation of wild Leguminosae in Israel. *Bull. Indep. Biol. Lab. (Ramatayim Israel)* 10 (2). 8 pp.

Jackson, H. V. 1910. Cultivation of coffee. *Philipp. Agric. Rev.* 3:512–524.

Jaensch, T. 1884. Zur Anatomie einiger Leguminosenhölzer. *Ber. Dtsch. Bot. Ges.* 2:268–292.

Jagoe, R. B. 1949. Beneficial effects of some leguminous shade trees on grassland in Malaya. *Malay. Agric. J.* 32:77–87.

Janse, J. M. 1897. Les endophytes radicaux de quelques plantes javanaises. *Ann. Jard. Bot. Buitenzorg* 14:53–201.

Jansen van Rensburg, H., and Strijdom, B. W. 1971. Stability of infectivity in strains of *Rhizobium japonicum. Phytophylactica* 3:125–130.

Jansen van Rensburg, H., and Strijdom, B. W. 1972. A bacterial contaminant in nodules of leguminous plants. *Phytophylactica* 4:1–8.

Janzen, D. H. 1966. Coevolution of mutualism between ants and acacias in Central America. *Evolution* 20 (3):249–275.

Janzen, D. H. 1967. Interaction of the bull's-horn *Acacia (Acacia cornigera* L.) with an ant inhabitant (*Pseudomyrmex ferruginea* F. Smith) in eastern Mexico. *Univ. Kans. Sci. Bull.* 47 (6):315–558.

Jaretzky, R., and Axer, B. 1934. Pharmacognosy of *Sarothamnus scoparius* Koch and parallel drugs. *Arch. Pharm.* 272:152–167.

Jeník, J., and Mensah, K. O. A. 1967. Root systems of tropical trees 1. Ectotrophic mycorrhizae of *Afzelia africana* Sm. *Preslia (Prague)* 39:59–65.

Jeník, J., and Kubíková, J. 1969. Root system of tropical trees. 3. The heterorhizis of *Aeschynemone elaphroxylon* (Guill. et Perr.) Taub. *Preslia (Prague)* 41:220–226.

Jensen, H. L. 1942. Nitrogen fixation in leguminous plants. I. General characters of root-nodule bacteria isolated from species of *Medicago* and *Trifolium* in Australia. *Proc. Linn. Soc. N. S. W.* 67 (1–2):98–108.

Jensen, H. L. 1963. Relations de la plante hôte avec les *Rhizobium* du groupe *Lotus-Anthyllis*. *Ann. Inst. Pasteur* 105:232–236.

Jensen, H. L. 1964. On the relation between hosts and root-nodule bacteria in certain leguminous plants [in Danish, English title and summary]. *Tidsskr. Planteavl* 68:1–22.

Jensen, H. L. 1967. Mutual host plant relationships in two groups of legume root nodule bacteria (*Rhizobium* spp.).*Archiv. Mikrobiol.* 56:174–179.

Jensen, H. L., and Hansen, A. 1968. Observations on host plant relations in root nodule on the *Lotus-Anthyllis* and the *Lupinus-Ornithopus* groups. *Acta Agric. Scand.* 18:135–142.

Jepson, W. L. 1936. *A Flora of California*, 2. 684 pp. San Francisco: California School Book Depository.

Jimbo, T. 1927. Physiological anatomy of the root nodule of *Wistaria sinensis*. *Proc. Imp. Acad. (Japan)* 3:164–166.

Jimbo, T. 1930. On the serological classification of the root-nodule bacteria of leguminous plants. *Bot. Mag. (Tokyo)* 44:158–168.

Joachim, A. W. R. 1929. Green manuring with particular reference to coconuts. *Dep. Agric. Ceylon Leafl.* 57. 10 pp.

John, C. M., Venkatanarayana, G., and Seshadri, C. R. 1954. Varieties and forms of groundnut (*Arachis hypogaea* Linn.). Their classification and economic characters. *Indian J. Agric. Sci.* 24 (3):159–193.

Johns, S. R., Lamberton, J. A., and Sioumis, A. 1966. Alkaloids of the Australian Leguminosae. VI. Alkaloids of *Petalostylis labicheoides* var. *casseoides*. *Aust. J. Chem.* 19 (5):893.

Johnson, M. D., and Allen, O. N. 1952a. Cultural reactions of rhizobia with special reference to strains isolated from *Sesbania* species. *Antonie van Leeuwenhoek J. Microbiol. Serol.* 18 (1):1–12.

Johnson, M. D., and Allen, O. N. 1952b. Nodulation studies with special reference to strains isolated from *Sesbania* species. *Antonie van Leeuwenhoek J. Microbial. Serol.* 18 (1):13–22.

Johnston, A. 1949. Vesicular-arbuscular mycorrhiza in sea island cotton and other tropical plants. *Trop. Agric. (Trinidad)* 26:118–121.

Johnston, I. M. 1924. Taxonomic records concerning American spermatophytes. I. *Parkinsonia* and *Cercidium*. *Contrib. Gray Herb.* 70:61–68.

Johnston, I. M. 1941. New phanerogams from Mexico. IV. *J. Arnold Arboretum Harv. Univ.* 22:110–124.

Jones, H. A. 1942. *A List of Plants Reported to Contain Rotenone or Rotenoids*. U. S. Dep. Agric. Bureau of Entomology and Plant Quarantine E-571. 14 pp.

Jones, K. H., Sanderson, D. M., and Noakes, D. N. 1968. Acute toxicity data for pesticides (1968). *World Rev. Pest Control* 7:135–143.

Jordan, D., and Opoku, A. A. 1966. The effect of selected soil covers on the establishment of cocoa. *Trop. Agric. (Trinidad)* 42:155–166.

References

Jordan, D. C., and Grinyer, I. 1965. Electron microscopy of the bacteroids and root nodules of *Lupinus luteus*. *Can. J. Microbiol.* 11:271–725.

Joshi, N. V. 1920. Studies on the root nodule organism of the leguminous plants. *India Dep. Agric. Mem.* Bact. Ser. 1:247–276.

Kamerman, P. 1926. The toxic principle of *Erythrophleum lasianthum*. *S. Afr. J. Sci.* 23:179–184.

Kaplan, L. 1965. Archeology and domestication in American *Phaseolus* (Beans). *Econ. Bot.* 19:358–368.

Kariyone, T., Atsumi, K., and Shimada, M. 1923. Constituents of *Millettia taiwaniana* Hayata. *J. Pharm. Soc. Japan* 500:739–746. *Abstr. Jour. Chem. Soc.* Part 1:251–252 (1924).

Keay, R. W. J., Onochie, C. F. A., and Stanfield, D. P. 1964. *Nigerian Trees*, 2. 495 pp. Ibadan, Nigeria: Federal Department of Forest Research.

Kellerman, K. F. 1910. Nitrogen-gathering plants. *U. S. Dep. Agric. Yearbook* 1910:213–218.

Kellerman, K. F., and Robinson, T. R. 1907. Conditions affecting legume inoculation. *U. S. Dep. Agric. Bur. Plant Ind. Bull.* 100 (8):73–83.

Kerharo, J., and Bouquet, A. 1950. *Plantes Médicinales et Toxiques de la Côte-d'Ivoire-Haute-Volta*. Paris: Vigot Frères. 296 pp.

Keuchenius, A. A. M. N. 1924. Botanische kemmerken en cultuurwaarde als groenbemester van een 60-tal nieuwe soorten van Leguminosen. *Dep. Landb. Nivj. Handel, Meded. Proefsta. voor Thee* 90:44 pp.

Keuren, R. W. van, and Marten, G. C. 1972. Pasture production and utilization. *In* Hanson, C. H., ed. *Alfalfa Science and Technology*. Agron. Ser. 15:641–658. Madison, Wis. American Society of Agronomy.

Kew Bulletin. 1890. The Weather Plant. *Kew Bull. Misc. Inf.* 37. 28 pp.

Khaidarov, K. K. 1963. A new spasmolytic preparation, triacanthin [in Russian]. *Med Prom-st'. SSSR* 17 (6):53–54. *Chem. Abstr.* 65:11215 (1966).

Kharbush, S. S. 1928. Recherches sur les tubercules radicaux de quelques Papilionacées Alpines. *Bull. Soc. Bot. Fr.* 75:674–696.

Khudairi, A. K. 1957. Root-nodule bacteria of *Prosopis stephaniana*. *Science* 125:399.

Khudairi, A. K. 1969. Mycorrhiza in desert soil. *BioScience* 19:598–599.

Kies, P. 1951. Revision of the genus *Cyclopia* and notes on some other sources of bush tea. *Bothalia* 6 (1):161–176.

Kinch, D. M., and Ripperton, J. C. 1962. Koa haole, production and processing. *Hawaii Agric. Exp. Stn. Bull.* 129. 58 pp.

King, G., Duthie, J. F., and Prain, D. 1901. A second century of new and rare Indian plants. *Ann. R. Bot. Gard. (Calcutta)* 9. (1) 80 pp. 93 plates.

Kingsbury, J. M. 1964. *Poisonous Plants of the United States and Canada*. Englewood Cliffs, N.J.: Prentice-Hall. 626 pp.

Kinman, C. F. 1916. Cover crops for Porto Rico. *Porto Rico Agric. Exp. Stn. Bull.* 19. 32 pp.

Kirchner, O. 1895. Die Wurzelknöllchen der Sojabohne. *Beitr. Biol. Pflanzen Cohn's* 7:213–223.

Kitagawa, M., and Tomiyama, T. 1929. A new amino compound in the Jack bean and a corresponding new enzyme. *J. Biochem. (Japan)* 11:265–271. *Chem. Abstr.* 24:1131 (1930).

Klebahn, H. 1891. Ueber Wurzelanlagen unter Lenticellen bei *Herminiera elaphroxylon* und *Solanum dulcamara*. Nebst einem Anhang über die Wurzelknöllchen der ersteren. *Flora* 74:125–139.

Klein, G., and Farkass, E. 1930. Die mikrochemische Nachweis der Alkaloide in der Pflanze. XIV. Die Nachweis von Cytisin. *Österreichische Bot. Z.* 79:107–124.

Knaap-van Meeuwen, M. S. 1962. Reduction of *Afrormosia* to *Pericopsis* (Papilionaceae). *Bull. Jard. Bot. État Brux.* 32 (2):213–219.

Knaap-van Meeuwen, M. S. 1970. A revision of four genera of the tribe Leguminosae-Caesalpinioideae-Cynometreae in Indomalesia and the Pacific. *Blumea* 18 (1):1–52.

Knight, W. A., and Dowsett, M. M. 1936. *Ceratoniae gummi*: Carob Gum: an inexpensive substitute for gum tragacanth. *Pharm. J.* 136 (3767):35–36 (ser. 4. vol. 82).

Kodama, A. 1967. Cytological studies on root nodules of some species in Leguminosae II [in Japanese, English summary]. *Bot. Mag. Tokyo* 80:92–99.

Koeppen, R. C. 1963. Observations on *Androcalymma* (Cassieae, Caesalpiniaceae). *Brittonia* 15 (2):145–150.

Koeppen, R. C. 1967. Revision of *Dicorynia* (Cassieae, Caesalpiniaceae). *Brittonia* 19 (1):42–61.

Koeppen, R., and Iltis, H. H. 1962. Revision of *Martiodendron* (Cassieae, Caesalpiniaceae). *Brittonia* 14:191–209.

Koller, D., and Negbi, M. 1955. Germination regulating mechanisms in some desert seeds. V. *Colutea istria* Mill. *Bull. Res. Counc Isr.* sect. D. 5:73–84.

Komarov, V. L., ed. 1965. *Flora of U.S.S.R.* Leguminosae: *Astragalus*, 12. 681 pp. Akad. Nauk SSSR. Translation from Russian by Israel Program for Scientific Translations.

Komarov, V. L., ed. 1971. *Flora of the U. S. S. R.* Papilionatae, Caesalpinoideae, Mimosoideae, 11. 327 pp. Akad. Nauk SSSR. Translation from Russian by Israel Program for Scientific Translations.

Komarov, V. L., ed. 1972. *Flora of the U. S. S. R.* Leguminosae, 13:292–294. Akad. Nauk SSSR. Translation from Russian by Israel Program for Scientific Translations.

Kosower, S. N., and Kosower, E. M. 1967. Does 3,4-dihydroxyphenylalanine play a part in favism? *Nature* 215:285–286.

Kostermans, A. J. G. H. 1952. Notes on two leguminous genera from eastern Indonesia. *Reinwardtia* 1 (4):451–457.

Kostermans, A. 1954. A monograph of the Asiatic, Malaysian, Australian and Pacific species of Mimosaceae, formerly included in *Pithecolobium* Mart. *Bull. Org. Sci. Res. Indonesia* 20 (11):1–122.

Krapovickas, A., and Rigoni, V. A. 1957. Neuvas especies de *Arachis* vinculadas al problema del origin del mani. *Darwiniana* 11:432–455.

Krauss, F. G. 1911. Leguminous crops for Hawaii. *Hawaii Agric. Exp. Stn. Bull.* 23. 30 pp.

Krauss, F. G. 1927. Improvement of the pigeon pea. *J. Hered.* 18:227–232.

Krishnaswamy, N. R., and Seshadri, T. R. 1962. Naturally occurring phenylcoumarins. *In* Gore, T. S., Joshi, B. S., Sunthankar, S. V., and Tilak,

B. D., eds. *Recent Progress in the Chemistry of Natural and Synthetic Colouring Matters and Related Fields*, 235–253. New York: Academic Press.

Kriss, A. E. 1947. Microorganisms of the tundra and arctic desert soils of the Arctic [in Russian, English summary]. *Mikrobiol.* 16 (5):437–448.

Kriss, A. E., Korenyako, A. I., and Migulina, V. M. 1941. Root-nodule bacteria in arctic regions [in Russian, English summary]. *Mikrobiol.* 10:67–73.

Krüger, R. 1914. Beiträge zur Artenfrage der Knöllchenbakterien einiger Leguminosen. Inaugural Dissertation, Leipzig. 56 pp.

Krukoff, B. A. 1939a. Preliminary notes on Asiatic-Polynesian species of *Erythrina*. *J. Arnold Arbor. Harv. Univ.* 20:225–233.

Krukoff, B. A. 1939b. The American species of *Erythrina*. *Brittonia* 3 (2):205–337.

Krukoff, B. A. 1970. Supplementary notes on the American species of *Erythrina*. IV. Field studies of Central American species. *Mem. N. Y. Bot. Gard.* 20 (2):159–177.

Krukoff, B. A., and Barneby, R. C. 1974. Conspectus of species of the genus *Erythrina*. *Lloydia* 37:332–459.

Kryn, J. M. 1956. Courbaril. U.S. For. Serv. Madison, Wis.: Forest Products Lab. *Foreign Wood Ser. Rep.* 1942 (revised). 5 pp.

Kubo, H. 1939. Über Hämoprotein aus den Wurzelknöllchen von Leguminosen. *Acta Phytochim.* 11:195–200.

Kuhlmann, J. G. 1949. O genero "Etabalea." *Lilloa* 17:57–60.

Kukachka, B. F. 1960. Kokrodua (*Afromosia elata* Harms). U.S. For. Serv., Madison, Wis.: Forest Products Lab. *Foreign Wood Ser. Rep.* 1978 (revised). 7 pp.

Kukachka, B. F. 1961a. Ishpingo, *Amburana acreana* (Ducke) A. C. Smith, Leguminosae (Papilionaceae). U.S. For. Serv., Madison, Wis.: Forest Products Lab. *Foreign Wood Ser. Rep.* 1915 (revised). 4 pp.

Kukachka, B. F. 1961b. Agba, *Gossweilerodendron balsamifera* (Verm.) Harms, Leguminosae (Caesalpinioideae). U.S. For. Serv., Madison, Wis.: Forest Products Lab. *Foreign Wood Ser. Rep.* 2024 (revised). 7 pp.

Kukachka, B. F. 1964. Angelique — *Dicorynia guianensis* Amsh. U.S. For. Serv., Madison, Wis.: Forest Products Lab. *Res. Note* FPL–071 (Rep. 1787 revised). 8 pp.

Kukachka, B. F. 1970. Properties of imported woods. U.S. For. Serv., Madison, Wis.: Forest Products Lab. *Res. Pap.* FPL–125. 67 pp.

Kukachka, B. F., and Kryn, J. M. 1965. Cativo, *Prioria copaifera* Gris. U.S. For. Serv., Madison, Wis.: Forest Products Lab. *Res. Note* FPL–095. 8 pp.

Kurz, S. 1877 (reprint 1974). Leguminosae. *In Forest Flora of British Burma*, 1:330–431. Calcutta: Office of the Superintendent of Government Printing. (Reprint. Delhi: M/S Periodical Experts.)

Lachmann, J. 1858. Ueber Knöllchen der Leguminosen. *Landwirtsch. Mitt. Z. K. Lehranst., Vers. Stn. Poppelsdorf (Bonn).* 37 pp.

Lachmann, J. 1891. Über Knollen an den Wurzeln der Leguminosen. *Biedermann's Zentralbl. Agric.* 20:837–854.

Lackey, J. A. 1977a. A revised classification of the tribe Phaseoleae (Leguminosae: Papilionoideae), and its relation to canavanine distribution. *Bot. J. Linn. Soc.* 74:163–178.

Lackey, J. A. 1977b. *Neonotonia*, a new generic name to include *Glycine wightii* (Arnott) Verdcourt (Leguminosae; Papilionoideae). *Phytologia* 37:209–212.

Lackey, J. A. 1978. New combinations in *Kerstingiella* Harms (Leguminosae; Papilionoideae). *Phytologia* 38:229.

LaForge, F. B., Haller, H. L., and Smith, L. E. 1933. The determination of the structure of rotenone. *Chem. Rev.* 12:181–214.

Lainz, M., S. J. 1969. In Floram Europaeam animadversiones. *Candollea* 24:253–262.

Lamprecht, H. 1956a. *Pisum sativum* L. oder *P. arvense* L. Eine nomenklatorische Studie auf genetischer Basis. *Agric. Hort. Genet.* 14:1–4.

Lamprecht, H. 1956b. Artberechtigung von *Pisum elatius* Stev. und *jomardi* Shrank. *Agric. Hort. Genet.* 14:5–18.

Lange, R. T. 1958. Symbiotic affinities between introduced legumes and native *Rhizobium* in Western Australia. *Rhizobium Newsl.* 3 (2):9–10.

Lange, R. T. 1959. Additions to the known nodulating species of Leguminosae. *Antonie van Leeuwenhoek J. Microbiol Serol.* 25:272–276.

Lange, R. T. 1961. Nodule bacteria associated with the indigenous Leguminosae of south-western Australia. *J. Gen. Microbiol.* 61:351–359.

Lange, R. T. 1962. Susceptibility of indigenous south-west Australian legumes to infection by *Rhizobium*. *Plant Soil* 17:134–136.

Lange, R. T., and Parker, C. A. 1960. The symbiotic performance of lupin bacteria under glasshouse and field conditions. *Plant Soil* 13 (2):137–146.

Lange, R. T., and Parker, C. A. 1961. Effective nodulation of *Lupinus digitatus* by native rhizobia in south-western Australia. *Plant Soil* 15:193–198.

Larisey, M. M. 1940a. Monograph of the genus *Baptisia*. *Ann. Mo. Bot. Gard.* 27:119–244.

Larisey, M. M. 1940b. A revision of the North American species of the genus *Thermopsis*. *Ann. Mo. Bot. Gard.* 27:245–258.

Laurent, É. 1891. Recherches sur les nodosités radicales des légumineuses. *Ann. Inst. Pasteur* 5:105–139.

Lawrence, G. H. M. 1949. Discussions in botanical names of cultivated plants. *Gentes Herb.* 8 Fasc. 1:3–76.

Lawson, G. W. 1966. *Plant Life in West Africa.* Oxford Tropical Handbooks. London: Oxford University Press. 150 pp.

Lechtova-Trnka, M. 1931. Étude sur les bactéries des légumineuses et observations sur quelques champignons parasites des nodosités. *Botaniste,* 23:301–530.

Lecomte, H. 1894. Les tubercules radicaux de l'Arachide (*Arachis hypogea* L.) *C. R. Acad. Sci.* 119:302–304.

Ledingham, G. F. 1957. Chromosome numbers of some Saskatchewan Leguminosae with particular reference to *Astragalus* and *Oxytropis*. *Can. J. Bot.* 35:657–666.

Ledingham, G. F. 1960. Chromosome numbers of *Astragalus* and *Oxytropis*. *J. Can. Génét. Cytol.* 2:119–128.

Ledingham, G. F., and Rever, B. M. 1963. Chromosome numbers of some southwest Asian species of *Astragalus* and *Oxytropis* (Leguminosae). *J. Can. Génét. Cytol.* 5 (1):18–32.

References

Lee, A. T. 1948. The genus *Swainsona*. *Contrib. N. S. W. Natl. Herb.* 1 (4):131–271.

Lee, A. T. 1973. A new genus of Papilionaceae and related Australian genera. *Contrib. N. S. W. Natl. Herb.* 4 (7):412–430.

Leifson, E., and Erdman, L. W. 1958. Flagellar characteristics of *Rhizobium* species. *Antonie van Leeuwenhoek J. Microbiol. Serol.* 24:97–110.

León, H., and Alain, H. 1951. *Flora de Cuba*. Dicotiledoneas: Casuarinaceae a Meliaceae, 2. 456 pp. Havana: *Contrib. Ocas. Mus. Hist. Nat. Col. Salle* 10.

León, J. 1966. Central American and West Indian species of *Inga* (Leguminosae). *Ann. Mo. Bot. Gard.* 53 (3):265–359.

Léonard, J. 1949. Notulae systematicae IV (Caesalpiniaceae-Amherstieae africanae americanaeque). *Bull. Jard. Bot. État Brux.* 19:383–408.

Léonard, J. 1950. Note sur les genres paléotropicaux *Afzelia, Intsia* et *Pahudia* (Légum.-Caesalp.). *Reinwardtia* 1 (1):61–66.

Léonard, J. 1951. Notulae systematicae XI. Les *Cynometra* et les genres voisins en Afrique tropicale. *Bull. Jard. Bot. État Brux.* 21:373–450.

Léonard, J. 1952. Cynometreae et Amherstieae. IV. *In* Robyns, W. *Flore du Congo Belge et du Ruanda-Urundi*, 3:279–495. Brussels: Publ. Inst. Nat. Étude Agron. Congo Belge (I.N.É.A.C.).

Léonard, J. 1954a. Hedysareae, *In* Robyns, W. *Flore du Congo Belge et du Ruanda-Urundi*, 5:176–359. Brussels: Publ. Inst. Nat. Étude Agron. Congo Belge. (I.N.É.A.C.).

Léonard, J. 1954b. Notulae systematicae XV. Papilionaceae-Hedysareae Africanae (*Aeschynomene, Alysicarpus, Ormocarpum*). *Bull. Jard. Bot. État Brux.* 24:63–106.

Léonard, J. 1954c. Notulae systematicae XVI. *Paramacrolobium* genre nouveau de Caesalpiniaceae d'Afrique tropicale. *Bull. Jard. Bot. État Brux.* 24:347–348.

Léonard, J. 1955. Notulae systematicae XVII. Les genres *Anthonotha* P. Beauv. et *Pellegriniodendron* J. Léonard en Afrique tropicale. *Bull. Jard. Bot. État Brux.* 25:201–203.

Léonard, J. 1957. Genera des Cynometreae et des Amherstieae africaines. *Mém. Acad. R. Belg. Classe Sci.* 30 Fasc. 2:1–314.

Leonard, L. T. 1923. Nodule-production kinship between the soybean and the cowpea. *Soil Sci.* 15:277–283.

Leonard, L. T. 1925. Lack of nodule formation in a subfamily of the Leguminosae. *Soil Sci.* 20:165–167.

Leonard, L. T. 1934. The nodule organisms of *Mimosa pudica* L. *J. Bacteriol.* 27:55.

Leonard, L. T. 1939. Bacteria associated with *Gleditsia triacanthos* L. *Trans. 3d Comm. Intern. Soc. Soil Sci., New Brunswick, N. J., 1939*. A:64–70.

Leonard, L. T., and Dodson, W. R. 1933. The effects of nonbeneficial nodule bacteria on Austrian winter peas. *J. Agric. Res.* 46:649–663.

Leonard, L. T., and Reed, H. R. 1930. A comparison of some nodule forming and non-nodule forming legumes for green manuring. *Soil Sci.* 30:231–236.

Leonard, N. J. 1953. Lupin alkaloids. *In* Manske, R. H. F., and Holmes, H. L., eds. *The Alkaloids, Chemistry and Physiology*, 3:119–199. New York: Academic Press.

Leppik, E. E. 1971. Assumed gene centers of peanuts and soybeans. *Econ. Bot.* 25 (2):188–194.

Li, H.-L. 1944. On *Flemingia* Roxburgh (1812), non Roxburgh (1803) versus *Moghania* J. St.-Hilaire (1813). *Am. J. Bot.* 31:224–228.

Liebmann, F. 1854. *Dipogon*, Leibm. — Nov. genus Papilionacearum. *In Index Seminum in Horto Academico Hauniensi. Ann. Sci. Nat.* ser. 4. Bot. 2:370–375.

Lim, G. 1977. Nodulation of tropical legumes in Singapore. *Trop. Agric. (Trinidad)* 54 (2):135–141.

Lim, G., and Ng, H. L. 1977. Root nodules of some tropical legumes in Singapore. *Plant Soil* 46:317–327.

Link, K. P. 1945. The anticoagulant 3,3′-methylene-bis (4-hydroxycoumarin). *Fed. Proc.* 4:176–182.

Link, K. P. 1959. The discovery of dicumarol and its sequels. *Circulation* 19:97–107.

Lipman, C. B., and Fowler, L. W. 1915. The isolation of *Bacillus radicicola* from soil. *Science* 41:256–259.

Little, E. L., Jr. 1945. Miscellaneous notes on nomenclature of United States trees. *Am. Midl. Nat.* 33:495–513.

Little, E. L., Jr., and Wadsworth, F. H. 1964. *Common Trees of Puerto Rico and the Virgin Islands.* U. S. Dep. Agric. Forest Service, Agriculture Handbook 249. 548 pp.

Little, V. A. 1931. A preliminary report on the insecticidal properties of Devil's shoe-string, *Cracca virginiana* Linn. *J. Econ. Entomol.* 24:743–753.

Loden, H. D., and Hildebrand, E. M. 1950. Peanuts — especially their diseases. *Econ. Bot.* 4:354–379.

Logan, M. D. 1960. The carob crusade. *Am. For.* 66 (5):18–19; 63–65.

Löhnis, F., and Hansen, R. 1921. Nodule bacteria of leguminous plants. *J. Agric. Res.* 20:543–556.

Löhnis, F., and Leonard, L. T. 1926. Inoculation of legumes and non-legumes with nitrogen-fixing and other bacteria. *U. S. Dep. Agric. Farmers Bull.* 1496. 27 pp.

Long, H. C. 1917. *Plants Poisonous to Live Stock.* Cambridge: Cambridge University Press. 119 pp.

Lopatina, G. V. 1931. Investigations of the nodule-forming bacteria. II. Observations on the nodule formation in leguminous plants [in Russian, English summary]. *Bull. State Inst. Agric. Microbiol. USSR* 4 (3):105–110.

Lyman, C., Rotar, P. P., and Brown, T. A. 1967. Koa haole. *Univ. Hawaii Coop. Ext. Serv. Leafl.* Conserv. Ser. 110. 2 pp.

Lyman, R. L., Bickoff, E. M., Booth, A. N., and Livingston, A. L. 1959. Detection of coumestrol in leguminous plants. *Arch. Biochem. Biophys.* 80:61–67.

Lynch, D. L., and Sears, O. H. 1950. The nitrogen-fixing efficiency of strains of *Lotus corniculatus* nodule bacteria. *Soil Sci. Soc. Am. Proc.* 14:168–170.

References

Maassen, and Müller, A. 1907. Über die Bakterien in den Knöllchen der verschiedenen Leguminosen arten. *Mitt. K. Biol. Anst. Land. Forstw.* 4:42–44.

McCawley, E. L. 1955. Cardioactive alkaloids. *In* Manske, R. H. F., ed. *The Alkaloids, Chemistry and Physiology*, 5:79–107. New York: Academic Press.

McComb, J. A., Elliott, J., and Dilworth, M. J. 1975. Acetylene reduction by *Rhizobium* in pure culture. *Nature* 256:409–410.

McCoy, E. F. 1929. A cytological and histological study of the root nodules of the bean, *Phaseolus vulgaris* L. *Centralbl. Bakt.* ser. 2. 79:394–412.

McDougall, W. B. 1921. Thick-walled root hairs of *Gleditsia* and related genera. *Am. J. Bot.* 8:171–175.

McEwan, T. 1964. Isolation and identification of the toxic principle of *Gastrolobium grandiflorum*. *Queensl. J. Agric. Sci.* 21 (1):1–14.

McIntyre, A. C., and Jeffries, C. D. 1932. The effect of Black Locust on soil nitrogen and growth of catalpa. *J. For.* 30:22–28.

McIntyre, A. R. 1947. *Curare — Its history, nature, and clinical use.* Chicago: University of Chicago Press. 240 pp.

McKee, R., and Ritchey, G. E. 1947. Lupines — New legumes for the South. *U. S. Dep. Agric. Farmers Bull.* 1946. 10 pp.

McKee, R., Schoth, H. A., and Stephens, J. L. 1931. Monantha vetch. *U. S. Dep. Agric. Circ.* 152. 13 pp.

Mackie, W. W. 1943. Origin, dispersal and variability of the lima bean, *Phaseolus lunatus*. *Hilgardia* 15:1–29.

McKnight, T. 1949a. Efficiency of isolates of *Rhizobium* in the cowpea group, with proposed additions to this group. *Queensl. J. Agric. Sci.* 6 (2):61–76.

McKnight, T. 1949b. Non-symbiotic nitrogen-fixing organisms in Queensland soils. *Queensl. J. Agric. Sci.* 6:177–197.

McLeod, R. W. 1960. Compatibility of summer legumes with local commercial inoculum strains. *Rhizobium Newsl.* 5 (2):63–64.

McLeod, R. W. 1962. Natural nodulation of tropical legumes in New South Wales. *Agric. Gaz. N. S. W.* 73 (8):419–421; 431.

MacNeish, R. S. 1960. Agricultural origins in Middle America and their diffusion into North America. *Katunob* 1 (2):25–29.

MacTaggart, A. 1937. *Stylosanthes. J. Counc. Sci. Ind. Res. Canberra* 10 (3):201–203.

Maeda, K. 1973a. The recent botanical taxonomic system of peanut cultivars. *Jap. Agric. Res. Q.* 7 (4):228–232.

Maeda, K. 1973b. Floral morphology and its application to the botanical classification of the peanut cultivars, *Arachis hypogaea* L. *Mem. Fac. Agric. Kochi Univ. Nankoku, Japan* 23. 55 pp.

Maheshwari, J. K. 1967. The genus *Bakerophyton* Hutch. (Fabaceae). *Taxon* 16:238.

Maire, R. 1919. Un nouveau genre de Papilionacées de la flore nord-africaine. *Bull. Soc. Hist. Nat. Afr. Nord* 10:22–26.

Maire, R. 1924. *Benedictella* nov. gen. *Papilionacearum Lotearum*. Contributions à l'étude de la Flore de l'Afrique du Nord. *Bull. Soc. Hist. Nat. Afr. Nord* 15:383–384.

Maire, R. 1949. Contributions à l'étude de la flore de l'Afrique du Nord. *Bull. Soc. Hist. Nat. Afr. Nord* 39 (7–8):129–137.

Majot-Rochez, R., and Duvigneaud, P. 1954. *Erythrina* L. *In* Robyns, W., ed. *Flore du Congo Belge et du Ruanda-Urundi*, 6:113–126. Brussels: Publ. Inst. Nat. Étude Agron. Congo Belge (I.N.É.A.C.).

Majumdar, D. N., and Paul, G. B. 1954. *Mucuna pruriens.* IV. Alkaloidal constituents and their derivatives. *Indian Pharm.* 10:79–84. *Chem. Abstr.* 48:8994 (1955).

Majumdar, D. N., and Zalani, C. D. 1953. *Mucuna pruriens.* III. Isolation of water-soluble alkaloids and a study of their chemical and physiological characteristics. *Indian J. Pharm.* 15:62–65. *Chem. Abstr.* 49:9881 (1955).

Malan, C. E. 1935. Richerche sui tubercoli radicali e sulle micorrize delle leguminose della zona alpina del Faggio, Abete, e Larice. *Nuovo G. Bot. Ital.* 42:475–476.

Malan, C. E. 1938. Root nodules and mycorrhiza of Alpine pastures [in Italian]. *Ann. Bot. (Rome)* 21:465–494.

Malme, G. O. A. 1905. *Dahlstedtia*, eine neue Leguminosen-Gattung. *Ark. Bot.* 4(9):1–7.

Malpighi, M. 1679. *Anatome Plantarum.* London: J. Martyn. 93 pp.

Mann, H. H. 1906. The renovation of deteriorated tea. *Agric. J. India* 1:83–96.

Mannetje, L. 't. 1969. *Rhizobium* affinities and phenetic relationships within the genus *Stylosanthes. Aust. J. Bot.* 17:553–564.

Mantell, C. L. 1947. *The Water-Soluble Gums.* New York: Reinhold. 279 pp.

Marassi, A. 1939. Della *Cordeauxia edulis* (Yebb nuts). *Agric. Colon.* 33:613–626.

Maréchal, R., and Baudet, J. C. 1977. Transfert du genre africain *Kerstingiella* Harms à *Macrotyloma* (Wight & Arn.) Verdc. (Papilionaceae). *Bull. Jard. Bot. Nat. Belg.* 47:49–52.

Marion, L. 1952. The *Erythrina* alkaloids. *In* Manske, R. H. F., and Holmes, H. L., eds. *The Alkaloids, Chemistry and Physiology,* 2:449–511. New York: Academic Press.

Marker, R. E., Wagner, R. B., Ulshafer, P. R., Goldsmith, D. P. J., and Ruof, C. H. 1943. Sterols CLIV. Sapogenins. LXVI. The sapogenin of *Trigonella foenum-graecum. J. Am. Chem. Soc.* 65:1247.

Marker, R. E., Wagner, R. B., Ulshafer, P. R., Wittbecker, E. L., Goldsmith, D. P. J., and Ruof, C. H. 1947. New sources for sapogenins. *J. Am. Chem. Soc.* 69:2242.

Markley, K. S., ed. 1950–1951. *Soybeans and Soybean Products.* New York: Interscience Publishers. 2 vols.

Marsh, C. D., and Clawson, A. B. 1920. *Daubentonia longifolia* (coffee bean), a poisonous plant. *J. Agric. Res.* 20 (6):507–513.

Martin, W. P. 1948. Observations on the nodulation of leguminous plants of the southwest. U. S. Dep. Agric. Soil Conserv. Serv. Region 6. Albuquerque, Regional Bull. 107. Plant Study Ser. 4. Mimeographed, 10 pp.

Mascré, M. 1937. Le leucaenol, principe défini retiré des graines de *Leucaena glauca* Benth. (Legumineuses Papilionacées). *C. R. Acad. Sci.* 204:890–891.

References

Masefield, G. B. 1952. The nodulation of annual legumes in England and Nigeria: preliminary observations. *Emp. J. Exp. Agric.* 20 (79):175–186.

Masefield, G. B. 1957. The nodulation of annual leguminous crops in Malaya. *Emp. J. Exp. Agric.* 25 (98):139–150.

Masefield, G. B. 1961. Root nodulation and agricultural potential of the leguminous genus *Psophocarpus. Trop. Agric. (Trinidad)* 38 (3):225–229.

Matsuoka, T. 1920. Fiber of *Sesbania aegyptiaca.* Japanese patent 25,952. March 11. *Chem. Abstr.* 15:1406 (1921).

Mattoon, W. R. 1930. Growing black locust trees. *U. S. Dep. Agric. Farmers Bull.* 1628. 13 pp.

Matyukhina, L. G., and Ryabinin, A. A. 1964. Structure of spherophysine and its derivatives [in Russian]. *Z. Obshch. Khim.* 34 (11):3854–3855. *Chem. Abstr.* 62:6525e (1965).

Maxwell, R. H. 1970. The genus *Cymbosema* (Leguminosae): notes and distribution. *Ann. Mo. Bot. Gard.* 57:252–257.

Maxwell, R. H. 1977. A resume of the genus *Cleobulia* (Leguminosae) and its relation to the genus *Dioclea. Phytologia* 38:51–65.

Mayer, F. 1943. *The Chemistry of Natural Coloring Matter.* New York: Reinhold. 354 pp.

Mazur, Y., Danieli, N., and Sondheimer, F. 1960. The synthesis of steroidal sapogenins. *J. Am. Chem. Soc.* 82:5889–5908.

Mears, J. A., and Mabry, T. J. 1971. Alkaloids in the Leguminosae. *In* Harborne, J. B., Boulter, D., and Turner, B. L., eds. *Chemotaxonomy of the Leguminosae,* 73–178. New York: Academic Press.

Medsger, O. P. 1939. *Edible Wild Plants.* New York: Macmillan. 323 pp.

Meeuwen, M. S. van. 1962. Reduction of *Afrormosia* to *Pericopsis* (Papilionaceae). *Bull. Jard. Bot. Brux.* 32:213–219.

Meeuwen, M. S. van, Steenis, C. G. G. J. van, and Stemmerik, J. 1961. Preliminary revisions of some genera of Malaysian Papilionaceae II. *Reinwardtia* 6 (1):85–108.

Mehta, J. C., and Majumdar, D. N. 1944. Indian medicinal plants. V. *Mucuna pruriens* I. *Indian J. Pharm.* 6:92–95. *Chem. Abstr.* 40:3227 (1946).

Mendes, A. J. T. 1947. Estudos citologicos no genera *Arachis. Bragantia* 7:257–267.

Menninger, E. A. 1962. *Flowering Trees of the World — For Tropics and Warm Climates.* New York: Heathside Press. 336 pp.

Menninger, E. A. 1967. *Fantastic Trees.* New York: Viking Press. 304 pp.

Merck Index — An Encyclopedia of Chemicals and Drugs. 1968. Stecher, P. G., and others, eds. 8th ed. Rahway, N.J.: Merck. 1,713 pp.

Merkenschlager, F. 1934. Migration and distribution of red clover in Europe. *Herb. Rev.* 2:88–92.

Merlis, V. M. 1952. Alkaloids of *Eremosparton flaccidum* [in Russian]. *Z. Obshch. Khim.* 22:347–350. *Chem. Abstr.* 46:7289 (1952).

Merrill, E. D. 1910. An enumeration of Philippine Leguminosae, with keys to the genera and species. *Philipp. J. Sci.* ser. C. Bot. 5:1–136.

Merrill, E. D. 1916. The systematic position of the "rain tree," *Pithecolobium Saman. J. Wash. Acad. Sci.* 6:42–48.

Merrill, E. D. 1935. A commentary on Loureiro's *Flora Cochinchinensis. Trans. Am. Philos. Soc.* n.s. 24 (2):1–445.

Merrill, E. D., and Chen, L. 1943. The Chinese and Indo-Chinese species of *Ormosia. Sargentia* 3:77–117.

Meusel, H., and Jager, E. 1962. Über die Verbreitung einiger Papilionaceen-Gattungen. *Kulturpflanze* 3:249–262.

Meyer, T. M. 1946. The insecticidal constituents of *Pachyrrhizus erosus* Urban. *Rec. Trav. Chim.* 65:835–842. *Chem. Abstr.* 41:2739 (1947).

Milne-Redhead, E. 1937. The genus *Cordyla* Loureiro. *Feddes Repert. Specierum Nov. Regni Veg.* 41:227–235.

Milne-Redhead, E. 1951. Tropical African Plants XXI. *Kew Bull.* 1950:335–384.

Milne-Redhead, E. 1954. *Zornia* in tropical Africa. *Bol. Soc. Broteriana* ser. 2. 28:79–104.

Milovidov, P. F. 1926. Über einige neue Beobachtungen an den Lupinenknöllchen. *Centralbl. Bakt.* ser. 2. 68:333–345.

Milovidov, P. F. 1928a. Ein neuer Leguminosen-knöllchenmikrob (*Bacterium radicicola* forma *carmichaeliana*). *Centralbl. Bakt.* ser. 2. 73:58–69. 2 plates.

Milovidov, P. 1928b. Recherches sur les tubercules du lupin. *Rev. Gen. Bot.* 40:1–13.

Misra, R., Pandey, R. C., and Dev, S. 1964. Chemistry of the oleoresin from *Hardwickia pinnata*. A series of new diterpenoids. *Tetrahedron Lett.* 1964 (49):3751–3759.

Mitchell, W. 1956. Liquorice and glycyrrhetinic acid. *Manuf. Chem.* 27:169–172.

Moffett, M. L., and Colwell, R. 1968. Adansonian analysis of the Rhizobiaceae. *J. Gen. Microbiol.* 51:245–266.

Mohlenbrock, R. H. 1957. A revision of the genus *Stylosanthes. Ann. Mo. Bot. Gard.* 44:299–355.

Mohlenbrock, R. H. 1961. A monograph of the leguminous genus *Zornia. Webbia* 16:1–141.

Mohlenbrock, R. H. 1962a. Tribe Hedysareae, subtribe Stylosanthinae (Leguminosae), of Central America and Mexico. *Southwest Nat.* 7 (1):1–22.

Mohlenbrock, R. H. 1962b. The leguminous genus *Riedeliella* Harms. *Webbia* 16:643–648.

Mohlenbrock, R. H. 1962c. A revision of the leguminous genus "Ateleia." *Webbia* 17:153–186.

Mohlenbrock, R. H. 1963a. Reorganization of genera within Tribe Ingeae of the mimosoid Leguminoseae, *Reinwardtia* 6 (4):429–442.

Mohlenbrock, R. H. 1963b. Subgeneric categories of *Pithecellobium* Mart. *Reinwardtia* 6 (4):443–447.

Mohlenbrock, R. H. 1963c. A revision of the genus *Sweetia. Webbia* 17:223–263.

Mohlenbrock, R. H. 1966. A revision of *Pithecellobium* Sect. "Archidendron." *Webbia* 21:653–724.

References

Moldenke, H. N., and Moldenke, A. L. 1952. *Plants of the Bible*. Waltham, Mass.: Chronica Botanica. 328 pp. 95 plates.

Molhuysen, J. A., Gerbrandy, J., de Vries, L. A., de Jong, J. C., Lenstra, J. B., Turner, K. P., and Borst, J. G. G. 1950. A liquorice extract with deoxycortone-like action. *Lancet* 2:381–386.

Molisch, H. 1926. Über die Periodizität der Wurzelknöllchen bei *Wistaria sinensis*. *In Pflanzenbiologie in Japan auf Grund eigener Beobachtungen*, 227–228. Jena: Gustav Fischer.

Moore, J. A. 1936. The vascular anatomy of the flower in the papilionaceous Leguminosae. I. *Am. J. Bot.* 23:279–290.

Moore, R. H. 1938. Investigations of insectical plants. *Puerto Rico (Mayaguëz) Agric. Exp. Stn. Rep.* 1939:71–93.

Moore, R. J. 1965. Colchicine tetraploid *Caragana arborescens*. *Can J. Genet. Cytol.* 7:103–107.

Morck, D. 1891. Über die Formen der Bakteroiden bei den einzelnen Spezies der Leguminosen. *Inaugural Dissertation, Leipzig*. 44 pp.

Morimoto, H., and Matsumoto, N. 1966. Alkaloids. VI. Components of *Lespedeza bicolor* var. *japonica* [in German]. *Ann. Chem.* 692:194–199. *Chem. Abstr.* 65:3921 (1966).

Morimoto, H., and Oshio, H. 1965. Alkaloids. V. Components of *Lespedeza bicolor* var. *japonica*. I. Lespedamine, a new alkaloid [in German]. *Ann. Chem.* 682:212–218. *Chem. Abstr.* 62:14740 (1965).

Morrison, J. H., and Neill, K. G. 1949. The isolation of *d*-sparteine from *Hovea* species. *Aust. J. Sci. Res.* 2A:427–428.

Morse, W. J. 1947. The versatile soybean. *Econ. Bot.* 1:137–147.

Morton, C. V. 1944. Taxonomic studies of tropical American plants. *Contrib. U. S. Natl. Herb.* 29:1–85.

Moshkov, B. S. 1939. Symbiosis of leguminous plants with assimilating bacteria as determined by photoperiodism [in Russian]. *Dokl. Akad. Sci. SSSR.* 22:187–188.

Mostert, J. W. C. 1955. Observations on the nodulation of some leguminous species. *Farm. S. Afr.* 30:338–340.

Mothes, K., and Pietz, J. 1937. Zur Physiologie der Leguminosensymbiose. *Naturwissenschaften.* 25:201–202.

Moustafa, E. 1963. Peroxidase isozymes in root nodules of various leguminous plants. *Nature* 199 (4899):1189.

Moxon, A. L., Olson, O. E., Searight, W. V., and Sandals, K. M. 1938. The stratigraphic distribution of selenium in the Cretaceous formations of South Dakota and the selenium content of some associated vegetation. *Am. J. Bot.* 25:784–810.

Muenscher, W. C. 1939. *Poisonous Plants of the United States*. Rural Science Series. New York: Macmillan. 266 pp.

Muller, H. R. A., and Frémont, T. 1935. Observations sur l'infection mycorhizienne dans le genre *Cassia* (Caesalpinaceae). *Ann. Agron.* 5:678–690.

Munk, W. J. de. 1962. Preliminary revisions of some genera of Malaysian Papilionaceae III. A census of the genus *Crotalaria*. *Reinwardtia* 6 (3):195–223.

Munns, E. N. 1922. Reproduction and nitrogen. *J. For.* 20:497–498.

Munson, W. M. 1899. The acquisition of atmospheric nitrogen. Soil inoculation. *14th Annu. Rep. Maine Agric. Exp. Stn.* 1898:208–212.

Nagai, K. 1902. Über Rotenon, ein wirksamer Bestandteil der *Derris* Wurzel [in Japanese]. *J. Tokyo Chem. Soc.* 23:740. (Reviewed in *Biochem. Z.* 157 (2):1925).

Nakai, T. 1923. Genera nova Rhamnacearum et Leguminosarum ex Asia orientali. *Bot. Mag. Tokyo* 37 (435):29–36.

Namboodiripad, C. P. 1966. Certain seed oils. *Indian Oil Soap J.* 31 (10):286–289.

Narayana, C. S., and Rangaswami, S. 1955. Chemical examination of plant insecticides. Part X. Seeds of *Mundulea suberosa* Benth. *J. Sci. Ind. Res. (India)* 14B:105–107.

Narayana, H. S. 1963a. Morphology and anatomy of the root nodules in Angiosperms. *Mem. Indian Bot. Soc.* 4:208–218.

Narayana, H. S. 1963b. A contribution to the structure of root nodule in *Cyamopsis tetragonoloba* Taub. *J. Indian Bot. Soc.* 42 (2):273–280.

Narayana, H. S., and Gothwal, B. D. 1964. A contribution to the study of root nodules in some legumes. *Proc. Indian Acad. Sci.* 59:350–359.

Natelson, B. H. 1969. Beans — a source of L-dopa. *Lancet* (Sept. 20) 640–641.

National Academy of Sciences. 1979. *Tropical Legumes: Resources for the Future*. Washington, D.C. National Academy of Sciences. 331 pp.

Naudin, C. 1894. Quelques mots sur les nodosités. *J. Agric. Prat.* 58:453–454.

Naudin, C. 1897a. Nouvelles recherches sur les tubercules et les nodosités des légumineuses et sur leurs rapports avec ces plantes. *J. Agric. Prat.* 61 (1): 491–495.

Naudin, C. 1897b. Nouvelles recherches sur les tubercules et les nodosités des légumineuses et sur leurs rapports avec ces plantes. *J. Agric. Prat.* 61 (1): 807–811.

Naudin, C. 1897c. Nouvelles recherches sur les tubercules et les nodosités des légumineuses et sur leurs rapports avec ces plantes. *J. Agric. Prat.* 61 (1): 842–846.

Naudin, C. 1897d. Nouvelles recherches sur les tubercules et les nodosités des légumineuses et sur leurs rapports avec ces plantes. *J. Agric. Prat.* 61 (2): 46–51.

Neal, M. C. 1948. *In Gardens of Hawaii*. Honolulu: Bernice P. Bishop Mus. *Spec. Publ.* 40. 850 pp.

Neal, W. M., Rusoff, L. L., and Ahmann, C. F. 1935. The isolation and some properties of an alkaloid from *Crotalaria spectabilis* Roth. *J. Am. Chem. Soc.* 57 (2):2560–2561.

Nelson, G. H., Nieschlag, H. J., Daxenbichler, M. E., Wolff, I. A., and Perdue, R. E. 1961. A search for new fiber crops. III. Laboratory-scale pulping studies. *TAPPI (Tech. Assoc. Pulp Pap. Ind.)* 44:319–325.

Newton, W. E., and Nyman, C. J., eds. 1974. *Proceedings of the 1st International Symposium on Nitrogen Fixation*, 2:313–717. Seattle: Washington State University Press.

Newton, W., Postgate, J. R., and Rodriquez-Barrueco, C. 1977. *Recent Developments in Nitrogen Fixation*. New York: Academic Press. 622 pp.

References

Nicholas, D. B. 1971. Genotypic variation in growth and nodulation in *Glycine wightii. J. Aust. Inst. Agric. Sci.* 37 (1):69–70.

Nicholas, D. B., and Haydock, K. P. 1971. Variation in growth and nodulation of *Glycine wightii* under controlled environment. *Aust. J. Agric. Res.* 22:223–230.

Nicolls, O. W., Provan, D. M., Cole, M. M., and Tooms, J. S. 1964–1965. Geobotany and geochemistry in mineral exploration in the Dugald River area, Cloncurry district, Australia. *Inst. Min. Metall. Trans.* 2:695–799.

Nikonov, G. K. 1959. Alkaloids of *Maackia amurensis* [in Russian]. *Tr. Vses. Nauch.-Issled. Inst. Lekarstv. Aromat. Rast.* 11:38–45. *Chem. Abstr.* 55:18893 (1961).

Nobbe, F., and Hiltner, L. 1896. Improvements relating to the inoculation of soil for the cultivation of leguminous plants. British Patent No. 11,460, April 25, 1896.

Nobbe, F., and Hiltner, L. 1899. Über die Wirkung der Leguminosenknöllchen in der Wasserkultur. *Landwirtsch. Vers. Stn.* 52:455–465.

Nobbe, F., Hiltner, L., and Schmid, E. 1895. Versuche über die Biologie der Knöllchenbakterien der Leguminosen, insbesondere über die Frage der Arteinheit derselben. *Landwirtsch. Vers. Stn.* 45:1–27.

Nobbe, F., Richter, L., and Simon, J. 1908. Weitere Untersuchungen über die wechselseitige Impfang verschiedener Leguminosengattungen. *Landwirtsch Vers. Stn.* 68:241–252.

Nobbe, F., Schmid, E., Hiltner, L., and Hotter, E. 1891. Versuche über die Stickstoff-Assimilation der Leguminosen. *Landwirtsch. Vers. Stn.* 39:327–359.

Norman, A. G. 1931. Studies on the gums. II. Tragacanthin — the soluble constituent of gum Tragacanth. *Biochem. J.* 25:200–204.

Norman, A. G., ed. 1963. *The Soybean — Genetics, Breeding, Physiology, Nutrition, Management.* London: Academic Press. 239 pp.

Normand, D. 1947. Note sur les bois de *Zingana* et autres Césalpiniées africaines à très petites folioles. *Rev. Int. Bot. Appl. Agric. Trop.* 27:139–150.

Norris, D. O. 1956. Legumes and the *Rhizobium* symbiosis. *Emp. J. Exp. Agric.* 24 (96):247–270.

Norris, D. O. 1958a. Lime in relation to the nodulation of tropical legumes. *In* Hallsworth, E. G., ed. *Nutrition of the Legumes* 164–182. New York: Academic Press.

Norris, D. O. 1958b. A red strain of *Rhizobium* from *Lotononis bainesii* Baker. *Aust. J. Agric. Res.* 9 (5):629–632.

Norris, D. O. 1959a. *Rhizobium* affinities of African species of *Trifolium. Emp. J. Exp. Agric.* 27 (106):87–97.

Norris, D. O. 1959b. The role of calcium and magnesium in the nutrition of *Rhizobium. Aust. J. Agric. Res.* 10 (5):651–698.

Norris, D. O. 1959c. Legume bacteriology in the tropics. *J. Aust. Inst. Agric. Sci.* 25:202–207.

Norris, D. O. 1964. Techniques used in work with *Rhizobium. In Some Concepts and Methods in Sub-tropical Pasture Research, Aust. C.S.I.R.O. Commonw. Bur. Pastures Field Crops (Brisbane) Bull.* 47:186–198.

Norris, D. O. 1965. *Rhizobium* relationships in legumes. *Proc. 9th Int. Grassl. Congr.* São Paulo, 1965. 2:1087–1092.

Norris, D. O. 1967. The intelligent use of inoculants and lime pelleting for tropical legumes. *Trop. Grassl.* 1 (2):107–121.

Norris, D. O. 1969. Observations on the nodulation status of rainforest leguminous species in Amazonia and Guyana. *Trop. Agric. (Trinidad)* 46:145–151.

Norris, D. O., and Mannetje, L. 't. 1964. The symbiotic specialization of African *Trifolium* spp. in relation to their taxonomy and their agronomic use. *East Afr. Agric. For. J.* 29:214–235.

Norris, D. O., Lopes, E. S., and Weber, D. F. 1970. Incorporacão de matéria orgânica ("mulching") e aplicação de péletes de calcário ("pelleting") para testar estirpes de *Rhizobium* em experimentos de campo sob condiçoes tropicais. *Pesqui. Agropecu. Bras.* 5:129–146.

Norton, L. B. 1943. Rotenone in the yam bean *(Pachyrrhizus erosus)*. *J. Am. Chem. Soc.* 65:2259–2260.

Norton, L. B., and Hansberry, R. 1945. Constituents of the insecticidal resin of the yam bean *(Pachyrrhizus erosus)*. *J. Am. Chem. Soc.* 67:1609–1614.

Nutman, P. S. 1956. The influence of the legume in root-nodule symbiosis. A comparative study of host determinants and functions. *Biol. Rev.* 31:109–151.

Nutman, P. S., ed. 1976. *Symbiotic Nitrogen Fixation in Plants*. Int. Biol. Progr. 7. 584 pp. Cambridge: Cambridge University Press.

Ochse, J. J., Soule, M. J., Jr., Dijkman, M. J., and Wehlburg, C. 1961. *Tropical and Subtropical Agriculture*, 2 Part 2:761–1446. New York: Macmillan.

Offutt, M. S., and Baldridge, J. D. 1973. The lespedezas. *In* Heath, M. E., Metcalfe, D. S., and Barnes, R. F., eds. *Forages*, 189–198. 3d ed. Ames, Iowa: Iowa State University Press.

Ogimi, C. 1964. Tests of the seed inoculation with *Rhizobium* on some species of genus *Acacia* (Preliminary report) [in Japanese, English summary]. *Div. Agric. Home Econ. Engin. Univ. Ryukyus, Sci. Bull.* 11:109–119. 2 plates.

Ohashi, H. 1973. Euchresteae, a new tribe of the family Leguminosae [in Japanese, English summary]. *J. Jap. Bot.* 48:225–234.

Ohashi, H., and Sohma, K. 1970. A revision of the genus *Euchresta* (Leguminosae). *J. Fac. Sci. Univ. Tokyo* Sec. 3. Bot. 10 (11–13):207–231.

Ohwi, J. 1965. *Flora of Japan*. Washington, D.C.: Smithsonian Institute. 1067 pp.

Ohwi, J., and Ohashi, H. 1969. Adzuki beans of Asia. *J. Jap. Bot.* 44:29–31.

Okon, Y., Eshel, Y., and Henis, Y. 1972. Cultural and symbiotic properties of *Rhizobium* strains isolated from nodules of *Cicer arietinum* L. *Soil Biol. Biochem.* 4:165–170.

Oldeman, R. A. A. 1964. Primitiae Africanae IV. Revision of *Didelotia* Baill. (Caesalpiniaceae). *Blumea* 12 (2):209–239.

Oliver, D. 1871. *Flora of Tropical Africa*. Leguminosae to Ficoideae, 2. 613 pp. London: L. Reeve.

Ollis, W. D. 1968. New structural variants among the isoflavonoid and neoflavonoid classes. *In* Mabry, T. J., ed. *Recent Advances in Phytochemistry* 1:329–378. New York: Appleton-Century-Crofts.

References

Orrù, A., and Fratoni, A. 1945. A toxic principle in the seeds of *Canavalia ensiformis*. I, II [in Italian, English summary]. *Boll. Soc. Ital. Biol. Sper.* 20:200–201; 201–203.

Ottinger, R., Chiurdoglu, G., and Vandendris, J. 1965. Determination of the structure of ivorine [in French]. *Bull. Soc. Chim. Belg.* 74 (3–4):198–199.

Oxley, T. 1848. Some account of the nutmeg and its cultivation. *J. Indian Archipel./East. Asia* 2 (10):641–660.

Pachter, I. J., Zacharias, D. E., and Ribeiro, O. 1959. Indole alkaloids of *Acer saccharinum* (the silver maple), *Dictyoloma incanescens*, *Piptadenia colubrina* and *Mimosa hostilis*. *J. Org. Chem.* 24:1285–1287.

Pagan, J. D., Child, J. J., Snowcroft, W. R., and Gibson, A. H. 1975. Nitrogen fixation by *Rhizobium* cultured on a defined medium. *Nature* 256:406–407.

Palacios, G., and Bari, A. 1936a. The physiology of Indian nodule bacteria. *Proc. Indian Acad. Sci.* sec. B. 3:334–361.

Palacios, G., and Bari, A. 1936b. A new micro-organism associated with the nodule-bacteria in *Cajanus indicus*. *Proc. Indian Acad. Sci.* sec. B. 3:362–365.

Palafox, A. L., and Quisenberry, J. H. 1948. Koa haole as a substitute for alfalfa meal. *Univ. Hawaii Agric. Exp. Stn. Rep.* 1946–1948:145–147.

Palmer, E., and Pitman, N. 1961. *Trees of South Africa*. Capetown: A. A. Balkema. 352 pp.

Palmer, E. J. 1931. Conspectus of the genus *Amorpha*. *J. Arnold Arbor. Harv. Univ.* 12:157–197.

Pammel, L. H. 1911. *A Manual of Poisonous Plants*. Cedar Rapids, Iowa: The Torch Press. 977 pp.

Paris, R. 1948. Sur un *Erythrophleum* d'Indochine: le "lim" (*E. fordii* Oliver). *Ann. Pharm. Fr.* 6:501–506.

Paris, R. R., and Bézanger-Beauquesne, L. 1956. Sur la constitution du swartziol: identité avec le kempférol. *C. R. Acad. Sci.* 242: 1761–1762.

Parker, C. A. 1968. On the evolution of symbiosis in legumes. *In Festskrift til Hans Laurits Jensen*, 107–116. Lemvig: Gadgaard Nielsens Bogtrykkeri.

Parker, C. A., and Oakley, A. E. 1963. Nodule bacteria for two species of serradella—*Ornithopus sativus* and *Ornithopus compressus*. *Aust. J. Exp. Agric. Animal Husb.* 3:9–10.

Parker, C. A., and Oakley, A. E. 1965. Reduced nodulation of lupins and serradella due to lime pelleting. *Aust. J. Exp. Agric. Animal Husb.* 5 (17):144–146.

Parker, D. T., and Allen, O. N. 1952. The nodulation status of *Trifolium ambiguum*. *Soil Sci. Soc. Am. Proc.* 16:350–353.

Parker, R. N. 1933. Leguminosae and root nodules. *Indian For.* 59:232–239.

Parks, H. B. 1937. Valuable plants native to Texas. *Tex. Agric. Exp. Stn. Bull.* 1937. 551 pp.

Pärsim, E. 1966. Occurrence and morphology of the nodules on leguminous plants growing in the Estonian SSR [in Russian]. *Izv. Akad. Nauk. Est. SSR.* ser. Biol. Nauk 1:16–28.

Pärsim, E. 1967. Acid production by fermentation of carbohydrates by rhizobia [in Russian, English summary]. *Eesti Nsv Tead. Akad. Toime.* XVI Biol. 2:163–174.

Pate, J. S. 1958a. Nodulation studies in legumes. II. The influence of various environmental factors on symbiotic expression in the vetch (*Vicia sativa* L.) and other legumes. *Aust. J. Biol. Sci.* 11 (4):496–515.

Pate, J. S. 1958b. Studies of the growth substances of legume symbiosis using paper chromatography. *Aust. J. Biol. Sci.* 11 (4):516–528.

Paul, W. R. C. 1949. On the value of some legumes. *Trop. Agric.* 105:109–116.

Peirce, G. J. 1902. The root-tubercles of bur clover (*Medicago denticulata* Willd.) and of some other leguminous plants. *Calif. Acad. Sci. Proc.* ser. 3 Bot. 2:295–328.

Pellegrin, F. 1924. *Augouardia* Pellegrin, genera nouveau de Césalpiniées du Congo. *Soc. Bot. Fr.* 22A:309–311.

Pellegrin, F. 1933. De quelques Légumineuses de l'Afrique occidentale. *Bull. Bot. Fr.* 80:463–467.

Pellegrin, F. 1948. *Les Légumineuses du Gabon.* Mémoires de l'Institut d'Études Centrafricaines, Brazzaville (A. E. F.). Paris: Librairie Larose. 284 pp.

Pellett, K., ed. 1969. World soybean production. *Soybean Digest Blue Book Issue of the American Soybean Association.* 29 (6):48–49.

Pereira Forjaz, A. 1929. Contribution to the Muntz process of nitrification [in French]. *C. R. Acad. Sci.* 189:585–589.

Perkin, W. H., Ray, J. N., and Robinson, R. 1926–1928. Experiments on the synthesis of brazilin and haematoxylin and their derivatives. *J. Chem. Soc.* 1926 Part I:941–953; 1927 Part II:2094–2100; 1928 Part III:1504–1513.

Perkins, J. 1907. The Leguminosae of Porto Rico. *Contrib. U. S. Natl. Herb.* 10 (4):133–220.

Perrier, H. de la Bathie. 1938. Un nouveau genre malgache de Caesalpiniacees. *Bull. Soc. Bot. Fr.* 85:493–496.

Perrin, D. R., and Bottomley, W. 1962. Studies on phytoalexins. V. The structure of pisatin from *Pisum sativum* L. *J. Am. Chem. Soc.* 84:1919–1921.

Pertchik, B., and Pertchik, H. 1951. *Flowering Trees of the Caribbean.* New York: Rinehart. 125 pp.

Pettit, A. S. 1895. *Arachis hypogaea. Mem. Torrey Bot. Club* 4:275–296.

Peyronel, B., and Fassi, B. 1957. Micorrize ectotrofiche in una cesalpiniacea del Congo Belga. *Atti Accad. Sci. Torino* 91:569–576.

Peyronel, B., and Fassi, B. 1960. Nuovi casi di simbiosi ectomicorrizica in leguminose della famiglia delle cesalpiniacee. *Atti Accad. Sci. Torino* 94 (1):36–38.

Pfeiffer, P., Quehl, K., and Tappermann, F. 1930. Verbindungen der *a*-phenoxy-*a*-phenyl-aceton-Reihe (10. Mitteil. Zur Brasilin-und-Hämatoxylin-Frage). *Ber. Dtsch. Chem. Ges.* 63:1301–1308.

Phillips, E. P. 1917. The genus *Calpurnia* E. Mey. (Leguminosae). *Ann. S. Afr. Mus.* 9:475–481.

Phillips, E. P. 1926. A preliminary list of the known poisonous plants found in South Africa. *Bot. Surv. S. Afr. Mem.* 9. 30 pp.

Phillips, E. P. 1951. *The Genera of South African Flowering Plants. Dep. Agric. Div. Bot. Plant Path. Bot. Surv. Memo.* 25. 2d ed. Pretoria: Government Printer. 923 pp.

Phillips, E. P., and Hutchinson, J. 1921. A revision of the African species of *Sesbania. Bothalia* 1:40–56.

Phillips, J. F. V. 1926. *Virgilia capensis* Lamk. ("Keurboom"): a contribution to its ecology and sylviculture. *S. Afr. J. Sci.* 23:435–454.

Pichi, P. 1888. Alcune osservazioni sui tubercoli radicali delle Leguminose. *Gabinetto Bot. R. Univ. Pìsa,* January 1888:45–47.

Pieters, A. J. 1927. *Green Manuring — Principles and Practice.* New York: John Wiley. 356 pp.

Pieters, A. J. 1934. *The Little Book of Lespedeza.* Washington, D.C.: Colonial Press. 94 pp.

Pieters, A. J., and Hollowell, E. A. 1937. Clover improvement. *U. S. Dep. Agric. Yearb. Agric.* 1937:1190–1214.

Piper, C. V. 1920. A new genus of Leguminosae. *J. Wash. Acad. Sci.* 10 (15):432–433.

Piper, C. V. 1924a. *Forage Plants and their Culture.* New York: Macmillan. 671 pp.

Piper, C. V. 1924b. The genus *Oxyrhynchus* Brandegee. *J. Wash. Acad. Sci.* 14 (2):46–49.

Piper, C. V. 1924c. A new genus of Leguminosae. *J. Wash. Acad. Sci.* 14 (15):363–364.

Piper, C. V. 1925. The American species of *Canavalia* and *Wenderothia.* Contrib. *U. S. Natl. Herb.* 20:555–588.

Piper, C. V. 1926. Studies of American Phaseolineae. *Contrib. U. S. Natl. Herb.* 22 (9):663–701.

Piper, C. V., and Dunn, S. T. 1922. A revision of *Canavalia. Kew Bull.* 1922:129–145.

Pittier, H. 1916a. New or noteworthy plants from Colombia and Central America. 5. *Contrib. U. S. Natl. Herb.* 18 (4):143–171.

Pittier, H. 1916b. Preliminary revision of the genus *Inga. Contrib. U. S. Natl. Herb.* 18 (5):173–224.

Pittier, H. 1917. The middle American species of *Lonchocarpus. Contrib. U. S. Natl. Herb.* 20:37–93.

Pittier, H. 1929. The middle American species of the genus *Inga. J. Dep. Agric.* 13 (4):117–177.

Poiteau, A. S. 1853. Note sur l'*Arachis hypogea. Ann. Sci. Nat.* ser. 3. 19:268–272.

Polhill, R. M. 1968a. *Argyrolobium* Eckl. & Zeyh. (Leguminosae) in tropical Africa. *Kew Bull.* 22 (1):145–168.

Polhill, R. M. 1968b. Miscellaneous notes on African species of *Crotalaria* L.: II. *Kew Bull.* 22 (2):169–348.

Polhill, R. M. 1974. Revision of *Pearsonia* (Leguminosae-Papilionoideae). *Kew Bull.* 29:383–410.

Polhill, R. M. 1976. Genisteae (Adans.) Benth. and related tribes (Leguminosae). *Bot. System.* 1:143–368.

Popov, M. G. 1929. The genus *Cicer* and its species [in Russian, English summary]. *Bull. Appl. Bot. Genet. Plant Breed.* 21:1–240.

Porsild, M. P. 1930. Giebt es Knöllchenbakterien auf Disko in Grönland? *Dan. Bot. Arkiv.* 6 (7):1–7.

Porter, C. L. 1956. The genus *Peteria* (Leguminosae). *Rhodora* 58 (696):344–354.

Porterfield, W. M. 1939. The yam bean as a source of food in China. *J. N. Y. Bot. Gard.* 40:107–108.

Posadas, S. S. 1941. A study of root-nodule bacteria of certain leguminous plants. *Philipp. Agric.* 30:215–226.

Pound, F. J. 1938. History and cultivation of the Tonka bean (*Dipteryx odorata*) with analysis of Trinidad, Venezuelan and Brazilian samples. *Trop. Agric. (Trinidad)* 15:4–9; 28–32.

Power, F. B. 1901. Bark of *Robinia Pseudacacia. J. Chem. Soc.* 80 (2):679.

Power, F. B., and Cambier, J. 1890. On the chemical constituents and poisonous principle of the bark of *Robinia Pseudoacacia*, Linn. *Pharm. Rundsch.* 8 (2):29–38.

Prain, D. 1908. A new species of *Butea*, with notes on the genus. *Kew Bull.* 1908 (9):381–388.

Prain, D., and Baker, E. 1902. Notes on *Indigofera. J. Bot.* 40:60–67.

Prillieux, E. 1879. Sur la nature et sur la cause de la formation des tubercules qui naissent sur les racines des légumineuses. *Bull. Soc. Bot. Fr.* 26:98–107.

Purchase, H. F., Vincent, J. M., and Ward, L. M. 1951. The field distribution of strains of nodule bacteria from species of *Medicago. Aust. J. Agric. Res.* 2:261–272.

Purseglove, J. W. 1968. *Tropical Crops. Dicotyledons I.* New York: John Wiley. 332 pp.

Quispel, A., ed. 1974. *The Biology of Nitrogen Fixation.* Front. Biol. 33. 769 pp. New York: American Elsevier.

Quisumbing, E. 1951. *Medicinal Plants of the Philippines.* Tech. Bull. 16. Manila: Department of Agriculture and Natural Resources. 1234 pp.

Raber, O. 1936. Shipsmast locust, a valuable undescribed variety of *Robinia pseudoacacia. U. S. Dep. Agric. Circ.* 379. 8 pp.

Raffauf, R. F. 1970. *A Handbook of Alkaloids and Alkaloid-Containing Plants.* New York: Wiley-Interscience. n.p.

Raju, M. S. 1936. Studies on the bacterial-plant groups of cowpea, cicer and dhaincha. I. Classification. *Centralbl. Bakt.* ser. 2. 94:249–262.

Raju, M. S. 1938a. Bacterial plant groups of dhaincha. *Science* 88:300.

Raju, M. S. 1938b. Studies on the bacterial-plant groups. IV. Variations in the fermentation characters of different strains of nodule bacteria of the cowpea, *Cicer* and dhaincha groups. *Centralbl. Bakt.* ser. 2. 99:133–141.

Rakhit, S., and Majumdar, D. N. 1956. *Mucuna pruriens.* V. Alkaloidal constituents and their characterization. *Indian J. Pharm.* 18:285–287. *Chem. Abstr.* 52:5748 (1958).

Rangaswami, G., and Oblisami, G. 1962. Studies on some legume root nodule bacteria. *J. Indian Soc. Soil Sci.* 10:175–186.

Ransohoff, J. W., ed. 1955. Abrin, lethal jewelry. *Arch. Ind. Health* 12:468–469.

Rassel, A. 1960. Le Voandzeu (*Voandzeia subterranea* Thouars) et sa culture au Kwango. *Bull. Agric. Congo Belge* 51. 26 pp.

Rasumowskaja, S. G. 1932. Zur Frage der Knöllchenbakteriophagen. *Arch. Sci. Biol. (Leningrad)* 32:304–314.

Rasumowskaja, S. G. 1934. Über die Knöllchenbakterien des *Cicer. Centralbl. Bakt.* ser. 2. 90:330–335.

Ratle, G., Monseur, X., Das, B. C., Yassi, J., Khuong-Huu, Q., and Goutarel,

References

R. 1966. Prosopine and prosopinine, alkaloids from *Prosopis africana* [in French]. *Bull. Soc. Chim. Fr.* 1966 (9):2945–2947. *Chem. Abstr.* 66:18779h (1967).

Rechinger, K. H. 1957. Leguminosae. *In* Køie, M., and Rechinger, K. H. *Symbolae Afghanicae*, 3:1–208. *Dan. Vidensk. Sel'sk. Biol. Skrift.* 9.

Record, S. J. 1940. Some new names for tropical American trees of the family Leguminosae. *Trop. Woods* 63:1–6.

Record, S. J., and Hess, R. W. 1943. *Timbers of the New World.* New Haven: Yale University Press. 640 pp.

Record, S. J., and Mell, C. D. 1924. *Timbers of Tropical America.* New Haven: Yale University Press. 610 pp.

Reed, E. L. 1924. Anatomy, embryology, and ecology of *Arachis hypogea. Bot. Gaz.* 78:289–310.

Rehder, A. 1940. *Manual of Cultivated Trees and Shrubs.* 2d ed. New York: Macmillan. 996 pp.

Reid, J. J., and Baldwin, I. L. 1937. The infective ability of rhizobia of the soybean, cowpea, and lupine cross-inoculation groups. *Soil Sci. Soc. Am. Proc.* (1936) 1:219.

Reithel, F. J. 1971. Ureases. *In* Boyer, P. D., ed. *The Enzymes,* 4:1–21. New York: Academic Press.

Rendle, B. J., comp. and ed. 1969. *World Timbers.* I. *Europe and Africa.* London: Ernest Benn. 191 pp.

Renz, J. 1936. Uber das Mimosin. *Z. Physiol. Chem.* 244:153–158.

Ressler, C., Redstone, P. A., and Erenberg, R. H. 1961. Isolation and identification of a neurotoxic factor from *Lathyrus latifolius. Science* 134:188–189.

Reuter, G. 1965. Biosynthesis of pharmacologically active guanidine derivatives with isoprenoid carbon chain [in German]. *Planta Med.* 13 (4):494. *Chem. Abstr.* 64:5459h (1966).

Riceman, D. S., and Powrie, J. K. 1952. A comparative study of *Pisum, Vicia, Lathyrus* and *Lupinus* varieties grown in Buckingham Sand in the Coonalpyn Downs, South Australia. *Aust. C.S.I.R.O. (Melbourne) Bull.* 269. 36 pp.

Richmond, T. E. 1926a. The nodule organism of the cowpea group. *J. Am. Soc. Agron.* 18:411–414.

Richmond, T. E. 1926b. Legume inoculation as influenced by stock and scion. *Bot Gaz.* 82 (4):438–442.

Ricker, P. L. 1934. The origin of the name *Lespedeza. Rhodora* 36:130–132.

Ricker, P. L. 1946. New Asiatic species of the legume genus *Campylotropis. J. Wash. Acad. Sci.* 36 (2):37–40.

Rickett, H. W., and Stafleu, F. A. 1959–1961. Nomina generica conservanda et rejicienda spermatophytorum. *Taxon* 8 (9):282–314.

Ridley, H. N. 1925. *The Flora of the Malay Peninsula,* 5. 470 pp. Ashford, Eng.: L. Reeve.

Riley, H. P. 1963. *Families of Flowering Plants of Southern Africa.* Lexington: University of Kentucky Press. 267 pp.

Rimington, C. 1935a. The occurrence of cyanogenetic glucosides in South African species of *Acacia.* II. Determination of the chemical constitution of acacipetalin. Its isolation from *Acacia stolonifera* Burch. *Onderstepoort J. Vet. Sci. Animal Ind.* 5 (2):445–464.

Rimington, C. 1935b. Acacipetalin, a new cyanogenetic glucoside from South

African species of *Acacia*. Its isolation from *Acacia stolonifera* Burch. and chemical composition. *J. S. Afr. Vet. Med. Assoc.* 6 (2):136–138.

Roark, R. C. 1932. A digest of the literature of *Derris* (*Deguelia*) species used as insecticides, 1747–1931. *U. S. Dept. Agric. Misc. Pub.* 120. 86 pp.

Roark, R. C. 1933. The chemical relationship between certain insecticidal species of fabaceous plants. *J. Econ. Entomol.* 26:587–594.

Roark, R. C. 1936. *Lonchocarpus* species (barbasco, cube, haiari, nekoe, and timbo) used as insecticides. *U. S. Dep. Agric. Bur. Entomol. Monogr.* E-367. 133 pp.

Roark, R. C. 1937. *Tephrosia* as an insecticide — A review of the literature. *U. S. Dep. Agric. Bur. Entomol. Plant Quar.* E-402. 165 pp.

Roark, R. C. 1938. *Lonchocarpus* (barbasco, cube, and timbo). A review of recent literature. *U. S. Dep. Agric. Bur. Entomol. Monogr.* E-453. 174 pp.

Roark, R. C. 1947. Some promising insecticidal plants. *Econ. Bot.* 1:437–445.

Robbins, W. J., Kavanaugh, F., and Thayer, J. D. 1947. Antibiotic activity of *Cassia reticulata* Willd. *Bull. Torrey Bot. Club* 74:287–292.

Roberty, G. 1954. *Xeroderris*. Notes sur la flore de l'Ouest-africain. *Bull. Inst. Fr. Afr. Noire* ser. A. Sci. Nat. 16 (2):321–369.

Robinson, B. 1968. Alkaloids of the Calabar Bean. *In* Manske, R. H. F., ed. *The Alkaloids, Chemistry and Physiology*, 10:383–400. New York: Academic Press.

Robinson, R. 1958. Chemistry of brazilin and hematoxylin. *Bull. Soc. Chim. Fr.* 1958:125–135. *Chem. Abstr.* 52:12852 (1958).

Robinson, R. 1962. Synthesis in the brazilin group. *In* Gore, T. S., and others, eds. *Recent Progress in the Chemistry of Natural and Synthetic Colouring Matters and Related Fields*, 1–11. New York: Academic Press.

Rock, J. F. 1920. *The Leguminous Plants of Hawaii*. Honolulu: Experiment Station of the Hawaiian Sugar Planters' Association. 234 pp.

Roderick, L. M. 1929. The pathology of sweet clover disease in cattle. *J. Am. Vet. Med. Assoc.* 74:314–326.

Roizin, M. B. 1959. The root nodules of leguminous plants of the Kola Peninsula [in Russian]. *Bot. Zh.* 44:467–474.

Rol, F. 1959. Locust bean gum. *In* Whistler, R. L., and BeMiller, J. N., eds. *Industrial Gums — Polysaccharides and their Derivatives*, 361–375. New York: Academic Press.

Rollins, R. C. 1940. Studies in the genus *Hedysarum* in North America. *Rhodora* 42 (499):217–239.

Rose, J. N. 1891. List of plants collected by Dr. Edward Palmer in Western Mexico and Arizona in 1890. *Contrib. U. S. Natl. Herb.* 1 (4):91–116.

Rose, J. N. 1897. Studies of Mexican and Central American Plants. *Contrib. U. S. Natl. Herb.* 5 (3):109–144.

Rose, J. N. 1899. Studies of Mexican and Central American Plants. No. 2. Synopsis of the North American species of *Nissolia*. *Contrib. U. S. Natl. Herb.* 5 (4):145–200.

Rose, J. N. 1903. Studies of Mexican and Central American Plants. No. 3. *Contrib. U. S. Natl. Herb.* 8:1–55.

Rose, J. N. 1906. Studies on Mexican and Central American Plants. No. 5. *Sphinctospermum*, a new genus. *Contrib. U. S. Herb.* 10 (3):79–132.

Rose, J. N. 1909. *Conzatti*, a new genus of Caesalpiniaceae. *Contrib. U. S. Natl. Herb.* 12:407–408.

References

Rosengarten, F., Jr. 1969. *The Book of Spices*. Wynnewood, Penn.: Livingston. 489 pp.

Rothmaler, W. 1941. Revision der Genisteen. I. Monographien der Gattungen um *Ulex*. *Bot. Jahrb.* 72:69–116.

Rothmaler, W. 1949. Revision der Genisteen. II. Die Gattungen *Erinacea, Spartium* und *Calicotome*. *Bot. Jahrb.* 74 (2):271–287.

Rothschild, D. I. de. 1963. Anatomia del nódulo radical de algunas leguminosas cultivadas. *Rev. Inst. Munic. Bot. Jard. Bot. "Carlos Thays"* 3 (1):1–32.

Rothschild, D. I. de. 1964. Morfologia y anatomia del nódulo radical en Leguminosas indigenas de América. Trab. *Comun. 1st Reunión Latino-americana sobre inoculantes para Leguminosas* Montevideo, Uruguay, 8. Extracto aparecido en la publicación respectiva, febrero de 1965.

Rothschild, D. I. de. 1967a. Anatomia del nódulo radical de origen bacteriano en *Adesmia* DC. (Leguminosae). *Rev. Mus. Argent. Cienc. Nat. "Bernardino Rivadavia" Inst. Nac. Invest. Cienc. Nat. B. Aires.* Bot. 3 (3):161–184.

Rothschild, D. I. de. 1967b. Estudio preliminar sobre la presencia de bacterias en la raiz de *Gleditsia amorphoides* (Gris.) Taub., y *G. triacanthos* L. *Comm. Mus. Argent. Cienc. Nat. "Bernardino Rivadavia" B. Aires.* Bot. 2 (3):9–14.

Rothschild, D. I. de. 1970. Nodulación en leguminosas subtropicales de la flora Argentina. *Rev. Mus. Argent. Cienc. Nat. "Bernardino Rivadavia" Inst. Nac. Invest. Cienc. Nat. B. Aires.* Bot. 3 (9):267–286.

Roti-Michelozzi, G. 1957. Adumbratio Florae Aethiopicae. 6. Caesalpiniaceae (excl. gen. *Cassia*). *Webbia* 13 (1):133–228.

Roux, E. R., and Marais, C. C. Q. 1964. Rhizobial nitrogen fixation in some South African acacias. *S. Afr. J. Sci.* 60:203–204.

Roux, E. R., and Warren, J. L. 1963. Studies in the aut-ecology of the Australian acacias in South Africa. II. Symbiotic nitrogen fixation in *Acacia cyclops* A. Cunn. *S. Afr. J. Sci.* 59 (6):294–295.

Rubinshtein, I. M. M., and Menshikov, G. P. 1944. The alkaloids of *Sphaerophysa salsula* [in Russian, English summary]. *Zh. Prikl. Khim.* [*J. Gen. Chem., U.S.S.R.*] 14:161–171. *Chem. Abstr.* 39:2291 (1944).

Rudd, V. E. 1954. *Centrolobium* (Leguminosae): validation of a specific name and a brief review of the genus. *J. Wash. Acad. Sci.* 44 (9):284–288.

Rudd, V. E. 1955. The American species of *Aeschynemone*. *Contrib. U. S. Natl. Herb.* 32 (1). 172 pp.

Rudd, V. E. 1956. A revision of the genus *Nissolia*. *Contrib. U. S. Natl. Herb.* 32 (2):173–206.

Rudd, V. E. 1958. A revision of the genus *Chaetocalyx*. *Contrib. U. S. Natl. Herb.* 32 (3):207–245.

Rudd, V. E. 1959. The genus *Aeschynemone* in Malaysia (Leguminosae-Papilionatae). *Reinwardtia* 5 (1):23–36.

Rudd, V. E. 1963. The genus *Dussia* (Leguminosae). *Contrib. U. S. Natl. Herb.* 32 (4):247–277.

Rudd, V. E. 1965. The American species of *Ormosia* (Leguminosae). *Contrib. U. S. Natl. Herb.* 32 (5):279–384.

Rudd, V. E. 1967. *Oxyrhyncus* and *Monoplegma* (Leguminosae). *Phytologia* 15:289–294.

Rudd, V. E. 1968a. A résumé of *Ateleia* and *Cyathostegia* (Leguminosae). *Contrib. U. S. Natl. Herb.* 32 (6):385–411.

Rudd, V. E. 1968b. Leguminosae of Mexico. Faboideae I. Sophoreae and Podalyrieae. *Rhodora* 70:492–532.

Rudd, V. E. 1969. A synopsis of the genus *Piscidia* (Leguminosae). *Phytologia* 18 (8):473–499.

Rudd, V. E. 1970a. Studies in the Sophoreae (Leguminosae) I. *Phytologia* 19:327.

Rudd, V. E. 1970b. Nomina conservanda proposita. *Taxon* 19:294–297.

Rudd, V. E. 1970c. *Etaballia dubia* (Leguminosae), a new combination. *Phytologia* 20 (7):426–428.

Ruffner, J. D., and Hall, J. G. 1963. Crownvetch in West Virginia. *W. Va. Univ. Agric. Exp. Stn. Bull.* 487. 19 pp.

Russell, A., and Kaczka, E. A. 1944. Fish poisons from *Ichthyomethia piscipula*. I. *J. Am. Chem. Soc.* 66:548–550.

Ryabinin, A. A., and Il'ina, E. M. 1954. Alkaloids of *Eremosparton* [in Russian]. *Zh. Priklad. Khim.* 27:221–223. *Chem. Abstr.* 49:3826 (1955).

Rydberg, P. A. 1919a. (Rosales) Family 24. Fabaceae: Psoraleae. *North Am. Flora* 24 (1):1–64.

Rydberg, P. A. 1919b. A genus of plants intermediate between *Petalostemon* and *Parosela*. *J. N. Y. Bot. Gard.* 20:64–66.

Rydberg, P. A. 1920. (Rosales) Fabaceae, Psoraleae. *North Am. Flora* 24 (2):65–136.

Rydberg, P. A. 1923. (Rosales) Fabaceae, Indigofereae, Galegeae (pars.) *North Am. Flora* 24 (3):137–200.

Rydberg, P. A. 1924. (Rosales) Fabaceae: Galegeae (pars.). *North Am. Flora* 24 (4):201–250.

Rydberg, P. A. 1928. (Rosales). Mimosaceae (conclusio). *North Am. Flora* 23 (3):137–194.

Rydberg, P. A. 1929. Astragalinae. *North Am. Flora* 24 (5–7):251–462.

Sadykov, A., and Lazur'evskii, G. 1943a. Dyes from *Ammothamnus lehmanni* Bge. [in Russian, English summary]. *Zh. Prikl Khim.* [*J. Gen. Chem. U.S.S.R.*] 13:309–313. Chem. Abstr. 38:1117 (1944).

Sadykov, A., and Lazur'evskii, G. 1943b. Alkaloids of *Ammothamnus lehmanni* Bge. [in Russian, English summary]. *Zh. Prikl. Khim.* [*J. Gen. Chem. U.S.S.R.*] 13:314–318. Chem. Abstr. 38:1240 (1944).

Salfeld, A. 1896. *Die Boden-Impfung zu den Pflanzen mit Schmetterlingsbluten in landwirtschaftlichen Betriebe.* Bremen: M. Heinsius. 100 pp.

Sánchez-Marroquín, A., García, L., and Méndez, M. 1958. Brazilin, antibacterial substance from *Haematoxylon brasiletto* Karst. *Rev. Latinoam. Microbiol.* 1 (3):225–232.

Sandberg, F. 1958. The saponin of *Swartzia madagascariensis*. *Pharm. Weekbl.* 93:5–7. *Chem. Abstr.* 52:5560 (1958).

Sandberg, F., Ahlenius, B., and Thorsen, R. 1958. The saponin of *Swartzia madagascariensis*. *Sven. Farm. Tidskr.* 62:541–553. *Chem. Abstr.* 53:2538 (1959).

References

Sandmann, W. P. L. 1970a. *Results of formal greenhouse tests (1963–1969) of strains of* Rhizobium *and information about some other trials.* Rhodesian Ministry of Agriculture. Government Printer. 39494-221. Mimeographed. 56 pp.

Sandmann, W. P. L. 1970b. *The Practical Aspects of* Rhizobium *Bacteriology at Grasslands Research Station, Marandellas, Rhodesia.* Rhodesian Ministry of Agriculture. Government Printer. 45661-55. Mimeographed. 56 pp.

Sandmann, W. P. L. 1970c. A note on nodules that are purple inside. *Rhizobium Newsl.* 12 (2):222–224.

Sando, C. E. 1932. The plant coloring matter, robinin. *J. Biol. Chem.* 94:675–680.

Sanlier-Lamark, R., and Cabezas de Herrera, E. 1956. Estudio de la infeccion del garbanzo por el genero *Rhizobium*. *An. Edafol. Fisiol. Veg.* 15 (12):943–953.

Sanna, A., and Chessa, G. 1927. The presence of sparteine in the flowers of *Spartium junceum* L. *Ann. Chim. Appl.* 17:283–284.

San Pietro, A., ed. 1972. *Photosynthesis and Nitrogen Fixation, Part B. Methods in Enzymology* 24. 526 pp. New York: Academic Press.

Santra, D. K., and Majumdar, D. N. 1953. *Mucuna pruriens*. II. Isolation of water-insoluble alkaloids. *Indian J. Pharm.* 15:60–61. *Chem. Abstr.* 48:8793 (1954).

Sargent, C. S. 1949. *Manual of the Trees of North America.* 2. 910 pp. 2d correct. ed. New York: Dover Publications.

Sarić, Z. 1959. Nodule bacteria with some plants of the *Vigna* group [in Yugoslavic, English summary]. *Letopisa naucnih radova Poljoprivrednog fakulteta u Novom Sadu* 3:100–106.

Sarić, Z. 1964. The nature of the virulence to peanuts of different strains of nodule bacteria [in Yugoslav and English publications]. *Arhiv Poljopriv. Nauke* 16 (53):90–101. *J. Sci. Agric. Res.* 16:83–96.

Sarkar, S. L. 1914. Colouring matter contained in the seed-coats of *Abrus precatorius*. *Biochem. J.* 8:281–286.

Satyanarayan, Y., and Guar, Y. D. 1965. Preliminary studies on the nodulation of arid zone legumes. *Curr. Sci.* 34:21–22.

Saubert, S., and Scheffler, J. G. 1967a. Strain variation and host specificity of *Rhizobium*. II. Host specificity of *Rhizobium trifolii* on European clovers. *S. Afr. J. Agric. Sci.* 10:85–94.

Saubert, S., and Scheffler, J. G. 1967b. Strain variation and host specificity of *Rhizobium*. III. Host specificity of *Rhizobium trifolii* on African clovers. *S. Afr. J. Agric. Sci.* 10:357–364.

Sauer, J. 1964. Revision of *Canavalia*. *Brittonia* 16 (2):106–181.

Schaaffhausen, R. v. 1963. *Dolichos lablab* or Hyacinth Bean: its uses for feed, food and soil improvement. *Econ. Bot.* 17:146–153.

Schaede, R. 1932. Das Schicksal der Bakterien in den Knöllchen von *Lupinus albus* nebst cytologischen Untersuchungen. *Zentralbl. Bakt. Parasitenkd.* ser. 2. 85:416–425.

Schaede, R. 1940. Die Knöllchen der adventiven Wasserwurzeln von *Neptunia oleracea* und ihre Bakteriensymbiose. *Planta* 31:1–21.

Schaede, R. 1962. *Die Pflanzlichen Symbiosen.* Stuttgart: Gustav Fischer. 238 pp.

Schiel, E. 1945. Ubicación de *"Trifolium polymorphum"* dentro de los grupos de leguminosas formadas sobre la base de la infeccion cruzada. *Rev. Argent. Agron.* 10:190–191.

Schiel, E., Olivero, E. G. de, and Ypes, M. 1960. Excipiente apto para inocular semillas de leguminosas en seco. *Rev. Invest. Agric.* 14 (1):48–108.

Schilling, E. D., and Strong, F. M. 1955. Isolation, structure and synthesis of a *Lathyrus* factor from *L. odoratus*. *J. Am. Chem. Soc.* 77 (10):2843–2845.

Schindler, A. K. 1912. *Kummerowia* Schindler *novum genus leguminosarum. Feddes Repert. Specierum Nov. Regni Veg.* 10:403–404.

Schindler, A. K. 1917. Desmodiinae novae. *Bot. Jahrb.* 54:51–68.

Schindler, A. K. 1924. Über einige kleine Gattungen aus der Verwandtschaft von *Desmodium* Desv. *Feddes Repert. Specierum Nov. Regni Veg.* 20:266–286.

Schindler, A. K. 1926. Desmodii generumque affinium species et combinations novae II. *Feddes Repert. Specierum Nov. Regni Veg.* 22:250–288.

Schlakman, I. A., and Bartilucci, A. J. 1957. A comparative study of commercially available guar gums. *Drug Stand.* 25:149–154.

Schneider, A. 1892. Observations on some American rhizobia. *Bull. Torrey Bot. Club* 19:203–218.

Schneider, A. 1893a. The morphology of root tubercles of Leguminosae. *Am. Nat.* 27:782–792.

Schneider, A. 1893b. A new factor in economic agriculture. *Ill. Agric. Exp. Stn. Bull.* 29:301–309.

Schneider, A. 1905. Contributions to the biology of rhizobia. IV. Two coast rhizobia of Vancouver Island, B.C. *Bot. Gaz.* 40:135–139.

Schofield, F. W. 1924. Damaged sweet clover: the cause of a new disease in cattle simulating hemorrhagic septicemia and blackleg. *J. Am. Vet. Med. Assoc.* 64:553–575.

Schofield, J. L. 1950. Serradella (*Ornithopus sativus*), a legume for light acid soils. *J. Br. Grassl. Soc.* 5:131–140.

Schollenberger, J. H., and Goss, W. H. 1945. Soybeans: certain agronomic, physical, chemical, economic and industrial aspects. *U. S. Dep. Agric. A.R.S. Bur. Agric. Ind. Chem. Bull. AIC* 74. 84 pp.

Schönberg, L. 1929. Untersuchingen über das Verhalten von *Bact. radicicola* Beij. gegenüber verschiedenen Kohlehydraten und in Milch. *Zentralbl. Bakt.* ser. 2. 79:205–221.

Schoth, H. A., and McKee, R. 1962. The vetches. *In* Hughes, H. D., Heath, M. E., and Metcalf, D. S., eds. *Forages*, 205–210. Ames, Iowa: Iowa State University Press.

Schramm, J. R. 1966. Plant colonization studies on black wastes from anthracite mining in Pennsylvania. *Trans. Am. Philos. Soc.* 56 Part 1. 194 pp.

Schreven, D. A. van. 1972. Note on the specificity of the rhizobia of crownvetch and sainfoin. *Plant Soil* 36:325–330.

Schroeter, J. 1886. Die Pilze Schlesiens. *In* Cohn, F., ed. *Kryptogamen-Flora von Schlesien*, 3 (9):134–135. Breslau: Kern Verlag.

Schubert, B. 1954. Genres *Desmodium* et *Droogmansia*. *In* Robyns, W. *Flore du Congo Belge et du Ruanda-Urundi*, 5:180–223. Brussels: Publ. Inst. Nat. Étude Agron. Congo Belge (I.N.É.A.C.).

Schwarz, J. S. P., Cohen, A. I., Ollis, W. D., Kaczka, E. A., and Jackman, L. M.

References

1964. The extractives of *Piscidia erythrina* L. I. The constitution of ichthynone. *Tetrahedron* 20:1317–1330.

Sealock, R. R. 1955. β-3,4,-dihydroxyphenyl-L-alanine. *Biochem. Prep.* 1:25–27.

Seaney, R. R. 1973. Birdsfoot trefoil. *In* Heath, M. E., Metcalfe, D. S., and Barnes, R. F., eds. *Forages*, 177–188. 3d ed. Ames, Iowa: Iowa State University Press.

Sears, O. H., and Clark, F. M. 1930. Non-reciprocal cross-inoculation of legume nodule bacteria. *Soil Sci.* 30:237–242.

Senn, H. A. 1938a. Cytological evidence on the status of the genus *Chamaecrista* Moench. *J. Arnold Arbor. Harv. Univ.* 19:153–157.

Senn, H. A. 1938b. Chromosome number relationships in the Leguminosae. *Bibliogr. Genet.* 12:175–336.

Senn, H. A. 1939. The North American species of *Crotalaria*. *Rhodora* 41:317–370.

Seppilli, A., Schreiber, M., and Hirsch, G. 1941. L'Infezione sperimentale "in vitro" delle colture di radici isolate di leguminose. *Ann. Acad. Bras. Cienc.* 13 (2):69–84.

Serrano, C. B. 1923. Prussic acid in *Phaseolus lunatus* and other beans. *Philipp. Agric.* 11 (6):163–176.

Shangraw, R. F., and Lynn, E. V. 1955. The saponins of *Pachyrrhizus erosus* (L.) Urban. *J. Am. Pharm. Assoc. Sci. Ed.* 44 (1):38–39.

Shaw, F. J. F., and Khan, K. S. A. R. 1931. Studies in Indian Pulses. (2) Some varieties of Indian gram (*Cicer arietinum* L.). *Mem. Dep. Agric. India. Bot. Ser.* 19:27–47.

Shelley, W. B., and Arthur, R. P. 1955a. Mucunain, the active pruritogenic proteinase of cowhage. *Science* 122:469–470.

Shelley, W. B., and Arthur, R. P. 1955b. Studies on cowhage (*Mucuna pruriens*) and its puritogenic proteinase mucunain. *Arch. Dermatol.* 72:399–406.

Shively, A. F. 1897. Contribution to the life history of *Amphicarpaea monoica*. *Contrib. Bot. Lab. Univ. Penn.* 1 (3):270–363.

Shunk, I. V. 1921. Notes on the flagellation of the nodule bacteria of Leguminosae. *J. Bacteriol.* 6:239–247.

Sillans, R., and Normand, R.-D. 1953. Sur le fruit et la structure du bois de *Neochevalierodendron stephanii*. *Rev. Int. Bot. Appli.* 33:565–570.

Silsbury, J. H., and Brittan, N. H. 1955. Distribution and ecology of the genus *Kennedya* Vent. in Western Australia. *Aust. J. Bot.* 3:113–135.

Simon, J. 1908. Die Wiederstandsfähigkeit der Wurzelbakterien der Leguminosen und ihre Bedeutung für die Bodenimpfung. *Jahresber. Ver. Angew. Bot.* (1907) 5:132–160.

Simon, J. 1914. Ueber die Verwandtschaftsverhältnisse der Leguminosen-Wurzelbakterien. *Centralbl. Bakt.* ser. 2. 41:470–479.

Simonitsch, E., Frei, H., and Schmid, H. 1957. The constitution of pachyrrhizin. *Monatsh. Chem.* 88:541–559. *Chem. Abstr.* 52:11033 (1958).

Sims, J. R., Muir, M. K., and Carleton, A. E. 1968. Evidence of ineffective rhizobia and its relation to the nitrogen nutrition of sainfoin (*Onobrychis viciaefolia*). *In* Cooper, C. S., and Carleton, A. E., eds. *Sainfoin Symposium*, 8–12. *Mont. Agric. Exp. Stn. Bull.* 627. 109 pp.

Skinner, C. M. 1911. *Myths and Legends of Flowers, Trees, Fruits, and Plants in All Ages and in All Climes*. Philadelphia: J. P. Lippincott. 302 pp.

Smith, A. C. 1940. Notes on the genus *Amburana* Schwacke & Taub. (*Torresea* Fr. Allem.) *Trop. Woods* 62:28–31.

Smith, A. W. 1963. *A Gardner's Book of Plant Names*. New York: Harper and Row. 428 pp.

Smith, C. A. 1932. *Chrysoscias*. In Burtt Davy, J. *A Manual of the Flowering Plants and Ferns of the Transvaal with Swaziland, South Africa*, Part 2:405–406. London: Longmans, Green.

Smith, F., and Montgomery, R. 1959. *The Chemistry of Plant Gums and Mucilages*. New York: Reinhold. 627 pp.

Smith, J. E. 1794. *Platylobium*. In *An Account of Two New Genera of Plants from New South Wales, Presented to the Linnean Society by Mr. Thomas Hoy, F. L. S., and Mr. John Fairbairn, F.L.S. Trans. Linn. Soc.* 2:350–352.

Smith, J. G. 1898. Leguminous forage crops. *U. S. Dep. Agric. Yearb.* 1897:487–508.

Smith, W. K. 1948. Transfer from *Melilotus dentata* to *M. alba* of the genes for reduction in coumarin content. *Genetics* 33:124–125.

So, L. L., and Goldstein, I. J. 1967. IV. Application of the quantitative precipitin method to polysaccharide-concanavalin A interaction. *J. Biol. Chem.* 242:1617–1622.

Soliman, G., and Mustafa, Z. 1943. The saponin of fenugreek seeds. *Nature* 151 (3824):195–196.

Soraru, S. B. 1973. Tres especies nuevas de Leguminosas: *Anarthrophyllum pedicellatum, A. capitatum* y *A. strigulipetalum. Darwiniana* 18:37–43.

Sornay, P. de. 1916. *Green Manures and Manuring in the Tropics*. London: John Bale, Sons and Danielsson. 466 pp.

Spratt, E. 1919. A comparative account of the root nodules of the Leguminosae. *Ann. Bot. Lond.* 33:189–199.

Stahmann, M. A., Huebner, C. F., and Link, K. P. 1941. The hemorrhagic sweet clover disease. V. Identification and synthesis of the hemorrhagic agent. *J. Biol. Chem.* 138:513–527.

Stalker, M. 1884. Crotalism — A new disease among horses. Ames, Iowa: Daily State Register, Sept. 24.

Standley, P. C. 1920–26. Trees and shrubs of Mexico. *Contrib. U. S. Natl. Herb.* 23 (2):170–515.

Standley, P. C., and Steyermark, J. A. 1943. Studies of Central American Plants — III. *Fieldiana Bot.* 23:3–28.

Standley, P. C., and Steyermark, J. A. 1946. Flora of Guatemala. *Fieldiana Bot.* 24 (5). 502 pp.

Stanton, W. R. 1966. *Grain Legumes in Africa*. Rome: Food and Agricultural Organization of the United Nations. 183 pp.

Stapf, O. 1912. A new ground bean (*Kerstingiella geocarpa*, Harms). *Kew Bull. Misc. Inf.* 5:209–213.

Staphorst, J. L., and Strijdom, B. W. 1972. Some observations on the bacteroids in nodules of *Arachis* spp. and the isolation of rhizobia from these nodules. *Phytophylactica* 4:87–92.

Stapp, C. 1923. Beiträge zum Studium der Bakterientyrosinase. *Biochem. Z.* 141:42–69.

References

Steenis, C. G. G. J. van. 1948. *Dansera* and *Uittienia. Bull. Bot. Gard. Buitenzorg* ser. 3. 17:414–419.

Steenis, C. G. G. J. van. 1960a. Malaysian Papilionaceae I. *Neocollettia* Hemsl. *Reinwardtia* 5:436–437.

Steenis, C. G. G. J. van. 1960b. Miscellaneous notes on New Guinea Plants. VI. *Nova Guinea Bot.* 3:13–19.

Štefan, J. 1906. Studien zur Frage der Leguminosenknöllchen. *Centralbl. Bakt.* ser. 2. 16:131–149.

Steiner, M. L. 1959. The Philippine Jadevine. *Natl. Hortic. Mag.* 38 (1):42–45.

Steinmann, 1930. De Bergcultures. *Orgaan Alg. Landbouw Synd.* 4:1099–1103.

Stephen, A. M. 1957. The isolation of 5-O-α-L-arabopyranosyl-L-arabinose from partly degraded gums of *Virgilia* species. *J. Chem. Soc. Lond.* 1957:1919–1921.

Stephens, E. L. 1913. A new species of *Haematoxylon* (Leguminosae-Caesalpineae) from Great Namaqualand. *Trans. R. Soc. S. Afr.* 3 (2):255–257.

Steyaert, R. 1952. Cassieae. *In* Robyns, W. *Flore du Congo Belge et du Ruanda-Urundi*, 3:496–545. Brussels: Publ. Inst. Nat. Étude Agron. Congo Belge (I.N.É.A.C.).

Steyaert, R. L. 1953. Étude sur les rapports entre les genres *Uittienia, Dansera* et *Dialium* (Legum.-Caesalp.) *Reinwardtia* 2 (2):351–355.

Steyermark, J. A. 1963. *Flora of Missouri*. Ames, Iowa: Iowa State University Press. 1725 pp.

Steyn, D. G. 1929. Recent investigations into the toxicity of known and unknown poisonous plants in the Union of South Africa. *15th Rep. Dir. Vet. Serv. Union S. Afri.* 777–803.

Steyn, D. G. 1934. *The Toxicology of Plants in South Africa Together with a Consideration of Poisonous Foodstuffs and Fungi*. S. Afr. Agric. Ser. 13. 631 pp. Johannesburg: Central News Agency.

Steyn, D. G., and Rimington, C. 1935. The occurrence of cyanogenetic glucosides in South African species of *Acacia*. I. *Onderstepoort J. Vet. Sci. Anim. Ind.* 4 (1):51–73.

Stockman, R. 1929. Lathyrism. *J. Pharmacol. Exp. Ther.* 37:43–53.

Stockman, R. 1932. Historical notes on poisoning by leguminous foods. *Janus* 36:180–189.

Stone, B. C. 1971. America's Asiatic flora: the plants of Guam. *Am. Sci.* 59:308–319.

Strange, R. 1958. Preliminary trials of grasses and legumes under grazing. *East Afri. Agric. J.* 24 (2):92–102.

Stromberg, V. L. 1954. The isolation of bufotenine from *Piptadenia peregrina. J. Am. Chem. Soc.* 76:1707.

Strong, T. H. 1938. On the role of seed inoculation. *Aust. C.S.I.R.O. Bull.* 122:23–24.

Suessenguth, K., and Beyerle, R. 1935. Über Bakterienknöllchen am Spross von *Aeschynemone paniculata* Willd. *Hedwigia* 75:234–237.

Summerfield, R. J., and Bunting, A. H., eds. 1980. *Advances in Legume Science*. Royal Botanic Gardens, Kew, England: Her Majesty's Stationery Office. 667 pp.

Sumner, J. B. 1926a. The isolation and crystallization of the enzyme urease. *J. Biol. Chem.* 69:435–441.

Sumner, J. B. 1926b. The recrystallization of urease. *J. Biol. Chem.* 70:97–98.

Sumner, J. B., and Howell, S. F. 1936. The identification of the hemagglutinin of the Jack Bean with concanavalin A. *J. Bacteriol.* 32:227–237.

Sutton, A. W. 1911. Compte rendu d'expérience de croisements faite entre le pois sauvage de Palestine et les pois de commerce dan le but de découvrir entre eux quelque trace d'identité specifique. *4th Conf. Int. Génét.* 1911:358–367.

Sykes, R. L., Hides, E. A., and Simon, T. D. 1954. Vegetable tannins in East Africa. *East Afr. Agric. J.* 20:59–65.

Szafer, W. 1966. *The Vegetation of Poland.* Oxford: Pergamon Press. 738 pp.

Takahashi, M., and Ripperton, J. C. 1949. Koa haole (*Leucaena glauca*). Its establishment, culture and utilization as a forage crop. *Hawaii Agric. Exp. Stn. Bull.* 100. 56 pp.

Takeda, H. 1913. *Cladrastis* and *Maackia*. *Notes R. Bot. Gard. Edinb.* 8:95–104.

Tasaki, B., and Tanaka, U. 1918. Toxic constituents in the bark of *Robinia pseudoacacia*. *J. Coll. Agric. Tokyo* 3:337–356. *Chem. Abstr.* 13:2696 (1919).

Tattersfield, F., and Gimingham, C. T. 1932. The insecticidal properties of *Tephrosia macropoda* Harv. and other tropical plants. *Ann. Appl. Biol.* 19:253–262.

Taubert, P. 1889. Leguminosae novae v. minus cognitae austro-americanae. *Flora* 72:421–430.

Taubert, P. 1891. *Eminia*, genus novum Papilionacearum. *Ber. Dtsch. Bot. Ges.* 9:28–32.

Taubert, P. 1894. Leguminosae. *In* Engler, A., and Prantl, K. *Die Natürlichen Pflanzenfamilien*, Part 3, sect. 3. Leipzig: Engelmann. 396 pp.

Taubert, P. 1896. Leguminosae africanae I. *Bot. Jahr.* 23:172–196.

Theiler, A. 1918. Jaagsiekte in horses. *7th–8th Rep. Dir. Vet. Res. Union S. Afr.* 59–103.

Thompstone, E., and Sawyer, A. M. 1914. The peas and beans of Burma. *Dep. Agric. Burma Bull.* 12. 107 pp.

Thorne, D. W., and Walker, R. H. 1936. The influence of seed inoculation upon the growth of black locust seedlings. *J. Am. Soc. Agron.* 28:28–34.

Titmuss, F. H. 1965. *Commercial Timbers of the World.* 3d ed., enlarged, of *A Concise Encyclopedia of World Timbers* London: Technical Press. 277 pp.

Tiwari, R. D., and Yadava, O. P. 1971. Structural study of the quinone pigments from the roots of *Cassia alata*. *Planta Med.* 19 (4):299–305. *Biol. Abstr.* 52 (18):103364 (1971).

Topham, P. 1930. The genus *Brachystegia* in Nyasaland. *Kew Bull. Misc. Inf.* 8:348–364.

Toussaint, L. 1953. *Bowringia* Champ. *In* Robyns, W. *Flore du Congo Belge et du Ruanda-Urundi*, 4:6–7. Brussels: Publ. Inst. Nat. Étude Agron. Congo Belge (I.N.É.A.C.).

Towle, M. A. 1961. *The Ethnobotany of pre-Columbian Peru.* Viking Fund Publ. Anthropol. No. 30. Chicago: Aldine. 180 pp.

Toxopeus, H. J. 1936. Over Physiologische specialisatie bij knolletjesbacterien van kedelee op Java. *Vers. 16e Vergadering vereeniging Proefstation-personeel Djember. Oct. 1936*, 53–62. Surabya: H. Van Ingen.

Tracy, S. M., and Coe, H. S. 1918. Velvet beans. *U. S. Dep. Agric. Farmers Bull.* 962. 39 pp.

References

Trelease, S. F., and Beath, O. A. 1949. *Selenium*. New York: Published by the authors. 292 pp.

Trelease, S. F., and Trelease, H. M. 1938. Selenium as a stimulating and possibly essential element for indicator plants. *Am. J. Bot.* 25:372–380.

Trelease, S. F., and Trelease, H. M. 1939. Physiological differentiation in *Astragalus* with reference to selenium. *Am. J. Bot.* 26:530–535.

Treviranus, L. C. 1853. Ueber die Neigung der Hülsengewächse zu unterirdischer Knöllenbildung. *Bot. Ztg.* 11:393–399.

Trinick, M. J. 1965. *Medicago sativa* nodulation with *Leucaena leucocephala* root-nodule bacteria. *Aust. J. Sci.* 27:263–264.

Trinick, M. J. 1968. Nodulation of tropical legumes. I. Specificity in the *Rhizobium* symbiosis of *Leucaena leucocephala*. *Exp. Agric.* 4:243–253.

Trinick, M. J., and Galbraith, J. 1976. Structure of root nodules formed by *Rhizobium* on the non-legume *Trema cannabina* var. *scabra*. *Arch. Microbiol.* 108:159–166.

Tschirch, A. 1887. Beiträge zur Kenntniss der Wurzelknöllchen der Leguminosen. *Ber. Dtsch. Bot. Ges.* 5:58–98.

Tulasne, L.-R. 1843. Nova quaedam proponit Genera in Leguminosarum classe. *Ann. Sci. Nat.* ser. 2. Bot. 20:136–144.

Tummin Katti, M. C. 1930. Chemical examination of the seeds of *Caesalpinia Bonducella*, Flem. Part I. *J. Indian Chem. Soc.* 7:207–220.

Tummin Katti, M. C., and Puntambekar, S. V. 1930. Chemical examination of the seeds of *Caesalpinia Bonducella*, Flem. Part II. Fatty oil. *J. Indian Chem. Soc.* 7:221–227.

Turner, B. L. 1956. Chromosome numbers in the Leguminosae. *Am. J. Bot.* 43:577–581.

Turner, B. L. 1971. Implications of the biochemical data: a summing up. *In* Harborne, J. B., Boulter, D., and Turner, B. L., eds. *Chemotaxonomy of the Leguminosae*, 549–558. New York: Academic Press.

Turner, B. L., and Fearing, O. S. 1959. Chromosome numbers in the Leguminosae. II. African species including phylogenetic interpretations. *Am. J. Bot.* 46:49–57.

Turner, B. L., and Fearing, O. S. 1964. A taxonomic study of the genus *Amphicarpaea* (Leguminosae). *Southwest. Nat.* 9 (4):207–218.

Turner, B. L., and Harborne, J. B. 1967. Distribution of canavanine in the plant kingdom. *Phytochemistry* 6:863–866.

Turner, W. J., and Merlis, S. 1959. Effect of some indolealkylamines on man. *Arch. Neurol. Psychiatr.* 81:121–129.

Tutin, T. G., Heywood, V. H., Burger, N. A., Moore, D. M., Valentine, D. H., Walters, S. M., and Webb, D. A., eds. 1968. *Flora Europaea*. Rosaceae to Umbelliferae. 2. LXXXI. 455 pp. + 5 maps. London: Cambridge University Press.

Ulbrich, E. 1906. *Weberbauerella* Ulbrich *genus novum*. Leguminosae andinae. III. *Bot. Jahr.* 37:551–553.

University of Pretoria. 1954. Studies on the symbiotic nitrogen fixation in legumes of importance to southern Africa. *Rep. Nuffield Found. Grant. Plant Physiol. Res. Inst.* Mimeographed, 17 pp.

University of Pretoria. 1955–1956. Annual Report on the research activities of the Plant Physiological Research Institute. 45 pp.

University of Pretoria. 1957. Annual Report on the research activities of the Plant Physiological Research Institute. 61 pp.

University of Pretoria. 1958. Annual Report on the research activities of the Plant Physiological Research Institute. 81 pp.

Uphof, J. C. T. 1968. *Dictionary of Economic Plants.* 2d ed. Lehre, Germany: J. Cramer. 591 pp.

Urban, I. 1899. V. Species novae, praesertim portoricenses. *Symbol. Antillanae* 1:291–481.

Urban, I. 1900. Leguminosae novae vel minus cognitae. I. *Symbol. Antillanae* 2:257–335.

Urban, I. 1928. Plantae cubenses novae vel rariores a clo. Er. L. Ekman lectae IV. Leguminosae. *Symbol. Antillanae* 9:433–458.

Urban, I. 1929. Plantae Haitiensis et Domingensis novae vel rariores. V. *Arkiv. Bot.* 22A (8):1–98.

Varner, J. E. 1960. Urease. *In* Boyer, P. D., Lardy, H., and Myrbäch, K., eds. *The Enzymes,* 4:247–256. 2d ed. New York: Academic Press.

Vasil'chenko, I. T. 1965. The evolution of the genus *Oxytropis* DC. (Leguminosae) [in Russian, English summary]. *Bot. Zh.* 50:313–323.

Veen, A. G. van, and Hyman, A. J. 1935. Djenkolic acid, a new sulfur-containing amino acid. *Rev. Trav. Chim.* 54:493–501.

Verboom, W. C. 1966. The grassland communities of Barotseland. *Trop. Agric. (Trinidad)* 43:107–115.

Verdcourt, B. 1966. A proposal concerning *Glycine* L. *Taxon* 15:34–36.

Verdcourt, B. 1968a. The identities of *Dolichos trilobus* L. and *Dolichos trilobatus* L. *Taxon* 17:170–173.

Verdcourt, B. 1968b. Notes on *Dipogon* and *Psophocarpus* (Leguminosae). *Taxon* 17 (5):537–539.

Verdcourt, B. 1969. New combinations in *Vigna* Savi. (Leguminosae-Papilionoideae). *Kew Bull.* 23:464.

Verdcourt, B. 1970a. Studies in the Leguminosae-Papilionoideae for the *Flora of Tropical East Africa:* II. *Kew Bull.* 24 (2):235–307.

Verdcourt, B. 1970b. *Macrotyloma* (Wight & Arn.) Verdc. *Kew Bull.* 24 (2):322.

Verdcourt, B. 1970c. Studies in Leguminosae-Papilionoideae for the *Flora of Tropical East Africa:* III. *Kew Bull.* 24 (3):379–447.

Verdcourt, B. 1970d. Studies in the Leguminosae-Papilionoideae for the *Flora of Tropical East Africa:* IV. *Kew Bull.* 24 (3):507–569.

Verdcourt, B. 1971. Studies in the Leguminosae-Papilionoideae for the *Flora of Tropical East Africa:* V. *Kew Bull.* 25 (1):65–169.

Verdcourt, B. 1973. Studies in the Leguminosae-Papilionoideae-Hedysareae (*sensu lato*) for the *Flora Zambesiaca:* 3. *Kew Bull.* 28 (3):429–431.

Verdcourt, B. 1978. A new combination in *Oxyrhynchus* (Leguminosae, Phaseoleae). *Kew Bull.* 32 (4):779–780.

Verdcourt, B., and Trump, E. C. 1969. *Common Poisonous Plants of East Africa.* Church, Mrs. M. E., illus. London: Collins. 254 pp.

Viguier, R. 1949. Leguminosae Madagascariensis novae. *Not. Syst.* 13:333–369.

Viguier, R. 1950. Leguminosae Madagascariensis novae. Suite 1. *Not. Syst.* 14:62–74.

Viguier, R. 1952. Leguminosae Madagascariensis novae. Suite 2. *Not. Syst.* 14:168–187.

References

Vilkomerson, H. 1943. Chromosomes of *Astragalus*. *Bull. Torrey Bot. Club.* 70:430–435.

Vincent, J. M. 1941. Serological studies of the root-nodule bacteria. I. Strains of *Rhizobium meliloti*. *Proc. Linn. Soc. N. S. W.* 66 (3–4): 145–154.

Vincent, J. M. 1945. Host specificity amongst root-nodule bacteria isolated from several clover species. *J. Aust. Inst. Agric. Sci.* 11:121–127.

Vincent, J. M. 1970. *A Manual for the Practical Study of Root-Nodule Bacteria.* IBP Handbook No. 15. 164 pp. Oxford: Blackwell Scientific Publications.

Vincent, J. M., Whitney, A. S., and Bose, J. 1977. Exploiting the legume-*Rhizobium* symbiosis in tropical agriculture. *Univ. Hawaii Coll. Trop. Agric. Misc. Publ.* 145. 469 pp.

Vivian, L. A. 1959. The leguminous fodder "stylo" or "tropical lucerne" in Kelantan. *Malay. Agric. J.* 42:183–198.

Vlamis, J., Schultz, A. M., and Biswell, H. H. 1964. Nitrogen fixation by root nodules of Western Mountain Mahogany. *J. Range Manage.* 17:73–74.

Voorhoeve, A. G. 1965. *Liberian High Forest Trees.* Wageningen, The Netherlands: Centre for Agricultural Publications and Documentation. 416 pp.

Voss, E. G. 1957. New records of vascular plants from the Douglas Lake region (Emmet and Cheboygan Counties), Michigan. *Pap. Mich. Acad. Sci. Arts Lett.* 42:3–34.

Vuillemin, P. 1888. Les tubercules radicaux des Légumineuses. *Ann. Sci. Agron. Fr. Étrang. 5th Ann.* 1:121–212.

Vyas, S. R., and Prasad, N. 1959. Studies on *Rhizobium* of *Zornia diphylla*. *Proc. Indian Acad. Sci.* 49B (3):156–160.

Wade, B. L. 1937. Breeding and improvement of peas and beans. *In Better Plants and Animals*, Book 2:251–282. *U. S. Dep. Agric. Yearb. Agric.*

Waldron, R. A. 1919. The peanut (*Arachis hypogaea*): its history, histology, physiology, and utility. *Contrib. Univ. Penn. Bot. Lab.* 4:301–338.

Walker, R. H. 1928. Physiological studies on the nitrogen fixing bacteria of the genus *Rhizobium*. *Iowa Agric. Coll. Exp. Stn. Res. Bull.* 113:371–406.

Walker, R. H., and Brown, P. E. 1935. The nomenclature of the cowpea group of root-nodule bacteria. *Soil Sci.* 39:221–225.

Walsh, S. R. 1958. Tropical legumes for better pastures. *Queensl. Agric. J.* 84:527–536.

Walters, J. A. T. 1924. The Carob Bean (*Ceratonia siliqua*) in Rhodesia and other pod-bearing indigenous trees of value for stock feed. *Rhodesia Agric. J.* 21:567–572.

Ward, H. M. 1887. On the tubercular swellings on the roots of *Vicia faba*. *Philo. Trans. R. Soc. Lond.* ser. B. 178:539–562.

Warren, J. A. 1909. Notes on the number and distribution of native legumes in Nebraska and Kansas. *U. S. Dep. Agric. Bur. Plant Ind. Circ.* 31. 9 pp.

Warren, J. A. 1910. Additional notes on the number and distribution of native legumes in Nebraska and Kansas. *U. S. Dep. Agric. Bur. Plant Ind. Circ.* 70. 8 pp.

Wasicky, R. 1942. Pharmacological investigation of leaves of *Cassia fistula* L. and some Brazilian species of *Cassia*. *Bol. Acad. Nac. Farm.* 4:133–135. *Chem. Abstr.* 37:1562 (1943).

Watson, G. A. 1957. Nitrogen fixation by *Centrosema pubescens. J. Rubber Res. Inst. Malaya Commun.* 15:168–174. *Chem. Abst.* 17582 (1958).

Watt, G., 1908. *The Commercial Products of India.* London: John Murray. 1189 pp.

Watt, J. M., and Breyer-Brandwijk, M. G. 1962. *The Medicinal and Poisonous Plants of Southern and Eastern Africa.* 2d ed. Edinburgh: Livingstone. 1457 pp.

Webb, L. J. 1948. Guide to the medicinal and poisonous plants of Queensland. *Aust. C.S.I.R.O. Bull.* 232. 202 pp.

Webster, G. T. 1955. Interspecific hybridization of *Melilotus alba* x *M. officinalis* using embryo culture. *Agron. J.* 47:138–142.

Wendel, E. 1918. Zur physiologischen Anatomie der Wurzelknöllchen einiger Leguminosen. *Beitr. Allg. Bot.* 1:151–189.

West, E., and Emmel, M. W. 1950. Some poisonous plants in Florida. *Fl. Agric. Exp. Stn. Bull.* 468. 47 pp.

Westphal, E. 1975. The proposed retypification of *Dolichos* L.: a review. *Taxon* 24 (1):189–192.

Wheeler, L. C. 1955. The husks of the prodigal son. *Turtox News* 33 (10,11), Oct., Nov. 7 pp.

Wheeler, W. M. 1921. The *Tachigalia* ants. *Zoologica* 3:137–168.

White, E. P. 1943. Alkaloids of the Leguminosae. Parts. I.–V. *N. Z. J. Sci. Tech.* 25B:93–98; 103–105; 106–108.

White, E. P. 1944. Alkaloids of the Leguminosae. VIII. *N. Z. J. Sci. Tech.* 25B:137–138.

White, E. P. 1951. Alkaloids of the Leguminosae. XXII. *N. Z. J. Sci. Tech.* 33B:54–60.

White, F.. 1962. *Forest Flora of Northern Rhodesia.* London: Oxford University Press. 454 pp.

White, G. A., and Haun, J. R. 1965. Growing *Crotalaria juncea*, a multipurpose legume for paper pulp. *Econ. Bot.* 19:175–183.

Whitford, H. N. 1911. The forests of the Philippines. Part II. The principal forest trees. *Philipp. Dep. Inter. Bur. For. Bull.* No. 10. 113 pp.

Whiting, A. L., and Hansen, R. 1920. Cross-inoculation studies with the nodule bacteria of lima beans, navy beans, cowpeas and others of the cowpea group. *Soil Sci.* 10:291–300.

Whiting, A. L., Fred, E. B., and Helz, G. E. 1926. A study of the root-nodule bacteria of Wood's clover (*Dalea alopecuroides*). *Soil Sci.* 22:467–475.

Whitton, B. A. 1962. Forests and dominant legumes of the Amatuk region, British Guiana. *Caribb. For.* 23 (1):35–57.

Whyte, R. O., Nilsson-Leissner, G., and Trumble, H. C. 1953. *Legumes in Agriculture.* FAO Agric. Stud. No. 21. Rome: Food and Agriculture Organization of the United Nations. 367 pp.

Wiesner, K., MacDonald, D. M., and Bankiewicz, C. 1953. The structure of pithecolobine. *J. Am. Chem. Soc.* 75:6348–6349.

Wiesner, K., McDonald, D. M., Valenta, Z., and Armstrong, R. 1952. Pithecolobine, the alkaloid of *Pithecolobium saman* Benth. I. *Can. J. Chem.* 30:761–772.

References

Wight, W. F. 1907. The history of the cowpea and its introduction into America. *U. S. Dep. Agric. Bur. Plant Ind. Bull.* 102. Misc. Pap. 6:43–59.

Wilbur, R. L. 1963. The leguminous plants of North Carolina. *N. C. Agric. Exp. Stn. Tech. Bull.* 151. 294 pp.

Wilbur, R. L. 1965. The lectotype of the Mimosaceous genus *Leucaena* Benth. *Taxon* 14 (7):246–247.

Wilcox, E. V. 1899. Lupines as plants poisonous to stock. *Mont. Agric. Exp. Stn. Bull.* 22:37–45.

Wilczek, R. 1952. Dimorphandreae; Eucaesalpinieae; Bauhinieae; Amphimanteae. *In* Robyns, W., *Flore du Congo Belge et du Ruanda-Urundi*, 3:237–246; 247–264; 265–278; 546–549. Brussels: Publ. Inst. Nat. Étude Agron. Congo Belge (I.N.É.A.C.).

Wilczek, R. 1953. Papilionaceae Genisteae Congolanae novae (*Robynsiophyton, Crotalaria, Argyrolobium*). *Bull. Jard. État Brux.* 23:125–221.

Wilczek, R. 1954. Phaseolinae. *In* Robyns, W., *Flore du Congo Belge et du Ruanda-Urundi*, 6:260–409. Brussels: Publ. Inst. Nat. Étude Agron. Congo Belge (I.N.É.A.C.).

Wilkins, J. 1967. The effects of high temperatures on certain root-nodule bacteria. *Aust. J. Agric. Res.* 18:299–304.

Willaman, J. J., and Schubert, B. G. 1961. Alkaloid-bearing plants and their contained alkaloids. *U. S. Dep. Agric.-A.R.S. Tech. Bull.* 1234. 287 pp.

Williams, L. O. 1952. Beans, maize and cultivation. *Ceiba* 3 (2):77–85.

Willis, J. C. 1957. *A Dictionary of the Flowering Plants and Trees.* 6th ed. Cambridge: Cambridge University Press. 752 pp.

Willis, J. C. 1966. *A Dictionary of the Flowering Plants and Ferns.* 7th ed. Cambridge: University Press. 1214 pp.

Wilson, C. M., ed. 1945. *New Crops for the New World.* New York: Macmillan. 295 pp.

Wilson, J. K. 1939a. Leguminous plants and their associated organisms. *Cornell Univ. Agric. Exp. Stn. Mem.* 221. 48 pp.

Wilson, J. K. 1939b. Symbiotic promiscuity of two species of *Crotalaria. J. Am. Soc. Agron.* 31:934–939.

Wilson, J. K. 1939c. The relationship between pollination and nodulation of the Leguminoseae. *J. Am. Soc. Agron.* 31:159–170.

Wilson, J. K. 1944a. Over five hundred reasons for abandoning the cross-inoculation groups of the legumes. *Soil Sci.* 58:61–69.

Wilson, J. K. 1944b. The nodulating performance of three species of legumes. *Soil Sci. Soc. Am. Proc.* 9:95–97.

Wilson, J. K. 1945a. The symbiotic performance of isolates from soybean with species of *Crotalaria* and certain other plants. *Cornell Univ. Agric. Exp. Stn. Bull.* 267. 20 pp.

Wilson, J. K. 1945b. Another nonnodulating legume. *J. Bacteriol.* 50:123.

Wilson, J. K. 1946. Variation in seed as shown by symbiosis. *Cornell Univ. Agric. Exp. Stn. Mem.* 272. 21 pp.

Wilson, J. K., and Chin, C.-H. 1947. Symbiotic studies with isolates from nodules of species of *Astragalus. Soil Sci.* 63:119–127.

Wilson, J. K., and Choudhri, R. S. 1946. Effects of DDT on certain microbiological processes in the soil. *J. Econ. Entomol.* 39:537–538.

Wilson, J. K., and Westgate, P. J. 1942. Variation in the percentage of nitrogen in the nodules of leguminous plants. *Soil Sci. Soc. Am. Proc.* 7:265–268.

Wilson, P. W., Burton, J. C., and Bond, V. S. 1937. Effect of species of host plant on nitrogen fixation in *Melilotus*. *J. Agric. Res.* 55 (8):619–629.

Wipf, L. 1939. Chromosome numbers in root nodules and root tips of certain Leguminosae. *Bot. Gaz.* 101:51–67.

Wipf, L., and Cooper, D. C. 1938. Chromosome numbers in nodules and roots of red clover, common vetch and garden pea. *Proc. Natl. Acad. Sci. U.S.A.* 24:87–91.

Wipf, L. and Cooper, D. C. 1940. Somatic doubling of chromosomes and nodular infections in certain Leguminosae. *Am. J. Bot.* 27:821–825.

Wolf, W. J., and Cowan, J. C. 1971. *Soybeans as a Food Source.* Cleveland: CRC Press, A Division of the Chemical Rubber Company. 86 pp.

Woodroof, J. G. 1966. *Peanuts, Production Processing, Products.* Westport, Conn.: AVI Publishing Company. 291 pp.

Worsley, R. R. Le G. 1936. The insecticidal properties of some East African plants. II. *Mundulea suberosa* Benth. *Ann. Appl. Biol.* 23:311–328.

Worsley, R. R. Le G. 1937a. The insecticidal properties of some East African plants. III. *Mundulea suberosa* Benth. Part 2. Chemical constituents. *Ann. Appl. Biol.* 24:651–658.

Worsley, R. R. Le G. 1937b. The insecticidal properties of some East African plants. IV. *Mundulea suberosa* Benth. Part 3. Variability of samples. *Ann. Appl. Biol.* 24:659–664.

Worsley, R. R. Le G., and Nutman, F. J. 1937. Biochemical studies of *Derris* and *Mundulea*. I. The histology of rotenone in *Derris elliptica*. *Ann. Appl. Biol.* 74:696–702.

Wright, H. 1903. Nitrogenous plants. *Circ. Agric. J. R. Bot. Gard. Ceylon* 2 (4):77–81. (See also, Anonymous. 1903. *Agric. Bull. Straits Fed. Malay States* 2:288–290).

Wright, H. 1905. Green manures. *Circ. Agric. J. R. Bot. Gard. Ceylon* 3 (12):180–198.

Wurdack, J. J. 1958. Indian narcotics in southern Venezuela. *Gard. J.* 8 (4):116–119.

Wydler, H. 1860. Kleinere Beiträge zur Kenntniss einheimischer Gewächse. *Flora* 43:17–32; 51–63; 83–96.

Yakovleva, A. P., Proskurnina, N. F., and Utkin, L. M. 1959. Alkaloids of *Ammothamnus songoricus* [in Russian]. *Zh. Obshch. Khim.* 29:1042–1044. *Chem. Abstr.* 54:1577 (1960).

Yarnell, S. H. 1962. Cytogenetics of the vegetable crops. III. Legumes. A. Garden peas, *Pisum sativum* L. *Bot. Rev.* 28 (4):465–537.

Younge, O. R., Plucknett, D. L., and Rotar, P. P. 1964. Culture and yield performance of *Desmodium intortum* and *D. canum* in Hawaii. *Hawaii Agric. Exp. Stn. Univ. Hawaii Tech. Bull.* 59. 28 pp.

References

Youngman, V. E. 1968. Lentils — A pulse of the Palouse. *Econ. Bot.* 22 (2):135–139.

Zimny, H. 1962. Root nodulation of certain species of papilionaceous plants in forest and meadow associations at Bialowieźa (NE Poland) [in Polish, English abstract]. *Fragm. Florist. Geobot.* (Krakow) 8 (2):157–183.

Zimny, H. 1964. The nodulation of papilionaceous plants and habitat conditions. [in Polish, English abstract]. *Fragm. Florist. Geobot.* Ann. 10 (2):199–237.

Zohary, D. 1972. The wild progenitor and the place of origin of the cultivated lentil: *Lens culinaris. Econ. Bot.* 26 (4):326–332.

Zohary, M. 1968. *The Legumes of Palestine.* Offprint from *Flora Palestina*, Part 2. 224 pp. Jerusalem: Israel Academy of Sciences and Humanities.

Zuijderhoudt, G. F. P. 1967. A revision of the genus *Saraca* L. (Legum.-Caes.). *Blumea* 15 (2):413–425.

Illustration Credits

Brenan, J. P. M. 1967. *Flora of Tropical East Africa: Leguminosae, Subfamily Caesalpinioideae.* London: Crown Agents for Overseas Governments and Administrations. Reproduced by permission of the Director, Royal Botanic Gardens, Kew, England.

> Leaves: *Monopetalanthus richardsiae*; *Tamarindus indica*.
>
> Flowers: *Berlinia orientalis*; *Mezoneuron angolense*; *Tamarindus indica*.

Britton, N. L., and Brown, A. 1970. *An Illustrated Flora of the Northern United States and Canada*, 2. New York: Dover Publications, Inc.

> Leaves: *Lupinus perennis*; *Melilotus officinalis*; *Ulex europaeus*; *Vicia hirsuta*.
>
> Flower: *Astragalus hypoglottis*.
>
> Pods: *Astragalus bisulcatus*; *Medicago hispida*.

Brown, W. H., ed. 1921. Minor Products of Philippine Forests. *Dep. Agric. Nat. Res. Bur. For. Bull.* 22.

> Leaves and flower: *Intsia bijuga*.
>
> Pod: *Pterocarpus vidaliana*.

Burkart, A. 1943. *Las Leguminosas Argentinas-Silvestres y Cultivadas.* Buenos Aires: Acme Agency.

> Leaves: *Acacia melanoxylon*.
>
> Flower: *Calliandra brevicaulis*.

Chuang, C., and Huang, C. 1965. *The Leguminosae of Taiwan, for Pasture and Soil Improvement.* Taipei, Taiwan: Dep. Botany, Taiwan University and J.C.R.R.

> Leaves: *Arachis hypogaea*; *Astragalus sinicus*; *Cassia tora*; *Clitoria ternatea*; *Leucaena leucocephala*; *Medicago hispida*; *Pisum sativum*; *Trifolium pratense*.
>
> Flowers: *Arachis hypogaea*; *Cassia tora*; *Clitoria ternatea*; *Crotalaria juncea*; *Indigofera endecaphylla*; *Leucaena leucocephala*; *Lupinus luteus*; *Medicago hispida*; *Melilotus indica*; *Mucuna macrocarpa*; *Tephrosia noctiflora*; *Trifolium pratense*.
>
> Pod: *Pisum sativum*.

Cockburn, P. F. 1976. *Trees of Sabah*, 1. Sabah Forest Record 10. Kuching, Malaysia: Borneo Literature Bureau.

> Leaves: *Pithecellobium splendens*.
>
> Flowers: *Albizia falcataria*; *Sympetalandra borneensis*.
>
> Pods: *Parkia singularis*; *Sindora beccariana*.

Cowan, R. S. 1963. Correct Name of the Powder-puff Tree. *Baileya* 11.
Leaves: *Calliandra inaequilatera*.

Ducke, A. 1930. Plantes nouvelles ou peu connues de la région amazonienne. Part 4. *Arch. Jard. Bot. Rio de Janeiro* 5.
Pods: *Cedrelinga catenaeformis*; *Tipuana speciosa*.

Elias, T. S. 1974. The genera of Mimosoideae (Leguminosae) in the southeastern United States. *J. Arnold Arboretum* 55 (1).
Leaves: *Neptunia pubescens*.

Garcia Barriga, Hernando, and Forero Gonzales, Enrique. 1968. *Catálogo ilustrado de las Plantas de Cundinamarca*, 3. Bogota: Universidad Nacional.
Leaves: *Abrus precatorius*.

Hutchinson, J., and Dalziel, J. M. 1958. *Flora of West Tropical Africa*. 2d ed. rev. Keay, R. W. J. Volume 1, Part 2. London: Crown Agents for Overseas Governments and Administrations.
Leaves: *Swartzia madagascariensis*.
Flowers: *Entada africana*; *Swartzia madagascariensis*; *Tetrapleura tetraptera*.
Pod: *Tetrapleura tetraptera*.

Isely, D. 1951. The Leguminosae of the north-central United States: I. Loteae and Trifoleae. *Iowa State Coll. J. Sci.* 25 (3).
Pods: *Medicago sativa*; *Melilotus officinalis*.

Little, E. L., Jr., and Wadsworth, F. H. 1964. *Common Trees of Puerto Rico and the Virgin Islands*. Washington, D.C.: U.S. Dep. Agric. Forest Service, Agriculture Handbook 249.
Leaves: *Bauhinia monandra*; *Parkinsonia aculeata*.
Flowers: *Bauhinia monandra*; *Delonix regia*; *Inga laurina*.
Pods: *Cassia javanica*; *Leucaena leucocephala*; *Lonchocarpus latifolius*; *Parkinsonia aculeata*.

Medsger, O. P. 1939. *Edible Wild Plants*. New York: Macmillan. Copyright 1939 by Macmillan Publishing Co., Inc., renewed 1967 by Henry O. Medsger, Thomas A. Medsger, and Oliver P. Medsger, Jr.
Leaves: *Prosopis glandulosa*.

Purseglove, J. W. 1968. *Tropical Crops: Dicotyledons I*. New York: John Wiley and Sons, Inc. Reproduced by permission of Longman Group Ltd., Essex, England.
Leaves: *Pachyrrhizus erosus*.
Pods: *Calopogonium mucunoides*; *Cicer arietinium*; *Psophocarpus tetragonolobus*; *Tamarindus indica*.

Robyns, W., ed. 1952. *Flore du Congo Belge et du Ruanda-Urundi* 3. Brussels: Institut National pour l'Étude Agronomique du Congo Belge.
Leaves: *Acacia macrothyrsa*.
Flowers: *Acacia macrothyrsa*; *Brachystegia boehmii*; *Parkia bicolor*; *Xylia ghesquierei*.
Pods: *Berlinia grandiflora*; *Xylia ghesquierei*.

Voorhoeve, A. G. 1965. *Liberian High Forest Trees*. Wageningen, The Netherlands: Centre for Agricultural Publications and Documentation.
Inflorescence: *Parkia bicolor*.

White, F. 1962. *Forest Flora of Northern Rhodesia*. London: Oxford University Press. The figures listed here, skillfully executed by Mrs. Maureen E. Church (née Griffiths) and Mrs. Janet Dyer (née Chandler) have been reproduced by courtesy of the author, Mr. F. White, and the Oxford University Press.

> Leaves: *Colospermum mopane*; *Mezoneuron welwitschianum*; *Piliostigma thonningia*.
>
> Flowers: *Cordyla africana*; *Erythrophleum guineense*.
>
> Pods: *Acacia hebeclada*; *Acacia nilotica*; *Afzelia quanzensis*; *Alysicarpus zeyheri*; *Burkea africana*; *Caesalpinia sepiaria*; *Crotalaria natalita*; *Desmodium velutinum*; *Dichrostachys glomerata*; *Entada abyssinica*; *Erythrina baumii*; *Erythrophleum africanum*; *Lablab purpureus*; *Mimosa pigra*; *Mucuna puriens*; *Ormocarpum bibracteatum*; *Swartzia madagascariensis*.

Wilbur, R. L. 1963. The leguminous plants of North Carolina. *N. C. Agric. Exp. Stn. Tech. Bull.* 151.

> Leaves: *Albizia julibrissin*; *Cercis canadensis*.
>
> Flower: *Albizia julibrissin*.
>
> Pods: *Albizia julibrissin*; *Baptisia australis*; *Cassia obtusifolia*; *Cercis canadensis*.

Nodule drawings by Barbara Goodsitt, Illustrator, Botany Dept., University of Wisconsin, Madison.

Nodulated root systems (except for *Sesbania grandiflora*) courtesy of Dr. Joe C. Burton, The Nitragin Company, Milwaukee, Wisconsin.

Scanning electron photomicrographs of nodules by the Scanning Electron Microscope Facility of the Department of Entomology, University of Wisconsin, Madison.

All other photographs by the authors.

Index of Common Names

The common names listed here are applied to species within the genera indicated.

Index of Common Names

Index of Common Names

General Index

Alkaloids present in legumes, 18, 41, 45, 47, 48, 90, 129, 163, 185, 193, 211, 229, 244, 271, 275, 277, 280, 291, 295, 299, 315, 334, 371, 385, 408, 414, 438, 447, 504, 509, 519, 522, 524, 529, 549, 577, 591, 594, 611, 614, 617, 619, 638, 644, 655, 667, 671, 689

Animals. *See* Fish stupefiers; Food for animals; Maladies affecting livestock

Antimicrobial properties, 49, 142, 317, 527, 549, 584

Antiquity of legumes
  evidence from archeology and ethnology, 61, 156, 303, 342, 511–512, 527, 548
  evidence from geology, 120, 133, 159, 283, 362, 482

Ants: hosts for (myrmecophytes), 6–7, 64, 335, 466, 535, 640

Arrow poisons: sources of, 224, 275, 277, 344, 436, 525, 645

Bacteroids, xix, 63, 77, 128, 304, 375, 425, 429, 460, 469, 513, 528, 585, 594, 660, 667, 678–679

Balsam: presence of, 181, 219, 365, 452, 453, 488. *See also* Copals and resins; Gums

Betel nut substitute, 235

Biblical references, 7, 35 ,156, 159, 382, 577, 666

Bufotenine, 47, 229, 522. *See also* Folk medicines: hallucinogen

Canavanine, 134, 199

Canoes. *See* Dug-out canoes

Carpentry. *See* Wood: special uses for veneer, cabinetwork

Characteristics of Leguminosae, xiv–xv

Charcoal: sources of, 7, 116, 118, 124, 209, 237, 270, 278, 367, 388, 396, 526, 548, 577, 585, 605, 642

Chemotaxonomy: applied, 134, 271, 280, 291, 524

Chromosome lines in *Cassia*, xxii–xxiii

Cleistogamy. *See* Flowers chasmogamous and cleistogamous

Composition cork: source of, 275

Copals and resins, 45, 113, 177, 181, 219, 232, 234, 248, 311, 314, 322, 356, 415, 416, 477, 479, 518, 537, 618, 651, 657, 658. *See also* Balsam; Gums

Cordage (rope): plants used for, 95, 118, 127, 226, 237, 257, 312, 322, 494, 544, 552, 570, 605, 618

Coumarin, 39, 248, 301, 429, 453

Cover crops, 20, 22, 36, 44, 55, 65, 82, 89, 124, 133, 154, 165, 172, 174, 237, 251, 288, 308, 333, 388, 402, 407, 423, 437, 447, 475, 512, 551, 556, 570, 574, 579, 585, 605, 631, 646, 652, 673, 678, 683

Cowpea miscellany. *See* Cross-inoculation, promiscuous

Crocodile repellent, 450

Cross-inoculation: promiscuous (cowpea miscellany), xviii, 5, 8, 24, 32, 37, 44, 50, 57, 62, 90, 104, 107, 124, 128, 135, 136, 142, 154, 164, 165, 169, 170, 172, 174, 194, 203, 211, 220, 226, 229, 242, 252, 262, 269, 276, 281, 293, 295, 308, 321, 334, 338, 344, 353, 357, 358, 362, 371, 372, 378, 385, 397, 418, 419, 439, 441, 448, 451, 480, 495, 525, 526, 535, 544, 549, 558, 565, 571, 572, 580, 591, 594, 601, 614, 617, 619, 631, 633, 637, 644, 646, 650, 684–685, 688, 690

Cross-inoculation: restricted, selective, xvii–xviii, 166, 186, 191, 217, 254, 306, 375, 382, 399, 403, 425, 429, 439, 468, 472, 476, 483, 500, 508, 528, 586, 606, 629, 652, 660, 667, 680, 697

DESIGNED BY IRVING PERKINS
COMPOSED BY THE NORTH CENTRAL PUBLISHING COMPANY
SAINT PAUL, MINNESOTA
MANUFACTURED BY KINGSPORT PRESS, KINGSPORT, TENNESSEE
TEXT IS SET IN BASKERVILLE, DISPLAY LINES IN OPTIMA SEMIBOLD

Library of Congress Cataloging in Publication Data
Allen, Oscar Nelson, 1905–1976
The Leguminosae, a source book of characteristics, uses, and nodulation.
Bibliography: pp. 731–798.
Includes index.
1. Leguminosae. I. Allen, Ethel Kullmann, 1906–   joint author. II. Title.
III. Title: Nodulation.
QK495.L52A57   583'.32   80-5104
ISBN 0-299-08400-0